Quaternary of Scotland

THE GEOLOGICAL CONSERVATION REVIEW SERIES

The comparatively small land area of Great Britain contains an unrivalled sequence of rocks, mineral and fossil deposits, and a variety of landforms that span much of the earth's long history. Well-documented ancient volcanic episodes, famous fossil sites, and sedimentary rock sections used internationally as comparative standards, have given these islands an importance out of all proportion to their size. These long sequences of strata and their organic and inorganic contents, have been studied by generations of leading geologists thus giving Britain a unique status in the development of the science. Many of the divisions of geological time used throughout the world are named after British sites or areas, for instance the Cambrian, Ordovician and Devonian systems, the Ludlow Series and the Kimmeridgian and Portlandian stages.

The Geological Conservation Review (GCR) was initiated by the Nature Conservancy Council in 1977 to assess, document, and ultimately publish accounts of the most important parts of this rich heritage. The GCR reviews the current state of knowledge of the key earth-science sites in Great Britain and provides a firm basis on which site conservation can be founded in years to come. Each GCR volume describes and assesses networks of sites of national or international importance in the context of a portion of the geological column, or a geological, palaeontological, or mineralogical topic. The full series of approximately 50 volumes will be published by the year 2000.

Within each individual volume, every GCR locality is described in detail in a self-contained account, consisting of highlights (a précis of the special interest of the site), an introduction (with a concise history of previous work), a description, an interpretation (assessing the fundamentals of the site's scientific interest and importance), and a conclusion (written in simpler terms for the non-specialist). Each site report is a justification of a particular scientific interest at a locality, of its importance in a British or international setting, and ultimately of its worthiness for conservation.

The aim of the Geological Conservation Review series is to provide a public record of the features of interest in sites being considered for notification as Sites of Special Scientific Interest (SSSIs). It is written to the highest scientific standards but in such a way that the assessment and conservation value of the site is clear. It is a public statement of the value set on our geological and geomorphological heritage by the earth-science community which has participated in its production, and it will be used by the Joint Nature Conservation Committee, English Nature, the Countryside Council for Wales, and Scottish Natural Heritage in carrying out their conservation functions. The three country agencies are also active in helping to establish sites of local and regional importance. Regionally Important Geological/Geomorphological Sites (RIGS) augment the SSSI coverage, with local groups identifying and conserving sites which have educational, historical, research or aesthetic value, enhancing the wider earth science conservation perspective.

All the sites in this volume have been proposed for notification as SSSIs, the final decision to notify or renotify lies with the governing Councils of the appropriate country conservation agency.

Information about the GCR publication programme may be obtained from:

Earth Science Branch,
Joint Nature Conservation Committee,
Monkstone House,
City Road,
Peterborough PE1 1JY.

Titles in the series

1. **Geological Conservation Review**
   An introduction
2. **Quaternary of Wales**
   S. Campbell and D. Q. Bowen
3. **Caledonian Structures in Britain**
   South of the Midland Valley
   Edited by J. E. Treagus
4. **British Tertiary Volcanic Province**
   C. H. Emeleus and M. C. Gyopari
5. **Igneous Rocks of South-west England**
   P. A. Floyd, C. S. Exley and M. T. Styles
6. **Quaternary of Scotland**
   Edited by J. E. Gordon and D. G. Sutherland
7. **Quaternary of the Thames**
   D. R. Bridgland

Quaternary of Scotland

Edited by

J. E. Gordon
Scottish Natural Heritage,
Edinburgh, Scotland.

and

D. G. Sutherland
Edinburgh, Scotland.

GCR Editor: W. A. Wimbledon

CHAPMAN & HALL
London · Glasgow · New York · Tokyo · Melbourne · Madras

Published by Chapman & Hall, 2–6 Boundary Row, London SE1 8HN

Chapman & Hall, 2–6 Boundary Row, London SE1 8HN, UK

Blackie Academic & Professional, Wester Cleddens Road, Bishopsbriggs, Glasgow G64 2NZ, UK

Chapman & Hall Inc., 29 West 35th Street, New York NY10001, USA

Chapman & Hall Japan, Thomson Publishing Japan, Hirakawacho Nemoto Building, 6F, 1–7–11 Hirakawa-cho, Chiyoda-ku, Tokyo 102, Japan

Chapman & Hall Australia, Thomas Nelson Australia, 102 Dodds Street, South Melbourne, Victoria 3205, Australia

Chapman & Hall India, R. Seshadri, 32 Second Main Road, CIT East, Madras 600 035, India

First edition 1993

Typeset in ITC Garamond 10/12 by Columns Design & Production Services Ltd, Reading
Printed in Great Britain at the University Press, Cambridge

ISBN 0 412 48840 X

A catalogue record for this book is available from the British Library

Library of Congress Cataloging-in-Publication data

The Quaternary of Scotland / edited by J. E. Gordon, D. G. Sutherland. —
1st ed.
p. cm. – (The Geological conservation review series ; 6)
Includes index.
ISBN 0–412–48840–X
1. Geology, Stratigraphic–Quaternary. 2. Geology–Scotland.
I. Gordon, J. E. (John E.) II. Sutherland, D. G. (Donald G.)
III. Series.
QE696.Q3365 1993
551.7′9′09411–dc20 93–12124
CIP

Contents

Contents

Contents

Contents

Contributors

J. E. Gordon Scottish Natural Heritage, 2 Anderson Place, Edinburgh, EH6 5NP.

D. G. Sutherland 2 London Street, Edinburgh, EH3 6NA.

C. A. Auton British Geological Survey, Murchison House, West Mains Road, Edinburgh, EH9 3LA.

C. K. Ballantyne Department of Geography and Geology, University of St Andrews, Purdie Building, North Haugh, St Andrews, Fife, KY16 9ST.

J. Birnie Cheltenham and Gloucester College of Higher Education, Shaftesbury Hall, St George's Place, Cheltenham, Gloucestershire, GL50 3PP.

H. J. B. Birks Botanical Institute, University of Bergen, Allégaten 41, N–5007 Bergen, Norway.

C. J. Caseldine Department of Geography, University of Exeter, Amory Building, Rennes Drive, Exeter, EX4 4RJ.

R. A. Cullingford Department of Geography, University of Exeter, Amory Building, Rennes Drive, Exeter, EX4 4RJ.

P. R. Cundill Department of Geography and Geology, University of St Andrews, Purdie Building, North Haugh, St Andrews, Fife, KY16 9ST.

A. G. Dawson Department of Geography, Coventry University, Priory Street, Coventry, CV1 5FB.

K. J. Edwards Department of Geography, University of Birmingham, P.O. Box 363, Edgbaston, Birmingham, B15 2TT.

C. R. Firth Geography Section, West London Institute of Higher Education, Borough Road, Isleworth, Middlesex, TW7 5DU.

J. M. Gray Department of Geography, Queen Mary and Westfield College, Mile End Road, London, E1 4NS.

A. M. Hall Department of Geography, University of Edinburgh, Drummond Street, Edinburgh, EH8 9XP and Department of Geography and Geology, University of St Andrews, Purdie Building, North Haugh, St Andrews, Fife, KY16 9ST.

B. Huntley Department of Biological Sciences, University of Durham, Science Laboratories, South Road, Durham, DH1 3LE.

J. Jarvis Department of Geography and Geology, University of St Andrews, Purdie Building, North Haugh, St Andrews, Fife, KY16 9ST.

V. J. Jones Environmental Change Research Centre, Department of Geography, University College London, 26 Bedford Way, London, WC1H 0AP.

T. J. Lawson 12 Bonaly Grove, Colinton, Edinburgh, EH13 0PD.

J. J. Lowe Department of Geography, Royal Holloway and Bedford New College, Egham, Surrey, TW20 0EX.

Contributors

L. J. McEwen Cheltenham and Gloucester College of Higher Education, Shaftesbury Hall, St George's Place, Cheltenham, Gloucestershire, GL50 3PP.

J. W. Merritt British Geological Survey, Murchison House, West Mains Road, Edinburgh, EH9 3LA.

P. D. Moore Division of Life Sciences, Kings College London, Campden Hill Road, London, W8 7AH.

J. D. Peacock 18 Maclaren Road, Edinburgh, EH9 2BN.

W. Ritchie Department of Geography, University of Aberdeen, Elphinstone Road, Old Aberdeen, AB9 2UF.

D. E. Smith Department of Geography, Coventry University, Priory Road, Coventry, CV1 5FB.

A. C. Stevenson Department of Geography, The University, Newcastle-Upon-Tyne, NE1 7RU.

R. M. Tipping Nether Kidston Cottage, Nether Kidston Farm, By Peebles, EH4 8PJ.

M. J. C. Walker Department of Geography, St David's University College, Lampeter, Dyfed, SY48 7ED.

A. Werritty Department of Geography and Geology, University of St Andrews, Purdie Building, North Haugh, St Andrews, Fife, KY16 9ST.

G. Whittington Department of Geography and Geology, University of St Andrews, Purdie Building, North Haugh, St Andrews, Fife, KY16 9ST.

Acknowledgements

Work on this volume was initiated by the Nature Conservancy Council and has been seen to completion by the Joint Nature Conservation Committee on behalf of the three country agencies, English Nature, Scottish Natural Heritage and the Countryside Council for Wales. Each site description bears the name of its author(s), but we would like to acknowledge the many colleagues who gave help and advice. In addition to the named contributors, many members of the Quaternary community assisted with information or advice during site selection and documentation: Dr K. D. Bennett, Professor G. S. Boulton, Dr S. Campbell, Dr C. M. Clapperton, Dr R. Cornish, Dr E. A. FitzPatrick, Dr A. M. D. Gemmell, Dr B. A. Haggart, Dr W. G. Jardine, Dr R. P. Kirby, Dr J. K. Maizels, Dr J. H. Martin, Dr R. J. Price, Professor J. Rose, Dr J. B. Sissons, Professor D. E. Sugden, Dr J. A. T. Young and Professor W. B. Whalley. Dr G. P. Black and Dr W. A. Wimbledon also provided early stimulus for the project.

We are particularly grateful to Professor D. Q. Bowen and Professor J. D. Peacock for reviewing the entire text, and we also thank the following people for reviewing or revising parts of the text: Dr C. A. Auton, Dr C. K. Ballantyne, Dr S. Campbell, Dr C. R. Firth, Dr A. M. Hall, Dr J. W. Merritt, Professor J. Rose, Professor D. E. Sugden, Dr M. J. C. Walker and Professor W. B. Whalley.

It is a pleasure to acknowledge the Word Processing Department of the former Nature Conservancy Council for typing the original drafts, in particular Susan Bull, Shirley Drake, Kathy Harrison, Ann Murkett and Maureen Symons; also Peter Cann, Caroline Mee, Kevin Hayward and Tanya Jardine and the Library Services of the Nature Conservancy Council, Scottish Natural Heritage and the British Geological Survey (Edinburgh) for assistance with the bibliography and tracing references.

Special thanks go to the GCR publication production team: Dr D. O'Halloran (Project Manager), Neil Ellis (Publications Manager), Valerie A. Wyld (Sub-editor) and Nicholas D. W. Davey (Scientific Officer); and to Lovell Johns Ltd of Colwyn Bay for cartographic production.

The contributions by C. A. Auton and J. W. Merritt are published with permission of the Director of the British Geological Survey (NERC).

Access to the countryside

This volume is not intended for use as a field guide. The description or mention of any site should not be taken as an indication that access to a site is open or that a right of way exists. Most sites described are in private ownership, and their inclusion herein is solely for the purpose of justifying their conservation. Their description or appearance on a map in this work should in no way be construed as an invitation to visit. Prior consent for visits should always be obtained from the landowner and/or occupier.

Information on conservation matters, including site ownership, relating to Sites of Special Scientific Interest (SSSIs) or National Nature Reserves (NNRs) in particular counties or districts may be obtained from the relevant country conservation agency headquarters listed below:

Scottish Natural Heritage
12 Hope Terrace,
Edinburgh EH9 2AS.

Countryside Council for Wales,
Plas Penrhos,
Ffordd Penrhos,
Bangor,
Gwynedd LL57 2LQ.

English Nature,
Northminster House,
Peterborough PE1 1UA.

Foreword

One of the great insights of nineteenth century geology was the recognition that the environmental backdrop against which the drama of human evolution and history had been played was not static, as had been hitherto assumed, but had changed dramatically on relatively short timescales. The young Swiss geologist, Louis Agassiz, who played a major role in bringing about this change in thinking, visited Scotland in 1840 to advocate his new glacial theory, which suggested that the northern continents had suffered widespread glaciation in the recent past. Scottish geologists such as Lyell, Jamieson, MacLaren, Croll and Geikie were quick to pick up his ideas and, seeing ubiquitous evidence of change in their own country's dramatic landscape, led the world in exploring the implications of this revolutionary new concept. These pioneers established the flow patterns of the ice masses which had moulded the rock slabs of the Cuillin of Skye and had dispersed the rocks of Ailsa Craig into England; demonstrated rebound of crust after ice disappearance, which uplifted old shorelines around Oban and Mull high above modern sea level; and showed that great floods of meltwater from the decaying ice masses had produced the hummocky ridges on which many of Scotland's best golf courses are now built. They also showed that there had been rapidly alternating warm and cold periods in the past and that the ultimate drive for climatic change was the Earth's fluctuating orbit around the Sun.

Only recently, however, has the advent of techniques such as pollen analysis, uranium-series dating and radiocarbon dating been able to place a precise timescale on these events. They have revealed the dramatic overlap between an almost unimaginable geological past and a human present as reflected in prehistory and history, showing for example, that 11,000–10,000 years ago, when Jericho was a thriving city, the sites of many modern cities in Scandanavia, and many towns and villages in the Scottish Highlands, were overlain by more than a kilometre of ice.

It is only in the last two decades, with increasing public awareness of the fragility of the ecosystem, of the fact that well-protected botanical reserves appear to 'deteriorate', and of the vulnerability of the Earth's climate itself, that the significance of the geological record of Quaternary environmental change has been generally realised. The record tells us about the frequency and magnitude of natural change in the past, how mean annual temperatures have changed by as much as 5°C in the period of a human lifetime; how floral assemblages have changed rapidly, both in response to climatic change and without any apparent climatic drive; and how the composition of the atmosphere, including its 'greenhouse gases', has varied cyclically and in phase with the climate. The record also tells us about the frequency and magnitude of natural change that long-term dwellers on this Earth should expect, what the consequences of an increased atmospheric greenhouse gas concentration might be, and what processes should be taken into account in theories about future climate and environmental change.

This understanding is drawn from natural geological archives such as those represented by the sites described in this volume. Many of these archives have been well-read and understood: many others, no doubt, await new techniques or new insights before they yield up their secrets. Just as no civilised person would lightly destroy the books in an ancient library, no more should we lightly contemplate the destruction of this record of the past. However, roads need to be built, minerals need to be mined, food must be grown and people need to be housed, and Quaternary sediments are soft and easily destroyed or removed. Moreover, farmers, in their desire to improve their pastures, may wish to drain bogs containing superb records of past climate and ecological change, whilst elsewhere some of our finest surviving eskers are the most readily available source of sand and gravel for building. Clearly there are difficult decisions to be made about the balance between the need to preserve the geological archive and the need for us to use the land. Such decisions as these, which must be made as a result of debate involving the new natural heritage organisations, need information about the extent and nature of our heritage. This splendid volume is of fundamental importance in helping to define that heritage.

The Quaternary of Scotland documents the most important known Quaternary sites in Scotland and provides a basic factual archive, although there are, no doubt, other sites which are known which will prove to be equally important as a result of new insights and new methods, and others as yet undiscovered which will also join these ranks in the future. The site-by-site observational information described in this volume is associated with interpretations, which indicate the significance of each site in adding to our understanding. The site descriptions are incorporated into regional and Scottish syntheses, so that the role of the individual observations in determining the large-scale theoretical framework can be seen. So great is the amount of the data now available that few syntheses are able to go back to the primary observations, but are based on second and third hand sources. John Gordon and Donald Sutherland have not only done a great service to conservation but also to Quaternary geology in relating the facts to the interpretive framework. Much of the speculation may not survive changes in scientific fashion and theory, but the basic observations will.

The text has great clarity for such a complex subject and the quality of the illustrations is a reminder of that great lure to field science in Scotland: the beauty of the land.

Geoffrey Stewart Boulton FRS, FRSE
Regius Professor of Geology, The University of Edinburgh

Preface

STRUCTURE OF THE VOLUME AND TERMINOLOGY USED

This book contains scientific descriptions of 138 localities of national importance for Quaternary geology, geomorphology and environmental change in Scotland. It consists of two chapters that provide a general overview, followed by 16 regional chapters. The objective of the former chapters is to permit the reader to understand how the details of individual sites fit into the national scheme.

The locations of the regions are shown in Figure 1.1. Each of the regional chapters has a brief introduction which outlines the Quaternary geology and geomorphology and places the individual sites in their regional context. The individual site descriptions form the core of the book. In each chapter they are arranged, broadly, from oldest to youngest, although many of the sites cover significant periods of time. Each site report consists of a description of the evidence; interpretation of that evidence, with correlation, where relevant, with other localities; and assessment of the significance of the site in a regional, national or international context. Where sites form part of a wider network, then cross-reference is made to related sites to provide fuller understanding of the feature or period being discussed. In addition, where sites are of particular historical significance, then the history of study of the site is dealt with in detail.

There is at present no universally accepted system of terminology for the subdivision of Quaternary deposits in Britain. Mitchell *et al.* (1973) proposed a correlation scheme based on standard stages. Since that date, however, not only has there been a great increase in knowledge of the Quaternary succession so that the 1973 system is now incomplete, but also certain of the stage names proposed at that time have been questioned as to their suitability or even existence. To avoid confusion, therefore, Table 1 and Figure 2.7 have been compiled to show the terminology and approximate accompanying chronology that is used in this book; a simplified summary chart showing the position of each site in the chronology is given in Table 2. The basis of the chronology is the oxygen isotope signal recognized in deep-sea sediments. This signal has been shown to be a function of the Earth's orbital parameters (Hays *et al.*, 1976), and astronomical data have been used to 'tune' the geological time-scale (cf. Imbrie *et al.*, 1984; Prell *et al.*, 1986; Ruddiman *et al.*, 1986, 1989; Martinson *et al.*, 1987). For the period back to about 620 ka, the time-scale is that developed by Imbrie *et al.* (1984), which has been substantiated by later work (Prell *et al.*, 1986; Shackleton *et al.*, 1990). For the earlier part of the Quaternary, the revised time-scale of Shackleton *et al.* (1990) is adopted.

Preface

Table 1 Terminology used in the subdivision of the Late Pleistocene and Holocene

			Age (years BP)	OI Stage
Holocene (late) (middle) (early)				1
			10,000	
Late Devensian	Lateglacial	Loch Lomond Stadial		
			11,000	
		Lateglacial Interstadial		
			12,000	
			13,000	2
			24,000	
			26,000	
Middle Devensian				3
			59,000	
Early Devensian				4
			71,000	
				5a-5d
			116,000	
Ipswichian				5e
			128,000	

Where radiocarbon 'dates' (age estimates) are cited, they are quoted in radiocarbon years before present (AD 1950). It should be noted that the radiocarbon time-scale diverges from the calendrical one, and although calibration is available back to 9000 years in detail (cf. Pilcher, 1991) and to 30,000 years in outline (Bard *et al.*, 1990), the interpretation of radiocarbon measurements,

Table 2 Summary of stratigraphical positions of sites described in this volume. Sites appear more than once where they have multiple interests or interests of different ages. Sites with features pre-dating the Late Devensian are grouped together because of uncertainties over dating

		Shetland Western Isles	Orkney Caithness North-west Highlands	Inverness Area North-east Scotland Eastern Grampians	South-west Highlands Inner Hebrides	Western Highland Boundary Eastern Highland Boundary Fife and Lower Tay	Western Central Lowlands Lothians and Borders South-west Scotland
Holocene		Garths Voe Ronas Hill Borve Gleann Mór	Ward Hill Loch of Winless An Teallach Sgùrr Mór Loch Sionascaig Lochan an Druim Loch Maree	Dores Barnyards Munlochy Valley Ben Wyvis Findhorn Terraces Muir of Dinnet Philorth Valley The Cairngorms Abernethy Forest Loch Etteridge Allt na Feithe Sheilich Coire Fee Glen Feshie Morrone	Glenacardoch Point Kingshouse Pulpit Hill Loch Cill an Aonghais Eas na Broige Western Hills of Rum West Coast of Jura Gribun Loch an t-Suidhe Loch Ashik Loch Cleat Loch Meodal	South Loch Lomond Rhu Point Western Forth Valley Mollands Tynaspirit Dryleys Maryton Milton Ness Stormont Loch Carey Silver Moss Pitlowie Kincraig Point Black Loch	Dundonald Burn Tinto Hill Din Moss Newbie Loch Dungeon Round Loch of Glenhead
Late Devensian	Loch Lomond Stadial	Burn of Aith Ronas Hill	Ward Hill Loch of Winless Achnasheen An Teallach Baosbheinn Beinn Alligin Cnoc a'Mhoraire Coire a'Cheud-chnoic Creag nan Uamh Cam Loch Lochan an Druim	Coire Dho Fort Augustus Dores Barnyards Munlochy Valley Ben Wyvis Muir of Dinnet The Cairngorms Lochnagar Loch Etteridge Morrone Glen Feshie Coire Fee	Isle of Lismore Moss of Achnacree South Shian Glen Roy Pulpit Hill The Cuillin Beinn Shiantaidh Western Hills of Rum Northern Islay? West Coast of Jura? Loch an t-Suidhe Loch Ashik	Aucheneck Croftamie Gartness South Loch Lomond Rhu Point Western Forth Valley Tynaspirit Stormont Loch Black Loch	Loch Skene Beanrig Moss Dunbar Tauchers Bigholm Burn Redkirk Point
Late Devensian	Lateglacial Interstadial	Burn of Aith	Loch of Winless Cam Loch Lochan an Druim	Ardersier Findhorn Terraces The Cairngorms Lochnagar Loch Etteridge Abernethy Forest Morrone Glen Feshie	Glenacardoch Point South Shian Pulpit Hill West Coast of Jura Loch an t-Suidhe Loch Ashik	Croftamie Gartness South Loch Lomond Geilston Rhu Point Tynaspirit Dryleys Milton Ness North Esk & West Water Stormont Loch Inchcoonans & Gallowflat Kincraig Point Black Loch	Beanrig Moss Bigholm Burn Redkirk Point
Late Devensian		North-west Coast of Lewis Port of Ness Tolsta Head Glen Valtos	Den Wick? Mill Bay? Baile an t-Sratha? Drumhollistan? Leavad? Gairloch Moraine An Teallach Corrieshalloch Gorge	Clava Ardersier Struie Channels Kildrummie Kames Littlemill Torvean Findhorn Terraces Boyne Quarry Teindland Castle Hill Kippet Hills Muir of Dinnet Kirkhill Bellscamphie The Cairngorms Loch Etteridge Glen Feshie	The Cuillin Scarisdale Beinn Shiantaidh? Northern Islay West Coast of Jura	Croftamie Gartness Geilston Nigg Bay Burn of Benholm Almondbank Shochie Burn North Esk & West Water	Afton Lodge Nith Bridge Greenock Mains Carstairs Kames Clochodrick Stone Falls of Clyde Agassiz Rock Hewan Bank Keith Water Carlops Rammer Cleugh Port Logan Bigholm Burn
Pre-Late Devensian		Fugla Ness Sel Ayre North-west Coast of Lewis Tolsta Head	Den Wick? Mill Bay? Muckle Head & Selwick Baile an t-Sratha? Drumhollistan ? Leavad? Corrieshalloch Gorge? Creag nan Uamh	Clunas Dalcharn Allt Odhar Clava Windy Hills Moss of Cruden Pittodrie Hill of Longhaven Kirkhill Bellscamphie Boyne Quarry? Teindland Castle Hill? The Cairngorms	Tangy Glen Glenacardoch Point Isle of Lismore? Northern Islay West Coast of Jura	Nigg Bay Burn of Benholm Milton Ness Kincraig Point	Afton Lodge Dunbar

particularly during parts of the Late Devensian is additionally complicated (cf. Ammann and Lotter, 1989; Zbinden *et al.*, 1989).

The informal term 'Lateglacial' (equivalent to Devensian late-glacial) is well established in the Scottish Quaternary literature and is used throughout this volume following the definitions of Gray and Lowe (1977a). The terms Lateglacial Interstadial and Loch Lomond Stadial are also used. These are climate-stratigraphic, or climatostratigraphic, terms, and as such differ from chronostratigraphic, or time-stratigraphic terms. The latter are intervals of time based on a definition tied to a particular rock-sequence. Climate-stratigraphic terms, however, are based on climatic inferences drawn from rocks, either at a site, or from several sites. The terms Lateglacial Interstadial and Loch Lomond Stadial describe the inferred nature of the climate towards the end of the Devensian Stage. In general terms, the former relates to the time between approximately 13,000 and 11,000 years ago, a time of overall climatic improvement, whereas the latter refers to the time between approximately 11,000 and 10,000 years ago, which corresponds to a time of climatic deterioration.

Comparison with the nomenclature used in Europe shows that the Lateglacial Interstadial corresponds with the Oldest Dryas, Bølling, Older Dryas and Allerød events. The Loch Lomond Stadial corresponds with the Younger Dryas. Attention is drawn to the latter in particular, in view of its importance as an international term in studies seeking to understand the Earth's climate system.

Finally, where the usage of certain local terms for particular landforms or deposits is widely accepted in the literature, these have been retained in the present volume; for example, corrie (cirque) and carse (estuarine silts and clays). Where possible, modern names of marine mollusca are used, following Smith and Heppell (1991).

Chapter 1

Introduction

J. E. Gordon and D. G. Sutherland

RATIONALE FOR CONSERVATION AND SELECTION OF QUATERNARY SITES IN SCOTLAND

The most striking feature of the Scottish landscape is the wide variety of landforms represented in a relatively small geographical area. The rugged Highlands with their accentuated relief contrast with the surrounding lowlands and the more rolling hills of the Southern Uplands and the Midland Valley. Further variety is introduced in the distinctive landscapes of the western and northern island groups and in the rich diversity of scenery around Scotland's coasts. This varied topography largely reflects the interplay of geological controls, geomorphological processes and the effects of climatic change, most recently during the Quaternary. When combined with the prevailing climate, the geological legacy has produced a complex natural environment which incorporates a remarkable geographical diversity of plant communities, soils and geomorphological processes. By virtue of Scotland's position on the extreme Atlantic fringe of north-western Europe, allied with its geomorphological diversity, many aspects of the natural environment are unique and demand to be conserved.

Diversity is also a significant theme in a temporal sense: the present landscape is the product of a long history of evolution which reflects the interaction of geology, topography, climate, geomorphological processes and their changes through time. The study of the evolution of the modern Scottish environment during the Quaternary has revealed a fascinating sequence of events that range from the shaping of many of the major elements of the landscape by glacial erosion to the establishment of the present vegetation cover after the last period of glaciation. These events are of considerable intrinsic scientific interest, not only because they are a particularly significant part of recent Earth history, but also because it has been possible to establish that, at various periods during the Quaternary, environments and their accompanying plant communities existed that have no known modern analogues. Knowledge of Quaternary history is important also in that it provides direct evidence for the rate at which natural processes can occur, in particular the response of geomorphological processes and plant communities to both major climatic deteriorations and ameliorations. Such information may become increasingly important as attempts are made to predict the likely effect of future natural or man-induced climatic changes. It is also in this last context that Quaternary science has a unique value, because it is only with a detailed knowledge of the natural environment that had developed prior to human impact that the full extent of that impact can be assessed, whether in terms of prehistoric forest clearance or modern industrial pollution.

Site selection guidelines

For the Quaternary scientist concerned about conservation, the wealth of detail present in the Quaternary geological record or preserved in the landforms presents problems as well as opportunities. Decisions as to which sites should be conserved have been made on the basis of a number of guidelines which try to encapsulate the range of scientific interest. These guidelines are: uniqueness; classic examples; representativeness; being part of a site network; providing understanding of present environments; historical importance; and research potential and educational value. Individual sites may fall into one or more of the categories: other things being equal, sites with multiple interests were favoured.

Certain sites are unique. They are either the only known representatives of particular parts of the geological record or they may be known, as part of the landscape, to have no comparable counterparts in Scotland or even internationally. Examples of the former are Fugla Ness, Kirkhill, North-west Coast of Lewis and Tangy Glen. The latter are best exemplified by Glen Roy and the Cairngorms.

Some sites are nationally or internationally recognized as being classic examples of particular features and are quoted in standard textbooks. Examples include Glen Roy, Northern Islay, West Coast of Jura and Carstairs Kames.

Other sites are representative of important aspects of geomorphology, landscape evolution or environmental change during the Quaternary in Scotland. Certain sites have therefore been selected because they are the best studied, are the best preserved and/or have the most complete local representation of phenomena that are quite widespread. They are therefore important reference sites for the particular phenomena or area concerned. Examples are sites concerning glacial deposits or landforms such as end and hummocky moraines, meltwater channels, kames, kettle holes, and eskers; others include representative or informally recognized reference sites for Quaternary stratigraphy.

Where there is a strong geographical component in the scientific interest, a series or network of sites has been chosen to include different aspects of one general type of phenomenon that shows significant regional variations in its characteristics, for example in relation to factors such as geology, climate or relief. Such networks may comprise unique, classic or representative sites. Prime examples are in vegetational change where, for example, the timing of the spread of trees following the climatic amelioration at the end of the Late Devensian varied throughout the country, as did the pattern of vegetation which that spread produced over a period of 3000 to 5000 years. Clearly, no one site in any part of the country can encapsulate such an aspect of the Quaternary history. Examples of the phenomena to which this guideline has been applied are mountain-top periglacial deposits and landforms, Lateglacial and Holocene vegetational and environmental change and Lateglacial and Holocene changes in sea level.

Certain sites are of particular importance because of the light they throw on the development of the present ecological landscape. An example is the occurrence of the arctic–alpine plant refuges on certain mountains, the nearest neighbours of which may be in Norway. Sites such as Coire Fee and Morrone demonstrate that these plants have survived on the Scottish mountains during the mild climate of the Holocene since the cold climate of the Late Devensian, when they were much more widespread in their occurrence. A very different example is that of the degradation of the environment by modern pollutants. The full extent of acidification of lochs by industrial emissions can only be known if the pH levels are known prior to the start of the pollution. A further question is the sensitivity of certain parts of the environment to environmental change, including human impact. This is exemplified by upland geomorphological processes and certain sites have provided basic information as to the rates of change that have occurred during the Holocene. Comparison of modern rates and past rates reveals whether the former are anomalous and give cause for concern.

If the above is a justification for conserving sites that preserve specific scientific evidence of certain Quaternary events, then in Scotland the history of the development of Quaternary science is a further important reason for the conservation of many Quaternary sites. The glacial theory was more widely accepted during the middle of the last century by the Earth science community in Scotland than in Britain in general, and over the 50-year period between 1840 (when the glacial theory was introduced by Agassiz) and 1890 many of the concepts related to glacial landforms, sediments and the interaction between solar radiation, climate, ice-sheets and sea-level change were elaborated with respect to the Scottish glacial deposits and landforms. Principal among these ideas may be set the advocacy of glacio-eustasy (MacLaren, 1842a), glacio-isostasy (Jamieson, 1865), multiple ice-sheet glaciation (Croll, 1870b, 1875; Geikie, 1894) controlled principally by variations in the Earth's orbit around the Sun (Croll, 1867, 1875, 1885) and the idea that the ocean currents were the principal agency for transporting heat from the tropics to the polar regions and were hence a fundamental control on the climate of the world (Croll, 1875). In addition, one of the legacies of this period has been the adoption of various Scots words as formal glacial-geological and geomorphological terms such as 'till', 'kame', 'kettle' and 'drumlin'. Certain of the Quaternary sites in Scotland are therefore of major significance in the history of geology and deserve to be conserved on this basis alone. Other sites have had a long history of research and have played a fundamental role in the development of new ideas and interpretations of Quaternary events, chronology or landscape processes.

A final justification of many sites is the interpretation or interpretations, frequently controversial, that have been placed upon them. Such sites may illustrate the development of scientific thinking on the subject of landscape history and, indeed, the debates, for example about process or chronology, that characterize certain areas of Quaternary science. It is important that such sites continue to be available for further study and to stimulate active scientific debate. Other sites provide fundamental baselines, for example in dating or as stratigraphical markers, and must remain accessible for reference. There are many outstanding questions not yet resolved in the Quaternary of Scotland, as this volume will show, but in many respects this is a strength and not a weakness of the science and will hopefully stimulate further generations to enquire in depth about the evolution of Scotland during the

Quaternary. Although new sites will become available, it is important that existing sites are maintained for the application of new research techniques. The long-term research potential of many sites is therefore a key factor in their selection. Finally, the educational importance of many of the sites should not be overlooked, and in total the coverage provides a history of the evolution of the Scottish landscape as recorded in its constituent landforms and sediments.

The present work is a compilation of all the sites in Scotland that merit conservation for their significance to Quaternary science (Figure 1.1). Coastal and fluvial geomorphology, in the sense of modern process studies, and large-scale mass movement features are reviewed in their own thematic volumes of the GCR. Site selection has been based on identifying the minimum number of sites necessary to represent the diverse history and form of the Scottish landscape; direct duplication of interests has therefore been minimized. Extensive consultations on site assessment were carried out with the appropriate specialists in Quaternary science.

Structure of the volume

In the chapters that follow, the sites are arranged in broad geographical areas (Figure 1.1), each with a brief introduction giving an overview of the presently understood Quaternary history of the area and highlighting the particular aspects of that history which are scientifically important. The individual site reports include syntheses of the currently available scientific documentation, and the interpretations of the site. A key part of each report is the assessment, which explains why the site is important. Where sites form part of a wider network, cross-reference is made to related sites to provide a fuller understanding of the event or period being discussed. In addition, where sites are of particular historical significance, the history of study of the site is dealt with in detail. Chapter 2 provides an overview of the Quaternary history of Scotland as understood from the available evidence.

INTRODUCTION TO THE QUATERNARY

Character of the Quaternary

The Quaternary is the portion of the late Cenozoic Era of geological time that spans approximately the last 1.6 Ma; the greater part of it is known as the Pleistocene and the last 10 ka as the Holocene. In a strict geological sense the base of the Quaternary, at the Pliocene–Pleistocene boundary, is defined in a section at a type locality at Vrica in Italy and dated at about 1.64 Ma (Aguirre and Pasini, 1985). The terms Pleistocene and Quaternary have often been used synonymously with 'Ice Age'. However, it is now known from many parts of the world that glacier advances occurred before the start of the Pleistocene *sensu stricto*, and recent studies of ocean-floor sediments have indicated significant climatic deterioration and initiation of major Northern Hemisphere glaciation at around 2.4 Ma (Shackleton *et al.*, 1984; Ruddiman and Raymo, 1988). This abrupt onset of the late Cenozoic ice ages is, as yet, unexplained, although changes in insolation associated with orbital periodicities (eccentricity of the Earth's orbit, tilt of the Earth's axis and precession of the Earth's axis) are now established as the driving force of changes in the Earth's climatic system (Hays *et al.*, 1976; Imbrie *et al.*, 1984; Berger, 1988). The ocean-floor sediments have also revealed that the late Cenozoic was characterized by many fluctuations in climatic conditions, with up to 50 major 'warm' and 'cold' oscillations recognized during the last 2.4 Ma (Ruddiman and Raymo, 1988). The cold phases are usually described as ice ages or glaciations, and the warmer interludes as interglacials. The ice ages were not unbroken in their frigidity, however, since the exceptionally cold phases (stadials) were interrupted by warmer interludes (interstadials) lasting for a few thousand years.

Factors controlling Quaternary climatic change

Both the duration and the intensity of the climatic cycles have varied through time (Ruddiman and Raymo, 1988; Ruddiman *et al.*, 1989). During the period 2.4–0.9 Ma, the climate was dominated by 41 ka cycles (corresponding to

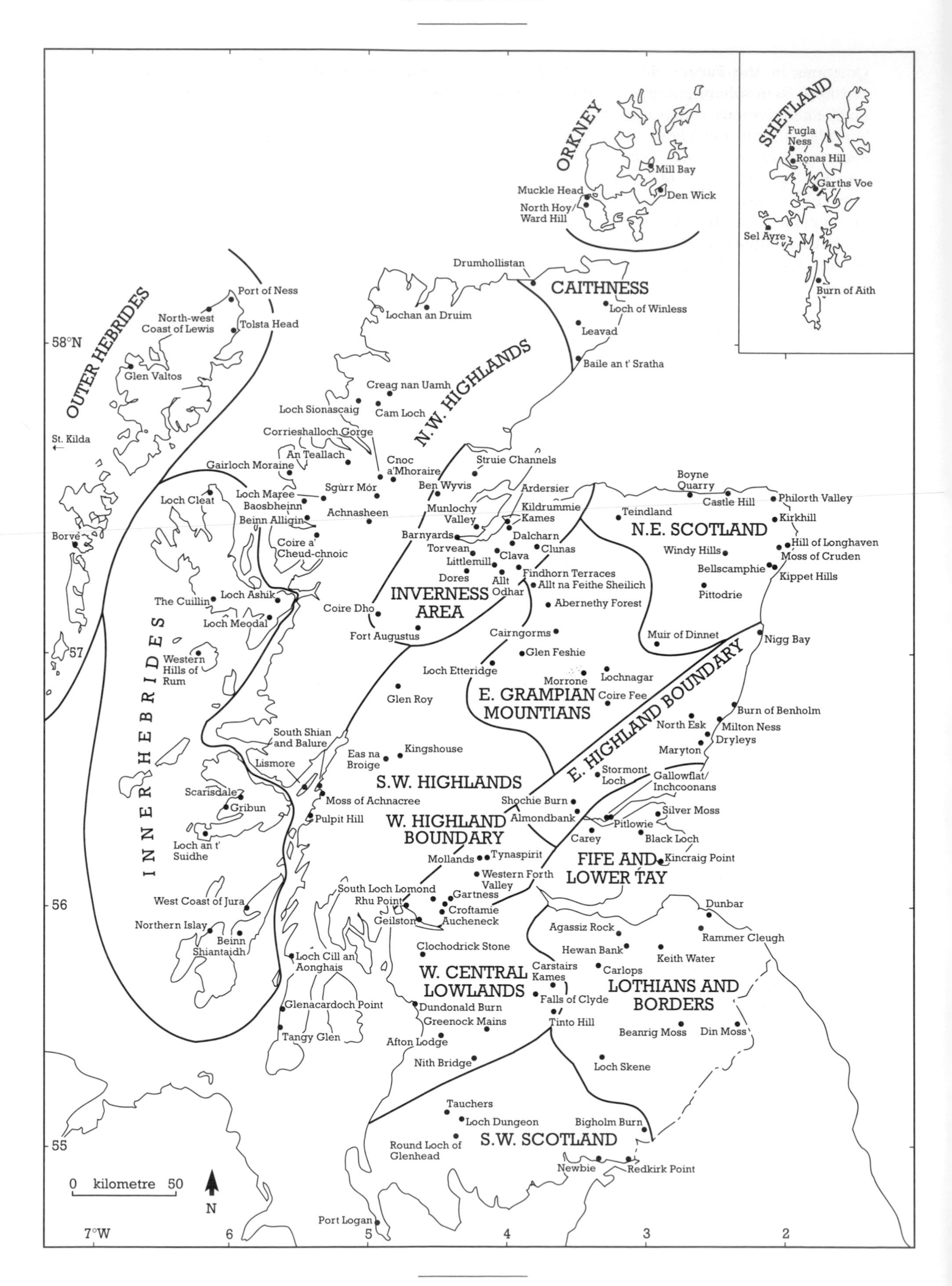
ORKNEY
SHETLAND
Mill Bay
Den Wick
Muckle Head
North Hoy/ Ward Hill
Fugla Ness
Ronas Hill
Garths Voe
Sel Ayre
Burn of Aith
OUTER HEBRIDES
Port of Ness
North-west Coast of Lewis
Tolsta Head
Glen Valtos
St. Kilda
Borve
58°N
57
56
55
7°W
6
5
4
3
2
Drumhollistan
CAITHNESS
Loch of Winless
Leavad
Baile an t' Sratha
Lochan an Druim
N.W. HIGHLANDS
Creag nan Uamh
Loch Sionascaig
Cam Loch
Corrieshalloch Gorge
An Teallach
Gairloch Moraine
Cnoc a'Mhoraire
Struie Channels
Loch Cleat
Loch Maree
Sgùrr Mór
Ben Wyvis
Ardersier
Baosbheinn
Achnasheen
Munlochy Valley
Kildrummie Kames
Beinn Alligin
Coire a' Cheud-chnoic
Barnyards
Torvean
Dalcharn
Clunas
Littlemill
Clava
Dores
Allt Odhar
Findhorn Terraces
Allt na Feithe Sheilich
Abernethy Forest
The Cuillin
Loch Ashik
Loch Meodal
Coire Dho
INVERNESS AREA
Fort Augustus
Cairngorms
Glen Feshie
Loch Etteridge
Morrone
Lochnagar
Coire Fee
E. GRAMPIAN MOUNTIANS
INNER HEBRIDES
Western Hills of Rum
Glen Roy
South Shian and Balure
Eas na Broige
Kingshouse
Lismore
S.W. HIGHLANDS
Scarisdale
Gribun
Moss of Achnacree
Pulpit Hill
W. HIGHLAND BOUNDARY
Loch an t' Suidhe
Mollands
Tynaspirit
Western Forth Valley
South Loch Lomond
Rhu Point
Gartness
Croftamie
Geilston
Aucheneck
West Coast of Jura
Northern Islay
Beinn Shiantaidh
Loch Cill an Aonghais
Clochodrick Stone
Glenacardoch Point
Tangy Glen
W. CENTRAL LOWLANDS
Dundonald Burn
Greenock Mains
Afton Lodge
Nith Bridge
Boyne Quarry
Castle Hill
Philorth Valley
Teindland
Kirkhill
N.E. SCOTLAND
Windy Hills
Hill of Longhaven
Moss of Cruden
Bellscamphie
Kippet Hills
Pittodrie
Muir of Dinnet
Nigg Bay
E. HIGHLAND BOUNDARY
Burn of Benholm
North Esk
Milton Ness
Dryleys
Maryton
Stormont Loch
Gallowflat/ Inchcoonans
Shochie Burn
Almondbank
Silver Moss
Pitlowie
Carey
Black Loch
FIFE AND LOWER TAY
Kincraig Point
Dunbar
Agassiz Rock
Rammer Cleugh
Hewan Bank
Keith Water
Carstairs Kames
Carlops
LOTHIANS AND BORDERS
Falls of Clyde
Tinto Hill
Beanrig Moss
Din Moss
Loch Skene
Tauchers
Loch Dungeon
Bigholm Burn
Round Loch of Glenhead
S.W. SCOTLAND
Newbie
Redkirk Point
Port Logan
0 kilometre 50
N

variations in the Earth's tilt). After 0.9 Ma, the Northern Hemisphere glaciation intensified, while a 100 ka periodicity (corresponding to variations in orbital eccentricity) became stronger and then dominant after 0.45 Ma. Also, during the last 0.45 Ma, the influence of the 41 ka and 23 ka (corresponding to variations in precession) cycles, superimposed on the longer term fluctuations, has been identified in the geological record. The driving forces for these patterns, linking the relatively small orbitally controlled variations in solar radiation with major climatic change and ice-sheet growth, undoubtedly reflect a complex set of interactions and feedbacks involving the atmosphere, oceans, cryosphere and global tectonics. It is clear that the climatic record is not simply a linear function of astronomical forcing (e.g. Broecker and Denton, 1989; Overpeck *et al.*, 1989; Rind *et al.*, 1989). Several potentially key links have been identified, although a unifying theory awaits further research. The case for global tectonics, particularly the impact of uplift in the American south-west and the Tibetan Plateau on Northern Hemisphere atmospheric circulation patterns, has been argued by Ruddiman and Raymo (1988) and Ruddiman and Kutzbach (1990). In the model of Broecker and Denton (1989, 1990) changes in the world's oceans and their links with the atmosphere provide a potential means of coupling the orbital changes with ice-sheet fluctuations. According to Broecker and Denton, orbitally induced changes in the intensity of the seasons influence water vapour transfer and the salinity structure of the oceans, producing major reorganizations in global ocean circulation and hence in climate. Accompanying changes in the interchange of carbon dioxide between the oceans and atmosphere may provide a means of amplifying the orbital forcing through the ocean–atmosphere system (cf. Saltzman and Maasch, 1990). It has been shown that changes in the atmospheric content of that gas follow the orbital periodicities (e.g. Barnola *et al.*, 1987), but precede changes in ice volume (Shackleton *et al.*, 1983; Shackleton and Pisias, 1985). However, it is apparent from modelling experiments that reductions in atmospheric carbon dioxide alone are inadequate to produce the inferred magnitude of global cooling (Broccoli and Manabe, 1987; Broecker and Denton, 1989). In addition, there are other factors which may modulate the growth and decay of ice-sheets, involving non-linear interactions between ice, climate, bedrock and sea level (cf. Sugden, 1987, 1991). Examples proposed have included the inherent instability of marine-based ice-sheets (e.g. Hughes, 1987; van der Veen, 1987; Jones and Keigwin, 1988), the effects of topography on ice-sheet growth (e.g. Payne and Sugden, 1990b), and the combined effects of climatic warming and isostatic depression in producing accelerated ice-sheet melting (Hyde and Peltier, 1985).

Figure 1.1 Location of sites and areas described.

Evidence for subdivision of the Quaternary

One of the major developments in Quaternary studies in the last 20 years has been the recovery of sediment cores from the floors of the world's oceans and the resolution of the climatic and other environmental records which they contain (Bowen, 1978; Imbrie and Imbrie, 1979). In contrast to the fragmentary terrestrial records, those from the ocean floors have a major advantage in being longer and more continuous, and for the first time have allowed a comprehensive picture of climatic variation during the Quaternary. Three indices have been employed to reconstruct this climatic record: variations in oxygen isotope ratios in the shells of marine microfossils (which in large part reflect changes in ice-sheet volume (Shackleton and Opdyke, 1973)), variations in sea-surface temperatures inferred from assemblages of these microfossils, and variations in the percentages of $CaCO_3$ (higher during warmer episodes) and ice-rafted or wind-blown continental detritus (higher during colder episodes) (Ruddiman *et al.*, 1986). Complementary records of climatic fluctuations have been obtained from studies of the oxygen isotope ratios in polar ice-sheets (Johnsen *et al.*, 1972; Robin, 1983; Jouzel *et al.*, 1987, 1990; Oeschger and Langway, 1989). Although these latter relate to relatively short time-scales, over the last 160 ka, and their interpretation may be subject to more constraints (see Paterson, 1981), they are nevertheless a key data source for Quaternary scientists. One particularly important conclusion to emerge is the rapidity of climatic change; for example, Greenland ice-cores indicate a warming of 7°C in about 50 years at the end of the Younger Dryas (Dansgaard *et al.*, 1989).

The deep-sea sediments, and in particular the

variations in oxygen isotope ratios with depth which they contain, have also provided the foundation for a subdivision of Quaternary time that has global applicability (see Bowen, 1978). This subdivision is based on the recognition of a series of oxygen isotope stages (cf. Figure 2.7), beginning with the Holocene (Stage 1) and numbered back in time with even numbers for glacial periods and odd numbers for interglacials. These stages are dated in relation to the established boundaries defining changes in the direction of polarity of the Earth's magnetic field and tuned according to the time-scale of orbital periodicities (Imbrie *et al.*, 1984; Martinson *et al.*, 1987; Shackleton *et al.*, 1990). The problem remains, however, to relate the fragmentary terrestrial sedimentary record to the oxygen isotope time-scale, particularly back beyond the last interglacial.

In Britain the extent of the area covered by ice varied during different glaciations. During the last (Late Devensian) glaciation ice extended as far south as the north Midlands and just impinged on the north coast of East Anglia. During some earlier glaciations ice extended farther south but probably never beyond the Thames Valley. However, the sedimentary record of the Quaternary in Britain is incomplete. The most comprehensive sequences of deposits occur in East Anglia and the Midlands, where a series of glacials and interglacials has been recognized. These are individually named, sometimes after particular reference localities, and provide the type sites (stratotypes) for stratigraphic correlation in Britain. Over the remainder of the country, the period during and since the last (Ipswichian) interglacial has been reconstructed in greatest detail, reflecting the more widespread preservation of younger sediments.

Quaternary environments

A fundamental feature of the Quaternary is the predominance of extreme climatic change, and as the climate fluctuated, so too did environmental conditions. Particular types of environmental change have left a strong imprint in the landforms, fossils and recent sedimentary deposits of Britain. By studying these features and by making comparisons with modern analogues, Quaternary scientists have made considerable progress in unravelling the past. They have shown that a wide range of landforms and sediments produced by ice erosion and deposition distinguish the glaciations. During the melting of the ice-sheets the liberation of vast volumes of meltwater produced an equally characteristic suite of water-lain glaciofluvial landforms and deposits. In the areas that lay beyond the ice-sheets, and also during less severe cold phases when glaciers were either restricted in their distribution or absent altogether, periglacial conditions prevailed. These are characterized by frost-assisted processes and by a range of frost and ground-ice generated landforms and deposits. Mass wasting (downslope movement of soil on both large and small scales) and increased wind action were prevalent and also produced a range of diagnostic features. In parts of upland Britain periglacial processes are still active today. As reflected in the fossil record, the flora and fauna of the cold periods show restricted diversity of species and, not surprisingly, the predominance of cold-tolerant types.

Interglacials have a sedimentary record characterized by the absence of glacial, periglacial and glaciofluvial features. They are often distinguished by periods of chemical weathering, soil formation or the accumulation of organic material. Changes in the amounts and types of pollen grains preserved in organic deposits have been used to define systems of pollen zones, each zone being characterized by particular vegetation types, allowing climatic and environmental conditions to be inferred. Progressive changes in vegetation through time can be summarized by sequences of pollen zones. In addition, environmental conditions were different in different interglacials, and hence the presence of particular types of pollen can be diagnostic of particular interglacials. Similarly, different mammal faunas occurred in different glacials and interglacials, and some species are diagnostic of particular glacials or interglacials. Both terrestrial and marine molluscs and Coleoptera (beetles) are also useful in reconstructing past environmental conditions by analogy with their present-day environmental tolerances and geographic ranges.

Running parallel with the growth and decay of ice-sheets, significant changes have occurred in the coastal zone of Britain associated with a complex interplay of changing land and sea levels. World sea level has varied according to the volume of water locked up within the ice-sheets, being lower during glacials than interglacials. The level of the land has also varied, sinking under the weight of advancing ice-sheets and rising up again when they melted. Such changes are evident in beaches, shore platforms and marine sediments now raised above present sea level. Some of the more important examples have been given particular names, and some have been dated or assigned to particular subdivisions of the Quaternary. Submerged shoreline features and drowned valleys also point to relatively lower sea levels in the past. Changes in river courses and channel patterns have followed changes in discharge, sediment supply and sea level. There have been times when rivers built up large thicknesses of glacially derived debris on their floodplains, and others when they eroded down into their floodplains. The resulting effects on the landscape are 'staircases' of terraces in many river valleys. In some cases the more important terraces have been individually named and dated with reference to a range of fossil materials.

Change through time is a fundamental aspect of Quaternary studies. Very often traces of successive environments are recorded in layers of sediment preserved on top of one another; for example, glacial deposits may overlie interglacial beach deposits and in turn be succeeded by periglacial slope deposits and later sand dunes. Sites with such sequences can provide particularly revealing perspectives on the Quaternary. Although a temporal theme has been emphasized, it is also the case that Quaternary environmental changes and associated processes have not been uniform in their operation throughout Britain, and allied with the variety of the geology, this has produced a remarkable regional diversity in surface landforms and deposits. This has been a key factor in compiling the national network of GCR sites representing the Quaternary.

Approaches to Quaternary science

The fundamentals of Quaternary science are explained in a range of texts (for example, West, 1977; Bowen, 1978; Birks and Birks, 1980; Lowe and Walker, 1984; Bradley, 1985; Dawson, 1992). In brief, key aims are (1) to establish the stratigraphy of the deposits and to effect their correlation and classification, and (2) to interpret and explain the evolution of the landscape during the Quaternary, the history of geomorphological events and processes, climatic and environmental change, and the development of the flora and fauna. It is also important to understand the relationships between each of these aspects and how they have varied both geographically and through time. Quaternary science is therefore multidisciplinary, involving the combined efforts of geologists, geographers, geomorphologists, botanists, zoologists and archaeologists. The basis of the geological approach is stratigraphy, that is correlating between individual localities using sediments or fossils. Other approaches can involve reconstruction of environmental change or spatial analysis, for example of landforms, ice movements or vegetation patterns. Certain sites are recognized, either formally or informally, to be important reference localities because they demonstrate, for example, particular events, sequences of environmental changes, types of sediment, or contain datable materials. Such sites provide the standards for future studies. Sites with organic or fossil remains are also highly valued as rich sources of information about past environments and yield material for dating.

Chapter 2

The Quaternary in Scotland

D. G. Sutherland and J. E. Gordon

PRE-GLACIAL LANDFORM INHERITANCE AND THE EFFECTS OF GLACIAL EROSION

The broad outlines of the Scottish landscape owe their origins to a combination of geological and tectonic controls and geomorphological processes in pre-Quaternary time. However, there are few on-land deposits that pre-date the Quaternary and the great majority of the relict features in the landscape are erosional and therefore difficult to date. Consequently there has been a range of interpretations and reconstructions concerning the existence, ages and origins of erosion surfaces, drainage systems and weathered bedrock (Linton, 1951b; Walton, 1964; George, 1965, 1966; Godard, 1965; Sissons, 1967a, 1976b; Haynes, 1983; Hall, 1991). Nevertheless, detailed regional studies, together with consideration of the sediments deposited on the adjacent continental shelves, have allowed progress to be made in recent years (Hall, 1983, 1986, 1991; Le Coeur, 1988) in understanding the longer-term evolution of the present landscape and its large-scale features.

Study of deep-sea cores recovered from the floor of the North Atlantic has suggested that the first mid-latitude ice-sheet glaciation during the present sequence of ice ages occurred around 2.4 Ma (Shackleton *et al.*, 1984; Loubere and Moss, 1986). Although no direct evidence has been found for glaciation of Scotland or the neighbouring shelves at that time, the relatively minor decline in temperature that would be necessary for glaciers to develop in Scotland suggests glaciation is likely to have occurred at least in the Highlands.

The general character of the topography on which the earliest glaciers developed may not have been too different from that of today, with a mountainous western Highland zone with a range of relief not dissimilar to that of the present and fringing lowlands, particularly on the eastern side of the country. The principal watershed was likely to have been located in the west, with major rivers draining to eastern estuaries. The North Sea was probably smaller than today, and the low-lying eastern coastal fringe consequently wider. The Orkney Islands were probably then joined to the mainland, but the western island groups may have been broadly the same as at present, with the exception that those islands close to the mainland (Skye, Mull, Scarba–Jura–Islay), which owe their isolation to glacial erosion, would then have been peninsulas. These inferences can be made from the known distribution of pre-glacial weathered bedrock profiles and erosion surfaces (Linton, 1959; Godard, 1965; Sissons, 1967a, 1976b; Hall 1985, 1986, 1987), the occurrence of Neogene sediments on neighbouring continental shelves (Andrews *et al.*, 1990), and consideration of the tectonic and erosional history of the Highland block during the Mesozoic and Tertiary (Watson, 1985; Hall, 1991).

Although areas of weathered bedrock occur locally in the west (Godard, 1961, 1965; Le Coeur, 1989), pre-glacial relics are most abundant in the Buchan area of north-east Scotland, where they exist in most notable form in deposits of fluvial and/or marine gravels at an altitude of around 150 m OD (see Moss of Cruden and Windy Hills) (Flett and Read, 1921; Koppi and Fitzpatrick, 1980; McMillan and Merritt, 1980; Kesel and Gemmell, 1981; Hall, 1982, 1984c) and in frequent occurrences of chemically decomposed bedrock (see Hill of Longhaven, Clunas and Pittodrie) (Phemister and Simpson, 1949; FitzPatrick, 1963; Hall, 1984b, 1985, 1986, 1987; Hall and Mellor, 1988; Hall *et al.*, 1989a). These features are of considerable interest, recording aspects of the Neogene geological history of Scotland. They occur in what was probably the most extensive on-land area of north-west Europe where such old features have been preserved despite a history of glaciation. Additionally, the distribution of the decomposed bedrock outcrops provides an indication of the intensity of glacial erosion in north-east Scotland (Hall and Sugden, 1987; Sugden, 1989) and illustrates graphically the localized nature of this process, even though certain of the weathered bedrock profiles may be truncated (cf. Godard, 1989).

Elsewhere in Scotland the effects of glacial erosion give a distinct aspect to the landscape, albeit one that varies from one part of the country to another (Figures 2.1 and 2.2). In the eastern Highlands, the landforms of glacial erosion, whether corries fretting the edges of the major mountain plateaux (see the Cairngorms and Lochnagar), or spectacular through valleys such as the Lairig Ghru or Glen Tilt, are clearly superimposed on a pre-existing topography, parts of which have survived in a largely unmodified form (Figure 2.3). In such areas glacial erosion has operated selectively, although to striking effect, in creating glacial troughs, breaches and

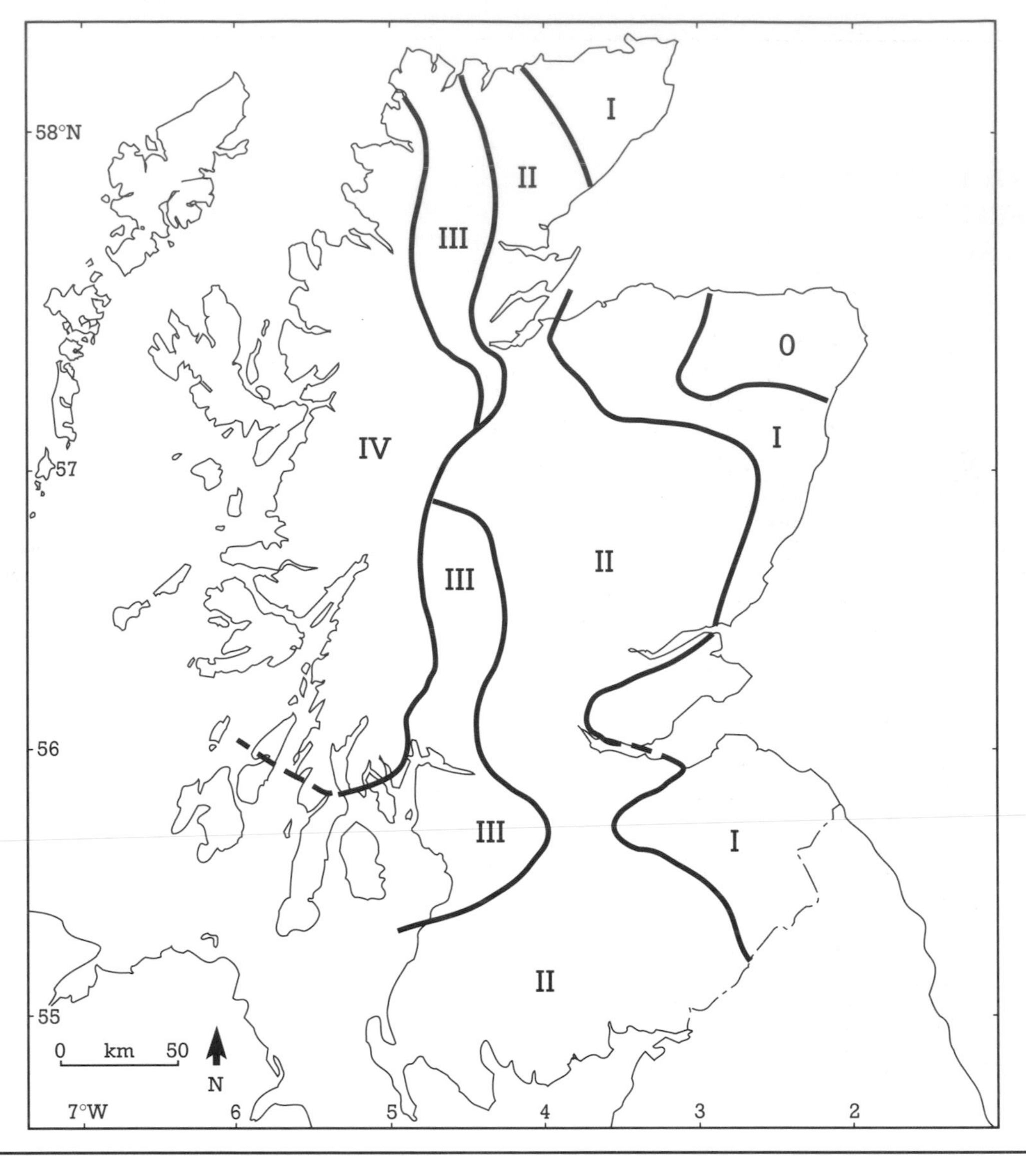

Zone	Lowlands	Uplands
O	No erosion Head on weathered rocks and slopes Outwash in concavities Rare occurrences of till on weathered rock	No erosion Outwash on valley floors Solifluction deposits on slopes Boulder fields and tors on divides
I	Ice erosion confined to detailed or subordinate modifications Concavities drift mantled but convexities may show some ice moulding Occasional roches moutonnées Ice-scoured bluffs in favourable locations	Ice erosion confined to detailed or subordinate modifications Suitable valley slopes ice steepened Entrenched meanders and spurs converted to rock knobs Interfluves still commonly Zone O
II	Extensive excavation along main flow-lines so that concavities may be drift free or floored by outwash or post-glacial deposits. Isolated obstacles may be given ovoid or cutwater forms if of soft rock, or crag-and-tail with associated scour troughs if hard Margins of larger masses converted to ice-scoured bluffs or planar slopes	Conversion of pre-glacial valleys to troughs common, but usually confined to those of direction concordant with ice flow Some diffluence; transfluence rare Interfluves may be Zone I or even Zone O, and separated from troughs by well-marked shoulders
III	Pre-glacial forms no longer recognizable but replaced by tapered or bridge interfluves with planar slopes on soft rocks, and by rock drumlins and knock-and-lochan topography on hard rocks	Transformation of valleys to troughs comprehensive giving compartmented relief with isolated plateau or mountain blocks. Zone O may still persist on interfluves at sufficiently great heights
IV	Complete domination of streamlined flow forms even over structural influences	Ice moulding extends to high summits Upland surfaces given knock-and-lochan topography (sometimes of great amplitude at lower levels) Lower divides extensively pared or streamlined

Figure 2.1 Zones of glacial erosion (from Clayton, 1974).

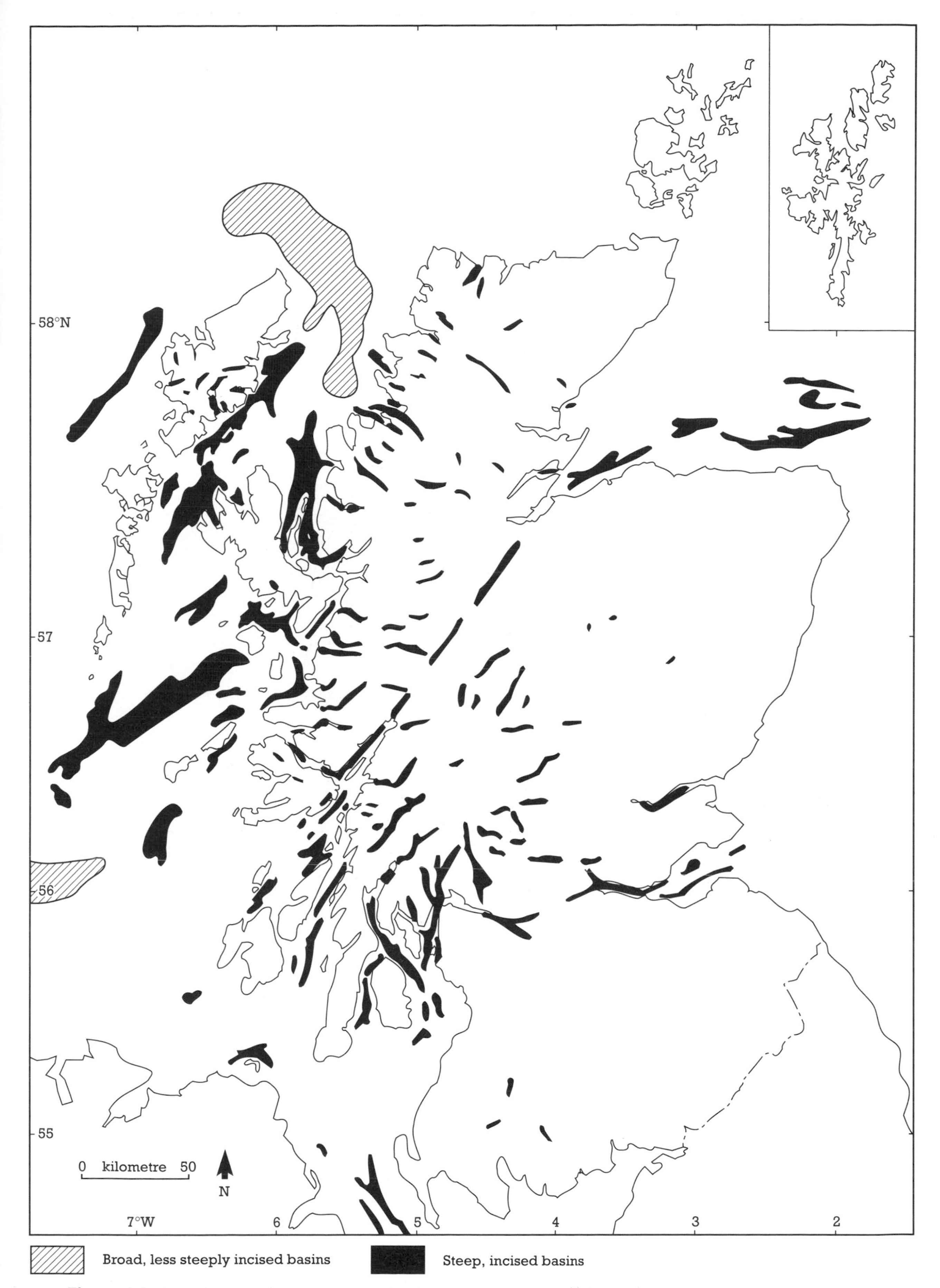

Figure 2.2 Distribution of rock basins on the Scottish mainland and neighbouring shelves.

Figure 2.3 The western Cairngorms, with Ben Macdui (centre) viewed from the east, showing the diversity of the geomorphology of the eastern Highlands. Rounded plateau surfaces with tors (on Beinn Mheadhoin, lower left), solifluction lobes and weathered bedrock contrast strikingly with landforms of glacial erosion, including corries, the glacial troughs of Loch Avon (foreground) and the Lairig Ghru (middle distance), and the truncated spur of the Devil's Point (top, left of centre). Locally, parts of the plateau are ice scoured, as between Ben Macdui and Loch Avon. 'Hummocky moraine' can also be seen at the head of Loch Avon. (Cambridge University Collection: copyright reserved.)

diversions of drainage (Linton, 1951a, 1963; Sugden, 1968). In the western Highlands and adjacent mountainous islands, however, erosion has been much more intense (see the Cuillin), as is shown by the frequency of corries, sharp ridges, pyramidal summits, overdeepened valleys, rock basins and fjords which often dominate the scenery (Linton, 1951a, 1957, 1959, 1963; Linton and Moisley, 1960; Sissons, 1967a, 1976b; Clayton, 1974; Haynes, 1977a; Dale, 1981). The lowlands, too, have received an erosional imprint from glaciation, whether in the form of knock-and-lochan topography (areal scouring) so typical of the Lewisian gneiss terrain of north-west Sutherland (Figure 2.4) and the Outer Hebrides (Linton, 1959; Godard, 1965; Gordon, 1981) or as the streamlined hills and crags and tails of the Midland Valley (Linton, 1962; Burke, 1969).

It has been argued (Sutherland, 1984a) that as the ice-flow directions at the maximum of the last ice-sheet glaciation (during the Late Devensian, see below) were, in places, at a high angle to the direction of ice movement that was responsible for the formation of major erosional landforms,

Figure 2.4 Suilven, with Canisp in the background (left), shows the striking relationships between geology and relief in north-west Scotland. These isolated mountains of Torridonian sandstone rise above the intensively ice-scoured surface of Lewisian gneiss. (© British Crown copyright 1992/MOD reproduced with the permission of the Controller of Her Britannic Majesty's Stationery Office.)

then those features must have originated during earlier glaciations. It is most probable that in Scotland there were numerous periods of either partial or ice-sheet glaciation during the Quaternary. During the former, erosional forms such as corries would have been progressively fashioned, and during the latter, major glacial breaches, overdeepened valleys and lowland ice-moulding would have formed. It may be further argued that as there would very probably have been extensive areas of deeply weathered bedrock throughout Scotland at the onset of glaciation, the initial glacial phases were likely to have had the greatest erosional impact, with later glaciations chiefly inheriting and only slightly modifying erosional landforms.

Two principal factors external to the relief and geological structure have influenced the distribution and degree of development of landforms of glacial erosion in Scotland. Major features (such as glacial breaches, overdeepened valleys and zones of areal scouring; Figures 2.1 and 2.2) were controlled principally by the dynamics and thermal regimes of the ice masses that were responsible for their development (cf. Sugden and John, 1976). They show regional patterns as a consequence (Linton, 1959; Clayton, 1974; Boulton *et al.*, 1977; Haynes, 1977a; Gordon, 1979; Hall and Sugden, 1987), not only with respect to the major ice-sheet that centred on the western Highlands but also around the minor ice centres such as that on the Outer Hebrides. In contrast, small features such as corries, which formed during periods of partial glaciation, have regional distributions and characteristics more closely related to palaeoclimate (Linton, 1959; Sissons, 1967a; Sale, 1970; Robinson *et al.*, 1971; Dale, 1981) and only secondarily or locally to pre-existing relief and geological controls (Thompson, 1950; Haynes, 1968; Sugden, 1969; Gordon, 1977).

QUATERNARY EVENTS PRIOR TO THE LATE DEVENSIAN ICE-SHEET GLACIATION

Deposits attributed to the Late Devensian ice-sheet glaciation can be recognized throughout most of Scotland and hence form a useful datum. Prior to this event, the onshore Quaternary record is fragmentary and, as a consequence, correlation between sites with evidence of earlier events is difficult. Offshore, particularly in the North Sea Basin, much longer sequences of Quaternary deposits are known and these provide invaluable information for interpreting the onshore record. Even offshore, however, there are major unconformities and difficulties of correlation.

Marine deposits and landforms

Pre-Late Devensian marine deposits and landforms are known from localities around almost all the Scottish coasts (Figure 2.5). Possibly the earliest marine event for which there is evidence onshore is represented by the ice-transported mollusc shell fragments in the Kippet Hills area of Aberdeenshire. These shell fragments are contained within glacial deposits of presumed Late Devensian age, but their derivation from an Early Pleistocene deposit, originally suggested on faunal grounds by Jamieson (1882a), has been confirmed by amino acid analysis (D. G. Sutherland, unpublished data). The shell fragments are probably derived from marine deposits close to the Aberdeenshire coast which correlate with the Aberdeen Ground Formation of the North Sea sequence (Stoker *et al.*, 1985).

The Aberdeen Ground Formation accumulated during the whole of the Early Pleistocene and the early Middle Pleistocene and the greater part of its thickness is comprised of silts, clays and fine sands deposited in a relatively shallow (around 50 m) sea as part of a prograding delta sequence (Stoker and Bent, 1987). Along the western margin of the central North Sea Basin, however, a number of beds of coarser sands and gravelly sands have been found in the Early Pleistocene sediments and these were deposited as channel lags in a nearshore environment with water depths of less than 15 m. These sediments are now at depths greater than 150 m below present sea level, which may in part be due to basin subsidence. However, such subsidence would also apply to the deeper-water sediments with which they are interbedded and hence the nearshore sediments indicate low relative sea levels during the Early Pleistocene which seem most probably to be the consequence of glacio-eustatic lowering of sea level. The uppermost of these beds occurs between the top of the Jaramillo palaeomagnetic event (*c.* 990 ka) and the Brunhes–Matuyama palaeomagnetic boundary (780 ka) and may broadly correlate with the first evidence found for glacial sedimentation in the central North Sea by Sejrup *et al.* (1987). Such a correlation would support the attribution of the other beds of nearshore sediment in these Early Pleistocene deposits to glaciations. For much of the Early Pleistocene, however, no glacial sediments have been recognized in the North Sea basin. Apart from the low sea levels mentioned above, the approximate position of the Scottish coastline during this period cannot be ascertained because of the erosion of the western margin of the Aberdeen Ground Formation during Middle and Late Pleistocene glaciations. Certain of the wide erosional platforms described between altitudes of −30 m to −70 m off the east coast (Stoker and Graham, 1985) may have been formed at this time.

Onshore, a distinct set of marine deposits are the so-called 'high-level' shell beds which occur at altitudes of up to 150 m OD (Figure 2.5). The genesis of these deposits has been debated since the discovery of the first sites in the last century (see Clava, Tangy Glen and Afton Lodge) (Bell, 1893a, 1893b, 1895a; Horne *et al.*, 1894, 1897; Eyles *et al.*, 1949; Holden, 1977a; Sutherland, 1981a; Merritt, 1990b). Their origin and chronology are poorly understood, but whether they are *in situ* or are large, ice-transported erratics, they clearly represent marine invasion of the Scottish

Figure 2.5 Distribution of pre-Late Devensian marine deposits and high rock platforms.

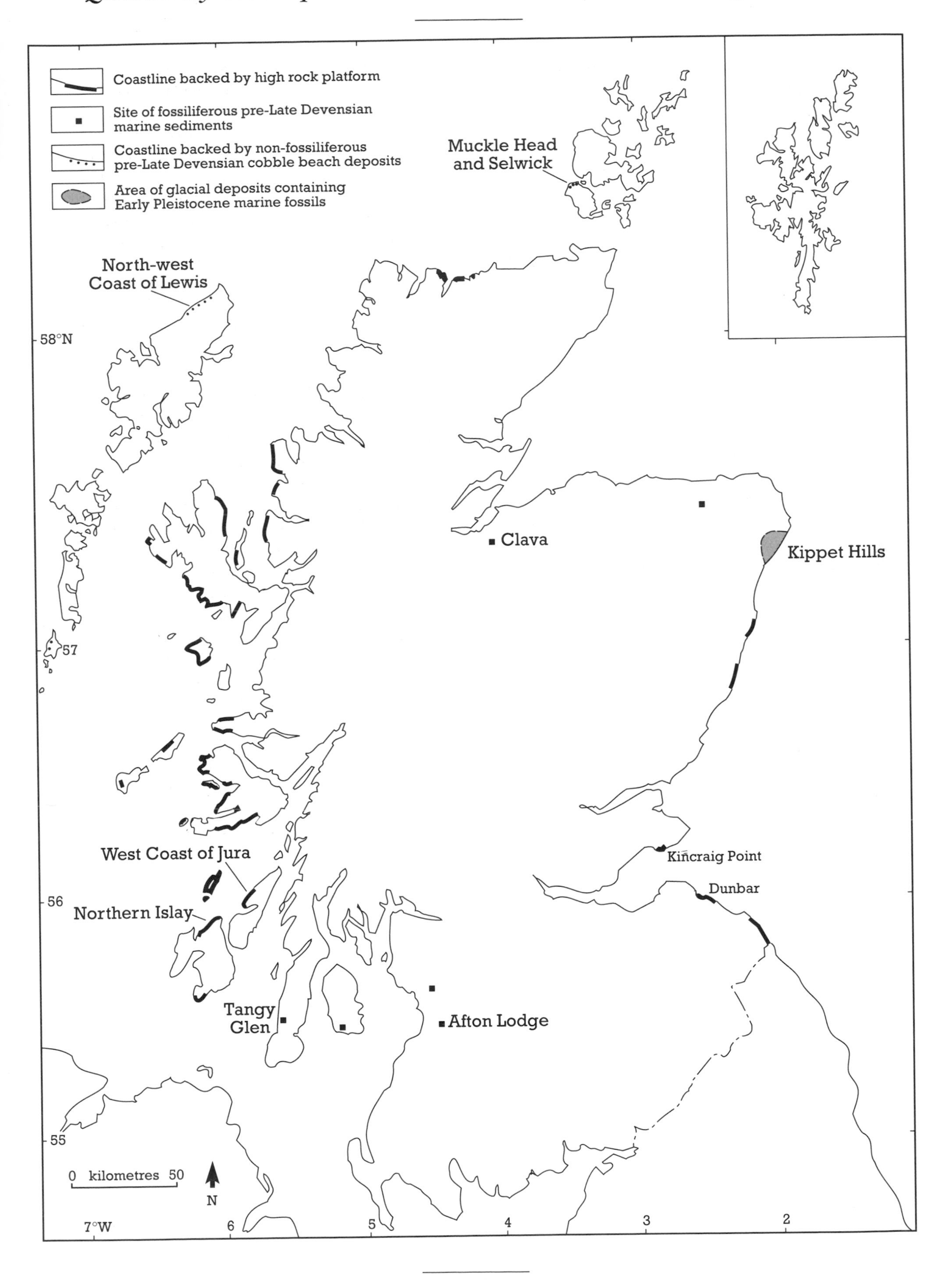
Coastline backed by high rock platform
Site of fossiliferous pre-Late Devensian marine sediments
Coastline backed by non-fossiliferous pre-Late Devensian cobble beach deposits
Area of glacial deposits containing Early Pleistocene marine fossils
Muckle Head and Selwick
North-west Coast of Lewis
58°N
Clava
Kippet Hills
57
West Coast of Jura
Kincraig Point
Dunbar
56
Northern Islay
Tangy Glen
Afton Lodge
55
0 kilometres 50
N
7°W
6
5
4
3
2

coastal zone at a time or times prior to the Late Devensian. Sutherland (1981a) argued that they may have been deposited during the Early Devensian during a period of ice-sheet expansion. The faunas contained in the silts and clays are high boreal to arctic in character (Horne *et al.*, 1894, 1897; Munthe, 1897; Jessen, 1905), implying the sediments were deposited during a glacial rather than an interglacial. At such a time, isostatic downwarping would be necessary for the sea to have access to the coastal zone. Amino acid analysis of mollusc fragments from various of the shell beds indicates that they were deposited in more than one event, however. Those at Clava and King Edward were deposited during the Middle or Early Devensian (Miller *et al.*, 1987; Merritt, 1990b), whereas those at Tangy Glen were deposited during the Middle Pleistocene (Gray, 1985; D. G. Sutherland, unpublished data), possibly between 250 ka and 300 ka. The fossiliferous clays at Burn of Benholm have amino acid ratios (Bowen, 1991) that suggest they were deposited, either as a till (Hall and Connell, 1991) or as a marine deposit (Sutherland, 1981a) during the Middle Pleistocene, possibly between 420 ka and 480 ka.

As with the Early Pleistocene marine event inferred from the presence of mollusc fragments in more recent glacial deposits in Aberdeenshire, so other marine events are implied by the age of shells contained in tills elsewhere around the Scottish coast. One such event appears to have occurred during the Middle Devensian. This has been suggested by amino acid ratios on mollusc fragments in till in Ayrshire (Jardine *et al.*, 1988) and amino acid ratios and radiocarbon dates on mollusc fragments from glacial deposits on Lewis (Sutherland and Walker, 1984).

A further group of marine deposits pre-dating the last ice-sheet is the non-fossiliferous cobble beaches reported from the Outer Hebrides and the Orkney Islands (see North-west Coast of Lewis and Muckle Head and Selwick) (Wilson *et al.*, 1935; Gailey, 1959; McCann, 1968; Peacock, 1984a; Sutherland and Walker, 1984; Selby, 1987). These occur at between 5 m and 15 m above present sea level and frequently rest upon eroded rock or, occasionally, till surfaces. The age of the deposits is not known, nor has it been established whether they represent one or more periods of marine deposition. By analogy with beach deposits in southern Britain, it has been suggested (von Weymarn, 1974) that the beaches are interglacial, but there is no direct evidence for this.

Whereas the marine depositional record is fragmentary in both time and space, marine erosional features, often spectacularly developed, can be traced for many kilometres around the coast of Scotland (Figure 2.5). Even more extensive marine erosion surfaces have been identified on the continental shelves to depths in excess of 100 m (Flinn, 1964, 1969; Sutherland, 1984c, 1987d; Stoker and Graham, 1985). On the coast, these erosional features typically occur as rock platforms and backing cliffs, the surfaces of which are often ice moulded or mantled with glacial deposits (Wright, 1911; Sissons, 1967a, 1982b; McCann, 1968). The glaciated platforms occur at altitudes ranging from the present intertidal zone to over 35 m OD and there is a pattern in their distribution that appears to relate to the centres of ice dispersal. Thus wide intertidal platforms or platforms only a few metres above present sea level are found around almost all of the Scottish coast (see North-west Coast of Lewis, Milton Ness, Dunbar, Kincraig Point and Northern Islay) except in areas of intense glacial erosion close to the former centres of ice dispersal, this pattern being apparent both in areas such as the Outer Hebrides and Skye as well as on mainland coasts. High rock platforms have a more restricted distribution, being most outstandingly developed around the Inner Hebridean islands and on parts of the neighbouring mainland coast (see Northern Islay). They also occur along parts of the east coast (see Dunbar). These high rock platforms do not occur in areas peripheral to the former ice centres, where glaciated platforms at between 5 m and 10 m are encountered (see North-west Coast of Lewis, Port Logan and Glenacardoch Point).

The times of formation of these erosional features remain speculative, and there is a very strong possibility that they have been occupied by the sea on more than one occasion, as have the platforms in the present intertidal zone. Sissons (1981a, 1982b) has suggested that the high platforms were formed when the Scottish ice-sheet was quite extensive, ice cover thus precluding platform formation towards the centres of ice dispersal. On this hypothesis, isostatic depression at the time of platform formation and subsequent rebound would explain the altitude of the features. As a consequence of isostatic tilting these shorelines would pass below present sea level in the areas peripheral to the centres of ice dispersal, thus explaining the apparent absence of the high platforms in these areas. The lower, glaciated platforms in the more peripheral areas may be of some antiquity, that in North-west Lewis, for example, being overlain by deposits of two distinct glacial phases as well as by the possible interglacial deposit of Toa Galson (McCann, 1968; Peacock, 1984a; Sutherland and Walker, 1984). Although many rock platforms and some marine deposits have been described as having been formed during interglacials, no unequivocal evidence is yet known from Scotland to support such an interpretation.

Glacial deposits

The early glacial record is only slightly better known than the marine record. As with the latter, the offshore Pleistocene deposits provide a fuller sequence of events and better chronological control than the onland deposits, but from neither area can the complete series of glaciations be inferred. The earliest evidence for glaciation has been found in the central North Sea Basin, where glaciomarine sediments occur between the top of the Jaramillo geomagnetic event (990 ka) and the Brunhes–Matuyama geomagnetic boundary (780 ka) (Sejrup *et al.*, 1987). These glaciomarine deposits are overlain by silts and clays containing a marine fauna indicative of interglacial conditions, above which the sediments indicate a return to a cold, glacial environment. Sejrup *et al.* (1987) correlated this glacial/interglacial/glacial sequence with Oxygen Isotope Stages 22/21/20, implying that the first glaciation occurred between 860 ka and 840 ka and the second cold period between 810 ka and 780 ka. It has not been established whether the glaciomarine sediments of the first glaciation were deposited from Scandinavian or Scottish ice (or both). If the correlation suggested earlier with the shallow-water marine sediments off the Scottish coast is correct, then this would favour deposition from Scandinavian ice.

The first appearance of glacial sediments in the North Sea Basin at this time is significant, for the oxygen isotope record, which is primarily an indicator of former ice volumes (Shackleton and Opdyke, 1973; Shackleton, 1987), indicates that there was a marked change in the magnitude and frequency of glacial events at approximately 750–800 ka (Ruddiman *et al*, 1986, 1989; Ruddiman and Raymo, 1988). Prior to this time, glaciations had been of shorter duration (with a dominant periodicity of 41,000 years) and of lesser magnitude. Subsequently, glacial cycles had a dominant periodicity of about 100,000 years and total ice volume at the glacial maxima was greater than previously. As will be discussed below, subsequent to this first glaciation the North Sea Basin was invaded by Scottish ice on at least five separate occasions. Correlation of the Early Pleistocene low sea-level events discussed above with periods of more limited glaciation on land (and by implication in the Highlands of Scotland) suggests at least three periods of such limited glaciation during the Early Pleistocene.

The earliest dated glacial deposits derived from Scotland also occur in the North Sea Basin, where they have been encountered in boreholes in the Forth Approaches. There, towards the top of the Aberdeen Ground Formation, a sedimentary sequence has been interpreted as having been deposited from a grounded ice-sheet in the west, with glaciomarine sediments occurring eastwards of the presumed maximum extent of the glaciation (Stoker and Bent, 1985). These deposits occur immediately above the Brunhes–Matuyama palaeomagnetic boundary, which suggests that the glaciation may have occurred during Oxygen Isotope Stage 18, between approximately 770 ka and 730 ka. Andrews *et al.* (1990) state that the till in the west-central North Sea Basin, which was assigned by Sejrup *et al.* (1987) to the Saalian glaciation, lies within the upper part of the Aberdeen Ground Formation and hence could be correlated with this earlier glacial event. No glacial deposits on land have been correlated with this glaciation.

The top of the Aberdeen Ground Formation is a major erosional surface which has been attributed to extensive glaciation of the North Sea Basin (Cameron *et al.*, 1987; Andrews *et al.*, 1990) by both Scottish and Scandinavian ice. Overlying this surface are interglacial marine deposits which are considered to correlate with the Holsteinian (Hoxnian), thus suggesting that the glaciation that produced the erosion surface was probably the Anglian, which may have occurred during Oxygen Isotope Stage 12, between 480 ka and 420 ka.

A number of glacial deposits onshore have been correlated with this glaciation, although none of these correlations rests on firm dating evidence. At Kirkhill in Buchan, basal fluvial sands and gravels contain erratics which were derived from the west and were probably transported by the same ice that deposited the basal till (Leys Till) in the nearby Leys quarry (A. M. Hall, this volume). These sediments are undated, but on the basis of likely minimum ages for the overlying glacial and interglacial deposits, they have been tentatively assigned to the Anglian glaciation (Connell and Hall, 1984a; Hall and Connell, 1991). The basal till at Bellscamphie (Hall, this volume) may also have been deposited at this time. Other evidence of early glaciation that may also correlate with the Anglian occurs at Fugla Ness in Shetland. If the interglacial deposits there are *in situ* and of Hoxnian age, as has been suggested by Birks and Ransom (1969), then the till described by Chapelhowe (1965) as underlying the interglacial sediments would be Anglian or older. At Dalcharn, 15 km east of Inverness, a lowermost till (Dearg Till) and overlying outwash gravels (Dearg Gravels) have been assigned to the Anglian (Merritt, 1990a), as the Dalcharn Biogenic Complex, which is partly developed in the Dearg Gravels, contains a pollen assemblage similar to that found at Fugla Ness.

In the North Sea Basin Quaternary succession, the Aberdeen Ground Formation is overlain by the Ling Bank Formation and then, unconformably, by the Fisher Formation. Within these last two formations there is evidence of two glacial phases separated by a marine event (Andrews *et al*., 1990). During both these phases, ice in the west-central North Sea Basin originated in the Scottish Highlands. The glacial and glaciomarine sediments deposited during the glaciations are underlain and overlain by interglacial deposits. The lower interglacial has been tentatively correlated with the Holsteinian (Hoxnian) and the upper interglacial correlated with the Eemian (Ipswichian), and hence the glacial sediments are considered to have been deposited during two phases of the Saalian ('Wolstonian'), between 180 ka and 130 ka. However, the occurrence of two quite distinct phases to the Saalian glaciation has also been noted in northern Europe (Ehlers *et al*., 1991), where it has been suggested that they may represent two quite distinct glaciations, the former in Oxygen Isotope Stage 8, between 300 ka and 245 ka, and the latter in Oxygen Isotope Stage 6, between 180 ka and 130 ka. Resolution of the number and age of glaciations in this time interval awaits further work.

A number of glacial deposits have been assigned to this period between the Hoxnian and Ipswichian. At Kirkhill, a till, which is underlain by deposits containing a soil horizon tentatively correlated with the Hoxnian (Hall, 1984b; Hall and Connell, 1991), has undergone interglacial weathering, presumably during the Ipswichian. This till may be correlated with other weathered tills in north-east Scotland, possibly at Boyne Bay (Peacock, 1966) and with the middle till at Bellscamphie (Hall, this volume). In the Inverness area, the Dalcharn Lower Till overlying the Dalcharn Biogenic Complex, the Suidheig Till underlying the Odhar Peat, and the Cassie Till at the base of the Clava succession have all been tentatively assigned to this period (Merritt, 1990a). More widely, the event represented by these deposits may correlate with the early phase of glaciation on the North-west Coast of Lewis which pre-dates the possible interglacial site of Toa Galson (Sutherland and Walker, 1984) and which may have been responsible for the early transport of erratics on to the outer islands such as St Kilda (Sutherland, 1984a; Sutherland *et al*., 1984). The glacial deposits on the western shelf edge identified by Stoker (1988) together with the earlier moraine system described by Stoker and Holmes (1991) may also date from this period, as may the basal till at Leavad in Caithness.

These glacial deposits all pre-date at least one interglacial or are overlain by sediments attributed to the Late Devensian glaciation. It has been suggested that rather than only one phase of ice-sheet glaciation post-dating the Ipswichian (i.e. during the Late Devensian) there was a build-up of an ice-sheet in Scotland during the Early Devensian (Sutherland, 1981a). There is as yet no unequivocal evidence for glaciation in Scotland during the Early Devensian (Bowen *et al*., 1986), although a number of glacial deposits have been tentatively attributed to this period. These include the uppermost till at Kirkhill (Hall, 1984b; Hall and Connell, 1991), the Moy Till and the lower member of the Dalcharn Upper Till in the Inverness area (Merritt, 1990a) and the basal till at Sourlie in Ayrshire (Jardine *et al*., 1988). Results of amino acid analysis of shells contained in the till in Caithness and Orkney have been interpreted as supporting the hypothesis of Early Devensian glaciation (Bowen and Sykes, 1988; Bowen, 1989, 1991), but no stratigraphic evidence has been found in these areas for such an event (cf. Hall and Whittington, 1989; Hall and Bent, 1990).

Periglacial deposits

Pre-Late Devensian periglacial deposits have not been frequently described, although it seems very likely that in the lowlands, in particular, cold non-glacial periglacial climatic regimes occurred in Scotland during much of the Pleistocene. The earliest such deposits known are those at the base of the Kirkhill sequence; and periglacial deposits also occur overlying the lower palaeosol at this site, formed prior to deposition of the weathered till. Periglacial reworking has also affected the Dalcharn Biogenic Complex sediments, which contain a pollen assemblage that may correlate with the Hoxnian. Above the weathered till at Kirkhill, but under the uppermost glacial deposits, there are further periglacial sediments (Connell and Hall, 1984a, 1987). These may have formed during the Early Devensian, if the attribution of the weathering of the underlying till to the Ipswichian is correct. Other putative Early Devensian periglacial deposits are those immediately overlying the possible Ipswichian palaeosol at Teindland, those at the base of the Castle Hill sequence at Gardenstown (Sutherland, 1984b), the sediments of the Moy Paraglacial Complex overlying the Allt Odhar interstadial peat (Merritt, 1990c) and those overlying the possible Ipswichian peat at Toa Galson on the North-west Coast of Lewis (Sutherland and Walker, 1984). Pre-Late Devensian soliflucted deposits are also present at Sel Ayre and Fugla Ness on Shetland, where they underlie inferred Late Devensian tills (Mykura and Phemister, 1976; A. M. Hall and J. E. Gordon, unpublished data).

Non-marine organic sediments

Non-marine organic sediments that pre-date the Late Devensian have only been located at a small number of localities in Scotland (Figure 2.6). They comprise peats, lacustrine sediments and palaeosols, and a small number have been found in association with mammal remains. They are rarely of any great thickness and almost all have been truncated by erosion. Some are considered to have been deposited during interglacials, others during interstadials.

The best preserved interglacial deposit is possibly also the oldest, a peat at Fugla Ness in Shetland. This has a clear thermophilous vegetational succession and macrofossils of trees have been recovered from the peat. It has been correlated with the Hoxnian (Birks and Ransom, 1969), but such an attribution must remain uncertain (Lowe, 1984), not least because of the considerable distance of Shetland from the reference sites for this interglacial in south-east England and the extent of vegetational differences between these areas today. The vegetational record from Fugla Ness shows a strong oceanic influence and certain species are strongly suggestive of a Middle Pleistocene age. Comparison with the oceanic climatic record for the North Atlantic reveals only two periods in the last 600 ka that were as warm or warmer than the Holocene, the Ipswichian and a period at around 380 ka (Oxygen Isotope Stage 11) (Ruddiman and McIntyre, 1976). Thus the Fugla Ness peat may have been formed during that earlier period (Sutherland, 1984a).

At Dalcharn, near Inverness, organic sediments comprising the Dalcharn Biogenic Complex (Merritt and Auton, 1990) have been shown to contain a pollen assemblage rather similar to that found at Fugla Ness. The presence of significant quantities of *Ilex* (holly) pollen, in particular, implies an interglacial origin, and its abundance at Dalcharn, which is close to the present northern limit of *Ilex*, may suggest a climate somewhat warmer than today (Walker, 1990a). On these grounds, tentative correlation may be made with the Hoxnian.

The lower palaeosol at Kirkhill and the immediately overlying organic sediments were originally taken to have been formed during an interglacial (Connell *et al.*, 1982), but more recent work has shown that the palaeosol is most akin to a cold-water gley which could develop today in Scotland above 900 m OD (Connell and Romans, 1984). Hence, at Kirkhill (altitude, 50 m OD), the palaeosol probably formed during an interstadial. However, detrital organic material contained in the overlying sands was eroded from a land surface with a temperate soil and vegetation cover (Connell, 1984a; Lowe, 1984) and thus, although no *in situ* interglacial deposits are to be observed at Kirkhill, an interglacial period can be inferred to have occurred prior to the development of the palaeosol. From its stratigraphic position, this interglacial has been correlated with the Hoxnian.

Other interglacial (or presumed interglacial) deposits have proved difficult to correlate. The principal reason is that the pollen spectra at these sites have generally revealed contemporaneous vegetational communities which were dominated

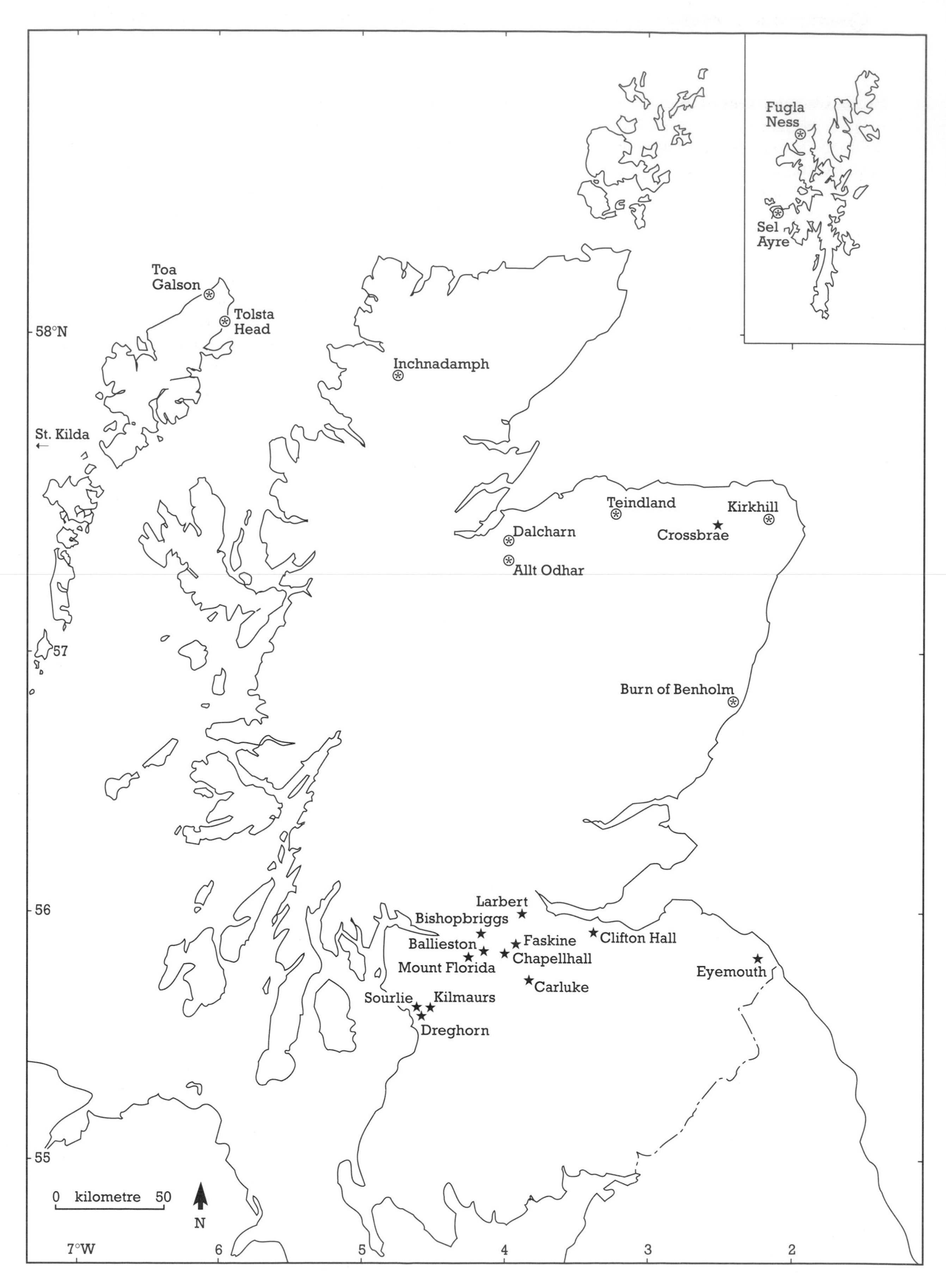

★ Site with terrestrial deposits and/or faunal remains pre-dating the Late Devensian

⊛ GCR site

by grasses and sedges, with acid heath development at certain sites. Tree pollen, which has been used to recognize and differentiate interglacial deposits in more southerly areas of Britain, is present in these pollen spectra in only very low amounts and can be attributed to long-distance transport. Thus the organic sediments at Sel Ayre (Birks and Peglar, 1979), Teindland (FitzPatrick, 1965; Edwards *et al.*, 1976) and Toa Galson (North-west Coast of Lewis) (Sutherland and Walker, 1984), each of which may have been deposited, at least in part, during the Ipswichian, cannot be attributed to that interglacial with any confidence (Lowe, 1984). The truncated palaeosol developed in the lower, weathered till at Kirkhill has also been assigned to the Ipswichian.

Pre-Late Devensian interstadial deposits, though still rare, are relatively more abundant than presumed interglacial deposits. With the exception of the interstadial represented by the lower palaeosol at Kirkhill discussed above, all the interstadial deposits have been attributed to the Devensian. They may be divided into two groups: those that are beyond the range of radiocarbon dating and which are likely to be Early or early Middle Devensian, and those that yield apparently finite radiocarbon dates from the Middle Devensian.

The most detailed record of a presumed Early Devensian interstadial site is that from the Odhar Peat at Allt Odhar in Inverness-shire. Pollen analysis of the peat (Walker, 1990b) has demonstrated a vegetational succession from birch woodlands with willow and juniper scrub interspersed with open grassland, to heathland and, finally, to an open vegetation dominated by species-poor grass and sedge communities. Pollen of taxa of northern or montane affinities are present throughout the deposit. Radiocarbon dating has indicated that the peat was formed before 51,000 BP and probably earlier than 62,000 BP (Harkness, 1990). An initial uranium-series date on the peat gave an age of 124 ± 13 ka (Heijnis, 1990) but the revised age estimate is 106 + 11/−10 ka, which places the peat in an Early Devensian interstadial, equivalent to Oxygen Isotope Substage 5c (Walker *et al.*, 1992).

Figure 2.6 Location of sites where pre-Late Devensian fossils or organic non-marine sediments have been found. Details of non-GCR sites are given in Sutherland (1984a).

An Early Devensian interstadial origin has been proposed for the peat lens in till at Burn of Benholm. This has a pollen spectrum of low floristic diversity dominated by grasses and sedges. It has a radiocarbon age of >42,000 BP (Donner, 1960, 1979). The limited information on this deposit precludes attempts at wider correlation.

Several sites with radiometric ages between approximately 26,000 BP and 35,000 BP appear to relate to an interstadial in the later part of the Middle Devensian. Various mammal remains (reindeer, woolly rhinoceros) have also been dated from this period. The most detailed record to have been published to date is that from Tolsta Head in Lewis (von Weymarn and Edwards, 1973; Birnie, 1983), for which a relatively cool climate and unstable soil conditions are suggested. Deposits at Sourlie (Jardine *et al.*, 1988) also date from this period, as do reindeer bones recovered from the caves at Creag nan Uamh (Lawson, 1984) and the woolly rhinoceros bone found in glacial sediments at Bishopbriggs (Rolfe, 1966). Although none of them have been radiocarbon dated, it is likely that at least some of the mammoth remains found in central Scotland are also of this age. Further Middle Devensian interstadial sites are those at Crossbrae Farm in north-east Scotland (Hall, 1984b) and possibly Abhainn Ruaival on St Kilda (Sutherland *et al.*, 1984). Interstadial pollen spectra from Teindland have also been attributed to this phase (Edwards *et al.*, 1976), but this has been disputed (Sissons, 1981b; Lowe, 1984) and an alternative explanation of an Early Devensian age advanced (Sissons, 1982c). During this interstadial, speleothem deposition occurred in caves in Sutherland (Atkinson *et al.*, 1986) and, as noted earlier, there was marine invasion of the Scottish coast.

In summary, the pre-Late Devensian Pleistocene record is only partially known and its fragmentary nature in the absence of radiometric dates makes correlation between sites a matter of some speculation. The available evidence can be synthesized into a scheme involving three and possibly five periods of ice-sheet glaciation (Figure 2.7) and certain other periods of limited glaciation during the Early Pleistocene may also be inferred. Marine sedimentation along the inner coasts has taken place on two (and probably more) occasions, though the greater part of the marine record, consisting of erosion platforms and non-fossiliferous beaches, remains as yet undated. The non-marine interglacial sites can be

Chrono-stratigraphy	Terrestrial: Organic	Terrestrial: Periglacial	Marine: Inner coast	Glaciation	OI Stage	Age ka

Holocene

Devensian

Ipswichian

'Wolstonian'

Hoxnian

Anglian

?

?

1
12
2
24
3
59
4
71
5
128
6
186
7
245
8
303
9
339
10
362
11
423
12
480
13
14
15
16
17
730
18
770
19
Brunhes-Matuyama boundary
780
20
810
21
840
22
860

interpreted as representing two separate interglacial stages, although these correlations are tentative. An interstadial succeeding the earlier of those interglacials is known from Kirkhill, and other interstadial sites have been assigned to the Devensian. At least one Early Devensian interstadial deposit has been proposed and deposits at a number of sites throughout the country appear to be from an interstadial towards the end of the Middle Devensian. No sites have been found that have been correlated with the Upton Warren interstadial in England. Lowland periglacial deposits have been identified from at least three periods. By comparison with the sequence of Pleistocene climatic change derived from both deep-sea cores (Ruddiman and McIntyre, 1976; Zimmerman *et al.*, 1984) and the terrestrial record in Holland (Zagwijn, 1975, 1986) and elsewhere in Britain (Bowen *et al.*, 1986; Ehlers *et al.*, 1991), it is apparent that there are very many events not represented in the present synthesis of the Scottish evidence. This implies that much remains to be discovered and, further, that the above synthesis is probably an over simplification. It is therefore clear that in the apparent absence of so much evidence the scientific value of the known sites is at a premium.

LATE DEVENSIAN ICE-SHEET GLACIATION

Ice-sheet chronology, flow patterns and dimensions

The locations of the sites at which late Middle Devensian interstadial deposits have been found indicate that much of lowland and coastal Scotland was ice-free during the period between about 35,000 BP and 26,000 BP. The climate, however, appears to have been more severe than at present and hence there is a strong possibility that glaciers would have existed in the Highlands at this time. Subsequently, there was a major expansion of ice, initially from the main ice centre of the western Highlands and the neighbouring islands such as Skye and Mull and later, as the glaciation progressed, from the ice centre in the Southern Uplands (Figure 2.8).

The chronology of this period of glaciation is not known in detail, but a radiocarbon date on ice-transported mollusc fragments, if correct, suggests that the independent Outer Hebridean ice-cap only expanded beyond the present coast after 23,000 BP (Sutherland and Walker, 1984). In the North Sea Basin, the Marr Bank Formation was deposited in a shallow-water glaciomarine environment east of the ice-sheet margin (Sutherland, 1984a; Stoker *et al.*, 1985) and a radiocarbon date of 17,730 ± 480 BP (SRR–625) indicates that the ice was close to its maximum at this time. Glaciomarine sediments at St Fergus in Aberdeenshire, deposited during ice retreat, have been radiocarbon dated to 15,320 ± 200 BP (Lu–3028) (Hall and Jarvis, 1989). Farther south, five raised shorelines between Stonehaven and Fife Ness (Cullingford and Smith, 1980) which were formed in succession during ice retreat (see Dryleys, and Milton Ness) can be dated to between 16,000 BP and 14,000 BP (Sutherland, 1991a) on the basis of a calculation of the change in shoreline gradient with shoreline age (Andrews and Dugdale, 1970). Radiocarbon dating of basal sediments in deglaciated areas (see Cam Loch, Loch Etteridge, and Loch an t-Suidhe), although potentially subject to error (Sutherland, 1980), together with similar dating of marine molluscs in deposits laid down upon retreat of the ice-sheet (see Geilston), implies that by around 13,000 BP almost all the ice had retreated to within the Highland boundary. Thus the whole period of lowland glaciation, both advance and retreat, may have lasted only around 10,000–12,000 years.

Not all of Scotland was covered by ice during this period as ice-free areas have been identified in both North-west Lewis (Sutherland and Walker, 1984) and around Crossbrae Farm near Turriff (Hall, 1984b), although the precise limits of the latter ice-free area are unclear (see Castle Hill and Boyne Quarry) (Sutherland, 1984a; Hall, 1984b; Hall and Bent, 1990). It has also been suggested that much of Caithness (see Baile an t-Stratha and Drumhollistan) and Orkney (see Den Wick and Mill Bay) were not glaciated at this time (Sutherland, 1984a), but there is no direct evidence for this, Hall and Whittington (1989) and Hall and Bent (1990) arguing for complete glaciation during the Late Devensian.

Figure 2.7 Summary of the principal glacial, periglacial, marine and terrestrial depositional events recognized in the Quaternary record in Scotland. Note that the time-scale is not linear. Only those events that can be related to specific deposits on land or on the adjacent continental shelves are plotted.

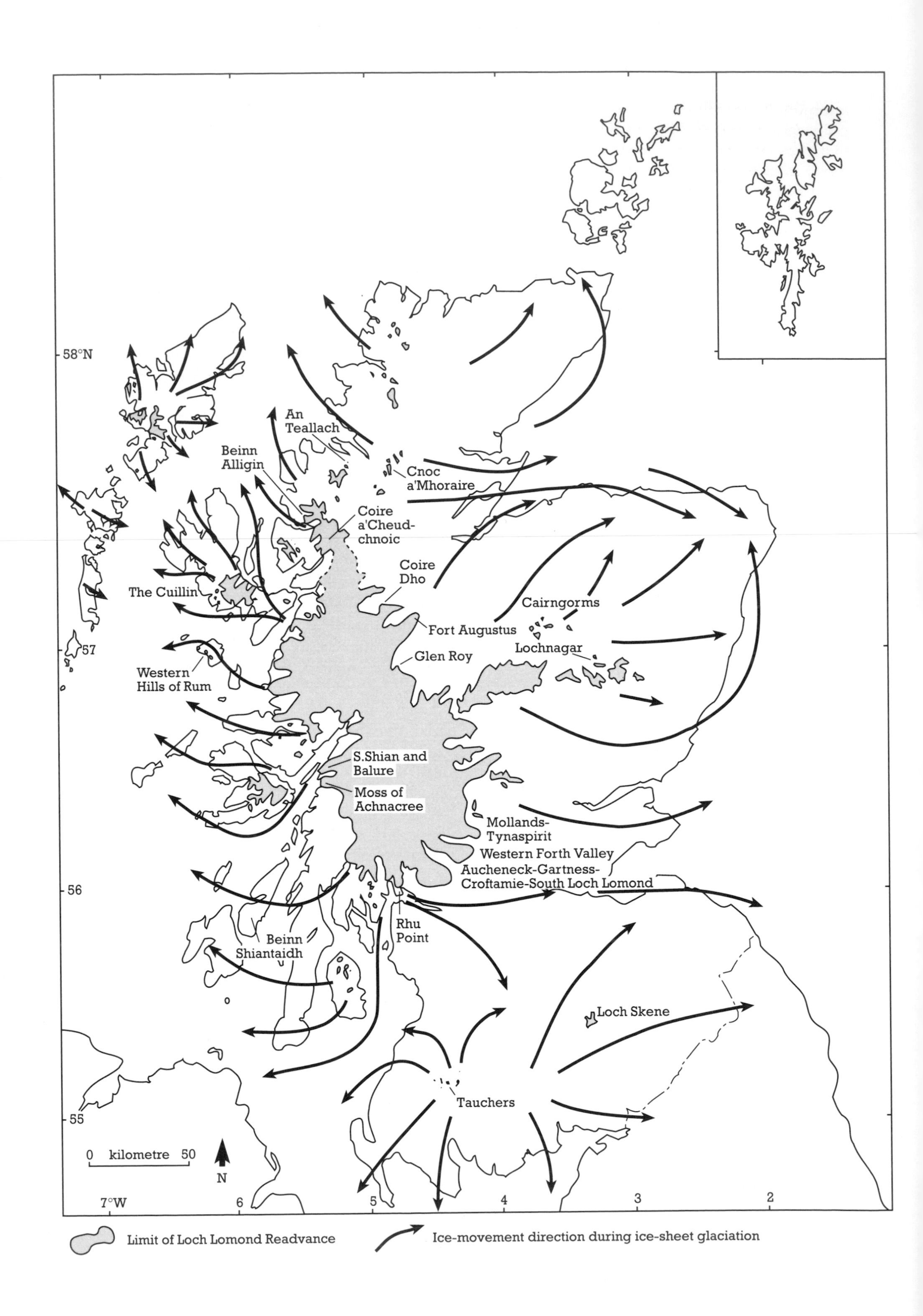

58°N
57
56
55
7°W
6
5
4
3
2
An Teallach
Beinn Alligin
Cnoc a'Mhoraire
Coire a'Cheud-chnoic
Coire Dho
The Cuillin
Fort Augustus
Cairngorms
Lochnagar
Glen Roy
Western Hills of Rum
S.Shian and Balure
Moss of Achnacree
Mollands-Tynaspirit
Western Forth Valley
Aucheneck-Gartness-Croftamie-South Loch Lomond
Rhu Point
Beinn Shiantaidh
Loch Skene
Tauchers
0 kilometre 50
N
Limit of Loch Lomond Readvance
Ice-movement direction during ice-sheet glaciation

At the maximum of the glaciation two major lobes of ice appear to have extended into the North Sea Basin, one emerging from the Midland Valley and the other from the inner Moray Firth. Both terminated in end-moraine complexes fronting into a shallow arctic sea (Sutherland, 1984a; Cameron *et al.*, 1987; Hall and Bent, 1990), with a large part of the central North Sea being dry land. Off the west coast, the ice-sheet terminated to the north of Lewis (Sutherland, 1991a), with a larger lobe extending towards the shelf edge to the south of the Hebridean islands (Selby, 1989; Peacock *et al.*, 1992). Low sea levels on the outer western continental shelf apparently also imply extensive areas of dry land at this time (Sutherland *et al.*, 1984; Sutherland, 1984c, 1987d). By 15,250 BP, the St Kilda Basin was free of glacier ice (Peacock *et al.*, 1992). During the Late Devensian, Shetland appears to have been covered by an independent ice-cap (Sutherland, 1991b). Comparison of models of glacio-isostatic rebound (Lambeck, 1991a, 1991b), with observed patterns of raised shorelines, also suggests that the ice in Scotland was less extensive about 18,000 years ago than suggested in some ice-sheet models (Boulton *et al.*, 1977, 1985; Denton and Hughes, 1981).

The presence of ice-free areas along parts of the northern fringes of the ice-sheet, its relatively modest dimensions (the flow-line distance from ice shed to ice margin in the North Sea Basin was about 200 km) and the coexistence of ice-caps nurtured on relatively small mountain masses all imply that the main Highland ice mass was not particularly thick at its maximum (see also Lambeck, 1991a, 1991b). Striations and erratics have been reported at up to 1100 m OD in the Ben Nevis range and, although they cannot be dated, they may be attributed to the last ice-sheet, suggesting that it overtopped the mountains in that area (Sissons, 1967a; Thorp, 1987). Similarly, erratics found on the top of Merrick, in the western Southern Uplands, suggest that the ice there also overtopped the highest hills (Charlesworth, 1926a; Cornish, 1982).

In the Northern Highlands, however, it is probable that the highest mountains stood above the ice-sheet as nunataks (see An Teallach and Ben Wyvis) (Godard, 1965; Ballantyne, 1984; Sutherland, 1984a; Ballantyne *et al.*, 1987; Reed, 1988), and at that time some of the major periglacial features on these mountain summits may have been formed. Ice-surface altitudes of around 700 m OD on An Teallach rising to over 800 m OD in the Fannich mountains have been suggested (Ballantyne *et al.*, 1987). On the Trotternish Peninsula on Skye, Ballantyne (1990) has mapped a periglacial trimline considered to represent the upper limit of the Late Devensian ice-sheet. This trimline descends from approximately 600 m to around 450 m OD in 24 km from south to north along the peninsula. In northern Lewis, the moraine that Sutherland and Walker (1984) identified as marking the edge of the last ice-sheet descends from over 110 m to around 50 m OD in 6 km.

Within the limits of the ice-sheet, till was deposited, this being almost continuous in the lowland areas, where it frequently forms drumlins. In different parts of the country the till contains specific suites of erratics, largely derived from the immediate locality but also containing material from much further afield. The distribution of these erratics (see Clochodrick Stone), together with the orientation of ice-streamlined landforms and striations (see Agassiz Rock), gives a graphic picture of the ice-flow patterns (Figure 2.8) and allows the relative strengths of the ice flow from different ice centres to be assessed.

The major ice mass was centred in the western Highlands, the ice shed extending in an approximately north–south alignment from Sutherland to the Cowal Peninsula in Argyll (Sissons, 1976b; Boulton *et al.*, 1991). Ice flowed outwards from this area, transporting erratics of Highland aspect throughout almost the whole of the Central Lowlands and even into the northern Southern Uplands. As the glaciation progressed, however, the Southern Uplands ice expanded and exerted sufficient pressure to force back the Highland ice throughout the southern Central Lowlands. This resulted in the classic two-fold stratigraphy in this 'debatable ground' of a till containing Southern Upland erratics overlying a till containing Highland erratics (see Nith Bridge, Hewan Bank and Port Logan). The Highland ice also abutted against ice flowing away from the independent ice domes that developed on Skye (see the Cuillin), Mull, the Outer Hebrides, the Cairngorms and the south-east Grampians but did not overwhelm any of them.

Figure 2.8 Ice-sheet flow patterns and Loch Lomond Readvance glaciers.

Ice-sheet deglaciation

The retreat of the last ice-sheet was accompanied by the release of abundant meltwater, which reworked debris both beneath and held within the ice. Intricate sequences of meltwater channels were often cut across spurs and other areas of positive relief, these being recognizable by their independence from the present drainage system, their anastomosing pattern in plan, their frequent up-and-down long profiles (resulting from formation under hydrostatic pressure below the ice) and, at times, their very large size (see Rammer Cleugh, Carlops, Corrieshalloch Gorge, Glen Valtos, the Cairngorms and Muir of Dinnet).

In topographic depressions and valley bottoms, the depositional counterparts of the meltwater channels are extensive sequences of bedded sands and gravels, often with very characteristic morphologies. Where deposited by rivers flowing through channels below or within the ice they possess an elongate, sinuous morphology, and frequently groups of these eskers form complex anastomosing patterns. The most noted of such features occur at Carstairs, but there are also outstanding examples at Kildrummie, Torvean, Littlemill and Kippet Hills. Other areas of bedded sands and gravels have a locally more complex morphology with frequent rounded, steep-sided hollows where individual masses of 'dead' ice became surrounded by sand and gravel, the hollows (kettle holes) being produced when the 'dead' ice melted (see Muir of Dinnet and the Cairngorms). The intervening mounds of sand and gravel are frequently flat-topped and, where banked against the sides of a valley, have a clear terrace form. These flat-topped kames and kame terraces can often be traced into outwash sands and gravels deposited in front of the decaying ice, this relationship indicating the role of meltwater drainage through or around the 'dead'-ice masses in forming these deposits. Outstanding examples illustrating these relationships are found at Torvean and Moss of Achnacree (although the latter example was formed during retreat of a Loch Lomond Readvance glacier, see below), and an excellent series of outwash terraces occurs along the North Esk near Edzell.

The various meltwater landforms and deposits did not form randomly in the landscape and when regional patterns of meltwater channels and bedded sands and gravels are mapped they reveal the major drainage routes across the country of the decaying ice-sheet (Figure 2.9) (Sutherland, 1984a, 1991a). Thus meltwater drainage from the hills of south-central Ayrshire was eastwards across the central Clyde valley and, via the Midlothian Basin, into the North Sea. Similarly, meltwaters from the Tay Valley initially drained north-eastwards through Strathmore to reach the coast near Montrose.

Corresponding to the coarse-grained sand and gravel deposits found inland are fine sands, silts and clays deposited along the coasts, most especially in the eastern estuaries and in the offshore zone during this period of ice-sheet retreat (Figure 2.10). Along much of the east coast these fine-grained deposits have long been famous for the high-arctic character of the marine molluscs, ostracods and Foraminifera contained within them (Jamieson, 1865; Brown, 1868; Brady *et al.*, 1874; Davidson, 1932; Graham and Gregory, 1981; Paterson *et al.*, 1981). They have been studied in most detail in the Inchcoonans and Gallowflat area beside the Tay estuary, where the Errol Clay Pit has given the sediments their informal name of the Errol beds (Peacock, 1975c). The offshore equivalents in the North Sea are termed the St Abbs Formation in the Forth Approaches (Stoker *et al.*, 1985) and the Fladen Member of the Witch Ground Formation in the central North Sea Basin (Long *et al.*, 1986). Radiocarbon dating of the North Sea sediments, together with the close association of the Errol beds with the sequence of raised shorelines formed as the ice retreated, indicates that the Errol beds were deposited from some time prior to 16,000 BP until approximately 13,000 BP. Deposition of sediments containing high-arctic faunas during a large part of the retreat of the last ice-sheet implies that the climate was still very cold and therefore that the retreat of the ice was not a consequence of warmer temperatures. It has therefore been inferred that a reduction in precipitation was responsible for the initial period of ice decay (Sutherland, 1984a).

The estuarine sediments generally coarsen towards the rivers that supplied the sediments and at these river mouths complex sedimentary sequences often developed. Sea-level fall in response to isostatic uplift was a major control on these sequences, with coarse-grained fluvial or outwash sediments interdigitating with or being deposited on top of the estuarine sediments as

Figure 2.9 Distribution of glaciofluvial deposits.

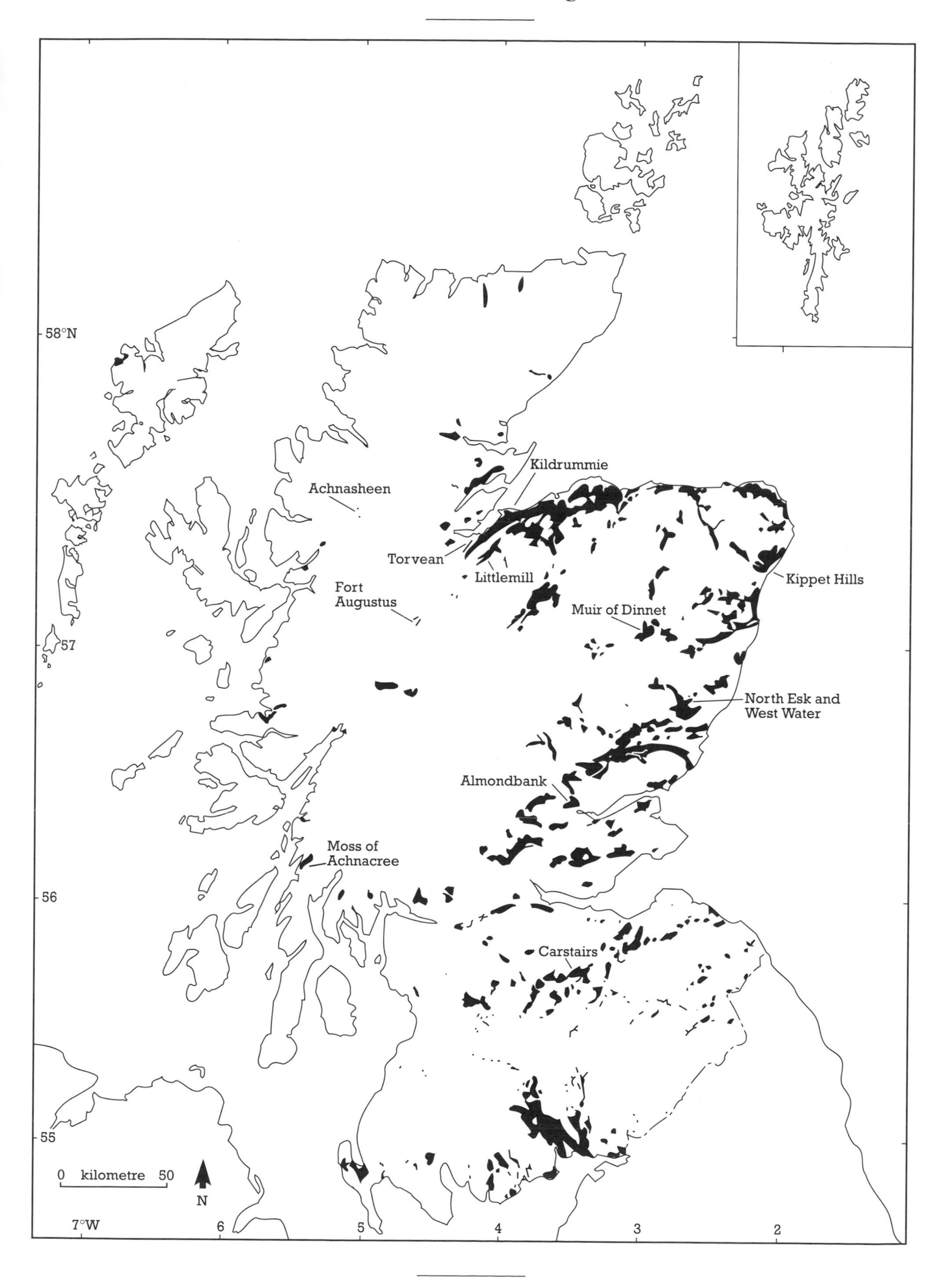

58°N
57
56
55
7°W
6
5
4
3
2
Kildrummie
Achnasheen
Torvean
Littlemill
Fort
Augustus
Kippet Hills
Muir of Dinnet
North Esk and
West Water
Almondbank
Moss of
Achnacree
Carstairs
0 kilometre 50
N

the shoreline retreated. In the Perth area, for example, such a sequence of outwash gravels overlying estuarine silts was originally thought to represent a readvance of the ice-sheet (the so-called Perth Readvance – Simpson, 1933; Sissons, 1963a). Paterson (1974), however, has demonstrated that the sequence need not represent a readvance but simply results from the deposition of outwash sediments across the estuarine silts as sea level fell (see Almondbank).

One of the major consequences of ice-sheet glaciation was glacio-isostatic depression of the Earth's crust, such depresssion being greatest where the ice was thickest in the western Highlands. This depression of much of Scotland was significantly greater than the glacio-eustatic lowering of world sea level due to the contemporaneous expansion of the North American and Scandinavian ice-sheets, so that the rebound upon removal of the ice load has resulted in a sequence of raised shorelines around much of the coasts of mainland Scotland and the Inner Hebrides (Sissons, 1967a; Gray, 1985). It is only around the coasts of the outer islands that the eustatic rise of sea level has been greater than any isostatic effect and where, as a consequence, shorelines that date from the deglacial phase are now submerged.

These raised shorelines frequently form prominent elements of the coastal scenery, none more so than the 'staircases' of white, unvegetated, quartzite shingle ridges along the West Coast of Jura. The finer-grained sediments in the low-energy estuaries of the east coast formed extensive sequences of low-gradient terraces, which are well displayed at Munlochy Valley, Dryleys and Inchcoonans and Gallowflat. The progressive decline in the rate of isostatic uplift since deglaciation has meant that terraced sequences of distinct shorelines formed, each younger shoreline occurring in a given area at a lower altitude and with a lower gradient than its predecessor. Particularly along the east coast where individual shorelines have been mapped, the younger features are found to extend progressively farther west, reflecting a period of continous retreat of the ice-sheet (Cullingford and Smith, 1966, 1980; Sissons *et al.*, 1966; Firth, 1989a), during which much of the Central Lowlands were deglaciated.

Although various halts or even readvances have been postulated during the period of ice-sheet retreat, almost all of these have now been discounted. An important exception occurs in the Northern Highlands where a major end moraine (see Gairloch Moraine) has been identified as marking the limit of a readvance that interrupted the later part of ice retreat (Robinson and Ballantyne, 1979; Sissons and Dawson, 1981; Ballantyne *et al.*, 1987). This readvance has been associated with a distinct shoreline which is partly cut in rock and it is possible that the readvance dates from the period around 13,500–13,000 BP when the climate changed from arctic cold to a more boreal regime at the opening of the Lateglacial Interstadial. Another halt in the ice retreat on the west coast may be marked by an end moraine complex identified on Islay (see Northern Islay) (Dawson, 1982) which Sutherland (1991a) has suggested formed at around 14,500 BP. Near the mouths of numerous sea lochs along the west coast there are marked drops in the marine limit (Peacock, 1970a; Sutherland, 1981b) which could be the result of glacier readvance but are equally likely to reflect topographic control on the rate of calving of the glaciers flowing down the sea lochs. Minor local readvances of ice may also be represented in the Perth area at the Shochie Burn site and near Inverness at Ardersier.

Errol beds sediments with their high-arctic faunas are absent along much of the west coast of Scotland, only being recorded in the North Minch (Graham *et al.*, 1990) and, possibly, around Stranraer (Brady *et al.*, 1874). Immediately upon deglaciation along the west coast, marine sands, silts and clays containing boreo-arctic faunas (the Clyde beds; see Geilston) were deposited. The absence of Errol beds has led to the suggestion that these areas were ice covered until the climate ameliorated (Sissons, 1967a; Peacock, 1975c). A corollary to this hypothesis is that a large ice dome existed across the Firth of Clyde until a very late stage of deglaciation, which may explain the easterly and westerly to north-westerly flow of ice out of this area at that time. Holden (1977a) has postulated a late readvance of ice from the Clyde area into central Ayrshire to explain a tripartite sedimentary sequence near Greenock Mains, and late-stage flow from the south-west across Jura is indicated by the Scriob na Caillich medial moraine on the west coast of

Figure 2.10 Areas flooded by the sea during retreat of the Late Devensian ice-sheet (1), and location of ice-dammed lakes formed during deglaciation of the Late Devensian ice-sheet and the Loch Lomond Readvance (2).

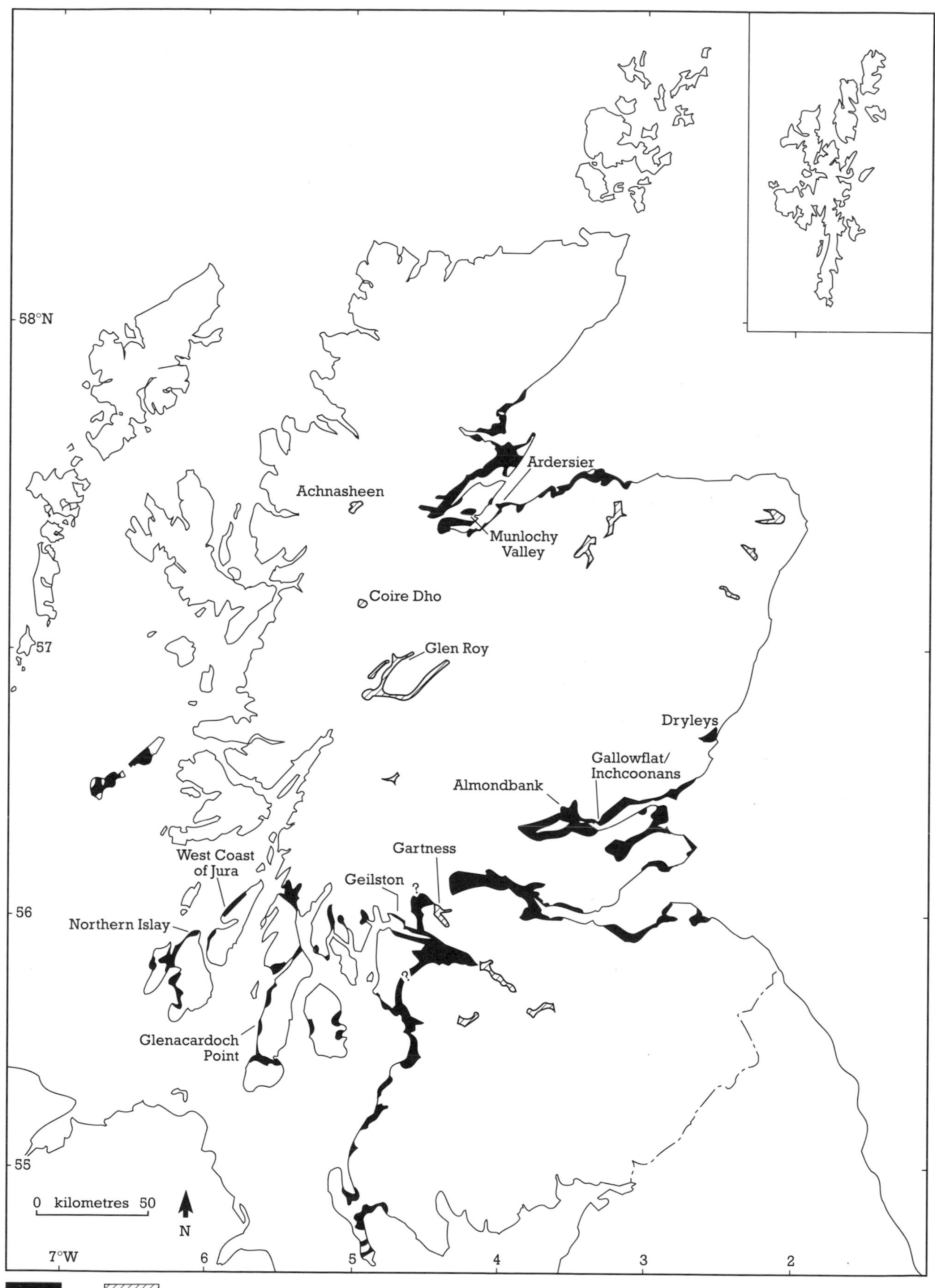

58°N
57
56
55
7°W
6
5
4
3
2
Ardersier
Achnasheen
Munlochy
Valley
Coire Dho
Glen Roy
Dryleys
Gallowflat/
Inchcoonans
Almondbank
Gartness
West Coast
of Jura
Geilston
Northern Islay
Glenacardoch
Point
0 kilometres 50
N
1.
2.

that island (Dawson, 1979b). It has been suggested (Sutherland, 1984a) that rapid deglaciation of the Firth of Clyde occurred slightly before 13,000 BP, as radiocarbon dates on marine shells from the head of the firth and the Clyde estuary imply deglaciation by that date (Sutherland, 1986).

LATEGLACIAL INTERSTADIAL

The Lateglacial Interstadial was a period of mild climate which opened around 13,000 BP. As the oceanic polar front in the North Atlantic migrated to the north of Scotland (Ruddiman and McIntyre, 1973, 1981a; Ruddiman *et al.*, 1977), North Atlantic Drift waters reached the Scottish coast (Peacock and Harkness, 1990) and both marine and atmospheric temperatures rose dramatically to close to present-day values (Bishop and Coope, 1977; Peacock, 1981b, 1983a, 1989b; Atkinson *et al.*, 1987). The change in the nearshore marine environment from high arctic to boreal may have occurred in only 50 years (Peacock and Harkness, 1990).

This rapid change in climate was not immediately registered in the nature of the vegetation because of the time necessary for soil development and plant expansion. Plant communities, as indicated by pollen records at sites throughout Scotland (such as Loch an t-Suidhe, Loch of Winless, Cam Loch, Din Moss, Loch Etteridge, Burn of Aith and Loch Cill an Aonghais) (Figure 2.11), were characterized by open-habitat taxa typical of pioneer vegetation on recently deglaciated terrain. A much more rapid response to climatic change has been revealed both by Coleoptera in the terrestrial environment and by marine faunas. It is only in southern Scotland that beetle assemblages from the Lateglacial Interstadial have been studied (see Redkirk Point and Bigholm Burn) and they have revealed that around 13,000 BP summer temperatures were close to present-day values (Bishop and Coope, 1977). Analysis of marine faunal assemblages has also suggested that water temperatures at around 12,800 BP were within 1–2°C of those of the present (Peacock, 1981b, 1983a; Peacock and Harkness, 1990).

As the interstadial progressed, plant succession led to the development of a closed vegetation cover throughout the lowlands and in the Highland valleys (Walker, 1984b; Tipping, 1991a). There was considerable geographical diversity in the interstadial plant communities, reflecting the varied topographic, edaphic and climatic controls on plant growth. At a number of sites, however, there was an apparent brief halt in the progressive establishment of the vegetation, shown by an increase in the representation in the pollen spectra of open-habitat taxa, a reduction in woody plants and, in places, an increase in the inwashing of mineral matter (see Stormont Loch, Pulpit Hill, Cam Loch, Loch an t-Suidhe and Loch Sionascaig). This phase, which has not been observed at all sites investigated (Tipping, 1991a, 1991b), has been correlated with the Older Dryas chronozone (12,000–11,800 BP) of the north-west European sequence (Pennington, 1975b). However, accurate radiocarbon dating at a number of sites to establish the age of the event and whether it does indeed correlate with the Older Dryas has not been carried out. On the assumption that it is a single event, Walker and Lowe (1990) have suggested that its registration at some sites but not at others may be a reflection of changes in seasonal temperatures, as Atkinson *et al.* (1987) reported that at about this time in the interstadial there were periods when winter temperatures decreased although summer temperatures remained broadly unchanged.

Subsequently, there developed throughout the country a complex mosaic of vegetational communities during the main part of the interstadial. In south-east and eastern Scotland open birch woods with subordinate juniper occurred (see Beanrig Moss, Stormont Loch and Muir of Dinnet) and along the western seaboard as far north as Mull (Bigholm Burn, Loch Cill an Aonghais, Loch an t-Suidhe) grassland communities were dominant, with local occurrence of juniper and *Empetrum* (crowberry) heaths and tree birch. In the north and west (Loch Sionascaig, Cam Loch) dwarf-shrub heaths were dominant, interspersed with grasslands, whilst in the central Highlands (Loch Etteridge, Abernethy Forest) a shrub tundra with *Empetrum*, *Betula* (birch) and some *Salix* (willow) occurred. In the valleys of the

Figure 2.11 Distribution of sites containing lacustrine or terrestrial organic sediments attributable to the Lateglacial Interstadial. (1) Lateglacial Interstadial site confirmed by palynology or coleopteran analyses, many with supporting radiocarbon dating; (2) GCR site; (3) Lateglacial Interstadial site confirmed only by radiocarbon dating.

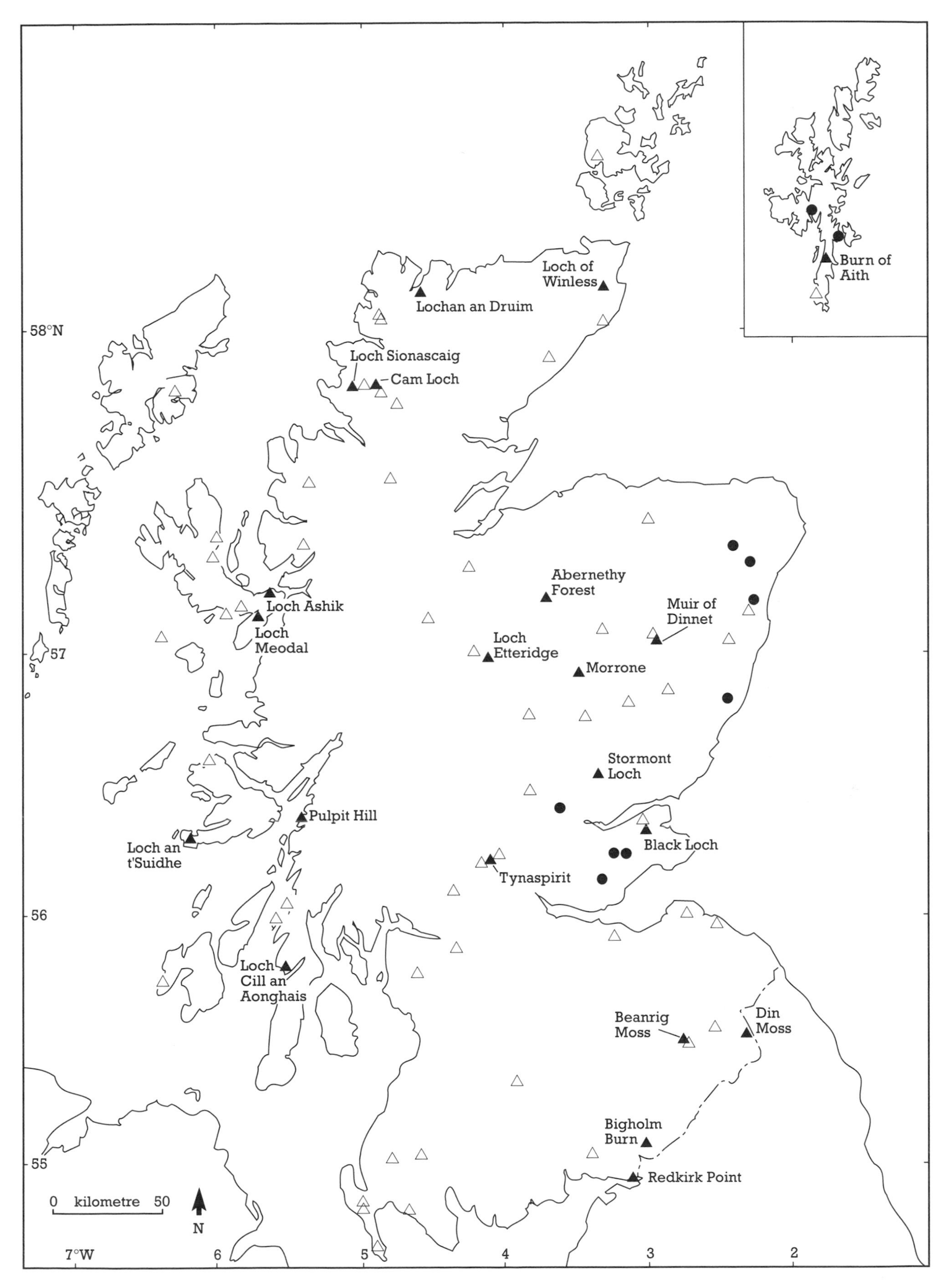
Loch of Winless
Lochan an Druim
Burn of Aith
58°N
Loch Sionascaig
Cam Loch
Abernethy Forest
Muir of Dinnet
Loch Ashik
Loch Meodal
Loch Etteridge
Morrone
57
Stormont Loch
Pulpit Hill
Loch an t'Suidhe
Black Loch
Tynaspirit
56
Loch Cill an Aonghais
Beanrig Moss
Din Moss
Bigholm Burn
Redkirk Point
55
0 kilometre 50
N
7°W
6
5
4
3
2
1.
2.
3.

south-east Grampians a closed grassland developed with juniper, willow, dwarf birch and, in sheltered areas, tree birch. Moss heaths and floristically poor grasslands were present on upper slopes.

Tree development during this part of the interstadial was limited to birch and pine, which occurred only in particularly favoured localities (Gray and Lowe, 1977b). Tree birch had a northern limit on the west coast in southern Skye (Loch Ashik, Loch Meodal) but it was also present along the east coast into Aberdeenshire and possibly even Caithness (Loch of Winless). Pine appeared only locally in the eastern central lowlands and Aberdeenshire (Black Loch, Muir of Dinnet).

The broad pattern of vegetation that developed during the interstadial shows vegetational zones which have a geographical expression similar to the vegetation zones that developed under the different climatic conditions of the middle Holocene (Birks, 1977; Bennett, 1989) and which have been identified in the relict vegetation of today (McVean and Ratcliffe, 1962).

The vegetational succession would appear to indicate that, for the greater part of the interstadial, the climate became milder (with a possible brief 'revertance' phase). As with the opening of the interstadial, however, environmental indicators with a faster response time to climatic change than plants suggest otherwise and in both the coleopteran and marine faunal records the main part of the interstadial has been found to be significantly cooler (by 2–3°C) than its opening phase shortly after 13,000 BP. It may be speculated that since such a drop of temperature would be compatible with glaciation of the highest mountains (Manley, 1949), then glaciers would have been likely to exist in the Highlands through the Lateglacial Interstadial. Sutherland (1987c) has suggested that the laminated sediments deposited throughout the interstadial at Loch Droma in Wester Ross (Kirk and Godwin, 1963) may be evidence for the continued existence of glaciers in the adjacent mountains.

Towards the end of the interstadial (between approximately 11,500 BP and 11,000 BP) both the vegetational and the coleopteran records indicate that a decline in temperature was under way as a precursor of the Loch Lomond Stadial. Interestingly, the nearshore marine faunal record indicates that there was a brief period at about 11,250–11,000 BP of warmer sea-surface temperatures (Peacock, 1981b, 1987, 1989b; Peacock and Harkness, 1990). It has been suggested (Peacock, 1987, 1989b) that this marine warming was due to a brief strengthening of the North Atlantic Drift as the oceanic polar front began to move southward in the later part of the interstadial. It can be speculated that this conjunction of a relatively warm sea and a cooling land (cf. Ruddiman and McIntyre, 1979; Ruddiman *et al.*, 1980) would have been particularly favourable for glacier development (Sutherland, 1984a).

During the early part of the Lateglacial Interstadial, sea level around most of the Scottish mainland coast, and particularly around the Highland margins, was falling rapidly. In the more peripheral areas, sea level was below its present level throughout this period and freshwater deposits accumulated within (and presumably below) the present intertidal zone (see Redkirk Point). In the later part of the interstadial, sea level everywhere around the Scottish coast was either below its present level or below the level attained following the Main Postglacial Transgression (see below), so that the course of sea-level change can only be deduced indirectly. Thus, from inferences based on the depth of water in which marine micro- and macrofaunas would have lived, Peacock *et al.* (1977, 1978) found that in the area of the sea lochs at the head of the Firth of Clyde, sea level was relatively stable during this period.

In summary, the Lateglacial Interstadial is one of the most intensively studied periods during the Scottish Late Pleistocene and a wealth of data have been accumulated on the natural environment and its evolution during the 2000 years after climatic amelioration at around 13,000 BP. The prime driving force of environmental change has been established to be the movement of the oceanic polar front in the North Atlantic. The studies that have been carried out have been rewarded with insights into the response of different parts of the environment to a common, major climatic stimulus followed by a period of mild but cooler climate than at present.

THE LOCH LOMOND STADIAL

At around 11,000 BP, the climatic decline that had become apparent towards the end of the Lateglacial Interstadial intensified and there was a return to arctic conditions during the Loch

Lomond Stadial. This period of severe climate, which lasted approximately 1000 years, has been registered by all palaeoenvironmental indicators that have been studied. It corresponds with a period of glacier readvance, a return to tundra and open-habitat plant communities throughout the country, the destruction of the soil profiles that had developed during the preceding interstadial and accompanying inwashing and solifluction of fine sands, silts and clays into lakes and enclosed basins, a return of polar waters to the neighbouring seas and the formation or reactivation of large-scale periglacial landforms and deposits (Sissons, 1979e).

Glaciation

The precise chronology of the glacier readvance (Loch Lomond Readvance) that reached its maximum during the climatic deterioration at the end of the Lateglacial has not been established. It is likely that glaciers began to readvance during the decline in climate at the end of the Lateglacial Interstadial. Radiocarbon dates on marine molluscs and terrestrial organic sediments that were transported or overridden by outlet glaciers of the western Highland ice-cap (see Western Forth Valley, Gartness, Croftamie, Rhu Point and South Shian and Balure of Shian) as well as on the Isle of Mull have confirmed the general age of the readvance to be between 11,000 and 10,000 BP (Sissons, 1967b; Peacock, 1971c; Gray and Brooks, 1972; Rose, 1980c, 1980e; Browne and Graham, 1981; Sutherland, 1981b; Browne *et al.*, 1983; Rose *et al.*, 1988; Peacock *et al.*, 1989; Merritt *et al.*, 1990). A similar inference that the readvance maximum occurred during the Loch Lomond Stadial can be made from a comparison of sediment sequences in enclosed basins 'inside' and 'outside' the readvance limit (Donner, 1957; Sissons *et al.*, 1973), the former typically containing only sediments from the end of the stadial and the Holocene, whereas the latter frequently have complete Lateglacial profiles in addition to Holocene sediments (see Mollands, Tynaspirit, Pulpit Hill and Loch Etteridge).

Details of the chronology of ice advance and retreat and the date of the maximum extent of the readvance are not well known, and locally, at least, it is possible that the glaciers did not all reach their maximum positions at the same time (Bennett, 1990). At Croftamie, Rose *et al.* (1988) have dated organic sediments underlying deposits of the Loch Lomond glacier, which have indicated that the readvance culminated after 10,500 BP in this area. This chronology is in agreement with the date of the readvance maximum in the Western Forth Valley which may be inferred from the shoreline sequence in that area (Sissons, 1983a; Sutherland, 1984a). Dates on ice-transported molluscs from South Shian and Balure of Shian (Peacock *et al.*, 1989) have also indicated that the Creran glacier reached its maximum late in the stadial, possibly as late as 10,000 BP.

The timing of the retreat and final disappearance of the glaciers has been studied using both pollen analysis and radiocarbon dating of the basal sediments in enclosed basins in the deglaciated areas (Walker and Lowe, 1981, 1982; Lowe and Walker, 1981, 1991). Radiocarbon dating of such deposits has produced inconsistent results (Sutherland, 1980; Walker and Lowe, 1980) but may suggest that deglaciation of the west Highland ice-field occurred between 10,600 BP and 10,200 BP (see Mollands and Kingshouse). A complementary approach to the establishment of the chronology of ice decay has been to examine the pollen assemblages from basal sediments in enclosed basins within the readvance limits. Such assemblages, when compared with those of regional pollen zonations for the end of the stadial and the early Holocene imply progressive ice retreat (Lowe and Walker, 1981, 1991; Walker and Lowe, 1985). These studies indicate that final deglaciation in the south-west Grampians, on Mull and on Skye was somewhat earlier than the maximum of *Juniperus* (Juniper) pollen values in the early Holocene (see below). However, this widely recognized feature of pollen diagrams is not well dated itself (Tipping, 1987) and could be as late as 9800 BP, which may imply final deglaciation of Scotland in the early Holocene.

The greatest extent of Loch Lomond Readvance glaciers was in the west Highland icefield (Figure 2.8) (Sissons *et al.*, 1973; Sissons, 1980b; Thorp, 1986, 1991a; Gray and Coxon, 1991) and significant subsidiary ice-caps and ice-fields existed on Mull (Gray and Brooks, 1972), Skye (Walker *et al.*, 1988; Ballantyne, 1989a; Ballantyne and Benn, 1991) and in the south-east Grampians (Sissons and Grant, 1972; Sissons, 1974b). In addition, numerous smaller glaciers developed in climatically favourable localities in the northern Highlands (Sissons 1977a), the eastern Grampians (Sissons, 1972a, 1979f) and the Southern Uplands (Sissons, 1967a; Cornish, 1981). Over

200 individual ice bodies have been identified (Sutherland, 1984a). Such a pattern probably reflects a combination of climatic factors and the form of the topography that controlled glacier initiation and subsequent development (Sissons and Sutherland, 1976; Payne and Sugden, 1990a, 1990b).

The limit of the readvance is marked in many areas by prominent end moraines. These may occur as massive accumulations of ice-pushed drift as at Menteith (see Western Forth Valley) and South Shian and Balure of Shian, as partly stratified deposits laid down in a glaciolacustrine environment (see Gartness) or as cross-valley ridges formed in a similar environment (see Glen Roy and Coire Dho), as bouldery ridges fronting corrie glaciers (see Lochnagar and the Cairngorms or as massive ridges of ice-transported debris (see Cnoc a'Mhoraire, Tauchers and An Teallach).

The Loch Lomond Readvance has also left a significant imprint on the landscape of many Highland valleys in the form of hummocky moraines; possibly the most outstanding example occurs at the Coire a'Cheud-chnoic in Glen Torridon. Another notable feature of the readvance was the creation of ice-dammed lakes as at Coire Dho (Sissons, 1977b), possibly (Ballantyne *et al.*, 1987) Achnaseen (Sissons, 1982a), Loch Tulla (Ballantyne, 1979) and, most famous of all, at Glen Roy (Jamieson, 1863, 1892; Sissons, 1978, 1979a, 1979c, 1981d; Peacock and Cornish, 1989). At all these localities former lake shorelines occur, but they are most clear and extensive at Glen Roy, where it has been demonstrated that they were cut mainly in rock and were faulted or warped, possibly due to rapid isostatic adjustments near the readvance limits (Sissons and Cornish, 1982a, 1982b). Sissons (1979a, 1979c) has also reconstructed the complex changes in lake drainage and river terrace development that occurred as the ice retreated from its maximum in Glen Roy and Glen Spean and hence has argued that there was one major and several more minor *jökulhlaups* (floods) during this deglaciation sequence. He suggested that a large fan of sand and gravel at Fort Augustus was formed by the major *jökulhlaup* and that the volume of water released was sufficient to raise the level of Loch Ness by several metres (see Dores) and also to build a large alluvial fan at Inverness blocking much of the Beauly Firth (Sissons, 1979c, 1981c).

The variations in the size of the glaciers and the altitude of their equilibrium lines allow inferences to be made about the climate during the period of glacier expansion. Mean summer temperatures of 7–9°C below those of the present have been deduced, and the pattern of precipitation was one of heavy snowfall in the western Highlands and along the Highland boundary, with markedly lower snowfall in the central and eastern Grampians and in the north of the country (Sissons, 1974b, 1980b; Sissons and Sutherland, 1976).

Terrestrial non-glacial environments

The severe climatic conditions of the stadial were reflected in the plant communities, as revealed by pollen analysis. A tundra-type vegetation occurred at low altitudes, with widespread presence of plants typical of unstable ground, although many of the plant communities that existed during this period appear to have no direct modern analogues (Walker, 1984b). Very high values of *Artemisia* pollen have been particularly noted in the eastern central Grampians, with declining frequencies to the west and south (see Abernethy Forest, Loch Etteridge, Stormont Loch and Loch an t-Suidhe) (Birks and Mathewes, 1978; Macpherson, 1980; Tipping, 1985). As certain species of *Artemisia* are chionophobus, the pollen values of that taxon have been interpreted as an indicator of precipitation during the stadial (Walker, 1975b; Birks and Mathewes, 1978; MacPherson, 1980; Lowe and Walker, 1986a). As such, the reduced levels of precipitation implied for the eastern central Highlands in general and Strathspey in particular correspond to the pattern of precipitation inferred from the distribution of contemporaneous glaciers (Sissons, 1980b). Certain pollen profiles also show variations in *Artemisia* pollen values during the stadial, with an early phase of low frequencies followed by a later phase of significantly higher frequencies (see Stormont Loch). Following the above line of argument, this change in *Artemisia* values has been interpreted as reflecting an early wet phase followed by a drier period (Caseldine, 1980a; MacPherson, 1980; Tipping, 1985, 1991a).

In southern Scotland knowledge of the stadial environment has been augmented by study of fossil Coleoptera (see Bigholm Burn), from which summer temperatures about 7°C below those of the present have been inferred (Bishop and Coope, 1977; Atkinson *et al.*, 1987).

Under the severe conditions, non-glacial geomorphological processes were particularly active. The most dramatic landforms produced at this time were in the mountains, where a number of protalus ramparts formed (Ballantyne and Kirkbride, 1986), including the remarkable feature on Baosbheinn in Wester Ross (Sissons, 1976c) which may have been partly formed by a landslip (Ballantyne, 1986a). The distribution and altitude of protalus ramparts in the Highlands shows a pattern remarkably similar to that of the contemporaneous glacier equilibrium line altitudes; they are lowest in the west of the country and increase in altitude towards the east (see the Cairngorms) (Ballantyne, 1984; Ballantyne and Kirkbride, 1986). It may be inferred that the precipitation variations that were a major control on the glacier distribution also influenced the formation of protalus features. The protalus ramparts also give information about the rate of rock-fall activity during the stadial (Ballantyne and Kirkbride, 1987), from which it is apparent that this process operated at a rate that is approximately an order of magnitude greater than at the present time and equivalent to some of the highest rates observed anywhere in the world today. Talus slopes formed during the Lateglacial are also of a size and maturity quite distinct from those that have accumulated during the Holocene (see An Teallach) (Ballantyne and Eckford, 1984).

The intense rock-fall activity also contributed to the development of a number of rock glaciers (see Beinn Shiantaidh and the Cairngorms), although the most spectacular of these, at Beinn Alligin in Wester Ross, is most probably the product of a landslip on to the surface of a decaying glacier (Sissons, 1975a; Ballantyne, 1987c). Landslip activity in general may have been much more intense during the stadial than during the late Holocene and Holmes (1984) has highlighted a relationship between a large proportion of such features and the limits of Loch Lomond Readvance glaciers. He suggested that raised porewater pressures in slopes adjacent to the glaciers played a crucial role in the weakening of the rock, with subsequent landslipping at some stage after ice retreat.

Mountain-top periglacial processes can be inferred to have been particularly active during the stadial from the mutually exclusive relationship in many mountain areas between Loch Lomond Readvance limits and large-scale periglacial features (see Lochnagar, Western Hills of Rum, An Teallach and the Cairngorms). Although the mountain-top detritus comprising the periglacial features may have been derived from periods of intense frost weathering during Late Devensian ice-sheet deglaciation or even, on certain summits, during the glaciation itself, the relict periglacial landforms that can be observed today were most probably fashioned during the Loch Lomond Stadial. The lowest altitude at which these features occur on mountains declines westwards and northwards, implying some form of regional climatic control on their formation (Ballantyne, 1984). The specific types of landform developed on any particular summit, however, are a product of the available topography and the type of regolith derived from frost weathering of the underlying bedrock, in addition to climatic variables (Ballantyne, 1984). Thus the summits underlain by granites (see the Cairngorms, Lochnagar, Western Hills of Rum and Ronas Hill) display different periglacial landforms from those underlain by schistose rocks (see Ben Wyvis and Sgùrr Mór), with further differences on summits underlain by sandstones (see An Teallach and Ward Hill).

At low altitudes, slope processes resulted in the destruction of the soil profiles that had developed during the preceding interstadial. In a number of areas interstadial peats or organic remains have been found underlying slumped or soliflucted material at the base of drift-covered slopes (see Bigholm Burn) (Donner, 1957; Gray, 1975b; Dickson *et al.*, 1976; Clapperton and Sugden, 1977). In enclosed basins the stadial is marked by a distinct layer of mineral sediment washed in from the surrounding slopes and frequently containing low concentrations of organic detritus derived from the interstadial soils (Lowe and Walker, 1977, 1986a; Pennington, 1977a; Walker and Lowe, 1990). River activity at low altitudes also appears to have been enhanced during the stadial, with large alluvial fans being built across interstadial sediments in the Lochwinnoch Gap (Ward, 1977), at the foot of the Ochil Hills (Kemp, 1971) and near Corstorphine in Edinburgh (Newey, 1970; Sissons, 1976b). A number of fossil frost wedges at low altitudes have been assigned to the stadial, suggesting permafrost conditions and mean annual temperatures of below -5°C (Rose, 1975; Sissons, 1977a, 1979e), but as none of these has been unequivocally dated to the stadial, climatic inferences must remain tentative.

Marine environments and sea-level change

The severe climate of the stadial was also registered in the contemporaneous marine fauna. Boreal species present during the interstadial disappeared and were replaced by high-arctic species such as *Portlandia arctica* (Gray) (Peacock *et al.*, 1978; Peacock, 1981b, 1983a, 1987, 1989b; Graham *et al.*, 1990). Nearshore marine summer temperatures of more than 10°C below present levels are implied. In offshore sediments, recognition and correlation of deposits from the stadial have been aided by the occurrence of shards of volcanic ash (the Vedde Ash) which has been dated to approximately 10,600 BP (Long and Morton, 1987; Graham *et al.*, 1990). This ash layer is considered to be derived from a volcano in Iceland and has been found widely along the west coast of Norway (Mangerud *et al.*, 1984) but not on-land in Scotland. Its occurrence in marine sediments off both the east and west coasts of Scotland may, however, have been due to redistribution by sea ice, depending on the time of the year when the volcano was in eruption.

In the Forth valley and along the Firth of Forth, a distinct erosional surface has been identified cutting across Lateglacial marine sediments, till and bedrock. It is immediately overlain by early Holocene sediments (Sissons, 1969, 1976a) and has been accordingly inferred to have been formed towards the end of the Lateglacial. The inner margin of the surface marks a clear, isostatically tilted shoreline termed the Main Lateglacial Shoreline. Its distinctive erosional nature in a low-energy environment which was normally characterized by fine-grained estuarine deposition, together with its stratigraphic position, led Sissons (1976b) to conclude that the feature was formed mainly during the severe climatic conditions of the stadial. A similar surface has been found in the inner Beauly Firth (see Barnyards) (Sissons, 1981c; Firth, 1984).

Around the coasts of the south-west Highlands and the neighbouring islands there is also a distinctive marine erosional feature (see Isle of Lismore, Glenacardoch Point and West Coast of Jura), the Main Rock Platform (Gray, 1974a; 1978a; Dawson, 1980b, 1988a; Gray and Ivanovich, 1988), which is tilted to the south and west in conformity with the local pattern of isostatic deformation. It also has a gradient similar to that of the Main Lateglacial Shoreline of the east coast. These similarities led Sissons (1974d) to conclude that the Main Rock Platform had also been produced by particularly active frost erosional processes in the intertidal zone. Such a hypothesis received support from those who subsequently worked on the platform (Gray, 1978a; Dawson, 1980b; Sutherland, 1981b) and the concept of rapid erosion of bedrock in a littoral environment during the stadial appeared to be confirmed by evidence from the Parallel Roads of Glen Roy. More recently, however, certain evidence seems to suggest that the platform may have been in part inherited from a pre-existing feature or features (Browne and McMillan, 1984; Gray and Ivanovich, 1988), thus diminishing (though not excluding) the need to postulate such active erosional processes during the stadial.

It is possible that the ice masses that developed during the readvance were of sufficient magnitude to re-depress the crust isostatically (Sutherland, 1981b; Firth, 1986, 1989c; Boulton *et al.*, 1991), and in the Western Forth Valley the maximum of the readvance coincides with a marine transgression which followed the formation of the Main Lateglacial Shoreline. At the peak of this transgression a distinct shoreline (the High Buried Shoreline) was formed seawards of the Menteith Moraine, but it is not found within the limits of the readvance (Sissons, 1966, 1983a). This rise of sea level, which was at least 8 m, coincided with the readvance maximum and is most easily explicable as the result of renewed isostatic depression, as there does not appear to be a significant coincidental eustatic rise in sea level that would otherwise explain it. Using a calculation of shoreline gradient against age, it is possible to infer that the High Buried Shoreline was formed between 10,000 BP and 10,500 BP (Sutherland, 1984a), thus placing a limit on the time of the maximum of the readvance in the Forth valley.

In summary, the Loch Lomond Stadial was a period of particularly marked environmental change and it was also the last period when glaciers existed in Scotland. In contrast to the preceding interstadial and subsequent early Holocene, the stadial demonstrates the impact of a severe, albeit short-lived (in geological terms), climatic deterioration and shows the sensitive and marginal nature of the Scottish environment to such a change. An interesting footnote is the possibility that Man was present in Scotland at this time, as has been inferred from a collection of reindeer antlers in the Creag nan Uamh caves in Sutherland (Lawson and Bonsall, 1986a, 1986b).

THE HOLOCENE

The boundary between the Late Devensian and the Holocene is conventionally placed at 10,000 BP, although the actual climatic amelioration at the end of the Loch Lomond Stadial may have occurred somewhat prior to this. The change in climate coincided with the return once more of North Atlantic Drift waters to the Scottish coasts. Peacock and Harkness (1990) have indicated that the warming occurred in two phases, with a period between approximately 10,100 BP and 9600 BP when marine temperatures were similar to those during the Lateglacial Interstadial, that is 2–3°C below the present level. The greater part of the warming that occurred slightly prior to 10,100 BP may have taken only 40 years (Peacock and Harkness, 1990). Similar extremely rapid temperature changes have been inferred from fossil coleopteran assemblages, and atmospheric temperature changes could have been of the order of 1°C per decade (Coope, 1977; Atkinson *et al.*, 1987).

The brief period of 'interstadial' conditions at the beginning of the Holocene that has been deduced from the nearshore marine faunas has not been noted in the vegetational record as inferred from pollen analytical studies, possibly because of the slow response time of plants to the major climatic amelioration. However, at two sites on Mull (see Gribun) and on Skye, which have long early Holocene sequences and hence permit very detailed analysis of vegetation changes, brief vegetation 'revertance' phases have been noted in the early Holocene (Dawson *et al.*, 1987a; Lowe and Walker, 1991).

Vegetational development

Immediately after the amelioration of climate, the vegetation throughout most of Scotland was dominated by open-habitat taxa, these being succeeded, as during the early part of the Lateglacial Interstadial, by shrub and scrub vegetation. Unlike the opening of the earlier interstadial, however, when an initial mild phase was soon followed by deterioration in climate with temperatures declining to somewhat lower than at present, in the Holocene the mild climate was maintained and hence the pattern of vegetational development was quite distinct from that of the interstadial. The dwarf shrub and scrub phase saw successive dominance in much of the country of *Empetrum* and then *Juniperus* (this phase being particularly marked). Subsequently there followed expansion of mixed deciduous woodland.

The pattern and timing of the spread of tree species into Scotland during the early to middle Holocene may broadly be conceived in terms of expansion from the south, but when viewed in detail this is clearly an oversimplification (Huntley and Birks, 1983; Birks, 1989). *Betula* was the first tree to spread across the country, which it did between 10,000 BP and 9000 BP. Earliest expansion was in the south-east and east, which may partly be explained by spreading from areas of dry land in the south and central North Sea Basin. *Corylus* (hazel) shows a completely different spreading pattern, with very early, rapid migration along the western coastal fringe and the Inner Hebrides, the northern coast being reached by 9500 BP. Thereafter, by 9000 BP, it had expanded across southern and central Scotland and, by 8500 BP, into the central Highlands and the north-east. The initial spread along the west coast may have been due to water transport of hazel nuts combined with the occurrence (by analogy with today) of infrequent frosts along the western coastal fringe.

Ulmus (elm) and *Quercus* (oak) both showed a broadly similar south-to-north spreading pattern. *Ulmus* arrived first, expanding across southern and central Scotland and into the Highlands between 9000 BP and 8500 BP. This rapid expansion, at around 500 m a^{-1}, thereafter slowed and the tree did not reach the north-east until 8000 BP and Caithness until 6500 BP. *Quercus* spread through southern and central Scotland and into the south-west Highland coastal fringe between 8500 BP and 8000 BP and had reached its northern limits in southern Skye and on the southern shore of the Moray Firth by 6000 BP.

The spread of *Pinus* (pine) had perhaps the most curious pattern (Bennett, 1984; Birks, 1989). Its earliest occurrence was in the northern Highlands at around 8500–8000 BP (Birks, 1972b) (see Loch Maree and Loch Sionascaig) and there was subsequent spread into the eastern Highlands at around 7500 BP (see Abernethy Forest) and throughout much of the central Grampians by 5000 BP. Pine was not a significant element in the woodlands at any time through

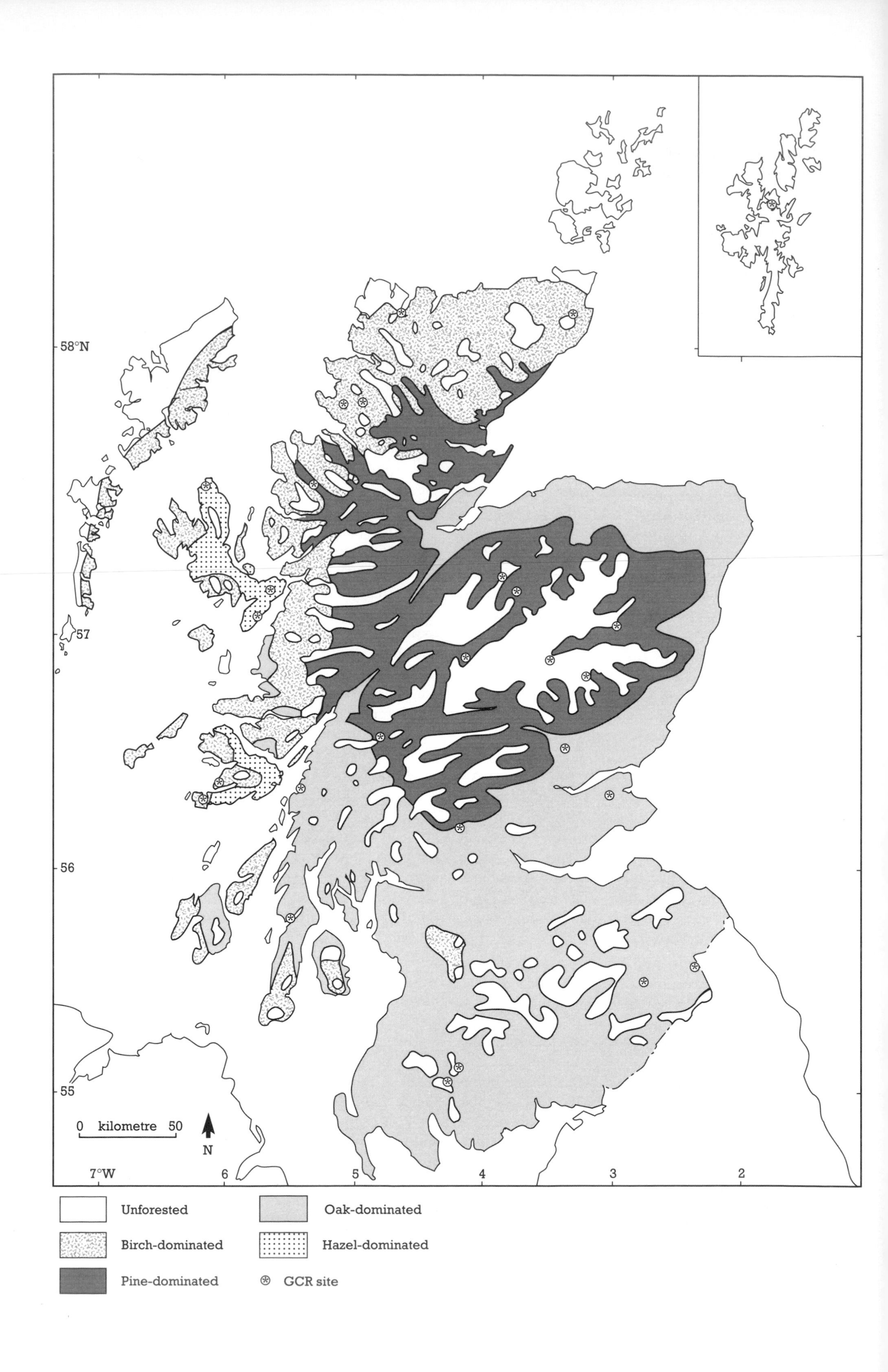
58°N
57
56
55
7°W
6
5
4
3
2
0 kilometre 50
N
Unforested
Birch-dominated
Pine-dominated
Oak-dominated
Hazel-dominated
GCR site

almost the whole of south-central Scotland and the south-west Highland coastal fringe (Bennett, 1984), an exception occurring in the western Southern Uplands (Birks, 1972a, 1975) (see Loch Dungeon), where expansion from Ireland at around 7500–7000 BP occurred (Birks, 1989). This unusual pattern is as yet unexplained, but the northern Scots Pine may be genetically distinct from that which occurred in southern Britain (Kinloch *et al.*, 1986), suggesting expansion from a different source area.

Alnus (alder) also exhibits an unusual pattern of spread (Birks, 1989; Bennett and Birks, 1990). It expanded through Scotland between 9000 BP and 5000 BP, but its occurrence was sporadic in both space and time, apparently due to its specific wet, mildly basic habitat requirements.

By the later part of the middle Holocene, the Scottish forests had reached their greatest extent and degree of diversity (Figure 2.12) (Bennett, 1989). In the very north-west of Sutherland and the northern and western island fringes there was little forest development, but birch and pine stumps have been recovered from peat profiles (see Garths Voe), suggesting scattered occurrences of woodland in favoured locations during the middle to early part of the late Holocene. After 5000 BP, a change to a cooler, moister climate together with human impact led to a reduction in tree cover and a corresponding expansion of blanket peat, heaths and grasslands.

The impact of Man was first noted slightly prior to 5000 BP, with a marked decline in elm pollen throughout the country (see Din Moss, Black Loch, Loch Cill an Aonghais, Loch Ashik and Loch Maree). This decline was accompanied or followed by increases in grasses, heaths and species that are considered typical of cultivation. In more recent times, sites such as Black Loch (Whittington *et al.*, 1991a), which have rapid sedimentation rates, have allowed identification of the timing of the introduction of exotic species as well as changes in agricultural practice during historical times. Of particular interest are the detailed studies that have been carried out using pollen and diatom analyses on the acidity of waters in enclosed basins. The most detailed such study, at Round Loch of Glenhead in the western Southern Uplands (Jones *et al.*, 1989), has reconstructed the acidity of the basin waters since the Lateglacial and demonstrated clearly the impact of industrial pollution of the area in the last 200 years.

Figure 2.12 Forest zones at approximately 5000 BP (from Bennett, 1989), and GCR sites with palynological data for the Holocene.

Sea-level change

Holocene changes of sea level were quite different along those parts of the coast strongly influenced by glacio-isostatic effects of the last ice-sheet (and probably also by the Loch Lomond Readvance ice-field) and the more peripheral regions where isostatic effects were minimal. In these latter regions, the world-wide glacio-eustatic rise of sea level during the early to middle Holocene was the dominant influence.

The most detailed studies of early Holocene sea-level change have been made along the coasts and the estuaries of eastern Scotland. The isostatic effects in this region were greatest in the Western Forth Valley and near the head of the Beauly Firth (see Barnyards) and declined eastwards, being least in north-east Scotland (see Philorth Valley). In the areas most influenced by isostatic uplift, relative sea level in the early Holocene was initially high, with a prominent shoreline (Main Buried Beach) being formed in the Western Forth Valley (Sissons, 1966, 1972b) at around 9600 BP. This shoreline is probably also present in the Tay–Earn valley (see Carey) (Cullingford *et al.*, 1980) and in the Beauly Firth area (see Barnyards). Following the abandonment of this shoreline, sea level fell and a lower beach (Low Buried Beach) was formed at around 8600 BP. Then sea level again dropped, reaching its lowest level at between 8500 BP and 8300 BP in the Forth, Tay and Beauly Firth areas (Sissons and Brooks, 1971; Cullingford *et al.*, 1980; Haggart, 1989). Studies on the west coast of Scotland, along the Solway Firth and the Firth of Clyde, also suggest that this period was one of relatively low sea level, with peat formation at altitudes close to present sea level (see Newbie, Redkirk Point and Dundonald Burn) (Jardine, 1975, 1977, 1980b; Bishop and Coope, 1977; Sutherland, 1981b; Boyd, 1982).

The early Holocene fall of sea level in those areas near to the centre of isostatic uplift was a consequence of the rate of uplift exceeding the rate of world sea-level rise. Subsequently, however, this pattern was reversed and a transgression (the Main Postglacial Transgression) occurred everywhere around the Scottish coast. This relative sea-level rise resulted in the deposition of estuarine and marine sediments on top of early Holocene peats that had formed during the period of low sea level. Thus the progress of the transgression is marked by the time-transgressive nature of the overlap of the marine or estuarine sediments on to the terrestrial deposits. This process has been clearly registered at sites around the Solway Firth and the Firth of Clyde (see Redkirk Point, Newbie and Dundonald Burn), in the Western Forth Valley, along the Tay estuary and neighbouring parts of the east coast (see Carey, Silver Moss and Maryton) and at the head of the Beauly Firth (see Barnyards).

At a number of sites along the east coast the apparently simple stratigraphy described above is interrupted by a distinct thin bed of fine sand which cuts across the estuarine sediments, tapering out in the adjoining terrestrial peats. This bed contains marine diatoms and has been radiocarbon dated to around 7000 BP (see Barnyards, Maryton, Silver Moss and Western Forth Valley). It has been interpreted as either the product of a major storm surge in the North Sea Basin (Smith *et al.*, 1985a; Haggart, 1988b) or the result of a tsunami caused by a major submarine slide on the edge of the Norwegian continental shelf to the north of the North Sea (Dawson *et al.*, 1988; Long *et al.*, 1989a). It is an important time-stratigraphic marker in the east coast marine and estuarine sediments, the variations in altitude of which, for example, provide precise information on the regional pattern of isostatic uplift in the last 7000 years (Long *et al.*, 1989a).

The maximum of the transgression appears to have been reached at different times in different parts of the country, being earliest in the Western Forth Valley (at around 6800 BP: Sissons, 1983a) and generally later in areas farther from the centre of isostatic uplift (see Silver Moss and Philorth Valley) (Smith *et al.*, 1983; Cullingford *et al.*, 1991). A particularly detailed study of the period close to the maximum of the transgression at Pitlowie in the Tay estuary (Smith *et al.*, 1985b) has revealed that sea level, after rising rapidly to near its maximum altitude, stayed relatively stable for approximately 1000 years. This relatively stable period may be registered as a double peak to the transgression in areas of greater isostatic uplift such as Loch Lomond (see South Loch Lomond), which was an arm of the sea at this time (Dickson *et al.*, 1978; Stewart, 1987; Browne and McMillan, 1989). A particularly prominent depositional shoreline, the Main Postglacial Shoreline, was formed at this time and it has been shown to be isostatically tilted away from a centre of uplift in the south-west Highlands (Sissons, 1983a; Cullingford *et al.*, 1991) (see Western Forth Valley, West Coast of Jura, Munlochy Valley and Dryleys).

The phase of transgression was the result of a rapid rise in world sea level that terminated at around 6000 BP (by which time both the Scandinavian and the North American ice-sheets had melted) and, as isostatic uplift was still continuing around most of the Scottish coast, albeit at a slower rate than in the early Holocene, a fall in sea level ensued. This fall is not well dated, but a series of isostatically tilted shorelines were formed (see Munlochy Valley, Dryleys and West Coast of Jura), possibly during halts in the general regression.

In the more peripheral areas there is little direct information on sea levels during the early Holocene, which appear everywhere to have been much lower than at present. By the middle Holocene, relative sea level was rising to close to the present level in the Outer Hebrides (see Borve) (Ritchie, 1966, 1985) but was apparently significantly lower than this in Shetland (Hoppe, 1965). The middle to late Holocene in these areas appears to have been a time of relatively stable or slightly rising sea level and, in contrast to the glacio-isostatically affected areas, no Holocene raised shorelines have been found.

Geomorphological processes

The establishment of soil and vegetation covers during the early Holocene together with a milder climate resulted in a marked diminution in the intensity of geomorphological activity compared with that of the Loch Lomond Stadial. However, the Late Devensian left a legacy in the form of large volumes of unconsolidated debris and oversteepened slopes. Consequently, during the early Holocene there appears to have been a considerable number of slope failures (Watters, 1972; Holmes, 1984; Ballantyne, 1986d, 1991c) as well as frequent reworking of the available debris by streams and debris flows to produce, respectively, river terraces and debris cones in many Highland valleys (see Eas na Broige and Glen Feshie). This activity appears to have slowed during the middle to late Holocene, but there has been a resurgence in the last 200–300 years (cf. Ballantyne, 1991c), possibly due to the effect of the climatic deterioration termed the 'Little Ice Age' (Brazier and Ballantyne, 1989) and possibly also under the influence of increased grazing pressure in the upland areas (Innes, 1983b; Brazier, 1987; Brazier *et al.*, 1988). Apart from in a few cases, however, the precise causes remain uncertain (Innes, 1983a; Ballantyne, 1991a).

Periglacial activity has continued on hill tops and mountain summits throughout the Holocene (see Ward Hill, Western Hills of Rum, An Teallach, Sgùrr Mór, Ben Wyvis, the Cairngorms and Tinto Hill), but, as at lower altitudes, the intensity of the processes has been markedly less than during the Loch Lomond Stadial (Ballantyne, 1987a, 1991c). Small-scale forms are typical of the Holocene and they show the same dependence of type on the underlying regolith as described above for the large-scale periglacial features formed during the colder periods. There also occurs a marked diminution northwards and westwards in the altitude at which given features form and, as noted above for the larger examples, this again implies an overall regional climatic control.

Where weathering of the underlying bedrock has released considerable quantities of sand, certain hills have areas covered in sand sheets (see Ronas Hill, Ward Hill and An Teallach) which contain organic layers. Dating of the latter suggests that during the early Holocene there was a period of sand accumulation followed, during much of the middle and late Holocene, by relative inactivity. However, in recent centuries there has been reactivation of sand movement, possibly due to grazing pressure (Ball and Goodier, 1974; Ballantyne and Whittington, 1987). As vegetation cover also has a major role in stabilization of the regolith, reduction in vegetation cover due to grazing pressure and/or climatic deterioration seems to have been responsible for a recent increase in the area of patterned ground on Tinto Hill (Miller *et al.*, 1954) and rapid solifluction on the Fannich mountains in the last few centuries (Ballantyne, 1986c).

CONCLUSION

There has been an upsurge in studies of the Quaternary during the last 20 years and, with the application of dating techniques as well as micropalaeontological analyses to both the marine and the terrestrial records, the foundations have been laid for a greatly improved understanding of the Earth's climatic system and the factors that govern climatic change. It has become apparent that relatively minor changes in the Earth's orbital parameters are amplified by 'feedback loops' in the atmosphere–ocean–cryosphere–biosphere system to produce the major climatic changes that have characterized the Quaternary. The North Atlantic oceanic circulation is a particularly important element in the Earth's climatic system, for not only is that ocean the source of the greater part of the precipitation that nurtured the principal northern hemisphere ice-sheets during ice ages but the North Atlantic is also, at present, the principal area of formation of oceanic deep water. Formation of deep water releases large amounts of heat to the North Atlantic surface waters and helps maintain the equable climate experienced in north-west Europe. During glaciations, however, deep water formation is retarded or halted, with consequent impact on the climate of the surrounding land masses.

The position of Scotland, jutting into the North Atlantic and with the west coast washed by the North Atlantic Drift, has therefore produced a climate and natural environment that are particularly dependent on and sensitive to changes in the circulation of the North Atlantic. Consequently, during the Quaternary Scotland has at different times been completely covered by ice, partially glaciated with accompanying intense non-glacial geomorphological activity beyond the ice margins, or, as at present, has experienced a mild, temperate climate. Coincident with the changes between the different environments there have been great fluctuations in sea level around the Scottish coasts and on the neighbouring continental shelves from more than 40 m above to less than 120 m below present level.

The challenge presented by these environmental changes is to establish their pattern and chronology from the evidence around Scotland and to understand how the different parts of the environmental system respond. In a global context, Scotland is significant in certain respects. First, the climatic record of its onshore and offshore deposits should mirror the changes in the circulation of the North Atlantic. Second, the relatively small size of the ice masses that developed in Scotland implies that they would have had a faster response time to climatic change than the Scandinavian and North American ice-sheets. Consequently, it may be anticipated that the glaciers and accompanying environmental fluctuations in Scotland were a sensitive monitor of global climatic change, particularly those changes of higher frequency and lower magnitude. At a national scale, knowledge of the mechanisms controlling change in the natural environment provides the necessary baselines against which human impacts on the environment can be assessed and differentiated from 'natural' environmental change. Also, by providing a temporal perspective, it allows a better understanding of the diversity of the natural environment and the potential rates of change arising from natural perturbations or human interference.

As is apparent from this review of current understanding of the Scottish Quaternary, much work remains to be done before even a reasonably complete record of Quaternary events is established. However, the existing information has offered important insights, at a variety of scales, into environmental stability and response times as well as the rates of change possible in different parts of the environment once thresholds are crossed and change initiated. The present review therefore concludes with three examples of response to environmental change which highlight the importance of thresholds in the natural environment and the small changes that may be sufficient to switch natural systems rapidly from stable to unstable states or from one level of dynamic equilibrium to another.

During the Quaternary, the evidence of both temperature and ice-sheet volume changes derived from deep-sea sediments indicates that Scotland is likely to have been glaciated on numerous occasions. However, in this review evidence has been found for only four (less certainly, six) major expansions of ice-sheets across the lowlands and on to the adjacent continental shelves. Even accepting that the record is incomplete, the implication is that major ice-sheet expansion is relatively uncommon. In addition, it has been suggested above that the expansion of the Late Devensian ice-sheet from the Highlands to its maximum and subsequent retreat to within the Highlands once more may only have taken approximately 10,000–12,000 years, with average rates of advance and retreat of the ice margin of 30 m a^{-1}. Such rates are of a similar order of magnitude to those observed at the North American and Scandinavian ice-sheet margins and, as noted by Ehlers *et al.* (1991), imply ice velocities comparable to the fastest Antarctic and Greenland ice streams, which are difficult to comprehend for an ice-sheet as a whole. In addition, almost the whole of the retreat of the last Scottish ice-sheet from its maximum position to within the Highlands occurred during a time of cold climate. The behaviour of the Scottish ice-sheet therefore appears to be one in which it is confined within the mountain zone until a certain threshold is crossed, whereupon it expands and then retreats rapidly. The ice-sheet modelling experiments of Payne and Sugden (1990b) may provide insight into one of the mechanisms responsible for this behaviour, for they have observed that the intra-mountain basins (such as Rannoch Moor) provided a threshold for ice-sheet expansion. They noted that only a very slight change was necessary in the climatic parameters controlling glacier development for a stable intra-mountain ice-field to change to a rapidly growing ice-sheet and that this occurred when the intra-mountain basins changed from being areas of ablation to areas of accumulation.

A quite different example is the rates of change in the biosphere that resulted from the major climatic ameliorations during the Lateglacial and at the opening of the Holocene, which have been found to be quite distinct between different species. The climatic changes, of the order of 5–10°C or more increase in mean annual temperatures, have been estimated to have taken place in only several decades and certain species of Coleoptera as well as of marine micro- and macro-faunas have been able to respond within these time periods. However, there was a delayed response by most plant species, not least because many have complex habitat requirements. At the opening of the Lateglacial Interstadial, however, climatic conditions similar to those of today were only maintained for 100–300 years before temperatures fell a few degrees. This decline was sufficient to preclude the spread of trees into Scotland during the interstadial, with the exception of tree birch in the south and east and possibly pine in favoured areas. In contrast, at the beginning of the Holocene the mild climate was maintained and created conditions that were suitable for trees to spread and through the southern and central parts of the country rates of expansion were very high, between 150 and 500 m a^{-1} on average. Towards the limits of their ranges in the north of the country expansion was much slower, typically less than 100 m a^{-1}. These rates of expansion are far higher than can be understood in terms of propagation directly from trees and additional mechanisms such as water or bird transport need to be invoked.

Finally, it is apparent from the studies of Holocene geomorphological processes that there is a very fine balance between stability and erosion in Scottish upland areas and that during the last 200–300 years this balance has been tilted in favour of erosion. This has been noted in increased debris-flow and alluvial fan activity, erosion of blanket peat, renewed activity of mountain-top aeolian sands and, at least locally, rapid mass movement of frost-derived debris on mountain summits. Upland blanket peats have accumulated during the Holocene at average rates of between 1.5 and 5 cm $100\,a^{-1}$. Today, almost throughout the upland areas, peat erosion, in places to depths of over 1 m, is occurring. This phase of erosion commenced 200–300 years ago and is apparently far more intense than occasional brief periods of erosion during the late Holocene (Stevenson *et al.*, 1990). The precise mechanism that has induced this erosion is not yet understood but may be the result of increased burning or grazing pressure or it may be a response to the climatic downturn known as the 'Little Ice Age'.

These three examples illustrate at both regional and local scales the rapidity with which the natural environment can change. All the changes are, or would have been, appreciable on a 'human' time-scale measured in decades. The causes of the changes were principally natural, although the role of human impact cannot yet be fully separated from natural changes in the third example. In all the examples there appears to be a rather narrow dividing line between an essentially 'stable' condition and rapid change. Future Quaternary studies will lead to elucidation of the factors involved and also a better understanding of landscape sensitivity, which is essential in developing sustainable environmental management and use of natural resources.

Chapter 3

The Shetland Islands

INTRODUCTION

D. G. Sutherland and J. E. Gordon

The Shetland Islands (Figure 3.1) are located approximately equidistant from Norway and the mainland of Scotland, and at the junction of three distinct oceanic areas: the North Sea to the east and south-east, the Norwegian Sea to the north, and the Atlantic Ocean to the west. Their situation is thus critical both for the study of the extent of the Scandinavian and Scottish ice-sheets and for the derivation of terrestrial palaeoclimatic information that may be compared with the contrasting marine records from the surrounding seas. The importance of Shetland's location was recognized in the last century by Croll (1870a, 1875), who considered it possible that the islands had been glaciated by ice originating in Scandinavia rather than in Scotland. Peach and Horne (1879) sought to test Croll's hypothesis in the field. They interpreted an initial ice-sheet glaciation from an external source, possibly Scandinavia, followed by a phase of local ice-cap glaciation. These ideas have formed the basis of subsequent studies of the glaciation of Shetland (Sutherland, 1991b). Shetland is also notable for the preservation of two probable interglacial deposits which suggest that the local Quaternary stratigraphy may cover, albeit sporadically, approximately the last 400 ka (Sutherland, 1984a).

Organic deposits that are older than 13,000 BP and overlain by till are uncommon in Scotland (Sissons, 1981b; Lowe, 1984; Sutherland, 1984a; Bowen *et al*., 1986). Two such deposits, probably interglacial in origin and probably dating from different Pleistocene stages, occur on the Mainland of Shetland. These two deposits, at Fugla Ness and Sel Ayre, have been studied in detail for both pollen and plant macrofossils (Birks and Ransom, 1969; Birks and Peglar, 1979). The plant assemblages contain several taxa that are now either extinct on Shetland or extinct in the British Isles. Even though the pollen assemblages are dominated by dwarf-shrub and herb pollen, comparisons with Holocene assemblages from Murraster, on the Mainland (Johansen, 1975), suggest that the Fugla Ness and Sel Ayre assemblages are interglacial (*sensu* Jessen and Milthers, 1928) in character. Numerical comparisons of the pollen spectra at Fugla Ness and Sel Ayre indicate that they differ in composition and that they are thus probably of different ages.

The site at Fugla Ness contains the earliest known Quaternary sequence in Shetland as well as the best documented, most northerly and one of the oldest interglacial deposits in Scotland. On the basis of the pollen content and plant macrofossil remains, Birks and Ransom (1969) suggested that the Fugla Ness peat may date from the Hoxnian, but the basis for this correlation is not secure (Lowe, 1984). Accepting that certain aspects of the pollen record imply a Middle Pleistocene age and that the thermophilous nature of the macrofossils indicates correlation with a period of very mild oceanic climate, Sutherland (1984a) suggested that the Fugla Ness peat may correlate with the event recorded in deep-sea cores at around 380 ka.

A further peat bed underlying till has been reported from Shetland, at Sel Ayre (Mykura and Phemister, 1976; Birks and Peglar, 1979). On the basis of its pollen content this peat has been tentatively assigned to the Ipswichian (Birks and Peglar, 1979), but both its interglacial status and its correlation with the Ipswichian may be questioned (Lowe, 1984). There is no information from Sel Ayre about glaciation prior to the deposition of the peat. However, the overlying till contains erratics of local sandstone as well as rare basic lavas which have been derived from the south-east (Mykura and Phemister, 1976). The ice that deposited this till may therefore be correlated with local ice-cap glaciation during the Late Devensian.

Investigations of the glacial history of Shetland have generated different interpretations of events (Coque-Delhuille and Veyret, 1988). Early studies (Hibbert, 1831; Peach, 1865a; Croll, 1870a; Helland, 1879) recorded the presence of striations and erratics, but until recently the definitive work on the glaciation of Shetland was that of Peach and Horne (1879) which was based on extensive field observations relating to roches moutonnées, striations, distributions of erratics and lithological variations in till composition. This evidence was interpreted as indicating ice movement from the North Sea towards the Atlantic, ice of Scandinavian origin possibly being deflected to the north-west by the presence of a Scottish ice-sheet further south. Evidence from striations and till lithologies on the eastern side of Shetland, together with morainic deposits in the valleys, was also used to demonstrate that local glaciers moved off the higher ground and into the valleys after the recession of the ice-sheet. Subsequently, in response to critical comments from Milne Home (1880b, 1881a, 1881b, 1881d), Peach and Horne (1881a, 1881b) elaborated both

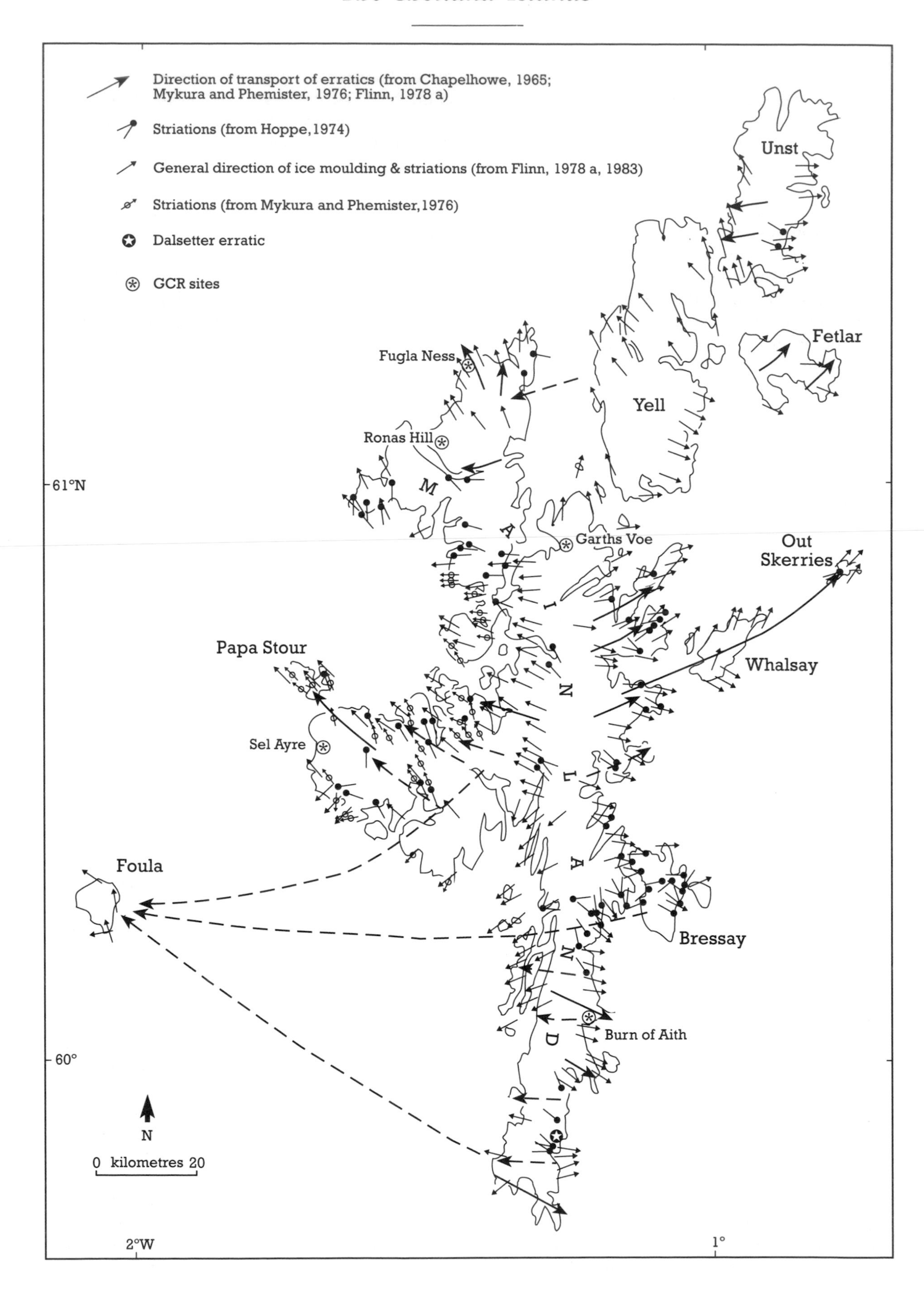

Direction of transport of erratics (from Chapelhowe, 1965; Mykura and Phemister, 1976; Flinn, 1978 a)
Striations (from Hoppe, 1974)
General direction of ice moulding & striations (from Flinn, 1978 a, 1983)
Striations (from Mykura and Phemister, 1976)
Dalsetter erratic
GCR sites
Unst
Fetlar
Yell
Fugla Ness
Ronas Hill
61°N
M
A
I
N
L
A
N
D
Garths Voe
Out Skerries
Whalsay
Papa Stour
Sel Ayre
Foula
Bressay
Burn of Aith
60°
N
0 kilometres 20
2°W
1°

on the field evidence and their interpretation of it.

Since the paper by Peach and Horne (1879) the debate as to the nature of the last glaciation of Shetland and the relative importance of Scandinavian and local ice has been concerned almost exclusively with two types of evidence: striated and ice-moulded rock surfaces and the distribution of erratics (Figure 3.1). In assessing the influence of Scandanavian ice, Hoppe (1974) placed emphasis on an older set of striations occurring on and to the west of Bressay, which he considered to have been formed by ice flowing from the north-east. However, this interpretation of the sense of direction of some of these striations has been contradicted by Flinn (1977, p. 141). Hoppe inferred that during an early stage of glaciation Shetland was overridden by an ice-sheet from the east, probably from Scandinavia; during a later stage, a local ice-cap formed over the islands. Further, since the patterns of striations suggested progressive change in the ice movements, Hoppe concluded that both stages were of Late Devensian age. Following the ice maximum, rapid calving of the margin of the Scandinavian ice in the northern part of the North Sea led to the isolation of the local Shetland ice-cap.

Flinn (1977, 1978a) argued that the only evidence for the glaciation of Shetland by Scandinavian ice was in the southern part of the islands. There, westward transport of certain local rock types across the reconstructed ice shed of the later local ice-cap, at least as far as Foula, indicated that Scandinavian ice crossed the area during either a previous glaciation or an early phase of the last glaciation. It is also in this area, at Dalsetter, that the only Scandinavian erratic that has been found in Shetland occurs (Finlay, 1926). However, the history of transport of a single erratic may be complex (Hoppe, 1974). The westward transport of erratics has also been described from certain of the northern isles and here too the influence of Scandinavian ice has been invoked (Mykura, 1976).

The evidence for Scandinavian ice reaching Shetland is therefore rather weak and relates solely to the margins of the island group. In particular, it may be noted that no shelly tills or North Sea Basin erratics have been reported from Shetland, a situation in marked contrast to the eastern Orkney Islands. It may be concluded that if Scandinavian ice did cross the Shetland Islands, the evidence for this has been largely obliterated by the subsequent local glaciation(s). It is also clear that any such ice movements pre-dated the Late Devensian, since offshore studies have shown that the floor of the central North Sea was unglaciated at that time (Jansen, 1976; Flinn, 1978a; Cameron *et al.*, 1987; Sejrup *et al.*, 1987).

Figure 3.1 Location map and principal features of the glaciation of Shetland, including patterns of striations, directions of transport of erratics and general directions of ice moulding (from Sutherland, 1991b).

The distribution of most erratics and ice-moulded surfaces can be explained by a local ice-cap having covered the Shetland Islands, with ice flowing outwards from an ice shed located along the long axis of the islands. Since westerly-directed ice moulding and erratic transport on the western side of the islands is compatible with local as well as Scandinavian glaciation, more significance has been given to the evidence for eastward movement of ice on the eastern side of the islands. Such evidence includes, for example, the transport of erratics from the Mainland as far east as the Out Skerries (Mykura, 1976) and has been found along the length of the eastern Mainland and other islands.

All workers since Peach and Horne (1879) agree that Shetland has been glaciated by a local ice-cap, but the age and extent of this ice-cap have been the subject of debate. Although it is broadly agreed that the last phase of local ice-cap glaciation dates from the Late Devensian, direct evidence for this is lacking (Sissons, 1981b). If the Sel Ayre peat is indeed of Ipswichian age and the overlying till relates to the latest, local ice-cap, then this only indicates Devensian glaciation. It is generally accepted that the ice-cap extended on to the adjacent shelf, but in the north of the islands there is some recent evidence that the ice margin during the last glacial maximum may have been located relatively close inshore (Flinn, 1983; Long and Skinner, 1985).

Similarly, little is known about the mode and timing of retreat of the local ice-cap. A 'moraine belt' has been described from the western island of Papa Stour (Mykura and Phemister, 1976) but its significance is unclear. The absence of raised shorelines limits the possibilities for studying ice-cap retreat: Hoppe (1974), on the basis that striation patterns reflect calving during deglaciation, has suggested that the contemporaneous sea level was only 20–25 m lower than it is today, but this figure must be regarded as speculative. A minimum age for the termination of the local ice-cap is provided by radiocarbon dates on basal organic sediments resting on the glacial deposits. The oldest such date is 13,680 ± 110 BP (at Burn of Aith), but this is not critically linked to glacier retreat.

The only evidence for the vegetation history of Shetland during the Lateglacial comes from the detailed work of Birnie (1981) at the Burn of Aith. Her results suggest that a relatively mild phase occurred during the early Lateglacial Interstadial. During the middle part of the interstadial juniper and possibly some birch scrub was present on the islands prior to a return during the Loch Lomond Stadial to soil instability and a vegetation pattern dominated by open habitat species. Minor moraines have been described on some of the hills (Charlesworth, 1956; Mykura, 1976; Mykura and Phemister, 1976; Flinn, 1977) and attributed to the Loch Lomond Readvance, but which of these features are in fact moraines has been disputed among the authors quoted. Whichever interpretation is adopted, however, it is clear that the Loch Lomond Readvance was of very limited extent on Shetland.

In contrast, the Holocene vegetational history has been studied at a number of sites (Hawksworth, 1970; Johansen, 1975, 1978, 1985; Hulme and Durno, 1980; Birnie, 1981; Bennett *et al.*, 1992). Despite the early records of fossil wood within the peat on Shetland (Lewis, 1907, 1911), it was widely thought that Shetland had remained essentially treeless during the Holocene, a view supported by the relative scarcity of tree pollen (Erdtman, 1924). However, radiocarbon dating and pollen evidence from Garths Voe (Birnie, 1981, 1984) show beyond all doubt that during the middle Holocene there was a distinct phase when willow, birch or hazel developed widely. On Foula it seems probable that trees were restricted to sheltered areas (Hawksworth, 1970).

Shetland is also notable for certain periglacial landforms. Although the uplands seldom exceed 300 m in altitude, the highest area, Ronas Hill (450 m OD), has a notable range of both fossil and active periglacial features (Ball and Goodier, 1974). The original mountain-top detritus is likely to have been produced during the Late Devensian, but present activity, particularly influenced by the wind, is producing stripes, sand sheets ('dunes') and various types of terrace (Veyret and Coque-Delhuille, 1989). Elsewhere in Shetland, small, active stone stripes are present at 60 m OD on Keen of Hamar on Unst (Spence, 1957; Carter *et al.*, 1987).

The coastline of Shetland is essentially one of submergence (Flinn, 1974), but there have been few detailed studies of relative sea-level change. Submerged peat deposits are common (Finlay, 1930; Flinn, 1964, 1974; Chapelhowe, 1965; Birnie, 1981), and radiocarbon dates from Whalsay suggest relative sea level at least 9 m below present before 5500 BP (Hoppe, 1965). Birnie (1981) reported radiocarbon dates of 5840 ± 50 BP (SRR – 1796) and 4586 ± 40 BP (SRR – 1795) from intertidal peats at Leebotten and the Houb, respectively, and although these dates confirm the trend of submergence, they are not related to specific sea-level altitudes.

FUGLA NESS

H. J. B. Birks

Highlights

The sequence of deposits in the coastal section at Fugla Ness includes two tills and an interglacial peat, ascribed to the Hoxnian. These deposits provide critical evidence for interpreting the Quaternary history of Scotland in an area peripheral to the main centres of glaciation.

Introduction

Fugla Ness is a promontory on the north-west coast of North Roe on the Mainland of Shetland. A broad platform slopes gently from about 30 m OD at the base of the Beorgs of Uyea to form a small sea-cliff about 10 m high. A drift-filled geo (at HU 312913) within this marine-eroded cliff reveals a succession of Pleistocene deposits. These are of great scientific interest, particularly since they include one of the oldest known interglacial peat deposits in Scotland. The stratigraphy of the succession is described by Chapelhowe (1965), and the vegetation history has been reconstructed by Birks and Ransom (1969) on the basis of the fossil assemblages of pollen and plant macrofossil remains in the peat.

Description

The following sequence of deposits is revealed in the cliff section (Figure 3.2) (Chapelhowe, 1965; Birks and Ransom, 1969):

Figure 3.2 Sediment sequence at Fugla Ness, Shetland, showing interglacial peat (lower left) overlain by slope deposits, with till at the top. (Photo: J. E. Gordon.)

6. Sandy, slightly organic topsoil 0.14 m
5. Reddish till containing granite pebbles average size 0.12 m, and some boulders of 0.9 m with long axes orientated parallel to local striations 2.05 m
4. Grey-brown till, horizontally stratified; clasts mainly of local origin with some granite; average size of clasts 0.005–0.03 m 3.17 m
3. Compacted, structureless peat, with much compressed wood and pine cones and frequent lenses of silt and clasts 0.50 m
2. Compacted, structureless peat with some wood and large clasts 1.05 m
1. Grey, cemented till similar to bed 4.

The peat (beds 2 and 3), which is exposed over a distance of at least 20 m, thins and becomes discontinuous towards its edges. In places it is slightly distorted, probably by the weight of the overlying deposits. The peat beds dip at about 20° towards the sea and may have been formed on a sloping surface. Alternatively, the peat may be an erratic block and not be *in situ*. Seven finite radiocarbon dates ranging in age from 34,800 +900/−800 BP (T–1092) to 47,500 +2900/−2100 BP (GrN–7634) and one infinite age (>33,300 BP SRR–666) have been obtained from the peat (Page, 1972; Harkness and Wilson, 1979; Harkness, 1981). The finite dates, like finite radiocarbon dates obtained from several English interglacial deposits (see also Kirkhill), are probably erroneous owing to sample contamination (Sissons, 1981b).

There have been no detailed sedimentological studies of beds 1, 4 and 5. However, work in progress (A. M. Hall and J. E. Gordon, unpublished data) suggests that bed 4 is a head deposit and contains bands of reworked peat.

Interpretation

The organic deposit appears to have formed at the edge of a small oligotrophic pond and consists of plant debris derived from the surrounds of the pond mixed with both *in situ* material and drift material. The deposit probably formed within a *Juncus*-dominated shoreline community with a variety of fen and damp-

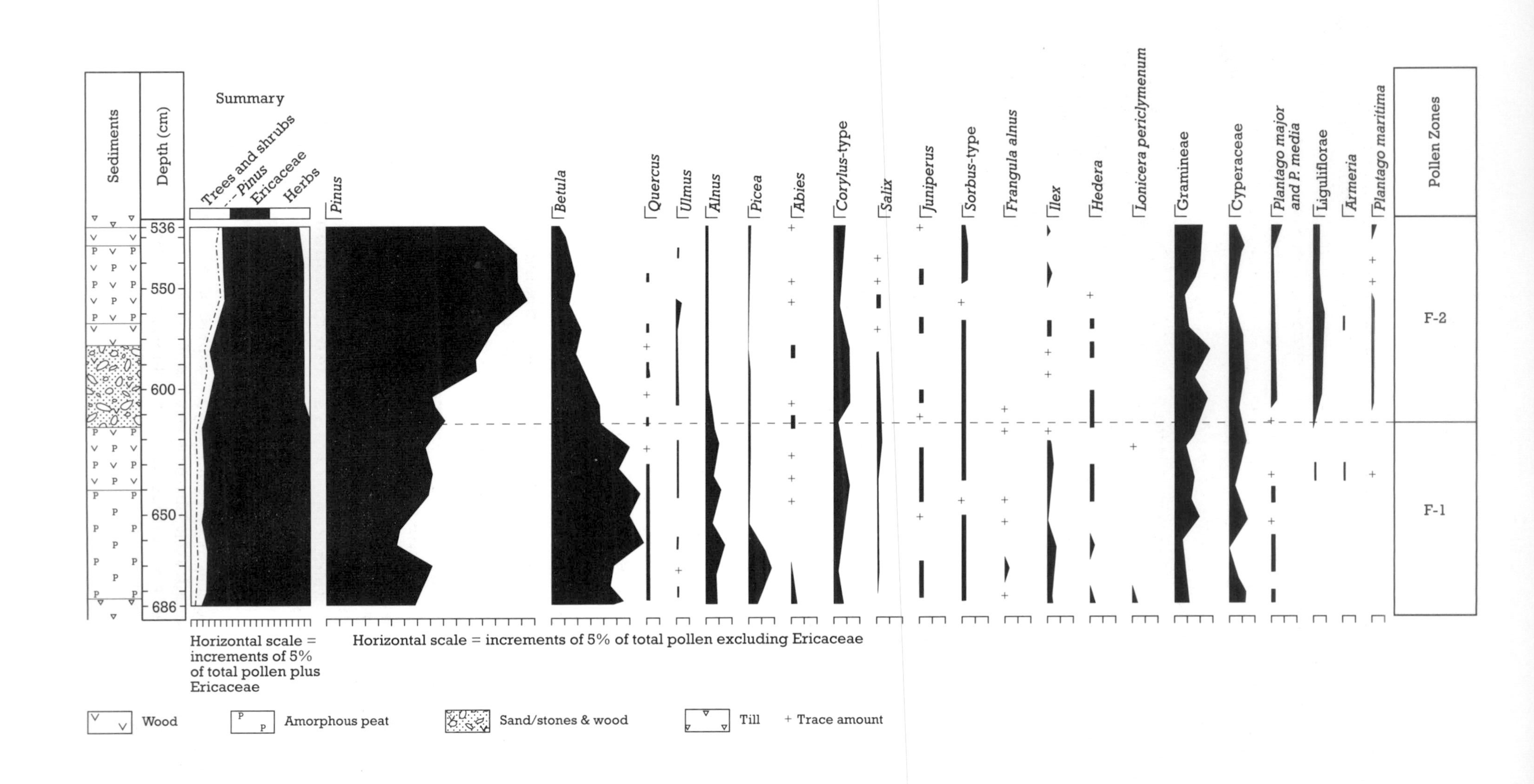
Sediments
Depth (cm)
Summary
Trees and shrubs
Pinus
Ericaceae
Herbs
536
550
600
650
686
Pinus
Betula
Quercus
Ulmus
Alnus
Picea
Abies
Corylus-type
Salix
Juniperus
Sorbus-type
Frangula alnus
Ilex
Hedera
Lonicera periclymenum
Gramineae
Cyperaceae
Plantago major and P. media
Liguliflorae
Armeria
Plantago maritima
Pollen Zones
F-2
F-1
Horizontal scale = increments of 5% of total pollen plus Ericaceae
Horizontal scale = increments of 5% of total pollen excluding Ericaceae
Wood
Amorphous peat
Sand/stones & wood
Till
+ Trace amount

ground herbs and ferns. The occasional pollen and macrofossils of floating-leaved and submerged aquatic plants probably originate from aquatics in the pool. On the better-drained areas around the site the vegetation represented in the lower pollen zone (F–1, Figure 3.3) appears to have been dominated by ericaceous dwarf-shrub heaths with *Bruckenthalia spiculifolia* (originally identified as the extinct taxon *Erica scoparia* var. *macrosperma* by Birks and Ransom, 1969), a plant now confined to the mountains of the Balkans and Asia Minor. *Juniperus*, *Lycopodium annotinum* and *Jasione montana* probably grew in these heaths, and *Betula*, *Pinus*, *Abies*, *Picea*, *Ilex* and *Sorbus* may have occurred locally in sheltered areas.

In the succeeding pollen zone (F–2, Figure 3.3) the composition of the heaths changed, with an expansion of *Erica tetralix*, *E. mackaiana* (now confined in the British Isles to Connemara and Donegal), *Empetrum nigrum* and *Daboecia cantabrica* (now restricted in the British Isles to Connemara), and of grassland communities rich in herbs such as *Jasione montana*, *Centaurium erythraea* and *Plantago* spp. Plants of biogeographical interest present in this zone include the 'peat-alpines' *Rubus chamaemorus* and *Chamaepericlymenum suecicum*, both of which are absent from Shetland today.

The change in vegetation from zone F–1 to zone F–2 (Figure 3.3) was interpreted by Birks and Ransom (1969) to be the result of climatic deterioration, with a decrease in annual temperature and an increased frequency of frosts.

If the Holocene pollen spectra from Murraster (Johansen, 1975) are viewed as representative of present 'interglacial' conditions on Shetland Mainland, it is clear that the low values of tree pollen and the high frequencies of dwarf-shrub pollen at Fugla Ness are of interglacial character (*sensu* Jessen and Milthers, 1928). The strongly calcifuge character of the plant assemblages at Fugla Ness suggests that the sequence reflects the oligocratic phase of an interglacial cycle (*sensu* Andersen, 1966, 1969) when acid, humus-rich soils and peats were widespread. Given the proximity of the site to present sea level, the absence of obvious 'littoral' pollen is noteworthy.

The age of the organic deposits at Fugla Ness cannot be established by radiocarbon dating. However, several features of the plant assemblages suggest correlation with the Hoxnian or Gortian (Birks and Ransom, 1969). These are:

1. the complete absence of so-called 'Tertiary' pollen types such as *Tsuga* and *Pterocarya*, suggesting that the deposits are not of Early Pleistocene age (West, 1980);
2. the absence of any pollen of *Carpinus betulus*, the abundance of which is considered by some to be characteristic of the later phases of the Ipswichian in England at sites at least as far north as County Durham (Beaumont *et al.*, 1969; Phillips, 1974; West, 1977, 1980);
3. the presence of *Abies* pollen, suggesting a Middle Pleistocene age (West, 1977, 1980);
4. the presence of the heaths *Daboecia cantabrica*, *Erica mackaiana* and *Bruckenthalia spiculifolia*, suggesting close floristic affinities with the closing phases of the Gortian in Ireland (Watts, 1967).

The plant assemblages from the Fugla Ness deposits may thus reflect a northern, oceanic variation of the vegetation of the Hoxnian and Gortian (but see Lowe, 1984). The mild climate implied by the reconstructed plant assemblages at Fugla Ness invites comparison with the palaeoclimatic record of the deep-sea sediments (Ruddiman and McIntyre, 1976). Only two periods are apparent in the deep-sea record of the last 600,000 years when the north-east Atlantic oceanic climate was as mild as, or milder than, at present. These were 125 ka (the Ipswichian) and 380 ka. Given the likely Middle Pleistocene age of the Fugla Ness peat, then a correlation with the oceanic event at 380 ka in Oxygen Isotope Stage 11 seems most likely (Sutherland, 1984a).

There are two possible interpretations of the glacial history represented in the deposits at Fugla Ness (Sutherland, 1991b). On present information, it is possible that the till units (beds 1 and 4) that enclose the interglacial deposits are from the same glaciation, with the peat occurring as an erratic block (see Birks and Ransom, 1969). If this is so, then the site indicates only two periods of glaciation, both post-dating the peat and represented by beds 1–4 and bed 5, respectively. Alternatively, if the tills in beds 1 and 4 are from separate glacial events, then three

Figure 3.3 Fugla Ness: relative pollen diagram showing selected taxa based on sum of total pollen excluding Ericaceae (from Birks and Ransom, 1969; Lowe, 1984).

periods of glaciation are recorded at this site, the earliest of which pre-dates the interglacial peat and hence is possibly of Middle or Early Pleistocene age. If bed 4 is a head, as proposed by Hall and Gordon (unpublished data), then possibly three cold episodes, including two periods of glaciation, are represented at the site. The uppermost till unit (bed 5) contains granite erratics and is apparently derived from the south-east (Chapelhowe, 1965), which conforms with the direction of movement of the last local ice-cap glaciation of probable Late Devensian age. Fugla Ness therefore has significant potential for further research to amplify knowledge of the glacial sequence in Shetland.

Fugla Ness is of considerable scientific importance, not only because of its remarkable fossil flora, but also because it represents perhaps the oldest interglacial deposit known in Scotland and the northernmost known interglacial sequence in the British Isles. Although its implications in terms of glacial history remain to be elucidated in detail, it is a site of great potential importance.

Conclusion

The sequence at Fugla Ness includes the most northerly and one of the oldest known interglacial deposits in Scotland. Analysis of the pollen and larger plant remains preserved in the interglacial peat indicates a period of mild, oceanic climate and the occurrence of trees, including pine and fir, in sheltered areas. The interglacial deposits have been ascribed to the Hoxnian Stage of the Pleistocene (about 380,000 years ago), but they have not been dated directly. The sediments at Fugla Ness also provide evidence for at least two separate periods of glaciation in Shetland.

SEL AYRE

H. J. B. Birks

Highlights

The sequence of deposits in the coastal section at Sel Ayre includes an interglacial peat, assigned to the Ipswichian, which is overlain by a succession of slope deposits and a till. These sediments have contributed significantly to the understanding of the Quaternary history of Scotland in an area well to the north of the main centres of glaciation.

Introduction

Sel Ayre [HU 176540] is located on the west coast of the Walls Peninsula on Shetland Mainland. It is a site of considerable scientific importance for the succession of Pleistocene deposits infilling a channel or gully on the cliff top. The deposits include peat that appears to have formed during an interglacial period, probably the Ipswichian. As the peat contains a different pollen assemblage from that recorded at Fugla Ness, the peats at the two sites are probably of different ages and represent the closing phases, at least, of two different interglacials. The sequence of deposits at Sel Ayre is described by Mykura and Phemister (1976), and the vegetation history has been reconstructed from pollen analysis of the peat by Birks and Peglar (1979).

Description

The following sequence of deposits is revealed in the cliff section (Mykura and Phemister, 1976):

10.	Peat	0.3 m
9.	Till, with clayey matrix and angular to subangular pebbles and boulders of Walls Sandstone (Old Red Sandstone) and rare basic lavas	2.7 m
8.	Gravel, well-bedded with predominantly angular pebbles set in a silty to clayey matrix	1.8 m
7.	Sand, pale brown with patchy brown iron staining; sparse pebbly bands, up to 0.23 m thick, more common at the sides of the channel	1.4 m
6.	Sand, black to dark brown, limonite-impregnated and peaty	0.025 m
5.	Sand, pale ochre-brown, pebbly, locally bleached white; patchily ochre-stained at base	0.4 m
4.	Peat with scattered round sand grains and some sandy lenses	0.038 m
3.	Soft, pale clayey sand with thin laminae of clay and peat	0.23–0.26 m
2.	Peat with scattered sand grains in lower part; passes laterally into sand with thin peat bands and thickens to 1 mm	0.45 m
1.	Sand with scattered pebbles, base not seen	0.4 m

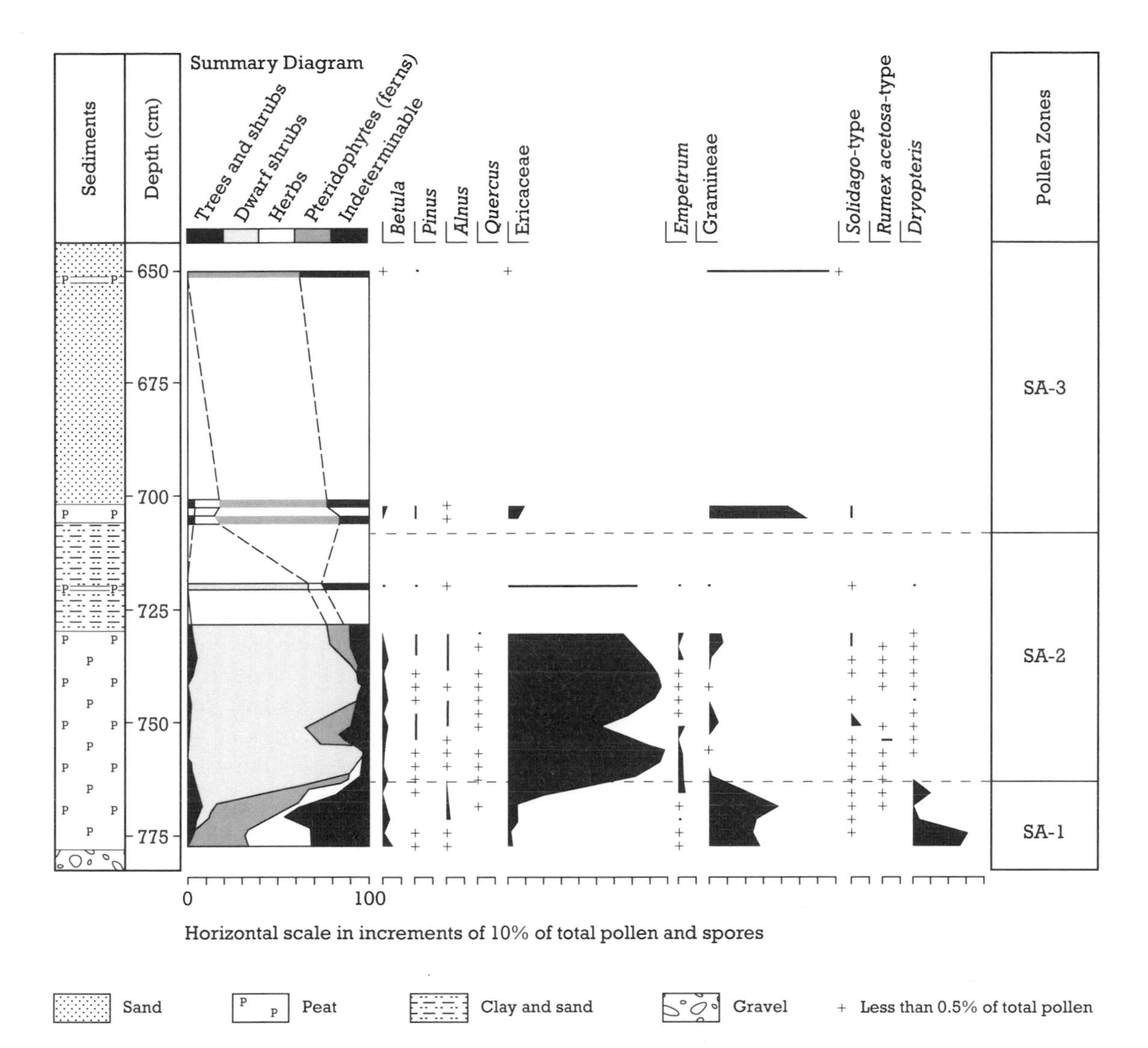

Figure 3.4 Sel Ayre: relative pollen diagram showing selected taxa as percentages of total pollen and spores (from Birks and Peglar, 1979; Lowe, 1984).

The deposits lie within an ENE-trending channel. There have been no detailed sedimentological studies. However, work in progress (A. M. Hall and J. E. Gordon, unpublished data) suggests that beds 1 and 4–8 comprise a series of slope deposits infilling the channel. The main peat bed (bed 2) splits laterally into a series of bands of sandy peat interspaced with sand lenses, often with ripple cross-laminations, as the section is traced from the centre of the channel to its edges. A radiocarbon date of 36,800 +1950/ −1960 BP (SRR–60) (Mykura and Phemister, 1976) has been obtained from the peat bed. This date is probably erroneous (Birks and Peglar, 1979; Sissons, 1981b).

The peat may have accumulated in a valley or basin mire within the channel or it may be highly compressed acid mor humus that accumulated within the channel. In the absence of any identifiable plant macrofossils, it has not been possible to distinguish between these two alternative origins. The pollen stratigraphy at Sel Ayre has been divided into three pollen assemblage zones (Figure 3.4) (Birks and Peglar, 1979).

Interpretation

The vegetation at the time of the lowest pollen zone (SA–1, lower part of bed 2) appears to have been dominated by fern and herb-rich grasslands, possibly similar to the ungrazed communities rich in ferns and tall-herbs that are confined to cliff-ledges and islands in lochs in Shetland today.

Such vegetation implies moderately fertile brown-earth soils.

Ericaceae dominated the vegetation of the succeeding pollen zone (SA–2, upper part of bed 2 and lower part of bed 3), suggesting acid humus-rich soils. Because of poor pollen preservation and the limitations of pollen morphology, it is not possible to say which taxa within the Ericaceae were abundant in the vegetation. The Balkan *Bruckenthalia spiculifolia* was present, however, along with a variety of calcifuge herbs and pteridophytes. Tree pollen values are low, as they are at Fugla Ness (Birks and Ransom, 1969) and in the Holocene sediments at Murraster (Johansen, 1975). In contrast to Fugla Ness, pollen of *Carpinus betulus*, *Quercus* and *Ulmus* are present at Sel Ayre. Also pollen of *Picea* and *Abies* are present in lower frequencies than at Fugla Ness.

The third pollen zone (SA–3) is restricted to the peat lenses (upper part of bed 3, bed 4 and bed 6) within the sands that overlie the main peat bed. The vegetation appears to have been open and grass-dominated with a variety of herbs characteristic of skeletal mineral soils.

In view of the stratigraphic setting of the Sel Ayre peat beneath 7.3 m of till, gravel and sand, and of the relatively low tree pollen values in Holocene pollen assemblages at Murraster (Johansen, 1975), even prior to any human disturbance, the Sel Ayre peat probably formed during an interglacial phase (but see Lowe, 1984). Numerical comparisons of the Fugla Ness, Sel Ayre and Murraster pollen assemblages (Birks and Peglar, 1979) indicate that the three sequences differ in their pollen composition, suggesting that they were formed in three different interglacials. The presence of *Carpinus betulus* pollen in the Sel Ayre profile suggests that the sequence was formed during the Ipswichian, as high *Carpinus* values are characteristic of its later phases in England (Phillips, 1974; West, 1977, 1980). The vegetational changes recorded at Sel Ayre suggest that the lowermost pollen zone (SA–1) reflects the mesocratic phase (*sensu* Andersen, 1966, 1969) of the Ipswichian, with fertile brown-earth soils. Pollen zone SA–2 represents the oligocratic phase (*sensu* Andersen, 1966, 1969), with acid podsols and peaty soils, and pollen zone SA–3 the cryocratic phase (*sensu* Iversen, 1958) of the Early Devensian, with skeletal mineral soils. The Sel Ayre sequence may thus represent an almost complete record of the Ipswichian on Shetland Mainland.

As in the case of Fugla Ness, the deposits at Sel Ayre merit detailed investigation to elucidate the glacial history of the area. The provenance of the sands and gravels interbedded with the peat has not been established and there is no information on glaciation of the site before deposition of the peat. The sequence of sands and gravels suggests a period of slope instability in the catchment prior to deposition of the overlying till (bed 9). The latter contains erratics of Walls Sandstone and rare basic lavas derived from the south-east (Mykura and Phemister, 1976), and may be correlated with the local ice-cap glaciation (Sutherland, 1991b). The latter is generally assigned to the Late Devensian.

If the correlation of the Sel Ayre sequence with the Ipswichian is correct (but see Sissons, 1981b; Caseldine and Edwards, 1982; Lowe, 1984) and the correlation of Fugla Ness with the Hoxnian and Gortian (or earlier stages) is correct, it would appear that there was very considerable regional variation in the vegetation of the British Isles in previous interglacials, just as there is in the vegetation of the present interglacial. The composition of the vegetation was, however, very different from one interglacial to another, a fact of considerable importance to plant ecologists and plant geographers.

Sel Ayre may be the most complete record of the Ipswichian currently known in Scotland and the most northerly such record in the British Isles. It is of considerable importance, not only for the information it contains on vegetation history and diversity, but also because it provides a limiting date on the last glaciation of Shetland, probably by a local ice-cap (Mykura and Phemister, 1976; Flinn, 1978a; Sutherland, 1984a).

Conclusion

The sequence at Sel Ayre includes an interglacial (temperate climate) peat deposit, the most northerly in Britain that has been ascribed to the Ipswichian Stage (about 125,000 years ago). The pollen preserved in this deposit provides a valuable record of vegetational history, showing a succession from grassland, through heathland with trees, to open, grass-dominated vegetation. The deposits also indicate an episode of later glaciation (probably Late Devensian), preceded by a cold phase during which slopes in the catchment were unstable.

BURN OF AITH

J. Birnie

Highlights

The sediments which infill a deep basin at the Burn of Aith comprise lake clays and fen peats. The pollen and diatoms contained in these sediments provide a detailed record, supported by radiocarbon dating, of environmental changes in Shetland during the Lateglacial and early Holocene.

Introduction

The Burn of Aith site [HU 441295] comprises a flattish area of lake and fen infill sediments approximately 1 km^2 in extent. It is located immediately inland from the east coast of Shetland Mainland near Cunningsburgh. The infill is exceptionally deep and contains a Lateglacial and Holocene sequence that is representative of lowland Shetland. Two cores have been analysed in detail, giving pollen, diatom and sedimentary characteristics (Birnie, 1981), and four radiocarbon dates have been obtained for the Lateglacial sequence in one core (J. Birnie, unpublished data). This is the earliest detailed information on the Lateglacial environment of Shetland.

Description

The infill sediments, lake clays and fen peats occupy a basin which exceeds 11 m in depth. It is formed in either till or bedrock and shallows gradually in the direction of the present outlet to the sea at Aith Voe. Despite the present surface being less than 5 m above sea level, all the infill sediments are of freshwater origin, so that the basin configuration must have prevented marine inundation, even with the rising Holocene sea level. The bedrock in the vicinity of the basin is Old Red Sandstone, although the adjacent upland catchment, drained by the Burn of Laxdale, is in phyllite and spilitic lavas (Mykura, 1976).

The stratigraphy at the coring site consists of the following sequence (Birnie, 1981, Aith II site):

6.	Poorly humified organic material with clay matrix and wood fragments	1.41 m
5.	Yellow clay with gritty layers and with fibrous vegetation	0.68 m
4.	Fibrous organic deposit (fen peat) with clay content decreasing upwards	0.58 m
3.	Light-grey inorganic clay	0.42 m
2.	Organic clay with plant remains	0.3 m
1.	Light-grey inorganic clay	0.11 m

Pollen frequencies from the different beds have been analysed (Figure 3.5). Radiocarbon dates from bed 2 were obtained from a separate core (Birnie, unpublished): lower part 13,680 ± 110 BP (SRR–2286), middle part 12,700 ± 80 BP (SRR–2285) and 12,670 ± 80 BP (SRR–2284) and upper part 12,190 ± 80 BP (SRR–2283).

Interpretation

The lowest clay (bed 1) is barren of both pollen and diatoms. Analysis of the overlying organic clay (bed 2) indicates a fairly alkaline lake, rich in diatoms and higher plant species, with pollen and macrofossils suggesting a land vegetation of open-ground herbs including *Rumex*, *Salix herbacea* and *Koenigia islandica* and thus substrate instability. However, pollen from the middle part of the bed indicates a phase when *Juniperus* and possibly some *Betula* shrubs were also present (Figure 3.5). In the overlying inorganic clay (bed 3) diatoms are relatively rare. Those present still suggest alkaline conditions, but they are much reduced in variety. The pollen record consists of *Salix herbacea*, Umbelliferae and Compositae, with *Rumex* not returning until the top of the bed, where organic content begins to increase once again. The continuous organic sedimentation above this level (beds 4, 5 and 6) was assumed to be Holocene (Birnie, 1981), although the pollen record shows that an initial phase of tall herb growth was followed by a return of open-ground herbs including *Artemisia* and *Rumex*, with *Lycopodium selago*. Stable ground conditions were not finally achieved until a level dated, on the basis of the appearance of *Corylus* and *Ulmus* in the long-distance pollen, to post–9600 BP. Details of the Holocene environmental history are described by Birnie (1981).

The radiocarbon dates from the Lateglacial organic clay appear to be relatively older than might be expected (see Cam Loch). Either they reflect errors arising from the 'hard-water effect' or they indicate a record of early Lateglacial Interstadial warming in Shetland, in comparison to the mainland of Scotland. Further dates are

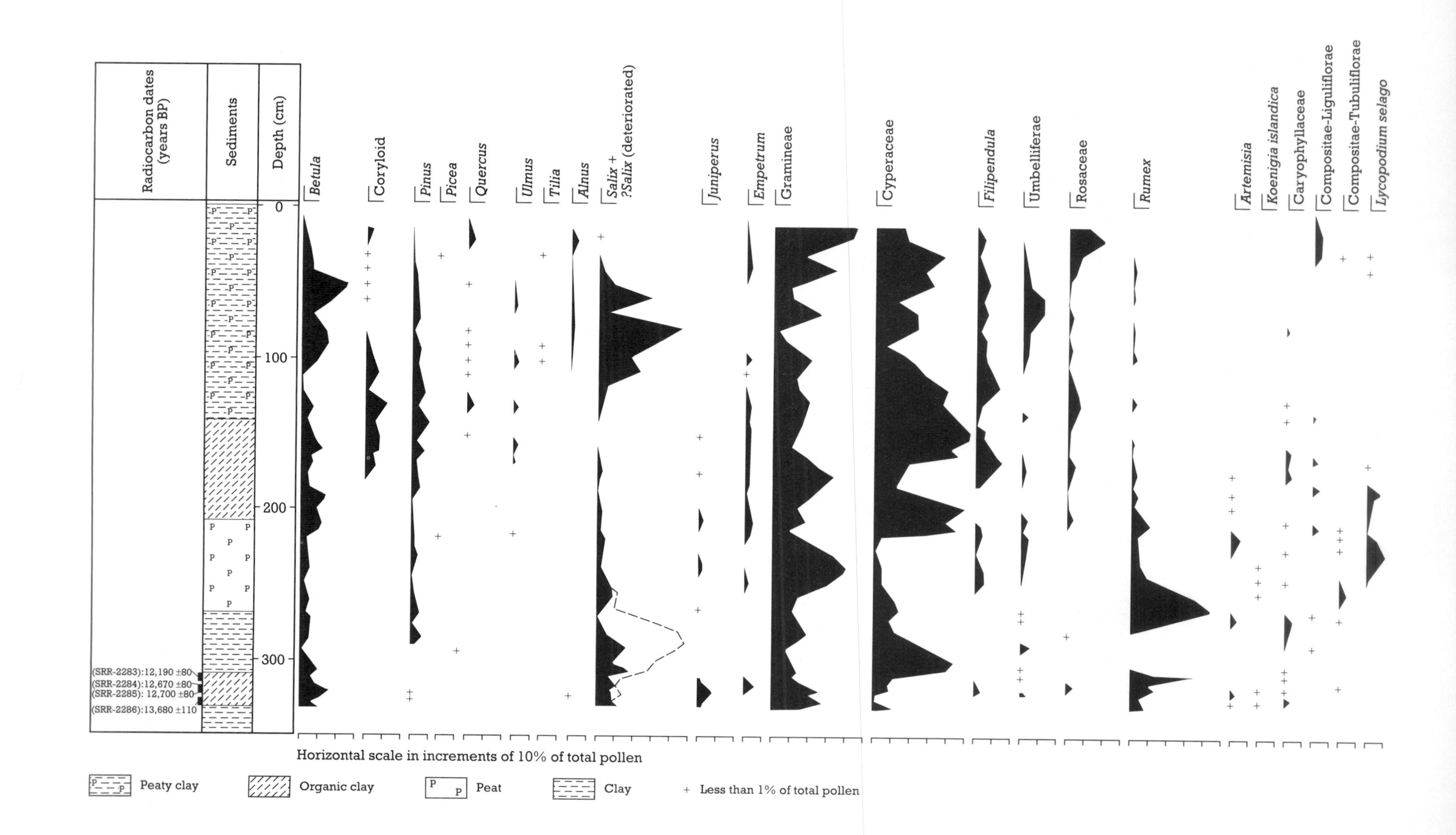

Radiocarbon dates (years BP)
Sediments
Depth (cm)
0
100
200
300
(SRR-2283):12,190 ±80
(SRR-2284):12,670 ±80
(SRR-2285): 12,700 ±80
(SRR-2286):13,680 ±110
Betula
Coryloid
Pinus
Picea
Quercus
Ulmus
Tilia
Alnus
Salix +
?Salix (deteriorated)
Juniperus
Empetrum
Gramineae
Cyperaceae
Filipendula
Umbelliferae
Rosaceae
Rumex
Artemisia
Koenigia islandica
Caryophyllaceae
Compositae-Liguliflorae
Compositae-Tubuliflorae
Lycopodium selago
Horizontal scale in increments of 10% of total pollen
Peaty clay
Organic clay
Peat
Clay
+ Less than 1% of total pollen

needed on the lower part of bed 4, which, with its indications of unstable ground conditions in the pollen record, may be part of the Lateglacial sequence, rather than the Holocene as had been assumed.

There is little published literature on the Holocene environment of Shetland, and nothing on the Lateglacial, apart from a moss identification (Hulme, 1979) and two radiocarbon dates on organic material within inorganic clays beneath Holocene peat. These were 12,090 ± 900 BP (St–1640) at Loch of Clickhimin, and 11,135 ± 135 BP (St–1714) on peat lenses within a minerogenic deposit at Tresta (Hoppe, 1974). The beginning of Holocene organic sedimentation has been dated to around 10,400 BP at two sites (Johansen, 1975; Hulme and Durno, 1980). Otherwise the stratigraphy from the Burn of Aith valley, described here, together with that of Spiggie Loch, which is undated, provides the only detailed information on the Lateglacial environment of Shetland (Birnie, 1981).

During the Lateglacial and early Holocene the oceanic circulation around the Shetland Islands was apparently rather different from that of today and this provides an instructive comparison with the evidence for environmental change at Burn of Aith. Following the last period of ice-sheet glaciation when arctic waters extended well to the south of the British Isles, an interstadial marine circulation with a weak North Atlantic Drift became established off western Scotland and in the Norwegian and North seas by 12,800 BP (Jansen and Bjørklund, 1985; Peacock and Harkness, 1990). The presence of the warmer waters off the British coast corresponds with the opening of the Lateglacial Interstadial in the terrestrial records from mainland Britain (Atkinson *et al.*, 1987), but contrasts with the very early date on the interstadial sediments from the Burn of Aith. Climatic deterioration during the Loch Lomond Stadial is clearly recorded both in the marine and the Burn of Aith records. Of particular interest at the Burn of Aith site, however, is the apparent delay in the onset of warm conditions at the Lateglacial–Holocene boundary, for the marine record along the Scandinavian coast suggests continuing colder conditions until approximately 9600 BP (Peacock and Harkness, 1990). At this time, therefore, the climate of Shetland may have been more closely akin to that of Scandinavia than the Scottish mainland.

Figure 3.5 Burn of Aith: relative pollen diagram showing selected taxa as percentages of total pollen (from Birnie, 1981).

Conclusion

Burn of Aith provides the only detailed record so far of the environmental history of Shetland during the Lateglacial (approximately 13,000–10,000 years ago). The pollen preserved in the sediments indicates a period of relatively mild climate with the development of herbs and shrubs during the Lateglacial Interstadial, about 13,000–11,000 years ago, followed by a return to more severe conditions with open-habitat vegetation during the Loch Lomond Stadial (about 11,000–10,000 years ago). At the start of the Holocene, the onset of stable ground conditions was delayed in comparison with sites elsewhere in Scotland, and shows more similarity in timing with events in Scandinavia. Burn of Aith is therefore an important reference site not only for studies of environmental history in Shetland, but also for establishing wider regional patterns of environmental change.

GARTHS VOE

J. Birnie

Highlights

Pollen and wood remains from peat exposed in the stream section at Garths Voe provide a record of vegetational change in Shetland during the Holocene. They are important in dating a period when trees in Shetland were more widespread than today. The peat also contains a thin layer of sand which appears to represent a major coastal flood about 6000 years ago.

Introduction

The site is a roadside section [HU 409741] at the head of Garths Voe, an eastern arm of Sullom Voe. It was revealed at the mouth of a small burn draining the Hill of Garth by road building for the Sullom Voe oil terminal. At present it is one of only two sites in Shetland where radiocarbon dates have been obtained on wood within

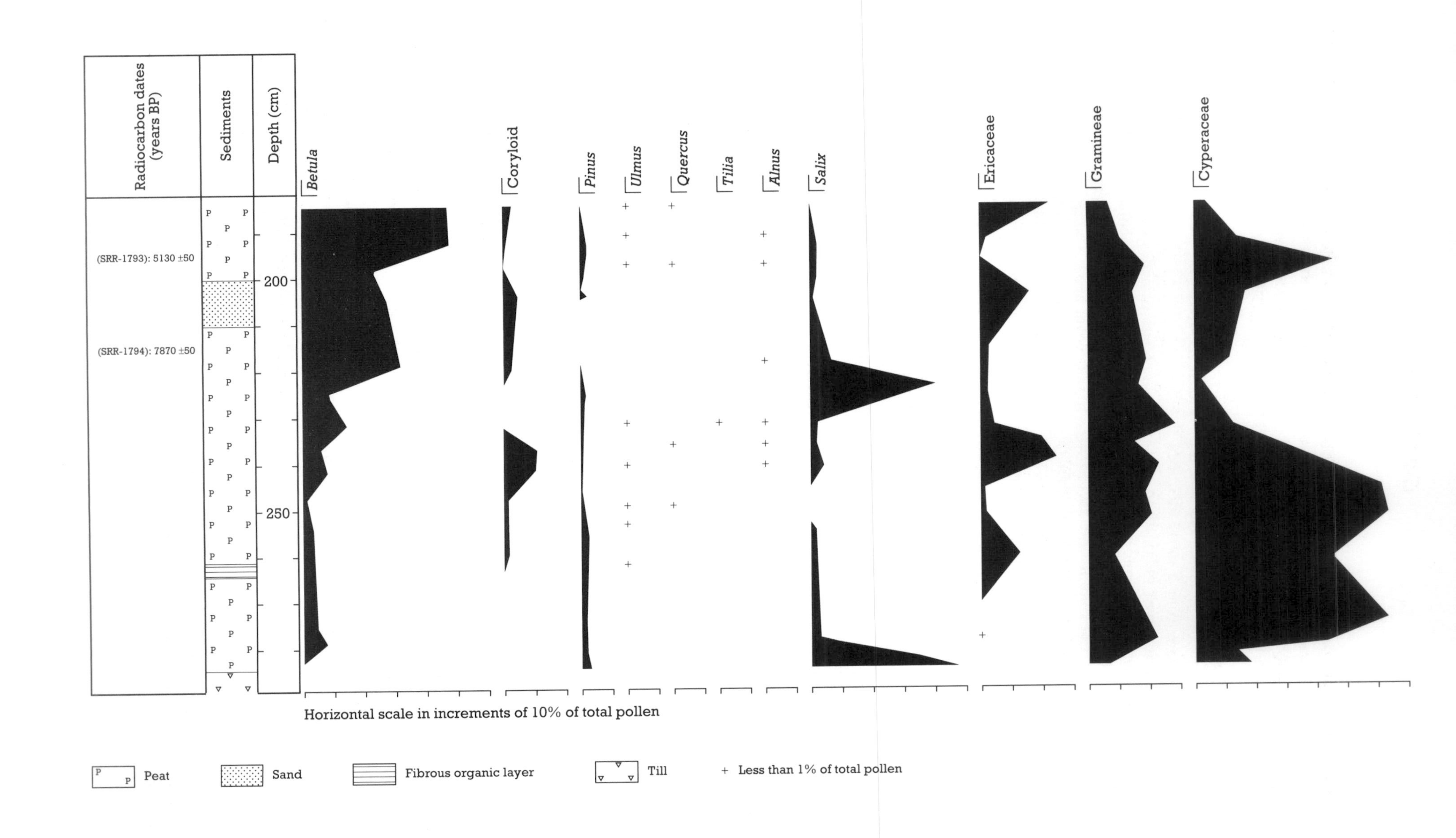

Radiocarbon dates (years BP)
Sediments
Depth (cm)
(SRR-1793): 5130 ±50
(SRR-1794): 7870 ±50
200
250
Betula
Coryloid
Pinus
Ulmus
Quercus
Tilia
Alnus
Salix
Ericaceae
Gramineae
Cyperaceae
Horizontal scale in increments of 10% of total pollen
Peat
Sand
Fibrous organic layer
Till
+ Less than 1% of total pollen

Table 3.1 Radiocarbon dates relating to the buried sand horizon at Garths Voe and Voe of Scatsta

Location	Material	Radiocarbon date	Laboratory number
Garths Voe	0.02 m thick layer of peat above sand	5315 ± 45 BP	SRR–3839
Garths Voe	0.02 m thick layer of peat below sand	5765 ± 45 BP	SRR–3838
Voe of Scatsta	0.02 m thick layer of peat above sand	3815 ± 45 BP	SRR–3841
Voe of Scatsta	0.02 m thick layer of peat below sand	5700 ± 45 BP	SRR–3840

Holocene peat deposits, with associated pollen analysis of the peat giving a record of regional vegetation history. Hence it indicates the past status of trees and shrubs in this northernmost part of the British Isles, which is currently treeless. The pollen stratigraphy of the site is described by Birnie (1981).

Description

The sequence, traced along both sides of the burn for approximately 8 m, is as follows (Birnie, 1981):

6.	Blanket peat	2.0 m
5.	Sand, unconsolidated	0.1 m
4.	Peat, less-humified, containing twigs and small wood fragments, and including branches of 0.05 m diameter in the upper part of the bed	0.52 m
3.	Discrete, fibrous organic layer of detrital vegetation remains	0.03 m
2.	Peat, well-humified, with a few plant remains	0.2 m
1.	Till	>0.1 m

Small fragments of wood are present in the lowest 0.5 m of the blanket peat (bed 6), but above this it is composed only of sedge, *Sphagnum* and ericaceous remains. *Salix* wood in bed 4, from below the sand layer, was dated to 7870 ± 50 BP (SRR–1794), and *Betula* wood from above the sand, in bed 6, to 5130 ± 50 BP (SRR–1793).

Figure 3.6 Garths Voe: relative pollen diagram showing selected taxa as percentages of total pollen (from Birnie, 1981).

Interpretation

Pollen analysis (Figure 3.6) indicates the presence of open-ground herbs in the lowest organic layer (bed 2), being replaced by Cyperaceae, and then covered with inwashed detritus of *Sphagnum*, Cyperaceae and Ericaceae comprising the fibrous layer. Autochthonous deposition then recommenced, with Cyperaceae and some *Sphagnum* locally and Ericaceae nearby. The site then became a willow fen, with grassland, ferns and tall herbs associated. This is represented by the lower woody peat (bed 4), with *Salix* constituting over 40% of the pollen total. Prior to the deposition of the sand, birch also appeared at the site, with *Betula* pollen values reaching 33% of the total. The pollen and macrofossils together show that there was open woodland or scrub at the site at around 7900 BP. Following deposition of the sand, peat accumulation recommenced, with *Betula* pollen reaching its maximum representation of 50% and *Salix* virtually absent. The birch was initially associated with ferns, but then heaths appeared at the site, and at some time after 5130 BP (bed 6) shrubs or trees disappeared and heath- and sedge-dominated blanket peat communities predominated, as at present. Birnie (1981) interpreted the sand layer (bed 5) as reflecting either erosion of minerogenic soils in the stream catchment or encroachment of beach sediments landwards from Garths Voe. However, recent investigations (D. E. Smith, unpublished data) suggest that the sand is part of a widespread deposit in Shetland, and may possibly represent a major marine flood with a run-up of several metres. Smith *et al.* (1991a) obtained radiocarbon dates on peat at the upper and lower contacts of the sand layer at Garths Voe and the adjacent Voe of Scatsta (Table 3.1). Two possible explanations for the origin of the sand layer are that it is a tsunami deposit rather like that recorded from *c.* 7000 BP in eastern Scotland (see Maryton), since a third Storegga slide is known at *c.* 6000 BP (Jansen *et al.*, 1987), or that it is the deposit of a storm surge of unusual magnitude.

There are written records of wood in Shetland peat from at least the early 19th century, for example Brewster (1829) and Bryden (1845), and Lewis (1907, 1911) published accounts of wood and other plant macrofossils as part of a study of peat in the whole of Scotland. Lewis wrote that the 'Forest Bed' (principally of birch, hazel and willow) was remarkably widespread in Shetland' and concluded 'these trees do not represent copses growing in sheltered valleys away from the coast but ... are just as well developed in the most exposed situations' (1911, p. 808). Hawksworth (1970) described the distribution of wood remains in peat on Foula, and concluded that on this small island they were restricted to the lower and more sheltered areas. Johansen (1975, 1978), examining the pollen record of a lake infill site on the Mainland, found that tree and shrub pollen did not exceed 50% of the sum of the land pollen at any one time, and so he disputed Lewis's term 'Forest' for what he interpreted as birch and hazel scrub. Birnie (1984) has examined wood and pollen records from a number of sites in Shetland and concluded that there was a distinct phase of widespread willow, birch or hazel development, with the radiocarbon dates from Garths Voe providing the only means of dating that phase at present – between 8000 and 5000 BP. The Garths Voe site is therefore an example of the vegetation record described by Lewis as representing forest, and by Johansen as scrub. It records the most advanced level of Holocene vegetation development in Shetland and as such will be significant in any interpretation of the climatic optimum, the timing and causes of environmental deterioration, and such issues as species dispersal and colonization rates.

There could, potentially, be very many sites in Shetland which demonstrate a similar Holocene stratigraphic record to that at Garths Voe (cf. Bennett *et al.*, 1992). Garths Voe is at present, however, the only site in which Holocene vegetation development has been examined by means of pollen and macrofossil analyses, with radiocarbon dates obtained from the wood. It appears to represent vegetation sequences described by Lewis in the early part of this century, and with the presence of macrofossils there is potential for more detailed examination of the age and relative importance of trees and shrubs contributing to the deposit. This would lead to a better understanding of the nature of former woodland cover in the Scottish Islands – an issue presently unresolved in the Western Isles and Orkney, let alone Shetland.

Conclusion

The deposits at Garths Voe provide a representative record of the vegetational history of Shetland during the Holocene (the last 10,000 years), based on pollen analysis and radiocarbon dating. A phase of open-habitat vegetation was followed by the development of willow, birch or hazel between about 8000 and 5000 years ago, but was subsequently replaced by blanket peat with heath and sedge communities. As a reference site for Shetland, Garths Voe and its record are also important for further investigation of the timing of the Holocene climatic optimum and the causes and wider patterns of the subsequent deterioration. In addition, the deposits at Garths Voe include a sand layer formed just after 6000 years ago, which may represent a tsunami (tidal wave) or storm-surge event in the North Sea.

RONAS HILL

J. E. Gordon

Highlights

The periglacial landforms and deposits at Ronas Hill include a range of active and fossil features formed by wind- and frost-related processes. On account of its northern location and relatively low altitude, Ronas Hill is a key locality for the study of periglacial activity in Scotland.

Introduction

Ronas Hill [HU 316835] is the highest summit (453 m OD) in Shetland. It is important for periglacial geomorphology and demonstrates an outstanding assemblage of both active and fossil patterned-ground landforms which are developed at relatively low altitudes. Combined wind and frost effects are particularly striking and include turf-banked terraces, wind-blown sand deposits and vegetated stripe features. The landforms of Ronas Hill have been described principally by

Ball and Goodier (1974), Goodier and Ball (1975) and Veyret and Coque-Delhuille (1989).

Description

Ronas Hill is formed of granophyre, part of the Ronas Hill Granite intrusion of Devonian age, which is itself cut by a series of north–south trending felsitic dykes (Mykura, 1976). Galloway (1958, 1961a) recorded the presence of blockfields and boulder terraces at 365–427 m OD on Ronas Hill similar to those on the Cairngorms at altitudes between 1067 and 1219 m OD. He posed the question whether the blockfields in such an oceanic environment as Shetland might derive in part from deep chemical weathering of the granite.

Ball and Goodier (1974) later identified four main groups of periglacial landforms developed on the granite regolith of Ronas Hill: turf-banked terraces, wind stripes, 'hill dunes' (wind-blown sand deposits) and composite stripe-terrace features. They also noted, but did not describe, large, fossil boulder terraces and blockfields, and small, active stone circles.

Turf-banked terraces include features aligned parallel to the ground surface contours, reflecting solifluction processes operating in a relatively mobile surface soil layer. A second type of terrace aligned obliquely to the contours probably reflects the combined influence of wind and solifluction activity. Wind stripes take the form of narrow strips of vegetation, with steeper windward and gentler leeward faces related to the effective wind direction. They occur both as continuous stripes and as fragmented crescents of vegetation. 'Hill dunes' appear as vegetated areas of sand rising up to 1 m above adjacent deflation surfaces. They are relics of a formerly more extensive soil and vegetation cover, but the presence of buried humic horizons indicates episodic accumulation. Composite features occur where terraces intersect wind stripes. Overall, however, it is important to emphasize the total assemblage of wind, frost and mass-movement features present and their spatial interactions according to local conditions of slope and exposure (Veyret and Coque-Delhuille, 1989); for example, terrace treads support wind stripes and bare areas between vegetated stripes show the effects of frost sorting.

Interpretation

Ball and Goodier (1974) considered that the large-scale relict landforms are probably of Late-glacial age and that the wind-blown material accumulated during the early Holocene. They also argued that erosion of the 'hill dune' vegetation cover and development of the turf-banked terraces occurred possibly during the Little Ice Age between about AD 1550 and 1750. Although possible wider correlations are premature in the absence of firm dating, it is worth noting that elsewhere in Scotland radiocarbon dating has indicated that there was active solifluction during the late Holocene (Sugden, 1971; Mottershead, 1978; Ballantyne, 1986c). Elsewhere, aeolian sand and silt deposits are represented on Ward Hill (see below) and mountain summits of north-west Scotland (Sissons, 1976b; Birse, 1980; Pye and Paine, 1984; Ballantyne, 1987a). On An Teallach (see below), Ballantyne and Whittington (1987) established that accumulation of niveo-aeolian sands began during the early Holocene, before about 7900 BP, and was reactivated more recently, possibly either during the 17th and 18th centuries when the Little Ice Age weather conditions were most severe in Scotland, or during the 19th century following overgrazing. In County Donegal in Ireland, Wilson (1989) identified two periods of Holocene sand-sheet accumulation, with the most recent erosion commencing before the late nineteenth century. Recognition and correlation of widespread climatic or anthropogenic causes, however, must await more detailed site-specific studies of sediments, palynology and dating. Ballantyne (1991a), in particular, has sounded a note of caution in ascribing late Holocene erosion to climatic deterioration, as the connection is based entirely on inferred coincidence of timing.

The assemblage of wind- and frost-related, patterned ground features has developed at a relatively low altitude on Ronas Hill under the subarctic, oceanic climatic conditions of the area (for details see Spence, 1957, 1974; Birse, 1971, 1974). Similar features typically occur at much higher altitudes in the mountains of mainland Scotland (see An Teallach, Sgùrr Mór, Ben Wyvis, and the Cairngorms) (Crampton, 1911; Peach *et al.*, 1913a; Crampton and Carruthers, 1914; Galloway, 1958, 1961a; Godard, 1965; Kelletat, 1970a, 1972; King, 1971b; Goodier and Ball, 1975; Ballantyne, 1981, 1987a; Pye and Paine,

1984). Ronas Hill is also distinguished by the range and quality of the features present in a relatively compact area, and Ball and Goodier (1974) noted that the interaction of wind and frost effects was more clearly demonstrated than at any other site in Britain known to them. Further, the combined effects of wind and frost can be readily investigated and evaluated in an area of uniform geology and with a range of slopes and aspects. Although Ward Hill in Orkney is similar in its range of landforms to Ronas Hill, active frost processes there are less evident, so that the two hill masses essentially complement each other in their periglacial interests.

In a national context Ronas Hill forms a northern end member of a network of sites representing past and present periglacial activity. It is a key site for studies of spatial variations in periglacial processes in Britain and the essential controlling variables (see Ballantyne, 1987a). Considerable potential exists for studies of current periglacial processes, landform history and dating of the buried soils. As yet no investigation has been undertaken of the larger relict landforms, their palaeoclimatic significance and their relationships to the active features. Ball and Goodier (1974) highlighted two important results that would arise from a comprehensive study of the wind- and frost-related features on Ronas Hill together with those on Ward Hill – 'on the one hand, it should help to elucidate the complex post-glacial and recent history of changing stability, wind erosion, and frost-induced disturbance and movement in the northern climatically stressed hill areas and, on the other, provide a sensitive long term monitor of regional climatic change'.

The fossil periglacial features form part of a network of such sites in Scotland (for example, An Teallach, the Cairngorms) which have attracted recent attention for their possible significance in delimiting the vertical dimensions of the last ice-sheet (Ballantyne *et al.*, 1987; Reed, 1988; Ballantyne, 1990). Those on Ronas Hill have not been studied in detail but merit investigation particularly in relation to their age or ages of origin and possible relationships to the limit of the last ice-sheet (Veyret and Coque-Delhuille, 1989). There appear to be three possibilities: such features may pre-date the last ice-sheet but were preserved beneath cold-based or inactive ice; they may have developed if Ronas Hill remained as a nunatak above the surface of the Shetland ice-cap; or they may have been formed or reactivated during the Loch Lomond Stadial.

Conclusion

Ronas Hill is outstanding for its assemblage of landforms developed under periglacial (cold climate) conditions, particularly those formed by the combined effects of wind and frost activity. It represents the northernmost occurrence of such features in Scotland and is notable for their development at relatively low altitudes. Ronas Hill has significant potential for research both on contemporary periglacial processes and on the history of upland geomorphological changes during approximately the last 13,000 years.

Chapter 4

The Orkney Islands

INTRODUCTION

D. G. Sutherland and J. E. Gordon

Perhaps surprisingly, in view of the position of the Orkney Islands (Figure 4.1) between the North Sea and the North Atlantic, the Quaternary deposits and landforms there have been little studied. Early work addressed whether Orkney had been glaciated or not (A. Geikie, 1877; Laing, 1877), but it was Peach and Horne (1880) who first established the essential outlines of the glacial history of the islands. Following up the ideas of Croll (1870a, 1875), Peach and Horne proposed that at the period of maximum glaciation, the North Sea had been inundated by ice from both Scandinavia and Scotland and that the combined ice masses flowed into the Atlantic Ocean in a westerly or north-westerly direction across the northern isles of Scotland. Hence the principal themes of studies of the glaciation of Orkney have been the direction of ice movement, the types of erratics found in the glacial deposits and, in particular, whether these erratics were derived from Scotland or from Scandinavia. Only minor attention was paid to the possibility of local glaciation, and it is only in recent years that a small amount of palynological information has become available on the history of the vegetation of the islands.

No organic interstadial or interglacial deposits have been reported from Orkney. The stratigraphically earliest Quaternary deposit is a raised cobble beach exposed in the north of Hoy (see Muckle Head and Selwick). The beach occurs at 6–12 m above present sea level and rests on the inner margin of a marine abrasion ramp. No erratic material and no fossils have been observed in the beach, which is overlain by a head deposit and then a till deposited during the last regional glaciation (Wilson *et al.*, 1935; D. G. Sutherland, unpublished data). The age of the beach is unknown.

A fundamental theme in studies of the glaciation of Orkney has been the movement of ice from the floor of the North Sea westwards and north-westwards across the islands. This model derived initially from studies on the Scottish mainland in Caithness (Jamieson, 1866; Peach, 1868 (cited by Croll, 1870a); Croll, 1870a, 1875) and was later substantiated by detailed mapping on the Orkney Islands themselves (Helland, 1879; Peach and Horne, 1880; Wilson *et al.*, 1935; Mykura, 1976; Rae, 1976). The glacial features in Orkney relate closely to those in Caithness in terms of ice-movement patterns, the presence of shelly tills and inferred chronology. The main glacial deposits of Orkney, like those of Caithness, have generally been ascribed to the last glaciation, but recent interpretations suggest that they may relate to Early Devensian glaciation and that Orkney was largely unglaciated during the Late Devensian (Sutherland, 1984a; Bowen, 1989, 1991). However, the issue is far from resolved and recent work by Hall and Whittington (1989) and Hall and Bent (1990) provides support for the former model.

In an early account, Laing (1877) argued that Orkney had not been glaciated. This view was challenged in a reply by A. Geikie (1877), who adduced the presence of roches moutonnées, striations, till and moraines, and concluded that the ice moved from south-east to north-west. From their investigations on the Shetland Islands, Peach and Horne (1879) concluded that Shetland had been glaciated by ice originating from Scandinavia and that during the glacial maximum convergent ice streams from the Scottish mainland and Scandinavia moved north-west across the floor of the North Sea and the Orkney Islands. Subsequently (Peach and Horne, 1880), they elaborated on this model and in support provided detailed field evidence from Orkney for the pattern of ice-sheet and, later, local mountain glaciation; Helland (1879) independently reached similar conclusions. Striations and roches moutonnées showed conclusively that ice moved to the north-west and north-north-west, from the North Sea to the Atlantic. Further evidence of such ice movement was provided by the presence of shell fragments in the till at numerous locations and systematic changes in the lithological composition of the till on several islands. Erratics included a number of rock types foreign to the islands: limestones (including chalk), igneous lithologies, and the famous Saville Boulder, which has been argued to be of Scandinavian origin (Heddle, 1880; Flett, 1898; Wilson *et al.*, 1935). Other supposedly Scandinavian erratics were noted by Saxton and Hopwood (1919), although Rae (1976) considered that they could equally well have been derived from the Scottish mainland. Peach and Horne (1893c) later provided further details of shelly till and erratics on North Ronaldsay. They concluded that some of the material could have been derived from Aberdeenshire, the eastern Highlands and Fife. Peach and Horne's work remained the definitive study of

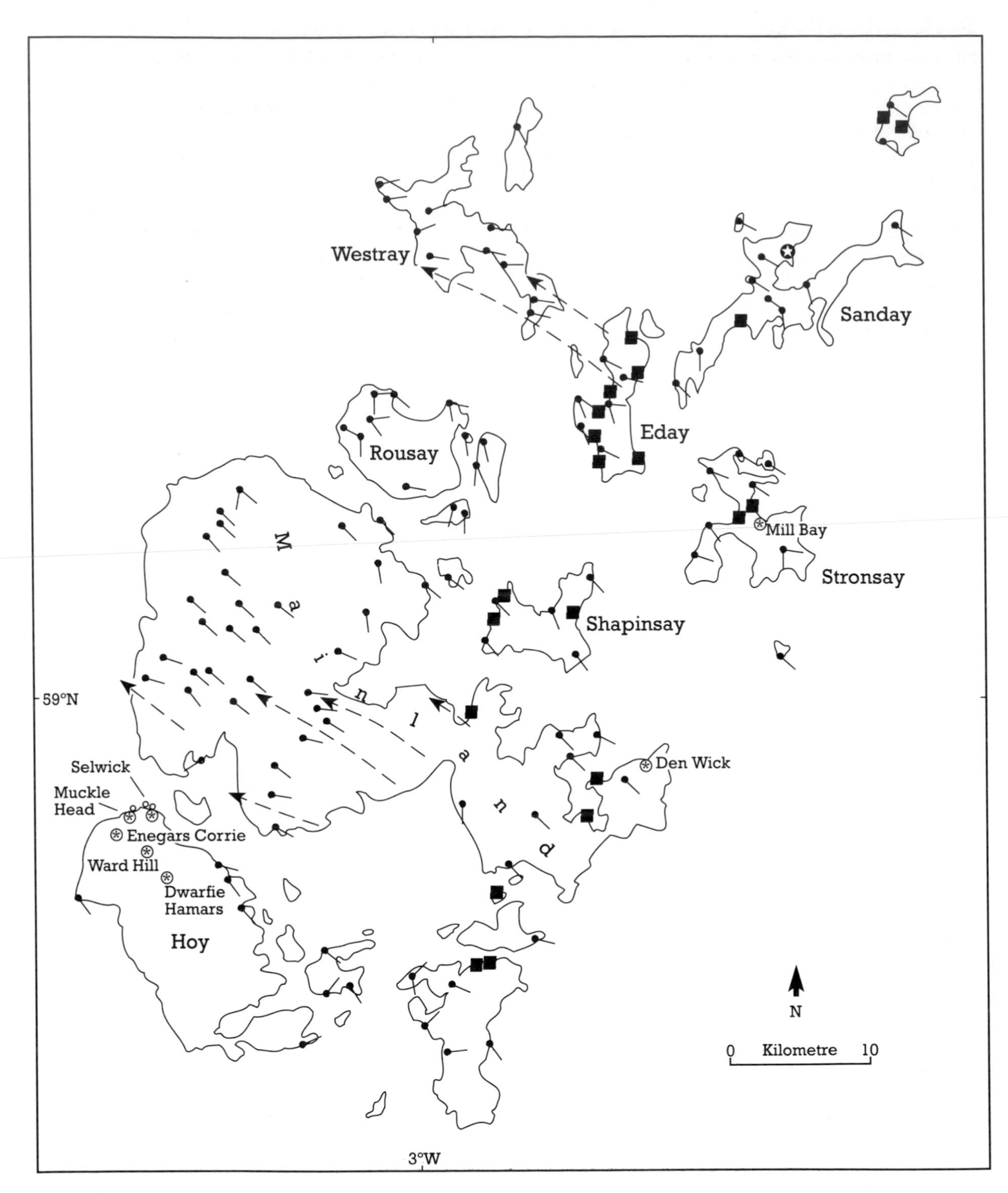

Striations from Wilson *et al.* (1935).

Transport of erratics recorded in Peach and Horne (1880).

Shelly till localities (after Peach and Horne, 1880; Rae, 1976; Sutherland, unpublished data).

Saville erratic.

Raised beach in north Hoy.

GCR sites.

Orkney for nearly a century and it has provided a basis for most regional syntheses.

Wilson *et al.* (1935) elaborated on Peach and Horne's results, suggesting that Orkney was crossed initially by Scandinavian ice moving from east to west and then by Scottish ice flowing at first towards the north-west and then towards the north. Charlesworth (1956) placed the limit of his Stage A (Highland Glaciation) across the Mainland, based on the distribution of morainic mounds. However, there is no convincing evidence that these are ice-marginal deposits and the current view is that they relate to ice-sheet wastage (Rae, 1976). Godard (1965) followed the interpretation of Wilson *et al.* (1935), but believed that the last ice-sheet movement probably correlated with Charlesworth's Scottish Readvance, a concept that is no longer tenable.

The most detailed recent study of the glaciation of Orkney is that of Rae (1976). He concluded that the earliest ice flow originated from between south and south-south-east. Later ice flowed from the south-east and finally from between east-south-east and east. Although he recognized two distinctive till units and cross striations, Rae concluded that there was no substantive evidence for more than one major ice-sheet glaciation on Orkney (see Den Wick). The age of this glaciation was, however, unconfirmed, but on the basis of an infinite radiocarbon date from Mill Bay, Rae believed that it could have occurred during the Early Devensian.

From his work in Shetland, Orkney, Foula and Fair Isle, Flinn (1978a) proposed two possible models for the glaciation of the islands. The first involved Scandinavian ice crossing the North Sea early in the last glaciation and deflecting Scottish ice across Orkney. Later, as both Scottish and Scandinavian ice waned, an independent ice-cap developed in Shetland. The second model, which Flinn considered more acceptable, placed the Scandinavian ice in an unspecified earlier stage.

In the absence of any known organic interglacial or interstadial deposits on Orkney, dating of the glacial sequence has remained conjectural, with most opinions assigning the main glacial features to the last ice-sheet (Late Devensian) or the ice maximum. In recent years, however, the extent of the last glaciation in Britain has been questioned and attention has re-focused on the possible existence of unglaciated areas in the more peripheral parts of the country. A fundamental problem in interpreting the glacial sequence in Orkney centres on explaining the westerly movement of the Scottish ice. This has generally been attributed to its deflection by Scandinavian ice in the central North Sea. However, offshore studies have shown that during the last glaciation the floor of the central North Sea was unglaciated (Cameron *et al.*, 1987; Sejrup *et al.*, 1987). Therefore, unless the deflection of the Scottish ice relates to an alternative, but as yet unidentified, mechanism, the glaciation of Orkney must have been earlier. Thus, although there is no conclusive evidence as yet, the current weight of opinion favours the interpretation that Orkney was last glaciated by an external ice-sheet during the Early Devensian and was not covered by the last Scottish ice-sheet (Rae, 1976; Synge, 1977a; Sissons, 1981b; Sutherland, 1984a; Bowen *et al.*, 1986). Recently, further support for Early Devensian glaciation has come from the preliminary results of amino acid dating of shells from the tills of Caithness and Orkney (Bowen and Sykes, 1988; Bowen, 1989, 1991). If this hypothesis is substantiated, then it places important constraints on wider palaeoclimatic and glacier reconstructions and also on the nature of landscape change and the impact of periglacial processes on Orkney during the Late Devensian. However, Hall and Whittington (1989) have adduced several lines of evidence from Caithness (see below) to support more extensive glaciation during the Late Devensian and a possible ice limit as far north as the Orkney – Shetland Channel (see also Hall and Bent, 1990). Subsequently, Hall and Bent (1990) have suggested that the north-westerly ice movement across Orkney and Caithness might relate to the development of a divergent flow pattern in the Moray Firth, where the ice-sheet moved from a rigid to a deformable bed.

Following A. Geikie (1877), Peach and Horne (1880) recognized a number of moraines on Hoy and the more elevated parts of the Mainland which they inferred to be associated with local valley glaciation. From later work, however, it is apparent that only on Hoy is there clear geomorphological evidence for local glaciation (Wilson *et al.*, 1935; Godard, 1965; Rae, 1976). Rae (1976) concluded that apart from Enegars Corrie, there was no evidence for corrie glaciers in any

Figure 4.1 Location map and principal features of the glaciation of Orkney, including patterns of striations, directions of transport of erratics and shelly till localities (from Sutherland, 1991b).

of the locations suggested by Charlesworth (1956), and in other localities in the valley of South Burn and the Ford of Hoy it was not clear whether the glacial deposits were associated with local ice or with the decay of the ice-sheet.

A number of arcuate end moraines occur, however, for example at Dwarfie Hamars and Enegars Corrie (D. G. Sutherland, unpublished data). Although not directly dated, these possibly relate to the Loch Lomond Readvance and represent the last minor phase of glaciation on Orkney. These small glaciers formed in favoured localities and had equilibrium line altitudes of approximately 150 m OD.

The hills of northern Hoy are the only ones in Orkney to give rise to significant areas of ground above 350 m OD and they are especially notable for the development of periglacial features (see Ward Hill) (Goodier and Ball, 1975). It seems probable that the periglacially weathered detritus that mantles the summits of these hills was produced during the Late Devensian (and conceivably earlier), but the periglacial features observed today were all apparently produced during the Holocene. The sand sheets ('dunes') on Ward Hill, in particular, seem to preserve a record of changing Holocene environments in the form of buried soil horizons.

Despite the pioneering work of Erdtman (1924), there have been few detailed pollen analytical studies of the vegetational development of Orkney. The only study of Lateglacial profiles is that of Moar (1969a), who indicated that the vegetation at that time was very restricted in its development. During the early to middle Holocene a dwarf-shrub vegetation developed with possibly only isolated patches of birch and hazel scrub (Moar, 1969a; Keatinge and Dickson, 1979). From the middle Holocene onwards there has been considerable human impact on the vegetation (Moar, 1969a; Caseldine and Whittington, 1976; Davidson *et al.*, 1976), and it may be that grazing pressure contributed to the spread of blanket bog in the later Holocene (Keatinge and Dickson, 1979).

MUCKLE HEAD AND SELWICK

D. G. Sutherland

Highlights

Coastal sections at Muckle Head and Selwick provide a rare example in Scotland of raised beach deposits which are overlain by till. These deposits pre-date at least the last glaciation on Orkney, and are important for interpreting the history of sea-level changes and glaciation in the northern isles.

Introduction

Muckle Head (HY 213053) and Selwick (HY 225055) are coastal sections located on the north of Hoy. They provide important exposures in raised beach deposits overlain by till. The raised beach deposits pre-date at least the last glaciation and indicate a former sea level slightly above that at present. They are important not only for interpreting the Pleistocene history of Orkney, but also in the wider context of those parts of Scotland which were peripheral to the centres of glaciation. The only published description of the sites is in Wilson *et al.* (1935).

Description

Wilson *et al.* (1935) recorded a raised beach at Muckle Head and Selwick at between 20 ft (6 m) and 40 ft (12 m) above present sea level and resting on a clear rock notch or platform. They described the beach as being up to 5 m in thickness and consisting of a coarse, rounded gravel with numerous rounded boulders. The beach deposits, which at both localities were cemented by calcite, contained no erratic clasts. Wilson *et al.* (1935) did not specifically record the beach deposits as being overlain by till (and a cross-section of the Selwick beach and rock platform does not show any glacial deposits) but they concluded that the beach was probably 'pre-Glacial' in age, on the grounds of the lack of erratic material.

Recently, the best exposure has been that at Muckle Head (Figure 4.2) (D. G. Sutherland, unpublished data). There, approximately 3.5 m of beach sediments rest in the angle of a rock notch at the back of a ramp cut in conformity with the bedding of the Old Red Sandstone flagstones. The beach consists of a fining upwards sequence of rounded cobbles and pebbles at the base to pebbles at the top, with large subangular flagstone boulders scattered throughout. Immediately overlying the beach is a head deposit, about 1.5 m thick, consisting of angular to very angular platy

Figure 4.2 Section at Muckle Head, Orkney. The raised beach gravel, which rests on a rock platform, incorporates large flagstone blocks and is overlain in turn by a head deposit and till. (Photo: D. G. Sutherland.)

clasts derived from the immediately adjacent bedrock. Resting on this head deposit is approximately 1 m of brown till with a well-developed clayey-sand matrix. The upper 0.3 m of this till has been periglacially disturbed and subject to solifluction, and the section is capped by 0.15–0.3 m of structureless sand with occasional gravel clasts scattered throughout.

The Selwick deposit is less accessible and the stratigraphy less clear than at Muckle Head. Also the beach does not show the fining upward sequence noted at Muckle Head, being composed of rounded cobbles and pebbles, with a scatter of angular boulders throughout. There was no clear contact observed with the overlying deposits, but exposures in slumps revealed a yellow-brown till.

Interpretation

It is apparent from the local stratigraphy that the beach pre-dates the last glaciation of Orkney. The age of this glaciation remains to be established, however. The stratigraphic sequence at Muckle Head can be interpreted as evidence for a period of high sea level, followed, when sea level had fallen, by cold periglacial conditions and then invasion of the area by ice. Subsequent to ice retreat there was a further period of periglaciation. The simplest chronological interpretation would be that this sequence of events represents the last interglacial–glacial cycle, but until direct dating evidence is available such a conclusion is speculative.

There are relatively few localities in Scotland where raised beach sediments can be observed to be overlain by till. Similar stratigraphic sequences are relatively common in the southern British Isles, where amino acid dating has established that there are several generations of beach deposits, each apparently being correlated with a separate interglacial period (Bowen *et al.*, 1986). In Scotland, beach deposits overlain by till are only known in those areas peripheral to the centres of glaciation, such as in the north of Lewis (see North-west Coast of Lewis), on the islands of Barra (Peacock, 1980a, 1984a; Selby, 1987) and North Rona (Gailey, 1959), and at two localities in the north of Hoy, at Muckle Head and Selwick. In none of these areas have marine shells been recovered from *in situ* beach sediments and hence there has been no opportunity to date the beaches other than with reference to the local stratigraphic sequence. It is therefore not known whether the various exposures of beach deposits are contemporaneous or not.

The raised beach deposits in the north of Hoy are two of the very few similar deposits in Scotland that record a period (or periods) of sea level only slightly above that of the present, but prior to at least the last glaciation. The Muckle Head section, in particular, demonstrates a sequence of sediments which apparently were deposited during a period of climatic cooling and glacial expansion, but the age of these sediments remains to be established. The site is an important element not only in the Pleistocene history of Orkney, but also as part of a pattern of interrelated sea level and glacial events in those areas of Scotland that were peripheral to the areas of build-up of the Scottish ice-sheets.

Conclusion

Muckle Head and Selwick are important in showing raised beach deposits overlain by ice-deposited sediment (till), a succession which indicates that sea level was slightly above that of the present during a period, as yet undated, before the last glaciation. This area of north Hoy is one of only a few locations in Scotland, all located near the periphery of ice-sheet glaciation, where such successions occur. It is therefore critical not only in establishing the history of sea-level change and glaciation in Orkney, but also in helping to reconstruct their wider regional patterns.

DEN WICK

J. E. Gordon

Highlights

The sequence of sediments in the coastal section at Den Wick is representative of the multiple till deposits of Orkney. The lithological contents and clast fabrics of the two tills provide evidence for changing ice-flow patterns during the last glaciation of Orkney.

Introduction

Den Wick (HY 576088) is a coastal section on the Deerness peninsula in the eastern part of the

Mainland; it shows one of the best examples of a multiple till sequence in Orkney. Two tills are present and have been described by Rae (1976). Their lithological characteristics provide significant evidence for reconstructing former ice-flow directions, and they demonstrate important general relationships between till properties and the source bedrock lithologies.

The first suggestion of multiple drift units on Orkney was made by Wilson *et al.* (1935), who noted the occurrence of a grey or yellowish clayey rubble below the more usual grey, red or purple till of sandy-clay composition, but they concluded that the former was little more than fractured and locally transported bedrock. Subsequently, Rae (1976) formally established and defined the existence of more than one till on the basis of colour, lithology and sedimentological properties. He argued that the till characteristics at a site reflect the bedrock composition up-ice and that contrasts between individual till units at a site relate to changing ice-movement patterns. Rae (1976) recorded twenty multiple till sections in Orkney.

Description

At Den Wick, Rae (1976) noted the following sequence:

3.	Red till	1.5–>3 m
2.	Brown till	>5.5 m
1.	Striated bedrock	

The contact between the two tills was sharp and undulating, and locally a thin sand and gravel horizon, 0.1–0.15 m thick, intervened between the tills. However, Rae considered that at this and all the other sites, there was no significant evidence to support more than one glacial episode, and therefore the multiple tills were deposited by a single ice-sheet.

From the orientations of striae on the bedrock beneath the brown till, the lithological composition of the till (a relatively high percentage of Lower Eday Sandstones) and clast fabric measurements, Rae concluded that the ice movement associated with the lower till was from between south and south-east. Ice following such a flow pattern would have traversed the Lower Eday Sandstone most of the way across the Deerness peninsula and avoided the Middle Eday Sandstones.

The overlying red till has a relatively higher content of red sandstone clasts similar to the Middle Eday Sandstone. Since no onshore outcrop of this lithology appeared to fit with the pattern of striations observed beneath the red till elsewhere in Deerness where it rests directly on bedrock, Rae postulated an offshore outcrop, an interpretation supported by the presence of chalk erratics in the red till but not in the brown till. The red till was therefore associated with ice flow from a more easterly azimuth than the underlying brown till. Shell fragments have been recovered from both till units (A. M. Hall, unpublished data).

Interpretation

Den Wick is an important section representing one of the key elements of the till stratigraphy of Orkney. The multiple till sequences, of which Den Wick is a particularly good example, are important in several respects. First, their lithological and sedimentological characteristics demonstrate close relationships with the source bedrock traversed by the ice. Second, the superimposition of the different till units indicates changes in ice-flow direction, thus supporting the inferences based on striation patterns. Third, the contacts between the different units at individual sites provide no conclusive evidence for deposition during more than one glacial period. Fourth, the superimposition of the multiple till units in a consistent fashion (Rae, 1976) and the inferred shifts in the direction of ice flow provide important evidence for significant changes in the wider regional dynamics of the ice-sheet and its driving forces. Further study at both regional and national scales, coupled with ice-sheet modelling, is required to elaborate the origins of these changes and the extent to which they may relate to climatic factors and other variables that determine ice-sheet flow patterns. Multiple till sequences, like that at Den Wick, will be an important source of field evidence to provide constraints on the appropriate mathematical models.

Conclusion

The sediments at Den Wick are representative of the multiple till deposits of Orkney. They comprise two superimposed tills, each containing distinctive rock fragments derived from the bedrock over which the ice that deposited the tills had flowed. The tills provide evidence, in the form of their stone orientations and rock and shell contents, for former ice-flow directions. The till succession indicates a change in the flow pattern during the course of a single glaciation, suggesting a significant change in ice-sheet dynamics.

MILL BAY

J. E. Gordon

Highlights

The coastal section at Mill Bay is the best locality for demonstrating the shelly till that is characteristic of the eastern part of Orkney. This deposit shows that the ice moved across the sea floor and then onshore in a west-north-west direction. The presence of the shells also provides an important opportunity to date the last main glaciation of the islands.

Introduction

A coastal section at Mill Bay (HY 665256), on Stronsay, provides the best available exposures of the shelly till of Orkney and has been described by Peach and Horne (1880) and Rae (1976). As well as demonstrating the characteristics of this important till unit, Mill Bay is particularly significant for the dating potential of the shells in the till.

Description

The section at Mill Bay forms a continuous cliff 6–10 m high over a distance of nearly 1 km. Peach and Horne (1880) recorded a reddish-brown, gritty clay containing striated stones. In composition the till mostly includes material from the adjacent flagstones and siltstones, but a range of exotic rock types is also present, including igneous and metamorphic lithologies, fossiliferous limestones, chalk, flints and fossil wood. Peach and Horne inferred that the erratics were derived from the Scottish mainland. Numerous fragments of shells, including *Arctica islandica* (L.), *Mytilus* and *Mya truncata* (L.), are also present in the till, and according to Peach and Horne they appear smoothed and striated. Striations on bedrock at Mill Bay are aligned W15°–35°N. Peach and Horne also recorded large blocks of what appeared to be petrified wood in the till. Rae (1976) has provided additional sedimentological and lithological details of the till at Mill Bay, and noted the presence of a lens of grey till incorporated in the red till. Rae (1976) also obtained an infinite radiocarbon date of >44,300 BP (T–1152) from shell fragments in the till.

Interpretation

The lithological composition and the erratic and shell content of the Mill Bay till clearly indicate that ice moved onshore from an easterly direction (Peach and Horne, 1880; Rae, 1976) in accordance with the general pattern established for the Orkney Islands. Although, strictly, the radiocarbon date is inconclusive, Rae argued that on the basis of probability it suggests an Early Devensian age for the till. More recently, preliminary results from amino acid epimerization analyses suggest that shells in the till are no younger than the last interglacial, which again lends support to the hypothesis that the maximum age of the till is Early Devensian (see Bowen and Sykes, 1988; Bowen, 1989, 1991).

Mill Bay is an important reference site for the shelly till of Orkney, and it represents the best exposure of its type on the islands. Although shelly till is exposed at several other localities (see Peach and Horne, 1880; Rae, 1976), predominantly in the eastern half of the Orkney Islands (Figure 4.1), Mill Bay is one of the few sites where shell fragments are relatively abundant and it has also yielded a relatively wide range of erratic types. Mill Bay therefore demonstrates particularly well the general direction of ice movement onshore and towards the west-north-west. In a wider context, the Mill Bay till forms part of a lithostratigraphic unit that extends across Caithness and Orkney. This formation is thought to be the product of a single glacial episode in which ice from the Scottish mainland moved north-west from the Moray Firth Basin and adjacent North Sea Basin. As yet the age of the glaciation is not securely established, but Mill Bay provides significant dating potential. Preliminary results from amino acid analyses, taken in conjunction with those from Caithness and with other evidence from the central North Sea, provide some support for glaciation during the Early Devensian. If substantiated, this would be

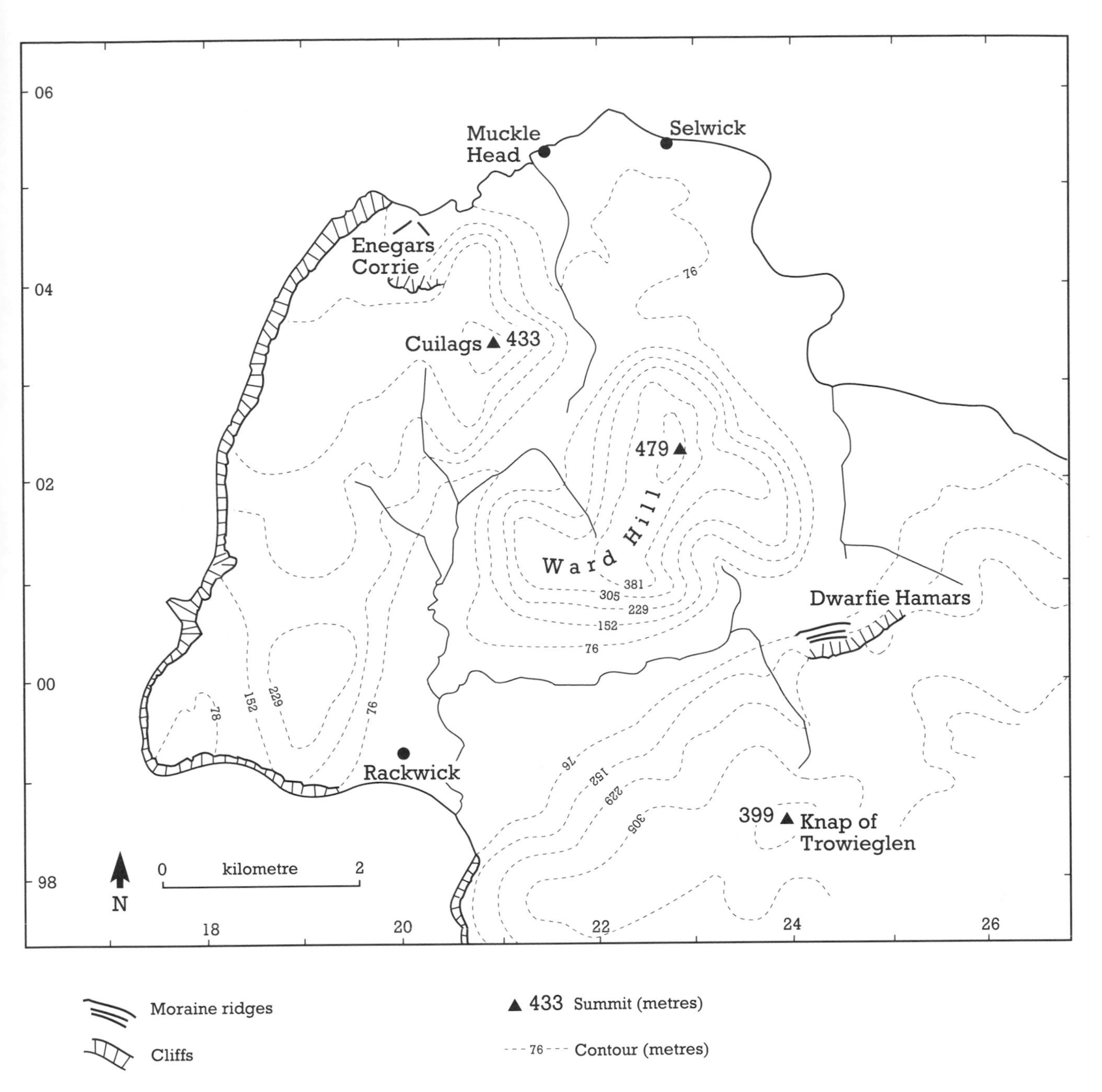

Figure 4.3 Location map of North Hoy.

the first positive evidence for Early Devensian glaciation in Scotland. On the other hand, there is evidence from Caithness (see below) that the shelly till may in fact date from the Late Devensian. Sites such as Mill Bay therefore have a major role to play in resolving this key issue of the limits of the last ice-sheet in northern Scotland and the adjacent shelf.

Conclusion

Mill Bay provides the best available exposures in the shelly till (sediments deposited by ice) of Orkney. The presence of the shells of marine molluscs in the till, together with other evidence, indicates the movement of ice across the sea bed and then onshore and towards the west-north-west. Not only is Mill Bay an important reference site for demonstrating this pattern of ice movement, it is also significant for the potential provided by the shells for dating (by geochemical techniques) the last major glaciation of the islands.

WARD HILL, ENEGARS CORRIE AND DWARFIE HAMARS

D. G. Sutherland

Highlights

This site is notable for its assemblage of periglacial and glacial interests. The periglacial landforms and deposits include a range of wind- and frost-related features on Ward Hill that provide a record of slope activity and soil movements both at the present day and earlier during the Holocene. The end moraines at Enegars Corrie and Dwarfie Hamars are thought to have formed during the Loch Lomond Stadial and provide important evidence for palaeoclimatic reconstruction.

Introduction

Ward Hill (HY 229023) is located in the north of the island of Hoy. It is the highest hill in Orkney at 479 m OD and is separated from the immediately neighbouring hills, Cuilags (433 m) and Knap of Trowieglen (399 m), by two valleys each over 200 m deep (Figure 4.3). The summit area of the hill forms a broad, approximately north–south aligned ridge some 1.6 km long with secondary ridges extending eastwards and westwards at its northern and southern ends. Over 2.5 km^2 of ground lie above 300 m OD, of which slightly under 1 km^2 is at over 400 m OD. The hill is fully exposed to winds from the open sea.

Ward Hill is important for periglacial geomorphology and demonstrates a fine assemblage of active and fossil landforms. By virtue of its northern location, it is a prime site in a network of mountain summits for studying the distribution of upland periglacial features in Scotland. It is particularly noted for landforms associated with wind activity, described by Goodier and Ball (1975). Additional interest in the hills of North Hoy is provided by end moraines at Enegars Corrie (HY 200043) on Cuilags and at Dwarfie Hamars (HY 245005) on Knap of Trowieglen, which are tentatively ascribed to the Loch Lomond Readvance. The moraine at Enegars Corrie has been described by A. Geikie (1877), Peach and Horne (1880), Wilson *et al.* (1935) and Rae (1976).

Description

The hills of northern Hoy are underlain by Hoy Sandstones of the Upper Old Red Sandstone (Mykura, 1976). These Devonian sandstones are medium-grained and generally well sorted, in keeping with their fluvial origin. They are near-horizontally bedded, but slight variations in lithology are reflected in their resistance to weathering. The harder beds form prominent crags and steps on the hillsides, whereas the softer beds weather to a loose sand.

Four main types of periglacial feature have been described from Ward Hill by Goodier and Ball (1975). These they termed turf-banked terraces, wind stripes, hill dunes and composite stripe/terrace features. Although they do not include them as a separate category, it is apparent from the descriptions that deflation surfaces are also well-developed.

The turf-banked terraces (Figure 4.4) have typical dimensions of 2–3 m across the 'treads', with slopes of about 15°, and the 'risers' have widths of 3–5 m and slopes of 25–35°. The 'treads' are vegetation-free, and the long-axis of the terraces is either parallel to the contours or aligned obliquely across the slope. According to Goodier and Ball (1975), in the former case, the principal influence in the formation of the terraces is frost action, whereas in the latter, the action of wind on the vegetation has orientated the features towards the effective wind direction, forming composite stripe/terraces.

Wind stripes are strikingly developed on Ward Hill. Goodier and Ball (1975) distinguished three separate types. The first comprises regular, continuous stripes in which the vegetated parts are generally straight-sided and parallel; the second, regular continuous stripes with non-uniform widths and curving margins; and the third, scattered crescentic vegetated areas on deflation surfaces. The first type occurs on the exposed southern face of the hill, and here the stripe alignment is generally parallel to the contours. The vegetated stripe widths in this area are from 0.2 to 1.7 m and the inter-stripe zones are

Figure 4.4 Solifluction terraces on Ward Hill, Hoy. The turf-banked terraces are orientated approximately parallel with the contours. (Photo: J. E. Gordon.)

0.52–0.78 m wide. The wind-cut stripe faces average 0.14 m in height. These regular features merge along-slope with wind-formed vegetation waves where vegetation cover is complete, but elongate zones of growing *Calluna vulgaris* alternate with zones of dead *Calluna* stems and lichens (see Bayfield, 1984).

The features termed hill dunes by Goodier and Ball (1975) consist of areas in which eroded sand-sheets capped by vegetation stand as remnants above deflation surfaces. The name is therefore somewhat misleading, as the 'dunes' are not constructional features. The sands have a distinct stratification: a surface horizon of about 0.1 m of grey sand overlies a yellowish-brown sand to a depth of 0.7 m, with a buried surface horizon between 0.7 m and 0.84 m, a second bed of brown sand to 0.94 m and then a further buried surface horizon at 0.9 m–0.95 m, with underlying sand to 1.2 m. Below the sands is a diamicton consisting of weathered sandstone clasts in a sandy clay matrix. A similar diamicton also underlies the deflation surfaces. The stratification of the sand deposits is indicative of alternating periods of stability and sand movement, but no studies have been carried out on the age of the deposits. Similar sand-sheets and interbedded soil horizons have been described on Ronas Hill (Ball and Goodier, 1974) and on An Teallach (Ballantyne and Whittington, 1987). At the latter locality it was demonstrated that sand deposition began in the early Holocene but was much reduced by the establishment of vegetation cover during the Holocene. However, recent disruption of the vegetation due either to climatic deterioration or to overgrazing has resulted in a renewed phase of sand erosion and redeposition. A similar history may apply to the Ward Hill sands, with the lower slopes of Ward Hill also providing clear evidence of former slope activity in the form of gullied drift and fan deposits.

The slopes of Ward Hill are covered in a debris mantle and there is no clear evidence within the corrie-like recesses flanking the hill for local glaciation post-dating the last period of ice-sheet glaciation. However, to the north in Enegars Corrie and to the south below Dwarfie Hamars, clear end moraines can be observed. The Enegars moraine is a single arcuate ridge, up to 6–8 m high, descending to approximately 100 m OD and associated with a former glacier with a north-east aspect. The Dwarfie Hamars landforms comprise at least three distinct arcuate moraines, the outermost reaching down to about 50 m OD. The famous Dwarfie Stone is an erratic boulder resting on one of the moraines. The age of these moraines has not been established, but there is a notable difference in the degree of development of screes and slope debris mantles within the moraines compared with those on the adjacent slopes outside them. This suggests that the small glaciers that formed the moraines developed in favoured localities during the Loch Lomond Stadial. If this attribution is correct, it suggests that the debris mantles on the slopes of Ward Hill developed in major part during, or prior to, the Loch Lomond Stadial.

Interpretation

Periglacial deposits occur on the summits of most Scottish mountains. It is possible to divide these deposits into two broad age-groups: those formed during the cold phases of the Late Pleistocene and those formed under the milder conditions of the Holocene (Ballantyne, 1984, 1987a). The processes responsible for the formation of the latter periglacial features are normally still operative on Scottish mountains. The types of both fossil and active periglacial landforms and sediments that may be encountered on particular summits are related to the underlying bedrock and the climatic conditions, especially the temperature regime and the degree of exposure. As the above conditions vary throughout the country, understanding of the development of the different types of periglacial deposits is dependent upon the study of summits in different areas. In this context, Ward Hill occupies a critical position because of its location in the Orkney Islands and the range of periglacial features developed on it. Initial studies have emphasized the role of wind in the landform development, a feature that Ballantyne (1981) considers important in understanding the unique periglacial environment of Scottish mountain tops. There is considerable potential for the further study of the sand deposits on Ward Hill. Their present erosion, together with the evidence from their stratification of episodic stability in the past, indicates the fragile nature of the balance between formation and disruption of these hill-top deposits.

The moraines of Enegars Corrie and Dwarfie Hamars are significant in a national context. If they are of Loch Lomond Stadial age, then they represent some of the northernmost glaciers at this time in Britain and therefore have a significant bearing on reconstructing the palaeoclimate of the stadial. In particular, they suggest glacier equilibrium line altitudes of about 150 m OD, which compares with values of 319 m for Skye (Ballantyne, 1989a) and 357 m OD for Rum (Ballantyne and Wain-Hobson, 1980) on the western seaboard of Scotland (see also Sissons, 1979d).

Conclusion

This area in the north of Hoy is important for its assemblage of landforms created by periglacial and glacial processes. It is particularly noted for a series of deposits formed principally by wind action, but also includes others modified by the combined action of wind and frost. These periglacial deposits provide a vital record from this northern locality of the history of past episodes of slope stability and erosion. There is significant potential for further research to establish the timing and causes of the erosion. The interest of the site also includes moraines believed to have been formed by glaciers, about 11,000–10,000 years ago, during the cold period known as the Loch Lomond Stadial. As such, they are the northernmost features of their kind in Britain and therefore are significant for reconstructing the climate of the stadial.

Chapter 5

Caithness

INTRODUCTION

D. G. Sutherland

Geographically, Caithness (Figure 5.1) forms a link between the Orkney Islands and the Northern Highlands. The north-east part of the district is a gently undulating erosion surface *c.* 90–180 m OD developed on the underlying sandstone bedrock, whereas to the south, conglomerates and quartzites form the upstanding hill masses of Morven (705 m), Maiden Pap and Scaraben. The effects of glacial erosion are limited by comparison with the north-west Highlands, and in studies of the Quaternary of Caithness, the glacial deposits have been the principal focus of research. Particular attention has centred on patterns of ice movement, the characteristics of a distinctive shelly till, the interaction between ice of local origin and the external ice that deposited the shelly till, the number and ages of separate glaciations represented in the stratigraphic record and the question of whether or not part of north-east Caithness was ice-free during the Late Devensian.

During the last century the glacial deposits were the subject of numerous papers particularly on account of their content of marine fossils (Busby, 1802; Peach, 1858, 1863a, 1863b, 1863c, 1865b, 1867; Jamieson, 1866; Crosskey and Robertson, 1868c; Brady *et al.*, 1874; Dick, *in* Smiles, 1878; Peach and Horne, 1881c). The abundance of these fossils led to a considerable debate as to whether the deposits were indeed the product of land ice (Croll, 1870a; J. Geikie, 1877) or whether they had been deposited in the sea from icebergs (Cleghorn, 1850, 1851; Peach, 1863a, 1863c; Jamieson, 1866). Caithness also became noted for the number and size of the Mesozoic and Cenozoic erratics found there (Peach, 1859, 1860; Peach and Horne, 1881c; Tait, 1908, 1909; Carruthers, 1911; Crampton and Carruthers, 1914). These erratics were thought to have come from the Moray Firth or the eastern coast of Sutherland and hence supported the interpretation that the last ice movement over most of Caithness was from the south-east towards the north-west (Croll, 1870a; J. Geikie, 1877; Peach and Horne, 1881c) and not, as had been suggested by Jamieson (1866), in the opposite direction.

During this early phase of study of the glacial deposits, the main outlines of the glacial stratigraphy of Caithness were also established. The sequence consisted of a lower till containing erratics derived from immediately inland (Jamieson, 1866; Peach and Horne, 1881c; Crampton and Carruthers, 1914), an overlying till containing the marine fossils and erratics already referred to, and, in the southern and western parts of the county, a local till together with morainic and glaciofluvial deposits (Peach and Horne, 1881c; Crampton and Carruthers, 1914). The superposition of the lower and the shelly tills can be seen at localities such as Baile an t-Sratha (Dunbeath) and Drumhollistan, and an additional lower till may be present at the site of the largest known Mesozoic erratic at Leavad. Significant observations were also made on the periglacial features found on the hills in the south of Caithness (Carruthers, 1911), and the absence of raised shorelines from much of the coast was also noted (Crampton and Carruthers, 1914).

Despite this early interest, little further work has been carried out on the glacial deposits in Caithness until recent years. Various summaries and syntheses of the published evidence were produced (for example, Bremner, 1934a; Charlesworth, 1956; Phemister, 1960; Godard, 1965), but little original work was carried out until that of Omand (1973). He systematized the glacial stratigraphy and named the lower, local till (the Dunbeath Till), the shelly till (the Lybster Till), and the upper local till (the Reay Till); the moraines and glaciofluvial deposits were assigned to the latest Bower stage. No break was noted in deposition between the Dunbeath and Lybster tills, the first being deposited by ice flowing from the south-west and the second by ice from the south-east. This sequence of events is broadly similar to that interpreted for the glacial deposits of Orkney by Rae (1976) (see Chapter 4), which strongly suggests contemporaneity of the glacial phases, as does the sedimentary succession (Sutherland, 1984a). This glacial phase was therefore characterized by initial unimpeded expansion of Scottish ice, but subsequently the ice turned to flow towards the north-west, presumably due to the influence of the Scandinavian ice in the North Sea Basin as originally suggested by Croll (1870a, 1875) (but see below). There is some debate as to whether the Reay Till and the Bower stage represent significantly later glacial events (Peach and Horne, 1881c; Smith, 1968, 1977; Omand, 1973; Flinn, 1981; Sutherland, 1984a) or are essentially retreat phenomena of the major glaciation (Crampton and Carruthers, 1914; Charlesworth, 1956; Sis-

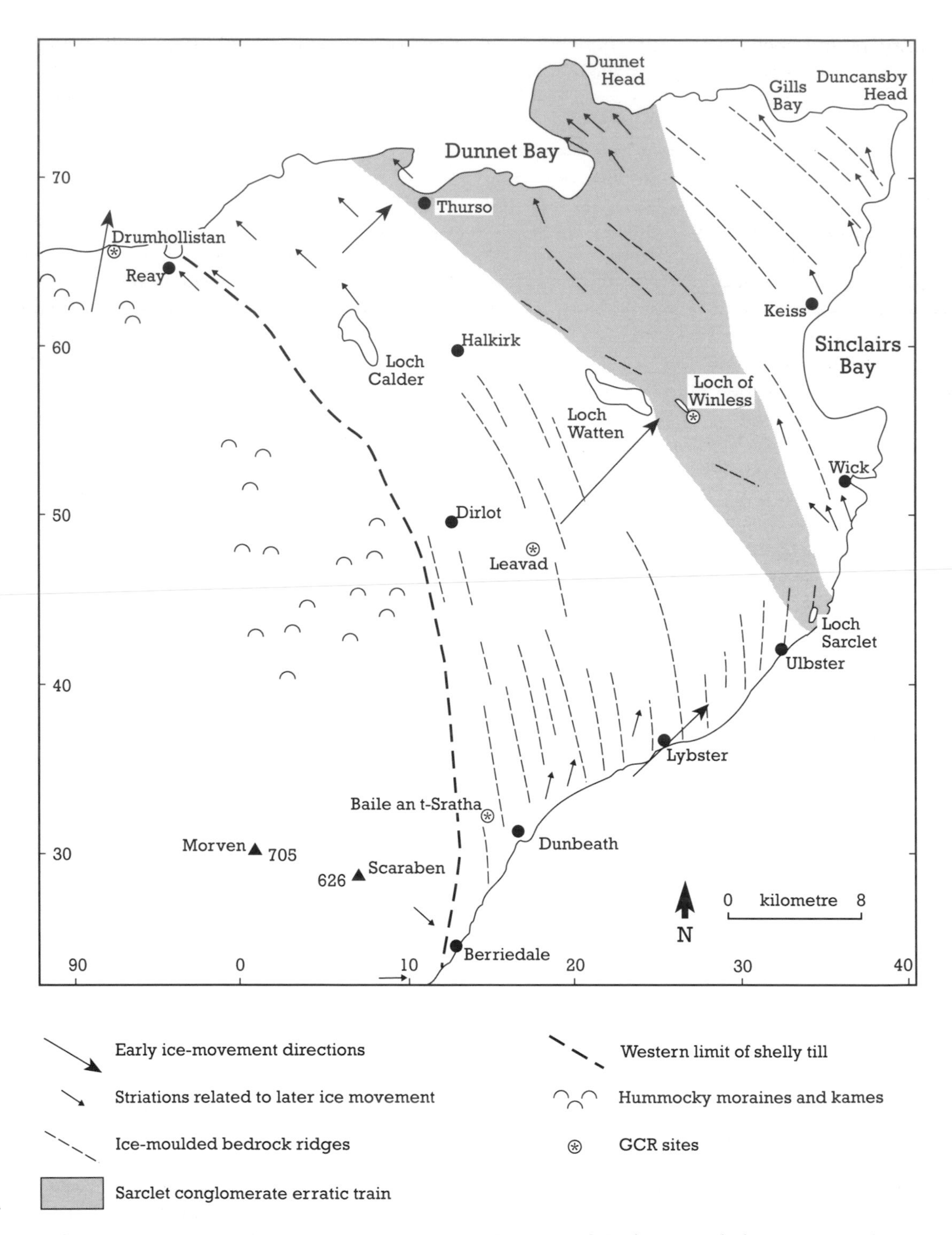

Figure 5.1 Location map and principal features of the glaciation of Caithness, including patterns of striations, ice-moulded landforms, distribution of erratics and shelly till (from Peach and Horne, 1881c; Sissons, 1967a; Sutherland, 1984a).

sons, 1967a). This outline stratigraphy has been confirmed and amplified in southern Caithness by Hall and Whittington (1989).

There is no direct evidence that bears on the age of the glacial deposits in Caithness. No *in situ* organic sediments have been found below or interstratified with the glacial deposits. Radiocarbon dating of the shells within the till has produced only a 'greater than' age (Omand, 1973) and amino acid analyses suggest that the majority of the shells have been derived from deposits of last interglacial age or older (Bowen and Sykes, 1988; Bowen, 1989). Various workers (Smith, 1977; Synge and Smith, 1980; Flinn, 1981; Sutherland, 1984a) have proposed that the shelly till may date from earlier than the Late Devensian and that only the overlying morainic and glaciofluvial deposits in the west and south of the county are of Late Devensian age. Hall and Whittington (1989) and Hall and Bent (1990), however, have argued that there is no evidence for a hiatus between sediments dated to the Lateglacial Interstadial and the underlying glacial deposits, thus inferring a Late Devensian age for the latter.

Overlying the glacial sediments are thicknesses of head and soliflucted till (Omand, 1973; Smith, 1977; Futty and Dry, 1977; Hall and Whittington, 1989). These have been interpreted as supporting the concept of this area being ice-free during the Late Devensian, the area being presumed to be subject to severe periglacial processes at that time. However, the mass-movement sediments may equally reflect early deglaciation as well as further activity during the Loch Lomond Stadial (see Hall and Whittington, 1989). In the south of the county the summits of the quartzite and conglomerate hills above about *c.* 450 m OD are notable for a range of active and fossil periglacial features, including wind and frost-related effects in the hill-top detritus and vegetation (Crampton, 1911; Crampton and Carruthers, 1914).

The earliest known deposits post-dating the last glaciation of Caithness are those at the Loch of Winless (Peglar, 1979), where the initial deposition of organic sediment has been dated to after 13,000 BP. This site has the most detailed published Lateglacial Interstadial pollen diagram in this region of Scotland and it shows certain differences from the north-west Highlands, with the possible local development of tree birch. Further Lateglacial Interstadial pollen records have been described by Hall and Whittington (1989) and by Charman (1990) from the Caithness–Sutherland border.

Apart from the information on the sparse local vegetation from the Loch of Winless, there is little information available on the environment during the Loch Lomond Stadial. It may be presumed, on the basis of inference from neighbouring areas, that certain of the periglacial features at both low level and on the mountain summits in the south of the county were formed or modified at this time, and this is confirmed by the work of Hall and Whittington (1989).

The vegetational development during the Holocene is important in this region for an understanding of the forest history of Scotland and the development of the blanket peat of the Flow Country. Lewis (1906) and Crampton (1911) provided details of vegetation and forest changes represented in the macrofossils preserved in peat sections, including the extensive occurrence of pine stumps in the central and western parts of Caithness. However, apart from the early work of Durno (1958) and the investigation of peat mounds by Robinson (1987), the only detailed pollen analytical work is that from the Loch of Winless. Caithness during the Holocene appears to have been largely treeless, although local, sheltered patches of birch–hazel woodland had developed by 9300 BP. Fossil pine stumps occur within blanket bog in western Caithness (Birks, 1977; Gear and Huntley, 1991) but in the east of the county the sampled pollen values for pine are so low that it seems unlikely that the tree ever grew there. Radiocarbon dating of the pine stumps has shown that the pine forest in Caithness was short-lived, expanding and retreating within a period of about 400 years around 4000 BP (Gear and Huntley, 1991). Clearance of the limited woodland cover in Caithness by Man appears to have started at around 3000 BP.

THE GLACIATION OF CAITHNESS

J. E. Gordon

The main features of the glaciation of Caithness are illustrated in Figure 5.1.

The glacial deposits of Caithness vary considerably in their distribution and thickness: typically they attain their greatest thickness in pre-existing valleys, topographic depressions and coastal re-entrants, and are relatively thin on higher ground and interfluves (Cleghorn, 1851; Jamieson, 1866; Peach and Horne, 1881c; Crampton and Carruthers, 1914; Godard, 1965; Omand, 1973; Hall and Whittington, 1989).

Three main suites of glacial deposits are characteristic of the area (Jamieson, 1866; Peach and Horne, 1881c; Crampton and Carruthers, 1914; Omand, 1973; Hall and Whittington, 1989). These comprise a locally derived till generally occurring west of a line from Reay to Berriedale (Figure 5.1) and deposited by ice originating in the hills of southern Caithness and Sutherland, a distinctive shelly till (often weathered in its upper layers) to the east and derived from offshore in the Moray Firth, and hummocky glacial deposits which locally overlie both till sheets in the main valleys.

Long recognized as a distinctive deposit on account of its abundant shell content, the shelly till, has received considerable documentation. The first known report of marine shells in the Caithness drift was by Busby (1802; cited by Crampton *et al.*, 1914, p. 118). Later, in the course of his energetic travels around Caithness, Robert Dick identified the widespread occurrence of shelly boulder clay, which he explained as a marine deposit associated with sea ice and glaciers (Smiles, 1878). He collected and identified many shells, some well preserved, from the deposits and also recorded the presence of Chalk flints and oolitic limestone clasts. Shells collected by Cleghorn were listed by Smith (1850a) and their arctic character noted (Smith, 1850b). Cleghorn (1850, 1851) thought the till to be a marine deposit and attributed the fragmentary nature of the shells to the feeding habits of catfish! Miller (1847 – see Miller, 1858) first publicized Dick's and Cleghorn's observations, and later exhibited some of Dick's specimens (Miller, 1851).

C. W. Peach added greatly to the knowledge of the fossil content of the till (Peach, 1858, 1859, 1860, 1863a, 1863b, 1863c, 1865b, 1867), listing the remains of 142 species, including molluscs, Foraminifera, bryozoa, sponges and algae. Further details of Foraminifera were provided by Crosskey and Robertson (1868c) and 'Entomostraca' by Brady *et al.* (1874). The most abundant molluscs are *Arctica islandica* (L.), *Spisula solida* (L.), *Mya truncata* L., *Turritella communis* Risso, *Tridonta elliptica* (Brown) and *Tridonta borealis* Schumacher (Crampton and Carruthers, 1914).

Interpretations of the glacial sequence in Caithness have centred notably on the patterns of ice movement represented by the different deposits and, more recently, on questions concerning the age(s) of the deposits and the extent of the Late Devensian glaciation.

The first regional synthesis of the glaciation of Caithness was put forward in an important study by Jamieson (1866). He believed the shelly drift to be a marine deposit disturbed and mixed into its present state by drift-ice moving north-west to south-east across the county, which he inferred from striations, the apparent distribution of stoss and lee slopes and the overlap patterns of drift on bedrock in different areas. Chronologically he related the shelly drift to the glacial-marine period or submergence, which he believed followed the main glaciation (see Clava and Nigg Bay) on the basis of the sub-arctic character of its fauna, which he did not appreciate was a derived assemblage. Jamieson (1866) also described up to 20 ft (6.1 m) of reddish-brown clay with boulders (sandstone, quartz, mica-schist and granite) with few or no shells resting on the shelly till at Wick and Keiss.

C. W. Peach reached a different conclusion about the origin of the Caithness till. Revising his earlier ideas of glaciomarine processes (Peach, 1863a, 1863c), he informed Croll in 1868 (Croll, 1870a) that the drift had been formed by land ice moving from the Moray Firth to the Atlantic. Croll (1870a, 1875) supported and elaborated this view, arguing that Scottish ice in the Moray Firth was deflected across Caithness by Scandinavian ice in the North Sea and carried with it shales, Chalk flints and Chalk fossils from the sea bed in its boulder clay. This interpretation was also favoured by J. Geikie (1877), who noted that the striation and till patterns cited by Jamieson for a south-east ice movement were equally explicable in terms of a north-west movement, and the stoss and lee forms were somewhat indistinct. Moreover, an ice movement towards the south-east did not take account of the Chalk and other foreign material in the till. Geikie concluded that the Caithness till was simply an ordinary glacial boulder clay, but with shells dispersed through it.

Using as evidence the patterns of striations, roches moutonnées and the distribution of indicator stones, Peach and Horne (1881c) confirmed the existence of two separate ice streams: local ice moving east-north-east and north-north-east, deflected by external ice moving north-west. They described the characteristics of the till units associated with the separate ice masses and presented a revised species list for the shelly till, noting the general deep-water character of the faunal assemblage. They concluded that both tills were deposited by land ice. Peach and Horne (1881c) also presented the first detailed account of 'moraines' and gravel in Caithness, although examples near Dirlot had been recorded earlier by Dick (Smiles, 1878). The features largely comprise ridges and mounds of gravel resting on both the local and shelly tills. They are largely confined to the central and north-west part of the county and are generally absent from the east, as noted by Jamieson (1866). Peach and Horne suggested that they were formed by local ice moving out from the hilly ground to the west and reaching a limit near Dirlot, long after the retreat of the Scandinavian ice.

Further confirmation of onshore ice movement was provided by Tait (1908, 1909) and by the location of the famous Leavad and other erratics (Tait, 1909, 1912; Carruthers, 1911; Crampton and Carruthers, 1914; Sutherland, 1920).

Crampton and Carruthers (1914) later showed the moraines and gravels to be more extensive than was previously recognized. From the existence of cross striations away from the zone of confluence of the two main ice streams and the presence of a local till beneath the shelly till at Leavad, they thought it highly probable that local ice advanced across part of lowland Caithness before the external ice reached its full extent. After the latter had waned, the moraines and gravels were deposited by an advance of local ice across part of the area covered by the shelly till with lobes extending to near Thurso and reaching the sea at Wick and Sinclair's Bay.

Bremner (1934a) explained the shelly till of Caithness as the ground moraine of his 'Second Ice Sheet' moving north-west from the Moray Firth to the Atlantic. He suggested that some of the erratic material recognized by Crampton and Carruthers (1914) was carried south-east from the northern Highlands into the Moray Firth by the 'First Ice Sheet', then moved into Caithness by the second. As evidence for a pre-shelly till glaciation he quoted striae patterns and the presence of local till beneath the Leavad erratic. A third ice sheet was indicated by the surface moraines and gravels, and although less extensive than the previous one, it covered 'quite half of the Caithness plain north-east of a line from Lybster to Reay'. Here Bremner was following Crampton and Carruthers (1914). Charlesworth (1956, 1957) largely echoed Bremner's views although he placed the limit of the third ice sheet (his Highland Glaciation) further north than Peach and Horne, in Orkney.

Phemister (1960) reconstructed two separate ice-sheet movements in Caithness from striae orientations. The earlier one, from south-east to north-west, deposited the shelly boulder clay. The later one was to the north as the Scandinavian ice barrier withdrew eastwards and although it must also have transported a shelly till, two separate shelly tills had not been identified. Phemister correlated the northerly ice movement with the so-called Strathmore Glaciation of north-east Scotland and mapped it as extending beyond the Caithness coast. He also noted an early, local ground moraine beneath the shelly till.

Godard (1965) identified three superimposed till sheets in Caithness, reflecting the changing predominance of mainland and Scandinavian ice, and interpreted them in terms of Bremner's triple ice-sheet model (Bremner, 1934a). A lower local ground moraine was overlain by a shelly till deposited by ice moving north-west from the Moray Firth; above was an upper local ground moraine associated with ice from a more southerly direction which extended into Orkney.

Sissons (1965, 1967a) considered that during the last glaciation Scottish ice bifurcated in the Moray Firth due to the presence of Scandinavian ice to the east. One stream moved north-west across Caithness and Orkney, depositing the shelly till. Subsequently, during the so-called Aberdeen–Lammermuir Readvance, inland ice moved north-east across Caithness to a limit of till, morainic mounds and kames resting on the shelly till near Dirlot, as identified by Peach and Horne (1881c).

Smith (1968, 1977) proposed a rather different chronology for glacial events in Caithness. He thought that the moraines at Dirlot and near Lothbeg represent the limit of the Weichselian (Late Devensian) glaciation in Caithness and that the shelly till was of Saale ('Wolstonian') age. He noted great depths of shattered bedrock, extensive colluvium at the base of slopes and lack of 'fresh' forms over large parts of the county. However, the one radiocarbon date from the shelly till at Gills Bay (>40,800 BP (Birm–179); Shotton and Williams, 1971) does not necessarily support his argument – as he states (Smith, 1977), it is inconclusive.

From his investigation of the field evidence Omand (1973) suggested there had been three main phases in the glaciation of Caithness. The first is represented in a number of sections, as at Baile an t-Sratha, where a local till (Dunbeath Till) underlies the shelly till, indicating early local glaciation. During the second phase, ice from the Moray Firth was diverted across Caithness by Scandinavian ice in the North Sea Basin and deposited the shelly till (Lybster Till). This ice stream merged with and deflected local inland ice, which deposited an upper local till (Reay Till) identified by Omand west of the shelly till limit. However, the distinction between the two local tills is unclear and, so far as is known, they are not seen together in any section. Probably they are both part of the same till unit, its distribution reflecting variations in the dominance of different ice masses. During the final phase (Bower stage), local ice deposited moraines and gravels as it wasted back more or less continuously in the eastern part of the county. Omand concluded there was no evidence for a separate Lateglacial readvance or a separate till associated with the Bower stage. From the periglacial modification of the tills by solifluction and cryoturbation and from the radiocarbon date cited above, Omand proposed either that the glacial maximum in Caithness was early in the Devensian, or the deposits in Caithness were older, possibly 'Wolstonian' in age as suggested by Smith (1968). He implied that the latter was more probable since evidence elsewhere in the country supported a Late Devensian ice maximum.

Studies in the central North Sea Basin apparently indicate that Scottish and Scandinavian ice did not coalesce during the Late Devensian (Jansen, 1976; Sutherland, 1984a; Cameron *et al.*, 1987; Sejrup *et al.*, 1987; Nesje and Sejrup, 1988) (see also Lambeck 1991a, 1991b). Thus, as Flinn (1981) and Sutherland (1984a) noted, if the flow of the ice that deposited the shelly till was to the north-west as a consequence of deflection by Scandinavian ice, then the till must pre-date the Late Devensian. However, the assumption that Scandinavian ice is required to produce such a flow pattern may not be warranted; for example, a strong outflow of ice from the Moray Firth might also produce such a pattern (see for example, Hall and Bent, 1990). On the basis of a small number of striations, Flinn (1981) proposed that subsequent to deposition of the shelly till, local ice flowed outwards across the Caithness coast, northwards to Orkney and south-eastwards into the Moray Firth. In contrast, Sutherland (1984a) suggested that much of northern and eastern Caithness may have been ice-free during the Late Devensian. Some support for such a proposal comes from preliminary results of amino acid analyses of shells from the Caithness till, which suggest that the shelly till may relate to glaciation during the Early Devensian (Bowen and Sykes, 1988; Bowen, 1989, 1991). The results also indicate a Late Devensian age for the shelly till at Latheronwheel. This raises questions of whether there might be more than one shelly till or whether sampling elsewhere might have missed younger shells.

Hall and Whittington (1989), working on the stratigraphy of south-east Caithness, have continued the debate. From the field evidence they conclude that the inland till and the shelly till relate to a single glacial episode and, following retreat of the ice in the Moray Firth, there was a late minor advance of the inland ice. Furthermore, from an assessment of several lines of relative age evidence, including the degree of weathering of the tills, periglacial deposits and comparisons with Buchan (see Chapter 8: North-east Scotland), Hall and Whittington infer that this episode was during the Late Devensian. They conclude that the maximum limit of the Late Devensian ice must therefore have been as far north as the Orkney–Shetland Channel, assuming that the shelly till of Caithness and Orkney is a single lithological unit. Subsequently, Hall and Bent (1990) have elaborated on the model citing additional evidence from a wider area of the Moray Firth and Buchan. They propose that the last Scottish ice-sheet extended northwards across Caithness and Orkney to a limit in the Orkney–Shetland channel, and eastwards across the adjacent shelf to a limit at the Bosies' Bank moraine (Bent, 1986).

Hall and Whittington (1989) also report the presence in south-east Caithness of peat deposits radiocarbon-dated to the Lateglacial Interstadial, which are overlain by Loch Lomond Stadial gelifluction deposits. This evidence is important in demonstrating the significance of periglacial mass wasting in Caithness during the stadial.

It is clear from the above review that several important questions remain to be answered about the glacial sequence in Caithness. In particular these concern:

1. the position of the Late Devensian ice limit;
2. the direction, age and extent of the last ice movement;
3. the age of the shelly till and its weathering and periglacial modifications;
4. the origin of the shelly till: was it deposited by land-based ice or as a glaciomarine deposit?
5. the age of the local till below the shelly till and its stratigraphic relationships to local till elsewhere;
6. the significance of the moraines and gravels above both the local and shelly tills;
7. the origin and significance of the reddish-brown, stoney clay on top of the shelly till at Wick and Keiss described by Jamieson (1866).

The answers to these questions have an important bearing on a wider scale concerning the extent of the last ice-sheet in northern Scotland and adjacent areas of the North Sea Basin, Orkney and Shetland. Three sites have been selected to represent the main units and features of the glacial deposits of Caithness: Baile an t-Sratha, Drumhollistan and Leavad.

BAILE AN T-SRATHA

J. E. Gordon

Highlights

The stream sections at Baile an t-Sratha demonstrate the stratigraphical relationship between the main glacial deposits recognized in south-east Caithness. Local till derived from inland is overlain by shelly till deposited by ice moving across the sea floor and then onshore towards the north-west. These deposits provide important evidence for interpreting the successive patterns of ice movement.

Introduction

Baile an t-Sratha (ND 142307) is a stream section in the valley of the Dunbeath Water; it is a key locality demonstrating the two main till units of Caithness, the local inland till characteristic of the western part of the county, and the shelly till characteristic of the eastern part (Peach and Horne, 1881c; Omand, 1973; Hall and Whittington, 1989). It is particularly important since it occurs in the zone of overlap of the two tills and illustrates their stratigraphic relations and contact. The deposits in the valley of the Dunbeath Water are described by Hall and Whittington (1989). The valley of the Dunbeath Water has long been a notable locality for glacial deposits. Dick (see Smiles, 1878), Jamieson (1866) and Peach and Horne (1881c) all referred to fine examples of shelly till, although they did not describe exposures. More recently, Omand (1973) noted that Dunbeath was one of only four places in Caithness where he found shelly till overlying a local till.

Description

The site indicated by Omand's (1973) grid reference is slumped and vegetated at present, and the best section in 1980 was at Baile an t-Sratha. There, the Allt an Learanaich, a tributary of the Dunbeath Water, has excavated sections up to 40 m high in a till infill on the south side of the valley. The exposures show two superimposed till units (Hall and Whittington, 1989: Balantrath and Balcraggie sites):

3. Grey, shelly till, weathered near the surface to a maximum depth of 2.7 m; clast lithologies dominated by Devonian sandstones, but also including distinctive pebbles of Mesozoic sedimentary rocks.
2. Brown to dark reddish-brown till of local provenance; Devonian sandstones dominant.
1. Devonian mudstone bedrock with striations orientated W–E.

Interpretation

Hall and Whittington (1989) interpreted the lower till (the Balantrath Till member in their lithostratigraphy), as a lodgement till emplaced by inland ice moving down the valley. Locally the contact with the overlying shelly till (their Forse Till member) is planar, sharp and erosional. Elsewhere at the contact, glaciotectonic folding was noted, and the presence of local structures and interlaminated brown and grey muds and diamicts is inferred by Hall and Whittington to indicate that the two respective ice masses were contemporaneous, and that the shelly till was deposited as the inland ice receded. According to them several features indicate that the shelly till is the product of land-based ice and not a glaciomarine deposit; first it rests on striated bedrock; second, it contains locally eroded bedrock and reworked local till; and third, it reaches an elevation of 110 m OD, which appears to be too high for glaciomarine deposition, even allowing for glacio-isostatic depression (cf. Sutherland, 1981a).

The lower till at Baile an t-Sratha is part of a suite of local till units, occurring west of a line from Reay to Berriedale and described by Jamieson (1866), Peach and Horne (1881c) and Crampton and Carruthers (1914). Currently, good sections occur in the valleys of the Dunbeath Water, Langwell Water and Berriedale Water, and also at the coast near Reay and Berriedale (for details see Omand, 1973; Hall and Whittington, 1989). The upper till at Baile an t-

Sratha is part of the classic shelly till unit of Caithness occurring east of a line from Reay to Berriedale, with many exposures along the east coast and in stream valleys (Omand, 1973; Hall and Whittington, 1989). The age of the tills is currently a matter of debate (see above).

The significance of Baile an t-Sratha is that it provides one of the best current exposures clearly showing local and shelly tills, their respective characteristics and their stratigraphic relationships in the zone of interaction of the two ice streams that converged in Caithness. As such it is an important reference locality demonstrating a key part of the glacial succession in Caithness. Similar sections showing the superimposition of the two till units occur at Latheronwheel, Watten and Drumhollistan (Omand, 1973), but the stratigraphic relationships are most clearly seen at Baile an t-Sratha. In a wider context the multiple till sequence at Baile an t-Sratha is similar in several respects to those, for example, at Den Wick, Boyne Quarry, Nigg Bay and Nith Bridge where the interaction of ice masses from separate sources also produced distinctive, superimposed tills. Further study of such sites will contribute important field evidence to test and constrain reconstructions and mathematical models of regional ice-sheet dynamics and the factors controlling changes in ice-sheet flow patterns.

Conclusion

Baile an t-Sratha demonstrates the succession of glacial (ice-deposited) sediments in south-east Caithness: a local till derived from inland is overlain by a distinctive shelly till deposited by ice moving onshore and towards the north-west. The importance of the site lies in its value as a reference locality for establishing the relationships between the two tills and the respective patterns of ice movement that they indicate. There is currently debate about the age of the deposits and whether they were formed during the Early or Late Devensian (approximately 65,000 and 18,000 years ago, respectively).

DRUMHOLLISTAN

J. E. Gordon

Highlights

Sections in the stream gully at Drumhollistan show the till succession on the north coast of Caithness and allow interpretation of the associated patterns of ice movement.

Introduction

Drumhollistan (NC 920654) is located on the north coast of Caithness, west of Reay, where stream erosion has dissected a drift plug in a cliff-top depression. The sections exposed are important in demonstrating the succession of Pleistocene deposits on the north coast of Caithness. They show a multiple till sequence locally overlain by head deposits, and complement the interests at Baile an t-Sratha and Leavad. The only account of the site is that by Omand (1973).

Description

The deposits are exposed along a narrow stream valley that deepens to over 30 m at its western end. The flanks of the valley are heavily dissected by gullies separated by narrow, intervening ridges. The deposits were recorded by Omand (1973) and comprise the following composite sequence:

4.	Head	up to 1 m
3.	Brown till with shell fragments near the base	up to 21 m
2.	Sand and gravel; boulders up to 0.75 m length in the lower 1.5 m, fining upwards	up to 3.5 m
1.	Grey till	> 3 m

The lower till (bed 1) is exposed only at the western end of the gully. It is lighter in colour than the upper till and the matrix appears gritty and sandy. No shells have been found in it. A stone count of 50 clasts by Omand (1973) revealed that 94% of the sample consisted of gneiss or migmatite, probably derived from the area to the south and west of the site. At the western end of the gully on the south side, the lower till is overlain by unbedded sand and gravel with a layer of sand at the top. The upper till (bed 3) in its lower part is a darker, chocolate-brown colour with a sandy/silty matrix. Of 50 clasts sampled by Omand, 76% were from the Old Red Sandstone group and probably derived from the area to the south-east of the site. The upper part of the till is weathered to a depth of 5–6 m and has a reddish appearance. A few shell fragments were found in this till, but they are scarce and considerably comminuted in comparison with those at Dunbeath and other sites on the east coast of Caithness. On the north side of the gully the brown till (bed 3) rests directly on the grey till (bed 1). The lower layers of the former are bouldery with lenses of sand and gravel in the lower 3 m.

In the middle part of the gully on the south side, a layer of sand occurs at the base of the upper till. Possibly it corresponds with the sand and gravel at the western end of the section, but the continuity of the beds could not be traced because of slumping. Bands of the upper till are interbedded with the sands which here have a rusty, weathered appearance, and the two units interdigitate.

A stone layer occurs at the surface of the upper till beneath a cover of peat. On the northern side of the gully at the western end, up to 1 m of head locally covers the upper till.

An additional feature of the site is the fine example it provides of active gullying and incipient earth pillar formation along the corrugated and dissected flanks of the main gully.

Interpretation

The general characteristics and distribution of the glacial deposits of Caithness, the theories proposed to explain them and the outstanding problems have already been discussed. The evidence at Drumhollistan appears to confirm two distinct ice movements on the north coast followed by a period or periods of weathering and head formation. The lower till, associated with a northerly ice movement occupies a similar stratigraphic position to the lower till at Dunbeath, but it is still open to question whether it represents an early local glaciation or was broadly contemporaneous with the shelly till as at Dunbeath (Hall and Whittington, 1989). The upper till is part of the widely recognized shelly till of Caithness deposited by ice moving northwest. The weathering of the upper part of the sequence at Drumhollistan is typical of many parts of Caithness (Jamieson, 1866; Peach and Horne, 1881c; Omand, 1973; Hall and Whittington, 1989), as is the overlying head (Omand, 1973; Futty and Dry, 1977; Hall and Whittington, 1989).

In south-east Caithness, Hall and Whittington (1989) noted weathering and mottling of the tills, with decalcification to a depth of 2.7 m. They concluded that this was compatible with Holocene weathering. Although the weathering at Drumhollistan extends to a greater depth of 5–6 m, this could reflect local factors and is still compatible with depths of Holocene weathering in eastern England (Madgett and Catt, 1978; Eyles and Sladen, 1981). At present there is no evidence that the weathering in Caithness might reflect a long period of ice-free conditions during the Devensian. The head deposits at Drumhollistan are undated, but in south-east Caithness comparable deposits have been ascribed to the Loch Lomond Stadial (Hall and Whittington, 1989). Again, there is no evidence that they might relate to periglacial episodes earlier in the Devensian, although this cannot be completely ruled out without further work.

Drumhollistan has not been investigated in detail, and (as is also the case for Baile an t-Stratha) several intriguing questions remain unanswered about the deposits, their wider relationships with other sites and the implications they carry for the Pleistocene history of Caithness. These concern:

1. the age(s) of the lower and upper tills and whether or not they are broadly contemporaneous;
2. the significance of the sand and gravel layer between the till beds and whether it is part of a single, complex melt-out sequence (cf. Nigg Bay, Boyne Quarry and Hewan Bank) or represents a significant break in glacial sedimentation;
3. the age and degree of the weathering in the upper till;
4. the age of the overlying head;
5. the relationships of the tills at Drumhollistan to the local till exposed at the surface along the Sandside Burn at Reay, 4 km to the east (Omand, 1973).

Drumhollistan is an important reference site for the succession of Pleistocene deposits on the north coast of Caithness. It demonstrates the western extent of the shelly till and the northern extent of local till, together with evidence for subsequent head formation and weathering. Comparable sites to Drumhollistan on the south-east coast of Caithness that show shelly till superimposed on local till, occur at Dunbeath (Baile an t-Stratha), Leavad and Latheronwheel (Omand, 1973; Hall and Whittington, 1989).

Baile an t-Stratha is notable for better exposure

of local till and better preservation of shells in the upper till, whereas at Leavad the main interest is a large erratic of soft strata incorporated in the shelly till. Exposures at Latheronwheel are poor and have not been recorded in detail. Together, therefore, Drumhollistan, Baile an t-Sratha and Leavad demonstrate key aspects of the Pleistocene deposits of Caithness.

The contemporary erosion and gullying of the till are also of interest and are representative of a type of phenomenon that is known from a number of localities in Scotland, for example at Rosemarkie Dens (Miller, 1858; Geikie, 1901), Fochabers (Hinxman and Wilson, 1902; Geikie, 1903) and in the Nairn valley near Clava. The narrow ridges illustrate an incipient form of earth pillar formation, although the absence of large boulders has inhibited the development of classic earth pillars (cf. Whalley, 1976b).

Conclusion

The deposits at Drumhollistan demonstrate the glacial history of the north coast of Caithness. They show two (ice-deposited) tills, locally separated by a layer of sand and gravel. The value of the site is as a reference locality for establishing the glacial succession and pattern of ice movements in this area. There are different interpretations of the age of these deposits and whether they were formed during the Early or Late Devensian (approximately 65,000 and 18,000 years ago, respectively).

LEAVAD

J. E. Gordon

Highlights

The principal interest at Leavad comprises a large raft of Lower Cretaceous sandstone believed to have been transported from the sea floor off Caithness by ice moving onshore. The erratic mass is an outstanding example of its kind and lies within a sequence of glacial deposits which includes three tills. The latter provide important evidence for interpreting the glacial sequence of Caithness and the patterns of ice movement.

Introduction

The principal interest at Leavad (ND 174462) in eastern Caithness is a large (nearly 800 m long) erratic of Lower Cretaceous sandstone believed to have been transported by ice over a distance of at least 15 km from the sea floor off the Caithness coast. The erratic occurs in a sequence of deposits, including three tills, so that Leavad is potentially an important reference site for the Pleistocene stratigraphy of Caithness. The deposits at Leavad were formerly exposed in a quarry and have also been investigated by boreholes (Tait, 1908, 1909, 1912; Carruthers, 1911; Crampton and Carruthers, 1914).

Description

The first accounts of the Leavad Quarry (Tait, 1908; Carruthers, 1911) described a decomposed calcareous sandstone with more resistant concretions, quite unlike any other sandstone in the neighbourhood. Tait (1908) initially considered it to be a small outlier of Jurassic rocks resting unconformably on Old Red Sandstone, but was unable to confirm this because no contact was seen. He recognized similarities between the large concretions in the sandstone and egg-shaped stones dredged from Wick harbour, transported there by ice from a source near Brora. Subsequently, Tait (1909) reported fissures of boulder clay with Highland metamorphic rocks between the concretions and, within the latter, fossils identified as a Lower Cretaceous (Neocomian) assemblage by Kitchin (Lee, 1909). Above the calcareous sandstone was a boulder clay, coarser in texture and with more stones of local origin (Old Red Sandstone). Tait raised the possibility that the Cretaceous sandstone might be a glacially transported mass, similar to that at Moreseat in Aberdeenshire (see Moss of Cruden).

To settle the question of the origin of the Leavad sandstone a series of boreholes was put down in 1910. The findings were described by Carruthers (1911), Tait (1912) and summarized by Crampton and Carruthers (1914). The deposits containing the sandstone mass occupied a buried channel of the Little River to a depth of at least 24.1 m below the floor of the quarry. They comprised the following sequence (Tait, 1909, 1912):

15. Sandy till with stones of local

	origin (exposed in the quarry)	0.6 m
14.	Sand and sandstone with Cretaceous fossils (5.2 m exposed in the quarry)	7.9 m
13.	Dark-green, bedded clay with shells and Foraminifera	1.5 m
12.	Dark clayey sand	0.6 m
11.	Dark-green clay with shells	0.4 m
10.	Clay and sand	0.8 m
9.	Greenish clay with shell fragments	0.2 m
8.	Dark shelly till with striated stones	0.6 m
7.	Yellowish-green, sandy till with stones of local origin	5.0 m
6.	Gravel	0.8 m
5.	Yellowish sandy clay and gravel	0.6 m
4.	Gravel	1.5 m
3.	Yellowish sandy clay and gravel	8.0 m
2.	Brown sandy clay	0.6 m
1.	Clean sand and gravel	>0.7 m

The boreholes confirmed that the sandstone (bed 14) is an erratic block, and its size was estimated at 878 m long, 549 m wide and 7.9 m thick.

Tait noted that beds 3, 5 and 7, which he interpreted as till, with interbedded gravel layers (beds 4 and 6), were quite distinctive from the dark, shelly till of bed 8, being more sandy in composition, lighter in colour and containing stones largely of local Old Red Sandstone origin. Lee (in Carruthers, 1911) identified the fossil assemblage in the dark-green clay (beds 9, 11, 13) as post-Cretaceous, but noted differences with that found elsewhere in the shelly till of Caithness. He suggested that the clay might belong to some part of the Crag, but the dominance of thermophilous forms, typical of the Miocene, indicates an earlier age (Hall, unpublished data). If the dark-green clay does indeed represent an erratic mass of Miocene clay then its occurrence at Leavad is of great interest, for no sediments of this age are known from the Moray Firth west of 1°W (Andrews *et al.*, 1990).

Interpretation

Carruthers (1911), Tait (1912) and Crampton and Carruthers (1914) all considered that the Leavad erratic was transported to its present position from the floor of the Moray Firth by the same ice moving onshore which deposited the shelly till of Caithness (Peach and Horne, 1881c; see Baile an t-Sratha). The nearest Lower Cretaceous sediments to Leavad occur just offshore (Andrews *et al.*, 1990) implying a minimum distance of transport of about 15 km. A similar sandstone has also been encountered about 1 km north of Leavad (Crampton and Carruthers, 1914). This may also be presumed to be an erratic.

On account of its size the Leavad erratic is frequently quoted as a spectacular example of a glacially transported mass (Charlesworth, 1957; Sissons, 1967a; Embleton and King, 1975a). Similar features in Scotland include the Moreseat erratic (Jamieson *et al.*, 1898), the Comiston boulder (Campbell and Anderson, 1909), the Kidlaw erratic (Kendall and Bailey, 1908) and the Plaidy erratic (Jamieson, 1859, 1906; Read, 1923; Pringle, 1936). Others elsewhere in Britain are described by Charlesworth (1957), Sparks and West (1972) and Embleton and King (1975a). Possible mechanisms of entrainment of such masses have been discussed by Weertman (1961) and Boulton (1972a) (see also Clava).

If Tait's descriptions of the section and boreholes are valid, the Leavad erratic is also of considerable stratigraphic interest as the only known site where the three main till units represented in Caithness are superimposed: a till of local origin appears to overlie a shelly till derived from offshore, which in turn overlies a lower local till. Leavad may therefore be an important reference site demonstrating the succession of Pleistocene deposits in Caithness, the interpretation and chronology of which are still debatable. At present no exposures exist in the original sandpit, although along the Burn of Tacher (ND 176465) a dark till overlies a lighter brown one. Possibly these represent the shelly till and lower local till, respectively. The upper local till is not exposed.

Conclusion

Leavad is notable for a large mass of sandstone incorporated within a sequence of till deposits. It is a well-known example of a large erratic and was transported by an ice-sheet from its original location in the bedrock offshore and deposited within the glacial sediments at Leavad. The occurrence of three tills in the sequence provides potentially important evidence for the pattern of successive ice movements across the area and the interaction of different ice masses. As yet, the ages of these deposits are not firmly established.

LOCH OF WINLESS

H. J. B. Birks

Highlights

Pollen grains preserved in the layers of sediment which infill the bed of this loch provide a detailed record of the vegetational history of Caithness during the Lateglacial and Holocene. This information contributes significantly towards the understanding of the environmental history of this ecologically unique area.

Introduction

Caithness is an area of unique ecological interest. In the north-east, the area of shelly till, the so-called Caithness Plain, is today mainly crofting and farm land. To the west and south, the area of non-calcareous, sandy till is dominated by vast areas of patterned blanket-bog or 'flows'. This is the largest area of continuous peatland in Britain and is of international importance to ecologists and palaeoecologists (Stroud *et al.*, 1987; Lindsay *et al.*, 1988). Caithness is today one of the most extensive treeless areas in Britain (excluding recent plantations of exotic conifers).

Despite its importance ecologically, very little is known of the Late Quaternary vegetational and ecological history of Caithness. Loch of Winless (ND 293545) is the only site from which a detailed Lateglacial and Holocene pollen diagram, with radiocarbon dates, has been published (Peglar, 1979). Loch of Winless lies approximately 8 km west-north-west of Wick at an elevation of 9 m OD. It is a small lochan within a basin now almost completely infilled with fen peat and lake sediments up to 6 m depth. The palynological results from this site are thus of particular significance in the context of Holocene vegetational history of Britain and of Scotland (see Birks, 1977, 1989).

Description

The sediment infill at Loch of Winless is *Sphagnum* peat overlying coarse detritus (organic) muds, marl, and silts (Figure 5.2). Ten radiocarbon dates are available (Figure 5.2) and suggest that the basal sediments may be as old as 12,800 BP. However, the earlier dates are likely to be subject to hard-water errors because of the calcareous nature of the surrounding boulder clay and underlying Old Red Sandstone. Peglar (1979) proposed that the basal sediments may only extend to about 12,500 BP. The pollen record has been divided into five local pollen assemblage zones (Figure 5.2). Zones LW1 and LW2 are assigned to the Lateglacial; zones LW3–LW5, to the Holocene.

Interpretation

The pollen record shows that the Lateglacial vegetation was grassland and tall-herb communities, with abundant Cyperaceae and a variety of open-ground herbs including some arctic–alpine species which today are restricted to high elevations in north-west Scotland. In contrast to Lateglacial sequences from the north-west Highlands and the Inner Hebrides (see Pennington, 1977a; Lowe and Walker, 1986a) dwarf-shrubs such as *Empetrum* appear to have been unimportant throughout at Loch of Winless and on Orkney (Moar, 1969a), perhaps as a result of the higher fertility of soils on the Old Red Sandstone.

The early and middle Holocene vegetation (10,000–6000 BP) was probably a mosaic of species-rich, tall-herb and fern communities, birch, hazel, and willow scrub, and grassland. The pollen assemblages clearly indicate that oak and pine were never major components of the Holocene vegetation. Indeed it is likely that they were absent from the Loch of Winless area throughout the Holocene (Birks, 1989). The pollen data also suggest that the area was never extensively forested at any time during the Holocene, with tree pollen reaching a maximum of 50% and averaging around 25%. The most likely reconstruction is of birch, willow, and hazel scrub occurring locally in sheltered areas, and of widespread grassland and fern and tall-herb dominated stands on moist, well-drained, and fertile soils. With nutrient depletion by leaching, soils deteriorated and *Calluna* heath and moorland began to expand at about 6000 BP. *Alnus* may have been present in low amounts after about 5500 BP. Despite the abundance of

Figure 5.2 Loch of Winless: relative pollen diagram showing selected taxa as percentages of total pollen (from Peglar, 1979). Note that the data are plotted against a radiocarbon time-scale.

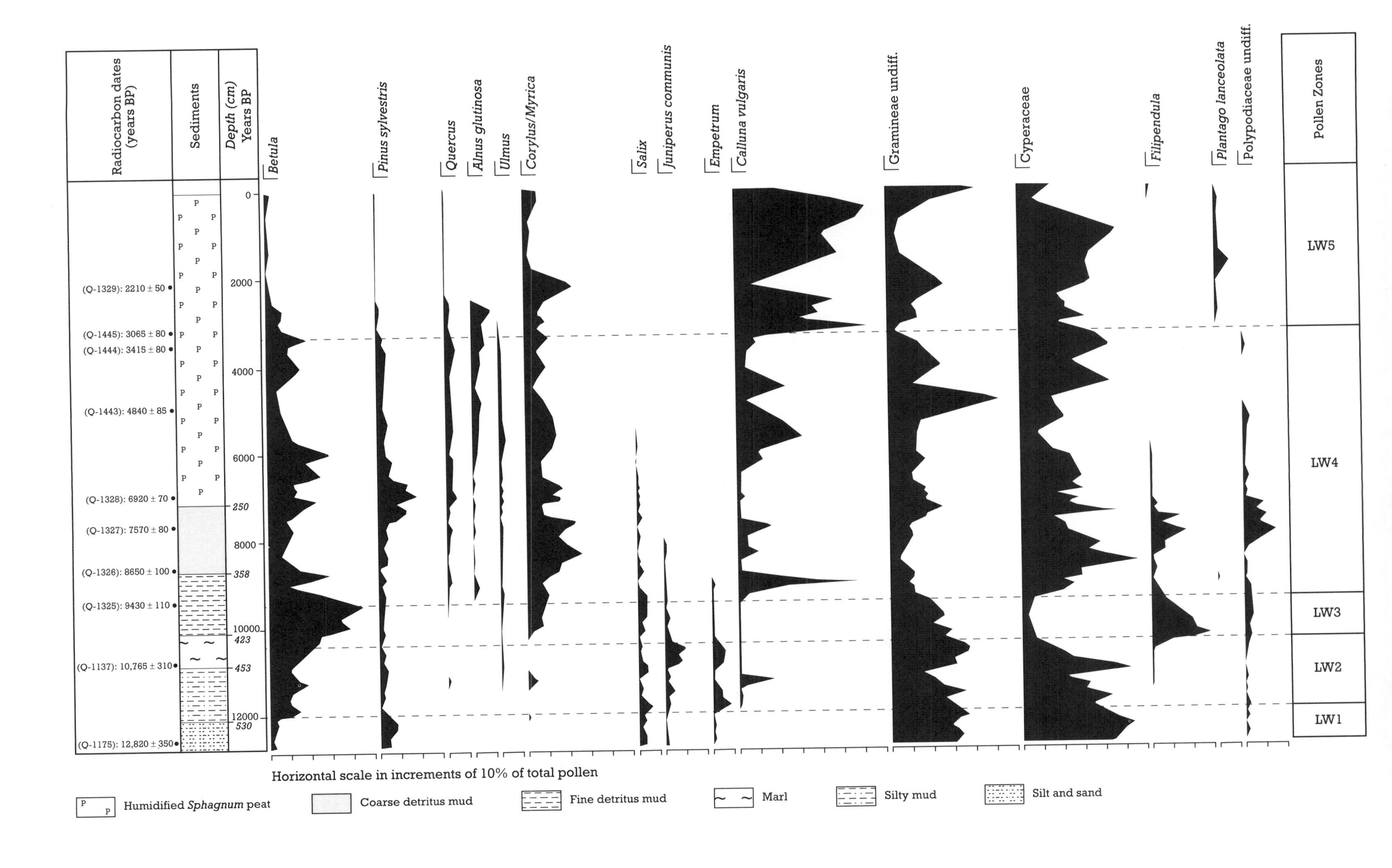
Radiocarbon dates (years BP)
Sediments
Depth (cm)
Years BP
(Q-1329): 2210 ± 50
(Q-1445): 3065 ± 80
(Q-1444): 3415 ± 80
(Q-1443): 4840 ± 85
(Q-1328): 6920 ± 70
(Q-1327): 7570 ± 80
(Q-1326): 8650 ± 100
(Q-1325): 9430 ± 110
(Q-1137): 10,765 ± 310
(Q-1175): 12,820 ± 350
0
2000
4000
6000
250
8000
358
10000
423
453
12000
530
Betula
Pinus sylvestris
Quercus
Alnus glutinosa
Ulmus
Corylus/Myrica
Salix
Juniperus communis
Empetrum
Calluna vulgaris
Gramineae undiff.
Cyperaceae
Filipendula
Plantago lanceolata
Polypodiaceae undiff.
Pollen Zones
LW5
LW4
LW3
LW2
LW1
Horizontal scale in increments of 10% of total pollen
Humidified Sphagnum peat
Coarse detritus mud
Fine detritus mud
Marl
Silty mud
Silt and sand

archaeological sites in north-east Caithness, there is very little unambiguous palynological evidence for human impact on the vegetation until about 2500 BP (see Huntley, in press).

Over a wider area this general pattern of Holocene vegetation is reflected in the results of Durno (1958) and Robinson (1987). Fossil pine stumps in the blanket peat of south-west Caithness (Lewis, 1906; Gear and Huntley, 1991) indicate that pine grew locally during the mid-Holocene, but perhaps for only a short time (Gear and Huntley, 1991).

The site and its pollen record are important for several reasons. First, they demonstrate that north-east Caithness was never extensively forested and that the area was the least wooded lowland area of mainland Britain during the Holocene (Birks, 1988). Pine and oak were absent, and birch, willow and hazel scrub probably occurred locally in sheltered areas, possibly with some elm and alder. By extrapolation, the data from Loch of Winless suggest that the flow country to the west and south was similarly never extensively wooded (Birks, 1988).

Second, Loch of Winless provides the youngest dates for the first arrival of *Ulmus* and *Alnus* during the Holocene in mainland Britain (Birks, 1989). These dates demonstrate that elm took 3000 years to complete its spread through Britain from south-east England (*c.* 9500 BP) to Caithness (6500 BP). Alder pollen is present in low amounts from 8500 BP, but it does not attain its maximum values until 5500 years later (Bennett and Birks, 1990), presumably because of the extreme climatic severity of this exposed northern area.

Third, the site provides a reference Holocene pollen profile for the ecologically unique area of Caithness. There are, at present, no other continuous, well-dated pollen profiles published from the area (Birks, 1989).

Fourth, the pollen data suggest, rather surprisingly, that human impact on vegetation has never been intense despite the abundance of archaeological sites nearby. Alternatively, the pollen data in such an extreme, exposed, and naturally open or lightly wooded area, are poor reflectors of human impact. There is support for such an explanation from charcoal evidence from further west (Charman, 1992) and south (Robinson, 1987), which suggests that human impact on the landscape may have been significant from as early as 7500 BP.

Fifth, it raises several intriguing and important palaeoecological hypotheses about the extent and control of tree growth in the Holocene (Gear and Huntley, 1991) and about the apparent steepness of vegetational gradients in northernmost Britain (Birks, 1988). The testing of these hypotheses can only be attempted by new investigations in the nearby flow-bog country of central and western Caithness.

Loch of Winless and its pollen record clearly show that the vegetational history of north-east Caithness has been different from anywhere else at low elevations on mainland Britain. It has a unique ecological history and suggests that forest growth this far north was always extremely limited during the last 10,000 years. This palaeoecological lesson from the past is an ecological indicator of the area's unsuitability for productive forestry and of the ecological uniqueness and conservation importance of this part of northern Scotland (Stroud *et al.*, 1987; Lindsay *et al.*, 1988; Huntley, 1991).

Conclusion

The deposits at Loch of Winless provide a detailed record of the vegetational history of Caithness during the Lateglacial and Holocene, during approximately the last 13,000 years, based on pollen analysis and radiocarbon dating. The results show that the area has never been extensively afforested and that it was the least-wooded lowland region of the mainland during the Holocene. Loch of Winless contributes important information towards establishing the pattern of spread of tree species during the Holocene and is a key reference site for reconstructing the vegetational and environmental history of this area of the northern mainland.

Chapter 6

North-west Highlands

INTRODUCTION
D. G. Sutherland

The north-west Highlands (Figure 6.1) contain some of the most spectacular scenery in the country, glaciation having resulted in valley overdeepening, watershed breaching and corrie formation in a landscape that already had considerable pre-glacial relief. There is relatively little known about pre-Late Devensian events in this region. The cave systems of Sutherland have recently been found to contain fossil material and other sediments that pre-date the Late Devensian, and published uranium-series disequilibrium dates on speleothems imply ice-free conditions around 122 ka (the Ipswichian) and again between approximately 38,000 BP and 26,000 BP (Lawson, 1981a; Atkinson *et al.*, 1986). The latter period of ice-free conditions is also substantiated by radiocarbon dates on reindeer antlers from Creag nan Uamh (Lawson, 1984) of approximately 25,000 BP. Thus there appears to have been an interstadial period towards the end of the Middle Devensian.

Late Devensian ice-sheet glaciation then ensued, although there is no site in the north-west Highlands that unequivocally demonstrates this. The ice shed during the period of maximum glaciation can be reconstructed from the transport of erratics, and it appears to have lain somewhat to the east of the present watershed (Peach and Horne, 1893a; Sissons, 1967a; Lawson, 1990). As some of the mountains of the region are capped by extensive areas of frost-shattered debris which show distinct down-slope limits (see An Teallach), it has been suggested that these mountains were not overtopped by the Late Devensian ice but existed as nunataks (Godard, 1965; Ballantyne *et al.*, 1987). If this is correct, then the occurrence of high-level erratics on certain of these summits (Ballantyne *et al.*, 1987) must be attributed to a pre-Late Devensian period of ice-sheet glaciation.

The period of deglaciation following the last ice-sheet maximum is noted in the north-west Highlands for the formation of end moraines at particular retreat stages. The most extensive of these moraines is that first described by Robinson and Ballantyne (1979) and attributed to the Wester Ross Readvance (see Gairloch Moraine); further moraines relating to this phase have been subsequently described by Sissons and Dawson (1981), Ballantyne (1986a) and Ballantyne *et al.* (1987) (Figure 6.1). The moraines apparently mark a specific phase in the deglaciation of the region and they have been related to a raised shoreline which is partly cut in bedrock (Sissons and Dawson, 1981). This has led to the supposition that the Wester Ross Readvance dates from the period at around 13,500 BP to 13,000 BP, when the climate changed from arctic cold to boreal as the oceanic polar front migrated north of Scotland. Other moraines have been reported from Easter Ross (see Achnasheen) (Sissons, 1982a) which may correlate with the Wester Ross Readvance (Sutherland, 1984a), although such an attribution is uncertain and a Loch Lomond Readvance age can also be maintained on the basis of the available field evidence (Ballantyne *et al.*, 1987).

Deglaciation of much of the lowland part of the region by 13,000 BP is suggested by the radiocarbon dates on basal organic sediments (see Cam Loch), but continued presence of ice in the high mountains during the Lateglacial Interstadial has been argued for on palaeoclimatic grounds (Ballantyne *et al.*, 1987). Subsequently there was an expansion of shrubs and heaths, such as *Empetrum* and *Juniperus*, and an accompanying increase in the acidity of the soils. This vegetational development, however, was briefly interrupted, with a re-expansion of indicators of disturbed soils and a decline in shrub cover shown by lower frequencies in pollen spectra, as well as an increase in the amount of mineral matter being washed into enclosed basins. This brief phase has been correlated with the Older Dryas (see Cam Loch and Loch Sionascaig) (Pennington, 1975b, 1975c). Following this, shrubs and heaths became re-established and throughout much of the region the vegetation of the main part of the Lateglacial Interstadial was characterized by ericaceous dwarf-shrub heathland or tundra. The high representation of pollen of *Empetrum* suggests a strong oceanic influence.

The Loch Lomond Stadial was a period of glacier resurgence throughout the region. A large ice-field, with outlet glaciers descending close to present sea level in the west, occupied an area from the mountains in the south to as far north as Torridon and there were numerous individual valley and corrie glaciers in the mountains farther to the north (Figure 6.1) (Peacock, 1970a; Robinson, 1977; Sissons, 1977a; Ballantyne, 1981; Boulton *et al.*, 1981; Wain-Hobson, 1981; Lawson, 1986; Ballantyne *et al.*, 1987). Major end-moraine systems (see Cnoc a'Mhoraire and An

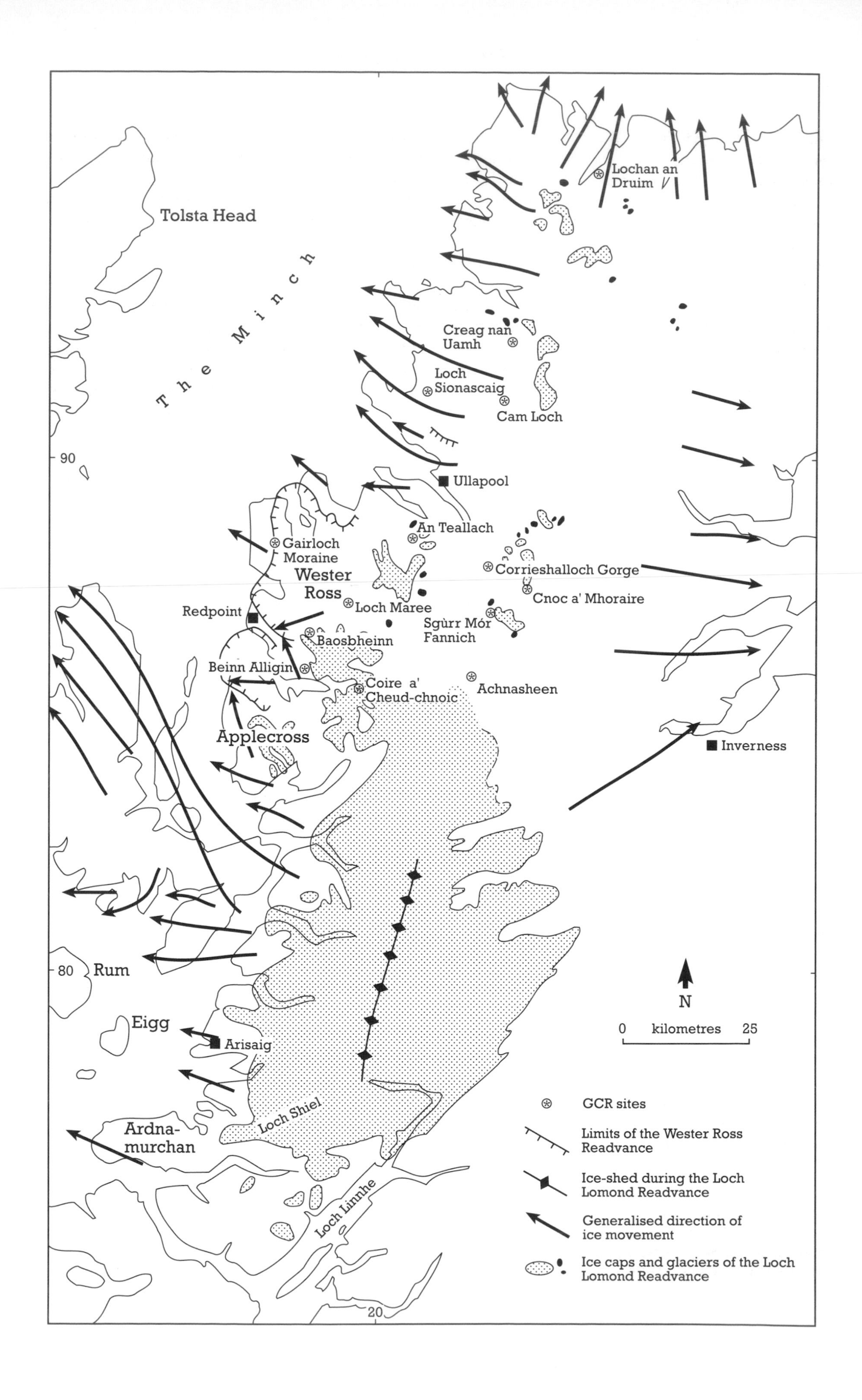

Tolsta Head
The Minch
Lochan an Druim
Creag nan Uamh
Loch Sionascaig
Cam Loch
90
Ullapool
An Teallach
Gairloch Moraine
Wester Ross
Corrieshalloch Gorge
Cnoc a' Mhoraire
Loch Maree
Redpoint
Sgùrr Mór Fannich
Baosbheinn
Beinn Alligin
Coire a' Cheud-chnoic
Achnasheen
Applecross
Inverness
80
Rum
Eigg
Arisaig
N
0 kilometres 25
Loch Shiel
Ardna-murchan
Loch Linnhe
20
GCR sites
Limits of the Wester Ross Readvance
Ice-shed during the Loch Lomond Readvance
Generalised direction of ice movement
Ice caps and glaciers of the Loch Lomond Readvance

Teallach) and abundant hummocky moraines (see Coire a'Cheud-chnoic) were deposited by these glaciers and are prominent elements of the modern scenery of the valleys of the north-west Highlands. Periglacial processes were particularly active in the unglaciated areas at this time, and the major periglacial landforms on the mountain summits received their final fashioning, even though the frost-weathered detritus on certain summits was derived from earlier periods of periglaciation. Two of the most spectacular landforms of the region, the Baosbheinn protalus rampart (Sissons, 1976c) and the Beinn Alligin rock glacier (Sissons, 1975a), formed at this time, with landslides being important in the formation of each (Ballantyne, 1986a).

The vegetation during this period was characteristic of open tundra, with large areas of disturbed ground. At Creag nan Uamh a cache of reindeer bones, which has been radiocarbon dated to the stadial, raises the possibility that Man was present in the area at the time, despite the severity of the climate (Lawson and Bonsall, 1986a, 1986b).

The change in climate at the end of the stadial was marked by a regular plant succession from dwarf-shrub tundra, through a juniper-dominated phase to, at around 9000 BP, the expansion of first birch and then birch–hazel woodland (see Loch Maree and Loch Sionascaig). Latterly in the southern part of the region, communities of oak and elm became established in favourable localities (Moore, 1977; Williams, 1977), but by around 8300 BP pine appeared, apparently earlier at certain sites in Wester Ross and farther north (see Loch Maree) compared with sites farther south, and even compared with neighbouring sites in the region (Loch Clair, Pennington *et al.*, 1972). In the northern part of the region there was probably only a brief phase of pine expansion in what was predominantly a birch-forest zone (see Lochan an Druim).

Reduction in the forest cover began around 5000 BP to 4000 BP, with accompanying expansion of blanket bog, possibly due to a climatic change to cooler and moister conditions. The role of Man in this process, although apparent farther south, remains to be clearly demonstrated in this region. Extensive clearance of birch forest in the last 1500 years can be more directly attributed to human activity.

Sea-level changes around the coasts have not been studied in detail. Following the phase of high sea level accompanying ice-sheet decay, there appears to have been a long period when sea level was below that of the present, and the only evidence for later higher sea levels is the sequence of beaches formed at the maximum of the Main Postglacial Transgression and subsequently. None of these shorelines has been dated, although by comparison with elsewhere in Scotland they can be presumed to have been formed during approximately the last 6000 years.

During the Holocene the mountain summits have continued to be the focus of periglacial activity, albeit at an intensity much reduced from the Loch Lomond Stadial. Small-scale patterned-ground features occur on many summits, and down-slope movement of detritus is recorded by buried organic horizons under solifluction lobes and terraces (White and Mottershead, 1972). A recent increase in such activity may be indicated by the soil horizon buried near Sgùrr Mór in the Fannich mountains in the last few hundred years (Ballantyne, 1986c). On An Teallach a notable feature is the extensive deposit of wind-blown sand with interstratified organic horizons. Following deposition of much of the sand during the early Holocene there has been a recent resurgence in sand blowing, possibly as a consequence of grazing pressure (Ballantyne and Whittington, 1987).

Figure 6.1 Location map and principal glacial features of the north-west Highlands (modified from Johnstone and Mykura, 1989).

GAIRLOCH MORAINE

C. K. Ballantyne

Highlights

This site demonstrates the best representative assemblage of landforms associated with the Gairloch Moraine, a feature formed by the Wester Ross Readvance of the Late Devensian ice-sheet.

Introduction

The Gairloch Moraine (NG 792815) extends over a distance of about 10.5 km in a north–south direction across the peninsula between Loch Gairloch and Loch Ewe. It provides important

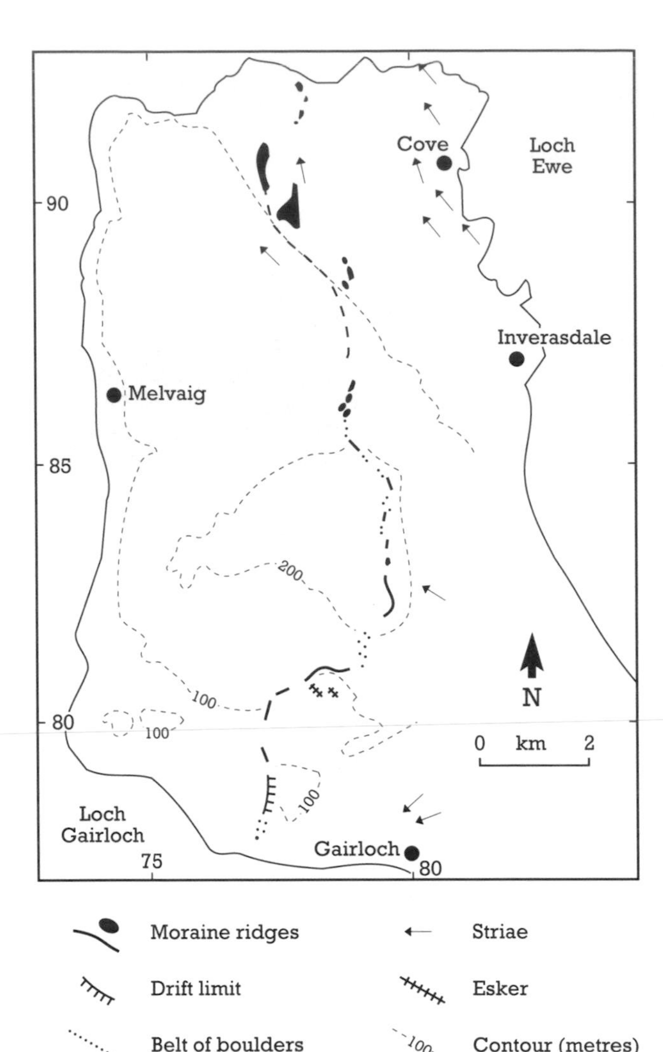

Figure 6.2 The Gairloch Moraine and associated landforms (from Robinson and Ballantyne, 1979).

geomorphological evidence demonstrating a readvance of the Late Devensian ice-sheet in north-west Scotland, the Wester Ross Readvance, which has been identified so far over an area extending from Applecross to north of Loch Broom (Robinson and Ballantyne, 1979). The only published description of the site is by Robinson and Ballantyne (1979).

Description

The Gairloch Moraine marks the western limit of a former glacier nearly 25 km across, which occupied Loch Gairloch and Loch Ewe and the intervening low ground (Robinson and Ballantyne, 1979). The moraine (Figure 6.2) can be traced northwards from a point (NG 771780) 3 km west-north-west of Gairloch, where the former ice margin is marked by a belt of boulders and a drift limit. This is continued northwards and westwards by a discontinuous moraine ridge up to 4 m high. A pair of beaded eskers up to 10 m high terminates just east of this ridge. From NG 797821 to NG 789866, the moraine takes the form of a well-marked, though discontinous, boulder ridge 4 km long (Figure 6.3). At the foot of the Loch Maree fault-line scarp the former glacier limit is represented by broad till ridges (for example, near NG 778901), although the moraine resumes its bouldery character 1 km from the coast, at NG 780915. The total length of the former ice margin delimited by the Gairloch Moraine is about 10.5 km. The form of the moraine is very similar to that of the moraines associated with the Inland Ice near Søndre Strømfjord in West Greenland (see Ten Brink and Weidick, 1974; Ten Brink, 1975). Like these, it runs across country for long distances over low undulating topography, with loops extending down the main valleys.

Local equivalents of the Gairloch Moraine, comprising boulder ridges, moraine ridges and drift limits, have been identified on the Redpoint and Applecross peninsulas near Aultbea, on An Teallach (Robinson, 1977, 1987b; Robinson and Ballantyne, 1979; Sissons and Dawson, 1981) and on the flank of Ben Mór Coigach (Sutherland, 1984a, figure 10). Robinson (1977) and Robinson and Ballantyne (1979) record that some of the individual features were first noted by officers of the Geological Survey on their manuscript maps, although their significance was not then recognized (but see Wright, 1937). However, a whole system of ice-marginal features has now been traced out (see Sutherland, 1984a) and interpreted on geomorphological grounds as marking the maximal extent of a glacial readvance, the Wester Ross Readvance (see Figure 6.1), that interrupted the retreat of the Late Devensian ice-sheet (Robinson and Ballantyne, 1979). This event is tentatively dated to 13,500–13,000 BP (Ballantyne *et al.*, 1987), the approximate time when the oceanic polar front migrated northwards of the west coast of Scotland (Ruddiman and McIntyre, 1973). Sissons and Dawson (1981) argued on glaciological grounds that a readvance is more probable than a stillstand. A readvance is also indicated by changes in the orientation of striae on either side of the Redpoint Moraine

Figure 6.3 The Gairloch Moraine, in the valley of the River Sand north-west of Gairloch, comprises a low ridge of boulders. (Photo: J. E. Gordon.)

(Robinson and Ballantyne, 1979), but it is as yet unconfirmed by stratigraphic evidence and is of unknown magnitude. The presence of an end moraine, however, demonstrates active retreat of the last ice-sheet in this area.

Interpretation

In the past, numerous readvances of the Late Devensian ice-sheet in Scotland have been proposed (see the reviews in Sissons, 1967a, 1974c and 1976b; and also Charlesworth, 1926b, 1956; Synge, 1966, 1977b; Synge and Stephens, 1966; Smith, 1977; Synge and Smith, 1980), but the evidence for most of these has now been reinterpreted (see reviews in Sissons, 1974c, 1976b; Gray and Sutherland, 1977; Sutherland, 1984a). The evidence from Wester Ross, however, and particularly the continuity of the ice-marginal features across large tracts of country, appears to substantiate a readvance. Significantly, also, the marine limit outside the moraine represents a broadly synchronous shoreline formed at approximately the same time as the moraine (Sissons and Dawson, 1981).

Relationships between the Wester Ross Moraine and former ice limits elsewhere are uncertain (Ballantyne *et al.*, 1987). For example, Smith (1977) and Synge (1977b) described a putative readvance limit at Ardersier (but see Firth, 1989b), and D. J. Balfour (unpublished data) has identified former ice limits near the mouths of a number of valleys on the north coast of Sutherland. Sissons and Dawson (1981) considered the possibility that the Wester Ross Readvance might relate to former ice limits associated with sharp drops in the marine limit at Stirling (Sissons *et al.*, 1966) and Otter Ferry (Sutherland, 1981b) and to a stillstand or readvance suggested by Peacock (1970a) in Inverness-shire. However, the evidence is conflicting (Sutherland, 1984a). On the one hand, the gradient of the Main Wester Ross Shoreline associated with the moraine is similar to that of the Main Perth Shoreline which terminates inland near Stirling. In contrast, the gradual drop in the marine limit as the ice retreated from the Wester Ross Moraine appears to argue against such correlations (Sissons and Dawson, 1981). Ballantyne (1988) has also suggested that the readvance may be represented by the 'Strollamus Moraine' in southern Skye, but subsequently reinterpreted the latter as a medial moraine deposited at the convergence of two ice streams (Ballantyne and Benn, 1991).

Similarly, it is not possible to substantiate correlations with the end moraines described in Easter Ross (Sissons, 1982a; Sutherland, 1984a; Ballantyne *et al.*, 1987) for it is as yet unclear which of those moraines relate to ice-sheet deglaciation and which relates to the Loch Lomond Readvance.

The continuity and extent of the Wester Ross Moraine is of considerable importance in providing the clearest geomorphological evidence yet for a readvance of the Late Devensian ice-sheet in Scotland. This evidence is particularly well developed in the Gairloch area where the features marking the ice limit are all clearly demonstrated and seen in close geographical association – drift limits, boulder ridges and till ridges. Here too, eskers occur as part of the landform assemblage. The Gairloch Moraine site may therefore be regarded as the single most important locality demonstrating key geomorphological features of the Wester Ross Readvance. Elsewhere, other aspects of the readvance are represented at An Teallach, notably the relationship with Loch Lomond Readvance moraines.

Conclusion

This site demonstrates an end moraine and other landforms formed by a readvance (the Wester Ross Readvance) of the Late Devensian ice-sheet, about 13,500–13,000 years ago. It includes the best assemblage of landforms that mark the former limit of the ice and is therefore an important reference locality for the geomorphological expression of the event.

ACHNASHEEN

J. E. Gordon and D. G. Sutherland

Highlights

The landforms and deposits at Achnasheen are outstanding examples of glaciofluvial outwash and delta terraces formed by meltwater deposition in an ice-dammed lake during the Loch Lomond Stadial. They are also important for studies of sedimentation in a glacial lake environment.

Introduction

The Achnasheen site (NH 160575) covers an area of 4 km^2 at the western end of Strath Bran at its junction with the through valleys leading to Strath Carron and Glen Docharty. It is important for its particularly fine suite of outwash delta terraces and ice-marginal landforms, which are related to a former ice-dammed lake and associated glacier limits in Strath Bran. It is also important in demonstrating two contrasting styles of sedimentation in the ice-dammed lake. Early accounts of the site were given by Nicol (1844), Campbell (1865), Milne Home (1878), Lucy (1886), Morrison (1888), Geikie (1901) and Peach *et al.* (1913b); more recently it has been studied by Sissons (1982a), Sutherland (1987a) and Benn (1989a). The interest of the site and the formation of the terraces is also summarized by Benn (1992).

Description

The terraces extend eastwards from Loch a'Chroisg and north-eastwards from Loch Gowan to the vicinity of Achnasheen (Figure 6.4). The most prominent features occur to the east of the Abhainn a'Chomair and between the Abhainn a'Chomair and the Abhainn Loch Chroisg (see figure 79 in Geikie, 1901; and plate 8 in Peach *et al.*, 1913b). In both areas the terraces comprise very conspicuous high-level surfaces and varying numbers of lower fragments down to the present river floodplains (Figures 6.4 and 6.5). At their maximum height the terraces are up to 25–30 m above the floodplains. Sissons (1982a) mapped and levelled the terraces in detail, showing the terrace north-east of Loch Gowan to descend from 191 m to 175 m OD over a distance of about 850 m (gradient of 19 m km^{-1}) and the terrace east of Loch a'Chroisg to fall from 185 m to 176 m OD in about 500 m (gradient of 18 m km^{-1}). Kettle holes occur on the proximal parts of the highest terraces, and the western and southern margins of those terraces near Loch a'Chroisg and Loch Gowan, respectively, are demarcated by sharply defined ice-contact slopes. Irregular drift mounds occur on the valley sides above the ice-contact features and Sissons (1982a)

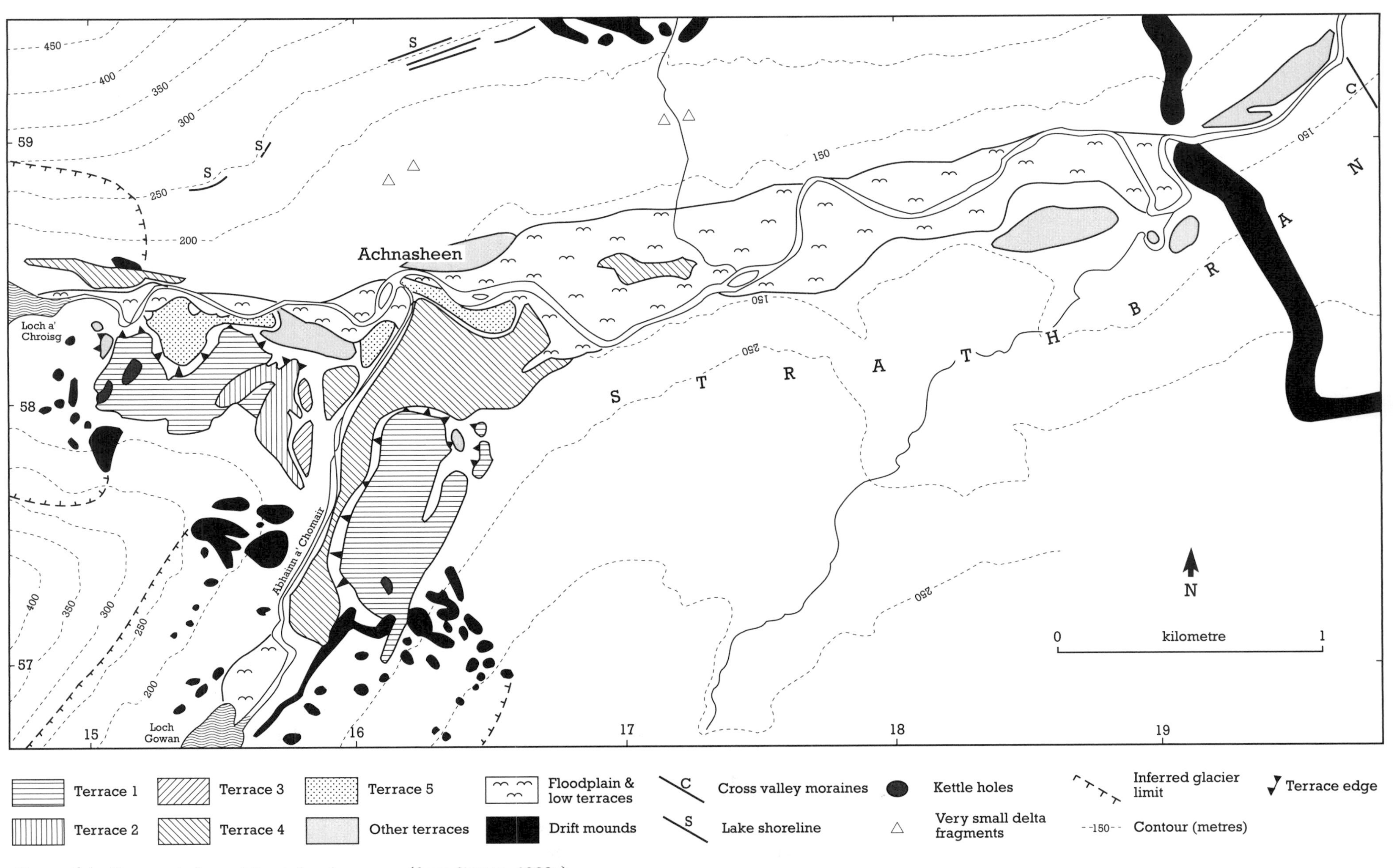

Figure 6.4 Geomorphology of the Achnasheen area (from Sissons, 1982a).

Figure 6.5 Glaciofluvial terraces at Achnasheen. (British Geological Survey photograph B915.)

interpreted these as marking the margins of glaciers contemporaneous with terrace formation.

There are few good contemporary sections in the terraces. Small exposures at NH 159577 show rhythmically bedded silts and clays, and recent stream erosion at NH 156581 has revealed a section with horizontally bedded sands at the bottom overlain by well-sorted sands and gravels dipping eastwards in foreset-type bedding (Benn, 1989a). Former sections described by Lucy (1886) and Peach *et al.* (1913b) showed similar composition and bedding.

About 2.5 km east of Achnasheen an impressive drift limit crosses Strath Bran (Figure 6.4). On its west side the drift margin is up to 20 m high and is continued eastwards on the south side of the valley by a ridge and mounds (Sissons, 1982a). To the east of the limit in the valley bottom, there are two further asymmetric cross-valley moraine ridges (Figure 6.4). Benn (1989a) described the facies geometry of sediments in three exposures at the former ice limit, noting the presence of alternating diamictic and laminated associations. The former comprise laminated, matrix-supported diamicton with thin sand interbeds, laminated silts with dispersed clasts, and clast-supported diamicton, while the latter comprise multiple sand–silt couplets and rippled sands. Overlying these sediments is a much coarser deposit of coarse diamicton with a sand-rich matrix, poorly sorted gravel and deformed sand with abundant clasts. Numerous small exposures of laminated lake-floor sediments with occasional drop stones occur in the valley between Achnasheen and the Strath Bran drift limit. Further evidence for a former ice-dammed lake is the presence of shoreline fragments identified by Sissons (1982a) at 255 m, 245 m and 237 m OD on the side of the valley north of Achnasheen (Figure 6.4) and overflow channels along the southern margin of the Strath Bran ice lobe (Sissons, 1982a).

Interpretation

On account of their striking landscape form, the Achnasheen terraces have been noted in the scientific literature for about 140 years. In early accounts, debate centred on whether they were of fluvial, marine or glaciolacustrine origin. In one of the first descriptions of the site Nicol (1844) considered that the terraces were formed by a great river flowing from the west. Campbell (1865), however, thought them to be of marine origin, a view supported by Milne Home (1878). Lucy (1886) proposed a lacustrine theory in which glacier-transported debris blocked the outlets of Lochs Gowan and Chroisg. Outflow of water from these dammed lakes then cut down into the barriers and levelled the terraces. He envisaged three separate episodes of damming and overspill to cut the three main terraces.

Morrison (1888) explained the terraces as the beaches of a former lake in Strath Bran, formed at a time when vast quantities of meltwater were released on to low ground by melting glaciers. Geikie (1901, p. 294) developed this interpretation of what he described as a 'remarkable group of ancient lake terraces' (p. 508), suggesting that they were formed in a lake ponded in Strath Bran by the build up of snow and ice to the east. Peach *et al.* (1913b) concluded that the finer-grained deposits in the lower part of the sequence were laid down as a lake delta (an explanation which they acknowledge was first proposed by A. Penck on a visit to the site in 1895), then subsequently covered by coarse glaciofluvial gravels. The lake in Strath Bran was impounded to the east by a lobe of ice from the Fannich mountains which extended west into the valley. The valleys to the west (now occupied by Lochs Gowan and Chroisg) were also occupied by glaciers, which acted as sediment sources for the terraces, initially providing fine-grained material at the advancing delta fronts, then coarser glaciofluvial gravels.

In a detailed study, Sissons (1982a) has confirmed and amplified the conclusions of Peach *et al.* (1913b). He mapped moraines and meltwater features that defined the margin of the large lobe of ice from the Fannich mountains, which produced the ice-dammed lake. The lake had a maximum depth of at least 125 m adjacent to the ice lobe. Sissons (1982a) also identified a moraine related to a small glacier that descended the northern side of Strath Bran, as well as the glaciers west and south-east of Achnasheen, at the margins of which the large outwash terraces formed. The height relationships of the terraces and the lake shorelines indicated that the terraces formed during deglaciation as the lake level fell to about 175 m OD, this level being controlled by a meltwater channel identified farther down Strath Bran (Sissons, 1982a). Sissons noted that minor deltas occur beside various side streams at this altitude. Subsequently, the lake level fell rapidly to about 140 m OD, the major proglacial terraces were dissected and the terraces close to the valley bottom formed.

Benn (1989a) drew attention to the contrasting landform and sediment associations related to, on the one hand, the Strath Bran glacier lobe in the east and, on the other hand, to the ice tongues in the valleys of Loch a'Chroisg and Loch Gowan to the west. The Strath Bran glacier produced a large, asymmetric drift ridge and drift limit, with cross-valley moraines on its proximal side. The latter have been interpreted as sublacustrine in origin (Sissons, 1982a), possibly associated with a floating ice ramp (Benn, 1989a). Outwash terraces and deltas, however, are absent. From the sediment characteristics and facies variations, Benn (1989a) inferred gravity-flow sedimentation at the ice margin, including deformation and remobilization of previously deposited materials. In contrast, the western glacier margins are marked by the large ice-contact terraces described above, which formed as prograding deltas built out into the lake. Benn (1989a) considered three factors, which together could explain the observed contrast between the 'glacier-contact' glaciolacustrine sedimentation of the Bran ice lobe and the 'glacier-fed' glaciolacustrine sedimentation at the western glacier margins: variations in ice thickness and water depth, meltwater discharge and glacier fluctuations. He concluded from both field evidence and glaciological theory that the contrast reflected relatively high meltwater discharge at relatively stable, grounded ice margins in the west, compared with lower meltwater discharge at a fluctuating and periodically calving margin of the Bran ice lobe which directly controlled the level of the ice-dammed lake.

The age of the Achnasheen landforms and sediments is not firmly established. Clapperton (1977, p. 31) reproduced a map of the terraces by A. M. D. Gemmell and suggested that they might mark the limits of 'the local ice cap during the Late-glacial'. However, Robinson (1977) and Sissons (1977a) mapped the limits of the Loch Lomond Readvance some distance to the west suggesting an earlier age for the events at Achnasheen (cf. Sutherland, 1984a). Subsequently, Sissons (1982a) was unable to locate any evidence that would refute the hypothesis that the Achnasheen ice limits and terraces were related to the Loch Lomond Readvance. Similarly, Ballantyne *et al.* (1987) concluded that there was no evidence to discount a Loch Lomond Stadial age for the deposits; of particular relevance was the apparent absence of Lateglacial Interstadial deposits within the ice limits.

The Achnasheen terraces are the best known example of a suite of outwash delta terraces in Scotland. They are one of the classic landform localities in the country (cf. Benn, 1992) and are important from both educational and scientific viewpoints in demonstrating terrace morphology and sedimentology. Achnasheen is also notable for a wider assemblage of glacial and glaciolacustrine landforms and sediments represented in a relatively compact area. These include end moraines, drift limits, cross-valley moraines, lake shorelines, ice-contact slopes and kettle holes. Recent work has further highlighted the value of the area for studies of glaciolacustrine sedimentation; two quite distinctive styles are represented and these provide a valuable opportunity to demonstrate and evaluate their relationships to reconstructed glacier, meltwater and lake-level controls. Achnasheen is therefore a key locality for glacial lake landforms and sediments.

The interests of Achnasheen complement those at several other sites in Scotland. The most comparable glacial lake outwash deltas are in Glen Spean (see Glen Roy and the Parallel Roads of Lochaber), but the terrace forms there are less clearly displayed than at Achnasheen, and the sediments have not been studied in detail; the origin of the terraces in Glen Roy (e.g. at Glen Turrett) is currently a matter of debate (see below). Detailed sedimentological studies and comparisons between the features at Achnasheen, in Lochaber and at Gartness (see below) would contribute towards a better understanding of glacial lake sedimentary environments in Scotland. In some cases individual landforms are equally or better developed elsewhere (for example, lake shorelines at Glen Roy, cross-valley moraines at Coire Dho), but Achnasheen is outstanding first, for the quality of its delta terraces; second, for a clear demonstration of some of the controls on glaciolacustrine sedimentation; and third, for the range of landforms and deposits developed in a relatively compact area. In their glacial lake associations, the Achnasheen landforms also differ genetically from the subaerial or marine-related terrace systems, for example, at Moss of Achnacree, Glen Feshie, Corran Ferry, Port a'Chuillin, Gruinard Bay, Glen Einich and Kilmartin Valley.

Conclusion

Achnasheen is a classic locality for a series of impressive outwash and delta terraces formed by glacial meltwater rivers that flowed into a lake dammed by glaciers in Strath Bran and adjacent valleys, probably during the Loch Lomond Stadial (about 11,000–10,000 years ago). The deposits are also significant for sedimentological studies, in particular for reconstructing the patterns and processes of sedimentation in glacial lakes and the factors that control them.

AN TEALLACH

C. K. Ballantyne

Highlights

An Teallach is a site of great importance for its assemblage of glacial and periglacial landforms. The interest includes a suite of moraines formed during different episodes of Late Devensian glaciation and outstanding examples of periglacial features, most notably deflation surfaces and mountain-top sand deposits.

Introduction

An Teallach (NH 038860) is located on the south side of Little Loch Broom, about 11 km south of Ullapool. It is one of the most spectacular mountains in north-west Scotland, rising to 1062 m OD, and supports many outstanding examples of glacial and periglacial features within a relatively small area (about 35 km^2). The landforms developed on and around An Teallach are typical of those of the Torridonian sandstone mountains, and several types are among the finest examples known (Ballantyne, 1984, 1987a, 1987b). The principal publications on the site are those of Peach *et al.* (1913a), Godard (1965), Sissons (1977a), Robinson and Ballantyne (1979), Ballantyne (1981, 1984, 1985, 1986b, 1987a, 1987b), Ballantyne and Eckford (1984), Ballantyne and Whittington (1987) and Benn (1989b).

Description

Abundant striae and chattermarks on both Torridonian sandstone and Cambrian Quartzite indicate that, during the Late Devensian ice-sheet glaciation, ice moving from the east and south-east was deflected around the flanks of An Teallach. The upper slopes and plateau areas of the mountain are mantled by a thick cover of frost-weathered detritus, which contrasts with the ice-moulded surfaces of the lower slopes, and as the transition from one type of surface to the other (at around 700 m OD) is apparently too abrupt to be explained by a climatic difference, it may be that the upper part of the mountain was not glaciated at the time of the last ice-sheet maximum (Reed, 1988). However, erratics are found at altitudes of up to 900 m OD (Peach *et al.*, 1913a), including a remarkable high-level train of Cambrian Quartzite erratics on the northern plateau. Such erratics indicate glaciation from the east-south-east, and may relate to an earlier period of more extensive ice-sheet glaciation (Ballantyne, 1987b).

A sequence of three end or lateral moraines is found in the broad valley west of the massif (Figure 6.6: a,b,c); this provides evidence that the retreat of the last ice-sheet was interrupted by local stillstands or readvances. These have been tentatively correlated with the Wester Ross Readvance by Robinson and Ballantyne (1979). A massive drift ridge over 1 km long at the north-east end of the valley (Figure 6.6: d) has been interpreted as a medial moraine deposited when ice in this valley parted from that occupying Little Loch Broom to the north.

During the Loch Lomond Readvance six separate corrie glaciers existed on An Teallach (Figure 6.6) (Sissons, 1977a; Ballantyne, 1987b), their limits being marked by striking end moraines, rising by as much as 30 m above the adjacent bare Torridonian bedrock in one example (at NH 092846) and, in the case of the former Coire Toll an Lochain glacier (Figures 6.6: 6, and 6.7), by a superb drift limit. On the north-west side of the mountain, below Mac is Màthair, one glacier (Figure 6.6: 2) formed a sequence of five nested boulder moraines, the outermost truncating a moraine associated with the Wester Ross Readvance (Robinson and Ballantyne, 1979; Ballantyne, 1987b). Controls on moraine asymmetry and debris transport in two of the eastern corries are discussed by Benn (1989b): moraine asymmetry strongly correlates with the distribution of source cliffs which principally reflects climatic controls.

Periglacial landforms are particularly well-developed on An Teallach (Figure 6.8). The frost-weathered detritus of plateau and summit areas underlain by Torridonian sandstone rarely exceeds 1 m in depth and consists of slabs embedded in a coarse sandy matrix. The slabs were produced by frost wedging along pseudobedding planes during the Late Devensian, the sand by the granular disintegration of rock surfaces. The latter process continues today. In contrast, on summit outcrops of Cambrian Quartzite, openwork blockfields and blockslopes are composed of more angular detritus, since quartzite is generally less susceptible to granular disintegration. In places (for example, NG 085845) Cambrian Quartzite blockslope material has moved downslope under the action of frost creep, completely covering the underlying Torridonian sandstone strata. The periglacial features of the upper slopes and plateaux of the massif may be subdivided into relict and active. The stability of the relict features under present conditions is illustrated by the strong contrast in the roundness of exposed and buried clasts, the former being rounded and the latter angular. There are relatively few relict periglacial landforms, though boulder lobes averaging 10 m in width occur in the southern part of the massif (for example, NG 059852), shallow, though well-developed solifluction lobes occur downslope, and dubious examples of sorted circles are found on the northern plateau (Ballantyne, 1981). An example of a Loch Lomond Stadial protalus rampart exists at the foot of a relict talus slope in the north-western part of the massif at NH 054875 (Ballantyne and Kirkbride, 1986).

Among the active periglacial features are probably the finest assemblage of high-level aeolian and niveo-aeolian features in Great Britain. Deflation of the extensive plateau surfaces with their sandy regolith has its counterpart in the accumulation of sand-sheets on flanking slopes (Godard, 1965; Ballantyne and Whittington, 1987). The deposits are up to 4 m in thickness at the crest of lee slopes, although they rarely exceed 1.5 m elsewhere. A formerly more widespread sand cover is indicated by the occurrence of isolated sand 'islands' on high-level cols. Interbedded with the sands are organic horizons, and analysis of these for pollen content together with radiocarbon dating has indicated that the deposits began to accumulate in the early Holocene (Ballantyne and Whittington, 1987), but that

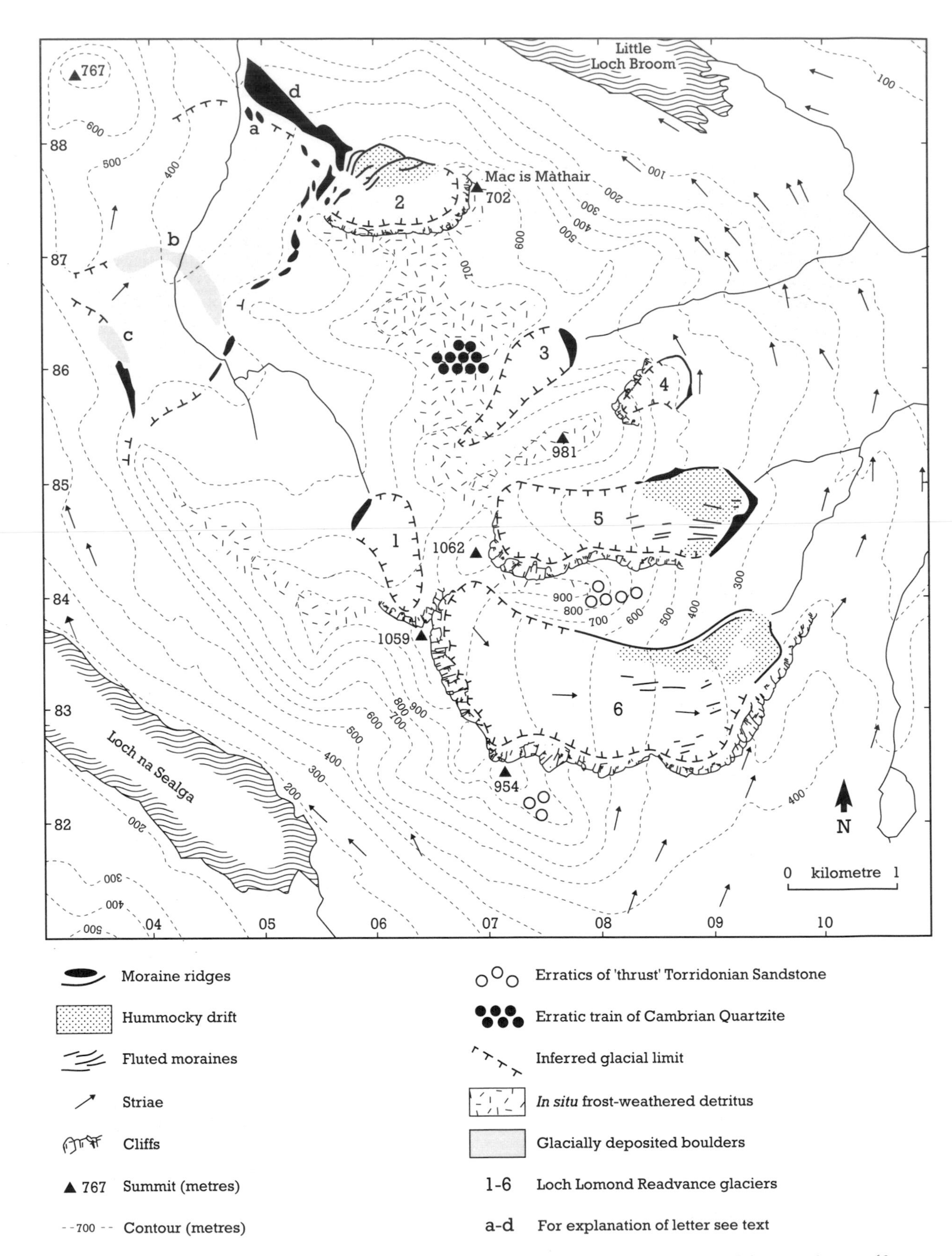

Figure 6.6 The An Teallach area, showing principal glacial features and the limits of former glaciers (from Ballantyne, 1987b).

Figure 6.7 The east flank of An Teallach. A clear lateral moraine and drift limit (centre) mark the extent of the Loch Lomond Readvance glacier that occupied the corrie of Toll an Lochain. The ice-scoured Torridonian sandstone of the lower slopes contrasts with the frost-shattered slopes of the mountain ridge and the quartzite blockslopes on Glas Mheall Liath. (British Geological Survey photograph D2102.)

deposition virtually ceased in the late Holocene following establishment of a vegetation cover on the plateau areas upwind. Recent disruption of the vegetation has resulted in renewed accumulation of wind-blown sand on lee slopes.

Modification to the sand deposits has occurred by both selective eluviation by nival meltwater and scarp erosion, the former process producing large sand hummocks like thúfur (earth hummocks) (Ballantyne, 1986b), and the latter producing small nivation hollows. However, Ballantyne (1985) found that the effects of nivation were confined to localized erosion and transport of the unconsolidated niveo-aeolian sands. Despite some contrary evidence from northern England (Tufnell, 1971; Vincent and Lee, 1982), the present geomorphological role of nivation in the British uplands appears to be limited or at least localized (Ballantyne, 1985, 1987a).

Frost creep is active on unvegetated debris slopes on the Torridonian sandstone, producing an average downslope movement of surface

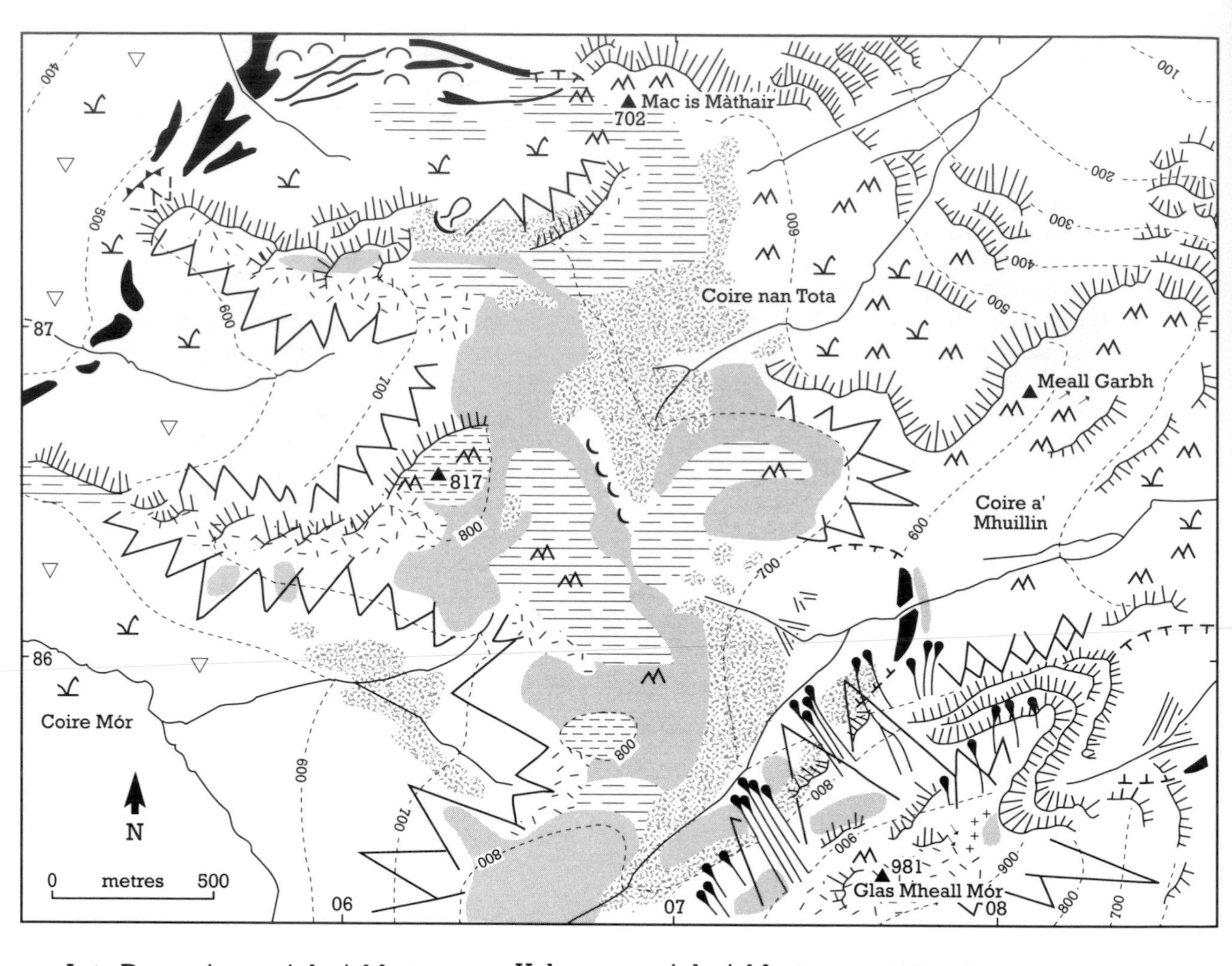

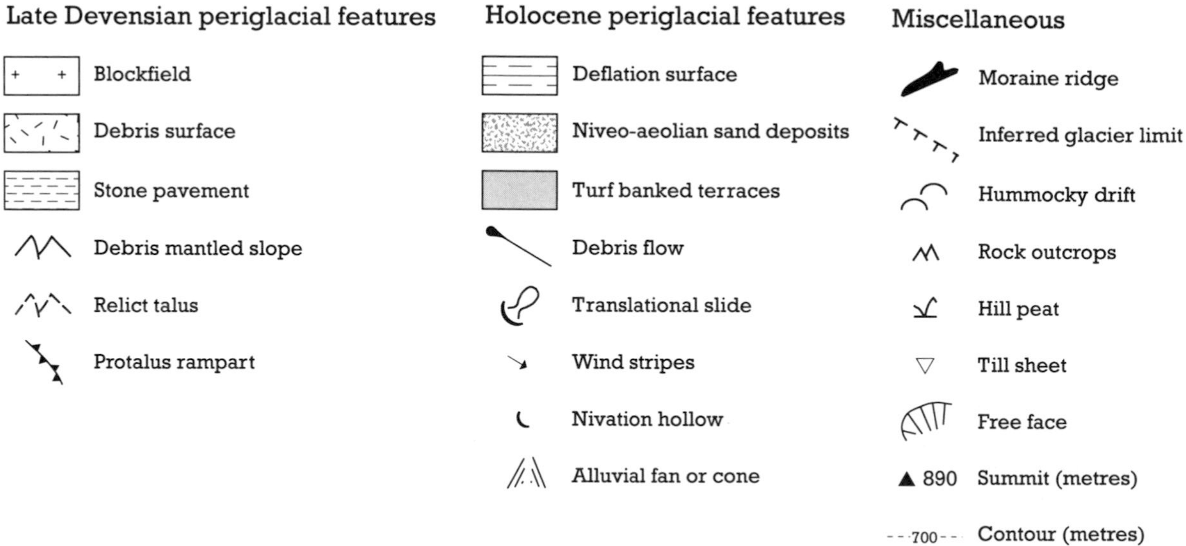

Figure 6.8 Periglacial landforms and deposits on the northern plateau of An Teallach (from Ballantyne, 1984).

stones around 0.01 m a^{-1} on the 30° slopes of Glas Mheall Mór (NG 076854). Such rates are, however, insufficient to account for the distance travelled downslope by Cambrian Quartzite clasts from summit outcrops, implying that rates of creep have been greater in the past. A combination of wind action and frost creep has resulted in the formation of outstanding examples of oblique and horizontal turf-banked terraces on slopes of between 5° and 25° (see also Ward Hill). The sparsely vegetated 'treads' range in width from 0.5 m to over 5 m downslope and from 2 m to over 100 m in length, size being closely correlated with the size of clasts in the 'riser'. On north- and south-facing slopes such terraces dip westwards, into the direction of the dominant winds; on steep slopes such as the south slope of Glas Mheall Mór, this has produced a series of remarkable oblique terraces.

Within the corries the most active processes today are rockfall and debris flow. Stonefalls from the glacially steepened cliffs of the eastern corries are fairly frequent, and a rockfall during the winter of 1976–77 on to a well-developed active talus cone (NG 845075) displaced boulders over 100 kg in mass. Talus slopes produced by rockfall during the Holocene (that is those within the limits of the Loch Lomond Readvance) show less mature profiles than the relict talus slopes formed during the Lateglacial (Ballantyne and Eckford, 1984) although rockfall activity is, after solute transport, the process responsible for the highest rate of erosion on the mountain today (Ballantyne, 1987b). Innes (1983b) has shown that there has been an increase in debris flow activity in recent centuries, which he has attributed to the effects of overgrazing or possibly burning of vegetation to improve the quality of rough pasture.

Interpretation

An Teallach is a site of great importance for Quaternary geomorphology. It not only supports a remarkable range of high-level periglacial phenomena developed on Torridonian sandstone, but has also been the focus of the most detailed investigation of upland periglaciation in Great Britain (Ballantyne, 1981). Furthermore, individual types of periglacial feature are exceptionally well developed, particularly in the case of aeolian and niveo-aeolian features and oblique terraces. The scientific work that has been carried out on An Teallach has also demonstrated the fragility of the mountain summit environment and how it has been disturbed in recent years. Such disturbance may have been due partly to climatic change (the Little Ice Age), but Man's activities, particularly the introduction of sheep and consequent degradation in the vegetation, has more probably had the major impact.

The assemblage of periglacial phenomena on An Teallach is outstanding not only in its own right, but also because it complements features of interest at a range of other sites on different rock types and in different mountain environments (see Ronas Hill, Ward Hill, Ben Wyvis, the Cairngorms, Sgùrr Mór, and the Western Hills of Rum), which together represent most major facets of the periglacial geomorphology of upland Scotland. An Teallach demonstrates better than any other mountain the range of landforms developed on a sandstone substrate.

An Teallach is of significant scientific interest in a second respect. It is exceptional in demonstrating, in a small area, evidence relating to several glacial phases. This includes high-level erratics that reflect the former passage of an ice-sheet over much of the mountain, periglacial trimlines that may represent the upper level of the last (Late Devensian) ice-sheet, and moraines of both the Wester Ross Readvance and the Loch Lomond Readvance. It provides key evidence showing Wester Ross Readvance moraines truncated by Loch Lomond Readvance moraines, thereby establishing their stratigraphic relationship (see also Ballantyne, 1986a). In at least two cases, the Loch Lomond Readvance moraines and drift limits are outstanding examples of their type.

Conclusion

An Teallach is outstanding for its assemblage of glacial and periglacial landforms. The former include an exceptional range of features, including deposits and moraines associated with the Late Devensian ice-sheet at its maximum, the Wester Ross Readvance and the Loch Lomond Readvance. The latter include both fossil and active forms, most notably a series of features formed by wind erosion, which is probably the finest of its kind in Britain. An Teallach is a key reference site for periglacial landforms developed on Torridonian sandstone mountains and is the most intensively studied site in upland Britain.

BAOSBHEINN

D. G. Sutherland

Highlights

Baosbheinn is important for a protalus rampart of

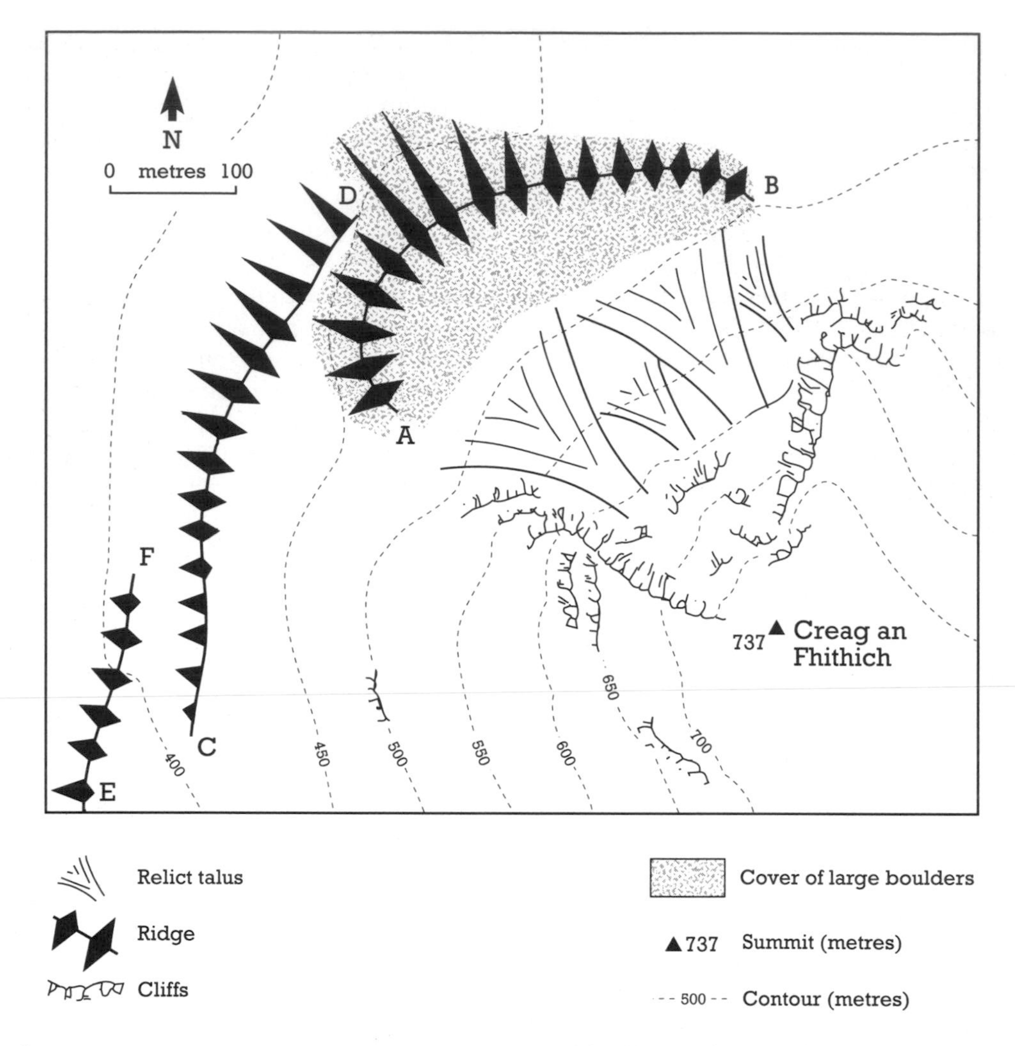

Figure 6.9 The Baosbheinn protalus rampart and associated landforms, showing the upper boulder ridge (AB) and lower ridges (CD and EF) (from Ballantyne, 1986a).

exceptional size, in part formed by large rock-slides. It is important in demonstrating slope-process activity during the Loch Lomond Stadial.

Introduction

Baosbheinn (NG 855676) is the site of the largest and most impressive protalus rampart in Great Britain (Sissons, 1976c). The rampart is located on the north-west end of the Baosbheinn ridge at an altitude of approximately 450 m OD beneath the cliffs of Creag an Fhithich. Its morphology and characteristics were originally described by Sissons (1976c) and further details were provided by Ballantyne (1986a). A number of protalus ramparts have been described in recent years (see Ballantyne and Kirkbride, 1986), but none of them rivals the Baosbheinn rampart in size. This has led Ballantyne (1986a, 1987d) to suggest that both landsliding and deformation by internal ice may have been involved in the production of this remarkable geomorphological feature.

Description

The protalus rampart consists of two distinct ridges (Figures 6.9 and 6.10): a massive, open-work, upper ridge (AB) composed of Torridonian sandstone boulders and a lower, vegetated ridge (CD) with few boulders. The arcuate upper ridge is separated from the base of the mountain by a depression approximately 70 m wide and up to

Figure 6.10 Protalus rampart on the north-west flank of Baosbheinn. The view also shows the rock avalanche scar, the two lower Wester Ross Readvance moraine ridges and a Loch Lomond Readvance moraine intersecting the latter at the base of the mountain. (Cambridge University Collection: copyright reserved.)

6 m deep, and contains two minor, enclosed hollows. This ridge is over 450 m long and its distal slope is, at maximum, 55 m high. The second, lower ridge was examined in detail by Ballantyne (1986a), who demonstrated that it is composed of a diamicton which, on sedimentological grounds (clast size, angularity, roundness, form, hardness and granulometry), is similar to the material comprising certain parts of the Wester Ross Moraine, but quite different to the upper boulder ridge. A third ridge (EF) is also present (Figure 6.9).

Interpretation

Sissons (1976c) interpreted both the main ridges (AB and CD) as part of a protalus rampart complex, the lower ridge forming during an early part of the Loch Lomond Stadial and the upper ridge subsequent to it, after a change in climate as the stadial progressed. Ballantyne (1986a), although agreeing with Sissons as to the age of the upper ridge, argued convincingly that both the lower ridge and ridge EF are lateral moraines formed during the Wester Ross Readvance at some time prior to 13,000 BP. No climatic inferences can therefore be based on the contrast in the nature of the two ridges.

Interpretation of the upper ridge as a 'conventional' protalus rampart formed by individual rockfall events was also questioned by Ballantyne (1986a). He pointed out that not only is the rampart particularly large by British and even world standards (see Washburn, 1979; Ballantyne and Kirkbride, 1986), but its formation also implies rock-wall retreat of the overlooking cliffs of an order of magnitude greater than could be inferred from other similar landforms formed during the Loch Lomond Stadial (Ballantyne and Kirkbride, 1987). He also observed that the cliffs upslope take the form of a major rockslide scar, and suggested that the rampart formed in response to one or more rockslides from the backing cliffs across a former snowbed. The occurrence of small enclosed depressions also suggested to Ballantyne the presence of buried snow, firn or ice at the time of formation of the rampart and this, in turn, raised the possibility that there had been some forward movement of the debris due to deformation of buried ice. The feature may, therefore, be regarded as a protalus or valley-wall rock glacier and has affinities with similar features on Beinn Shiantaidh on Jura (Dawson, 1977) and Coire Beanaidh in the Cairngorms (Sissons, 1979f; Chattopadhyay, 1984). Many protalus rock glaciers in the Scottish Highlands occur at sites of rock-slope failures (Ballantyne, 1987d; Maclean, 1991).

In recent years protalus ramparts have been described from a number of localities in Scotland (Sissons, 1977a, 1979f; Rose, 1980d; Sutherland *et al.*, 1982, 1984; Ballantyne and Kirkbride, 1986), as well as more widely in the United Kingdom (Sissons, 1980a; Colhoun, 1981; Gray, 1982a; Gray and Coxon, 1991). Of the known protalus ramparts, however, none is as spectacularly developed as that at Baosbheinn. The countrywide distribution of protalus ramparts, all of which are thought to have formed during the Loch Lomond Stadial, shows a pattern of decline in altitude from the eastern Grampian Highlands to the western Highlands and Islands (Ballantyne, 1984; Ballantyne and Kirkbride, 1986) that is remarkably similar to the variation in equilibrium line altitudes for Loch Lomond Readvance glaciers (Sissons, 1980b). This similarity is considered to be of significance in palaeoclimatic terms, according with a pronounced eastwards and northwards decline in precipitation in Scotland during the Loch Lomond Stadial (Ballantyne and Kirkbride, 1986). The Baosbheinn rampart is thus part of a wider sequence of landforms of significance in the understanding of the environment of the Loch Lomond Stadial.

Conclusion

Baosbheinn demonstrates an exceptional example of a protalus rampart, a ridge formed at the base of a snowbank through the accumulation of rock debris which fell from cliffs above. Its large size appears to reflect the incorporation of material from a lateral moraine and the supply of debris from large rockslides. Baosbheinn is important in illustrating the nature of slope processes, between about 11,000 and 10,000 years ago, during the Loch Lomond Stadial. It forms part of a network of related sites, the distribution of which shows the interplay of debris supply and climatic factors during this intensely cold phase.

BEINN ALLIGIN

J. E. Gordon

Highlights

The principal interest at Beinn Alligin is a large rockslide which occurred on to the surface of a glacier during the Loch Lomond Stadial. It is the largest and clearest such feature in Britain and illustrates the geomorphological effects of high-magnitude slope failure in a glacial environment.

Introduction

Beinn Alligin (NG 870600) is located on the north side of Loch Torridon in a highly dissected landscape of glacial erosion and narrow mountain ridges. On its north-west and south-east sides the mountain is indented by corries, which acted as ice source areas for Loch Lomond Readvance glaciers (Sissons, 1977a). The most impressive corries are those of Toll a'Mhadaidh Mór and Toll a'Mhadaidh Beag, which together with Coire Mhic Nòbuil demonstrate a fine range of glacial and mass-movement landforms, including most notably an extensive area of glacially transported rock slope failure (rockslide or rock avalanche) debris that has been interpreted as constituting the largest fossil rock glacier in Scotland. Its characteristics and origin are described in detail by Sissons (1975a) and discussed by Sissons (1976d, 1977d), Whalley (1976a) and Ballantyne (1987c). In addition, lateral, medial and hummocky moraines are well represented.

Description

The rock slope failure debris in Toll a'Mhadaidh Mór (Figures 6.11 and 6.12) comprises a massive accumulation of Torridonian sandstone blocks, some exceeding 5 m in length, that form a debris tongue 1.2 km long, up to 15 m high and tapering in width from 400 m at the head to 200 m near the toe. At the surface, the debris mass displays both longitudinal and transverse ridges and depressions. The lowermost 300–400 m of the debris tongue is much less thick than the rest (Figure 6.12). The source of the debris is marked by a large scar and cleft high on the corrie headwall below the summit of Sgùrr Mhór. In part, the scar is structurally defined by two intersecting faults.

Beinn Alligin is also notable for fine examples of double lateral moraines and a medial moraine of Loch Lomond Stadial age (Figures 6.11 and 6.12). The lateral moraines are well-developed on both sides of Coire Mhic Nòbuil and mark the limit of a Loch Lomond Readvance glacier which extended just offshore into Loch Torridon (Sissons, 1977a). In places double ridges are present. Sections in the western lateral moraine along the Diabaig road reveal boulders and cobbles in a sandy-gritty matrix; upslope from here the lateral moraine comprises boulder ridges. The medial moraine in Coire Mhic Nòbuil is clearly seen as a line of boulders running south-west from the northern limb of the Beinn Alligin ridge. On its western side a zone of hummocky moraine and fluted moraine completes the landform assemblage.

Interpretation

Beinn Alligin provides a particularly striking illustration of a major rock slope failure apparently associated with glacier ice, although there are varying interpretations of the resulting deposits.

Sissons (1975a) considered that the debris accumulation in Toll a'Mhadaidh Mór was not simply a landslide deposit, since several aspects of its morphology are quite unlike those of other rockslides in the Highlands, notably its sharply defined lateral margins and long travel distance. Certain of its morphological characteristics, however, are similar to those of rock glaciers elsewhere: plan shape, transverse ridges and closed depressions. Sissons considered that the feature represented reactivation of a small decaying glacier in the corrie by a rockslide during the Loch Lomond Stadial.

As an alternative explanation, Whalley (1976a) put forward the hypothesis that the feature was simply a rockslide deposit, with only a morphological resemblance to a rock glacier from which any ice had melted (cf. Whalley and Martin, 1992). He argued that its characteristics are typical of many large features of this kind and could relate to either flotation on a cushion of air (Shreve, 1968a, 1968b) or the development of particular flow properties that aid momentum transfer (Hsü, 1975; Eisbacher, 1979; Cruden and Hungr, 1986). In reply, Sissons (1976d) contended that topographic conditions at Beinn Alligin did not favour airborne flow and that the form of the debris tongue differed in certain respects from rockslides that apparently moved in this way. In particular, it lacks the highly distinctive lateral and distal rims and surface fluting typical of many large-scale rock avalanches or flowslide deposits (Shreve, 1966, 1968a, 1968b; Marangunić and Bull, 1968; Reid, 1969; Hsü, 1975; Gordon *et al.*, 1978).

Ballantyne (1987c) accepted that the Beinn Alligin deposit is a glacially transported rockslide or rock avalanche, but noted that it lacked the steep front and high terminal ridges typical of rock glaciers and that the debris thinned downslope. He therefore suggested that the glacier may have been larger than envisaged by Sissons (1975a) and that a rock glacier in the normal sense may not have formed (i.e. through the deformation of an ice core or interstitial ice within the debris). Accordingly, the Beinn Alligin feature may resemble several modern instances of rockslides or rock avalanches on to glaciers (Marangunić and Bull, 1964; Reid, 1969; Gordon *et al.*, 1978).

Glacially transported rock slope failure deposits have been reported elsewhere in Scotland, for example on Eigg (Peacock, 1975d) and in Gorm Coire, Ben Hee (Haynes, 1977b). None of these, however, rivals the Beinn Alligin feature in terms of size, or clarity of the relationship between source and transported debris. Holmes (1984) has established that the great majority of rock slope failures in the Highlands occur within a short distance of the former limits of Loch Lomond Readvance glaciers, and he suggested that excess pore water pressures developing in oversteepened slopes during deglaciation may have contributed significantly to subseqent slope failure (cf. Whalley, 1974; Whalley *et al.*, 1983).

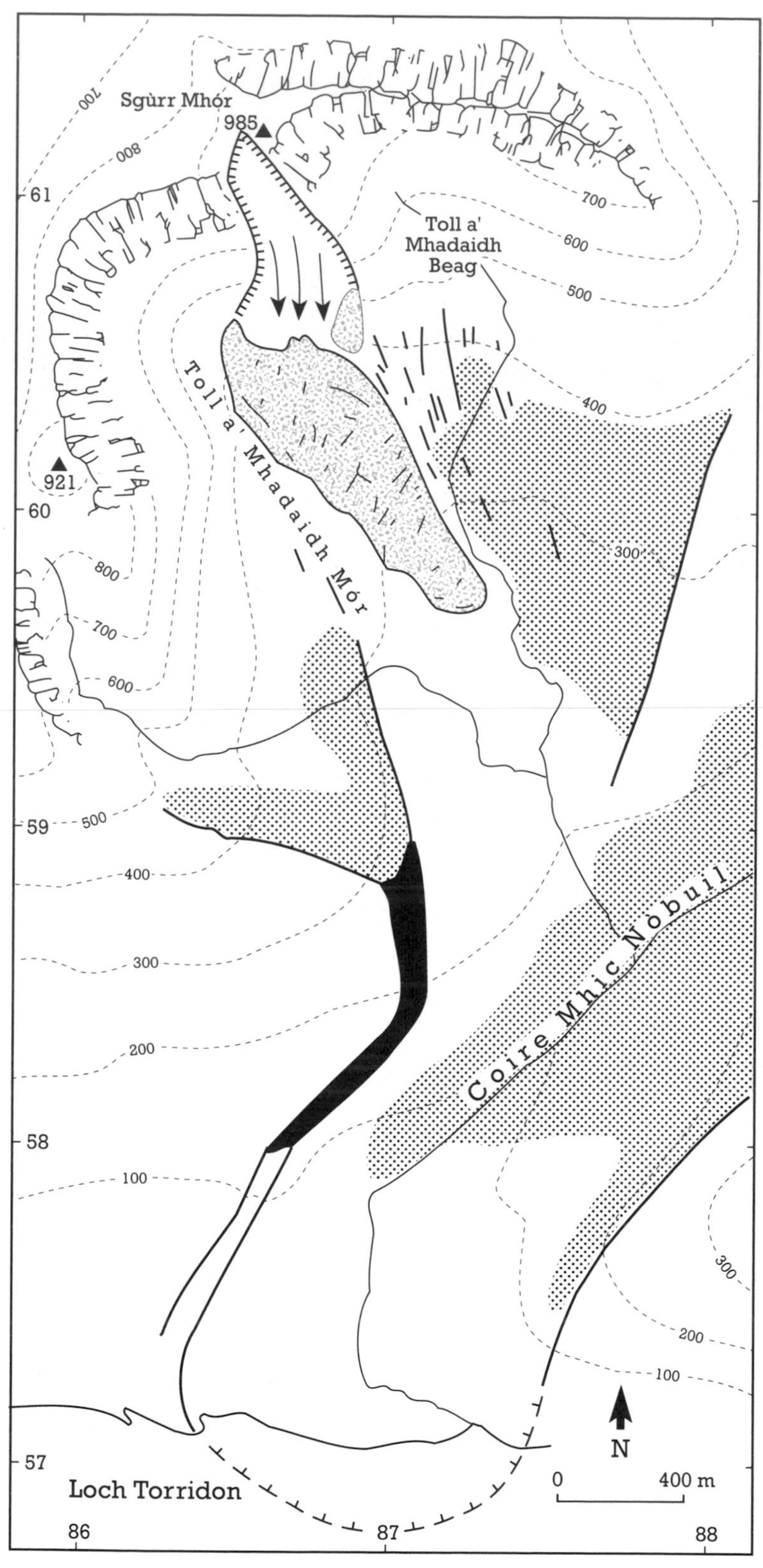

Sgùrr Mhór
985
Toll a' Mhadaidh Beag
Toll a' Mhadaidh Mór
921
Coire Mhic Nòbuil
Loch Torridon
N
0
400 m
700
600
500
400
300
200
100
800
61
60
59
58
57
86
87
88
Cliffs
Rock slope failure scar and track
Rock slope failure debris; lines indicate surface ridges
Moraine ridge
Fluted moraines
Hummocky moraines
Inferred glacier limit
▲921 Summit (metres)
100 Contour (metres)

Figure 6.12 Beinn Alligin, a Torridonian sandstone mountain in Wester Ross, rises above an ice-scoured surface of Lewisian gneiss to the west. A lateral moraine (centre left) marks the limit of a Loch Lomond Readvance glacier and is succeeded on the lower slopes by hummocky moraine; fluted drift can also be seen in the centre and to the right of the photograph. A large rock avalanche scar is prominent below the summit of Sgùrr Mhór. The resulting deposit on the corrie floor includes a low tongue of boulders extending beyond the main part of the deposit. (Cambridge University Collection: copyright reserved.)

Elsewhere in Scotland, fossil rock glaciers have been described from Beinn Shiantaidh on Jura (Dawson, 1977) and the Cairngorms (Sissons, 1979f; Ballantyne, 1984; Chattopadhyay, 1984). They are all smaller features, however, and differ in their morphology and inferred mode of formation, as they take the form of protalus lobes (see Wahrhaftig and Cox, 1959; Liestøl, 1961; Outcalt and Benedict, 1965; White, 1976; Lindner and Marks, 1985; Martin and Whalley, 1987; Whalley and Martin, 1992) that have apparently developed as a result of deformation of ice within rockfall talus accumulations. The only other features in Britain that may be of similar origin to that at Beinn Alligin are at Moelwyn Mawr in North Wales (Campbell and Bowen, 1989) and Beinn an Lochain in Argyll (Holmes, 1984; Maclean, 1991).

Figure 6.11 Geomorphology of Toll a'Mhadaidh, Beinn Alligin, showing rockslide debris and principal glacial landforms (from Sissons, 1977d).

Although good examples of the different types of Loch Lomond Readvance moraine represented at Beinn Alligin occur both individually and in various combinations at other sites in the Highlands, the Beinn Alligin landforms provide a particularly fine assemblage which enhances the overall geomorphological value of the site. Further, the medial moraine in Coire Mhic Nòbuil was described by Sissons (1977a) as the 'best individual example' in the area of the northern Highlands that he had mapped.

Conclusion

Beinn Alligin is noted for a large rockslide transported by a glacier. Although it has been interpreted as a fossil rock glacier, the largest such feature in Scotland, several lines of evidence suggest that it comprises a rockslide from the corrie headwall, which accumulated on the surface of a glacier during the intensely cold phase known as the Loch Lomond Stadial (about 11,000–10,000 years ago). The site is important in illustrating the geomorphological impact of high-magnitude slope processes during the stadial, and it also shows the problems of interpreting the origins of fossil landforms from morphological evidence alone. The wider landform assemblage also includes good examples of different types of moraines.

CNOC A'MHORAIRE

J. E. Gordon

Highlights

Cnoc a'Mhoraire is a representative example of a large end-moraine ridge formed during the Loch Lomond Stadial. Its importance is enhanced by the availability of indirect dating evidence from nearby deposits at Loch Droma.

Introduction

Cnoc a'Mhoraire (NH 284755) is a large end moraine located at the mouth of the Allt Lair valley where the latter enters the Dirrie More, one of the major glacial breaches cut through the watershed of the north-west Highlands (Linton, 1951a; Dury, 1953). It is important as one of the largest and most accessible end moraines associated with Loch Lomond Readvance glaciers in northern Scotland and has been described by Geikie (1901), Peach *et al.* (1913a), Kirk and Godwin (1963), Kirk *et al.* (1966), Sissons (1977a), Smith (1977), and Synge and Smith (1980). It is also significant through its proximity to the Lateglacial pollen site at Loch Droma (Kirk and Godwin, 1963).

Description

The end moraine is an impressive landform up to 25 m high, 200 m wide and 800 m long. It dams Loch a'Gharbhrain and was formed by a Loch Lomond Readvance valley glacier fed from sources in the Beinn Dearg massif to the north (Sissons, 1977a). The lateral limits of this glacier are demarcated along both sides of Loch a'Gharbhrain by lateral moraines, hummocky moraines and a notably fine drift limit (Sissons, 1977a).

Interpretation

In an early reference, Geikie (1901) mentioned the striking moraines of the Dirrie More, referring in particular to the terminal feature at the southern end of Loch a'Gharbhrain. Later, Peach *et al.* (1913a) also made special reference to the moraine, noting its great height and conspicuous rampart when viewed from the Garve–Ullapool road. They related its formation to their valley glacier stage following ice-sheet glaciation and an episode when glaciers were confluent. Surprisingly, Charlesworth (1956) did not specifically refer to the Cnoc a'Mhoraire moraine in his extensive exposition of glacial retreat stages in the Highlands. However, he placed the limit of his Stage M glaciation, later related by Donner (1957) to Pollen Zone III (Loch Lomond Stadial) of the Lateglacial, well to the east, in the vicinity of Contin.

The Cnoc a'Mhoraire moraine assumed considerable significance following the discovery of Lateglacial organic deposits nearby at Loch Droma (Kirk and Godwin, 1963). These deposits contained Lateglacial Interstadial pollen and provided a radiocarbon date of 12,810 ± 155 BP (Q–457). Together with the sedimentary record of the site, this evidence had profound implications for the deglaciation chronology of the area. It apparently demonstrated that no active ice had passed down the Dirrie More after about 12,800 BP and that deglaciation of the area had taken place much earlier than previously suspected. The features to the east mapped by Charlesworth as Stage M were therefore older than the Loch Lomond Stadial.

Although comparison of the pollen stratigraphy with other radiocarbon-dated sites in northern Scotland (for example, Cam Loch) suggests that the Loch Droma date may be as much as 1000 years too old (Sutherland, 1987c), the main conclusion of Kirk and Godwin (1963) remains valid. Of additional note is that, at Loch Droma, laminated sediments were deposited throughout the Lateglacial, suggesting that glaciers were present in the surrounding mountains during this time. A coarsening upwards sequence in the sediments deposited in the basin during the Loch Lomond Stadial would accord with ice readvance closer to the Loch Droma site (Sutherland, 1987c).

The evidence for local glaciation following decay of the ice-sheet is confined to the tributary glens of the Dirrie More, notably at Loch a'Gharbhrain and the northern side of the Fannich mountains (Kirk and Godwin, 1963; Kirk *et al.*, 1966). Kirk and Godwin (1963) and Kirk *et al.* (1966) termed this readvance 'the Gharbhrain stage', and the latter authors inferred it to be the local equivalent of the Zone III (Loch Lomond) Readvance. This has subsequently been supported by Sissons (1977a) following his systematic mapping of Loch Lomond Stadial glaciers in the northern Highlands and by Ballantyne *et al.* (1987) and Reed (1988). Smith (1977) and Synge and Smith (1980), however, have retained the local name of the Gharbhrain stage.

The Cnoc a'Mhoraire moraine is distinguished from the many others produced during the Loch Lomond Readvance in the northern Highlands (Sissons, 1977a) by its great size and relative ease of accessibility. The most comparable feature is probably in Strath Oykell, but it is partly afforested. In addition, Cnoc a'Mhoraire is significant for its close association with the Loch Droma site. Historically, the latter was a critical locality demonstrating early ice-sheet deglaciation of the northern Highlands and providing a maximum date for the Loch Lomond Readvance represented by the Cnoc a'Mhoraire moraine.

Conclusion

This site comprises a large end-moraine ridge formed by a glacier during the Loch Lomond Stadial (about 11,000–10,000 years ago). It is a well-known and representative example of its type. The age of the moraine has been inferred indirectly from a sequence of Lateglacial deposits nearby at Loch Droma.

COIRE A'CHEUD-CHNOIC

J. E. Gordon and D. G. Sutherland

Highlights

This outstanding geomorphological locality is noted for its assemblage of hummocky moraines formed during the Loch Lomond Stadial. These deposits provide important evidence for the processes of glacier activity and wastage.

Introduction

Coire a'Cheud-chnoic (Valley of a Hundred Hills) (NG 955550) in Glen Torridon, is a classic geomorphological locality displaying one of the best and most accessible examples of hummocky moraine in Scotland. The moraines were first mentioned by Geikie (1863a) in a footnote (p. 157) reporting 'a vast but unrecorded accumulation of glacier mounds blocking up a glen between Loch Torridon and Loch Maree', and a plate of the valley, first published in the Geological Survey memoir (Peach *et al.*, 1913b), has been reproduced in standard texts such as Wright (1937) and Sissons (1967a). Despite the clarity of development of the moraines little research was carried out on the area until recent years when Robinson (1977) mapped their distribution and Hodgson (1982, 1987) carried out sedimentological and morphological studies to ascertain the genesis of the features.

Description

Coire a'Cheud-chnoic drains into Glen Torridon from the south-east and is underlain by Torridonian sandstone and north-east aligned belts of Cambrian Quartzite. The greater part of the valley floor is covered by hummocky moraines which reach 8 m in height. Areas of bedrock crop out, implying that there is no great thickness of till under the moraines. Although they have a chaotic appearance, mapping of the moraines by Hodgson (1982, 1987) has shown that, in the lower valley, the majority of the moraines are elongated in a roughly north–south direction (in conformity with striations on the neighbouring bedrock surfaces) (Figure 6.13). Many of the moraines are asymmetric in long profile, with the highest part at the northern end, which gives them a conical appearance when viewed from Glen Torridon (Figure 6.14).

Constituent materials of the hummocky moraine are revealed in a recent roadside quarry at NG 954567. The section shows till at the bottom, consisting of angular clasts and boulders in a sandy and gritty matrix. Its upper surface is deformed, and an overturned fold partly encloses a large lense of poorly bedded sands and gravels. Above, layers of sands and gravels form the core of the upper section of the mound. The depression adjacent to it is underlain by a semicircular lens of angular clasts in a gravel and grit matrix, and is capped by a layer of poorly sorted sands and gravels. It is not possible to assess how representative this section is of the hummocky moraine as a whole, although Hodgson (1982, 1987) showed that the moraines elsewhere are composed of till and that there is little vertical variation in either particle size or erratic content. He found that the majority of the moraine mounds were elongated in a north–south direction and that the clast fabrics had a primary orientation mode parallel to the long axis of the moraines. Although quartzite underlies all the steep ground above the valley, it represents only a small proportion of the surface boulders on the moraines, implying that supraglacially derived debris from the valley sides was not a significant source of the debris. Instead, a significant portion of the erratic content of the moraines is material that was introduced into the area during the Late Devensian ice-sheet glaciation. The mapping of Robinson (1977, 1987a) and Sissons (1977a) has demonstrated that the moraines in Coire a'Cheud-chnoic were formed during the Loch Lomond Readvance.

Interpretation

Hodgson (1982, 1987) concluded that, since the moraines were dominantly composed of exotic material but had been formed by glaciers that were confined to the local valley system, then the moraines were the result of a Loch Lomond Readvance glacier overriding and partially deforming pre-existing debris into a crude type of fluted moraine. He considered that such an explanation could be applied more widely to hummocky moraine in the Highlands of Scotland, although he cited only one particular study, that of Donner and West (1955) on Skye (see the Cuillin), which apparently described similar moraines. Detailed studies elsewhere in Torridon supported his contention that the fluted moraine had formed by subglacial deformation of the sediments (Hodgson, 1982, 1986).

Hummocky moraine occurs widely in the Highlands, Inner Hebrides and parts of the Southern Uplands (see, for example, Geikie, 1863a; Geikie, 1901; Bailey *et al.*, 1924; Peacock, 1967; Sissons, 1967a, 1977a; Gray and Brooks, 1972; Sissons *et al.*, 1973; Bennett, 1991), but its best development in the British Isles is in north-west Scotland (Sissons, 1976b). The Torridon area contains some particularly notable examples, especially to the north of Beinn Eighe and Liathach (Sissons, 1977a; Hodgson, 1982). The moraines of Coire a'Cheud-chnoic are especially significant, both as a representative example of these landforms in the Torridon area and as an area on which detailed research has been carried out into their mode of formation. The locality is also well-known because of its use as a text-book illustration.

Despite considerable documentation of the location of hummocky moraine and some controversy over its implications for the reconstruction of former ice margins (see the Cairngorms), there have been few detailed studies of its origin until recently. Three main explanations have been proposed (see also Loch Skene and the Cuillin). First, it may be a type of chaotic, 'dead-ice' topography formed by rapid stagnation of ice that carried an extensive cover of supraglacial debris (Sissons, 1967a; Thompson, 1972). Second, it may be produced by controlled or uncontrolled deposition by actively retreating glaciers (Eyles, 1979, 1983; Day, 1983; Horsfield, 1983; Benn, 1990, 1991; Bennett, 1990, 1991; Bennett and Glasser, 1991) (see the Cairngorms). Third, as demonstrated by Hodgson (1982, 1987) for the Coire a'Cheud-chnoic features, it may be a subglacial deposit formed by deformation of pre-existing till (see also the Cuillin; Ballantyne, 1989a, Benn, 1991). Probably all three types exist, often in a single area (as in the Cuillin and at Loch Skene), but their relative importance is generally unknown. As the focus of a detailed

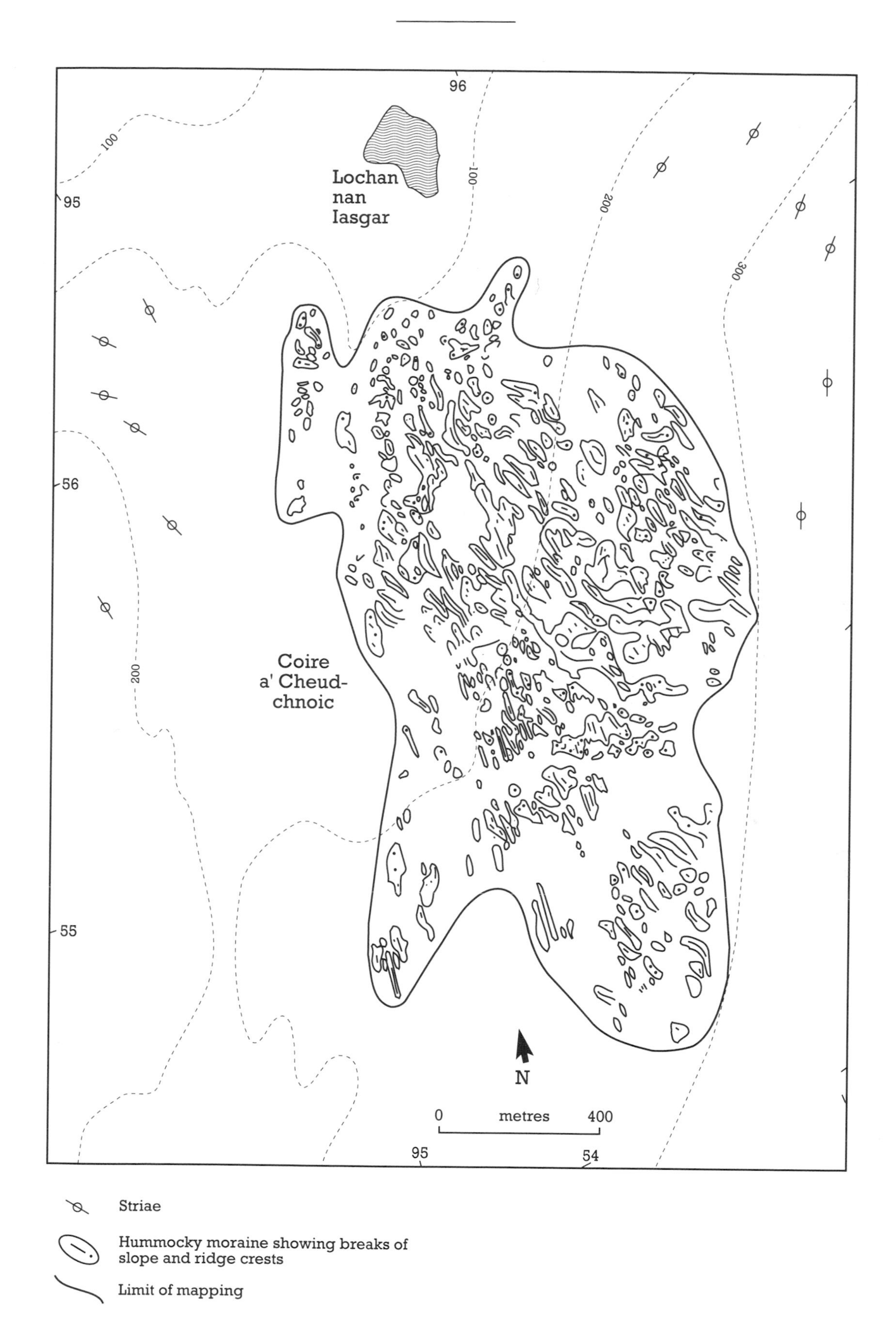

Figure 6.13 Geomorphology of Coire a'Cheud-chnoic (from Hodgson, 1987).

Figure 6.14 Hummocky moraine at Coire a'Cheud-chnoic in Glen Torridon. A clear drift limit on the valley side demarcates the Loch Lomond Readvance deposits below, from the ice-scoured bedrock slopes above. (British Geological Survey photograph D2737.)

case study, Coire a'Cheud-chnoic is therefore an important reference locality for wider comparative investigations of the genesis of the different landform types that appear to comprise hummocky moraine. The results of such work may have a significant bearing on interpreting styles of deglaciation and recognition of diagnostic landform and sediment facies assemblages (see Eyles, 1979, 1983; Sharp, 1985; Evans, 1989; Benn, 1990, 1991; Benn *et al.*, 1992; Bennett, 1991). The distribution patterns of hummocky moraine types, when compared with reconstructed glacier margins, flowlines and debris sources, should provide important information on sediment transfer controls and glacier processes during the Loch Lomond Readvance (cf. Benn, 1989b for end moraines). Further, if the subglacial origin of the Coire a'Cheud-chnoic deposits is confirmed, then the site can provide important field evidence for inputs to models of glacier dynamics for deformable bed materials (see Boulton and Jones, 1979; Alley *et al.*, 1986; Boulton and Hindmarsh, 1987).

Conclusion

Coire a'Cheud-chnoic is a classic geomorphological locality for an assemblage of hummocky moraines formed by a glacier during the Loch Lomond Stadial (about 11,000–10,000 years ago). It is one of the best locations in Scotland showing this type of glacial deposit, the form and distribution of which are important for interpreting the mode and pattern of ice wastage.

CORRIESHALLOCH GORGE

L. J. McEwen

Highlights

The gorge at Corrieshalloch is an excellent example of a gorge formed by glacial meltwaters; its form closely reflects the effects of bedrock controls.

Introduction

Corrieshalloch Gorge (NH 203782) occurs in the step of the valley profile formed where the Dirrie More glacial breach joins the Loch Broom glaciated trough. It is an impressive, steep-sided, slot gorge cut by glacial meltwaters. The gorge and its formation have been described by Peach *et al.* (1913a), Kirk *et al.* (1966), Whittow (1977), Ferguson (1981) and Sutherland (1987b).

Description

Corrieshalloch Gorge is exceptional in its length (*c.* 1.25 km), depth (60 m) and width (as narrow as 10 m at the lip). It is cut in undifferentiated Moine schists (psammite). Its form is determined by steeply dipping or vertical joints trending NW–SE and NE–SW (Peacock, unpublished data). The gorge can be subdivided into at least two parts, separated by the major waterfalls at Falls of Measach. In addition, there are several minor falls, for example, near Braemore junction. The modern channel is thus characterized by a stepped profile over these falls, which are interspaced with deep, boulder-filled pools. Ferguson (1981) observed that these are being progressively extended upstream by waterfall recession, at rates enhanced by the rock jointing.

Interpretation

The present form of Corrieshalloch Gorge can only be explained by a past period of extreme erosive activity. Although Whittow (1977) cites an earlier theory that the gorge reflects post-glacial entrenchment, he emphasizes that the scale of the feature makes this explanation unviable as a complete explanation. The currently favoured hypothesis is that the gorge was cut by meltwaters from the direction of Dirrie More (Peach *et al.*, 1913a; Kirk *et al.*, 1966; Sutherland, 1987b), although it is not established whether these were subglacial or proglacial. However, since the valley was probably a major meltwater discharge route during both ice-sheet deglaciation and the Loch Lomond Readvance, both may have contributed to the formation of the gorge (Sutherland, 1987b). Although there is no evidence that the gorge was ever re-occupied by ice after its incision, it would not be surpising if the gorge was a polycyclic feature, and that a significant channel was already cut into the valley shoulder at this locality prior to the onset of the last glaciation. Due to its glacial legacy, the present River Droma thus undergoes a dramatic change in controls (increased slope, increased confinement, and reduced sediment supply) as it enters the gorge.

Corrieshalloch Gorge is the most impressive of a number of gorges in this area (see, for example, the gorge along the lower Abhainn Cuileig), the form of which is determined by bedrock controls in the flaggy Moine schists (Peach *et al.*, 1913a). It is a classic landform and is recognized as the best-known example of a steep-sided slot gorge in Scotland. The gorge was cut by meltwaters during the last phases of glaciation of the area as well as, probably, during earlier glacial events. The scale of the feature is particularly impressive, and in terms of its size and the presence of waterfalls, it is a more striking and varied feature than the other gorges in the area and also the dramatic, but less well-known gorge at Black Rock of Novar in Easter Ross (Miller, 1887; Peach *et al.*, 1912).

Conclusion

Corrieshalloch Gorge is a classic example of a gorge formed by glacial meltwaters. It is notable for its length and depth and for showing, particularly well, the effects of geological controls on gorge formation. Its development probably reflects erosion by meltwaters during several glacial episodes.

CREAG NAN UAMH

T. J. Lawson

Highlights

Creag nan Uamh is of great importance for cave deposits and the fossil remains which they contain. Together these provide a record of environmental conditions and faunal changes dating back to the Ipswichian. The record of Late Devensian mammal faunas is particularly important, being the most detailed, diverse and best dated in Scotland.

Introduction

The white dolomite Creag nan Uamh ('Crag of the Caves') (NC 268170) in Sutherland possesses three main caves, known locally as the 'Bone Caves', and a number of niches and rock-shelters (Young, 1988). Situated 45 m above the normally dry stream bed of the Allt nan Uamh, the Bone Caves (Badger Cave, Reindeer Cave and Bone Cave, from west to east) have wide, semicircular entrances and large entrance chambers. Both Badger Cave and Reindeer Cave have inner chambers largely choked with fine-grained sediments; Bone Cave possibly also has one, but access is prevented by deposits in the back of the entrance chamber.

The caves at Creag nan Uamh are of considerable interest for Quaternary research because of the paucity of cave sites with datable sediments in Scotland as a whole, as well as their position so far north. They have yielded a Late Devensian and Holocene fauna that shows how the area acted as a refuge for sub-arctic species at the end of the last glaciation (Peach and Horne, 1893b, 1917; Callender *et al.*, 1927; Cree, 1927; Ritchie, 1928; Lawson, 1981b, 1983, 1984). The stratigraphy that has been reconstructed may extend back to before the Ipswichian (Lawson, 1981a), and recent research suggests that the caves are an important archaeological site (Lawson and Bonsall, 1986a, 1986b; Morrison and Bonsall, 1989).

Description

Bone Cave was first excavated by B. N. Peach and J. Horne in 1889 (Peach and Horne, 1893b, 1917). They described a six-layered stratigraphy and were of the opinion that the deposits related to the '... Lateglacial time, or at least to a period before the final disappearance of local glaciers in that region' (Peach and Horne, 1917, p. 327).

Badger Cave and Reindeer Cave were excavated in 1926 by J. G. Callender, J. E. Cree and J. Ritchie (Callender *et al.*, 1927; Cree, 1927; Ritchie, 1928; Lawson, 1981b). The results of work in Badger Cave were largely insignificant, but a more complicated stratigraphy in Reindeer Cave, together with rich faunal remains, prompted a reinvestigation of the deposits in Bone Cave in 1927 in order to correlate the stratigraphy of the two caves. However, this latter investigation uncovered largely undisturbed deposits with a stratigraphy somewhat different from that previously published, casting doubts on Peach and Horne's stratigraphic description (Lawson, 1981b, pp. 9 and 16).

The Creag nan Uamh caves are the truncated remains of large, high-level phreatic passages forming part of a cave system that once extended across the area now occupied by the Allt nan Uamh Valley (Lawson, 1983). Remnants of other abandoned phreatic passages occur at approximately the same altitude as the Bone Caves (330 m OD) in Allt nan Uamh Stream Cave and Uamh an Claonaite (entrances at NC 27461713 and NC 27091659, respectively), and these are possibly parts of the same former cave system. Uranium-series disequilibrium dating of a speleothem from Uamh an Claonaite gave an age of 122 ± 12 ka, indicating ice-free conditions at that time and placing the flowstone block in the Ipswichian (Lawson, 1981a). Three additional uranium-series dates of 181 +24/−18 ka, 143 +13/−16 ka and 192 +53/−39 ka, also from Uamh an Claonaite, indicate that part of the master cave system, at least, is considerably older (Lawson, 1983).

A recent survey of the caves showed that nearly all of the clastic deposits in the outer chambers were removed during the 1926 and 1927 excavations (Lawson, 1983). The only *in situ* deposits in outer Badger Cave are a series of yellow sands and silts occupying rock ledges on the western side. Much of the original silty sand in the inner chamber remains, overlain by breakdown slabs. Friable, red 'cave earth' comprising small dolomite splinters in a red or sandy-brown matrix containing numerous small faunal remains, especially amphibian bones, occupies ledges around the cave walls; in places it overlies or is indurated with flowstone.

No *in situ* deposits remain in the entrance chamber of Reindeer Cave or in the fissure or shaft connecting it to the inner chamber. The 1926 excavations uncovered a gravel, rich in erratic lithologies, deposited by a stream that had entered the cave by way of the side passage from Bone Cave and then plunged down the shaft into the inner cave. This was overlain by an angular dolomite-rich gravel containing an arctic mammal fauna, which was in turn overlain by red 'cave earth'. Deposits in the inner cave appear to have been largely untouched since the 1927 excavation. A number of different strata are present. The lowest layer visible is a silty clay containing breakdown clasts which have been intensively weathered to form areas of grey clay. A lens of dark-stained gravel, 0.08–0.15 m thick, with a sharp, irregular upper surface, separates this layer from reddish-yellow silts (0.33 m thick) containing a few fallen roof-stones. A concentration of these angular dolomite cobbles separates the latter from an overlying stratum of structureless, pale-yellow silty-sand, 0.45 m thick. This layer is devoid of breakdown material. In places where the intervening marker-horizon of breakdown fragments is absent, the reddish-yellow silts merge imperceptibly into the pale-yellow silty sand, suggesting that they are all part of a single depositional unit. The whole profile is capped by slabs of dolomite breakdown. Clarification of the sedimentary history of this inner chamber is complicated by the presence of a band of bedded deposits, coarsening upwards from laminated sand to rounded pebbles in a silty sand matrix, which cuts through the other deposits on the eastern side of the chamber. The relationship between this gravel layer and the one found in the outer chamber and shaft in 1926 in not known.

In Bone Cave few of the original deposits remain *in situ*. In the passage connecting it with Reindeer Cave a section through cyclically bedded stream gravels occurs. They are not as well sorted nor as rounded as the gravels just described in the inner Reindeer Cave. Foreset bedding on the leeward side of a rock bar indicates that they were deposited by water flowing from Bone Cave to Reindeer Cave. These gravels most probably represent those described in 1927 (Lawson, 1981b) as also forming the lower gravel stratum in the outer Reindeer Cave. The gravel is capped by a thin flowstone layer and overlies yellow-brown clayey sediments occupying pockets in the cave floor. A brown silty mud containing amphibian bones occurs in the fissure on the eastern side of the cave; it is unlike other 'cave earths' in these caves, and probably represents the uppermost layers of the clay underlying red 'cave earth' described here in 1927. A cemented breccia at the cave entrance cannot be related to the former sedimentary fill as no evidence of the adjoining layers has been preserved.

Interpretation

Figure 6.15 is an attempt to reconstruct the basic composite stratigraphy of the Creag nan Uamh caves. The relationship of the silts and sands to the 'cave earth' in the outer chamber of Badger Cave is entirely speculative; it assumes a correlation between these silts and sands and the yellow-brown silts of the inner cave, which is impossible to demonstrate in view of the disturbed nature of the latter. The silts of the inner recesses of Badger Cave also probably correlate with one or more of the fine-grained deposits in the inner Reindeer Cave; a smoke test in 1926 proved the two caves were linked (Lawson, 1981b), but no penetrable passages have yet been found.

Samples of sediment taken at the time of the 1926–7 excavations (and hitherto preserved in the Royal Museum of Scotland, Edinburgh), together with samples from the various lithostratigraphic units still preserved in the cave have been analysed (Lawson, 1983) in order to ascertain their provenance and mode of deposition. Five radiocarbon dates have been obtained on selected faunal remains from two different layers in Reindeer Cave (Figure 6.15) (Lawson, 1984). Table 6.1 gives a composite faunal list after analysis of the material excavated in 1926–7. A synthesis of all the various evidence currently available allows the following stratigraphy to be proposed for the Creag nan Uamh caves.

The earliest deposit presently visible is the silty clay containing weathered breakdown material in the inner Reindeer Cave, which may correlate with the fine deposits occupying pockets in the floor of the outer chamber of Reindeer Cave (silts and sands) and Bone Cave (silty clay). It may represent a 'wash' deposit, carried into the cave by groundwater percolating through the rock above. The intense weathering of this layer perhaps suggests that the deposit is the oldest in the caves. At least one stream channel extended over this material. The small grain size of this deposit implies low energy conditions, and the staining of the overlying gravel (possibly manganese) is similar to that which affects cave and peaty surface streams elsewhere in the area at present, suggesting an interglacial or interstadial origin for the chemical precipitation. The sharp, irregular contact with the overlying gravel layer reflects a break in sedimentation of unknown duration and subsequent erosion of these deposits.

The overlying reddish-yellow silt and sand containing fallen roof-stones, which gives way vertically to structureless, pale-yellow silty sand, again implies low-energy conditions. It is thought that these sediments are the finer fractions of the surficial glacial deposits washed into the caves by way of fissures in the dolomite (Lawson, 1983). Deposition to the roof of the inner Reindeer Cave and Badger Cave, that is to levels higher than the present cave entrances, requires those entrances to have been blocked and the caves to have been flooded. In the absence of material evidence of such barriers, it is presumed that the cave entrances were blocked by glacier ice. The faunal assemblage obtained from the surface of this stratum of silt and silty sand is dominantly arctic or sub-arctic in character (Table 6.1). It implies that the land surface around the caves was free of glacier ice, although the climate was still cold and the possibility of glacier ice existing elsewhere in the area cannot be excluded. Two individual reindeer antler fragments yielded radiocarbon dates of 25,360 +810/−740 BP (SRR–2103) and 24,590 +790/−720 (SRR–2104) BP, and give a minimum age for the fine-grained deposits in the inner Reindeer Cave (Lawson, 1984). These dates

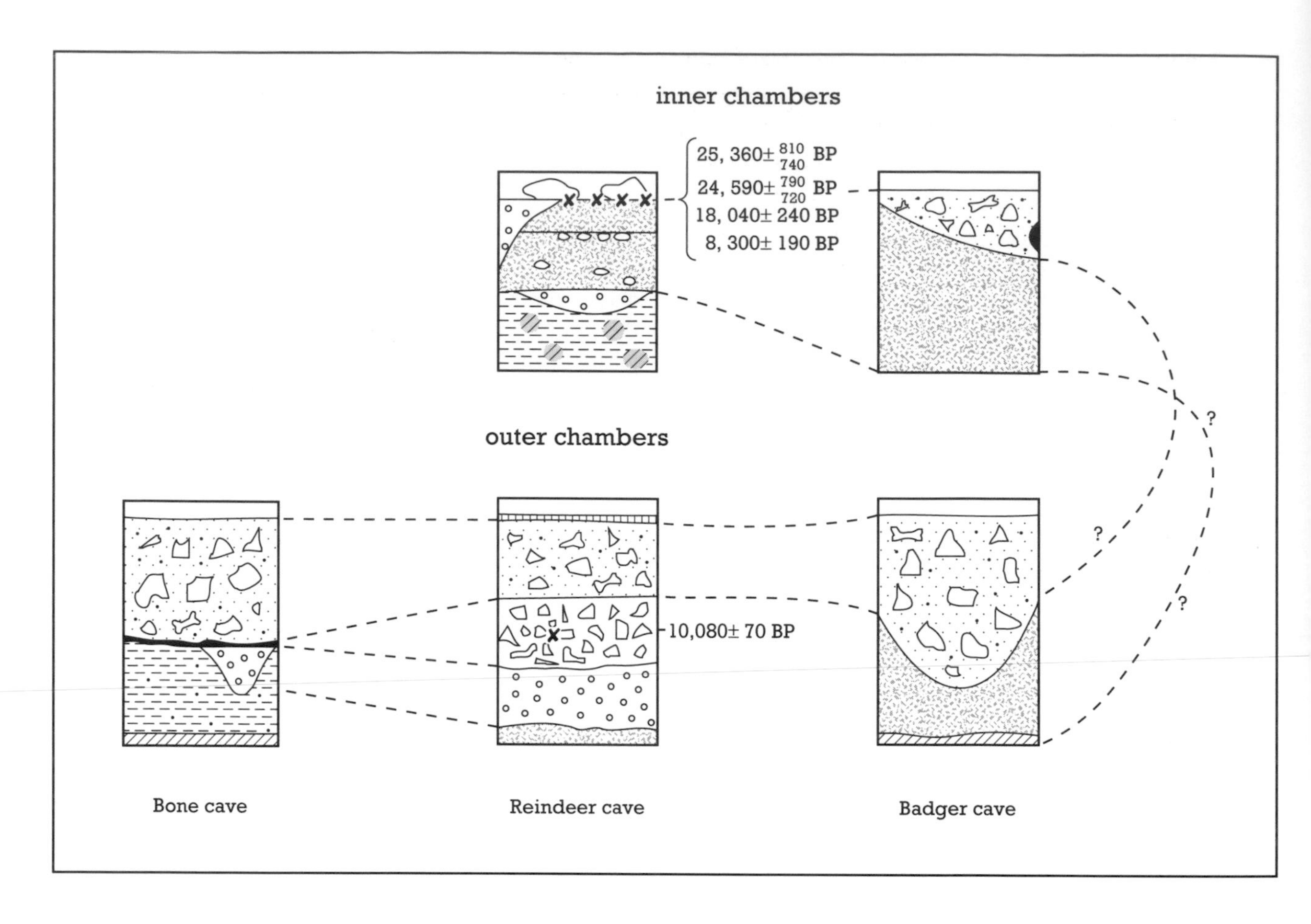

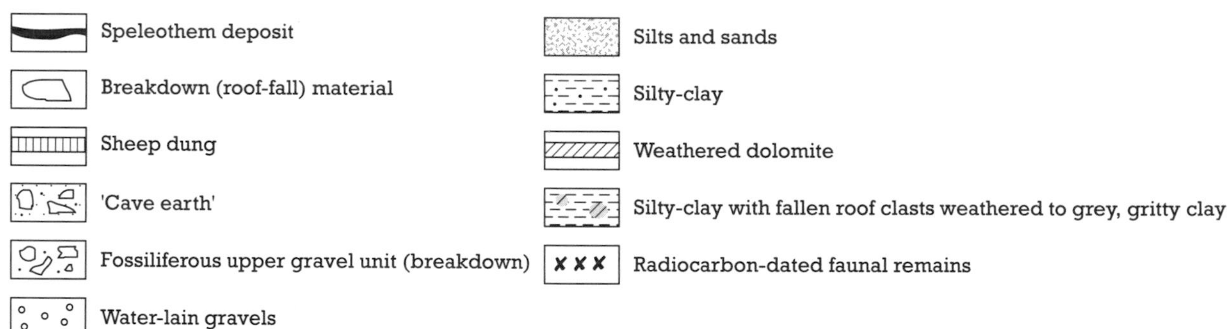

Figure 6.15 Diagrammatic reconstruction of the lithostratigraphy of the Creag nan Uamh caves, showing proposed relationships between certain of the layers (from Lawson, 1983).

correspond well with uranium-series disequilibrium dates on speleothem samples from other caves in the general area (Atkinson *et al.*, 1986), which suggests that this part of north-west Scotland experienced climatic conditions allowing groundwater recharge in a Middle Devensian interstadial approximately 38,000 to 26,000 BP, prior to the onset of full glacial conditions during the Late Devensian. A radiocarbon date of 8300 ± 190 BP (SRR–2105) on a leg bone of a juvenile reindeer from the same stratigraphic horizon shows that faunal material of a mixture of ages is present in the inner cave. This casts serious doubt on the reliability of a date of 18,040 ± 240 BP (SRR–1789) obtained from a bulked sample of antler fragments collected from this same level.

The stratigraphic relationships of the two main fluvially-deposited gravels are as yet incompletely understood. They are lithologically distinct, which

Table 6.1 Faunal assemblages from the Creag nan Uamh caves.

(a) Surface of fine-grained deposits in the inner Reindeer Cave		
Carnivora	*Ursus arctos* L.	Brown bear
	Alopex lagopus (L.)	Arctic fox
	Lynx lynx (L.)	Northern lynx
	Canis lupus L.	Wolf
Artiodactyla	*Rangifer tarandus* L.	Reindeer
Rodentia	*Dicrostonyx torquatus* (Pallas)	Arctic/collared lemming
	Microtus cf. *agrestis* L.	Field vole
	Microtus cf. *oeconomus* (Pallas)	Northern vole
	Arvicola terrestris L.	Water vole
	Apodemus sylvaticus (L.)	Wood mouse
(b) Dolomite-rich, upper gravel unit, outer Reindeer Cave		
Carnivora	*Ursus arctos* L.	Brown bear
Artiodactyla	*Rangifer tarandus* L.	Reindeer
Rodentia	*Dicrostonyx torquatus* (Pallas)	Arctic/collared lemming
	Microtus gregalis (Pallas)	Tundra vole
	Microtus sp.	Vole
(c) 'Cave earth' in Badger Cave (1), Reindeer Cave (2), Bone Cave (3)		
Insectivora	*Sorex araneus* L.	Common shrew (1)
	Sorex sp.	Shrew (2)
Primates	*Homo sapiens* L.	Man (1,2)
Carnivora	*Felis silvestris* Schreber	Wildcat (1)
	Mustela erminea L.	Stoat (3)
	Meles meles (L).	Badger (1)
	Canis lupus L.	Wolf (3)
	Vulpes vulpes (L).	Common fox (1,3)
	Ursus arctos L.	Brown bear (1,2,3)
Artiodactyla	*Sus* sp.	Pig (1)
	Cervus elaphus L.	Red deer (1,2)
	Capreolus capreolus (L.)	Roe deer (1)
	Rangifer tarandus L.	Reindeer (1,2,3)
	Ovis sp.	Sheep (1,2)
	Bos sp.	Ox (1,3)
Lagomorpha	*Oryctolagus cuniculus* (L.)	Rabbit (1,3)
	Lepus sp.	Hare (1)
Rodentia	*Arvicola terrestris* L.	Water vole (1)
	Microtus cf. agrestis L.	Field vole (1)
	Microtus oeconomus (Pallas)	Northern vole (1)
	Dicrostonyx torquatus (Pallas)	Arctic/collared lemming (2)

Continued overleaf

Table 6.1 *continued*

A variety of different unidentified bird species and fish, gastropods (including *Cepaea nemoralis* (L.) (2) and *Patella* sp. (2)) and amphibians (mainly frogs and toads) were also retrieved from this layer (c). E.T. Newton (in Peach and Horne, 1917) identified the following additional species from Bone Cave, presumably from the 'cave earth' stratum of the revised stratigraphy: weasel (*Mustela nivalis* L.), otter (*Lutra lutra* (L.)), rat vole (*Microtus ratticeps* (Keyserling and Blasius)), bank vole (*Clethrionomys glareolus* (Schreber)), ?chaffinch (*Fringilla coelebs* L.), barnacle goose (*Branta leucopsis* (Bechstein)), ?mute swan (*Cygnus olor* (Gm)), ?mallard (*Anas platyrhynchos* (L.)), teal (*Anas crecca* L.), wigeon (*Anas penelope* L.), tufted duck (*Aythya fuligula* (L.)), long-tailed duck (*Clangula hyemalis* (L.)), eider duck (*Somateria mollissima* (L.)), common scoter (*Melanitta nigra* (L.)), ptarmigan (*Lagopus mutus* (Montin)), red grouse (*Lagopus lagopus scoticus* (Latham)), golden plover (*Pluvialis apricaria* (L.)), grey plover (*Pluvialis squatarola* (L.)), little auk (*Alle alle* (L.)), puffin (*Fratercula arctica* (L.)), frog (*Rana temporaria* L.), toad (*Bufo bufo* L.), natterjack toad (*Bufo calamita* (Laurenti)), salmon or trout (*Salmo* sp.).

can be attributed to different transport paths prior to deposition. The erratic-rich gravel in the outer cave was traced by the previous excavators from the entrance of Bone Cave, through the connecting passageway to Reindeer Cave; the exposed section through this gravel in the connecting passageway attests to water flowing in that direction. Both the lack of faunal material within it and its situation in the cave are consistent with meltwater drainage from glacier ice occupying the Allt nan Uamh Valley. In contrast, the higher percentage of dolomite clasts and the greater degree of roundness of the pebbles in the gravel that cuts through the sediments on the eastern side of the inner Reindeer Cave, suggest a longer transport route through a cave system prior to deposition. Imbrication of the pebbles, the position of the gravel close to the steeply sloping cave roof and wall and the limited spread of this material over the yellow silty sand, suggest that it was deposited by a stream entering the chamber from below under hydrostatic pressure, and hence probably under phreatic conditions. These gravels are therefore also attributed to a glacial phase, and are post-dated by the faunal remains in the inner Reindeer Cave noted above.

The upper gravel unit found in the outer chamber of Reindeer Cave represents extensive frost shattering and breakdown of the cave roof and walls. Why similar layers were not found in either Badger Cave or Bone Cave is unknown. This layer yielded a faunal assemblage of distinctly 'northern' character (Table 6.1). The tundra vole (*Microtus gregalis* (Pallas)) is thought to have become extinct during the Lateglacial period (Bramwell, 1977; Stuart, 1982; A. Currant, unpublished data). A radiocarbon date of 10,080 ± 70 BP (SRR–1788) on reindeer antler fragments (Lawson, 1984), in accord with the cold-climate fauna, indicates a Loch Lomond Stadial age for the deposit. The caves lie outside the area affected by glacier ice at that time (Lawson, 1986). The faunal assemblage is peculiar in that over 830 shed reindeer antler burrs were excavated from the 0.5 m thick layer, but only two reindeer antler bones were unearthed. An analysis of the age and sex characteristics of the antler material suggests that Man may have been responsible for its introduction into the cave, which would make Reindeer Cave the oldest archaeological site yet found in Scotland (Lawson and Bonsall, 1986a, 1986b; Morrison and Bonsall, 1989).

'Cave earth' comprises the uppermost deposits in the entrance chambers of all the caves, and includes varying amounts of small breakdown flakes in a red-brown silty matrix containing abundant faunal material. These sediments are clearly of Holocene age, but they do not appear to be actively forming at present. The presence of a thin flowstone deposit low down in this layer probably reflects a temporarily moister climatic regime. The 'cave earth' contains species that are still present in Scotland today (Table 6.1). Other species no longer present (for example, brown bear, wolf, and reindeer) probably owe their extinction to Man rather than to the changing Holocene climate. The location of the Creag nan Uamh caves helps to explain the presence of other animals usually associated with more northern climes, as a region so far north is likely to have remained a refuge area for many arctic species trapped there as climate ameliorated. This may account for the presence of arctic lemming (*Dicrostonyx torquatus* (Pallas)). The presence of *Cepaea nemoralis* (L.) is interesting as this land snail is no longer found farther north than southern Skye (Kerney and Cameron, 1979, distribution map 272). The numerous frog bones in these deposits are due to the death of many of these amphibians whilst aestivating in pockets of mud.

A reassessment of the previous excavations in the caves, together with selective analysis of various *in situ* deposits, has allowed the construction of a lithostratigraphy covering a large part of the Late Quaternary. Radiocarbon dates indicate that the fine-grained sediment (silty clay, reddish-yellow silts and pale-yellow silty sand) of the inner Reindeer Cave, at least in part probably deposited beneath an ice-sheet, pre-dates a Middle Devensian interstadial; further work may show that the lowest stratum (silty clay with weathered clasts) dates from before the Ipswichian. Other stratigraphic units can be demonstrated to be of Lateglacial and Holocene age on sedimentological, geomorphological and faunal grounds. The faunal assemblages from these most recent deposits are unique in Scotland and are important in palaeobiological terms, showing how certain species with sub-arctic affinities survived into the Holocene in this remote part of northern Britain. The increasing evidence that the site is of considerable archaeological significance, being temporarily occupied by Late Upper Palaeolithic Man around 10,000 years ago, also adds to the importance of the Creag nan Uamh caves as a Quaternary site.

Conclusion

The unique assemblage of deposits and fossil animal remains preserved in the caves at Creag nan Uamh provides an important record of environmental conditions and changes in Scotland during at least the last 125,000 years. Particularly important is the detailed evidence for the range of sub-arctic mammalian species present in the area at the end of the last ice age (about 10,000 years ago).

SGÙRR MÓR

C. K. Ballantyne

Highlights

The example of solifluction terraces on Sgùrr Mór is one of the best in Scotland and one of the first to have been described.

Introduction

Sgùrr Mór (NH 204715), the highest summit (1110 m OD) in the Fannich Mountains, is important for one the best examples in Scotland of a suite of solifluction terraces. The landforms were first described by Peach *et al.* (1913a), who included a photograph showing 'plateau frost debris ... arranged in parallel lines or terraces due to soil creep aided by the movement of snow' on the ridge south-east of Sgùrr Mór. This is possibly the earliest reference to solifluction features on the mountains of Scotland. Subsequently these impressive features were used by Sissons (1967a, plate 23A) as an illustration of mass movement of frost-weathered debris. The most recent investigation of the site is that of Ballantyne (1981), later summarized in Ballantyne (1987e).

Description

Sgùrr Mór is composed of Moine pelitic schists that have weathered to produce a shallow, frost-susceptible regolith comprising slabby, angular clasts embedded in a matrix of silt and fine- to medium-grained sand ('Type 3' regolith of Ballantyne (1981, 1984)). On the south-east flank of the mountain, from 1050 m to 950 m OD, on slopes of around 13° there is an unbroken sequence of broad steps which range from 6.6 m to 12.7 m in width (downslope) and 33 m to 86 m in length (across slope). The junction between adjacent sheets is marked by a steep, vegetated 'riser', 0.5 m to 1.0 m high, interrupted in several places by the formation of small vegetation-covered solifluction lobes.

The vegetation cover of the 'treads' ranges from 0–70%, the bare areas being covered with slabs of local mica schist embedded in a deflation surface of coarse sand and granules. Excavation of one sheet revealed a typical solifluction deposit consisting of slabs up to 0.3 m across embedded in a dark-brown, structureless, sandy soil.

Interpretation

Ballantyne (1981) interpreted the features as stepped solifluction sheets (terraces) and considered that they were probably still active. Late Holocene solifluction activity on the Fannich Mountains was subsequently established by Ballantyne (1986c) at a site at 840 m altitude on the flank of Sgùrr nan Clach Geala, approximately 2 km to the west of Sgùrr Mór. Excavations at this site revealed an almost undisturbed soil horizon underlying a solifluction lobe. A series of radiocarbon dates indicated a very recent, very rapid advance of the lobe, which Ballantyne considered may have been due either to the Little Ice Age climatic deterioration in the 17th and 18th centuries AD or to vegetational disturbance due to overgrazing. Elsewhere in Scotland, Holocene solifluction has been established by radiocarbon dating of organic material from under or within solifluction lobes or sheets in the Cairngorms (Sugden, 1971) and Ben Arkle in Sutherland (Mottershead, 1978).

Frost-weathered regolith occurs widely on the mountain tops in Scotland and supports a range of solifluction deposits and landforms (see Ronas Hill, Ward Hill, An Teallach, Ben Wyvis and the Cairngorms). Sgùrr Mór is probably the most outstanding example of these landforms in Scotland, both for the clarity of the individual features and the extent of their development. The site is also of historical interest as being probably the first recorded example of solifluction features on Scottish mountains.

Conclusion

Sgùrr Mór is important for periglacial landforms developed on Moine schist. It demonstrates one of the best, and earliest described, examples of solifluction terraces in Scotland, formed by the slow downslope movement of frost-weathered debris. Some of the terraces may have been active during the last few hundred years.

CAM LOCH

H. J. B. Birks

Highlights

The sediments on the floor of Cam Loch contain a valuable record of environmental changes during the Lateglacial. This record has been studied in great detail using pollen, diatom and chemical methods, together with radiocarbon dating. Cam Loch is an important reference site for the Lateglacial in north-west Scotland.

Introduction

Cam Loch (NC 209135) is a large (2.6 km^2), irregularly shaped loch situated at 124 m OD just north of Elphin in west Sutherland. It has a large catchment (*c.* 87 km^2) with major inflowing and outflowing rivers. The sediments preserved on the floor of the loch are important for reconstructing the environmental history of the Lateglacial between about 13,000 and 10,000 BP. Evidence for rapid and marked climatic change during this interval is strongly represented in Scotland, both as landforms and as sedimentary, geochemical and biostratigraphical records. Cam Loch is one of the most intensively studied Lateglacial sites in Scotland and is of international importance because of the numerous interdisciplinary studies that have been made on its Lateglacial sediments by members of the Freshwater Biological Association (Pennington, 1975a, 1975b, 1975c, 1977a, 1977b; Pennington and Sackin, 1975; Haworth, 1976; Cranwell, 1977). It is the major reference site for the stratigraphy of the Lateglacial in north-west Scotland, and it provides critically important palaeoecological comparisons with sites elsewhere in the British Isles and in north-west Europe.

Description

The maximum water depth of Cam Loch is 37 m and the mean depth, 12 m. A sediment core has been obtained from a water depth of 10 m. The basal 0.65 m of sediment in this core comprise a succession of clay, silt and gyttja deposits (Figure 6.16) and were formed during the Lateglacial. These sediments have been intensively studied for their inorganic (Pennington and Sackin, 1975; Pennington, 1975a, 1977a) and organic geochemistry (Cranwell, 1977; Pennington, 1977a), their pollen stratigraphy (both as relative and absolute data) (Pennington, 1975a, 1975b, 1977a, 1977b; Pennington and Sackin, 1975) and their diatom stratigraphy (Haworth, 1976; Pennington, 1977a). As a result Cam Loch provides an extremely detailed environmental history for this time period. Seven radiocarbon dates (SRR–247 to SRR–253) have been obtained for consecutive sediment samples for the period 13,000–10,000 BP (Figure 6.16). The pollen record has been divided into seven local pollen assemblage zones (zones Ca–Ch) and three regional pollen assemblage subzones (subzones NWS A, B, C) described from the nearby sites of Loch Sionascaig and Loch Borralan. Correlation has been made with the Late Weichselian chronozones of

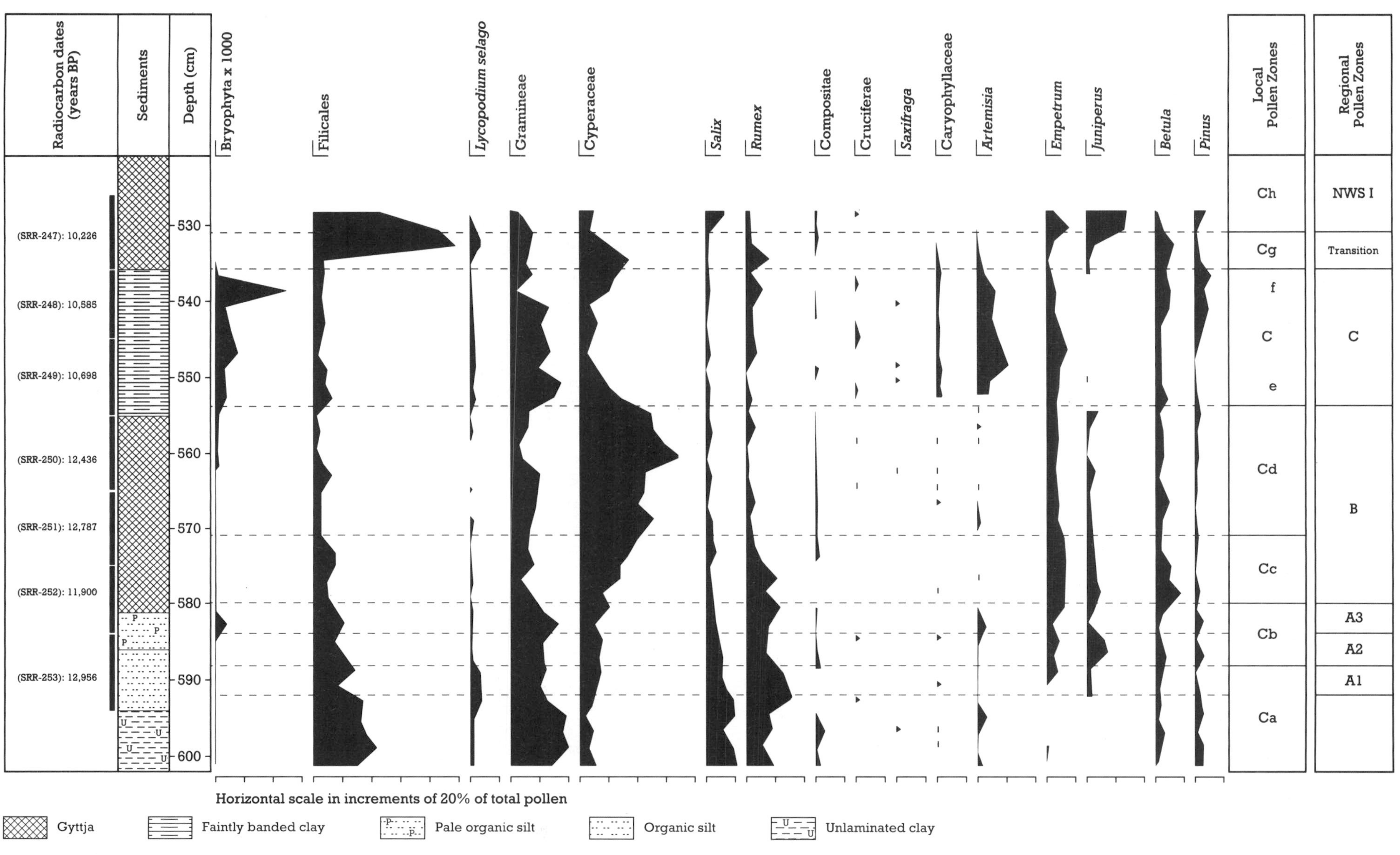

Figure 6.16 Cam Loch: relative pollen diagram, showing selected taxa as percentages of total pollen (from Pennington, 1977a).

Mangerud *et al.* (1974). The presence at Cam Loch of fossiliferous organic sediments that pre-date 13,000 BP suggests that the site and its surrounds were ice-free before that time.

Interpretation

The basal sediments (Figure 6.16) are barren clays and silts with high sodium, potassium, calcium, and magnesium values, indicating erosion and inwash of unleached skeletal mineral soils. Immediately above these sediments there are pollen assemblages deposited prior to 13,000 BP (zone Ca) that are characterized by *Salix* (including *S. herbacea*), *Rumex acetosa* and *Oxyria digyna* pollen, and by *Lycopodium selago* spores. The vegetation was probably a mosaic of very open, pioneer vegetation with *Rumex acetosa*, *Oxyria digyna* and *Lycopodium selago*, and with extensive areas of prolonged snow-lie dominated by *Salix herbacea* (Pennington, 1980). The climate was probably oceanic with much snow and comparatively little periglacial erosion because of the extensive snow cover. At about 13,000 BP there was an increase in the carbon, nitrogen, and iodine contents of the sediments and the calcium and sodium contents fell, followed by a decrease in the potassium and magnesium values. These chemical changes suggest some stabilization of the landscape and the onset of soil leaching, humus accumulation, and chemical weathering. The pollen assemblages (lower part of zone Cb) are characterized by a rise, both in percentages and absolute values, of *Juniperus communis* and *Empetrum nigrum* pollen, suggesting that the vegetation was a dwarf-shrub heath, possibly analogous in physiognomy to the shrub tundra of coastal south-west Greenland today (Pennington, 1977a). This phase is correlated with the Bølling Chronozone. Between 12,000 and 11,800 BP there was a temporary fall in the values of *Juniperus communis* pollen and a rise in the values of *Artemisia* pollen (upper part of zone Cb). Chemical changes suggest that this was a phase of soil disturbance and erosion, and the diatom flora includes inwashed terrestrial diatoms. This short-lived phase is correlated, on the basis of the radiocarbon dates, with the Older Dryas Chronozone of Fennoscandia. The phase probably reflects a climatic recession of comparatively small magnitude. Interestingly, the geochemical data suggest that although some soil erosion occurred, no periglacial erosion of mineral soils is detectable, such as occurred prior to 13,000 BP and between 11,000 and 10,400 BP.

In the sediments deposited from 11,800 to 11,000 BP (correlated with the Allerød Interstadial and Chronozone) pollen of *Empetrum*, Cyperaceae and *Betula* attain their highest values (zones Cc and Cd). The absolute *Betula* values do not, however, indicate that tree birches grew near the site at this time. The carbon content of the sediments attains its highest Lateglacial values during this period. The organic geochemistry indicates that much of the organic component of the Allerød Chronozone sediments was derived from *in situ* lake biota and lake productivity, suggesting a high trophic status at this time. The reduced state of the iron and manganese in the sediments presumably resulted from oxygen depletion in the hypolimnion, a characteristic of productive eutrophic lakes today. The diatom assemblages similarly indicate that the loch was productive at this time. The soils within the catchment appear, however, to have become leached during the interstadial, presumably because of high rainfall and the generally calcifuge nature of the dwarf-shrub and sedge-dominated vegetation. Soil erosion began during the interstadial, probably as a result of climatic deterioration. It resulted in the redeposition of organic material and some apparent anomalies in the radiocarbon dates.

Sediments deposited between 11,000 and 10,400 BP contain pollen assemblages that are characterized by high values of *Artemisia* (including *A. norvegica*), Cruciferae, Compositae, Caryophyllaceae and Gramineae pollen and of *Lycopodium selago* spores (zones Ce and Cf). The assemblages date from the first part of the Younger Dryas Stadial and Chronozone, the time during which small corrie glaciers became re-established on the mountains of north-west Scotland (Sissons, 1977a). The sediments are minerogenic and reflect the onset of intense periglacial erosion of mineral soils by frost disturbance. They resemble the lowermost sediments in their high calcium, magnesium, potassium and sodium content. Terrestrial diatom species are frequent, again indicating extensive erosion and inwashing of terrestrial material. The vegetation was probably predominantly open and species-rich, with abundant sedges and grasses, *Artemisia*, and other herbs (Pennington, 1980). The climate was probably more continental and with less snowfall than the climate pre-13,000 BP.

This phase is part of the Loch Lomond Stadial and reflects a period of very marked climatic deterioration, with extensive erosion and inwashing of mineral soils.

At about 10,400 BP the sediments become more organic, the carbon content sharply rises to values higher than at any time in the Lateglacial, and elements such as sodium, potassium, calcium and magnesium decrease equally rapidly. These chemical changes indicate humus accumulation, soil stability, and high lake productivity. The pollen assemblages are dominated by Cyperaceae and *Rumex* pollen (zone Cg) and then by *Juniperus communis* and *Empetrum* pollen (zone Ch). The pollen stratigraphy clearly reflects the rapid vegetational changes that occurred at the end of the Lateglacial and the beginning of the Holocene.

The numerous geochemical and palaeontological variables that were studied in the Cam Loch sediments all indicate that a major and rapid climatic amelioration occurred at about 13,000 BP; a minor climatic recession occurred between 12,000 and 11,800 BP, and a major climatic deterioration occurred between 11,000 and 10,400 BP. These phases are correlated by Pennington (1975b, 1977a, 1977b) with the Bølling, Older Dryas, and Younger Dryas Chronozones, respectively, of Fennoscandia.

Cam Loch is of national importance because of its detailed and diverse stratigraphical record of the environmental history of the Lateglacial. It is one of the most intensively studied sites of Lateglacial age in Scotland, with detailed and integrated pollen, diatom and inorganic and organic chemical analyses, thereby providing valuable independent evidence for environmental change during the Lateglacial. Its palaeoclimate record shows marked differences with sites elsewhere in Britain (for example, Pennington, 1977b; Walker, 1984b) and mainland Europe (for example, Watts, 1980). Such spatial differences even within north-west Scotland (Birks, 1984) indicate the complex vegetational and hence environmental patterns that may have existed during the Lateglacial, especially in extreme, marginal areas. Cam Loch provides convincing evidence from various data sources for a minor climatic recession between about 12,000 and 11,800 BP. Increasing interest is now focused on such short-lived, often rather abrupt, climatic recessions during the Lateglacial (Tipping, 1991b). Because of its wealth of palaeoecological and palaeolimnological data from the Lateglacial, Cam Loch is an important reference site for studies of Late Quaternary vegetational and climatic history.

Conclusion

The sediments at Cam Loch are important for reconstructing the environmental history of north-west Scotland from about 13,000 to 10,000 years ago, during the Lateglacial. Pollen, diatom and chemical analyses, together with radiocarbon dating, show that the climatic amelioration at the end of the last glaciation was twice interrupted by colder phases, with corresponding changes in vegetation patterns and soil stability. The first interruption was a relatively short-lived event; the second corresponds with the Loch Lomond Stadial (about 11,000 to 10,000 years ago) and brought about a change from a stable ground cover of dwarf-shrub vegetation to unstable soils with open-habitat sedges, grasses and herbs. As one of the most intensively studied Lateglacial sites in north-west Scotland, Cam Loch is a key reference locality for this area.

LOCH SIONASCAIG

H. J. B. Birks

Highlights

The sediments on the floor of Loch Sionascaig and in a bog on Eilean Mór provide detailed pollen records, supported by radiocarbon dating, of vegetational changes in Scotland during the Lateglacial and Holocene. The Holocene record is particularly important for the environmental changes it demonstrates.

Introduction

Loch Sionascaig (NC 120140) is a large (6.1 km^2), deep (maximum depth 184 m, mean depth 70 m), oligotrophic (alkalinity 6 ppm CaC0, pH 6.4–6.6) and nutrient-poor loch within the Inverpolly National Nature Reserve, west of Elphin in Sutherland. The reserve is primarily moorland and blanket-bog developed over Lewisian gneiss, interspaced by numerous small lochs and wet and soligenous valley mires, with Loch Sionascaig and its wooded islands dominating the central part. The area is mainly treeless, although small areas of birch wood persist locally on steep, rocky slopes and on islands in the larger lochs. It is 'a good example of this kind of submontane moorland and wetland complex so characteristic of west Ross and west Sutherland' (Ratcliffe, 1977).

Detailed palaeoecological and palaeolimnological studies of the sediments of Loch Sionascaig (Pennington *et al.*, 1972) and of a small bog (0.42 m deep) (NC 120139) on Eilean Mór (Kerslake, 1982), its largest woodland island, provide important evidence on the Lateglacial and Holocene ecological history of this internationally biologically important and unique landscape.

Description

The shore of Loch Sionascaig is mainly bare Lewisian gneiss and sediments are absent over much of the loch floor. In the southern arm of the loch, where the bedrock is Torridonian sandstone, up to 5.5 m of sediments occur overlying glacial clays and sands. These comprise a succession of clay, silt, sand and gyttja deposits (Figure 6.17).

Pollen analyses of the Sionascaig sediments (Pennington *et al.*, 1972) show that the sequence represents a complete Lateglacial and Holocene record. This has been divided into three Lateglacial regional pollen assemblage zones (A, B, C) and six Holocene regional pollen assemblage zones (NWS I–NWS VI) (Figure 6.17). Seven radiocarbon dates (SRR–12 to SRR–15 and Y–2362 to Y–2364) (Figure 6.17) indicate a remarkably constant sediment accumulation rate for the Holocene.

Interpretation

The Lateglacial pollen stratigraphy is typical for north-west Scotland, with a pre-Lateglacial Interstadial assemblage dominated by *Rumex*, grasses and sedges, including a wide variety of other herbs characteristic of open, pioneer vegetation. During the interstadial closed-heath vegetation with *Empetrum* and *Juniperus communis*, developed on humus-rich acid soils. The overlying Loch Lomond Stadial deposits are characterized by increased values of *Artemisia* (including *A. norvegica*), Cruciferae, and Caryophyllaceae, suggesting open disturbed soils.

The boundary between the Devensian and the Holocene is marked by a series of rapid changes in the pollen record, with successive peaks of *Rumex*, *Lycopodium selago*, Gramineae, *Empetrum* and *Juniperus*, representing the characteristic vegetational succession from stadial to interglacial conditions. At about 9500 BP *Betula* woodland, with some *Corylus avellana* and abundant ferns, developed. *Pinus sylvestris* expanded at about 8000 BP, early by comparison with England and elsewhere in Scotland, with the exception of the Loch Maree area (Birks, 1972b; Birks, 1989). Open pine-dominated woodland, with abundant *Pteridium aquilinum* and some *Betula* and *Calluna vulgaris*, was the major regional vegetation from about 7500 BP to 4500 BP (Pennington, 1986). Mires were locally present in the catchment, presumably in waterlogged areas. *Alnus glutinosa* expanded locally about 6000 BP, but, as elsewhere in the north-west Highlands, it was never an important forest component (Bennett and Birks, 1990). *Ulmus* and *Quercus* were also present, but in low numbers throughout the Holocene. At about 4500 BP pine underwent a major decline here, as elsewhere in north-west Scotland (Birks, 1972b, 1975; Birks, 1977, 1989; Bennett, 1984; Gear and Huntley, 1991), probably as a result of rapid climatic change with a shift towards wetter and windier 'oceanic' conditions. Such a shift would, on the leached, acid soils of the area, inhibit the natural regeneration of pine and lead to the widespread development of blanket-bog on flat and gently sloping areas. The decline of *Pinus* pollen in the sediments at Loch Sionascaig is accompanied by rises of Cyperaceae, Gramineae and *Calluna* pollen and of *Sphagnum* spores, all suggestive of peat development and the spread of blanket-mires. *Pinus sylvestris* died out locally sometime in the last 4000 years, as it did in much of north-west Scotland (Birks, 1989). Its former widespread occurrence is strikingly evidenced by the abundant pine stumps within the blanket-peats of the Inverpolly area.

Sediment–chemical analyses provide independent evidence for soil changes within the Sionascaig catchment (Pennington *et al.*, 1972). There is a close correspondence between the pollen-record and sediment–chemical changes, suggesting important vegetation–soil relationships over the last 13,000 years (Pennington, 1986). During the Lateglacial there was progressive soil development until the later part of the interstadial with humus accumulation, nutrient leaching and chemical weathering leading to clay-mineral formation. Soils became less organic and more nutrient-rich

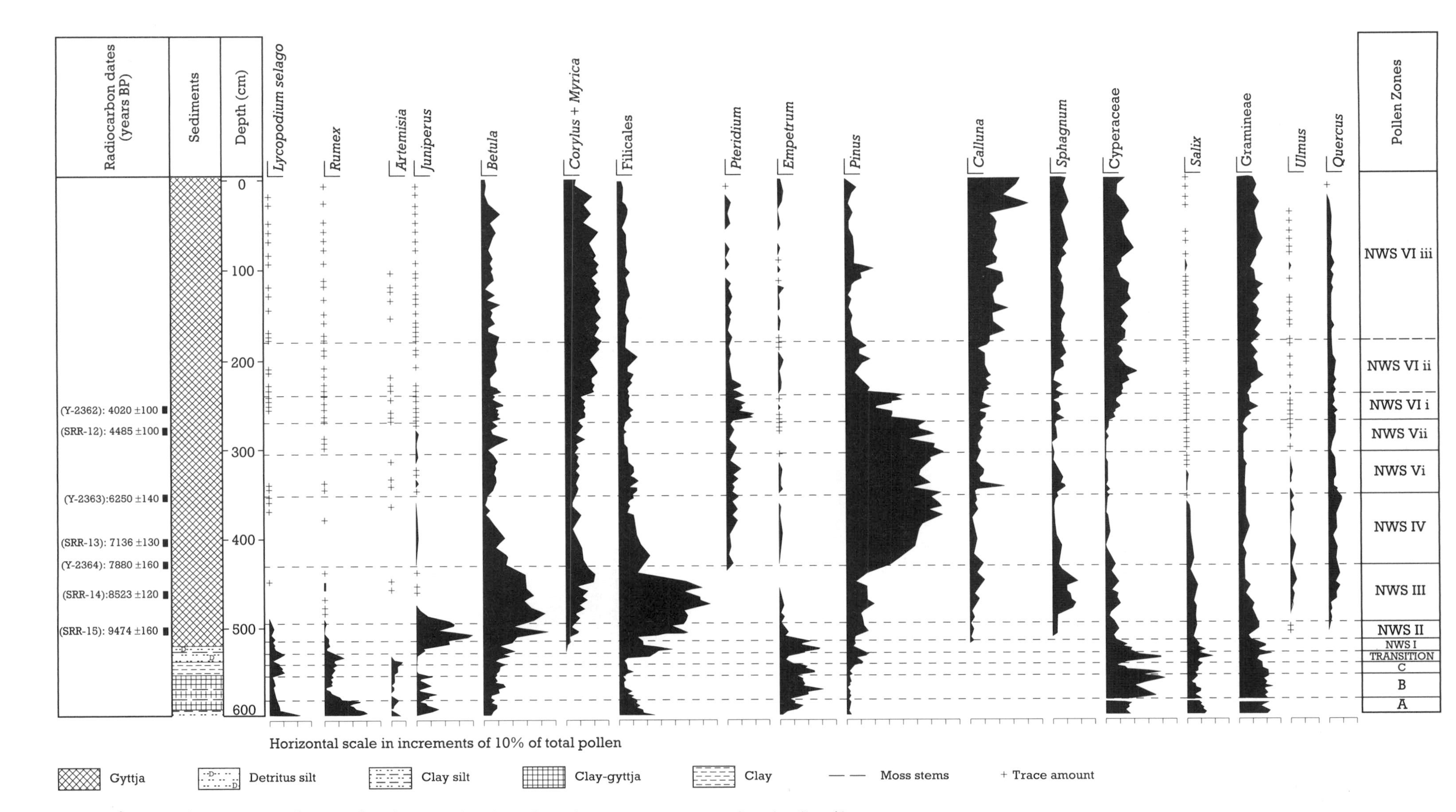

Figure 6.17 Loch Sionascaig: relative pollen diagram, showing selected taxa as percentages of total pollen (from Pennington *et al.*, 1972).

in the Loch Lomond Stadial. However, the soils were quickly leached and were forming acid humus before the expansion of birch about 9500 BP, as shown by electron spin-resonance studies of the humic acids preserved in the early Holocene sediments.

Chemical analyses suggest that between 8000 and 6000 BP the soils were acid but well-drained. At about 6000 BP the soils appear to have become increasingly waterlogged, as evidenced by changes in the iron and manganese contents of the sediments. By 5000 BP blanket peat was common, and by 4000 BP it was widespread.

Diatom analyses provide data about the loch's water chemistry and nutrient status (Haworth, 1976). The Lateglacial diatom flora is unusual within the British Isles in that it contains numerous planktonic taxa (Pennington *et al.*, 1972). The pre-interstadial assemblage suggests a high base status, with many alkaliphilous taxa, including some species that occur near glaciers today. In the interstadial there was a decline in these alkaliphilous taxa and an increase in taxa characteristic of acid water, suggesting a decline in lake nutrient status. Loch Lomond Stadial sediments contain terrestrial diatoms, suggesting inwashing and erosion of terrestrial soils. At the Devensian–Holocene boundary, the diatoms indicate a return to nutrient-rich conditions, presumably as a result of soil rejuvenation during the stadial. During the early Holocene there is a marked change in the diatoms, indicating a major and rapid reduction in the base status of the loch. By the time of the expansion of pine at about 8000 BP, the loch was acid, with very few alkalibiontic and alkaliphilous taxa, and large numbers of acidophilous and acidobiontic taxa. This trend towards loch acidification stabilized at about 4500 BP and the loch remained weakly acid until the present day. There is no evidence that the major decline of pine at about this time had any effects on the loch's water or sediment chemistry. There is also no evidence for recent lake acidification in Loch Sionascaig (Atkinson and Haworth, 1990), suggesting that it is still a 'pristine' lake ecosystem.

Pollen assemblages from a large loch such as Sionascaig are derived from a large source area, the size of which is not known. They provide a record of regional vegetational history. In contrast, pollen spectra from small lakes and bogs (<100 m radius) reflect a local, spatially restricted vegetational history. Kerslake's (1982) pollen diagram based on samples from a small bog on Eilean Mór provides a record of local vegetational history on the largest island in Loch Sionascaig. The basal sediments from the bog have yielded a radiocarbon date of 7400 BP. A series of 11 dates provides an internally consistent chronology for the sequence. The pollen record is of particular interest in that it suggests that Eilean Mór has been dominated by *Betula* with some *Corylus*, *Salix* and, from 6000 BP, *Alnus* for much of the Holocene. *Pinus* has never been a major component of the island's vegetation. From about 2500 BP alder disappeared and a series of alternating phases of birch dominance and *Calluna* dominance occurred, suggesting a natural alternation between birch woodland and heathland. The Eilean Mór profile is so strikingly different from the Loch Sionascaig pollen diagram and from a diagram from Lochan Dubh to the west of Sionascaig (Kerslake, 1982), that it is clear that there may have been major vegetational mosaics within the Inverpolly area through much of the Holocene.

Loch Sionascaig is one of the most intensively studied Lateglacial and Holocene sites in Scotland, with pollen analysis, diatom analysis, sediment geochemistry, local studies of peat stratigraphy and radiocarbon dating. These various lines of evidence provide important insights into the ecological history of this extreme oceanic landscape. The palaeoecological record shows very clearly the rapidity of soil leaching and accumulation of acid humus in the earliest Holocene, a feature of the extreme oceanic situation of the site. It illustrates the delicate balance between open pine woodland and blanket bog and how this balance was changed by rapid and abrupt climatic change about 4200 BP. It also shows the close interrelationship between vegetation, soils and sediment chemistry in a landscape where human disturbances have been relatively minimal compared with the extreme dominance of climate and geology. Its sedimentary record provides clear and unambiguous evidence of how the remarkable blanket-bog and loch landscape of north-west Scotland came into being. As blanket bog is better developed and more abundant in the British Isles than elsewhere in Europe, the environmental history of Loch Sionascaig and its catchment provides a detailed historical background for understanding the history and development of a landscape dominated by blanket-bog. It also demonstrates that the loch has not been affected by recent increased atmospheric acidification and is thus still in a 'pristine' condition; such lochs are becoming fewer in northern and western Britain. Loch Sionascaig is thus a site of great palaeoecological and palaeolimnological importance.

Conclusion

Loch Sionascaig is a reference site for reconstructing the environmental history of north-west Scotland, during approximately the last 13,000 years, that is, in Lateglacial and Holocene times. Particularly important is the intensively studied Holocene record contained in the loch bed sediments and in a peat bog on Eilean Mór. The demise of pine around 4200 years ago, following rapid climatic change to wetter and windier conditions, and the subsequent development of blanket bog are clearly demonstrated.

LOCHAN AN DRUIM

H. J. B. Birks

Highlights

The pollen and plant macro-fossils preserved in the sediments which infill the basin at Lochan an Druim provide an important record of vegetational history and environmental changes during the Lateglacial and Holocene in the extreme north-west of Scotland.

Introduction

Lochan an Druim (NC 435568) is a small lochan at an altitude of 25 m OD in the shallow valley between the A838 road and the ridge of An Druim that runs northwards from Eriboll. It contains an important Lateglacial vegetational sequence and it has a Holocene vegetational history that is unique to the extreme north-west Scottish Highlands (Birks, 1977, 1980).

The local bedrock is Durness Limestone and, where it crops out to the east of the lochan, there are botanically interesting *Dryas octopetala* heaths, with associated arctic–alpine plants such as *Carex capillaris*, *C. rupestris, Polygonum viviparum* and *Saxifraga aizoides*. A pollen diagram for the site was included in Birks (1980), and Birks (1984) gives full details of the site and its environmental setting.

Description

An 8.9 m long core collected from the west side of the lochan showed a succession of silty sand, silty mud, clay and detritus mud sediments (Figure 6.18). Eleven radiocarbon dates (SRR–776 to SRR–785, SRR–866) provide a chronology for the observed pollen stratigraphy (Figure 6.18), although some or all of the dates may be subject to hard-water errors owing to the incorporation of ^{14}C-deficient carbon from the surrounding limestone and calcareous drift. The vegetational history of the site has been studied in detail by Birks (1984 and unpublished data; pollen diagram in Birks, 1980).

Interpretation

The Lateglacial part of the sequence (7.85–8.91 m) has been studied by Birks (1984) for both its pollen and spore assemblages and its plant macrofossils. The sequence begins at about 12,500 BP, implying that deglaciation of the area had occurred by that time. An open, pioneer vegetation of grasses and sedges with *Salix*, *Empetrum nigrum*, *Dryas octopetala* and *Saxifraga* spp. was initially present. During the Lateglacial Interstadial a vegetation dominated by dwarf-shrubs, such as *Empetrum nigrum*, *Dryas octopetala* and *Salix herbacea*, with some juniper was widespread. Leaves of the obligate snow-bed moss *Polytrichum sexangulare* are present, indicating that areas of late snow-lie persisted at low altitudes near the lochan, even in interstadial times.

Inwashing of mineral material occurred during the Loch Lomond Stadial, suggesting unstable soils and discontinuous vegetational cover. The vegetation contained an abundance of open ground, arctic–alpine herbs, such as *Artemisia norvegica*, *Astragalus alpinus*, *Arenaria norvegica*, *Cherleria sedoides*, *Minuartia rubella*, *Saxifraga oppositifolia* and *S. cespitosa*. This assemblage is suggestive of present-day, high-arctic vegetation, with scattered plants forming a sparse cover on open mineral soils, and with long-lasting snow beds and associated meltwater runnels. At this time small corrie glaciers reformed on the higher mountains in north-west Scotland (Sissons, 1977a).

The opening of the Holocene is clearly marked by the expansion initially of dwarf-shrub heaths, dominated by *Empetrum* and *Juniperus*, followed

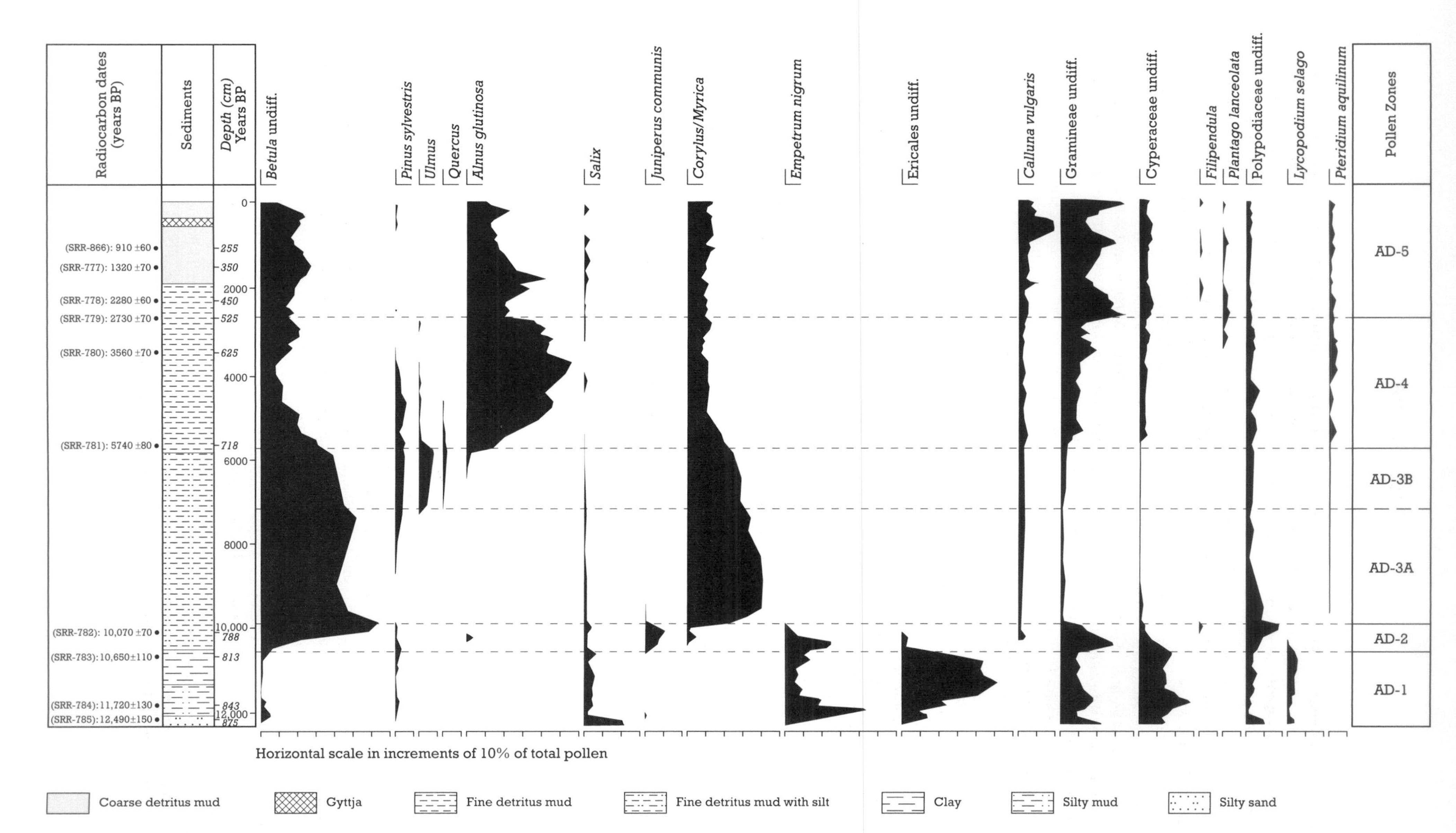

Figure 6.18 Lochan an Druim: relative pollen diagram, showing selected taxa as percentages of total pollen (from Birks, 1980). Note that the data are plotted against a radiocarbon time-scale.

by the development of open birch woods with *Populus tremula* and *Salix* spp. *Corylus avellana* expanded rapidly at about 9800 BP along with *Sorbus aucuparia*. By 9500 BP the landscape around the lochan would have been a mosaic of birch and hazel woods with aspen, rowan and willows, and with an abundance of ferns and tall herbs. Small treeless areas may have persisted where soils were shallow, and where there were natural rock outcrops.

At about 7200 BP *Ulmus* arrived in the area, but it never became an important component of the local forest vegetation. There is no evidence from the pollen stratigraphy at this site to suggest that pine or oak ever grew this far north (Birks, 1977, 1989), even though pine stumps occur locally in the Eriboll area. The natural woodland cover of this part of Scotland appears to have been primarily birch and hazel woods. At about 5800 BP *Alnus* migrated into the area (Birks, 1989; Bennett and Birks, 1990) and expanded locally in wet sites near the lochan.

Destruction of the birch and hazel woods began at about 5000 BP, resulting in the expansion of grasslands and, to a lesser extent, of heathland. There is palynological evidence for arable cultivation, presumably on the fertile limestone soils, from about 2500 BP. By this time extensive forest clearance had occurred, resulting in the virtually treeless landscape of the Loch Eriboll area today.

The site is of national importance because of its detailed and well-dated Lateglacial and Holocene pollen stratigraphy and for the co-ordinated study of pollen and plant macrofossils in the Lateglacial. These palaeoecological data provide important insights into the Lateglacial environment at low altitudes in the extreme north-west of Scotland; there is no other site that has been studied in such detail from this part of Scotland. These insights are as follows: (1) Even during the Lateglacial Interstadial, snow beds occurred at or near the site, thereby providing an interesting British parallel for the extremely open, chionophilous vegetation of south-west Norway during the Lateglacial (H.H. Birks, unpublished). (2) Major north–south and west–east floristic contrasts existed during the Lateglacial of northern Scotland and south-western Norway, as revealed by plant macrofossil studies (Birks, 1984; Birks and Mathewes, 1978; H. H. Birks, unpublished). (3) Lateglacial flora and vegetation at this far north-western site were similar to low- or mid-alpine situations today in western Norway, suggesting a considerably cooler climate than at contemporaneous sites further south in western Scotland. (4) The Holocene vegetational history from Lochan an Druim is particularly important because of the apparent lack of pine and oak. Its pollen record is thus intermediate between sites further south in western Scotland (Birks, 1980) and sites further west on Skye (Birks and Williams, 1983) and the Outer Hebrides (Bennett *et al.*, 1990). Lochan an Druim thus represents a uniquely important site for the reconstruction of Quaternary vegetational history and past environments.

Conclusion

Pollen and plant remains in the sediments from Lochan an Druim provide a record of the environmental history of the far north-west of Scotland during the Lateglacial and Holocene (approximately the last 12,500 years). They show that conditions remained extreme even during the relative climatic warming in the Lateglacial Interstadial. Later, during the Holocene, birch and hazel woodland developed, but oak did not extend this far north and pine was probably only locally present. Lochan an Druim provides valuable comparisons with other areas and is important as part of the network of sites that show the wider geographical variations in the patterns of vegetation development since the end of the last ice age.

LOCH MAREE
H. J. B. Birks

Highlights

The pollen records preserved in the sediments on the floor of Loch Maree and in the bogs on its islands provide a valuable record of Holocene vegetational changes in an area of high ecological importance. In particular, they allow important insights into the development of the native pinewoods.

Introduction

Loch Maree is a long (20 km), narrow (1.6–3.7 km) and deep (in excess of 110 m) loch situated north-west of Kinlochewe in Wester Ross. The sediments preserved on the floor of the loch and in several bogs and small lochans on the islands of Eilean Subhainn and Eilean Dubh na Sroine provide important pollen records of the Holocene vegetation history of this area. The Loch Maree woods, including Coille na Glas-Leitire on the south side of Loch Maree, the islands at the north-west end of Loch Maree, and the Letterewe oakwoods on the north side of Loch Maree are internationally important (Ratcliffe, 1977) because of the abundance of *Pinus sylvestris*, a tree that today is characteristic of the eastern Highlands and is rare or absent in much of north-west Scotland. The Loch Maree pine populations are distinct in terms of their monoterpene and isoenzyme loci and 'show little genetic affinity between contemporary Scottish and continental European populations' (Kinloch *et al.*, 1986). This suggests that these pine populations may have had a history that differs from those elsewhere in Scotland. Further, the Letterewe woods are also significant as the northernmost extensive semi-natural oakwood in Scotland. The vegetation history of this ecologically unique area is thus an important and integral part of its overall conservation importance, and has been investigated through pollen analytical studies of a sediment core from Loch Maree Hotel Bay (NG 919709) by Birks (1972b).

Kerslake (1982) has also studied three pollen profiles from islands in the loch. These islands support fine stands of pine woodland, alternating with a range of mire communities and small lochans, and provide a 'natural experiment' in vegetational history, in that by being isolated and difficult to reach, they are less likely to have been influenced by grazing and by human disturbance than the mainland. The site on Eilean Dubh na Sroine (NG 909720) is a small lochan on a rocky island that supports almost continuous pine woodland today. Subhainn Lochan (NG 923721) lies within a mosaic of pine woodland and blanket mires on Eilean Suhainn; Subhainn Bog (NG 922726) represents an area of deep peat and was studied to elucidate the history of peat development on the same island as Subhainn Lochan. An interesting feature of these island sites is the occurrence of Lateglacial sediments, in contrast to their absence in Loch Maree itself. However, these Lateglacial sediments have not been studied in detail.

Description

From the site in Loch Maree Hotel Bay, Birks (1972b) recovered a 5.42 m long core from below 29 m of water. The sediments comprised 4.67 m of organic muds overlying clays and silts (Figure 6.19). Six radiocarbon dates were obtained (Figure 6.19) and these form an internally consistent series. The pollen record has been divided into five local pollen assemblage zones (Figure 6.19).

Interpretation

The radiocarbon dates indicate that organic sedimentation began at about 9500 BP. The pollen stratigraphy shows a dominance of *Juniperus communis* with some *Empetrum* and herbs, such as *Rumex acetosa*, in the early Holocene. This was replaced at about 9000 BP by *Betula* and, to a lesser extent, *Corylus*. At the 8300 BP level, small but consistent amounts of *Quercus* and *Ulmus* pollen are found suggesting the local occurrence of these trees, perhaps on the Letterewe side of the loch. The most important feature, however, is the rapid and early expansion of *Pinus sylvestris* at about 8250 BP. This is one of the earliest occurrences of pine known in Scotland (Birks, 1989). The source of this early arrival is unknown (Kinloch *et al.*, 1986; Birks, 1989). At that time pine was present in southern Ireland and parts of southern and central England. Glacial survival or long-distance seed dispersal are the most plausible hypotheses (Birks, 1989). The pollen record provides strong support for the unique history of the Loch Maree pine populations that has been inferred by Kinloch *et al.* (1986) from biochemical evidence.

After 8000 BP *Pinus sylvestris* formed extensive open woodlands with abundant *Pteridium aquilinum*, other ferns and *Calluna vulgaris*. However, the extent of these began to decline from about 6500 BP, perhaps because of climatic change or soil degradation (Birks, 1972b). At about this time *Alnus* expanded, presumably in wet areas by streams and rivers and around the loch. As elsewhere in north-west Scotland (Birks, 1977, 1988; Bennett, 1984; Gear and Huntley, 1991), *Pinus* underwent a major decline in the Loch Maree area at about 4250–4000 BP, probably because of increased oceanicity and moisture, which restricted its survival and regeneration to well-drained, steep, blocky slopes

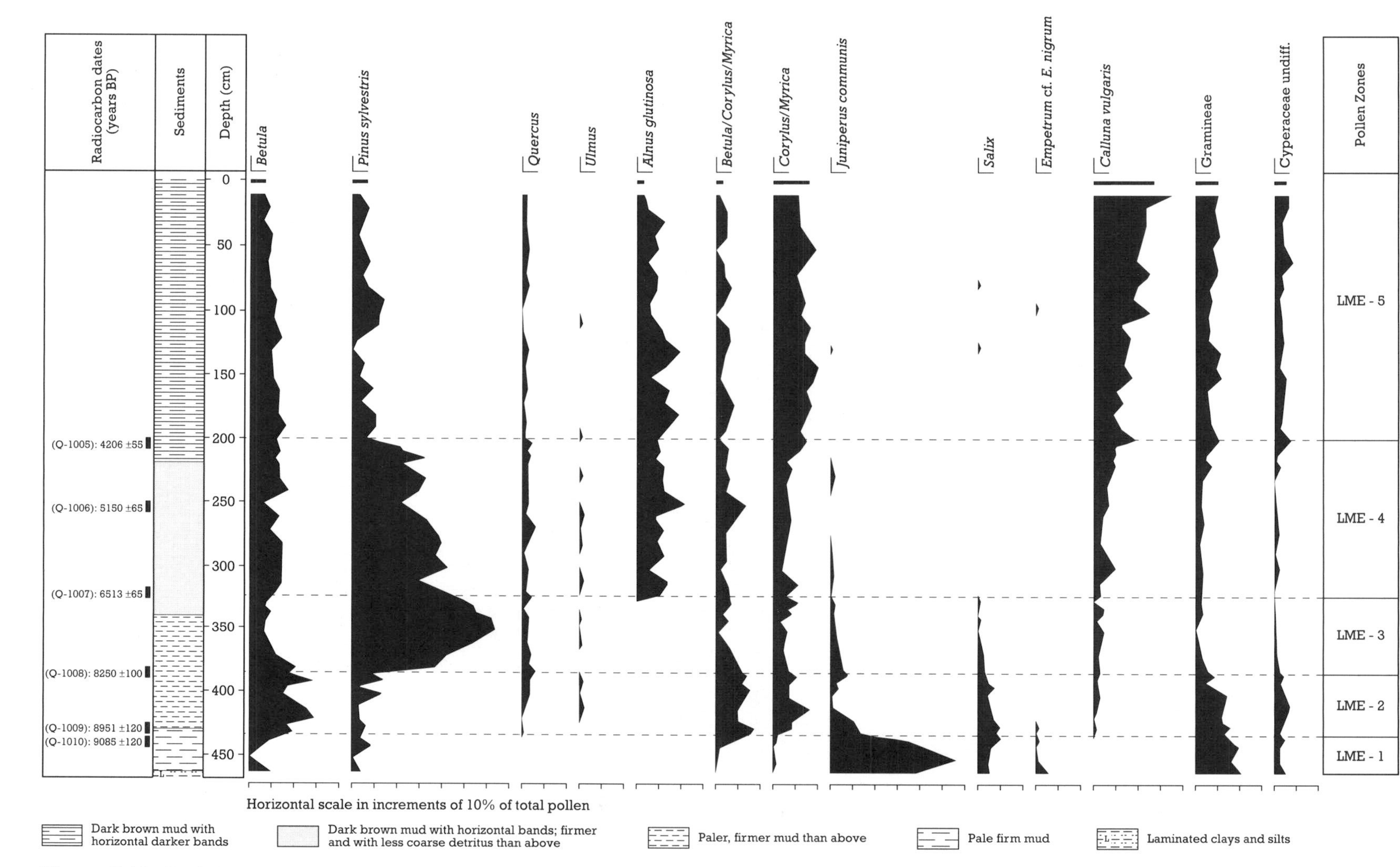

Figure 6.19 Loch Maree: relative pollen diagram, showing selected taxa as percentages of total pollen (from Birks, 1972b).

where blanket bog could not develop. The pollen record suggests that the present mosaic of pine, birch and oak woods, all spatially and ecologically separated today, of blanket bogs and moorland, and of alder stands had largely developed by about 4000 BP. Indeed there have been virtually no regional pollen-compositional changes in the last 4000 years. Anthropogenic impact appears to have been low in the area throughout the last 5000 years.

The Eilean Dubh na Sroine profile (4.28 m of sediment, seven radiocarbon dates) shows an early Holocene expansion of *Betula* and *Corylus* at about 9000 BP, along with *Pinus sylvestris*. This is the earliest known occurrence of pine in Scotland (Birks, 1989) and is about 800 years earlier than suggested by the Loch Maree regional profile. Almost no change occurred on the island throughout the Holocene, except for the arrival and expansion of *Alnus glutinosa* at about 6500 BP and the arrival of *Myrica gale* at about 3800 BP and its expansion at 3200 BP. Alder was never important on the island, whereas *Myrica gale* became locally prominent in damp areas within the pine forests, in soligeneous stream-side mires, and in moist areas near the lochan. The vegetation of the island has largely been pine–birch–juniper woodland for over 8000 years.

The sequence from Subhainn Lochan (7.38 m of sediment, nine radiocarbon dates) shows a later arrival and expansion of *Pinus* at about 7800 BP. The early and middle Holocene vegetation was *Betula* woodland with some *Corylus*, *Juniperus*, and *Salix*. *Pinus* was never dominant, and it declined to modern values at about 4000 BP, paralleled by an expansion of *Myrica gale* at 3800 BP, suggesting the spread of blanket and soligenous mires at this time.

The Subhainn Bog profile (6.25 m of sediment, nine radiocarbon dates) shows that the site was originally a shallow lake and that *Sphagnum* peat began to form there at about 8000 BP. Early Holocene vegetation consisted of open birch–hazel–willow communities. Pine expanded at about 8000 BP and was locally important until 3800 BP, when it declined markedly. At this time plants of wet blanket-mires expanded (for example *Menyanthes trifoliata*, *Rhynchospora alba*, *Narthecium ossifragum*), suggesting an increase in surface wetness of the bog. As pine was growing on the bog until this time, this profile provides direct evidence that increased waterlogging was the cause of the decline in pine. Interestingly, at this time *Myrica gale* expanded on the gentle slopes around Subhainn Lochan, presumably in response to regional hydrological changes. However, on Subhainn Bog *Myrica* did not expand until about 1700 BP.

The available pollen data from Loch Maree and its islands provide a detailed picture of the regional and local history of the pinewoods in the area. Pine grew on a range of soil types, including acid mor humus and peats overlying lake sediments. Over the region as a whole, pine declined at about 4250 BP with the onset of a steady rise in *Calluna*, grasses and sedges, suggesting the regional spread of blanket-bog and soligenous mires. At a local scale there has been considerable variation in the onset of peat development depending on local topography and hydrological thresholds. The Loch Maree area provides strong support for the hypotheses that blanket-bog development was an entirely natural process in this extreme oceanic environment and that the widespread and spectacular pine decline of north-west Scotland was a response to a major climatic change at about 4250–4000 BP (Birks, 1988; Gear and Huntley, 1991).

In view of the wealth of palaeoecological information concerning the history of *Pinus sylvestris* in the area, Loch Maree and its islands are sites of great importance in our understanding of the Holocene vegetational and environmental history of Scotland. They provide insights into the history and status of the north-west Scotland races of *Pinus sylvestris*, populations that are of considerable international importance.

Conclusion

The pollen grains contained in the sediments from Loch Maree and its islands provide an important record of the vegetation history of an area that is ecologically significant for its pine and oak woodlands. The pollen data, together with radiocarbon dating, show the rapid expansion of pine around 8250 years ago, followed by its decline after about 4250 years ago. The present pattern of pine, birch and oak woodlands has been in existence for the last 4000 years. Loch Maree is therefore a reference site for the vegetation history of north-west Scotland during the Holocene, and in particular for understanding the development of the native pine forest.

Chapter 7

Inverness area

INTRODUCTION

D. G. Sutherland and J. E. Gordon

The Inverness area comprises the lowlands along the Moray Firth coast from the Dornoch Firth to east of Nairn, the upland areas of the hinterland and the glaciated valleys extending to the west and south-west, including the Great Glen (Figure 7.1). The principal focus of research on this area has centred on the evidence for the last ice-sheet, the pattern of deglaciation and the changes in relative sea level that both accompanied and followed the ice wastage (Auton, 1990a; Firth, 1990a). Until recently, only one deposit was known to pre-date the Late Devensian, the high-level shelly clay at Clava. However, the discovery of possible Hoxnian deposits at Dalcharn and probable Early Devensian interstadial deposits at a site on the Allt Odhar, both associated with multiple till successions, together with detailed reinvestigation of the Clava succession, has enabled the development of a provisional stratigraphy extending back to the Anglian (Merritt, 1990a).

The Inverness area was glaciated during the Pleistocene by one of the major ice streams flowing out from the Highlands, receiving ice from the mountains both to the north and the south of the Great Glen. The pattern of striations indicates that the ice generally moved towards the north-east during the later phases of glaciation (Merritt, 1990a) and was channelled into the inner Moray Firth, beyond which it diverged and flowed back on to the land in both Caithness (Chapter 5) and in Buchan (Chapter 8). As discussed for other regions, such an ice-flow pattern may have occurred on more than one occasion during the Pleistocene. On the higher ground between the Rivers Nairn and Findhorn, there is evidence for an earlier southwards ice-movement in the form of striations aligned north–south and the transport of Middle Old Red Sandstone erratics from source areas around Loch Ness and the inner Moray Firth (Horne, 1923; Merritt, 1990a).

Early work in the area recorded the presence of many glacial features including striations, erratics, moraines, glaciofluvial deposits and terraces (Jamieson, 1865, 1874, 1882b, 1906; Fraser, 1877, 1880; Aitken, 1880; Milne Home, 1880a; MacDonald, 1881; Cameron, 1882a, 1882b; MacDonald and Fraser, 1881; Wallace, 1883, 1898, 1901, 1906; Mackie, 1901; Peach *et al.*, 1912; Horne and Hinxman, 1914; Hinxman and Anderson, 1915; Horne, 1923), as well as a high-level (*c.* 150 m OD), shelly clay at Clava (Fraser, 1882a, 1882b; Horne *et al.*, 1894). In the Geological Survey Memoirs (Peach *et al.*, 1912; Horne and Hinxman, 1914; Hinxman and Anderson, 1915; Horne, 1923), three phases of glaciation were recognized: a period of maximum glaciation in which the whole landscape was covered by ice moving out from the mountains to the west and south, a period of confluent valley glaciers as the higher hills became ice-free and a final valley-glacier period accompanied by the formation of moraines. In the area south of the Moray Firth, ice-dammed lakes formed as the ice retreated (Horne, 1923). Later, Bremner (1934a, 1939c) and Charlesworth (1956) developed similar ideas on ice recession and glacial lakes. The effects of glacial erosion have been variable across the area, generally being greater in the west than in the east. To the east of the Great Glen, pockets of deeply weathered bedrock have survived glaciation, as at Clunas, providing a link with the more extensive occurrences of this phenomenon in north-east Scotland (see Chapter 8). More recent studies of glaciofluvial landforms in mid-Strathdearn (Young, 1980) and in the adjacent Dulnain and Spey valleys (Young, 1977b, 1978) indicate that the ice-sheet wasted largely *in situ*. The pattern locally, however, may be quite complex. In the middle Findhorn valley (see below), Auton (1990b) identified six stages involving both downwasting and recession and also the formation of an ice-dammed lake.

In the inner Moray Firth area and in the valleys to the west, a series of ice-sheet recessional stages was proposed by Kirk *et al.* (1966), J. S. Smith (1966, 1968, 1977) and Synge and Smith (1980). Readvances or stillstands were recognized at Ardersier, Alturlie, Kessock, Englishton, Muir of Ord, Balblair-Contin and Garve. However, reassessment of the evidence by Firth (1984, 1989a, 1989b, 1990a) led him to propose an alternative model of progressive ice recession without significant interruptions.

Evidence for pre-Late Devensian deposits occurs at three sites in the Inverness area. At Dalcharn, an organic deposit containing pollen of interglacial affinity forms part of a complex glacigenic succession observed in two separate exposures (Merritt and Auton, 1990; Walker *et al.*, 1992). The organic deposit, which contains lenses of compressed peat of infinite radiocarbon age and appears to have been cryoturbated and glaciotectonically disturbed, overlies weathered outwash

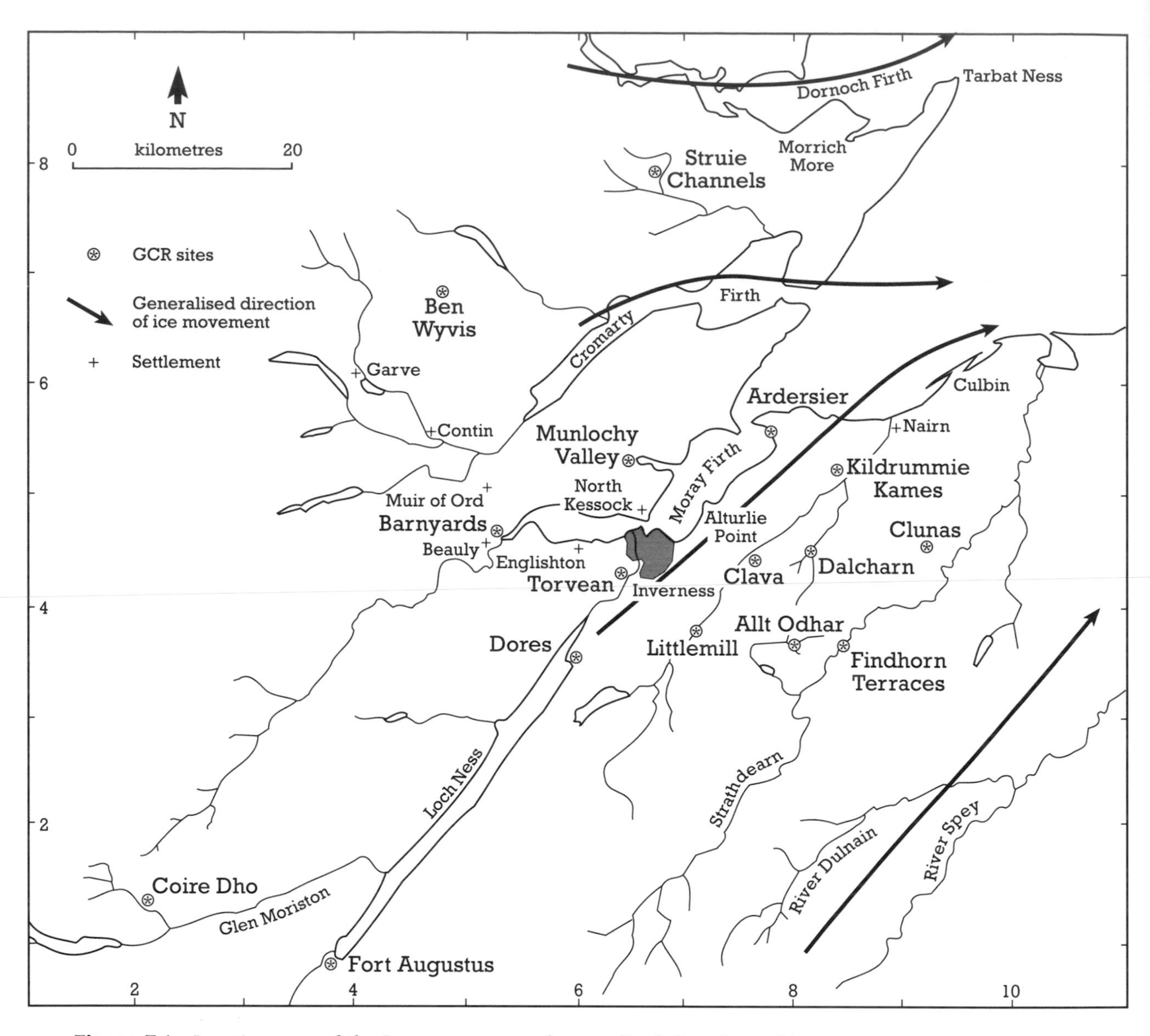

Figure 7.1 Location map of the Inverness area and generalized directions of ice movement.

gravels, which in turn rest on till. Above the organic deposit are two distinct formations of lodgement till, each comprising three separate members. Pollen analysis of the peat (Walker, 1990a) indicates the development of pine forest with birch, alder and holly, which appears to reflect an interglacial episode. The pollen record is similar to that from Fugla Ness in Shetland, which has been assigned to the Hoxnian, although such a correlation is now considered insecure (Lowe, 1984). At present, therefore, it is not possible to provide a firm age for the Dalcharn deposits (Walker, 1990a; Whittington, 1990; Walker *et al.*, 1992).

The second pre-Late Devensian deposit occurs in a section along the Allt Odhar, where a layer of compressed peat containing pollen of interstadial affinity forms part of a succession of glacigenic deposits (Merritt, 1990c; Walker *et al.*, 1992). The peat rests on gravel, which in turn overlies a weathered till. Above the peat is a succession of interbedded paraglacial deposits and till. Pollen analysis (Walker, 1990b) indicates a vegetation–climate cycle involving a succession from birch woodland with juniper and willow, through a phase of expansion of grassland and heathland, to an open landscape of species-poor grass and sedge communities. The peat has yielded an infinite radiocarbon date (Harkness, 1990) and a uranium series date of 106 ka +11/−10 ka, the latter placing the deposit in Oxygen Isotope Substage 5c (Walker *et al.*, 1992). The pollen and insect evidence also support an interstadial rather than an interglacial origin for the deposit (Walker *et al.*, 1992).

At Clava there is a high-level shelly clay. Originally discovered by Fraser (1882a, 1882b), this deposit was subsequently examined in detail by a committee of the British Association (Horne *et al.*, 1894) in the context of the debate about the existence of a major interglacial submergence of the British Isles. The majority of the committee considered the deposit to be in place and to result from such a submergence, but this was disputed (Bell, 1895a, 1897a). Later publications tended to emphasize the likelihood of the shell bed being a glacial erratic (Sissons, 1967a; Peacock, 1975b). More recently Sutherland (1981a) has suggested that the deposit is indeed *in situ* and results from a marine transgression consequent upon isostatic depression in front of an expanding Scottish ice-sheet. However, re-investigation of the site led Merritt (1990b) to conclude that the deposit is an erratic of marine sediments, transported by the Late Devensian ice-sheet from the Loch Ness basin.

The direction of movement of the last ice-sheet is well-illustrated by the transport of erratics from within the Highlands to the lower ground of the shores of the Moray Firth (Horne, 1923; Sissons, 1967a; Smith, 1977; Synge and Smith, 1980). Notable among these is the Inchbae augen-gneiss, fragments of which can be traced eastwards from its outcrop to the north-west of Ben Wyvis across the Black Isle and into the coastlands of Moray (Mackie, 1901; Peach *et al.*, 1912; Sutherland, 1984a). It is of interest that there is a distinct upper limit to these erratics on Ben Wyvis (Peach *et al.*, 1912), raising the possibility that the summit of this mountain was ice-free at the time of the maximum extent of the last ice-sheet. The major flow of ice emerging from the Great Glen is also reflected in the transport of erratics in an easterly direction (Horne and Hinxman, 1914; Horne, 1923; Bremner, 1934a; Peacock *et al.*, 1968).

The most striking landforms in the Inverness area relate to the period of deglaciation of the last ice-sheet. Particularly notable in this connection are the deposits at Ardersier and the sequence of glaciofluvial sediments that extends from Torvean at the mouth of the Great Glen at Inverness, through Littlemill to Kildrummie (Harris and Peacock, 1969; Smith, 1977; Synge and Smith, 1980; Firth, 1984). The deposits at Ardersier have been associated with a readvance of the last ice-sheet (J. S. Smith, 1968, 1977; Synge, 1977b; Synge and Smith, 1980), but this interpretation has been questioned by Firth (1984, 1989b). The esker and kame terraces at Torvean are outstanding landforms, among the largest examples of their type of Scotland, whilst the Flemington Esker at Kildrummie is one of the longest continuous such features in Scotland. Associated with such glaciofluvial deposits are meltwater channel systems, particularly on the higher ground and across spurs and interfluves. Young (1977b, 1980) has mapped the intricate sequences of channels in the valleys and hills to the south of Inverness, and a complex sequence of channels crosses the watershed between the Dornoch Firth and Cromarty Firth in upper Strathrory. Further elements in the suite of glaciofluvial deposits that resulted from the last deglaciation are outwash terraces, which are found in many of the valleys of the region. Along the River Findhorn, near the Streens Gorge, there is an extremely complex sequence of terraces (Young, 1980). This sequence was initiated during deglaciation and evolved still further during the Lateglacial and Holocene (Auton, 1990b).

Around the coasts the progressive ice wastage resulted in the formation of a series of raised shorelines (see Ardersier and Munlochy Valley). These are isostatically tilted towards the north-east, each successively younger shoreline extending farther west and having a lower gradient than its predecessor (Synge and Smith, 1980; Firth, 1984, 1989a). Shorelines also formed around Loch Ness at this time (see Dores and Fort Augustus), and were considered by Synge (1977b) and Synge and Smith (1980) to be marine, and part of the coastal sequence of shorelines. Firth (1984, 1986) has re-mapped these features and considers that there was no connection with the sea and that the shorelines are lacustrine in origin. As such, they are among the clearest former lake shoreline deposits in Scotland.

At the time of deglaciation, and continuing into the Lateglacial Interstadial, fossiliferous marine sediments were laid down in the sea lochs and along the coastal fringe. These have been studied in boreholes (Peacock, 1974a, 1977a; Peacock *et al.*, 1980) and, to date, no equivalents of the Errol beds (see Chapter 15 below), otherwise widely distributed on the east coast of Scotland, have been found in the inner firths of the Inverness area (Peacock, 1975c). This may suggest relatively late deglaciation of this area, although there is no dating evidence available to support such an idea.

There has been little investigation in the Inverness area of the Lateglacial Interstadial terrestrial environment. Pennington *et al.* (1972) and Pennington (1977a) provided details of pollen and other analyses carried out at Loch Tarff in the extreme south-west of the region. Following an initial phase of pioneer vegetation with species such as *Rumex* and *Lycopodium selago*, which are characteristic of disturbed, skeletal soils, there occurred a period of dwarf-shrub tundra with increased representation of *Empetrum*. A brief period of reduced *Empetrum* values then occurred, which may be correlated with the Older Dryas period of increased climatic severity between 12,000 and 11,800 BP. Thereafter there was a marked expansion of *Empetrum*, which was accompanied immediately prior to the onset of the Loch Lomond Stadial by a peak in juniper pollen. This period of *Empetrum* dominance corresponds to the major part of the Lateglacial Interstadial and represents a time of soil stability and increasing acidity as is also confirmed by the diatom assemblages contained in the sediments.

The events of the Loch Lomond Stadial left a strong imprint on the scenery of this area, most particularly in the south-west, which was invaded by glaciers flowing out from the ice-field of the western Highlands. Major outlet glaciers flowed along the Great Glen to terminate near Fort Augustus (see below) and along lower Glen Moriston. This latter glacier flowed across the exit to Coire Dho (see below) damming a lake and resulting in the production of an outstanding suite of landforms relating both to the ice-dammed lake (shorelines and cross-valley moraines) and to the drainage of the lake (water-swept bedrock and meltwater channels) (Sissons, 1977b).

Sissons (1979c) also suggested that certain of the deposits at Fort Augustus related to a *jökulhlaup*, when the ice-dammed lake in Glen Spean and Glen Roy (see Chapter 10) drained catastrophically during the decay phase of the Loch Lomond Readvance. He suggested that this raised the level of Loch Ness several metres and that the resultant flood and corresponding erosion of the glaciofluvial sediments in the Ness valley resulted in the construction of a major fan at Inverness, producing the narrows between Inverness and Kessock (Sissons, 1981c). Firth (1984, 1986) accepted the broad outline of Sissons' hypothesis, although he suggested modifications to details, such as the position of the ice front at the time of the supposed *jökulhlaup* and the precise nature of the changes in the level of Loch Ness accompanying the flood. In proposing this dramatic sequence of events, Sissons (1981c) also suggested that a marine erosion surface, which he identified underlying Holocene deposits at the head of the Beauly Firth and reaching an altitude a few metres above present sea level, was the equivalent of the Main Lateglacial Shoreline of the Forth valley.

Periglacial processes would have been active, particularly on the mountains during the Loch Lomond Stadial, and the large-scale, sorted ground features on Ben Wyvis may have received their final fashioning at this time, although the cover of frost-weathered debris on the summit area of that mountain may date from the early phases of deglaciation or, if the mountain was not in fact covered by the last ice-sheet, from the fully glacial part of the Late Devensian and earlier (Ballantyne, 1984; Ballantyne *et al.*, 1987).

There has been little study of the changes in the Holocene terrestrial environment of this region, although considerable work has been done on the evolution of the coastline. In the inner firths, Haggart (1982, 1986, 1987, 1988b) and Firth and Haggart (1989, 1990) have shown, by study of sites such as that at Barnyards, that during the early Holocene sea level initially fell, reaching a low some time after 9000 BP, then subsequently rose during the Main Postglacial Transgression to culminate, between 7100 BP and 5800 BP, in the formation of the Main Postglacial Shoreline. The subsequent fall in sea level to its present level is not securely dated, but several distinct shorelines were formed during this period, as is illustrated by the sequence of estuarine flats at Munlochy Valley (Firth, 1984). Elsewhere in the area, Holocene raised shoreline features are well-developed (Ogilvie, 1914, 1923; J.S. Smith, 1968, 1977; Comber, 1991; Hansom, 1991) and include shingle ridges (Dornoch Firth, Tarbat Ness, Spey Bay and Culbin) and sand beach ridges (Morrich More).

CLUNAS

A. M. Hall

Highlights

The stream section at Clunas illustrates the

effects of differential weathering processes in a Devonian conglomerate. The survival of the weathered bedrock also highlights the relatively low intensity of glacial erosion in this area.

Introduction

The site at Clunas (NH 907446) is a stream exposure of weathered conglomerate at an altitude of 210 m OD on the Muckle Burn, 12.5 km south of Nairn. It is important for studies of deeply weathered bedrock, which is unusually widespread in both sedimentary and crystalline rocks to the east of Inverness and more especially in north-east Scotland (FitzPatrick, 1963; Hall, 1985, 1986; Hall *et al.*, 1989a). In particular, Clunas is a good example of deep weathering of a Devonian conglomerate, which is part of a small outlier of Middle Old Red Sandstone age (Horne, 1923). The site also gives important insights into the processes of differential weathering. The conglomerate contains large boulders of a variety of rock types and allows study of the chemical alteration of different rock types under identical environmental conditions. The only detailed study of the site is by Wilson *et al.* (1971).

Description

The conglomerate is exposed to a depth of up to 5 m and is overlain by up to 2 m of glaciofluvial gravel of probable Devensian age. The conglomerate contains large rounded boulders up to 1 m in diameter in various stages of weathering. Boulders of banded metaquartzite, metamorphosed grit and silicified volcanic rocks remain fresh, whereas boulders of granite and quartz–biotite-granulite are more or less decomposed to a clayey gritty sand.

The clay mineralogy of the weathered boulders has been studied in detail by Wilson *et al.* (1971). Feldspars, with the exception of microcline, are altered to montmorillonite. Kaolinite, derived from muscovite, is also present in smaller amounts. The presence of carbonate minerals lining microfractures in the weathered boulders indicates that these transformations occurred under a relatively closed, alkaline weathering system.

Interpretation

Chemically weathered rock is preserved at many sites in north-east Scotland and this reflects the relatively low degree of glacial erosion in the region. Deep weathering of Devonian sedimentary rocks is relatively rare (Hall, 1986), although several sites do occur around Elgin (Peacock *et al.*, 1968), Turriff (Hall, 1983) and between New Aberdour and Pennan (J. D. Peacock, unpublished data). At Clunas, the low degree of chemical alteration of the granite and granulite boulders is characteristic of the grus weathering type recognized by Hall (1985), which developed under humid temperate environments prior to the first regional glaciation, as well as during interglacial periods. The precise age of the weathering at Clunas, however, is presently unknown. The possibility that weathering of the boulders may have started soon after deposition in the Devonian has not yet been investigated, but is suggested by the presence of carbonate minerals infilling microfractures.

The site has interest for both regional pre-Pleistocene and Pleistocene geomorphology and also for its clay mineralogy. The survival of weathered bedrock demonstrates that the last, and probably the earlier, ice-sheets have failed significantly to lower the bedrock surface in the Clunas area. If the weathering is of pre-Pleistocene age, then minimal Pleistocene erosion has occurred (see Pittodrie) and the form of the preglacial landsurface can be reconstructed. The dominance of montmorillonite and the evidence of a relatively closed weathering system suggests that only the poorly drained base of the former weathering profile has been preserved. Elsewhere in north-east Scotland alteration of feldspars in granitic rocks generally gives kaolinitic clays (see Hill of Longhaven and Pittodrie) (Hall, 1983; Hall *et al.*, 1989a) and the abundance of montmorillonite at Clunas is therefore unusual. Its coexistence with small amounts of kaolinite, derived from alteration of muscovite, is also noteworthy as it demonstrates the importance of small-scale equilibria in clay mineral genesis (Wilson *et al.*, 1971).

Conclusion

Clunas forms part of a network of sites showing deeply weathered bedrock, one of the principal features of the geomorphology of north-eastern

Scotland. The example at Clunas is particularly interesting as the bedrock is a conglomerate in which pebbles of different lithologies show different degrees of alteration. Not only does the site provide insights into the processes of rock weathering, it also indicates minimal erosion by ice, a characteristic of this area during the Quaternary glaciations.

DALCHARN

C. A. Auton

Highlights

The sequence of sediments exposed in the stream sections at this site includes interglacial organic deposits which are both underlain and overlain by till. Although the deposits are undated, the sequence is remarkable for the detail of information it has yielded on the Quaternary history of the Inverness area and the potential it holds for providing further elaboration of this record.

Introduction

Sediments containing compressed and disseminated biogenic matter are exposed beneath a thick sequence of tills in a river cliff of the Allt Dearg at Dalcharn (NH 815452), some 6 km south-west of the village of Cawdor, near Nairn. The organic deposits, which lie at an altitude of *c.* 200 m OD, have been cryogenically and glaciotectonically disturbed, but contain pollen of full interglacial affinity reflecting the middle and later stages of an interglacial cycle. The overlying till sequence provides evidence of at least two separate glacial episodes, and although the age of the interglacial material cannot be firmly established at present it is probable that it pre-dates the Ipswichian.

The organic deposits occur near the bleached top of a deeply weathered gravel. The base of the gravel is not exposed at the section containing the organic material, but the gravel can be seen to overlie an older till at the base of another cliff section some 200 m to the north-east. Various aspects of these sections have been described by Bloodworth (1990), Merritt and Auton (1990), Walker (1990a), Whittington (1990) and Walker *et al.* (1992). Dalcharn provides the first evidence that the northern Grampian Highlands were covered by pine forest during at least one interglacial stage of the Middle or Late Quaternary.

Description

Exposures in the cliffs of the Allt Dearg, east of Dalcharn Cottages, display a succession of Quaternary sediments *c.* 25 m in thickness (Figure 7.2). The lithological subdivisions used in this account follow those of Merritt and Auton (1990) and Walker *et al.* (1992), and are based on a composite log of three sections: Dalcharn East (NH 81574537), Dalcharn West–Section A (NH 81464521) and Dalcharn West–Section B (NH 81434516), shown in Figure 7.3. The recognized sequence is as follows:

7.	Humic soil	*c.* 0.3 m
6.	Glaciofluvial gravel	up to 2.5 m
5.	Dalcharn 'upper till formation'	8.5–10.0 m
4.	Dalcharn 'lower till formation'	8.5–9.5 m
3.	Dalcharn 'biogenic formation'	1.3–1.6 m
2.	Dalcharn 'gravel formation'	up to 3.0 m
1.	Dearg 'till formation'	at least 1.0 m

The Dearg 'till formation' (unit 1) is a moderate yellowish brown, very stiff, massive diamicton, with abundant clasts of Devonian sandstone and is exposed beneath the Dalcharn 'gravel formation' at the Dalcharn East Section.

The Dalcharn 'gravel formation' (unit 2) is a poorly sorted, matrix-rich gravel, bleached in its upper part, containing a high proportion of decomposed and unsound clasts. The clay mineralogy of the matrix of this deposit, in which vermiculite occurs as a product of subaerial weathering, has been described by Bloodworth (1990).

The Dalcharn 'biogenic formation' (unit 3) is subdivided into an upper unit, the Dalcharn 'biogenic member' (0.5–0.6 m) and a lower unit, the Dalcharn 'cryoturbate member' (0.8–1.0 m). The uppermost 0.1–0.2 m of the 'biogenic member' comprises compact, laminated olive grey sandy and clayey silt with discontinuous wisps of pebbly sand and disseminated peaty matter. This overlies compact carbonaceous sandy silt and diamicton containing fibres and lumps of dark peaty material as well as discrete lenses, up to 0.05 by 0.01 m, of compressed sandy peat; an infinite radiocarbon date (>41,300 BP (GU–2340)) has been obtained from compressed peat close to the base of the 'biogenic member'.

Figure 7.2 Section at Dalcharn showing the Dalcharn 'gravel formation' and the Dalcharn 'biogenic formation' (bottom left), overlain by a sequence of tills (right). (Photo: D. G. Sutherland.)

The 'cryoturbate member' consists of massive, matrix-supported clayey gravel diamicton, with a matrix of light grey to white silty fine-grained sand. Small fragments of organic material are sparsely disseminated throughout the deposit. Clasts within the diamicton are mainly of yellowish grey coarse-grained sandstone, many with white weathering rinds. Five pollen assemblage zones (Figure 7.4) have been recognized within the Dalcharn 'biogenic formation' exposed at Section A – Dalcharn West (Walker, 1990a; Walker *et al.*, 1992). The pollen record appears to show that closed pine forest with birch, alder and holly (D–1) was followed by a pine and heathland episode (D–2). This was succeeded by a gradual disappearance of the pine forest, which was initially replaced by birch (D–3) and later by heath and open grassland (D–4 and D–5). No plant macrofossils or insect remains have been found in the biogenic deposits.

The Dalcharn 'lower till formation' (unit 4) is subdivided into upper (*c.* 3.0 m), middle (*c.* 3.1 m) and lower (4–5 m) 'members', which all comprise reddish brown sandy diamicton, characterized by abundant clasts of Devonian sandstone. The upper and lower 'members' are massive and matrix-supported; the middle 'member' is stratified and friable.

The Dalcharn 'upper till formation' (unit 5) is divided into upper and lower 'members', which both comprise brown, massive, matrix-supported diamicton with clasts predominantly comprising psammite, semipelite and pink and grey granite. The upper 'member' (3.0–3.5 m) is separated from the lower by a sharp subhorizontal planar discontinuity and is characterized by a strongly developed clast fabric indicating former ice movement towards N034°; a deformed mass of claybound gravel occurs close to its base. The lower 'member' (5.5–6.5 m) contains a smaller proportion of clasts of metamorphic rock types and a larger proportion of pink granitic clasts than the upper 'member'; it is characterized by a clast fabric indicating former ice movement towards N097°.

The glaciofluvial gravel (unit 6) comprises orange stained, poorly sorted, clast-supported cobble gravel showing poorly developed horizontal stratification and an imbrication indicating a north-easterly palaeocurrent.

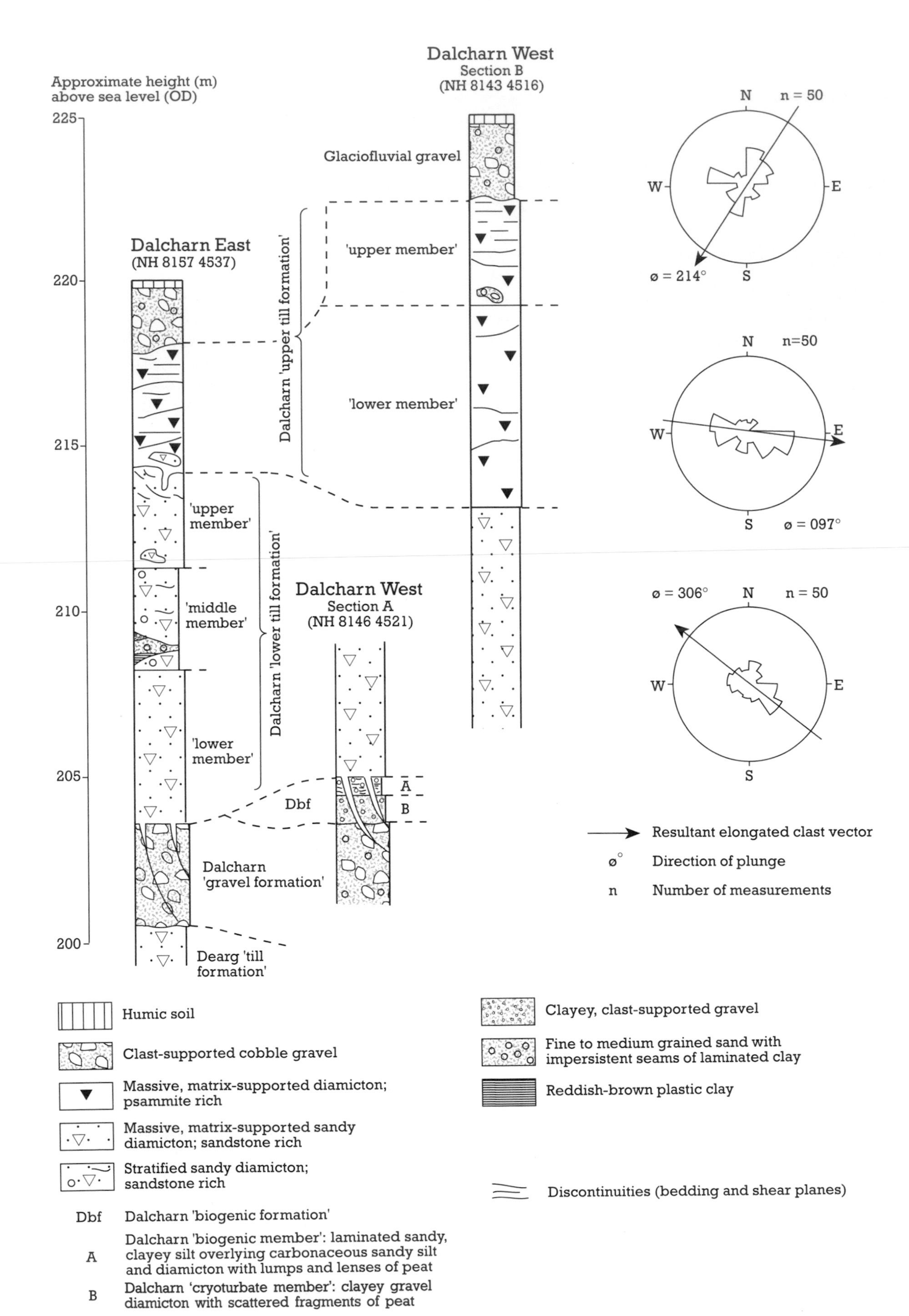
Approximate height (m)
above sea level (OD)
225
220
215
210
205
200
Dalcharn West
Section B
(NH 8143 4516)
Glaciofluvial gravel
'upper member'
'lower member'
Dalcharn 'upper till formation'
Dalcharn East
(NH 8157 4537)
'upper
member'
'middle
member'
'lower
member'
Dalcharn 'lower till formation'
Dalcharn West
Section A
(NH 8146 4521)
A
B
Dbf
Dalcharn
'gravel formation'
Dearg 'till
formation'
N
n = 50
W
E
ø = 214°
S
N
n=50
W
E
S
ø = 097°
ø = 306°
N
n = 50
W
E
S
Resultant elongated clast vector
ø°
Direction of plunge
n
Number of measurements
Humic soil
Clast-supported cobble gravel
Massive, matrix-supported diamicton;
psammite rich
Massive, matrix-supported sandy
diamicton; sandstone rich
Stratified sandy diamicton;
sandstone rich
Dbf
Dalcharn 'biogenic formation'
A
Dalcharn 'biogenic member': laminated sandy,
clayey silt overlying carbonaceous sandy silt
and diamicton with lumps and lenses of peat
B
Dalcharn 'cryoturbate member': clayey gravel
diamicton with scattered fragments of peat
Clayey, clast-supported gravel
Fine to medium grained sand with
impersistent seams of laminated clay
Reddish-brown plastic clay
Discontinuities (bedding and shear planes)

Interpretation

The occurrence of brown till with few sandstone erratics, overlying reddish brown till with abundant sandstone clasts is a common feature of many of the sequences of Quaternary deposits which mantle the high ground flanking the coastal lowlands of the Moray Firth between Inverness and Nairn. This stratigraphic relationship, which was first recognized by Fraser (1880) in Strathnairn and subsequently by Horne and Hinxman (1914) and Horne (1923) during the primary geological surveys of the surrounding districts, is clearly seen in the cliffs of the Allt Dearg and those of its tributaries. At Dalcharn, the recognition of discontinuities between the various tills, the change in composition of their clasts and in the orientation of their fabrics support the contention that the two till formations, which overlie the organic sediments and weathered gravel, are the products of at least two distinct glacial episodes.

Both 'members' of the Dalcharn 'upper till' formation contain flat-iron shaped cobbles and elongate clasts with striations parallel to their longer axes, and the matrices of both units are penetrated by subhorizontal fissures and sharp concavo-convex discontinuities. These features, together with the very poor sorting and overconsolidation of the diamictons, are considered to be characteristic attributes of lodgement tills (Dreimanis, 1989). The clast fabric of the upper 'member' indicates former ice movement towards the north-east, which is parallel to the general alignment of glacially streamlined features near the Dalcharn site. The fabric of the lower 'member' indicates former ice movement towards the east.

The relative abundance of clasts of Devonian sandstone, together with a weakly developed fabric suggesting former ice movement towards the south-east, serves to distinguish the Dalcharn 'lower till formation' from the overlying diamictons. This south-eastward direction of ice movement corresponds to the orientation of some striae on bedrock observed at a few sites on the high ground between Loch Moy and Loch Ness (see Merritt 1990a, fig. 1). The poorly sorted and overconsolidated nature of the upper 'member' of the 'lower till formation', the presence of striated cobbles and discontinuity surfaces suggest that it is probably a lodgement till, whereas the presence of winnowed horizons and discrete lenses of sand and gravel, particularly within the middle 'member' of the 'lower formation', suggests that these lower parts of the deposit may have been formed by basal meltout rather than by lodgement processes.

Figure 7.3 Sediment logs and stratigraphy at Dalcharn (from Merritt and Auton, 1990).

The highly decomposed nature of the gravel underlying the biogenic deposits at Dalcharn indicates that it has been subjected to prolonged weathering under warm humid conditions, and suggests that the gravel and the associated organic material is of considerable antiquity. The pollen recorded from the organic horizons suggests that the weathering occurred during at least one interglacial episode prior to the Devensian.

It is also apparent that the biogenic deposits have been affected by severe post-depositional (and probably also syn-depositional) cryoturbation, as shown by the fragmentation of the peaty material and the mixing of bleached clasts from the underlying gravel into the biogenic sediments. The penetration of fissures, lined with silt and orange sand, from above the base of the overlying till, through the biogenic deposits and into the underlying gravel indicates that both the lower units have also been affected by glaciotectonic disturbance.

The origins of the biogenic deposits are uncertain and the pollen diagram may not reflect complete sequential vegetation development (Walker, 1990a; Whittington, 1990; Walker *et al.*, 1992). Nevertheless, the sequence of pollen zones appears to reflect a consistent pattern of vegetation development during an interglacial. Pollen data from sites elsewhere in Scotland indicate that pine woodland was the climax forest of the north-central Grampians during the Holocene (Pennington *et al.*, 1972; O'Sullivan, 1974a, 1976; Walker, 1975c) and, if these records can be used as an analogue for previous interglacials, then the Dalcharn sequence probably reflects a warm episode of interglacial rather than interstadial status.

That temperatures comparable with, or even higher than, those of today may have prevailed during the accumulation of the Dalcharn 'biogenic formation' can be inferred from the relatively high counts for *Ilex* pollen. Holly is known to be intolerant of winter cold, the limiting mean temperature of the coldest month being −0.5°C while that of the warmest is 12–13°C (Iversen,

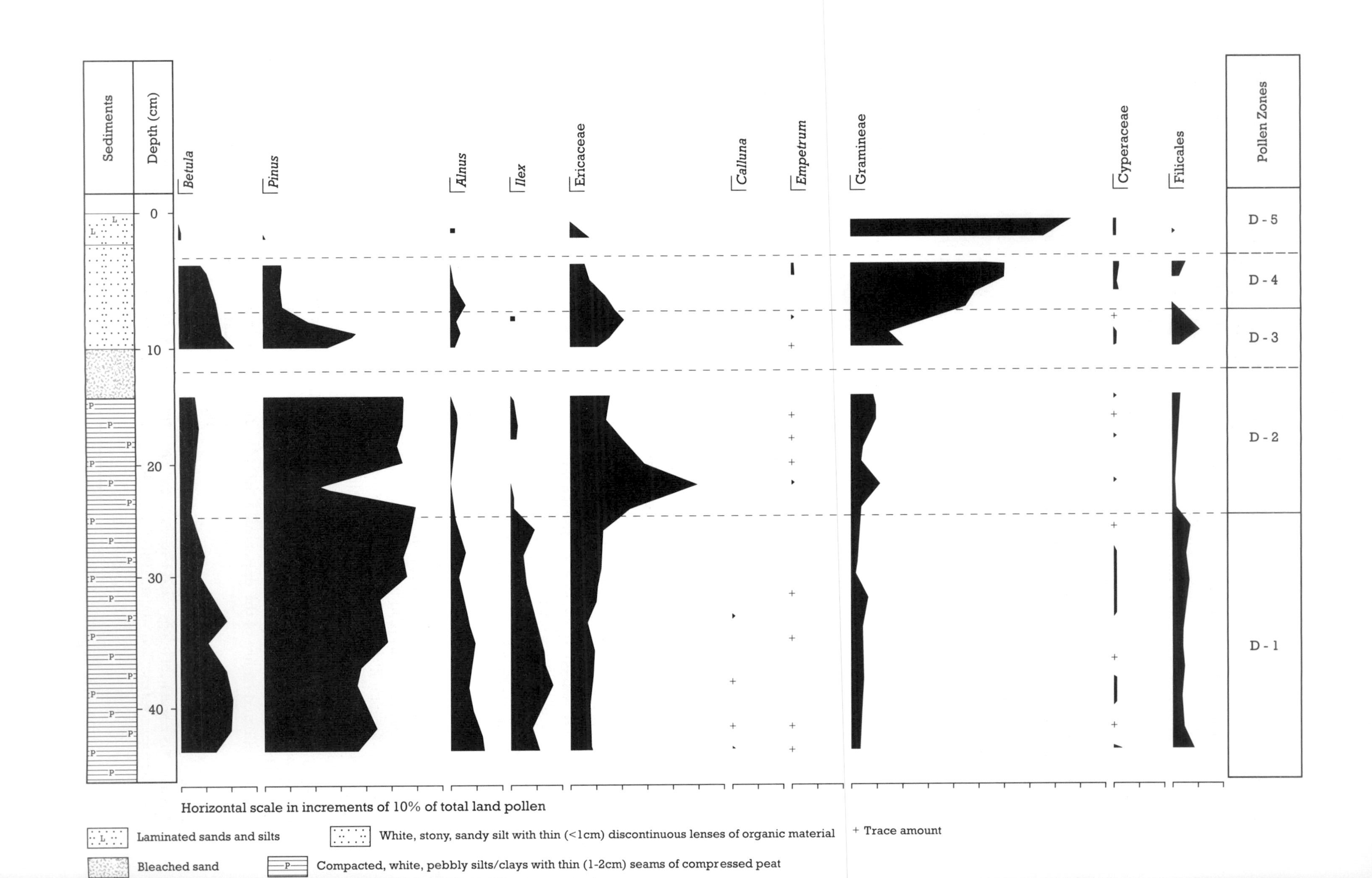
Sediments
Depth (cm)
0
10
20
30
40
Betula
Pinus
Alnus
Ilex
Ericaceae
Calluna
Empetrum
Gramineae
Cyperaceae
Filicales
Pollen Zones
D - 5
D - 4
D - 3
D - 2
D - 1
Horizontal scale in increments of 10% of total land pollen
Laminated sands and silts
White, stony, sandy silt with thin (<1cm) discontinuous lenses of organic material
+ Trace amount
Bleached sand
Compacted, white, pebbly silts/clays with thin (1-2cm) seams of compressed peat

1944). The Dalcharn site lies near to the present northern limit of *Ilex* in Britain (Godwin, 1975), and hence the relative abundance of *Ilex* pollen in Zone D–1 of the Dalcharn profile almost certainly reflects a climate somewhat warmer than that of today. The decline in *Ilex* at the D–1/D–2 boundary and its subsequent disappearance from the pollen record may therefore be seen as a response to deteriorating climatic conditions. Overall, the pollen record may represent the middle and later phases of an interglacial vegetation cycle, corresponding with the mesocratic, oligocratic and initial cryocratic phases of Iversen (1958) and Andersen (1966) (see also Birks, 1986).

The pollen assemblage from Dalcharn is similar in some respects to that described from Fugla Ness on Shetland by Birks and Ransom (1969), who equated the latter with the Gortian of Ireland, and hence the Hoxnian of southern England; although a Cromerian origin for the Fugla Ness record was not excluded. However, on present evidence it is not possible to firmly attribute either the Dalcharn or the Fugla Ness record to a particular interglacial within the Middle or Late Quaternary. Similarly, correlations with other interglacial or interglacial/interstadial sites in Scotland cannot be made. The pollen records from Sel Ayre on Shetland (Birks and Peglar, 1979), Toa Galson in north-west Lewis (Sutherland and Walker, 1984) and Abhainn Ruaival on St Kilda (Sutherland *et al.*, 1984) are characterized by open grassland or heathland vegetation, and there are difficulties in establishing correlations with the palaeosols at Teindland (Edwards *et al.*, 1976) and Kirkhill (Connell *et al.*, 1982) in north-east Scotland (Lowe, 1984; Walker, 1984b).

The pollen assemblage at Dalcharn represents the first record of an undoubted interglacial deposit beneath tills in the Moray Firth area, and is the first to provide unequivocal evidence of interglacial pine forest and its history. The presence of till beneath the weathered gravel at Dalcharn is also the first reported occurrence of a glacial deposit formed prior to at least one interglacial of the Middle or Late Quaternary on this part of the Scottish mainland. This recognition of pre-Late Devensian glacial and interglacial sediments has critical implications for the interpretation of multiple till sequences throughout northern Britain, which have hitherto been attributed to variations in the direction of movement within a single Late Devensian ice-sheet, but which may in fact represent successive earlier glacial episodes.

Figure 7.4 Relative pollen diagram for the Dalcharn 'biogenic formation', showing selected taxa only as percentage of total land pollen (from Walker, 1990a).

Conclusion

The sequence of deposits at Dalcharn is of considerable importance for the evidence it provides for the climatic and glacial history during the Quaternary. Although dating has yet to be firmly established, the length and detail of the record are exceptional, including evidence for an interglacial (temperate climate) and episodes of multiple glaciation and periglacial conditions. The interglacial deposits are significant in providing the first clear record of interglacial pine forest development in Scotland. The site has outstanding potential for elucidating further the glacial history of Scotland.

ALLT ODHAR

J. W. Merritt

Highlights

The stream section at Allt Odhar contains a bed of peat preserved within a sequence of glacial deposits. Analysis of pollen, plant-macrofossil and beetle remains, preserved in the peat, has allowed a detailed reconstruction of environmental conditions during an Early Devensian interstadial, the only one so far in Scotland to be unequivocally dated. The deposits also have significant potential for establishing a detailed glacial history of the area.

Introduction

The site (NJ 798368) is a river cliff located 16 km south-east of Inverness. It lies at *c.* 370 m OD immediately upstream of the confluence of the Allt Odhar and the Caochan nan Suidheig. A deposit of compressed peat was found towards the base of the section in 1988, during the systematic resurvey of Sheet 84 (Fortrose) by the

British Geological Survey. The peat, which contains pollen, insect remains and plant macrofossils of interstadial affinity, occurs above a weathered till and there is at least one till higher in the sequence (Merritt, 1990c). The precise age of the peat is in some doubt, but there is a convergence of evidence suggesting that it accumulated during an Early Devensian interstadial. The close proximity of this site to the Dalcharn interglacial site (see above) is of major significance in Scottish Quaternary research. They are the first sites from the mainland of Scotland to provide evidence of wooded conditions during both an interstadial and an interglacial period of the Middle or Late Quaternary. The Allt Odhar deposits provide the most detailed record yet published from a Scottish site of vegetational change during a pre-Late Devensian interstadial (Walker, 1990b; Walker *et al.*, 1992).

Description

The lithostratigraphy in the vicinity of the Allt Odhar site, given below, is that established by Merritt (1990c), with minor modifications after Walker *et al.* (1992):

6.	Blanket Peat	up to 2 m
5b.	Sheet-wash Gravel	up to 1.5 m
5a.	Carn Monadh Gravel	up to 10 m
4.	Moy Formation:	
c	Upper Till Member	up to 10 m
b	Lower Till Member	up to 6 m
a	Paraglacial Member	up to 2.2 m
3.	Odhar Peat	up to 0.6 m
2.	Odhar Gravel	up to 1.5 m
1.	Suidheig Till	at least 1.5 m

The lithostratigraphy is based on several sections because no single exposure reveals the complete sequence. At the Allt Odhar section (Figure 7.5) only units 1 to 4b are present.

The Suidheig Till (bed 1) is only recognized unequivocally at the type section, where it comprises a very stiff, light brown to moderate yellowish brown, massive, matrix-supported diamicton. Many of the clasts are decomposed and have orange weathering rinds. The nature and composition of the diamicton is similar to that of the lower till member (bed 4b) of the Moy Formation, but it is more deeply weathered.

The Odhar Gravel (bed 2) is a dense, poorly sorted, cobble gravel with a ferruginous pan towards the base. Pink granite is the dominant lithology, many clasts being unsound. The less abundant clasts of gneiss and schist are commonly decomposed. The deposit is fluvial in origin, possibly glaciofluvial.

The Odhar Peat (bed 3) (Figure 7.6) lies within a shallow depression at the top of the underlying gravel. Four distinct beds are apparent:

(i)	pebbly, peaty sand	0.2–0.3 m
(ii)	black amorphous peat with sand wisps	0.15–0.3 m
(iii)	compressed, felted, fibrous peat	0.35 m
(iv)	interlaminated sand and peat	0.2 m

The sand is generally bleached and the deposit as a whole most probably accumulated in a soligenous mire (Walker, 1990b). The results of pollen analysis on the Odhar Peat are reported by Walker (1990b) and Walker *et al.* (1992). Three pollen assemblage zones are recognized (Figure 7.7). A small number of plant macrofossil types were also recovered (cf. *Campanula* sp. (p), *Carex* sp. (p), *Cenococcum geophilum* (Fr.), *Montia fontana* ssp. *fontana* L., *Selaginella selaginoides* (L.) Link and *Viola* sp. (p.) (Walker *et al.*, 1992).

The lower part of the deposit (beds iii and iv) has yielded the remains of 31 taxa (23 species) of fossil insect (Coleoptera) (Walker *et al.*, 1992). The species generally show a preference for humus-rich or peaty soils or damp habitats (e.g. *Patrobus assimilis* Chaud., *Pterostichus diligens* (Sturm) and *Diacheila polita* Fald.), including deciduous woodland (*Pterostichus niger* (Schall.)) and thinly wooded environments with drier soils (*Calathus melanocephalus* (L.)). Five are no longer present in Britain (*Diacheila polita* Fald., *Helophorus* cf. *glacialis* Villa, *Olophrum boreale* (Payk.), *Euconecosum norvegicum* Munst. and *Boreaphilus henningianus* Sahlb.) but occur today in Fennoscandia. *Diacheila polita* Fald. is characteristic of tundra environments, but also occurs on the northern margins of the boreal forest.

Radiocarbon dating of samples from near the base of bed (iii) and from near the top of bed (ii) in both cases gave age estimates >51,100 BP (SSR–3677 and SSR–3678), indicating that the materials are older than the upper limit of radiocarbon dating (Harkness, 1990). A uranium-series disequilibrium age estimate of 124 ka ± 13 ka was initially obtained on a sample of peat from bed (iii) (Heijnis, 1990). Subsequently, based on additional measurements, a revised

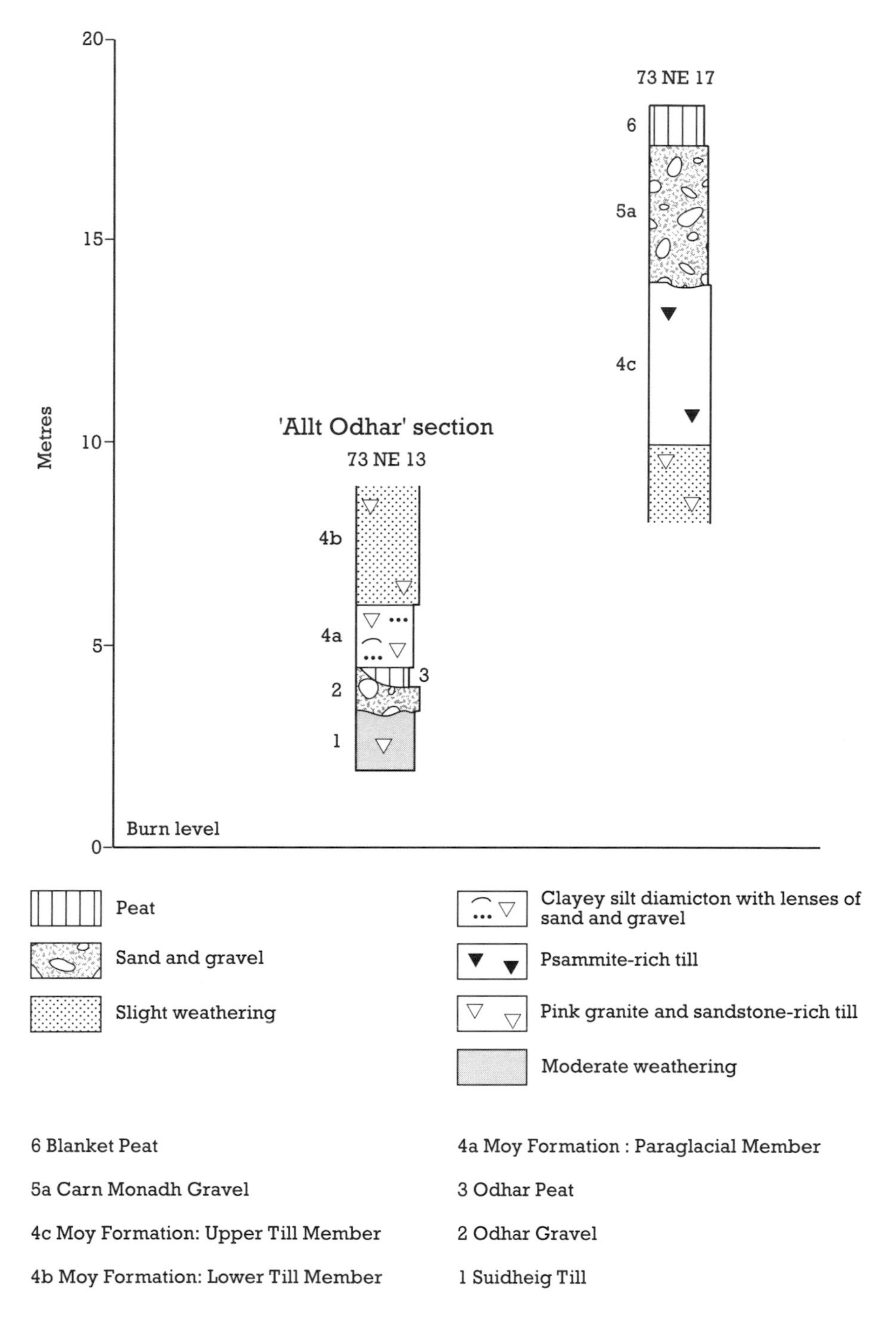

Figure 7.5 Sediment logs of the Allt Odhar and adjacent sections (from Merritt, 1990c).

estimate of 106 ka +11/−10 ka was obtained (Walker *et al.*, 1992).

The Moy Formation comprises two till members and a paraglacial member. The paraglacial member (bed 4a) is an extremely compact unit of pebbly clayey silt diamicton and silty sand with lenses of sand and gravel. The upper and lower contacts of the member are gradational, and there is evidence that the deposits have been consolidated and sheared subglacially. The unit was probably originally deposited by debris-flow processes, either proglacially or in periglacial conditions prior to the arrival of glacier ice.

Throughout the area, a psammite-rich till (bed 4c) overlies a pink granite-rich and sandstone-

Figure 7.6 Section at Allt Odhar showing the Odhar Peat resting on the Odhar Gravel and overlain by the Paraglacial Member of the Moy Formation. (Photo: D. G. Sutherland.)

rich one (bed 4b). Both tills are lodgement tills as defined by Dreimanis (1989). The upper till member (bed 4c) comprises very stiff, olive grey to pale olive grey, massive, matrix-supported diamicton. The lower till member (bed 4b) is more sandy, its colour varies from moderate yellowish brown to pale olive grey and it contains clasts with orange weathering rinds. There is no unequivocal evidence that the upper and lower members formed in more than one glacial episode, but the generally greater degree of weathering of the latter may indicate that it is the product of an earlier glaciation.

The Carn Monadh Gravel (bed 5a) appears to be restricted to the valley of the Allt Odhar. It mainly comprises thinly bedded silty sandy gravel with distinct planar subhorizontal stratification and it was probably deposited in ice-marginal fans as the higher ground became free of ice during deglaciation. The Sheet-wash Gravel (bed 5b) is a coarse, poorly sorted deposit that caps most of the river cliffs in the area. The overlying blanket peat (bed 6) contains pine stumps near the base.

Interpretation

The results of pollen analysis on the Odhar Peat (Walker, 1990b; Walker *et al.*, 1992) reveal that a landscape of birch woodland, with juniper and willow scrub interspersed with open grassland (pollen zone AD–1), was replaced first by grassland and heathland (zone AD–2) and then by an open landscape dominated by species-poor grass and sedge communities (zone AD–3). The pollen record reflects an episode of climatic amelioration, followed by a decline in temperature accompanied, perhaps, by a shift to wetter climatic conditions, and finally to a markedly more severe climatic regime. The scarcity of pine pollen, relatively low arboreal pollen counts, the absence of thermophilous taxa and the presence of herbaceous taxa with northern or montane affinities all indicate a climatic regime markedly cooler than that of a full interglacial. This conclusion is strongly supported by analysis of the fossil Coleoptera, which suggests a cool to cold climate similar to that occurring today in the birch zone of the Scandinavian mountains, with

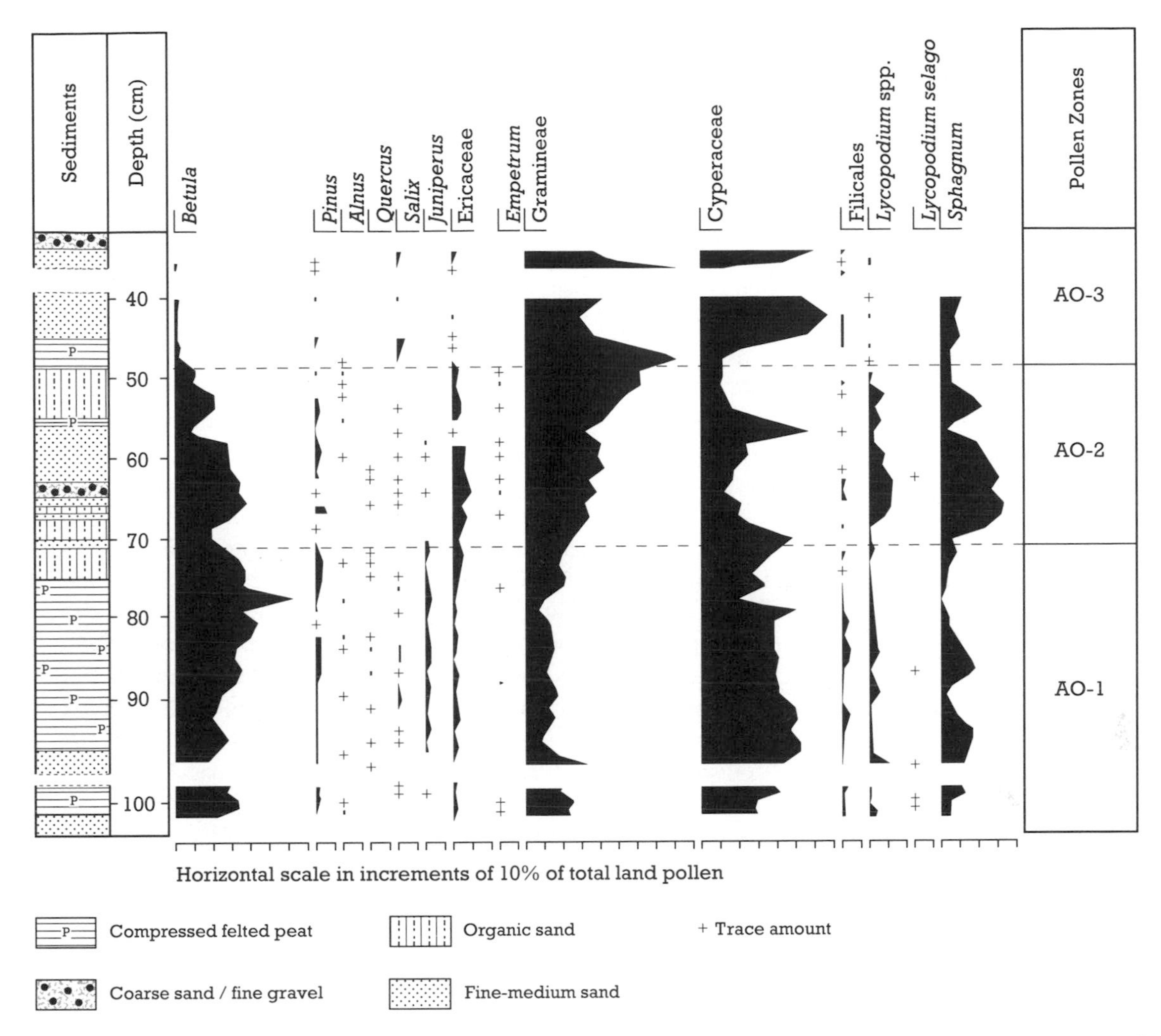

Figure 7.7 Relative pollen diagram for the Odhar Peat, showing selected taxa as percentages of total land pollen (from Walker, 1990b).

mean July temperatures a little above 10°C and colder winters than at present (Walker *et al.*, 1992). The pollen and insect data taken together strongly indicate that the Odhar Peat is more likely to have formed during an interstadial than an interglacial.

The radiocarbon dates from near the base and the top of the Odhar Peat indicate that the deposit is older than Middle Devensian (Harkness, 1990). Hence, it pre-dates organic remains from a number of previously published Scottish Devensian interstadial sites, including Tolsta Head on Lewis (von Weymarn and Edwards, 1973), Sourlie near Glasgow (Jardine *et al.*, 1988) and Crossbrae Farm near Turriff in north-east Scotland (Hall, 1984b; Hall and Connell, 1991), from which finite dates in the range 22,000–29,000 BP have been obtained. The relatively high frequencies of birch pollen in the spectra from Allt Odhar also suggest that the Odhar Peat is different in age from organic horizons at Teindland (Edwards *et al.*, 1976), Burn of Benholm (Donner, 1979) and Abhainn Ruaival on St Kilda (Sutherland *et al.*, 1984), where the radiocarbon dates were either infinite or best regarded as minimal (Sutherland, 1984a), and where the pollen shows evidence only of open grassland.

The uranium series date of *c.* 106 ka on the Odhar Peat places it firmly in the Early Devensian (Walker *et al.*, 1992). The interstadial episode may therefore be the terrestrial equivalent of Oxygen Isotope Substage 5c of the ocean record which has been dated using the technique of 'orbital tuning' to 103.29 ± 3.41 ka (Martinson *et al.*, 1987). On the basis of the uranium series date and the pollen and insect evidence, Walker *et al.* (1992) have argued that the nearest correlative of the Odhar Peat is the interstadial deposit at Chelford in Cheshire (Simpson and West, 1958; Coope, 1959). This deposit has a thermoluminescence (TL) date in the range 90–100 ka (Rendell *et al.*, 1991), and on the basis of amino acid geochronology, the Chelford Interstadial is considered to correlate with Substage 5c (Bowen, 1989). Correlation of Allt Odhar peat with Substage 5c also allows wider comparisons with pollen records from similar interstadial sites on the European mainland (Walker *et al.*, 1992). It appears that open birch forest predominated along the north-west margin of Europe, with boreal forest of pine, spruce and birch to the south and east.

The date from Allt Odhar is broadly in agreement with uranium series dates from Chelford, but other provisional TL and OSL dates at both Chelford and Allt Odhar are significantly younger than indicated by the uranium series dating (H. McKerrell, unpublished data). More work is clearly required to resolve this apparent anomaly, which has far-reaching implications for Quaternary geochronology.

Psammite-rich till containing few sandstone erratics commonly overlies diamictons with abundant clasts of sandstone and flagstone over the high ground flanking the coastal lowlands of the Moray Firth between Inverness and Nairn. This relationship was first recognized by Fraser (1880) in Strathnairn and it was substantiated during the primary geological survey of the Inverness area (Horne, 1923). It is clear that material derived from outcrops of Old Red Sandstone along the Moray Firth coast and around Loch Ness was transported eastwards and upwards on to the highest ground in the area (Sissons, 1967a). The final movement of ice, however, was towards the north-east, as indicated by the orientation of glacial striae, streamlined landforms and the clast composition of the youngest, psammite-rich tills (Merritt, 1990a). The precise ages and interrelationship of the psammite-rich and sandstone-rich tills are still unclear, but it is now certain at the Allt Odhar site that both are Devensian in age. The greater degree of weathering of the sandstone-rich till suggests that it could be an Early Devensian deposit, and the psammite-rich till a Late Devensian deposit. The Suidheig Till is almost certainly pre-Devensian in age. The Suidheig Till, together with the Dearg Till (?Anglian) at Dalcharn, are the only definite pre-Devensian tills known in mainland northern Scotland (cf. Worsley, 1991). The sections in the vicinity of the Allt Odhar site and those at Dalcharn offer particularly good opportunities to test new methods of dating on glacigenic sediments.

The Allt Odhar site is a critical reference site for Quaternary studies in Scotland. First, it contains the most detailed pollen record so far for vegetation change during a pre-Late Devensian interstadial in Scotland. Second, it includes the only organic deposits dated to the Early Devensian and therefore provides a unique record of environmental conditions at that time. Third, the pollen record is the first to demonstrate unequivocally that birch woodland, as opposed to grass and heathland, occurred in Scotland during a Devensian interstadial. Fourth, Allt Odhar has also provided the first pre-Lateglacial insect fauna from Scotland. Fifth, the uranium series date on the peat is the first from such a deposit that lies beyond the age of radiocarbon dating. Sixth, the correlation of the organic sediments with those from Chelford in England and with interstadial sites on the European mainland makes the site internationally important for establishing wider patterns of vegetation and climate during Oxygen Isotope Substage 5c. Seventh, Allt Odhar is notable for its sequence of glacial deposits. It is one of only two sites in mainland northern Scotland where pre-Devensian tills can be demonstrated. In addition, there is significant potential for further research, which may allow the establishment of Early Devensian glaciation in northern Scotland.

Conclusion

Allt Odhar is a site of great importance for Quaternary studies in Scotland. A bed of peat within a sequence of glacial deposits has yielded pollen and beetle remains that provide, so far, unique evidence for environmental conditions during the Early Devensian (around 106,000 years ago). They show a climatic deterioration and also that a landscape of birch woodland gave way to a more barren one with grass and sedge communities as the climate deteriorated. The site is also notable for older, pre-Devensian, glacial deposits, one of only two sites in northern Scotland where such deposits are known, and also a further till at the locality may be of Early Devensian origin.

CLAVA

J. E. Gordon

Highlights

The glacial deposits at Clava are famous for a shelly clay which has been interpreted either as an *in situ* marine deposit or a glacially transported raft of sediment. These deposits are Early or Middle Devensian in age. The site also has evidence for ice-sheet glaciation both older and younger than the shelly clay.

Introduction

The site at Clava (NH 766442), located at an altitude of *c.* 150 m OD, 9 km east of Inverness, comprises a long-disused claypit and a series of sections along the lower Cassie Burn and the lower Finglack Burn, both tributaries of the River Nairn. The deposits, proved in the sections and in boreholes, form a complex glacigenic succession and include the famous, 'arctic' shelly clay. The latter was first described by Fraser (1882a, 1882b) in an old claypit excavated into a broad terrace feature (see Fraser, 1880). The shelly clay excited considerable controversy in the last century, concerning whether it was *in situ*, thereby representing a great submergence of the country during the Pleistocene, or whether it was transported by ice from offshore. This controversy was recently revived (Sutherland, 1981a), but subsequent detailed investigation by Merritt (1990b) has provided support for a glacially transported origin for the deposit. The site and its significance have been widely discussed in the literature, but the principal references are by Fraser (1882a, 1882b), Horne *et al.* (1894), Sissons (1967a), Peacock (1975b), Holden (1977a), Sutherland (1981a) and Merritt (1990b, in press).

Description

Fraser (1882a, 1882b) noted shelly clay at the bottom of the claypit (now disused and overgrown), where it was overlain by fine, stratified sand, which was, in turn, overlain by boulder clay, soil and gravel. Chemical analysis of the shelly clay by W. I. Macadam showed it to be similar to clays derived from 'mixed gneiss and schist districts in the Highlands'. Although fragile, some of the shells were intact, with their periostraca preserved, and Fraser inferred that they occurred *in situ* in a former sea-bed deposit and at a similar altitude to the one supposed to exist at Chapelhall, near Airdrie (Smith, 1850a, 1850b). Fraser provided lists of mollusc shells, Foraminifera, ostracods and barnacles identified in the Clava deposits by T. F. Jamieson and D. Robertson, and considered that several species were diagnostic of arctic and shallow water marine conditions.

In view of the controversy over the Clava marine clay and its designated keystone role in the whole concept of glacial submergence (see below), a Committee of the British Association for the Advancement of Science was convened to carry out further investigations on the deposits. In the vicinity of the disused claypit (the 'Main Pit'), they excavated two pits and sank seven boreholes, which contributed greatly to the detailed knowledge of the site (Horne *et al.*, 1894). They established that the complete sequence of deposits was (Figure 7.8):

6.	Surface soil and sandy boulder clay	13.1 m
5.	Fine sand	6.1 m
4.	Shelly blue clay with stones in lower part	4.9 m
3.	Coarse gravel and sand	4.6 m
2.	Brown clay and stones	6.6 m
1.	Old Red Sandstone bedrock	—

Bed 4 was confirmed to be a marine deposit extending for at least 170 m and reaching a maximum thickness of 4.9 m. It was essentially horizontal and had well-defined contacts with the adjacent beds above and below. There was little sign of disturbance, although cracks and fissures were noted. Silt and clay were the main constituents, with a small number of clasts at the base. Of the latter, 59% comprised micaceous gneiss and 17% only of the local Old Red Sandstone. One piece of supposed Jurassic grit was found, the nearest source being *c.* 20 km to the north. Organic remains in the deposit were identified by D. Robertson, supplementing Fraser's original faunal list. They were shallow-water species representing an arctic or sub-arctic faunal assemblage. However, the variety of species was poorer than that recorded in the Lateglacial marine clays of the Clyde estuary or the east coast. The shells were not striated and were generally well-preserved with the periostraca intact, although some were partially crushed.

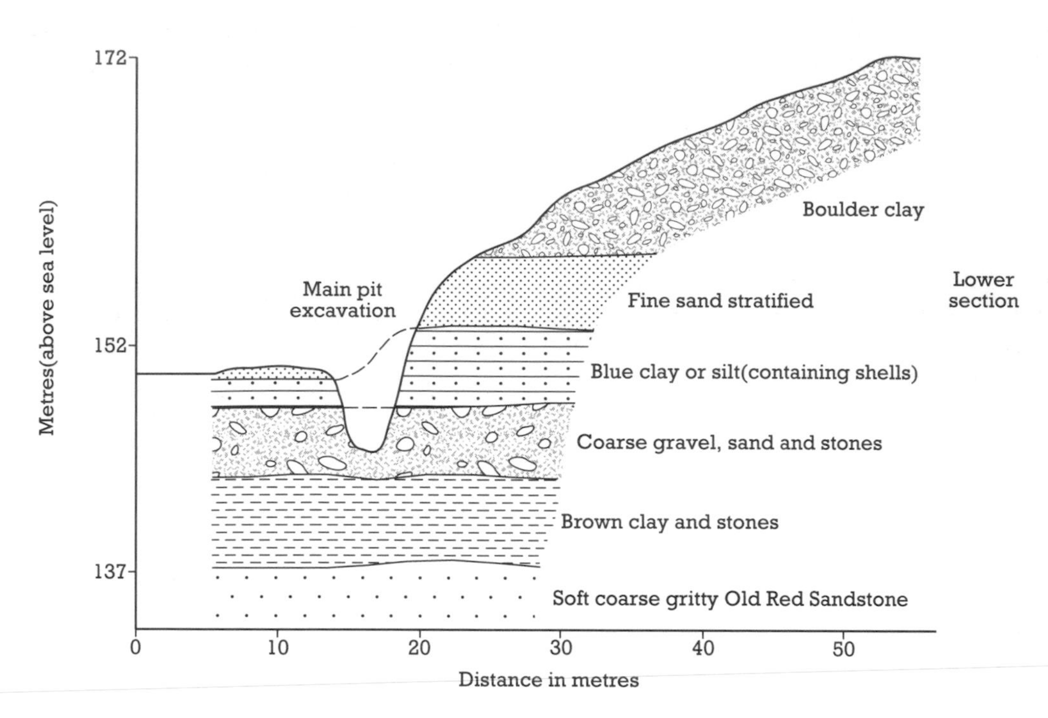

Figure 7.8 Clava: lithological succession at the 'Main Pit' (from Horne *et al.*, 1894).

Compressed annelid burrows were also observed in the marine clay. Clasts in beds 3 and 6 were predominantly derived from the local bedrock, up to 76% in the case of the latter.

Subsequently the deposits were reinvestigated by Peacock (1975b). Although the original sections were no longer exposed, fresh ones had appeared along the Cassie Burn a few hundred metres to the south-west. Here, Peacock described three main sedimentary units. At the base was a till varying in composition from a stiff, silty till to almost stoneless soft sand and silt, and with marine shells, including *Portlandia arctica* (Gray), which had not previously been recorded at Clava. Above the till was a bed of poorly sorted sand and gravel, and on top a bed of silty till interbedded with discontinuous layers and streaks of finely laminated silt, sand and fine-grained gravel.

By interpolation, Peacock correlated these beds with beds 4, 5 and 6 respectively in the succession reported by Horne *et al.* (1894), although he did not relocate the shelly clay of Horne *et al.* (1894). Subsequently, from a detailed reinvestigation of the various sections along the Cassie Burn and the Finglack Burn, Merritt (1990b, in press) has proposed a composite succession. The full lithostratigraphy recognized by him and the provisional correlations with the successions of the British Association Committee (Horne *et al.*, 1894) are as follows:

	British Association bed numbers
12. Diamictic gravel	–
11. Finglack Till (flow-till facies)	–
10. Finglack Till (melt-out facies)	–
9. Finglack Till (lodgement facies)	6
8. Finglack Till (resedimented)	–
7. Glaciofluvial Ice-contact Deposits	–
6. Clava Sand	5
5. Clava Shelly Clay	4
4. Clava Shelly Till	–
3. Drummore Gravel	3
2. Cassie Till	2
1. Bedrock	1

Merritt interpreted the Cassie Till (bed 2), known only from the British Association boreholes, as a lodgement or basal melt-out till. The

Drummore Gravel (Cassie Gravel of Merritt, 1990b) is exposed along the lower Finglack Burn, where it comprises principally a stratified, matrix-supported, gravelly, silty sand diamicton with a gravel composition of mainly sandstone and flagstone, with some metamorphic clasts; many of the gneiss and schist clasts are weathered. Merritt interpreted this unit as comprising subaerial sediment gravity flows that accumulated in an ice-marginal or supraglacial environment. From the relatively greater degree of weathering of this deposit, compared with the overlying bed, he inferred that it formed during an earlier glacial episode.

The Clava Shelly Till is exposed on the west side of the Cassie Burn and is described by Peacock (1975b) and Merritt (1990b). It comprises a stiff, matrix-supported diamicton with a matrix of silty, fine-sandy clay and contains clasts mainly of metamorphic rocks, but with some of granite and sandstone. Graham (1990) has confirmed the presence of *Portlandia arctica* in the deposit and listed a sparse microfossil assemblage. The presence of this bivalve provides evidence for a fully arctic environment (Peacock, 1975b; Graham, 1990) in contrast to the Clava Shelly Clay.

The Clava Shelly Clay is not presently exposed, and the main source of information has therefore been from the British Association investigation. Re-excavation of a temporary section at the site of the Main Pit, for a Quaternary Research Association field meeting in 1990, confirmed the principal findings, with particular attention drawn to the heavily sheared nature of the sediments. Graham (1990) has recently updated the taxonomic lists of the macrofauna and microfauna of the shelly clay (Table 7.1). He noted the absence of exclusively arctic forms and concluded that the assemblage was indicative of a high-boreal or colder environment. Amino acid analysis of museum specimens of *Littorina littorea* (L.) yielded a mean ratio (D-alloisoleucine: L-isoleucine) of 0.04, suggesting that the shells may be Middle Devensian in age (D. Q. Bowen, unpublished data). Radiocarbon dating of a museum specimen of *Littorina littorea* also has yielded a Middle Devensian age (43,800 ± 3,300 BP, OxA–2483), but this date should be regarded as a minimum age, being so close to the limit of radiocarbon dating. Further radiocarbon dating of field specimens collected in 1990 has yielded infinite age estimates: *Astarte sulcata* (da Costa), >41,200 (OxA–2483) and *Littorina littorea*,

Table 7.1 The macrofauna of the Clava Shelly Clay (from Graham, 1990)

Gastropoda
Boreotrophon clathratus (Ström)
Buccinum undatum Linné
Littorina littorea Linné
Littorina sp.
Littorina saxatilis (Olivi) *rudis* (Maton)
Lunatia pallida (Broderip and Sowerby)
Lunatia sp.
Margarites helicinus (Fabricius)
Margarites groenlandicus (Gmelin)
Neptunea antiqua (Linné)
Oenopota scalaris (Moeller)
Oenopota trevelliana (Turton)
Oenopota turricula nobilis (Moeller)
Oenopota turricula (Montagu)
Oenopota sp.
Omalogyra atomus (Philippi)
Rissoa parva? (da Costa)
Bivalvia
Astarte sulcata (da Costa)
Cerastoderma edule (Linné)
Lepton nitidum Turton
Macoma balthica (Linné)
Macoma calcarea (Gmelin)
Macoma sp.
Mytilus edulis (Linné)
Nicania montagui (Dillwyn)
Nucula sp.
Nuculoma tenuis (Montagu)
Nuculana pernula (Müller)
Nuculana sp.
Thyasira flexuosa (Montagu)*
Tridonta elliptica (Brown)
Tridonta sp.
Yoldiella lenticula s.l. (Müller)
Yoldiella sp.
bivalve indet.
mytilacean fragments
unidentifiable bivalve fragments
Cirripedia
Balanus balanoides Linné
Balanus crenatus? Bruguière
Balanus sp. plates
Decapoda
crustacean claw

* probably a misidentification of one of the colder water species *Thyasira gouldi* (Philippi) or *Thyasira sarsi* (Philippi).

>43,000 (OxA–2876) (Merritt, in press).

The Clava Sand at the Main Pit shows poorly defined subhorizontal lamination and a well-developed system of clastic veins, which Merritt (1990b) has interpreted as the product of brittle fracture while the material was frozen and either overridden by, or transported within, glacial ice. Evidence of shearing of the deposit is also present near the junction with the overlying till. Clava Sand is also found overlying Clava Shelly Till in the Cassie Burn section (Merritt, 1990b), where it is folded as well as being cut by microfaults and shear planes. The Clava Shelly Till has been folded also.

The Glaciofluvial Ice-contact Deposits unit (bed 7), exposed on the west bank of the Cassie Burn, is a part waterlain and part mass-flow deposit, which is cut by a series of shear planes (Merritt, 1990b).

The Finglack Till is exposed in sections along the lower Finglack Burn and on the west side of the Cassie Burn (Peacock, 1975b; Merritt, 1990b). The clasts consist mainly of sandstone and flagstone, and the unit comprises a succession of lodgement, melt-out and flow-till facies. The origin of the overlying Diamictic gravel is uncertain (Merritt, 1990b).

Interpretation

In the period following its discovery, the significance of the Clava Shelly Clay was considered by both protagonists and antagonists of the glacial submergence theory of the time (cf. Davies, 1968a). Richardson (1882) referred to Clava as one of a number of high-level arctic shell beds in the British Isles indicating an extensive submergence during the glacial period. He also correlated these high-level deposits with shelly clays at many lower-level sites around the coast of Scotland.

Jamieson (1882b) thought the Clava deposit implied a similar amount of submergence to that which he inferred from the quartz and flint gravels at Windy Hills. Wilson (1886) associated the shelly clay at Clava with the so-called interglacial beds of Aberdeenshire (for example at Kippet Hills) and ascribed them to the same submergence. Crosskey (1887) re-examined the Clava shell-bed and supported Fraser's conclusion that it was a true *in situ* sea-bed deposit, citing its lack of disturbance and mixing with other debris, its sharp junction with the overlying sands and the preservation of the distinctive arctic shell assemblage.

Bell (1893a, 1893b, 1895a), however, argued that because of their limited extent and the lack of marine organisms in the overlying deposits, both the Clava and Chapelhall shell beds had been transported to their present positions by land ice. He considered that if subsequent glaciation had removed all traces of high-level marine deposits except at a few localities, it was difficult to explain why there were no marine remains in the overlying till. Significantly, also, clasts in the marine clay at Clava were not derived from local Old Red Sandstone rocks. In the case of Clava, Bell suggested that ice issuing from the Great Glen was deflected eastwards by an ice barrier in the Moray Firth and that it crossed part of the sea floor before reaching Clava. He also argued that the high-level shelly deposits in North Wales, Ireland and at Chapelhall did not provide substantive evidence for a 'great submergence' during the glacial period (Bell, 1891a, 1893b).

The majority of the British Association Committee concluded that the marine deposits were *in situ*, indicating former submergence of the land up to about 150 m OD. As evidence they cited the assemblage of organic remains, their mode of occurrence, the extent of the deposits and their apparently undisturbed character. A minority of the Committee (Bell and Kendall), however, argued that there was insufficient evidence to reach a firm conclusion and, moreover, doubted that there was any substantial evidence at all in Scotland for a great submergence. They questioned the widespread absence of shell beds and other traces of submergence and the lack of marine organisms in the overlying till. Although acknowledging certain difficulties, notably the extent of the deposit and the good preservation of the shells, they favoured an ice-rafted origin for the shelly clay, with a source area in Loch Ness, judging from ice-movement patterns inferred from striae and erratics.

In view of the lack of unanimity on the conclusions to the British Association Report, it was not surprising that debate on the origin and implications of the Clava shell bed continued. Indeed, Clava assumed even greater significance following the failure of a similar British Association Committee to relocate the shell bed at Chapelhall (Horne *et al.*, 1895). Further shell beds, however, were found in Kintyre (Horne *et al.*, 1897). In the years immediately following the

British Association investigations, Bell (1896a, 1897a, 1897b) continued to argue against the Clava and other similar deposits being *in situ* and resulting from an extensive marine submergence. He received support from Lamplugh (1906). In contrast, Reade (1896), Smith (1896a) and Jamieson (1906) maintained the view that the deposits were the product of a major marine submergence.

In the Geological Survey Memoir for the area, Horne (1923) summarized the findings of the British Association Committee, but produced no new evidence. Later, Bremner (1934a) speculated that the Clava marine beds might be preglacial, since there was no boulder clay beneath them. Bourcart (1938), however, suggested a possible correlation of the Clava marine clays with the Tyrrhenian marine transgression (Mindel–Riss interglacial). Charlesworth (1956) also favoured the view that the deposits were *in situ*, representing an interglacial submergence.

In contrast, Sissons (1967a) considered that the Jurassic erratic, the low content of local rocks and the marine fossils themselves all suggested that the material had been ice-rafted from the north prior to the last ice movement, from south-west to north-east.

Despite the inconclusiveness of the British Association Report and the significant implications for the Pleistocene history of Scotland if the deposits were *in situ*, there was no reinvestigation of the area until the work of Peacock (1975b). He argued that the shelly clay at Clava was part of a till unit comprising reworked sea-floor material. It was probably an autochthonous melt-out till (cf. Boulton, 1968) rather than a lodgement or flow till, since the preservation of annelid burrows precludes resedimentation. The existence of such intact erratics of marine sediments, often with well-preserved fossils and structures, has been widely documented (Jamieson *et al.*, 1898; Lamplugh, 1911; Debenham, 1919; Read, 1923; Eyles *et al.*, 1949; Peacock, 1966, 1971a), and possible mechanisms of entrainment have been proposed (Moran, 1971; Boulton, 1972a; Banham, 1975). Peacock related the characteristics of the sand and gravel bed (bed 5 in the succession of Horne *et al.*, 1894) along the Cassie Burn to deposition in a high-energy, fluvial environment and he explained the overlying beds (see above) as a succession of flow tills deposited in an environment similar to that of modern ice margins in Svalbard (Boulton, 1968).

Holden (1977a) considered that the last ice movement in the area was from the Great Glen and that it was moving, therefore, in the wrong direction to transport sea-floor materials inland. He argued that the Clava Shelly Clay was similar in characteristics and location to those at Tangy Glen in Kintyre and at Afton Lodge in Ayrshire. From his investigation of the Afton Lodge deposits, and in view of the apparent problems entailed in the ice-rafting explanation at Clava and Kintyre, he concluded that all three were *in situ* and represented a pre-Devensian sea-level stand at between 115 m and 150 m OD. Sutherland (1981a) also accepted an *in situ* origin for the high-level shell beds and presented a model relating these deposits to glacio-isostatically induced submergence in front of an expanding ice-sheet. However, the deposits at Clava fit only partly into the overall distribution pattern of the high-level shell beds and are apparently at too great an altitude to be fully explained by the model.

More recently, the detailed work of Merritt (1990b) has provided a significant advance in resolving the origin of what has been one of the most enigmatic Pleistocene deposits in Scotland. He summarizes the evidence for and against the shelly clay being *in situ*. The arguments in support have included the good state of preservation of the shells, the large size of the bed and its near horizontal form; counter arguments have included the occurrence elsewhere of large ice-rafted deposits with well-preserved shell contents, the lack of other evidence of sea levels at a comparable altitude to the Clava deposits and the distinctive composition of the shelly clay compared with the overlying and underlying deposits. In considering the evidence, Merritt (1990b) noted similarities with large, glacially transported rafts or 'megablocks' in North America (Moran *et al.*, 1980; Aber, 1985, 1989), particularly concerning the presence of glaciotectonic deformations. He correlated the Clava Sand overlying the Clava Shelly Clay at the Main Pit with lithologically similar sand overlying the Clava Shelly Till in the Cassie Burn sections, the latter in turn overlain unconformably by the glaciofluvial sand and gravel unit described by Peacock (1975b). He interpreted the deformation structures in the shelly till and the overlying sands, not simply as the result of overriding by the ice that deposited the Finglack Till, but as structures that were already in place and therefore formed as part of the transport and emplacement processes of these deposits. Peacock (1975b) considered that shelly clay graded into the shelly till, in a similar manner to the rafts of marine clay in the till at Boyne Quarry, near Portsoy (see below); such a pattern would conform with glaciotectonic deformation of megablocks during transportation (cf. Aber, 1985). However, as the shelly till contains a fauna indicative of colder conditions than the main raft of shelly clay, the two formations are not simply derived one from the other; the till is formed from a part of the sequence not preserved at the Main Pit (Merritt, in press). The reconstructed pattern of ice movement suggested that the source of the shelly clay was the Loch Ness Basin, which was inferred to have been an arm of the sea prior to the ice advance. The process of rafting involved the development of high porewater pressures within interbedded sands and clays within the semi-enclosed basin of Loch Ness. The results of amino acid analysis and radiocarbon dating indicate that the shelly clay is Middle or Early Devensian in age and that it was transported by the Late Devensian ice-sheet.

Conclusion

Clava is best known for a bed of shelly clay that has had a significant bearing on interpretations of the Pleistocene history of Scotland. Although it has been suggested that the deposit is an *in situ* marine clay and reflects a phase of high sea level, either before or at the time of the build-up of the last (Late Devensian) ice-sheet, current interpretations indicate that it was transported *en masse* to its present location by the last ice-sheet (approximately 18,000 years ago). Clava is not only a site of historical importance, but is also recognized to be significant for studies of glacial sedimentation.

ARDERSIER

J. E. Gordon and J. W. Merritt

Highlights

The interest at Ardersier comprises ice-contact deposits of glaciomarine origin and a sequence of Lateglacial and Holocene raised shorelines. These features provide important evidence for interpreting the pattern of wastage of the Late Devensian ice-sheet, including a possible readvance, and the changes in relative sea level that both accompanied and followed the period of ice-melting.

Introduction

The site (NH 780562) is located on the Ardersier peninsula on the east coast of the Moray Firth, between Inverness and Nairn. It forms part of a suite of glaciomarine ice-contact deposits and raised shorelines and includes an area of high ground consisting of contorted silts, sands and clays, trimmed on the north and west sides by a series of Lateglacial and Holocene raised shorelines. These features provide significant evidence for interpreting both the pattern of ice-sheet deglaciation during the Late Devensian and subsequent changes in relative sea level. The landforms and deposits at Ardersier have been described by Jamieson (1874), Wallace (1883), Ogilvie (1914), Horne (1923), J.S. Smith (1968, 1977), Small and Smith (1971), Synge (1977b), Synge and Smith (1980) and Firth (1984, 1989b, 1990b) and have featured in most reconstructions of the Pleistocene history of the area. They have been interpreted as demonstrating a major readvance of the Late Devensian ice-sheet ('Ardersier Readvance') (J.S. Smith, 1968, 1977; Synge, 1977b; Synge and Smith, 1980), although this interpretation has recently been challenged (Firth, 1984, 1989b).

Description

The deposits at Ardersier (Figure 7.9) were first described by Jamieson (1874), who recorded, near Kirkton, a small exposure (NH 793561) of grey clay containing shells of arctic molluscs, which was either overlain by, or incorporated within a brownish deposit of gravel and silt. Wallace (1883) later reported further details of the shelly deposit and noted that specimens examined by Jamieson included *Nuculana pernula* (Müller), *Macoma calcarea* (Gmelin) and *Tridonta elliptica* (Brown). Robertson (in Wallace, 1883) identified *Astarte sulcata* (da Costa) and several species of ostracod and Foraminifera. An updated and corrected faunal list by D. K. Graham is presented in Firth (1990b). Although the shells were largely fragmented, Robertson noted that many of the pieces were in a natural position. Horne (1923) subsequently provided additional information on the stratigraphy of the

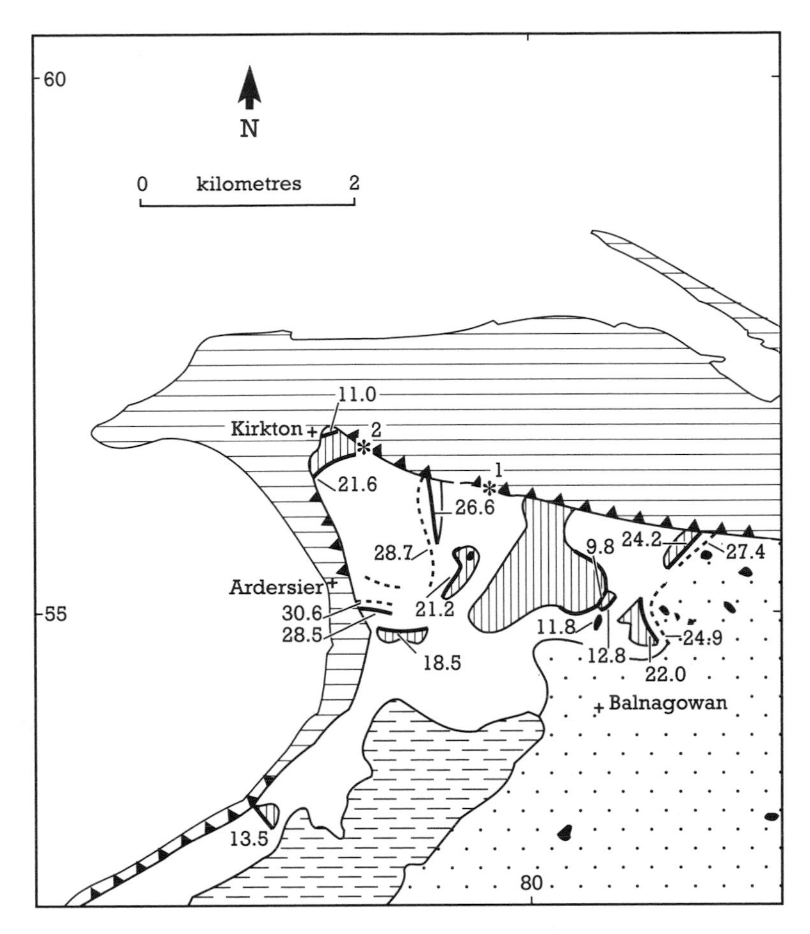

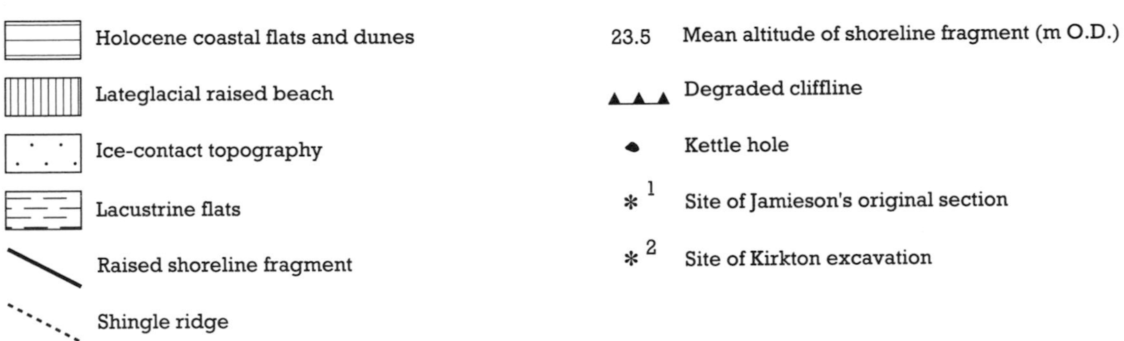

Figure 7.9 Geomorphology of the Ardersier area (from Firth, 1989b).

Kirkton section:

3.	Sand and clay deposit with some stones	1.2 m
2.	Stratified sand	1.8 m
1.	Grey, shelly clay	–

J.S. Smith (1968, 1977) referred to additional sections in the bluff behind the Ardersier village, which 'revealed beds of sand and silt which were folded, faulted and thrust, with inliers of blue clays' (Smith, 1977, p. 74).

Firth (1984, 1989b) described the following sequence in the bluff (NH 783565) east of Kirkton, but not, as he records, at the site of Jamieson's original section:

3. Horizontally bedded sands with well-rounded clasts.
2. Finely laminated sands interbedded with massively bedded silts.
1. Massively bedded, grey clay.

This section was re-exposed in September 1990. The uppermost bed of pebbly sand was not exposed, but the remainder of the sequence was as follows:

4. Obscured 4 m
3. Thinly interbedded clay, very fine sand and silt occurring as graded couplets stacked into discrete units, possibly varves 1.5 m
2. Thinly interbedded and interlaminated clay, silty clay and very fine sand; graded bedding 5 m
1. Silty clay, medium grey, mainly massive but with some graded beds 1.9 m

Although the sequence was conformable, it was seen to be gently dipping and possibly glaciotectonized to a small degree. No faunal remains were recovered from the deposits.

In sections exposed in the fossil cliffline north of Ardersier village, Firth recorded two formations:

2. Undeformed sands and fine gravels (Hillhead Beds).
1. Deformed and/or tilted, massive silts interbedded with layers of clay, laminated fine-grained sands and lenses of gravel (Ardersier Silts).

The Ardersier Silts show convolute bedding and disturbance by load casts and water escape structures. Palaeocurrent directions are towards the north-east. Recent re-examination of the section at Hillhead has revealed a stack of thrust slices.

Re-excavation of the original section of Jamieson (1874), east of Kirkton, in September 1991 revealed the following succession:

3. Sandy diamicton, including mainly finely stratified, but irregular shaped masses of massive stony diamicton; clasts up to 0.5 m size; laminated sand, silt and clay at base above planar contact with bed 2 2.0 m
2. Sand, comprising a stack of thrust-bound slices (0.2–0.4 m thick), intercalated with thin (<0.15 m), sheared silty clay seams 3.2 m
1. Sand, poorly bedded, with steeply dipping thrusts lined by sheared silty clay and folded seams of silty clay <0.2 m thick 3.4 m

No shells were recovered in any of the beds.

The deposits at Ardersier rise to an altitude of about 40 m OD. They were trimmed by the sea during the Lateglacial and Holocene, producing a series of raised shorelines on the north and west side of the peninsula and an extensive relict cliffline (Ogilvie, 1914, 1923; J. S. Smith, 1968, 1977; Synge, 1977b; Synge and Smith, 1980; Firth, 1989b, 1990b). The highest marine features are well-developed Lateglacial shingle ridges at 28–31 m OD (Figure 7.9). Below, Lateglacial shoreline fragments occur at altitudes of 28.5 m, 26.6 m, 21–21.6 m, 18.5 m and 11 m OD (Figure 7.9). Later Holocene changes in relative sea level (see Firth and Haggart, 1989) are demonstrated by the raised beach deposits in front of the relict cliffline. These include the prominent shingle ridges noted by Synge and Smith (1980) near Kirkton, which are associated with the development of the distinctive coastal foreland and spit (Ogilvie, 1914).

Interpretation

Jamieson believed that the marine clay was the remnant of a more extensive deposit which had been destroyed during a later glacier advance. J. S. Smith (1968, 1977) interpreted the deposits, together with complementary features on the north side of the Moray Firth at Fortrose, as a readvance moraine of the last ice-sheet. The shelly, marine clay was translocated from the floor of the firth during the readvance and glaciotectonically deformed with the other deposits to produce the high ground at Ardersier. As supporting evidence for an ice readvance, Smith (1977) also adduced a significant drop in the marine limit west of Ardersier. The concept of a readvance at Ardersier was later reaffirmed by Synge (1977b) and Synge and Smith (1980); these authors recognized it as the first of a series of retreat stages or readvances in their model of deglaciation of the inner Moray Firth and its hinterland.

Synge (1977b) and Synge and Smith (1980) integrated the Ardersier evidence into a general model of Lateglacial shoreline changes and ice-sheet retreat stages in the Moray Firth and Loch Ness areas. Briefly, deglaciation of the area was accompanied by high relative sea level at about 38–42 m OD as the ice retreated from near Nairn to Inverness. Subsequently the ice readvanced to Ardersier, and deglaciation was interrupted by further halt stages (see Chapter 6, Introduction). As ice recession continued west and south-west from Inverness, shorelines formed at 28–34 m, represented at Ardersier by a raised shingle ridge at 28–31 m and raised beach terraces at 28 m OD (Synge and Smith, 1980). Relative sea level continued to fall to a low position and then rose again, extending westwards along Strath Conon where deltas formed at Balblair and Contin at 26 m OD, and into Loch Ness where a clear shoreline developed; at Ardersier a prominent shoreline formed at 24 m OD. Sea level then fell before the Holocene transgression, and its associated shoreline development, which is represented by the raised beach terrace and cliffline at 8–9 m OD at Ardersier and the raised spits on which Kirkton stands.

Aspects of this model have been seriously questioned by Firth (1984, 1986, 1989a, 1989b) (see also Dores, Fort Augustus and Torvean) who presented a detailed reconstruction of relative sea-level changes and ice limits based on instrumental levelling of shorelines combined with geomorphological mapping. From the sedimentary and shoreline evidence both at Ardersier and over the wider area, Firth (1984, 1989b) concluded that the Ardersier Readvance could not be substantiated. He interpreted the Ardersier Silts, which extend up to an altitude of 37.8 m OD, as characteristic of subaqueous outwash deposition (cf. Anderson *et al.*, 1983; Eyles *et al.*, 1983; Eyles *et al.*, 1985; Powell, 1983; Benn and Dawson, 1987), and probably glaciomarine in view of the marine fauna at Kirkton. The Hillhead Beds, which extend up to about 40 m OD, were probably deposited in a high-energy marine environment. Firth considered that the deformation structures in the Ardersier Silts were not indicative of a major ice advance, but probably reflected loading, slumping or minor ice-front movement. It was also significant that major thrust structures and lodgement till, typical of large-scale glaciotectonics (cf. Moran, 1971; Banham, 1975), were apparently absent. Furthermore, analysis of the shoreline data showed no significant drop in the marine limit at Ardersier. The highest in the sequence of ten Lateglacial shorelines identified by Firth includes the highest shingle ridges at Ardersier (30.6 m OD) and is associated with an ice limit near Inverness. The glaciomarine sediments at Ardersier indicate relatively higher sea level when they were deposited at an ice margin at Ardersier, so that sea level dropped from at least 37.8 m to about 30 m OD while the ice retreated from Ardersier to Inverness. However, there is no evidence that this occurred while the active ice front was at Ardersier, as required by the model of Synge (1977b) and Synge and Smith (1980). In contrast, Firth's reconstructed shoreline sequence demonstrated a progressive fall in sea level as the ice-sheet retreated westwards (Firth, 1984, 1989a). Since the shorelines are truncated by the Main Lateglacial Shoreline in the area (Sissons, 1981c; Firth, 1984), which is believed to have formed during the Loch Lomond Stadial (Sissons, 1974d, 1981c), they must have been formed sometime before about 11,000 BP (Firth, 1989a).

The folds and thrust planes in the Ardersier Silts and in the sands at Kirkton, together with the presence of the diamicton at the top of the succession, originally described by Jamieson (1874) and confirmed in the recent excavation, have a significant bearing on the question of ice readvance. The diamicton may be interpreted either as a subaerial mass-flow or glaciomarine deposit, but in either case appears to require the close proximity of ice following glaciotectonism of the underlying finer-grained sediments. A readvance of the ice front is clearly indicated, although the shoreline evidence presented by Firth (1989b) appears to preclude an event of the magnitude suggested by Smith and Synge (J. S. Smith, 1968, 1977; Synge, 1977b; Synge and Smith, 1980).

The Quaternary geomorphology and sediments at Ardersier are important in several respects. First, the juxtaposition of glacial and marine features is particularly significant; it has provided important evidence for interpreting the pattern of ice-sheet recession and relative sea-level change in the inner Moray Firth at the end of the last glaciation. As outlined above, two different reconstructions have been proposed. J. S. Smith (1968, 1977), Synge (1977b) and Synge and Smith (1980) believe that the deposits at Ardersier represent a readvance of the last ice-sheet. Elsewhere in eastern Scotland similar readvances are now largely discounted (Sissons, 1974c; Sutherland, 1984a), apart from the Elgin Oscillation (Peacock *et al.*, 1968), although in north-west Scotland the Wester Ross Readvance is based on clear geomorphological evidence (Robinson and Ballantyne, 1979) (see Gairloch Moraine). The work of Firth (1984, 1989b), however, suggests a different interpretation for the Ardersier evidence, which does not require a major ice readvance but rather progressive ice recession. Ardersier is therefore one of a number of key sites in Scotland for interpreting the mode of deglaciation of the last ice-sheet.

Second, Ardersier is important for the sediments that comprise the core of the higher ground. These have a significant bearing on the interpretation of the mode of deglaciation and the contemporary sedimentary processes and environments. In the model of J.S. Smith (1968, 1977), Synge (1977b) and Synge and Smith (1980) these sediments represent ice-transported and deformed marine deposits; in the model of Firth (1984, 1989b), they represent a glaciomarine sequence deposited in front of a stationary ice margin. The Ardersier deposits therefore provide key evidence for reconstructing deglacial events in the inner Moray Firth and have significant potential for further sedimentological investigation, particularly at the original section described by Jamieson (1874) at Kirkton.

Third, Ardersier is notable for its shell-bearing deposits (Jamieson, 1874; Wallace, 1883). Although these have not been relocated, they nevertheless offer significant potential for establishing firm dating control on the deglaciation of the inner Moray Firth for the first time.

Fourth, Ardersier provides important morphological and stratigraphic evidence for relative sea-level changes during deglaciation. The Ardersier shorelines provide important datum points on the shoreline diagram constructed by Firth (1984, 1989a) for the Lateglacial Interstadial, complementing the evidence from Munlochy Valley and Barnyards. According to Firth (1984, 1989b) the deposits indicate that relative sea level attained an altitude of about 40 m OD at Ardersier and that it subsequently fell from at least 37.8 m to about 30 m OD as the ice retreated to Inverness. As the ice retreated farther west, the progressive fall in relative sea level continued.

Fifth, the assemblage of features at Ardersier is completed by Holocene raised beach deposits, including raised shingle ridges. Although these add to the diversity of interests of the site, the Holocene sequence in the area is better demonstrated at Barnyards and Munlochy Valley, and raised shingle ridges, both Lateglacial and Holocene, are most spectacularly developed at Tarbat Ness and Spey Bay (Ogilvie, 1923; Smith, 1977).

Conclusion

The landforms and deposits at Ardersier are important for interpreting the pattern of wastage of the last (Late Devensian) ice-sheet (about 14,000–13,000 years ago) and the changes in the relative position of sea level that accompanied and followed deglaciation. The deposits may represent a readvance of the last ice-sheet, although such an event and its magnitude are still a matter of debate. Beach deposits and shorelines show that relative sea level was as high as 40 m above present as the ice melted, and their presence has contributed towards understanding the wider regional patterns of sea-level changes during the Devensian (Lateglacial) and the Holocene (approximately the last 13,000 years).

STRUIE CHANNELS

J. E. Gordon

Highlights

This site provides a good example of a meltwater channel system formed during the melting of the Late Devensian ice-sheet; such systems are relatively rare in northern Scotland.

Introduction

The site (NH 670790) is located in Strathrory on the south side of a col at *c.* 208 m OD between the Dornoch Firth to the north and the Cromarty Firth to the south. It provides a good example of a glacial meltwater channel system in the northern Highlands. The Struie Channels were first recorded by Peach *et al.* (1912) and subsequently have been mapped and described by J. S. Smith (1968) and Leftley (1991); otherwise they have attracted little published comment despite the relative scarcity of well-developed meltwater channel systems north of the Great Glen.

Description

The interest comprises a series of subparallel meltwater channels. The largest is up to 33 m deep, 89 m wide and 2.5 km long (Figure 7.10). In plan form, the channels show anastomosing

Figure 7.10 Geomorphology of the Struie meltwater channels, Strathrory (from J. S. Smith, 1968; Leftley, 1991).

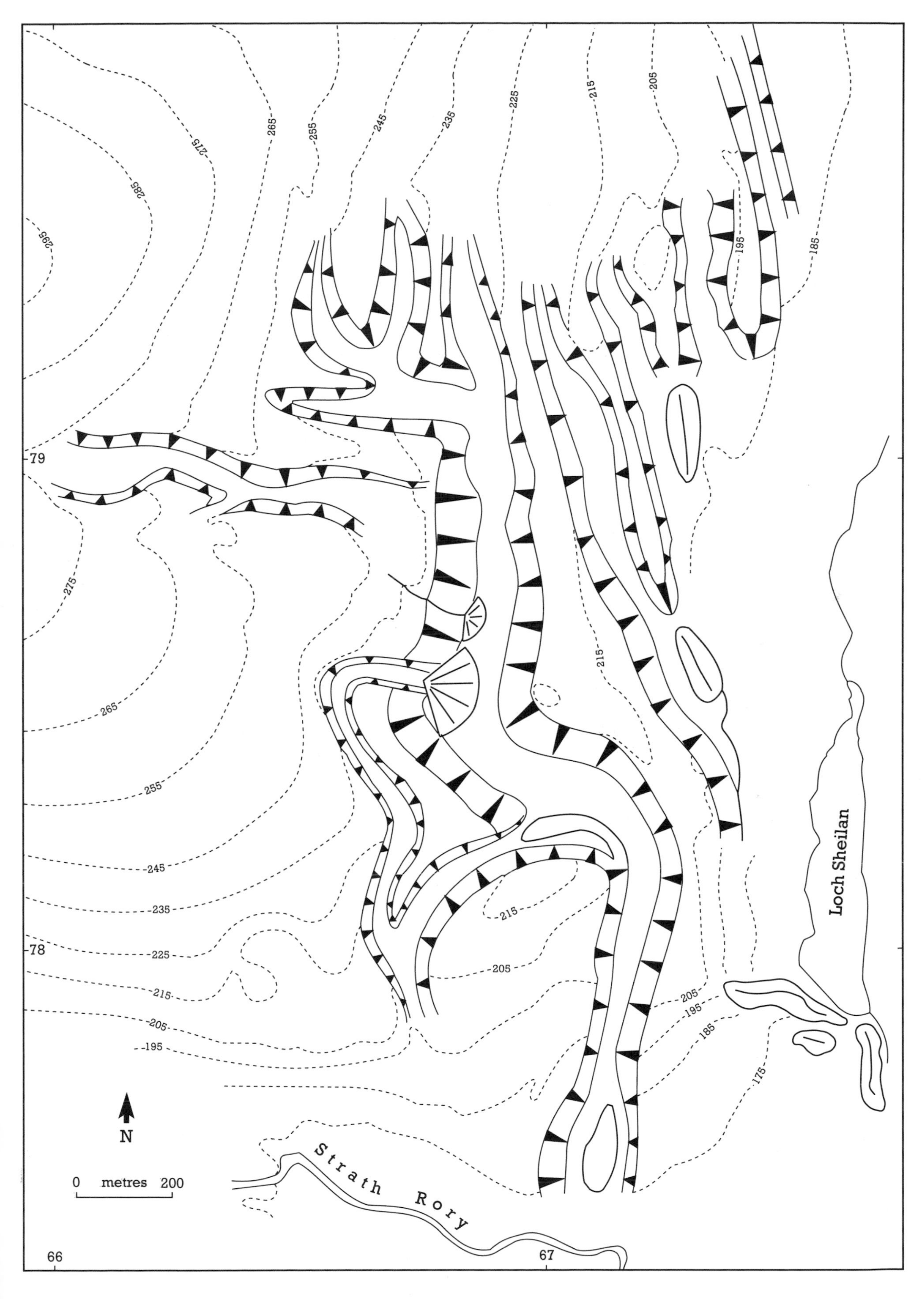

 Large meltwater channels

 Glaciofluvial ridges

 Alluvial fans

- - 215 - - Contours (metres)

and branching patterns, as well as parallel forms, and locally small cut-off loops lie perched above the main channel (Figure 7.10). According to J. S. Smith (1968), the channels originate at the lowest point of the col, but in fact they begin several hundred metres on the lee side and some 30–40 m above the lowest point. J. S. Smith (1968) also noted that another channel runs south from the next col to the west, and a cross channel extends eastwards from it to link with the Strathrory system. The channels are also associated with glaciofluvial deposits, including an esker at the south end of Loch Sheilan (NH 676780) and gravel terraces to the west in Strathrory, which have been partially quarried (Harris and Peacock, 1969; Mykura *et al.*, 1978), revealing glaciolacustrine deposits (Leftley, 1991).

Interpretation

Peach *et al.* (1912) interpreted the Struie Channels as representing ice-marginal drainage at successive levels along the edge of a wasting glacier which occupied the valley of the Allt Dearg immediately to the north of Strathrory. However, on account of their apparent relationship to the col, J. S. Smith (1968) considered that the channels were of subglacial origin and were associated with his Fortrose stage of deglaciation, when they carried meltwater south from the Dornoch Firth to the Cromarty Firth across the low col into Strathrory and then towards Scotsburn (NH 720763). Leftley (1991) also interpreted the channels as subglacial in origin, formed during a late stage in the deglaciation of the last ice-sheet, when a lobe of ice extended across the col from the north into Strathrory; this occurred penecontemporaneously with the development of a series of ice-dammed lakes in Strathrory.

In their anastomosing forms and location on the lee side of a col, the Struie Channels are similar to the superimposed subglacial forms described by Clapperton (1968) from the Cheviots. Such channels are considered to reflect regional hydraulic gradients associated with active ice (cf. Sugden and John, 1976; Shreve, 1985a, 1985b). However, in other aspects, particularly their parallel forms, the Struie Channels resemble many of the channel systems described from lowland Scotland (Sissons, 1960, 1961a) and Scandinavia (Mannerfelt, 1945, 1949), which have been interpreted as marginal or submarginal in origin. The Struie site therefore provides an interesting assemblage of meltwater channel features that would benefit from further detailed investigation.

Meltwater channel systems are relatively rare in the northern Highlands of Scotland, but Struie is a particularly good example. It demonstrates many of the typical features of meltwater channels in Scotland (see Carlops and Rammer Cleugh), including a combination of subglacial and marginal/submarginal characteristics. In the wider context of the Moray Firth area, the Struie Channels complement the interests of the depositional, glaciofluvial landform assemblages at Torvean, Kildrummie Kames and Littlemill.

Conclusion

The Struie channels were eroded by glacial rivers during the melting of the last (Late Devensian) ice-sheet, between approximately 14,000 and 13,000 years ago. They form part of a network of sites showing glacial meltwater landforms formed at this time, and are notable as one of the few well-known examples of a system of meltwater channels in northern Scotland.

KILDRUMMIE KAMES

J. E. Gordon and C. A. Auton

Highlights

This site demonstrates an outstanding example of a system of braided eskers formed by the Late Devensian ice-sheet. It shows particularly clearly the morphology of the landforms and is also important for interpreting the development of glacial drainage during the wastage of the Late Devensian ice-sheet.

Introduction

The Kildrummie Kames (also known as the Flemington Kames or more properly as the Flemington Eskers) extend over a distance of about 10 km to the south-west of Nairn (from approximately NH 783502 to NH 874540). They are probably the best example of large, braided eskers and one of the longest continuous esker systems in the country that remains essentially unmodified by sand and gravel extraction. The Kildrummie Kames form part of an extensive system of glaciofluvial deposits, including eskers, kames and kame terraces that occupies the low ground on the south side of the Moray Firth from the north end of Great Glen to near Elgin. They were formed by glacial meltwaters during the wastage of the Late Devensian ice-sheet and have been mapped and described in various publications (Jamieson, 1866, 1874; Horne, 1923; Ogilvie, 1923; Gregory, 1926; J. S. Smith, 1968, 1977; Harris and Peacock, 1969; Small and Smith, 1971; Synge, 1977b; Mykura *et al.*, 1978; Synge and Smith, 1980; Firth, 1984, 1990b). The eskers are illustrated in Sparks and West (1972, plate 15). The most detailed maps available of the eskers are those recently published by the British Geological Survey (1:10,000 Sheets NH 85 SW, SE and 75 SE). Although the Kildrummie landforms are clearly eskers, they are widely known as 'kames' following the historical usage of the latter term.

Description

The Kildrummie Kames comprise a series of up to eight braided ridges, 5–10 m high, with intervening kettle holes often partially infilled by peat or waterlogged silt and sand which is several metres deep in places. The braided forms occur in three distinct groups, linked together by a single discontinuous ridge (Figures 7.11 and 7.12). To the west of the B9090 the form of the esker ridges is less distinct and their exact morphology is masked by thick coniferous woodland. However, the system extends almost unbroken as far as Culaird (NH 782500). A number of eskers, separated from, but aligned with the main group, occur to the south of Tornagrain (NH 769499). According to Small and Smith (1971) these are 'fed' by a series of meltwater channels at the western end of the system, between High Wood and Balnabual (NH 776490). To the north-east of Meikle Kildrummie (NH 856539), the esker system terminates in a broad, flat-topped and steep-sided ridge, which is pitted by small kettle holes and slopes regularly towards the east (Figure 7.13). To the north-west of the ridge, terraces slope down northwards into a large depression almost entirely enclosed by stagnant-ice terrain. This depression drains northwards through a channel south of Tradespark (NH 869568) (Figure 7.13).

Mapping of the eskers by staff of the British Geological Survey has shown that although the ridges are principally composed of sandy, well-rounded coarse gravel, lenses of claybound gravel and brown sandy diamicton are also present, notably within exposures to the east of Bemuchlye (NH 827531). These show up to 8 m of diamicton, which appears to overlie finely interlaminated sand and silt, suggesting that this part of the esker system may have been laid down in a body of standing water ponded within or beneath the ice.

Interpretation

Most accounts recognize the Kildrummie Kames as classic features. As early as 1866 Jamieson described them as a 'remarkable series of ridges' and the 'finest of all' the gravel hills in the Moray Firth area. He again referred to them in 1874, believing they were moraines of the last glacial episode in Scotland. However, despite their striking landscape appearance and classic lines, the Kildrummie Kames feature only infrequently in published literature, generally in a descriptive context or in discussions of relative sea-level change.

Horne (1923) recognized the deposits as glaciofluvial and described them briefly as part of the 'kame series' of the area, noting the anastomosing forms and composition of sand and well-rounded gravel. Ogilvie (1923) in his descriptive account of the physiography of the Moray Firth coast presented a topographic map of the eskers east of Loch Flemington. He noted that the eskers terminated abruptly in what might be a sea cliff, cut during the maximum submergence of the land following deglaciation. To the north and east of the eskers he recognized a zone of kames that had been washed and trimmed by the sea.

Gregory (1926), in his review of similar features throughout Scotland, described a section which he considered to show marine trimming of the esker and also beach deposits banked against it. He also recorded sections showing beds of coarse cobbles and smaller pebbles, and coarse gravel overlying sand and gravel layers, with coarse gravel again at the base.

J. S. Smith (1968, 1977) and Small and Smith (1971) referred to the eskers in the context of the extensive suite of meltwater channels and glaciofluvial deposits associated with the melting

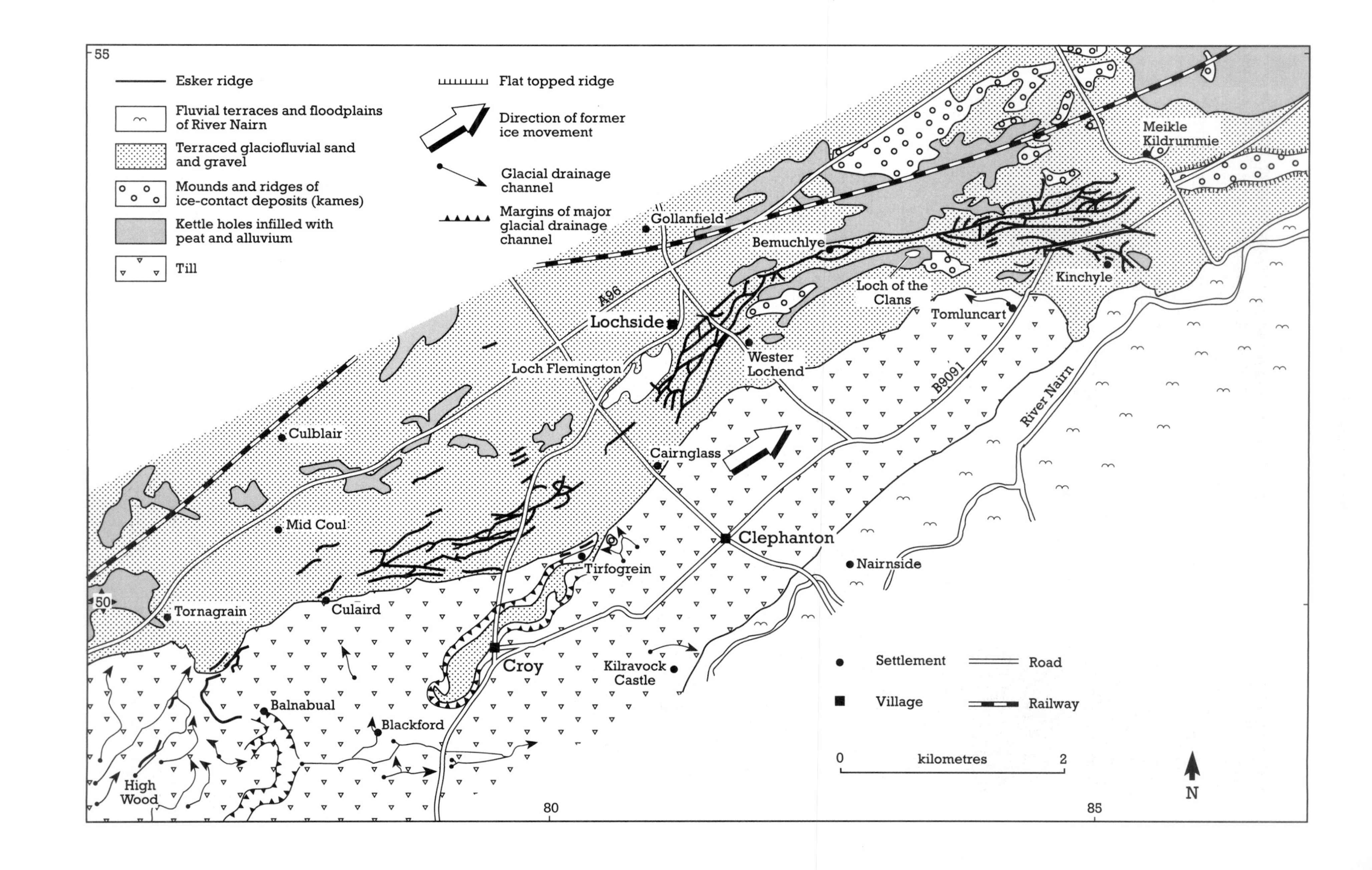
55
Esker ridge
Fluvial terraces and floodplains of River Nairn
Terraced glaciofluvial sand and gravel
Mounds and ridges of ice-contact deposits (kames)
Kettle holes infilled with peat and alluvium
Till
Flat topped ridge
Direction of former ice movement
Glacial drainage channel
Margins of major glacial drainage channel
Meikle Kildrummie
Gollanfield
Bemuchlye
Kinchyle
Loch of the Clans
Tomluncart
A96
Lochside
Wester Lochend
Loch Flemington
B9091
River Nairn
Culblair
Cairnglass
Mid Coul
Clephanton
Nairnside
Tirfogrein
50
Tornagrain
Culaird
Croy
Kilravock Castle
Balnabual
Blackford
High Wood
Settlement
Road
Village
Railway
0
kilometres
2
N
80
85

Figure 7.12 The Kildrummie Kames esker system viewed towards the east. Two areas of braided ridges (right foreground and centre distance) are linked by a single ridge. (Cambridge University Collection: copyright reserved.)

of the Late Devensian ice-sheet on the south side of the Moray Firth. The landforms indicate easterly flow of subglacial meltwater, controlled by the ice-surface gradient, and demonstrate a continuous phase of ice-sheet downwasting (Smith, 1977). Small and Smith (1971) noted that the esker system had been washed on its seaward side near Gollanfield (NH 815533) by the Lateglacial marine transgression. This latter theme was developed more fully by Synge (1977b) and Synge and Smith (1980) (see also Synge, 1977a). According to these authors, the highest Lateglacial shoreline in the area ('Kildrummie Shoreline' at 37–42 m OD) is represented by the flat-topped ridge east of Meikle Kildrummie at 36–38 m OD. This ridge was interpreted as a form of marine delta deposited in association with decaying ice, when the roof of the subglacial esker tunnel collapsed. The steep eastern ends of the individual esker ridges west of Meikle Kildrummie were thought to reflect wave trimming and to represent a lower shoreline at 33 m OD, part of the 'Culcabock Shoreline' at 28–34 m OD. Beach ridges associated with this shoreline were also

Figure 7.11 Geomorphology of the Kildrummie Kames esker system between High Wood and Meikle Kildrummie (from mapping by C. A. Auton for the British Geological Survey 1:50,000 Geological Sheet 84W (Fortrose), in press).

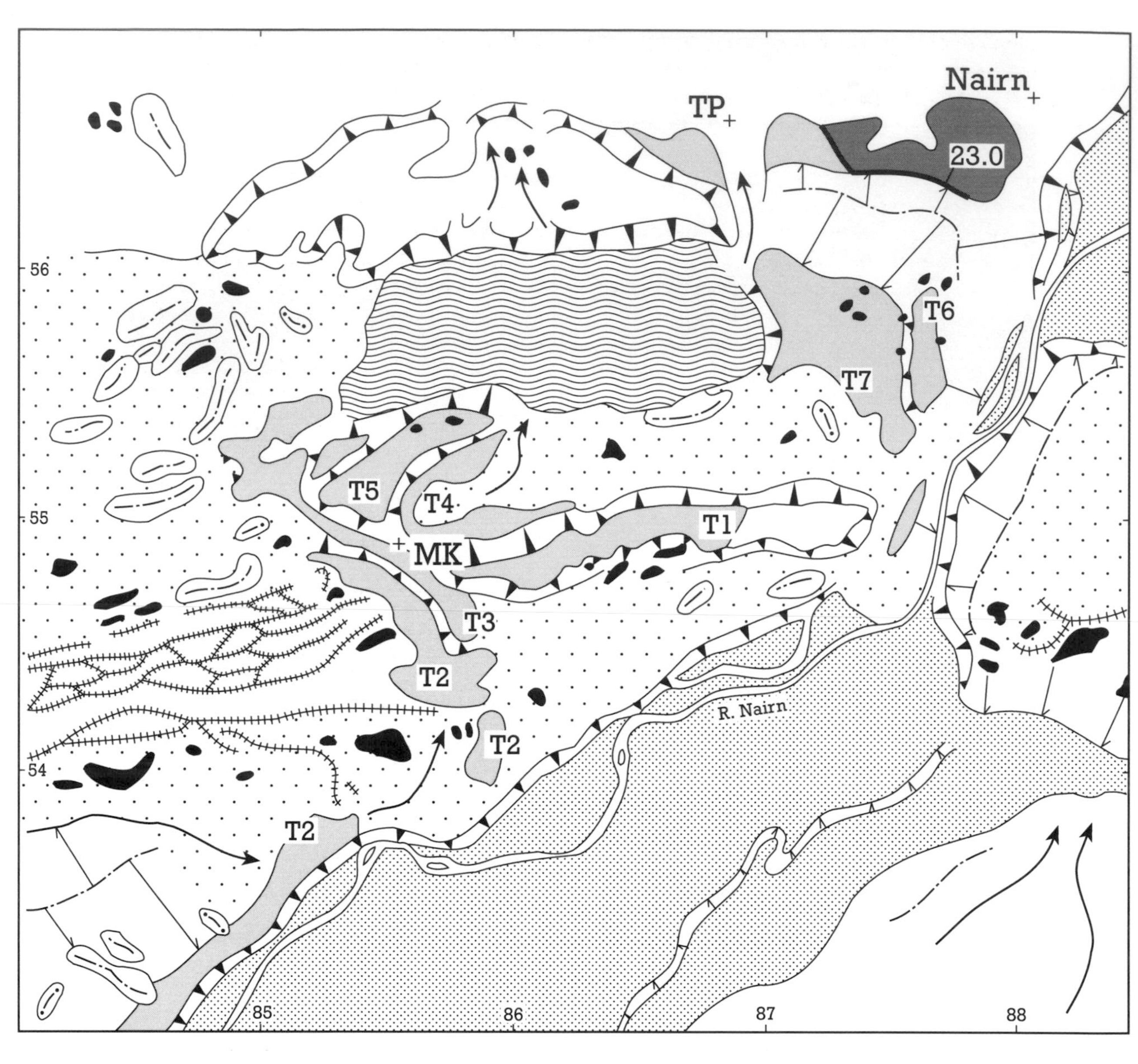
Nairn
TP
23.0
56
55
54
T1
T2
T2
T2
T3
T4
T5
T6
T7
MK
R. Nairn
85
86
87
88

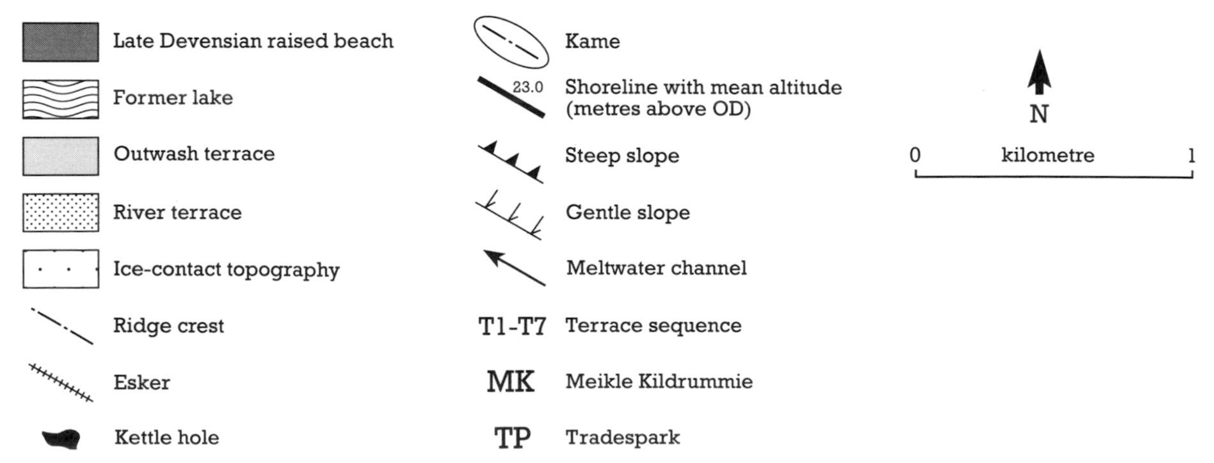
Late Devensian raised beach
Former lake
Outwash terrace
River terrace
Ice-contact topography
Ridge crest
Esker
Kettle hole
Kame
23.0
Shoreline with mean altitude (metres above OD)
Steep slope
Gentle slope
Meltwater channel
T1-T7 Terrace sequence
MK Meikle Kildrummie
TP Tradespark
N
0 kilometre 1

identified between the ridge ends (Synge and Smith, 1980). Below the altitude of these shorelines, the steep slopes of the eskers and kettle holes were degraded by the marine transgression.

Firth (1984) mapped in detail the eastern part of the Kildrummie Kames (Figure 7.13), the adjacent glaciofluvial deposits and the terraces of lower Strathnairn. During the downwasting of the ice-sheet, marginal and submarginal glacial meltwaters formed a series of kame terraces and meltwater channels on the southern slopes of Strathnairn; at the same time subglacial meltwaters drained eastwards forming the Kildrummie esker system. As deglaciation progressed, meltwater from the kame terraces drained into the eskers, and as the ice thinned at the eastern end, the main tunnel roof collapsed forming an open crevasse in which the large flat-topped ridge east of Meikle Kildrummie formed. Firth's (1984) interpretation of the terrace sequence indicates that, while the lower valley remained blocked by stagnant ice, meltwater in Strathnairn drained both across the esker system, forming a gap and terraces between the main esker and the flat-topped ridge, and around the eastern end of the latter and into a proglacial lake occupying the depression to the north of Meikle Kildrummie. Drainage from the proglacial lake was northwards past Tradespark (Figure 7.13), where the highest shoreline fragment in the area is at 23 m OD. Firth (1984) found no evidence that either the esker system or the glaciofluvial deposits immediately to the north had been washed by the sea. Ice remained in the area until relative sea level fell below 21 m OD.

Kildrummie Kames are important in several respects. They are an outstanding example of a braided esker system, one of the finest and largest in Britain. They have largely escaped sand and gravel extraction and other large-scale modifications and therefore demonstrate landform morphology in a particularly clear fashion. Kildrummie Kames offer significant potential for further research on subglacial hydrology and the controls on meltwater routes and sedimentation (see Shreve, 1985a). Recently Shaw *et al.* (1989) have suggested that anastomosing channel patterns, similar to that indicated by the Kildrummie Kames, reflect major subglacial floods. Alternatively, each braided area may represent a series of channels on an ice-cored fan surface developed in front of a receding ice margin (cf. Jenkins, 1991); the total assemblage of landforms therefore represents three successive stages in the ice recession. The well-preserved landforms of Kildrummie Kames offer good opportunity for testing such ideas and for applying the theories of glacier physics and hydrology to reconstruct Late Devensian ice-sheet characteristics and drainage conditions.

In their braided forms, the Kildrummie Kames share morphological similarities with the Carstairs Kames (see below). Proposed origins for the latter have included subglacial and proglacial processes, and recent work (Jenkins, 1991) has suggested that the ridges formed on the surface of buried ice or in englacial tunnels near the ice margin under conditions of high energy flows (large floods). Detailed comparative investigation of the two sites should help clarify the respective origin of their landforms and their implications for patterns of deglaciation. Morphologically, Kildrummie Kames differ from the system of parallel esker ridges at Littlemill and the assemblage of single eskers and kame terraces at Torvean. The interpretation of such differences and their implications for glacier hydrology during deglaciation await resolution.

Conclusion

The Kildrummie Kames represent a classic site for geomorphology, showing a large system of braided esker ridges formed by meltwater rivers during the wastage of the last (Late Devensian) ice-sheet (approximately 14,000–13,000 years ago). The landforms are largely intact and display particularly clearly the surface forms of the eskers. The Kildrummie Kames have significant potential for developing an understanding of glacial drainage systems and patterns of ice decay.

Figure 7.13 Geomorphology of the eastern part of the Kildrummie Kames in the vicinity of Meikle Kildrummie (from Firth, 1984).

LITTLEMILL

J. E. Gordon

Highlights

The landforms at Littlemill provide an excellent example of a system of large, parallel eskers

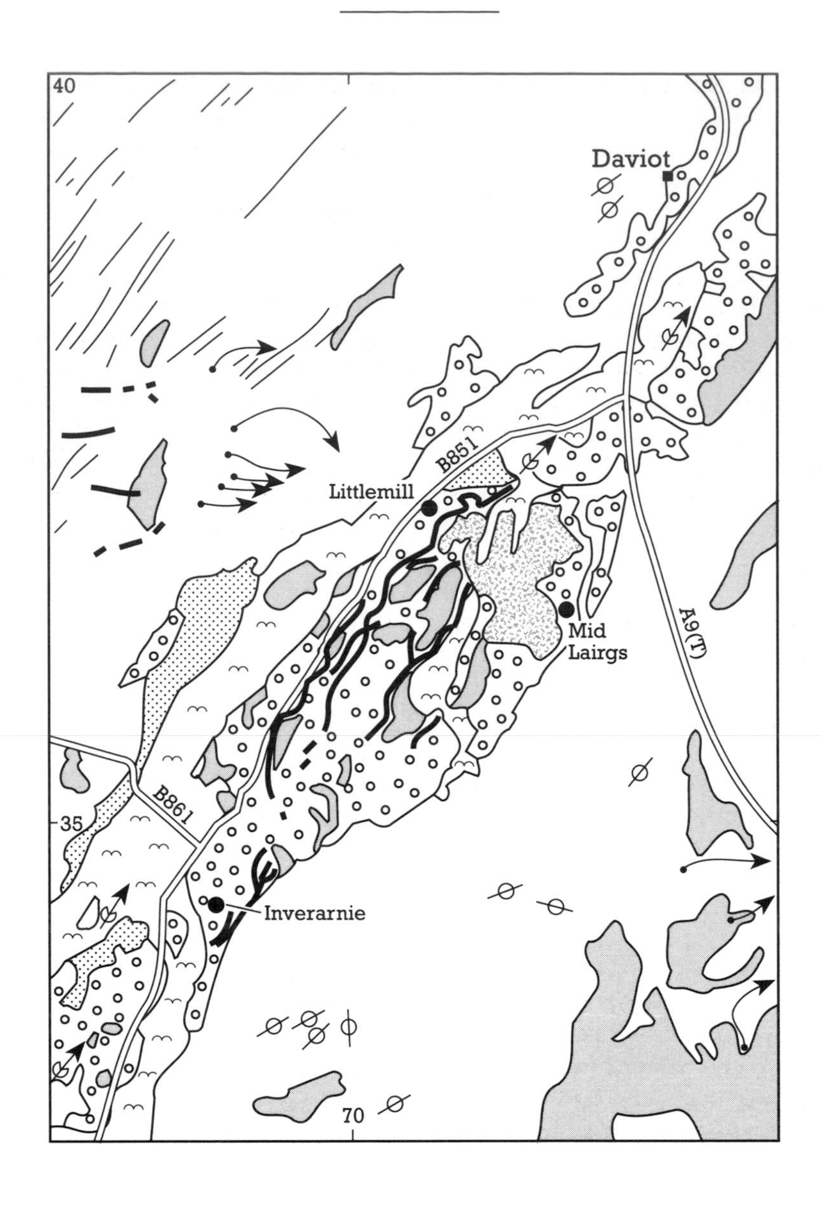
40
Daviot
B851
Littlemill
Mid
Lairgs
A9(T)
B861
35
Inverarnie
70

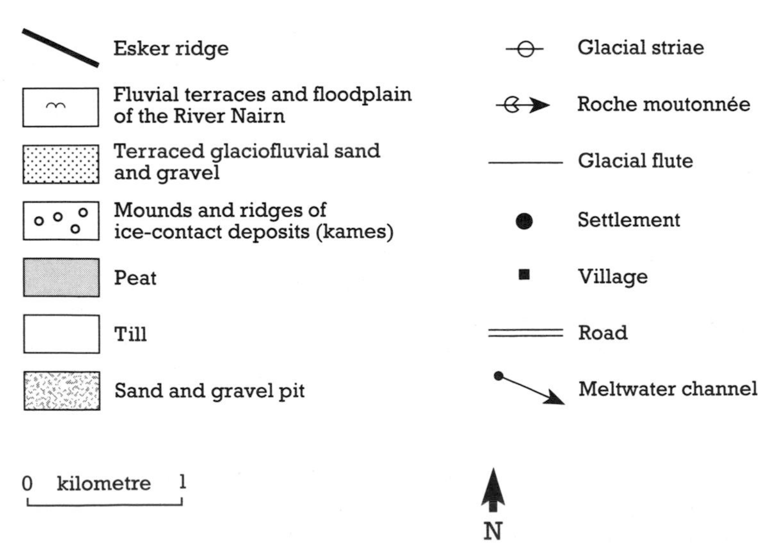
Esker ridge
Fluvial terraces and floodplain of the River Nairn
Terraced glaciofluvial sand and gravel
Mounds and ridges of ice-contact deposits (kames)
Peat
Till
Sand and gravel pit
Glacial striae
Roche moutonnée
Glacial flute
Settlement
Village
Road
Meltwater channel
0 kilometre 1
N

formed during the melting of the Late Devensian ice-sheet.

Introduction

The site at Littlemill (NH 695352–NH 717377) extends over a distance of *c.* 3.5 km south-west of Daviot in Strathnairn and includes a system of subparallel esker ridges, kames and kettle holes which together comprise an outstanding assemblage of glaciofluvial landforms. They relate to the wastage of the Late Devensian ice-sheet, forming part of a more extensive suite of related deposits in Strathnairn between Brinmore and Daviot (Harris and Peacock, 1969; Mykura *et al.*, 1978). There have been no detailed studies of the Littlemill landforms. Early descriptions were published by Fraser (1880) and Cameron (1882a, 1882b). In addition, the site has been figured in publications by Sissons (1967a, plate XIIIA) and Gray (1991, figure 315).

Geikie described the main characteristics of the Littlemill eskers, commenting that 'the observer will find one of the most remarkable groups of glacial ridges in the north of Scotland' (1901, p. 497). In the Geological Survey Memoir to the area Horne (1923) referred to a number of boulders scattered over the ridges at their northern end and suggested that they might be partly morainic in origin. Gregory (1926) included a note on them in his survey of Scottish kames, and they have also been recognized and mapped by Harris and Peacock (1969), Small and Smith (1971), Smith (1977), Mykura *et al.* (1978) and Synge and Smith (1980).

Description

The key features occur in a topographic embayment on the east side of the valley and comprise four major and several minor, subparallel esker ridges (Figure 7.14). Typically they are about 15 m high but in places reach as much as 40 m. Their maximum length is almost 2 km. North-east of Littlemill Farm (NH 365372) the eskers converge in a broad, flat-topped mound. A second large mound immediately to the north displays hummocky and linear forms and kettle holes on its surfaces. Fine kettle holes also occur between the main esker ridges, whose striking topographic expression is spectacularly displayed in Sissons (1967a, plate XIIIA). In the north-east part of the site, excellent exposures have been revealed in eskers and associated mounds in the large sand and gravel quarry at Mid-Lairgs, with both transverse and longitudinal sections present. There is considerable variability between sections in the type of material present, ranging from entirely coarse gravel to bedded sands and gravels (see also Harris and Peacock, 1969; Mykura *et al.*, 1978). Large-scale arch bedding was formerly displayed in one transverse section through an esker ridge (1980). At the south-west end of the system, sections in a forestry pit (NH 698354) reveal gravel and sand with boulders up to 0.5 m in size.

Additional interest in the site includes an excellent example of roches moutonnées at Scatraig (NH 713376).

Figure 7.14 Geomorphology of the Littlemill esker system between Inverarnie and Daviot (from mapping by J. W. Merritt for the British Geological Survey 1:50,000 Geological Sheet 84W (Fortrose), in press).

Interpretation

There are few specific references in the literature to the Littlemill landforms. Fraser (1880) interpreted the ridges as lateral moraines, although he noted that they differed from other moraines in their composition of sand and gravel. He suggested that the material accumulated along a glacier margin where it was reworked and stratified by glacial meltwater. Cameron (1882a) noted sand and gravel mounds and terraces north of Littlemill, which he explained in terms of marine submergence. In a subsequent paper (Cameron, 1882b), he extended his observations to include the Littlemill landforms, which he described as kames. He noted sections showing large rounded gravel and, in places, sand layers. Cameron disagreed with 'some geologists' who had proposed a subglacial origin for the ridges, arguing instead that extensive deposits, formed during marine submergence, were later dissected by overspill and drainage from lakes which they dammed. This latter explanation, however, is now discounted in favour of the former: the deposits were probably formed either by an englacial or subglacial meltwater drainage system.

The Littlemill eskers are particularly notable for their size and extent. Although part of the system has been extensively quarried and part is afforested, the intact ridges show clearly defined esker morphology. The individual ridges are generally similar in form to single eskers elsewhere in the area; for example at Torvean, Edderton, Dornoch, Alness and Brora. The key additional interest that distinguishes the Littlemill features centres on the system of forms there and their spatial arrangement: they are one of the finest examples of a system of subparallel esker ridges in the country. In so far as the key interest lies in the complete network of forms, the Littlemill eskers provide an interesting contrast with the system of braided eskers at Kildrummie Kames (see above). These two sites complement each other, illustrating different types of former glacial drainage pattern. Tentatively, the morphological differences between these two sites may also reflect different origins. The Littlemill landforms appear to be more typical of eskers formed in a subglacial or near subglacial position, whereas those at Kildrummie may have formed in braided channels incised into the surface of buried ice (cf. Carstairs Kames). The quarry at Mid-Lairgs, together with the small forestry gravel pit at the southern end of the site, provides an excellent opportunity to relate esker morphology and sedimentology. Apart from the study of Middle Mause esker by Terwindt and Augustinus (1985), there are no published modern studies of the detailed sedimentology and palaeohydrology of Scottish eskers.

Conclusion

Littlemill is an important landform site, providing a particularly good example of large esker ridges aligned in a parallel arrangement. They were formed by meltwaters during the wastage of the last (Late Devensian) ice-sheet (approximately 14,000–13,000 years ago) and allow an interesting comparison with the braided eskers of Kildrummie Kames, the two sites illustrating different types of glacial drainage system and possibly differences in the details of origin of the landforms.

TORVEAN

J. E. Gordon

Highlights

The landform assemblage at Torvean includes an excellent range of glaciofluvial features formed during the melting of the Late Devensian ice-sheet. It illustrates particularly well the evolution of a glacial drainage system and associated sedimentary environments during deglaciation.

Introduction

The Torvean site (centred on NH 630420) covers an area of *c.* 4.1 km^2 and forms part of an extensive zone of glaciofluvial deposits that extends more than 5 km south-west from the outskirts of Inverness on the west side of the River Ness. It contains an excellent assemblage of glaciofluvial landforms, including one of the best examples in Britain of a suite of kame terraces and one of the highest eskers. It is also important in demonstrating the development of an integrated marginal, submarginal and subglacial drainage system during the wastage of the Late Devensian ice-sheet. Several brief accounts of the Torvean landforms have appeared in the literature (Jamieson, 1865; Aitken, 1880; Horne and Hinxman, 1914; Ogilvie, 1923; Small and Smith, 1971), and they have been mapped in detail by Synge and Smith (1980) and Firth (1984).

Description

Torvean includes a suite of six major and five minor kame terraces, eskers, kames, river terraces and kettle holes (Figure 7.15) (Synge and Smith, 1980; Firth, 1984). The kame terraces form a striking 'staircase' below 120 m OD on the valley side (Figure 7.15, T161–T171). The higher-level terraces merge eastwards with the massive Torvean and Tomnahurich eskers (over 68 m high); the lower terraces descend below the level of the Torvean esker. Below the lowest terrace (T171), at 60–70 m OD a series of eskers, kames and outwash terraces (T172–T174) are truncated abruptly to the south-east by an extensive bluff. Below is a suite of river terraces (T175–T181) with four separate levels (Firth, 1984), the

Figure 7.15 Geomorphology of the Torvean area (from Firth, 1984). The terrace fragments include kame terraces (T161–T171), Lateglacial outwash and river terraces (T172–179, T159), and Holocene river terraces (T180–181).

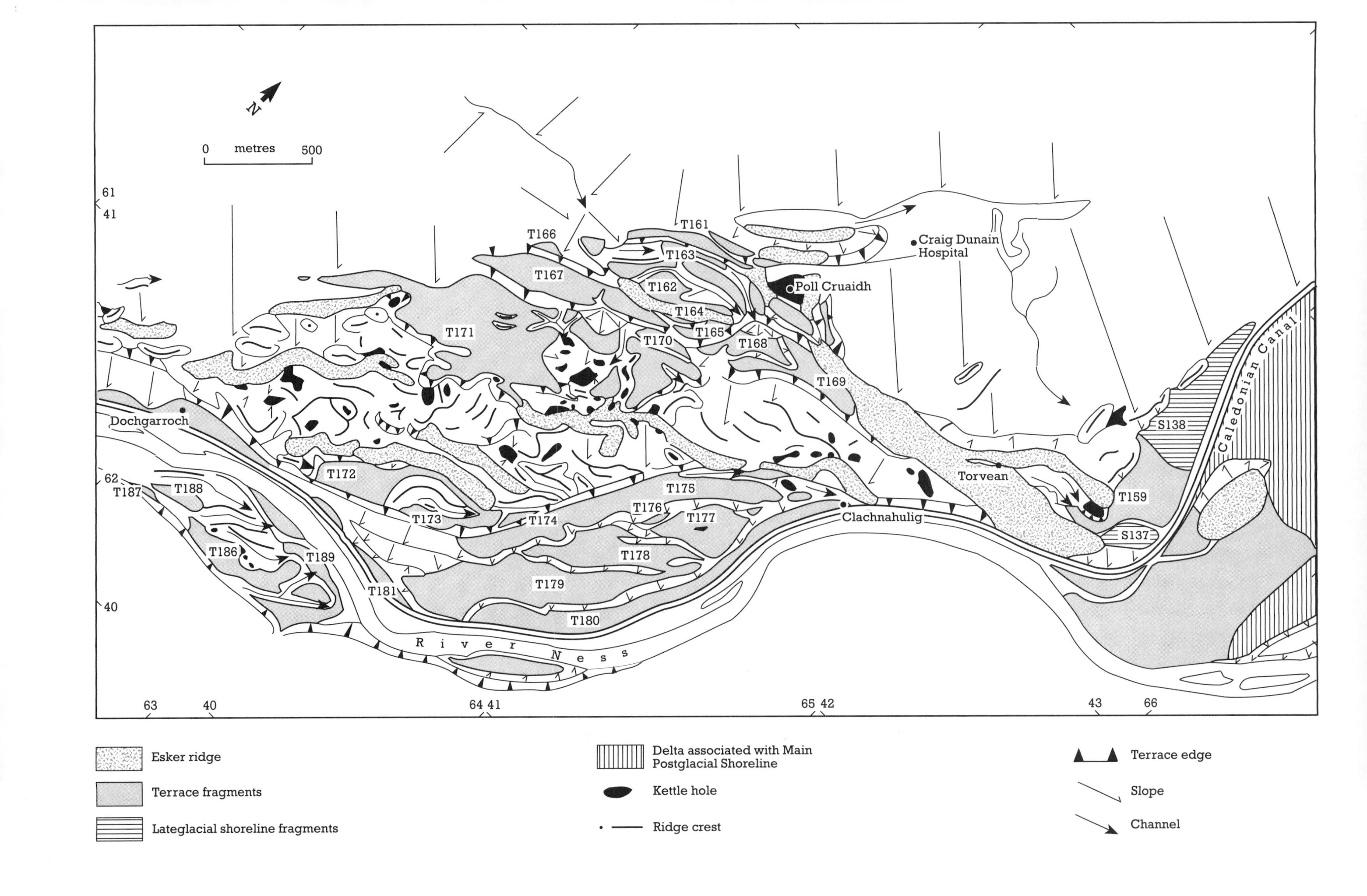
N
0 metres 500
61
41
62
40
63
40
64 41
65 42
43
66
Dochgarroch
Craig Dunain Hospital
Poll Cruaidh
Torvean
Clachnahulig
Caledonian Canal
River Ness
T159
T161
T162
T163
T164
T165
T166
T167
T168
T169
T170
T171
T172
T173
T174
T175
T176
T177
T178
T179
T180
T181
T186
T187
T188
T189
S137
S138
Esker ridge
Terrace fragments
Lateglacial shoreline fragments
Delta associated with Main Postglacial Shoreline
Kettle hole
Ridge crest
Terrace edge
Slope
Channel

highest three being Lateglacial in age and the lowest (T180 and T181) being Holocene in age; dating is based on the relationships observed between the terrace fragments and marine shoreline fragments (see Firth, 1984). In addition, there are terrace fragments (T178, T179 and T188) above the Holocene terrace which merge into kettled and channelled topography. At the eastern end of the site, terrace fragment T159 grades into a raised shoreline (S138) at 13.7 m OD. Kettle holes are frequent in the glaciofluvial deposits, including the impressive example at Poll Cruaidh. Well-bedded sands and gravels have been exposed in the quarry in the Torvean esker (NH 646432), but there have been no detailed sedimentological studies. Harris and Peacock (1969) recorded poorly-sorted gravel in a matrix of silty sand or silt, and also arched bedding in conformity with the topographic form of the ridge. The core of the esker, where it was formerly exposed, was formed of large boulders (J. D. Peacock, unpublished data). Near Dochgarroch (NH 620408) on the north-west bank of the Caledonian Canal, till has been recorded beneath the sands and gravels of the glaciofluvial deposits (Horne and Hinxman, 1914; Small and Smith, 1971).

Interpretation

The individual landforms at Torvean are all closely related and demonstrate the development of an integrated marginal, submarginal and subglacial drainage system when the Great Glen acted as a major route for meltwaters draining north-eastwards during the wastage of the Late Devensian ice-sheet. Two main stages in the development of the drainage system can be recognized. Initially, ice-marginal drainage associated with downwasting ice is indicated by the kame terraces. To the north-east, the drainage became subglacial, confined in an ice-walled tunnel in which the Torvean and Tomnahurich eskers were deposited. The lower kame terraces represent successive ice-marginal positions as ice wastage continued. The extensive kame and kettle topography and lower lying eskers then formed as the ice downwasted *in situ* on the valley floor, possibly in a manner similar to that described by Boulton (1972b) or Price (1973) in modern glacier environments. Meltwater continued to drain along the lowest kame terrace at the margin of the decaying ice mass and also through the ice via a system of eskers that lead north-eastwards to Clachnahulig (NH 644428).

In the lower part of the valley, between Loch Dochfour (NH 605387) and Clachnahulig, an extensive terrace system records the subsequent history of subaerial drainage development. The terraces form two groups according to their altitudes (Firth, 1984). The lower group occurs within a narrow, steep-sided valley, which includes the floodplain of the River Ness. Some of the lower terrace fragments have been interpreted as marine features, representing higher former sea levels in the Ness valley and Loch Ness. Horne and Hinxman (1914) reported that the '25-foot' raised beach extended along the Ness valley to a height of 50 ft (16 m) at Tomnahurich and Torvean.

According to one interpretation the sea also penetrated along the Ness valley and into Loch Ness during the Lateglacial (Small and Smith, 1971; Synge, 1977b; Synge and Smith, 1980). Synge (1977b) and Synge and Smith (1980) inferred that two shorelines were present at 33 m and 25–28 m OD. The higher shoreline was represented at Clachnahulig in the form of a breached and infilled kettlehole. The lower shoreline was a major erosional feature, extending from Clachnahulig to Dunain, and it truncated both the glaciofluvial deposits and a river terrace graded to a lower sea level. It therefore represented a marine transgression. Synge (1977a, 1977b, 1980) placed these inferred shorelines in wider regional and national perspectives. However, from a detailed study of the Ness valley and the shores of Loch Ness, Firth (1984, 1986) concluded there was no geomorphological evidence for any marine incursion into Loch Ness. In particular, the gradients of the terraces indicate that they are glaciofluvial or fluvial features. Firth therefore reinterpreted the shoreline fragments identified in the Ness valley by earlier workers as outwash terraces, formed as the ice downwasted and retreated progressively into the Loch Ness Basin. As the ice margin retreated from Inverness to Loch Ness, relative sea level fell from *c.* 35 m to 13.8 m OD. Firth suggested that this fall occurred after 13,000 BP.

The Ness valley and Torvean deposits are also of interest in relation to the drainage of the ice-dammed lakes in Glen Roy and Glen Spean during the Loch Lomond Stadial. Jamieson (1865) first recognized a possible connection. He speculated whether the Torvean glaciofluvial deposits themselves might relate in some way to catastrophic drainage of the lakes. Although this specific link is now seen to be incorrect, Sissons (1979c, 1981c) developed the basic idea. Drainage of the Glen Roy and Glen Spean lakes via the Great Glen raised the level of Loch Ness. Overspill of water at the north-east end of the loch enlarged the valley of the River Ness and the floods of water deposited an extensive fan of bouldery gravel out into the Beauly Firth (see also Peacock, 1977a). Firth (1984) has elaborated on some of the details of this event, reconstructing the sequence of Lateglacial and Holocene changes in the level of the loch and demonstrating the links between the evidence around Loch Ness, in the Ness valley and in the Beauly Firth (see Fort Augustus and Dores). Specifically, Firth concluded that the terrace fragments above the Holocene terrace at Torvean were formed during the flood associated with the drainage of the Glen Roy and Glen Spean ice-dammed lake.

Torvean is important in several respects. First, it contains what is believed to be the highest esker in Britain, with a height of over 68 m (Clapperton, 1977). Second, it contains what is one of the finest examples of a suite of kame terraces in Britain. Although individual examples of such landforms are not uncommon (see Moss of Achnacree and the Cairngorms), an assemblage such as that at Torvean is exceptional in terms of the size, extent and number of terraces present.

Third, the total landform assemblage at Torvean illustrates particularly well the relationships between different types of glaciofluvial deposits and landforms and the development of a marginal, submarginal and subglacial drainage system associated with the decay of the last ice-sheet. Torvean is exceptional among sites of this type in illustrating the relationships of kame terraces and eskers; at most other sites, drainage systems are represented by meltwater channels (see Struie Channels and Rammer Cleugh), kames or eskers (see Carstairs Kames), or by some combination of such landforms.

Fourth, the lower terraces at Torvean have a bearing on the interpretation of the Lateglacial history of Loch Ness, including the question of a marine incursion and the effects of the catastrophic drainage of the Glen Roy and Glen Spean ice-dammed lakes.

Fifth, Torvean is important in the wider regional context of glaciofluvial landform development in the Inverness area. Individually, Torvean, Kildrummie Kames and Littlemill provide classic examples of individual glaciofluvial depositional landforms; together they demonstrate within a relatively short distance what is arguably the finest group of such landforms in Britain.

Conclusion

Torvean is notable for glacial geomorphology, containing an outstanding range of landforms and deposits formed by the meltwaters of the last (Late Devensian) ice-sheet, between approximately 14,000 and 13,000 years ago. Not only are there particularly good examples of individual landforms (kame terraces, eskers, kames, kettle holes and river terraces), but the arrangement of the landforms and the relationships between different features (deposited beneath, in front of or along the side of the glacier) illustrate the development of the glacial drainage system as ice wastage progressed.

FINDHORN TERRACES

L. J. McEwen and A. Werritty

Highlights

This site demonstrates a particularly good assemblage of glacial outwash and river terraces formed respectively during and following the melting of the Late Devensian ice-sheet.

Introduction

The site (NH 845366) is located on the southern side of the middle River Findhorn, within the Streens Gorge, 20 km south of Nairn near the settlement of Ballachrochin. It is notable for a series of glaciofluvial and fluvial terraces (Figure 7.16), which occupy the lower part of the north-west facing slope of Carn Torr Mheadhoin (543 m OD) and are cut into the extensive glacial and glaciofluvial deposits found throughout the Streens Gorge (Young, 1980). The area is described by Horne (1923), Young (1980) and Auton (1990b).

Description

Horne originally identified eleven terrace levels. More recent mapping by Auton (1990b) has

Figure 7.16 The Findhorn terraces at Ballachrochin. (British Geological Survey photograph C1415.)

shown that there are thirteen terraces, of which the lowest five occur at 245–275 m OD and exhibit downvalley gradients of 35–50 m km^{-1}. These terraces locally abut terrace 6 at 285 m OD. In section this flat-topped feature comprises 1.0 m of clast-supported, well-rounded gravel underlain by 1.5 m of a horizontally laminated, low-angle cross-bedded, silty, fine-grained sand. This sand in turn passes down into 2.0 m of finely interlaminated sandy silt and clay with dropstone cobbles and sparse interbeds of diamicton. Above terrace 6, terraces 7 to 11 extend from 287–310 m OD, with the terrace at 305 m OD containing a small steep-sided circular kettle hole 5 m deep. By contrast, terrace 12 (at 340 m OD) is cut into bedrock. The sequence ends at 365 m OD with a small outwash fan on the western side of the Allt a'Choire Bhuidhe.

Interpretation

Horne (1923) interpreted the terraces as fluvial features, although accepting that some of the higher levels were probably glaciofluvial in origin. Young (1980) regarded them as eskers. Auton (1990b), in the most recent investigation, interpreted the landforms as kame terraces and thus of glaciofluvial origin, being closely related to the downwasting of an isolated mass of stagnant ice.

A key part of the sequence in Auton's interpretation of the site is terrace 6 (at 285 m OD). This he considers to be the remains of a glaciolacustrine delta, since the sedimentary sequence closely resembles that of the lower part of the Malaspina delta in Alaska, as described by Gustavson *et al.* (1975). Such an interpretation is not new, having already been anticipated in part by Horne (1923). However, this reconstruction clearly requires the presence of a temporary glacial lake. Young (1980) claimed that the higher terraces are eskers and as such do not require the existence of a glacial lake within the valley, as suggested earlier by Bremner (1939c) and Charlesworth (1956). Auton rejected Young's interpretation and developed a model in which most of the landforms in this middle part of the Findhorn Valley are of paraglacial origin, that is they were formed by 'non-glacial processes that are conditioned by glaciation' (Church and Ryder, 1972). In particular, he considered that the terrace sequence at Ballachrochin developed in reponse to a stagnating ice mass in the Streens Gorge, which steadily downwasted during the Late Devensian and in so doing created local, temporary glacial lakes. Successive ice margins have been reconstructed by Auton at 460 m, 400 m, 380–350 m, 340–300 m (310 and 305 m benches cut at this stage), 300–260 m (benches between 255 and 287 m cut at this stage) and 250 m OD (final benches cut after this stage).

All the major river valleys in upland Scotland possess sets of terraces which are of fluvial and glaciofluvial origin. It is unusual, however, to find staircases of terraces which extend 80 m above the valley floor and possess 13 identifiable benches. This site on the River Findhorn is notable on both accounts.

The flight of terraces is one of the highest and most remarkable in Scotland, the sequence of 13 levels being related to a complex pattern of deglaciation in this part of the middle Findhorn valley. Although the site has recently been investigated in considerable detail in term of its glacial history, the Holocene development of the lower, fluvial, terraces has yet to be attempted. Only when this has been completed will the full significance of the site be disclosed.

Conclusion

The principal landforms at this site comprise a sequence of glacial outwash and river terraces. They are remarkable for the number of levels present and their altitudinal extent. Their development reflects the complex pattern of melting and wastage (deglaciation) of the last (Late Devensian) ice-sheet in the area (approximately 14,000–13,000 years ago). The site represents a striking example of terraces formed by glacial meltwater and river processes during and following deglaciation.

COIRE DHO

J. E. Gordon

Highlights

The assemblage of landforms and deposits at Coire Dho provides an excellent illustration of the development and sudden drainage of an ice-dammed lake during the Loch Lomond Stadial. The cross-valley moraines and bedrock surfaces washed by meltwater floods are the most outstanding examples of these features in Britain.

Introduction

Coire Dho (NH 193142) is a *c.* 10 km^2 area located in upper Glen Moriston, 55 km south-west of Inverness. It provides an excellent assemblage of glacial, glaciofluvial and glacio-lacustrine landforms and sediments and is particularly noted for the best example in Britain of cross-valley moraines. The landform assemblage is associated with a Loch Lomond Readvance glacier that impounded a lake in the upper part of the valley. The former presence of the lake is also recorded by shorelines, lacustrine sediments, overspill channels and areas of washed bedrock where the water drained in a succession of floods. The geomorphology of the site was noted briefly by Milne Home (1878) and has been investigated in detail by Sissons (1977b) and also summarized by him (Sissons 1977d).

Description

Although in an early account Milne Home (1878) reported the presence of 'remarkable terraces' and sand and gravel deposits in Coire Dho, the site subsequently attracted little attention until the work of Sissons (1977b). Sissons produced a comprehensive description and interpretation of the geomorphology of Coire Dho and the wider assemblage of landforms extending a further 15 km to the east in Glen Moriston. During the Loch Lomond Readvance, a glacier flowed eastwards along the Cluanie valley and down Glen Moriston to a limit several kilometres east of Dundreggan (Peacock, 1975a; Sissons, 1977b). The glacier transported distinctive erratics from the Cluanie granodiorite intrusion into Coire Dho, a tributary of Glen Moriston. As a lobe of ice extended up Coire Dho, it dammed a lake in the upper part of the valley. The formation of this lake and its subsequent periodic drainage, together with a fluctuating ice margin, produced a remarkable and varied assemblage of landforms and sediments in a relatively small area.

The limit of the ice in Coire Dho is marked by

a low end moraine which merges upslope with a large lateral moraine (Figure 7.17). Inside the end moraine and extending downvalley for a distance of 3 km is a sequence of at least 30 cross-valley moraines with individual ridges up to 7–8 m high and 20–40 m wide. They run more or less straight down the valley sides and in some cases across the valley floor. The composition of the ridges varies from boulders and unstratified sand to till comprised of grit, sand and stones. In places the ridges consist of lake-floor sediments with coarser material deposited on top.

Although not included in the GCR site, two of the tributary valleys at the head of Coire Dho were also occupied by Loch Lomond Readvance glaciers, which produced end moraines and hummocky moraines.

Glaciolacustrine and glaciofluvial interests include lake shorelines, lake sediments, washed bedrock, meltwater channels and river terraces. Up to seven former lake shorelines have been identified on the north-east side of Coire Dho (Sissons, 1977b). Like similar shorelines in Glen Roy they are partly erosional and partly depositional features. The highest shoreline at 406 m OD is the clearest and most extensively developed, reaching a maximum width of 15–20 m and extending over a distance of 2.6 km (Figure 7.17). It is associated with a col at 407 m OD, suggesting topographic control on the lake level. Several of the lower shorelines are also at the same altitudes as adjacent cols, indicating that drainage through these cols controlled the level of the corresponding ice-dammed lake.

Lake-floor deposits occur extensively in Coire Dho. Sissons (1977b) recorded several sections which show bedded sands overlying laminated silts, sands and clays. Drop stones are locally present, while penecontemporaneous slumping is indicated by low-angle normal faults and contorted bedding. In places the lacustrine sediments have been incorporated into cross-valley moraine ridges (Sissons, 1977b). They extend up to an altitude of 450 m OD, implying lake levels well above the highest shoreline. At the head of Coire Dho the lake sediments have been gullied by water erosion and now take the form of ridges.

Two suites of terraces are present in Coire Dho. Those at higher levels are formed in lake deposits and are erosional features associated with falling lake levels. The lower terraces comprise coarse-gravel aggradational features unrelated to lake levels. These terraces were probably formed by outwash from the Loch Lomond Readvance glaciers at the head of Coire Dho.

An important and unusual interest in Coire Dho and eastwards into Glen Moriston is the presence of extensive areas of water-washed bedrock. The individual areas of washed bedrock are up to 1.5 km long and up to 0.4 km wide. Together with interconnecting meltwater channels they form two distinct belts over 9 km long. Sissons (1977b, 1977d) considered the water-washed bedrock to be associated with sudden drainage of the ice-dammed lake. Such drainage stripped away the former drift cover, often leaving a well-defined washing limit. Bedrock erosional marks comparable to those ascribed to subglacial floods by Shaw (1988) and Sharpe and Shaw (1989) have not been recorded at Coire Dho. The meltwater channels associated with the washed bedrock occur as two main types. The first occupy pre-existing valleys or depressions and have flat floors up to 0.35 km wide, infilled with glaciofluvial sediments which sometimes demonstrate braided channel surfaces or terraces. These channels are associated with catastrophic lake drainage. The second are smaller, narrower channels typical of subglacial meltwater erosion.

Interpretation

There is a considerable body of literature on the geomorphology of moraines that lie transverse to ice flow (cf. Sugden and John, 1976). Sissons (1977b) considered the Coire Dho features to be sublacustrine cross-valley moraines as described by Andrews (1963a, 1963b) from Baffin Island. Various processes have been proposed to explain the formation of such moraines, including the squeezing of material into basal crevasses (Andrews, 1963b), squeezing or pushing of debris at the ice front (Andrews and Smithson, 1966) and accumulation at the grounding margin of floating ice ramps (Holdsworth, 1973; Barnett and Holdsworth, 1974). Sissons (1977b) favoured an ice-marginal origin for the Coire Dho ridges, arguing that their bouldery composition and close spacing were incompatible with squeezing of material into subglacial crevasses. However, as acknow-

Figure 7.17 Geomorphology of Coire Dho showing landforms associated with the former ice-dammed lake (from Sissons, 1977b).

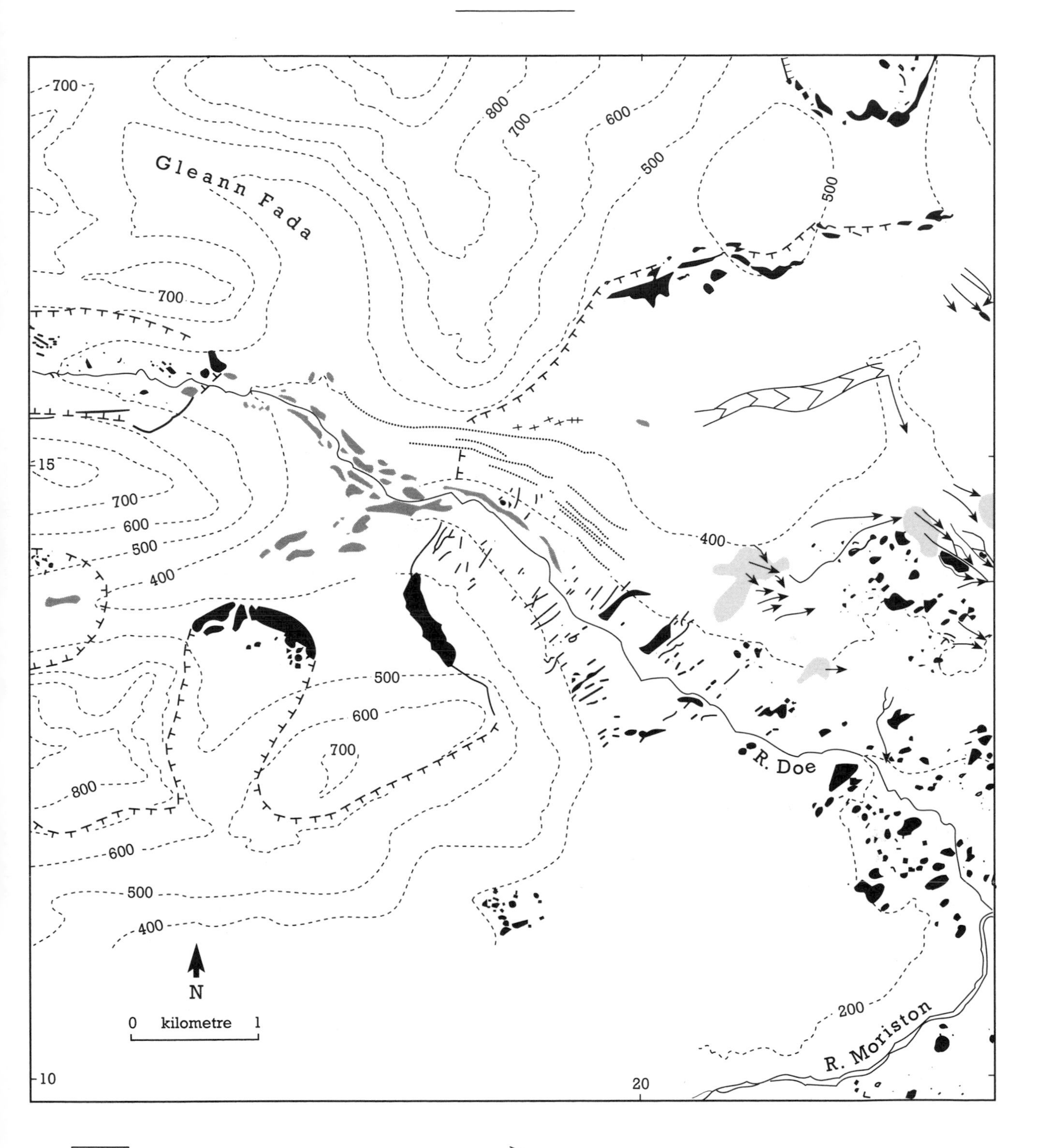
700
800
700
600
500
500
Gleann Fada
700
15
700
600
500
400
400
500
600
700
800
600
500
400
R. Doe
200
R. Moriston
N
0 kilometre 1
10
20

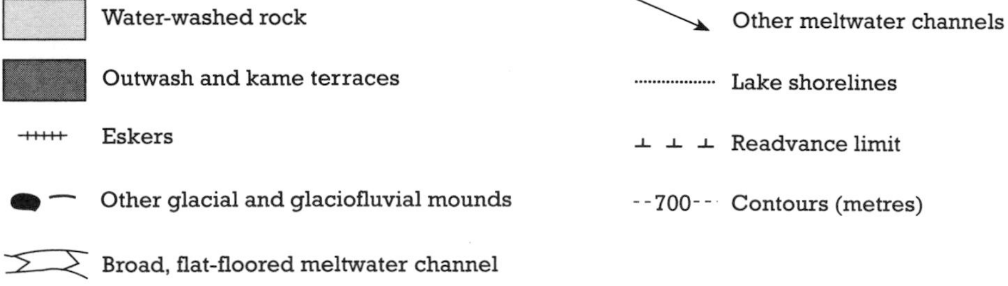
Water-washed rock
Outwash and kame terraces
Eskers
Other glacial and glaciofluvial mounds
Broad, flat-floored meltwater channel
Other meltwater channels
Lake shorelines
Readvance limit
--700-- Contours (metres)

ledged by Sissons, the precise mechanism or mechanisms requires detailed study, which could usefully focus on the sedimentary characteristics of the ridges (cf. Dardis, 1985; Benn, 1989a).

The drainage of the ice-dammed lake is considered in some detail by Sissons (1977b). The characteristics and the spatial arrangement of meltwater channels and washed bedrock suggest a series of major floods (*jökulhlaups*) following two major marginal, submarginal and subglacial drainage routes towards the snout of the glacier in Glen Moriston, in the manner of contemporary glacial lake drainage (see for example, Sugden *et al.*, 1985). Following these periodic floods, final drainage of the lake was probably through the gorge now occupied by the present River Doe.

Coire Dho is important in several respects. First, it provides the best example in Britain of cross-valley moraines. These are normally associated with ice-dammed lakes and the only other localities in Scotland where they have been reported are at Glen Spean (Sissons, 1979c; see Glen Roy and the Parallel Roads of Lochaber), Achnasheen (Sissons, 1982a; Benn, 1989a), Gartness (Rose 1980e) and possibly Loch Muick (Lowe *et al.*, 1991). The Coire Dho moraines are exceptional for the number of ridges present, their clarity of development and the clear spatial relationships between the moraines, lake sediments, shorelines and lake drainage features that can be demonstrated in a relatively compact area. In Glen Spean the cross-valley moraines are partly afforested, but provide an important element in interpreting ice-front positions during the later stages of the sequence of ice-dammed lakes in Glen Roy and Glen Spean. At Achnasheen the ridges are a relatively minor component in the landform and sediment assemblage; at Gartness they are associated with a considerably greater volume of glaciolaustrine sedimentation and additional features of stratigraphic interest.

Second, Coire Dho is unique in Britain for the clear geomorphological evidence it provides for glacial lake drainage routeways. Areas of water-washed bedrock and meltwater channels show two integrated systems of marginal, submarginal and subglacial water movement, which can clearly be related to the evidence for an ice-dammed lake. No comparable reconstructions can be made in such detail elsewhere in Britain. In this aspect of its geomorphology Coire Dho differs from Glen Roy, where steeper topography confined lake drainage to the subglacial valley floors and hence restricted the development and survival of comparable landforms.

Third, Coire Dho is unique in Britain for the extent of the areas of water-washed bedrock associated with glacial lake drainage.

Fourth, Coire Dho is important in illustrating the characteristics of unequivocal glacial lake overspill channels and channels associated with glacial lake drainage. During the last few decades the hypothesis of subglacial meltwater erosion has prevailed in the interpretation of meltwater channel formation in Britain. However, as Sissons (1977b) concludes, ice-dammed lakes may have been more common than recently supposed during the deglaciation of upland Britain and all the field evidence should be critically evaluated in interpreting meltwater channels elsewhere. Since this evidence may be incomplete or only partly preserved, the geomorphology of Coire Dho may provide an invaluable template against which to compare other locations.

Fifth, Coire Dho is important for sedimentological studies, providing a good range of exposures in both cross-valley moraines and lake sediments. In some examples the sections in the fine-grained lake sediments are more clearly exposed than those in Glen Roy.

Conclusion

Coire Dho is exceptional for its range of landforms and deposits associated with a lake dammed by glacier ice during the Loch Lomond Stadial (approximately 11,000–10,000 years ago) and its sudden drainage. It includes the best examples in Britain of cross-valley moraines and washed bedrock. It also provides a particularly good example of integrated ice marginal, submarginal and subglacial meltwater routeways clearly related to the drainage of the lake. Coire Dho is a key site for studies of the geomorphology and sedimentology of former ice-dammed lakes.

FORT AUGUSTUS

C. R. Firth

Highlights

The landforms and deposits at Fort Augustus include kame and kettle topography, glacier flood deposits and lake shorelines. They provide important information for interpreting the geomorphological changes that occurred in the landscape during the Loch Lomond Stadial. Particularly significant is the evidence for re-depression of the Earth's crust by the build-up of glaciers in the west Highlands during the stadial.

Introduction

The site at Fort Augustus consists of three areas at Borlum (NH 383084), the north shore of Loch Ness (NH 386105) and Auchteraw (NH 366090). These demonstrate an assemblage of landforms and deposits including drift limits, outwash gravels and raised shoreline fragments. The interpretation of these features has varied considerably between authors. Charlesworth (1956) and J. S. Smith (1968) proposed that the drift limits represented a stillstand in the decay of the Late Devensian ice-sheet, and Synge (1977b) and Smith (1977) suggested that the highest shoreline terraces were marine. In contrast, Sissons (1979b, 1979c) and Firth (1984, 1986) proposed that all the features were of Loch Lomond Stadial age and that all the shoreline terraces were lacustrine. Sissons (1979c, 1981c) also suggested that the extensive outwash terrace in the area was a *jökulhlaup* deposit formed by the drainage of the former ice-dammed lake in Glen Spean. Firth (1986) largely agreed with this interpretation and further proposed that the morphological evidence indicated glacio-isostatic depression of the Earth's crust during the Loch Lomond Stadial.

Description

The low ground at the southern end of Loch Ness and the surrounding slopes of the Great Glen are mantled by extensive glacial and glaciofluvial deposits (Figure 7.18). On the eastern side of the Great Glen above Fort Augustus, Charlesworth (1956) and Synge (1977b) recorded a lateral moraine which rose from the shores of Loch Ness along the Allt an Dubhair to an altitude of 100 m OD. Firth (1984) has indicated that there is no true lateral moraine at this site, only a drift limit. A similar drift limit has been identified by Sissons (1979b) on the western side of the valley.

At Borlum, inside the drift limit on the eastern side of the valley, a number of Late Devensian features are present. To the east of the River Tarff an area of ice-decay topography (Figure 7.18) consists of kame terraces in the south, which lead into a meltwater channel through kame and kettle topography. The channel descends to an outwash terrace, which in turn grades to a shoreline fragment at 32.4 m OD. Synge (1977b) and Smith (1977) proposed that this shoreline is of marine origin. However, Sissons (1979b, 1979c) and Firth (1984, 1986) maintained that it was a lacustrine feature.

The ice-decay topography is truncated to the north and west by an erosional bluff, which is fronted by a series of terraces. Synge and Smith (1980) suggested that both the bluff and the terraces were of fluvial origin. In contrast, Sissons (1979c) and Firth (1986) have proposed that the terraces and bluff to the east are fluvial in origin, but that the bluff to the north was produced by lacustrine processes, forming a raised shoreline fragment at 22.4 m OD. Further north, directly adjacent to the shore of Loch Ness, is a raised shingle ridge at 17.9–18.8 m OD, also interpreted as a lacustrine feature.

On the lower slopes of the Great Glen, above the northern shores of Loch Ness, there is a distinctive bench at 29.0–29.5 m OD (Figure 7.18). It is up to 10 m wide, and stream sections indicate that it is an erosional feature formed in both bedrock and drift deposits. In places the bench is backed by degraded cliffs cut in bedrock. Firth (1984, 1986) recorded the bench on both sides of the glen but only outside the limit of the Loch Lomond Readvance glacier at Fort Augustus.

The area between Auchteraw and the River Oich contains glaciofluvial outwash deposits (Figure 7.18). To the south of Auchteraw there is an extensive area of kame and kettle topography which is continued to the north by a large outwash terrace. Sissons (1979b, 1979c) believed that these features were associated with an ice margin that lay 4 km inside the Loch Lomond Readvance limit in the area. Deposits in sections (NH 073355) exposed in the terrace comprise poorly sorted materials ranging in size from sand to boulders. The nature of the deposits and the absence of major erosional contacts appears to be consistent with high-energy fluvial deposition, and they have been tentatively interpreted as *jökulhlaup* (large flood) deposits (A. J. Russell, unpublished data): given the probable deltaic depositional environment, the deposits may be related to below-peak flows.

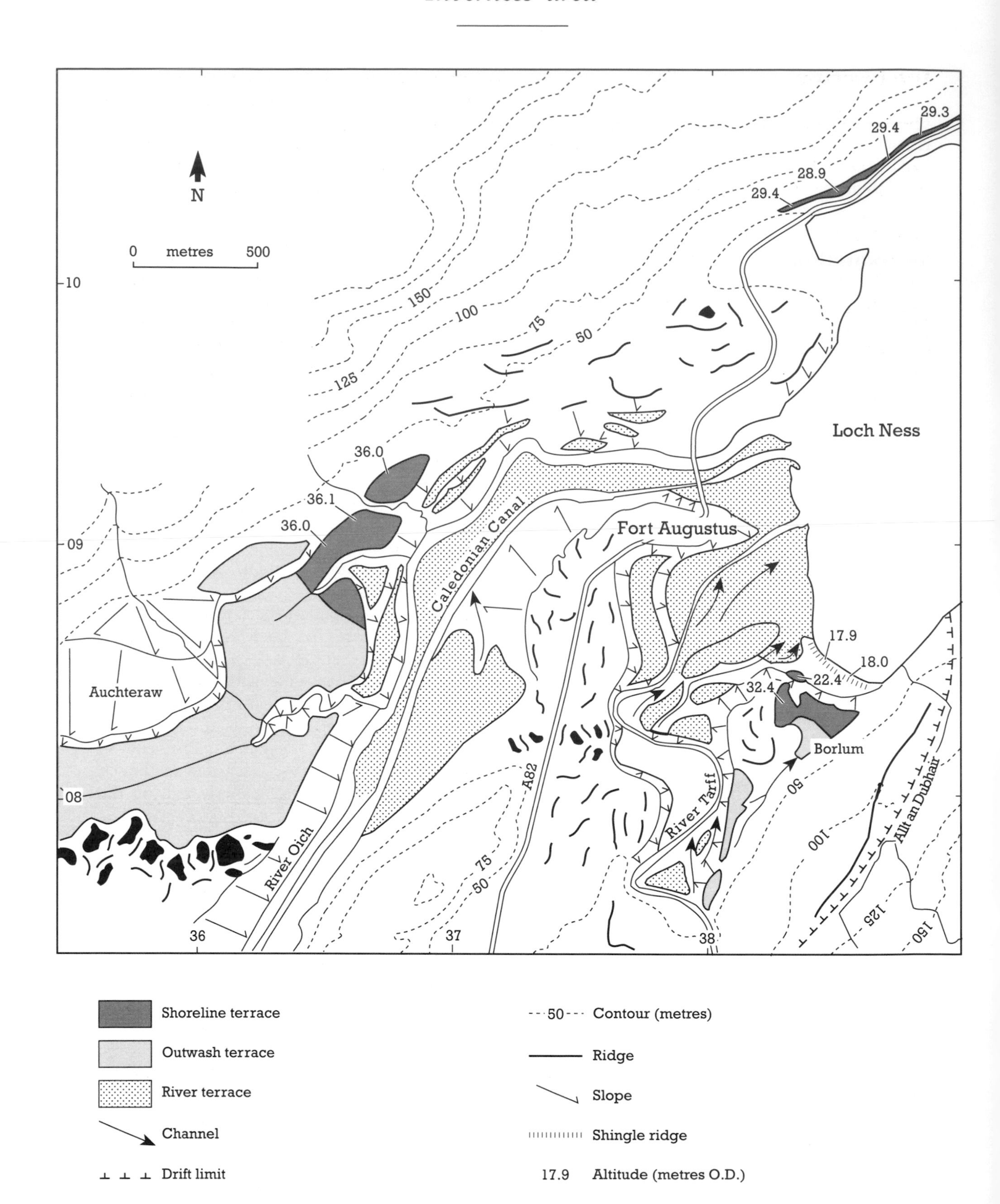

Figure 7.18 Geomorphology of the Fort Augustus area (from Firth, 1984).

Interpretation

The drift limits identified on the slopes of the Great Glen were initially considered to mark a halt in the retreat of the Late Devensian ice-sheet (Charlesworth, 1956; J. S. Smith, 1968). Mapping by Synge (1977b), Sissons (1979b) and Firth (1984) and palaeoenvironmental evidence from Loch Tarff (NH 425100) and Loch Oich (NH 330020) by Pennington *et al.* (1972) implied that the drift limits were deposited during the Loch Lomond Readvance. The limits mark the northernmost extent of readvance ice in the Great Glen and as such are important in the reconstruction of ice coverage during this glacial phase.

Synge (1977b) and Synge and Smith (1980) proposed that the shoreline bench at 29.0–29.5 m OD was a marine feature formed during the decay of the Late Devensian ice-sheet, when the sea penetrated into Loch Ness around 12,800 BP. They considered that marine terraces occurred throughout the River Ness valley, near Inverness, providing a former link between Loch Ness and the sea. However, Firth (1984) reinterpreted the landforms in the Ness valley and concluded that there was no evidence there to support a marine incursion into Loch Ness (see Torvean). Instead he suggested that all the terraces around the shores of the loch were lacustrine in origin. He argued that the erosional feature at Fort Augustus would not have formed in an area of such limited fetch during the period of deglaciation (Firth, 1984, 1986) and proposed that it could only have been produced by periglacial processes during the Loch Lomond Stadial. As such, it is a lacustrine feature and thus comparable to the Parallel Roads of Glen Roy. Dawson *et al.* (1987b) have indicated that such features could easily have formed in an area of limited fetch under periglacial conditions. Firth used the bench to reconstruct the Loch Lomond Stadial lake-level changes of Loch Ness, and it provided important evidence for his idea that the Earth's crust was glacio-isostatically re-depressed at this time (Firth 1986, 1989c). In particular, it demonstrated a rise in loch level from 29 m to 32 m OD between the formation of the erosional terraces at Fort Augustus and the high-level delta at Borlum. Such a rise is present only at the south-west end of the loch and not at the north-east end (see Dores). According to Firth, the simplest explanation of these observations is that the build-up of glaciers in the western Highlands during the stadial was sufficient to re-depress the crust in the area at the south-west end of the loch, resulting in a lake transgression there.

Sissons (1979b, 1979c, 1981c, 1981d) suggested that many of the glaciofluvial features within the Loch Lomond Readvance limits were produced by the catastrophic drainage of the ice-dammed lake in Glen Roy and Glen Spean. He ascribed the outwash terrace at Auchteraw to such an origin and also proposed that it formed part of a more extensive surface including the terrace at 32 m OD at Borlum. Sissons (1981c) suggested that the flood of water into Loch Ness during the *jökulhlaup* was so great that the water level of the loch was temporarily raised by 8.5 m from 22.5 to 31 m OD. He suggested that the loch level then fell to its pre-*jökulhlaup* level. Firth (1984, 1986) agreed that the Auchteraw terrace was the product of a Loch Lomond Stadial *jökulhlaup* but proposed that it descended towards the north-east to grade into a raised shoreline fragment at 36.0–36.1 m OD, and was not associated with the terrace at Borlum. Alternatively, he suggested that the latter was formed during the initial retreat of the Loch Lomond Readvance glacier and thus is distinct from the terraces south-west of Fort Augustus. Firth (1984) interpreted the terrace at Borlum as a delta associated with the 32 m OD loch level and, as a result, he suggested that the *jökulhlaup* flood temporarily raised the loch level by only 4 m from 32 m to 36 m OD.

The field evidence in the Fort Augustus area therefore suggests a sequence of events which spans the period of the Loch Lomond Stadial (Firth, 1984, 1986). During the early part of the stadial, loch level stood at 29 m OD and the erosional benches were formed. The Loch Lomond Readvance then reached its maximum extent; during the initial retreat, loch level stood at 32 m OD. After the ice-front had retreated 4 km, the ice-dammed lake in Glen Spean drained catastrophically to produce a large outwash spread related to a temporarily high loch level at 36 m OD. Subsequently, the level of the loch fell to 22.5 m OD in response to erosion of the outlet of Loch Ness produced by the floodwaters.

The Quaternary landforms and deposits in the Fort Augustus area are important in a number of respects. The features at Borlum are important in determining the Late Devensian and early Holocene evolution of Loch Ness. Synge (1977b) and Synge and Smith (1980) interpreted the 32.4 m OD feature as a marine terrace. In contrast,

Sissons (1979c) suggested that it represented a fragment of the *jökulhlaup* deposit. More recently, Firth (1984, 1986), has proposed that the terrace formed during the Loch Lomond Stadial prior to the *jökulhlaup* and thus provides key evidence for the level of Loch Ness before the event. Firth (1984, 1986, 1989c) also used the features in this area to identify re-depression of the Earth's crust during the Loch Lomond Stadial. The evidence from Fort Augustus, together with that from other sites along the Great Glen (see Dores), is therefore important in reconstructing variations in the regional pattern of isostatic uplift in northern Scotland.

The features at Auchteraw represent the first *jökulhlaup* deposit identified in Great Britian (Sissons, 1979c, 1981d). The evidence indicates that the floods had a considerable impact on Loch Ness, being of sufficient volume to temporarily raise the level of the loch by 4 m and of sufficient erosive power eventually to lower the loch exit by 9.5 m (Firth, 1984). The landforms at Auchteraw also demonstrate the nature of early deglaciation of the Loch Lomond Readvance glacier at Fort Augustus, indicating active retreat over a distance of 4 km before the *jökulhlaup* event.

The erosive benches on the north side of Loch Ness are important in the context of the Late Devensian evolution of the area: they provide one of the key lines of evidence relating to changes in water level. They also represent an erosive feature formed by periglacial lacustrine processes at low altitude and thus complement the higher altitude shorelines of Glen Roy and the low altitude marine features of the Main Rock Platform of western Scotland.

Conclusion

The assemblage of landforms and deposits at Fort Augustus provides evidence for a range of geomorphological processes in the Loch Ness area during the Loch Lomond Stadial (about 11,000–10,000 years ago). They include deposits that indicate the drainage pathway of the meltwaters from the Glen Roy ice-dammed lakes and a series of benches and terraces that show the changes in level of Loch Ness and how the build-up of the Loch Lomond Readvance glaciers in the western Highlands was sufficient to affect the isostatic recovery of the Earth's crust (the weight of the ice on the land had depressed the level of the ground surface) following the melting of the Late Devensian ice-sheet (approximately 14,000–13,000 years ago). Together, the features at Fort Augustus contribute significantly to the understanding of key events and changes in the landscape that occurred during the stadial.

DORES

C. R. Firth

Highlights

The landforms at Dores comprise an exceptional suite of raised shingle ridges of lacustrine origin. These shorelines provide important evidence for interpreting geomorphological changes that occurred in the Loch Ness area during the Loch Lomond Stadial.

Introduction

The site at Dores (NH 598354), located on the eastern side of Loch Ness 12.5 km south-west of Inverness, is notable for a series of raised shorelines. In early accounts, Horne and Hinxman (1914) and Ogilvie (1923) commented on the clarity of the '100 ft raised beach' (a marine terrace) around the northern shores of Loch Ness, and the former suggested that fragments of three raised shorelines were present near Dores. J. S. Smith (1968, 1977), Synge (1977b) and Synge and Smith (1980) identified three shingle ridges and proposed that the highest feature was marine in origin. In contrast, Firth (1984) identified four additional shoreline terraces at this locality and suggested that all the features were lacustrine in origin.

Description

The low ground beween Dores and Inverness was noted by Horne and Hinxman (1914) for its extensive glaciofluvial deposits that merge eastward into till. They suggested that the deposits adjacent to Loch Ness up to an altitude of 33 m (100 ft) OD, had been modified by wave action, with three distinct levels of wave activity being recognized near Dores.

J. S. Smith (1968, 1977), Synge (1977b) and Synge and Smith (1980) also identified three

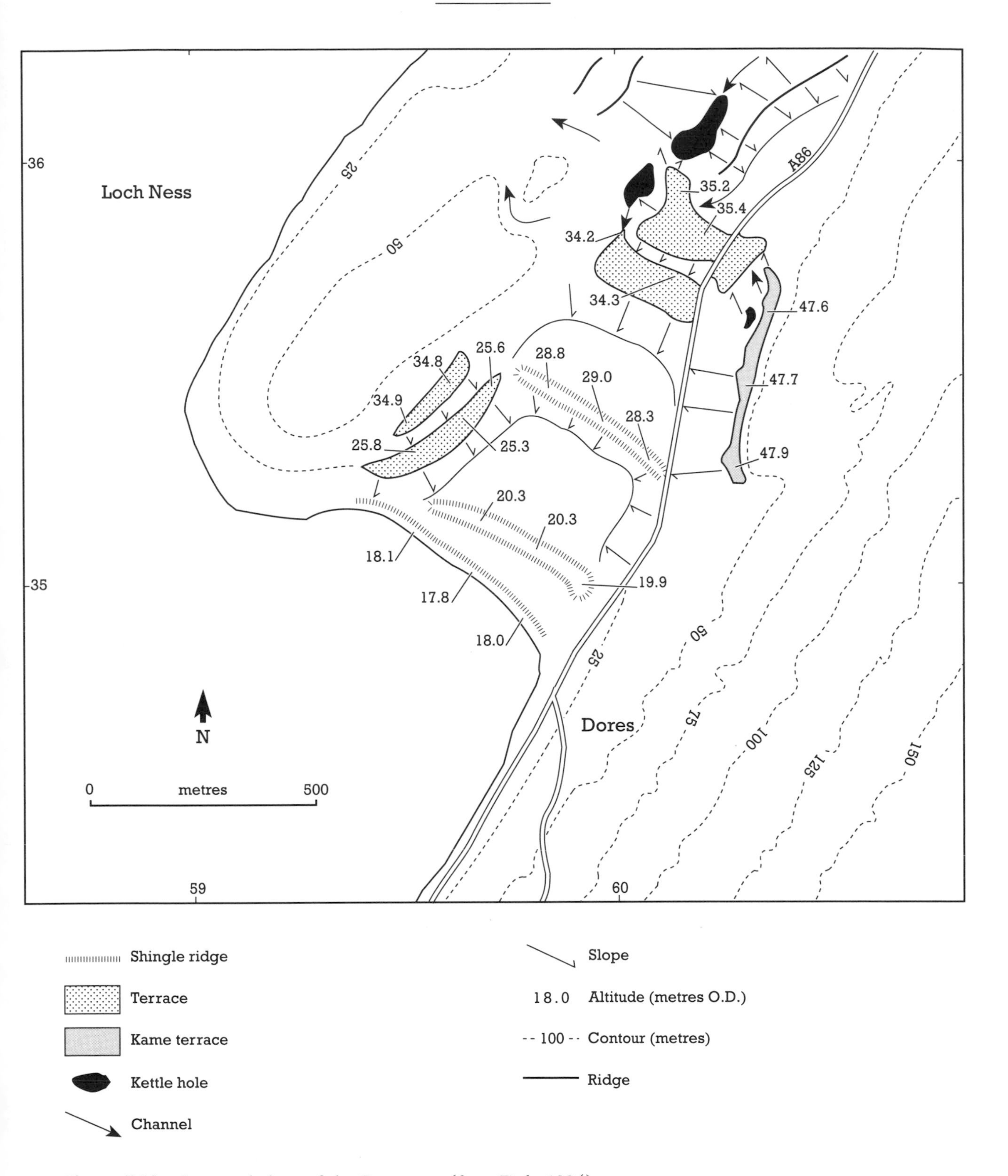

Figure 7.19 Geomorphology of the Dores area (from Firth, 1984).

distinct levels in the deposits at Dores, each level represented by a shingle ridge. The lowest ridge at 17.9–18.1 m OD is related to present loch level (16 m OD), whereas the features at 19.9–20.3 m OD and 28.3–29.0 m OD are indicative of higher water levels (Figure 7.19).

In addition to the three shingle ridges, Firth (1984) identified a series of outwash and shoreline terraces at Dores (Figure 7.19). On the slopes to the east of the highest ridge there is a kame terrace at 47.6–47.9 m OD, and to the north-east of the shingle ridges, shoreline terraces at 35.2–35.4 m OD and 34.2–34.3 m OD are present. These terraces lie above adjacent kame and kettle topography and have meltwater channels descending to them. It is inferred that these higher terraces were formed while ice remained in the area. Poorly developed shoreline terraces at 34.8–34.9 m OD and 25.3–25.8 m OD have also been identified on the slopes west of the shingle ridges (Firth, 1984). Firth (1984) proposed that the 25.3–25.8 m OD terrace was formed at the same time as the 28.3–29.0 m OD shingle ridge and provided a better indication of former loch level.

Cores of sediments have been obtained from Loch Ness in Dores Bay (Pennington *et al.*, 1972; Haggart, 1982). Pennington *et al.* (1972) identified grey microlaminated glaciolacustrine clay, but no organic lake sediments. They inferred that the absence of the latter was due to strong currents. Analysis of the cores for total halide content (Pennington *et al.*, 1972) and diatoms (Haggart, 1982) suggested that the sediments were of freshwater origin, although one sample had a total halide content that was double that found in other layers.

Interpretation

It is generally agreed that the lower shoreline features (17.9–18.1 m OD and 19.9–20.3 m OD shingle ridges) in the Dores area are lacustrine in origin (Synge, 1977b; Firth, 1984). However, there is debate over the origin of the higher shoreline fragments. J. S. Smith (1968, 1977), Synge (1977b) and Synge and Smith (1980) proposed that the 28.3–29.0 m OD shingle ridge is a Lateglacial feature formed during a marine incursion into Loch Ness dated to 12,800 BP. Synge and Smith (1980) proposed this marine origin for several reasons. First, they argued that there was no rock bar or evidence of a former outlet to suggest the presence of former lake at this high level. Second, they considered the sediments from Dores Bay and, in particular, the layer with a high halide content as being evidence in support of a marine incursion. Third, they identified marine terraces at a corresponding altitude throughout the Ness valley, thus providing a former link between Loch Ness and the sea.

In contrast Horne and Hinxman (1914) and Firth (1984) proposed that the highest shingle ridge (28.3–29.0 m OD) was of lacustrine origin. Firth (1984) reinterpreted the landforms in the Ness valley and concluded that there is no evidence to support a marine incursion into Loch Ness and that consequently all the terraces around the shores of Loch Ness were lacustrine in origin. This view was supported by Pennington *et al.* (1972) and Haggart (1982) in their interpretation of the sediments from Dores Bay. Firth (1984, 1986) suggested that the ridge was of Loch Lomond Stadial age. He proposed that no outlet associated with a lake at this level had been recognized because it was destroyed during a catastrophic flood which resulted from the drainage of the former ice-dammed lake in Glen Spean and Glen Roy (Sissons, 1979c).

The shoreline terraces at 34–35 m OD occur at a higher elevation than the adjacent kame and kettle topography, and hence Firth (1984) concluded that they were apparently formed while Late Devensian ice remained in the area. However, there was insufficient evidence from around the shores of Loch Ness to determine the extent of the former lake that formed this shoreline, and Firth considered it possible that the 34–35 m lake may have been localized to the northern end of Loch Ness.

The evidence at Dores indicates that only a single lacustrine shoreline occurs at 25–26 m OD in contrast to more southerly sites in the Great Glen (see Fort Augustus) where two shorelines are associated with this level. This pattern is central to the interpretation that the build-up of Loch Lomond Readvance glaciers was sufficient to halt and reverse the isostatic rebound at the south-west end of Loch Ness (Firth, 1986, 1989b).

The sequence of changes in lake level may be summarized as follows (Firth, 1984). During the decay of the Late Devensian ice-sheet, a lake was formed at the northern end of Loch Ness at an altitude of around 35 m OD. During the Loch Lomond Stadial, lake level stood at 25–26 m OD in the Dores area and subsequently fell to 18–19 m OD (associated shingle ridge at 19.9–20.3 m OD) after a catastrophic flood resulted in the erosion of the outlet along the Ness Valley. During the Holocene, lake level fell to 15 m OD as a result of further downcutting at the outlet, only to be raised to the present day 16 m OD level as a result of the construction of the Caledonian Canal.

The landforms and deposits at Dores provide a key record of Lateglacial and Holocene changes in the level of Loch Ness near its outlet. In particular, the site is of considerable importance in the determination of regional patterns of isostatic uplift in northern Scotland. The evidence from Dores and other sites in the Great Glen (see Fort Augustus) indicates that the build-up of Loch Lomond Readvance glaciers was sufficient to halt and reverse the isostatic rebound at the south-west end of Loch Ness (Firth, 1986, 1989b).

Although marine shingle ridges are arguably better developed on the West Coast of Jura and Northern Islay (see below), the three shingle ridges at Dores are noteworthy due to their lacustrine origin. As raised lacustrine features they are the only landforms of this type identified in Scotland.

Conclusion

The raised shorelines at Dores indicate the changing water levels of Loch Ness during the Lateglacial and Holocene (approximately the last 13,000 years). They provide important evidence that shows how the level of the loch changed when the floodwaters from the ice-dammed lakes in Glen Roy discharged into it. Evidence from the shorelines also contributes towards understanding the wider pattern of isostatic rebound (the result of the release of pressure as the ice melted, following the depression of the land surface by the weight of the ice-sheet on it) at the end of the last glaciation, and the interruption of isostatic rebound during the Loch Lomond Stadial (approximately 11,000–10,000 years ago).

BARNYARDS

C. R. Firth

Highlights

The sub-surface deposits at Barnyards include a sequence of estuarine sediments and peat. These provide a detailed record, supported by radiocarbon dating, of the changes in relative sea level that occurred in the Beauly Firth area during the Lateglacial and Holocene.

Introduction

The Barnyards site (NH 531470) is an area of carseland (flat, low-lying area of silty clay deposits of estuarine origin) located 0.5 km north of Beauly. The Beauly Firth area provides important evidence for reconstructing Lateglacial and Holocene sea-level changes in northern Scotland. Until recently the Beauly carse deposits have received little detailed study. Early researchers, such as Horne and Hinxman (1914), Ogilvie (1923) and J.S. Smith (1968), noted only that Holocene raised beaches at '25 ft' and '15 ft' were present, and they made no detailed assessment of the carse deposits. In contrast, recent studies by Sissons (1981c), Haggart (1982, 1986, 1987), Firth (1984, 1989a) and Firth and Haggart (1989) have indicated that evidence of relative sea-level changes which date back to the Loch Lomond Stadial is present within the area. From stratigraphical and morphological studies these authors have identified seven former marine levels, two of which lie buried beneath later deposits. Barnyards is a key locality for interpreting the stratigraphic evidence.

Description

Horne and Hinxman (1914) first identified raised marine deposits in the Beauly area. They reported that the village of Beauly stands on a 'wide tract of marine alluvium, with a mean level of about 25 ft'. Ogilvie (1923) also recognized a '25 ft' raised beach in the Beauly carselands and he suggested that a lower '15 ft' beach was present. These views were reiterated by J. S. Smith (1968), who suggested that the carse deposits were Holocene in age. He also noted the presence of a degraded cliff landward of the carse and he considered that this erosional feature was also of Holocene age.

Eyles and Anderson (1946) provided brief details of the deposits in the Barnyards area, recording 3.4 m of carse clay overlying 0.3 m of peat. More detailed investigations of the stratigraphy of the Beauly Firth carselands have been undertaken subsequently by Sissons (1981c), Haggart (1982, 1986, 1987), Firth (1984) and Firth and Haggart (1989). Sissons (1981c) confined his investigations to the carselands south of the Beauly River and in the basin of Moniack Burn. He noted that the clay–silt carse deposits are underlain by an extensive gravel layer. Sissons

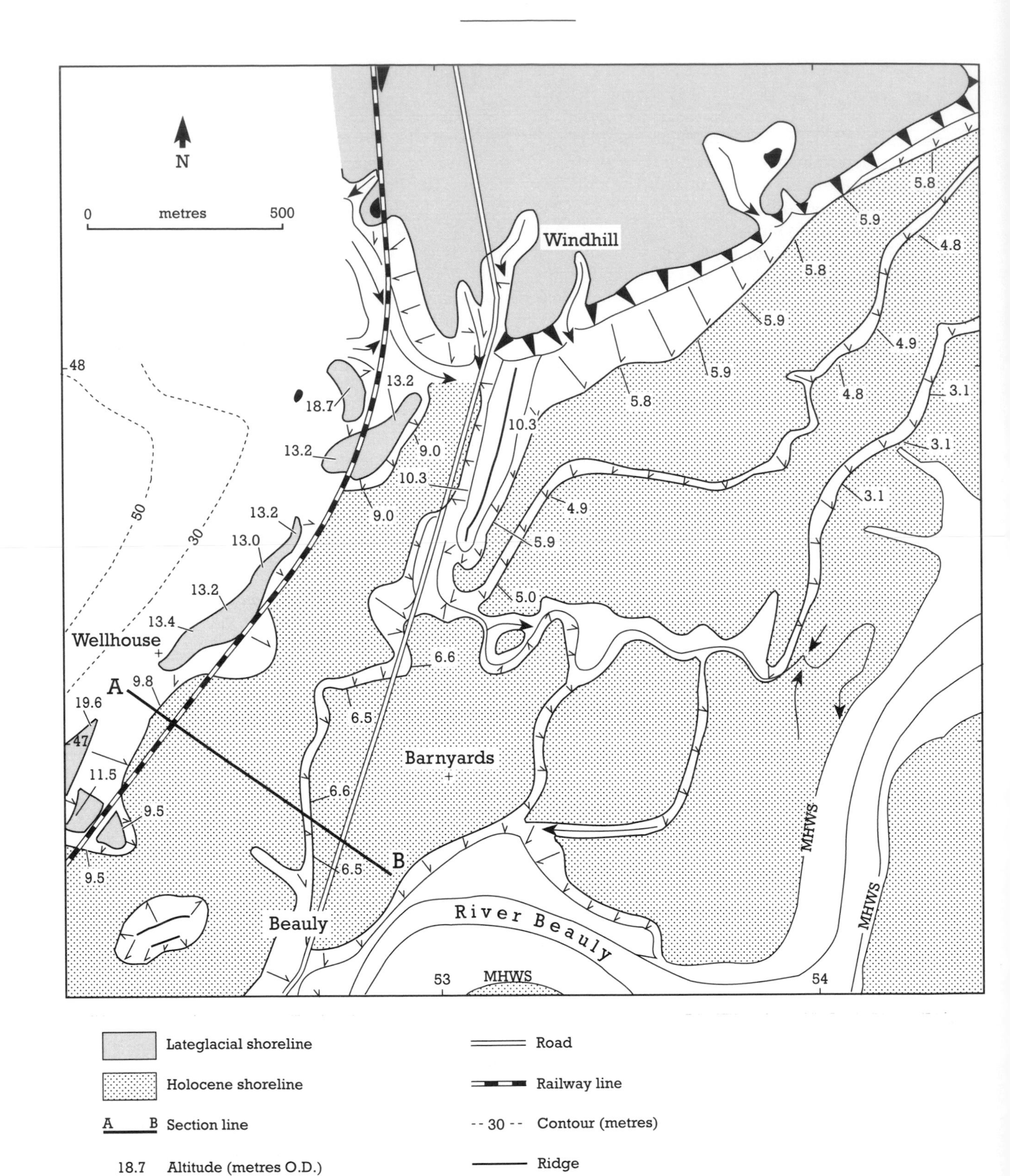

Figure 7.20 Geomorphology of the Beauly carselands (from Firth, 1984).

(1981c) indicated that the gravel layer rises landward from below −1 m OD to a maximum altitude of 2 m OD. He noted that it rests on Lateglacial marine sediments and that it terminates landward near the base of the degraded cliff that backs the carselands. Between the cliff and the buried gravel layer he identified a steeply sloping surface of erosion, which rises to 6–6.5 m OD at the base of the cliff. Sissons (1981c) concluded that these buried features were marine in origin.

Similar stratigraphical investigations were conducted by Firth (1984) on the carselands north of the Beauly River (Figure 7.20). Firth (1984) also identified a buried gravel layer rising to near 2 m OD (for example, south-east of Windhill, Figure 7.20) and a steeply sloping erosion surface which terminated at the base of a degraded cliff, at an altitude of about 8 m OD. Firth (1984) also suggested that these features were marine.

Haggart (1982, 1986, 1987) undertook stratigraphical investigations of the carselands between Wellhouse and Barnyards (Figure 7.20, A-B). In this area he identified a sequence of marine and terrestrial deposits (Figure 7.21). At the base of the sequence till is overlain by light-grey, silty clay. The latter is truncated by a layer of silt, sand and gravel, which is succeeded by grey, clayey silt rising to an altitude of 7.6 m OD. This layer is overlain by peat, the base of which has been dated to 9610 ± 130 BP (Birm–1123). Haggart (1986) has interpreted this clayey silt/peat contact as a regressive overlap caused by a fall in relative sea level. An additional radiocarbon date from a buried peat layer farther seaward indicates that relative sea level was still falling after 9200 BP. Lying above the peat in the sequence is a grey, marine silty clay (carse deposits), which rises to a maximum altitude of 9.5 m OD. This in turn is overlain by another peat whose base has been dated to 5510 ± 80 BP (Birm–1122) (Haggart, 1982, 1986). Haggart (1982) also identified a light-grey, micaceous silty sand within the carse deposits. He noted that this sand layer rises to a maximum altitude of 9.3 m OD and is marine in origin. A similar deposit at Moniack, 4 km to the south-east, was dated to 7270 ± 90 BP (Birm–1126) (Haggart, 1982). A similar sequence of marine and terrestrial deposits was identified by Firth (1984) in the carseland north of Barnyards.

The surface morphology of the Beauly carselands has been mapped in detail by Firth (1984). He noted that the degraded cliff truncates Lateglacial marine deposits and extends as far west as Windhill (Figure 7.20). Directly south of Windhill, a broad sand ridge with a maximum altitude of 10.3 m OD has been identified (Haggart, 1982; Firth, 1984). Both Haggart (1982) and Firth (1984) have interpreted this ridge as a Holocene sand spit. Firth (1984) has also indicated that five separate terrace levels are present in the carselands. The highest level at 9.0–9.8 m OD occurs west of the sand ridge (Figure 7.20). Lower levels are present at 7.5–6.6 m, 5.8–5.9 m, 4.8–5.0 m and 3.1 m OD, each lower level extending farther east. On the slopes west of the carselands Firth (1984) identified a series of Lateglacial marine terraces at 19.6 m, 18.7 m, 13.4–13.0 m, and 11.5 m OD (Figure 7.20); to the north of the carselands a kettled outwash surface grades into a Lateglacial shoreline fragment at 27 m OD.

Interpretation

The stratigraphical and morphological investigations of the carse deposits in the Beauly area provide a detailed sequence of Late Quaternary relative sea-level movements for northern Scotland. It is agreed by all authors that the carse deposits are of Holocene age, a fact reinforced by the palaeoenvironmental evidence presented by Haggart (1982, 1986). In contrast, some debate exists over the age of the degraded cliffline. The early researchers (Horne and Hinxman, 1914; Ogilvie, 1923; J. S. Smith, 1968) proposed that the cliff was Holocene in age, being formed at the culmination of the Main Postglacial Transgression. In contrast, Sissons (1981c) and Firth (1984) suggested that the cliff was formed in association with the buried gravel layer and was a Lateglacial feature. This view was advanced for three reasons. First, the cliff, the steeply sloping surface of marine erosion and the buried gravel layer are continuous. Second, the cliff is partly buried by a variety of Holocene deposits. Third, the deposition of the Holocene silty clays (carse deposits) could not have occurred at the same time as the erosion that produced the cliff.

Sissons (1981c) and Firth (1984) concluded that the buried gravel layer was formed by marine erosion during a slow transgression that occurred during the Loch Lomond Stadial. Both authors correlated the inner margin of the buried gravel layer at about 2 m OD with the Main Lateglacial Shoreline identified in the Forth

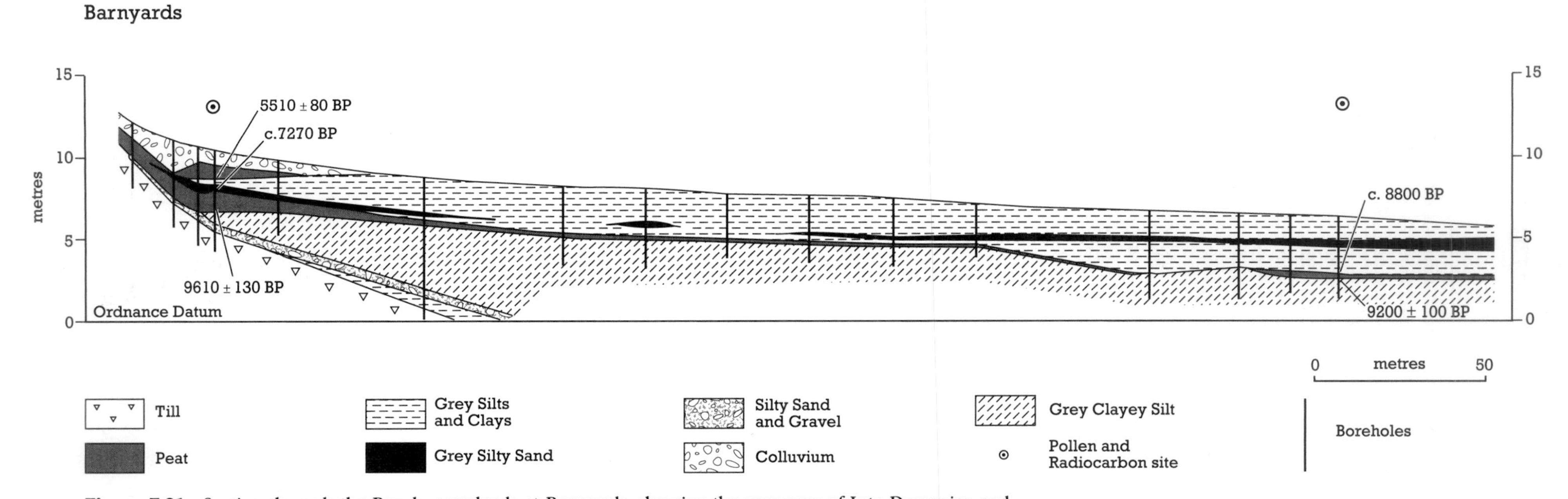

Figure 7.21 Section through the Beauly carselands at Barnyards, showing the sequence of Late Devensian and Holocene deposits (from Haggart, 1986).

estuary (Sissons, 1969, 1976a). Sissons (1981c) suggested that the marine transgression must have continued at a faster rate in order to produce the marine erosion surface which separates the buried gravel layer from the degraded cliff. This transgressive event culminated at 7–8 m OD. Similar steeply sloping surfaces of marine erosion have been identified in the Forth estuary (Sissons, 1976a).

Haggart (1982, 1986) suggested that relative sea level then fell at the beginning of the Holocene, with marine-estuarine, grey, clayey silt being deposited in the Barnyards area up to an altitude of 7.6 m OD until about 9600 BP. Haggart (1986) proposed that these buried marine deposits were equivalent to the Main Buried Beach identified in the Forth and Tay estuaries (see Western Forth Valley and Carey) (Sissons, 1966; Cullingford *et al.*, 1980).

Haggart (1982, 1986) presented evidence that relative sea level continued to fall during the early Holocene, until around 8800 BP, when a major transgression started, during which the marine, grey silty clay was deposited. Interruption of the deposition of the silty clays occurred around 7200 BP (Haggart, 1982, 1986, 1987, 1988b) with the formation of a grey, micaceous, silty fine sand layer. This deposit was thought to represent either a period of increased marine transgression or a storm surge event (Haggart, 1982, 1986, 1987, 1988b). It has been correlated with similar deposits identified throughout eastern Scotland (see Silver Moss and Maryton) (Smith *et al.*, 1985a; Dawson *et al.*, 1990), and is now ascribed to a tsunami associated with the second Storegga Slide on the Norwegian continental slope (Dawson *et al.*, 1988; Long *et al.*, 1989a).

The transgressive event which deposited the carse silty clays culminated at about 6500 BP with the formation of the highest Holocene shoreline fragment in the area, at 9.5 m OD (Haggart, 1982; Firth, 1984; Firth and Haggart, 1989). Firth (1984) correlated this shoreline fragment with the Main Postglacial Shoreline, which has been identified throughout eastern Scotland (Sissons and Smith, 1965b; D.E. Smith, 1968; Morrison *et al.*, 1981). Relative sea level then fell to its present level via intermediate shorelines at 7.5 m, 5.9 m, 4.9 m and 3.1 m OD. These shoreline fragments formed the basis of the Holocene shorelines diagram produced by Firth (1984) and Firth and Haggart (1989), which incorporated six tilted shorelines declining in altitude towards N20°E.

The sequence of deposits at Beauly represents a key stratigraphic record of Late Quaternary environmental and geomorphological changes in northern Scotland. In this respect it complements the interest at Munlochy Valley, where the evidence is principally morphological. It demonstrates that a slow transgression during the Loch Lomond Stadial culminated in the formation of the Main Lateglacial Shoreline, and illustrates the subsequent rapid transgression that formed the marine erosion surface. The area also reveals the early Holocene regression, during which the Main Buried Beach was formed. Similarly the area records the Holocene transgression that culminated in the formation of the Main Postglacial Shoreline, as well as a possible storm surge deposit dated to 7200 BP. The four, lower, Holocene shorelines demonstrate interruptions in the subsequent fall of relative sea level to its present level.

Similar changes in relative sea level have been identified in other areas of Scotland, namely the Forth and Tay estuaries (Morrison *et al.*, 1981) and more recently at Creich in the Dornoch Firth (Smith *et al.*, 1991b). However, the Beauly carse deposits are important because they contain a wealth of stratigraphical and morphological evidence within such a limited area. Haggart (unpublished data) has suggested that further stratigraphical investigations may identify an equivalent to the Lower Buried Beach of the Forth Valley (Sissons, 1966).

The area is an integral member of a national network of Quaternary sites which illustrate the changes in relative sea level in Scotland (see Silver Moss, Western Forth Valley, Carey, Dryleys, Maryton and Philorth Valley). The area also provides details which are important in the determination of regional and national patterns of isostatic movements in the British Isles (Haggart, 1989; Shennan, 1989). For example, Firth and Haggart (1989) noted apparent shifts in the isobases of different shorelines, implying, if confirmed, shifts in the centre of isostatic uplift between the formation of different shorelines (see also Gray, 1983, 1985). Furthermore, the gradients of shorelines in the Moray Firth area are steeper than expected from existing isobase maps (Sissons, 1967a, 1983a; Jardine, 1982) suggesting that the pattern of isostatic uplift in Scotland may not be represented by a simple ellipsoid (Firth and Haggart, 1989). In addition, the complexity of isostatic uplift is attested from comparison of sea-level curves from the Western Forth Valley, Lower Strathearn, the eastern Solway Firth and the inner Moray Firth, which suggests apparent differences in relative rates of uplift between the different areas (Haggart, 1989).

Conclusion

The sediments at Barnyards contain a detailed record of relative sea-level changes in the Beauly Firth area during the Lateglacial and Holocene (approximately the last 13,000 years). The record shows several significant fluctuations in relative sea-level position, providing valuable comparisons with the changes that are documented at reference sites elsewhere in Scotland. Together, the records from these sites allow the wider patterns of isostatic movements (movements of the Earth's crust triggered by ice-sheet growth and loading, and melting and unloading) and sea-level changes to be established.

MUNLOCHY VALLEY

C. R. Firth

Highlights

Munlochy Valley is notable for a series of raised shorelines. In conjunction with a succession of estuarine and peat deposits buried beneath the valley floor, these provide a detailed record of coastline changes during the Lateglacial and Holocene.

Introduction

The site (NH 645528) lies 0.5 km south-west of Munlochy and comprises an area on the north-west side of Munlochy Valley and part of the valley floor. It is the most representative area for a series of raised shoreline fragments at the head of Munlochy Bay. Munlochy Valley has long been recognized for the detailed morphological evidence it provides of changes in relative sea level. Horne and Hinxman (1914), Ogilvie (1923) and J. S. Smith (1966, 1968) all noted the highest raised marine shoreline in the valley at 90–100 ft (27–30 m) OD. However, only Ogilvie (1923) proposed that this feature was formed in close association with a downwasting ice-sheet. These same authors also indicated that other raised marine features are present lower in the valley, although the only description provided was by Horne and Hinxman (1914), who identified raised shoreline fragments at 50 ft (15 m) and 25 ft (8 m) OD. In contrast, Firth (1984) has identified eleven raised marine levels, six of these being Lateglacial in age and five Holocene. Firth (1984, 1989a) also suggested that the highest shorelines in the valley were formed in close association with a downwasting ice mass.

Description

Within Munlochy Valley and on the slopes above Munlochy Bay there are a series of raised marine shoreline fragments and glaciofluvial features (Firth, 1984, 1989a) (Figure 7.22). The highest and most distinctive of the marine terraces occurs at an altitude of 28.9–29.4 m OD and extends for 2 km along the northern slope of the valley. The lower marine terraces are only poorly developed, occurring in a 'staircase', one feature above another. They indicate four marine levels at 27.0 m, 24.6 m, 17.2–17.5 m and 14–15 m OD.

The floor of Munlochy Valley is composed of grey, silty clays up to 2 m thick, which contain shell fragments and overlie sands and gravels. Towards the head of the valley the silty clay deposits become peaty. These sediments underlie a series of horizontal surfaces, former salt marshes, linked by gently sloping ramps, former mudflats. The horizontal surfaces are interpreted as raised marine shoreline fragments and they occur at five distinct levels (8.0 m, 7.7–6.8 m, 5.3–5.5 m, 4.2–4.3 m, 3.0–3.2 m OD).

Around the shores at Munlochy Bay there is a degraded cliffline, against which, for the most part, raised shingle beaches are deposited. West of Munlochy, however, fine-grained estuarine deposits lie adjacent to the cliffline. Stratigraphical investigations by Firth (1984) indicate that the cliff is fronted by a steeply sloping surface, which descends from *c.* 5 m to 0 m OD. Beyond this there is an extensive planar surface which can be traced throughout the bay and is interpreted as a platform of marine erosion.

Above the highest marine terrace the slopes comprise kame and kettle topography and are dissected by meltwater channels indicative of a downwasting ice mass. The meltwater channels can be traced westwards, either to kame and

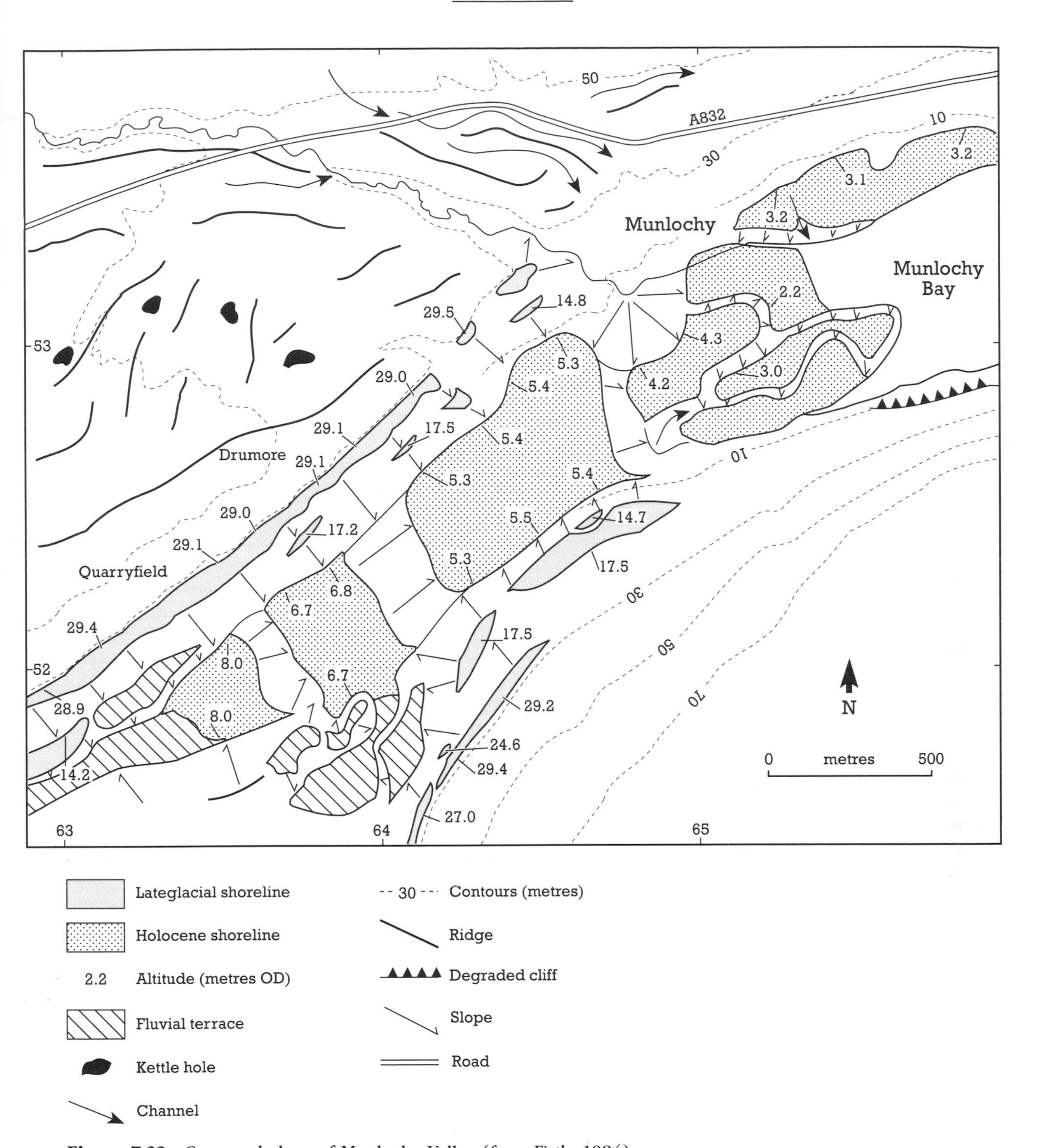

Figure 7.22 Geomorphology of Munlochy Valley (from Firth, 1984).

kettle topography or to cols that separate the Munlochy Valley drainage basin from the Beauly Firth. The clearest of these meltwater channels descends from the col near Ashley (NH 633502) past Bogalian Church (NH 635505) to an altitude of 30 m on the southern slopes of Munlochy Valley.

Interpretation

The glaciofluvial features which mantle the upper slopes of Munlochy Valley testify to the decay of the Late Devensian ice-sheet. Many of the meltwater channels associated with these deposits ultimately lead into Munlochy Valley, but only one channel, which descends from the col at Ashley, has been directly linked with the raised marine features (Firth, 1984, 1989a). Firth (1984) proposed that while relative sea level stood at 29 m OD in Munlochy Valley, meltwater flowed across the Ashley col into the valley from the Beauly Firth. This implies that ice must have occupied the Beauly Firth up to an altitude of 55 m while the highest marine terrace was being formed.

The marine terraces, down to 14 m OD, were produced as relative sea level fell and after the flow of meltwater into the valley had ceased (Firth, 1984). The occurrence of the shoreline fragments as steps, one below the other on the hillside, has facilitated identification of altitudinally close but chronologically distinct Lateglacial shorelines within the inner Moray Firth area.

The grey, silty clay deposits present on the floor of the valley were considered by Horne and Hinxman (1914) to be part of the '25 ft raised beach'. These are similar to deposits found at Beauly (see Barnyards) and in the carselands of Scotland. Such deposits are considered to be estuarine in origin and Holocene in age, a view supported by Haggart (1982, 1986, 1987, 1988b) in his study of the carse clays at Beauly. Firth (1984) proposed that the 8 m OD surface in these silty clay deposits is equivalent to the Main Postglacial Shoreline which has been identified throughout eastern Scotland (Sissons and Smith, 1965b; Morrison *et al.*, 1981; Sissons, 1983a). The four lower marine levels in the estuarine deposits were produced as relative sea level fell in response to continued isostatic uplift. The inferred storm surge or tsunami deposit, about 7200 BP, is also represented in the succession at Munlochy Valley (Firth and Haggart, 1989) (see also Barnyards, Maryton, Silver Moss and Western Forth Valley; Smith *et al.*, 1985a; Dawson *et al.*, 1988; Haggart, 1988b; Long *et al.*, 1989a).

The degraded cliffline which borders the shores of Munlochy Bay was originally considered to have formed when relative sea level stood at 8 m OD (J. S. Smith, 1968). In contrast, Firth (1984) proposed that the erosional feature was produced at the same time as the extensive surface of marine planation that occurs throughout Munlochy Bay, and which rises to 0 m OD. This erosional feature has been equated to the Main Lateglacial Shoreline and is thought to have formed during the Loch Lomond Stadial (Sissons, 1981c; Firth, 1984; Firth and Haggart, 1989).

Limited investigations of the estuarine deposits in Munlochy Valley indicate that buried marine deposits, which possibly date from the early Holocene, may be present and worthy of further investigation (Firth and Haggart, 1989).

From the morphological and stratigraphical evidence in Munlochy Valley, Firth (1984) interpreted the following sequence of events. As the Late Devensian ice-sheet retreated, the sea flooded into Munlochy Valley to a maximum altitude of 29.4 m OD. While the sea stood at this level, ice occupied the Beauly Firth and meltwaters flowed over the watershed into Munlochy Valley. Subsequently, relative sea level fell and formed Lateglacial marine depositional terraces at 27.0 m, 24.6 m, 17.5 m and 14–15 m OD, and then continued to fall to some unknown level below 0 m OD. During the Loch Lomond Stadial there was a slow marine transgression combined with extensive marine erosion which formed the Main Lateglacial Shoreline at 0 m OD. There followed a more rapid rise in relative sea level that culminated at about 6 m OD. Subsequently, relative sea level fell to an unknown level. The evidence indicates that during the Holocene there was another marine transgression between 7100 BP and 5510 BP, which culminated at 8.0 m OD with the formation of the Main Postglacial Shoreline. Since that time relative sea level has fallen to its present level via intermediate shorelines at 7.7–6.8 m, 5.3–5.5 m, 4.2–4.3 m and 3.0–3.2 m OD.

The landforms and deposits in Munlochy Valley provide a key record of Lateglacial and Holocene relative sea-level changes in northern Scotland. The site demonstrates the best Lateglacial shorelines in the Beauly Firth area and provides morphological representation of all the Holocene shorelines. It contains five Lateglacial beaches at different levels, five distinct Holocene beaches, including the Main Postglacial Shoreline, and the buried Main Lateglacial Shoreline. In providing detailed morphological evidence for relative sea-level changes in the inner Moray Firth area, Munlochy Valley therefore complements the stratigraphic record represented at Barnyards (see above). The features indicate a close relationship between Lateglacial raised marine terraces and ice-sheet decay and illustrate the fall in relative sea level associated with deglaciation in Scotland. The area also provides evidence of a period of marine erosion during the Loch Lomond Stadial and of a major marine transgression during the Holocene, which culminated in the formation of the Main Postglacial Shoreline. The four lower Holocene shorelines demonstrate temporary stillstands in the fall of relative sea level to its present position.

The features in Munlochy Valley are important for a number of reasons. First, the relatively high number of raised marine levels recorded in the valley are of regional significance in determining the number of shorelines and patterns of isostatic uplift in the inner Moray Firth. Second, the area is a key reference site demonstrating changes in relative sea level during the Lateglacial and Holocene in northern Scotland. Third, the area is an integral member of a national network of Quaternary sites which together represent relative sea-level movements in Scotland, and as such demonstrate national patterns of isostatic uplift (see for example Barnyards, Milton Ness, Dryleys, Western Forth Valley and Glenacardoch Point) (see Barnyards for further discussion of the wider significance of the inner Moray Firth area in this context, and also Firth, 1989a; Firth and Haggart, 1989; Haggart, 1989; Shennan, 1989).

Conclusion

Munlochy Valley provides an important geomorphological record of sea-level changes during the Lateglacial and Holocene (approximately the last 13,000 years). The evidence comprises a combination of both shoreline terraces and buried estuarine and peat sediments. In particular, the site is noted for the high number of raised marine levels, the majority represented as clear landscape features. Such a detailed geomorphological record complements the sedimentary record at Barnyards and makes this a key reference area in the network of localities for studies of sea-level change.

BEN WYVIS

C. K. Ballantyne

Highlights

Ben Wyvis is outstanding for an assemblage of periglacial landforms developed on Moine schist, most notably non-sorted, patterned ground features and solifluction sheets and lobes.

Introduction

The Ben Wyvis massif in Easter Ross, 30 km north-west of Inverness, is composed of Moine schists and gneisses and rises to 1046 m OD at the summit of Glas Leathad Mór (NH 463684). The summit ridge and upper slopes of the mountain are notable for a range of periglacial landforms formed during the Lateglacial and Holocene. These were first studied by Galloway (1958, 1961a) in the course of his pioneering work on the periglaciation of Scotland, and his map of the solifluction phenomena in this area, although somewhat misleading, has been reproduced in several texts (for example, Sissons, 1965; Embleton and King, 1968, 1975b; Curtis *et al.*, 1976). The features described by Galloway have subsequently been investigated in detail by Ballantyne (1981, 1986b).

Description

The periglacial landforms developed on Ben Wyvis include solifluction lobes and sheets, 'ploughing boulders', nivation hollows, turf hummocks and non-sorted stripes (Figure 7.23). During the Late Devensian glaciation, ice from the west flowed around the flanks of the mountain, but there is no direct evidence to indicate whether the main ridge, which generally exceeds 900 m, was ice-covered (Peach *et al.*, 1912). Romans *et al.* (1966) concluded that the smooth, vegetation-covered regolith that mantles most of the massif is till of pre-Devensian age. The presence of silt droplets in the soils of the main ridge was interpreted by Romans *et al.* (1966) and Romans and Robertson (1974) as evidence of former permafrost and indicative of periglacial modification of the regolith. However, Ballantyne (1981, 1984) has argued that the whole thickness of the regolith, which varies in depth from 0.7 m on the plateau to over 2 m on the slopes, was produced by frost weathering. Where underlain by schists and gneisses, the regolith consists of pebbles and boulders embedded in a mica-rich, silty sand matrix, this being typical of the frost-susceptible soils developed on schistose mountains in the Highlands. As indicated by Ballantyne (1981, 1987a), the nature of the regolith represents a primary control on the range of periglacial landforms that develop on a particular mountain summit.

Lobes and sheets of debris are outstandingly well-developed on the steep (25°–30°) north-west flank of Ben Wyvis (Figure 7.23), where Galloway described 'stone-banked' (actually turf-banked) and smaller 'turf-banked' (actually vegetation-covered) lobes. He explained the distribu-

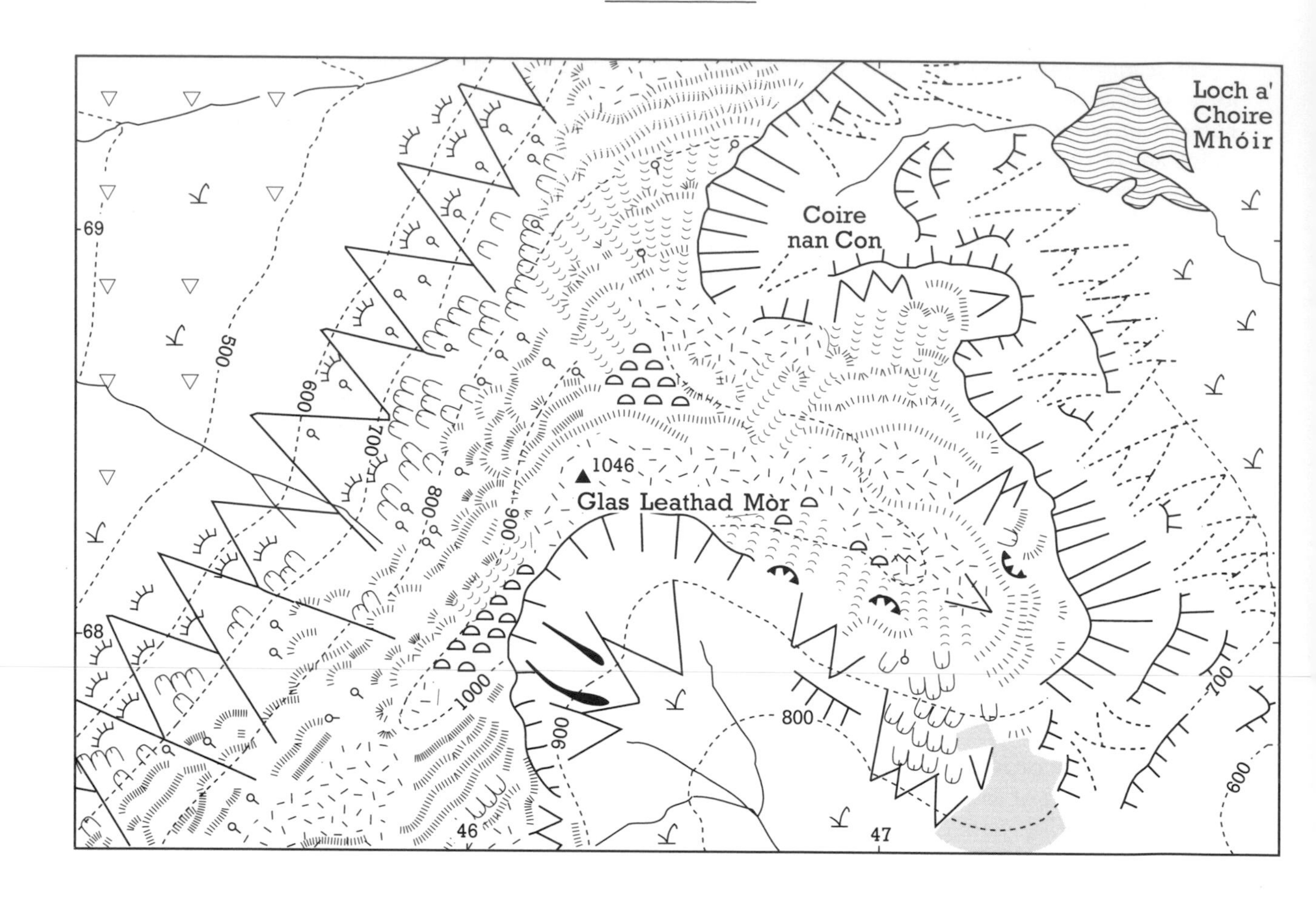

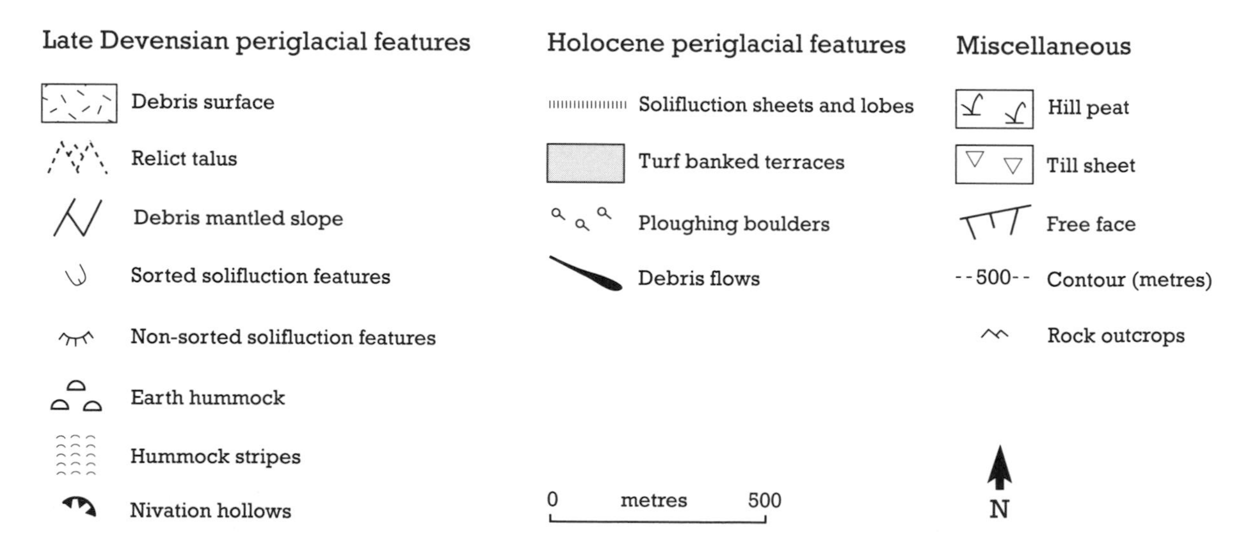

Figure 7.23 Periglacial landforms and deposits on the summit ridge of Ben Wyvis (from Ballantyne, 1984).

tion of the two types in terms of lithological differences. Ballantyne (1981) found that the vegetation-covered features had encroached on the turf-banked lobes, burying the relict block-slopes on which the latter were developed. He concluded that the fossil turf-banked lobes, which show pronounced evidence of vertical frost sorting, had moved downslope by a combination of frost creep and gelifluction, probably under permafrost conditions. The vegetation-covered features show a transition from sheets with straight-fronted scarps on gentle slopes near the crest of the slope, through lobe-fronted sheets, to individual lobes that descend to an altitude of about 750 m OD. These forms are non-sorted and have developed by gelifluction *sensu stricto*, and are currently moving downslope at a rate of a few millimetres per year. The vertically sorted, turf-banked lobes are apparently inactive, apart from the washing out of interstitial fine material, evident in places in the form of spreads of sand in front of lobe 'risers'. However, in one excavation a buried podsol was found containing Holocene pollen assemblages (Ballantyne, 1984), implying movement under climatic conditions not dissimilar from those of today. Both types of feature are also developed on the east side of the massif, as are relict, vegetation-covered boulder lobes, active ploughing boulders and, on sheltered slopes, active turf-banked terraces produced by frost creep (Figure 7.23). Nivation benches on the higher eastern slopes of the mountain are relict features (Ballantyne, 1985).

On the plateau area and the surrounding gentle slopes, Galloway (1961a) identified 'ring' and 'stripe' patterns in the vegetation. The former take the form of turf hummocks 0.06–0.44 m high and 0.28–1.27 m in diameter; on very gentle slopes these tend to form lines downslope which, in turn, grade into a well-developed, non-sorted ridge-and-furrow stripe pattern of similar dimensions (Ballantyne, 1986b). Transitions between these types occur at 1–6° and 6–11°, respectively and relief stripes are poorly developed on slopes greater than 20° and absent from slopes above 25°. The hummocks are clearest to the north of the main summit and the stripes are superbly developed south-east of Tom a'Choinnich (NH 464700). Excavation of these patterned ground features revealed mature podsols underlying both the up-raised portions and the depressions, but no evidence of lateral frost sorting (Ballantyne, 1986b). The origin of these features is uncertain. Galloway (1958, 1961a) interpreted them as being inherited from fossil sorted polygons and stripes formed under climatic conditions more severe than at present (see also Chattopadhyay, 1982). Ballantyne (1986b) offered the alternative explanation that the features are the product of modification by mass displacement of non-sorted, vegetation-defined patterns similar to those found in the high arctic at present (see Washburn, 1979, pp. 151–153).

Interpretation

Ben Wyvis is an important locality for its range of types and degree of development of relict and active periglacial features. In particular, the massif supports the most extensive known area of non-sorted relict stripes in Scotland, together with some of the finest examples of active non-sorted solifluction sheets and lobes. The presence of the undisturbed podsols within the inactive features suggests that they are of Lateglacial age (Ballantyne, 1986b). The frost-susceptible nature of the regolith derived from the Moine schist and gneiss bedrock is a fundamental control on the development of these features, and hence the periglacial landforms found on Ben Wyvis contrast with those found on mountain summits underlain by different lithologies and therefore regoliths (see Lochnagar, An Teallach, Ward Hill, Ronas Hill and the Cairngorms). Particularly striking is the contrast between the assemblage of periglacial features on Ben Wyvis and that on An Teallach, a nearby mountain of Torridonian sandstone of similar altitude and relief. Active solifluction features, 'ploughing boulders', earth hummocks and non-sorted stripes are absent from the coarser regolith of the latter, which supports instead a wide range of wind-related forms, such as deflation surfaces and niveo-aeolian sand deposits (Ballantyne, 1984, 1987a).

Conclusion

Ben Wyvis is important in the network of localities for periglacial landforms, providing an exceptionally good range of active and fossil features developed on Moine schist. It is particularly noted for the fine development of an assemblage of patterned ground forms that comprise vegetated hummocks and vegetated ridges and furrows; these originally formed under more severe climatic conditions than at present. Ben Wyvis also displays excellent examples of sheets and lobes of debris, moved downslope by slow mass movement of the soil; some of these features are still active.

Chapter 8

North-east Scotland

INTRODUCTION

D. G. Sutherland and J. E. Gordon

North-east Scotland covers the mainly lowland area between lower Strathspey and Aberdeen (Figure 8.1). Geologically, it is underlain by the same Dalradian metamorphic rocks and younger granitic intrusions as the central and south-west Highlands, and hence its lowland character indicates a distinctive geomorphological evolution. Since the early Tertiary, north-east Scotland has been a 'hinge zone' between the mountain zone to the west, which has been uplifted and subsequently dissected to produce spectacular mountain scenery, and the North Sea Basin, which has undergone continuing downwarping (Hall, 1987). This relative stability has resulted in the preservation of Tertiary gravel deposits and extensive areas of deeply weathered bedrock.

The Tertiary gravels occur in two distinct groups, a western quartzite gravel group, best exemplified at Windy Hills, and an eastern flint gravel group, most extensively preserved at the Moss of Cruden (Figure 8.1). The western gravels are generally accepted to be fluvial in origin (McMillan and Merrit, 1980; Hall, 1987), deposited by a precursor of the present River Ythan, but the origin of the flint gravels is disputed, possibly being marine (McMillan and Merritt, 1980), possibly fluvial (Hall, 1982) or possibly glacial and glaciofluvial (Kesel and Gemmell, 1981).

Two distinct types of weathered bedrock have been recognized in north-east Scotland (Hall, 1985, 1986; Hall *et al.*, 1989a). An older, less widespread type is characterized by more intensive alteration to clay minerals (Pittodrie), whereas the younger, more widespread type is less chemically altered, having typically disintegrated to a granular sand, with the great proportion of the original minerals preserved (Hill of Longhaven Quarry; see also Clunas, Chapter 7). The alteration has taken place under humid temperate conditions; the older weathering type is probably Miocene in age, and the younger type of late Tertiary to early Pleistocene age. Depths of weathering vary considerably, up to a maximum of over 50 m. Local controls on the development and preservation of the weathered bedrock have been relief, bedrock type and limited glacial erosion (Hall, 1986; Hall and Sugden, 1987; Sugden, 1989).

The low intensity of glacial erosion of north-east Scotland that is implied by the preservation of the Tertiary sediments and the weathering profiles is matched by a longer record of Quaternary glacial (and non-glacial) events than has been found anywhere else in Scotland. The most outstanding site is at Kirkhill. There, and at the neighbouring Leys Quarry, evidence has been found for at least three periods of ice-sheet glaciation, separated by possibly two interglacials, the earlier of which was succeeded by an interstadial period. Deposits of four periods of periglacial activity are interstratified with the glacial and interglacial/interstadial sediments (Connell *et al.*, 1982; Hall, 1984a; Connell and Hall, 1987). Unfortunately, the ages of the Kirkhill deposits have not been established, although certain correlations have been proposed: Hall and Connell (1991) have tentatively correlated the glacial episodes with the Early Devensian, the 'Wolstonian' and the Anglian.

A weathered till at Kirkhill, which represents the middle of the three glaciations, may correlate with weathered tills at Boyne Bay (see below) (Peacock, 1966; Hall, 1984b) and Kings Cross, Aberdeen (Synge, 1963). The weathering of the till has been considered to have occurred under interglacial conditions (Hall, 1984a) and may correlate with the palaeosol preserved at Teindland (FitzPatrick, 1965; Edwards *et al.*, 1976) in lower Strathspey. This weathered till may also correlate with the Bellscamphie Middle Till, suggesting correlation of the Bellscamphie Lower Till with the earliest known glaciation at Kirkhill and Leys. The grey till reported by Jamieson (1906) as overlying the equivalent of the Bellscamphie Middle Till, would then be correlated with upper till at Kirkhill. However, some apparently weathered tills (e.g. at Moreseat – Hall and Connell, 1982) may reflect the incorporation of weathered bedrock rather than prolonged *in situ* weathering (A. M. Hall, unpublished data).

The occurrence of tills superposed, at certain localities, upon the above weathered tills, and possibly correlative palaeosols, indicates subsequent ice-sheet glaciation, and three distinct drift sheets are recognized: the Inland 'Series', the Red 'Series' and the Blue Grey 'Series' (Hall, 1984b) (Figure 8.1). If the weathering episode occurred during the Ipswichian, then Devensian glaciation is implied, but it is uncertain whether there was one or two periods of glaciation of north-east Scotland during the Devensian and, indeed, whether the whole region was glaciated during the Late Devensian (Clapperton and Sugden, 1977; Hall, 1984b; Sutherland, 1984a; Hall and

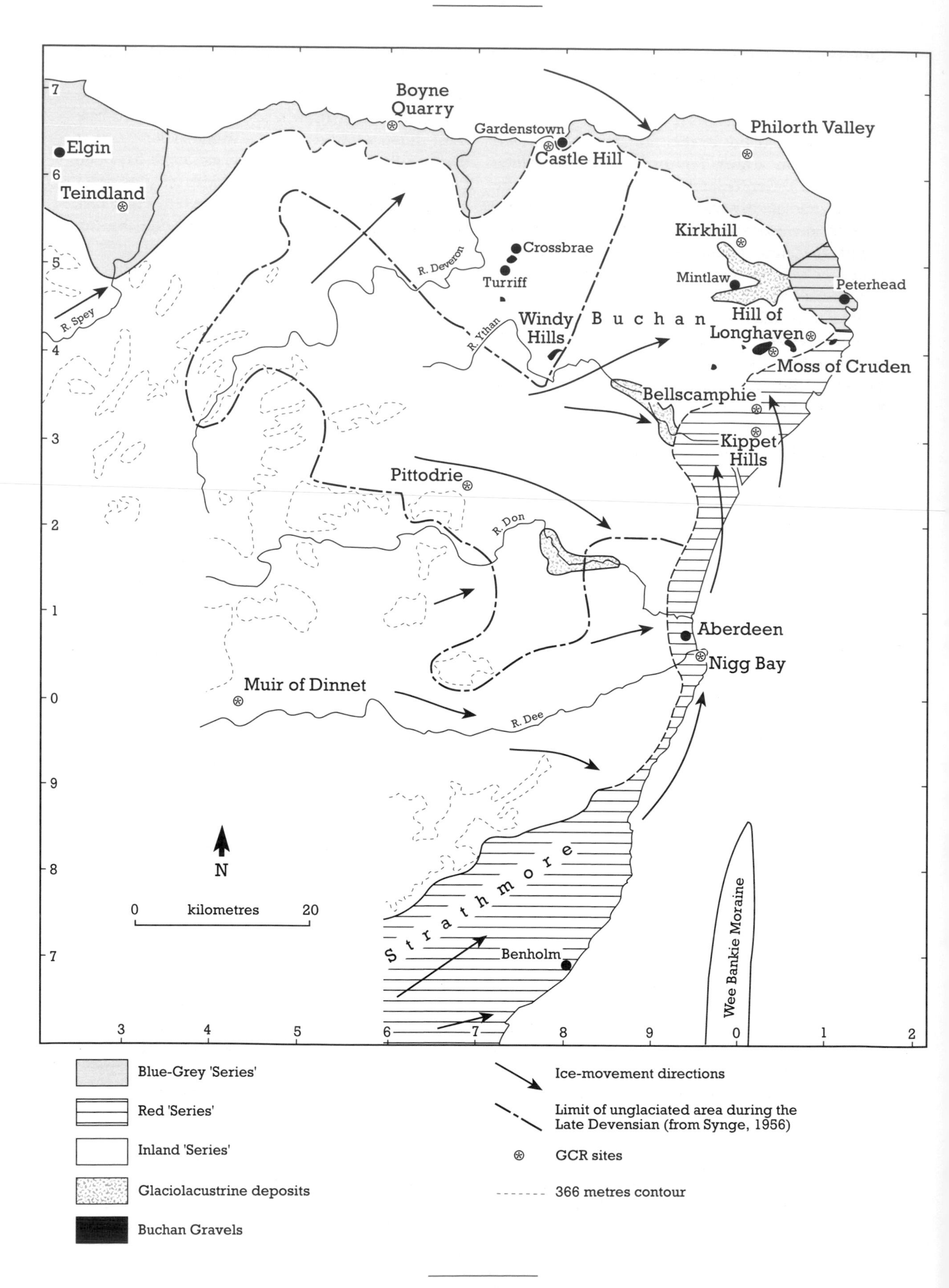
Boyne Quarry
Gardenstown
Castle Hill
Philorth Valley
Elgin
Teindland
Kirkhill
Crossbrae
Turriff
R. Deveron
Mintlaw
Peterhead
R. Spey
Windy Hills
Buchan
Hill of Longhaven
R. Ythan
Moss of Cruden
Bellscamphie
Kippet Hills
Pittodrie
R. Don
Aberdeen
Nigg Bay
Muir of Dinnet
R. Dee
N
0 kilometres 20
Strathmore
Benholm
Wee Bankie Moraine
Blue-Grey 'Series'
Red 'Series'
Inland 'Series'
Glaciolacustrine deposits
Buchan Gravels
Ice-movement directions
Limit of unglaciated area during the Late Devensian (from Synge, 1956)
GCR sites
366 metres contour

Connell, 1991). For example, at King Edward (NJ 722561) till and glaciofluvial gravels overlie an apparently *in situ* shell bed at *c.* 46 m OD (Jamieson, 1866; Horne in Read, 1923; Sutherland, 1984b). If the shell bed is indeed *in situ*, then its radiocarbon and amino acid age (Miller *et al.*, 1987) implies Early Devensian glaciation. Further, the nearby Middle Devensian interstadial deposit at Crossbrae Farm, unaffected by subsequent glaciation (Hall, 1984b), suggests that at least part of the north-east was ice-free at the time of the last glacial maximum.

Also uncertain is the status of deposits found along the coasts, such as at Boyne Bay, Castle Hill and the Red 'Series' drift deposits that occur inland of the east coast southwards from Peterhead (see Bellscamphie). These deposits consist of interstratified tills, glaciolacustrine, glaciomarine and glaciofluvial sediments (Murdoch, 1977; Hall, 1984b; Peacock, 1984b). They are coeval with lacustrine deposition along valleys, such as those of the Ugie Water and River Ythan (McMillan and Aitken, 1981; Merritt, 1981), and this implies that for at least part, if not all, of the period when the glacial sediments were being deposited, the interior of Buchan was ice-free. Glaciomarine sediments at St Fergus have been radiocarbon dated to approximately 15,000 BP (Hall and Jarvis, 1989). These sediments rest on glacial deposits derived from the Moray Firth but their stratigraphic relationship to the Red 'Series' deposits remains to be established. The presence of pre-Quaternary palynomorphs in radiocarbon-dated material from the Red 'Series' at Errolston indicates that the date obtained is of no practical value in establishing the age of these deposits (Peacock, 1984b; Connell *et al.*, 1985). Other possible interglacial or interstadial sites in north-east Scotland, at Tipperty Brickworks and Balmedie Village, have not been confirmed (cf. Bremner, 1938, 1943a; Peacock, 1980b; Edwards and Connell, 1981).

During final ice retreat, glaciofluvial deposition resulted in the formation of major esker and kame systems parallel to the coast (Kippet Hills), in places associated with glacial lakes (Merritt, 1981; Thomas, 1984; Aitken, 1991); lakes also formed along valleys inland (Aitken, 1990). Elsewhere, sequences of large outwash terraces were constructed along some principal valleys, the relationships of these to both ice-decay features and raised sea levels being particularly well exemplified in lower Strathspey. In topographically suitable hollows, stagnant ice masses became isolated from the main retreating ice mass, resulting in the formation of complex sequences of kames, kettle holes and eskers (Muir of Dinnet).

Environmental change during the Lateglacial has been relatively little studied in the north-east, although radiocarbon-dated Lateglacial Interstadial profiles have been described from Loch Kinord (Muir of Dinnet), Loch of Park and Garral Hill (Donner, 1957; Vasari and Vasari, 1968; Vasari, 1977). These sites indicate that an initial phase of open-habitat vegetation was succeeded by closed heath and scrub vegetation. Interestingly, two pine needles were recovered from Lateglacial deposits at Loch Kinord, suggesting local occurrence of pine in this area during the interstadial. Locally, tree birch may also have developed.

The Loch Lomond Stadial resulted in a return to tundra vegetation with marked slope instability and inwashing of sediment into closed basins. Sediment cores show a typical Lateglacial 'tripartite' sequence (Gunson, 1975), and radiocarbon dates have been obtained from peat beneath solifluction deposits at several sites (Clapperton and Sugden, 1977; Hall, 1984b; Connell and Hall, 1987). It is also possible that at this period some of the fossil frost polygon networks found in north-east Scotland were formed, although formation during the retreat of the last ice-sheet is also possible (Gemmell and Ralston, 1984, 1985; Armstrong and Paterson, 1985). Whatever the timing, the impact of periglacial processes on the soils of north-east Scotland has been widespread (FitzPatrick, 1956, 1958, 1969, 1972, 1975a, 1987; Galloway, 1961b, 1961c; Connell and Hall, 1987).

During the Holocene, sea level was below that of the present day for considerable periods. In the middle Holocene, the Main Postglacial Transgression resulted in marine invasion of the lower parts of the river valleys, with deposition of estuarine silts, fine sands and clays on terrestrial peats and fluvial muds. The transgression reached its maximum after 6100 BP in the Ythan valley (Smith *et al.*, 1983) and between 6300 and 5700 BP in the Philorth valley (Smith *et al.*, 1982). The subsequent regression was accompanied by renewed terrestrial sedimentation in

Figure 8.1 Location map and principal features of the Quaternary geomorphology of north-east Scotland (from Hall and Connell, 1991).

the valley mouths.

Holocene vegetation history has been investigated at several localities (Durno, 1956, 1957), but there are few well-dated pollen profiles. The most detailed studies have been at sites in the Dee Valley at Loch Kinord (see Muir of Dinnet), Loch Davan, Loch of Park and Braeroddach Loch (Vasari and Vasari, 1968; Edwards, 1978). During the middle Holocene, pine forest developed in the western part of the area, whereas birch–hazel forest was predominant to the east and on the coastal lowlands (Vasari and Vasari, 1968; Gunson, 1975; Birks, 1977). The impact of Man on the landscape is apparent in the pollen and sediment records before 5000 BP (Edwards, 1978, 1979b; Edwards and Rowntree, 1980).

WINDY HILLS

J. E. Gordon and D. G. Sutherland

Highlights

Windy Hills is a locality of outstanding importance for a suite of quartzite gravels deposited by a pre-Quaternary river. The sedimentary characteristics of the gravels and their subsequent, post-depositional modifications provide unique evidence for interpreting long-term landscape evolution in north-east Scotland.

Introduction

The Windy Hills site lies about 12 km south-east of Turriff. It covers two areas (total 0.43 km^2) between NJ 786392 and NJ 805402 on the top of a low ridge orientated south-west to north-east, overlooking the River Ythan to the south-east. The crest of the ridge, at about 120 m OD, is some 90 m above the valley bottom. Windy Hills is important for the unique, so-called 'Pliocene' gravels of Buchan. Long recognized as lithologically distinct deposits because of their very high proportions of quartzite and flint (Christie, 1831), the gravels were discussed in early papers by Ferguson (1850, 1855, 1857, 1877, 1893), Jamieson (1858, 1865, 1874, 1882b, 1906) and Wilson (1886), and they were described in some detail in an important paper by Flett and Read (1921). More recently there has been renewed interest in these deposits (Clapperton, 1977; Gemmell and Kesel, 1979, 1982; Koppi and FitzPatrick, 1980; McMillan and Merritt, 1980; Kesel and Gemmell, 1981; McMillan and Aitken, 1981; Merritt, 1981; Merritt and McMillan, 1982; Hall, 1982, 1983, 1984c).

The gravels (termed the Buchan Gravels by McMillan and Merritt (1980) and Buchan Gravels Group by Hall (1984c)) occur in a restricted area of Buchan (Figure 8.1) and comprise two lithologically distinct groups of well-rounded, water-worn pebbles discontinuously capping a number of hilltops at altitudes between 75 m and 150 m OD: a western quartzite-dominated group in the Windy Hills and Turriff areas (termed the Windy Hills Gravels by McMillan and Merritt (1980) and Windy Hills Formation by Hall (1984c)), and an eastern flint-dominated group on the summits of a broad ridge running north-east from the Hill of Dudwick (NJ 979378) to Stirling Hill (NK 125413) (termed the Buchan Ridge Gravels by McMillan and Merritt (1980) and Buchan Ridge Formation by Hall (1984c)). It has been assumed in the past that these two distinct groups are of the same age although there is no evidence that this is so.

Description

The most extensive occurrence and best exposures of the quartzite gravels are at Windy Hills. Detailed accounts of their stratigraphic relations and sedimentary character have been given by Flett and Read (1921), Read (1923), FitzPatrick (1975a, 1975b), Clapperton (1977), McMillan and Merritt (1980) and Kesel and Gemmell (1981). At Windy Hills over 10 m of predominantly quartzite gravels interbedded with white quartz sand overlie deeply weathered and kaolinized pelitic schist (Koppi, 1977; Koppi and FitzPatrick, 1980; Kesel and Gemmell, 1981; Hall *et al.*, 1989a). The quartzite pebbles are comparatively fresh and unweathered and some show chatter marks, in contrast to occasional cobbles of granite and schist, which are nearly always decomposed to a kaolinitic sand. A small number of flint pebbles is also present, some weathered with a dull grey or white rind, and the very rare presence of chert of Lower Cretaceous age is also recorded (Flett and Read, 1921; Hall, 1987). Jamieson (1865) reported the presence of (?)Late Cretaceous fossils typical of the Chalk associated with the flints at Windy Hills, but gave no specific identifications; his observations have not been confirmed. However, among the flints at Delgaty, near Turriff, Christie (1831) found fossils, principally of sponges or alcyonaria. Details of fossils from other sites are given by Salter (1857) and Ferguson (1857).

Clast imbrication and rare cross-bedding indicate that the gravels were deposited by water flowing from approximately west-south-west to east-north-east (McMillan and Merritt, 1980; Kesel and Gemmell, 1981). Preservation of such sedimentary structures implies little disturbance of the gravels since deposition, but the upper 1 m has been cryoturbated. This shows several features associated with periglacial modification: ice-wedge casts up to a metre across, vertically aligned clasts, an indurated horizon, clasts with silt cappings and an increase in clast concentration at the surface (FitzPatrick, 1975a, 1975b, 1987). The gravels are only overlain by patches of till (Bremner, 1916b; Kesel and Gemmell, 1981), but unweathered erratics have been incorporated into the upper layer of the gravels, probably by frost churning (Clapperton, 1977).

Interpretation

The quartzite of the gravels was considered by Flett and Read (1921) to be unlike that of any known quartzite outcrop in north-east Scotland but was comparable to the quartzite of Scaraben in Caithness. Koppi (1977) and Kesel and Gemmell (1981), however, examined the quartzite in thin section and concluded that it was most probably derived from Banffshire quartzites. This conclusion was supported by the heavy-mineral assemblage associated with the gravels. Hall (1987) has suggested the further possibility that much of the quartzite debris was recycled from Devonian conglomerates.

Jamieson (1858, 1865) originally interpreted the gravels to be locally derived and of pre-glacial marine origin, but later suggested glacial derivation from the floor of the Moray Firth (Jamieson, 1906). Wilson (1886) thought that they were residual deposits from a denuded chalk cover and had been glacially reworked. Flett and Read (1921) concluded that the gravels were remnants of formerly more extensive marine deposits resting on an old platform and were of Tertiary, possibly Pliocene, age.

Recent interpretations are agreed that the Windy Hills gravels are fluvial in origin (McMillan and Merritt, 1980; Kesel and Gemmell, 1981; Hall 1982, 1983, 1987), although Kesel and Gemmell (1981) also considered a glaciofluvial origin as a possibility in view of certain grain-surface textures, identified on a scanning electron microscope, suggestive of glacial transport.

An important facet of the Windy Hills site is that the gravels, together with the associated deep weathering, glacial and periglacial phenomena, hold important clues about landscape evolution and environmental change in north-east Scotland during the late Tertiary and Pleistocene (see also Moss of Cruden):

1. The weathering characteristics of both the gravels and the underlying bedrock led Hall (1983, 1987) to conclude that the gravels were probably Neogene in age, being deposited along a proto-Ythan valley. Subsequent surface lowering has resulted in topographic inversion with the gravels now occupying hill-top positions.
2. The presence of flint has been used as evidence of Late Cretaceous marine transgression (Wilson, 1886; Flett and Read, 1921; Hall, 1983, 1987).
3. Hall (1983, 1985, 1987) associated the kaolinitic alteration in the gravels and the underlying bedrock with his clayey grus weathering type. The latter is older (probably Miocene in age) than the gruss weathering type (Pliocene to Pleistocene in age) based on a greater degree of alteration (see Hall *et al.*, 1989a).
4. The occurrence of deep-weathering profiles and the Windy Hills and Buchan Ridge gravels also testify to the limited and selective nature of glacial erosion in the Buchan area (Clayton, 1974; Hall, 1982, 1986; Hall and Sugden, 1987), despite the evidence for repeated ice-sheet invasion during the Quaternary (see Kirkhill).

Windy Hills is the most important locality for the quartzite gravels of probable Tertiary age of north-east Scotland. The site is complementary to that of Moss of Cruden, where flint gravels of broadly similar age are preserved, although the origin of the two gravel bodies may not be the same. The gravels provide a rare insight into the middle to late Tertiary environment of north-east Scotland and their occurrence, together with that of contemporaneous deep-weathering profiles, provides an important reference level for the extent of glacial erosion in this region.

Conclusion

Windy Hills is the type area for the famous quartzite gravels of Buchan. These deposits are now agreed to have been formed by a pre-Quaternary river; subsequent erosion has lowered the adjacent landscape, leaving the gravels in their present hill-top location. The gravels show evidence of weathering and frost-disturbance and are locally overlain by glacial deposits. They are important for the unique evidence they provide about the long-term evolution of the landscape in north-east Scotland, both before and during the Quaternary ice ages.

MOSS OF CRUDEN

A. M. Hall

Highlights

The Moss of Cruden is a key locality for a suite of flint gravels of pre-Quaternary origin. Like the quartzite gravels at Windy Hills, the syn- and post-depositional sedimentary characteristics of the flint gravels are a unique source of evidence for interpreting the history of landscape evolution in north-east Scotland during the late Tertiary and Quaternary.

Introduction

The Moss of Cruden site (NK 028403) occupies an area (0.85 km^2) at the top of a broad ridge orientated south-west to north-east, approximately 10 km south-west of Peterhead. The ridge, which reaches a maximum altitude of 139 m OD is the most important locality for the Buchan Ridge Formation, part of the Buchan Gravels Group (McMillan and Merritt, 1980). These gravel deposits are notable for:

1. the presence of Chalk flints;
2. a highly distinctive lithology of flint and quartzite clasts with a matrix of kaolinitic sand;
3. the advanced degree of post-depositional alteration.

Flint gravels in Buchan were first described by Christie (1831) and subsequently generated considerable scientific interest (Ferguson, 1850, 1855, 1857, 1877, 1893; Salter, 1857; Jamieson, 1858, 1865, 1874, 1882b, 1906; Wilson, 1886; Flett and Read, 1921). Despite much recent work (Koppi and FitzPatrick, 1980; McMillan and Merritt, 1980; Kesel and Gemmell, 1981; McMillan and Aitken, 1981; Merritt, 1981; Gemmell and Kesel, 1982; Hall, 1982, 1983, 1984c; Merritt and McMillan, 1982; Saville and Bridgland, 1992), the age and origin of the Buchan Gravels remain controversial, but there is no doubt that these deposits are of key importance for understanding Tertiary and early Pleistocene environments in north-east Scotland. Recent and continuing excavations at Moss of Cruden (Hall *et al.*, unpublished data) have added to the interest of the site by the discoveries of a small outlier of sandstone of probable Devonian age adjacent to the flint gravel margin and two masses of weathered Lower Cretaceous sandstone, apparently *in situ* and underlying the Buchan Ridge gravels.

Description

The Buchan Ridge Formation comprises deposits of flint gravel found at altitudes of 75 m to 150 m OD discontinuously capping the summits of a broad ridge running south-west from Den of Boddam (NK 115415) to Hill of Dudwick (NJ 979378). These gravels probably reach their maximum extent at Moss of Cruden, where a thickness of at least 25 m has been recorded (McMillan and Aitken, 1981). Present exposures, however, are poor and infrequent.

The deposit at Moss of Cruden comprises white, clay-bound, coarse gravels with minor sandy and silty units. The gravel ranges from granule to boulder size and is composed mainly of flint with metaquartzite and vein quartz. Flint and metaquartzite clasts are generally well-rounded and bear numerous chatter marks. The deposits originally contained small numbers of less siliceous clasts, which have decomposed to balls of white, kaolinitic, sandy clayey silt (McMillan and Merritt, 1980).

Sandy units are composed of quartz and flint with seams of muscovite. Both sand and gravel units are bound and, in places, supported by white, sandy clayey silt consisting of well-ordered kaolinite with minor illite (Hall, 1982). The base of the gravels rests on kaolinized granite and gneiss (McMillan and Merritt, 1980; Hall, 1983, 1987; Hall *et al.*, 1989a).

Recent excavations (A. M. Hall *et al.*, unpublished data) on the north-western edge of the

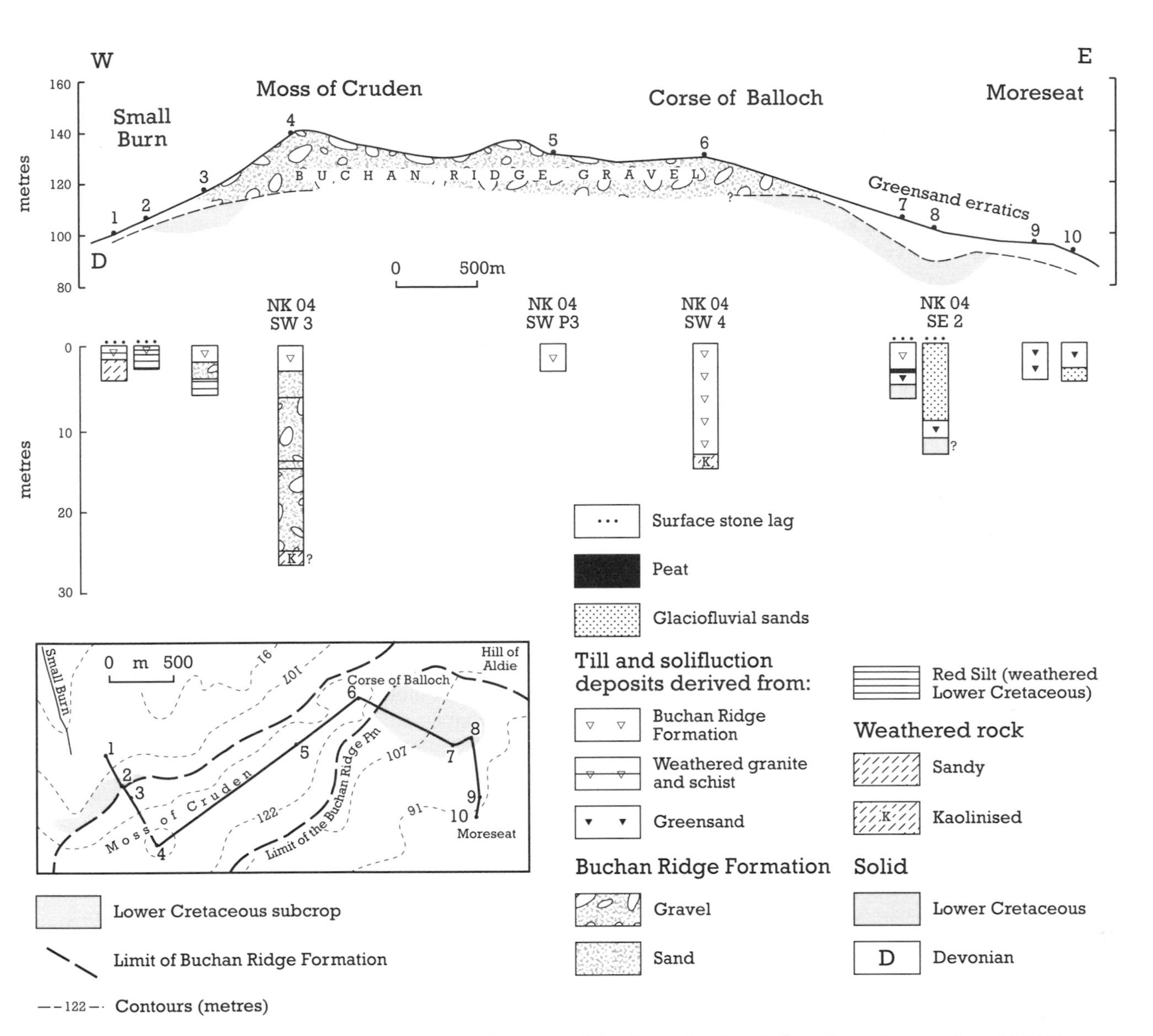

Figure 8.2 Schematic cross-section through the Moss of Cruden ridge. Borehole and pit data are from McMillan and Aitken (1981), Hall and Connell (1982) and A. M. Hall *et al.* (unpublished data).

ridge have shown that the flint gravel margin approaches to within a few hundred metres of a small, concealed outlier of red-brown arkosic sandstone of probable Devonian age. The Moreseat locality on the south-eastern margin of the ridge is famous for the occurrence of large masses of Lower Cretaceous greensand, previously interpreted as erratics transported to the site by ice from the Moray Firth (Jamieson *et al.*, 1898; Hall and Connell, 1982). Recent work has revealed, however, that the Lower Cretaceous sandstones are more extensive beneath the lower slopes of the Moss of Cruden than previously thought, and that these rocks are probably *in situ*. Evidence for this interpretation includes the manner in which mottled, red silts, representing highly weathered Lower Cretaceous sediments, pass below the north-west margin of the Buchan Ridge gravels north of Smallburn (NK 019407).

On the crest of Moss of Cruden, the Buchan Ridge Formation is overlain by a variable thickness of white to grey, gravelly till incorporating occasional erratics of fresh Peterhead granite (Figure 8.2). To the north of the ridge, a sheet of soliflucted flint gravel, up to 2 m thick, extends well downslope on to sandy, weathered granite. North of Moreseat, recent excavations have shown that the gravels are locally overlain by two tills, which are separated by a peat of interglacial or interstadial origin.

Interpretation

The presence of flint gravels in the Buchan area was first described by Christie (1831). Ferguson (1850, 1855, 1857, 1877, 1893) noted the rounded character of the clasts and suggested an origin as a beach, an idea initially supported by Jamieson (1858, 1882b). Jamieson (1865, 1874) thought the gravels to be pre-glacial, but later suggested deposition by ice moving south from the Moray Firth (Jamieson, 1906). More detailed study of petrology and sedimentology led Flett and Read (1921) subsequently to favour earlier interpretations of the gravels as pre-glacial beach deposits, possibly of Pliocene age.

Several studies have added much new information about these distinctive deposits (Koppi and FitzPatrick, 1980; McMillan and Merritt, 1980; McMillan and Aitken, 1981; Merritt, 1981; Kesel and Gemmell, 1981; Hall, 1983), but their origins remain highly controversial (Merritt and McMillan, 1982; Gemmell and Kessel, 1982; Hall, 1982, 1984c). Three main areas of recent discussion can be identified: first, the provenance of the gravel constituents; second, the mechanism or mechanisms of transport; and third, the age of the deposits.

The three main components of the gravels are flint, quartzite and kaolinitic silts and sands. The origin of each of these components has been disputed. The two sources of flint that have been proposed are: Chalk outcrops beneath the Moray Firth, transported southwards as glacial erratics (Jamieson, 1906; Kesel and Gemmell, 1981); or Tertiary weathering of a former Chalk cover (Wilson, 1886), with concentration into later fluvial (Hall, 1982, 1983) or marine gravels (Koppi and FitzPatrick, 1980; McMillan and Merritt, 1980). The recent discovery of little-worn, nodular flint in gravels at the Moss of Cruden indicates that some of the flints are not far-travelled and is consistent with reworking of a nearby remanié flint deposit. This interpretation is also consistent with the discovery that the gravels locally rest on Lower Cretaceous sandstone.

According to Flett and Read (1921), the quartzite clasts are distinct in character from the Dalradian quartzites of north-east Scotland, but recent mineralogical study has demonstrated many similarities (Kesel and Gemmell, 1981). An overlooked possible additional source is the Old Red Sandstone (Hall, 1984c), whose basal conglomerates contain large amounts of well-rounded Dalradian quartzite cobbles (Read, 1923; Peacock *et al.*, 1968).

The kaolinitic sand matrix has been interpreted as the result of secondary infilling of an openwork sandy gravel by kaolinitic fines due to alteration and breakdown of some clasts in the gravel (McMillan and Merritt, 1980; Merritt and McMillan, 1982), or glacial transport and deposition of clay-rich facies of the gravels as till (Kesel and Gemmell, 1981; Gemmell and Kesel, 1982).

The origin of the Buchan Ridge Formation has been variously interpreted. Evidence adduced for a beach origin includes the presence of chatter-marked clasts (Flett and Read, 1921; Koppi and FitzPatrick, 1980), the presumed original openwork character of the deposit (McMillan and Merritt, 1980) and the geomorphological setting of the deposit (Flett and Read, 1921). In contrast, a fluvial origin has been invoked to explain the sedimentary structures, clast rounding and quartz grain surface textures, and the association with a deeply weathered land surface (Hall, 1982, 1983, 1986, 1987). The third proposed origin is as a glacial or glaciofluvial deposit. Supporting evidence includes the presence of matrix-supported beds of gravel and the breakage of previously well-rounded quartz grains (Kesel and Gemmell, 1981).

The presence of kaolinized clasts throughout the entire known thickness of the Moss of Cruden gravels and the evidence of topographic inversion since deposition, have allowed agreement that the flint gravels are older than all known Pleistocene deposits in Buchan. Following on from their interpretation of the gravels as early glacial or glaciofluvial deposits, Kesel and Gemmell (1981) suggested a Pliocene or early Pleistocene age. Flett and Read (1921) also originally proposed a Pliocene age based on long-range height correlations with marine deposits in southern England. McMillan and Merritt (1980) suggested that the degree of post-depositional weathering indicates prolonged alteration under warm climates in the middle to late Tertiary; Hall (1982, 1983, 1985) proposed a Neogene age based on comparisons with types of deep-weathering cover recognized by him in Buchan. However, firm dating of the flint gravels is not yet possible from the evidence available.

The Buchan Ridge Formation is a unique deposit in Scotland. The review above reveals that many questions about the provenance, origin and age of these gravels remain unanswered, and further advances may await the opening of deep sections. However, it is clear that the flint gravels have a bearing on several important problems of Scottish pre-Pleistocene and Pleistocene landscape history including:

1. the former extent of Cretaceous cover in north-east Scotland;
2. the nature of Tertiary weathering environments in Buchan and landscape evolution in the region (Hall, 1985, 1986);
3. the timing of the onset of regional glaciation in Scotland (Hall, 1984c; Sutherland, 1984a).

The Moss of Cruden site sheds important new light on long-term rates of denudation in this part of Scotland. The Peterhead granite, which partly underlies the Moss of Cruden, is of Caledonian age; it intrudes Dalradian metasediments. It was unroofed by Devonian times and later buried by Devonian sediment. This Old Red Sandstone cover was almost completely removed at the site, apart from a thin remnant now represented by the small outlier of arkosic sandstone, prior to marine transgression and deposition of sandstone in the Early Cretaceous. Further burial by Late Cretaceous Chalk is demonstrated by the presence of nodular Chalk flints in the basal layers of the Buchan Ridge Gravels. The Cretaceous cover was probably largely stripped in response to early Tertiary uplift, and the survival of the small remnants of Devonian and Cretaceous sediments at Moss of Cruden are undoubtedly a result of their subsequent burial by the Buchan Ridge Gravels in the (?)late Tertiary. The granite and metasediments at Moss of Cruden have therefore been protected from erosion for the last *c.* 350 Ma. It is unlikely, however, that these igneous and metamorphic rocks were ever deeply buried and this allows the possibility of a highly complex weathering history in these rocks.

The only outcrop of the Buchan Ridge Formation of comparable dimensions to the Moss of Cruden deposit lies beneath the Hill of Aldie (NK 059414), but exposure is extremely poor. Fieldwork there suggests the existence of a flint gravel deposit extending over an area of about 1 km^2 and reaching depths of at least 17.8 m (McMillan and Aitken, 1981). Smaller deposits occur at Whitestones Hill (NJ 979389), Sandfordhill (NK 115416) and Den Muir (NK 105406), but the possibility of glacial and periglacial disturbance is much greater, particularly at the last two localities.

The Windy Hills Formation consists of quartzite gravels with occasional flints and shows many similarities with the Buchan Ridge Formation (see Windy Hills). However, these two gravel bodies are lithologically distinct and may be of different origin and age (McMillan and Merritt, 1980; Kesel and Gemmell, 1981; Hall, 1983).

Conclusion

Moss of Cruden is the type area for the unique flint gravels of Buchan. These deposits include weathered material and are locally overlain by till. Although acknowledged to be pre-Quaternary in age (see Windy Hills above), their origin is still arguable and has been ascribed to marine, river and glacial processes. The Moss of Cruden gravels are different in their composition from those at Windy Hills and the two deposits may be of different age and origin. Like the gravels at Windy Hills, those at Moss of Cruden provide unique evidence for interpreting the long-term evolution of the landscape during the Quaternary ice ages and earlier.

PITTODRIE

A. M. Hall

Highlights

At Pittodrie, a small pit shows an exposure of decomposed granite bedrock which has undergone relatively advanced chemical weathering during pre-glacial times. It provides valuable information for interpreting the longer-term geomorphological evolution of north-east Scotland and shows the limited effects of glacial erosion in this area.

Introduction

The site at Pittodrie (NJ 693245), 30 km north-west of Aberdeen, is a small exposure at *c.* 180 m OD at the foot of Bennachie, a granite hill with good examples of tors at its summit (518 m OD). It shows kaolinized granite containing a hematite/layer-silicate clay mineral, macaulayite, known only from this locality (Wilson *et al.*, 1981, 1984). The weathering of the granite probably occurred during the late Tertiary (Hall, 1985; Hall *et al.*, 1989a). Additional interest in the site is provided by the excellent examples of downslope flaring of bedrock structures as a

Figure 8.3 Weathered granite at Pittodrie overlain by soliflucted deposits. (Photo: J. E. Gordon.)

result of mass movement under former periglacial conditions.

Description

The section is up to 4 m high and shows 3–4 m of friable weathered granite overlain by 1–2 m of stoney and sandy solifluction deposits, on which a podsolic soil profile has developed (Figure 8.3).

The parent rock at the site is a leucogranite consisting of quartz, K-feldspar and subsidiary mica. The rock has generally weathered to a friable, clayey, silty sand in which bedrock structures, such as joints and quartz veins, are still clearly visible. The weathered granite is usually white to light brown, but also shows striking zones of bright red mottling. Kaolinite and illite are the dominant clay minerals and the rock is strongly depleted of more soluble bases but retains approximately 3.0% K_2O owing to the survival of partially altered grains of K-feldspar. Occasional patches of harder rock coincide with zones of relatively quartz-rich granite.

In situ weathered rock is overlain by about 0.5 m of banded growan (cf. Brunsden, 1964) in which vertical bands of white, pink and red weathered granite have been bent or flared downslope in response to former solifluction. This layer is overlain by up to 1.5 m of dark-brown, soliflucted till. The soliflucted till has a high content of basic igneous clasts, indicating original deposition by ice impinging on the hill from the west.

Interpretation

The Pittodrie site was mentioned as an example of weathered rock of pre-glacial age by FitzPatrick (1963) and Glentworth and Muir (1963). However, the only detailed descriptions of the exposure are those by Wilson *et al.* (1981) and Hall (1983).

Wilson *et al.* (1981, 1984) demonstrated the existence of a previously undescribed swelling hematite/layer-silicate complex, now formally named as macaulayite, in rubefied zones in the weathered granite. They suggested that the mineral formed in pre-glacial or interglacial times, when iron was complexed at higher, subsequently eroded, levels in the weathering profile and passed down joint planes to be oxidized at depth.

Hall (1983) and Hall *et al.* (1989a) described the site in more general terms and gave data on granulometry, geochemistry and clay mineralogy. Hall allocated the Pittodrie site to his 'clayey grus weathering type', a type of intense kaolinitic weathering found at only a few sites in north-east Scotland. Comparisons with the mineralogy of North Sea drill holes (Karlsson *et al.*, 1979) suggests that the clayey gruses are of Miocene to middle Pliocene age (Hall, 1985; Hall *et al.*, 1989a).

The possibility that the observed weathering features are a result of hydrothermal alteration has still to be fully considered. This origin is suggested by the brief reference of Wilson and Hinxman (1890) to a 'segregation vein' in the granite in the vicinity of Pittodrie.

The intensity of weathering at Pittodrie is far in excess of that usually found in weathered granites in north-east Scotland. Comparable kaolinitic alteration is currently exposed at only a small number of other sites in Buchan, where it is generally confined to quartz schist parent rocks (Hall, 1983; Hall *et al.*, 1989a).

Deep weathering of granite and other rocks is widespread in north-east Scotland (Phemister and Simpson, 1949; FitzPatrick, 1963; Hall, 1985, 1986, 1987; Auton and Crofts, 1986; Munro, 1986; Hall *et al.*, 1989a) and is generally ascribed to a pre-glacial or interglacial origin. The degree of preservation of deep weathering despite multiple glaciation is remarkable and is matched in only a few other formerly glaciated areas around the North Atlantic (Hall, 1984b, 1985). The weathering has considerable geomorphological significance:

1. The survival of weathered rock has been used as evidence of minimal glacial erosion in north-east Scotland (Clayton, 1974) and to identify local variations in the intensity of such erosion (Hall, 1983, 1985; Hall and Sugden, 1987; Sugden, 1989). Such studies have significant potential for elaborating the basal processes and dynamics of former mid-latitude ice-sheets (Sugden, 1989).
2. Analysis of the mineralogy of the weathered rock has provided important information on former weathering environments (Basham, 1974; Wilson and Tait, 1977; Hall, 1983; Hall *et al.*, 1989a).
3. The characteristics and distribution of the weathering have been used to investigate long-term landscape evolution in the region (Hall, 1983, 1986, 1987, 1991).

North-east Scotland, particularly the Buchan district, has therefore become an important area for the study of weathering and landform development in formerly glaciated regions (see Hill of Longhaven Quarry). Pittodrie provides a rare example of rubefied and kaolinitic weathered granite and it has supplied important information on pre-glacial or interglacial weathering environments. Pittodrie is also the only known locality for the clay mineral macaulayite.

Conclusion

Pittodrie is a reference site for deeply weathered bedrock, one of the characteristic features of the geomorphology of north-east Scotland. It shows a relatively intense type of weathering in granite represented at only a few sites in the area, and it is the only known location for a particular type of clay mineral. The decomposed bedrock, considered to be the product of pre-glacial weathering, is important for interpreting geomorphological processes during the landscape evolution of north-east Scotland. Its preservation is also significant in demonstrating a relatively low degree of glacial erosion in this area during the Quaternary ice ages.

HILL OF LONGHAVEN QUARRY

A. M. Hall

Highlights

The disused quarry at Hill of Longhaven shows an exposure of granite bedrock that has been heavily decomposed through granular disaggregation and limited chemical weathering. It is a particularly good example of this type of phenomenon which is widespread in north-east Scotland.

Introduction

Hill of Longhaven Quarry (NK 083424) is excavated at an altitude of 110 m OD, near to the summit of a broad west–east ridge which extends almost to the North Sea coast. It provides good exposures of weathered granite typical of the granular disintegration to grus accompanied by a relatively low degree of chemical alteration that is found at many locations in north-east Scotland (Hall, 1985; Hall *et al.*, 1989a). The nature of the weathering contrasts with that, for example, at Pittodrie (see above), where the alteration to clay minerals has been much greater. The only account of the Longhaven site is by Hall (1983), who gave data on granulometry, geochemistry and clay mineralogy (see also Hall, 1985; Hall *et al.*, 1989a).

Description

The lithological succession at the site is simple and continuous along 100 m of quarry face. Extending down from the surface and through a thin humic soil is a layer of cryoturbated till up to 0.3 m thick and containing abundant flint clasts. Beneath this lies up to 5 m of weathered granite, which is locally incorporated into the overlying deposit. The parent rock is typical pink, coarse-grained Peterhead granite, which has disintegrated to a uniform gravelly sand in which original rock structures, such as joint systems and thin quartz veins, are perfectly preserved (see Chorley *et al.*, 1984, plate 8). Corestones up to 1.5 m in diameter are found on the quarry floor, but no fresh rock, apart from occasional aplite veins, occurs in the quarry faces. The basal surface of weathering is locally exposed at the quarry floor by several whaleback-shaped rock 'risers'.

Weathering elsewhere in the Peterhead granite is described by Edmond and Graham (1977) and Moore and Gribble (1980). In an engineering study for Boddam power station, Edmond and Graham (1977) note that weathering in the Peterhead granite:

1. reaches a depth of 56 m in a fault zone;
2. decreases in intensity with depth, thereby indicating that subaerial, rather than hydrothermal alteration is the main cause of weathering;
3. includes thin seams of white and red clay along vertical joint planes, which may reflect minor hydrothermal activity.

Moore and Gribble (1980) gave geochemical data for a 10 m deep weathering profile in Stirling Hill Quarry (NK 123415), approximately 4 km east of the Longhaven site. Kaolinite and illite are the main clay minerals present and the amount of geochemical change is shown to decrease with depth, again supporting the idea of subaerial weathering.

At Hill of Longhaven, the content of fines is unusually low (<6%) and the granite has disintegrated to a gravelly sand (Hall, 1983). Average losses of CaO, Na_2O and MgO are also low (5.9%). Samples relatively enriched in clay, taken from joint planes, are dominated by illite with kaolinite and halloysite. Chlorite is present as a green coating on certain joint surfaces and may reflect slight alteration by hydrothermal solutions. Hall noted that the incorporation of previously weathered granite into the overlying cryoturbated till demonstrates that the period of rock weathering pre-dates at least one glaciation.

Interpretation

A puzzling feature of this site is that an apparently very low degree of chemical alteration has produced such deep and thorough disintegration of the rock. Hall (1983) suggested that some form of mechanical disintegration in response to buttressed expansion (Folk and Patton, 1982), or even ice-sheet loading and unloading (Carlsson and Olsson, 1982), may have been involved. The site has considerable research potential in this respect.

Hill of Longhaven Quarry provides a large exposure in weathered granite showing many features typical of weathering in coarse-grained granites in north-east Scotland. The site also has potential for research into the causes of rock breakdown in the initial stages of chemical alteration.

Other good exposures in similar weathered granites in north-east Scotland are found at Mill Maud (NJ 566067), Glen Cat (NJ 574949) and East Den (NK 082443). The degree of weathering at these sites is slightly greater than at Hill of Longhaven, but other features are broadly similar. Deep weathering in finer-grained granites is well-exposed at Redhouse (NJ 577203), Cairnlea (NJ 901537) and Cairngall (NK 053471).

Deep weathering of granite and other rocks is widespread in north-east Scotland and its geomorphological significance is outlined above (see Pittodrie). The weathering is widely regarded as pre-glacial or interglacial in origin (FitzPatrick, 1963; Basham, 1974; Wilson and Tait, 1977; Hall, 1983, 1985; Hall *et al.*, 1989a). In view of the low degree of alteration at the Longhaven site it is likely that this profile is of interglacial origin (see Hall, 1985; Hall *et al.*, 1989a). Its survival beneath a relatively exposed hilltop site is evidence of the inefficiency of glacial erosion in the Peterhead area, a characteristic feature of the geomorphology of this part of Scotland (Hall, 1983, 1985; Hall and Sugden, 1987; Sugden, 1989). It is also interesting to note that stripping of the weathered bedrock would reveal a hummocky bedrock topography resembling certain types of ice-moulded terrain. The site may thus have further research potential in elucidating the origins of certain classic landforms of glacial erosion, such as roches moutonnées and 'knock-and-lochan' topography (areal scouring) (Linton, 1959).

Conclusion

Hill of Longhaven Quarry demonstrates a particularly good example of weathered granite. It is representative of a type of weathering that shows granular disaggregation of the rock and a relatively low degree of chemical modification. In contrast to the more intense (?Tertiary age) chemical alteration seen at Pittodrie, the weathering at Hill of Longhaven Quarry may have occurred during one or more interglacial climatic phases in Pleistocene times. It is important for interpreting the geomorphological processes that have shaped the landscape of north-east Scotland.

KIRKHILL

A. M. Hall and J. Jarvis

Highlights

The sequence of deposits formerly exposed at Kirkhill Quarry, and proved by boreholes and pits to extend beneath adjacent fields, provides the longest and most complete record of Quaternary events in Scotland. It is a unique locality of the very highest importance for Quaternary studies, providing evidence of multiple glacial, interglacial and periglacial episodes.

Introduction

Kirkhill Quarry (NH 012529) is located 14 km north-west of Peterhead. The interest of the site lies in the variety of environments recorded by a succession of sediments which spans a large part of Middle and Late Pleistocene time. The succession includes:

1. Two interglacial soils of possible Ipswichian and Hoxnian age.
2. Organic sands of possible late Hoxnian age.
3. Tills which record at least three separate phases of glaciation in this part of Buchan.
4. Evidence for at least four separate periglacial phases, two of which pre-dated the last interglacial and three of which involved soil development.

The site represents the most complete Middle Pleistocene sequence known in Scotland and is of key importance in the controversy over the extent and timing of Devensian glaciation in north-east Scotland (Hall, 1984b; Sutherland, 1984a). The sediments have been described in a series of papers by Connell, Hall and co-workers (Connell *et al.*, 1982; Connell, 1984a, 1984b; Connell and Romans, 1984; Hall, 1984a; Connell and Hall, 1987; Hall and Connell, 1991); Lowe (1984) has critically reviewed the available pollen data. The quarry is now filled, but the succession has been proved beneath the fields to the south and east (Hall *et al.*, 1989b) and at Leys Quarry, 0.6 m south-west of Kirkhill (Hall and Connell, 1986).

Description

The sediment succession is up to 10 m thick and occurs in a number of bedrock channels or basins. The lithostratigraphy is summarized in Figure 8.4. Terminology is after Connell and Hall (1987).

Deposits below the Lower Buried Soil

The base of the sequence is composed of blocky talus and gelifluctate deposits derived from frost shattering of the channel margins. These sediments comprise Gelifluctate Complex 1 and are overlain by, and interstratified with, up to 4.5 m of bedded sand and gravel of fluvial or glaciofluvial origin. The gelifluctate and the base of the sand and gravel contain occasional erratic pebbles, which suggest, together with the existence of bedrock channels probably carved by meltwater, an early phase of glaciation at this site.

Recent work at nearby Leys Quarry indicates that the Kirkhill Lower Sands and Gravels rest on,

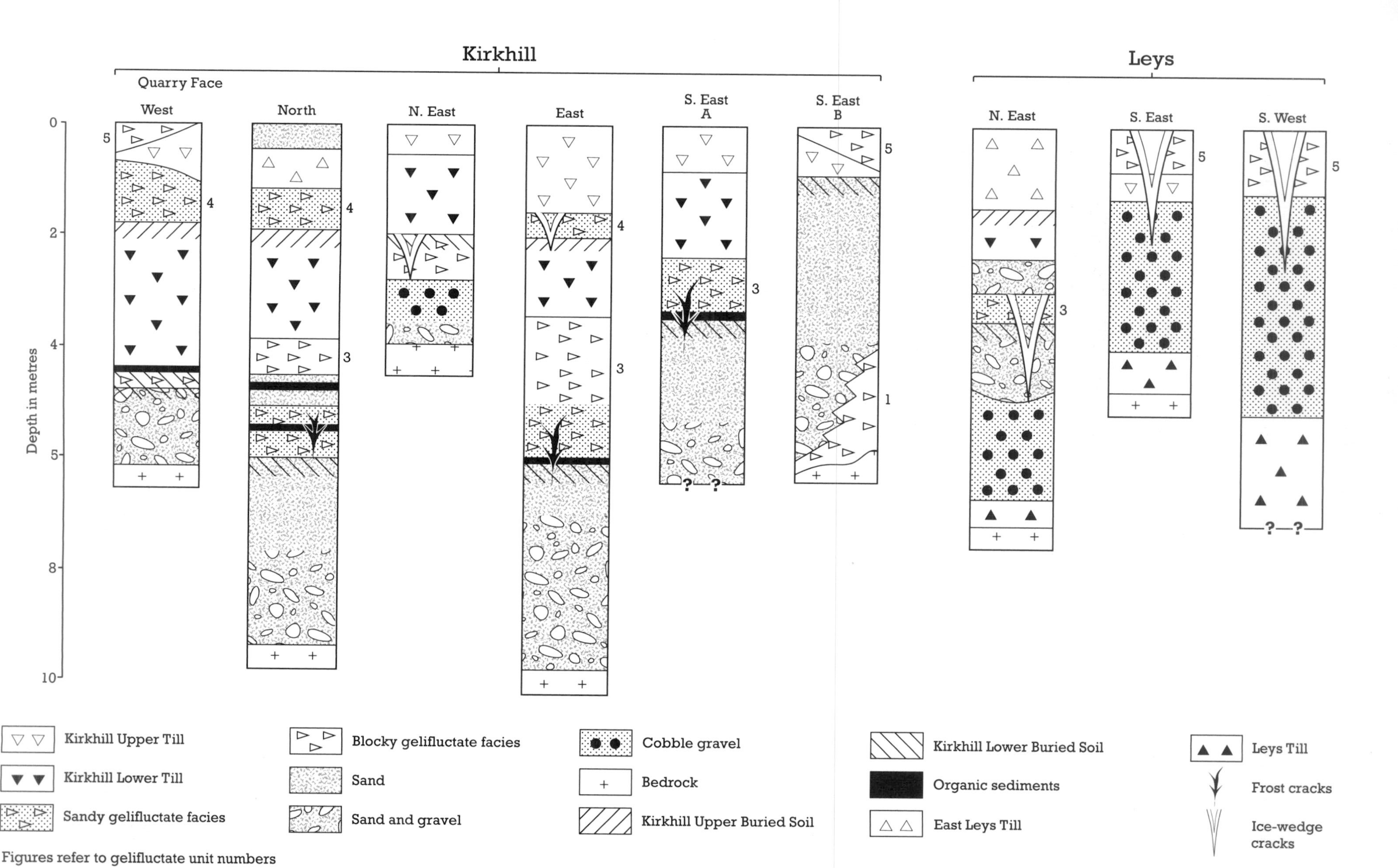

Figure 8.4 Kirkhill and Leys: lithostratigraphy.

or are a facies of the bouldery Leys Gravels, carried by meltwater from the east (Hall and Connell, 1986). Beneath the latter and resting on bedrock lies a further till, the Leys Till, derived from the west. The erratics in the Lower Sands and Gravels are presumably derived from the Leys Till.

The Lower Buried Soil

The lower deposits are truncated by an erosion surface on which is developed the Lower Buried Soil. The soil comprises a distinctive, upper, light-grey, bleached horizon and a lower, grey-brown, mottled horizon, with a basal iron pan (Figure 8.5). Analytical data are given in Connell and Romans (1984). The Lower Buried Soil is of podsolic type and it developed under humid temperate conditions (Connell *et al.*, 1982), but micromorphological evidence of silt droplet fabrics indicates subsequent climatic deterioration (Connell and Romans, 1984).

Deposits between the Lower and Upper Buried Soils

The Lower Buried Soil is truncated and draped by a 0.02 m thick layer of laminated organic mud, deposited in shallow ponds. The mud contains pollen of Gramineae, together with a marked arboreal component of mainly *Pinus* and *Alnus* (Connell *et al.*, 1982). Initial radiocarbon dating of the organic mud gave finite radiocarbon dates for three samples of 45,630 +1740/−1430 BP (SRR–1635), 44,900 +1580/−1320 BP (SRR–1637) and 33,810 +630/−590 BP (SRR–1636), but contamination was suspected (Connell *et al.*, 1982). Later dating of a much larger sample (Connell, 1984b) gave a date of >47,360 BP (SRR–2416) and confirmed that the sediments are beyond the range of radiocarbon dating (Hall, 1984b).

The organic mud is succeeded by 0.1–0.7 m of poorly-organic sands containing frost cracks. Pollen analysis shows a reduction in arboreal pollen and an increase in grasses and *Calluna*, possibly reflecting the establishment of an open, treeless environment. The organic sediments indicate erosion and recycling of organic deposits, possibly including the upper horizons of the Lower Buried Soil, from neighbouring slopes under deteriorating climatic conditions. The organic sands are overlain by sediments belonging to Gelifluctate Complex 3 and then by the Lower Till, containing erratics transported from the north-west.

The Upper Buried Soil

The Upper Buried Soil is developed on the Lower Till. Diagnostic features of soil development include mottling, clay translocation, alteration of clasts and down-profile changes in colour and clay mineralogy (Connell and Romans, 1984). The Upper Buried Soil bears many similarities with Holocene gleyed brown-earth profiles in north-east Scotland and is therefore considered to have formed during an interglacial. Countable pollen dominated by *Alnus* was recovered from one sample in the weathered till but its significance is uncertain.

Deposits above the Upper Buried Soil

The A horizon of the Upper Buried Soil is missing due to erosion. The truncated soil surface is cryoturbated and overlain by Gelifluctate Complex 4. This depositional surface then stabilized, ice wedges developed and periglacial soil formation began. Renewed glaciation then took place across the permafrost surface, with deposition of the brown Upper Till, derived from the west. Of interest is the discovery of a single Norwegian erratic in the Kirkhill Upper Till. In the northern part of the quarry the Upper Till underlies a large erratic mass of dark grey silty clay containing Jurassic and Cretaceous dinoflagellate cysts, which was originally derived from the Moray Firth. West of the quarry, excavations have shown that this material grades into dark tills at least 5 m thick in places and containing fragments of shell (the East Leys Till). The Kirkhill Upper Till and the East Leys Till are overlain by glaciofluvial sands and gravels and by periglacial slope deposits of Kirkhill Gelifluctate 5. This is the last depositional event recorded at the site prior to Holocene soil development.

Interpretation

Kirkhill Quarry contains an extraordinary range of sediments, which provide an unrivalled record of Quaternary environmental change in Scotland. The site is considered to include evidence of two interglacials, at least three glacial phases and at least four periglacial phases. No firm dates, however, exist for sediments in the Kirkhill

Figure 8.5 Kirkhill Quarry, south-east face B (1984). Developed in the Kirkhill Lower Sands and Gravels is the Kirkhill Lower Buried Soil, with its striking bleached horizon. The soil is truncated and overlain by laminated organic muds and sands, and then by periglacial slope deposits of Kirkhill Gelifluctate 3. These rubble layers have been partly reworked to form the Kirkhill Upper Till. The section is about 6 m high. (Photo: J. Jarvis.)

sequence. The simplest view is that the Upper Buried Soil represents the Ipswichian and that the Lower Buried Soil represents the Hoxnian. On this basis, the glacial and periglacial sediments above the Upper Buried Soil would be Devensian, there was one period of regional glaciation preceded by periglacial conditions between the Ipswichian and the Hoxnian, and the latter interglacial was preceded by one period of glaciation as well as periglacial conditions of unknown duration.

Kirkhill Quarry clearly has great potential as a regional reference site, but this potential has not yet been realized owing to the shortage of age determinations and the lack of firm stratigraphic correlations with Pleistocene sediments elsewhere in north-east Scotland. Sands and gravels below the Lower Buried Soil have been tentatively correlated with gravels at Leys Quarry (Hall and Connell, 1986), but no other deposits of comparable age are known in Buchan (Hall and Connell, 1991). Equally, the Lower Buried Soil currently stands alone in the regional stratigraphy as a presumed pre-Ipswichian deposit. The Lower Till may correlate with other weathered pre-Devensian tills of inland derivation at Kings Cross, Aberdeen (Synge, 1963) and Boyne Quarry, Portsoy (see below) (Peacock, 1966). If these correlations are justified, then these tills represent a major advance of inland ice (Hall and Connell, 1991). Acceptance of an Ipswichian age for the Upper Buried Soil invites correlation with the weathered tills above and with the buried interglacial soil at Teindland (FitzPatrick, 1965; Sutherland, 1984a).

The deposits below the Upper Buried Soil point to a complex sequence of pre-Devensian events in Buchan. An early, possibly Anglian ice-sheet transported the erratics in the basal Gelifluctate Complex 1 and deposited the basal till at Leys. Subsequently, ice-sheets covered this area at least in two later, separate periods, probably during a pre-Ipswichian cold period and during the Devensian. In addition, the sediments record at least two pre-Devensian periglacial phases, and the thickness of periglacial sediments in the sequence suggests important periglacial slope modification in the Middle Pleistocene (Connell and Hall, 1987). Finally, the Lower Buried Soil and the overlying organic sediment appear to represent an interglacial phase and a subsequent climatic deterioration. This interglacial was probably the Hoxnian. The only other organic sediments in Scotland for which a Hoxnian age has been claimed are from Fugla Ness, Shetland (Birks and Ransom, 1969; Lowe, 1984) (see above), and Dalcharn near Inverness (Merritt, 1990a) (see above).

The Upper Till has been correlated with other tills of inland derivation, the Inland 'Series' (Hall, 1984b), which represent the last glaciation of central Buchan. The Upper Till at Kirkhill provides an upper limiting age of post-Ipswichian for the Inland 'Series'. At Crossbrae, Turriff, a peat containing pollen of interstadial character and with radiocarbon dates of 26,400 ± 170 BP (SRR–2041a) and 22,380 ± 250 BP (SRR–2041b) rests on till correlated with the Inland 'Series' and is overlain by periglacial solifluction deposits (Hall and Connell, 1991). The combined evidence from Kirkhill and Crossbrae indicates that the last glaciation of central Buchan was during the Early or Middle Devensian and that an ice-free area existed there in the Late Devensian (Sutherland, 1984a; Connell and Hall, 1987). The Kirkhill site thus has crucial bearing on the problems of the timing, number and extent of Devensian glaciations in Scotland (Clapperton and Sugden, 1977; Hall, 1984b; Sutherland, 1984a; Bowen *et al.*, 1986; Hall and Connell, 1991).

The details of the stratigraphic succession at Kirkhill seem assured after over a decade of sediment logging. The quarry is now infilled with refuse, but geophysical and trial pit investigations (Hall *et al.*, 1989b) indicate that the sequence extends to the east of the former quarry, and recent work at Leys Quarry has further expanded the basic stratigraphic sequence established at Kirkhill. Several problems remain outstanding, however: the status of the Lower Buried Soil is not yet resolved (Connell and Romans, 1984) and published pollen data are scanty (Lowe, 1984). Crucially, also, the sediments at Kirkhill are not yet firmly dated. None the less, the Kirkhill succession is the longest known in Scotland and offers potential for correlations, both on land and offshore.

Conclusion

Kirkhill is a unique site of the highest importance for Quaternary studies in Scotland. Its sequence of deposits provides evidence for two interglacial phases, three separate glaciations and at least four periglacial episodes. Kirkhill has the longest succession of Quaternary deposits in Scotland (extending back over approximately 450,000 years) and is therefore a critical reference site for establishing a better understanding of the sequence and timing of glacial and non-glacial climatic episodes.

BELLSCAMPHIE

A. M. Hall and J. Jarvis

Highlights

Deposits exposed in the disused railway cutting at Bellscamphie include three distinct tills, two of which are believed to pre-date the Late Devensian. The sequence provides important evidence for interpreting the glacial history of north-east Scotland and the patterns of ice-movement associated with different glacial events.

Introduction

The section at Bellscamphie (NK 019338) lies 7 km north-east of Ellon in a railway cutting, now disused, and was described by Jamieson (1906) from his observations during the original excavation for the railway in the late 19th century. The site is important for a sequence of three superimposed tills, which include the 'indigo boulder clay' of Jamieson (1906). The tills, the Bellscamphie Upper, Middle and Lower Till Members, may mark three separate phases of glaciation, two of which pre-date the Late Devensian. The Bellscamphie Middle Till (the indigo boulder clay of Jamieson) was originally described at three sites, but is now accessible only at Bellscamphie. The Bellscamphie Lower Till has not been previously described in the literature and has no known correlative deposits in the Ellon area. The Bellscamphie site is therefore of great importance in the study of the Pleistocene history of Buchan. The only published work on the site is that of Jamieson (1906), but it has recently been reinvestigated by Hall (1989b).

Description

Jamieson (1906) described the following succession which had a total thickness of 16 ft (4.9 m):

6. Red clay, a few feet thick, of the Red Clay 'Series'.
5. Gravel, containing pebbles of yellow limestone and sandstone and shell fragments of Crag affinities. Similar in lithology to the gravels forming the Late Devensian Kippet Hills esker, 2.5 km to the south.
4. Sand, fine, clean-washed.
3. Red Clay, similar to bed 6.
2. Indigo boulder clay, with small fragments of schist and granite. No more than a few feet thick.
1. Bedrock, schist.

No other descriptions were produced as the section was soon obscured by vegetation, but numerous references to the site and Jamieson's description exist in the literature (for example, Bremner, 1916b, 1928, 1934a; Synge, 1956; Hall, 1984b; Sutherland, 1984a).

Jamieson (1906) also described the indigo boulder clay at two other sites:

A. Craigs of Auchterellon (NJ 952308). The indigo boulder clay was discovered at a depth of 14 m in wells dug east of the rock knoll of Craigs of Auchterellon, apparently preserved in the lee of the bedrock high. The site is now covered by housing.
B. 'A railway cutting a few miles south of Ellon'. Jamieson's description is vague and the location of the site was not precisely recorded. Consequently, this site has not been relocated.

Excavations in 1988 (Hall, 1989b) revealed a different stratigraphy at Bellscamphie (Figure 8.6) to that described by Jamieson (1906):

5. Grey gravels and sands (Bellscamphie Gravels and Sands).
4. Interbedded red diamictons (Bellscamphie Upper Till).
3. Dark grey diamicton – the indigo boulder clay (Bellscamphie Middle Till).
2. Brown diamicton (Bellscamphie Lower Till).
1. Dalradian psammites.

The Bellscamphie Lower Till is a brown, sandy, silty, massive and overconsolidated diamicton. At the time of the original railway excavations this unit would have lain largely concealed below the cutting floor and this is probably why the unit was not described by Jamieson (1906). The Bellscamphie Lower Till is dominated by clasts of local metamorphic rocks. It rests in shallow hollows on the surface of the Dalradian.

The Bellscamphie Middle Till is a dark grey, clayey silty, massive and overconsolidated diamic-

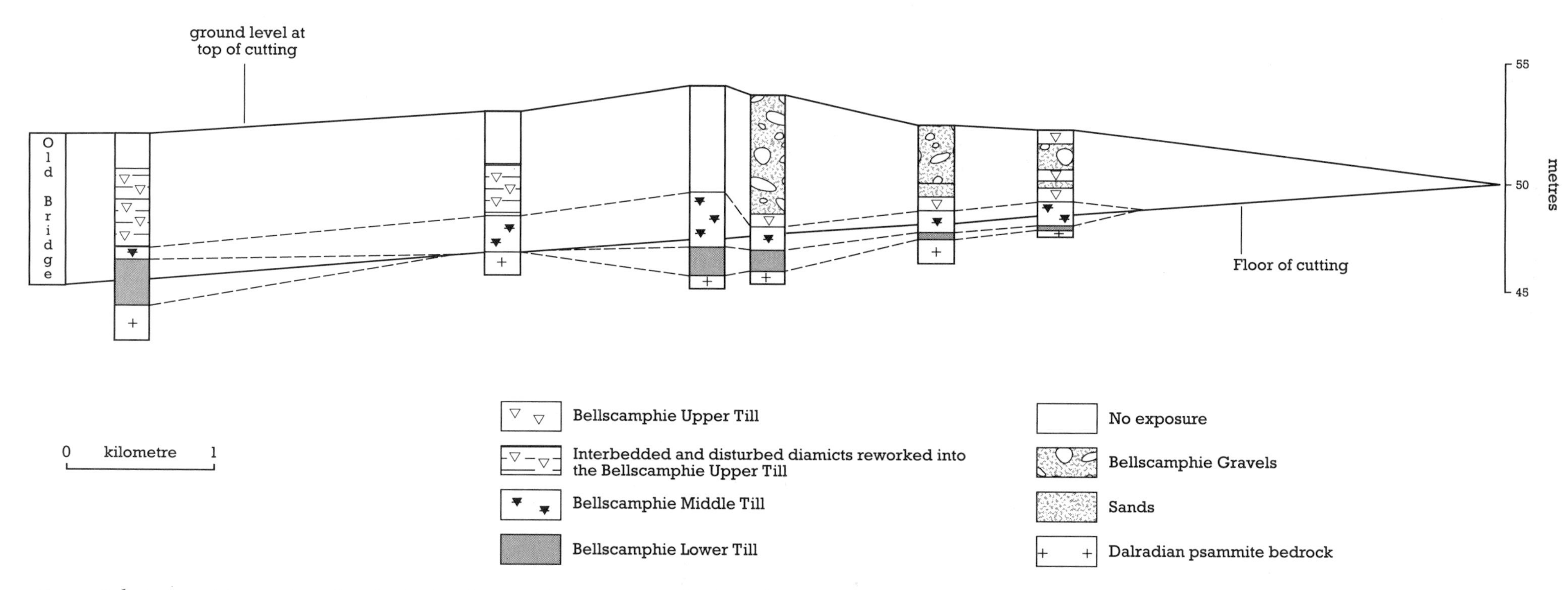

Figure 8.6 Sediment logs and stratigraphy in the disused railway cutting at Bellscamphie.

ton with sparse, small shell fragments. In the deposit at Craigs of Auchterellon, Jamieson (1906) recognized fragments of shells including *Arctica islandica* (L.) and *Astarte arctica* (Gray). At Bellscamphie, both species are also present. The Middle Till reaches a maximum observed thickness of 2.5 m and shows planar and erosive upper and lower contacts, except that in one pit it was seen to have incorporated small bodies of the underlying Bellscamphie Lower Till. The till incorporates palynomorphs of Kimmeridge Clay affinity (W. Braham and E. R. Connell, unpublished data), and its overall lithology is consistent with transport by ice from the Moray Firth, as Jamieson (1906) suggested.

Overlying the Bellscamphie Middle Till are interbedded red diamictons (Bellscamphie Upper Till) and grey gravels and sands (Bellscamphie Gravels and Sands). The red diamictons are, in places, crudely bedded and washed and may represent deposition into water bodies beneath glacier ice. The Bellscamphie Upper Till closely resembles in its lithology other diamictons in the Ellon area, which belong to the Red 'Series' (Hall, 1984b). The Bellscamphie Gravels and Sands are poorly sorted, horizontally bedded and rich in shell debris. Lithological comparisons indicate that the Bellscamphie Gravels and Sands form the western extension of the body of shelly gravel, with clasts of late Tertiary (?)limestone and mudstone, which extends from the coast to the Kippet Hills and to Bellscamphie (Jamieson, 1906; Merritt, 1981; Hall, 1984b). Both units are to be correlated with the Late Devensian Red 'Series' deposits found southwards from Peterhead along the North Sea coast and deposited by ice moving from the south or south-east. In the north-east part of the cutting, excavations revealed a complex sequence of interbedded and interdigitating diamictons of varied colour and lithology, which probably represents reworking of older diamictons by the ice that deposited the Bellscamphie Upper Till.

Interpretation

Analysis of the deposits at Bellscamphie continues and interpretation of their age and significance is preliminary. Strong lithological similarities leave little doubt that the Bellscamphie Upper Till and Bellscamphie Gravels and Sands belong to the Red 'Series' formation and date from the Late Devensian.

The Bellscamphie Middle Till is correlated with Jamieson's (1906) indigo boulder clay, for which a pre-Late Devensian age has been widely accepted (for example, Synge, 1956; Sutherland, 1984a), although this has yet to be demonstrated. At Craigs of Auchterellon, however, originally the key locality for the indigo boulder clay, there is a grey till derived from the west which underlies the Red 'Series' and overlies the indigo boulder clay. This inland till is known from several other sites in the Ellon area and is probably of Early Devensian or early Late Devensian age (Hall, 1984b; Hall and Connell, 1991). This till is not represented at Bellscamphie, where the Middle Till is covered only by deposits of the Red 'Series'. Bremner (1916b, 1928, 1934a) regarded the indigo boulder clay as part of the ground moraine of a Scandinavian ice-sheet and suggested correlation on lithological grounds with dark shelly tills at Burn of Benholm and Aberdeen. Such correlation is dubious, however, because (1) palynological work indicates that the Benholm shelly till is derived from the North Sea, whereas other dark shelly tills in Buchan are derived from the Moray Firth (Hall and Connell, 1991) and (2) there is only one doubtful record of the recovery of a Scandinavian erratic from these dark shelly tills (Hall and Connell, 1991).

As indicated above, transport of the Bellscamphie Middle Till from the Moray Firth seems likely. Dark clayey tills and Mesozoic erratics of probable Moray Firth provenance occur at numerous sites in Buchan (Hall, 1984b, figure 3) and are thought originally to have been part of a more extensive till-sheet, the 'Mesozoic Drift' (Hall, 1984b). The ice movement or movements which formed these deposits probably pre-dated the Late Devensian, as inland tills of Early or early Late Devensian age incorporate masses of this material at a number of localities. Amino acid analysis of shell fragments from the Middle Till is awaited and may help to establish the age of this deposit more precisely.

No new sites for the indigo boulder clay have been reported since the turn of the century. However, in 1988 drainage ditches at Pitlurg Station (NK 021344) exposed a diamicton of very similar lithology to the Bellscamphie Middle Till underlying Red 'Series' deposits. A dark grey till underlying red till was also logged by the Department of Geology, University of Aberdeen, in pipeline excavations at Eastertown of Auchleuchries (NK 015364). Dark grey till underlies red-brown till and gravel in British Geological

Survey (BGS) borehole NK 03 SW 5 at Pitlurg (NK 02633312) and underlies brown till and gravels in BGS borehole NK 03 NW 1 at Hill of Auchleuchries (NK 00573649) (Merritt, 1981). Some or all of these occurrences may correlate with the Bellscamphie Middle Till, but detailed lithological comparisons are required.

Jamieson regarded the indigo boulder clay as the oldest glacial deposit in the Ellon area, but the discovery of an underlying till, the Bellscamphie Lower Till, forces the recognition of an earlier ice movement from the west or north-west. The Lower Till has not been recognized elsewhere in the Ellon area, but may correlate with the weathered till derived from inland at Kirkhill (Connell *et al.*, 1982; Hall, 1984b).

Deposits of the Red 'Series', which are directly comparable to the Upper Till and Gravels and Sands at Bellscamphie, are exposed at numerous localities in the Ellon area, notably around the Kippet Hills.

In summary, the Bellscamphie Lower Till has so far been recognized only at the type site. The Bellscamphie Middle Till can only be easily re-exposed at Bellscamphie. Moreover, the superimposition of three tills probably reflecting separate phases of glaciation is matched at only one other site in the Ellon area, Craigs of Auchterellon, and this site has been fossilized by house building.

The sequence of deposits at Bellscamphie provides a key stratigraphic record of the glacial history of the Ellon area and it is the only accessible site which shows three superimposed till units. The Bellscamphie Upper Till and interbedded Gravels and Sands were deposited by ice moving from the south during the Late Devensian and are well represented at other sites. The Middle Till (Jamieson's indigo boulder clay) was deposited by ice moving out of the Moray Firth, probably prior to the Late Devensian. The Lower Till is known only from this site and represents the oldest period of glaciation known in this part of Buchan. Bellscamphie is one of only two sites known in Buchan where two superimposed tills of probable pre-Late Devensian age occur; the other is at Leys Quarry (Hall and Connell, 1986; see Kirkhill). As Buchan contains the most complete pre-Devensian sequence known on land in Scotland (Hall, 1984b; Sutherland, 1984a; Hall and Connell, 1991), the site is also of national importance for the study of Pleistocene stratigraphy.

Conclusion

The deposits at Bellscamphie provide evidence for three separate phases of ice movement in north-east Scotland, two of which pre-date the last (Late Devensian) glaciation. Such a sequence is rare in Scotland, so that Bellscamphie is a valuable reference site for studies of Quaternary history and for interpreting the patterns of ice movement across the landscape. It is also a type locality for two of the individual ice-deposited till units.

BOYNE QUARRY

J. E. Gordon

Highlights

The superficial deposits exposed at Boyne Quarry include a sequence of multiple tills and ice-marginal lake deposits. These provide important evidence for interpreting the glacial history of the Moray Firth coastal area and the associated depositional environments.

Introduction

Boyne Quarry (NH 613659), *c.* 7.5 km west of Banff, is important for a succession of deposits demonstrating the glacial stratigraphy of the southern coast of the Moray Firth. Two, or possibly three tills occur (Peacock, 1966), and the basal one has been correlated with pre-Devensian weathered tills known elsewhere in north-east Scotland. The exposure is a comparatively recent one and does not appear in the extensive early literature on the glaciation of the Moray Firth coast; the only published references are by Peacock (1966, 1971a) and Connell and Hall (1984b). However, as one of the few localities showing an extended succession of deposits which help to clarify the controversial stratigraphic sequence in the region, it is an important reference site.

Description

The Boyne Quarry section was described by Peacock (1966) and comprises the following sequence:

6. Peaty soil 0.6 m
5. Horizontally bedded silt, sand and gravels 0–1.2 m
4. Grey, silty till with gabbro boulders derived from the Huntly basic complex 1.5 m
3. Dark grey, clayey till with shell fragments 3.0–4.6 m
2. Dark grey, shelly silt and clay, in part till-like and in part laminated, with pockets of shelly gravel especially at the base 1.5–3.7 m
1. Light-brown till 0–3.7 m

The till of bed 1 is considerably decomposed in comparison with the deposits above it. It comprises subrounded pebbles and cobbles of quartzite, with lesser amounts of granite, calc-schist and gabbroic rocks, in a gritty, clayey matrix. Apart from the quartzites, most of the clasts are decomposed. According to Connell and Hall (1984b), the weathering profile is truncated and the till surface covered by a thin (0.02 m) layer of brown silt, and by fine sand and gravel infilling shallow depressions. Subsequent examination (J. D. Peacock, unpublished data) has, however, shown that the till also includes fresh cobbles and boulders of gabbro and that the weathered appearance may partly reflect the incorporation of weathered bedrock. Such weathered bedrock was formerly seen below bed 1 on the south side of the quarry.

Further work has shown that the clay (bed 2) contains Foraminifera, and that the crushed mollusc shells reported earlier are single valves of *Nuculana pernula* (Müller) (J. D. Peacock, unpublished data). Connell and Hall (1984b) have suggested that bed 2 can be subdivided into a lower unit of sheared and interbedded layers of red and brown silt and fine sand, and an upper unit of dark grey, clayey till. Peacock (1966) also noted that the site was apparently unaffected by solifluction.

Interpretation

The glacial deposits of the Moray Firth coast have a long history of investigation, which provides the context for interpreting the sequence at Boyne Quarry. Early accounts drew attention to the presence of anomalous beds of clay containing Jurassic fossils (Christie, 1830; Prestwich, 1838a, 1840; Brickenden, 1851) and exposures of shelly stratified sands, gravels and clays, notably at Gamrie (see Castle Hill) and King Edward (Prestwich, 1838b, 1840; Chambers, 1857).

Martin (1856) described many of the characteristics of the glacial drift of the area and attempted to explain it in terms of debris transported by sea ice during a great submergence. From his observations on the distribution of erratics and patterns of striae, Mackie (1901) deduced ice movement initially from the interior towards the north-east, but it gradually became diverted towards the east and south-east by ice in the Moray Firth, in the first instance by ice from the Northern Highlands then by ice from Scandinavia.

The most significant early contribution, however, was that of T. F. Jamieson, who, in the course of his investigations of the glacial phenomena of Scotland, described many of the exposures and characteristics of the Moray Firth drift and pieced together a stratigraphic sequence for the area (Jamieson, 1858, 1865, 1866, 1874, 1882b, 1906, 1910). His more important observations included:

1. detailed descriptions of the King Edward and Castle Hill sections and identification of arctic shell assemblages (Jamieson, 1858, 1865, 1866, 1906);
2. confirmation of Chamber's (1857) report of a dark grey boulder clay overlying the shelly silts and sands at King Edward (Jamieson, 1866, 1906);
3. identification of a dark, bluish-grey shelly clay between Cullen and Banff, with erratics derived from westerly sources and Jurassic material from the Moray Firth sea floor (Jamieson, 1866, 1882b, 1906);
4. the eastwards continuation of the dark shelly clay to Peterhead and its interdigitation with the red clay of the east coast of Aberdeenshire (Jamieson, 1866, 1906);
5. the inference, from erratics and striae, that ice moved from west to east (Jamieson, 1866, 1906).

From these observations, and work elsewhere in Scotland, Jamieson (1865, 1906) deduced the glacial succession in the Moray Firth area as: (1) early glaciation from the north and west (based on evidence in the Ellon area only), (2) main glaciation with ice from the west succeeding ice moving down Strathspey, (3) as the ice melted away, submergence of the coast and deposition of the dark shelly clay and shelly sands contemporaneous with the red deposits of eastern Aberdeenshire, and (4) less extensive glaciation from the west, responsible for boulder clays (on top of the shelly clays and sands), and equivalent to the Aberdeen moraines (see Nigg Bay).

Unfortunately, Jamieson's interpretations were constrained by a rigid conceptual model requiring all 'clays' with shells, including those which he definitely recognized as boulder clays, to be deposited in glaciomarine or glaciolacustrine conditions during a great submergence of the coast at the end of the main glaciation. This seems to have arisen from a failure to distinguish between true marine clays and shelly boulder clays. Although Bell (1895a, 1895b, 1895c, 1895d) had clearly questioned the evidence for submergence to explain many of the deposits, his arguments went unheeded in Jamieson's later work (Jamieson, 1906, 1910).

In the Geological Survey Memoir, Read (1923) published a comprehensive review of the field evidence and proposed a fourfold division of the glacial stratigraphy of north Banffshire and north-west Aberdeenshire (sheets 86 and 87):

4. Lateglacial sands and gravels.
3. Upper or northerly drift deposited by ice moving out from Central Banffshire as the south-easterly ice withdrew.
2. The 'Coastal Deposits' (see Castle Hill) comprising a suite of sands, clays and gravels, including those at King Edward and Castle Hill, formed in a lake as the south-easterly ice withdrew.
1. Lower or south-easterly drift moving onshore out of the Moray Firth and including the shelly drift and the drift with Jurassic fossils.

This succession, which was based on stratigraphic evidence, striae and erratics, hinged on stratigraphic interpolation since no single exposure showed all the elements of the succession to be superimposed. An important facet of Read's interpretation was that the entire sequence related to a single period of glaciation.

Bremner (1916b) noted the pattern of cross-striations on the south coast of the Moray Firth. Later he agreed with Read's interpretation of two separate ice movements, from the north-west and south, but argued that they represented two distinct glacial periods (his first and second glaciations) (Bremner, 1928, 1934a, 1938, 1943a). Moreover, he proposed the existence of a third ice-sheet, moving from the north-west after deposition of the northerly drift. In support he cited supposedly marginal meltwater channels, evidence from two sections near Rothes and one near Cullen, and glaciofluvial deposits resting on the northerly till and extending from Inverness to Buchan. Bremner (1928) also disputed Jamieson's correlation of the blue shelly till with the red till of the east coast of Aberdeenshire, relating them to separate glaciations.

Charlesworth (1956) supported Bremner's view that the last ice in the Moray Firth area came from the north-west (his Highland Glaciation), and related the 'Coastal Deposits' to the final stages of retreat of this ice.

Synge (1956) also accepted that the last ice came from the north-west. However, a problem with Bremner's scheme was that this ice did not leave an extensive till cover, which Synge thought was improbable. Therefore, he proposed that Read's sequence of deposits was inverted and that the lower drift in fact related to the last ice-sheet. Read's upper drift might not be *in situ* but could have been soliflucted on to the 'Coastal Deposits' in the very few sections where it was seen to overlie them. Synge suggested that the 'Coastal Deposits' might be the equivalent of the Lateglacial gravels (of Read, 1923), noting that the two were never seen in a section together. He concluded, therefore, that the drifts of Banffshire could all be explained in terms of a single glaciation, the last or Moray Firth – Strathmore Glaciation. Erratics from the south were probably incorporated from an earlier drift.

The deposits at Boyne Quarry have an important bearing on these earlier interpretations. Peacock (1966) considered that the weathered till (bed 1) was the relic of a very early glacial episode and that it had undergone prolonged subaerial weathering before deposition of the overlying sediments. It may be of pre-Devensian age (Jardine and Peacock, 1973) and may correlate with the weathered tills reported from Kirkhill (Connell *et al.*, 1982; Hall, 1984b) and Kings Cross, Aberdeen (Synge, 1963). However, further work is required to establish the extent to which the apparent weathering reflects the incorporation of previously decomposed bedrock into the till, rather than being *in situ* weathering of the till itself (J. D. Peacock, unpublished data). Peacock (1966) interpreted bed 2 as an erratic and considered it to be the source material for bed 3. Bed 3 appeared to correspond to the shelly till of Banffshire and the 'lower or south-easterly drift' of Read (1923). The till of bed 4 was probably part of Read's 'upper and northerly drift'. Bed 5 formed part of the 'Coastal Deposits'.

The Boyne Quarry section as recorded by Peacock (1966), thus apparently established that Read's till succession was correct at least there, and not inverted as Synge suggested. However, 'Coastal Deposits' did not intervene between the tills, and Peacock (1971a) showed elsewhere that they were probably deposited in freshwater lakes during the melting of the last ice-sheet (see Castle Hill).

The apparent confirmation of Read's till succession was subsequently rejected by Peacock (1971a), who reinterpreted the uppermost till unit (bed 4) as simply a separate facies of the immediately underlying shelly till; a view supported by the gradational contact between the two units. Peacock (1971a) favoured the interpretation that the gabbro boulders in the top till had been incorporated from an earlier till-sheet derived from the south, so that the sequence of deposits above the basal weathered till, comprising beds 2 to 5, was the product of the last ice-sheet flowing from the north-west. An early north to south or south to north ice movement is shown by striations formerly seen on the bedrock surface, but their relationship to the strata in the adjacent drift section could not be ascertained (J. D. Peacock, unpublished data).

The age of the deposits at Boyne Quarry is uncertain. Peacock (1966) considered that the uppermost till (bed 4) and the overlying 'Coastal Deposits' (bed 5) were the product of Late Devensian ice-sheet advance and retreat, and the reinterpretation of Peacock (1971a) indicated that all the deposits except the weathered till (bed 1) were of Late Devensian age. This was also implied by Clapperton and Sugden (1975, 1977) in their consideration of the glaciation of north-east Scotland. Sutherland (1984a), however, suggested that the Late Devensian ice-sheet may have terminated to the west of the site and that hence the last glaciation of this area was pre-Late Devensian. This view was not accepted by Hall (1984b), who preferred a Late Devensian age for the last glaciation of the southern Moray Firth coast. Amino acid ratios on shells from sand and gravel horizons immediately overlying the weathered till and from the shelly till imply a Devensian age for the glaciation, but do not yet allow a fuller resolution of the chronological problems (D. G. Sutherland, unpublished data).

Boyne Quarry is a key stratigraphic site demonstrating the much-debated Pleistocene succession on the south coast of the Moray Firth, including the two till units and the 'Coastal Deposits' which have formed the basic field evidence for reconstructions of the glacial history of the area. Current interpretations suggest that most of the deposits were produced during a single glacial episode of Devensian age. Boyne Quarry is also particularly significant for one of the few exposures in Scotland of a till which has been considered to be of pre-Devensian age. From a sedimentological viewpoint, Boyne Quarry is also notable in providing a good illustration of the complex sequence of deposits which may be associated with a single glaciation, including a raft of marine clay within shelly till. Finally, Boyne Quarry has significant research potential (Connell and Hall, 1984b). This relates to:

1. the pedological characteristics of the weathered till (bed 1), their origins and the possible correlations with weathered tills elsewhere in north-east Scotland (cf. Kirkhill Quarry),
2. the sedimentary characteristics of the deposits in bed 2 and the process responsible for their origin,
3. the significance of the gradational contact between the tills of beds 3 and 4, and the interpretation of these deposits.

Conclusion

Boyne Quarry is a reference site for the ice-deposited sediments of the south coast of the Moray Firth. The sequence includes two, or possibly three tills, one of which may pre-date the Devensian, and demonstrates the main deposits that have been described from the area. Boyne Quarry is therefore important for establishing the pattern of ice movements across the area and also for studies of the formation of the glacial sediments.

TEINDLAND QUARRY

D. G. Sutherland

Highlights

The sequence of deposits at Teindland Quarry includes a palaeosol which has yielded pollen of both interglacial and interstadial affinites. Sites which preserve such evidence are rare, and

Teindland is a key locality for interpreting the Quaternary history of Scotland.

Introduction

Teindland Quarry (NJ 297570) is located in Teindland Forest in lower Strathspey, 5 km south-west of Fochabers at an altitude of approximately 100 m OD. It is one of a few known sites on the Scottish mainland where organic deposits older than the Late Devensian have been both radiocarbon dated and analysed for pollen. Since its original description by FitzPatrick (1965), the site and its interpretation have proved controversial (Edwards *et al.*, 1976; Romans, 1977; Sissons, 1981b, 1982c; Caseldine and Edwards, 1982; Lowe, 1984). Despite the significance of the site, no detailed description has yet been published of the full succession.

Description

The section originally described by FitzPatrick (1965) showed 1.8–2.4 m of sandy till and outwash gravel overlying an iron podsol developed on glaciofluvial outwash (Figure 8.7). A radiocarbon date of 28,140 +480/−450 BP (NPL–78) was obtained from the soil.

More data were provided by Edwards *et al.* (1976), particularly with respect to the pollen content of the sediments. They described a more complex succession than was recognized by FitzPatrick (1965) (cf. Figure 8.8):

5. Humified modern soil layer. *c.* 0.56 m
4. Yellow sand. *c.* 1.11 m
3. Sandy till. *c.* 0.74 m
2. Interbedded sequence of grey and yellow sands and black organic layers. *c.* 0.31 m
1. Fossil podsol developed in a sequence of yellow and grey sands with an iron-rich horizon *c.* 0.17 m below the top of the unit. at least 0.56 m

Six pollen assemblage zones (Figure 8.8) were recognized by Edwards *et al.* (1976), although the upper two (T–5 and T–6) related to the present soil horizon (bed 5) and a single pollen spectrum (T–4) was from a sandy horizon in the till which was considered to have been reworked from the sands beneath (bed 1). Within the basal sands below the iron-rich horizon, Edwards *et al.* identified a pollen assemblage zone (T–1) that contained high proportions of *Alnus*, Gramineae and *Plantago lanceolata*. Also present were low frequencies of *Betula*, *Quercus* and *Corylus* pollen. One sample had up to 50% Filicales spores. The iron-rich horizon yielded a single pollen spectrum (T–2), markedly different from that in the sand below (zone T–1) with little *Alnus*, high values of *Pinus* pollen, a significant level of *Calluna vulgaris* pollen and a reduced percentage of Gramineae. Above this, and coinciding with the interbedded sequence of sands and organic horizons, pollen counts (zone T–3) were dominated by Gramineae and there was a low representation of tree pollen and an erratic component of *Calluna vulgaris*.

Romans (1977) added the further observations that the A horizon at the top of the buried soil was composed of heavily charred organic material together with coniferous charcoal and that the deposition of this material was contemporaneous with the development of frost cracks in the soil.

Figure 8.7 Fossil podsol and overlying organic horizon at Teindland. (Photo: D. G. Sutherland.)

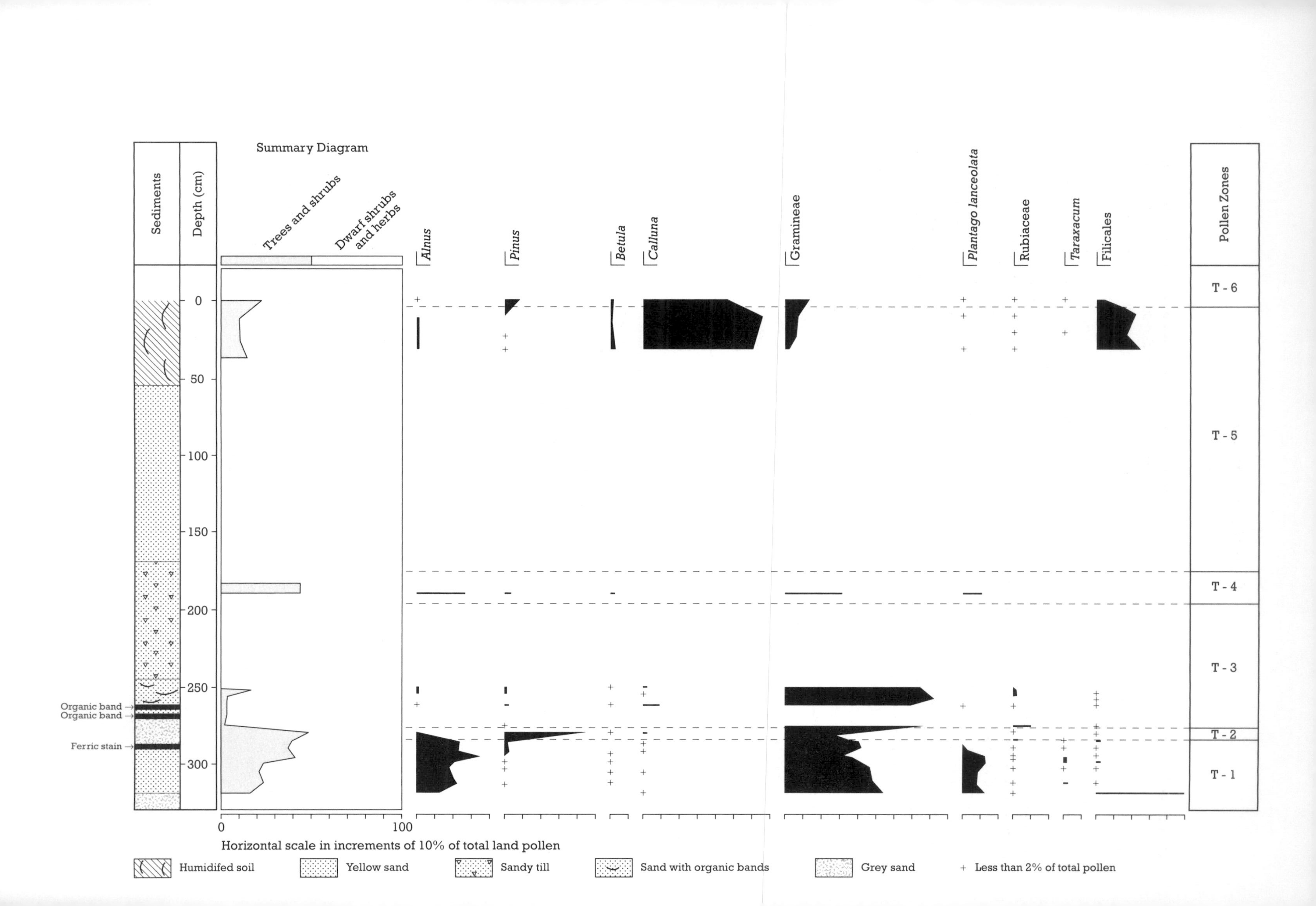

Summary Diagram
Sediments
Depth (cm)
Trees and shrubs
Dwarf shrubs and herbs
Alnus
Pinus
Betula
Calluna
Gramineae
Plantago lanceolata
Rubiaceae
Taraxacum
Filicales
Pollen Zones
T - 6
T - 5
T - 4
T - 3
T - 2
T - 1
0
50
100
150
200
250
300
Organic band
Organic band
Ferric stain
0
100
Horizontal scale in increments of 10% of total land pollen
Humidifed soil
Yellow sand
Sandy till
Sand with organic bands
Grey sand
+ Less than 2% of total pollen

Interpretation

FitzPatrick (1965) considered that the degree of soil development was the product of interglacial conditions and cautioned that the radiocarbon date on the soil should be regarded as minimal because of the lack of alkali pre-treatment. The site therefore apparently revealed evidence for a period of glaciation followed by an interglacial (?Ipswichian) then periglaciation and further glaciation (?Late Devensian).

Interpretation of the pollen data was made difficult by the possibilities of reworking during erosion of the palaeosol, differential pollen preservation in soils and downwashing of pollen grains through the sandy sediments. These difficulties notwithstanding, Edwards *et al.* (1976) suggested that zone T–1 represented the vegetation of an interglacial, and drew attention to the similarities in the pollen spectra dominated by *Alnus* and *Plantago lanceolata* and pollen assemblages from the later part of certain interglacial sequences assigned to the Ipswichian in East Anglia (see Phillips, 1976). The sole spectrum comprising zone T–2 was considered most probably to represent the closing phase of the interglacial represented by zone T–1, possibly coinciding with the period of podsol formation.

Zone T–3 was more complex and more difficult to interpret. Edwards *et al.* thought that the basal spectrum in this zone might either indicate cold climatic conditions immediately following the interglacial, or it was part of the sequence of interbedded sand and organic horizons that produced the remainder of the samples in zone T–3. These last were considered to relate to an interstadial phase and, accepting the radiocarbon date of FitzPatrick (1965), this interstadial was assigned a Middle Devensian age. The overlying sediments (bed 3) were considered to be the product of a glacial episode, the fabric of the sandy till indicating ice movement from S60°E to S61°E.

This last interpretation was disputed by Romans (1977), who considered that there was no till exposed at Teindland, but rather that the sediments overlying the buried soil were emplaced by solifluction. He therefore argued that the site had not been glaciated during the Devensian, and inferred a very cold climate during the final phase of soil development from his observations of frost cracks in the soil.

The problem of using the radiocarbon date to infer a Middle Devensian age for part of the sequence was raised by Sissons (1981b). In response, Caseldine and Edwards (1982) provided new radiocarbon age estimates of 40,710 ± 2000 BP (UB–2121) and 38,400 ± 1000 BP (UB–2209), the material for which had been subjected to both acid and alkali pre-treatment. However, as with other interstadial or interglacial sites (for example, Kirkhill and Fugla Ness), finite radiocarbon dates are not on their own sufficient evidence to assign a deposit to the Middle Devensian (Sissons, 1982c). Sissons went on to suggest that, if indeed the Teindland site did contain evidence for interglacial and subsequent interstadial conditions, then it would be possible to interpret the interstadial as having occurred during the Early Devensian during the period immediately after the last interglacial, a suggestion that had previously been partially advanced by Edwards *et al.* (1976) for the basal sample in their pollen assemblage zone T–3.

Lowe (1984) also raised a number of questions concerning the interpretation of the pollen data from Teindland, considering it necessary to obtain more information before either an interglacial or an interstadial interpretation could be unequivocally upheld. In addition, the published sections for the site are greatly simplified (Sutherland, 1984a, unpublished data) and a complex sequence of sediments overlies the soil horizon.

Despite the controversy surrounding the interpretation of the Teindland Quarry sediments, they hold great potential for providing information on Devensian and possibly earlier environments. The site is one of the very few accessible interstadial or interglacial sites on the Scottish mainland (see for example, Clapperton, 1977). It is one of only a small number of sites in Scotland with deposits to which an interglacial origin has been ascribed (see Kirkhill, Sel Ayre, Fugla Ness, Toa Galson in North-west Coast of Lewis and Dalcharn) and the only site at which both interglacial and interstadial deposits have been interpreted as occurring in superposition. Further research should result in clarification of the stratigraphy and it is apparent that the site should provide important evidence bearing on the question of both Early and Late Devensian glaciation of Scotland (Sutherland, 1984a).

Figure 8.8 Teindland: relative pollen diagram showing selected taxa as percentage of total land pollen (from Edwards *et al.*, 1976; Lowe, 1984).

Conclusion

Teindland Quarry is a site of great importance for Quaternary studies in Scotland. The sequence includes organic deposits which pre-date the last (Late Devensian) glaciation. They appear to comprise sediments formed in both interstadial and interglacial climatic phases (that is, respectively, in a warmer interlude within an ice age (glacial) and a warmer phase between two separate glacials), but their ages are not yet unequivocally established. As one of only a few sites where such deposits are accessible, Teindland is a key locality for further research to amplify the Quaternary history of Scotland. The origin of the sediments overlying the organic deposits is also controversial and has a bearing on the argument about whether parts of north-east Scotland were not covered by the last ice-sheet.

CASTLE HILL

D. G. Sutherland

Highlights

Deposits exposed in the coastal section at Castle Hill include a complex sequence of glacial, glaciolacustrine and possibly marine sediments. Organic remains, including shells of marine molluscs and pollen, are also present in some of the beds. Although the importance of the site for interpreting the Quaternary history and changing sedimentary environments of the Moray Firth coast is firmly established, the full details of the record await further investigation.

Introduction

Castle Hill (NJ 794643), a coastal section at Gardenstown, demonstrates an important sequence of Pleistocene sediments, the interpretation of which is fundamental to an understanding of the timing and mode of the last glaciation of the north coast of Buchan. The site has been described in most detail by Jamieson (1906), Read (1923) and Peacock (1971a), and Jamieson (1865) provided a list of marine shells collected at the locality. A progress report of current research was given in Sutherland (1984b). There is, however, no overall agreement as to the stratigraphy of the site, nor is the chronology of the events recorded at Castle Hill established.

Description

Jamieson (1906) recorded a basal layer of partly worn clasts mixed with earthy debris that was overlain by a sequence of shelly sands and clays, which he considered to be a marine deposit. He did not report any till in the sections. Read (1923), however, noted a sequence comprising a basal layer of 'gravelly material', red till and, above, a series of sands and clays . Peacock (1971a) recorded a basal till overlain by a sequence of shelly sands, gravels and clays, the last being pebbly and till-like in one bed. Overlying the shelly sediments was a bed of dark grey silt and the section was capped by around 17 m of fine-grained non-fossiliferous, micaceous, yellow sand. A radiocarbon date of >39,500 BP (Birm–191b) was reported by Peacock (1971a) from the shelly horizons, but this failed to resolve the age of the deposits.

The full succession was not observed by Sutherland (1984b), but he identified the following sequence:

11.	Silts, sands and minor clays.	*c.* 17 m
10.	Dark grey, laminated silts and fine sands.	?
9.	Red-brown till.	?
8.	Gap.	10 m
7.	Bedded, medium sands and silts.	?
6.	Shelly gravel.	0.1 m
5.	Irregularly bedded and disturbed, yellow, medium sands and massive, grey clays.	*c.* 2 m
4.	Dark grey, massive clay.	*c.* 1.5 m
3.	Organic sand.	0.2–0.6 m
2.	Angular, platy clasts of Dalradian metasediments in a loose sandy matrix with occasional bedded sand lenses.	0.8–2.0 m
1.	Weathered Old Red Sandstone bedrock.	

The clasts in bed 2 comprise Dalradian metasediments indicating derivation from the slopes immediately to the west of Castle Hill. The upper part of bed 5 was thought to be slumped. Sutherland was uncertain as to whether the organic sand (bed 3) was part of the slumped material or whether it represented an interstadial deposit. The bedding in bed 7 indicated deposition of the sands by a west to east current. No clearly *in situ* beds were reported by Sutherland in the gap in the section, but minor exposures revealed very shelly gravels, sands and massive, grey clays. Bed 9, the only till encountered at Castle Hill, formed a bench around the hill and was similar to till on the neighbouring hill slopes. One sample from bed 10 had an organic content of 14% and a sparse pollen assemblage, dominantly of *Pinus* grains; the upper part of the silts was involuted and cracked. Cross-bedding in the lower 2–3 m of bed 11 indicated deposition by a current flowing from N35°W, and in the upper part, dewatering structures were present.

Sutherland (1984b) also provided additional information on the age of the deposits: amino acid analyses (D-alloisoleucine : L-isoleucine ratios) of two shell fragments of *Arctica islandica* (L.) from the shelly gravel below the till suggested that the shells were of Ipswichian to Devensian age. The details are:

Laboratory No.	Ratio	
	Free	Total
BAL–309B	0.333	0.097
BAL–309C	0.366	0.176

Information on the Castle Hill fossil shells has been summarized as follows by J. D. Peacock (unpublished data). The Castle Hill locality was examined by Miller (1859, p. 333) and shells of marine molluscs in the Hugh Miller Collection in the Royal Museum of Scotland labelled 'Glacial shell beds, Gamrie, Banff' were almost certainly collected from this site (J. D. Peacock, unpublished data). They include paired valves of *Macoma balthica* (L.), *M. calcarea* (Chemnitz) and *Tridonta montagui* (Dillwyn) as well as fragments of *Timoclea ovata* (Pennant). These are partly enclosed in a matrix of coarse-grained, slightly clayey sand. *Arctica islandica* (L.) occurs only as umbonal fragments. The bed from which the shells were collected probably corresponds to bed 7 of Jamieson (1906), bed 5 of Peacock (1971a) and bed 6 of Sutherland (1984b). Jamieson (1906) listed 25 species of mollusc from this locality, many occurring as entire shells. The fauna as a whole is of high-boreal to low-arctic aspect and the excellent preservation of the paired valves suggests that the bed itself is either *in situ* or was transported and deposited by ice *en masse* without significant deformation (see Boyne Quarry).

Interpretation

Read (1923) assigned the sands and clays that he described to his Banffshire 'Coastal Deposits', which he considered to pre-date the last glaciation (from the south) of this coastal area. The shelly deposits described by Peacock (1971a) were considered by him to be the product of the last glaciation (from the north-west) of the area, and the upper sequence of grey silts and yellow sands he assigned to the Banffshire 'Coastal Deposits' which, he argued, had been deposited in an ice-dammed lake during the retreat of the last (Late Devensian) ice-sheet. This lake drained southwards along the Afforsk meltwater channel (NJ 790623).

The Castle Hill sequence is important for resolving the problems of the last glaciation of the eastern Moray Firth coast. Most workers are agreed that the last till to be deposited along this coastline was the product of glaciation from the north-west (Jamieson, 1906; Synge, 1956; Hall, 1984b; Sutherland, 1984a; Hall and Connell, 1991). Both Read (1923) and Bremner (1928, 1934a) advocated later glaciation to explain the movements of erratics and the sequence of meltwater channels but, with the exception of a few sections, no till has been ascribed to this later glaciation. Those sections originally considered to show an upper till have been reinterpreted as either due to complex glacial deposition (Simpson, 1955) or solifluction (Synge, 1956; Peacock, 1971a). The till (bed 9) at Castle Hill identified by Sutherland (1984b) is therefore considered to have been deposited during this period of glaciation from the north-west.

Debate, however, surrounds the date of the glaciation that deposited the till at Castle Hill. The majority opinion is that the last glaciation of this area was during the Late Devensian (Synge, 1956; Peacock, 1971a; Clapperton and Sugden, 1977; Hall, 1984b), but Sutherland (1984a) has suggested that the Late Devensian ice limit may have lain to the west and that the last glaciation here was an Early Devensian event (see also Sutherland, 1981a). The available dating evidence from Castle Hill does not allow these two conflicting interpretations to be resolved.

The sediments overlying the till at Castle Hill

(beds 10 and 11) are most probably glaciolacustrine sediments, as envisaged by Peacock (1971a), but the possibility of an organic cryoturbated horizon in these deposits, as reported by Sutherland (1984b), may indicate a more complex history than the brief existence of an ice-dammed lake during a period of deglaciation. The sediments underlying the till are of uncertain origin. They may represent a period of lacustrine sedimentation during ice advance or possibly, in part, marine sedimentation, as suggested by Sutherland (1981a).

Castle Hill has one of the thickest and most complex sequences of Pleistocene sediments along the north Buchan coast. The site is important for elucidation of the glacial chronology of this area, and with possibly two organic horizons within the sequence and abundant marine mollusc shells in certain strata, including one bed possibly *in situ*, there is considerable potential for dating the events represented by the sediments. More widely, the site will ultimately provide evidence relevant to the debate on the extent of the Late Devensian ice-sheet and the possibility of ice-free areas remaining in north-east Scotland at the maximum of that glaciation.

Conclusion

Although the sequence of deposits at Castle Hill has a long history of research, the ages and origins of the different beds are not yet fully established. The site is nevertheless important because of the dating potential provided by the organic contents of the deposits. Further research at Castle Hill should help significantly to resolve the question of the timing of the last glaciation of the eastern Moray Firth coast and the detailed sequence of geomorphological changes that occurred both before and after that event.

KIPPET HILLS

J. E. Gordon

Highlights

The landforms and deposits at Kippet Hills include an excellent assemblage of glaciofluvial features formed during the melting of the Late Devensian ice-sheet. They are also noted for their erratic content and shells of Early Pleistocene marine molluscs, derived from sources offshore. The locality provides important evidence for interpreting ice movements and the pattern of deglaciation in the coastal lowlands north of Aberdeen.

Introduction

The Kippet Hills (NK 030315) are located 7 km east of Ellon and comprise an esker ridge and an area (*c.* 10 km^2) of hummocky, ice-contact glaciofluvial deposits associated with the onshore movement of ice during the Late Devensian. Several kettle holes are present, including the impressive example now occupied by Meikle Loch. In addition, the deposits are noted for an assemblage of fossil shells similar to that of the Red Crag of East Anglia, believed to have been picked up from the floor of the North Sea and incorporated into the deposits by the ice. The locality has featured frequently in the literature on glacial studies in north-east Scotland, notably in the work of Jamieson (1858, 1860a, 1865, 1882a, 1906), and considerable attention has been focused on the origin of the deposits and the source of the shells (Wilson, 1886; Bell, 1895b, 1895c; Hull, 1895; Gregory, 1926; Simpson, 1955). The most recent work on the site is that of Gemmell (1975), Merritt (1981), Cambridge (1982) and Smith (1984).

Description

Extensive sand and gravel deposits in the Kippet Hills area form a striking assemblage of ice-contact glaciofluvial landforms (Gemmell, 1975; Merritt, 1981; Smith, 1984). These comprise a 3 km long esker ridge up to 15 m high (the Kippet Hills) and associated kames, kettle holes and terraces (Figure 8.9). The esker ridge continues northwards into an undulating, triangular-shaped feature (*c.* 0.3 km^2 in area), interpreted by Smith (1984) as a kettled delta. On the west side of the esker, the basin occupied by Meikle Loch forms a large kettle hole some 800 m long. Exposures in Whitefields Sand Pit (NK 032321) have revealed up to about 12 m of

Figure 8.9 Geomorphology of the Kippet Hills (from Smith, 1984).

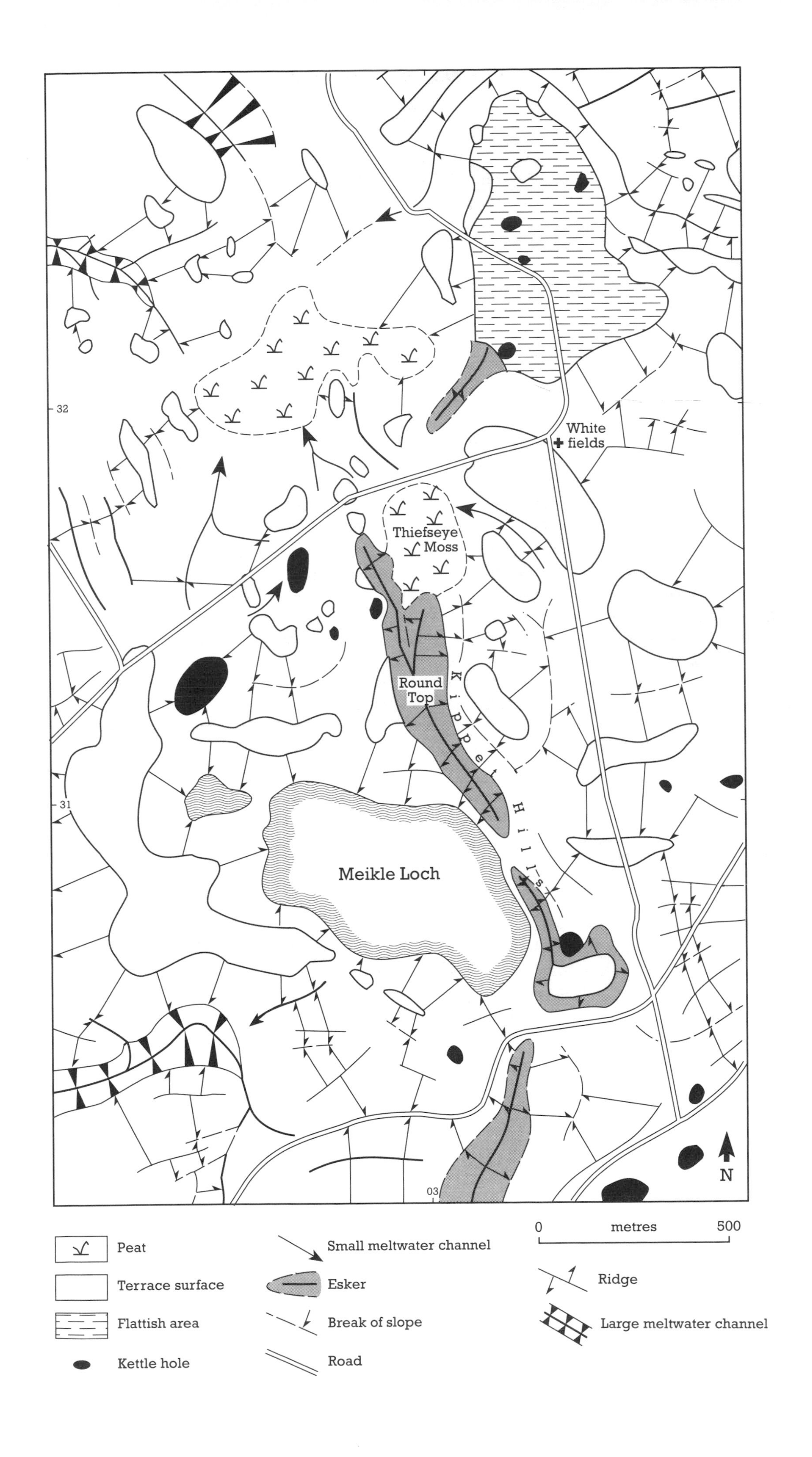
32
31
White fields
Thiefseye Moss
Round Top
Kippet Hills
Meikle Loch
03
N
0 metres 500
Peat
Terrace surface
Flattish area
Kettle hole
Small meltwater channel
Esker
Break of slope
Road
Ridge
Large meltwater channel

sands and gravels overlain by discontinuous red till (Gemmell, 1975; Merritt, 1981; Smith, 1984). Correlative shelly gravels and sands also occur at Bellscampie.

In a series of papers on the glaciation of north-east Scotland, Jamieson described the Kippet Hills deposits in some detail (Jamieson, 1858, 1860a, 1865, 1882a, 1906). He noted their sinuous, esker-like form and composition of local lithologies, together with significant amounts of red and grey sandstone, limestone and calcareous shale which he thought might be Jurassic or Permian. He identified a number of mollusc shells mixed through the gravels and an imprint of a fossil fish in a piece of limestone (Jamieson, 1858). He published species lists (Jamieson, 1860a, 1882a), noting the similarity of the assemblage to the Red Crag of Norfolk, although certain shells were anomalous. Jamieson also recorded that red clay with glacial characteristics was draped over the ridges and hollows, but there were apparently no traces of glacial deposits beneath the sands and gravels at Kippet Hills (Jamieson, 1858, 1865).

The most recent work has confirmed many of the previous observations. Merritt (1981) considered there to be a broad, threefold stratigraphic sequence in the area comprising a basal, dark clayey till, a widespread brown till, and a complex series of interbedded sands, gravels, clays and till. This last element in the sequence was found in the Kippet Hills. A borehole near the highest part of the esker revealed 3.6 m of red till overlying 21 m of well-sorted sand and gravel (Merritt, 1981), conforming with the stratigraphy at Whitefields Sand Pit. Merritt (1981) carried out a stone count of 10–14 mm gravel at the Whitefields pit. This showed that 41% of the clasts were ?Jurassic limestone and calcareous siltstone. Pebbles of yellow, shelly sandstone, possibly of Pliocene age were also found at this locality.

Cambridge (1982) was unable to confirm the shell identifications of Jamieson (1882a), but amino acid (D-alloisoleucine : L-isoleucine) ratios on *Arctica islandica* (L.) reported by Smith (1984) implied an Early to Middle Pleistocene age for those specimens:

Laboratory No.	Ratio	
	Free	Total
BAL–90A	1.097	0.537
BAL–90B	1.084	0.586

Interpretation

From his observations, Jamieson (1858, 1865) at first concluded that the Kippet Hills were *in situ* Crag deposits of marine origin formed during the late Tertiary, and were part of the same sequence of sand and gravel beds exposed in coastal sections at Collieston. Later he argued that in contrast to the latter, which were undisturbed, the Kippet Hills had been ploughed up by land ice into their present topographic form (Jamieson, 1865). However, in the course of further work Jamieson (1882a) noted grey till with local erratics underneath the deposits at Collieston and recognized that both they and the Kippet Hills sands and gravels were glacigenic deposits.

Jamieson (1906) observed that the red clay occurred both above and below the sand and gravel mounds in the Kippet Hills area, concluding that both were part of the same series of red deposits characteristic of the Aberdeenshire coast (Jamieson, 1882b). The latter he associated with his 'Second Glaciation', during which local ice from the west gave way to ice moving north along the coast from Strathmore. Originally he speculated about a local source in the Ythan estuary for the shell beds (Jamieson, 1882a), but later believed the material to have been transported to the Scottish coast by Scandinavian ice then entrained by the Strathmore ice (Jamieson, 1906). The red clay was then deposited during the wastage of the Strathmore ice in a glaciolacustrine or marine environment. It is a feature of Jamieson's work that although he recognized the glacial derivation of the red deposits and in places their boulder clay composition, he did not satisfactorily distinguish the till lithologies from the water-laid lithologies and therefore interpreted the whole succession in terms of the latter.

In the Geological Survey Memoir accompanying Sheet 87, Wilson (1886) recognized the same sequence of deposits as Jamieson, but reversed the interpretation: the Kippet Hills and Collieston gravels were interglacial marine beds, equivalent to those supposedly at Clava and King Edward, and the overlying red clays were a true glacial deposit. Bell (1895b) reviewed the respective interpretations of Jamieson and Wilson and concluded that both the shelly gravels and the till were in fact glacial in origin. Hull (1895) disagreed with Bell's interpretation, invoking submergence to explain both the sands and gravels and the clays. Bell (1895c), however, effectively rebutted Hull's arguments.

Bremner did not discuss the Kippet Hills in detail in his interpretation of the glacial sequence in Aberdeenshire. However, as part of his 'Second Glaciation' (Bremner, 1934a, 1938), he envisaged, first, deposition of basal grey till by local ice, second, deposition of the sands and gravels with bands of red clay in advance of Strathmore ice encroaching onshore, and third, deposition of the red clay as the latter ice decayed (Bremner, 1916b).

Gregory (1926) described the Kippet Hills as a 'pseudo-kame' or residual ridge following denudation, by unspecified processes, of a sheet of glaciofluvial drift. He also mentioned raised beach deposits in the upper part of one exposure, but these have not been recorded by other authors.

Simpson (1955) interpreted the Kippet Hills as part of a suite of glaciofluvial landforms extending from south of Nigg Bay (see below) northwards along the Aberdeenshire coast and deposited by the last ice-sheet. The content of shells and erratics in the sediments suggested that the ice had crossed the sea floor. Subsequent detailed studies on the mineralogy of the deposits in the area support this hypothesis (Glentworth *et al.*, 1964). Further evidence for ice moving onshore in north-east Scotland is provided by the presence of Upper Cretaceous erratics in the Belhelvie area (Gibb, 1905; Hill, 1915). Smith (1984) studied the morphology of the Kippet Hills, concluding that they were deposited by a subglacial stream where it emerged from the decaying ice into a standing body of water; the red clay was then deposited by glaciolacustrine sedimentation. A similar explanation was invoked for deposits at Strabathie to the south of the Kippet Hills by Thomas (1984).

The abundant occurrence of late Tertiary to Early Pleistocene erratics in a limited area around the Kippet Hills led Sutherland (1984a) to suggest that deposits of that age were likely to occur *in situ* in the immediate vicinity. Hall and Connell (1991) have suggested derivation of the shells from the Early to Middle Pleistocene Aberdeen Ground Formation, which occurs offshore (Stoker *et al.*, 1985).

Kippet Hills is important both for glacial landforms and stratigraphy. Although much of the interest in the site has centred on the sediments, Kippet Hills does provide a particularly fine assemblage of ice-contact glaciofluvial landforms. These form part of the Red 'Series' deposits in north-east Scotland (see Hall, 1984b), associated with ice moving northwards along the North Sea coast and also inland during the Late Devensian. They are important in contributing to the geomorphological and sedimentological evidence for interpreting the pattern of glaciation and deglaciation of the area during the Late Devensian. In particular, it appears from the evidence at Kippet Hills and other sites (see Hall, 1984b) that the inland ice had receded at the time of expansion and wastage of the coastal ice (Hall, 1984b) and that ice-marginal lakes were present (see Bremner, 1916b; Murdoch, 1977; Merritt, 1981; Hall, 1984b; Thomas, 1984; Thomas and Connell, 1985). Kippet Hills is also notable for an assemblage of fossil shells unique in Scotland that were derived from the floor of the North Sea in the immediate vicinity and reworked into glaciofluvial deposits as ice moved onshore during the last glaciation.

Conclusion

Kippet Hills is important for a series of glacial meltwater landforms and deposits formed by the last (Late Devensian) ice-sheet (approximately 17,000–16,000 years ago). These include an esker ridge, kames, terraces and kettle holes. The deposits have a long history of research and have featured in most studies of the glaciation of north-east Scotland. In addition to their geomorphological interest, the deposits at Kippet Hills are noted for their content of fossil molluscan shells and erratic material (stones of rock types not occurring locally) derived apparently from the sea floor and carried onshore by the ice. The landforms and deposits at Kippet Hills contribute significant evidence for interpreting the pattern of glaciation and deglaciation in this part of Scotland.

MUIR OF DINNET

J. E. Gordon

Highlights

The landforms at Muir of Dinnet include an assemblage of meltwater channels, eskers and related deposits. These are important in demonstrating the mode of deglaciation of the Late Devensian ice-sheet and the effects of topographic controls on the pattern of ice wastage. In addition, pollen and plant macro-fossils preserved in the sediments that infill the floor of Loch Kinord provide a detailed record of vegetational history and environmental changes during the Lateglacial and Holocene.

Introduction

The Muir of Dinnet site (centred on NJ 430000) occupies an area (*c.* 22.9 km^2) in the south-west corner of the Howe of Cromar, one of a number of major topographic basins in north-east Scotland, and part of the River Dee valley east of Ballater. It is important on three main accounts: first, for its fine assemblage of glaciofluvial landforms demonstrating the progressive downwastage of the Late Devensian ice-sheet (Clapperton and Sugden, 1972; Sugden and Clapperton, 1975); second, in a historical context as the site of the supposed Dinnet Readvance ice limit (Synge, 1956); and third, for the Lateglacial and Holocene pollen and plant macrofossil records preserved in the sediments of Loch Kinord (Vasari and Vasari, 1968; Vasari, 1977). The geomorphology of the area has a long history of investigation (Jamieson, 1860b, 1874; Barrow *et al.*, 1912; Bremner, 1912, 1916a, 1920, 1925a, 1931; Charlesworth, 1956; Synge, 1956; Sissons, 1967a; Clapperton and Sugden, 1972; Sugden and Clapperton, 1975; Maizels, 1985).

Glacial and glaciofluvial landforms

Description

The main glacial landforms of the area were described by Bremner (1931), Clapperton and Sugden (1972) and Sugden and Clapperton (1975). They include eskers and meltwater channels in the River Dee valley between Milton of Tullich and Cambus o'May and on the eastern flank of Culblean Hill, an extensive area of dead-ice topography comprising kames and kettle holes around Lochs Davan and Kinord, and spreads of outwash gravels extending eastwards from Cambus o'May across the Muir of Dinnet and eastwards from the two lochs (Figure 8.10). Additional landforms of note are the roches moutonnées on Cnoc Dubh, and river terraces are present along the margins of the River Dee. Palaeochannels are extensively developed on terrace surfaces (Bremner, 1931) and also down-valley to the east of Dinnet (Maizels, 1985; Maizels and Aitken, 1991). Various sections are present in small pits in the glaciofluvial deposits but have not been described in detail, and in sections along the River Dee the glaciofluvial deposits are seen resting on till.

Interpretation

Jamieson (1860b) first described gravel mounds in the Dee Valley east of Ballater and a great spread of water-rolled gravel extending across the Muir of Dinnet. Initially he explained the deposits as the product of marine processes (Jamieson, 1860b), but later reinterpreted them as moraines and an outwash spread representing a halt stage in the gradual retreat of the last ice-sheet from a limit on the coast at Aberdeen (see Nigg Bay) into the centre of the Cairngorms (Jamieson, 1874).

Barrow *et al.* (1912) noted Jamieson's observations and the gravelly nature of the moraines. They related the moraines and the associated outwash spread to a valley glacier debouching from the constriction of the Dee Valley at the Muir of Dinnet during a phase of local valley glaciation after the ice maximum episode.

Bremner (1912, 1920, 1931) subsequently described the deposits between Ballater and Dinnet in greater detail, concluding that they represented a distinct stage of valley glaciation equivalent to Geikie's Fourth Glacial Stage (Geikie, 1894). He based this conclusion on the interpretation of certain landforms, such as lateral moraines and marginal meltwater channels along the valley sides, moraines in the valley floor that were fresher than those down the valley and contrasts in the lithology and composition of two superimposed 'tills'. However, there was no conclusive evidence to suggest complete deglaciation or an interglacial period immediately prior

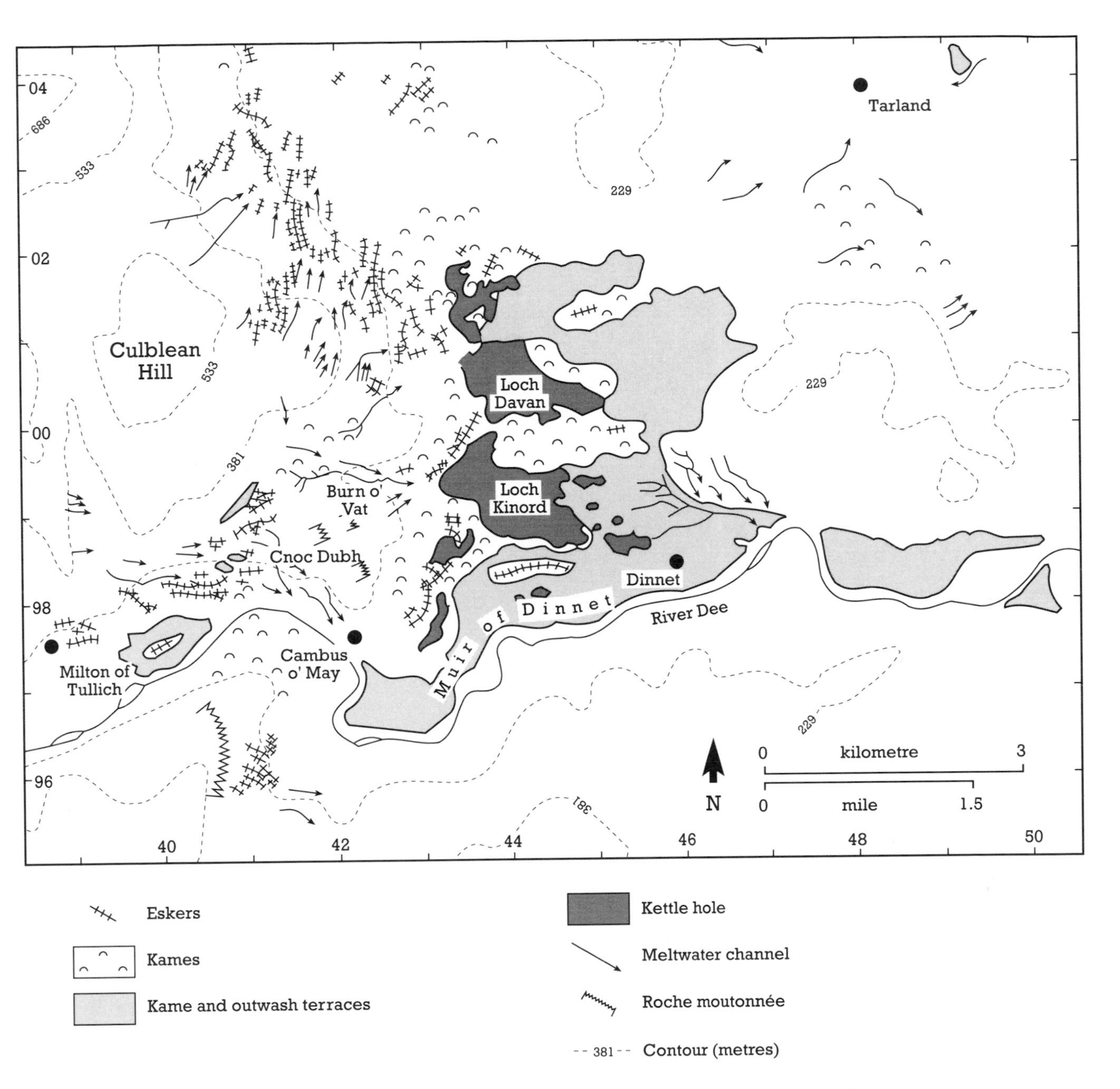

Figure 8.10 Geomorphology of the Muir of Dinnet (from Clapperton and Sugden, 1972).

to the valley glaciation.

Charlesworth (1956) believed the Dinnet deposits and landforms marked the local limit of his Stage M, a Lateglacial ice readvance equivalent to the Loch Lomond Readvance of Simpson (1933). Synge (1956), too, reached the conclusion that 'a massive terminal moraine' at Dinnet marked the limit of what he called the Dinnet Readvance, the local equivalent of the Loch Lomond Readvance. Sissons (1965, 1967a), however, correlated the Dinnet deposits with the Perth Readvance. He suggested they reflected the rapid downwasting of ice leading to stagnation and the formation of dead-ice topography following this readvance.

Subsequent detailed mapping in the Dinnet area by Clapperton and Sugden (1972) led them to dismiss the idea of an ice readvance in the Dee Valley. Instead, they related the assemblage of landforms and deposits to meltwater drainage in a progressively downwasting ice-sheet, concluding that their remarkable concentration near Dinnet was explained by the topography of the area, which allowed a large mass of ice to become isolated from the main ice-sheet in the Tarland

Basin and to downwast *in situ*. This interpretation is supported by several lines of evidence, subsequently summarized in Sugden and Clapperton (1975). First, many of the features formerly described as moraines are in fact of glaciofluvial origin; they form complex, interlinked systems of eskers, kames, kettle holes and meltwater channels. Second, many of the meltwater channels formerly interpreted as ice-marginal features display typical characteristics of subglacial channels. Third, the outwash spread east of Cambus o'May is not associated with a terminal moraine but appears to reflect extensive glaciofluvial deposition around stagnant ice blocks contemporaneously with the formation of eskers and kames within the ice west of Loch Kinord. Fourth, and most importantly, there is a progressive change from ice-directed channels and deposits on the higher slopes to topographically directed channels at lower levels. The subglacial ice-directed features follow the former regional ice-surface gradient, which trends north and north-east out of the Dee valley and are best displayed on the northern and higher eastern flanks of Culblean Hill (Figure 8.10). At lower levels on the east side of Culblean Hill they trend across and downslope into the Tarland Basin, reflecting increasing topographic influence on meltwater drainage as the ice downwasted. On the south side of Culblean Hill, eskers and channels at higher levels indicate meltwater flow up through the col with Cnoc Dubh (NJ 421991) and down on to the floor of the basin, notably via the Burn o'Vat channel (NJ 435996) with its spectacular pothole (Bremner, 1912, 1916a, 1925a). At lower levels, however, they follow the Dee Valley. During a later stage of ice decay an ice mass became stagnant in the lee of the Cnoc Dubh spur, with subsequent formation of kames, kame terraces, kettle holes and outwash gravels. Lochs Davan and Kinord are two particularly impressive kettle holes associated with this final stage of ice decay, being about 0.6 km and 1 km in diameter, respectively. Complete deglaciation had occurred at least by 11,520 ± 220 BP (HEL–418), the oldest date obtained from a core taken from Loch Kinord by Vasari (1977), and probably much earlier, since the Loch Builg area farther to the west is inferred to have been deglaciated before 12,000 BP (Clapperton *et al*., 1975).

In the context of this latest interpretation of events in the Dinnet area it should be noted that Bremner (1920) had, much earlier, recognized the evidence for downwasting ice and the presence of a residual mass of dead ice in Loch Kinord isolated by the form of the topography. This pattern of ice-sheet decay in the Late Devensian is typical of many parts of Scotland and northern England (see the Cairngorms) but is particularly well-exemplified at Muir of Dinnet where the relationships between the assemblage of landforms are clearly seen within a relatively compact area. Not only are individual landforms well-developed (meltwater channels, eskers, kames, kettle holes, terraces), but the overall continuum of features makes Muir of Dinnet an outstanding area for geomorphology. The site also illustrates particularly well the evolution of a glacial drainage system during ice-sheet downwastage, demonstrating clearly the pattern of glacial and topographic controls. The close association of meltwater channels and eskers is also of significant interest, and offers opportunities for detailed study and reconstruction of glacier dynamics and hydrological characteristics (see Shreve, 1972, 1985a, 1985b).

Lateglacial and Holocene vegetation history

Description

The pollen and plant macrofossil sequences from Loch Kinord are of particular interest for the almost complete record they provide of the Lateglacial and early Holocene vegetation history of the area. Vasari and Vasari (1968) described the sediments (a sequence of organic lake muds and silts) and organic contents of cores from the loch, and correlated the sequence of pollen zones (Figure 8.11) they identified with the Jessen–Godwin scheme. Recognizing that the respective pollen zones might not be synchronous because of regional variations in environmental factors, Vasari (1977) obtained radiocarbon dates (HEL–174, and HEL–418 to HEL–421) to provide a geochronometric framework for the Lateglacial stratigraphy of the site. On this basis he was able to correlate the zonation of the Loch Kinord pollen record with the conventional British scheme and also with the continental sequence of chronozones.

Figure 8.11 Loch Kinord: relative pollen diagram showing selected taxa as percentages of total land pollen (from Vasari, 1977).

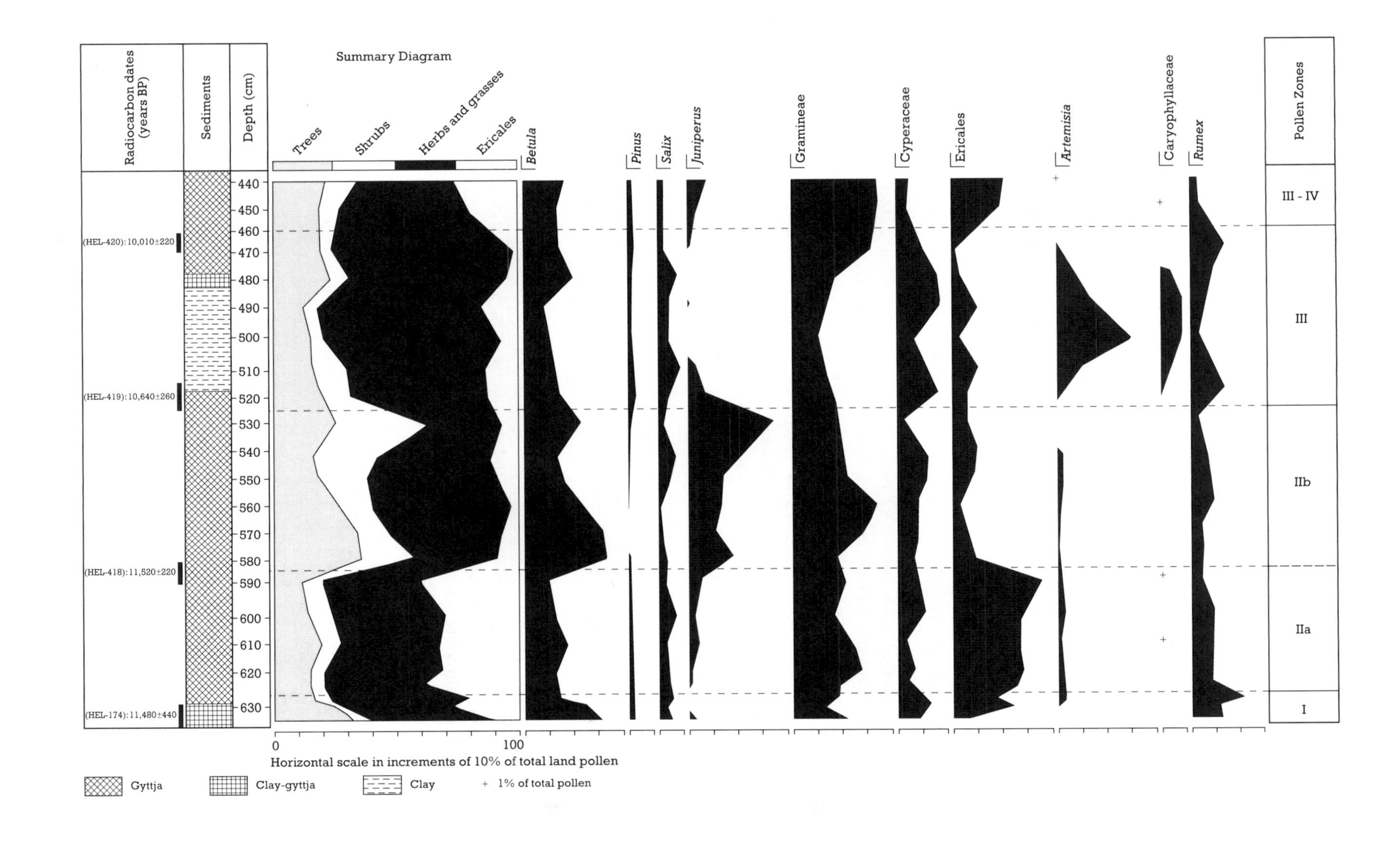

Radiocarbon dates (years BP)
Sediments
Depth (cm)
Summary Diagram
Trees
Shrubs
Herbs and grasses
Ericales
Betula
Pinus
Salix
Juniperus
Gramineae
Cyperaceae
Ericales
Artemisia
Caryophyllaceae
Rumex
Pollen Zones
III - IV
III
IIb
IIa
I
(HEL-420): 10,010±220
(HEL-419): 10,640±260
(HEL-418): 11,520±220
(HEL-174): 11,480±440
440
450
460
470
480
490
500
510
520
530
540
550
560
570
580
590
600
610
620
630
0
100
Horizontal scale in increments of 10% of total land pollen
Gyttja
Clay-gyttja
Clay
+ 1% of total pollen

Interpretation

The main features of the vegetation succession at Loch Kinord are as follows (Vasari and Vasari, 1968; Vasari, 1970, 1977). Open vegetation (Zone I) dominated by *Rumex* is followed by more closed vegetation (Zone II) in which a succession from *Rumex–Empetrum* to *Juniperus–Betula* assemblages occurs. Climatic deterioration is then thought to be reflected in the dominance of *Juniperus* over *Betula* (latter part of Zone II), which, as it progressed, led to an impoverished flora (Zone III; Loch Lomond Stadial) with *Salix*, *Artemisia* and *Rumex* prominent, although tree birch remained present. A transitional zone (III–IV; Loch Lomond Stadial–early Holocene) shows successive maxima of *Empetrum*, *Juniperus* and *Betula*. Light birch forests then succeeded park tundra (Zone IV), followed by an expansion of *Corylus* (Zone V). In a marked change *Pinus* becomes the dominant tree pollen type in the latter part of Zone VI, and *Ulmus* appears for the first time. *Alnus* then increases in frequency (Zone VIIa) and *Ulmus* declines (Zone VIIb). From the latter part of Zone VI to Zone VIII, pine–birch–alder forest assemblages prevail.

Vasari (1977) obtained a date of 11,520 ± 220 BP (HEL–418) for the Zone IIa/IIb boundary and speculated that the Zone I/II boundary might correlate with the Older Dryas/Allerød chronozone boundary (11,950–11,800 BP). The Zone II/III boundary was dated at 10,640 ± 260 BP (HEL–419), and although younger than the Allerød/Younger Dryas chronozone boundary (11,000 BP), this date is comparable to dates from similar stratigraphic horizons at sites in Scotland and northern England (Vasari, 1977). The Zone III/Zone III–IV boundary was dated at 10,010 ± 220 BP (HEL–420) and was placed at the rise of *Empetrum* at the start of the Holocene. A further date of 9,820 ± 250 BP (HEL–421) was obtained for the Zone III–IV/Zone IV boundary between the early Holocene juniper and birch maxima.

The vegetation sequence described from Loch Kinord has been discussed in its wider regional context by Vasari and Vasari (1968), Vasari (1970, 1977), Gunson (1975) and O'Sullivan (1975). The Lateglacial pollen record at Loch Kinord broadly parallels that found elsewhere in north-east Scotland, although the colder phase interrupting the Lateglacial Interstadial at several other sites is absent at Loch Kinord. In north-east Scotland as a whole radiocarbon dates suggest that the climatic deterioration of the Loch Lomond Stadial started later than in central and western Scotland, whereas the stadial phase was of much shorter duration (Vasari, 1977).

The Holocene vegetation history recorded in the deposits at Loch Kinord was reconstructed by Vasari and Vasari (1968) using the Jessen–Godwin scheme of pollen zonation. In transition Zone III–IV, the pollen diagram shows successive peaks in *Empetrum* and *Juniperus*, and in Zone IV, in *Betula*, indicating a development from open park-tundra to birch forest. In Zone V *Corylus* spread into the area and reached its maximum; *Quercus* and *Ulmus* also appear in the pollen spectra in this zone. During the earlier part of Zone VI, birch–hazel forest continued to predominate, but later *Pinus* became the dominant tree species. At the start of Zone VII, *Alnus* expanded, although pine, particularly, and birch continued to dominate the tree pollen. In Zone VIII, *Alnus* expanded further at the expense of birch and pine. Overall, the Holocene vegetation sequence at Loch Kinord shows greater affinity with that developed in upper Deeside and Strathspey (see Abernethy Forest) than with lowland Aberdeenshire, particularly in the expansion and subsequent predominance of pine in the middle Holocene (Vasari and Vasari, 1968; Gunson, 1975; O'Sullivan, 1975; Birks, 1977; Edwards, 1978). Evidence from nearby Loch Davan and Braeroddach Loch indicates human impact on the vegetation of the area starting around 5300 BP, followed by a series of clearance and regeneration episodes (Edwards, 1978, 1979b; Edwards and Rowntree, 1980).

Conclusion

Muir of Dinnet is noted for its assemblage of glacial meltwater landforms (notably meltwater channels and eskers). These were formerly interpreted in terms of a valley glacier readvance, but are now thought to relate to the pattern of deglaciation of the last ice-sheet (approximately 14,000–13,000 years ago). The landforms illustrate clearly how the glacial drainage system developed and, particularly, how it was increasingly influenced by the form of the underlying topography as ice wastage progressed. In addition to this geomorphological interest, the site is important for the pollen and larger plant remains preserved in the sediments of Loch Kinord. These record an almost complete sequence of the vegetation history and environmental changes in this area of north-east Scotland during the Lateglacial and the Holocene (approximately the last 13,000 years). Muir of Dinnet is therefore an important reference area for interpreting the patterns of landscape change both at the end of, and following, the last glaciation.

PHILORTH VALLEY

D. E. Smith

Highlights

The sub-surface deposits in the Philorth Valley include a sequence of estuarine sediments and peat. These provide important evidence for interpreting the pattern of relative sea-level changes during the Holocene. Because the Philorth Valley is located towards the margin of the area of isostatic uplift, the deposits there have preserved a more detailed record of coastal changes than sites elsewhere.

Introduction

The Water of Philorth is a small stream draining through a landscape of glacial deposits in the district of Buchan, north-east Scotland. The area (NK 011635) of significant interest lies at the northern end of the valley, west of the farm of Milltown, 3.5 km south of Fraserburgh. In this area, Smith *et al.* (1982) have identified a sequence of deposits that record changes in relative sea level during the middle and late Holocene, including a transgressive episode not found elsewhere in Scotland. This record is important because of its location near the periphery of the area of isostatic uplift in Scotland.

Description

The Philorth Valley is, for the most part, narrow and unremarkable, but in the final 3 km of its course it opens out and an extensive flat area occurs before the stream cuts through a rampart of sand dunes to reach Fraserburgh Bay. The surface of the flat area is largely composed of a brown, silty clay with some areas of sand. A sharp break of slope occurs where this surface meets the surrounding rising ground.

The area was studied by Smith *et al.* (1982). They mapped the surface deposits and found that the brown, silty clay surface lies at a consistent altitude of between +2.2 m and +3.2 m OD, but rises to over +5 m OD where it becomes restricted at its southern margin, near The Neuk (NK 002624). Boreholes across the area proved a succession of sands and gravels overlain by peat, then brown, silty clay, and discontinuous sand above (Figure 8.12). Within the peat two minerogenic layers occur, an upper layer of micaceous sandy silt, which tapers up-valley and a lower layer of grey sand, irregularly distributed (Figure 8.12). The surface of the micaceous sandy silt was found to be relatively consistent in elevation at +1.22 m to +2.26 m OD, but the grey sand was found to vary between −1.15 m and +1.80 m OD; at Mains of Philorth, two layers of grey sand were recorded near the side of the valley.

The sequence of deposits is best represented in the area between Milltown and Philorth Home Farm. Here Smith *et al.* (1982) undertook pollen analysis of the deposits at one site and radiocarbon dates on part of the sequence at two sites. Radiocarbon dates were also obtained from a site further up-valley (Figure 8.12 and Table 8.1).

At the pollen site, Smith *et al.* (1982) found that through the basal peat and intervening minerogenic horizons to 0.4 m below the surface of the brown, silty clay, where sampling ended, the vegetational sequences span the early to middle Holocene. The basal peat, grey sand, and peat below the micaceous sandy silt are associated with early Holocene sequences indicating scattered stands of *Betula* and *Pinus*, with *Corylus* and *Salix* in the general area; the valley floor being subject to a fluctuating water table, with sedges, grasses, and a variety of aquatic communities including *Lemna*, *Potamogeton* and *Typha angustifolia*. The grey sand layer is associated with a temporary decline in aboreal pollen. The top of the peat above the grey sand layer, together with the overlying micaceous sandy silt and much of the peat above it, are associated with increasing values of *Quercus* and particularly of *Alnus*; the silt is associated with high values of *Pinus* and *Quercus*, together with a concentration of *Plantago maritima* and significant representation of freshwater aquatics, notably *Lemna* and *Potamogeton*. The top of the peat and the overlying brown, silty clay yielded pollen indicating *Betula–Quercus* woodland,

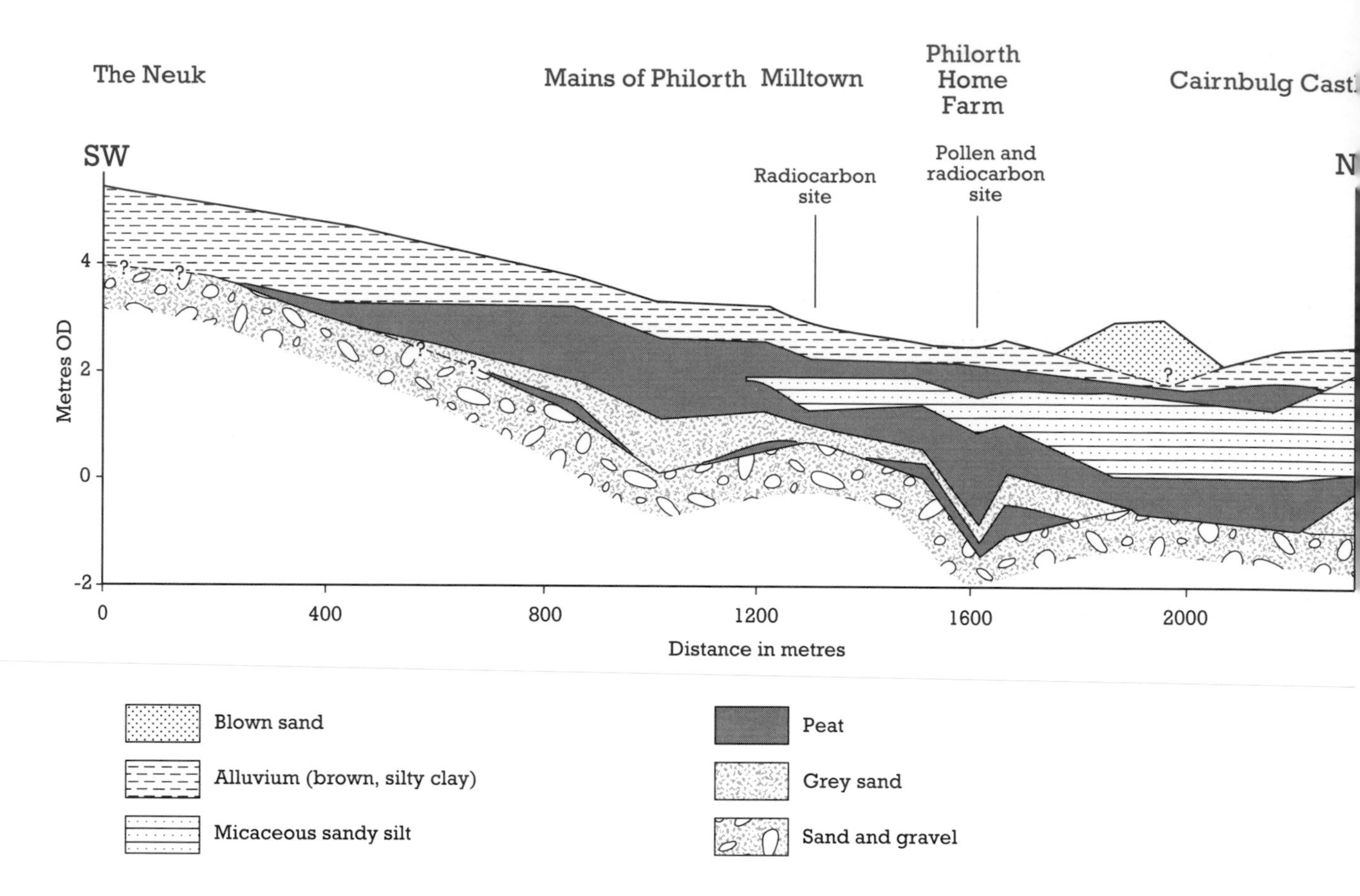

Figure 8.12 Section along the length of the Lower Philorth Valley showing the sequence of sediments (from Smith *et al.*, 1982).

with *Alnus* and local freshwater aquatic communities.

At the pollen site (Philorth Home Farm) and Milltown, peat above and below the micaceous, sandy silt was dated. At Mains of Philorth the sampling site was beyond the up-valley limit of the micaceous, sandy silt, and dates were obtained on peat from below the brown, silty clay and above and below two layers of grey sand (Table 8.1).

Interpretation

Both the pollen analysis and radiocarbon dates of Smith *et al.* (1982) indicate that the deposits which fill the lower end of the Philorth Valley accumulated during the Holocene. It seems likely that the grey sand layer is the deposit of a series of tsunami waves that struck the east coast of Scotland some 7000 years ago (e.g. Dawson *et al.*, 1988). In recent work (1990) the layer has been traced up-valley as far as The Neuk, where it tapers out at an altitude of 2.4 m OD. The two grey sand layers dated at Mains of Philorth lie at the side of the valley and may include colluvial material. One of them (the upper one) may equate with the extensive grey sand layer, but their provenance is uncertain.

The age and provenance of the micaceous sandy silt are better known. Both the pollen and the radiocarbon dates concur in indicating a middle Holocene age. The association of *Plantago maritima* pollen with the layer, together with the increase in representation of *Pinus* and *Quercus*, characteristic of a littoral depositional environment (Traverse and Ginsberg, 1966), indicate that the layer is of marine–estuarine origin, and Smith *et al.* maintain that this is the local expression of the Main Postglacial Transgression. They interpret the radiocarbon evidence as showing that the transgression was under way in the area by 6300 ± 60 BP, that it culminated after 6096 ± 75 BP, and that peat growth on the surface of the resulting deposit (the micaceous sandy silt) had commenced by 5700 ± 90 BP.

Table 8.1 Radiocarbon dates from sites in the Philorth Valley (after Smith *et al.* 1982)

Location	Details of sample	Altitude (metres OD) of sample at contact with minerogenic layer	14C date (years BP)	Laboratory number
Philorth Home Farm	Bottom 2 cm of peat above micaceous sandy silt	1.48	5700 ± 90	SRR–1660
Philorth Home Farm	Top 2 cm of peat below micaceous sandy silt	0.82	6300 ± 60	SRR–1661
Milltown	Bottom 2 cm of peat above micaceous sandy silt	1.82	5140 ± 60	SRR–1686
Milltown	Top 2 cm of peat below micaceous sandy silt	1.11	6095 ± 75	SRR–1687
Mains of Philorth	Top 1cm of peat below brown silty clay	2.59	4760 ± 60	SRR–1655
Mains of Philorth	Bottom 2 cm of peat above grey sand	1.51	6150 ± 250	SRR–1656
Mains of Philorth	Top 2 cm of peat below grey sand	1.47	6885 ± 90	SRR–1657
Mains of Philorth	Bottom 2 cm of peat above grey sand	1.40	7510 ± 120	SRR–1658
Mains of Philorth	Top 2 cm of peat below grey sand	1.34	8465 ± 95	SRR–1659

The brown silty clay present at the surface of the area is today often partially inundated during high tides, and would be even more so affected were it not for dykes along the lower Philorth. Smith *et al.* (1982) suggest that this deposit began to accumulate at about 4760 ± 60 BP as the result of a rise in relative sea level, and that it is still accumulating in places. The silty clay is essentially an estuarine deposit which becomes increasingly alluvial up-valley.

The deposits of the lower Philorth Valley contain evidence of tsunami activity dated at *c.* 7000 BP, and two major marine incursions, the Main Postglacial Transgression and the later one, after 4700 BP, that formed the surface mudflats. The earlier marine incursion in which the micaceous, sandy silt was deposited, culminated between 6096 ± 75 BP and 5700 ± 90 BP. It appears to have been the Main Postglacial Transgression, but is somewhat younger than that event further south, where ages of about 6200 BP in the Tay estuary (see Pitlowie) (Smith *et al.*, 1985b) and around 6800 BP in the Western Forth Valley (see below) (Sissons, 1983a) have been proposed. The Philorth dates, therefore, may be evidence for a time-transgressive shoreline (Smith *et al.*, 1983). Such diachroneity would accord with theories on the formation of relict shorelines in isostatically affected areas. Altitudes on the surface of the micaceous sandy silt (about +2 m OD) demonstrate the decline in altitude of the Main Postglacial Shoreline from the maximum altitudes of over 14 m OD at the head of the Forth Valley, near the centre of isostatic uplift (see Sissons, 1976b). In a recent publication, Cullingford *et al.* (1991) have identified detailed isobase patterns for the Main Postglacial shoreline in eastern Scotland. They place the Philorth Valley below the 2 m isobase and therefore close to the margins of Holocene isostatic uplift.

The second marine incursion identified in this area, in which the brown, silty clay was deposited, took place about 4760 ± 60 BP. It is unlikely to be found in many areas of Scotland. The pace of isostatic uplift over most of Scotland during the late Holocene would probably have exceeded regional sea-level rise. Only areas towards the periphery of the uplifted area would register the more minor fluctuations of the sea surface after the Main Postglacial Transgression. The Philorth Valley site is therefore uniquely valuable for studies of Holocene relative sea-level change in Scotland, and will repay further scientific enquiry.

Conclusion

The sediments in the Philorth Valley record sea-level changes in north-east Scotland during the Holocene. They show that a major coastal flood occurred about 7000 years ago and that there were two subsequent episodes when sea level was relatively higher than at present. This evidence allows comparisons with sites elsewhere in Scotland and contributes towards establishing the wider pattern of sea-level changes during the Holocene. The particular significance of the Philorth Valley lies in its location towards the margin of the area of isostatic uplift (the recovery of the Earth's crust following its depression by the weight of the ice-sheet); as such it preserves a more sensitive record of sea-level change than more central areas that have undergone greater uplift (where the ice was thicker) following the melting of the last ice-sheet.

Chapter 9

Eastern Grampian Mountains

INTRODUCTION

D. G. Sutherland

The eastern Grampian Mountains are considered here as the highland areas to the east of the Tay–Tummel–Truim–Spey through valley (Figure 9.1). This valley separates the western mountain areas, characterized by intense glacial erosion, and the eastern mountain plateau country, where glacial erosion has produced only specific features superimposed on an easily recognizable pre-existing landscape. There are three principal mountain groups in this area, the Gaick Plateau, the Cairngorms and the south-east Grampians around Lochnagar. Each of these is characterized by high-level plateau surfaces, which are most impressively developed in the Cairngorms. These surfaces are widely acknowledged as having formed prior to glaciation (Fleet, 1938; Linton, 1949b, 1959; Sissons, 1967a; Sugden, 1968; Hall, 1983) and carry apparently relict pre-glacial features, such as tors (Linton, 1950a, 1955) and decomposed bedrock (Barrow *et al.*, 1913; Sugden, 1968; Hall and Mellor, 1988).

The presence of these apparent pre-glacial features has resulted in certain authors suggesting that parts of the Cairngorms may never have been glaciated, but such an idea was effectively refuted by Sugden (1968). However, the eastern mountains do demonstrate eloquently the selectivity of glacial erosion, for the plateaux are frequently flanked by spectacular glacial breaches, such as the Lairig Ghru and Glen Tilt, or are bitten into by corries as on Lochnagar and on the northern flanks of the Cairngorms. The form of the corries and glacially eroded rock walls has been studied by Haynes (1968), Sugden (1969) and Dale (1981), the last author demonstrating a lower frequency and amplitude of rock walls in the eastern Grampians than in the western mountain groups. The altitude of the corries and the base of rock walls is also higher in the eastern mountains than the west and this has been related to the precipitation distribution in Scotland (today and, by inference, in the past) by Linton (1959).

With the exception of the landforms of glacial erosion, which have developed during multiple periods of both local and ice-sheet glaciations, the Quaternary history of the eastern Grampians, as presently known, relates only to the Late Devensian and the Holocene. During the Late Devensian ice-sheet glaciation, the western part of the area was covered by ice emanating from the Rannoch Moor area, and erratics of Rannoch granite can be found on the flanks of the Gaick plateau (Barrow *et al.*, 1905, 1913) and into the Truim Valley (Barrow *et al.*, 1913). Ice from the west also carried schistose erratics on to the flanks of the Cairngorms to an altitude of up to *c.* 840 m OD (Hinxman and Anderson, 1915; Sugden, 1970). Within the principal mountain masses, however, external erratics occur only sporadically, and it is probable that local ice masses developed which were sufficiently powerful to exclude ice from western sources. Erratics from these local areas were carried to the north-east, east or south-east (Sutherland, 1984a).

Throughout most of the valleys of the region there are abundant glaciofluvial landforms and deposits, as in Glen More at the foot of the Cairngorms (Sugden, 1970; Young, 1974), at the mouth of Glen Feshie (Young, 1975a) and along the Dee Valley (Sugden and Clapperton, 1975) (see Muir of Dinnet, Chapter 7). These areas of ice-decay deposits are typically accompanied by sequences of meltwater channels on the adjacent slopes (Sugden, 1968; Young, 1974) and, in places, extensive outwash terraces as in Glen Feshie (Young, 1976; Robertson-Rintoul, 1986b).

The timing of ice-sheet wastage has not been established in detail, but the occurrence of a considerable number of enclosed basins that are known to contain lacustrine sediments deposited during the Lateglacial Interstadial has demonstrated that much of the area was deglaciated prior to 13,000 BP. Important among these sites are those of Loch Etteridge (Sissons and Walker, 1974) and Abernethy Forest (Vasari, 1977; Birks and Mathewes, 1978). The vegetational succession in the eastern Grampians during the Lateglacial Interstadial showed some differentiation between the valleys of the northern and central parts and those of the south-east. After an initial phase in both areas of pioneer grass- and sedge-dominated communities, in the central mountain area there developed a shrub tundra dominated by *Empetrum* with stands of birch and willow (Walker, 1975b; Birks and Mathewes, 1978). In the south-eastern valleys there was a grassland with juniper, dwarf birch and willow and, in more sheltered areas, tree birch (Walker, 1975b, 1977, 1984b; Lowe and Walker, 1977). Higher-level sites, such as Morrone, reveal a sparser, less differentiated vegetation on the upper slopes with the development of moss heaths and grasslands (Huntley, 1976, 1981).

The Lateglacial Interstadial was terminated by a

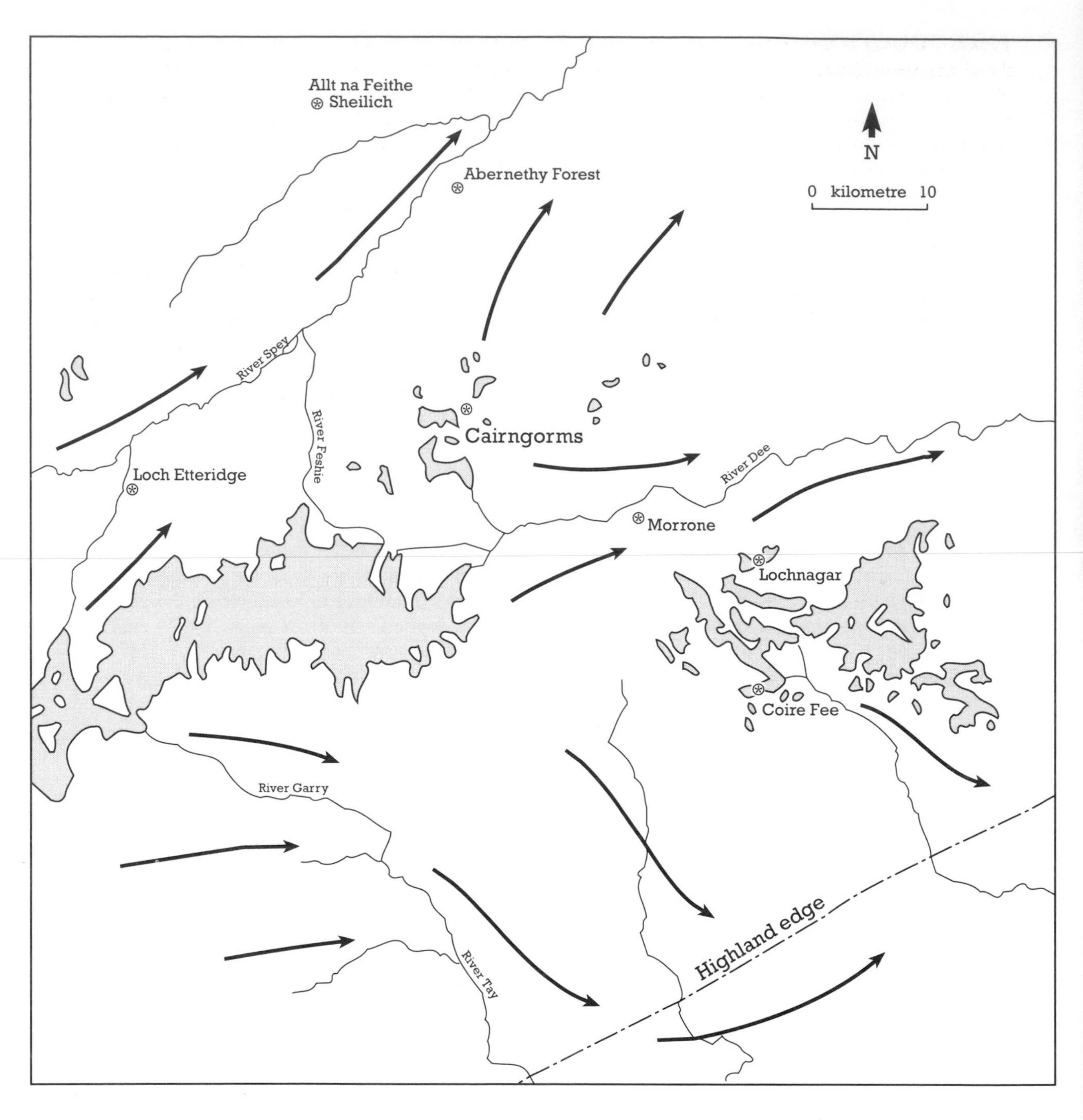

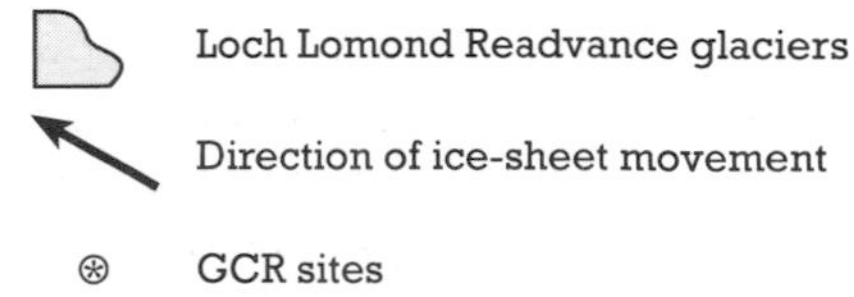

Figure 9.1 Location map of the eastern Grampian Mountains. The limits of the Loch Lomond Readvance glaciers are from Sissons (1972a, 1974b, 1979f) and Sissons and Grant (1972).

return to severe climatic conditions during the Loch Lomond Stadial. There was a recrudescence of glaciers in the high valleys and corries and the development of ice-caps on the Gaick Plateau and in the south-east Grampians between Glen Muick and Glen Clova (Figure 9.1) (Sissons, 1972a,

1974b, 1979f; Sissons and Grant, 1972). Although there has been discussion as to the exact extent of these glaciers (Sissons, 1973a; Sugden, 1973a, 1980), many of the landforms which they produced are clear, such as the end moraines in the corries of Lochnagar (Sissons and Grant, 1972; Clapperton, 1986) and the Cairngorms (Sissons, 1979f; Rapson, 1985).

It was probably during the stadial that the large-scale periglacial features that mantle many of the mountain tops received their final form. Most impressive of these features are the boulder sheets and lobes that occur on the granite of the Cairngorms and Lochnagar (King, 1972; Shaw, 1977). Additional periglacial features that formed at this time are the protalus ramparts and rock glaciers found in the Cairngorms (Sissons, 1979f; Ballantyne, 1984; Ballantyne and Kirkbride, 1986; Chattopadhyay, 1984). Enhanced fluvial activity associated with seasonal regimes was probably responsible for terrace development in areas such as Glen Feshie.

Vegetation at this time was dominated by open-habitat species and there were particularly high *Artemisia* pollen values in the sites investigated in Strathspey such as Abernethy Forest (Birks and Mathewes, 1978). These high values and their contrast with lower values in the valleys of the south-east Grampians (Walker, 1975b), have been interpreted as relating to variations in the snow-cover (Walker, 1975b; MacPherson, 1980), with Strathspey receiving very low precipitation during this period (Birks and Mathewes, 1978). Similar inferences have been made (Sissons and Sutherland, 1976; Sissons, 1980b) on the basis of variations in Loch Lomond Readvance equilibrium line altitudes throughout the region.

Following amelioration of the climate at the end of the Loch Lomond Stadial, the last glaciers melted and vegetation rapidly changed from open habitats through dwarf shrub and scrub communities (including a distinct phase of juniper dominance at around 10,000 BP or slightly later) to birch and hazel woodland in the valleys, with grass and heathland on the upper slopes by about 9,000 BP. Thereafter development of the woodlands again shows significant differences between the southern part of the region, where oak and elm dominated the forests in the valleys, and in the northern area, where pine became the principal tree species during the middle Holocene (see Allt na Feithe Sheilich, Abernethy Forest and the Cairngorms) (Birks, 1975; Birks, 1977).

One interesting aspect of vegetation development is the occurrence today of arctic–alpine species in certain of the mountains. Their presence raises the question of whether they have survived since the Lateglacial in certain favourable refuge habitats, or whether they died out in the early Holocene only to be reintroduced at a later date. The high-level pollen sites at both Coire Fee and Morrone provide direct evidence for the survival of these species throughout the Holocene.

Geomorphological activity has also continued during the Holocene, with small-scale periglacial features, such as turf-banked terraces and lobes, 'ploughing' boulders and patterned ground, forming at high altitudes (Chattopadhyay, 1986; Ballantyne, 1987a), and fluvial activity resulting in the formation of river terraces, debris cones and alluvial fans in the valley bottoms, as in Glen Feshie (Brazier, 1987; Robertson-Rintoul, 1986a, 1986b).

THE CAIRNGORMS

J. E. Gordon

Highlights

The Cairngorms is an area of outstanding importance for geomorphology. The interest comprises an exceptional assemblage of pre-glacial, glacial, glaciofluvial and periglacial features. Together these provide a great wealth of information for interpreting landscape evolution and environmental change in the uplands during the Quaternary.

Introduction

The Cairngorms massif, extending from Glen Feshie (NN 850960) in the west to Glen Builg (NJ 185055) in the east and from Glen More (NH 970100) in the north to Glen Dee (NN 970870) in the south, includes the largest area of high-level ground in Britain. It is one of the most outstanding mountain areas in Britain for its range of glacial, periglacial and pre-glacial landforms and deposits (Figure 9.2). Among the principal publications relating to the geomorphology and Quaternary history of the Cairngorms are those by Jamieson (1908), Barrow *et al.* (1912, 1913), Hinxman and Anderson (1915), Bremner (1929), Linton (1949a, 1950a, 1951a, 1954, 1955), Baird and Lewis (1957), Galloway

10
05
95
90
85
90
95
100
Loch Morlich
River Spey
300
Coire Laogh Beag
Strath Nethy
River Feshie
Glen Einich
Coire an Lochain
Cairngorm
Glen Avon
Loch Avon
Braeriach
Loch Einich
Glen Derry
Glen Feshie
Devil's Point
Derry Lodge
Glen Eidart
500
700
900
1100

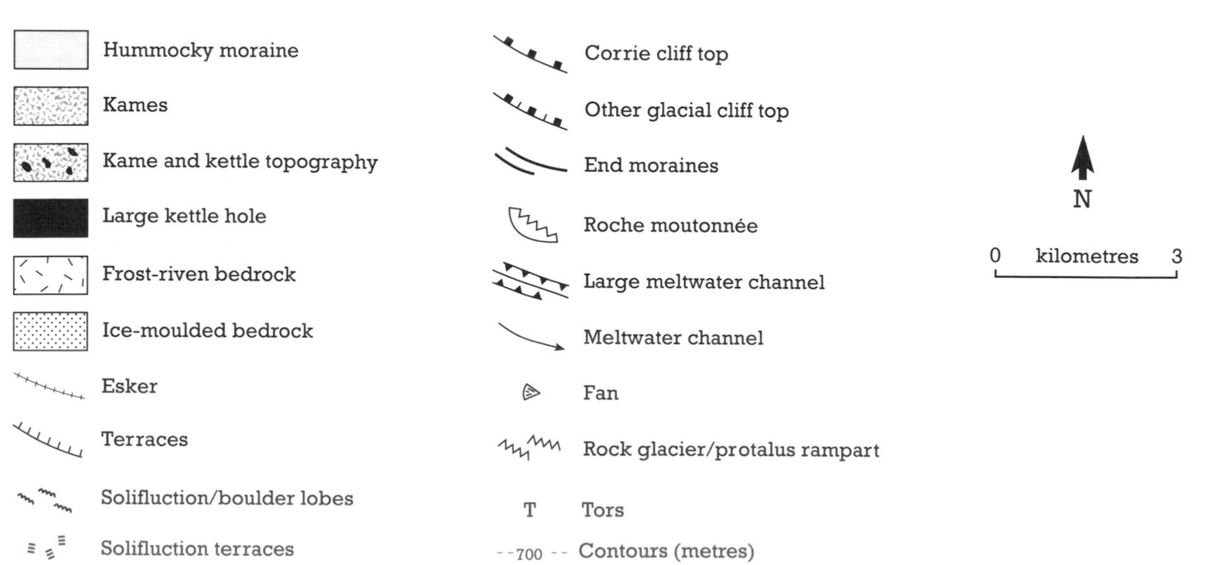

Hummocky moraine
Kames
Kame and kettle topography
Large kettle hole
Frost-riven bedrock
Ice-moulded bedrock
Esker
Terraces
Solifluction/boulder lobes
Solifluction terraces
Corrie cliff top
Other glacial cliff top
End moraines
Roche moutonnée
Large meltwater channel
Meltwater channel
Fan
Rock glacier/protalus rampart
T
Tors
700
Contours (metres)
N
0
kilometres
3

(1958), Pears (1964, 1968), Sugden (1965, 1968, 1969, 1970, 1971, 1977, 1983), Sissons (1967a, 1979f), King (1968, 1971a, 1971b, 1972), Birks (1969, 1975), Kelletat (1970a, 1970b, 1972), Young (1974, 1975a, 1975b), Clapperton *et al.* (1975), Dubois and Ferguson (1985), Rapson (1985), McEwen and Werritty (1988) and Bennett and Glasser (1991).

Description

Geology and pre-glacial landform elements

The Cairngorms massif largely comprises a granite pluton intruded during the Caledonian orogeny, in late Silurian–early Devonian times, into Precambrian Moine metamorphic country rocks. Recent work has shown the granite to be a discordant, stock-like mass and to consist of at least two intrusions, the more extensive Main Granite which is largely even-grained, and the Porphyritic Granite which occurs on the Carn Bàn Mór–Geal-charn ridge west of Loch Einich (Harry, 1965). The most detailed accounts of the solid geology of the area appeared in the early Geological Survey Memoirs (Hinxman, 1896; Hinxman and Anderson, 1915; Barrow *et al.*, 1912, 1913).

The broad outlines of the Cairngorms massif are characterized by two major morphological elements: undulating plateau surfaces and precipitous cliffs of corries and glacial troughs (Linton, 1950a; Sugden, 1968). The former are part of a suite of surfaces rising in 'steps' inland from the coast of north-east Scotland (Fleet, 1938; Walton, 1963). Two main breaks of slope at 760 m and 910 m OD are represented on the margins of the Cairngorms, and the higher summits rise gently above a third at 1070–1220 m OD (Sugden, 1968). Similar surfaces have long been recorded throughout the Highlands and Southern Uplands, either as vast 'tablelands' or as summit accordances (see Geikie, 1901; Peach and Horne, 1930; Godard, 1965). Although the origin of the surfaces and intervening breaks of slope is a matter of debate (Sissons, 1976b), they are generally held to be of pre-glacial, probably late Tertiary age. The hypothesis that the Cairngorm summits are part of a sub-Cenomanian surface, as proposed by Linton (1951b), has been effectively refuted by George (1966).

The essentially pre-glacial aspect of the surfaces is emphasized by a number of relict features of non-glacial origin associated with them, namely tors, decomposed granite, fluvial forms and 'pseudobedding' (sheet jointing) forms in the bedrock.

The Cairngorm tors are the finest in Scotland and are best seen in the north and east of the massif, on Beinn Mheadhoin (NJ 025017), Bynack More (NJ 042063) (notably the 30 m high Barns of Bynack) and Ben Avon (NJ 132018), where a series of forms occurs across the summit plateau (Figure 9.3). On Ben Avon, fantastically sculpted potholes occur, and other weathering forms can be found on Clach Bhàn (NJ 162054) (Hinxman, 1896; Hinxman and Anderson, 1915; Alexander, 1928). First described by Hinxman (1896) and Hinxman and Anderson (1915), the Cairngorm tors were interpreted by Linton (1955) in terms of a two-stage model as residuals of deep weathering during the Tertiary, subsequently exhumed during the Quaternary by solifluction or meltwater. Sugden (1968, 1974c) concurred with this interpretation, which is apparently supported by the presence of isolated pockets of deeply weathered bedrock on the plateaux surfaces (see below). The most detailed investigation and description of the tors is that of King (1968), who found that they occurred primarily in coarser-grained granite on plateaux and gentle to moderate slopes between 710 m and 1240 m OD. He concluded that they were formed by subsurface decomposition then exhumed and modified by periglacial frost action; chemical weathering later modified the surface forms and produced undercutting of the tors. Linton (1949b, 1950b, 1950c, 1952, 1954, 1955) believed that the tors had survived because the Cairngorm plateau had escaped at least the last glaciation, a view shared by Galloway (1958). However, Sugden (1965) found evidence of ice moulding on some of the tors, and metamorphic erratics surrounding the Argyll Stone (NH 905040), a small granite tor in the western Cairngorms. It is now believed that the entire Cairngorms were ice-covered during the last glaciation (see below), and Sissons (1976b) suggested that the preservation of the large tors in the north and east of the massif resulted from their relatively sheltered

Figure 9.2 Principal geomorphological features of the Cairngorm Mountains (sources include Sugden, 1968, 1970; Young, 1974, 1975a; Sissons, 1979f; J. E. Gordon, unpublished data).

Figure 9.3 Summit plateau of Ben Avon in the Cairngorms showing well-developed tors which appear to have survived glaciation. The adjacent slopes have been affected by periglacial processes and the development of solifluction lobes. (Photo: J. E. Gordon.)

location in relation to ice moving from the south-west.

In some cases an exclusively periglacial or cold-climate origin has been advocated for the development of tors similar to those on the Cairngorms (Palmer and Radley, 1961; Demek, 1964; Martini, 1969; Derbyshire, 1972); in others, a polygenetic origin has been advocated (Caine, 1967; Fahey, 1981; Söderman *et al.*, 1983). Tors preserved in glaciated areas have been reported in Britain from the Cheviots (Common, 1954b; Clapperton, 1970; Clark, 1970, 1971) and Pembrokeshire (John, 1973; Battiau-Queney, 1984), and from abroad in Tasmania (Caine, 1967), northern Finland (Kaitanen, 1969; Söderman *et al.*, 1983), Norway (Dahl, 1966), Somerset Island (Dyke, 1976, 1983), Ellesmere Island (Watts, 1981, 1983) and Baffin Island (Boyer and Pheasant, 1974; Sugden and Watts, 1977). Although some authors have argued that tors and the surfaces on which they stand have either not been glaciated themselves or lay beyond the limits of the last glaciation (Boyer and Pheasant, 1974; Dyke, 1976, 1983), Sugden and Watts (1977) suggested that tors could survive glaciation where slow-moving, cold-based ice was unable to exert sufficient tractive force to overcome the resistance of the intact, massive bedrock outcrops, and they drew analogies with the process of subglacial lodgement (cf. Boulton, 1975). Sugden (1983) subsequently reviewed the evidence for a pre-glacial origin for the Cairngorm tors and advocated such a hypothesis to explain their preservation. Although Dubois and Ferguson (1985) expressed a contrary view, the balance of evidence (see below), however, appears to favour a long period of landscape evolution extending back into the Tertiary, so that the present surfaces and slopes of the Cairngorm plateaux reflect the complex interaction of pre-glacial weathering, periglacial weathering and mass movement, and limited glacial action. Models such as those of Linton and Sugden appear to be valid in addressing the main elements of this polygenetic evolution, although they are probably oversimplified in terms of their spatial and temporal resolution. Elsewhere in Scotland tors occur on the granites of Lochnagar, Mount Keen, Broad Cairn, Bennachie and Ben Rinnes; on the syenites of Ben Loyal; the gabbros of the Insch basic intrusion at Cabrach; and the conglomerates of Morven in Caithness. However, there has been no thorough investigation of their ages or processes of formation, and not all may be exhumed features.

A second pre-glacial feature of the plateau surfaces is the decomposed granite. This is best exposed in a stream section near the head of Coire Raibert (NJ 001038) but can also be seen in a gully at the top of the Glen Avon cliffs (NH 997022). King (1968) reported examples in the headwalls of Coire an t-Sneachda (NH 995032) and Coire Bhrochain (NN 955995); other sites occur in stream sections on the Mòine Mhór, above the head of Gleann Einich, and on the flanks of Beinn Bhrotain (NN 455923) (A. M. Hall, unpublished data). Preliminary mineralogical investigations have revealed the presence of kaolinite in the Coire an t-Sneachda and Coire Bhrochain exposures (King, 1968), and kaolinite, gibbsite and hematite in the Coire Raibert exposure (Hall, 1983). The granular disintegration (arenization) of the rock and the assemblage of secondary minerals are similar to examples widely reported from deeply weathered bedrock, both in the Gaick area to the south (Barrow *et al*., 1913; Hall and Mellor, 1988) and on lower ground in north-east Scotland (see Hill of Longhaven), where such features have been generally ascribed to a long period of pre-glacial weathering (FitzPatrick, 1963; Basham, 1974; Hall, 1985, 1986; Wilson, 1985; Hall *et al*., 1989a). From a wider study of soil profiles on Scottish mountains, Mellor and Wilson (1989) concluded that the presence of gibbsite is a feature that pre-dates the last glaciation, but its precise time of formation is uncertain; it could have formed under humid, warm-temperate to subtropical conditions during the Tertiary and/or during Pleistocene interglacials and survived under cold-based ice with limited erosional capacity. The status of gibbsite, however, as an indicator of former warm environments is uncertain, as this mineral is also believed to form at the initial stages of rock breakdown (Hall *et al*., 1989a). Much further work is needed on the mineralogy and other characteristics of the weathering profiles in the Cairngorms before their significance can be properly assessed. In particular, more investigation is needed of the possible contribution of hydrothermal alteration to the breakdown of, and clay mineral genesis within, the granite (Hall, 1983). Additional study is also required of the possible development of clay minerals during chemical weathering beneath snow patches before a recent origin for such grusification can be ruled out (A.M. Hall, unpublished data).

The plateau surfaces of the Cairngorms are not flat but comprise smooth, rolling slopes and shallow fluvial valleys; for example Coire Raibert, Coire Domhain (NH 995023) and the valley of the Feithe Buidhe (NH 990015) on the Cairngorm–Ben Macdui plateau, and the valley of Caochan Dubh (NN 895947) on the Mòine Mhór plateau. Although such valleys probably owe part of their form to periglacial processes, they have clearly been little modified by the passage of ice and are abruptly truncated by cliffs at the plateaux margins. Such landscapes are comparable with parts of the Canadian Arctic archipelago (Sugden, 1978), although on a much smaller scale.

A fourth characteristic feature of the plateau surfaces is the 'pseudobedding' or sheet jointing present in the upper layers of the granite (Hinxman and Anderson, 1915; King, 1968; Sugden, 1968). It is particularly well seen in the tors and at the tops of many of the glacial cliffs; for example in Coire an Lochain (NH 985026) and above the Saddle (NJ 015033). Sugden (1968) noted that the spacing of the sheet joints increased with depth and that everywhere it lay parallel with the slope of the ground. He therefore concluded that since the sheet jointing conformed extensively with the detailed surface form of the plateaux, the pre-glacial surface over much of the Cairngorms was 'faithfully preserved', and he was able to complete a tentative reconstruction of its form. Sheet jointing parallel to pre-glacial surfaces has also been described from Dartmoor (Waters, 1954), New England (Jahns, 1943) and Maine (Chapman and Rioux, 1958) and is generally held to relate to stress unloading (Ollier, 1969). However, in the context of the Cairngorms, Sissons (1976b) has suggested an alternative possibility of intensive frost action under periglacial conditions.

Landforms and patterns of glacial erosion

Despite the presence of pre-glacial features it is now accepted that the entire Cairngorms massif was ice-covered during the Late Devensian glaciation, contrary to the views of Linton and Galloway outlined above. From the lack of metamorphic erratics in the central Cairngorms Bremner (1929) and Sissons (1976b) concluded that the massif was the site of an independent ice dome, albeit constrained by external ice streams on all but its north-eastern side. A similar view was put forward by Sugden (1970) for at least an early phase in the deglaciation of the mountains, although he thought that at an earlier time (not necessarily the Late Devensian), the entire massif had been overwhelmed by ice from the south-west, in order to explain the pattern of glacial troughs.

Whatever the source of the ice, successive glaciations have resulted in impressive, selective erosion of the plateaux (Linton, 1950a, 1951a; Sugden, 1968), producing the classic 'Icelandic' glacial troughs of Glen Einich and Glen Avon and their rock basins; the glacial breaches of the Lairig Ghru, Glen Feshie, Inchrory and Glen Avon–Glen Derry among others; the diffluent breach of Strath Nethy and the dramatic truncated spur at the Devil's Point (Figure 9.2). Many of these breaches are associated with abrupt changes in drainage direction and were traditionally explained in terms of pre- or post-glacial river capture (Hinxman, 1901; Gibb, 1909; Peach and Horne, 1910; Bremner, 1912, 1915, 1919, 1921, 1942, 1943b). However, Linton (1949a, 1951a, 1954) clearly demonstrated the role of watershed breaching by ice and glacial diversion of drainage. Roches moutonnées and ice-moulded bedrock are also well-displayed (see figure 3 in Sugden, 1968).

Linton (1950a, 1951a) interpreted the troughs and breaches as the product of erosion by local glaciers. Sugden (1968), however, argued strongly that they were cut by ice streams within an ice-sheet, drawing analogies with modern ice-sheets and glacierized areas, notably East Greenland. The Cairngorms massif as a whole may be described as a landscape of selective linear erosion (Sugden, 1968; Sugden and John, 1976) in which the glacial troughs and breaches contrast sharply with the little-modified plateau surfaces.

Such selectivity of erosion is widely represented in the eastern Grampians (Linton, 1963; Clayton, 1974; Haynes, 1977a), but the range of the older features, including the particularly fine development of the tors, and their close association with the glacial landforms, makes the Cairngorms by far the most outstanding illustration of this type of landscape in Britain and comparable, albeit on a much smaller scale, with examples from Baffin Island (Boyer and Pheasant, 1974; Sugden and Watts, 1977), the Torngat Mountains of Labrador (Ives, 1958, 1978); East Greenland (Bretz, 1935; Sugden, 1974a) and the Finger Lakes area of North America (Clayton, 1965). However, not all apparent examples from the Canadian Arctic may be directly analagous because of the role of tectonics there (England, 1987). The most satisfactory hypothesis to explain such landscapes relates to variations in glacier thermal regime. According to this hypothesis, relatively thin, slow moving, cold-based ice on the Cairngorm plateaux effected minimal erosion, whereas thicker and faster flowing outlet glaciers formed by ice converging on the troughs were warm based and therefore capable of more effective glacial erosion. In support, there is a body of theoretical and observational evidence, although other factors, such as topography, geology and glacier dynamics, may also interact (Sugden, 1974a, 1978; Sugden and John, 1976; Gordon, 1979; Andrews *et al*., 1985; Gellatly *et al*., 1988; Kaitanen, 1989).

In contrast to ice-sheet erosion, local mountain glaciers have been responsible for the formation of classic examples of corries, particularly on Braeriach, Cairn Toul, Cairn Lochan, and Beinn a'Bhuird (Westoll, 1942). The corries are notable for their regular geometric form (Sugden, 1969), which may reflect their plateau-edge location and the relatively uniform bedrock. Locally, however, structural influences can be seen in the 'schrund-line' long-profile form of Coire an Lochain (Haynes, 1968). In many cases, for example on the northern margin of the massif and in Glen Dee and Glen Lui, Sugden (1969) showed that the location, altitude and size of the corries closely relate to the form of the pre-glacial relief, in particular the pre-glacial valley heads: other corries occur on the flanks of the Glen Einich, Glen Derry and Glen Dee glacial troughs, while some form secondary basins within the larger corries; for example in Coire an t-Sneachda and Coire an Lochain. Corrie sizes are considerably greater than the volumes of debris in their respective moraines, indicating that their formation spans several periods of local glaciation. Indeed, on the basis of their size and position Sugden (1969) identified three generations of corries. Whether they can be correlated with specific glacial episodes, as he tentatively suggested, is questionable. Also open to question is the extent to which the corries may have been modified by ice-sheet glaciation (cf. Sugden, 1969; Holmund, 1991), either through direct erosion of bedrock or removal of screes and moraine debris. The location and distribution of the Cairngorm corries in a broader national context are considered by Linton (1959), Sale (1970), Sissons (1976b) and Dale (1981). On the northern flanks of the Cairngorms there is an interesting transition as the corries become increasingly shallower eastwards from Coire an Lochain to Coire Laogh Beag (NJ 013073).

Landforms and patterns of deglaciation

The deglaciation of the Cairngorms and the nature and extent of Lateglacial events have engendered considerable debate. Traditionally, the drift and meltwater channels which characterize many of the lower slopes and margins of the massif were explained in relation to a series of valley glaciers receding into the glens and corries following the last ice maximum (Hinxman, 1896; Jamieson, 1908; Barrow *et al.*, 1912, 1913; Hinxman and Anderson, 1915; Bremner, 1929; Charlesworth, 1956). Both Barrow *et al.* (1913) and Hinxman and Anderson (1915), however, acknowledged that the corrie moraines might indicate a subsequent recrudescence of glaciers. In Glen More on the northern flanks of the Cairngorms (Figure 9.4), Hinxman and Anderson (1915) identified a series of landforms (lateral moraines, terraces and meltwater channels) which they interpreted as ice-marginal retreat features of a lobe of ice derived from the south-west. According to Hinxman and Anderson this lobe of ice also advanced into the lower parts of Glen Einich and the Lairig Ghru, which were already ice-free, forming moraines there.

However, from detailed mapping and considerations of the relationships between the drift landforms and the meltwater channels, Sugden (1965, 1970) proposed an alternative model of an ice-sheet largely downwasting *in situ*, in similar fashion to the pattern of deglaciation proposed for parts of Scandinavia. He identified two main stages of landform formation, an ice-directed phase and a valley-controlled phase. To the first he attributed a series of meltwater channels and deposits which run across the northern flanks of the Cairngorms (Figure 9.4); these are discordant with the form of the underlying topography and are part of a regional meltwater drainage system mapped by Young (1974, 1975a, 1975b), extending from west of Glen Feshie north-eastwards to Abernethy Forest. In general, the altitude of the highest channels falls from west to east, reflecting a former ice-surface gradient. Many channels are cut across cols at right angles to north–south orientated spurs and several have up-and-down long profiles. The channels are similar in form, location and relationships to the superimposed englacial channels described from the south of Scotland (Price, 1960, 1963a) and the Cheviots (Clapperton, 1968, 1971a, 1971b), and are probably of similar origin. Particularly fine examples occur south-east of Creag a'Chalamain (NH 965053), south of Airgiod-meall (NH 965066) and south of Stac na h-Iolaire (NJ 015086) and on the north flank of Carn Eilrig (NH 938053) (Young, 1974). The most spectacular feature, interpreted as a glacio-fluvial deposit, is the flat-topped ridge extending east from Airgiod-meall, which is over 30 m thick and supports fine sections exposed along the Allt Mór (Sugden, 1970, figure 3). Its precise origin, however, is uncertain and it may in part be an ice-marginal feature, as may some of the ridges and bouldery deposits upslope that cross the outer slopes of Coire an Lochain and Coire an t-Sneachda. A question not addressed by Sugden is whether some of the ice-directed channels and deposits might represent successive ice margin positions of the downwasting glacier in Glen More (cf. Hinxman and Anderson, 1915). In the southern Cairngorms, easterly ice-directed meltwaters have cut channels across the summit area of Carn a'Mhaim (NN 995452), in the Meirleach col (ND 000936) and at the spectacular Clais Fhearnaig (NO 070935). The Water of Caiplich gorge at the Castle (NJ 123110) is another impressive meltwater channel, reflecting glacial diversion of drainage (Linton, 1954).

In contrast to the ice-directed landforms, the valley-controlled features show progressive conformity downslope with topography and have been described in detail by Young (1974, 1975a) for Glen More and Glen Feshie. On the northern flanks of the massif, whereas the higher channels and deposits are orientated towards the east, those at lower altitudes trend more towards the north-east; for example the channels on Creag nan Gall (NJ 015103) and the suite of channels and deposits in Glenmore aligned with the Ryvoan gap (NH 001105). At lower levels still, the features have a more northerly alignment with the An Slugain gap (NH 945130) reflecting increasing topographic control as the ice progressively downwasted. Finally, widespread stagnation of the ice is indicated by kame and kettle topography to the north and east of Loch nan Eilein (NH 895075) and west of Loch Morlich (NH 965093); masses of residual ice formed the

Figure 9.4 Glen More and the northern flank of the Cairngorms. The assemblage of landforms in this area includes the Cairngorm plateau and adjacent slopes extensively modified by solifluction lobes and terraces (Lurchers Gully – top centre), corries cut into the upper slopes of the massif, the striking glacial breach of the Lairig Ghru (top right), a system of ice-directed meltwater channels (including open-walled features – centre) and partly wooded glaciofluvial deposits in the valley bottom. (© British Crown copyright 1992/MOD reproduced with the permission of the Controller of Her Britannic Majesty's Stationery Office.)

large kettle holes now occupied by these lochs. During the latter stages extensive terraces were developed in western Glen More from the outlets of the Lairig Ghru and Gleann Einich towards Strathspey at Aviemore (Figure 9.2).

The overall pattern of a downwasting ice-sheet with progressive topographic control on meltwater discharge is one that is representative of Late Devensian ice-sheet decay in many parts of Scotland and northern England; for example see Sissons (1958b, 1961b), Stone (1959), Price (1963b), Clapperton (1971a, 1971b), Clapperton and Sugden (1972) and Young (1975b, 1977a, 1977b, 1978).

In the central Cairngorms, Sugden (1970) described channels and deposits that are concentrated, particularly, near the valley heads. Many of the deposits have the appearance of hummocky moraine, but have distinct linear alignments when viewed from the air (see below). Typically they have a scatter of surface boulders but largely comprise sand and gravel, a characteristic noted much earlier for those in Glen Derry by Jamieson (1860b) and Bremner (1929). Frequently channels run downslope above the ridges, but many of these clearly relate to post-glacial gullying of the valley-side drift cover, particularly by debris-flow activity. Sugden interpreted the total assemblage of landforms as valley- or topographically-controlled glaciofluvial features associated with an ice-sheet downwasting *in situ*. Moreover, he concluded that true moraines occurred only in a few of the corries and in one or two other localities. This led him to preclude a separate Loch Lomond Readvance valley glaciation in the Cairngorms as previously suggested by Sissons (1967a) for a large area of the Highlands, including the Cairngorms. In contrast, Sugden (1970) considered three possibilities for glaciation of the Cairngorms during the Loch Lomond Stadial, suggesting that it was represented by (1) a few corrie moraines; (2) a few moraines marking a stage in the wastage of the main ice-sheet, or (3) by an early stage in the deglaciation of the main Scottish ice-sheet, which persisted in the Cairngorms and Strathspey until the end of the Lateglacial, as in Scandinavia. Sugden initially (1970) favoured the third hypothesis.

In the debate that ensued, the key issues were whether complete deglaciation occurred in the Cairngorms during the Lateglacial Interstadial, and the location and extent of the Loch Lomond Readvance glaciers (Sugden, 1973a, 1973b, 1974b; Sugden and Clapperton, 1975; Sissons, 1972a, 1973a, 1973b, 1974a, 1975b; Sissons and Grant, 1972). Sissons argued for complete deglaciation in the interstadial followed by a fresh build-up of ice in the corries, parts of the plateaux and in some of the upper valleys, a pattern which he had established generally for the eastern Grampians and elsewhere in Scotland, based on evidence including the down-valley limits of fresh hummocky moraine, occasional end moraines, the distribution of solifluction lobes and pollen stratigraphy (Sissons, 1967a; Sissons *et al.*, 1973). He indicated that similar evidence occurred in the Cairngorms.

Sugden, however, argued that the evidence was inconclusive for complete deglaciation in the Cairngorms during the interstadial. He asserted that the hummocky moraine could be alternatively explained and was not a definitive characteristic of the Loch Lomond Readvance glaciers, as evidence at Loch Builg (NJ 187035) demonstrated (Clapperton *et al.*, 1975). Moreover, the formation of solifluction lobes could be diachronous and their distribution related to a wide range of local variables.

An important piece of evidence bearing on the debate came from a core from Loch Etteridge (see below). Not only did it contain a full Lateglacial pollen sequence, but the basal layers provided a radiocarbon date of about 13,150 BP, clearly indicating that Strathspey was already deglaciated early in the Lateglacial Interstadial (Sissons and Walker, 1974). In the light of this discovery, Sugden rejected his second two hypotheses in favour of the first; that the Loch Lomond Readvance in the Cairngorms was confined to a few corries.

Subsequently, Sissons published a map and detailed account of his interpretation of the Loch Lomond Readvance in the Cairngorms based on the criteria and arguments which he had applied elsewhere (Sissons, 1979f). In addition to identifying a number of important periglacial features, it differs in one major respect from that of Sugden: not only did small glaciers develop in some of the corries, but larger valley glaciers also existed at the head of Loch Avon, in Glen Eidart, in Glen Geusachan extending into the Dee Valley and in An Garbh Choire extending into the Lairig Ghru.

As yet no radiocarbon dates or pollen records are available to confirm either interpretation. However, it is clear that regardless of which interpretation one accepts, the extent of the Loch Lomond Readvance in the Cairngorms was somewhat less than in the eastern Grampians and the Gaick area to the south, as mapped by Sissons (1972a, 1974b). This pattern is thought to reflect the prevalence of snow-bearing winds from the south-east during the Loch Lomond Stadial, creating a precipitation shadow effect and a snowline rising towards the Cairngorms (Sissons and Sutherland, 1976; Sissons, 1980b).

The boulder moraines of Coire Lochan Uaine (NO 001981), Coire an Lochain (NH 945006), Coire Bhrochain (NN 960995) and Coire na Ciche (NO 103983), and the abrupt terminations of boulder spreads in Coire an Lochain (NH 981033) and Coire an t-Sneachda (NH 995034) (Figure 9.5), are particularly fine examples of Loch Lomond Readvance ice limits. Hummocky moraine associated with Sissons' Loch Lomond

Figure 9.5 Loch Lomond Readvance boulder moraine in Coire an t-Sneachda in the Cairngorms. The outer part of the moraine comprises several clearly defined ridges of boulders. (Photo: J. E. Gordon.)

Readvance glaciers is well-displayed at the head of Loch Avon (NJ 005016), and there are clear down-valley limits to its extent in Glen Eidart (NN 918922) and Glen Dee (NN 988921). Other fine examples of hummocky moraine beyond the readvance limits occur in Glen Derry (NO 033995) and at Loch Builg (NO 190025). In several areas the hummocky moraine comprises linear alignments of ridges and mounds, for example in Glen Eidart (Figure 9.6) and Glen Geusachan. The deposits in Glen Geusachan have been mapped in detail by Bennett and Glasser (1991) and interpreted by them as a series of ice-marginal landforms, the pattern of which implies active recession of the glacier towards the head of the glen. Bennett and Glasser (1991) also speculated that the relatively large size of the inferred Loch Lomond Readvance glacier in Glen Geusachan could be explained by the survival of ice during the Lateglacial Interstadial.

Multiple end moraines or boulder ridges are a characteristic feature of a number of the corries, for example Coire an t-Sneachda, Coire an Lochain (Cairn Lochan) and Coire an Lochain (Braeriach) (Sugden and Clapperton, 1975; Sissons, 1979f). Their significance in the context of climatic change during the Loch Lomond Stadial has not been fully evaluated. However, as MacPherson (1980) has shown, there is evidence at least for a generalized pattern of a decline in precipitation in the area after the start of the stadial, which is likely to have had a significant effect on the mass balance of small corrie glaciers, followed by increased precipitation and therefore possibly by glacier expansion towards the end of the stadial. Sugden (1977) questioned whether the innermost ridges might relate to renewed glaciation during the Little Ice Age of the 16th–19th centuries (see below). However, the presence of middle and late Holocene pollen profiles on the ice-proximal (inner) sides of the moraines in Coire an Lochain (Braeriach) and Coire an Lochain Uaine (Ben Macdui) demonstrates that the sediments pre-date the Little Ice Age and that active glacier ice could not have existed there at this time (Rapson, 1985) (see also Lochnagar).

Periglacial landforms

A wide range of periglacial landforms is developed on the slopes and plateau surfaces of the

Figure 9.6 Loch Lomond Readvance 'hummocky moraine' on the east flank of Glen Eidart in the Cairngorms. The deposits have a clear upper limit on the valley side and show well-defined lineations, which may mark successive ice-front positions of an actively retreating glacier. (Photo: J. E. Gordon.)

Cairngorms. Sissons (1979f) mapped and described several large-scale features of Loch Lomond Stadial age, including rock glaciers (protalus lobes), protalus ramparts and spreads of boulders which he inferred were deposited at the downslope margins of former snow patches. The rock glaciers, which take the form of protalus lobes (for discussion of terminology and origins, see for example Wahrhaftig and Cox, 1959; Outcalt and Benedict, 1965; Lindner and Marks, 1985; Martin and Whalley, 1987; Barsch, 1988; Whalley and Martin, 1992), occur north of Loch Etchachan (NJ 007009) and in Coire Beanaidh (NH 956006 and NH 954016). The latter was described by Chattopadhyay (1984). The boulder spreads are essentially similar features (Ballantyne and Kirkbride, 1986) and bear striking comparison to the 'talus terraces' described by Liestøl (1961) in Svalbard. The largest example, some 2 km long, is in Strath Nethy (NJ 020045), and a smaller example occurs at the northern end of the Lairig Ghru (NH 962037). Protalus ramparts occur, for example, below the Devil's Point (NN 978946) and in Lairig Ghru (NH 964028). A ridge in Coire an t-Sneachda (NH 995038), described as the largest protalus rampart in the Cairngorms (Sissons, 1979f), includes a bedrock outcrop with a quartz dyke; it is considered to be either a landslide deposit or a residual rock ridge isolated by marginal meltwater channels (C. K. Ballantyne, unpublished data).

Elsewhere in Scotland fossil rock glaciers have been reported from Jura (see Beinn Shiantaidh; Dawson, 1977) and Wester Ross (see Beinn Alligin; Sissons, 1975a; Ballantyne, 1987c), and protalus ramparts notably from Wester Ross (see Baosbheinn; Sissons, 1976c; Ballantyne, 1986a) and other parts of the Highlands (Sissons, 1977a; Ballantyne and Kirkbride, 1986). The Cairngorms, however, are particularly notable for the variety of features present within the massif, ranging from simple protalus ramparts to protalus lobes.

On the summits of Ben Macdui and Derry Cairngorm extensive boulder fields or 'felsenmeer' comparable to those of other mid-latitude mountains and parts of the Arctic (cf. Dahl, 1966; Boyer and Pheasant, 1974; Nesje, 1989) have developed through frost disruption of the underlying granite bedrock (cf. Sugden, 1971; Paine, 1982). The susceptibility of the granite to weathering has been ascribed to the mechanical weakness of the rock (Hills, 1969), but may also relate to chemical weathering of the latter (Innes, 1982). In places *in situ* joint blocks can be seen below displaced blocks on the surface. On Creag an Leth-choin (NH 968033) and other locations shattered rock outcrops and a blockfield on the summit are succeeded downslope by blockslopes and bouldery solifluction lobes.

Throughout the Cairngorms, periglacial mass movement has resulted in the widespread development of sheets, terraces and lobes of frost-weathered detritus. These were first investigated by ecologists working in the area (Watt and Jones, 1948; Metcalfe, 1950) and later by Galloway (1958), who considered them to be relict features immobilized by eluviation of fines. The most abundant and impressive examples are massive sheets of large boulders that terminate in risers up to 3 m high (as at NH 960028). These become increasingly lobate in plan form as the slope steepens (as at NH 963041). Such features have been variously described as stone-banked and vegetation-covered lobes (King, 1968, 1972), solifluction lobes (Sugden, 1971), 'blockloben' (Kelletat, 1970a) and boulder lobes (Sissons, 1979f). Sugden (1971) suggested that they may have survived the passage of the last ice-sheet, or may have formed during the Little Ice Age. King (1968, 1972) distinguished between vegetation-covered and stone-banked lobes. He believed that the former developed during the early Holocene, the latter during the Little Ice Age. This interpretation, however, was constrained by King's acceptance of Sugden's (1970) reconstruction of the timing of deglaciation in the area. Sissons (1979f) noted that boulder sheets and lobes are entirely absent inside the limits that he mapped for the Loch Lomond Readvance. This strongly suggests a Lateglacial origin, or at least reactivation, of such forms. Both Galloway (1958) and King (1968) maintained that some of the stone-banked lobes at higher altitudes are currently active; for example on Carn Bàn Mór. If so, it seems probable that they were active also during the Little Ice Age climatic deterioration. However, stone-banked lobes appear to occur nowhere within the limits of Loch Lomond Readvance glaciers, which casts doubts on any significant activity during the Holocene. King (1968, 1972) concluded that both types of lobe formed by viscous flow but were influenced in the form and location of their fronts by bedrock joints. Particularly fine examples of suites of stone-banked lobes occur on Creag an Leth-choin (NH 970034), extensively on the western slopes of Carn Bàn Mór (NN 893974) and the Sròn na Lairige spur of Braeriach (NH 960028); suites of vegetation-covered lobes occur on Creag an Leth-choin (NH 979041), on the south-west slope of Sgòran Dubh Mór (NN 901999) and between Coire Bogha-cloiche and Coire an Lochain (NH 993003).

In addition to the large-scale features described above, smaller-scale solifluction lobes and sheets occur in the Cairngorms. R. M. G. O'Brien (cited in Sugden, 1971) obtained radiocarbon dates of 4880 ± 140 BP (N–622) and 2680 ± 120 BP (N–623) for organic material under such features, and Sugden (1971) thought that the most likely periods of formation of the latter were the cold phases of about 2500 BP and the 16th–18th centuries AD. Similar conclusions were reached by White and Mottershead (1972) and Mottershead (1978) concerning solifluction terraces overlying organic material dated at between 5440 ± 55 BP (SRR–723) and 3990 ± 50 BP (SRR–724) on Arkle in Sutherland. However these dates may simply reflect the time of burial of the organic material and need not indicate a relationship between solifluction and climatic deterioration (Ballantyne, 1991a). There is also evidence that some of the solifluction features in the Cairngorms are active at present (King, 1968, 1972; Kelletat, 1970b, 1972).

Other features which appear to be currently active are small, turf-banked terraces ('steps'), formed by the combined action of wind and frost-creep (King, 1968, 1971b). These frequently occur in association with deflation surfaces (see below) and in some instances have formed on the 'treads' of older boulder sheets and lobes. Recent mass movement is also indicated by 'ploughing' boulders (King, 1968; Kelletat, 1970b), and debris flows resulting from heavy rainfall have left many steep slopes scarred by gullies and debris chutes (Figure 9.7) (Baird and Lewis, 1957; Innes, 1983b; Kotarba, 1987; Ballantyne, 1991c; Luckman, 1992), notably in Glen Geusachan, at the northern end of the Lairig Ghru and in Coire an t-Sneachda. These flows are of the hillslope type (cf. Innes, 1983c), with levées of debris along the margins, and tongues of debris at the foot of the flow tracks. Debris-flow activity appears to have increased in the last 250 years, possibly as a response to land-use changes (Innes, 1983b), although in Glen Feshie natural processes

Figure 9.7 Active debris flows on the slopes above the Lairig Ghru. (Photo: J. E. Gordon.)

have been identified as the principal cause (Brazier and Ballantyne, 1989). Innes (1985) noted the importance of relatively large events in the Cairngorms compared with other areas of the Highlands (see also Kotarba, 1987). Other effects of heavy rainfall are seen in flood deposits; for example, gravel spreads in the Dee Valley (Baird and Lewis, 1957; Clapperton and Crofts, 1969) and along the Allt Mór where the ski access road has been washed away several times in recent years (Sugden and Ward, 1980; McEwen and Werritty, 1988).

Both King (1968, 1971a, 1971b) and Kelletat (1970a, 1970b, 1972) have described fine examples of patterned ground in the Cairngorms. In general, frost-sorted forms (circles and stripes) are abundant only on vegetation-free areas above 900 m OD. Some of the large, fine-grained stone circles appear to experience frost heave at present (King, 1968, 1971a). However, most of the features described by these authors are inactive, and their widths (often over 2 m) suggest that they were formed under permafrost conditions, probably during the Lateglacial, rather than the Little Ice Age, as suggested by King, although some reactivation at that later time is a possibility (C. K. Ballantyne, unpublished data). It is also possible that they are even older, since patterned-ground features are known to be preserved under cold-based ice in modern glacial environments (Whalley *et al.*, 1981). Good examples of large-scale circles occur on Carn Bàn Mór (NN 894973) and on Ben Macdui (NH 981004, NH 991012); stone stripes occur on Carn Bàn Mór, where they grade downslope into lobes, and on Geal-charn (NH 893000) and Creag Follais (NH 893043). Much smaller, active circles and stripes have been described by Kelletat

(1970a) and are well represented at an altitude of 1065 m OD near the highest point of the Lairig Ghru (NH 974023) and in Coire Raibert (NJ 005030) (C. K. Ballantyne, unpublished data). King (1968, 1971a) also described a form of wind-patterned ground which he termed denudation surfaces. These are deflation scars in the vegetation cover, typically 1 m wide and 2–4 m long; they occur extensively above an altitude of 450 m OD. According to King, they formed through a combination of needle-ice erosion and deflation. More extensive areas of deflation surface occur, for example, on Meall Gaineimh (NJ 167052) and Beinn Bhrotain (NN 955924). Pedogenesis and the influence of periglacial processes in the regolith of the Cairngorms have been considered by Romans *et al.* (1966) and Romans and Robertson (1974); in particular, the development of silt droplets in the soil profile is believed to reflect former permafrost conditions.

A further component of the periglacial landscape for which the Cairngorms are renowned is late-lying or semipermanent snowbeds (Manley, 1949, 1971; Green, 1968; King, 1968), the most famous and persistent being in An Garbh Choire (NN 942980) (Gordon, 1943). Sugden (1977) considered the intriguing question of whether such snowbeds might have expanded during the Little Ice Age to form small glaciers in a number of the corries. He argued that there was some supporting evidence, although not conclusive, from historical records, reconstructed snowlines, and lichen sizes on the innermost moraines of certain corries. However, radiocarbon dating and pollen analyses of organic sediments behind the moraines in Coire Bhrochain (Sugden, 1977) and Coire an Lochain (Braeriach) and Coire an Lochain Uaine (Rapson, 1985) have indicated that these corries have remained unglaciated during the Holocene. Nevertheless, the possibility remains open that the moraine in Garbh Choire Mór, lying well inside the Loch Lomond Readvance limit (Sissons, 1979f), formed during the Little Ice Age (Rapson, 1990).

Although the Cairngorms are almost high enough to support glacier ice at present (see Manley, 1949), Manley (1971) doubted that there had been sufficiently long unbroken sequences of cool summers (at least 20) to form glaciers in historical time, although there would have been decades during the 17th to the 19th centuries when persistent snowbeds occurred at lower levels than today.

Associated with a number of the persistent snowbeds are nivation hollows, one of the best examples being Ciste Mhearaid (NJ 012045) (see McVean, 1963b, figure 2). The location, size and site characteristics of these and a range of other mountain hollows, some perhaps better described as incipient glacial corries, are described by King (1968). Generally in the mountains of Scotland, current snow patch erosion is limited in its effects (Ballantyne, 1987a).

Snow avalanches have received increasing attention in the Cairngorms. The area, with its steep slopes and massive cornices built up by snow drifting off the plateau surfaces, is probably the most conducive in Britain for avalanche activity. Some aspects of the snow and weather conditions associated with avalanches were described by Langmuir (1970), and subsequent research has addressed the types of avalanche that occur, their frequency and magnitude, the factors governing their location and release, physical characteristics of the snowpack and a predictive model (Ward, 1980, 1981, 1984a, 1984b, 1985a; Ward *et al.*, 1985). Spectacular examples of snow avalanches occur each spring from the slabs on the headwall of Coire an Lochain (NH 984027) (see Langmuir, 1970, figures 4 and 5). The geomorpholgical effects of such avalanches are variable. Good examples of avalanche boulder tongues that are currently active occur in the Lairig Ghru, and there are excellent fossil features on the western slopes of Derry Cairngorm (Ballantyne, 1989b, 1991c; Luckman, 1992). Most of the current geomorphological activity is associated with reworking of debris flow deposits (Luckman, 1992); in most other areas the effects appear to be relatively minor (Ward, 1985b; Davison and Davison, 1987), and only on Ben Nevis in the western Highlands have features such as avalanche impact landforms been recorded (Ballantyne, 1989b).

Lateglacial and Holocene vegetation history

The Lateglacial and Holocene vegetation history of the Cairngorms area is represented in the sediments at Abernethy Forest (see below) and is summarized in part by Dubois and Ferguson (1985). Within the massif itself, biostratigraphic evidence in the form of pollen and plant macrofossil records has been described from sites at Eidart, Sgòr Mór and Carn Mór (Pears, 1964,

1968) and Loch Einich (Birks, 1969, 1975), with the particular aim of elucidating the forest history of the area. Pears also carried out a more extensive survey of macrofossils in the peat, and Dubois and Ferguson (1985) investigated climatic history using stable isotope analysis and radiocarbon dating of fossil pine stumps from the northern flanks of the massif. Further palaeoenvironmental information is available from a Holocene tufa deposit at Inchrory in Glen Avon, in the eastern Cairngorms (Preece *et al.*, 1984).

The most detailed pollen diagram compiled is that from Loch Einich. Here Birks (1969, 1975) described a lower layer of pine stumps and an upper layer of birch stumps embedded in an area of eroded, deep blanket peat. From these stumps and the pollen stratigraphy which extends over the *Betula–Corylus/Myrica*, *Pinus* and *Calluna–Plantago lanceolata* Holocene regional assemblage zones (Birks, 1970), she interpreted the following sequence. Peat from a *Juncus effusus–Sphagnum* mire community (McVean and Ratcliffe, 1962) began to accumulate in waterlogged hollows in the underlying glacial deposits. Birch then colonized the surface of the bog (as evidenced by wood remains) and coexisted with *Empetrum*, *Calluna* and *Sphagnum*. Pine initially spread on to the valley sides and subsequently on to the surface of the bog. Pine stumps dated at 5970 ± 120 BP (K–1418) are overlain by *Sphagnum* peat, indicating increased waterlogging and the demise of pine at the site, although pine trees continued to grow on the valley sides. During a drier phase birch subsequently colonized the bog surface and is represented by stumps dated at 4150 ± 100 BP (Q–883). The birch, too, was eventually overwhelmed by *Sphagnum*, and treeless conditions then prevailed until the present day. In the uppermost layers of the bog, which provide the reference site for the *Calluna–Plantago lanceolata* regional pollen assemblage zone (Birks, 1970), there is evidence of forest clearance on the valley sides near Loch Einich and recession of the treeline to near its present limit at about 400 m OD in Rothiemurchus Forest.

The pollen diagrams of Pears (1968) are less detailed than those of Birks and are zoned according to the Jessen–Godwin scheme, so that direct comparisons are not facilitated. However, it is clear that *Pinus* formed the dominant element in the forests, with *Betula* playing a subsidiary role. Moreover, the radiocarbon dates obtained by Pears (1970, 1975a) from Carn Mór, the former site of Jean's Hut, Sgòr Mór, Coire Laogh Mór, Meall a'Bhuachaille and Barns of Bynack and by Birks (1975) from Loch Einich demonstrate that the stumps in both the lower and upper wood layers in the peat are asynchronous. Therefore they cannot be assigned to particular climatic periods (Boreal and Sub-boreal) in the Blytt–Sernander scheme, as was assumed before radiocarbon dates were available (see discussion for the Allt na Feithe Sheilich site); their occurrence reflects instead local site topographical and hydrological factors (Pears, 1970, 1972). This applies also to tree stumps in peat investigated elsewhere in Scotland (Birks, 1975).

Dubois and Ferguson (1985) reported a series of 40 radiocarbon dates on pine stumps from the northern flanks of the Cairngorms. The oldest, 7350 ± 85 BP (IRPA–594), provides a minimum age for the establishment of pine in the Cairngorms. Dates obtained for the inception of blanket bog range between 5230 ± 260 BP (IRPA–361) and 6090 ± 300 BP (IRPA–362), which again may reflect the influence of local conditions (Pears, 1970, 1988). Dubois and Ferguson (1985) also investigated the stable isotope chemistry of wood cellulose extracted from the dated pine stumps. Assuming that the deuterium/hydrogen ratio of precipitation is related to local surface air temperature and to the intensity of the precipitation (see Dansgaard, 1964), and that the moisture taken up by the tree roots and utilized in the production of cellulose reflects the isotopic composition of the precipitation, then the isotopic composition of the wood potentially provides a valuable palaeoenvironmental record (but see Dubois, 1984; Siegenthaler and Eicher, 1986). On this basis Dubois and Ferguson (1985) identified four periods of increased rainfall distinguished by low deuterium/hydrogen ratios around 7300 BP, between 6200 BP and 5800 BP, between 4200 BP and 3940 BP and around 3300 BP. Pears (1988), however, cautioned against interpretation of the stable isotope results purely in terms of precipitation and argued that the low deuterium/hydrogen ratios could equally reflect locally high values of relative humidity at individual growth sites. In reply Dubois and Ferguson (1988) provided further evidence to support their case. The results of Bridge *et al.* (1990) from a detailed study of macrofossils in the Rannoch Moor area, supported by radiocarbon dating and pollen analysis, lend some support to the conclusions of Dubois and Ferguson (1985), but also emphasize both the complex relationships between climatic factors, site conditions, forest ecology, tree growth and preservation of macrofossils, and the need for further investigations.

Pears (1975b) estimated growth rates of peat in the Cairngorms for various periods covered by his radiocarbon dates between 6700 BP and the present. Values lie in the range 1.4 to 3.4 cm 100 years^{-1}. The rates are consistent for sites in the Cairngorms and also with Birk's (1975) estimate for peat growth at Loch Einich. They are also broadly similar, although slightly lower, than rates of peat growth in Deeside reported by Durno (1961). A significant conclusion reached by Pears (1975b) is that, due to the slow growth rates, even relatively minor peat erosion scars are unlikely to develop sufficient vegetation cover to heal themselves, particularly in view of the increased human pressures on the sites.

The present treeline in the Cairngorms is constrained by exposure to high winds and biotic stress (Pears, 1967) and only in one area, on Creag Fhiaclach (NH 898055), does it approach its natural level of 610–685 m OD (Pears, 1968). The maximum Holocene altitude of the treeline was 793 m OD, recorded by the highest stumps (*Betula pubescens*) found in the region in Coire Laogh Mór. These have been dated to 4040 ± 120 BP (Pears, 1975a).

Interpretation

Five key elements can be identified within the total assemblage of geomorphology and Quaternary interests in the Cairngorms:

1. The surviving elements of the pre-glacial landscape, comprising tors, weathered regolith, plateau surfaces and river valleys are exceptional for the assemblage of forms within a single area. Although individual elements, such as tors and deep weathering, are represented in other mountain areas, such as Lochnagar and the Gaick Plateau respectively, the Cairngorms are unsurpassed for the range, scale and quality of the features preserved. The Cairngorms therefore provide an invaluable insight into long-term processes of mountain landscape development in Britain. In this respect they differ significantly from western mountain areas such as the Cuillin, Lake District and North Wales, where the imprint of glaciation is dominant. Within Britain the Cairngorms provide potentially important comparisons with areas such as Dartmoor where many similar features occur, albeit in an unglaciated environment, and with parts of south-west Wales located close to the margins of the Pleistocene ice-sheets. On an international scale, the pre-glacial forms of the Cairngorms bear comparison with those that have survived glacierization in Norway and East Greenland, but the closest parallels are probably with parts of north Finland, the Canadian Arctic islands and Baffin Island.
2. The elements of glacial erosion provide an assemblage of landforms for which the Cairngorms is both nationally and internationally recognized. These include the glacial troughs, breaches, corries and large-scale diversions of drainage. Although individual examples of troughs and breaches are arguably as well developed in the northern and western Highlands, on Skye (Loch Coruisk), in the Lake District and Wales, few areas can demonstrate a range of features comparable to those of Loch Avon, Glen Einich, the Lairig Ghru and Strath Nethy. Moreover, unlike those in other areas, in the Cairngorms the features of glacial erosion are juxtaposed with pre-glacial landscape elements to form a classic landscape of selective glacial erosion. In this respect, the Cairngorms surpass other examples in the eastern Grampians (Loch Muick area and Glen Clova) and rank on an international scale along with examples from parts of East Greenland, Labrador, Norway and Baffin Island. In terms of the diversity of features, the closest comparison is with parts of eastern Baffin Island.

   Cut into the granite plateau surfaces, the corries display classic forms which are relatively simple in outline; their principal interest lies in the diversity they lend to the geomorphology of the Cairngorms.

   Glacial diversions of drainage are particularly well demonstrated in the Cairngorms area. While other examples are known from the eastern Grampians, there is probably no finer an assemblage of such forms in Britain than occurs in the Cairngorms.
3. A third element of the Cairngorms landscape is the evidence for patterns of deglaciation in a mountain area, as represented by meltwater channels, glaciofluvial deposits (eskers, kame terraces and dead-ice topography) and mo-

raines. The meltwater features are best developed on the northern flanks of the massif and are of interest both as individual landform examples and as an assemblage of landforms which demonstrates downwasting of the last ice-sheet and the accompanying changes from ice-directed to topographically-controlled meltwater flow patterns. Some of the individual landforms are notable examples of their type, although not as distinctive as, for example, the meltwater channels at Carlops or Rammer Cleugh or the deposits at Carstairs Kames, Torvean or Kildrummie Kames. However, they are distinguished by their clear spatial patterns and the evidence that they provide for evolution of the meltwater drainage system during deglaciation. In this respect, they provide an outstanding assemblage of landforms comparable only to that at Muir of Dinnet (see above).

Morainic landforms are principally associated with the Loch Lomond Readvance, although there are some notable exceptions, for example at Loch Builg. The boulder moraines in the corries and the hummocky moraines in the valleys include many fine examples of landforms that are widely represented elsewhere in the Highlands. The particular significance of the Cairngorms features is the diversity they add to the geomorphology of the massif.

4. The Cairngorms contain an outstanding range of periglacial landforms and deposits. Individual examples of many of the types present are equally well, or better represented elsewhere in the Highlands (see in particular, An Teallach, Ben Wyvis, Beinn Shiantaidh, Sgùrr Mór, Ward Hill and Ronas Hill), but it is the combination and range of features which distinguishes this element of the geomorphology of the Cairngorms. The high plateau surfaces provide the closest analogue in Britain to sub-arctic or montane fellfield landscapes, containing blockfields, deflation surfaces and large-scale patterned ground. Slopes below the plateaux support a variety of mass-movement features. These include excellent examples of relict, bouldery gelifluction lobes, protalus ramparts and rock glaciers, as well as debris flows and snow avalanche landforms of more recent origin. The Cairngorms contain the greatest number of fossil rock glaciers (protalus lobes) of any mountain range in Britain. The gelifluction lobes compare with examples on Lochnagar, Mount Keen and Creag Mheagaidh in terms of size and extent, and the debris flows are characteristic of similar features elsewhere in the Highlands, for example, in Glen Coe and Drumochter Pass. The small-scale patterned ground, wind and active frost features form part of a network of upland sites of current periglacial activity ranging from Shetland to the Southern Uplands, the Cairngorm examples representing the conditions of the more continental high summits of the eastern Highlands. Overall, the assemblage of periglacial features in the Cairngorms is of national importance.
5. In terms of Holocene vegetation history and environmental change, the Cairngorms are important in several respects. The extensive peat deposits and bogs provide a record of both regional upland environmental changes and the role of local site-specific factors such as topography and drainage, in influencing changing vegetation patterns. The pine stumps extensively preserved in the peat provide a record of Holocene treeline changes and have significant potential for elucidating palaeoenvironmental conditions through the application of stable isotope analyses.

Key areas and individual sites have been identified in the descriptive sections above. However, it is important to stress the integrity of the total landform assemblage since this is an aspect of the geomorphology that is as important as each of the individual elements. Relationships between landforms and landform types are important and clearly demonstrated, and the scale of the site is such that spatial and altitudinal patterns can be distinguished. Thus, for example, the northern flanks of the massif from Cairngorm and Braeriach down to Glen More provide a transect from plateau surfaces with tors, weathered regolith and periglacial features, through corries with boulder moraines, protalus ramparts, rock glaciers and slope mass movements, leading downslope to ice-marginal features, meltwater channels, and finally eskers, dead-ice topography, outwash and river terraces.

In summary, the Cairngorms are exceptional for the range of particular landform elements and for the diversity of the total assemblage of features. Each of the five elements identified above ranks on a national scale of importance, while some rank on an international scale. Further, when the total range of interests is combined, the Cairngorms qualify as a site of international importance for geomorphology. The Cairngorms represent a striking demonstration of landscape evolution over a long time-scale, of the impact of successive geomorphological systems on the landscape and of the spatial variation of forms and processes within individual landform systems. They provide crucial field evidence for testing models of landscape evolution, patterns of glacial erosion, meltwater drainage evolution, periglacial and slope processes and Holocene environmental changes. Above all, they demonstrate the diversity of the geomorphology of glaciated mid-latitude mountains.

Conclusion

The Cairngorms is an area of the very highest importance for Quaternary geomorphology in Britain, providing an outstanding range of features for interpreting landscape evolution and environmental change during the Quaternary. The interest comprises five principal components. First, there are the planation surfaces, tors and pockets of deeply weathered bedrock that appear to have survived the effects of glaciation, and which illustrate aspects of longer-term landscape development. Second, a striking assemblage of landforms of glacial erosion demonstrates the powerful capacity of glaciers to modify the landscape, but in a selective fashion. Third, the Cairngorms display particularly well the landforms and deposits formed as the glaciers melted, including moraines, meltwater channels and meltwater deposits. Fourth, there is a range of periglacial landforms and deposits that illustrate the effects of cold climate conditions on the soil and its movement downslope. Fifth, the peat deposits and bogs provide a detailed record of environmental changes and vegetational history during the Holocene (the last 10,000 years). Many of these features are essential components of the wider site networks for their particular interests. Some are among the best examples of their kind in Britain, and others rank on an international scale for their clarity of development and interrelationships. However, it is the total assemblage of interests, developed in a relatively compact area, that makes the Cairngorms so remarkable.

LOCHNAGAR

J. E. Gordon and C. K. Ballantyne

Highlights

Lochnagar is important for its glacial and periglacial landforms, including corrie moraines and gelifluncted boulder lobes. These features formed during the Loch Lomond Stadial and provide a record of glacier dynamics and geomorphological processes active at that time.

Introduction

Lochnagar (NO 250860), a mountain massif rising to 1150 m OD and located *c.* 10 km south-east of Braemar, is important for its assemblage of glacial and periglacial landforms. It is noted for one of the best examples in Scotland of a suite of boulder lobes and terraces dating from the Loch Lomond Stadial. It also includes a fine example of a corrie and an excellent sequence of moraines formed during the Loch Lomond Readvance. The periglacial landforms have been investigated by Galloway (1958) and Shaw (1977), and the glacial landforms by Sissons and Grant (1972), Rapson (1985) and Clapperton (1986).

Description

In broad outline Lochnagar has the form of a residual granite massif rising above the plateau of the 'Grampian Main Surface' (Fleet, 1938). Landforms of glacial erosion are impressive and stand in sharp contrast to adjacent plateau surfaces and slopes that are essentially pre-glacial in their broad outlines, although modified in detail by periglacial processes. The glacial troughs of Glen Callater (NO 190835) to the south-west and Loch Muick (NO 290830) to the south-east are fine examples of selective linear erosion by ice-sheets and valley glaciers; on the northern slopes of the massif, corrie development has been predominant. These erosional landforms reflect the effects not only of Late Devensian glaciation, but also of earlier glaciation. Depositional landforms associated with both the Late Devensian ice-sheet and Loch Lomond Readvance glaciers are extensively developed and include hummocky moraines, fluted moraines and corrie-glacier moraines, together with eskers and other meltwater deposits (Sissons and Grant, 1972; Clapperton, 1986). However, it is the north-east corrie of Lochnagar and the higher southern slopes of the mountain that are of special interest for particular glacial and periglacial landforms, formed mainly during the Loch Lomond Stadial.

The north-east Corrie

The north-east corrie of Lochnagar is a striking example of a corrie with a steep headwall, an enclosed loch basin and a sequence of end- and lateral-moraine ridges (Figure 9.8). A total of nine moraine ridges is present, each less than 3 m high and comprising arcuate lines of boulders resting on a bouldery till substrate (Clapperton, 1986). The boulder moraines represent progressive recession of an active glacier into the corrie, but there are different interpretations of its former extent (Sissons and Grant, 1972; Clapperton, 1986).

The landforms and deposits in this area have a significant bearing on the interpretation of the Late Devensian history of the Cairngorms and adjacent parts of the eastern Grampian Highlands. Sissons and Grant (1972) mapped the geomorphological evidence for the last glaciers in the Lochnagar area and defined the limits of a series of corrie and valley glaciers associated with the Loch Lomond Readvance. This work and its subsequent extension in adjacent areas stimulated keen debate on the wider regional implications of the status of the Loch Lomond Readvance in the Cairngorms and vicinity (Sissons, 1972a, 1973a, 1973b, 1975b, 1979f; Sugden, 1973a, 1973b, 1980; Clapperton *et al.*, 1975; Sugden and Clapperton, 1975). In the absence of a locally established dating framework, interest centred on the extent of the Loch Lomond Readvance and indeed whether some of the small boulder moraines, including those on Lochnagar, could have formed during the Little Ice Age (Sugden, 1977). However, pollen analysis of cores taken from sites within the glacial limits defined by the corrie moraines on Lochnagar demonstrate that sedimentation has continued undisturbed from about 9700 BP (Rapson, 1985). A peat sample and pine stump located within these moraines have yielded radiocarbon dates of 7170 ± 80 BP (SRR–2272) and 6080 ± 50 BP (SRR–1808), respectively (Rapson, 1985). Together with similar evidence from the Cairngorms, these results preclude any significant Little Ice Age or earlier Holocene glacier development in the Lochnager corrie and imply that the boulder moraines are most probably of Loch Lomond Readvance age.

Using reconstructed glaciers as palaeoclimatic indicators, Sissons and Sutherland (1976) derived equilibrium line altitudes for the eastern Grampian Highlands during the Loch Lomond Stadial and inferred that precipitation and snow accumulation in this area were principally associated with winds from a southerly or south-easterly direction. Clapperton (1986), however, questioned this interpretation. His reconstruction of the glacier that formerly occupied the north-east corrie of Lochnager not only identified a slightly lower snowline, but also suggested that the dominant ice source was the south-west basin of the corrie. He therefore concluded that south-westerly winds were dominant at the time of glacier growth and that it was unnecessary to invoke unusual climatic conditions and south-easterly air flows.

Periglacial landforms

On the granite massif of Lochnagar there is a striking contrast between well-developed corries to the north and smooth slopes to the south. The latter are covered by frost-weathered debris on which have developed massive boulder lobes and terraces, considered by Galloway (1958) to be the finest in Scotland.

The boulder lobes and boulder sheets of Lochnagar are best developed on Broad Cairn around (NO 240818), Cac Carn Beag (NO 244861) and on the south-east slope of Cuidhe Cròm (NO 262848) (Figure 9.9). Galloway (1958) described those of Cuidhe Cròm in detail, and concluded that the lobes were stone-banked solifluction lobes of Lateglacial age, immobilized by eluviation of fine material, but that nearby boulder sheets and blockfields are still undergoing mass movement. Shaw (1977) carried out a very detailed study of the boulder lobes. He found that they occupy slopes of 10°–34° at altitudes of 640 m to 1110 m OD, and are most frequently developed on west-facing slopes. They range in thickness from 0.3 m to 5.9 m, in width (across slope) from 3.9 m to 33.3 m, and in length (downslope) from 2.1 m to 76.3 m. They are composed of openwork boulders with an average length of more than 0.7 m, with little interstitial finer material and a cover of peat up to 1.2 m thick over the 'treads'. Boulder terraces generally occupy gentler slopes (14°–22°) and

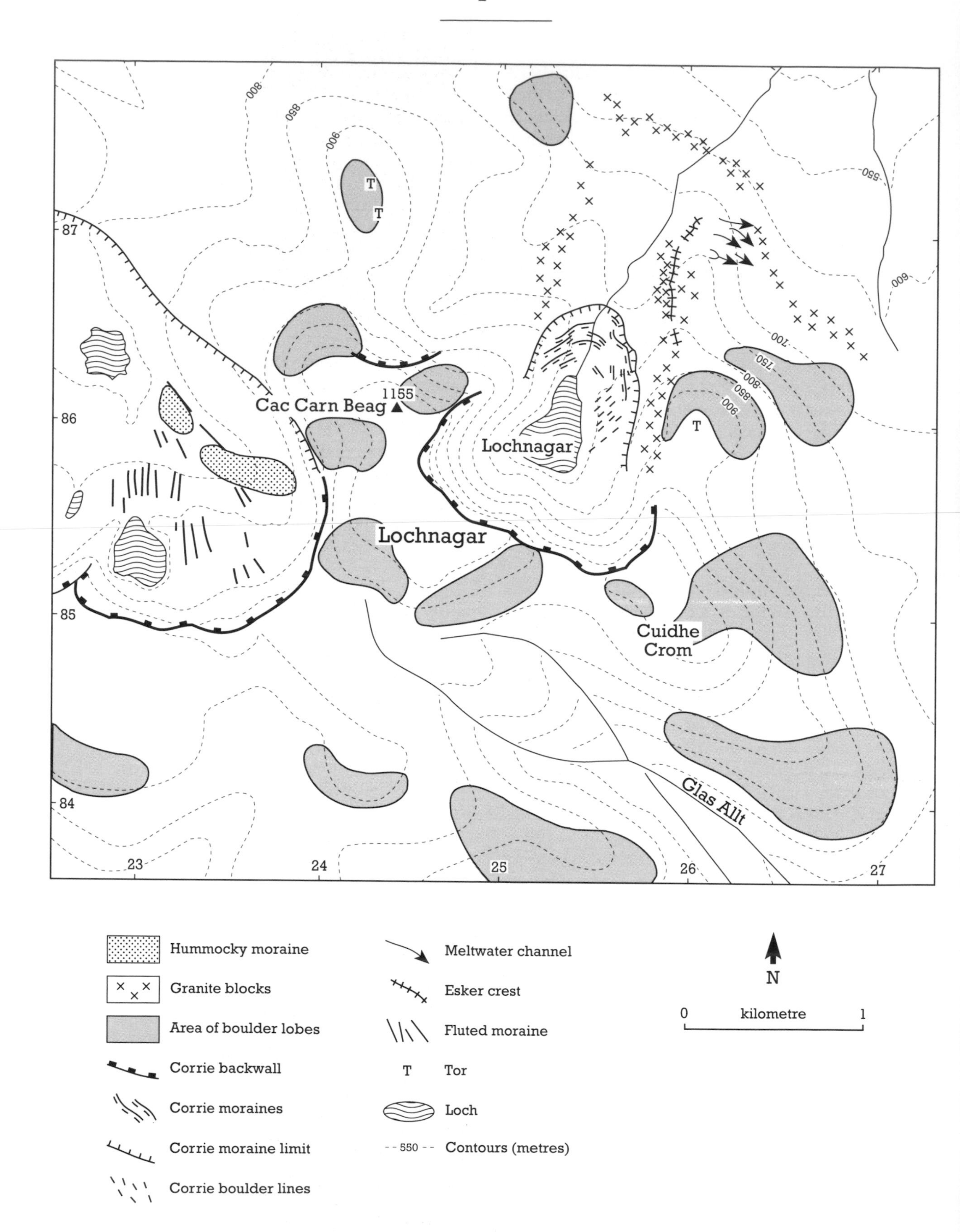

Figure 9.8 Geomorphology of the Lochnagar area (from Shaw, 1977; Clapperton, 1986).

Figure 9.9 Summit blockfield, blockslopes and boulder lobes on the south-east flank of Cuidhe Cròm, Lochnagar. (Photo: J. E. Gordon.)

are similar in composition and thickness. Shaw considered all of these features to be inactive, and concluded that they had crept downslope as a result of the deformation of interstitial ice. Strong evidence of a Lateglacial age for these features has been provided by Sissons and Grant (1972), who observed that they are entirely absent from the areas that were occupied by Loch Lomond Readvance glaciers, and indeed that near Loch Buidhe (NO 252827) the boulder features are apparently truncated by a lateral moraine deposited by the Glen Muick glacier.

Shaw (1977) also made observations on present mass-movement activity in this area. Ploughing blocks are common on slopes of 9°–38°; these range from 0.39 m to 2.4 m in length, and movement is marked by furrows 0.13 m to 3.07 m long and turf banks that have been pushed downslope by as much as 0.83 m above the adjacent ground. Current rates of movement do not exceed a few millimetres per year. Shaw also documented several rockfalls from the Lochnagar corries, and described avalanche tracks 100 m and 180 m wide cutting through woods on the slopes north-west of Loch Muick. These terminate in fan-like avalanche tongues under the surface of the loch. The effects of recent snow avalanche activity in the north-east corrie of Lochnager are relatively minor and restricted to occasional perched boulders, pits in the surface of talus and scratch marks (Ward, 1985b). Similar findings have been reported by Davison and Davison (1987) for an avalanche site in Glen Shee. Elsewhere, however, the geomorphological role of such avalanches is locally more important, for instance, in parts of the Cairngorms and on Ben Nevis (Ballantyne, 1989b; Luckman, 1992).

Interpretation

Lochnagar is important both for a range of individual glacial and periglacial landforms and also for the complete assemblage of features present. The north-east corrie of Lochnagar is a particularly fine example of this landform type, with steep, enclosed rock walls, a loch basin and suite of Loch Lomond Readvance moraines. The latter have had an important bearing on the debate concerning the extent of the Loch Lomond Readvance both locally and regionally. The moraines also demonstrate clearly the progressive active retreat of the glacier back into the corrie and offer scope for correlations with similar features in the Cairngorms and for interpreting the relationships between glacier fluctuations and climate during the readvance. Dating of organic deposits inside the moraines demonstrates conclusively that there was no recrudescence of glacier ice on the corrie floor during the Little Ice Age.

Lochnagar also provides particularly fine examples of periglacial features, notably relict boulder lobes that are among the finest in Scotland.

In many respects the geomorphology of Lochnagar is similar to that of the Cairngorms. Although the assemblage of features on Lochnagar is less complete, it is developed in a much more compact area. Lochnagar therefore provides an outstanding demonstration of key features of the geomorphology of the Loch Lomond Stadial and their spatial relationships, notably boulder lobes and end moraines.

Conclusion

Lochnagar is important for glacial and periglacial geomorphology. It provides a particularly good example of an assemblage of landforms that developed during the intensely cold climatic phase at the end of the Devensian stage and known as the Loch Lomond Stadial (approximately 11,000–10,000 years ago). These include periglacial boulder lobes formed by the slow mass movement of the soil downslope and a series of bouldery moraines formed by a corrie glacier. The landforms are developed in a relatively small area and their interrelationships are clearly demonstrated. Lochnagar forms part of a network of sites representing the geomorphology of the Loch Lomond Stadial.

LOCH ETTERIDGE

M. J. C. Walker

Highlights

Pollen grains preserved in the sediments on the floor of Loch Etteridge provide an important record, supported by radiocarbon dating, of glacial history, vegetational history and environmental change during the Lateglacial and early Holocene in the eastern Grampian Highlands. The site also includes an excellent assemblage of glaciofluvial landforms formed during the melting of the Late Devensian ice-sheet.

Introduction

Loch Etteridge (NN 688929) is located in Glen Truim, a tributary of upper Strathspey. The site occurs at an altitude of 300 m OD, approximately 5 km south-west of Newtonmore. It is important in the context of the Devensian Lateglacial in northern Britain. The sediments preserved on the floor of the loch contain a record of vegetational changes in the Grampian Highlands throughout the Lateglacial and early/middle Holocene, and the radiocarbon assay from the base of the sequence provides a minimum date for the wastage of the Late Devensian ice-sheet from the surrounding area. In addition, stratigraphic data from Loch Etteridge and from the nearby site of Drumochter form the basis for the establishment of a Late Devensian glacial chronology for this part of the Scottish Highlands. The central location of Loch Etteridge in the heart of the Grampian Highlands means that the lithostratigraphic, biostratigraphic and chronostratigraphic evidence from the site are of national significance in terms of both vegetational history and glacial chronology. The radiocarbon dates from Loch Etteridge are discussed in Sissons and Walker (1974), and the palaeoenvironmental data are described in Walker (1975a).

Description

Loch Etteridge lies in a large, dead-ice hollow and is surrounded by a complex system of kames, kame terraces and eskers (Young, 1978). The kame terraces, which slope from *c.* 310–240 m OD towards the north-east along the valley of the Milton Burn (Figure 9.10), contain small kettle holes and comprise rounded cobbles and boulders in a coarse sand matrix (Young, 1978). Kames and eskers occur principally at lower altitudes and often merge into the kame terraces; the eskers are steep, narrow ridges, ranging in height from 1 to 15 m (Young, 1978). According to Young (1978) the kame terraces formed at the margins of the melting ice-sheet.

The loch, which measures approximately 500 m in length and up to 150 m in width, is infilled at the south-west end, where over 7 m of sediment have accumulated. The lowermost sediments show a typical Lateglacial lithological sequence (Sissons *et al.*, 1973) comprising, from the base, gravels, grey silt/clay, green-grey silt/clay, green-brown gyttja, and light grey silt/clay (Figure 9.11). The last named deposits represent the Loch Lomond Stadial, while the underlying sediments are of Lateglacial Interstadial age. These Lateglacial deposits are overlain by Holocene lake muds and peats. Four radiocarbon dates (SRR–301 to SRR–304) were obtained from the Loch Etteridge sediments (Figure 9.11).

The base of the organic sediments in the profile was dated at 13,151 ± 390 BP (SRR–304), the Lateglacial *Empetrum* maximum at 11,290 ± 165 BP (SRR–303), the onset of the Loch Lomond Stadial at 10,674 ± 120 BP (SRR–302) and the close of the Loch Lomond Stadial at 9405 ± 260 BP (SRR–301); the last age determination now appears to be at least 1000 years too young when compared with dates on similar horizons at other Scottish sites (Lowe and Walker, 1984).

Interpretation

Five local pollen assemblage zones were identified in the Lateglacial sediments at Loch Etteridge (Figure 9.11). The lowermost (zone LE–1) is characterized by high values for *Rumex* and *Salix* (including *Salix herbacea*), with significant percentages of Gramineae, *Saxifraga* and *Artemisia*. These pollen spectra are indicative of an open-habitat landscape with a limited shrub component. In the next three zones (LE–2, LE–3, LE–4) there are higher counts of pollen of shrubby, woody plants. The dominant element is *Empetrum*, reflecting the widespread development of *Empetrum* heath in Strathspey during the Lateglacial Interstadial (see also Birks and Mathewes, 1978). There are two maxima in the *Empetrum* curve, one at the initial rise in values for the genus and a secondary peak at the close of the interstadial. Similar double maxima for *Empetrum* have been recorded in Lateglacial Interstadial deposits at Loch Tarff near the Great Glen (Pennington *et al.*, 1972), and at Tirinie in Glen Fender to the south of the Grampian watershed (Lowe and Walker, 1977). The episodes of *Empetrum* dominance are separated by a zone (LE–3) in which there are higher values of *Juniperus* and *Betula*. The majority of the birch grains appear to be from dwarf birch, however, and it seems unlikely that the regional birch treeline reached this area of the Grampians

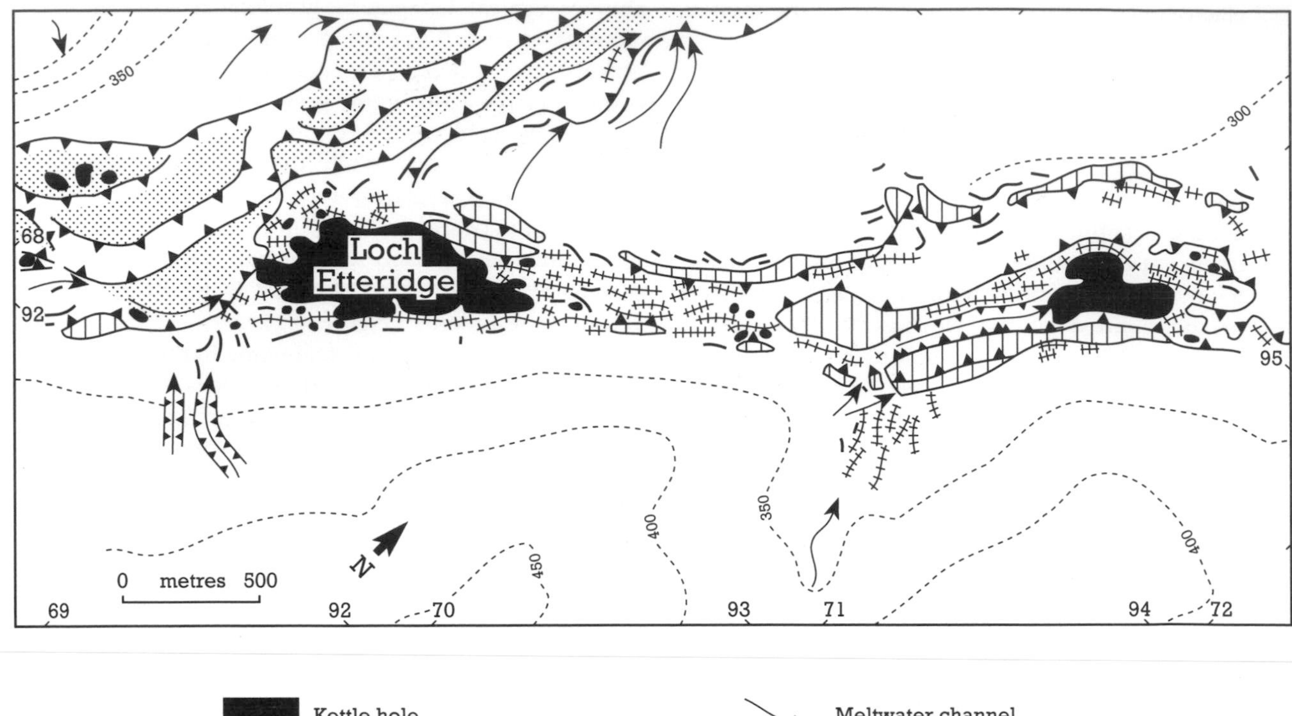

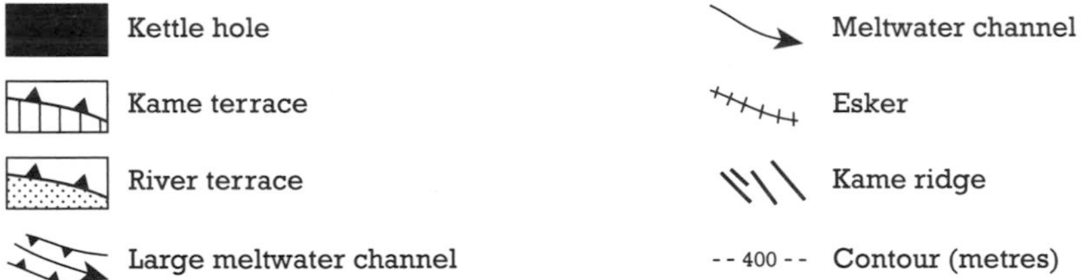

Figure 9.10 Geomorphology of the Loch Etteridge area (from Young, 1978).

during the Lateglacial (Walker, 1984b). Isolated stands of tree birch may have become established late in the interstadial, however, an inference supported by the discovery of tree birch macrofossils in late interstadial sediments at the nearby site of Abernethy Forest (Birks and Mathewes, 1978). *Salix* pollen is found at all levels representing the interstadial, reflecting the presence of shrub willows (probably such northern forms as *S. polaris*, *S. reticulata* or *S. glauca*) on the slopes around the Loch Etteridge basin. The occurrence of *Rumex* as an important element in the pollen spectra is indicative of the incomplete nature of the shrub-heath cover, and the continued presence of bare and disturbed ground throughout the Lateglacial Interstadial.

The Lateglacial Interstadial pollen records from Loch Etteridge and Abernethy Forest, which show the development of a shrub tundra dominated by *Empetrum* with *Betula nana* and *Salix*, are in marked contrast to those obtained from sites in the eastern and southern Grampian Highlands. In those areas the pollen data reflect an interstadial vegetation cover of moss heaths and poor grassland communities on the upper slopes, whereas a closed grassland with juniper, dwarf birch, willow, and stands of tree birch developed on the lower slopes and valley floors. Extensive copses of tree birch may have been found in more sheltered localities of the southern valleys (Walker, 1975b, 1977; Lowe and Walker, 1977; Lowe, 1978; Merritt *et al.*, 1990).

Sediments formed during the Loch Lomond

Figure 9.11 Loch Etteridge: relative pollen diagram showing selected taxa as percentages of total land pollen. The samples for radiocarbon dating were taken from comparable lithostratigraphic horizons in an adjacent core.

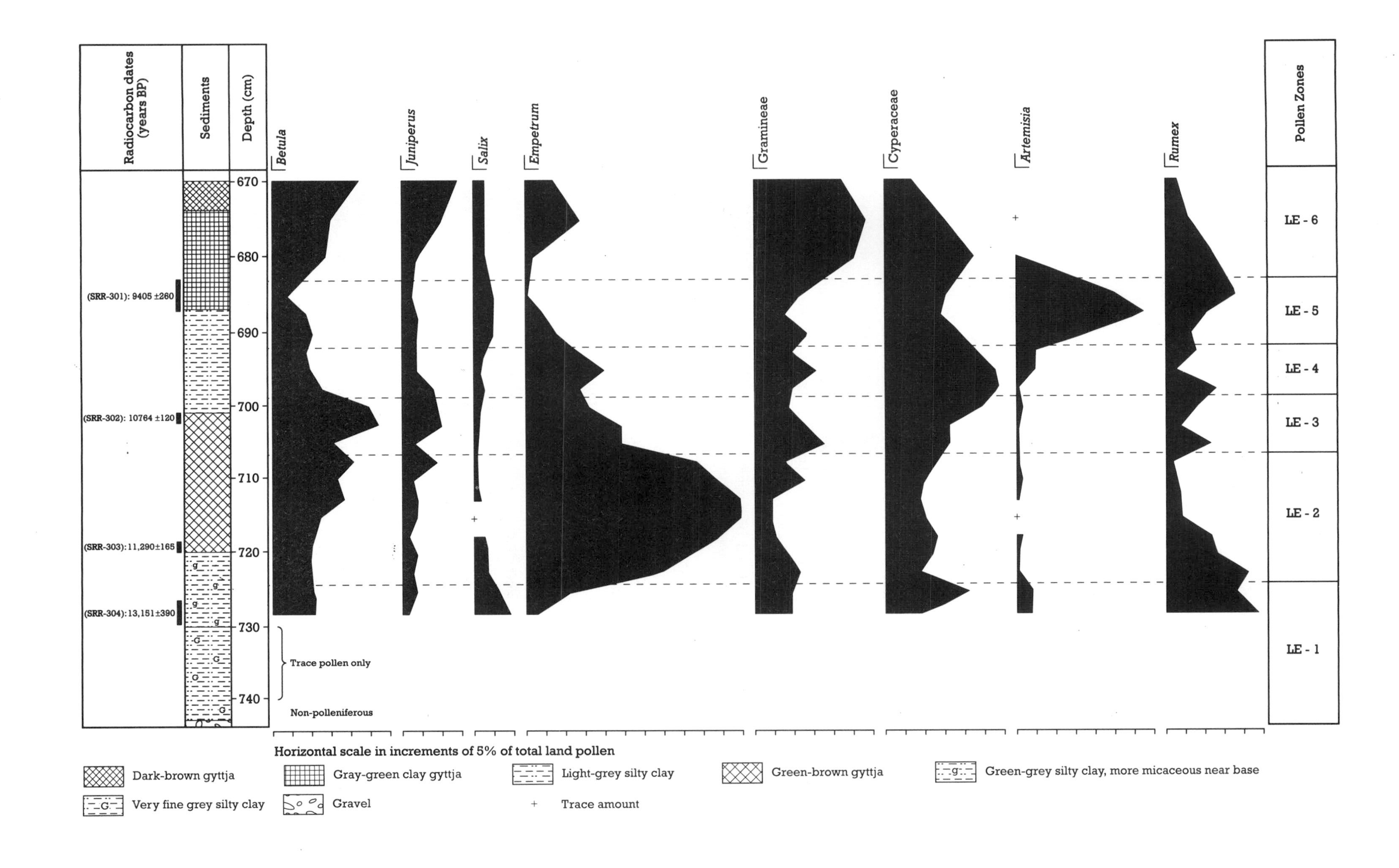
Radiocarbon dates (years BP)
Sediments
Depth (cm)
Betula
Juniperus
Salix
Empetrum
Gramineae
Cyperaceae
Artemisia
Rumex
Pollen Zones
(SRR-301): 9405 ±260
(SRR-302): 10764 ±120
(SRR-303):11,290±165
(SRR-304):13,151±390
670
680
690
700
710
720
730
740
LE - 6
LE - 5
LE - 4
LE - 3
LE - 2
LE - 1
Trace pollen only
Non-polleniferous
Horizontal scale in increments of 5% of total land pollen
Dark-brown gyttja
Gray-green clay gyttja
Light-grey silty clay
Green-brown gyttja
Green-grey silty clay, more micaceous near base
Very fine grey silty clay
Gravel
Trace amount

Stadial at Loch Etteridge are characterized by a relatively low pollen content, uniformly low counts for woody taxa, and the virtual absence of pollen of thermophilous plants. Open-habitat taxa dominate the pollen spectra (zone LE–5), notably *Artemisia* and *Rumex*, along with species of Caryophyllaceae, Chenopodiaceae and *Saxifraga*. *Betula* cf. *nana* and *Salix* cf. *herbacea* are also recorded. An arctic-alpine tundra landscape is indicated, with large areas of bare ground and skeletal soils, intermittent snowbeds, and widespread gelifluction. Of particular significance are the very high counts for *Artemisia* pollen. This is a characteristic feature of Loch Lomond Stadial sediments at other sites in the area (Birks and Mathewes, 1978; MacPherson, 1980), but is found in much lower frequencies at sites in the eastern and south-eastern Grampians (Walker, 1975b, 1977; Lowe and Walker, 1977). In view of the known chionophobous and xerix affinities of many species of *Artemisia*, this has led to the suggestion that a relatively arid climatic regime prevailed in upper Strathspey during the Loch Lomond Stadial (Birks and Mathewes, 1978), whereas in the south-eastern Highlands the spread of *Artemisia* was restricted by snowbeds and higher soil moisture levels (Walker, 1975b). This inference has been supported by palaeoclimatic reconstructions based on glaciological evidence which indicate heavier snowfall in the eastern Grampians brought by winds from the south-east (Sissons and Sutherland, 1976). The initially low frequencies for *Artemisia* pollen in the Loch Lomond Stadial sediments at Loch Etteridge, however, may also reflect changes in snow cover/precipitation levels during the stadial (MacPherson, 1980; Lowe and Walker, 1986a), with heaviest snowfall being implied during the early and later parts of the stadial, related to the initial southward and subsequent northward migration of the oceanic Polar Front (Sissons, 1979d; see also Tipping, 1985).

Of the radiocarbon age determinations, that on the basal organic sediments is the most significant. As the date was obtained from a bulk sample of organic lake mud, the possibility of an ageing effect produced by the hard-water factor or by contamination by older mineral carbon residues cannot be excluded (Sutherland, 1980; Walker and Harkness, 1990). However, the date is comparable with those obtained from the base of a number of Scottish Lateglacial profiles (Bishop, 1963; Kirk and Godwin, 1963; Pennington, 1975b; Lowe and Walker, 1977; Vasari, 1977; Walker and Lowe, 1982) and, if correct, supports the concept of climatic amelioration in the Lateglacial Interstadial at or before 13,000 BP (Coope, 1975, 1977; Atkinson *et al*., 1987). By that time, Late Devensian glacier ice had disappeared from upper Strathspey and, by implication, from much of the Grampian Highlands. Whether Scotland was completely deglaciated during the Lateglacial Interstadial (Sissons, 1974c, 1976b; Sissons and Walker, 1974), however, remains a matter for conjecture (Sutherland, 1984a).

The stratigraphy and pollen content of the sediments at Loch Etteridge and the nearby site of Drumochter have implications for the Late Devensian glacial sequence in this part of the Grampian Highlands. As Loch Etteridge contains a suite of Lateglacial sediments, the surrounding glaciofluvial landforms (Young, 1978) must be the product of the decay of the Late Devensian ice-sheet. To the south and east of the site, however, readvance limits have been identified (Sissons, 1974b) relating to outlet glaciers from a later ice-cap which developed on the Gaick Plateau. The most clearly defined of these limits occurs some 13 km to the south of Loch Etteridge, where a series of hummocky moraines terminates abruptly and is succeeded by outwash. A deep kettle hole within the hummocky moraines on the Drumochter Pass contains only Holocene sediments (Walker, 1975a). The pollen records from sediments 'inside' and 'outside' the glacier limits, therefore, suggest that these readvance limits date from the Loch Lomond Stadial (see Mollands and Tynaspirit).

Loch Etteridge contains a record of environmental change in the central Grampians throughout the Lateglacial period, the palynological data providing the basis for both vegetational and climatic reconstructions. The site occupies a critical position immediately to the north of the Highland watershed and is a key element in a network of sites that together provide a regional pattern of landscape change in the Grampian region throughout the Lateglacial. Loch Etteridge is also an important element in the establishment of a glacial chronology for the Grampian Highlands, the radiocarbon date from the base of the profile providing a minimum age for the disappearance of Late Devensian ice from much of the Scottish Highlands and, by implication, from the British Isles as a whole.

The site is also notable for glacial geomorphology. It provides an excellent assemblage of glaciofluvial landforms associated with the deglaciation of the Late Devensian ice-sheet. In a relatively compact area eskers, kame terraces, kames and kettle holes are all represented.

Conclusion

The sediments at Loch Etteridge provide a valuable record of the environmental history of the eastern Grampian Highlands during the Lateglacial (approximately 13,000–10,000 years ago), after the wastage of the last ice-sheet. Pollen grains preserved in the sediments show the development of open-habitat vegetation, followed by the spread of heath and shrubs with some tree birch. Open habitat species then became dominant again as climate deteriorated during the Loch Lomond Stadial (about 11,000–10,000 years ago). Loch Etteridge is an important member of the network of sites that record the pattern of landscape changes during the Lateglacial, providing valuable comparisons with sites in the southern and eastern Grampians. It is also significant for the evidence it provides for establishing a glacial chronology (time framework) for the area.

ABERNETHY FOREST

J. E. Gordon

Highlights

Pollen and plant macro-fossils preserved in the sediments which infill a topographic depression at Abernethy Forest provide an important record of Lateglacial and Holocene vegetational history. This record is particularly important in the context of the development of native Scots pine forest.

Introduction

The site known as Abernethy Forest (NH 967175) comprises a bog infill of a glacial channel or kettle hole between Loch Garten and Loch Mallachie at an altitude of 221 m OD. It is a key biostratigraphic site representing the sequence of Lateglacial and Holocene vegetation development in the western Cairngorms area. It is notable for the length and completeness of its stratigraphic record, the radiocarbon time-scale calibration of the latter, and the detailed studies which have been carried out on its pollen and plant macrofossils (Birks, 1969, 1970; Vasari, 1977; Birks and Mathewes, 1978). One particularly important aspect of the vegetational history preserved in the deposits is a record of the development and evolution of the native Scots pine forest, remnants of which still occur near the site.

The Late Devensian pollen and plant macrofossils and their dating have also been considered by Vasari (1977). O'Sullivan (1970, 1974a, 1974b, 1975) studied the Holocene vegetational history of nearby sites at Loch Garten and Loch a'Chnuic within the wider area of Abernethy Forest and also at Loch Pityoulish (O'Sullivan, 1976).

Description

The bog lies in an area of glaciofluvial deposits related to the downwastage of the Late Devensian ice-sheet (Young, 1977a). Cores from the bog have been described by Birks (1969, 1970) and Birks and Mathewes (1978). The sequence of sediments in a 5 m long core (Birks and Mathewes, 1978) comprises silt, sand, a series of detritus muds and peat (Figure 9.12). Seven radiocarbon dates (Q–1266 to Q–1272) were obtained from the sediments (Figure 9.12).

Interpretation

Birks (1969, 1970) defined five regional pollen assemblage zones in the sequence at Abernethy Forest (Figure 9.12). The Gramineae–*Rumex*–*Artemisia* assemblage zone at the base was subsequently confirmed to be of Late Devensian age by Birks and Mathewes (1978) and was subdivided by them into three parts. The first, (AFP–1) (Figure 9.12) dated between 12,150 and 11,650 BP, was characterized by a low pollen influx dominated by *Salix*, Cyperaceae, Gramineae and *Rumex acetosa* type, indicating tundra conditions and an open treeless vegetation of largely sedge and grass pioneer communities colonizing recently deglaciated moraine. Similar plant assemblages have been reported from Loch Etteridge (Walker, 1975a), Loch of Park and Loch Kinord (see Muir of Dinnet) (Vasari and Vasari, 1968) and Garral Hill (Donner, 1957). The second part (AFP–2), dated between 11,650–11,150 BP, was characterized by a higher pollen

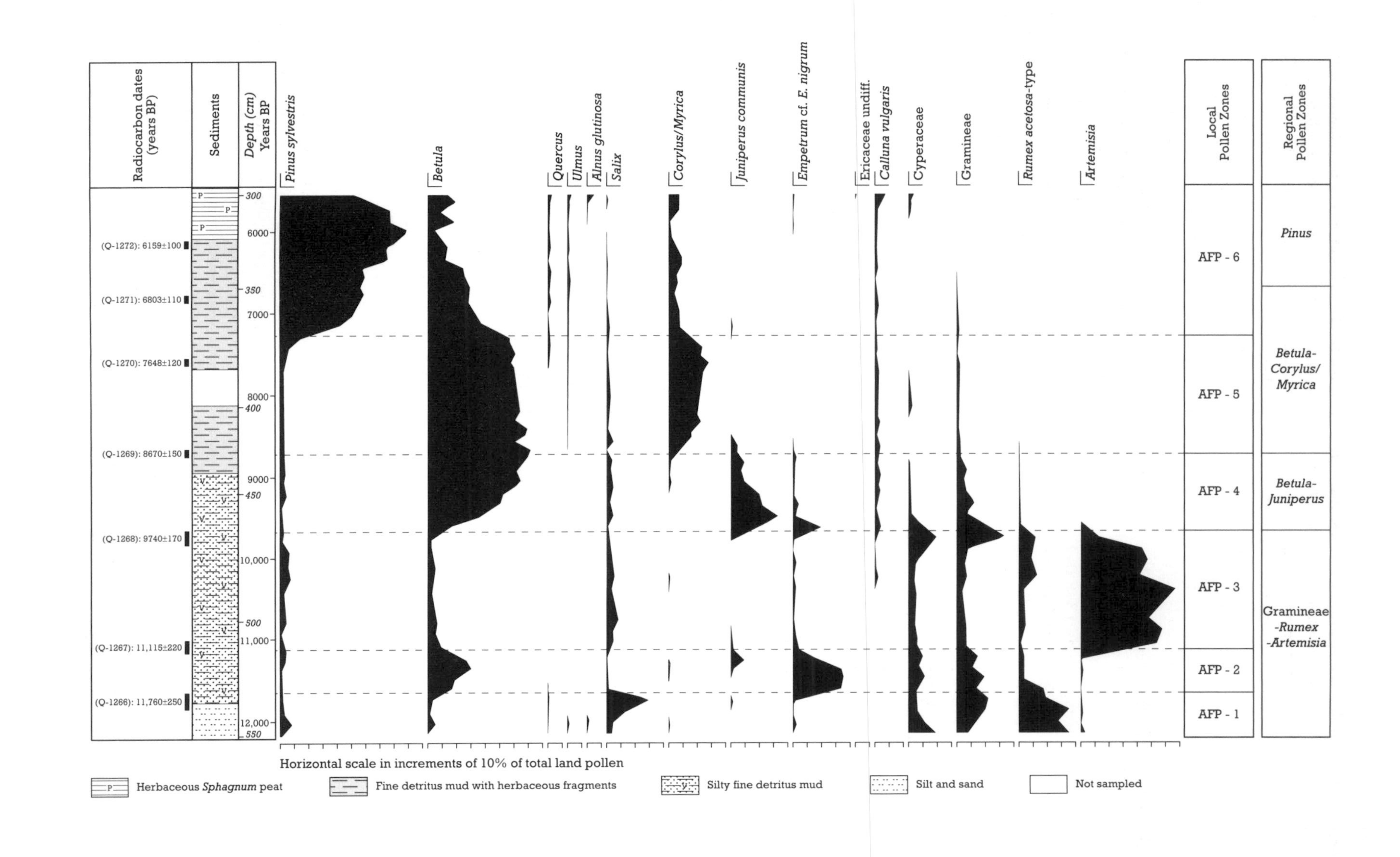

Radiocarbon dates (years BP)
Sediments
Depth (cm) Years BP
Pinus sylvestris
Betula
Quercus
Ulmus
Alnus glutinosa
Salix
Corylus/Myrica
Juniperus communis
Empetrum cf. E. nigrum
Ericaceae undiff.
Calluna vulgaris
Cyperaceae
Gramineae
Rumex acetosa-type
Artemisia
Local Pollen Zones
Regional Pollen Zones
(Q-1272): 6159±100
(Q-1271): 6803±110
(Q-1270): 7648±120
(Q-1269): 8670±150
(Q-1268): 9740±170
(Q-1267): 11,115±220
(Q-1266): 11,760±250
300
6000
350
7000
8000
400
9000
450
10,000
500
11,000
12,000
550
AFP - 6
AFP - 5
AFP - 4
AFP - 3
AFP - 2
AFP - 1
Pinus
Betula-Corylus/Myrica
Betula-Juniperus
Gramineae-Rumex-Artemisia
Horizontal scale in increments of 10% of total land pollen
Herbaceous Sphagnum peat
Fine detritus mud with herbaceous fragments
Silty fine detritus mud
Silt and sand
Not sampled

influx representative of arctic shrub-tundra dominated by *Betula nana* and *Empetrum*. Around 11,220 BP birch trees began to colonize at the site. Similar assemblages occur at Loch Etteridge (Walker, 1975a) and Loch Kinord (Vasari and Vasari, 1968), and, with slight variations, at Loch of Park (Vasari and Vasari, 1968) and Garral Hill (Donner, 1957). The third phase (AFP–3), dated between 11,190 and 9670 BP, showed a marked vegetational 'revertance' during the Loch Lomond Stadial. The low pollen influx is typical of exposed, unstable habitats and snowbed sites. The high percentage of *Artemisia* pollen, which is characteristic of Loch Lomond Stadial pollen assemblages at sites in Strathspey (for example at Tom na Moine (MacPherson, 1980), Loch Etteridge (Walker, 1975a) and Loch a'Chnuic (O'Sullivan, 1974a)), suggests a relatively arid climatic regime, whereas in lower Deeside (Vasari and Vasari, 1968), parts of the south-east Grampians (Walker, 1975b; Lowe and Walker, 1977) and western Scotland (Pennington *et al.*, 1972; Birks, 1973) less arid conditions appear to have prevailed. Birks and Mathewes (1978) consider that this pattern reflects a precipitation shadow effect in the Cairngorm area and Strathspey (see also Sissons, 1979d, 1980b; MacPherson, 1980; Walker, 1984b). After about 9910 BP there are signs of increasing vegetational stability and climatic improvement prior to the spread of *Betula* and *Juniperus* in the early Holocene which has been dated to *c.* 9670 BP.

Birks and Mathewes (1978) noted a broadly consistent Late Devensian pollen stratigraphy between sites in Strathspey (Abernethy Forest, Loch a'Chnuic and Loch Etteridge) and Deeside (Loch Kinord). However, the radiocarbon geochronology for Loch Etteridge (Walker, 1975a) suggests that the three corresponding pollen zones at that site are 300–500 years younger. Birks and Mathewes suggested that the higher altitude of the Etteridge site (300 m OD) might explain later climatic amelioration, but this did not satisfactorily account for the apparently later onset of the Loch Lomond Stadial.

Figure 9.12 Abernethy Forest: relative pollen diagram showing selected taxa as percentages of total land pollen (from Birks and Mathewes, 1978). Regional pollen assemblage zones are from Birks (1970). Note that the data are plotted against a radiocarbon timescale.

Vasari (1977) has also investigated the Lateglacial pollen and plant macrofossils of a core from the Abernethy Forest site. He subdivided the profile into conventional Lateglacial pollen zones following principles outlined in his earlier work (Vasari and Vasari, 1968). The succession was generally in agreement with that of Birks and Mathewes (1978), although there were some differences in the relative dating of particular events. Vasari obtained the following dates for zone boundaries of the Jessen–Godwin scheme: 12,710 ± 270 BP (Hel–424) (I/II), 11,260 ± 240 BP (Hel–423) (II/III), and 10,230 ± 220 BP (Hel–422) (III/III–IV). The middle date was similar to that of the corresponding boundary in the zonation scheme of Birks and Mathewes (1978). However, Vasari's dates implied that organic sedimentation began rather earlier than suggested by Birks and Mathewes. The date obtained on the lowermost level was earlier than that for the start of the Allerød chronozone (*sensu stricto*) and the corresponding boundary in Birks and Mathewes' scheme. The basal Holocene date was also somewhat earlier than that obtained by Birks and Mathewes. Vasari's Abernethy Forest dates were also consistently older than those he obtained for corresponding zones at Loch Kinord and Loch of Park, although the youngest is broadly comparable in the latter case. He raised the possibility of a hard-water error in the Abernethy Forest dates, but based his deductions on the assumption that they were correct. He inferred that climatic amelioration began relatively early and progressed in an uninterrupted manner until the middle of the Allerød.

Vasari's (1977) date of 10,230 ± 220 BP (Hel–422) for the zone III–zone III/IV boundary at Abernethy Forest is broadly comparable to those from a variety of sites, both in eastern and western Scotland (see Gray and Lowe, 1977b, table 1). If correct, this suggests some degree of regional synchroneity.

At Abernethy Forest, in the profile examined by Birks and Mathewes (1978), the Holocene part of the sequence commences with the *Betula–Juniperus* zone, dated between 9670 and 8740 BP. This zone records the replacement of open, unstable habitats by a stable, shrub-dominated vegetation, particularly *Juniperus* (which increased to a peak and then declined) and *Betula nana* scrub. Birch forest, probably open at the start of the zone, progressively increased in density as colonization by tree *Betula* and then

Corylus took place. Similar assemblage zones are recorded at Loch a'Chnuic (O'Sullivan, 1974a), Loch Etteridge and Drumochter (Walker, 1975a), Tom na Moine (MacPherson, 1980), Loch Kinord and Loch of Park (Vasari and Vasari, 1968), Roineach Mhor and Blackness (Walker, 1975b), and possibly Morrone (Huntley, 1976).

The succeeding *Betula* and *Corylus/Myrica* assemblage zone, dated between 8740–7230 BP probably reflects dense birch–hazel forest. *Quercus* and *Ulmus* pollen are recorded continuously for the first time, but these trees were probably very local in their distribution and confined to the valleys. This zone is also represented at Allt na Feithe Sheilich and Loch Einich (Birks, 1969, 1970), Loch Garten and Loch a'Chnuic (O'Sullivan, 1974a), Tom na Moine (MacPherson, 1980), and in a large number of profiles throughout north and west Britain (Birks, 1970).

In the core examined by Birks and Mathewes (1978), the *Pinus* assemblage zone at Abernethy Forest commenced at 7230 BP and lasted until 5520 BP. From considerations of pollen influx as well as the occurrence of *Pinus* macrofossils in the sediment, Birks and Mathewes inferred that pine first arrived in the area about 7165 BP, but that it did not become established close to the site until about 6800 BP. A similar *Pinus*-dominated pollen zone has been widely recorded in the Cairngorm area, Deeside and Speyside, at Allt na Feithe Sheilich and Loch Einich (Birks, 1969, 1970, 1975), Glen Eidart, Sgòr Mór and Carn Mór (Pears, 1968), Allachy Moss (Durno, 1959), Loch Kinord (Vasari and Vasari, 1968), Loch Etteridge and Drumochter (Walker, 1975a), Tom na Moine (MacPherson, 1980), Morrone (Huntley, 1976), and Loch Garten, Loch a'Chnuic and Loch Pityoulish (O'Sullivan, 1974a, 1975, 1976). Apparent discrepancies in the rate and timing of the pine expansion in the area were discussed by O'Sullivan (1975) and explained in terms of differential sediment accumulation rates, possibly during periods of low or falling lake levels. However, Birks and Mathewes' (1978) date of 7165 BP is comparable to the age determination of 7000 BP derived by Birks (1969, 1970) from pine stumps at Loch Einich and Allt na Feithe Sheilich and with a date of 7100 BP interpolated by Birks and Mathewes from O'Sullivan's (1975) results at Loch Pityoulish. It is also comparable with the oldest date of 7350 ± 85 (IRPA–594) obtained by Dubois and Ferguson (1985) from pine stumps on the northern slopes of the Cairngorms. In addition Birks and Mathewes point out that a date of 7585 ± 335 BP (UB–852) from Loch Garten (O'Sullivan, 1974a) accords with the others if its large standard deviation is taken into account.

Following its establishment, pine came to dominate the natural forest on the more acid, well-drained soils of the Cairngorm area up to a treeline altitude of 793 m OD (Pears, 1968, 1972), and formed an important unit of one of the major distinctive forest regions of Scotland (McVean and Ratcliffe, 1962). The pollen evidence shows that thermophilous deciduous trees, such as *Ulmus* and *Quercus*, were comparatively rare in the area during the Holocene, probably due to the relatively continental climate, particularly the severe winters.

Alnus also appears to have been relatively rare in the area during the Holocene, being confined to stream-sides and fens. The *Alnus* rise at Abernethy Forest was estimated at 5520 BP (Birks and Mathewes, 1978), which is in agreement with the date of 5548 ± 50 BP (SRR–459) from Loch Pityoulish (O'Sullivan, 1975, 1976). A slightly older date of 5860 ± 100 BP (UB–851) was obtained from Loch Garten but, like the date for the *Pinus* expansion, the difference is probably not significant in palaeoecological terms (Birks and Mathewes, 1978). On Deeside, also, the pine forest appears to have been established before the alder rise, which occurred there sometime after 6700 BP (O'Sullivan, 1975).

Birks and Mathewes (1978) did not sample the sediments corresponding with the *Calluna–Plantago lanceolata* zone, which had previously been reported by Birks (1969, 1970). The zone is characterized by a reduction in woodland cover and increased non-arboreal pollen frequencies, especially *Calluna*. The decline of woodland is more marked in upland sites, for example at the Loch Einich type locality for the zone (Birks, 1969, 1970), than at Abernethy Forest where the pine forest, although it may have thinned, never entirely disappeared. The evolution of the pine forest in the Abernethy area during the late Holocene, when anthropogenic effects intruded, has been considered by Steven and Carlisle (1959) and O'Sullivan (1970, 1973a, 1973b, 1974a, 1974b, 1976, 1977).

An important point made by Birks (1970) about the vegetation succession in the area following the climatic amelioration at the beginning of the Holocene is that it can be explained in terms of biological and environmental factors without invoking further climatic change. She concluded that the order of immigration of species reflected the distance from their refuge and their rate of migration, and that their relative abundance was a function of the regional climate of the area, soil factors, and competition among species. More recently human interference has been an additional factor. Such variables also account for the distinctive forest history and patterns of the Cairngorm area when seen in an overall national perspective (Birks, 1977).

The development of the hydrosere at Abernethy Forest through lake to bog communities was traced by Birks and Mathewes (1978), mainly from the macrofossil stratigraphy. In general the aquatic plant succession is similar to that at the other sites so far investigated in Scotland, at Loch of Park, Loch Kinord and Drymen (Vasari and Vasari, 1968), although there are some local differences at Abernethy Forest, including a delay to the typical early Holocene plant expansion. Birks and Mathewes note that Abernethy Forest is additionally interesting for the record it provides of hydroseral development from open water through a relatively brief poor-fen stage to a *Sphagnum*-dominated acid mire.

Abernethy Forest is an outstanding biostratigraphic locality and is particularly important in demonstrating the Lateglacial and Holocene vegetation history of the Strathspey and Cairngorm area. It is especially significant in the context of the development and history of the native pine forest. It has also been studied in greater detail than most other sites in terms of combined pollen and plant-macrofossil analyses. Furthermore, it provides important contrasts with sites further west in Scotland. In a wider context, the Loch Lomond Stadial in Scotland is more pronounced and more intensively recorded biostratigraphically than anywhere else in north-west Europe, and Abernethy Forest contributes significantly to the detail of this record.

Conclusion

Abernethy Forest is an important reference site for studies of environmental changes during the Lateglacial and Holocene, that is approximately the last 11,000 years. The pollen and larger plant remains preserved in the sediments have been studied in considerable detail and they allow valuable comparisons with records from sites in other areas. Abernethy Forest is an integral member of the network of sites recording the vegetational history of Scotland and its major regional variations. It also contributes significantly to understanding the development of the native pine forest.

ALLT NA FEITHE SHEILICH

H. J. B. Birks

Highlights

Stream sections eroded in blanket peat by the Allt na Feithe Sheilich have provided an important pollen and plant macro-fossil record, supported by radiocarbon dating, of vegetational history during the Holocene. This record is particularly significant for understanding the development of blanket bog and pine forest.

Introduction

This site (NH 850260) is located on the Monadhliath Plateau at an altitude of about 600 m near the summit of Carn nam Bain-tighearna and comprises a large area of blanket bog that is presently being eroded by the headwaters of the Allt na Feithe Sheilich. The site contains radiocarbon-dated pine stumps and birch-wood layers buried within the peat and is of considerable importance in understanding the local and regional Holocene vegetational history of the eastern Highlands of Scotland. The peat sequence is one of the oldest blanket-peat profiles in the British Isles, as it extends almost to the Late Devensian/Holocene boundary. The site was first described by Lewis (1906) (his Spey–Findhorn watershed site) in his study of buried tree-layers in Scottish peats, and subsequently by Samuelsson (1910). The stratigraphy, palaeobotany, and palaeoecology of the sequence have since been studied in detail by Birks (1975).

Description

Lewis (1906) recorded two layers of pine stumps with birch wood below in two sections, and one layer of pine stumps with birch wood below in a third section. Samuelsson (1910) reinvestigated the site and found only one layer of pine. Birks (1975) similarly found only one layer of stumps underlain by peat rich in birch-wood remains, but overlain by peat with an upper indistinct birch-wood horizon.

Birks (1975) prepared a detailed pollen diagram from a 3.25 m deep peat profile (Figure 9.13). The profile contained large pine stumps overlying 1.3 m of humified peat with frequent fragments of birch and willow wood. The pine stumps are overlain by 1.8 m of humified *Sphagnum* peat containing an indistinct layer of birch and *Calluna* twigs at 1.5 m depth. One pine stump yielded a radiocarbon date of 6960 ± 130 BP (K–1419) and the upper birch-wood layer is dated to 4425 ± 100 BP (Q–886). The profile is divided into eight local pollen zones that are correlated with the four, radiocarbon-dated, Holocene regional pollen-assemblage zones established for the Cairngorm area by Birks (1970) (see also Birks and Mathewes, 1978) (Figure 9.13).

Interpretation

The record of regional vegetational history preserved in the Allt na Feithe Sheilich deposits suggests that before 9400 BP open birch–willow–juniper scrub with a wide variety of herbs was widespread. From 9400 BP to 7500 BP birch woodland with some hazel and willow was predominant. At about 7500 BP pine migrated into this area of Scotland (Birks, 1977) and became the co-dominant tree of the region along with birch. The pollen assemblage of the top 0.5 m of the succession reflects the regional clearance of these pine and birch woods and the extensive development of *Calluna* moor.

The local vegetational history of the site suggests that wet mesotrophic birch–willow woods with abundant *Sphagnum* and *Empetrum* and some *Calluna vulgaris* developed in poorly drained hollows on the Carn nam Bain-tighearna plateau at about 9400 BP. By 9000 BP the field-layer composition of these woods changed to an abundance of grasses, sedges, and *Melampyrum* and a variety of fen herbs. Such communities may have resembled modern 'lagg' communities of bogs in central Sweden. This type of community is extremely rare in the British Isles today, owing to Man's drainage activities or, as at Allt na Feithe Sheilich, to subsequent natural burial by the growth of blanket peat, leaving little trace of these communities except for the wood of birch and willow at and near the base of many blanket peat profiles.

By 7500 BP the local vegetation changed to a drier, more acid *Calluna–Empetrum* bog with some birch, into which pine was able to establish itself, forming an open pine–birch bog with abundant dwarf shrubs. At 6900 BP the bog became wetter, leading to the death of the trees and the development of a *Sphagnum–Calluna* bog. At about 4400 BP the bog became drier again, allowing the local growth of birch, as reflected by the upper layer of birch and *Calluna* remains rich in carbonized fragments. Thereafter, *Eriophorum vaginatum* became dominant to form the characteristic ombrotrophic *Eriophorum–Calluna* blanket-bog community of plateau sites in the eastern Highlands today.

The local vegetational history of the site is of considerable importance because it records the earliest development of blanket bog in the British Isles and the establishment and subsequent demise of pine on the bog. Although Lewis (1906) assumed that buried wood layers in peats of different parts of Scotland were the same age, work by Birks (1975) clearly shows that pine stumps within an area such as the Cairngorms yield different radiocarbon ages. This asynchroneity of woodland development and the varied causes of death of the trees suggest that there were no simple, overriding, regional climatic factors controlling tree growth on peat here or in any other areas of Scotland (Bridge *et al.*, 1990). Tree establishment, growth, and death on blanket peat may have been controlled by small climatic fluctuations to which the vegetation at one site may have been sensitive at a particular time, whereas at other sites it was not, depending on local vegetational succession, aspect, altitude, hydrology and topography. Pine stumps in the Cairngorm area invariably occur at a level just after the peat became acid, but before the establishment of *Calluna–Eriophorum vaginatum* dominated ombrotrophic bog, a habitat which is unsuitable for pine regeneration today. The death of the pines seems to be due to various causes, the most common, as at Allt na Feithe Sheilich, being some increase in wetness of the peat surface. However, such an increase need not be due to any large regional climatic change because McVean (1963a) has shown that a single

Figure 9.13 Allt na Feithe Sheilich: relative pollen diagram showing selected taxa as percentages of the pollen sums indicated (from Birks, 1975).

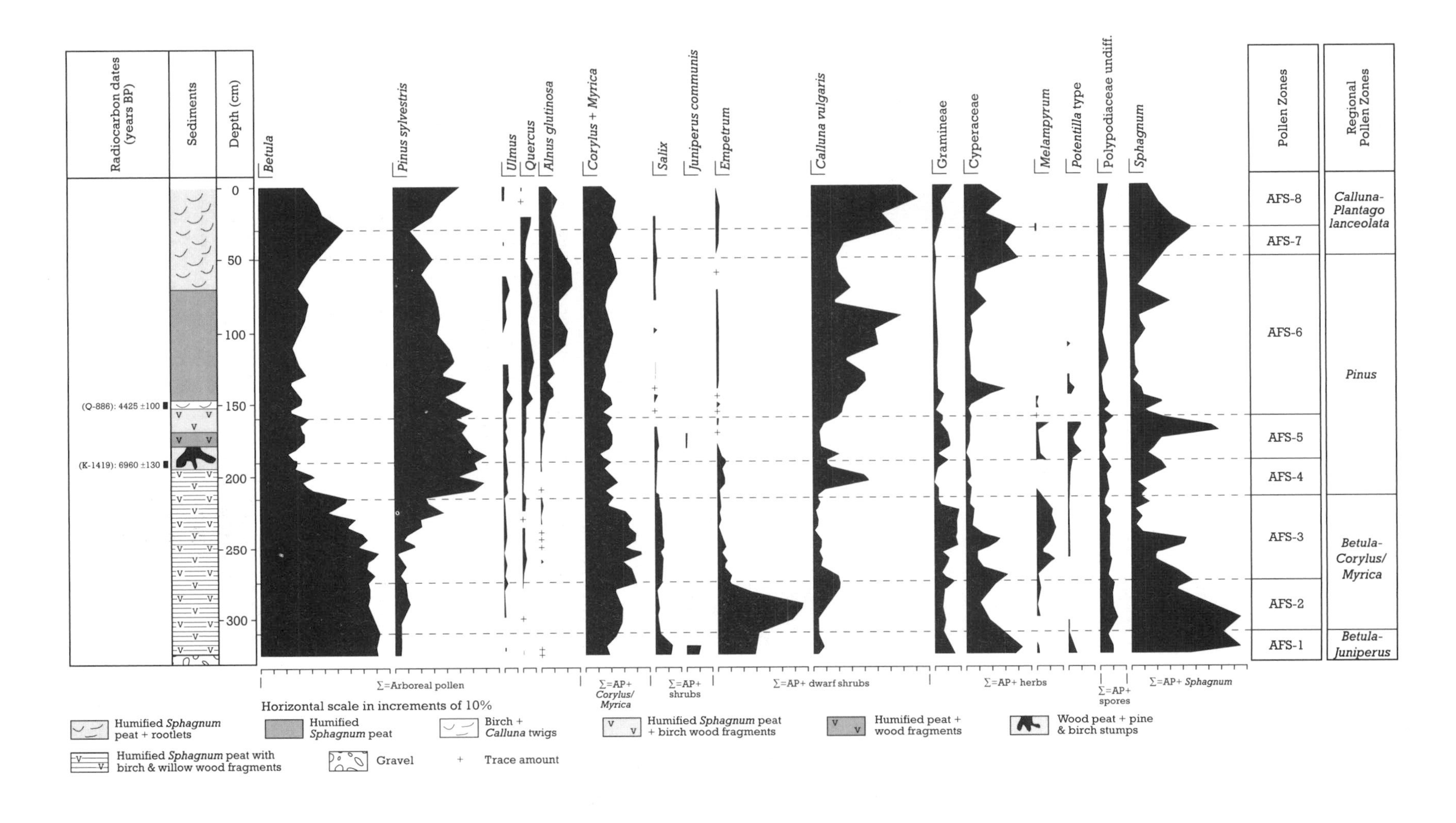

Radiocarbon dates (years BP)
Sediments
Depth (cm)
Betula
Pinus sylvestris
Ulmus
Quercus
Alnus glutinosa
Corylus + Myrica
Salix
Juniperus communis
Empetrum
Calluna vulgaris
Gramineae
Cyperaceae
Melampyrum
Potentilla type
Polypodiaceae undiff.
Sphagnum
Pollen Zones
Regional Pollen Zones
(Q-886): 4425 ±100
(K-1419): 6960 ±130
0
50
100
150
200
250
300
AFS-8
AFS-7
AFS-6
AFS-5
AFS-4
AFS-3
AFS-2
AFS-1
Calluna-Plantago lanceolata
Pinus
Betula-Corylus/Myrica
Betula-Juniperus
Σ=Arboreal pollen
Σ=AP+ Corylus/Myrica
Σ=AP+ shrubs
Σ=AP+ dwarf shrubs
Σ=AP+ herbs
Σ=AP+ spores
Σ=AP+ Sphagnum
Horizontal scale in increments of 10%
Humified Sphagnum peat + rootlets
Humified Sphagnum peat
Birch + Calluna twigs
Humified Sphagnum peat + birch wood fragments
Humified peat + wood fragments
Wood peat + pine & birch stumps
Humified Sphagnum peat with birch & willow wood fragments
Gravel
Trace amount

wet season may be sufficient to kill pines growing in marginal situations.

Allt na Feithe Sheilich is of considerable importance because of the wealth of available palaeobotanical and palaeoecological information about blanket bog development. Blanket bog is better developed in the British Isles than in any other country in Europe. The Holocene history and development of representative blanket bogs are thus topics of considerable importance. Allt na Feithe Sheilich is a well-studied site that is representative of blanket bog in the eastern Highlands. Moreover, it represents the earliest development of blanket peat known in the British Isles and it records the complex local vegetational changes that have occurred associated with the growth and subsequent death of pine and birch on the bog. Although pine stumps and birch wood are locally frequent in blanket peats in the Cairngorm area (Pears, 1964, 1968; Birks, 1970; Dubois and Ferguson, 1985), few sequences are as complete or have been studied in as much detail as Allt na Feithe Sheilich. Recent work (for example, Dubois and Ferguson, 1985; Birks, 1988; Gear and Huntley, 1991) on pine stumps preserved in blanket peat, suggests that such stumps are not only an important 'archive' of Holocene climatic information, but also may reflect abrupt, short-lived climatic perturbations that may be extremely important in understanding climatic change. The site is also important in the context of regional Holocene pollen stratigraphy and vegetational history, as it provides a record from a relatively high altitude (595 m OD), in contrast to the pollen records from the Strathspey lowlands (Birks, 1970; O'Sullivan, 1974a, 1976; Birks and Mathewes, 1978). It is thus one of the most important blanket-peat profiles in Scotland for the elucidation of the Holocene vegetational and environmental history of an upland area.

Conclusion

The peat deposits at Allt na Feithe Sheilich provide important information for interpreting the vegetational history of Scotland during the Holocene (the last 10,000 years). The pollen and tree remains show the early development of blanket bog (around 9000 years ago) and the subsequent growth and decline of pine and birch woodland. Allt na Feithe Sheilich is part of the network of sites demonstrating Holocene vegetational and environmental change and is particularly significant for understanding the history of blanket bog and pine forest development in the uplands.

COIRE FEE

H. J. B. Birks

Highlights

Pollen grains preserved in the sediments on the floor of Coire Fee provide a record, supported by radiocarbon dating, of Holocene vegetational history. This record is particularly significant for understanding the history of the montane species for which this area is of great botanical significance.

Introduction

The floor of Coire Fee (NO 250750) in Glen Doll, in Angus, contains an infilled basin at about 450 m OD located within a series of Loch Lomond Readvance moraines. The sediments in the basin provide a valuable record of Holocene vegetational history, particularly of the montane element in the British flora. This aspect has been investigated by Huntley (1976, 1979, 1981) and the glacial landforms by Sissons (1972a). The site is located within Caenlochan National Nature Reserve, an internationally important location for montane species (Ratcliffe, 1977).

Description

The glacial deposits at Coire Fee comprise an assemblage of hummocky moraines, lateral boulder moraines, fluted moraine, a medial moraine and recessional moraines all associated with a Loch Lomond Readvance glacier that flowed out from the corrie and became became confluent with a glacier in Glen Doll (Sissons, 1972a). The southern and western sides of the basin are flanked by the boulder-strewn slopes below the cliffs of Coire Sharroch and Coire Fee. The cliffs today support a range of rare montane vegetation types and species (Ratcliffe, 1977). The basin contains at least 16 m of stiff minerogenic lake sediments, but the base of the deposits has not been proved (Huntley, 1981). The sediments are very complex, with interbedding and intermixture of muds, silts, and sands. 'No convenient simplification is possible, except by regarding the whole core as one lithological unit within which there is irregular stratification and a slight trend towards lower levels of organic material towards the base' (Huntley, 1981, p. 196). Seven radiocarbon dates are available. The oldest, from the lowermost sediment samples, is 9460 ± 110 BP (Q–1424). The dates form an internally consistent series and permit the calculation of sediment accumulation rates and the dating, by interpolation, of changes in the pollen record. Four local pollen assemblages zones have been identified in the profile (Figure 9.14) (Huntley, 1979, 1981).

Interpretation

Detailed pollen analyses and identification of many rare herbaceous types (Huntley, 1981) provide a basis for reconstructing the vegetational and floristic history of the site. The pollen record suggests that the area was never densely wooded, as tree and shrub pollen values rarely exceed 50% of total pollen. Scrub and small areas of woodland of *Betula*, *Corylus*, *Salix* and *Juniperus* occurred locally. *Pinus* was probably never an important component, and *Ulmus*, *Quercus* and *Alnus* were always rare. Areas of grassland, heathland, and fern-rich vegetation were widespread, especially from about 6000 BP. Tall herbs, today confined to ungrazed ledges, were once more widespread (for example *Filipendula*, *Rumex acetosa*, *Valeriana officinalis*, *Trollius europaeus*). Comparatively few changes occur in the pollen stratigraphy over the period 7000–5000 BP. Botanically, the most important feature is the occurrence in low, but significant amounts of pollen grains of many montane species scattered through much of the sequence – *Oxyria digyna*, *Saxifraga stellaris*, *S. aizoides*, *S. oppositifolia*, *S. nivalis*, *Dryas octopetala*, *Silene acaulis*, *Sedum rosea* – thereby providing clear support for Piggott and Walter's (1954) hypothesis of long-term persistence through the Holocene. Vegetationally the important feature is the recent (100–200 years) decrease in trees and the associated expansion of grasslands (Huntley, 1981; Birks, 1988). The pollen record suggests that, for much of the Holocene, Coire Fee was little affected by human activities. However, in the last two centuries areas of scrub and woodland decreased here and in nearby Caenlochan Glen, probably as a result of excessive grazing by sheep and deer causing lack of tree regeneration and associated woodland decline (Huntley, 1981). Mountain glens such as this appear to have been some of the last areas in Britain to have been affected by human impact (Birks, 1988).

Coire Fee is an important Holocene site because it provides unique data on the history of the montane flora in Scotland and on the ecological history of Caenlochan National Nature Reserve. The history of the montane element in the British flora (that is, arctic–alpine, alpine, arctic–subarctic species) has long been an important topic in historical plant geography and Quaternary botany (Godwin, 1975). Debate has centred on possible reasons why concentrations of montane species are restricted to a few areas in the Scottish Highlands (for example, Ben Lawers, Inchnadamph, Caenlochan, Glen Clova), northern England (for example Upper Teesdale), and North Wales (for example, Cwm Idwal). One hypothesis (Pigott and Walters, 1954) proposes that these areas provide not only suitable ecological conditions today (basic soils, steep slopes, cliffs, etc.), but that in the past they also provided open areas where dense forest never developed and where shade-intolerant, slow-growing herbs of low competitive ability could persist from the Lateglacial. Tests of this hypothesis are few; in Scotland, the pollen record at Coire Fee provides an important test. The evidence from the pollen contained in the sequence indicates the presence of significant numbers of montane species in the catchment area and hence supports the hypothesis that the occurrence of these species in this (and other) areas is the result of their survival from Lateglacial times (see Morrone).

The site is also of interest for its assemblage of Loch Lomond Readvance moraines. Not only does it provide good representative examples of the range of moraine types formed in Scotland during the readvance, but also it has research potential for reconstructing the detailed pattern of ice wastage from the distribution of lateral, recessional and hummocky moraines (cf. the Cairngorms, Loch Skene, Coire a'Cheud-chnoic and the Cuillin).

Conclusion

The sediments at Coire Fee, and the pollen they contain, are important for the information they provide on vegetational development during

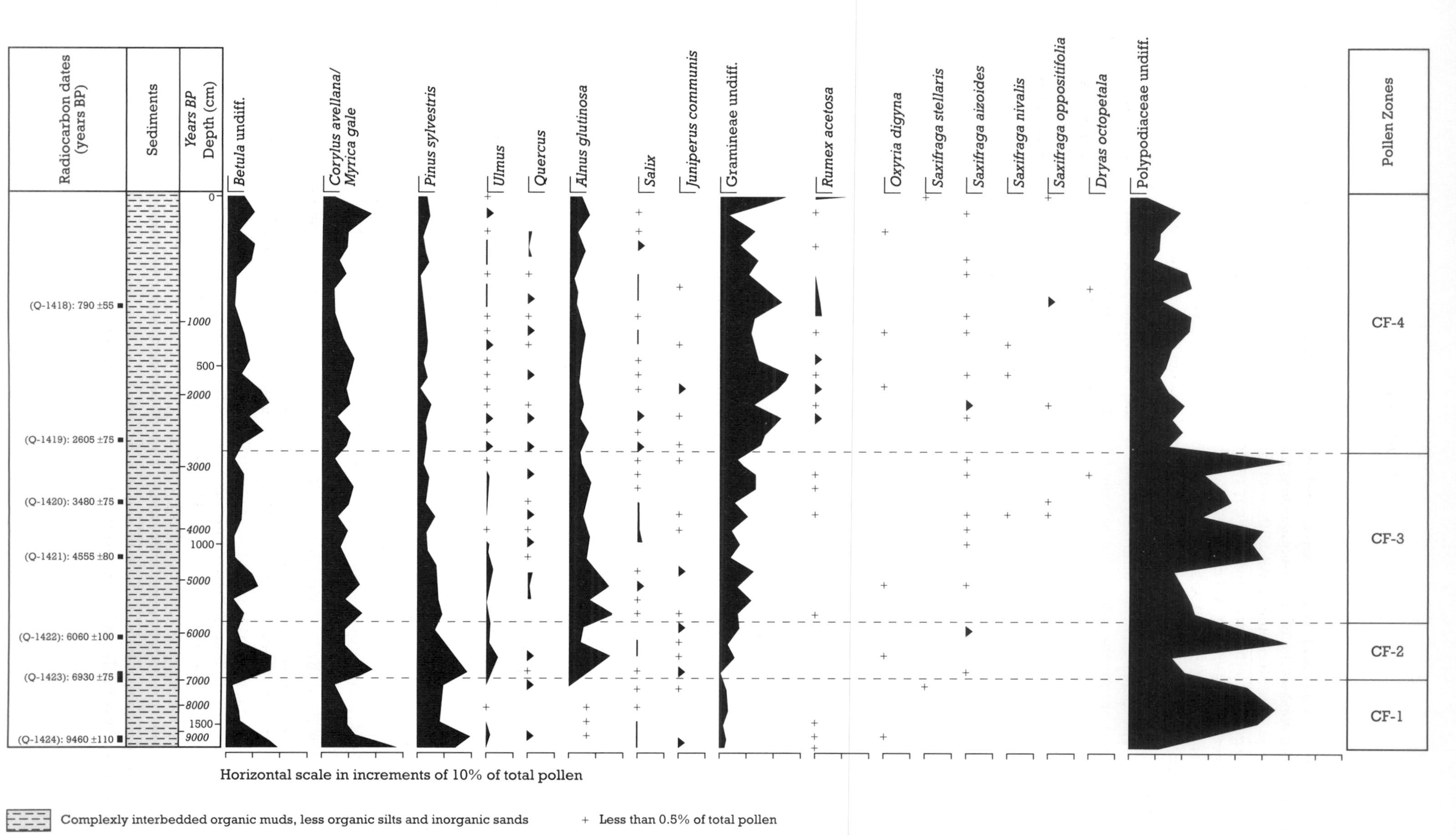

Figure 9.14 Coire Fee: relative pollen diagram showing selected taxa as a percentage of total pollen (from Huntley, 1981).

the Holocene (the last 10,000 years) and, in particular, on the history of the montane species for which the area is noted today. The pollen record supports the argument that these species have survived from preceding Lateglacial times. Coire Fee is one of the few sites in Britain where survival of these species can be demonstrated.

MORRONE

B. Huntley

Highlights

The sediments preserved in a topographic depression on the flanks of Morrone contain a wealth of palaeoecological information. Analysis of pollen and plant macrofossils, supported by radiocarbon dating, has allowed detailed reconstruction of vegetational history and environmental change during the Lateglacial and Holocene in an area of outstanding importance today for its arctic–alpine and northern–montane communities.

Introduction

The Morrone site (NO 135900) is situated at about 420 m OD on the north-facing slope of Morrone, a hill lying immediately to the south of the village of Braemar, on Deeside. The Morrone Birkwoods National Nature Reserve contains a complex of sub-alpine woodlands, scrub, flushes, mires, grasslands, tall herb and upland heath communities (Huntley, 1976; Ratcliffe, 1977; Huntley and Birks, 1979a, 1979b). The Quaternary interest at the site arises principally from the presence within the woodland area of a series of small, shallow, infilled basins which contain Lateglacial and Holocene sediments. Stratigraphic, pollen and plant macrofossil studies, and radiocarbon dating have been carried out upon a core from the deepest of these basins (Huntley, 1976 and unpublished data). These studies have revealed the history of the site since Lateglacial Interstadial times, and the plant macrofossil record for the Lateglacial period is particularly rich, allowing detailed palaeovegetation and palaeoenvironmental reconstructions to be made. The altitude of the site and its relative proximity to Loch Lomond Readvance ice limits, coupled to the known quality of the palaeoecological record that it preserves, make it unique in Scotland.

Description

The basin from which the core used in the palaeoecological studies was collected is located on a relatively gently sloping, till-covered area of the lower slopes of the hill. A small stream flows into the basin at its south-west corner, and there is also general seepage along the western side from a neighbouring basin some 350 m to the south-west. Another small stream drains from the basin on its eastern side.

Although it is primarily surrounded by heathland communities, birch woodland stands approach to within less than 300 m. The surface of the basin itself supports a soligenous mire community, and the inflow stream drains from gravel flushes and soligenous mires within the woodland area upslope of the site.

The core studied extends to a depth of 4.0 m, at which point the corer struck either bedrock or a large rock within the till. The sediments were described in detail by Huntley (1976) and comprise a sequence of organic silty mud, silts, silty muds, silts and fine sands, and gyttja (Figure 9.15).

Huntley (1976) also included a loss-in-weight upon ignition profile for the core which emphasizes the marked difference between the upper 1.5 m or so of often highly organic sediments and the predominantly minerogenic sediments of the remainder of the core.

Ten radiocarbon age determinations have been made on 0.05 m thick samples from the core (Q–1287 to Q–1291, Q–1344 and Q–2316 to Q–2319) (Figure 9.15).

The radiocarbon dates show that the lower 2.5 m or so of the core accumulated during the Lateglacial, with the first sediment accumulating in the basin about 12,600 BP. The dates Q–2319, Q–2316 and Q–1289 are indistinguishable at approximately 9800 BP, indicating a brief time of very rapid sediment accumulation during the early Holocene. In contrast, during the period between 9700 BP and 6600 BP only about 0.17 m of sediment accumulated, a very slow rate of accumulation. Although the uppermost part of the core contains a dating reversal, the results none the less indicate relatively rapid sedimentation once again during the last three to four millenia.

An absolute pollen stratigraphic study was

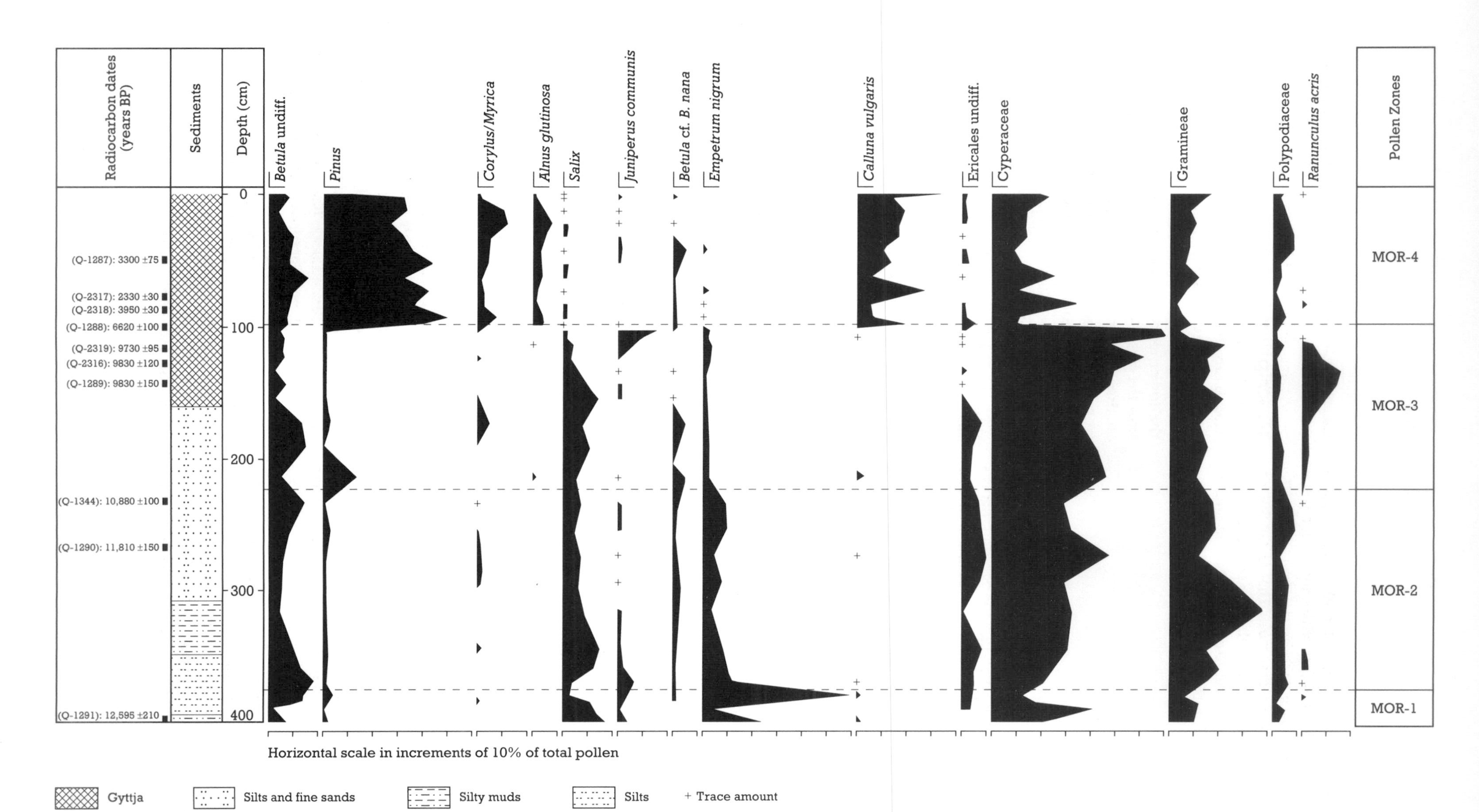

Figure 9.15 Morrone: relative pollen diagram showing selected taxa as percentages of total pollen (from Huntley, 1976).

performed on the core, and both relative pollen diagrams and diagrams of absolute pollen accumulation rates were prepared (Huntley, 1976). The relative pollen diagram (Figure 9.15) was divided into four local pollen assemblage zones (Huntley, 1976), using the results of numerical zonation techniques (Gordon and Birks, 1972) as a guide.

Interpretation

During the time represented by the first zone (MOR–1) dwarf-shrub heaths and grasslands were the predominant local vegetation types, although some scrub of *Juniperus* and *Salix* and even local stands of *Betula* were probably also present. A variety of pollen taxa together indicate the presence also of open, unstable soil areas. The evidence from the second zone (MOR–2) indicates a similar mosaic of vegetation including dwarf-shrub heaths, grasslands, scrub and occasional local stands of *Betula*, but with an increased representation of arctic–alpine and northern-montane taxa and of low-growing herbs, including taxa indicative of open soil areas. The coarse minerogenic sediments at this time, coupled to the evidence of more extensive open soil areas, combine to indicate severe environmental conditions. The environment subsequently stabilized during the third zone (MOR–3), with aquatic taxa becoming abundant in the small lake, increasingly organic sediment accumulating, and the development of extensive *Juniperus* scrub towards the end of the time represented. A vegetational mosaic continues to be present, none the less, although during this period the areas of open soil diminish. Major vegetational changes occurred at the time of the change to the fourth zone (MOR–4); *Pinus* forests became extensive on Deeside and surrounded the stands of *Betula* woodland, with their *Juniperus* understorey, at Morrone. These latter woodlands seem none the less to have persisted, perhaps favoured by the areas of better soils associated with the outcrop of calcareous rock and the base-rich tills at this site.

Although to a large extent the plant macrofossils at the site simply support the interpretation of the vegetational history based upon the pollen record, the macrofossil evidence adds an extra dimension to the palaeovegetational and palaeoenvironmental reconstruction, as well as documenting in much greater taxonomic detail the history of the flora of Morrone, especially during the Lateglacial. The identification of bryophyte macrofossils also provides information about a group of plants that leave no useful pollen and spore record, and yet are important, often dominant, components of many upland and arctic plant communities.

The macrofossil record is notable for its extreme species richness, and for the large numbers of arctic–alpine taxa represented. Most noteworthy perhaps are several taxa that are today absent from the British Isles but are found in Scandinavia (for example, *Papaver radicatum* and *Meesia triticha*). However, the abundant remains of *Polytrichum norvegicum* throughout much of the Lateglacial indicate the presence of long-lying snow patches near to the site, an inference which could not be reached from the pollen data alone, and complete shoots of *Saxifraga oppositifolia* and rosettes of *S. cespitosa* demonstrate that both species were growing in close proximity to the site and give a clearer picture of the nature of the open-soil communities of dwarf herbs present. The wealth of ecological detail provided by the macrofossil records allows an unrivalled picture to be assembled of the Lateglacial vegetation at the site.

Together, the pollen and macrofossil records show that this site, noted today for its rich flora and the abundance of arctic–alpine and northern-montane taxa that it supports, has had a long history of such floristic wealth and biogeographic character.

This site is of regional importance because of the radiocarbon-dated pollen and macrofossil records that are available from it and which are unmatched in terms of altitude and timespan of record within the region. The wealth of data that it has furnished, and the detailed vegetational history which has been reconstructed from these data combine to give the site national importance. The modern Morrone Birkwoods are unique within the British Isles, and share their closest ecological affinities with the sub-arctic birchwoods of Scandinavia. The pollen analytical and macrofossil evidence have documented the occurrence of non-British taxa at the site throughout the Lateglacial and the Holocene. The unusual modern vegetation of the area can therefore be understood in terms of the development of the local vegetational communities during the last 12,500 years.

Conclusion

The plant fossil contents of the deposits on the lower slopes of Morrone Hill and within the Morrone Birkwoods National Nature Reserve are of great importance for studies of vegetational history. Analyses of the distribution of pollen and larger plant remains in vertical sediment profiles, together with radiocarbon dating, have provided palaeoecological data of considerable value and have allowed the history of a nationally unique area of vegetation to be elucidated in great detail. The site is exceptional for the wealth of information on vegetational and environmental changes during the Lateglacial (about 13,000–10,00 years ago) that is available from the larger plant fossil remains.

GLEN FESHIE

A. Werritty and L. J. McEwen

Highlights

The landforms and deposits at Glen Feshie include outwash and river terraces, alluvial fans, palaeochannels and debris cones. This assemblage of features provides an outstanding record of valley-floor and valley-slope development during the Lateglacial and Holocene.

Introduction

The River Feshie is a right-bank tributary of the River Spey, draining a catchment area of 240 km^2 in the western Cairngorms. It is one of the most important sites in Britain for fluvial and Holocene geomorphology. As one of the most active gravel-bed rivers in the country it has attracted considerable research interest, particularly during the 1970s and 1980s (Young, 1976; Buck, 1978; Werritty and Ferguson, 1980; Ferguson, 1981; Ferguson and Werritty, 1983; McEwen, 1986; Robertson-Rintoul, 1986a, 1986b; Brazier, 1987; Brazier and Ballantyne, 1989; Werritty and Brazier, 1991). The River Feshie and the glen it occupies are particularly important in three respects: first, for the unique opportunity they provide for the study of present-day river processes and rates of channel and landform change in a large, highly active, gravel-bed river; second, for the record of such changes in the past, which are represented in documentary, geomorphological and stratigraphic evidence; third, for the unrivalled opportunity they allow to set the present-day river dynamics into a long-term perspective of geomorphological changes during the Lateglacial and Holocene. In scientific terms, these three aspects are closely interlinked, and it is the combination of all three, as well as each individual interest, which distinguishes the site. In this report, the emphasis is placed on Lateglacial and Holocene geomorphology and palaeohydrology, whereas the historical and present-day river dynamics and landforms are to be reviewed in the *Fluvial Geomorphology* volume of the Geological Conservation Review.

Three parts of the glen are important for Lateglacial and Holocene geomorphology and palaeohydrology: (1) a 3.3 km long reach extending from Allt Garbhlach (NN 850952) to north of Achlean (NN 850986); (2) the Allt Lorgaidh fan (NN 842908 to NN 846918); and (3) an area of debris cones extending over a distance of *c.* 0.8 km below Creag na Caillich (NN 853903).

Description

Most of the drainage basin lying between 700 m and 1000 m OD is underlain by Moine schist, but to the north-east the ground rises to 1265 m OD on the Cairngorm granite batholith. The basin is dissected by a steep-sided glacial trough (Linton, 1949a) through which the River Feshie flows westwards before turning north at about 400 m OD into the wider, lower valley cut into glacial tills and outwash, these glacigenic deposits being restricted to the valley floor and lower slopes. Bedrock on the plateau (600–800 m OD) is mantled by blanket peat; at the highest levels bare, frost-shattered regolith occurs. The lower course of the river is confined locally by bedrock outcrops, and more extensively by Lateglacial and Holocene terraces, but in three reaches (upper Glen Feshie, Lagganlia and at the confluence with the River Spey) the river is free to migrate laterally and is actively reworking its floodplain.

The geomorphological impact of Late Devensian ice-sheet wastage in lower Glen Feshie has been discussed in detail by Young (1975a). During the Loch Lomond Readvance, outlet glaciers descended from the Gaick ice-cap northwards into the upper valleys of the Feshie (Sissons, 1974b) and small glaciers occurred in the Cairngorms massif to the east (Sissons,

1979f). Glaciofluvial landforms are abundant in the lower part of the catchment and local accumulations of outwash materials are remarkably thick. Young (1976) identified three stages of terrace development, but more recently five terraces levels have been described in the lower part of the valley by Robertson-Rintoul (1986b). During the Holocene in certain parts of the valley the river has trimmed the distal margins of fans and cones (Brazier, 1987). Rates of channel change, which have resulted in extensive reworking of the floodplain over the past 200 years, are remarkably high for the British uplands (Werritty and Ferguson, 1980; McEwen, 1986).

Terraces and alluvial fans

Glen Feshie is typical of valleys in upland Scotland in containing large accumulations of glaciofluvial and fluvial sediments deposited as valley fills. The particular valley fill in Glen Feshie was created towards the end of the Late Devensian as the ice-sheet in the valley downwasted *in situ* (Young, 1975a). Three groups of landforms comprise the major geomorphological features of this valley fill:

1. kame and kettle landforms and an extensive associated palaeosandur;
2. an extensive suite of terraces;
3. tributary valley alluvial fans.

All of these features are extremely well exhibited in the reach from the Allt Garbhlach to north of Achlean. The dominant landform assemblage within this reach is the Allt Garbhlach fan and dissected palaeosandur (Figure 9.16). The latter is pitted with kettleholes, between which former braided channel networks can be traced. Partially buried by fan deposits, but projecting above the level of the fan are several kames. The fan deposits have also buried the ice-contact slopes between the sandur and the kame and kettle deposits upstream. The fan was probably built during the later phases of Late Devensian ice-sheet deglaciation about 13,000 BP (Robertson-Rintoul, 1986a).

Within this area lying some metres below the level of the 13,000 BP pitted outwash terrace, there is a group of three low-level terraces. The highest of the terraces, dated by soil stratigraphic methods at 10,000 BP, is about 5 m above present river level (Robertson-Rintoul, 1986b). The middle terrace, about 3 m above the present river, has been dated to 3600 BP; the lowest terrace, about 1.5 m above the present river, has been dated to approximately 1,000 BP. All of the more extensive terrace fragments possess well developed, braided palaeochannel networks on the terrace surfaces. A fifth terrace (late 19th century) is not represented within this site.

In this reach north of the Allt Garbhlach, discharges of the prior River Feshie around 13,000 BP were about 520% higher than present discharges. This earlier river was also considerably more braided and had much higher sediment transport rates than those of the present river. The 3,600 BP terrace surface was formed by a river which had discharges 100–120% greater than those of today (Robertson-Rintoul, 1986a). Again the stream was more braided and probably had higher rates of sediment transport than the present-day river. Discharges for the 1000 BP channel were about 8–34% higher than the present-day mean annual flood (estimated at 70–80 $m^3 s^{-1}$).

The river terraces in Glen Feshie are the product of temporal changes in the balance between fluvial transport capacity and sediment supply. The patterns of runoff and sediment production have fluctuated throughout the Holocene in response to climatic change and vegetational disturbance. This has resulted in at least five phases of incision within the main valley floor since 13,000 BP, locally these phases being accompanied by aggradation.

Allt Lorgaidh fan

In the upper braided reach of the River Feshie (*c.* NN 847917 to NN 845937), where the valley floor is almost 0.7 km wide, the terrace fragments become laterally very extensive. The high dissected palaeosandur is not represented in this reach, and the terraces comprise the three low-level late Holocene surfaces discussed above. At the upstream end of the upper braided reach is the tributary valley of the Allt Lorgaidh, which terminates in a complex alluvial fan. This fan probably owes its dimensions to debris provided by meltwaters from a tongue of the Gaick ice-cap which descended into Glen Feshie during the Loch Lomond Readvance (Sissons, 1974b). The eastern side of this fan comprises three units which correlate with the three low-level Holocene terraces in the upper braided reach. On the western side of the Allt Lorgaidh stream the low angle fan has been subjected to cut-and-fill processes. Local trenching of the fan by the

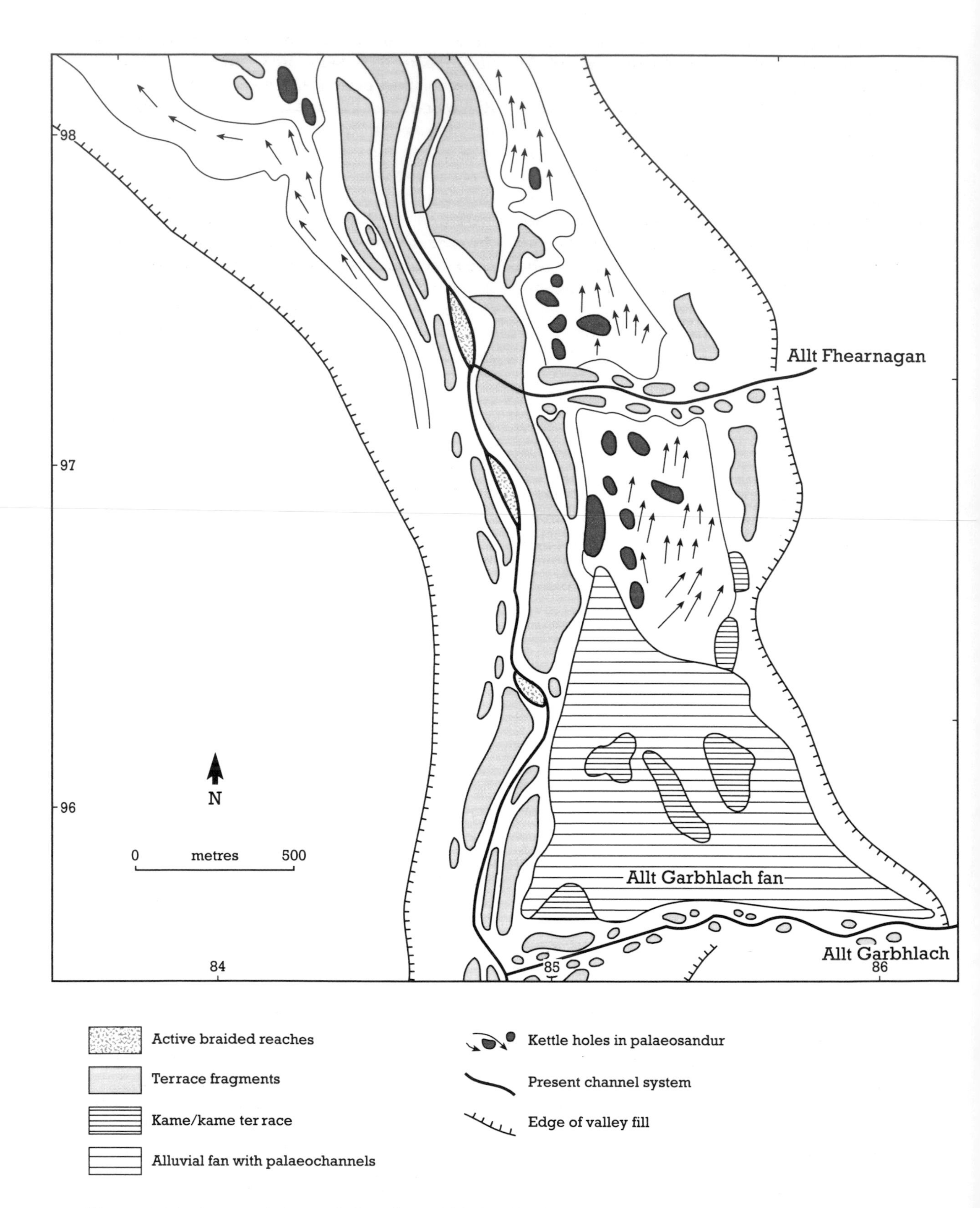

Figure 9.16 Geomorphology of the Allt Garbhlach–Allt Fhearnagan area of Glen Feshie (from Robertson-Rintoul, 1986a; Werritty and Brazier, 1991).

stream has exposed a buried podsol for which a radiocarbon date of 3620 ± 50 BP (Har–4535) has been obtained on charcoal found in the organic-rich layer. This podsol is buried beneath fluvial gravels which comprise the present upper surface of the fan, the latter forming the upper terrace surface in the tributary valley. The buried soil is traceable for some distance upstream in the Allt Lorgaidh and occurs in exposures on both banks of the tributary valley. The date of 3600 BP thus gives an approximate age for the initiation of a phase of late Holocene sediment aggradation in the tributary valley.

In the area of confluence between the River Feshie and River Spey the present alluvial fan is actively reworking a small part of a much larger fan formed during the Lateglacial. On this larger palaeofan a dendritic palaeochannel network can be identified from aerial photographs.

Glen Feshie debris cones

Three coalescing debris cones have built out from the steep gullied walls of the glacial trough in upper Glen Feshie at a mean altitude of 390 m OD. These gullies are cut into the Moine schist of Creag na Caillich from which sediment has been readily supplied into a set of coalescing cones. Basal erosion of these cones by the River Feshie has resulted in the exposure of an extensive section over 60 m long and in places up to 10 m high (Figure 9.17). The exposure consists almost entirely of coarse debris-flow deposits, with poorly sorted and dominantly angular clasts embedded in a coarse sandy matrix. The deposits extend down to the level of the river, with the exception of the northernmost cone which has buried a low river terrace. Stratification is largely absent, although when the section was freshly exposed in 1984 there were linear discontinuities that marked the boundaries between individual debris-flow units (Brazier, 1987). The flow units revealed comprise broad sheets of debris up to 1 m in thickness, which contrasts with open hillslope flows where the forms are narrower and delimited by levées (see the Cairngorms).

Radiocarbon dating of organic material (mainly woody roots) has been undertaken at four of the cones (Figure 9.17) in order to establish the timing of debris cone initiation and the subsequent periodicity in debris-flow activity at the site (Brazier and Ballantyne, 1989). The oldest age of 2090 ± 50 BP (SRR–2877) is from the base of the centre of cone 3. The other dates (320 ± 50 BP (SRR–2880) to 'modern' (SRR–2873, SRR–2874, SRR–2875 and SRR–2879) are too similar to permit any meaningful analysis of the periodicity of debris flows on the cones. It is, however, clear that the three upper debris-flow units have been deposited within the last 300 years. A minimum age for the river terrace buried by cone 1 is 270 ± 50 BP (SRR–2881).

Brazier and Ballantyne (1989) concluded that the aggradation of these debris cones in upper Glen Feshie was initiated by approximately 2000 BP. The site may then have remained stable for about 1700 years until, within the last 300 years, rapid and episodic debris-flow aggradation formed the bulk of the deposits visible in the stream-cut exposure.

It is also important to note the relationship between river undercutting and the source of the debris-flow sediments. The abundance of palaeochannels and the well-defined terraces preserved on the valley floor opposite the cones indicates that the formerly braided River Feshie has repeatedly migrated across the valley floor episodically reworking this area of the valley fill. Prior to 2000 BP the river may have eroded earlier slope deposits at the site currently occupied by the debris cones. Thus the stratigraphy of the present cones only provides evidence for debris-cone activity at this site for a maximum timespan of 2000 years (Brazier, 1987).

Debris flows and cones are a characteristic feature of the Holocene geomorphology of upland Britain (see Eas na Broige and the Cairngorms; Harvey *et al.*, 1981; Innes, 1983b, 1989; Ballantyne, 1986d; Brazier, 1987; Brazier *et al.*, 1988) and recent debris-flow events in these areas have all been triggered by heavy rainstorms (Common, 1954a; Baird and Lewis, 1957; Harvey, 1986; Carling, 1987; Jenkins *et al.*, 1988). Brazier and Ballantyne (1989) considered three possible hypotheses to explain the episodic nature of the Glen Feshie features and the marked increase in activity within the last few hundred years. First, as suggested by Innes (1983b), they may relate to changes in estate management practices and the introduction of systematic burning or overgrazing. Second, they may relate to secular climatic change and the known incidence of increased storminess during the period *c.* 2950–2250 BP and the Little Ice Age of the 16th–19th centuries. Third, they may be controlled by the dynamics of the River Feshie, debris cone formation occurring only when the river followed a course on the opposite side of its floodplain. Brazier and

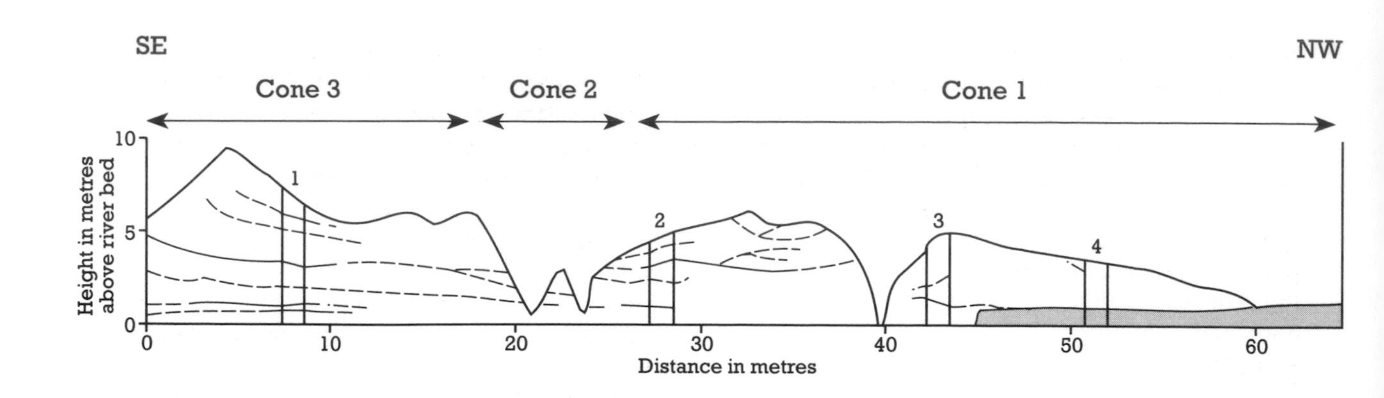

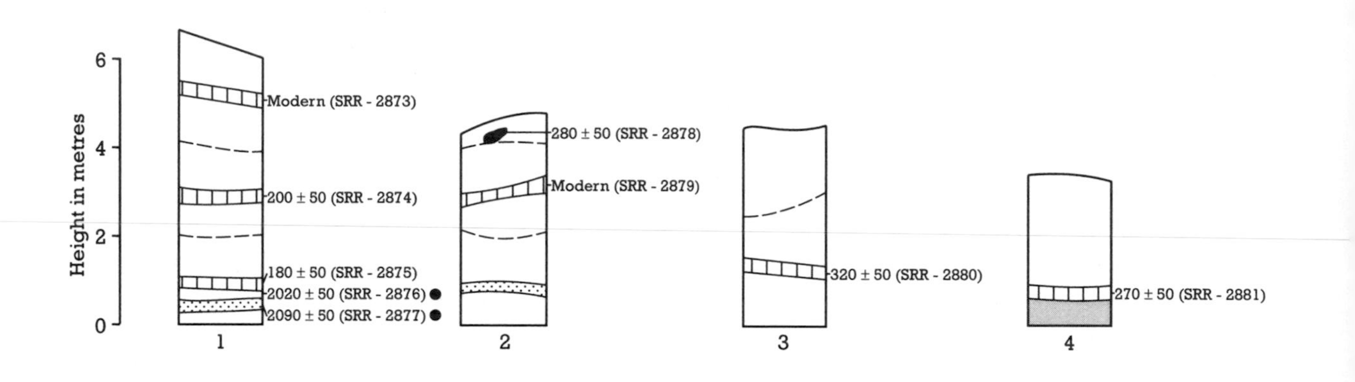

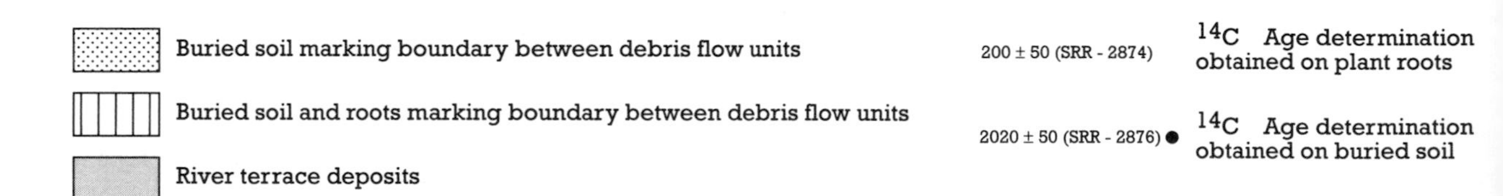

Boundary between debris flow units

Isolated root

1 to 4 Sampling sites

Figure 9.17 Top: surveyed section across the base of the Glen Feshie debris cones showing boundaries between individual debris-flow units. Bottom: detailed sections at sampling sites 1–4 (from Brazier and Ballantyne, 1989).

Ballantyne rejected the first hypothesis partly because the very steep and rocky nature of the source area would probably have precluded systematic forest clearance, and partly because of the absence of charcoal fragments in the sediments. Although the general coincidence in timing of debris-flow activity and known periods of climatic instability was notable, direct relationships were likely to be modulated by other variables. The major control was therefore attributed to the lateral migration of the river channel, although exceptional rainstorms were still required to trigger individual debris flows or periods of increased debris-flow activity.

In relating the Glen Feshie debris cones to natural processes, Brazier and Ballantyne (1989) also highlighted the contrast with debris cones elsewhere where anthropogenic effects had been significant in cone development (see Eas na Broige; Harvey *et al.*, 1981; Innes, 1983b; Brazier *et al.*, 1988). In terms of their more recent activity, the Glen Feshie cones also offer a further contrast with Eas na Broige, where the important control on debris cone formation is exercised by the inherited sediment supply. The geomorphological importance of debris-flow processes in upper Glen Feshie is indicated by the volume of material transported, which, when averaged over the last 300 years, represents an annual accumulation of about 50–60 m^3.

Interpretation

The geomorphological features in Glen Feshie described above are significant in a number of respects.

1. They provide a particularly good assemblage of fluvial and slope landforms and deposits, encompassing outwash and river terraces, alluvial fans, braided palaeochannels and debris cones.
2. Together these features provide one of the most detailed records of valley-floor development in Scotland during the Lateglacial and Holocene. Following deglaciation during the Late Devensian a pitted sandur surface was formed, inset within which a series of alluvial terraces developed during the Holocene. Large alluvial fans formed on the main valley floor at the confluence with tributary valleys. These formed in response to episodic release of large quantities of sediment from the steeper tributary streams. At other sites where slope processes have constructed debris cones directly on to low terraces and the adjacent floodplain, substantial debris-flow activity is reported over the last 300 years. Within the same time-scale, in response to a flashy runoff regime and a steep slope, the River Feshie has extensively reworked its valley floor in three major reaches.
3. The geomorphological and stratigraphic record provides a firm basis for setting the present-day river processes and geomorphological changes into a longer-term perspective.
4. The landforms and deposits in Glen Feshie demonstrate with remarkable clarity the complex nature of the coupling of slope and channel processes in the Scottish uplands and the highly episodic nature of fan and debris cone development.

Conclusion

Glen Feshie is outstanding for an assemblage of landforms and deposits that record the processes and patterns of valley-floor and valley-slope development during Lateglacial and Holocene times (approximately the last 13,000 years). Not only are the individual features particularly well developed, but also the total assemblage is one of the best of its kind in Scotland for the range of evidence provided and the detail of the record. This record is also fundamental to an understanding of the evolution of the present River Feshie, which is a key site for studies of active river processes.

Chapter 10

South-west Highlands

INTRODUCTION

D. G. Sutherland

The area termed the south-west Highlands in this chapter extends south of the Great Glen to the Highland boundary and from the central Grampians to the west coast, including the Kintyre peninsula (Figure 10.1). As elsewhere, there is a considerable range of environments in this region, which is reflected in its Quaternary history. Deposits older than the Late Devensian occur in southern Kintyre, but the main mountain area has been the major centre of successive episodes of ice accumulation and dispersion in Scotland (see Chapter 1), with the result that no deposits older than the Loch Lomond Stadial are known from the central part of the area. The often impressive features of glacial erosion, both in the mountains and in the fjord-like sea lochs, have developed through many periods of both ice-sheet and partial glaciation during the Pleistocene.

Only two glacial episodes are therefore clearly recognized, related to the Late Devensian ice-sheet and the Loch Lomond Readvance. Both are associated with ice flow principally to the west and south-west, radiating out along the troughs and valleys away from the central ice accumulation areas; during the Loch Lomond Readvance, the largest ice-field was centred over the south-west Highlands. The main themes of research in this area have focused on the geomorphology, chronology and vegetation history of the Lateglacial and the significance of the pre-Late Devensian landforms and deposits.

Around the southern Kintyre peninsula there are marine landforms and deposits that pre-date the last glaciation. These consist of rock platforms up to 10 m above present sea level (see Glenacardoch Point), and which are overlain by glacial deposits, and high-level shell beds at altitudes of as much as 45 m OD (see Tangy Glen). The rock platforms are thought to have been formed during interglacials (Gray, 1978a; Sissons, 1981a) although no direct evidence for this is available. The origin and age of the high-level shell beds have been controversial, there being uncertainty as to whether they are *in situ* or ice-transported (Horne *et al.*, 1897; Munthe, 1897; Jessen, 1905; Synge and Stephens, 1966; Sutherland, 1981a) and, if *in situ*, whether they were partly coeval with the last ice-sheet or were significantly older, or represent a glacial–interglacial–glacial cycle. Amino acid analysis of mollusc shells suggests that the deposits are older than the Ipswichian (Gray, 1985; D. G. Sutherland, unpublished data). Their exact position in the Scottish Pleistocene sequence awaits further work.

Studies of raised shorelines around the Scottish coast have indicated that the greatest isostatic depression resulting from the last ice-sheet was in the area of the south-west Highlands (Sissons, 1967a, 1976b, 1983a). That this was the area of thickest ice is also demonstrated by the transport of erratics such as Glen Fyne granite and Rannoch granite to both the east and the west (Sissons, 1967a; Sutherland, 1984a; Thorp, 1987).

Decay of the Late Devensian ice-sheet was accompanied by high relative sea levels (typically 30–40 m OD), marked by raised glaciomarine deltas, shingle ridges and beach terraces, such as occur at Glenacardoch Point. Particularly good examples of outwash terraces grading into raised shorelines occur at the mouth of Glen Scamadale and in the Ford–Kilmartin valley (Gray and Sutherland, 1977). During this period of ice retreat deposition of the fossiliferous Clyde beds began around the coasts (Peacock, 1975c), and dating of the included mollusc shells suggests that the greater part of the Clyde Sea area was free of ice by at least 13,000 BP (Sutherland, 1986). Analysis of the marine macro- and microfaunas in the Clyde beds has indicated that after an initial relatively mild phase at the beginning of the Lateglacial Interstadial, sea temperatures along the west coast of Scotland were 2–3°C cooler than the present for much of the interstadial (Peacock, 1981b, 1983a, 1989b). Interestingly, there is evidence in the inshore marine palaeotemperature record for a brief warmer phase at the end of the interstadial (Graham *et al.*, 1990; Peacock and Harkness, 1990).

The change in terrestrial environments during the Lateglacial Interstadial has been studied by pollen analysis at a number of sites along the western coastal fringe, as at Drimnagall and Pulpit Hill (Donner, 1957; Rymer, 1974, 1977; Birks, 1980; Tipping, 1984, 1986, 1989a, 1991b). During the interstadial these studies demonstrate that the vegetation was dominantly grassland interspersed with clumps of juniper, willow and *Empetrum*. As elsewhere in Scotland, certain sites suggest a brief period of climatic deterioration during the earlier or middle part of the interstadial.

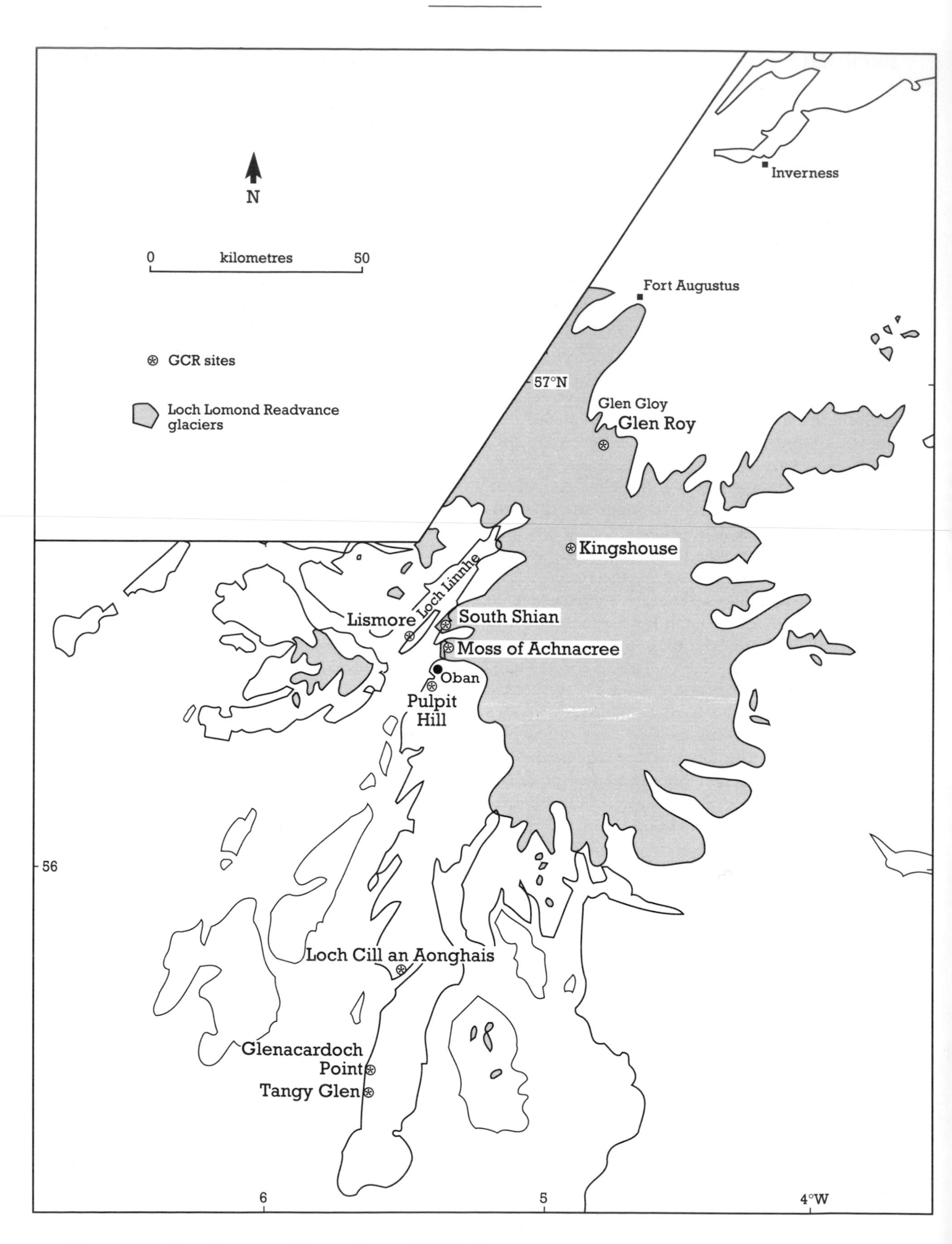

Figure 10.1 Location map of the south-west Highlands.

The effects of the Loch Lomond Stadial climatic deterioration were particularly pronounced in the south-west Highlands. The extent of deglaciation during the preceding interstadial is not known but it seems probable that at least all the sea lochs were ice-free (Sutherland, 1981b). Thereafter, a major readvance of the ice occurred, with the principal ice mass being centred on the hills to the west of Rannoch Moor (Thorp, 1986, 1991a, 1991b) and in the mountains north of the Cowal peninsula (Figure 10.1). Outlet glaciers extended to the mouths of many of the sea lochs, overriding or eroding fossiliferous marine sediments, which has allowed the dating of the readvance, as at South Shian (Peacock, 1971b, 1971c; Peacock *et al.*, 1989). Impressive sequences of ice-marginal features were also formed by these glaciers, as with the kame terraces and outwash plains at Moss of Achnacree at the mouth of Loch Etive (McCann, 1961b; Gray, 1975a). Two separate groups of features largely dating from this time are of particular prominence in the south-west Highlands and deserve special comment. They are the Main Rock Platform and the landforms of the Glen Roy area.

The Main Rock Platform occurs widely around the coasts of the south-west Highlands where it is a prominent feature up to 100 m (occasionally more) wide, with a backing cliff typically over 10 m high. The platform surface carries numerous stacks (as the Dog Stone by Oban) and the cliffline has many undercuts and caves (as on Lismore) (Gray, 1974a, 1978a). The platform is tilted, and declines in altitude away from the centre of isostatic uplift (Gray, 1978a; Sutherland, 1984a) from a maximum of over 11 m OD in the inner sea lochs to about present sea level along the Mull of Kintyre (see Glenacardoch Point). It was originally considered to be a Holocene marine feature (Bailey *et al.*, 1924) but subsequently was assigned an earlier interglacial origin (Gray, 1974a). Sissons (1974d), however, noted similarities with the erosional Main Lateglacial Shoreline of the south-east of Scotland and suggested, controversially, that the Main Rock Platform was formed towards the end of the Lateglacial, principally in the severe climate of the Loch Lomond Stadial. This view was accepted by those mapping the platform (Gray, 1978a; Dawson, 1979a, 1980b; Sutherland, 1981b, 1984a) although the possibility of inheritance from an earlier feature was pointed out (Peacock *et al.*, 1978; Sutherland, 1981b). This latter idea was also put forward by Browne and MacMillan (1984), and the uranium-series disequilibrium dating of speleothems from caves and undercuts in the cliff backing the platform on Lismore (Gray, 1987; Gray and Ivanovich, 1988) has also supported the concept (see also Dawson, 1989; Gray, 1989).

The Glen Roy area is quite outstanding for its assemblage of landforms and sediments related to the sequence of ice-dammed lakes when Loch Lomond Readvance glaciers blocked the mouths of Glen Spean, Glen Roy and Glen Gloy. These landforms have long been famous, the most detailed early work being that of Jamieson (1863, 1892); in recent years Sissons has added many details to the known sequence of events and the nature of the landforms (Sissons, 1978, 1979a, 1979b, 1979c, 1981d; Sissons and Cornish, 1982a, 1982b, 1983). Important results of Sissons' work are the clarification of the pattern of ice advance and retreat and the corresponding successive lake levels; the mode of formation of the shorelines; the fact that the shorelines in Glen Roy were formed principally during the rising lake sequence; the catastrophic drainage of the lakes; the warping and dislocation of the shorelines, particularly near ice-margins suggesting a glacio-isostatic influence; and the complex sequence of river terraces developed after lake drainage.

Glen Roy is not the only locality in the south-west Highlands where ice-dammed lake shorelines are of note, similar features having been formed by Loch Tulla during ice retreat at the end of the readvance (Ballantyne, 1979).

The severity of the climate during the Loch Lomond Stadial is reflected in a very reduced marine fauna of arctic affinity which inhabited the waters around the coasts of Scotland at that time (Peacock *et al.*, 1978; Peacock, 1981b; Graham *et al.*, 1990; Peacock and Harkness, 1990). Periglacial processes were particularly active on the mountain summits, resulting in a clear differentiation between the ice-moulded cols and mountain slopes buried by the readvance ice and the frost-riven and shattered bedrock surfaces that extended above the ice surface (Thorp, 1981a, 1981b, 1986, 1991a).

The decay of the readvance glaciers has been studied by examining the basal sedimentary sequences in enclosed basins inside the glacier limits (Lowe and Walker, 1981; Walker and Lowe, 1981; Tipping, 1988, 1989b). The basal pollen assemblages become progressively younger towards Rannoch Moor and this implies progressive deglaciation in that direction. A particularly valuable site is that at Kingshouse (Lowe and Walker, 1976; Walker and Lowe, 1977), where a radiocarbon date on moss fragments suggests that this area could have been deglaciated as early as 10,200 BP. The pollen zonation at the base of various enclosed depressions confirms deglaciation by the time of widespread expansion of juniper shrubs, but radiocarbon assays on this characteristic and widely recognized phase in the vegetation development are not sufficiently accurate to provide a clear limiting date on deglaciation (Tipping, 1987).

In the south-west Highland region two distinct zones of forest development are recognized during the Holocene. In the western coastal areas oak forest with birch developed (Loch Cill an Aonghais), whereas to the east, in the mountainous zone, pine became the major element of the forest during the middle Holocene (Birks, 1977; Bennett, 1984). The pine appears to have expanded after 8000 BP, following a phase of birch–hazel woodland dominance, and after about 4000 BP blanket bog expanded at the expense of pine. Human impact on the vegetation has been inferred from the time of the elm decline and possibly earlier in Kintyre (Nichols, 1967; Edwards and McIntosh, 1988).

Following low relative sea levels during the early Holocene, the Main Postglacial Transgression reached its climax, probably between 7000 and 6000 BP and a distinct, isostatically tilted shoreline was formed, equivalent to the Main Postglacial Shoreline of the east coast (Gray, 1974b; Sutherland, 1981b). Subsequently, as sea level fell towards its present level, four (Gray, 1974b) or five (Sutherland, 1981b) lower shorelines were formed.

The mountainous nature of much of the south-west Highlands has resulted in the rivers having a very 'flashy' regime and alluvial fans are common in many of the valleys. Although parts of these fans are active today, they typically have a complex history of development during the Holocene, as is illustrated by Eas na Broige in Glen Etive (Brazier *et al.*, 1988). The radiocarbon dating of the debris-flow and alluvial-fan deposits there indicates two periods of activity, one during the early to middle Holocene when glacial debris was available for reworking, and the other in recent times initiated by either climatic deterioration or overgrazing (Innes, 1983b; Brazier *et al.*, 1988).

TANGY GLEN

D. G. Sutherland

Highlights

The stream sections at this locality provide exposures in high-level, shelly clays. These are believed to be a unique occurrence of *in situ* marine sediments deposited during a Middle Pleistocene glaciation when relative sea level was higher than at present.

Introduction

The Tangy Glen site comprises stream sections along three burns on the west coast of Kintyre *c.* 13 km north-west of Campbeltown: Tangy Burn (NR 659279), Drumore Burn (NR 670329) and Allt a'Ghlaoidh (NR 670337). During the last century a number of localities were discovered in Scotland at which marine shells contained in sands or clays were overlain by glacial sediments (see Jamieson, 1858, 1866; Smith of Jordanhill, 1862; Crosskey and Robertson, 1873b; Fraser, 1882a, 1882b). These sites occurred at altitudes considerably in excess of the more widely distributed fossiliferous clays which post-dated the glacial deposits, and considerable debate ensued as to the nature of the high-level deposits and their place in the glacial sequence (J. Geikie, 1874; see also Clava). As a consequence a special committee of the British Association was convened to investigate these occurrences and one of the three areas they investigated was the sequence of shelly clays on the west coast of the Mull of Kintyre. These deposits were first reported by Crosskey and Robertson (1873b) at Tangy Glen and, although the main exposures of the clays are found in stream-cut sections to the north (Horne *et al.*, 1897), it is by the name of the original locality that the site is commonly known. The most detailed description of the stratigraphy of the shelly clays of Kintyre is that given by Horne *et al.* (1897). Subsequently the deposits have been investigated by Munthe (1897), Jessen (1905) and Gray (1985). Gray (1985) and Bowen and Sykes (1988) provided details of amino acid analyses of shells from the clays, and Sutherland (1981a) placed them in a wider context of related deposits in Scotland.

Description

The glaciation of the Kintyre peninsula was recorded by Horne *et al.* (1897) as being in a generally westerly direction based on the evidence of striae as well as the occurrence of erratics of Arran granite across the area.

Horne *et al.* (1897) reported clays occurring in three separate localities along approximately 4 km of coastline. In Tangy Glen the clays occur up to a height of about 135 ft (41 m) above present sea level, whereas to the north in Drummore Glen the top of the clays was taken to be 199 ft (60 m) and beside the Cleongart Burn (Allt a'Ghlaoidh), about 179 ft (55 m). Of these three exposures that at Cleongart was by far the most fossiliferous and it is the one that has been most closely investigated.

Horne *et al.* (1897) excavated two trenches across the top and base of the shelly clays in the Allt a'Ghlaoidh exposure as well as drilling three holes upstream and two in a line to the south in order to determine the continuity of the deposit in those directions. The stratigraphy they established is given below and illustrated in Figure 10.2.

3. Till, reddish-brown with abundant boulders most of which are local schist. No Arran granite erratics were observed in the section, but these occur frequently in the neighbourhood. 22.5 m
2. Shelly clay, stiff, dark bluish. The upper part is relatively stone-free, but fragments of schist occur in the lower part. Shells occur throughout. 8.4 m
1. Compact coarse sand and gravel. Unfossiliferous. >1.2 m

This succession has been confirmed in its broad details by subsequent workers (Munthe, 1897; Jessen, 1905; Gray, 1985) although more detailed descriptions have been given of the shelly deposits. Both Munthe (1897) and Gray (1985) identified a stoney layer in the clays, which Munthe compared to a 'veritable boulder clay', although Jessen (1905) disagreed that it could be compared to a glacial sediment.

Interpretation

The initial work of Crosskey and Robertson (1873b), Brady *et al.* (1874) and Horne *et al.* (1897) produced lists of species of marine molluscs, Foraminifera and ostracods, but did not attempt to establish whether there were any vertical variations in the species representation except for sequential analyses in the most southerly of the boreholes. No comment was made on the significance of any changes in the occurrence of different species at different levels in the clays. The most common molluscs recorded by Horne *et al.* (1897) were the bivalves *Arctica islandica* (Linné), *Astarte sulcata* (da Costa),

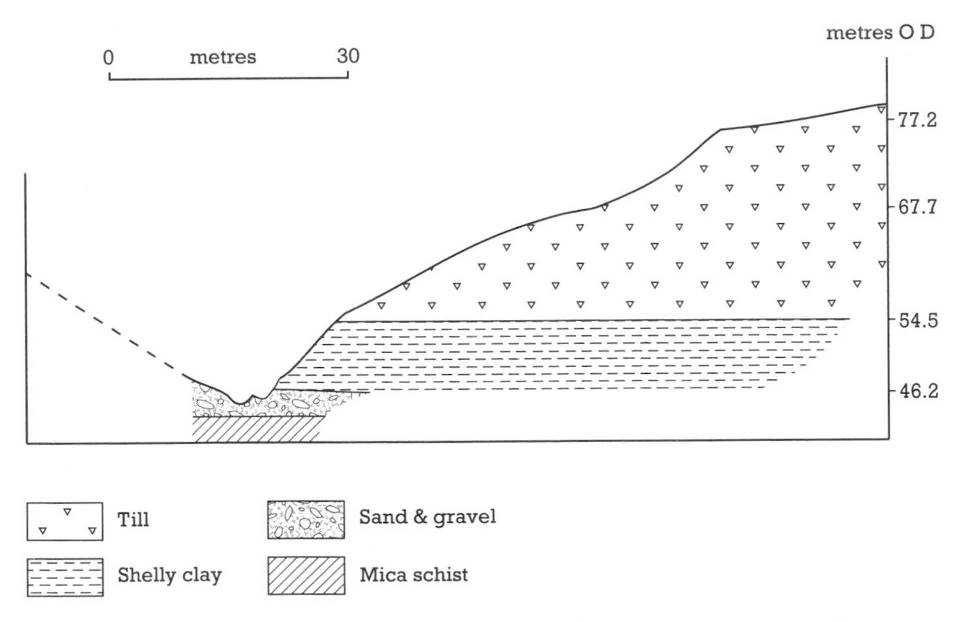

Figure 10.2 Tangy Glen: lithological succession at Cleongart (from Horne *et al.*, 1897).

Macoma calcarea (Gmelin), *Nicania montagui* (Dillwyn), and *Nuculana pernula* (Müller) and the gastropod *Turritella tenebra* (Linné). It was noted, however, that there was an apparent mixture of species deriving from both arctic and more southerly latitudes. It was also noted that the majority of the mollusc shells were broken.

In contrast to this initial work, Munthe (1897) examined twelve samples taken vertically through the deposit. On the basis of the fossil content of these samples Munthe concluded that the period of deposition of the clays coincided with a succession of cold to warm to cold climate. Munthe further interpreted the gravel at the base of the section (bed 1) as a glacial deposit, and hence argued that the warm phase in the middle portion of the clays was an interglacial. He considered that during the two cold phases water depth would have been at least 40 m (that is, upwards of 95 m above present sea level), but that during the warm phase a water depth of between 6 m and 25 m was more likely (between 51 m and 70 m above present sea level). Munthe accepted, however, that there were certain species which apparently implied different climatic conditions from those he interpreted and he suggested that these were reworked from previous deposits.

Jessen (1905) did not accept Munthe's (1897) subdivision of the deposit on the basis of its fauna. In contrast, he pointed out that the only mollusc shells to be found with valves together and in growth position were those indicative of an arctic climate. He suggested that since the shells of all the southerly indicators were in a fragmentary condition, they were derived. The deposition of the clays was envisaged by Jessen as taking place during a glacial phase when the sea level was about 90 m above that of the present.

During deposition of the clays in the lower part of bed 2 the ice front was at some distance from the site, but an advance brought the ice margin near to the adjacent shore and at this period a pre-existing deposit was eroded and the broken shells from this were redeposited in the middle portion of the clays. An increased quantity of gravel and boulders was deposited at this time because of the proximity of the ice front. Subsequently the glacier retreated and conditions similar to those during deposition of the basal clays resumed. At some later date ice advanced over the whole site depositing the till that caps the section. This last advance was considered by Jessen to be 'much younger' than the underlying clays. Both Munthe (1897) and Jessen (1905) accepted that during the glaciation(s) of the area ice movement was from east to west.

A quite distinct glacial history was offered for the southern Kintyre peninsula by Synge and Stephens (1966). They suggested that the Tangy Glen shelly clays were a till emplaced during an early glaciation with ice movement from north to south. There was a subsequent 'main' ice movement from east to west during which the overlying red till was deposited, but after this there occurred a final ice movement, again from the north, with an ice margin in the general area of Tangy Glen, where 'morainic accumulations' were referred to as being present. Striae oriented S10°E were cited as supporting evidence for this final ice movement. No further justification was advanced for regarding the shelly clays as being glacial in origin.

The shelly clays were regarded by Sutherland (1981a) as being *in situ* marine deposits and he argued that they had a common origin with the other similar deposits encountered around the Scottish coast, such as at Clava and Afton Lodge. He noted that the maximum altitude to which the high-level shell beds had been encountered decreased with increasing distance along glacier flow-lines, and presented a model explaining the distribution in terms of crustal depression in front of an expanding ice-sheet, with world sea level being relatively high during the initial glacial advance but subsequently falling as the large ice-sheets in the Northern Hemisphere expanded. The fauna of the shell beds was considered by Sutherland to indicate moderately arctic conditions and he argued that the conditions necessary to produce the deposits seemed to have occurred during the Early Devensian when, according to deep-sea core evidence, the last major period of glaciation was initiated (Shackleton and Opdyke, 1973; Ruddiman *et al.*, 1980).

Amino acid analysis of the contained mollusc shells at Allt a'Ghlaoidh (Gray, 1985; Bowen and Sykes, 1988; Sutherland, unpublished data) has cast doubt on the chronology proposed by Sutherland (1981a). On that basis, the Allt a'Ghlaoidh deposits are clearly older than the last interglacial and may pre-date the previous interglacial as well: the amino acid analyses suggest correlation with oxygen isotope stage 8, between approximately 300 ka BP and 250 ka BP (Bowen and Sykes, 1988).

The high-level shelly clays underlying till along the west coast of Kintyre are some of the best examples of a type of deposit encountered at only a few sites in Scotland. Their origin has been controversial since their discovery last century. Current published opinion favours the interpretation of the deposits as being *in situ*. The amino acid analyses imply that the deposit is of Middle Pleistocene age, although the length of period during which they were deposited is unclear. A Middle Pleistocene age distinguishes the Tangy Glen deposits from those at Clava (Devensian age) and would mean that the deposits are the only known representatives in Scotland of a marine event, possibly associated with ice-sheet glaciation, during oxygen isotope stage 8.

Conclusion

The clays containing shells of marine molluscs at Tangy Glen form part of a suite of such deposits that occurs in Scotland well above present sea level. Although they have a long history of research dating back to the last century, their respective ages and origins remain a source of scientific argument. Current work suggests that the high-level clays at Tangy Glen were laid down when the land was depressed by ice during a Middle Pleistocene glaciation (approximately 275,000 years ago), which would make them the only known deposit of their kind in Scotland. They therefore have an important bearing on interpreting this significant part of the Quaternary history of Scotland, about which little is otherwise known.

GLENACARDOCH POINT

J. M. Gray

Highlights

The coastal landforms at Glenacardoch Point comprise an assemblage of shore platforms and raised beach deposits. These include representative examples of all the principal landforms and deposits recognized in the south-west Highland region, that have resulted from sea-level change during the Quaternary.

Introduction

Glenacardoch Point lies on the west coast of the Kintyre peninsula. The site comprises a *c.* 2 km length of coastline (between NR 667384 and NR 659366), and extends 0.25–0.55 km inland. It is important in demonstrating two shore platforms and a sequence of raised beach deposits which provide a record of sea-level changes in the south-west Highlands. Little research has been carried out on the site. Apart from a few brief mentions in the early literature (Nicol, 1852; Hull, 1866; Sinclair, 1911), the only recent work is by Gray (1978a) who mapped and levelled the two shore platforms.

Description

Several accounts of former shorelines in Kintyre appeared in the early literature, drawing attention to raised and intertidal platforms with well-developed rock cliffs and caves (Nicol, 1852; Hull, 1866). Sinclair (1911) also made an important observation that raised sea stacks on some of the platforms were covered by till and therefore pre-dated a period of glaciation. The geomorphology of the site is shown in Figure 10.3, and Figure 10.4 is a generalized cross-profile. As can be seen from these figures, the site consists of a series of terraces of marine origin. Four main features or groups of features can be recognized. The oldest is a till-covered platform occurring over a distance of about 400 m at, and immediately south of Glenacardoch Point, though poorer fragments occur both to the north and south. Till is exposed in a section in the cliff behind the platform (at NR 660377), and at this point the rock platform can be seen clearly to extend inland below the till (Gray, 1978a, p. 155). The platform at this point lies at 13.1 m OD. It is cut by a number of geos, and its front slope forms the low backing cliff of a second, lower, intertidal platform at 0.6 m OD.

In several places the rocky coast immediately above sea level is overlain by a thin veneer of Holocene raised beach sediments, the clearest Holocene beach being at the extreme south of the site where a terrace, 100 m or more wide and a few metres above OD, is backed by the 20 m high till cliffline that runs through the site.

A higher terrace complex of raised beaches is present above this till cliffline. At least three levels occur immediately above Glenacardoch Point itself and, although these have not been levelled, the photogrammetrically determined contours on the 1:10,000 map of the area indicate that the three levels lie between about

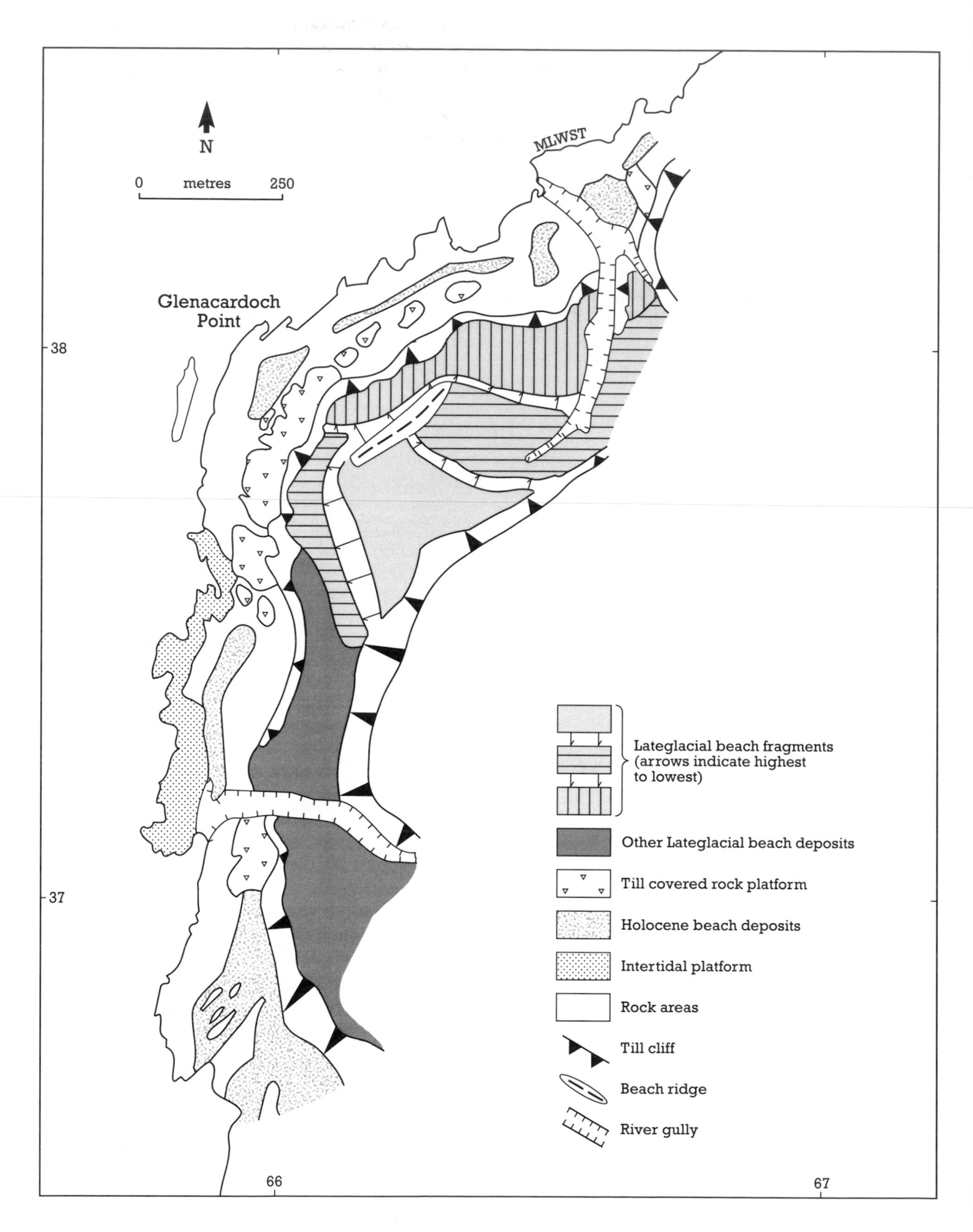

Figure 10.3 Geomorphology of the Glenacardoch Point area.

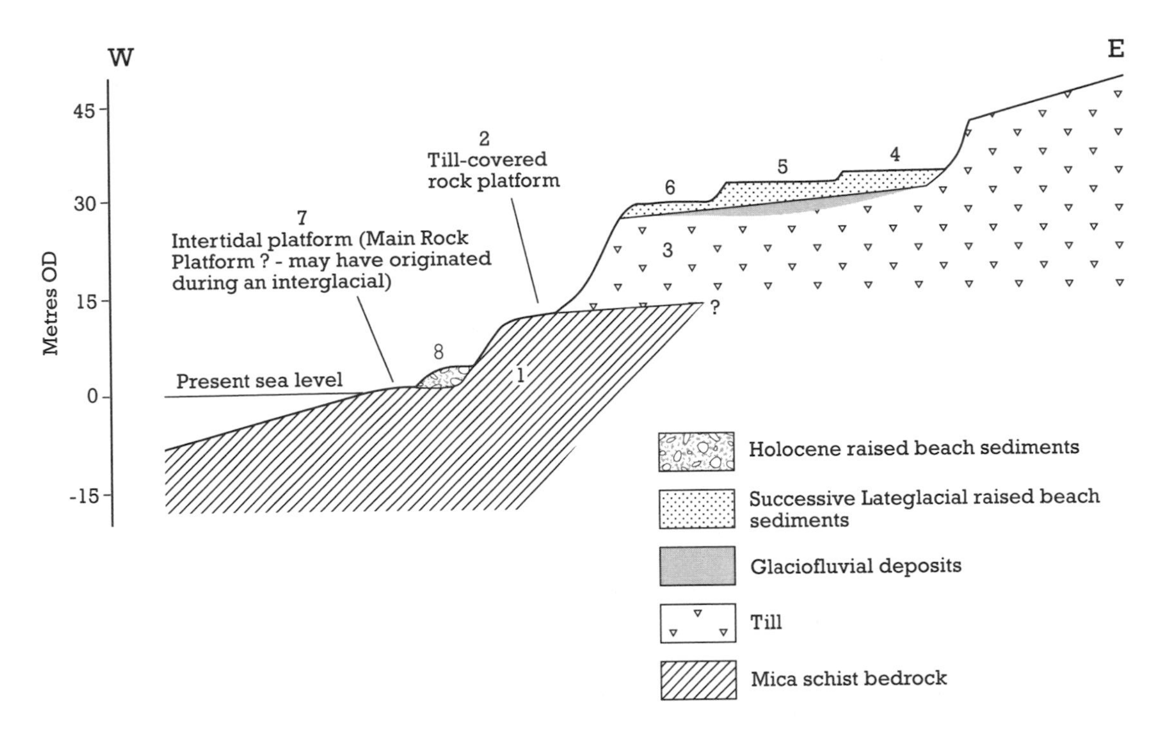

Figure 10.4 Coastal profile at Glenacardoch Point showing the relationships between the morphology and succession of the features and their probable sequence of formation (1–8).

28 m and 34 m OD. A possible storm ridge occurs at the front edge of the highest level, and at the back edge there is a further low till cliff. A section in the middle beach terrace (at NR 665378) reveals 2 m of well-bedded sand and gravel and (at NR 666381) sections in a stream valley show these beach sediments overlying till and glaciofluvial deposits.

Interpretation

The interpreted relationships between the different features are summarized in Figure 10.4, together with the probable sequence of formation. The oldest feature is the till-covered platform which pre-dates at least the main Late Devensian glaciation represented by the till. Following deglaciation, the till was trimmed by the sea and a sequence of Lateglacial beaches, now isostatically uplifted to *c*. 28–34 m OD, was formed. Subsequently, sea level fell to close to its present level and the platform in the present intertidal zone developed. From the overall platform distribution in Kintyre and neighbouring areas, Gray (1978a) suggested that the intertidal platform at Glenacardoch Point correlates with the Main Rock Platform of western Scotland either formed or last occupied during the Loch Lomond Stadial (Gray, 1974a, 1978a). During the Holocene, sea level once again rose, resulting in deposition of the beach gravels that occur below 10 m OD.

Glenacardoch Point is important in two main respects:

1. Till covered low-level platform. It is one of the few sites in Scotland where a low-level shore platform can clearly be seen to pass below till (see Port Logan and Dunbar). Similar situations are common in Ireland, but on the Scottish coast the Kintyre peninsula provides the best examples. Neighbouring equivalent sites a few kilometres to the south between Bellochantuy and Westport have been disturbed. The succession demonstrates that the platform pre-dates at least one glacial episode and although much work remains to be done on dating and correlating the platforms of western Britain, current opinion regards such low-level, till-covered platforms as having formed during interglacials. It is probably part of a suite of subhorizontal, low-level platforms in western Britain (Sissons, 1981a).
2. Sequence of sea-level changes. The site is also notable in preserving, within a compact area, evidence for several phases of sea-level

change. It is important for demonstrating the morphological and stratigraphical relationships between several Scottish raised beaches and shore platforms (see also Milton Ness and Kincraig Point). Particularly helpful in this respect is the till, since this clearly overlies a rock platform yet is overlain and partly eroded by raised beaches. Although the latter cannot be dated at this site, comparison with elsewhere in Scotland allows division of the raised beaches at Glenacardoch Point into Lateglacial (>10 m OD) and Holocene (<10 m OD). Similarly, the intertidal platform cannot be dated, but altitudinal comparisons with other platforms in Kintyre suggests that it may belong to the Main Rock Platform of western Scotland (see Isle of Lismore, Northern Islay and West Coast of Jura) which many authors over the last 15 years have suggested is Loch Lomond Stadial in age (but see Isle of Lismore).

Conclusion

The landforms and deposits at Glenacardoch Point are important since within a 2 km stretch of coast it is possible to establish and demonstrate most of the major elements of the recognized sequence of sea-level changes which occurred in the south-west Highlands during Quaternary times. Of particular interest is a very clear example of a low-level, coastal shore platform overlain by glacial deposits (till), indicating that the former pre-dates at least one glaciation. Glenacardoch Point is therefore both a representative site and a valuable component in the network of sites demonstrating sea-level changes.

ISLE OF LISMORE, THE DOG STONE AND CLACH THOLL

J. M. Gray

Highlights

These three coastal localities together demonstrate key features relating to the geomorphology and dating of the Main Rock Platform, one of the most distinctive Quaternary shorelines in western Scotland. The speleothem deposits on Lismore, in particular, hold great potential for the dating of this shoreline.

Introduction

This site consists of three separate parts which together demonstrate the key aspects of the Main Rock Platform, a striking feature of the geomorphology of the western seaboard of Scotland (Bailey *et al.*, 1924; Wright, 1928; McCann, 1968; Gray, 1974a, 1978a; Dawson, 1980b, 1984, 1988a; Rose, 1980b; Sutherland, 1981b, 1984a; Wain-Hobson, 1981; Gray and Ivanovich, 1988). The shore platform and its associated landforms are best developed along the coast of the Firth of Lorne (Gray, 1974a, figure 1). The Isle of Lismore site comprises two stretches of the north-west coastline of the island; the northern one at Port Ramsay (NM 872454) is 0.4 km long, the southern one (between NM 805395 and NM 831411), north-east of Achadun Bay, is 3 km long. Together these provide an excellent demonstration of the extensive development of the platform and its backing cliff, and also include the major speleothem sites that have a significant bearing on interpreting the age(s) of the platform (Gray and Ivanovich, 1988). On the adjacent mainland, two classic erosional features are associated with the platform. The Dog Stone (NM 853311) is a raised sea stack located immediately north of the promenade at Oban. Clach Tholl (NM 900448) is a raised natural arch located *c.* 1 km south-west of Port Appin.

Description

The Isle of Lismore provides an example of the continuity and excellent development of the shoreline over a long stretch of coast (Figure 10.5; see Gray and Ivanovich, 1988, figure 3), and as such is typical of the shoreline in the Firth of Lorn area. A cliffline 5–15 m high can be traced uninterrupted along virtually the whole length of the site, and a platform up to 100 m wide is also present, particularly in the south-west. Some of the classic features associated with the shoreline are also present at this site, including undercuts at the base of the cliffs (for example, around localities 14–17, Figure 10.5) and caves (such as Uamh na Cathaig – locality 18 – which is about 10 m deep; see Gray and Ivanovich, 1988, figure 4).

The presence and development of the shoreline on Lismore have been important factors in the evolution of ideas on the age and origin of the Main Rock Platform. In particular the sheltered

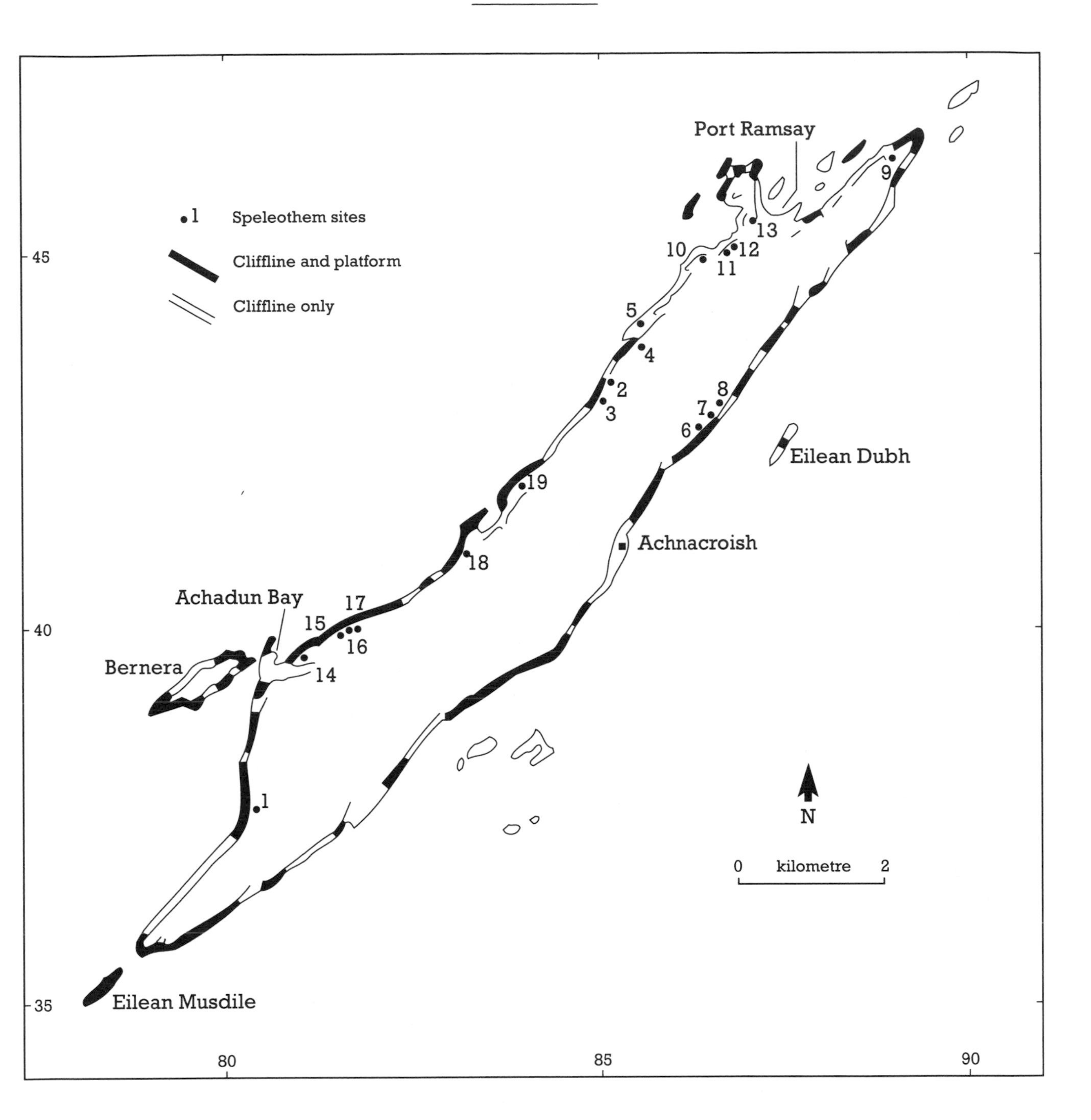

Figure 10.5 Distribution of the Main Rock Platform on the Isle of Lismore and localities mentioned in the text (from Gray and Ivanovich, 1988).

location of the island and its occurrence within the glacial trough of the Great Glen were both referred to by Sissons (1974d, p. 43) in his paper arguing for a Loch Lomond Stadial age for the shoreline.

Speleothem samples from four sites, undercuts and caves in the base of the cliffline, have been dated by uranium-series disequilibrium methods (Gray and Ivanovich, 1988). Five samples produced a range of ages: two from the Holocene, two from the Late Devensian and one from the Early Devensian. From analytical results, Gray and Ivanovich (1988) considered that the two samples that gave Late Devensian ages were unreliable and that they may actually be of Holocene age. However, the sample that gave the Early Devensian age (of 103.3 ka +28.4/−20.0 ka, HAR–3228) was considered reliable. Thus at least two periods of speleothem formation occurred.

The Dog Stone is an undercut, raised sea stack eroded in Devonian (Old Red Sandstone) conglomerate standing in front of the Main Rock Platform cliffline. Its name is derived from a legend that relates that this is where the giant, Fingal, tied up his dog, Bran, when he went hunting in the Hebrides. The real significance of the feature was first appreciated by the geologist Hugh Miller in 1857 (Miller, 1858) and since then it has been singled out for special mention by several authors (for example, Bailey *et al.*, 1924; Sissons, 1967a; Gray, 1974a, figure 6; Gray and Ivanovich, 1988, figure 2b). It has been selected here because of the absence of stacks from the Lismore site, the long history of description and its legendary associations.

Like the Dog Stone, Clach Tholl is an example of a special type of feature not present at the Lismore site. Clach Tholl is the clearest raised, natural arch associated with the Main Rock Platform in western Scotland, although another good example occurs below Gylen Castle on South Kerrera (NM 805265) (see Gray, 1974a, figure 5). It is developed along a dipping fault plane in a quartzite headland 1 km south-west of Port Appin. Its name is derived from the Gaelic words meaning 'hole in the rock'. It is widely known as a famous geomorphological landmark (see, for example, Price, 1976, figure 33; Gray and Ivanovich, 1988, figure 2a).

Interpretation

In the work of the Geological Survey of Scotland, undertaken at the end of last century and the beginning of the present one, the presence of the separate erosional shoreline that is now termed the Main Rock Platform was not recognized. Instead the surveyors included the erosional shoreline fragments with the '25 ft beach', of Holocene age (Kynaston and Hill, 1908; Bailey *et al.*, 1924), mainly due to the fact that the two sets of features broadly correspond in altitudinal range. Indeed, Wright (1928, p. 100) called the postglacial sea 'the cliff-maker *par excellence*'. McCallien (1937a), however, was not convinced that the rock platform and cliffs had been cut by the same sea that deposited the Holocene beach sediments. He believed (McCallien, 1937, p. 197) that 'since the Ice Age there has not been enough time for the cutting away of so much solid rock as is indicated in the raised platform around our coasts'. Instead he suggested that the platform was pre-glacial or interglacial in age, and believed it to be a coincidence that the Holocene sea had re-attained the altitude of the earlier platform.

Subsequently, an interglacial origin for the platform became widely accepted (McCann, 1966b, 1968; Synge, 1966; Sissons, 1967a; Gray, 1974a). Gray (1974a) undertook a detailed study of the platform in the vicinity of the Firth of Lorn, giving it the name Main Rock Platform. On the basis of 304 levelled altitudes on 106 platform fragments, he demonstrated a clear east–west tilt on the shoreline from about 11 m OD north of Oban to 4 m OD in mid-Mull, an overall gradient of 0.16 m km^{-1}. However, the gradient is not uniform throughout, for a number of bends and one possible fault were identified (see also Ringrose, 1989b). Gray argued that if the platform was formed during an interglacial episode, then the tilting is likely to imply tectonic instability of the area.

A major challenge to the idea that the Main Rock Platform was an interglacial feature came from Sissons (1974d). He was struck by the similarity between it and the Buried Gravel Layer, an erosional shoreline that he had identified earlier in the Firth of Forth. Since the latter is eroded into till and Lateglacial marine sediments, yet is overlain by Holocene marine and estuarine sediments, he argued that the shoreline must have been formed during the latter part of the Lateglacial. To explain the extent of erosion in such a short interval in the sheltered estuary of the Firth of Forth he suggested (1974d, p. 46) 'that the critical factor was the periglacial climate that characterized the stadial ... the erosion of unconsolidated sediments by the sea would be facilitated by slumping and flowing associated with seasonal thawing'. He argued that the Main Rock Platform could be correlated with the Buried Gravel Layer, thus explaining many characteristics of the former. For example, its tilt (due to differential glacio-isostatic rebound), its apparent lack of direct evidence of having been glaciated (due to its Lateglacial age), and its development in sheltered locations (due to periglacial frost shattering rather than wave action). Sissons used the term 'Main Lateglacial Shoreline' to refer to the correlation of the two features.

Fieldwork by Gray (1978a) in the area between the Firth of Lorn and the Firth of Clyde showed that the Main Rock Platform as well as having an east–west tilt also has a north–south tilt, reaching sea level in south Kintyre, south

Arran and south Ayrshire. He demonstrated that the shoreline does not correlate with the till-covered platforms of eastern Ireland; these also extend into Kintyre (see Glenacardoch Point) and other areas of south-west Scotland (see Port Logan) but at a higher level than the Main Rock Platform. Subsequent work by Dawson (1980b, 1988a), Rose (1980b, 1980f), Sutherland (1981b) and Wain-Hobson (1981) in other parts of western Scotland have added to the altitudinal information on the Main Rock Platform and have lent support to the correlation of the Main Lateglacial Shoreline. Studies in modern periglacial environments have also substantiated the hypothesis of rapid rates of shore platform formation (for example, Sissons, 1974a; Dawson, 1980b) and indicated the processes that may have been involved in Scotland (Hansom, 1983; Matthews *et al.*, 1986; Dawson *et al.*, 1987b; Shakesby and Matthews, 1987; see also review by Trenhaile, 1983).

Gray and Ivanovich (1988) reviewed the geomorphological arguments for and against a Lateglacial age: most of the evidence favours such an age, but some is contradictory. Fresh light on the problem has come recently from the uranium-series disequilibrium dates on the Isle of Lismore (Gray and Ivanovich, 1988). The samples giving Holocene ages are in accord with the hypothesis that the platform was eroded during the Lateglacial. However, if valid, the Early Devensian date throws doubt on the view that the shoreline was entirely formed during the Lateglacial. From these results and the contradictory geomorphological evidence for a Lateglacial age, Gray and Ivanovich (1988) were led to the conclusion that the Main Rock Platform may be polycyclic in origin. Browne and McMillan (1984) have also disputed a solely Lateglacial age, suggesting partial inheritance from an earlier platform which pre-dated at least the Late Devensian ice-sheet. Gray (1989) and Dawson (1989) have recently debated several aspects of the distribution and development of the Main Rock Platform in western Scotland.

The Main Rock Platform is important in several respects. First, it is one of the best-developed raised shorelines in Scotland and indeed in Europe. It is directly comparable with the 'Main Line' of northern Norway, also a rock-cut, glacio-isostatically tilted shoreline which was formed during the Younger Dryas (Marthinussen, 1960; Andersen, 1968; Sollid *et al.*, 1973). Second, recent ideas on rapid, periglacial formation of the platform have been widely adopted and incorporated into models of shore platform formation (see, for example, Trenhaile and Mercan, 1984). Third, the deformation of the platform with its tilt, bends and possible faults (Gray, 1974a, 1978a; Ringrose, 1989b) are relevant to understanding the crustal stability/instability of the area. It is particularly important that the age of the shoreline is understood so that the time-scale and origin of the deformations can be better appreciated. Such aspects are relevant to earthquake engineering (see, for example, Davenport and Ringrose, 1985; Davenport *et al.*, 1989). The Isle of Lismore site has great potential for clarifying the age and origin of the Main Rock Platform: the other two sites are outstanding examples of specific raised shoreline features associated with the platform.

Conclusion

These three sites together represent key features of the geomorphology of the Main Rock Platform, one of the most prominent fossil shorelines in western Scotland. The age of the platform is uncertain. It appears to have been formed, at least in part, during the Loch Lomond Stadial (about 11,000–10,000 years ago), but it may also be partly an older feature that has been reworked. Isle of Lismore demonstrates the shore platform and cliffline and also includes the critical cave sites where dating of deposits has been undertaken; the Dog Stone and Clach Tholl show additional landforms (stack and rock arch, respectively) associated with the shoreline.

MOSS OF ACHNACREE AND ACHNABA LANDFORMS

J. M. Gray

Highlights

This site demonstrates an outstanding assemblage of glaciofluvial landforms deposited by a Loch Lomond Readvance glacier.

Introduction

The site is located on the north side of Loch Etive and includes the major part (*c.* 4 km^2 in area) of an outwash plain underlying the Moss of Achnacree (NM 920358) and a 2.5 km length of adjoining kame terrace fragments and related glaciofluvial features to the east near Achnaba (NM 945365). It forms what is arguably the finest outwash and kame terrace system in Great Britain and its importance lies in the way that it is possible to demonstrate relationships between ice-marginal glaciofluvial features, and between ice-contact and proglacial drainage systems. Moss of Achnacree and Achnaba is also a particularly good representative of a landform assemblage associated with a number of Loch Lomond Readvance glaciers which terminated near to sea level in western Scotland. The area has been described by Kynaston and Hill (1908), McCann (1961b, 1966a), Synge (1966) and Gray (1972, 1975a).

Description

Moss of Achnacree

The Moss of Achnacree outwash plain (1, Figure 10.6) covers an area of almost 4 km^2. Most of the central part is covered by about 2 m of peat (the Moss of Achnacree), the edges having been cleared over the years for crofting and farming. Building construction and augering on these and other parts of the outwash plain have shown it to be underlain by sand and gravel. The feature forms a marked constriction at the entrance to Loch Etive. Here the loch is confined to a <0.5 km wide channel on the southern margin of the valley compared with a width of over 1.5 km a short distance to the east. When the outwash sediments were being deposited, the glacier snout lay immediately to the east and at that time the outwash plain would have been continuous across the valley. A large terrace behind Connel (NM 915342) is a likely remnant of the same outwash plain on the south side of the loch (Gray, 1972).

The outwash plain slopes from about 25 m OD near its eastern edge at Achnacairn (NM 927357) to about 12 m OD near the A828 Connel to Ballachullish road. This general gradient is, however, interrupted by a wide meltwater channel that runs westwards from Cairnbaan and by a number of deep, lochan-filled kettle holes (Figure 10.6). Bathymetric surveying of the three largest hollows has shown each to be complex. For example, Murray and Pullar (1910) noted that the Lochan na Beithe hollow consists of two major interlinked depressions. The deepest point lies 25 m below the terrace surface and 8 m below OD. Conacher (1932) found that Laga Beaga (NM 924357), west of Achnacairn, consists of four interconnected hollows with a deepest point 17 m below the terrace surface. The presence of these large kettle holes in the outwash plain has been seen by several authors as indicating that the glacier originally extended beyond Connel, and during retreat large blocks of ice were left behind, to be surrounded or covered by outwash sediments when the snout became stabilized at the eastern edge of the outwash plain.

As McCann (1961b) pointed out, the margins of the outwash plain have been modified by later marine action at a level of approximately 13 m OD. Along the eastern margin the ice-contact slope has been eroded so that there is a marked break of slope at this level. Westwards, along the southern edge of the terrace, the gradient of the outwash plain carries it down to the 13 m level at North Connel, and on the west side the sea has built a 2.5 km long north–south spit which rises to over 14 m OD. The sea level concerned is probably that of the Main Postglacial Shoreline. Because of the presence of this shoreline, it is difficult to decipher the relative sea level at the time the outwash plain was deposited, but it was certainly below the 13 m OD level. Following formation of the Main Postglacial Shoreline, relative sea level gradually fell to its present level and in doing so raised-beach sediments were deposited to the west of the spit around North Connel Airfield (NM 905353) and North Ledaig Caravan Site (NM 907369), as well as below the south-east corner of the outwash plain (around NM 928351).

Achnaba

The Moss of Achnacree outwash plain is continued eastwards by two main terrace fragments, one at Achnacreebeag (NM 933364) and the other 200 m to the south-east and partially built upon (2, Figure 10.6). At this locality the transition from a proglacial outwash plain to an ice-marginal kame terrace occurs. At its eastern end, where the kame terrace fragment rises to 26.4 m OD, it narrows into a short, sharp-crested ridge that is probably an esker. Thus it is possible to identify the locality at which a meltwater stream escaped from the ice and became an ice-

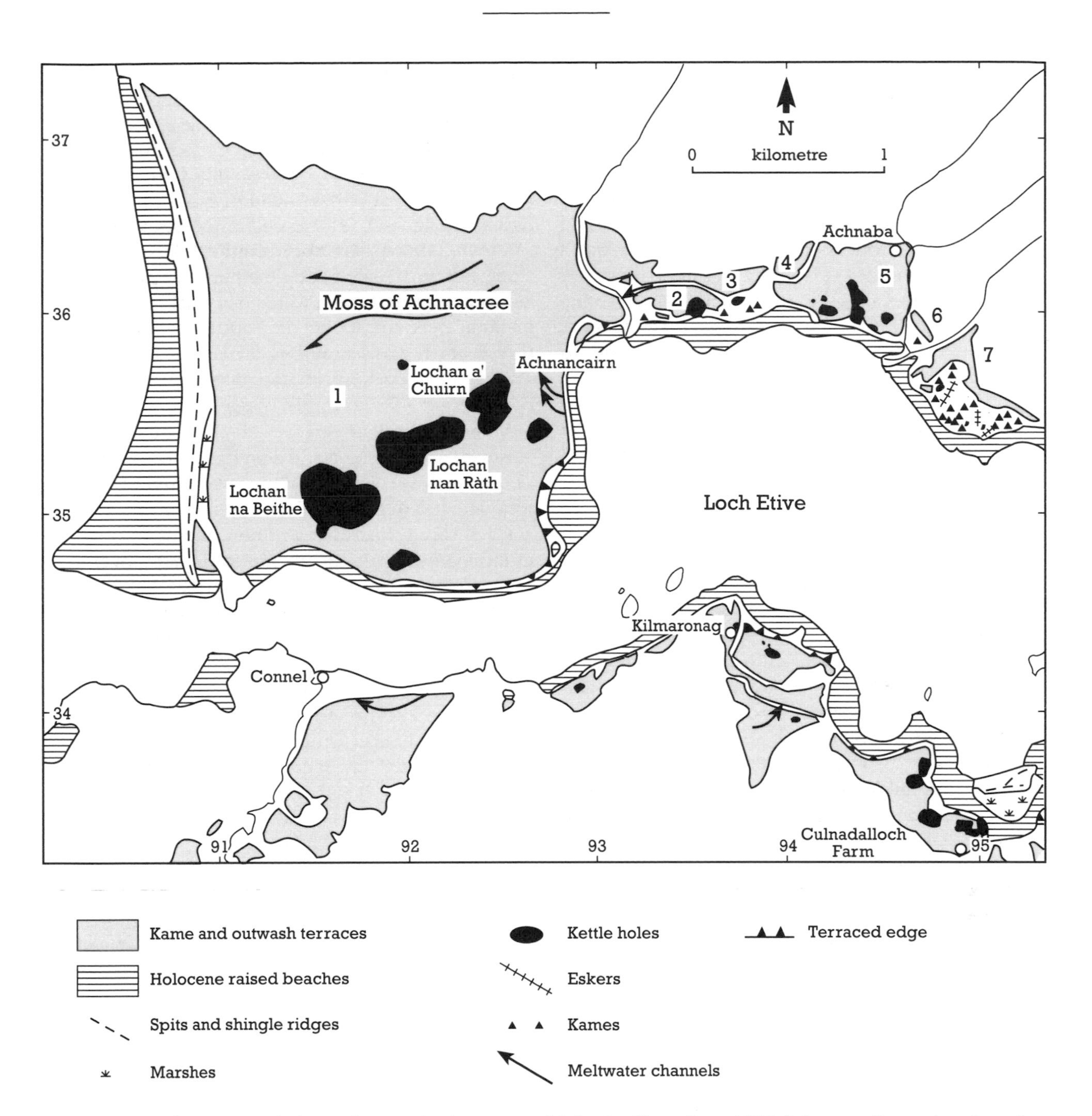

Figure 10.6 Geomorphology of Moss of Achnacree and Achnaba (from Gray, 1975a). See text for explanation of numbers.

marginal river contributing first to the deposition of the kame terrace and then to the outwash plain, west of Achnacreebeag.

As a glacier decays and contracts, the normal situation is for the earliest and highest kame terraces to occur along the valley sides and for later terraces to occur at lower elevations towards the valley centre. The situation at Achnacreebeag is different since fragment 2 is separated from the hill slope to the north by a channel/lower terrace (3, Figure 10.6) that is one of a slightly later group of kame terrace fragments. The explanation probably lies in the fact that in this case drainage was unable to flow along the front edge of fragment 2 because the glacier margin had not wasted away from it. Thus the drainage instead flowed along the back edge of the terrace and cut down into it.

The lower group of terraces is continued eastwards by terrace fragments 4, 5, 6 and 7

(Figure 10.6) extending to the western end of the site. The overall rise is from 19.3 m OD at the western end of fragment/channel 3 to 30.3 m OD at the eastern end of fragment 7. Heights along the back edges of the fragments are often anomalously high due to the presence of alluvial fans. At Achnaba, fragment 5 is 400 m wide and towards its front edge it is perforated by a remarkable series of kettle holes. These are striking features, being steep-sided hollows sometimes over 10 m deep located in a flat terrace. Many of them are complex in shape due to coalescing of hollows. As described above for the outwash plain, the front edges of the kame terraces were also eroded by the Main Postglacial sea at about 13 m OD and in places this sea washed into kettle holes, thus creating large embayments. A good example occurs 200 m north-west of the church (at NM 943362).

The stream that separates fragments 5 and 6 has cut the best sections in the kame terrace sediments. Although poorly bedded sand and gravel is predominant, there are also lenses of laminated sands, silts and occasionally clays, indicating that small ponds were present on the kame terraces, perhaps in abandoned drainage channels or at points of ice-melt subsidence.

Fragment 7 is mainly confined to a narrow strip along the valley side except at its eastern end where it extends out towards a bedrock area. Between fragment 7 and the loch the terrain is very irregular. Although this may in part be due to differential kame terrace subsidence following ice-melt, the presence of two steep-sided, sinuous eskers suggests that at least some of the landforms were deposited subglacially, and that some of the mounds and hollows are kames and kettles. The downslope trend of the eskers suggests that they are examples of the subglacially engorged type, probably formed by rivers that found a route down towards the valley floor from the kame terrace above.

Interpretation

The terraces at the entrance to Loch Etive and bordering the north and south sides of the lower part of the loch were originally interpreted as marine. Thus Kynaston and Hill (1908) described the Moss of Achnacree as resting on sands and gravels of the '50 ft beach' and assigned the terraces farther east to the '100 ft beach'.

McCann (1961b) was the first to appreciate the significance of the terraces. From the slope of the Moss of Achnacree feature, the presence of kettle holes in it and its similarity with the terraces at Corran, farther up Loch Linnhe, McCann concluded that it was an outwash plain marking a halt in the retreat of the Loch Etive glacier, which he later assigned to the Loch Lomond Readvance (McCann, 1966a). He also examined the terraces along the north side of the loch around Achnaba, and reinterpreted them as ice-marginal lacustrine infillings between ice to the south, the hill slope to the north and the outwash plain to the west. This general reinterpretation was confirmed by later authors (for example, Synge, 1966).

A more detailed survey of the Loch Etive outwash and kame terrace system was undertaken by Gray (1972, 1975a) who mapped the area and levelled all the terrace fragments. This work showed that the altitudes of the terrace surfaces on both the north and south sides of the loch generally fall within clearly defined sloping bands, with only a few surface altitudes failing to fit the scheme. The gradients of these bands (3.5–6.0 m km^{-1}) are much too steep to be due to differential glacio-isostatic rebound, and hence were interpreted mainly as original fluvial drainage gradients. Thus the kame terraces are now interpreted as ice-marginal fluvial rather than lacustrine features, while the different terrace groups are seen as being the result of contraction of the Loch Etive glacier following the maximum of the Loch Lomond Readvance bringing about changes in drainage patterns. The relationships of the kame terraces to other ice-contact landforms (kames, kettle holes, eskers) has enabled a detailed reconstruction of these changes in ice-marginal drainage, as outlined above.

The Loch Etive outwash plain and associated landforms (kame terraces, kettle holes, eskers, raised beaches) is arguably the finest such system in Britain. The system was formed during the early stages of wastage of the Loch Lomond Readvance glacier in the Loch Etive valley, and illustrates the meltwater drainage patterns and mode of ice decay at this time. Although individual outwash spreads and/or kame terraces are commonplace, it is the excellent development of the features at Loch Etive, the clarity of the relationships between features, and the size and completeness of the overall system that makes it exceptional. The system covers an area of 7 km by 4 km, but only part of this, including the outwash plain and adjoining 2.5 km stretch of kame terraces and related glaciofluvial features has been selected for conservation. Not only does this area include individual features of note (the Moss of Achnacree outwash plain, the kame terraces at Achnaba, the kettle holes on the outwash plain and at Achnaba), but it also demonstrates clearly the geomorphological relationships between kame terrace (ice-marginal drainage) and outwash (proglacial drainage), between kame terraces of different age, between kame terrace and ice-margin position, between kame terrace (ice-marginal drainage) and eskers (subglacial drainage) and between outwash plain and raised shorelines.

The Loch Etive landform assemblage is also an excellent representative example of a series of outwash deposits associated with a number of Loch Lomond Readvance glaciers on the western seaboard of Scotland, for example at Loch Creran, (see South Shian and Balure of Shian), Mull, Ballachulish, Corran, Loch Shiel, Loch Morar and Loch Torridon (McCann, 1961b, 1966a; Peacock, 1970a, 1971b; Gray, 1975a; Robinson, 1987a). Compared with these other sites, the Moss of Achnacree and Achnaba area stands out for the fine detail of the landform assemblage and the clarity of the geomorphological relationships.

Conclusion

This site is important for an assemblage of landforms produced by a Loch Lomond Readvance glacier, and its subsequent melting, during the Loch Lomond Stadial (approximately 11,000–10,000 years ago). Not only is the assemblage a particularly good representative example of its type, but many of the individual landforms are also exceptionally well developed. Relationships between individual landforms are clearly demonstrated and have allowed the pattern of glacier wastage to be reconstructed.

SOUTH SHIAN AND BALURE OF SHIAN

J. D. Peacock

Highlights

Deposits exposed in coastal sections at these two sites include fossiliferous marine sediments which have been deformed by glacier ice. These provide important evidence for establishing the timing of the Loch Lomond Readvance and the marine environmental conditions during the Loch Lomond Stadial.

Introduction

The sites at South Shian (NM 910420) and Balure of Shian (NM 896420) comprise two stretches of coast in western Banderloch, 13 km north of Oban, each *c.* 0.65 km in length and including exposures on the respective foreshores and in the adjacent backing cliffs. The terminal position of the former Loch Lomond Readvance Creran glacier is associated with a wide and impressive range of landforms and deposits, some of which show evidence of glaciotectonic structures associated with overriding ice. The deposits are particularly well exposed at South Shian and Balure of Shian where glacially disturbed Lateglacial marine clay (Clyde beds) and bedded ice-contact sediments, as well as Holocene raised beach gravels can be seen. The marine clays at South Shian have yielded a diverse assemblage of molluscan and ostracod shells. With part of the similar Rhu Point site now being concealed by sea defences, South Shian and Balure of Shian are probably the most accessible localities currently available in Scotland for the examination of glacially disturbed marine clays and their relationship to other glacial and marine deposits. The landforms and sediments at South Shian and Balure have been investigated by Kynaston and Hill (1908), McCann (1966a), Peacock (1971b, 1971c), Gray (1972, 1975) and Peacock *et al.* (1989).

Description

The deposits and landforms in Benderloch comprise end moraines, glaciofluvial outwash and mounds of ice-contact gravel, sand and silt: all have been modified by Holocene marine erosion and redeposition up to a level of about 13–14 m OD (Figure 10.7). At the west end of Loch Creran there is an arcuate end moraine which reaches over 30 m OD. It is formed chiefly of transported marine clay with minor sand and gravel (Peacock, 1971a and unpublished data). West and south of this ridge there is a peat-covered composite outwash fan that laps around rock ridges and mounds of ice-contact silt, sand and gravel (Gray,

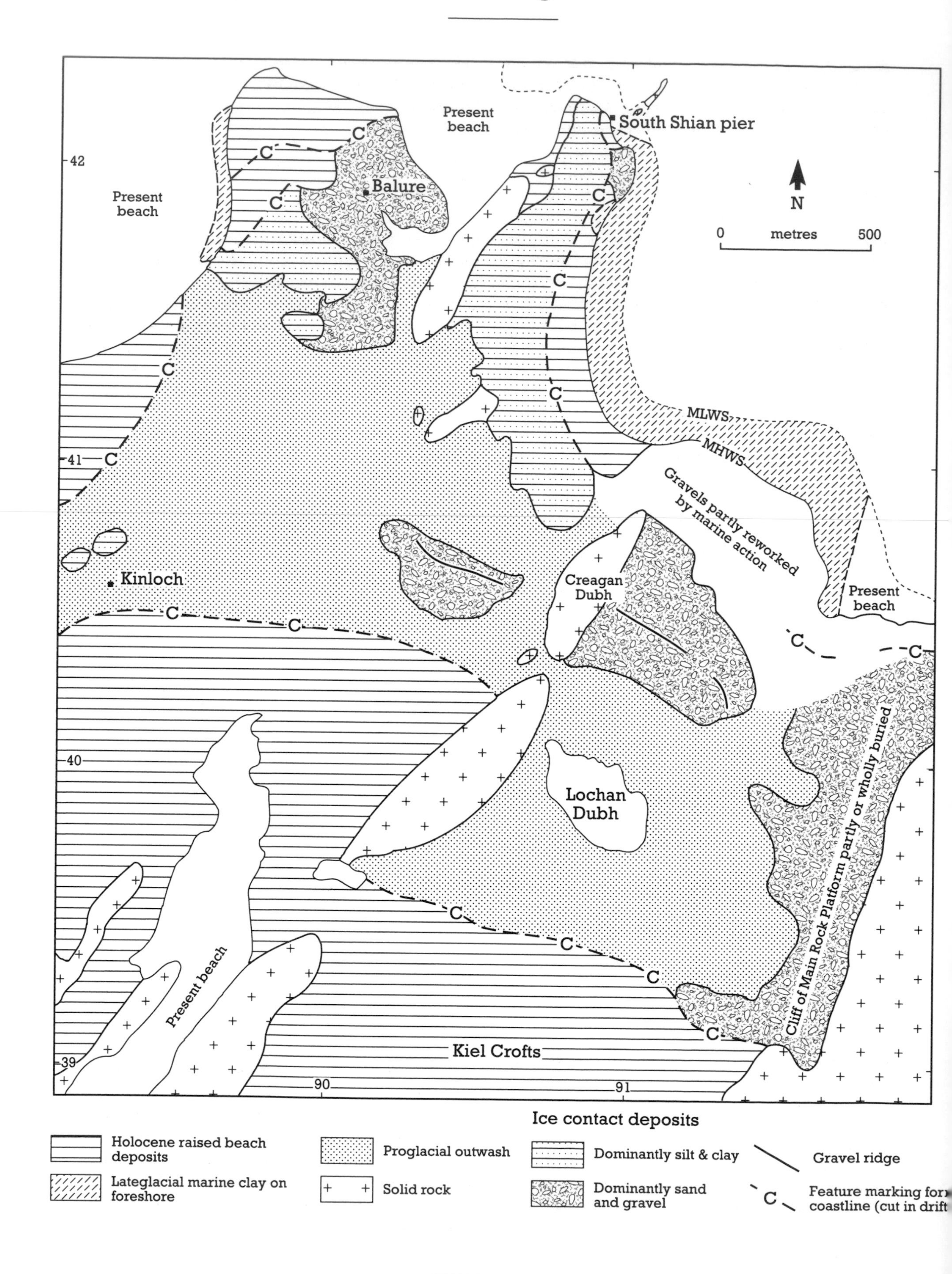
Present beach
South Shian pier
Balure
Present beach
N
0 metres 500
42
41
40
39
90
91
MLWS
MHWS
Gravels partly reworked by marine action
Kinloch
Creagan Dubh
Present beach
Lochan Dubh
Cliff of Main Rock Platform partly or wholly buried
Present beach
Kiel Crofts
Ice contact deposits
Holocene raised beach deposits
Lateglacial marine clay on foreshore
Proglacial outwash
Solid rock
Dominantly silt & clay
Dominantly sand and gravel
Gravel ridge
Feature marking for coastline (cut in drift

1972, 1975b). Moundy ice-contact gravels occupy a strip of country adjacent to the rock cliff of the Main Rock Platform east of Lochan Dubh and are terminated southwards by the back feature of the highest Holocene beach at about 12 m OD (Figure 10.7). Esker-like ridges formed of bedded gravel, which occur on both sides of the rock mound Creagan Dubh (NM 908407), were originally thought to have been produced by glacial disturbance within the outwash gravels (McCann, 1966a), but have been reinterpreted as the deposits of a subglacial stream (Peacock, 1971b). It is now considered that they could be minor end moraines. Peacock (1971b) concluded that, following the maximum of the readvance, the terminal part of the glacier stagnated while ice up-valley supplied material for deposition of outwash. This ice was probably that which formed the end moraines south of the site.

Immediately west of South Shian pier about 1.5 m of marine clay are exposed in a low cliff and are overlain by a small thickness of sand (Peacock, 1971b). The marine deposit is stiff, dark grey, brownish-weathering and silty and is streaked in places with black disseminated sulphide and organic matter. It is weakly laminated and contains scattered angular pebbles of red granite, black schist and phyllite. Similar clay crops out between tide-marks in the bay south of the pier southwards to, and beyond the southern limit of the site. Where lamination can be seen in the clay it is commonly contorted. Shells of about 40 species of mollusc have been recovered from the disturbed marine beds (Kynaston and Hill, 1908, p.168; Peacock, 1971b, p. 356 and revision in Table 10.1). These are chiefly cool-water taxa, but include the arctic bivalve *Portlandia arctica* (Gray), which in western Scotland is otherwise known only in beds attributed to the Loch Lomond Stadial (Peacock, 1977b). Radiocarbon dates on shells collected a few metres west of South Shian pier (calculated to $\delta^{14}C$) are as follows (Peacock, 1971c):

Chlamys islandica (Müller)	11,300 ± 300 BP (outer) 11,530 ± 210 BP (inner)	(IGS-C14/16)
Tridonta elliptica (Brown)	11,805 ± 190 BP	(IGS-C14/17)
Mixed shell debris	6705 ± 130 BP (outer) 11,430 ± 220 BP (inner)	(IGS-C14/18)

Figure 10.7 Quaternary deposits of the South Shian and Balure of Shian area, Benderloch (from Peacock, 1971a, unpublished data).

At Balure of Shian glacially disturbed Lateglacial marine clay with *Tridonta elliptica* (Brown) is exposed on the foreshore and in a low cliff, where it crops out below ice-contact silt, sand and clay as well as below glaciofluvial outwash gravel and storm beach gravel associated with the highest Holocene raised beach. In the sea cliff west of Balure, some 10 m of interbedded silt, clay and fine-grained sand capped by about 3 m of poorly sorted gravel were formerly to be seen in a temporary section exposed by marine erosion in the mound (NM 896418). Marine clay at the base of this section (McCann, 1966a) is folded up into the overlying beds (Peacock, 1971b). The clays contain entire hinged valves of *Portlandia arctica* (Gray) and small quantities of *Nuculoma belloti* (Adams) and *Yoldiella lenticula* (Müller) (Table 10.1). No microfauna was found (Peacock *et al.*, 1989).

Shells (*Portlandia arctica* (Gray)) from Balure of Shian have been radiocarbon dated using both conventional and accelerator methods. The results are as follows (Peacock *et al.*, 1989):

	Conventional age (^{14}C years BP ± 1σ)	Reservoir adjusted age (^{14}C years BP ± 1σ)
SRR–3182	10,510 ± 90 (outer)	10,105 ± 100
	10,320 ± 70 (inner)	9915 ± 80
SRR–3204	10,550 ± 100	10,145 ± 110
OxA–1345	10,960 ± 120	10,555 ± 130

Interpretation

Most of the glacial deposits in Benderloch were regarded as '100 ft Beach' sediments by the original geological surveyors (Kynaston and Hill, 1908) and are shown as 'Higher Beach' and 'Highest Beach' on the accompanying 'One-inch' map (Geological Survey of Scotland, 1907, Sheet 45). Charlesworth (1956) noted morainic deposits east of Lochan Dubh (NM 908400), but the possibility that the landforms and sediments were laid down near the terminus of a Loch Lomond Readvance glacier was first put forward by McCann (1966a) and independently by Synge (1966). McCann suggested that the 'high raised beach' deposits were in fact glaciofluvial outwash post-dating the marine clay, and Synge described the arcuate ridge south of South Shian as a terminal moraine. Both authors were of the opinion that the glacier terminated at the west end of Loch Creran (*c.* NM 915425), but Peacock (1971b) put forward evidence that the ice had extended considerably farther west and south (by as much as 2 km), a view supported by Gray (1972, 1975b) and Peacock *et al.* (1989).

The radiocarbon dates from South Shian confirm that the dated molluscan fauna flourished in the latter half of the Lateglacial Interstadial and that the Loch Lomond Readvance glacier in Loch Creran reached its maximum later than about 11,400 BP. This is a similar picture to that obtained from radiocarbon dating of glacially disturbed marine clays at Rhu Point, Loch Spelve (Mull) and Menteith, west of Stirling (Sissons, 1967b; Gray and Brooks, 1972; Rose, 1980c; Sutherland, 1984a).

The Balure of Shian site provides clear evidence for the stratigraphic position of the Lateglacial marine clay, that is, it antedates deposits laid down by the Loch Lomond Readvance ice (when this was near its maximum extent). Further work (Peacock *et al.*, 1989) suggests that the marine clays at Balure are less deformed than those at South Shian and contain an arctic marine fauna typical of glaciomarine conditions. The low-diversity fauna dominated by *Portlandia arctica* (Gray), is typical of that found seaward of the mouths of glacial rivers and tidewater glaciers in east Greenland and Spitsbergen today (Odhner, 1915; Ockelmann, 1958). The marine clays would thus be expected to immediately antedate the arrival of the Creran glacier at its maximum position.

The radiocarbon dates obtained at Balure of Shian allow a revised estimate of the age of the maximum extent of the Loch Lomond Readvance glacier in the Loch Creran Valley, to within, and possibly towards the end of, the period 10,500–10,000 BP (Peacock *et al.*, 1989). This agrees well with recent estimates of the age of the maximum extent of the Loch Lomond Readvance glaciers at Loch Lomond and in the Upper Forth Valley (see Croftamie and Western Forth Valley; Browne and Graham, 1981; Sissons, 1983a; Rose *et al.*, 1988), but contrasts with dates from organic lake sediments which suggest earlier deglaciation (see discussion for Croftamie).

The deposits and landforms in western Benderloch, including those at South Shian and Balure of Shian, provide a well-documented record of environmental changes during parts of the Late Devensian and during the Holocene. As such they are integral members of a national network of key sites demonstrating changing marine and terrestrial conditions during the latter part of the Quaternary. The disturbed marine clays contain faunas which can be confidently referred to both the cool water (interstadial) and cold water (arctic) parts of the Clyde beds (see Chapter 1 and Geilston). Deformation of these deposits by overriding Loch Lomond Readvance ice can be clearly demonstrated, as can their burial by glaciofluvial deposits attributed to the subsequent retreat of the Creran glacier from its nearby maximum position. The presence of erosional and depositional landforms associated with the high Holocene sea levels and their relationship to the deposits of the Loch Lomond Readvance lend additional significance, particularly to the Balure site. Further, the sites offer considerable potential for research into Lateglacial marine microfaunas, sedimentology and the physical properties of glacially deformed marine and glaciomarine deposits (see Peacock *et al.*, 1989). Finally, the radiocarbon dates from the site provide important evidence for interpreting the age of the maximum extent of Loch Lomond Readvance glaciers.

Conclusion

The deposits at South Shian and Balure of Shian provide important evidence for interpreting critical aspects concerning the geomorphology and timing of the resurgence of glacier ice associated with the Loch Lomond Readvance, and the prevailing environmental conditions. The sea formerly covered this area, and fossiliferous marine deposits were laid down under cool-water and then cold-water conditions; these were subsequently deformed as Loch Lomond Readvance glacier ice in the Creran valley advanced across them to its maximum position. Radiocarbon dates provide important age estimates for the timing of this event (between about 10,500 and 10,000 years ago). South Shian and Balure of Shian is therefore a key locality for studies of the Loch Lomond Readvance in western Scotland.

Table 10.1 Mollusca from South Shian and Balure of Shian

	1	2	3	4
Antalis entalis (L.)	**	*		
Boreotrophon clathratus (Ström)		*	*	
Boreotrophon truncatus (Ström)	**			
Buccinum undatum (L.)	*	*		
Gibbula cineria (L.)		*		
Lacuna parva (da Costa)			**	
Lacuna vincta (Montagu)	*			
Littorina sp	*			
Littorina saxatilis (Olivi)				*
Lora turricula (Montagu)		*	*	
Manzonia zetlandica (Montagu)	*			
Margarites costalis (Gould)	*			
Margarites helicinus (Fabricius)			*	
Moelleria costulata (Möller)	*			
Onoba aculeus (Gould)			**	
Onoba semicostata (Montagu)	*		**	
Polinices pallidus (Broderip and Sowerby)	*	*		
Puncturella noachina (L.)	*	*		
Rissoa interrupta (Adams)		*	**	
Skeneopsis planorbis (Fabricius)			**	
Tectonatica affinis (Gmelin)	*			
Velutina velutina (Müller)		*		
Abra sp.	*			
Abra alba (Wood)			**	
Acanthocardia echinata (L.)	*			
Arctica islandica (L.)	*	**		
Astarte sulcata (da Costa)		*		
Chlamys islandica (Müller)	***	***		
Heteranomia squamula (L.)	*			
Hiatella arctica (L.)	*	*		
Jupiteria minuta (Müller)	*		**	
Lyonsia arenosa (Möller)			*	
Macoma calcarea (Chemnitz)		*	**	
Mya truncata (L.)		*	**	
Nucula nucleus (L.)	*	*		
Nuculana pernula (Müller)	**	***	**	
Nuculoma sp.		*	**	
Nuculoma belloti (Adams)	*			**
Parvicardium ovale (Sowerby)	**	*		
Portlandia arctica (Gray)		*	*	****
Spisula subtruncata (da Costa)			*	
Thracia cf. *myopsis* (Möller)			*	
T. cf. *villosiuscula* (Macgillivray)	*			
Thyasira gouldi (Philippi)	*		**	
Tridonta elliptica (Brown)	***	***		
Tridonta montagui (Dillwyn)	**	**	*	
Yoldiella solidula (Warén)	**		***	
Yoldiella lenticula (Müller)	**	**	***	**

Continued overleaf

Table 10.1 *continued*

1. *Shore 10 m west of South Shian pier (NM 90834228).*
2. *Shore 10 m west of South Shian pier (NM 90834228), British Geological Survey collection.*
3. *Shore east of shellfish factory (NM 908416).*
4. *Shore west of Balure of Shian (glaciomarine bed) (NM 8962 4216).*

* *rare*
** *common*
*** *very common*
**** *dominant*

GLEN ROY AND THE PARALLEL ROADS OF LOCHABER

J. E. Gordon

Highlights

The area of Glen Roy and adjacent parts of Glen Spean and Glen Gloy is one of outstanding international importance for geomorphology. It is best known for the Parallel Roads, a series of ice-dammed lake shorelines which developed during the Loch Lomond Stadial. These form part of a much wider assemblage of glacial, glaciofluvial and glaciolacustrine features which provide unique evidence for the dramatic impact of geomorphological processes on the landscape during the stadial.

Introduction

The interest of this site extends across an area *c.* 146 km^2 east of Fort William, in Lochaber, covering parts of Glen Roy, Glen Gloy and Glen Spean. Glen Roy is a long-recognized site of international importance for its former ice-dammed lake shorelines, the 'Parallel Roads', which are the most extensive and best developed examples in Britain. The Parallel Roads, first documented by Thomas Pennant in 1771, have been the subject of some 70 scientific papers, and the site is widely regarded as being a classic example of former lake shorelines in standard texts on geomorphology and physical geology. Much of the original research on the Parallel Roads, which also occur in Glen Gloy and Glen Spean, was carried out during the 19th century when the landforms of this area were found to provide significant evidence for the former existence of glaciers in Scotland (Agassiz, 1842). The Parallel Roads were first recognized as the shorelines of ice-dammed lakes by Agassiz (1841b, 1842), an interpretation later confirmed in the definitive work of Jamieson (1863, 1892). More recently, in a series of papers Sissons (1977e, 1978, 1979a, 1979b, 1979c, 1981c, 1981d) has elucidated the formation of the Parallel Roads through detailed field observations and mapping and by setting them into the wider geomorphological context of contemporaneous events in Glen Spean and the Great Glen; additional evidence and details have been considered by Sissons and Cornish (1982a, 1982b, 1983), Peacock (1986, 1989a) and Peacock and Cornish (1989).

Of outstanding interest in their own right, the Parallel Roads also form part of a remarkable system of glacial, glaciofluvial and glaciolacustrine landforms extending from Loch Laggan west to near Fort William and north to the Great Glen (Figure 10.8). The total system and many of its individual elements are of considerable geomorphological interest both intrinsically and in their relationships to the Parallel Roads and the sequence of later events in Glen Roy. The scientific interest of the area therefore extends well beyond Glen Roy, and the site boundary is drawn to include not only the Parallel Roads of Glen Roy but also the wider landform system of which they are part.

Description

The landform assemblage and key localities

The geomorphology of the Glen Roy–Glen Spean area, including the form and location of the Parallel Roads, have been described extensively in the literature; the principal references are by MacCulloch (1817), Dick (1823), Darwin (1839), Maclaren (1839), Agassiz (1841b, 1842), Milne Home (1847b, 1849, 1876, 1879), Chambers

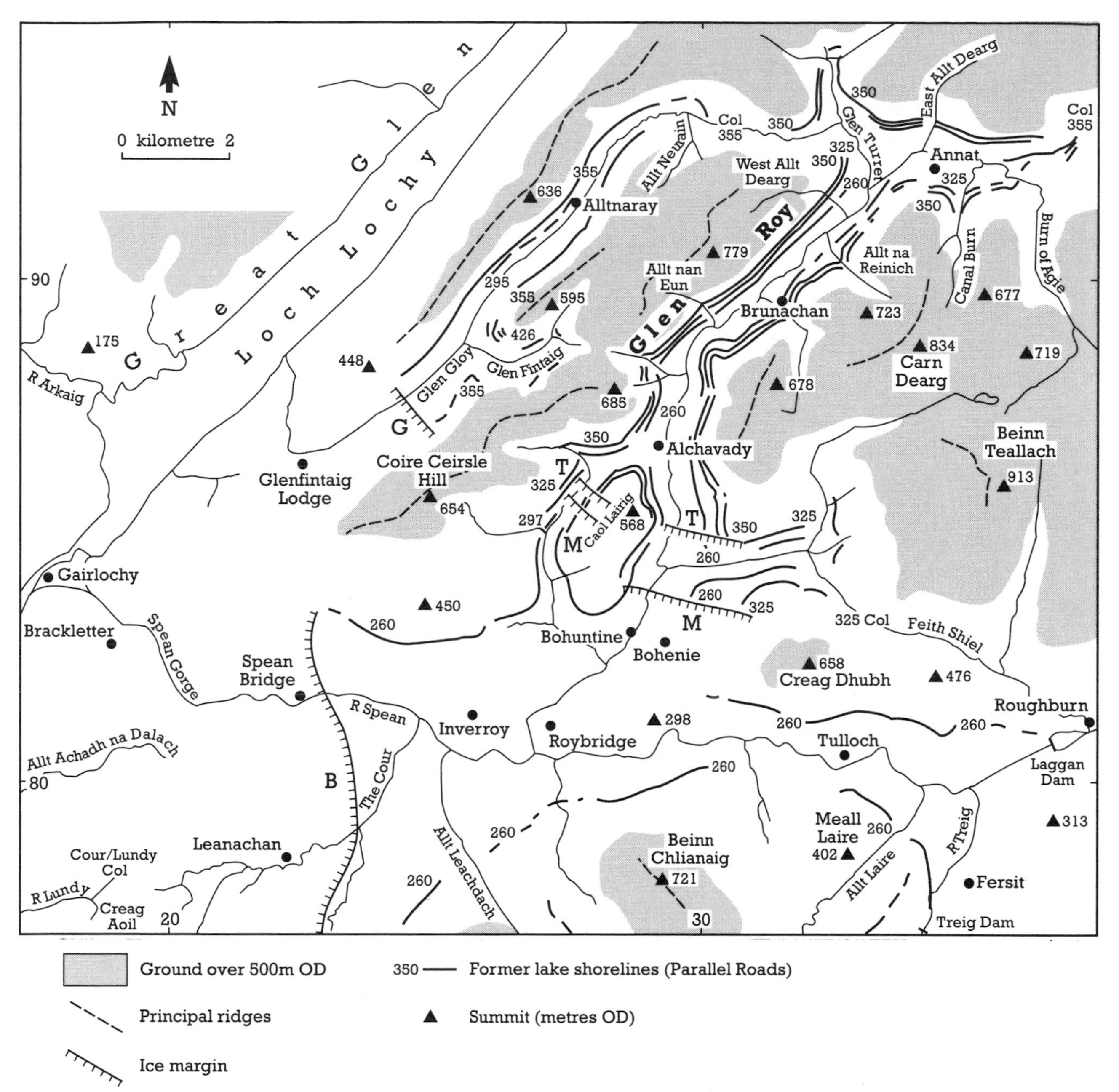

Figure 10.8 The Parallel Roads of Lochaber. The letters T, M, B and G identify the final positions of the ice-fronts damming the 355 m, 325 m, 260 m and Glen Gloy lakes respectively (from Peacock and Cornish, 1989).

(1848), Mackenzie (1848), Thomson (1848), Bryce (1855), Jamieson (1862, 1863, 1892), Rogers (1862), Mackie (1863), Watson (1866), Babbage (1868), Lubbock (1868), Nicol (1869, 1872), James (1874), Jolly (1873, 1880a, 1880b, 1886a, 1886b), Brown (1875), Campbell (1877), Dakyns (1879), Tyndall (1879), Livingston (1880, 1906), Prestwich (1880), Macfadzean (1883), Melvin (1887), Kinahan (1887), Wilson (1900), MacDonald (1903), Peacock (1970b) and Sissons (1978). The Parallel Roads are almost entirely former lake shorelines, although locally they occur as glaciofluvial terraces. Three main roads occur in Glen Roy at average altitudes of 350 m, 325 m and 260 m OD; one in Glen Gloy at 355 m and one in Glen Spean at 260 m OD (Figure 10.8). Typically they are cut in bedrock (Figure 10.9) and comprise an erosional floor and backslope and a depositional foreslope. Horizontal widths range from 1.6 to 63.6 m, and the backing cliff reaches a maximum height of 6 m (Sissons, 1978). To explain the formation of the

Figure 10.9 The Parallel Roads of Glen Roy on the east flank of the glen, north-east of the viewpoint, are cut in bedrock which can be seen clearly exposed in the gully in the foreground. (Photo: J.E. Gordon.)

features, Sissons (1978) invoked a combination of wave action and powerful frost disruption of the bedrock along each shoreline (see Matthews *et al.*, 1986; Dawson *et al.*, 1987b and Shakesby and Matthews, 1987 for discussion of possible modern analogues). Detailed levelling by Sissons and Cornish (1982a, 1982b) has shown that the shorelines are not uniformly tilted or warped, and that differential movements have occurred between blocks of the Earth's crust.

As noted above, the Quaternary landforms and deposits of the Glen Roy area are not only many and varied, but are also represented at a large number of key localities. Only the main features are summarized below, while additional details and sites are reported in Peacock (1989a).

Glen Roy

1. The important features in the uppermost part of Glen Roy are the Roy–Spey col (NN 410943) at 350 m, which controlled the level of the highest lake in the glen, and a suite of glacio-fluvial landforms extending from the Allt Chonnal across the lower valley slopes on the north side of the River Roy to the col. These deposits are crossed

by the highest Parallel Road and although their origin and relations have not been determined, they probably relate to the decay of the main Late Devensian ice-sheet. Palynological evidence from a bog on the col overflow channel shows that sediment began to accumulate there during the early Holocene (MacPherson, 1978; Lowe and Cairns, 1989, 1991), which supports a Loch Lomond Stadial age for the ice-dammed lakes.

2. Several sites demonstrate key aspects of the lake shorelines. The section of Parallel Road on the south side of Glen Roy north of the Burn of Agie (NN 369920) is one of the clearest examples of a shoreline cut in bedrock. It is associated with a prominent delta formed by the penecontemporaneous Burn of Agie. For a distance of about 300 m north of the burn the middle road is a rock-cut platform up to 12 m wide with a backing cliff up to 5 m high. Shorelines cut in bedrock are also well demonstrated at Braigh Bac (NN 306882) and Creagan na Gaoithe (NN 370925). In a gully on the east side of Glen Roy at (NN 307877) there is a good exposure showing the middle road cut in bedrock, and in a similar situation at (NN 304868) the top road is clearly cut across the structural grain of highly fractured bedrock. The susceptibility of the bedrock to weathering, demonstrated at the latter locality and elsewhere, is an important consideration in explaining the processes of formation of the Parallel Roads (Peacock and Cornish, 1989). Well-developed aggradational shorelines are represented in grid squares NN 3592 and NN 3692. Locally, additional Parallel Roads are present, for example at 334 m and possibly 344 m at Braigh Bac.

3. At the junction of Glen Roy and Glen Turret there is an important and controversial set of deposits comprising a fan with, at its northern end, an irregular, hummocky surface aligned with a series of subparallel mounds and terraces climbing obliquely up-valley on the east side of Glen Turret (Figure 10.10). Details of several important sections are described by Peacock (1986) and Peacock and Cornish (1989). Sections exposed in the south-east bluff of the fan (for example at NN 346924) show it to comprise coarse, poorly bedded gravels, and fine-grained lake sediments can be seen in scrapings on its surface. In a section in the fan (at NN 338919) Peacock and Cornish (1989) reported the following sequence (see also Peacock, 1986):

(3)	Well-bedded gravel, clast-supported, bouldery and cobbly (particularly towards the top), with a poorly-sorted, sandy matrix. Bedding subhorizontal, parallel to the fan surface, with beds less than 0.3 m thick. Local sand beds a few centimetres thick. Local imbrication.	21 m
(2)	Interbedded, hard, pebbly, laminated silt, and gravel.	5.0 m
(1)	Gravelly till.	1.5 m

Sections in the mounds at the back of the terrace (for example, at NN 339928)) reveal a variety of materials ranging from silts and clays to coarse, angular debris. The sedimentology of these fan deposits and their interpretation is critical in understanding the sequence of events (Sissons and Cornish, 1983; Peacock, 1986; Peacock and Cornish, 1989). Sissons (1977e) interpreted the fan as a delta, and later as a subaerial fan (quoted in Gray, 1978b). However, the association of the deposits, the terrace, the mounds on its surface and the lateral ridges up-valley closely resembles that of a former ice margin, and the north-west flank of the terrace closely resembles an ice-contact slope. Thus Rose (quoted in Gray, 1978b) interpreted the terrace feature as an outwash fan formed at an ice limit at some time during ice-sheet decay. Peacock (1986) concurred with this interpretation. Sissons and Cornish (1983), however, favoured outwash deposition into the 260 m lake of the rising sequence, at a time when the Gloy glacier extended across the col between Glen Gloy and Glen Roy. They suggested that the rise in lake level in Glen Roy resulted in ablation of the Gloy glacier and its retreat into Glen Gloy, which thus allowed the higher shorelines to form in Glen Turret. The absence of Lateglacial pollen from a sequence of organic deposits in a section and borehole at Turret Bank (NN 337925) suggested to Lowe and Cairns (1989, 1991) that Glen Turret was occupied by a Loch Lomond Readvance glacier. Although the pollen evidence on its own is inconclusive, Lowe and Cairns (1991) considered that this interpretation best fitted the wider pattern of landforms. However, Peacock (in Peacock and Cornish, 1989) considered that the commencement of organic sedimentation simply related to the drainage of the 260 m lake and not to the end of any glacial event. Further work on the Turret fan to resolve these outstanding issues is awaited (cf. Lowe and Cairns, 1991).

4. In the lower part of the valley of the Allt

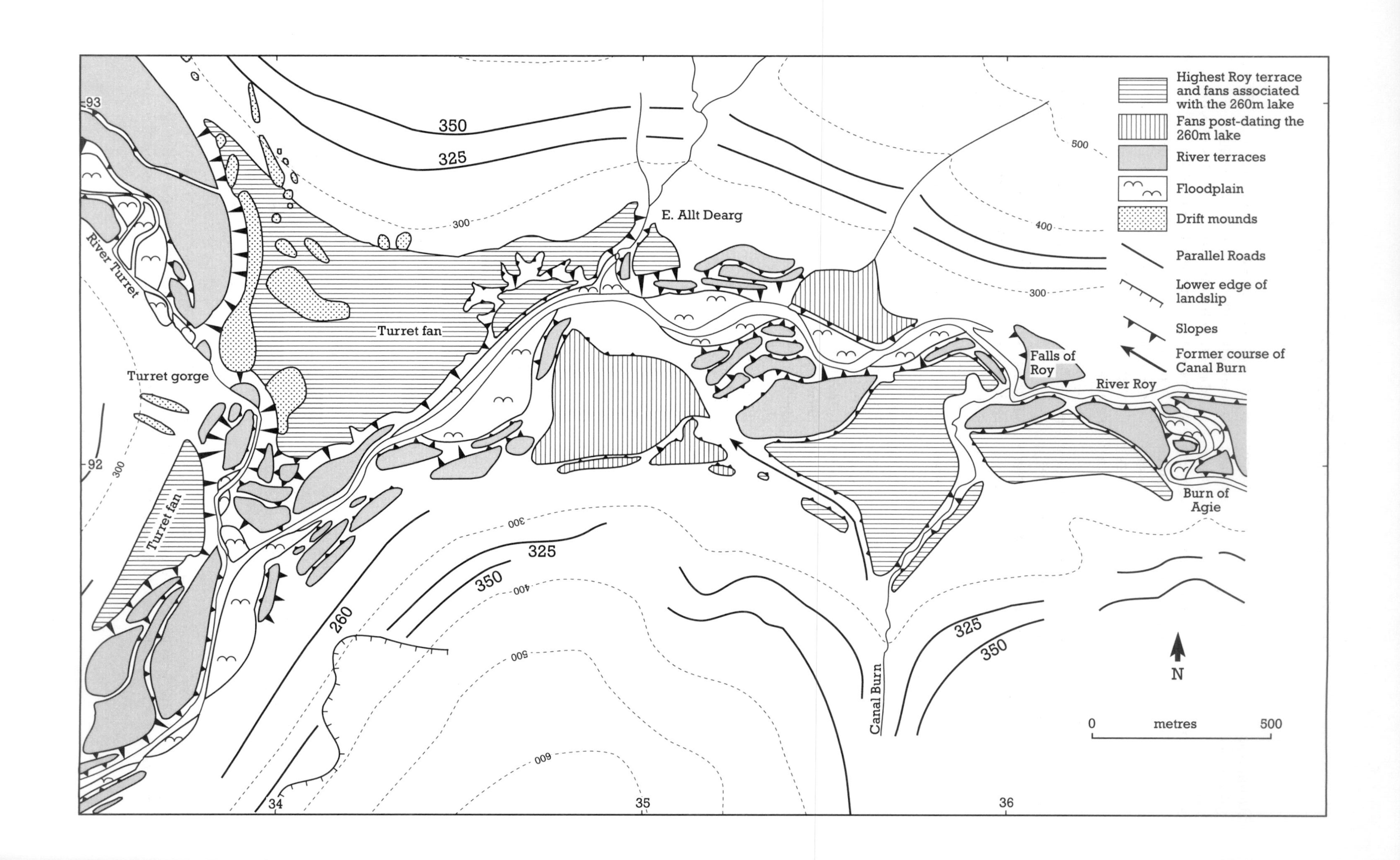
Highest Roy terrace and fans associated with the 260m lake
Fans post-dating the 260m lake
River terraces
Floodplain
Drift mounds
Parallel Roads
Lower edge of landslip
Slopes
Former course of Canal Burn
N
0
metres
500
350
325
300
400
500
600
260
E. Allt Dearg
Turret fan
Turret gorge
River Turret
Falls of Roy
River Roy
Burn of Agie
Canal Burn
93
92
34
35
36

a'Chomlain near its junction with Glen Turret (NN 330929) is a series of gravel mounds and deposits, with kettle holes, which have been terraced and dissected by the river. These were formed during the deglaciation of the area, although the precise details are unclear (Peacock and Cornish, 1989). Also in this area is a terrace which appears to be a delta of the 325 m lake (Peacock and Cornish, 1989).

5. Several superb examples of alluvial fans occur in Glen Roy (Sissons and Cornish, 1983; Peacock, 1986; Evans and Hansom, 1991, figures 1 and 2). On the east side (at NN 330907 and NN 318896) two large dissected fans extend across the valley floor from Coire na Reinich and Coire Dubh (the Reinich and Brunachan fans, respectively). Others are associated with the Burn of Agie, Canal Burn, the East Allt Dearg and the West Allt Dearg. Peacock (1986) described several sections in the fans, which principally comprise coarse gravels and sands, in places both overlain and underlain by laminated sediments. According to Sissons and Cornish (1983) these fans were deposited into the lowest lake of the rising sequence, but Peacock (1986) interpreted them as being older, subaerial features.

6. Thick drift deposits are present at the head of Glen Turret. In a gully section (NN 329944) there are up to 27 m of laminated silts, sands and gravels containing many angular clasts, which are overlain by up to 3 m of till. Sissons (1978) believed the source of the angular material to have been frost-riven debris transported from the lake shores by ice floes. Peacock (1986), however, considered the material to be waterlain till. East of the section a prominent fan appears to be graded to the level of the 325 m Parallel Road and may therefore be, in part, a delta (Peacock and Cornish, 1989).

7. In upper Glen Roy a particularly fine suite of river terraces, formed by the River Roy after drainage of the lowest lake, occurs on the south side of the River Roy between about NH 368920 and NH 345920 (Figure 10.10). Terraces also continue along the floor of the glen south-west from Braeroy Lodge.

8. Landslides are well represented (Sissons and Cornish, 1982a, 1982b; Holmes, 1984; Peacock and Cornish, 1989): a fine example occurs on the east side of Glen Roy (at NN 342915) and crosses the upper two roads (Figure 10.10). Another, which cuts across both the middle and lower roads, occurs 0.5 km down the valley from the viewpoint (at NN 295849). On the west side of Glen Roy, opposite Brunachan, Sissons and Cornish (1982a, 1982b) described a large landslide which they related to earthquake activity along a fault line activated by glacio-isostatic uplift. Ringrose (1987) (see also Davenport *et al.*, 1989; Peacock and Cornish, 1989), however, has suggested that the fault could have been activated by lateral movement along an adjacent fault line; it may therefore be only indirectly associated with glacio-isostatic uplift, if at all.

Figure 10.10 Geomorphology of the northern part of upper Glen Roy (from Sissons and Cornish, 1983).

9. A series of interesting landforms and deposits are represented in the Allt Bhreac Achaidh area (NN 298875) (Peacock and Cornish, 1989). These include ridges of laminated silt and gravel with liquefaction and other deformation structures (Ringrose, 1987, 1989c), river terraces underlain by laminated silt, and glacial and paraglacial landforms and deposits.

10. The viewpoint (NN 297853) affords the classic view of the Parallel Roads, which are strikingly displayed on both the west and east hillsides of Glen Roy. On the hillside north of the viewpoint, the limit of the Loch Lomond Readvance ice in the glen occurs at, or a little beyond, the northern end of a massive, dissected drift plug up to 80 m thick (approximately NN 298864–NN 300850) (Sissons, 1979b). The former ice margin is probably marked by a clear drift limit, while on the east side of the glen there is a landslide and drift ridge at the ice limit. Older moraine ridges occur beyond the readvance limit. Roadside sections near the top of the infill reveal glaciofluvial sands and gravels, and lacustrine silts and sands with drop stones and slump structures (NN 296858). Various gully exposures (see Peacock and Cornish, 1989) reveal further sands and gravels, and till near the base. These deposits form a glaciolacustrine delta with foreset and bottomset beds. A sequence of river terraces extends from the southern end of the drift plug to Roy Bridge and merges with the Glen Spean terraces. The former relate to the dissection of the drift plug by the waters of a remnant lake impounded by the plug following the drainage of the 260 m lake (Sissons, 1979a).

11. The Caol Lairig is an important site where a variety of glacial, glaciofluvial and glaciolacustrine landforms are easily accessible. The Loch Lomond Readvance ice limit is marked by an

arcuate moraine ridge 5 m high across the col (NN 288864) and its lateral extension can be traced along the hillslope to the west (NN 861276) as the upper limit of small meltwater channels (Sissons, 1979b). Four shorelines, in part lacustrine deltas, are present on the valley sides; the additional one at 297 m is related to the altitude of the Caol Lairig–Glen Roy col, and the lake overflow can be seen as a channel cut through the end moraine. Inside the latter, deltas and fans occur on the valley floor. Several sections occur in glaciolacustrine sediments (Peacock and Cornish, 1989), which include sedimentary structures that may relate to earthquake deformation (Ringrose, 1987, 1989c).

12. North of Bohuntine and Bohenie end moraine ridges on both sides of Glen Roy (at NN 291839 and NN 297836) mark the ice limit when the 325 m lake was formed.

13. Good sections in lake sediments are frequently exposed in cuttings along the public road in Glen Roy, and they provide a valuable source of sedimentary information. For example, Miller (1987) has identified two types of rhythmic deposit on the basis of their sediment characteristics and stratigraphic position. 'Group I laminates' (fine sands and silts) tend to cap major sediment bodies. They are typical of proximal glaciolacustrine deposits and they were probably deposited in the 350 m lake during the Loch Lomond Stadial. 'Group II laminates' (silts and clays) typically underlie major sediment bodies. They have characteristics of distal glaciolacustrine sediments, probably deposited during an early stage of the rising lake sequence.

Glen Gloy

1. Several mounds (at NN 280910) near Alltnaray are believed to mark the limit of the Loch Lomond Readvance ice in Glen Gloy (Peacock, 1970b; Sissons, 1979b), although this was not accepted by Sissons and Cornish (1983) (see also discussion of the Turret fan above). Inside this limit thick drift deposits are exposed along the forest road on the west side of the glen. They are attributed to debris flows and delta formation (Peacock and Cornish, 1989).

2. A second important site in Glen Gloy is the col at the head of the glen through which the waters of the 355 m lake spilled over into Glen Turret and Glen Roy. A small glaciofluvial terrace is present. Lowe and Cairns (1989, 1991) recorded 7.0 m of peat and lake sediments and showed that organic sedimentation began during the early Holocene.

3. At the Allt Neurlain (NN 303926) several features are of interest, including fault-controlled streams, possible recent movement along a fault (Ringrose, 1987), a delta at the 355 m road and sandy hummocks that possibly comprise a subglacial fan.

4. Glenfintaig (NN 265885) is important for an assemblage of landforms, comprising a sequence of up to eight shorelines (the clearest at 295 m, 355 m, 416 m and 426 m), a landslide, a drift limit possibly marking the maximum extent of the Loch Lomond Readvance glacier, and lake sediments and river terraces (Peacock and Cornish, 1989).

5. In addition to Glenfintaig, the main Parallel Road in Glen Gloy at 355 m is also well-developed at Allt Grianach (also 295 m road and delta) (NN 270905), Auchivarie (NN 287928) (partly cut in bedrock) and Allt Fearna (partly cut in bedrock) (NN 305935).

Glen Spean

1. The Roughburn area (Figure 10.11) is important for an assemblage of glacial and glaciolacustrine deposits. To the north of the A86 in the valley of the Feith Shiol a double end moraine marks the limit of the confluent Spean–Treig glacier, which impounded the 260 m lake in Glen Spean at the Loch Lomond Readvance maximum (Sissons, 1979b). The overflow from the 325 m lake in Glen Roy through the col at the head of Glen Glas Dhoire followed the valley of the Feith Shiol and breached the moraine ridges before entering the 260 m lake. At Roughburn a delta (NN 377813), comprising up to 10 m of coarse gravel in steeply dipping foreset beds on top of silty sands, records the torrential overspill into the lake (Jamieson, 1863; Peacock and Cornish, 1989). Eastwards along the north shore of Loch Laggan, fine-grained sediments of the distal part of the delta (bottomset or low-angle foreset beds) are well exposed (Peacock and Cornish, 1989).

2. The Inverlair–Fersit area north of Loch Treig (Figure 10.11) is important in several respects. It demonstrates an excellent example of a partly kettled delta formed in the 260 m lake as the Treig glacier receded back into the valley now occupied by Loch Treig (Peacock and Cornish, 1989). The delta extends from around Inverlair to south of the Treig dam and comprises an extensive area of sand and gravel, with

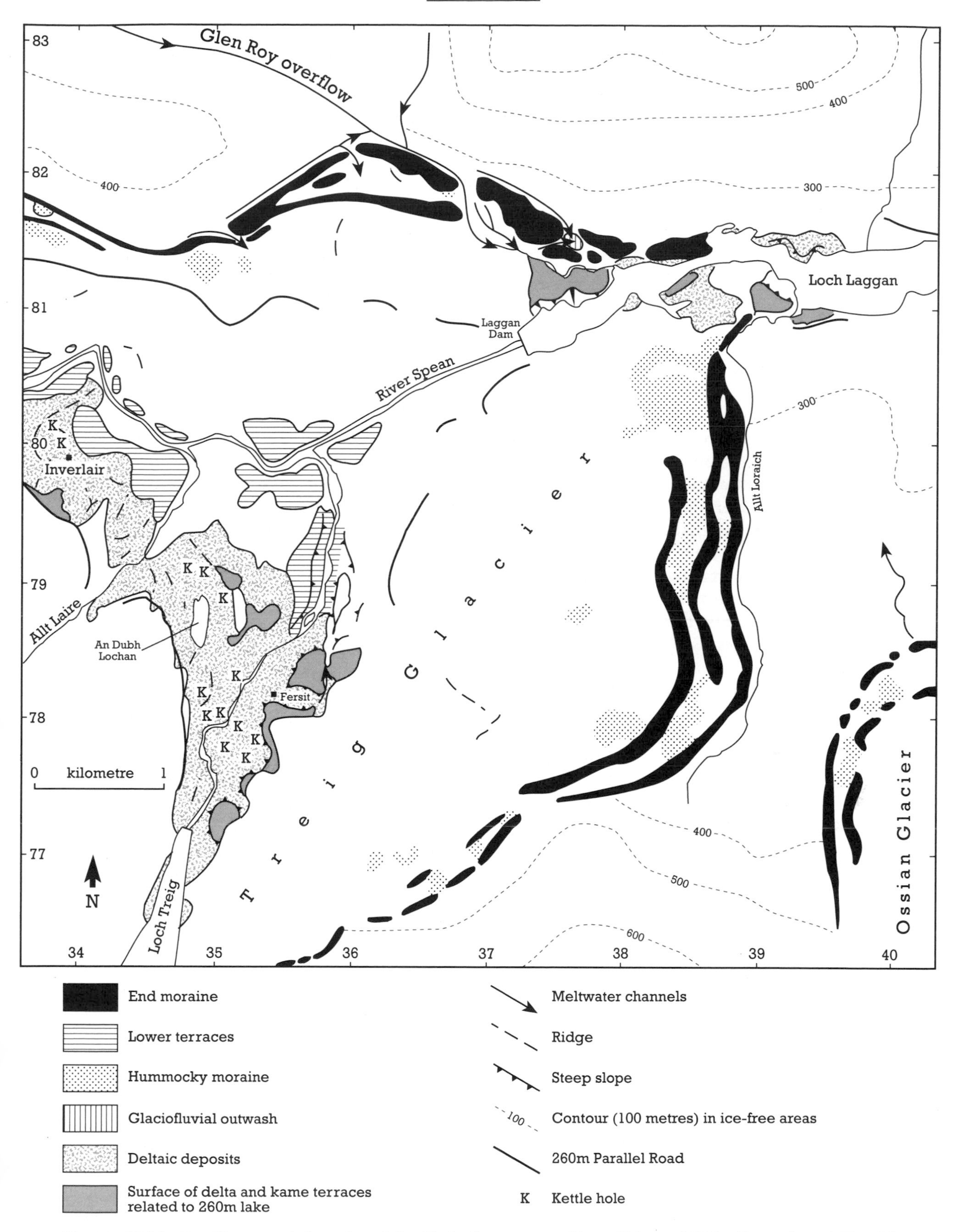

Figure 10.11 Landforms and deposits of the Treig–Laggan area (from Sissons, 1977e; Peacock and Cornish, 1989).

remnants of the original delta surface preserved, particularly around Fersit. A series of kame terraces lead from the delta southwards between Fersit and Loch Treig, notably on the east side of the valley. Foreset beds in the delta are exposed in the former gravel quarry at Fersit and in sections on the west side of Loch Treig. Following drainage of the lake, a series of outwash and river terraces formed in front of the receding glacier. These are represented on the east bank of the Treig (Peacock and Cornish, 1989), and younger terraces are particularly well seen to the south of Tulloch Station, where they continue down Glen Spean (Sissons, 1979a). Areas of water-worn bedrock and p-forms occur on the west shores of Loch Treig. The Fersit area also demonstrates relationships between the lowest Parallel Road and glaciofluvial landforms: south of about NN 345789, the 260 m shoreline merges with, and becomes a kame terrace. Spectacular kettle holes, up to 25 m deep, are present in deltaic deposits at Inverlair. Palynological investigations of several of these in the Inverlair–Fersit area have revealed that organic sedimentation began during the early Holocene (McPherson, 1978; Lowe and Cairns, 1989, 1991). Finally, the Inverlair–Fersit area is also of significant historical interest. The glacial features there greatly impressed Agassiz during his tour in 1840 (Agassiz, 1842), when he first recognized the former existence of glaciers in Scotland.

3. The valley of the Allt Leachdach provides important evidence for lake levels above 113 m (Peacock and Cornish, 1989). Near Loch a'Bhuic (NN 264788), which is dammed by an esker, a kame terrace grades into the 260 m shoreline and a 'collapsed' fan/delta is also associated with it. Lower down the valley, deltas and fans are associated with successively lower lake levels at about 143 m, 130 m, 122 m and 114 m. The last level corresponds to the 113 m lake discussed by Sissons (1979a). These levels provide significant evidence for interpreting the sequence of lakes that followed drainage of the 260 m lake. However, it is unclear whether they relate to the period of variable lake level following drainage of the 260 m lake (see Sissons, 1979a) or indicate an intermittent drop in lake level (Peacock and Cornish, 1989). Later terraces and Hjulström-type deltas in the Spean valley are also well demonstrated in this area, for example near Coirechoille (NN 250807).

4. Deltas, fans and high-level terraces elsewhere in Glen Spean provide important evidence for interpreting the sequence of events at the time of, and following, the 260 m lake:

(i) Kame terrace/delta at Achnacochine (NN 310807) associated with the 260 m Parallel Road and with retreat of the Spean glacier.
(ii) The 175 m delta of the River Spean at Tulloch (NN 330807).
(iii) Glaciolacustrine delta at Innis nan Seangan (NN 317794) above the level of the 260 m Parallel Road and with a good section showing internal composition.
(iv) Large outwash trains in the valley of the Allt nam Bruach (NN 314807), associated with the 260 m lake. Following the drainage of the lake, the outwash was dissected by the Allt nam Bruach and the material redeposited at the mouth of the valley as steeply sloping terraces which merge with those of Glen Spean (Sissons, 1979a). Near NN 309802 the lowest Parallel Road merges with a glaciofluvial terrace.
(v) High Spean terrace at Insch (NN 264802), with good sections in deltaic bottomset beds.

Many of these Spean valley deposits consist of delta topset beds overlying bottomset beds, without foreset beds, in contrast to the Roughburn and Treig deltas. They are thus probably of Hjulström type rather than Gilbert type (J. D. Peacock, unpublished data).

5. An important suite of river terraces recording the stages of valley infill and dissection after the drainage of the 260 m lake occurs between Roy Bridge and Spean Bridge (Sissons, 1979a). The upper terraces largely comprise sands (seen in section at NN 217819 and NN 274811), which overlie lacustrine silts and clays (Peacock, 1970b). The lower terraces are believed to be cut in lake sediments (Sissons, 1979a). East of Roy Bridge a higher-level terrace remnant is prominent (Peacock and Cornish, 1989). At Spean Bridge a sandpit (NN 217819) shows that the terrace in which it is excavated comprises laminated sands with ripple bedding and a small channel near the surface (Peacock and Cornish, 1989). On the south side of the Spean valley, Peacock and Cornish (1989) recorded a series of exposures in the terrace sequence between Insch and Spean Bridge.

6. In addition to Roughburn (see above) several sites are notable for landforms associated with the Loch Lomond Readvance limit:

(i) On the west side of the Allt nam Bruach the upper limit of hummocky moraine on the valley side (grid square NN 3178) marks the former ice limit, which is continued northwards by a series of lateral moraines (Sissons, 1979a, 1979b).
(ii) Lateral moraines (grid square NN 2979) below the ice limit suggest that the ice remained active during the early part of its retreat.

7. In the area of Murlaggan (NN 317812), in Glen Spean, a gap in the river terraces and the presence of kame and kettle topography records the position of a residual mass of stagnant ice, left after the active glacier had receded westwards to the vicinity of Spean Bridge and the 260 m lake had drained (Sissons, 1979a).

8. The Inverlair (NN 341806) and Monessie (NN 298811) gorges on the River Spean are of interest as features of fluvial erosion and, although utilized during Lateglacial times, are possibly older in origin. At the eastern end of the Monessie gorge several large and numerous small potholes are of note.

9. The 260 m Parallel Road is extensively developed in Glen Spean. Particular areas of note are: (i) at Creag Bhuidhe (NN 304803), where there is a well preserved stretch 10–13 m wide; and (ii) in grid square NN 2979 where it is cut in drift and demonstrates the original lakeward slope of the shore.

10. The cross-valley moraines that occur in the Spean and Allt Achadh na Dalach valleys west of Spean Bridge are an important assemblage of landforms (Figure 10.12, A–E). They comprise five sets of aligned ridge fragments made largely of till, although locally of sand and gravel. Peacock (1970b) described them in some detail and concluded that they were unlikely to be ice-marginal landforms. Sissons (1979c), however, interpreted them as end-moraine ridges of the Spean glacier and related their occurrence to the transfer of drainage from the Spean to the Lundy Gorge, when the calving ice front may have become lower and more stable after drainage of the 260 m lake. The ridges are similar in their form and lacustrine association to those of Coire Dho (see above) and to cross-valley moraines described from the arctic (Andrews, 1963a, 1963b; Andrews and Smithson, 1966; Holdsworth, 1973; Barnett and Holdsworth, 1974), but their processes of formation have not been fully investigated. The westernmost three ridges (Figure 10.12, F–H) in the valley of the Allt Achadh na Dalach comprise sand and gravel (for example in a section at Tom na Brataich (NN 179795)) and may have formed in crevasses parallel with the ice edge (Sissons, 1979c).

11. West of Spean Bridge the River Spean turns abruptly northwards to flow through a gorge, 3 km long and up to 30 m deep, into the Great Glen at Gairlochy, while the obvious continuation of the valley to the south-west is occupied by the misfit Allt Achadh na Dalach. The gorge functioned as a subglacial routeway for the catastrophic drainage of ice-dammed lakes in Glen Spean, but may have originated earlier (Sissons, 1979a). The relationships of river terraces to the gorge are discussed by Sissons (1979a, 1979c). In this area, around Brackletter and across the valley to the east, there is a varied and important assemblage of landforms (Figure 10.12):

(i) A sequence of cross-valley moraines associated with the Spean Glacier.
(ii) A Gilbert-type glaciolacustrine delta related to the 113 m lake (Figure 10.12, I). Good sections in topset, foreset and bottomset beds have been exposed in Brackletter sandpit.
(iii) Giant potholes in the gorge of the Allt a'Mhill Dhuibh (NN 197827), possibly formed subglacially by *jökulhlaup* discharge (Peacock and Cornish, 1989).
(iv) Glaciofluvial landforms including eskers, kames and kettles.

12. At the northern exit of the Spean Gorge and in the area around Gairlochy two suites of terraces relate to former higher levels of Loch Lochy (Peacock, 1970b; Sissons, 1979a, 1979c).

13. The meltwater channel between (NN 203831) and (NN 205837) (Figure 10.12, J) is an important landform in the sequence of events associated with the draining of the ice-dammed lakes in Glen Spean: it functioned as the overspill channel for the 113 m lake (Sissons, 1979a, 1979c).

14. The Lundy Gorge (Figure 10.12) is a large meltwater channel which functioned as an outlet for ice-dammed lakes in Glen Spean for a period after the drainage of the 260 m lake. As such it is an important element in the history of events in the area. Its role and relationships have been discussed in detail by Sissons (1979c). Recent sand and gravel extraction has exposed the rock-cut north wall of the gorge from beneath the

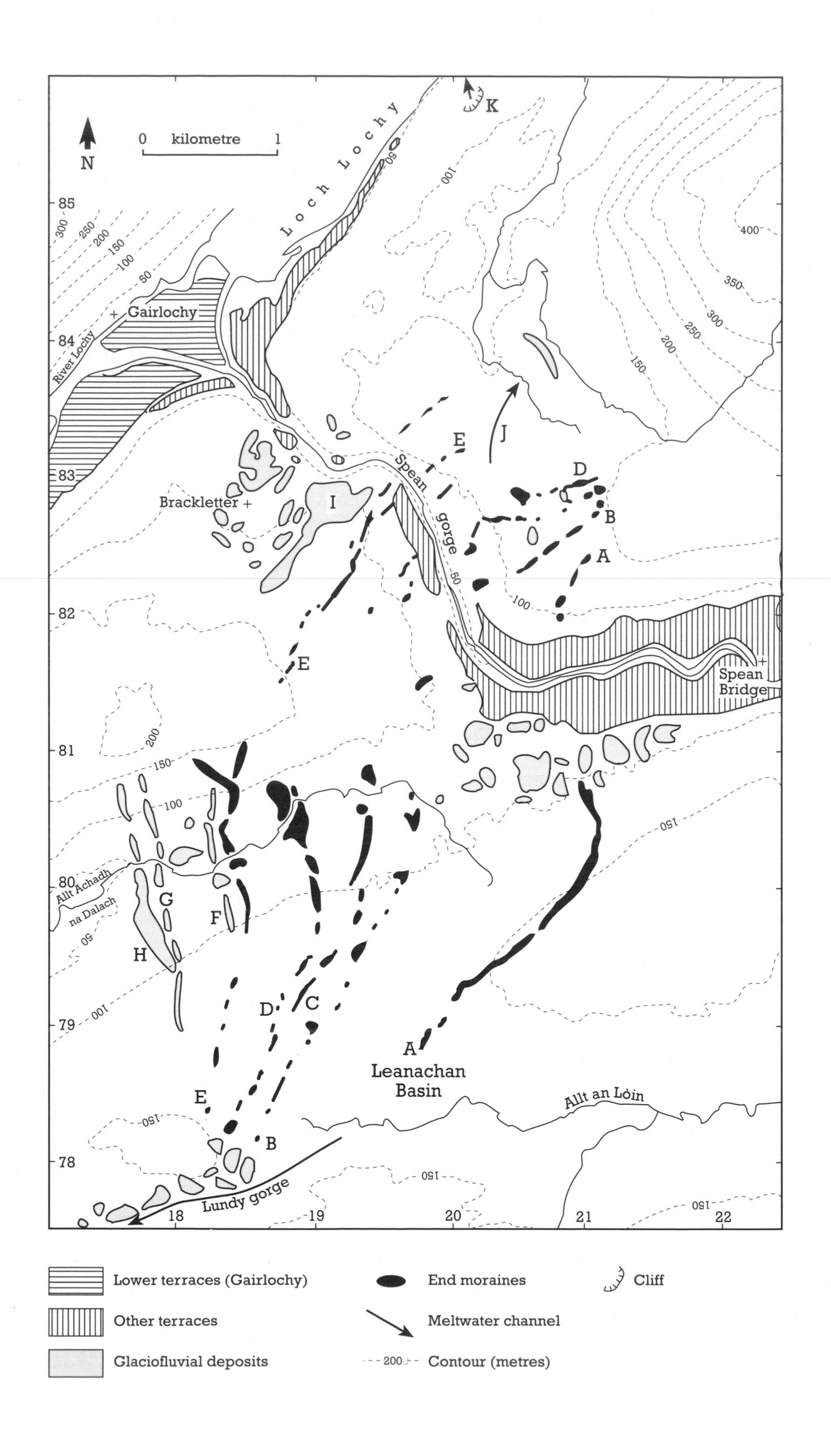

N
0 kilometre 1
K
Loch Lochy
Gairlochy
River Lochy
Brackletter
I
J
E
D
B
A
Spean gorge
Spean Bridge
Allt Achadh na Dalach
G
F
H
C
Leanachan Basin
Allt an Lòin
Lundy gorge
Lower terraces (Gairlochy)
Other terraces
Glaciofluvial deposits
End moraines
Meltwater channel
Contour (metres)
Cliff

kamiform sand and gravel deposits that extend to the north and north-east. There is a good section in these deposits at Tom na h-Iolaire.

15. An unusual, 'cirque-like' feature which leads into a meltwater channel on a hilltop south of Glenfintaig House (NN 201857) (Figure 10.12, K), has been interpreted by Sissons (1979c) as an abandoned waterfall site recording the final *jökulhlaup* of the ice-dammed lake in Glen Spean that had been periodically discharging through the Lundy Gorge.

Interpretation

The first published description of the Parallel Roads was by Thomas Pennant in 1771 in his work *A Tour in Scotland, 1769*. Although bad weather prevented him from visiting what he called the 'celebrated parallel roads', he noted the local belief that they had been constructed to facilitate hunting, a view later echoed by Rev. Thomas Ross in the *Old Statistical Account*. According to Ross (1796) the roads, or the 'Casan' as they were known locally, were 'one of the most stupendous monuments of human industry' (p. 549). Local tradition held that they were built either by the Kings of Scotland when they resided in the Castle at Inverlochy, or by the Gaelic mythical hero, Fingal, and his followers. In support of the latter explanation Ross noted that the features were locally called 'Fingalian roads'.

Historically Glen Roy played an important role in the development of geomorphological theories of landscape evolution. In addition, the search for a theory of formation of the Parallel Roads provides an instructive case study in the history and philosophy of science and the development of scientific ideas (Rudwick, 1974). In the 19th century various theories were proposed in the scientific literature to account for the origin of the Parallel Roads (Rudwick, 1974). These included aqueducts for irrigation (Playfair, cited by Jolly, 1880b), diluvial shorelines (Mackenzie, 1848; Rogers, 1862), lake shorelines (Greenough, 1805, cited by Rudwick, 1962; MacCulloch, 1817), marine shorelines (Darwin, 1839; Maclaren, 1839; Lyell, 1841b; Chambers, 1848; Watson, 1866; Nicol, 1869, 1872; Campbell, 1877; Macfadzean, 1883) and shorelines of debris-dammed (Dick, 1823; Milne Home, 1847b, 1849, 1876, 1879) or ice-dammed lakes (Agassiz, 1841b, 1842; Buckland, 1841b; Thomson, 1848; Jamieson, 1863, 1892; Lyell, 1863; Mackie, 1863; Geikie, 1865; Jolly, 1873, 1880a, 1880b, 1886a, 1886b; James, 1874; Brown, 1875; J. Geikie, 1877; Tyndall, 1879; Livingston, 1880; Prestwich, 1880). Several authors considered the shorelines to have formed by mass movements of slope debris (Jamieson, 1863; Lyell, 1863; MacCulloch, 1817; Babbage, 1868; Prestwich, 1880). Lubbock (1868) advocated redistribution of sediments by wave processes, while Melvin (1887) and Livingston (1906) believed that the roads were glacier-margin deposits. Dakyns (1879) made an important observation that the roads were locally cut in bedrock.

The marine school initially found strong proponents in both Charles Darwin and Charles Lyell. The former, in particular, was deeply impressed by Glen Roy. On 9 August 1838 he wrote to Lyell, 'I wandered over the mountains in all directions and examined that most extraordinary district. I think without any exception, not even the first volcanic island, the first elevated beach, or the passage of the Cordillera, was so interesting to me as this week. It is far the most remarkable area I ever examined. . . . I can assure you Glen Roy has astonished me' (Darwin, 1887, p. 293). At that time Darwin favoured a marine origin for the Parallel Roads. It was only 23 years later, in 1861, that he recanted in print this belief and accepted the fact that the roads represented the shores of a glacial lake (Barrett, 1973; Rudwick, 1974). However, it was Agassiz (1841b, 1842), a pre-eminent figure in the application of the glacial theory in Britain, who first identified the imprint of glacier ice and propounded the existence of former ice-dammed lakes in Glen Roy, following a visit there in 1840 with William Buckland. This idea was subsequently elaborated by Jamieson (1863, 1892). More recently, as outlined below, Sissons (1977e, 1978, 1979a, 1979b, 1979c, 1981c, 1981d) has refined the explanation of the Parallel Roads and established in some detail the sequence of events in their formation. His work also reveals the Parallel Roads to be part of a remarkable complex of glacial, glaciofluvial and glaciolacustrine landforms and sediments extending from Loch Laggan in the east through Glen Spean, Glen Roy and Glen Gloy, to near Fort William in the west, and north-

Figure 10.12 Geomorphology of the Spean Bridge–Gairlochy area (from Sissons, 1979c). See text for explanation of letters.

east to the Great Glen, Loch Ness and Inverness.

Current understanding of the sequence of events in the formation of the ice-dammed lakes and their subsequent drainage was summarized by Sissons (1981d), drawing on the details of his earlier papers (Sissons, 1977e, 1979a, 1979b, 1979c). Lakes in Glen Roy, Glen Gloy and Glen Spean were impounded by ice of the Loch Lomond Readvance from west of the Great Glen, coalescing with glaciers from the Ben Nevis range and from the ground to the south via the Laire and Treig breached valleys (Figure 10.13). Wilson (1900) and Peacock (1970b) established ice-movement patterns from striations and the distribution of erratics, and Sissons (1979b) has mapped and discussed the ice limits and related landforms, which include spectacular lateral moraines, end moraines and hummocky moraine. At its maximum extent the ice reached the western end of the present Loch Laggan and penetrated up-valley into Glen Roy and Glen Gloy (Figure 10.13). As it advanced, the ice ponded back a series of ice-dammed lakes, successively at 260 m, 325 m and 350 m OD (the rising sequence). The levels of these were controlled by the altitudes of the lowest ice-free cols on their perimeters (Jamieson, 1863; Sissons, 1977e). At the maximum extent of the ice the Glen Gloy lake overflowed through the col at 355 m on the Gloy–Turret watershed into the Glen Roy lake which attained maximum dimensions of 16 km in length and 200 m in depth. The level of the Glen Roy lake was controlled by the 350 m col leading into Strathspey at the head of the glen. The waters of a contemporary lake in Glen Glas Dhoire escaped to the east through a col at 325 m into an extensive lake at 260 m controlled by the Feagour col at the eastern end of the present Loch Laggan. As the ice retreated, lakes were formed at successively lower levels (the falling sequence). First in Glen Roy, the 325 m col became available as an outlet for the Roy lake, and the latter fell to its second major level. Subsequent decay and westward retreat of the ice margin to the vicinity of Spean Bridge allowed the Roy lake to fall to the level of the 260 m lake in Glen Spean, which at its maximum extent was 35 km long. In Glen Gloy the level of the lake remained constant, as the col at the head of the glen is the lowest in the watershed.

Drainage of the 260 m lake may be inferred by analogy with modern ice-dammed lakes in many parts of the world, which drain periodically by catastrophic subglacial flow of the ponded water (for example, Liestøl, 1956; Stone, 1963; Mathews, 1973; Dawson, 1983c; Clement, 1984; Shakesby, 1985; Russell, 1989); the resulting floods are commonly described by the Icelandic term *'jökulhlaup'* (glacier burst). From his detailed investigation of the field evidence, Sissons (1979c) proposed that the 260 m lake was drained by catastrophic subglacial flow through the Spean Gorge and northwards along the Great Glen to the Moray Firth. At Fort Augustus (see above) an extensive spread of sand and gravel is thought to have been deposited by the *jökulhlaup*, as is a large gravel deposit in the Beauly Firth at Inverness (Sissons, 1981c). Very perceptively, Jamieson (1865) first raised the possibility that gravel deposits in the Inverness area might be related to the final catastrophic drainage of the Glen Roy and Glen Gloy lakes, although those he possibly had in mind are eskers and kames (see Torvean). Subsequently, there was a period of oscillating lake levels and smaller *jökulhlaup* events through the Spean Gorge and later through the Lundy Gorge. Upon the abandonment of the latter route, drainage shifted back to the north-east, first in the form of a *jökulhlaup* along a now-abandoned waterfall and channel near Glenfintaig House then via an overspill channel from a later lake in Glen Spean at 113 m. Considerable fluvial infill took place in Glen Roy and Glen Spean after the drainage of the 260 m lake, and a complex series of over twenty terraces has been identified (Sissons, 1979a), some of which relate to a variety of lower lake levels in Glen Spean and other, later, ones to higher levels of Loch Lochy. Failure of the ice dam in Glen Spean led to final drainage through the Spean Gorge, dissection of the valley infill and terrace deposition in the Gairlochy area.

In upper Glen Roy, Sissons and Cornish (1983) mapped extensive fans of coarse gravel deposited in the lowest lake in the sequence of rising lake levels. The largest feature is associated with outwash from a glacier in Glen Turret. Sissons and Cornish (1983) suggested this glacier had flowed over the col from Glen Gloy. As the lake level rose, the glacier retreated and the gravels were mantled with lake sediments (clays and silts). However, following a re-examination of the

Figure 10.13 Loch Lomond Readvance ice limits and associated ice-dammed lakes in the Glen Roy–Glen Spean area (from Sissons, 1981d).

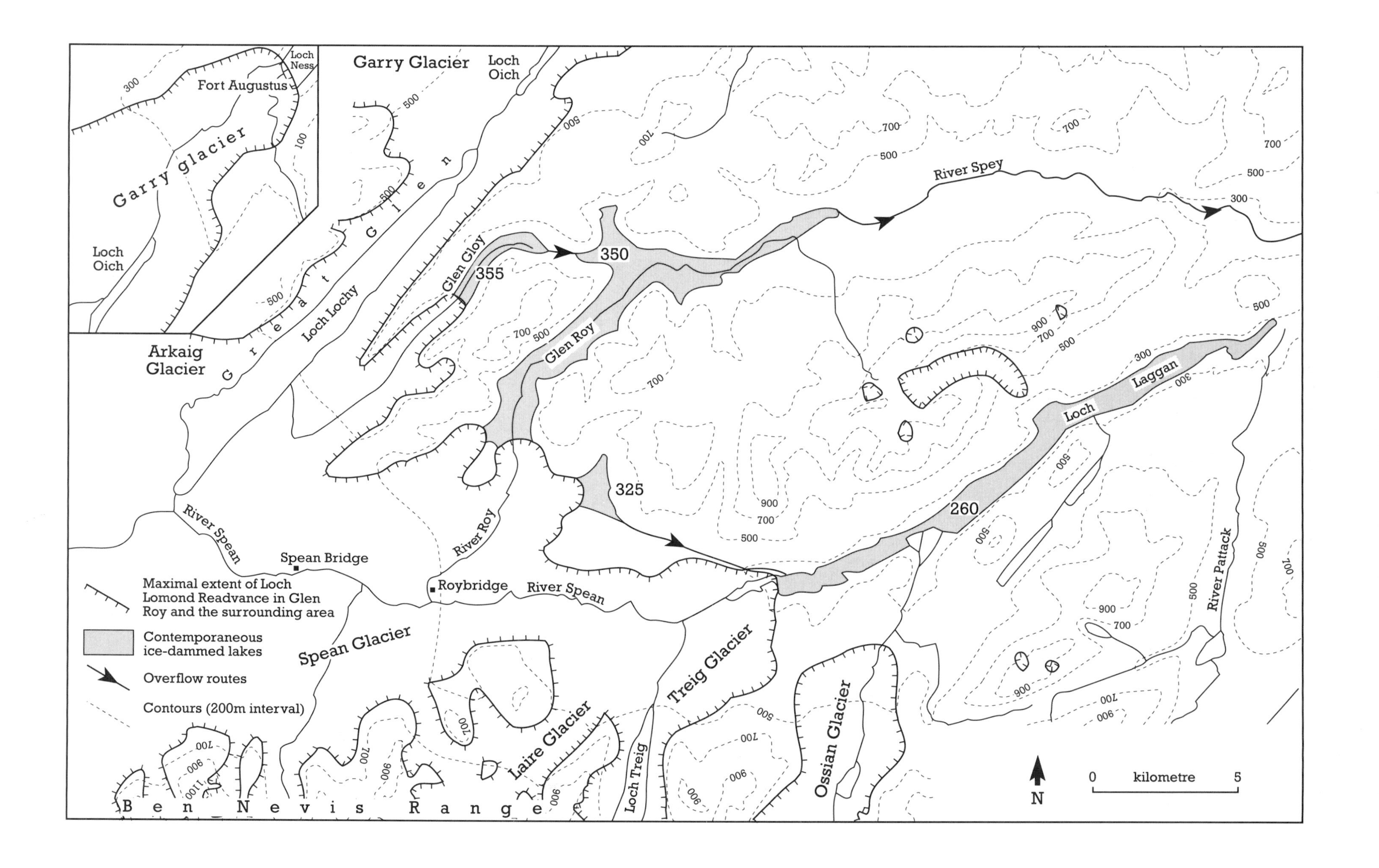

Garry Glacier
Loch Oich
Loch Ness
Fort Augustus
Garry glacier
Loch Oich
Arkaig Glacier
Great Glen
Loch Lochy
Glen Gloy
355
350
Glen Roy
325
260
Loch Laggan
River Spey
River Pattack
River Spean
Spean Bridge
Roybridge
River Roy
River Spean
Spean Glacier
Laire Glacier
Treig Glacier
Loch Treig
Ossian Glacier
Ben Nevis Range
Maximal extent of Loch Lomond Readvance in Glen Roy and the surrounding area
Contemporaneous ice-dammed lakes
Overflow routes
Contours (200m interval)
N
0 kilometre 5

sediments, Peacock (1986) considered that the fans were largely subaerial in origin and that they pre-dated the lakes. He suggested that the Turret outwash dated from the time of Late Devensian ice-sheet decay. Lowe and Cairns (1991) favoured Sisson's hypothesis, but the evidence is inconclusive and further investigation is required.

Detailed levelling of the Glen Roy shorelines has demonstrated differential glacio-isostatic uplift and dislocation of crustal blocks at the start of the Holocene (Sissons and Cornish, 1982a, 1982b). The dislocations, together with several associated landslips may have been triggered by stresses induced by the loading and unloading of the crust by the Loch Lomond Readvance glaciers and by the formation and catastrophic drainage of the lakes. This evidence raises the possibility that crustal dislocation at sites of ice limits and glacial lakes may be of wider significance than formerly recognized. Holmes (1984) observed a correlation in Glen Gloy, Glen Roy and Glenfintaig between the occurrence of landslips and possible Loch Lomond Readvance ice limits. Further evidence for palaeoseismicity has been recorded by Ringrose (1987, 1989a, 1989c) (see also Davenport and Ringrose, 1985, 1987; Davenport *et al.*, 1989) who inferred two deformation events from the pattern of liquefaction structures preserved in the lake sediments. The first was attributed to an earthquake which occurred before drainage of the 260 m lake in Glen Roy and the 355 m lake in Glen Gloy. The second was interpreted as a response to either a second earthquake or lake drainage.

Aspects of the vegetational history of the area and the chronology of lake drainage were studied by McPherson (1978) from pollen sites in Glen Roy and Glen Spean. She concluded that the highest lake existed until the time of the juniper pollen zone (transition from the Lateglacial to the Holocene), and that the lowest had drained by the start of the birch pollen zone (early Holocene). More detailed investigations by Lowe and Cairns (1989, 1991), however, suggest that some revision of MacPherson's chronology is necessary and that organic sedimentation began earlier, at the start of the Holocene. In addition, the absence of Lateglacial pollen from deposits in Glen Turret and on the Gloy–Turret col lends some support to the interpretation that these areas were glaciated during the Loch Lomond Stadial (Sissons and Cornish, 1983).

Glen Roy is a site of outstanding importance for geomorphology. It is unique in Britain not only for the extent, clarity and degree of development of its shorelines, but also for the remarkable assemblage of related landforms and deposits. These record geomorphological processes both during and following successive episodes of ice-dammed lake development and catastrophic drainage, and include glacier moraines, stagnant-ice deposits, kame terraces, meltwater gorges, lake-floor sediments, fans, Gilbert-type and Hjulström-type deltas, river terraces and landslides. Glen Roy and adjacent areas provide the clearest and most complete assemblage of morphological and sedimentological evidence in Britain for the formation and drainage of ice-dammed lakes. Moreover, variations in the altitudes of the shorelines have provided new and significant evidence concerning deformation and dislocation of the Earth's crust in glaciated areas. The pre-eminence of Glen Roy is also recognized historically when, particularly during the 19th century, Glen Roy played a significant role in the development of geomorphological ideas and models of landscape formation.

Scientific interest in Glen Roy, Glen Gloy and Glen Spean is therefore focused not only on individual or unique landforms, but also on the total assemblage of features, how they interrelate and together provide the evidence for interpreting the complex sequence of events recorded in the geomorphology and sediments of the area. The prime features of this interest are as follows:

1. The lake shorelines (the Parallel Roads, which are the best examples in Britain); their extent, altitudes, clarity of preservation, variations in form and nature (both erosional and depositional) and relationships to former ice fronts are all of major importance.
2. Landforms associated with former ice limits, including end moraines, drift limits, hummocky moraine, outwash fans and cross-valley moraines. Individual features, such as the Turret fan and the cross-valley moraines, are exceptional examples of their kind in Britain.
3. The alluvial fans in Glen Roy, which are among the best of their type in Britain, both as landform examples and for their potential for sedimentological studies.
4. The lake deltas, particularly at Inverlair–Fersit, Roughburn and Brackletter, which

are of key interest both for landforms and sedimentology, and are among the best examples of their kind in Britain; compared with Achnasheen (see above) they generally demonstrate much more extensive sediment collapse related to burial and melting of masses of glacier ice. The contrasting Gilbert-type and Hjulström-type deltas are essential elements in understanding the sedimentary processes during and following the time of the Parallel Roads lakes.

5. The river terraces in the lower Roy and middle and lower Spean valleys, which in their landforms and sediments preserve a detailed record of Holocene geomorphological change.
6. The numerous landslides, which are significant in relation to former ice-front positions, earthquake history and controls on release mechanisms.
7. The meltwater gorges, possibly related to catastrophic lake drainage.
8. The lake sediments with their potential for process studies and interpreting patterns of palaeoseismicity.
9. The periglacial slope deposits, which as yet have received little attention.
10. The organic sediments preserved in kettle holes and bogs, which have potential for elaboration of the chronology of lake drainage.
11. The total assemblage of features, which provides uniquely detailed evidence in Britain for catastrophic glacial lake drainage.
12. The archive of landforms and deposits clearly related to a particular geological datum, which provides unsurpassed potential for comparative studies of a whole range of geomorphological process magnitudes and rates during a period of extremely rapid environmental change.

Although ice-dammed lakes have been identified elsewhere in Britain (see Shotton, 1953; Straw, 1979; Gaunt, 1981), extensive shorelines are rarely preserved. They have been recognized in association with, for example, Lake Harrison (Dury, 1951 – but see Ambrose and Brewster, 1982) and Lake Humber (Edwards, 1937). However, it is in the Highlands of Scotland, in areas glaciated during the Loch Lomond Stadial, that examples of shorelines are best preserved, as in Coire Dho (Sissons, 1977b), at Loch Tulla (Ballantyne, 1979), at Achnasheen (see above) and, most remarkably of all, in Glen Roy. Beyond the limits of the last glaciers any shorelines will be considerably older and will therefore have undergone significantly greater modification, particularly through the activity of periglacial processes known to have been widespread in Britain during the Loch Lomond Stadial (Sissons, 1979e).

The Parallel Roads are comparable, for example, to Pleistocene lake shorelines in Scandinavia (Mannerfelt, 1945) and the Holocene shorelines in south-west Greenland recently described by Dawson (1983c); or even to some of the features associated with the Great Lakes of North America during the Wisconsin (last glaciation) (Spencer, 1890), although the latter occur on a vastly greater scale (Leverett and Taylor, 1915; Fulton, 1989). However, what distinguishes Glen Roy and the Parallel Roads as a locality of international importance for geomorphology is the total range of landforms, their clearly demonstrated relationships and the relatively compact extent of the whole assemblage.

Although the area has been studied for over two hundred years, it still has significant potential for further research, particularly on the sedimentology of the various deposits, the relationships between sediments, landforms and geomorphological processes, process rates and outstanding problems of landform genesis and chronology.

Conclusion

Glen Roy is one of the most famous landform landmarks in Britain and is internationally recognized as a classic locality for the shorelines of an ice-dammed lake, represented by the Parallel Roads, that formed during the period of glacier readvance known as the Loch Lomond Stadial (approximately 11,000–10,000 years ago). In their extent, continuity and degree of preservation, the Parallel Roads of Glen Roy and adjacent glens are unique in Britain. They are of outstanding geomorphological interest both in their own right, and as part of a remarkable system of glacial, glaciofluvial and glaciolacustrine landforms and deposits recording a complex sequence of landscape changes in Lateglacial and early Holocene times.

KINGSHOUSE

M. J. C. Walker

Highlights

Pollen preserved in the sediments that infill a topographic depression near Kingshouse provide an important record, supported by radiocarbon dating, of vegetational history and environmental changes during the early and middle Holocene. A radiocarbon date from the basal organic sediments provides a minimum date for the deglaciation of Rannoch Moor at the end of the Loch Lomond Readvance.

Introduction

The Kingshouse site (NN 282555) is situated in the north-western part of Rannoch Moor near the heads of Glen Etive and Glen Coe, and lies approximately 2.25 km north-east of the Kingshouse Hotel at an altitude of *c.* 340 m OD. It is the second of three sites in the Kingshouse area studied by Lowe and Walker (1976), and is known as Kingshouse 2. The importance of the site lies partly in the fact that it contains a detailed record of vegetational changes in the Rannoch Moor area of the west-central Grampian Highlands during the early and middle Holocene, but principally in the sequence of radiocarbon dates obtained from the lowermost sediments. The age determination on the basal organic sediments is one of the earliest from a site inside the Loch Lomond Readvance limits and, if correct, provides a minimum date for the deglaciation of Rannoch Moor. Details of the stratigraphy of the site can be found in Lowe and Walker (1976, 1980) and in Walker and Lowe (1977, 1980).

Description

The site is a small enclosed basin (maximum dimensions 40 m by 25 m) which has been buried beneath blanket peat. Almost 4.5 m of sediment have accumulated at the deepest point. The basal sediment sequence is complex (Figure 10.14): coarse gravels and sand are succeeded upwards by laminated silt and fine sand, clay, a thin band of gyttja, fine sand with abundant remains of the terrestrial moss *Rhacomitrium lanuginosum* and fine–medium-grained sand with occasional moss fragments. This sequence is overlain by over 4 m of limnic, telmatic and terrestrial organic sediments. The lower inorganic sediments are essentially barren of pollen, but a pollen diagram has been constructed for the lower 2.35 m of the overlying organic sediments, and a single pollen count was obtained from the thin gyttja layer near the base of the profile (Figure 10.14). The diagram was divided into local pollen assemblage zones and these were integrated with pollen data from the nearby sites at Kingshouse 1 and Kingshouse 3 to form a sequence of regional pollen asssemblage zones (R zones) for western Rannoch Moor (Walker and Lowe, 1977).

Three radiocarbon dates were obtained from the lowermost part of the Kingshouse 2 profile (Figure 10.14). The thin layer of gyttja within the basal sediments yielded a radiocarbon age of 10,520 ± 330 BP (Birm–723), and a date of 10,290 ± 180 BP (Birm–722) was obtained from fragments of moss, *Rhacomitrium lanuginosum*, in the fine sands above. The contact between these sands and the overlying organic sediments was dated at 9910 ± 200 BP (Birm–724) (Lowe and Walker, 1976).

Interpretation

The earliest vegetational records reflect a landscape immediately following deglaciation of *Empetrum* heath and juniper scrub with patches of ground covered by a moss carpet or by grasses. Indeed, the high *Empetrum* and *Juniperus* pollen frequencies, in conjunction with the macrofossil evidence of *Rhacomitrium lanuginosum*, suggest vegetational affinities with the '*Rhacomitreto–Empetretum*' (*Rhacomitrium–Empetrum* heath) and *Juniperetum nanae* (dwarf juniper scrub) associations which are common in parts of the western and northern Scottish Highlands today (McVean and Ratcliffe, 1962). Subsequently, the *Empetrum* heath communities were invaded by *Juniperus communis* and then by tree birch. Following the arrival of *Betula*, the landscape of the area appears to have been a mosaic of birch copses and heath and moorland communities, separated by areas of grassland on the steeper slopes around the Rannoch basin (see also Walker and Lowe, 1979, 1981). The juniper maximum in the Kingshouse 2 profile has been dated at 9910 ± 200 BP (Birm–724) and dates

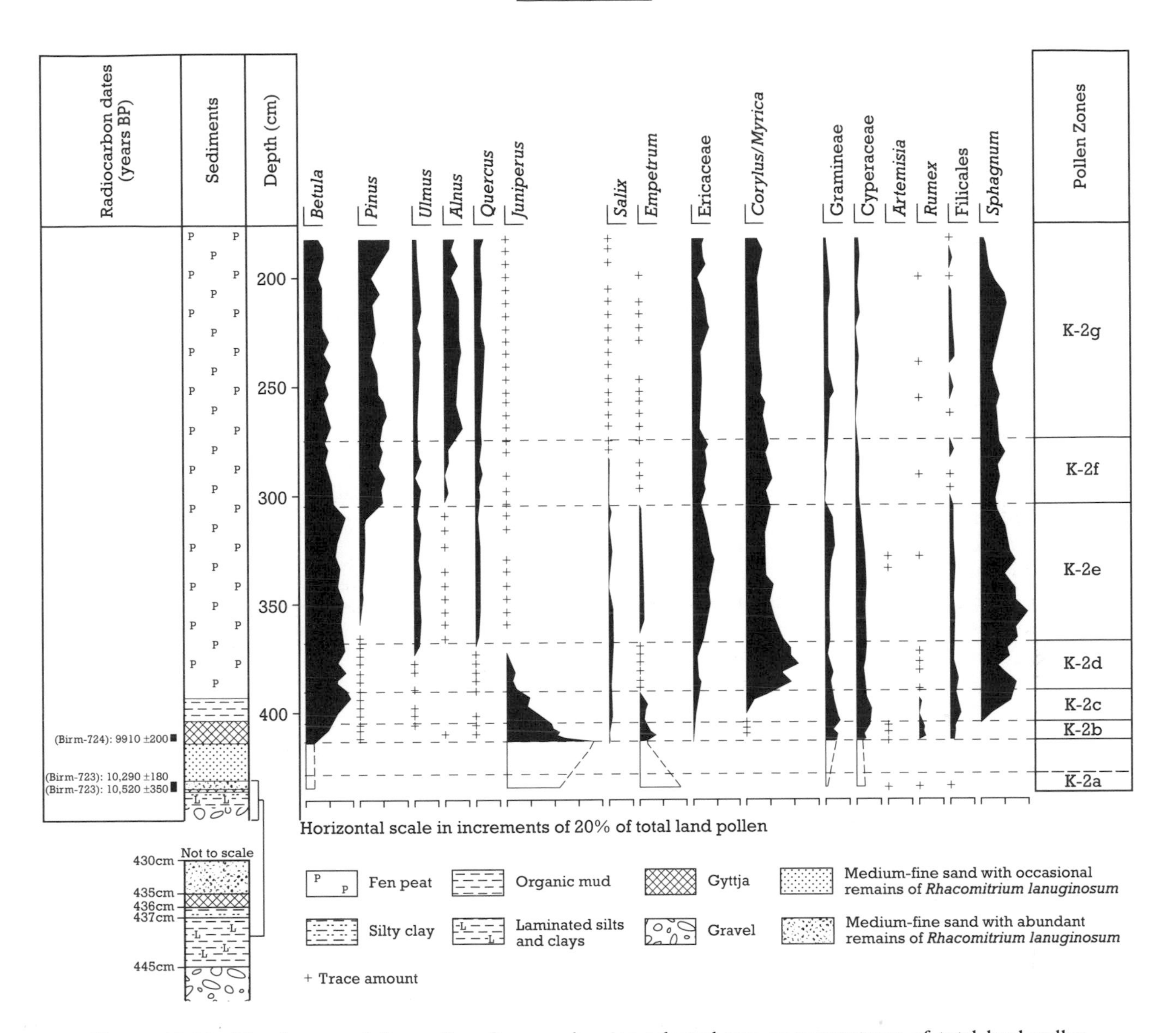

Figure 10.14 Kingshouse: relative pollen diagram showing selected taxa as percentages of total land pollen (from Walker and Lowe, 1977).

from other areas of Scotland suggest that tree birch would have been well established on western Rannoch Moor by 9000 BP (Birks, 1972b; Harkness and Wilson, 1979; O'Sullivan, 1976; Walker and Lowe, 1982, 1985).

The pollen records from the Kingshouse sites show that, following the period of open birch-woods, the landscape underwent further change with the arrival first of hazel (*Corylus avellana*) which expanded throughout Scotland early in the eighth millennium BP (Huntley and Birks, 1983), and subsequently with the immigration of woodland taxa including elm, oak, pine and, at a later date, alder. *Pinus sylvestris* and *Betula* appear to have formed the dominant elements of the woodland cover during the middle Holocene, whereas *Ulmus* and *Quercus* were more restricted in their distribution. The continuous representation of Ericaceae and Gramineae throughout the middle Holocene part of the Kingshouse 2 profile suggests that areas of heath and grassland communities may have existed between the woodland stands, and on the upper slopes and plateau surfaces above the regional treeline. Radiocarbon dates from the Cairngorms (Birks, 1970; O'Sullivan, 1975, 1976) and from north-west Scotland (Birks, 1972b; Birks and Williams, 1983; Pennington *et al.*, 1972; Williams, 1977) indicate that pine began to arrive in the Scottish Highlands soon after 8000 BP, that pine and birch forest was forest was widely established by 7000 BP and that the woodlands were further diversified by the immigration of alder between 7000 and 6000 BP.

The radiocarbon date (Birm–722) obtained from the terrestrial moss fragments is particularly important, for although errors arising from the 'hard-water effect' or from the incorporation of reworked mineral carbon residues may have influenced the dates obtained from organic mud/gyttja (Lowe and Walker, 1980; Sutherland, 1980; Walker and Harkness, 1990), such problems would not be encountered where moss constitutes the dating medium.

A more intractable problem arises from the discovery of 'plateaux' of essentially constant ^{14}C enrichment at around 10,000 BP (Ammann and Lotter, 1989; Zbinden *et al.*, 1989). This appears to have been caused by fluctuations in atmospheric ^{14}C production and clearly poses a major difficulty for the radiocarbon chronology at the Lateglacial/Holocene boundary. However, the date on the terrestrial moss sample from Kingshouse 2 falls outside the envelope of constant ^{14}C age on the curves of Ammann and Lotter (1989). Moreover, the fact that the three dates from the profile are internally consistent and that there is a broad measure of agreement between the dates from Kingshouse 2 and those from comparable biostratigraphic horizons at other sites on Rannoch Moor (Walker and Lowe, 1979, 1980) may be significant. If correct, the dates point towards deglaciation of Rannoch Moor well before 10,000 BP and probably before 10,200 BP by which time *Empetrum* heath and juniper scrub had become widely established locally.

The data from Kingshouse 2 are important in a wider context. The contrast between the dwarf-shrub-dominated pollen assemblage at the base of the profile and the largely herbaceous pollen assemblages found in the lowermost horizons of kettle hole basins in the valleys to the east of Rannoch Moor points towards a pattern of time-transgressive deglaciation following the Loch Lomond Readvance ice maximum (Lowe and Walker, 1981), although strictly the basal pollen relate only to the time of melting of buried ice and not necessarily to regional deglaciation. Basal organic sediments from a site located behind the Loch Lomond Readvance moraine at Callander (see Mollands) have been dated at 10,670 ± 85 BP (Lowe, 1978). If this age determination and those from Kingshouse 2 are correct, deglaciation from the Loch Lomond ice maximum may have been completed within around 300 years. However, a radiocarbon date of 10,560 ± 160 BP (Q–2673) on organic detritus beneath till of the Loch Lomond Readvance at Croftamie (see below) to the north of Glasgow (Rose *et al.*, 1988), suggests either marked spatial and temporal variations in the pattern of the Loch Lomond Readvance ice wastage, or significant errors in the available radiocarbon chronology. Nevertheless, studies have shown that the Rannoch basin was one of the major ice accumulation and dispersal centres in Scotland during the Loch Lomond Readvance and that, in view of the thick ice cover (over 400 m in places – Thorp, 1984, 1986), Rannoch Moor would have been one of the last localities in Scotland to be deglaciated following the readvance (see Sutherland, 1984a).

Kingshouse 2 is a site of considerable significance. In association with the neighbouring sites of Kingshouse 1 and Kingshouse 3, it provides a vegetational record for this area of the Grampian Highlands in the period immediately following the wastage of the last glaciers until the establishment of alder in the middle Holocene some 5000 years later. The site appears to lie near the western edge of former pine woodland and may prove to be useful in delimiting the ecotone between the middle Holocene birch–pine forests of the central Grampian Highlands and the birch and oak forests of the coastal lowlands to the west. Kingshouse 2 is most important, however, in the establishment of a chronology for deglaciation following the Loch Lomond Readvance. The radiocarbon date on the moss fragments is of particular significance and thus far is the only one obtained from plant macrofossil material from sites within the last glacier limits. In conjunction with the other age determinations from the site and from elsewhere on Rannoch Moor, it may imply much earlier deglaciation following the Loch Lomond Readvance than has previously been assumed (Sissons, 1979e).

The pollen stratigraphy indicates the complex and rapid vegetational changes that occurred following deglaciation, and suggests that the environmental changes associated with ice wastage were rapid and large. Nowhere else in north-west Europe is the evidence for rapid deglaciation and climatic change so clearly developed as on Rannoch Moor (H. J. B. Birks, unpublished data).

Conclusion

Kingshouse is one of a number of sites that contribute significant evidence for interpreting the environmental changes that occurred at the end of the last ice age (around 10,000 years ago). It is particularly important for establishing the timing of deglaciation at the end of the Loch Lomond Readvance (approximately 11,000–10,000 years ago), the evidence suggesting that this may have occurred relatively earlier than has previously been assumed. The pollen preserved in the sediments at Kingshouse also provide a valuable record of subsequent vegetational development in the west-central Highlands in the period after glacier ice had melted.

PULPIT HILL

R. M. Tipping

Highlights

The sedimentary infill in the topographic basin at Pulpit Hill contains an important and detailed pollen record of vegetational history and environmental change during the Lateglacial.

Introduction

The Pulpit Hill site (NM 852292) is a small (500 m^2) infilled basin on lower Devonian (Old Red Sandstone) lavas and conglomeratic sandstones, located 0.75 km south of Oban, Argyllshire. The basin contains a sequence of polleniferous, lacustrine sediments of Lateglacial and early Holocene age, important for its position in the historical development of Quaternary pollen studies in Scotland (Donner, 1955, 1957). Recent investigations (Tipping, 1984, 1991b) have demonstrated a number of environmental changes within the Lateglacial Interstadial, in particular, a climatic deterioration early in the interstadial and a later, but pre-Loch Lomond Stadial, climatic decline, which appear to have wider significance for the Scottish Quaternary, and which have also been reported from pollen sites on Skye (Walker and Lowe, 1990).

Description

At its deepest point the floor of the basin is 7.2 m below the surface peat. Donner (1955, 1957) first sampled the site (then designated 'Oban 2') in his study of pollen sites in central and western Scotland designed to delineate by biostratigraphic means the extent of Pollen Zone III ('Highland' or Loch Lomond Readvance) glaciation. His intention was to demonstrate a Lateglacial biostratigraphy at the site in order to show that the readvancing ice terminated at a limit (Donner, 1957) 2 km east of Oban. He failed, however, to obtain Lateglacial sediments, and there was no palynological evidence to support his contention that the basal sediments were deposited during the Lateglacial.

Tipping (1984, 1991b) reinvestigated the site, and was able to demonstrate a 0.8 m thick Lateglacial succession. Two cores within the basin showed closely similar sediment and pollen records. The findings confirm Donner's (1955, 1957) suggestion that the Loch Lomond Readvance limit lay to the east of Oban (see Gray, 1972, 1975b; Thorp, 1984, 1986).

The sediment infill in the basin comprises a succession of clay, clay/gyttja, detrital mud and peat deposits (Figure 10.15). Seven local pollen assemblage zones were identified, spanning the Lateglacial and the early Holocene (Figure 10.15).

Interpretation

The sediment and pollen records show several changes within the Lateglacial Interstadial. Following the establishment of organic-rich sedimentation in the basal pollen assemblage zone (A), a gradual change to pure clay (6.875–6.91 m depth) occurs with the reappearance of the earliest colonizing species (subzone Ab). The origin of the clay band is thought to lie in accelerated solifluction of material from the catchment at a time of short-lived climatic deterioration. Basin-edge collapse of pre-existing sediment is considered unlikely because of the diffuse boundary between the clay band and the underlying organic-rich clay (implying a gradual environmental change), the increasing pollen concentration values within the clay band (perhaps indicating that its deposition was not instantaneous) and the fact that the sedimentological, chemical (carbon/nitrogen) and palynological changes did not occur synchronously

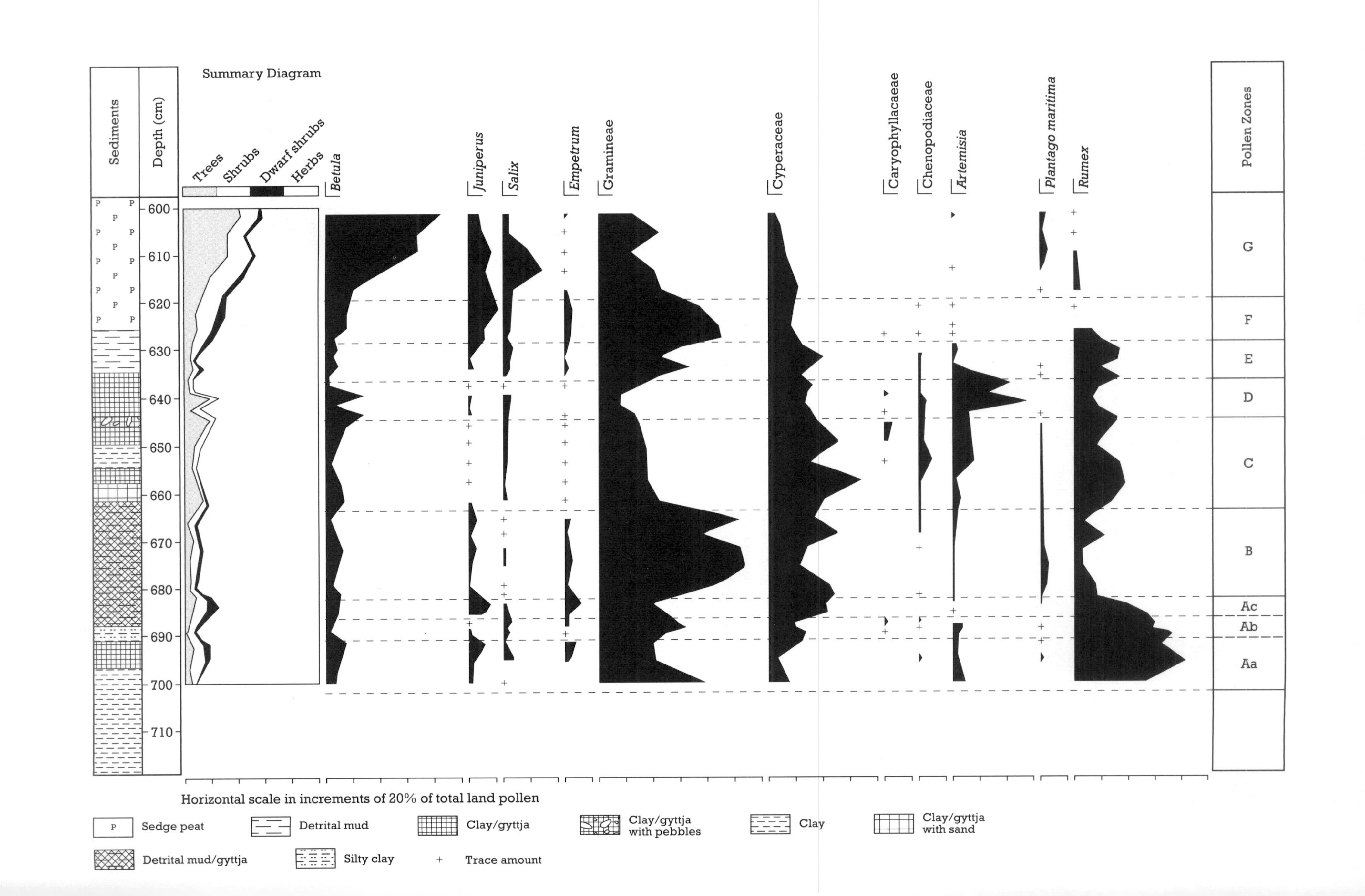

Summary Diagram
Sediments
Depth (cm)
Trees
Shrubs
Dwarf shrubs
Herbs
Betula
Juniperus
Salix
Empetrum
Gramineae
Cyperaceae
Caryophyllacaeae
Chenopodiaceae
Artemisia
Plantago maritima
Rumex
Pollen Zones
G
F
E
D
C
B
Ac
Ab
Aa
600
610
620
630
640
650
660
670
680
690
700
710
Horizontal scale in increments of 20% of total land pollen
Sedge peat
Detrital mud
Clay/gyttja
Clay/gyttja with pebbles
Clay
Clay/gyttja with sand
Detrital mud/gyttja
Silty clay
Trace amount

(Tipping, 1984, 1991b).

The *Juniperus–Empetrum* subzone (Ac) represents the mildest climatic phase of the Lateglacial Interstadial, which, characteristically for the west coast of Scotland, has no arboreal pollen taxa at percentages high enough to imply local growth. The suppression of tree growth by westerly winds stronger than those of the present day was suggested by Tipping (1984, 1991b) as the reason for this vegetational pattern, which contrasts sharply with the birch parkland found during this period in eastern Scotland (Lowe and Walker, 1977).

The Gramineae–*Plantago maritima* pollen zone (B) is thought to indicate a second, but more sustained, climatic decline and with no amelioration until after the Loch Lomond Stadial. This is recognized in the pollen record through the displacement by grassland of *Juniperus* and *Empetrum* associations, and in the chemistry of the sediments by declining carbon and nitrogen values. This climatic decline has become increasingly widely recognized in western Scotland (Tipping, 1984, 1991b), the Inner Hebrides (Lowe and Walker, 1986a; Walker and Lowe, 1990) and Ireland (Craig, 1978), and is radiocarbon dated to *c.* 12,000 BP.

Clay-dominated sedimentation did not recommence, however, until the Loch Lomond Stadial, when solifluction again introduced minerogenic sediment from catchment soils and produced several inwashed moss bands (cf. Birks, 1970). The pollen assemblages (zones C and D) from the stadial sediments are typical of tundra communities at the present day, with *Artemisia* a prominent taxon; in this respect the pollen record at Pulpit Hill conforms with the pattern at a great number of Lateglacial pollen sites in Scotland (Walker, 1984b; Tipping, 1985).

Pulpit Hill is of importance to the history of Lateglacial climatic and glacial geological studies, in that the site was examined by Donner (1957) in the first attempt to identify his 'Highland Readvance' (synonymous with the Loch Lomond Readvance) with the climatic deterioration recorded in Godwin Pollen Zone III.

More recent palynological investigations (Tipping, 1984, 1991b) have succeeded in clarifying the Lateglacial stratigraphy and have shown that the site is of major significance for present understanding of the climatic changes during the Lateglacial, as well as in the clarity with which several climatic changes are shown. Recent syntheses (for example, Gray and Lowe, 1977b) have suggested that there were no climatic fluctuations during the interstadial comparable to those recognized in north-west Europe by Mangerud *et al.* (1974). This now seems to be incorrect and Pulpit Hill shows, in some detail, all the major climatic changes now recognized in the Lateglacial.

These changes principally are:

1. A short-lived climatic deterioration prior to *c.* 12,000 BP, probably of a similar character to the later Loch Lomond Stadial, though of apparent less severity or duration. This feature is now recognized at a number of sites in western Scotland (Walker *et al.*, 1988; Walker and Lowe, 1990), but is perhaps best exhibited at Pulpit Hill, in the clarity of the stratigraphy, the high temporal resolution of the pollen counts and the geochemical assays. These show that the changes in certain indicators were not synchronous. The decline in organic content occurred before the appearance of pollen types of a more disturbed-ground community. The latter also occurred prior to indications in the sediments of soil instability. This pattern is most easily interpreted as climatic in origin.

   This climatic deterioration appears to have correlatives at, for example, Loch Sionascaig, Loch Borralan and Lochan an Smuraich (Pennington *et al.*, 1972), Corrydon (Walker, 1977), Stormont Loch (Caseldine, 1980a) and Loch Ashik, Slioch Dubh, Elgol and Druim Loch on Skye (Walker and Lowe, 1990). Correlations are not constrained by reliable radiocarbon dates at these sites, and accordingly, the synchroneity of this event cannot be demonstrated. The suggestion that this is correlated with the Older Dryas in north-west Europe (Pennington, 1975b) may not be justified on present evidence (Tipping, 1991b).
2. A brief period of climatic amelioration occurred following this phase before a further, and seemingly sustained, climatic deterioration set in at *c.* 11,800–12,000 BP (by correlation with dated pollen sites on Mull (Lowe and Walker, 1986a) and south-west Ireland (Craig, 1978)). The evidence from carbon/nitrogen contents and pollen stratig-

Figure 10.15 Pulpit Hill: relative pollen diagram showing selected taxa as percentages of total land pollen (from Tipping, 1991b).

raphy at Pulpit Hill is that this decline continued without a break to the markedly more intense deterioration of the Loch Lomond Stadial, and so accords closely with coleopteran evidence for this time period (Atkinson *et al.*, 1987).

3. Within the Loch Lomond Stadial a trend to increasing aridity is clearly seen in the pollen record. This feature has been noted at only a few other sites in Scotland (MacPherson, 1980), due perhaps to inadequate resolution of pollen counts within the clay sediments of the stadial. Should it be confirmed at other sites it would have clear significance for the age of maximum glaciation of the Loch Lomond Readvance.

Finally, Pulpit Hill is important in that it remains one of only a few pollen sites to have been examined by the analysis of more than one pollen core. At Pulpit Hill, two cores were analysed, and the major vegetational and climatic changes discussed above replicated. This approach clearly indicates that the fluctuations recognized are not localized perturbations induced through sedimentological disturbance, nor are they statistical artefacts, but are real indications of the complexity of Lateglacial climatic evolution.

Conclusion

The deposits at Pulpit Hill provide a valuable record of environmental changes in the south-west Highland area during the Lateglacial (about 13,000–10,000 years ago). In particular, detailed study of the pollen and sediments has revealed two separate phases of climatic deterioration, the later one corresponding to the intensely cold Loch Lomond Stadial (about 11,000–10,000 years ago), together with the accompanying vegetation and soil changes. The detail of information available at Pulpit Hill makes it a valuable reference site and allows comparisons with the climatic records of sites in other areas.

LOCH CILL AN AONGHAIS

H. J. B. Birks

Highlights

Pollen preserved in the sediments that infill the floor of this loch provide a valuable record, supported by radiocarbon dates, of the vegetational history of the south-west Highland region during the Holocene. It is particularly important for understanding the woodland history of the area.

Introduction

Loch Cill an Aonghais (NR 776618) is situated at an altitude of about 25 m OD on the north side of West Loch Tarbert, 12 km south-west of Tarbert in southern Knapdale. It is a Quaternary site of considerable importance in the study and reconstruction of the Holocene vegetational history of western Scotland (Birks, 1977). Its detailed and extensively radiocarbon-dated pollen record, elucidated by Sylvia M. Peglar (unpublished data), documents the forest history of the area and clearly illustrates that pine never grew in this part of Scotland (Birks, 1989). The pollen stratigraphy demonstrates the history and development of the western oak-forest type, a type of woodland that is restricted to the high rainfall areas of western Britain, such as Knapdale (Ratcliffe, 1977). Summary pollen diagrams for the site have been published by Birks (1977, 1980).

Description

Loch Cill an Aonghais occupies a small hollow within till. The underlying bedrock is Dalradian schist and quartzite. The loch is about 130 m long and 90 m wide and is bordered by western bryophyte-rich oak and birch woods on its northern and western edges. An extensive fen has developed on the southern and eastern sides of the loch. A 9.5 m core was obtained from this fen. It consists of 0.24 m of *Phragmites* peat overlying 7.0 m of fine-detritus organic muds. Below these muds there are 2.26 m of silty clays and sulphide-rich clays (Figure 10.16). Eight radiocarbon dates (Q–1410 to Q–1417) were obtained from the organic muds to provide a detailed chronological framework for the Holocene pollen stratigraphy of the site.

The basal clays contain a fully marine macrofauna, indicative of cool water conditions, similar to that found at Lateglacial Interstadial sites elsewhere in western Scotland (D. K. Graham, unpublished data).

Interpretation

It is clear that the site was an arm of the sea at a time when relative sea level was higher than today owing to isostatic depression. The pollen record (Figure 10.16) indicates that a treeless landscape prevailed prior to about 10,000 BP. The vegetation was probably a mosaic of *Salix* scrub, *Empetrum* heath, and species-rich grassland. Between 10,000 and 9600 BP fern-rich juniper scrub expanded, presumably in response to climatic amelioration at the onset of the Holocene.

Betula was the first tree to migrate into the area at about 9600 BP, at the expense of the shrub, dwarf–shrub, and grassland communities that were prevalent before 9600 BP. *Corylus avellana* expanded very rapidly between 9400 BP and 9100 BP (Birks, 1989). Hazel was quickly followed by the arrival of elm at about 9000 BP and its subsequent expansion at about 8500 BP. *Quercus* appears to have expanded gradually from about 8500 BP, but it was not locally frequent near the site until about 8000 BP. *Alnus glutinosa* expanded locally at about 7500 BP (Bennett and Birks, 1990). By about 6000 BP the forests in this part of Scotland were probably dominated by oak and birch on poorer soils and by hazel and elm on the richer sites. Alder was locally abundant in moist areas around the loch and within the forests.

At 5100 BP *Ulmus* pollen values fell dramatically and there was a small increase in the frequency of herbaceous pollen, suggesting the presence of small clearings within the forest. Further forest clearance occurred between 4600 BP and 2100 BP. These clearance activities resulted in accelerated rates of erosion and inwashing of clastic material from the catchment of the loch. Extensive forest clearance occurred at about 1300 BP. Cereals and flax appear to have been cultivated locally from 800 BP to 250 BP. Local fen development at the coring site obscures the regional vegetational history for the last 250 years.

The pine pollen values at Loch Cill an Aonghais are very low throughout the Holocene part of the record, indicating that pine was absent from this area of Scotland (Birks, 1989). This conclusion is unexpected, as it contrasts with the undoubted presence of pine further north (see Kingshouse and Loch Maree) and further south in the Galloway Hills (see Loch Dungeon).

Loch Cill an Aonghais is a site of national importance because of its very detailed and well-dated Holocene pollen record. It provides critical evidence for the timing of the arrival and subsequent expansion of the major forest trees of western Scotland during the Holocene. It also demonstrates the absence of pine in the southern Knapdale area and the history of forest clearance and agricultural land use over the last 5000 years. It is particularly important in the context of regional Holocene vegetational history as it provides a geological perspective for an area of Scotland where internationally important woodland types occur today (Ratcliffe, 1977). The Lateglacial part of the sequence demonstrates the transition from a marine to a non-marine environment during the fall in relative sea level over this period.

Conclusion

Loch Cill an Aonghais is a key site in the network of localities representing the pattern of vegetational development in Scotland during the Holocene (the last 10,000 years). In particular, the pollen preserved in its sediments provide a detailed record of the vegetation history of the western oak forest area, showing the pattern of forest development and the notable absence of pine.

EAS NA BROIGE DEBRIS CONE

A. Werritty and L. J. McEwen

Highlights

The deposits in the debris cone at Eas na Broige provide an important geomorphological and sedimentary record of slope processes during the Holocene. They show successive phases of debris-flow activity and alluvial fan development.

Introduction

The Eas na Broige debris cone (NN 192598) is located in Glen Etive at the base of a near-vertical south-facing rock gully (Dalness Chasm) which drains Stob na Broige (956 m OD). Debris cones are fan-shaped accumulations of poorly-sorted

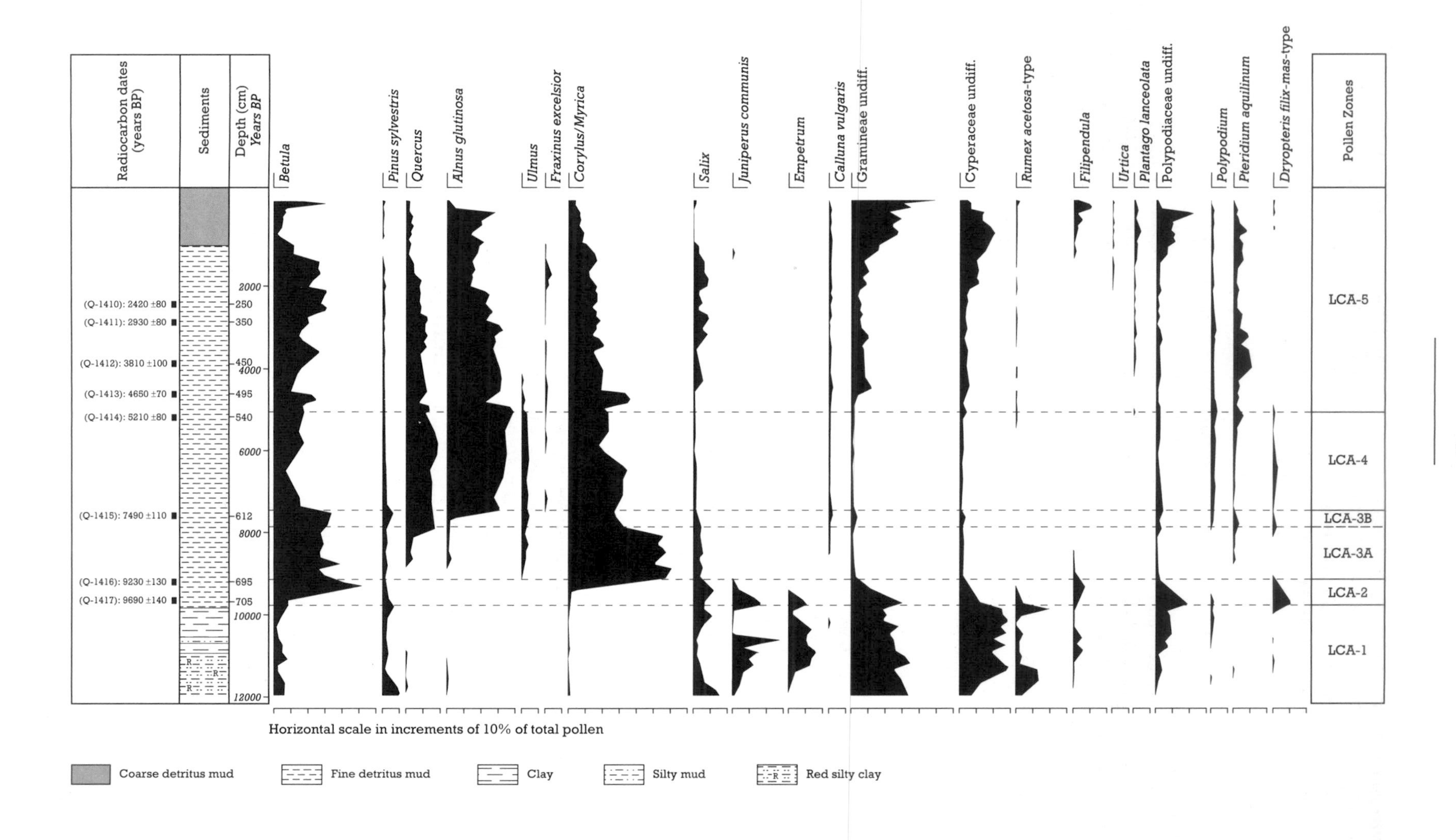
Radiocarbon dates (years BP)
Sediments
Depth (cm)
Years BP
Betula
Pinus sylvestris
Quercus
Alnus glutinosa
Ulmus
Fraxinus excelsior
Corylus/Myrica
Salix
Juniperus communis
Empetrum
Calluna vulgaris
Gramineae undiff.
Cyperaceae undiff.
Rumex acetosa-type
Filipendula
Urtica
Plantago lanceolata
Polypodiaceae undiff.
Polypodium
Pteridium aquilinum
Dryopteris filix-mas-type
Pollen Zones
LCA-5
LCA-4
LCA-3B
LCA-3A
LCA-2
LCA-1
(Q-1410): 2420 ±80
(Q-1411): 2930 ±80
(Q-1412): 3810 ±100
(Q-1413): 4650 ±70
(Q-1414): 5210 ±80
(Q-1415): 7490 ±110
(Q-1416): 9230 ±130
(Q-1417): 9690 ±140
2000
250
350
450
4000
495
540
6000
612
8000
695
705
10000
12000
Horizontal scale in increments of 10% of total pollen
Coarse detritus mud
Fine detritus mud
Clay
Silty mud
Red silty clay

debris formed by successive debris flows at the base of steep gullies. Such debris cones have developed extensively at the margins of valley floors in upland Scotland over the last 13,000 years and have formed in response to changes in sediment supply from adjacent gullies and slopes. Collectively these cones represent an important class of Lateglacial and Holocene landform found throughout upland Britain (Statham, 1976b; Harvey *et al.*, 1981; Innes, 1983b, 1985; Brazier *et al.*, 1988; Brazier and Ballantyne, 1989). The Eas na Broige cone is a particularly good example, and the deposits which comprise it have provided a detailed record of slope processes during the Holocene (Brazier *et al.*, 1988).

Description

The local bedrock comprises granite on the lower slopes with rhyolite lavas on the higher ground; the Dalness Chasm having been etched from a porphyritic dyke (Bailey and Maufe, 1916). During the Loch Lomond Stadial, glacier ice extended up to 650 m OD in upper Glen Etive (Thorp, 1981a, 1986). The debris cone is thus Holocene in age.

The Eas na Broige cone comprises two units: an upper debris cone with a concave long profile and mean gradient of 13.7°, and a lower alluvial fan with a mean gradient of 6.2°. Five discrete cone and fan surfaces can be identified within the lower alluvial surfaces which are inset into the steeper debris cone surfaces. The respective volumes of the debris cone and alluvial fan have been estimated at 170,000 m^3 and 100 m^3; the latter being entirely derived by incision and reworking of the former (Brazier, 1987).

The stratigraphy of much of the cone has been exposed by stream incision, and at two contrasting sites clearly distinguishable debris flow and fluvial sediments have been identified (Figure 10.17). At the apex of the alluvial fan (site 1) a coarse debris flow deposit lies beneath a distinctive and strongly podsolized palaeosol, radiocarbon dated at 550 ± 50 BP (SRR–2882). The overlying sediments comprise poorly-sorted alluvial gravels with only a weak soil development. Higher up the debris cone (site 2) a second palaeosol, radiocarbon dated at 4480 ± 300 BP (SRR–2884), separates two debris-flow units. The upper of these units is the continuation of the lower unit at site 1 and it appears to be the final debris flow unit which was deposited on this part of the cone (Brazier *et al.*, 1988). Pollen samples have been collected from site 1 in order to investigate possible vegetational changes associated with the onset of fluvial reworking. An initial cover of *Corylus*, *Alnus* and *Pinus* before 550 BP was replaced by Gramineae, *Plantago* and *Calluna* after that date. A strong presence of charcoal is also recorded in the unit above the palaeosol.

Figure 10.16 Loch Cill an Aonghais: relative pollen diagram showing selected taxa as percentages of total pollen (from Birks, 1980, after S. Peglar). Note that the data are plotted against a radiocarbon timescale.

Interpretation

Three major phases in the development of this fluvially modified debris cone can be identified. The debris cone initially developed during the first 6000 years of the Holocene, with aggradation ceasing about 4000 BP as a result of exhaustion of the sediment supply through the Dalness Chasm. A prolonged period of stability then ensued until 550 BP. This was followed by a final phase in which the incision into the debris cone produced the inset alluvial fan. The pollen evidence strongly suggests that fluvial activity was contemporaneous with changes in the vegetation cover caused by human interference. The removal of the tree cover destabilized the cone surface and triggered fluvial incision. This instability has continued until the present day.

This site has a potentially wider significance in terms of it being representative of the class of fluvially modified debris cones found throughout the Highlands (Brazier *et al.*, 1988) and other parts of upland Britain (Harvey *et al.*, 1981). First, the initial accumulation of debris-flow deposits involved the reworking of sediment deposited during deglaciation. This implies that this cone, like many others in upland Britain, is 'paraglacial' in origin (cf. Ryder, 1971; Church and Ryder, 1972), that is, its formation was dependent upon an abundant sediment supply following deglaciation. Once this was exhausted, aggradation on the cone ceased. Second, fluvial incision at this site is attributed to recent human

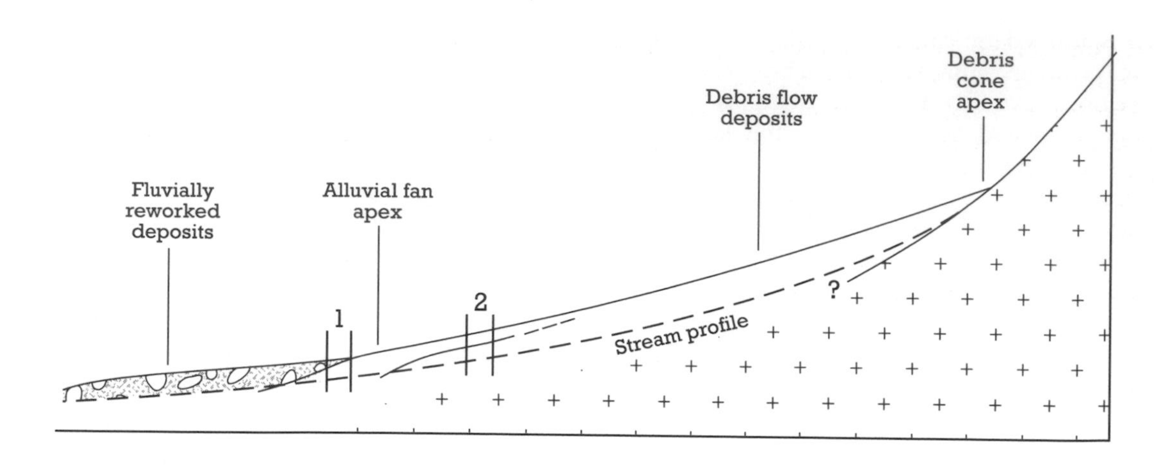

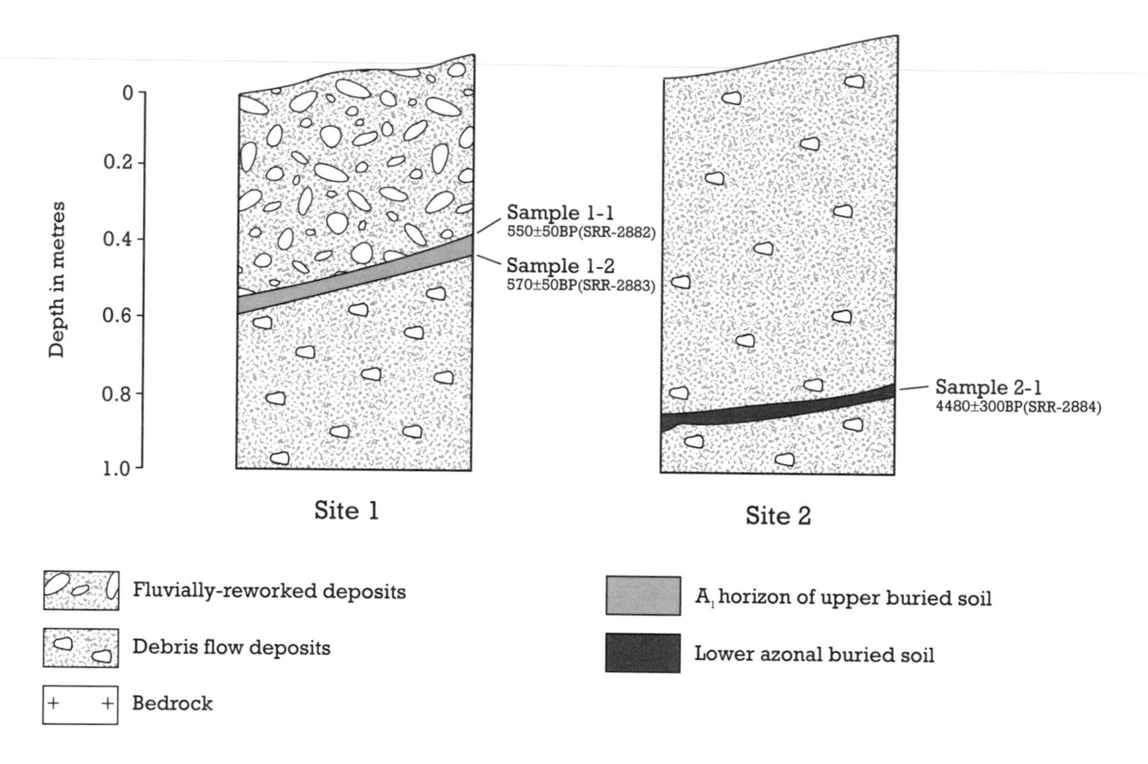

Figure 10.17 Top: schematic section along the length of the Eas na Broige debris cone. Bottom: detail of section at sampling sites (from Brazier *et al.*, 1988).

disturbance of the vegetation. There are many other fluvially modified debris cones in the Scottish Highlands and upland Britain where a similar anthropogenic trigger may have initiated the same change in the process regime on debris cones (see Statham, 1976b; Harvey *et al.*, 1981; Innes, 1983b). In this latter respect the Eas na Broige cone contrasts with those in Glen Feshie, where natural processes are considered to have been responsible for reactivation (Brazier and Ballantyne, 1989).

The Eas na Broige cone is a very good example of a Holocene debris cone that has been subject to fluvial modification. Although debris cones are ubiquitous throughout the Scottish Highlands, this site in Glen Etive is unique in that the date and extent of the fluvial reworking of the original cone have been precisely determined.

The Eas na Broige debris cone provides the most detailed record currently available of Holocene sedimentation at the margins of a major valley in Scotland. Following deglaciation at approximately 10,000 BP the debris cone developed over the next 4000 years. At this time the cone surface became stabilized until about 500 years ago, when fluvial reworking of the basal part of the cone arose in response to the removal of the forest cover by human activity. The cone thus represents a particularly good example of a fluvially modified debris cone in which the most recent phase of development has been in response to human settlement on the valley floor.

Conclusion

Eas na Broige debris cone is a fan-shaped accumulation of poorly sorted material (mixed particles of various sizes) formed by flows, from the slopes above, of rock and soil debris mixed with water. It provides an important record of slope processes during the Holocene and is representative of a type of landform and process system that occurs widely in the Highlands. In particular it shows two phases of development, the first reflecting high sediment supply following deglaciation (ice melting and retreat) and the second, the impact of forest clearance by Man. The site is not only a good landform example but has a well-documented history of development.

Chapter 11

Inner Hebrides

INTRODUCTION

D. G. Sutherland

The Inner Hebrides comprise the islands from Skye in the north to Islay in the south (Figure 11.1). By virtue of their position and topography, they give rise to a diverse series of environments. The Quaternary history of the islands can be understood in terms of the interplay of changes in the local environments resulting from climatic and sea-level variations and the, at times, dominant influences from the nearby mainland. Thus the history of glaciation is that of the interaction of locally nurtured ice caps and the invading mainland ice-sheet. The Holocene vegetational history is also one of local floristic diversity resulting from plant migration from the mainland, with a number of trees and shrubs reaching the northern or western limits of their ranges in the islands. No interglacial or interstadial sites are known from the Inner Hebrides and the established Quaternary history is thus relatively short, being confined to the Late Devensian. Although various shore platforms have been ascribed to interglacial or pre-glacial episodes, their origins remain uncertain. The area is a classic one for features of mountain glaciation, shore platforms and raised beaches. The principal themes of research have therefore focused on these aspects and, in addition, on the interaction of local and mainland ice, landforms of the Loch Lomond Readvance and Lateglacial and Holocene vegetation history.

The principal islands and mountain groups of the Inner Hebrides relate to the central Tertiary igneous complexes of Skye, Rum and Mull. Tectonic warping and accompanying erosion in the Tertiary would have fashioned a landscape with a magnitude of relief similar to that of the present by the time of the onset of glaciation, but subsequent glacial erosion has produced many of the familiar and dramatic details of the scenery. Thus it is likely that many of the islands, such as Skye, Mull, Scarba, Jura, and Islay, were initially attached to the mainland, with glacial erosion producing the intervening narrow stretches of sea (Sissons, 1983c). It may be noted that due to sea-level change since the last glaciation, Knapdale and Kintyre were, at different times, islands, whereas the Ardnamurchan peninsula, although never an island, has many affinities with the neighbouring islands.

Striations, roches moutonnées and the transport of erratics provide clear evidence of the interplay of the mainland ice-sheet and the glaciers and ice-caps developed during the last glaciation on the islands. The mainland ice overwhelmed almost all the islands, flowing in a westerly or north-westerly direction (Clough and Harker, 1904; Harker, 1908; Peach *et al.*, 1910b, 1911; Cunningham Craig *et al.*, 1911; Bailey *et al.*, 1924; Bailey and Anderson, 1925; Richey and Thomas, 1930), but local ice on both Skye (Harker, 1901) and Mull (Bailey *et al.*, 1924) was sufficiently powerful to maintain independent ice centres, and no mainland erratics are found in the central mountain areas of these islands. Surprisingly, in contrast to other areas of Scotland where the ice-sheet traversed part of the sea floor (see Mill Bay, Baile an t-Sratha, Boyne Quarry, Kippet Hills, Port of Ness, Nith Bridge and Port Logan), there are few reports of shelly glacial deposits in the Inner Hebrides. Only in western Islay have shelly tills (Synge and Stephens, 1966; Peacock, 1974b) and poorly fossiliferous glaciomarine sediments (Benn and Dawson, 1987) been described. A possible lateral limit of the last ice-sheet has been recognized on the Trotternish peninsula on Skye by Ballantyne (1990).

The most spectacular effects of glaciation occur in the Cuillin Hills of Skye. The Cuillin have the greatest frequency and magnitude of ice-eroded corries and rock walls in Scotland, and the summit ridge of arêtes and pyramid-shaped peaks is the product of headward erosion of the corries. Also in the Cuillin and particularly by Loch Coruisk is a quite outstanding development of ice-moulded and striated bedrock. Other small-scale features resulting from ice erosion that are of particular note in the Inner Hebrides are the bowls, channels and troughs (p-forms) reported by Gray (1981) at Scarisdale on Mull.

All the Inner Hebridean islands were close enough to the centre of the last mainland ice-sheet to have been strongly influenced by glacio-isostatic depression and rebound. This has resulted in a complex and often impressive development of raised marine erosional and depositional features around the coasts. Three distinct generations may be recognized; those that pre-date at least one period of glaciation; those that are contemporaneous with the retreat of the last ice-sheet; and those that post-date the disappearance of the last ice-sheet.

The first of these three groups has two separate elements: the first at close to present sea level and the second at high level, generally above

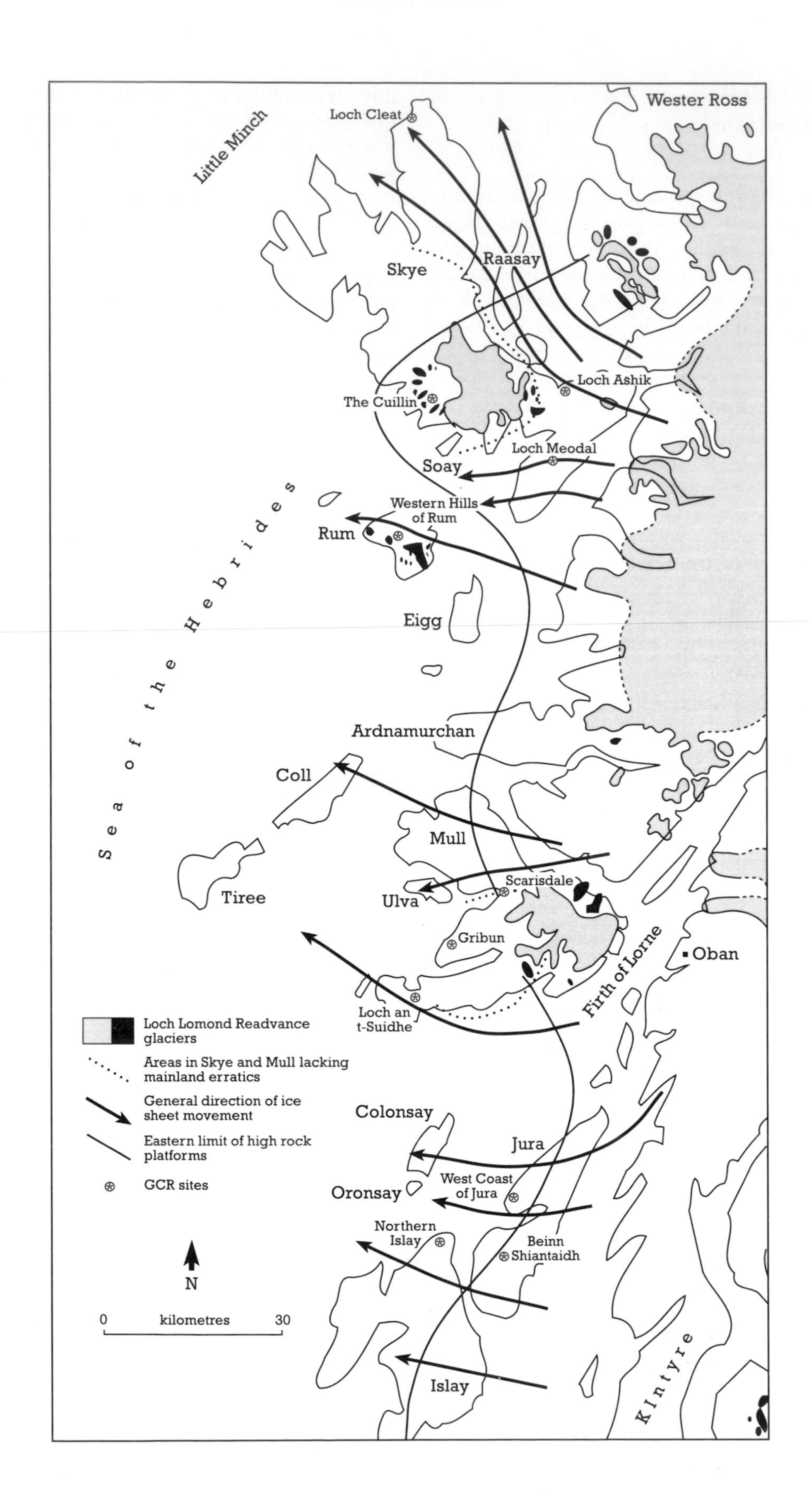
Wester Ross
Loch Cleat
Little Minch
Skye
Raasay
Loch Ashik
The Cuillin
Loch Meodal
Soay
Western Hills of Rum
Rum
Sea of the Hebrides
Eigg
Ardnamurchan
Coll
Mull
Tiree
Ulva
Scarisdale
Gribun
Oban
Firth of Lorne
Loch an t-Suidhe
Loch Lomond Readvance glaciers
Areas in Skye and Mull lacking mainland erratics
General direction of ice sheet movement
Eastern limit of high rock platforms
GCR sites
Colonsay
Jura
Oronsay
West Coast of Jura
Northern Islay
Beinn Shiantaidh
N
0 kilometres 30
Islay
Kintyre

18 m OD. Close to present sea level is a set of apparently horizontal rock platforms. Their surfaces are striated and the backing cliffs can, in places, be seen to disappear beneath glacial drift. On Northern Islay and the West Coast of Jura, Dawson (1980a) has termed this set of platforms the Low Rock Platform and it, or they (as there appear to be a number of levels in different parts of the Hebrides), have been suggested to be of interglacial origin (Dawson, 1980a; Sissons, 1981a, 1983c). There is no direct evidence for this, however.

These low platforms are in marked contrast to the second set of platforms comprising the so-called High Rock Platform, examples of which occur widely throughout the Inner Hebrides as well as on certain of the neighbouring peninsulas, such as Ardnamurchan and Applecross. These rock platforms are frequently spectacularly developed, as in Northern Islay, and range in altitude from 18 m to as much as 51 m OD. Originally thought to be 'preglacial' (Wright, 1911) because of the striations and glacial deposits found in places on the platform surfaces, their age is today considered much less certain. An interglacial origin was favoured until recently (McCann, 1964, 1968; McCann and Richards, 1969; Jardine, 1977), but has been supplanted in favour of the idea that the platforms formed contemporaneously with the last ice-sheet (Sissons, 1981a, 1982b; Sutherland, 1981a; Dawson, 1984; Gray, 1985). Sissons (1981a, 1982b) noted that not all fragments of the platform carried evidence of glaciation and, further, that there was a distinct line running through the Inner Hebrides to the east of which high rock platforms were either absent or only very poorly developed, whereas to the west they were typically very clear and extensive features (Figure 11.1). Sissons argued that this contrast was due to ice occupying the eastern area while the platforms were being formed to the west. Minor advances and retreats of the ice-sheet would have resulted in glaciation of certain of the platforms, but those formed during ice retreat would not be glaciated. A corollary of Sissons' hypothesis is that the platforms should be isostatically tilted. Unfortunately, the number of accurate altitudes available on the inner margin of the platform fragments is as yet too few to test this prediction. Sissons (1981a, 1982b) also suggested a chronology for the development of the rock platforms, envisaging initial expansion of the Scottish ice-sheet in the Early Devensian (see Sutherland, 1981a) and glaciation (and accompanying platform formation) continuing throughout the Devensian until final decay of the ice towards the end of the Late Devensian. There is as yet no direct evidence to support such a chronology.

Figure 11.1 Location map and principal Quaternary features of the Inner Hebrides (from Peacock, 1983b; Sissons, 1983c; Ballantyne and Benn, 1991).

During ice-sheet retreat in the Late Devensian, the most rapid deglaciation initially occurred in the deep-water channels between the islands, and the residual ice masses on the islands then flowed outwards into these ice-free areas. At a relatively late stage, however, a north-westerly ice flow was maintained across Islay and Jura, and as the Paps of Jura emerged from the ice as nunataks, the remarkable medial moraine of Scriob na Caillich (Dawson, 1979b) on the West Coast of Jura was deposited. At about this time the end moraine at Coir' Odhar in Northern Islay (McCann, 1964; Synge and Stephens, 1966; Dawson, 1982) was also formed. Sea level at this period was relatively high and the second generation of raised marine features was formed during the subsequent fall. These features consist of glaciomarine deltas as well as gravel spreads and shingle ridges. The last are most spectacularly developed along the West Coast of Jura (McCann, 1964; Dawson, 1979a, 1982), where staircases of over 20 distinct ridges have been mapped. In central Islay a moraine records the position of the ice-sheet margin while relative sea level fell by up to 12 m (Dawson, 1982).

The precise timing of deglaciation is unclear. A radiocarbon date of 16,470 ± 300 BP (SRR–118) (Harkness and Wilson, 1979) from marine sediments offshore from Colonsay (Binns *et al.*, 1974) may be too old due to contamination. The earliest dates from basal terrestrial sediments of 13,870 ± 150 BP (SRR–3121) at Loch Ashik on Skye (Walker *et al.*, 1988) and 13,140 ± 100 BP (SRR–1805) at Loch an t-Suidhe on Mull (Walker and Lowe, 1982; Lowe and Walker, 1986a) are also uncertain because of possible contamination (Sutherland, 1980). Lateglacial vegetational development has only been studied in detail on Mull and Skye (Birks, 1973; Birks and Williams, 1983; Walker and Lowe, 1982, 1990, 1991; Lowe and Walker, 1986a; Walker *et al.*, 1988). Throughout the Lateglacial Interstadial the vegetation of the islands was essentially treeless, and although tree birch occurred in the more southerly islands, its distribution was limited to those areas not exposed to westerly winds (Lowe and Walker, 1986a). It may have reached its north-westernmost limit at this period in south-east Skye (Birks and Williams, 1983). The early vegetation of the interstadial consisted primarily of open grass- and sedge-dominated communities, but a juniper scrub and then an *Empetrum* heath developed. As with the tree birch, exposure to westerly winds may have been the main factor in limiting plant diversity during this period (Walker *et al.*, 1988).

The Loch Lomond Stadial had a major impact on the Inner Hebrides. Glaciers readvanced or built-up anew on Skye (Walker *et al.*, 1988; Ballantyne, 1989a), Rum (Ballantyne and Wain-Hobson, 1980) and Mull (Gray and Brooks, 1972). The age of the readvance is established by the ice-transported shells at Loch Spelve on Mull, which gave a radiocarbon date of 11,330 ± 170 BP (I–5308) (Gray and Brooks, 1972), and the occurrence of only Holocene sediments in enclosed basins from within the readvance limits (Walker and Lowe, 1985). Radiocarbon dating and accompanying pollen analyses from such sites implies that the glaciers on Mull had largely disappeared by between 10,500 BP and 10,000 BP. On Skye comparisons of the pollen spectra from the basal sediments in enclosed basins within the readvance limits has indicated a diachronous retreat of the glaciers (Lowe and Walker, 1991). Initial retreat may have been underway by 10,200 BP and was apparently completed by the time of the juniper peak of the early Holocene vegetational succession, that is, no later than 9,600 BP.

During the latter part of the Lateglacial period a distinctive shoreline was eroded around the coasts of the southern islands. This shoreline, termed the Main Rock Platform, is isostatically tilted (Gray, 1978a; Dawson, 1980b; Sutherland, 1984a) and slopes away from the area of maximum isostatic depression in the south-west Highlands, such that it passes below present sea level in Northern Islay and around the coast of Mull. It has also been identified on the coast of southern Skye (Peacock, 1985). The evidence from the Isle of Lismore (see above) (Gray, 1987; Gray and Ivanovich, 1988) suggests that the platform could in part be a reoccupied, older feature.

The Loch Lomond Stadial vegetation reflected the harshness of the climate, with a dominance of open-habitat taxa and species characteristic of disturbed soils (Lowe and Walker, 1986a). Periglacial processes were particularly active (Ballantyne, 1991b), and the large-scale patterned-ground features found in the Western Hills of Rum (Ballantyne and Wain-Hobson, 1980; Ballantyne, 1984) were most probably fashioned at this period. On Jura the impressive protalus rock glacier of Beinn Shiantaidh (Dawson, 1977) was formed.

The early Holocene was initially characterized by rapid vegetational development as the climate ameliorated, and a plant succession from tundra heath through *Empetrum* heath and juniper scrub to hazel–birch woodland probably occurred in the first 1500 years. Details of this early phase are particularly well preserved at Gribun on Mull (Walker and Lowe, 1987). Subsequent development of the Holocene vegetation cover reflected the diversity of local environments, as is illustrated by the pollen sites in Skye at Loch Ashik, Loch Meodal and Loch Cleat (Birks, 1973; Birks and Williams, 1983). Around Loch Meodal, in the Sleat peninsula, during the middle Holocene the dominant vegetation was mixed birch–hazel–alder woods with some oak, elm, ash, rowan and holly. This was probably close to the northern limit of predominant oak at that time (Birks, 1977; Birks and Williams, 1983). Pine was absent, yet a short distance to the north around Loch Ashik pine flourished (Williams, 1977). In contrast again, at Loch Cleat in northern Skye only birch, hazel and willow scrub developed at this time (Williams, 1977; Birks and Williams, 1983). These three sites, considered together, demonstrate the diversity of vegetational development resulting from the geological, topographical and climatic variations on a single island in the Inner Hebrides.

Farther south than Skye, the vegetation of the middle Holocene consisted of birch–hazel scrub or woodland with some oak and elm (Andrews *et al.*, 1987; Walker and Lowe, 1987; Hirons and Edwards, 1990). Subsequent to 4000 BP woodland contracted and heathland and grassland expanded. A reduction in woodland cover at about this time has been widely noted in Scotland (see Allt na Feithe Sheilich and Loch Maree) and may be due to either an increase in storminess or human impact, or a combination of the two (Birks, 1987).

The final generation of raised marine features formed around the coasts of the inner Hebrides was produced at the maximum of, and subsequent to the Main Postglacial Transgression. As with the Lateglacial shorelines, the features are most impressively developed along part of the West Coast of Jura, particularly at Inver where as many as 30 distinct shingle ridges occur one above the other (Dawson, 1979). Radiocarbon dates on marine shells from Oronsay (Jardine, 1978, 1987) suggest that the maximum Holocene sea level was attained at around 6500–7000 BP. Thus the 'staircase' of shingle ridges at Inver was likely to have been formed in the last 6500 years.

Small-scale periglacial features are currently active on a number of mountains on Rum (Western Hills) and Mull (Godard, 1959).

THE CUILLIN

D. G. Sutherland

Highlights

The Cuillin is an outstanding area for glacial geomorphology. It is particularly noted for landforms of mountain glacier erosion, demonstrating an assemblage of classic features unmatched elsewhere in Britain. This interest is also complemented by an excellent range of moraine types formed by corrie and icefield glaciers of the Loch Lomond Readvance.

Introduction

The Cuillin site in southern Skye includes the main ridge of the Black Cuillin, extending *c.* 13 km from Gars-bheinn (NG 468187) in the south to Sgùrr nan Gillean (NG 472253) in the north, the slopes leading down to Glen Brittle to the west and the Bealach a'Mhàim to the north, the central trough of Loch Coruisk (NG 485205), and to the east, Glen Sligachan (Figure 11.2A). The Cuillin are arguably the most spectacular mountain range in Scotland. Taking the form of a semicircle concave to the south-east, the mountains rise abruptly from sea level to a maximum altitude of 993 m OD in Sgùrr Alasdair (NG 451208). Bitten into on all sides by corries, the central ridge to the mountains never drops below 750 m OD and is typically sharp and narrow. The individual peaks, overlooking corries on several sides, have the form of triangular pyramids and pinnacles and the spectacular glacial scenery is enhanced by the dominance of bare rock at the ground surface, there being little glacial drift within the area of the mountains (Figure 11.3). The Cuillin are of particular geomorphological interest as an outstanding example of the effects of glacial erosion, at both large and small scales, on a mountain massif (Forbes, 1846; Harker, 1901; Lewis, 1938, 1947; Haynes, 1968; Dale, 1981). They also contain evidence for two phases of glaciation during the Late Devensian: the main ice-sheet and Loch Lomond Readvance (Harker, 1901; Sissons, 1977c; Walther, 1987; Walker *et al.*, 1988; Ballantyne, 1989a; Ballantyne and Benn, 1991). The lower slopes and valleys are also important for depositional landforms associated with the Loch Lomond Readvance (Donner and West, 1955; Ballantyne, 1989a; Benn, 1991).

Description

The Cuillin occur on the western side of the Skye Tertiary central igneous complex (Richey, 1961; Emeleus, 1983). The principal rock in this part of the complex is a layered gabbro and the semicircular nature of the gabbro outcrop defines the mountain range. The layering dips steeply inwards at right angles to the trend of the hills, resulting in the steep western mountain front. The gabbro is cut by many minor intrusions including cone sheets and dykes of dolerite.

The intrusions of the igneous complex reached to a shallow depth in the Earth's crust and were unroofed and eroded while igneous activity continued, as is shown by the presence of clasts derived from the intrusive rocks contained in fluvial sediments interbedded with basaltic lavas to the west. By the time of cessation of volcanic activity the mountain mass had dimensions approximately similar to those of today, although it may be anticipated that the lavas abutted against the lower flanks of the hills and the major valleys had yet to be eroded.

Extensive erosion of the western Highlands and the Inner Hebrides during the Tertiary may be inferred from the relationship of the Tertiary dykes to the valley system, the disrupted nature of the once continuous lava flows and the distribution of erosion surfaces (Godard, 1965; George, 1966; Sissons, 1967a). Thus at the onset of Pleistocene glaciation the broad outlines of the topography were unlikely to have been significantly different from today, although without those features of detail, such as corries, overdeepened valleys and stripped and polished

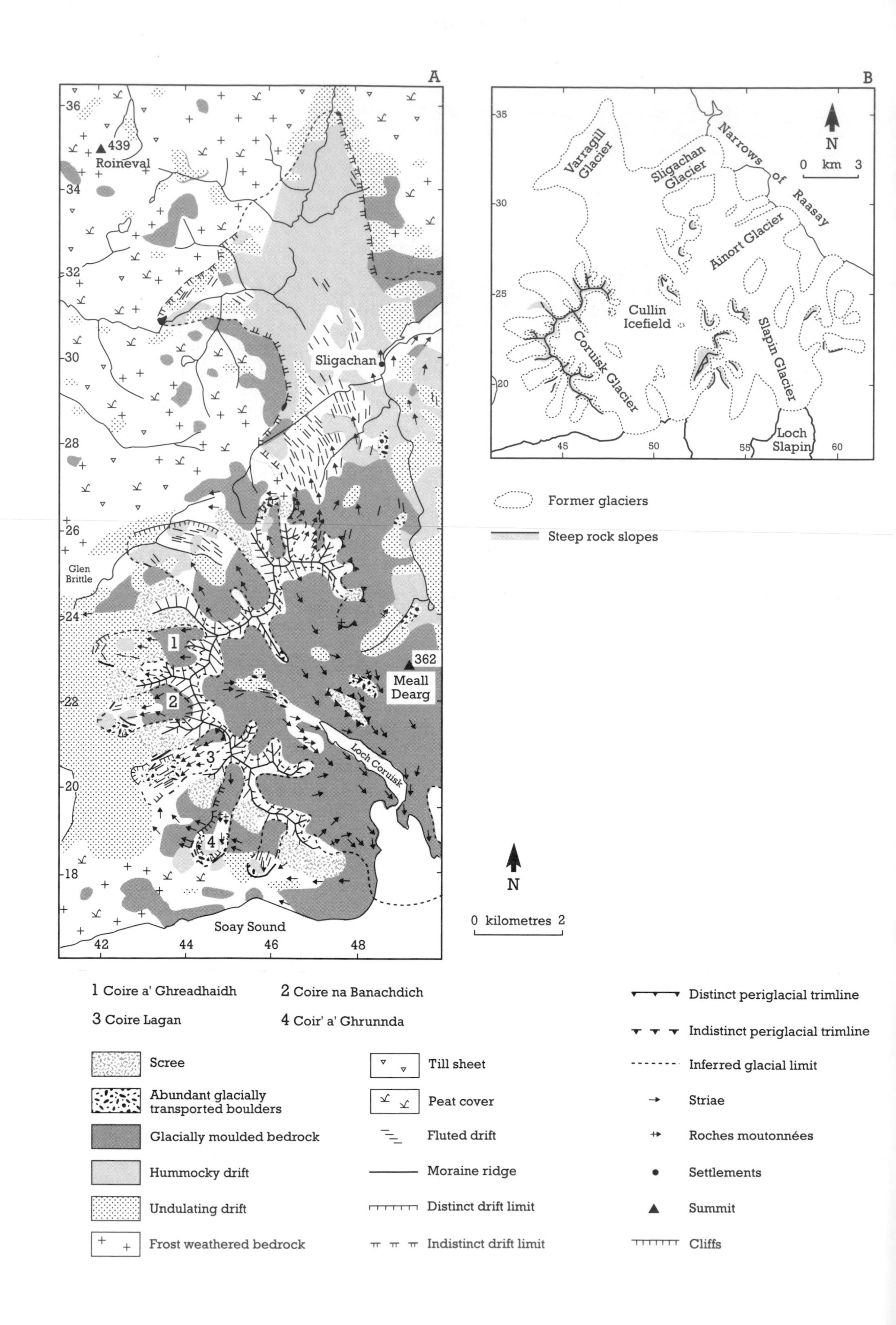

A
B
Roineval
439
Sligachan
Glen Brittle
Meall Dearg
362
Loch Coruisk
Soay Sound
N
0 kilometres 2
Varragill Glacier
Sligachan Glacier
Narrows of Raasay
Ainort Glacier
Cullin Icefield
Coruisk Glacier
Slapin Glacier
Loch Slapin
0 km 3
Former glaciers
Steep rock slopes
1 Coire a' Ghreadhaidh
2 Coire na Banachdich
3 Coire Lagan
4 Coir' a' Ghrunnda
Scree
Abundant glacially transported boulders
Glacially moulded bedrock
Hummocky drift
Undulating drift
Frost weathered bedrock
Till sheet
Peat cover
Fluted drift
Moraine ridge
Distinct drift limit
Indistinct drift limit
Distinct periglacial trimline
Indistinct periglacial trimline
Inferred glacial limit
Striae
Roches moutonnées
Settlements
Summit
Cliffs

Figure 11.3 Landforms of glacial and periglacial erosion are strikingly developed in the Cuillin of Skye. The serrated aspect of the main Cuillin ridge reflects intense periglacial weathering, whereas the lower slopes are heavily ice-scoured. (British Geological Survey photograph B168.)

bedrock surfaces, that can be ascribed to glacial action.

Within the area of the Cuillin there is evidence for only two glacial phases, both during the Late Devensian: one during the period of the last ice-sheet glaciation and the other during the Loch Lomond Readvance. However, it is only reasonable, given the evidence from surrounding areas in the British Isles (Bowen *et al.*, 1986), to presume that the hills have been repeatedly glaciated, perhaps from as early as 2.4 Ma BP (Shackleton *et al.*, 1984). The features of glacial erosion, which are so outstandingly developed in the Cuillin, can therefore be anticipated to have developed over a very long period.

The major features of glacial erosion centre on the main Cuillin ridge which has the form of a semicircular arête produced by the intersection of multiple corries (Figure 11.2). The floors of the corries display extensive areas of ice-moulded bedrock and rock steps, while taluses occur below the rock headwalls. At the centre of the massif there is a spectacular glacial trough with extensive areas of ice-scoured bedrock and the basin occupied by Loch Coruisk.

That the area showed evidence of glaciation was recognized at an early stage after the introduction of the glacial theory (Forbes, 1846), but despite occasional references (Geikie, 1863a, 1984; Bonney, 1871) it was not until the turn of the century that the glacial deposits and the effects of glacial erosion were discussed in detail (Harker, 1899a, 1899b, 1901; Clough and Harker, 1904). Harker (1901), in one of the first detailed systematic papers on glacial erosion, noted that the form of the valleys both in cross-profile and long profile was a direct result of glaciation. He pointed out that many of the corries 'hang' above the main valley and that this could be explained by glacial overdeepening of the main valley, or

Figure 11.2 (A) Principal glacial features of the Cuillin. (B) Reconstructed Loch Lomond Readvance glaciers in central Skye (from Ballantyne, 1989a).

the erosion of the valley sides. Harker attributed variations in the long profiles of the corries and the valleys, including valley steps as well as rock basins, to the tendency of ice to exaggerate any pre-existing marked inequalities in the profile: increased erosion under those parts of a glacier where the ice was thicker was considered the mechanism to explain these features. The form of the ridges and the summits was attributed to the backward erosion of the corrie heads which came close to intersecting, leaving only narrow arêtes and triangular, slightly concave-faced pyramids. An asymmetry was also apparent to Harker in the ridges of the Cuillin, with the northern slopes typically being steeper than the southern. A similar asymmetry was noted in the distribution of the corries, in particular among those towards the edges of the mountains, with the majority facing the north to north-east sector. These asymmetries Harker ascribed to periods of partial glaciation when the distribution of glaciers and their activity was influenced by their exposure (or lack of it) to the Sun.

In detail, Harker also noted the lack of discrimination by glacial erosion in the exploitation of small geological features in those areas where erosion was intense, the surface of dykes being planed smooth with the surrounding bedrock. This, as was evidenced by the mountain summits, was not typical of subaerial erosion. Harker thought, however, that the mechanism for the erosion of the dykes was distinct from that of the gabbro, for he recorded that there was a disproportionate amount of dyke rock in the cobble to boulder-sized material in the glacial deposits. He concluded that the dykes had been eroded by plucking and tearing away, whereas the gabbro had been ground down by debris lodged in the sole of the ice. It is of note that almost all of these observations and explanations on glacial erosion subsequently came to be accepted in standard texts (Flint, 1957; Sissons, 1967a; Andrews, 1975; Embleton and King, 1975a; Sugden and John, 1976).

W. V. Lewis also made significant contributions to the theory of glacial erosion, drawing on field examples from the Cuillin. He described the morphology of several of the Cuillin corries, noting the sharp contrast between the shattered headwalls and the ice-moulded floors, and such observations in part laid the foundations for his meltwater theory of corrie formation (Lewis, 1938). He also quoted notable examples of rock steps and roches moutonnées in support of his theories of glacial valley erosion (Lewis, 1947).

The morphology of the corries in the Cuillin has subsequently been investigated as a part of a larger study by Haynes (1968, 1969), who concluded that a large proportion of the long profiles of the corries did not fit the idealized mathematical shapes to which many other corries on the Scottish mainland closely approximated. This she attributed to the complex rock structure of the Cuillin Hills.

Dale (1981) studied the rock walls that comprise the corries and considered their development with respect to similar features in an east–west transect across the Scottish Highlands. She identified 28 individual rock walls in the Cuillin corries and demonstrated that the frequency of occurrence was among the highest measured. The size of the rock walls, as determined by both amplitude and area, was the greatest in the region studied; the average altitude of the base of the rock walls was the lowest, but this figure disguised two distinct altitudes of formation, one between 607 m and 649 m OD and the other between 470 m and 535 m OD. There was a strong northward component in the orientation of the rock walls, and even in corries that faced the west or southwest, the rock walls were best developed on their north- to north-west-facing sides. This conclusion mirrors the observations of Harker (1901) on the asymmetry of the ridges in the Cuillin Hills. Dale (1981) attributed the exceptional development of rock walls in the Cuillin to a combination of available relief, glacial history (the lack of inundation by the mainland ice-sheet) and by the mountains being particularly susceptible to glacier initiation, leading to relatively long periods of local glaciation.

Drawing on the then unpublished work of Harker, Geikie (1894) first reported that during the last glacial maximum the hills of Skye nurtured an independent ice-cap which had sufficient strength to prevent the main Scottish ice-sheet from overriding them. Harker (1901) and Clough and Harker (1904) amplified the ice-flow pattern, tracing the direction of flow of the coalescent ice-sheet and ice-cap both by the orientation of striae and by one of the earliest quantitive analyses of the erratic content of glacial drift. By such stone counts Harker (1901) traced the line of contact in Glen Sligachan between ice originating from the (granitic) Red Hills and ice from the (gabbroic) Cuillin Hills, and noted how the combined northwards ice flow was deflected to the west and then southwest by the mainland ice-sheet.

Subsequent to the period of ice-sheet glaciation, Harker (1901) recognized a distinct phase of local glaciation in which local corrie and valley glaciers flowed outwards from the hills. The impressive end moraine at the mouth of Coir' a'Ghrunnda (NG 445184) (Forbes, 1846) was considered to have been formed at this time. Various other authors have agreed on the existence of later glaciation, but its extent and age have remained in dispute. Charlesworth (1956) reconstructed a small ice-field over the mountains of southern Skye, with the corries on the west of the Cuillin feeding an outlet glacier in Glen Brittle. He correlated these glaciers with his Stage M glaciation (considered to be a readvance) and depicted various retreat stages. Anderson and Dunham (1966) argued that what they termed a 'late-glacial readvance' extended well beyond the limits of the hills.

A more complex scheme was proposed by Birks (1973) who interpreted an extensive local glaciation prior to 12,000 BP, followed by a phase of corrie glaciation in the Cuillin during the Loch Lomond Readvance. Sissons (1977c) also inferred limited glaciation during the Loch Lomond Readvance, identifying seven corrie and one valley (Coruisk) glacier in the Cuillin. This reconstruction was not accepted by Walther (1984, 1987) who envisaged a much greater extent of ice, reaching to north of Sligachan. He proposed certain retreat stages with a final glacial episode in the corries during the early Holocene, when block moraines were formed.

The most recent reconstruction (Walker *et al.*, 1988; Ballantyne, 1989a) favoured the development of a large ice-cap over the southern mountains of Skye during the Loch Lomond Readvance with contemporaneous corrie glaciers in the western Cuillin (Figure 11.2B). These last small glaciers were similar to those identified by Sissons (1977c).

The detailed geomorphological evidence for this reconstruction is outlined in Ballantyne (1989a). Ballantyne (1989a) also reconstructed the three-dimensional form of the Loch Lomond Readvance glaciers on Skye and derived a number of palaeoclimatic inferences. The area-weighted mean equilibrium line altitude of the glaciers conforms with the overall pattern of an eastwards rise in equilibrium line altitude reconstructed for the Loch Lomond Readvance glaciers in the Inner Hebrides and Western Highlands (Sissons, 1979d, 1980b). Local variations in the reconstructed equilibrium line altitude were inferred to reflect precipitation patterns associated with southerly winds (see Sissons, 1979d, 1980b). Mean July sea-level temperature for the stadial was estimated to be about 6°C. A notable feature of Ballantyne's glacier reconstruction is the north–south asymmetry of the Cuillin ice-field. Although the southern outlet glaciers descended steeply to Loch Scavaig, the main outlets to the north in Glen Varragill and Glen Sligachan were relatively more extensive and had relatively lower surface gradients. Such asymmetry could have reflected, in part, variations in the respective glacier mass budgets, but an intriguing possibility explored by Ballantyne is that it relates to differences in the nature of the beds over which the former glaciers flowed. The glaciers on the south side of the ice-field moved over ice-scoured bedrock and their reconstructed surface profiles are typical of those predicted for rigid beds from glacier physics theory (cf. Paterson, 1981). The northern glaciers, however, were associated with extensive deposits of fluted and hummocky moraine which are believed to have formed subglacially (see below), in effect forming a deformable glacier bed. Under such conditions, the predicted ice surface profile is lower than for the rigid bed case (cf. Boulton and Jones, 1979; Nesje and Sejrup, 1988).

Particularly good examples of depositional landforms associated with the Loch Lomond Readvance glaciers include the end and lateral moraines below Coir' a'Ghrunnda in Glen Brittle and the hummocky and fluted moraine at Sligachan (Figure 11.2A) (see Donner and West, 1955; Ballantyne, 1989a; Benn, 1991; Benn *et al.*, 1992). Benn (1989b) discussed aspects of the asymmetry of the end moraines in Coire na Banachdich (NG 430218) and Coire a'Ghreadhaidh (NG 430234). The origins of hummocky moraine are controversial and it seems probable that several landform types exist (see Coire a'Cheud-chnoic). Ballantyne (1989a) noted that large areas of moraine in Glen Varragill were fluted and streamlined, reflecting subglacial deformation comparable to that described from Coire a'Cheud-chnoic (Hodgson, 1982, 1987) and elsewhere in Torridon (Hodgson, 1986). As in the case of Coire a'Cheud-chnoic, the Sligachan and Glen Varragill areas provide potentially important evidence for further investigation of the genesis of hummocky moraine, and the different landform types, together with comparative studies of other sites and related landform assemblages, may allow not only the possibility of recognizing different styles of glaciation and deglaciation (see Eyles, 1979, 1983; Sharp, 1985; Evans, 1989), but also inputs to, and the testing of, theoretical models of glacier behaviour on deformable beds (cf. Boulton and Jones, 1979; Alley *et al.*, 1986; Boulton and Hindmarsh, 1987).

Small-scale erosional landforms are also spectacularly developed and owe their final detail of form to the Loch Lomond Readvance glaciers. Notable examples are in the corries above Glen Brittle (e.g. Coire Lagan – NG 444209) and around Loch Coruisk, where the ice-moulded and striated bedrock and roches moutonnées are among the best examples of their kind in Britain (Figure 11.4). These landforms, some described by Lewis (1947) and Haynes (1969), have significant potential for detailed studies of subglacial erosional processes, as in the types of investigation reported from Snowdonia in North Wales (Gray, 1982b; Gray and Lowe, 1982; Sharp *et al.*, 1989a).

The nature of the very steep and sharp ridges and summits of the Cuillin has precluded the extensive development of periglacial features (Ballantyne, 1991b). Harker (1899b, 1901) and Clough and Harker (1904) noted the extent of rock shattering on the summit areas and the spectacular development of scree slopes, the formation of which was considered to have commenced during the period of local glaciation. Harker (1899b, 1901) also commented on the contrast in the nature of the rock surfaces of the intensely glaciated areas and on the summits, and Ballantyne (1989a) noted that this contrast relates to the upper surface of the Loch Lomond Readvance glaciers. Ballantyne (1989a) used periglacial trimlines to delimit, in part, the limits of Loch Lomond Readvance glaciers. The assignment of the local glaciation to the Loch Lomond Readvance also implies that the accumulation of the scree in the corries must have occurred in the Holocene. Aspects of scree-slope development were investigated by Statham (1976a), who concluded that the debris movement and morphological characteristics accorded with a rockfall model of genesis. Modification of some of the taluses by snow avalanches may also have occurred (Benn, 1990). The Holocene talus accumulations may be the most outstanding

Figure 11.4 Detail of ice-moulded bedrock near Loch Coruisk showing glacially abraded and smoothed stoss slopes and localized joint-block removal. (Photo: D. G. Sutherland.)

features of their type in Britain (Ballantyne, 1991b).

Interpretation

The Cuillin is an area of spectacular scenery which owes its striking impact to the outstanding development of features of glacial erosion. It represents probably the single most intensive area of mountain glacier erosion in Britain. Within Scotland the Cuillin represent the classic development of corrie, arête, rock step, ice-moulded bedrock and glaciated-valley landforms typical of the zone of high-intensity glacial erosion of western Scotland (see Chapter 2), and in this respect rank ahead of even such areas as An Teallach, Rum and the mountains of North Arran, where excellent features of glacial erosion are also present. The intensity of erosion of the Cuillin contrasts markedly with that of the mountains in the central and eastern Highlands, such as the Cairngorms and Lochnagar, where more extensive elements of the pre-glacial landscape have survived and the dominant landform pattern is the equally distinctive one of selective glacial erosion. Elsewhere in Britain the area most comparable to the Cuillin is North Wales, where glacial erosion has left a strong imprint on the landscape (see Campbell and Bowen, 1989). Nevertheless, the Cuillin remain pre-eminent, particularly for the fine detail, compactness and overall impact of the geomorphology. Although the Lake District has fine landforms of glacial erosion, many classic in their own right, the overall intensity of glacial erosion there has been less.

Historically, the Cuillin have played a significant role in the study of glacial erosion, beginning with the work of Forbes (1846) and continued through that of Harker (1901), and Lewis (1938, 1947) to the recent investigations of Haynes (1968, 1969) and Dale (1981). These studies, which have provided textbook examples and laid the foundation for modern developments, have been essentially complemented by the important historical work on glacial erosion in North Wales (see Campbell and Bowen, 1989).

The Cuillin are also important in the evidence they provide for the development of local glaciers during the Loch Lomond Stadial. This evidence has allowed the reconstruction of a number of corrie glaciers and a central mountain ice-field, together with the associated palaeoclimatic controls. This has provided important insights into climatic conditions during the stadial and has raised questions about the nature of the glacier dynamics which have potentially wide-ranging implications (Ballantyne, 1989a): first, deformable-bed models of glacier dynamics may apply to relatively small glaciers and ice-caps as well as to large ice-sheets, and second, where such conditions pertain, they provide additional constraints on palaeoclimatic reconstructions derived from glacier–climate relationships.

Finally, the excellent assemblage of depositional landforms, particularly the fluted and hummocky moraine, has significant potential for studies of glacial sedimentation and styles of deglaciation (cf. Benn *et al.*, 1992), for example in comparison with the assemblages of hummocky moraine types at Coire a'Cheud-chnoic, the Cairngorms and Loch Skene.

Conclusion

The Cuillin is an area of outstanding importance for glacial geomorphology. It is a classic locality for landforms of glacial erosion and has played an important part in the development of ideas in this field. The features of erosion, many among the best of their kind in Britain, are notable in spanning a range of scales, from corries and arêtes to ice-moulded and striated bedrock surfaces. The area is also of great interest for a fine assemblage of morainic landforms produced by Loch Lomond Readvance glaciers (approximately 11,000–10,000 years ago).

SCARISDALE

J. M. Gray

Highlights

The bedrock surfaces exposed on the coast at Scarisdale demonstrate the best examples in Britain of p-forms, small-scale features produced by a combination of meltwater and glacial erosion.

Introduction

The Scarisdale site covers approximately a 3 km long stretch of coast on the southern shore of

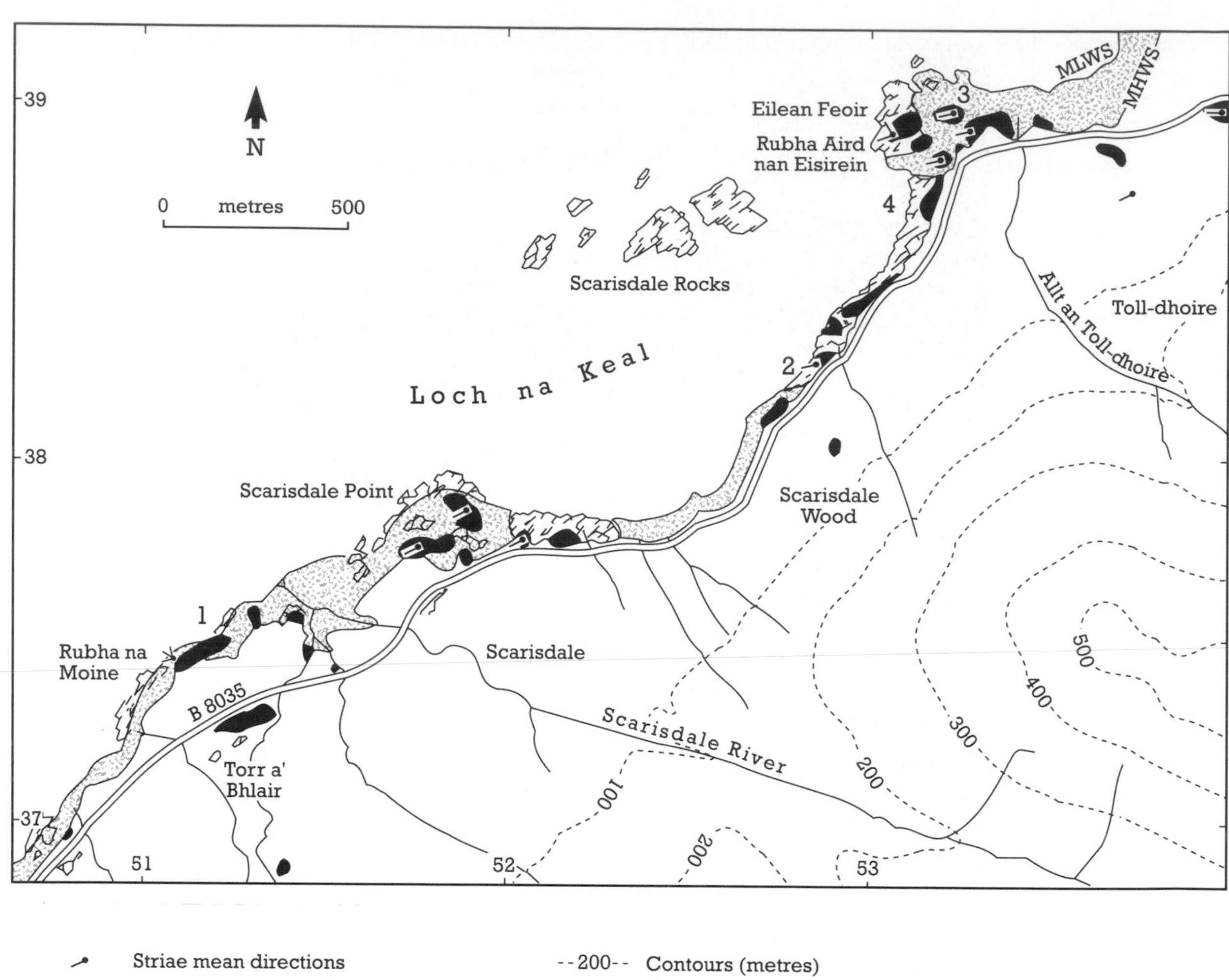

Figure 11.5 Scarisdale: localities with well-preserved p-forms (from Gray, 1981).

Loch na Keal on the Isle of Mull, between the head of the loch (NM 535389) and Rubha na Moine, about half way along the loch (NM 510372). The site does not have a long history of investigation. It was first noted by J. E. Richey in the Geological Survey Memoir for Mull (Bailey *et al.*, 1924, p.396) for a 'remarkable series of little striated hollows and winding grooves' eroded in the bedrock. Over 50 years later the site was reinvestigated by Gray (1981) who recognized the features as p-forms (cf. Dahl, 1965). He described them, mapped their distribution, and discussed their origin. He argued that no single genesis could explain all the characteristics of the features and instead suggested that they were formed by meltwater erosion, but later striated by active ice moving through them (see also Gray, 1984).

Description

The landforms are best seen between low water mark and a few metres above high water mark, probably mainly due to the absence of masking sediment, soil and vegetation. They comprise an assemblage of small-scale, smooth depressions eroded in the Palaeogene basalt and have the appearance of plastically sculptured forms (p-forms) (Dahl, 1965). The p-forms occur on

Figure 11.6 P-form channels at Scarisdale, Mull. (Photo: S. Campbell.)

flattish rock surfaces, but they have been cut irrespective of geological structure. Although p-forms occur along the entire length of the site there are areas where they are particularly common (Figure 11.5). The most impressive suites of features are at localities 1 and 2.

Although channels are by far the most abundant p-form type at Scarisdale, they are very variable in both size and morphology (Figure 11.6). The largest channel (at site 3) is about 3 m wide with steep side walls over 1 m high. It consists of a single curve about 12 m long with an undercut outer wall and smooth inner wall, both of which are covered with glacial striae that follow the curvature of the channel. At the other extreme, some channels are only 0.01 m or so deep and under 0.1 m wide. Occasionally, individual channels can be traced for over 20 m.

Channel sides vary from very gently sloping to vertical or even undercut. A particularly good example of undercutting (site 2) has a depth of 0.05 m, but the deepest undercut (site 3) is 0.1 m high. Asymmetrical cross-profiles are common, with the south (inland) slopes usually being the steeper. Most channels have rounded upper edges, although in several places sharp edges occur.

In plan, some channels are winding, others

describe single curves, while some are almost straight. Some curve around the flanks of abraded hillocks, though in some cases they run over the crests. Channels may bifurcate or join. In places overdeepened floor sections occur and sometimes facetting of rock surfaces suggests more than one phase of erosion (Gray, 1981, plate 2).

The approximate overall orientations of 142 channels reveals a clear pattern, with 96% orientated between 50° and 90°E of N. This is consistent with the trend of Loch na Keal and the main striae direction. Where a number of channels with similar orientations occur together, the rock may take on a furrowed appearance (Gray, 1981, plate 7).

Other p-form types also occur at Scarisdale. Bowls are fairly common, most being under 2 m in diameter and only a few centimetres deep. A few larger examples are also present. 'Sichelwannen' (sickle-shaped troughs) are quite rare, although two exceptionally large examples occur (at sites 1 and 4). All those discovered are concave to the west-south-west.

Interpretation

The three most favoured theories for p-form formation are (1) glacial abrasion (for example, Boulton, 1974), (2) movement of water-soaked till (Gjessing, 1965) and (3) meltwater moving under high velocities (for example, Dahl, 1965). The main characteristics of the Scarisdale channels are most successfully explained by meltwater flow, which accounts particularly well for the sinuousity, overdeepened floors, undercut lips, sharp edges, facets, and asymmetrical cross-profiles of the channels. None of the p-forms explained in the literature by glacial abrasion or till squeezing has all these characteristics. On the other hand, turbulent meltwater flow cannot account for the regular striae on the walls and floors of the channels. Thus it has been suggested that a two-stage origin is likely. First, the channels were cut by turbulent, high-velocity meltwater flow probably involving corrasion and/or cavitation associated with current vortices. Subsequently, active ice moved through the channels striating their floors and sides (Gray, 1981, 1984). This explanation had previously been proposed by J. E. Richey (*in* Bailey *et al.*, 1924) when first describing the Scarisdale site – 'the hollows are almost certainly potholes; the winding channels stream courses. Ice has been merely a modifying agent ...'. The two phases may have been closely related in time since subglacial tunnels kept open by meltwater flow would probably close if the flow diminished or shifted, allowing active ice to come into contact with the bed.

Although not accepted by all workers such ideas have gained wide acceptance in Canada, where research is leading to a realization of the importance of meltwater as a subglacial erosional agent (see Sharpe, 1987). Shaw (1988) and Sharpe and Shaw (1989) have described comparable features from Ontario and Quebec and emphasized the important role of turbulent subglacial meltwaters in their formation. They suggested that the glacier was decoupled from its bed during periodic subglacial floods, then subsequently reattached.

Other British p-form sites occur on the Isle of Islay (Gray, 1984), the Isle of Seil (J. M. Gray, unpublished data), the shore of Loch Treig (see Glen Roy and the Parallel Roads of Lochaber), and in Snowdonia (Gray and Lowe, 1982). However, the Scarisdale site represents the best assemblage of p-forms in Britain. The site is important since the characteristics of the features may be used to test the various hypotheses proposed to explain the formation of such smooth depressions. In particular, the wider importance of glacial meltwater as a subglacial erosional agent is suggested, especially in association with subglacial floods.

The significance of subglacial meltwater in understanding both subglacial erosion and glacier dynamics at both large and small scales has become increasingly apparent (for example Bindschadler, 1983; Kamb *et al.*, 1985; Drewry, 1986; Röthlisberger and Lang, 1987). Sites such as Scarisdale potentially provide important field evidence for reconstructing former subglacial drainage systems on bedrock and their hydrological characteristics (for example, see Hallet and Anderson, 1980; Sharpe and Shaw, 1989; Sharp *et al.*, 1989b). Such reconstructions would not only allow field testing of theoretical models of glacier hydrology, but would also provide valuable insights into the local dynamics of Pleistocene glaciers.

Conclusion

Scarisdale is the best example in Britain of an assemblage of small-scale features of erosion known as p-forms. These are smoothed grooves, channels and scalloping in the bedrock. The range of features present and their clarity of detail provides an unrivalled opportunity to test the different explanations proposed for their origin. The most likely of these is that they were formed by a combination of glacial meltwaters and moulding by overlying glacier ice, and therefore they may allow a reconstruction of aspects of glacier hydrology.

BEINN SHIANTAIDH
A. G. Dawson

Highlights

This site is of geomorphological interest for one of the best examples in Scotland of a fossil rock glacier formed at the base of a talus accumulation. It is believed to have been active during the Loch Lomond Stadial and provides information about slope processes and environmental conditions at that time.

Introduction

The site (NR 521749) is located on the island of Jura at the foot of the eastern slopes of Beinn Shiantaidh, one of the Paps of Jura. It is notable for one of the most spectacular fossil rock glaciers in Scotland. The only detailed account of the feature is given in Dawson (1977).

Description

The rock glacier consists of a lobate accumulation of poorly-sorted quartzite debris and has an area of 0.045 km^2, the maximum width along the foot of the hill being 380 m and the maximum length 180 m (Figure 11.7). It is located between 355 m and 400 m OD on the margin of the exposed col that separates Beinn Shiantaidh (755 m OD) from its neighbouring summit Corra Bheinn (569 m OD). The constituent boulders, many of which exceed 0.5 m in diameter, are arranged in a nested series of arcuate ridges and depressions which terminates in a sharply defined frontal margin. On the eastern margin of Beinn Shiantaidh, above the mass of debris, a talus of angular quartzite blocks rises by as much as 200 m towards the mountain summit.

The front edge of the rock glacier is represented by a ridge of unvegetated, angular boulders and slopes at about 20° towards the col surface. To the north, the continuity of the front margin is interrupted by numerous, small, transverse boulder hollows which are generally less than 1.5 m deep and 3 m wide. To the south, the outer rim becomes progressively more subdued. Here, there occurs a distinct outer ridge, the crest of which stands 5 m above the col surface. An arcuate depression flanks the inner edge of the ridge, but is replaced farther north by shallow hollows within a higher front ridge that stands 20 m above the col floor. The inner margin of this ridge descends to a semicircular depression that follows the ridge for most of its length. The radius of curvature of the ridge is 85 m and it represents the largest such feature upon the debris surface. Both ridge ends are overlain by taluses that slope consistently upwards at 35° towards the mountain summit.

Perhaps the most notable feature of the debris accumulation is the deep semicircular depression along the inner margin of the outermost ridge. The central area of the depression lies 6 m below the ridge crest and abuts an area of extremely large boulders which rises abruptly above the hollow at a gradient of 20–25°. The boulders, most of which exceed 0.5 m in diameter, comprise an upper surface slope which, measured from the base of the talus to the frontal ridge crest, is generally 10–16°.

Interpretation

Active rock glaciers are composed of coarse debris that is moved downslope by deformation of internal ice. Many are elongated and tongue-like in plan form; others are small and arcuate with low length-to-width ratios (Wahrhaftig and Cox, 1959; Barsch, 1969). The latter type is widely regarded as forming through the deformation of internal ice lenses or ice-rich frozen sediment and is unrelated to glaciers. The Beinn Shiantaidh feature is of this latter type (cf. Wahrhaftig and Cox, 1959; Outcalt and Benedict 1965; Lindner and Marks, 1985) and may be described as a protalus lobe (Martin and Whalley, 1987; Whalley and Martin, 1992).

During the formation and decay of the rock glacier, the persistence of snowbeds at the foot of the talus may have resulted in the accumulation

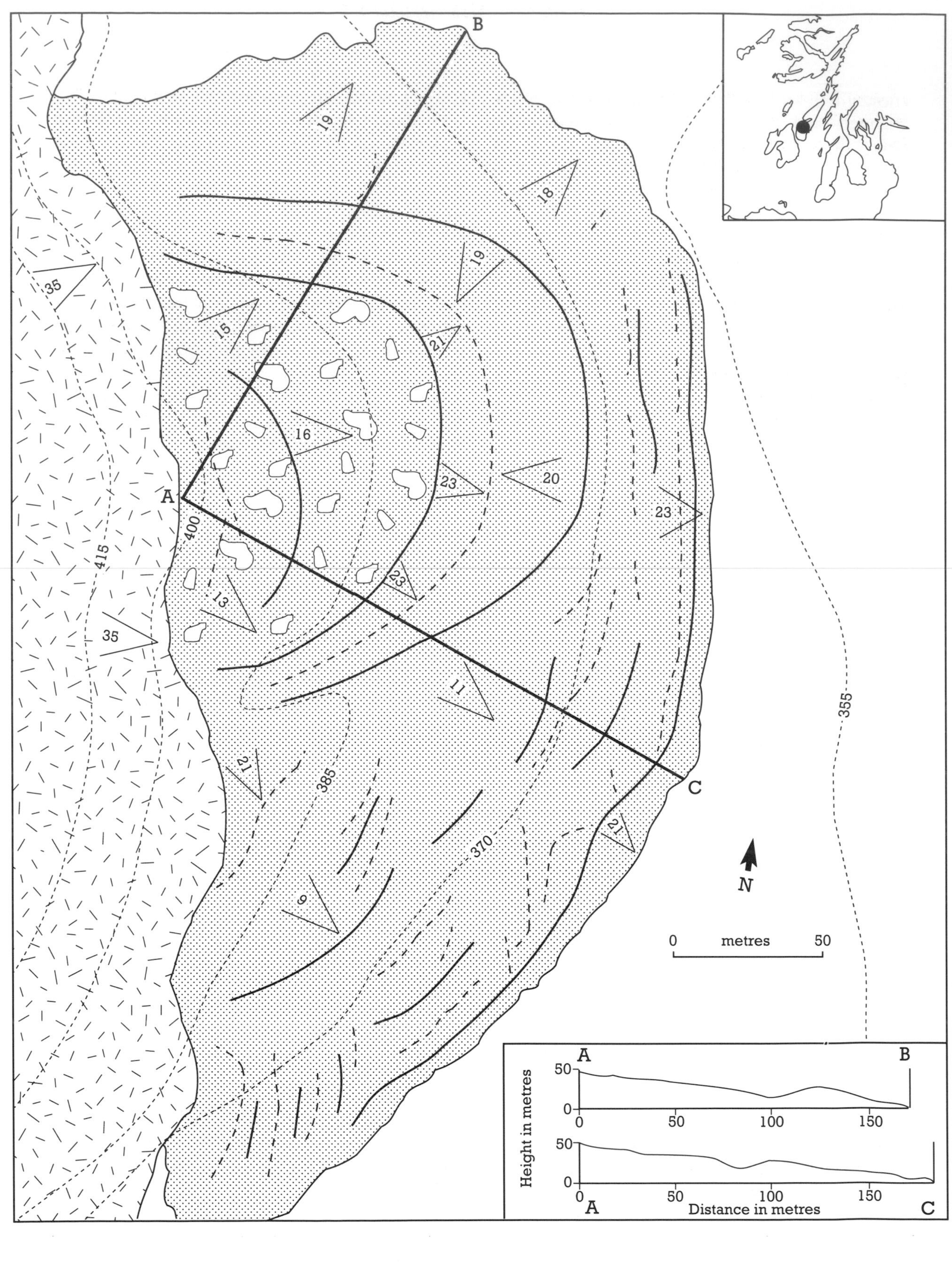

A
B
C
35
35
19
18
19
15
21
16
23
20
23
23
13
11
21
21
9
400
415
385
370
355
N
0
metres
50
Height in metres
50
0
0
50
100
150
Distance in metres

Talus
Rock debris
Ridge
Depression
Large boulders
21
Slope angle in degrees
400
Contours (metres OD)
Section line

of small protalus ramparts that were incorporated within the rock glacier. Indeed, in front of the talus slope that flanks the high north-facing buttress of Beinn Shiantaidh there is an arcuate ridge, 50 m long and composed of angular boulders, that was interpreted by Dawson (1977) as a fossil protalus rampart that formed contemporaneously with the rock glacier.

The east-north-east aspect of the fossil rock glacier would appear to have favoured the accumulation and persistence of snow and ice. Its development may have been assisted by the presence of permafrost which is thought to have last occurred in Scotland during the Loch Lomond Stadial (Sissons, 1974c). Although permafrost may not have been essential if the debris cover was sufficient to insulate the internal snow and ice, it is nevertheless reasonable to infer that the feature was formed (or at least last active) at that time.

Dawson (1977) estimated that the average rate of debris supply to the rock glacier from the cliffs upslope during the stadial was about 185 m^3a^{-1}, assuming a maximum duration of 1000 years for the period of formation. He also calculated that the average cliff retreat rate behind the rock glacier was approximately 9 mma^{-1}. This inferred cliff retreat rate is much larger than cliff retreat rates that can be inferred for other locations in Scotland during the Loch Lomond Stadial (Ballantyne and Kirkbride, 1987).

It is not known with complete certainty, however, that the material comprising the Bheinn Shiantaidh rock glacier was produced entirely during the cold climate of the Loch Lomond Stadial. For example, it is entirely possible (although not proven) that talus production on the slopes of Bheinn Shiantaidh may have commenced during ice-sheet deglaciation (cf. Chattopadhyay, 1984; Wilson, 1990a, 1990b). Under such circumstances the inferred rates of cliff retreat and talus production would be significantly lower than stated above.

The Bheinn Shiantaidh fossil rock glacier is one of the few such landforms in Great Britain and it represents a classic example of the lobate type (compare with Beinn Alligin). The almost complete lack of vegetation cover makes the detailed surface morphology resulting from former debris flowage particularly clear and impressive. The rock glacier is also potentially of palaeoclimatic significance, possibly indicating the former presence of permafrost in the mountains of the Inner Hebrides, probably during the Loch Lomond Stadial. If this is confirmed, it corroborates other lines of evidence of permanently frozen ground at low levels in western Scotland (Sissons, 1974c, 1976b). Similar landforms occur on granite in the Cairngorms (Sissons, 1979f; Chattopadhyay, 1984; Maclean, 1991), but at much higher altitudes (*c.* 800–1000 m OD). Although smaller, the Beinn Shiantaidh rock glacier also resembles the excellent examples on quartzite at altitudes of 150–400 m OD on Errigal Mountain and Muckish Mountain in County Donegal in north-west Ireland (Wilson, 1990a, 1990b). Together, such sites provide an opportunity to assess the relative roles of climate and debris supply factors in rock glacier (protalus lobe) formation.

Figure 11.7 The Beinn Shiantaidh rock glacier (from Dawson, 1977).

Conclusion

Beinn Shiantaidh provides one of the best examples in Scotland of a 'rock glacier' formed at the foot of a scree slope. This lobate landform developed through the slow deformation of ice that formed within the scree during the cold climatic conditions of the Loch Lomond Stadial (about 11,000–10,000 years ago). The feature is important in demonstrating geomorphological processes during the stadial, and its presence may support the suggestion that permanently frozen ground existed at relatively low altitudes at that time in western Scotland.

WESTERN HILLS OF RUM

C. K. Ballantyne

Highlights

This upland site is important for an assemblage of periglacial landforms developed on different rock types in an exposed maritime environment and includes both active and fossil features.

Introduction

The Western Hills of Rum, Sròn an t-Saighdeir (NM 323989), Orval (NM 334991) and Ard Nev

(NM 346986), do not exceed 571 m OD in altitude and occupy an area of only 5 km^2. However, they contain a remarkable assemblage of periglacial landforms, including some of the few examples of large-scale patterned ground known on Scottish mountains (Clark, 1962; Godard, 1965; Ryder, 1968, 1975; Ryder and McCann, 1971; Ballantyne and Wain-Hobson, 1980; Ballantyne, 1984). As the Late Devensian ice-sheet apparently overrode these hills (Harker, 1908; Charlesworth, 1956; Clark, 1962; Peacock, 1976), all of the features present post-date its downwastage, and the formation of large-scale periglacial forms can be attributed to the operation of periglacial processes during ice-sheet decay or the Loch Lomond Stadial. During the latter, two small glaciers occupied corries on the north side of these hills (Peacock, 1976; Ballantyne and Wain-Hobson, 1980). Small-scale periglacial forms including sorted stripes and circles are active at present.

Description

The Western Hills are underlain by acid igneous rocks and basalts and have broad, rounded outlines. Lithology has been of paramount importance in determining the nature of the frost-weathered regolith on the high ground. The microgranite of Sròn an t-Saighdeir has yielded the openwork, clast-supported, 'Type 1' regolith of Ballantyne (1981, 1984), whereas the basalt of Orval and, to some extent, the granophyre of Ard Nev have weathered to produce matrix-supported, 'Type 3' regolith. This latter type has a sufficient proportion of silt and fine sand to make the detritus frost-susceptible, so that periglacial features dependent upon ice-segregation for their formation occur on the last two hills.

Relict periglacial features are best represented on the Type 1 regolith of Sròn an t-Saighdeir, which is almost entirely covered by an openwork blockfield of large angular boulders (Figure 11.8). The blockslopes that surround the summit are partly vegetation-covered; they descend westwards to an altitude of only 270 m OD, where they terminate at sea cliffs. The lack of bedrock outcrops testifies to the susceptibility of the well-jointed microgranite to large-scale frost wedging, as does the remarkable cover of frost-shattered rocks in the corrie north of Sròn an t-Saighdeir. Former solifuction on these slopes has resulted in the formation of terraces and lobes of large boulders (Ballantyne and Wain-Hobson, 1980). On the summit plateau the blockfield detritus has been frost-sorted into circles 2–3 m in diameter and, on gentle slopes, into stripes of similar width. Sorted features of this size are generally regarded as indicative of permafrost conditions (Williams, 1975; Goldthwait, 1976).

The basalt and granophyre regoliths on Orval and Ard Nev support a completely different suite of periglacial features. On these hills, frost weathering has produced fine as well as coarse material and, where wind has stripped the vegetation cover, active sorted circles and polygons up to 0.5 m in diameter have developed under present conditions (Figure 11.8) (Ryder, 1975; Ballantyne and Wain-Hobson, 1980). Active sorted stripes 0.2 m in width are found on nearby slopes. Although the boulder sheets and lobes that occupy the gentler slopes around these hills are apparently inactive, the presence of 'ploughing' boulders (Figure 11.8) indicates that limited solifuction is still taking place.

Interpretation

There is a notable contrast between the debris-mantled slopes and blockslopes that fringe much of the Western Hills and the bedrock slopes partly covered by active talus within the limits of the two Loch Lomond Readvance glaciers in the corries on the northern face of Sròn an t-Saighdeir (Ballantyne and Wain-Hobson, 1980). This implies that the production of almost all of the frost-weathered debris pre-dates the Holocene. It is possible that much of the debris was formed during the decay of the Late Devensian ice-sheet, as the Rum hills may have been deglaciated at a time when the climate was still severe (see Sissons, 1983c; Sutherland, 1984a). However, the final morphology of the relict periglacial features developed on such debris probably reflects cryogenic activity during the Loch Lomond Stadial (Sissons, 1976b, 1983b).

Kotarba (1984) has also noted that major slope processes have been relatively inactive during the Holocene and that slow mass movements have been dominant on the slopes of the western Rum hills. According to Kotarba (1987) this contrasts with the situation in the Cairngorms, where high-magnitude processes have been more common.

The Western Hills of Rum support one of the most varied assemblages of fossil and active periglacial features of any Scottish mountain. The

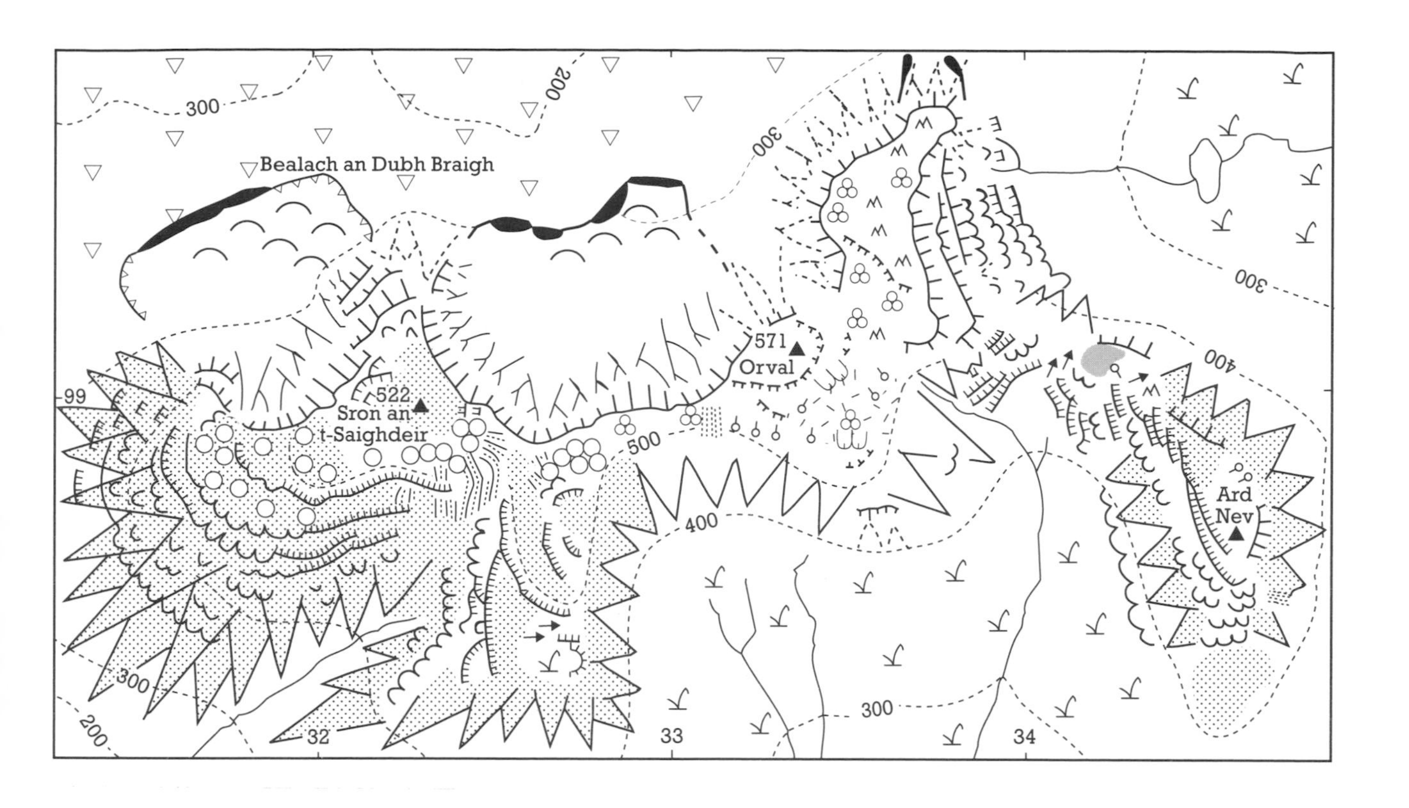

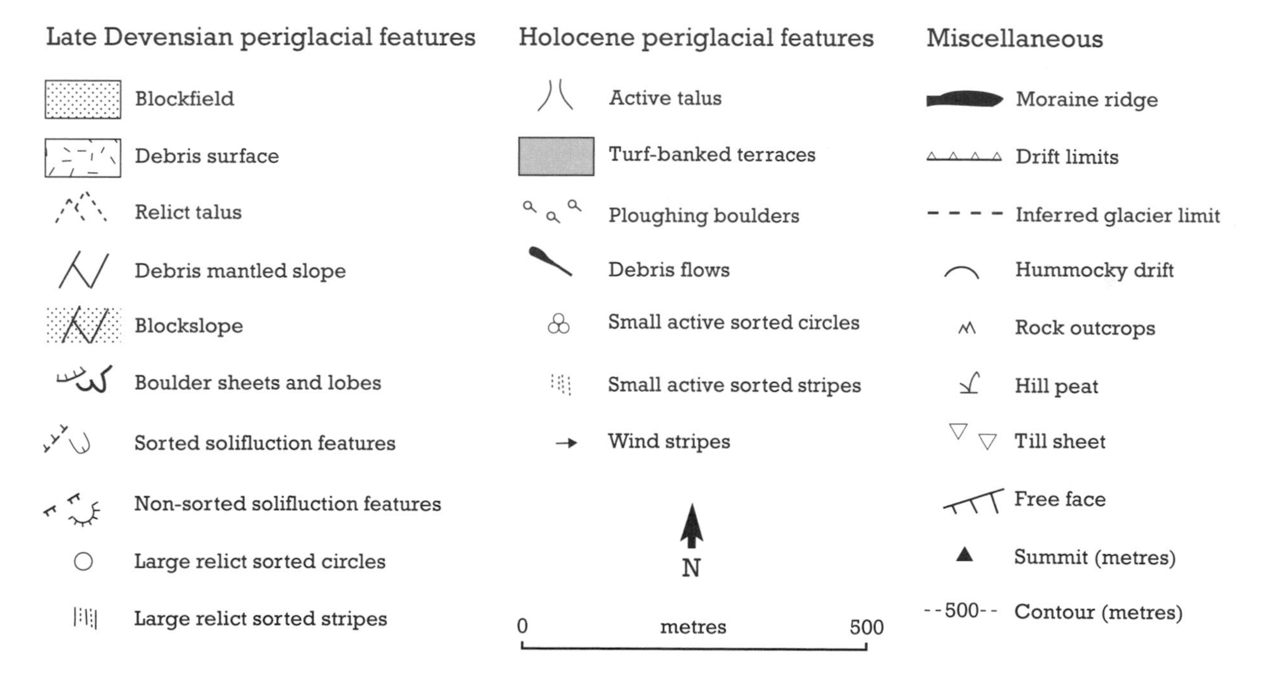

Figure 11.8 Periglacial features on the Western Hills of Rum (from Ballantyne, 1984).

types of landforms occur at apparently low altitudes by comparison with similar features on the mountains on the mainland (see An Teallach, Ben Wyvis and the Cairngorms) but, as Ballantyne (1984, 1987a) has demonstrated, this is part of a general pattern in the decline in altitude westwards of periglacial features across the Scottish Highlands and Islands. The reasons for such a decline relate to both past and present climatic variation and the limits of former glaciers, and emphasize the role of a network of national sites in understanding the genesis of periglacial landforms. In this context, the Western Hills of Rum are a particularly valuable site by virtue of their location as the most westerly site selected for periglacial features.

Conclusion

The Western Hills of Rum are important for periglacial geomorphology. In particular, the contrasting rock types on the different mountains illustrate clearly the importance of the lithology of the bedrock in controlling the characteristics and appearance of frost-weathered debris. The varied assemblage of features, for which the site is particularly noted, ranges from fossil blockfields and large stone circles (formed at the end of the last ice age), to actively forming stone circles and stripes. The landforms of the Western Hills are also representative of past and present periglacial conditions in the far west of Scotland.

NORTHERN ISLAY

A. G. Dawson

Highlights

Northern Islay is outstanding for its assemblage of fossil shoreline landforms, particularly shore platforms, and overlying glacial deposits. Most notable is the High Rock Platform, unparalleled elsewhere in its degree of development.

Introduction

This site, extending 7 km along the coastline of northern Islay between Rubha a'Mhàil (NR 428789) and Port Domhnuill Chruinn (NR 367769), is one of the classic areas of raised shoreline landscape in the British Isles. Photographs of this coastline have been published in several major texts (for example, Johnson, 1919; Sissons, 1967a), and the Quaternary geomorphology of this area has been extensively discussed (Wright, 1911; McCann, 1961a, 1964, 1968; Synge and Stephens, 1966; Dawson, 1979a, 1982, 1983a, 1983b).

The coastal zone of northern Islay is dominated by a spectacular, high, coastal rock platform and cliff (the High Rock Platform). In addition, two other shore platforms are present, the Low Rock Platform and the Main Rock Platform. The High Rock Platform was first described by Wright (1911) and later by McCann (1961a, 1964, 1968) who provided detailed descriptions of the raised shoreline features and Quaternary stratigraphy of this area. McCann (1964) also described a large end moraine that rests upon the High Rock Platform at Coir' Odhar (NR 400783). This feature was reinterpreted by Synge and Stephens (1966). Subsequent investigations by Dawson (1979a, 1982) have demonstrated a complex history of Quaternary sea-level changes and glacial events for this area.

The High Rock Platform

Description

In northern Islay a high rock platform eroded in Dalradian quartzite is almost continuous between Lòn na Cnuasachd (NR 405787) and west of Mala Bholsa (NR 378777) (Figure 11.9). East of Mala Bholsa the platform is spectacularly developed, having a maximum width of 650 m and backed by a cliff up to 60 m in height. (Figure 11.10). Along the entire length of the coastline, the cliff backing the platform is a degraded feature and is characterized by vegetated talus and slumped or soliflucted till, which blanket the rock face of the cliff and obscure the platform inner edge. The platform declines gently in altitude seaward at around 4° and its surface is free of stacks. Its front edge forms the backing cliff of a broad intertidal rock platform (the Low Rock Platform).

Between Aonan Port an-t-Sruthain (NR 385781) and Aonan na Mala (NR 375776) several exposures reveal accumulations of till that rest on the platform surface and which are, in turn, overlain by raised beach gravels. There, the distribution of the high raised beach gravels is limited to the seaward areas of the platform surface, generally below 27 m OD. Landward of these beach gravels the platform surface is overlain by till, and farther east along the coast the platform is overlain by the Coir' Odhar moraine (Figure 11.9).

Owing to the presence of drift deposits on the platform, its inner edge is only visible at six locations along stream channels and on the sides of geos. The altitudes (from 32.1 to 34.9 m OD) measured at these six locations indicate only minor variations in platform altitude and are similar to the values (32.1 m and 34.1 m OD) obtained for the west coast of Jura High Rock

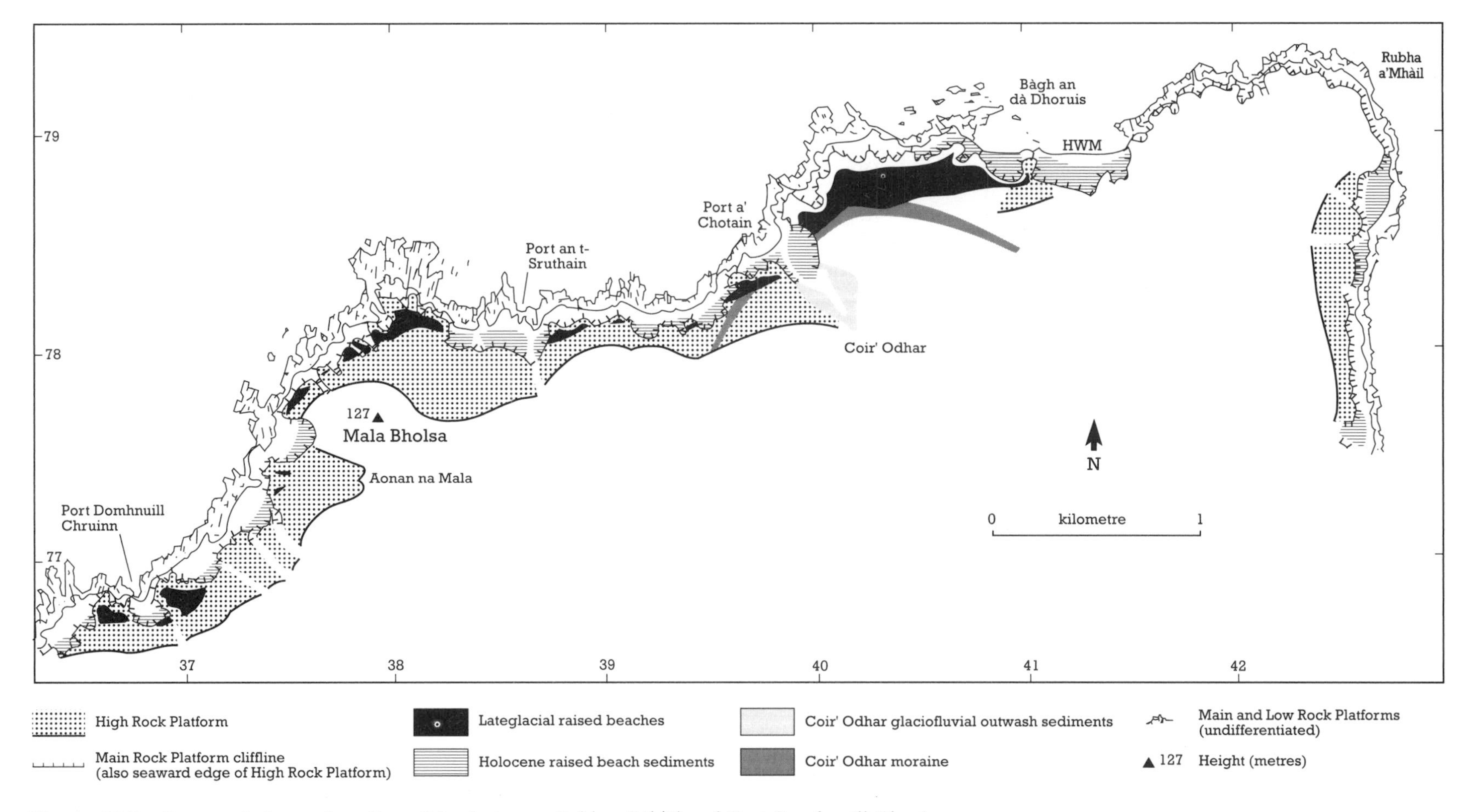

Figure 11.9 Geomorphology of northern Islay between Rubha a'Mhàil and Port Domhnuill Chruinn.

Figure 11.10 The coast of northern Islay, south of Rubha a'Mhàil, showing the High Rock Platform and its backing cliff. In the foreground the Main Rock Platform and its backing cliff are also clearly developed. (Photo: J. E. Gordon.)

Platform. One slightly higher altitude (35.4 m OD) has also been measured for the surface of the platform elsewhere in northern Islay. The similar nature of the Islay and Jura platforms, and their close proximity to each other implies that they are part of the same shoreline (the High Rock Platform).

Interpretation

Wright (1911) first described the High Rock Platform of northern Islay and discussed its age and origin. He considered that the till-covered feature was 'preglacial' in age and that it had been affected by subsequent tectonic activity. A similar view was expressed by McCann (1968, p. 24) although he proposed that the feature was 'interglacial' since glacial erosion '... must surely have resulted in more than the trifling amount of surface modifications of the platform ...'. McCann, however, considered that the 'till' described by Wright as overlying the platform was in fact soliflucted material.

The origin and age of the High Rock Platform, however, remain problematic, and depend largely on how the shoreline altitude data are interpreted (see also west coast of Jura). Sissons (1982b) suggested that the High Rock Platform in western Scotland represents a series of isostatically tilted shorelines produced during the last and previous glacials by frost action and wave action. However, the measured altitudes of the inner edge of the platform, as described above, appear to indicate that the platform is not glacio-isostatically tilted. Instead the pattern of measured altitudes indicates a generally horizontal platform surface that may be slightly warped.

Dawson (1983a) estimated that the widest High Rock Platform fragment in northern Islay would have required about 28,000 years of rapid periglacial shore erosion for its formation. The occurrence of such prolonged coastal erosion during a single period of cold climate is unlikely because of glacio-isostatic instability of the land surface and glacio-eustatic changes in sea level. It would therefore appear that the platform in northern Islay represents the product of several periods of shore erosion during the Pleistocene.

The Low and Main Rock Platforms

Description

On the foreshore beneath Rhuvaal lighthouse (NR 426792), two distinct rock platforms have been reported (Dawson, 1979a, 1980a). Both platforms occur in the intertidal zone but are markedly different not only in width, but also in morphology. The lower is the more conspicuous and forms an almost continuous feature along the northern Islay coastline. This platform is generally 100 m wide and in the Coir' Odhar embayment reaches a maximum width of almost 300 m. With the exception of the area near Rhuvaal lighthouse this platform is terminated landwards by quartzite cliffs, generally 30–35 m in elevation. In addition, its smooth ice-moulded surface and its considerable width strongly suggest that it forms part of the similar feature described in south-west Jura.

At Rhuvaal, however, the inner edge of the lower platform is separated from the main cliff by a second shore platform, which is 20–25 m wide. Between the two platforms is a 1–2 m high cliff. Unlike the lower platform, the surface of the higher platform (the Main Rock Platform) is characterized by protruding angular and inclined quartzite ridges. This platform can be traced intermittently for a considerable distance along the northern Islay coastline.

Interpretation

The fact that the lower set of platform fragments has been ice-moulded demonstrates that they were produced prior to the last glaciation. This platform, first noted by Wright (1911) as a '... preglacial plain of marine denudation ...' has been termed the Low Rock Platform by Dawson (1979a, 1980a). He noted that its presence as an ice-moulded intertidal feature, along many parts of the Scottish coastline, implied that it is interglacial in origin and unaffected by glacio-isostatic tilting. An alternative explanation was proposed by Sissons (1981a) who argued that the glaciated intertidal features represented a set of platform fragments of different ages that have been subject to glacio-isostatic deformation and which has been exhumed in the intertidal zone as a result of present marine activity. According to this hypothesis, these rock platform features were initially produced by cold-climate shore erosion processes. The higher platform fragments are considered part of the glacio-isostatically tilted Main Rock Platform (see Isle of Lismore), regarded as having been produced during the cold climate of the Loch Lomond Stadial (Dawson, 1979a, 1980b) (but see Isle of Lismore). This shoreline, owing to its glacio-isostatic deformation, is generally considered to pass below sea level west of Mala Bholsa (Dawson, 1980b).

The Coir' Odhar Moraine

Description

The Coir' Odhar moraine forms two distinct north-west-facing arcuate ridges which are separated by a small embayment 200 m wide (Figure 11.9). On both sides of the embayment the ridges rest on the High Rock Platform.

Exposures in the eastern ridge (at NR 400785) reveal angular quartzite blocks embedded in a matrix of stiff, orange clay. The deposits, together with the morphology of the feature, indicate clearly that it is a moraine. Rounded raised beach cobbles mantle the outer edge of the ridge and demonstrate that McCann's (1964) view that these gravels are incorporated within the moraine is invalid.

On both sides of the embayment the outer margin of the moraine is cliffed and the cliff forms the inner edge of a distinct raised shoreline at 26–27 m OD. This shoreline forms the marine limit in the area and is a clear feature along considerable stretches of the northern Islay coastline.

Inland of the moraine, the backing cliffs at the head of the embayment are composed of stratified sands and gravels which have been deeply incised by several streams. The surface of the stratified deposits descends seaward from over 42 m to 26–27 m OD, with an average gradient of 40 m km^{-1}; they have been interpreted as outwash deposits formed as the ice retreated from the moraine (Dawson, 1979a).

Interpretation

McCann (1964) first described a terminal moraine in northern Islay resting on the High Rock Platform and concluded that it represented the outer margin of a valley glacier that flowed seaward from a corrie located farther inland (McCann, 1964. p. 5). He considered that, since raised beach deposits were apparently incorporated within the moraine, a readvance of ice had occurred in northern Islay that was contemporaneous with the Highland (Loch Lomond) Readvance identified by Charlesworth (1956). McCann (1964, p. 5) stated that '... the outer face of the morainic ridge at 77 ft (23.5 m) above high water mark is unmodified by marine erosion showing that the sea must have fallen below this level before the onset of the readvance of the ice'.

In direct contrast, Synge and Stephens (1966, p. 107–8) concluded that the moraine was '... one of a series of drift ridges deposited by the general glaciation on this coast ... the seaward edge of this 'moraine' is the erosion scarp, or cliff, of the Lateglacial marine limit. An accumulation of rounded beach gravels occurs at the foot of this small cliff at 90–99 ft (29.3–30.2 m) ... (the) marine limit along this stretch of coast is uniform in height, and uninterrupted by any later glacial phase'.

Dawson (1979a) agreed with Synge and Stephens and established that the outwash deposits, formed as ice retreated from the moraine, were graded to the raised shoreline at an altitude of 26–27 m OD. The Coir' Odhar moraine was therefore formed during the retreat of the main Late Devensian ice-sheet and was not of Loch Lomond Readvance age as proposed by McCann (1964). The raised shoreline at *c.* 26–27 m OD at Coir' Odhar was considered by Dawson (1982) to be contemporaneous with the Central Islay moraine: no other moraine systems have been correlated with the Coir' Odhar moraine.

Following the first account of the High Rock Platform of northern Islay by Wright (1911), Johnson's (1919) description of the feature in his standard textbook, *Shore Processes and Shoreline Development*, focused international attention on this coastal zone as a superb example of a raised shoreline landscape (cf. Dawson, 1991). Stratigraphic studies and accurate levelling have since demonstrated the complexity of Late Quaternary sea-level changes (including glacio-isostatic shoreline deformation) and glaciation history in this area (Dawson, 1979a, 1982), and although a considerable amount of detailed information is now available, the origins and ages of certain geomorphological features still remain controversial. Thus the origin and age of the High Rock Platform are the subject of considerable disagreement (see Sissons, 1982b; Dawson, 1983a, 1984). Moreover, Sissons (1981a) and Dawson (1980a) maintain opposing views on the nature of the Low Rock Platform. Finally, the Coir' Odhar moraine, although now firmly established as having been produced during deglaciation of the last ice-sheet, appears to be related to a glacial episode for which there is only limited evidence in western Scotland. To a large extent, the detailed discussions of these features have arisen from their superb development in the Northern Islay landscape. The area is thus of outstanding scientific interest both for the classic development of shore platforms and for the associated geomorphological and stratigraphic evidence, which together provide an important record of Late Quaternary sea-level changes and glacier fluctuations in western Scotland.

Conclusion

Northern Islay is outstanding for Quaternary coastal geomorphology, displaying some of the finest examples of isostatically uplifted raised shoreline features in Europe. These have been raised to their present levels above the sea by the uplift that followed the depression of the Earth's crust by the weight of ice-sheets during the ice ages. It is a classic locality for raised shore platforms, most notably the High Rock Platform of western Scotland. The superb development of raised coastal terraces, together with the presence of a moraine formed by the last ice-sheet (approximately 15,000 years ago), makes the area quite unique in Scotland.

WEST COAST OF JURA

A. G. Dawson

Highlights

The coastal area of western Jura contains a remarkable assemblage of raised shoreline landforms. These include shore platforms and the best-developed spreads of Lateglacial shingle ridges in Britain which provide valuable information for understanding changes in relative sea level. The area is also noted for a medial moraine formed by the Late Devensian ice-sheet.

Introduction

This site comprises a *c.* 37 km long stretch of the west coast of Jura, between Glengarrisdale Bay

(NR 659985) in the north, and Rubha Aoineadh an Reithe (NR 448751) in the south, and also a small area (*c.* 0.2 km^2) at Inver (NR 442724). It is one of the classic localities in Great Britain for raised coastal landforms, most notably spectacular unvegetated spreads of Late Devensian and Holocene raised beach shingle. The area also includes excellent examples of three raised shore platforms, the High Rock Platform, Main Rock Platform and Low Rock Platform. Also represented in Sgriob na Caillich (NR 475765) is the finest example of a medial moraine in Great Britain.

There is only a limited amount of published information on the Quaternary features of this area. The raised shingle spreads were first described by officers of the Geological Survey (Wilkinson, 1900, 1907; Peach *et al.*, 1911), and Ting (1936, 1937) gave further details of the ridges. The first major study of the raised beaches was by McCann (1961a, 1964), who sought to describe and explain the origin of the western Jura shingle spreads and relate them to patterns of Lateglacial relative sea-level change. In a later paper, McCann (1968) extended his discussion to include the raised coastal rock platforms. More recently, the raised shorelines of western Jura have been investigated in detail by Dawson (1979a). The results of this research are published in a number of later papers (Dawson, 1980a, 1980b, 1982, 1983a, 1984, 1988b, 1991).

High Rock Platform

Description

A high rock platform and associated cliffline are almost continuous between Shian Bay (NR 530875) and Ruantallain (NR 505833). In this area the platform has an average width of 350 m and in places is as wide as 600 m; the backing cliffs are typically 5–15 m high although they reach a maximum of 50 m at Loch an Aoinidh Dhuibh (Figure 11.11). The inner edge of the platform is only visible in two stream sections and hence the altitude of the feature is only known for these locations: north of Shian Bay (NR 53858915) at an altitude of 34.1 m OD and 150 m north of Bhrein Port (NR 50948415) at 32.1 m OD. Along this stretch of coastline the platform possesses an average seaward slope of about 4°, its gently sloping surface having provided an environment favourable for the later deposition of the overlying Late Devensian shingle spreads.

The western Jura platform pre-dates one period of general glaciation, since at two locations it is separated from the overlying raised beach sediments by lodgement till. First, at Bhrein Port (NR 50688405) 1 km north of Ruantallain, a wedge of orange lodgement till is embedded in a platform depression between two inclined ridges of quartzite and is overlain by thick accumulations of shingle. Second, on the banks of a stream channel 150 m north of Bhrein Port (NR 50948415) the inner edge of the platform is choked by 2.5 m of creamy lodgement till beneath shingle. Additional evidence for glaciation is the occurrence of an ice-moulded and striated bedrock sea stack immediately north of Loch a'Mhile (NR 51398501). It is also possible that the exceptionally low altitude of the platform surface at Shian Bay may be due to the effects of glacial erosion.

The rock platform described above is considered to correlate with the classic High Rock Platform of northern Islay (Dawson, 1979a). The High Rock Platform has not been found, however, along the intervening south-west Jura coast probably due to the presence in that area of great thicknesses of till.

Interpretation

The High Rock Platform of the southern Inner Hebrides (see also Northern Islay) was first described by Wright (1911). Although he did not describe the Jura platform, Wright considered that the platform was 'pre-glacial' in age. Later, McCann (1968) provided the first description of the western Jura platform and suggested instead that it was 'interglacial'. Dawson (1979a, p. 161) accepted an interglacial origin for the western Jura (and also the northern Islay) platform but considered that the shoreline had been warped by neotectonic activity. A contrary view was expressed by Sissons (1982b), who proposed that the platforms that comprise the High Rock Platform were produced by cold-climate shore erosional processes and that the features exhibit glacio-isostatic tilting. On that view, the various platform fragments are part of a series of tilted shorelines. However, the available altitudes from the west coast of Jura and northern Islay do not demonstrate any tilt to the platform, which is

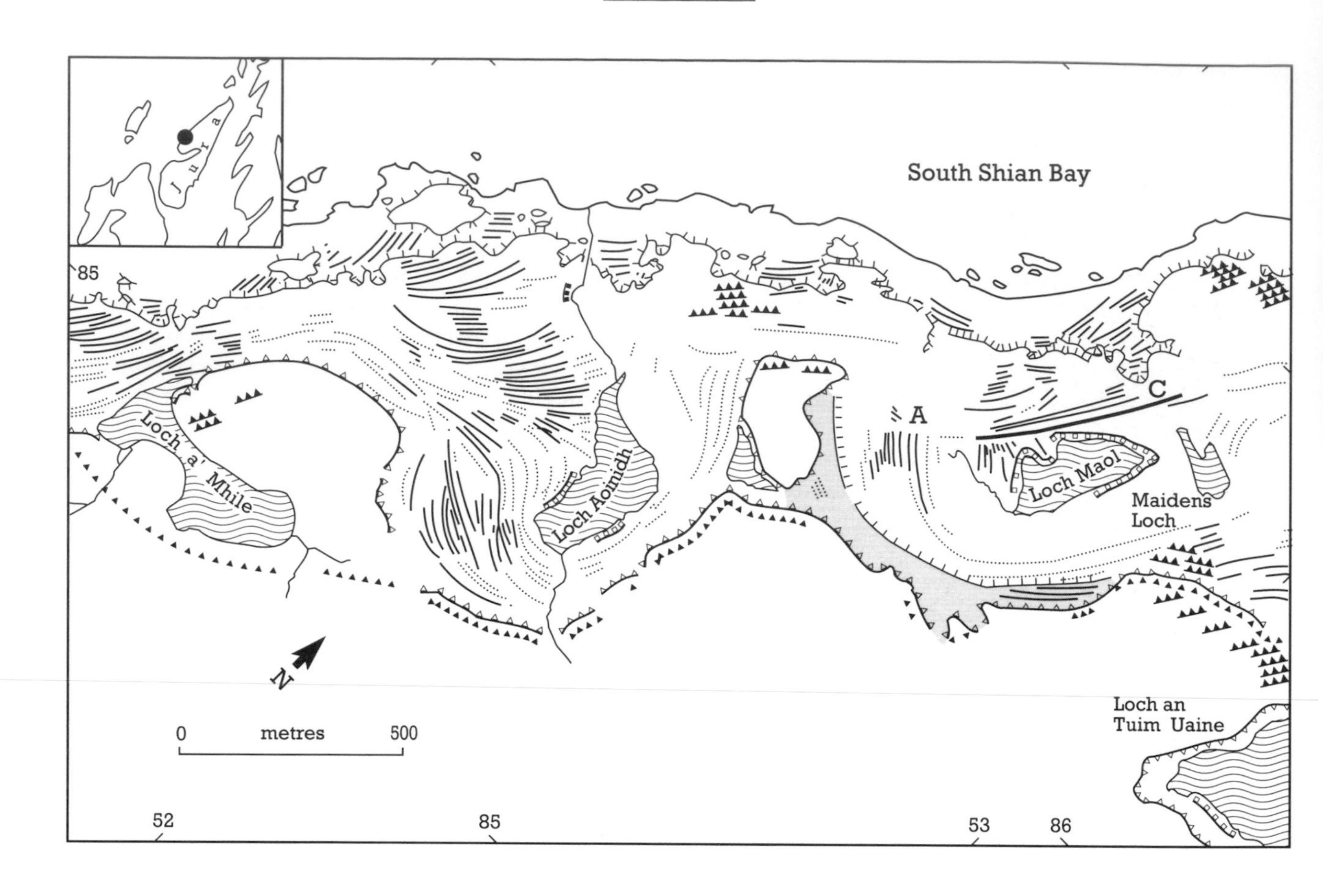

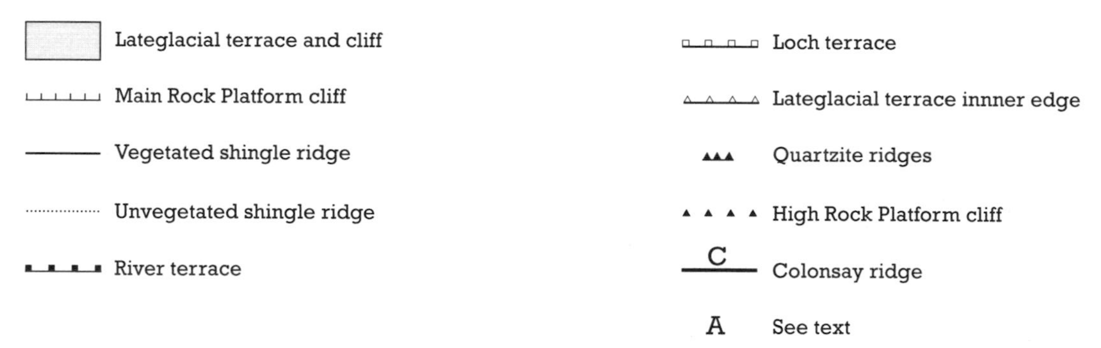

Figure 11.11 Geomorphology of western Jura in the area of South Shian Bay.

therefore regarded as a single feature. Thus there is at present no general agreement on their origin and age. Dawson (1983a) has argued on several grounds that formation of the western Jura platform by cold-climate shore erosion would have taken a minimum of 8000 years. Relative sea-level stability of such duration during a single period of cold climate is unlikely due to glacio-isostatic instability of the land surface and glacio-eustatic changes in sea level. It would therefore appear that the western Jura High Rock Platform represents the product of several periods of Pleistocene coastal erosion.

Main Rock Platform

Description

Between Shian Bay and Ruantallain the seaward edge of the High Rock Platform forms the cliffline of a lower platform 50–150 m wide (Figures 11.11 and 11.12). The inner edge of this platform occurs at 3–5 m OD. It is locally overlain by Holocene raised beach sediments and, along its length, the cliffline is indented by numerous raised sea caves. The platform is unglaciated and is characterized by inclined and jagged quartzite

Figure 11.12 View south along the west coast of Jura between Shian Bay and Ruantallain. Lateglacial shingle ridges extend across a high rock platform and to the west of Loch a'Mhile (centre). The loch was formerly a marine inlet prior to the deposition of the shingle ridges. Note also a prominent rock platform and backing cliff (the Main Rock Platform) seaward of the high shingle ridge 'staircases'. Holocene shingle ridges also cover the Main Rock Platform. (Photo: John Dewar Studios.)

ridges; the cliffs are typically crenulate and are usually 10–15 m high. The platform is also continuous between Shian Bay and Glendebadel Bay (NR 622951), where it is locally associated with cliffs up to 100 m high.

Interpretation

The platform constitutes part of a glacio-isostatically tilted shoreline that declines in altitude to the south-west, from 6 m OD in northern Jura to sea level in northern Islay (Dawson, 1980b). The shoreline gradient is $0.13\,\mathrm{m\,km^{-1}}$ and this, together with its rock-cut nature, general altitude and freshness of form, indicates correlation with the Main Rock Platform (Gray, 1974a, 1978a). McCann (1968) suggested that this feature exhibited evidence of glaciation, although Dawson

(1980b) did not report such evidence, and considered that McCann's evidence for glaciation related to the Low Rock Platform (see below). The origin and age of the Main Rock Platform in western Scotland have been discussed at great length in numerous publications (for example, McCann, 1968; Sissons, 1974d; Gray, 1978a, 1989; Dawson, 1979a, 1980b, 1983a, 1988a, 1989; Sutherland, 1984a). However, the occurrence of the regionally tilted Main Rock Platform in western Jura is of particular significance since it is in this area (and also northern Islay) that this platform merges with and crosses the regionally horizontal intertidal Low Rock Platform (Dawson, 1979a, 1980a) (see below).

Low Rock Platform

Description

In south-west Jura, low intertidal rock platform fragments are conspicuous along long stretches of coast. These are typically 100 m wide and are best developed on the foreshore between Rubh'-Aird na Sgitheich (NR 476793) and Allt Bun an Eas (NR 458763). At several locations, the continuity of the platform surfaces is interrupted by numerous Tertiary dolerite dykes. The platform surfaces are locally ice-moulded; throughout most of the area the platforms pass inland beneath till. Between Rubh'Aird na Sgitheich and Glenbatrick (NR 518801) the platform is overlain by considerable thicknesses (up to 15 m) of Late Devensian raised beach gravels.

Interpretation

Intertidal ice-moulded rock platforms also occur in north-west Jura, northern Islay and neighbouring Colonsay and were first described by Wright (1911) as representing a '... preglacial plain of marine denudation ...'. Dawson (1980a) referred to the Jura and Islay feature as the Low Rock Platform and explained its regional horizontality as having been produced by marine processes during interglacials. The shoreline is also of considerable significance since the glaciated platform fragments of south-west Jura, northern Islay and Colonsay were considered by McCann (1968) to demonstrate glaciation of the Main Rock Platform.

Late Devensian raised beaches

Description

The coastal zone of western Jura is dominated by conspicuous raised beach terraces and 'staircases' of unvegetated beach ridges, the widespread occurrence of which first attracted the attention of the Geological Survey (Wilkinson, 1900, 1907). Although discussed by Ting (1936, 1937), the most detailed studies of these raised coastal features are by McCann (1964, 1968) and Dawson (1979a, 1982).

In western Jura, raised coastal terraces can be traced almost continuously southward from Shian Bay as far as Inver (cf. Dawson, 1991, figure 5). Additional areas of raised coastal terrace occur at Corpach Bay (NR 568917) and Glendebadel. In most areas the raised marine deposits are ridges of unvegetated quartzite shingle (Figure 11.12). In western Jura the highest coastal terraces decline in altitude from north-east to south-west, from 40 m OD at Corpach Bay to 24.5 m OD at Inver. The raised beach terraces and shingle spreads were produced in association with the deglaciation of the Late Devensian ice-sheet in western Scotland and their altitudes reflect the effect of subsequent glacio-isostatic uplift.

Interpretation

Analysis of the regional altitude variations of the highest raised beach terraces on the western Jura coast suggest the existence of two shorelines. The older of these (shoreline L1) declines in altitude to the south-west, from 40 m OD at Corpach Bay to 34 m OD at Bàgh Gleann Righ Mór, 1.5 km east of Ruantallain. This raised shoreline is also thought to occur in northern Islay; it has a regional gradient of 0.56 m km^{-1} (Dawson, 1982). A separate and slightly younger shoreline (L2) is considered to be present in south-west Jura. This shoreline declines in altitude to the south-west, from 31 m OD at Glenbatrick to 24 m OD at Inver and has a regional gradient of 0.53 m km^{-1}. It was therefore inferred by Dawson (1979a, 1982) that south-west Jura remained ice-covered while shoreline L1 was formed between Corpach Bay and Shian Bay, and also on the northern Islay coastline. Dawson (1979a, 1982) also concluded that, owing to the drop in the marine limit between Corpach Bay and Glendebadel, north-west Jura was also ice-covered during this period. Deglaciation of south-west Jura took place at a slightly later date and was accompanied by the formation of shoreline L2.

The pattern of ridge-crest altitudes exhibited in the western Jura shingle 'staircases' indicates that, although stillstands may have occurred during the fall in the sea level from the marine limit, there were no major sea-level oscillations as relative sea level fell from near 35 m to 20 m OD. Most of the western Jura shingle spreads below this altitude terminate at the cliffline of the Main Rock Platform and consequently patterns of sea-level change lower than 20 m OD cannot be established. One exception, however, occurs at South Shian Bay, where owing to the exceptionally low altitude of the High Rock Platform, raised beach gravels descend to almost 11 m OD (Figure 11.11:A). McCann's (1964) proposal that a major sea-level oscillation is represented in this area by the Colonsay Ridge (a shingle spit) (Figure 11.11) may be correct and, if so, suggests that a pause in the overall fall in sea level occurred when relative sea level at South Shian Bay was at about 19 m OD (see Dawson, 1983a) (Figure 11.11). The other exception is at Lochan Maol an t-Sornaich (NR 547805) where a relative marine transgression is suggested, the sea rising from around 9 m to 14 m OD.

The presence on the west coast of Jura of extensive spreads of Late Devensian raised shingle is primarily due to the glacio-isostatic uplift of shoreline L1 and its altitudinal relationship with the till-covered High Rock Platform. Thus upon deglaciation, the maximum level of the sea along this stretch of coast (34–40 m OD) stood several metres higher than the inner edge of the High Rock Platform. Marine erosion of the till cover during the ensuing fall in relative sea level resulted in extensive shingle deposition. This process would have undoubtedly been promoted by the gentle seaward slope of the underlying platform surface and also by the exposure of the coastal zone to the effects of Atlantic waves.

The Sgriob na Caillich medial moraine

Description

On the western side of the Paps of Jura, there occurs perhaps the finest example of a fossil medial moraine in the British Isles (Dawson, 1979b). It is 3.5 km long and trends approximately NW–SE. It originates at 450 m OD at the western foot of Beinn an Oir (NR 498750) and descends gently to an altitude of 330 m OD before passing over the rock outcrop of Cnoc na Sgrioba (360 m OD). Seaward of this ridge, the boulder belts of the moraine lie on top of a thick till cover, until at 30 m OD, they are truncated by a low cliff and raised coastal platform, both of which are cut in till (Figure 11.13). At the junction of the till platform and the boulder belts is a small lochan (Loch na Sgrioba), impounded by a suite of raised shingle ridges which mantle the platform.

The boulder complex is composed in places of up to four parallel lines of angular blocks, each line rarely exceeding 27 m in width and 2.5 m in vertical thickness. The boulders in the belts, almost entirely of Dalradian quartzite (the local bedrock) though occasionally of slate and phyllite, bear no evidence of striation or ice moulding. They range from 0.2 m to 1.3 m in length and contrast markedly with the generally smaller quartzite blocks that are found in local till exposures. Additionally, the mean diameter of boulders measured at 500 m intervals along the feature decreases seaward by 0.07 m km^{-1}.

For most of its length the junction between each belt and the vegetation cover exhibits little variation in relief, though in places it is characterized by small boulder 'cliffs' up to 2 m in height. The main boulder belts are oriented parallel to each other; they are separate units which rarely merge. Coalescence of the belts is limited to the crest and flanks of Cnoc na Sgrioba, where the entire orientation of the feature changes slightly.

Interpretation

Dawson (1979b) argued that the medial moraine was produced during the waning of the Late Devensian ice-sheet. He suggested that the boulders of the moraine were deposited supraglacially from the Beinn an Oir nunatak on to a relatively thin, yet dynamically active, ice mass. Its preservation in the landscape as a series of unvegetated quartzite boulder belts is remarkable. Other medial moraines have been described from former Loch Lomond Readvance glaciers (see for example, Beinn Alligin). Only one similar feature, near Strollamus on Skye (Ballantyne, 1988; Benn, 1991), has been ascribed to the Late Devensian ice-sheet. No other medial moraines in Scotland

Figure 11.13 West coast of Jura. The Sgriob na Caillich medial moraine (centre) descends to the level of a high raised shoreline. (Photo: D. G. Sutherland.)

compare with Sgriob na Caillich in either size or complexity.

Holocene raised beaches

Description

Throughout western Jura, most Holocene beach accumulations mantle the rock surfaces of the Main and Low Rock Platforms. Relatively few raised coastal terraces are present and those that occur exhibit a gradual decline in altitude on Jura from near 10 m in the north-west to 8.5 m in the south-west. Most coastal areas, however, are characterized by banks of shingle and by shingle ridge 'staircases'. The most spectacular suite of Holocene shingle ridges occurs north of Inver, south-west Jura, where 31 unvegetated raised beach ridges descend from 12.3 m OD to the modern beach (Dawson, 1979a, 1991).

Interpretation

The highest ridge appears to have been produced during the culmination of the Main Postglacial Transgression. Hence the staircase of 31 ridges and intervening swales are likely to have been produced during the last 6000–7000 years, largely as a result of decreasing rates of glacio-isostatic uplift during this period.

Summary

The west coast of Jura is therefore outstanding for its assemblage of raised coastal landforms and deposits. Both the range of features and their extent and degree of development are exceptional. The interest includes not only examples of the three major rock platforms recognized in western Scotland, the High, Main and Low Rock Platforms, but also spreads of unvegetated Lateglacial and Holocene shingle beach ridges unparalleled elsewhere in Scotland for the length of their morphological record of sea-level changes. The latter features, in particular, distinguish the west coast of Jura from northern Islay (see above). Elsewhere in Scotland, there are notable sequences of raised shingle ridges at Spey Bay and Tarbat Ness on the Moray Firth coast (Ogilvie, 1923). Those at Spey Bay are comparable in their scale of development to the Jura features but occur in a different geomorphological process environment, being associated with a major river (the Spey) and significant longshore drift. Moreover, they have not been studied in comparable detail to the features on the west coast of Jura.

Conclusion

The coastline of western Jura is one of the classic localities in Britain for raised beaches (formed by isostatic uplift – see Northern Islay above). It is characterized by a variety of well-developed coastal landforms, of which the spectacular, unvegetated spreads of raised beach shingle (formed during the last 14,500 years) are without parallel in Britain and have allowed a detailed pattern of relative sea-level changes to be reconstructed. The area also includes excellent examples of raised platforms cut in bedrock as well as the finest example of a medial moraine in Britain, a ridge of boulders deposited by the last ice-sheet (approximately 15,000 years ago).

GRIBUN
M. J. C. Walker

Highlights

Pollen preserved in the sediments that infill a topographic basin at Gribun provide an unusually long and detailed record of Holocene vegetational history and environmental change. The sediments accumulated after the retreat of a Loch Lomond Readvance glacier, one of the lowest to have existed in Scotland during the Loch Lomond Stadial.

Introduction

The site (NM 450326), located at Gribun in western Mull, comprises a deep, infilled basin behind an arcuate end moraine (Bailey *et al.*, 1924; Dawson *et al.*, 1987a). The sediments in the basin have yielded a high-resolution pollen record which spans most of the Holocene and displays a level of detail seldom achieved in Scottish Holocene pollen profiles (Walker and Lowe, 1987). Hence Gribun is possibly the most important site for reconstructing the Holocene vegetational history in the Hebridean islands and adjacent areas of the west coast of Scotland.

Description

The coastal cliffs at Gribun are characterized by a series of extensive landslips resulting from failure of the Upper Cretaceous sedimentary rocks that underlie the Tertiary basalts in this area. One of the largest of the debris accumulations occurs to the south of Balmeanach Farm (NM 448329), where an impressive multiple-ridged, arcuate rampart composed of large boulders within a fine-grained matrix has developed at the foot of the Creag a'Ghaill escarpment. Deep, infilled basins are enclosed within the rampart complex. Although the arcuate ridge was originally described by the Geological Survey as '... a landslip of the completely disintegrated type, and accordingly might be claimed with some propriety as a moraine' (Bailey *et al.*, 1924, p. 414), a combination of geomorphological and sedimentary evidence confirms the view that the feature is an end moraine that formed as a consequence of glacier activity during the Loch Lomond Stadial (Walker *et al.*, 1985; Dawson *et al.*, 1987a). As such, it reflects the existence of one of the lowest glaciers in Scotland during the last cold phase, with an equilibrium line altitude of about 100 m. Of wider significance, however, is the biostratigraphical record contained within the sediments of the deep basin enclosed by the outer rampart.

Over 13 m of limnic and terrestrial sediment have accumulated in the largest basin within the morainic complex. The lowermost sediments (approximately 1.8 m) are minerogenic and consist of a generally upward-fining sequence of pebbles, grits, sands, silts and clays. Particularly distinctive is a series of over 90 silt/clay laminations which overlies the coarser basal beds. These sediments accumulated in a proglacial lake that developed behind the outer moraine following glacier recession. The lower gravels and sands are considered to reflect glaciofluvial deposition, whereas the laminated deposits are interpreted as glaciolacustrine varves. Overlying the basal minerogenic sediments are some 4 m of fine-grained gyttja (organic mud) and clay–gyttja, these limnic deposits being succeeded, in turn, by over 7 m of amorphous organic muds and peats (Figure 11.14). The pollen evidence (below) suggests that these sediments accumulated very rapidly, with rates of 0.2 m–0.3 m 100 years^{-1} being

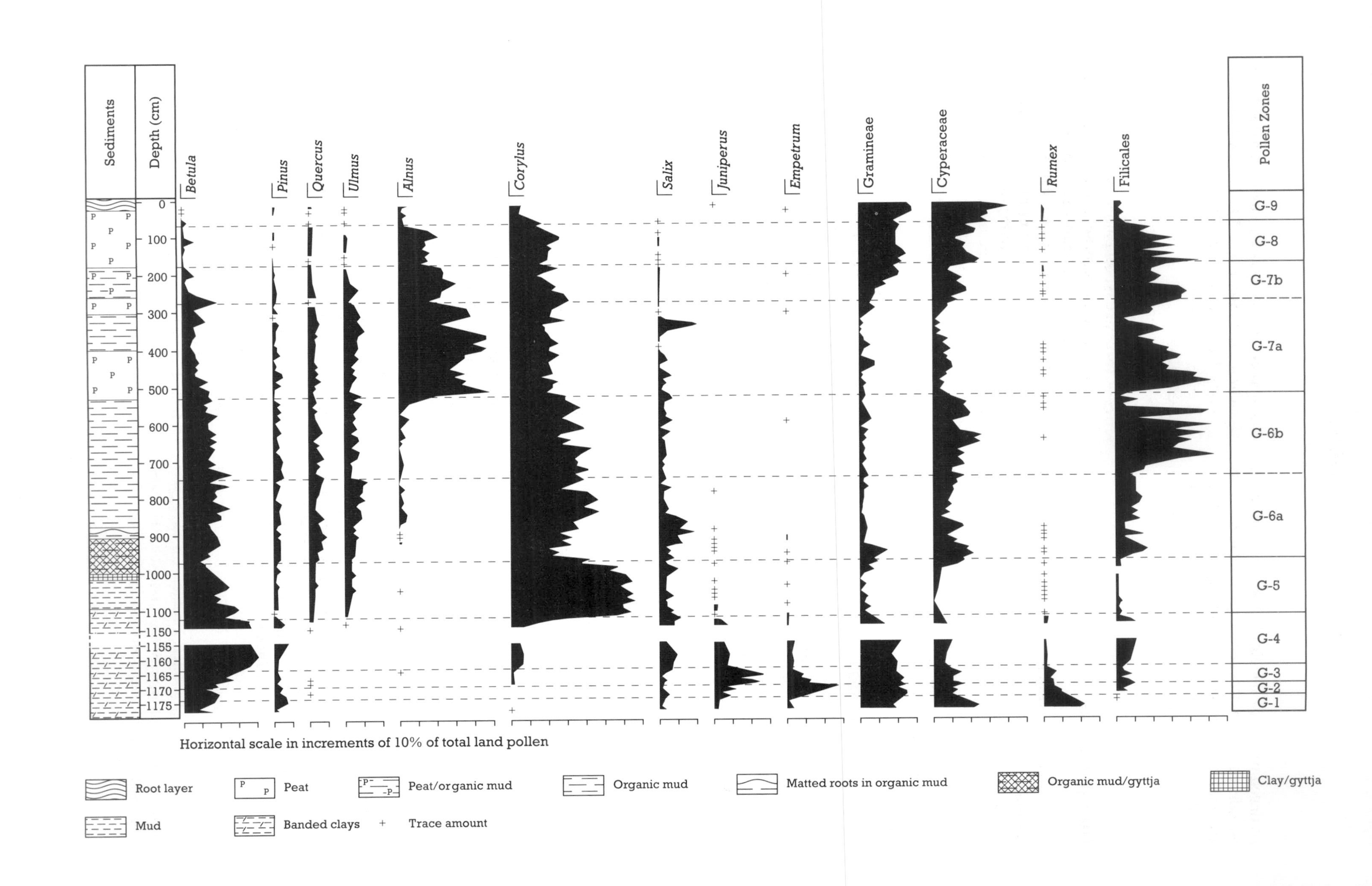
Sediments
Depth (cm)
0
100
200
300
400
500
600
700
800
900
1000
1100
1150
1155
1160
1165
1170
1175
Betula
Pinus
Quercus
Ulmus
Alnus
Corylus
Salix
Juniperus
Empetrum
Gramineae
Cyperaceae
Rumex
Filicales
Pollen Zones
G-9
G-8
G-7b
G-7a
G-6b
G-6a
G-5
G-4
G-3
G-2
G-1
Horizontal scale in increments of 10% of total land pollen
Root layer
Peat
Peat/organic mud
Organic mud
Matted roots in organic mud
Organic mud/gyttja
Clay/gyttja
Mud
Banded clays
Trace amount

recorded for the middle Holocene; and even during the early Holocene, organic sediment was accumulating at about 0.04 m 100 years^{-1}. These are significantly higher rates of deposition than are usually encountered in Holocene pollen sites in Scotland.

Interpretation

The minerogenic sediments at the base of the Gribun profile contained too little pollen for counting, but nine local pollen assemblage zones were identified in the overlying organic deposits (Figure 11.14). This sequence of pollen assemblage zones is broadly similar to that recorded at other Holocene sites on Mull (Walker and Lowe, 1985; Lowe and Walker, 1986b), although the early and late Holocene is more fully represented in the sediments from Gribun. On the wider scale, the sequence is comparable with Holocene pollen records from other parts of Scotland, including the north-west Highlands (see Cam Loch and Loch Sionascaig) (Pennington *et al.*, 1972), the Western Isles (Bennett *et al.*, 1990), Skye (see Loch Ashik) (Williams, 1977; Birks and Williams, 1983), Ardnamurchan (Moore, 1977), Argyll (see Loch Cill an Aonghais) (Rymer, 1974; Tipping, 1984) and the Rannoch Moor area of the Grampian Highlands (see Kingshouse) (Walker and Lowe, 1977, 1979, 1981). In terms of vegetational development, three distinct stages can be recognized.

1. An early Holocene succession in which open-habitat herbaceous vegetation was succeeded by heathland and ultimately by a landscape of trees and shrubs. This is reflected in the Gribun profile by a basal pollen assemblage zone (G–1) dominated by pollen of herbaceous plants including Gramineae, Cyperaceae, *Rumex* and *Artemisia*, along with spores of the clubmosses *Lycopodium selago* and *L. annotinum*. This initial pollen assemblage zone is succeeded by zones dominated by *Empetrum* with some *Salix*, *Juniperus* and *Betula* (G–2); *Juniperus*, *Salix* and *Betula* (G–3); *Betula* and *Salix* with a rising *Corylus* curve (G–4); and *Corylus* with *Betula* and *Salix* (G–5). No radiocarbon dates have been obtained from the Gribun profile, but there are profiles dated to the early Holocene available from other sites on Mull (Walker and Lowe, 1982). On the not unreasonable assumption that the early Holocene local pollen assemblage zones from the various profiles broadly correlate, then the earliest organic sediments at Gribun may be inferred to have accumulated prior to 10,200 BP, the *Empetrum* maximum may be dated to close to 10,000 BP, the phase of *Juniperus* expansion occurred around 9600 BP, the birch episode can be dated to near 9300 BP, and the *Corylus* rise began around 8800 BP.
2. A phase of middle Holocene woodland expansion and diversification following the establishment of *Corylus*, with *Quercus*, *Ulmus*, *Pinus* and *Alnus* forming the dominant elements. The first consistent pollen records for oak and elm are recorded in pollen zone G–5, but these taxa are better represented (along with *Pinus*) in pollen zone G–6. *Alnus* also appears during that zone and dominates the spectra along with *Corylus*/*Myrica* in pollen zone G–7. Radiocarbon dates from sites in western Scotland suggest that the *Alnus* expansion occurred around 6500 BP (Birks, 1972b; Pennington *et al.*, 1972; Williams, 1977). The generally low frequencies of arboreal taxa in the Gribun profile are found not only at other sites on Mull, but also in records from elsewhere in the Hebrides (Flenley and Pearson, 1967; Vasari and Vasari, 1968; Birks, 1973; Williams, 1977; Birks and Madsen, 1979; Birks and Williams, 1983), suggesting that only a scattering of oak, elm and pine woods developed on the islands of western Scotland even at the 'climatic optimum' of the Holocene. This reflects, above all, the effects of exposure to strong westerly winds, although generally thin soils which became rapidly leached during the course of the Holocene may also have significantly reduced tree vigour (Pennington *et al.*, 1972; Birks, 1975).
3. A period of woodland contraction and the replacement of woodland stands and tall-shrub-dominated communities by heathland and grassland. This phase is represented in the Gribun profile in pollen zones G–8 and G–9, throughout which there is a progressive

Figure 11.14 Gribun: relative pollen diagram showing selected taxa as percentages of total land pollen (from Walker and Lowe, 1987).

increase in pollen of Gramineae, Cyperaceae and other herbaceous taxa including *Potentilla*, *Rumex* and *Plantago*, and a marked reduction in woody plant pollen, a phenomenon which becomes particularly apparent in pollen zone G–9. Although these landscape changes largely reflect natural processes, namely progressive soil deterioration through accelerated leaching and increasingly stormy conditions along the western littoral, anthropogenic activity may be partly responsible for some of the inferred vegetational changes. There is abundant evidence on Mull to suggest a long history of human occupation (Morrison, 1980; Royal Commission, 1980), and hence the pollen changes that are apparent throughout zones G–8 and G–9 of the Gribun profile may reflect not only the decline of woodland stands through natural processes, but also the acceleration of that trend as a consequence of anthropogenic activity from the Neolithic period onwards.

Gribun is therefore a pollen site of major importance in the context of the Holocene in Scotland, for few profiles combine a length of stratigraphic record with such a high level of detail. The broad similarity between the local pollen assemblage zones in the Gribun diagram, and those from elsewhere on Mull and other sites in western Scotland suggests that the Gribun pollen assemblage zones have wider application and that they can constitute a basis for regional correlation of Holocene deposits. On the broader scale, the high degree of stratigraphic resolution in the Gribun profile and the relative scarcity of detailed pollen records from the islands of the Hebrides and nearby areas of the Scottish mainland, indicate that Gribun is possibly the most important site for the central stretch of the west coast of Scotland.

Conclusion

The sediments preserved in the infilled basin at Gribun provide an exceptionally detailed record of vegetational change during the Holocene (the last 10,000 years). Analysis of pollen contained in the sediments shows the development of open-habitat vegetation, heathland and trees during the early Holocene, a phase of woodland expansion during middle Holocene times and the subsequent contraction of the woodland and its replacement by heath and grassland. The pollen record from Gribun is particularly important because of its length and detail. Consequently Gribun is a key reference site for studies of vegetational history in the islands and the adjacent mainland of the central part of the west coast of Scotland.

LOCH AN T-SUIDHE

M. J. C. Walker

Highlights

The sediments that infill the floor of this loch contain a valuable pollen record, supported by radiocarbon dating, of vegetational history and environmental change on Mull during the Lateglacial.

Introduction

Loch an t-Suidhe (NM 370215) is a small lochan at an altitude of 30 m OD, approximately 1 km west-south-west of Bunessan on the Ross of Mull. Although a large number of pollen sites have now been investigated on the Scottish mainland (Walker, 1984b), until recently relatively little information was available about the Lateglacial on the islands of the Inner and Outer Hebrides. The publication (Lowe and Walker, 1986a) of two radiocarbon-dated pollen diagrams from sites on the Isle of Mull was therefore of significance in the context of the Scottish Lateglacial. Of the two sites investigated, Loch an t-Suidhe offered the better stratigraphic resolution and a coherent series of six radiocarbon dates was obtained from the profile (Walker and Lowe, 1982). The wealth of palaeoenvironmental evidence contained within the sediments of this basin make it one of the most important Lateglacial pollen sites so far described from western Scotland.

Description

The sedimentary sequence (Figure 11.15) near the southern shore of the lochan clearly resembles

Figure 11.15 Loch an t-Suidhe: relative pollen diagram showing selected taxa as percentages of total land pollen (from Lowe and Walker, 1986a).

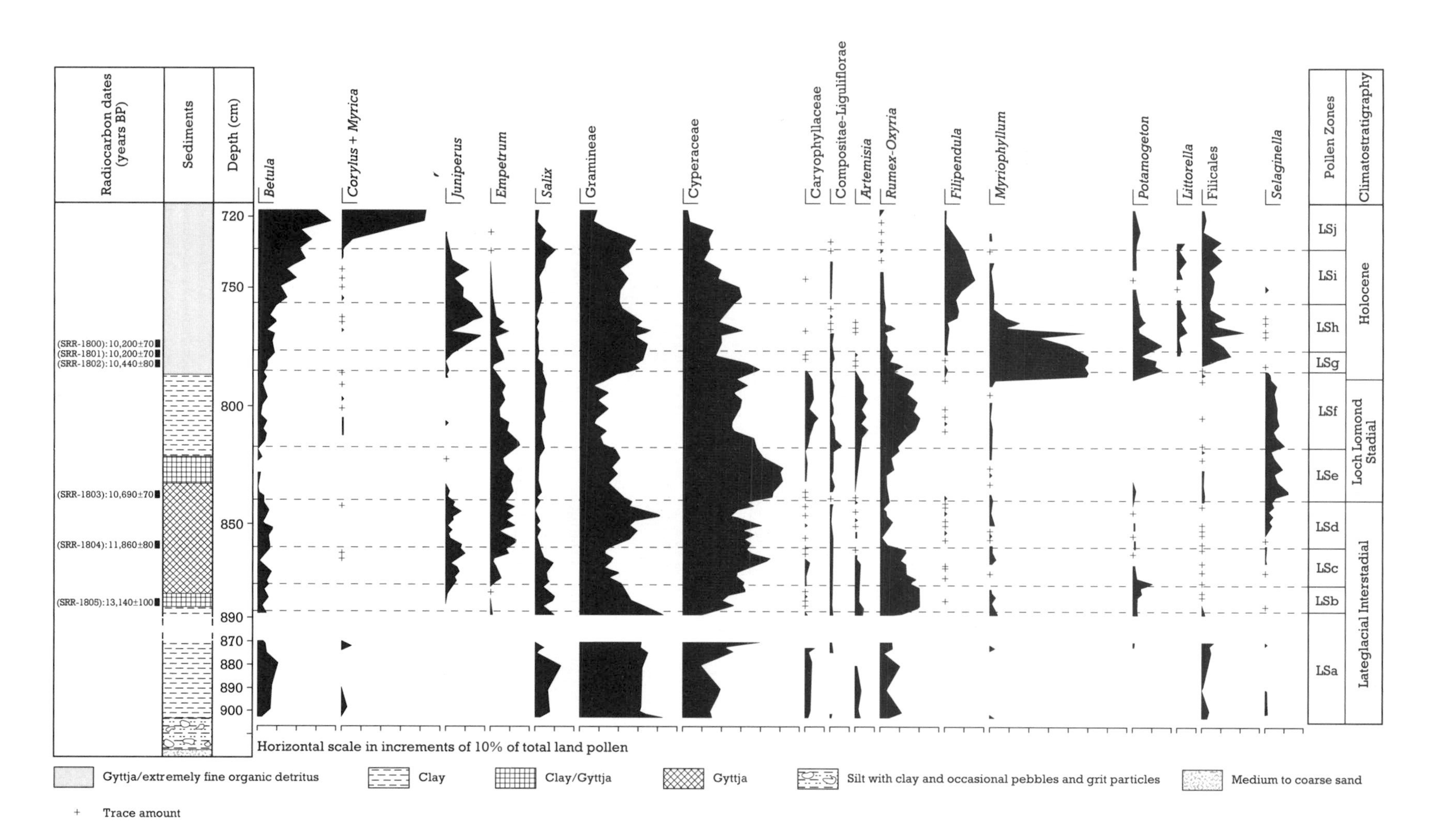
Radiocarbon dates (years BP)
Sediments
Depth (cm)
Betula
Corylus + Myrica
Juniperus
Empetrum
Salix
Gramineae
Cyperaceae
Caryophyllaceae
Compositae-Liguliflorae
Artemisia
Rumex-Oxyria
Filipendula
Myriophyllum
Potamogeton
Littorella
Filicales
Selaginella
Pollen Zones
Climatostratigraphy
(SRR-1800): 10,200±70
(SRR-1801): 10,200±70
(SRR-1802): 10,440±80
(SRR-1803): 10,690±70
(SRR-1804): 11,860±80
(SRR-1805): 13,140±100
720
750
800
850
890
870
880
890
900
LSj
LSi
LSh
LSg
LSf
LSe
LSd
LSc
LSb
LSa
Holocene
Loch Lomond Stadial
Lateglacial Interstadial
Horizontal scale in increments of 10% of total land pollen
Gyttja/extremely fine organic detritus
Clay
Clay/Gyttja
Gyttja
Silt with clay and occasional pebbles and grit particles
Medium to coarse sand
+ Trace amount

the tripartite lithostratigraphic Lateglacial succession commonly found in infilled Lateglacial lake sites in Britain (Sissons *et al.*, 1973). A lower minerogenic unit (approximately 8.8 m to base) is overlain by organic sediments (8.2–8.8 m) and these, in turn, are succeeded by a further minerogenic suite (7.9–8.2 m). The whole sequence is overlain by Holocene lake muds and peats. Details of the pollen stratigraphy of the Holocene sediments are contained in Lowe and Walker (1986b).

The pollen diagram from the Lateglacial and early Holocene deposits was divided into a series of local pollen assemblage zones based on fluctuations in the curves for the principal taxa (Figure 11.15). These show (from the base) a sequence of increasing pollen content and diversity accompanied by rising curves for woody plant pollen (zones LSa to LSd; the Lateglacial Interstadial), followed by a phase of reduced woody plant pollen and an increase in pollen from taxa indicative of bare or disturbed soils (zones LSe and LSf; the Loch Lomond Stadial). This, in turn, is followed by a series of pollen assemblage zones dominated by successive maxima in *Empetrum*, *Juniperus*, *Betula* and *Corylus* (zones LSg to LSj; the early Holocene).

Interpretation

In terms of regional landscape changes, the pollen record from Loch an t-Suidhe reflects initial vegetational colonization of freshly exposed substrates during the early Lateglacial Interstadial following the wastage of the Late Devensian ice-sheet. Subsequently, there developed a juniper scrub and *Empetrum* heath vegetation cover. This was succeeded by a marked vegetational 'revertance' phase (during pollen zones LSe and LSf) in response to the harsh climatic conditions of the Loch Lomond Stadial, when a tundra landscape developed. Finally, woody plants re-expanded as climate improved at the close of the Lateglacial and there occurred a vegetational succession from open heathland to *Betula–Corylus* woodland during the early Holocene.

Although this sequence is very similar to that recorded at Mishnish in northern Mull (Lowe and Walker, 1986a) and is comparable in broad outline with many other Lateglacial pollen successions from northern Britain (Pennington *et al.*, 1972; Gray and Lowe, 1977b; Pennington, 1977a, 1977b; Walker and Lowe, 1990), a number of features combine to make the Lateglacial and early Holocene record from Loch an t-Suidhe particularly distinctive. First, the data on deteriorated pollen provide independent evidence of episodes of increased geomorphological activity around the basin (for example, accelerated minerogenic inwash during the Loch Lomond Stadial). In addition, variations in levels of deterioration between individual pollen taxa made possible the differentiation between primary and secondary components in the pollen spectra, an aspect of pollen analysis that is becoming increasingly important in understanding the plant communities that developed during the Loch Lomond Stadial (Walker and Lowe, 1990). Second, following the work of Mackereth (1965, 1966) and Pennington *et al.* (1972), fluctuations in the curves for organic carbon and for the chemical elements Na, K, Mg and Ca, provide further evidence of the extent of mineral inwash into the basin and hence constitute an additional indirect record for landscape change around the site. Third, the fall in *Juniperus* pollen during the mid-interstadial and its replacement by *Empetrum*, in association with other evidence for disturbance in the vegetation cover, appears to reflect climatic deterioration some 1000 years before the onset of the Loch Lomond Stadial. This inference is in broad agreement with coleopteran records, which show evidence of a cooling trend, and particularly a fall in winter temperatures, from *c.* 12,500 BP onwards (Atkinson *et al.*, 1987). Fourth, systematic analysis of the record of *Artemisia* pollen reveals a change from oceanic conditions to a more continental climatic regime during the course of the Loch Lomond Stadial, which appears to be related to the southward migration of the oceanic and atmospheric polar fronts (Duplessy *et al.*, 1981; Ruddiman and McIntyre, 1981b; Bard *et al.*, 1987).

Finally, Loch an t-Suidhe is notable because of the internally consistent series of radiocarbon dates that was obtained from the profile. As with all age determinations on bulk samples of organic lake muds, however, contamination by older or younger carbon residues (Sutherland, 1980; Walker and Harkness, 1990) cannot be entirely excluded, and hence the dates must be treated with a degree of caution. The basal organic sediments were dated at 13,140 ± 100 BP (SRR–1805), an age determination which is in broad agreement with dates on comparable biostratigraphic horizons at sites on the Scottish mainland (Bishop, 1963; Kirk and Godwin, 1963; Sissons and Walker, 1974; Pennington, 1975b; Lowe and Walker, 1977; Vasari, 1977). It is also in agreement with the date inferred for the replacement of polar by warmer waters around the shores of western Britain (Duplessy *et al.*, 1981; Ruddiman and McIntyre, 1981b; Peacock and Harkness, 1990). If correct, the date from the basal sediments would support the view that climatic amelioration at the beginning of the Lateglacial Interstadial occurred around 13,000 BP (Coope, 1975; Atkinson *et al.*, 1987). The *Juniperus* decline in the interstadial was dated at 11,860 ± 80 BP (SRR–1804), an age determination that accords with the inference that climatic deterioration in western Britain began around 12,000 BP (Watts, 1977, 1985; Craig, 1978; Walker and Lowe, 1990). The onset of the Loch Lomond Stadial (as indicated by the sediment and pollen records) was dated at 10,690 ± 70 BP (SRR–1803), which is very close to the age determination (10,730 ± 60 BP; SRR–1807) on the comparable horizon at Mishnish in northern Mull (Walker and Lowe, 1982). Three dates of 10,440 ± 80 BP (SRR–1802), 10,200 ± 70 BP (SRR–1801) and 10,200 ± 70 BP (SRR–1800) were obtained from the early Holocene sediments which pre-date the expansion of *Juniperus*. These dates are in broad agreement with a number of dates from basal Holocene sediments from other Scottish sites (Walker and Lowe, 1979, 1980, 1985) and, if correct, they reinforce the suggestion (Lowe and Walker, 1976; Walker and Lowe, 1981) that climatic amelioration at the close of the Loch Lomond Stadial occurred well before 10,000 BP. It should be noted, however, that the recently discovered plateau of constant radiocarbon age around 10,000 BP, and which appears to reflect fluctuations in atmospheric radiocarbon production (Ammann and Lotter, 1989; Zbinden *et al.*, 1989), poses a major difficulty in the establishment of 'reliable' age estimates at the Lateglacial–Holocene boundary (see also Kingshouse).

Loch an t-Suidhe contains a wealth of data on the Lateglacial and early Holocene environmental history of the Isle of Mull. It forms a key element in a network of published sites from the mainland (Pennington *et al.*, 1972; Pennington, 1975b, 1977a; Rymer, 1977; Tipping, 1984) and from the Isle of Skye (Birks, 1973; Walker and Lowe, 1990) which now enable a regional picture to be established of environmental change along the western Scottish seaboard following the wastage of the last ice-sheet. However, Loch an t-Suidhe may also be important in a wider context. Although it is now generally recognized that a dominant influence on the British climate and environment has been oceanographic changes in the north-east Atlantic province (see, for example, Lowe and Walker, 1984), it has not always proved possible to establish clear links between the marine and terrestrial records, particularly over the relatively restricted timespan of the Lateglacial. The location of Loch an t-Suidhe on the maritime fringes of north-west Britain, coupled with the detailed stratigraphic evidence contained within the profile and the internally consistent radiocarbon chronology, make it a potentially valuable site for correlation between the marine and terrestrial records. In this respect, therefore, Loch an t-Suidhe may prove to be a site of both national and international significance.

Conclusion

Loch an t-Suidhe is a key site for interpreting Lateglacial environmental history in western Scotland during the Lateglacial, between about 13,000 and 10,000 years ago. The pollen record shows the pattern of vegetation colonization and development in the period after melting of the last ice-sheet (about 13,000 years ago). There is clear evidence for a subsequent return to tundra conditions during the Loch Lomond Stadial (approximately 11,000 and 10,000 years ago), followed by the development of open heathland and birch woodland during the early Holocene. Loch an t-Suidhe is particularly significant for the wealth of information it provides about this important time period, allowing detailed reconstruction of the palaeoenvironmental conditions.

LOCH ASHIK (LATEGLACIAL PROFILE)

M. J. C. Walker

Highlights

Pollen preserved in the sediments that infill the floor of Loch Ashik provide a detailed record, supported by radiocarbon dating, of the vegetational history and environmental changes on Skye during the Lateglacial.

Introduction

Loch Ashik (NG 691232) is located 4 km east of Broadford on the Isle of Skye. It lies at around 40 m OD and, at its maximum, is 175 m long and 125 m wide. In recent years the vegetational history of the island has undergone a major revision (Walker *et al.*, 1988; Walker and Lowe, 1990), and the pattern of Lateglacial and early Holocene environmental change that has emerged is more compatible with data from the Scottish mainland and other Hebridean islands than were the previously published interpretations from Skye (Erdtman, 1924, 1928; Blackburn, in Godwin, 1943; Vasari and Vasari, 1968; Birks, 1973; Vasari, 1977; Williams, 1977; Birks and Williams, 1983; Walther, 1984). A key element in this reinterpretation is the site of Loch Ashik, which contains an unequivocal Lateglacial pollen record and can now be regarded as the type Lateglacial profile for the Isle of Skye. Full details of the pollen record are contained in Walker and Lowe (1990), and a more concise description can be found in Walker and Lowe (1991).

Description

In an infilled embayment by an inflowing stream at the western end of the loch, a characteristic Lateglacial tripartite sediment sequence (cf. Sissons *et al.*, 1973) is preserved (Figure 11.16) (Walker and Lowe, 1990, 1991). This consists of basal minerogenic sediments (below 5.69 m) overlain by a unit of higher organic content (5.69–5.29 m) which, in turn, is succeeded by a further suite of minerogenic deposits (5.26–5.07 m). This Lateglacial sequence is overlain by over 5 m of Holocene lake muds and peats. Six radiocarbon dates (SRR–3116 to SRR–3121) have been obtained from the sediments (Figure 11.16).

The pollen diagram from the Lateglacial and early Holocene deposits (Figure 11.16) has been divided into eleven local pollen assemblage zones (LA–1 to LA–11) based on fluctuations in the curves for the principal taxa. These show an early Lateglacial pioneer vegetational stage dominated by open-habitat communities (LA–1 and LA–2), the expansion of woody plants, including birch and juniper (LA–3), and the subsequent establishment of *Empetrum* and *Erica* heaths (LA–5), the development of a grass-sedge tundra with some heathland stands during the Loch Lomond Stadial (LA–7), and finally the early Holocene vegetational succession from arctic–alpine communities to birch and hazel woodland (LA–8 to LA–11).

Interpretation

The sequence of vegetational changes represents a clear biotic response to climatic fluctuations at the last glacial–interglacial transition which began with rapid climatic amelioration around 13,000 BP, followed by gradual climatic deterioration during the Lateglacial Interstadial (from *c.* 12,000 BP onwards), the development of a climatic regime of arctic severity during the Loch Lomond Stadial (*c.* 11,000–10,200 BP), and a subsequent rapid rise in both winter and summer temperatures in the first five hundred years of the Holocene (Atkinson *et al.*, 1987).

Radiocarbon dating of the Loch Ashik sediments proved problematical, for although six age determinations were made on bulk samples of organic lake muds obtained from the site, the majority appear to be too old by comparison with radiocarbon dates on comparable biostratigraphic horizons from other sites in northern Britain. Indeed, only the date of 11,590 ± 160 BP (SRR–3118) on the late interstadial expansion of *Empetrum* appears to be consistent with the currently accepted radiocarbon chronology of Lateglacial biozones (Walker and Harkness, 1990). The measured ages most probably reflect the inwash of inert carbon residues into the lake basin, either in the form of mineral carbon from the local bedrock, or older organic carbon residues from soils around the lake catchment and/or from the inwashing of older carbon detritus (Olsson, 1979, 1986). Whatever the source of contamination, it is apparent that a reliable Lateglacial chronology cannot be established from the Loch Ashik sediments.

The Loch Ashik site is, nevertheless, significant in a number of respects. First, a high-resolution pollen record of the Lateglacial and early Holocene periods has been obtained from the basal sedimentary sequence, the percentage pollen counts being supported by pollen concentration, deteriorated pollen and sediment chemistry data (Walker and Lowe, 1990). Indeed the pollen

Figure 11.16 Loch Ashik: Lateglacial relative pollen diagram showing selected taxa as percentages of total land pollen (from Walker and Lowe, 1991).

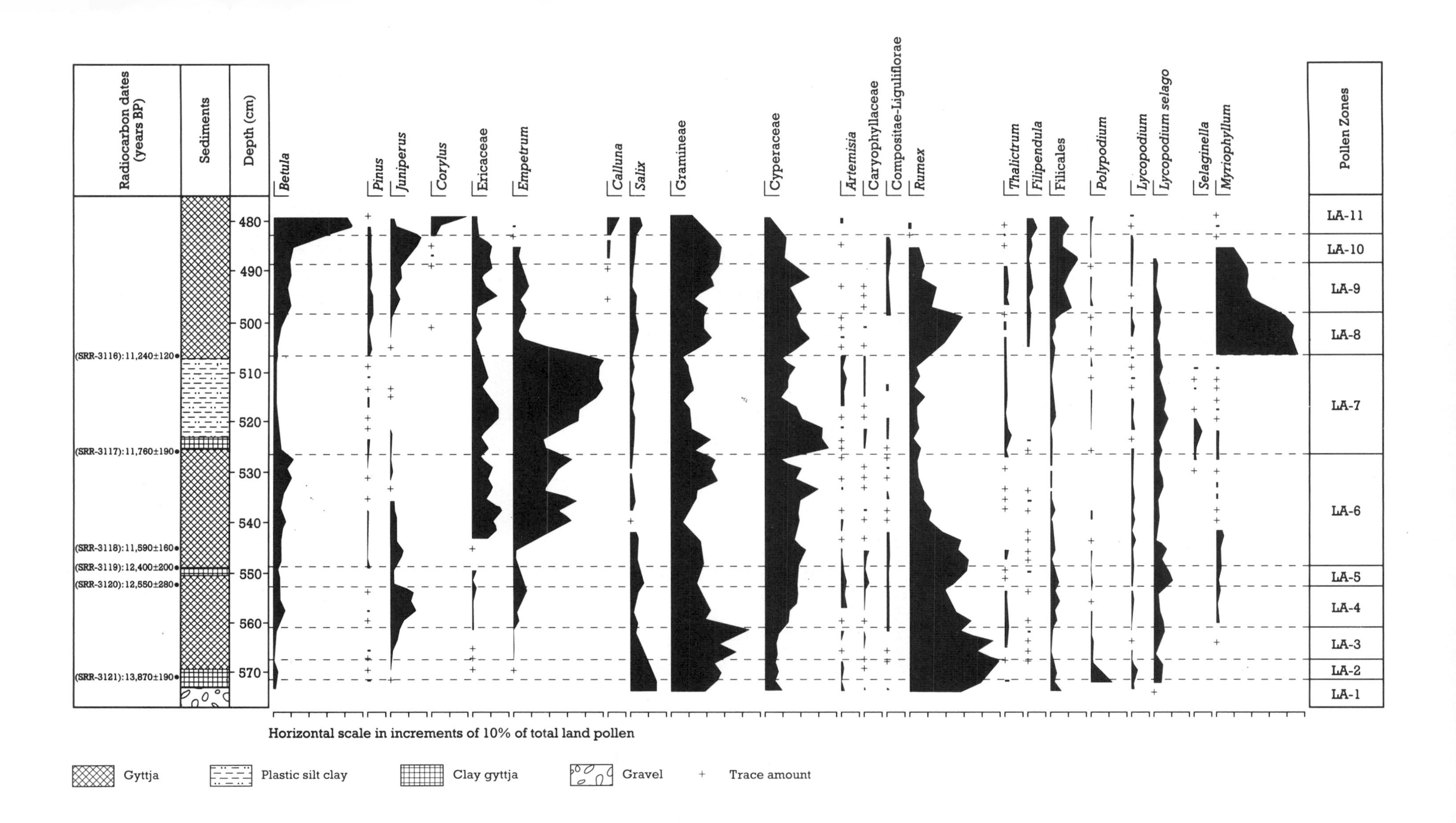

Radiocarbon dates (years BP)
Sediments
Depth (cm)
Betula
Pinus
Juniperus
Corylus
Ericaceae
Empetrum
Calluna
Salix
Gramineae
Cyperaceae
Artemisia
Caryophyllaceae
Compositae-Liguliflorae
Rumex
Thalictrum
Filipendula
Filicales
Polypodium
Lycopodium
Lycopodium selago
Selaginella
Myriophyllum
Pollen Zones
(SRR-3116):11,240±120
(SRR-3117):11,760±190
(SRR-3118):11,590±160
(SRR-3119):12,400±200
(SRR-3120):12,550±280
(SRR-3121):13,870±190
480
490
500
510
520
530
540
550
560
570
LA-11
LA-10
LA-9
LA-8
LA-7
LA-6
LA-5
LA-4
LA-3
LA-2
LA-1
Horizontal scale in increments of 10% of total land pollen
Gyttja
Plastic silt clay
Clay gyttja
Gravel
+ Trace amount

concentration diagram is the first to be published from a Lateglacial site in the Hebrides. Moreoever, the pollen record is directly comparable not only to those from sites on other Hebridean islands and adjacent areas of the Scottish mainland (Walker and Lowe, 1982, 1985, 1987; Tipping, 1984, 1986; Lowe and Walker, 1986a, 1986b; Robinson, 1987c), but also to those from other recently-investigated profiles on Skye (Walker and Lowe, 1990). It therefore demonstrates that the sequence of vegetational changes on Skye conforms with the pattern inferred from other sites in northern and western Scotland, a fact that was not apparent in previously published data from the island (Birks, 1973).

Second, the site is important for demonstrating the significance of deteriorated pollen analysis in palaeoecological reconstructions. The very high counts for *Empetrum* pollen in pollen assemblage zone LA–7 may be taken to indicate extensive heathland communities around the basin catchment during the Loch Lomond Stadial, and certainly the rising curve for this taxon is indicative of some local *Empetrum* presence. However, deteriorated pollen counts show that the majority of the *Empetrum* pollen exhibit signs of exine damage and hence are most likely to be of secondary derivation from eroding soils around the catchment (see also Loch an t-Suidhe).

Third, the profile shows unequivocal evidence of a mid-interstadial 'revertance' episode, reflected in both the sediment stratigraphy between 5.53 m and 5.49 m, and the pollen record changes from LA–3 to LA–5. The decline in *Juniperus* and *Empetrum* values in LA–4 is accompanied by increases in *Rumex*, Caryophyllaceae, *Salix* and *Lycopodium*, and also by a peak in the curve for deteriorated pollen. A similar lithological and biological oscillation in interstadial sediments has been noted at a number of sites in Scotland (Walker, 1984b; Tipping, 1991b) and has been widely interpreted as reflecting a break-up of the vegetation cover and increased soil erosion as a consequence of short-lived climatic deterioration. At other sites in Scotland and Ireland, a date of around 12,000 BP has been inferred for this event (Pennington, 1975b; Watts, 1985; Lowe and Walker, 1986a). Coleopteran evidence suggests a fall of almost 10°C in temperatures of the coldest months of the year from 12,300 to 11,800 BP after which winter temperatures rose by 4–5°C (Atkinson *et al.*, 1987), and it may be this climatic oscillation that is being reflected in the Loch Ashik profile.

Finally, Loch Ashik is located in a critical position relative to the mapped glacier limits in south-eastern Skye. It lies approximately 7 km east of the Loch Lomond Stadial glacier that developed in Coir Gorm in the Eastern Red Hills and just over 4 km north-west of the Loch Lomond Readvance limit in the Kyleakin Hills of eastern Skye. The site is therefore a key element in the establishment of a glacial chronology for this part of the Isle of Skye (Walker *et al.*, 1988; Ballantyne, 1989a; Ballantyne and Benn, 1991).

Loch Ashik is thus of considerable importance, and may justifiably be regarded as the main reference site for the Lateglacial on the Isle of Skye. In regional terms, it is a key element in establishing the spatial and temporal pattern of environmental change in western Scotland during the Lateglacial and early Holocene periods. In a wider context, it provides one of the few detailed Lateglacial pollen sequences from the maritime fringes of western Britain (see also Loch an t-Suidhe) which offers a basis for correlation between the nearshore marine evidence (e.g. Peacock, 1989b; Peacock and Harkness, 1990) and offshore records (e.g. Duplessy *et al.*, 1981; Ruddiman and McIntyre, 1981b; Bard *et al.*, 1987) during the last glacial–interglacial transition. As such, the comprehensive stratigraphic record from the Loch Ashik profile may have a much wider significance.

Conclusion

Loch Ashik is a key reference site for reconstructing the vegetational and environmental history on Skye during the Lateglacial, between approximately 13,000 and 10,000 years ago. Its full and detailed pollen record provides valuable insights into the changes that occurred during this period, including rapid climatic amelioration at the start of the Lateglacial Interstadial around 13,000 years ago, the development of intensely cold conditions during the Loch Lomond Stadial (about 11,000–10,000 years ago) and the rapid warming at the start of the Holocene (10,000 years ago). The pollen and sediments also reveal a brief climatic deterioration during the Lateglacial Interstadial (between about 13,000 and 10,000 years ago). Loch Ashik forms an integral part of a network of sites that demonstrate the geographical and temporal pattern of environmental change during the Lateglacial.

LOCH ASHIK, LOCH CLEAT AND LOCH MEODAL

H. J. B. Birks

Highlights

The detailed pollen records from the sediments in these loch basins provide evidence of the vegetational history of Skye during the Holocene and allow important insights into the pattern of woodland development. The latter shows major regional variations unique at the scale of the island.

Introduction

The Isle of Skye, on which these three sites are located, is unique within the Inner Hebrides today because of its great geological and topographical diversity, its botanical richness (Birks, 1973), its wide range of present-day plant communities (Birks, 1973), and its critical geographical position in relation to the boundaries of McVean and Ratcliffe's (1962) reconstructed Holocene potential woodland zones of Scotland (Williams, 1977; Birks and Williams, 1983). The sites of Loch Ashik, Loch Cleat and Loch Meodal are scientifically important because they provide detailed and extensively radiocarbon-dated Holocene pollen records and vegetational histories for three strongly contrasting ecological regions (*sensu* Birks, 1973) on Skye today. The regions are the Kyleakin area in the east (Loch Ashik, NG 691232), the Tertiary basalt country of northern Skye (Loch Cleat, NG 416742), and the sheltered Sleat peninsula in the south (Loch Meodal, NG 656113).

Although each site is important individually from the viewpoint of vegetational and environmental history, the three sites are of even greater scientific importance when their individual pollen records are compared (Williams, 1977; Birks and Williams, 1983). The three sites provide important evidence for marked vegetational differentiation within Skye throughout the Holocene. Nowhere else in Scotland can this remarkable range of forest history and vegetational differentiation be found within such a small area.

Because of the importance of the three sites when considered together, their combined importance is reviewed after the individual site accounts.

Loch Ashik

Description

Loch Ashik is situated at an altitude of 40 m OD, 4 km east of Broadford and occupies a small depression within the local Torridonian sandstone. It is 175 m long and 125 m wide and is surrounded by blanket bog and soligenous mires. The immediate area is treeless today, although birch woods with some hazel, rowan, oak and holly occur on steep slopes 5 km to the east. At Gleann na Beiste, 6.5 km to the north-east of Loch Ashik, fossil pine stumps occur within the blanket peat. These stumps have a radiocarbon date of 4420 ± 75 BP (Q–1309). This is the only known locality for dated pine stumps on Skye (Lewis, 1906; Birks, 1975), although stumps of comparable age also occur on western Lewis and Harris (Bennett, 1984; Wilkins, 1984).

The vegetational history of Loch Ashik and its surrounds was reconstructed from pollen analysis of cores obtained from the marginal fen on the western side of the loch. The stratigraphy (Figure 11.17) consists of 1.5 m of herbaceous sedge peat underlain by 2.85 m of fine-detritus (organic) mud. Below this is 0.55 m of silty, fine-detritus mud of early Holocene age underlain by 0.75 m of silts and silty muds of Lateglacial age (see above). Ten radiocarbon dates (SRR–804 to SRR–813) are available from the fine-detritus and silty, fine-detritus muds to provide a chronology for the Holocene pollen record of Williams (1977).

Interpretation

After an early Holocene phase of juniper scrub and grassland, birch and hazel expanded at about 9600 BP to form fern- and tall, herb-rich woods with willow and rowan. Elm and oak were probably present in small amounts after about 9000 BP. *Calluna* heath, species-poor grassland and bog appear to have also been present near Loch Ashik as early as 9000 BP, presumably as a result of podsolization and paludification of the predominantly acid soils derived from the underlying Torridonian sandstone. At 6300 BP alder expanded rapidly at the expense of willow and hazel (Bennett and Birks, 1990). In contrast to southern and northern Skye there is no palynological evidence at Loch Ashik for any human interference at 5000 BP. A second and very important contrast is the expansion of *Pinus*

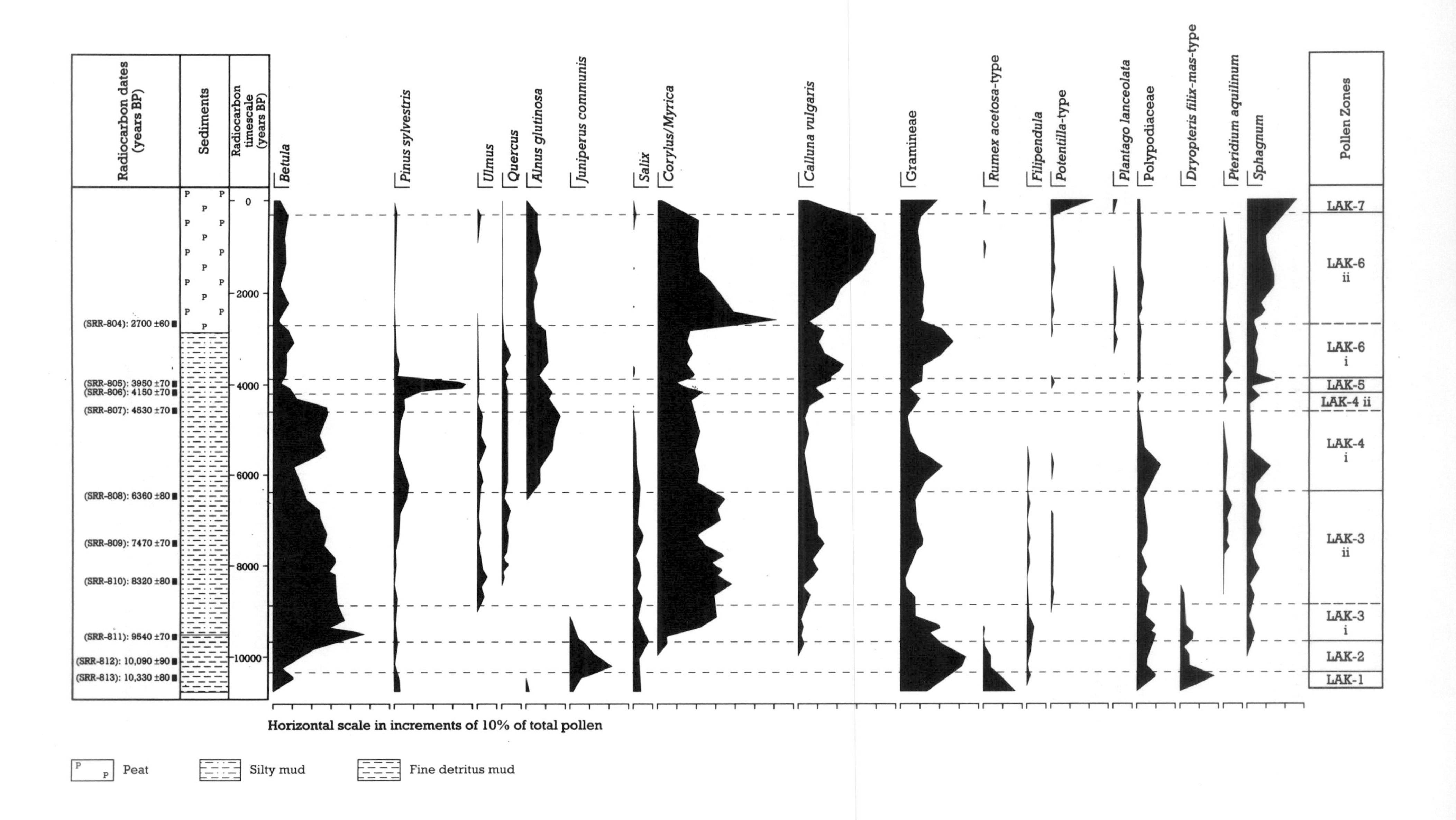

Radiocarbon dates (years BP)
(SRR-804): 2700 ±60
(SRR-805): 3950 ±70
(SRR-806): 4150 ±70
(SRR-807): 4530 ±70
(SRR-808): 6360 ±80
(SRR-809): 7470 ±70
(SRR-810): 8320 ±80
(SRR-811): 9540 ±70
(SRR-812): 10,090 ±90
(SRR-813): 10,330 ±80
Sediments
Radiocarbon timescale (years BP)
0
2000
4000
6000
8000
10000
Betula
Pinus sylvestris
Ulmus
Quercus
Alnus glutinosa
Juniperus communis
Salix
Corylus/Myrica
Calluna vulgaris
Gramineae
Rumex acetosa-type
Filipendula
Potentilla-type
Plantago lanceolata
Polypodiaceae
Dryopteris filix-mas-type
Pteridium aquilinum
Sphagnum
Pollen Zones
LAK-7
LAK-6 ii
LAK-6 i
LAK-5
LAK-4 ii
LAK-4 i
LAK-3 ii
LAK-3 i
LAK-2
LAK-1
Horizontal scale in increments of 10% of total pollen
Peat
Silty mud
Fine detritus mud

pollen between 4600 BP and 3900 BP at Loch Ashik (Birks, 1989). This may reflect the local growth of pine on dried peat surfaces in eastern Skye, a widespread phenomenon in north-west Scotland at that time (Birks, 1975; Birks 1988, 1989; Gear and Huntley, 1991), as at Gleann na Beiste.

The sharp decline in *Pinus* pollen at Loch Ashik correlates with the widespread demise of pine throughout north-west Scotland at about 4000 BP (Birks, 1972b, 1975; Birks, 1977, 1988; Bennett, 1984; Gear and Huntley, 1991) and the widespread development and expansion of blanket bog with *Calluna vulgaris*, *Sphagnum* and *Narthecium ossifragum*, of acid grasslands, and of heaths around Loch Ashik. The reasons for this widespread and dramatic decline of pine in north-west Scotland, eastern Skye, and parts of Lewis are not fully understood (Birks, 1988). A combination of rapid climatic change and human activity may have initiated the replacement of pine on flat and gently sloping ground by treeless blanket bog. There is, however, independent chemical evidence from the Inverpolly area (Pennington *et al.*, 1972) that suggests a major change to a more oceanic climate with increased precipitation and stronger winds at about 4000 BP. Such an abrupt change would have caused waterlogging, encouraged the expansion of blanket bog, and inhibited the regeneration of pine by reducing the number of good seed years (Birks, 1972b).

By 2700 BP bog and heath were widespread near Loch Ashik and woodland, mainly of birch, was rare and presumably restricted, as today, to steep slopes where blanket bog could not develop. This situation has continued to the present day, suggesting that the modern bog-dominated landscape is of considerable antiquity.

The Holocene pollen record of Loch Ashik (Williams, 1977; Birks and Williams, 1983) indicates that pine was locally abundant between 4600 BP and 3900 BP and that the vegetational history of eastern Skye has affinities with parts of Wester Ross, in McVean and Ratcliffe's (1962) 'predominant pine forest zone'. In contrast, pine appears to have been absent from southern and northern Skye during the Holocene. Loch Ashik is thus of considerable palaeoecological importance in illustrating the extremely localized geographical distribution of *Pinus sylvestris* during the Holocene and the local, but very rapid, extinction of pine close to the limits of its natural geographical range (see also Gear and Huntley, 1991).

Figure 11.17 Loch Ashik: Holocene relative pollen diagram showing selected taxa as percentages of total pollen (from Birks and Williams, 1983). Note that the data are plotted against a radiocarbon time-scale.

Loch Cleat

Description

Loch Cleat occupies a rock basin 200 m long and 100 m wide at about 40 m OD on the west side of the northern tip of the Trotternish peninsula near Duntulm. The solid geology is predominantly Palaeogene dolerite sills and Jurassic sedimentary rocks. To the south there are the westerly dipping, Palaeogene basalt lavas that form the impressive Trotternish ridge with its steep, east-facing scarp slope. The vegetation of the area near Loch Cleat today is predominantly grassland, meadow and bog. Small areas of birch, hazel, willow and rowan scrub occur locally on sheltered, steep, block-strewn slopes.

Organic sediments up to 9.4 m thick, underlain by 1.85 m of minerogenic sediments of presumed Lateglacial age, occur at the western edge of Loch Cleat (Figure 11.18). These organic sediments consist of 2 m of herbaceous sedge peat overlying 7.07 m of fine-detritus mud. There are 0.33 m of silty, fine-detritus mud overlying the basal minerogenic sediments. Ten radiocarbon dates (SRR–932 to SRR–941) are available from the organic sediments. A detailed pollen diagram for these sediments has been prepared by Williams (1977; see also Birks and Williams, 1983) (Figure 11.18).

Interpretation

The early Holocene (10,000–8900 BP) vegetation was juniper, willow and birch scrub with abundant grasses, ferns and tall herbs. This was replaced at about 8900 BP by birch, hazel and willow scrub with rowan and *Prunus padus*. Species-rich grasslands and tall herb communities continued to be locally frequent. Low pollen values of *Quercus*, *Ulmus* and *Pinus* throughout the Holocene at Loch Cleat and at other sites in northern Skye (Vasari and Vasari, 1968) indicate that none of these trees was ever an important component of the local vegetation (Birks, 1989),

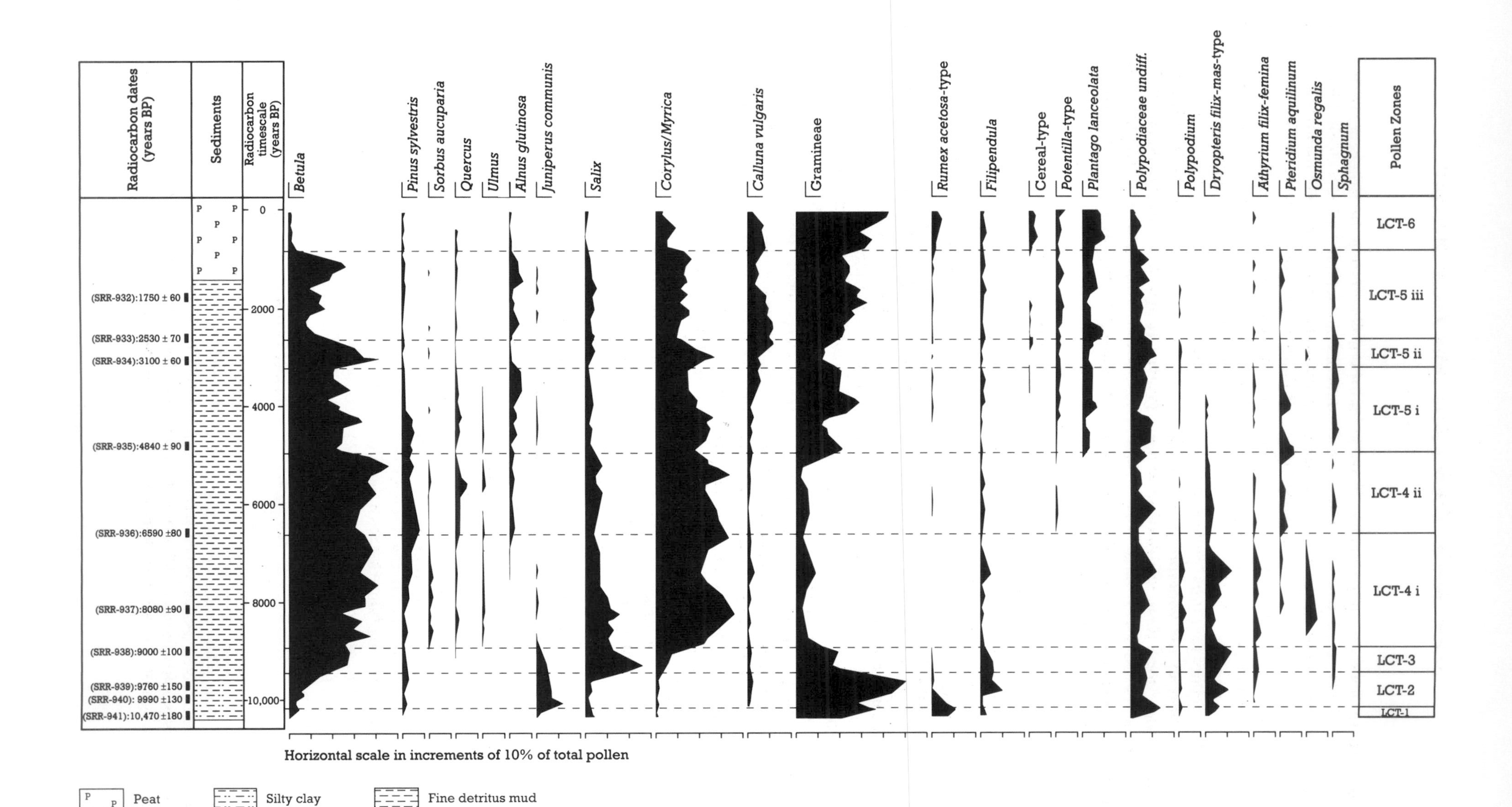

Radiocarbon dates (years BP)
Sediments
Radiocarbon timescale (years BP)
Betula
Pinus sylvestris
Sorbus aucuparia
Quercus
Ulmus
Alnus glutinosa
Juniperus communis
Salix
Corylus/Myrica
Calluna vulgaris
Gramineae
Rumex acetosa-type
Filipendula
Cereal-type
Potentilla-type
Plantago lanceolata
Polypodiaceae undiff.
Polypodium
Dryopteris filix-mas-type
Athyrium filix-femina
Pteridium aquilinum
Osmunda regalis
Sphagnum
Pollen Zones
(SRR-932):1750 ± 60
(SRR-933):2530 ± 70
(SRR-934):3100 ± 60
(SRR-935):4840 ± 90
(SRR-936):6590 ±80
(SRR-937):8080 ±90
(SRR-938):9000 ±100
(SRR-939):9760 ±150
(SRR-940): 9990 ±130
(SRR-941):10,470 ±180
0
2000
4000
6000
8000
10,000
LCT-6
LCT-5 iii
LCT-5 ii
LCT-5 i
LCT-4 ii
LCT-4 i
LCT-3
LCT-2
LCT-1
Horizontal scale in increments of 10% of total pollen
Peat
Silty clay
Fine detritus mud

in contrast to southern Skye (Loch Meodal) where elm and oak were frequent, and eastern Skye (Loch Ashik) where oak, elm, and pine were present locally. Alder arrived at Loch Cleat at about 6300 BP but, in contrast to eastern and southern Skye, it was never abundant in northern Skye (Bennett and Birks, 1990).

The pollen record at Loch Cleat reveals a marked increase in herbaceous pollen types (mainly grasses, *Potentilla* type, Chenopodiaceae and Cruciferae) at 5000 BP along with the first appearance of cereal type and *Plantago lanceolata* pollen. At the same time there is a large decrease in the pollen values of birch and hazel, suggesting clearance of scrub and the local development of arable and pastoral agriculture. The pollen spectra suggest that between 5000 BP and 700 BP the landscape of northern Trotternish was mainly treeless, with patches of scrub probably restricted to steep, rocky slopes that were difficult to clear. There are abundant Iron Age archaeological remains in the area, such as brochs and duns, the ages of which are unfortunately not known.

In the last 700 years there has been widespread clearance of the remaining areas of birch and hazel scrub to produce the virtually treeless landscape of northern Skye today. Cereal-type pollen is present in significant amounts, suggesting extensive cereal cultivation in this part of northern Skye. With its fertile soils, northern Skye was noted for its cereal crops and the parish of Kilmuir, in which Loch Cleat is situated, was referred to as the 'granary of Skye' (MacSween, 1959).

The pollen record from Loch Cleat provides the most detailed record of the Holocene vegetational history currently available for the basalt areas of northern Skye. The landscape of this area is virtually treeless today with a few stands of birch and hazel scrub confined to steep, sheltered, coastal cliffs and to ravines. McVean and Ratcliffe (1962) suggest that northern Skye lies within the 'predominant birch forest' zone. The pollen stratigraphy at Loch Cleat (Williams, 1977) confirms this suggestion and shows that only birch, hazel and willow scrub developed near the site, even in middle Holocene times. The pollen record also shows that oak, elm, pine and alder were never important components of the Holocene vegetation of northern Skye, in contrast to southern and eastern Skye (Williams, 1977; Birks and Williams, 1983). This history contrasts markedly with southern and eastern Skye. Loch Cleat is thus important because of its detailed and well-dated pollen record, its contribution to the reconstruction and understanding of Holocene forest history of western Scotland, and its record of land-use history over the last 5000 years.

Figure 11.18 Loch Cleat: Holocene relative pollen diagram showing selected taxa as percentages of total pollen (from Birks and Williams, 1983). Note that the data are plotted against a radiocarbon time-scale.

Loch Meodal

Description

Loch Meodal is located 5 km south-east of Ord in the Sleat peninsula, in a gently sloping area at an altitude of 105 m OD. The loch is 500 m long and 400 m wide. The landscape is one of variable relief, with the underlying Torridonian and Lewisian rocks frequently cropping out as rocky knolls. Gentle slopes and depressions are covered by blanket peat, *Calluna vulgaris* heath, or species-poor grassland. Areas of birch, hazel and oak woodland occur nearby in the Ord Valley, between Ord and Tokavaig, and by Loch na Dal (Birks, 1973). An extensive fen has developed at the northern end of Loch Meodal, where organic sediments up to 7.9 m have accumulated (Figure 11.19). These are underlain by 0.3 m of minerogenic sediments of possible Lateglacial age. The organic sediments consist of 1–2 m of herbaceous peat overlying 6.4–5.4 m of fine-detritus mud. There is a narrow (0.1 m) layer of sandy, fine-detritus mud overlying 0.4 m of silty, fine-detritus mud. The basal sediments are sand, silt and gravel, at least 0.3 m thick. The pollen record of these sediments has been studied in detail by Birks (1973; basal 1.20 m) and by Williams (1977; upper 7.70 m) (Figure 11.19). Eleven radiocarbon dates (Q–961, Q–1301 to Q–1310) have been obtained from the Loch Meodal sediments.

Interpretation

The presumed Lateglacial vegetational history, as reconstructed by pollen analyses of the basal 1.2 m of sediment (Birks, 1973) is as follows. The earliest vegetation was acid dwarf-shrub heaths with abundant *Betula nana* and herb-rich grass-

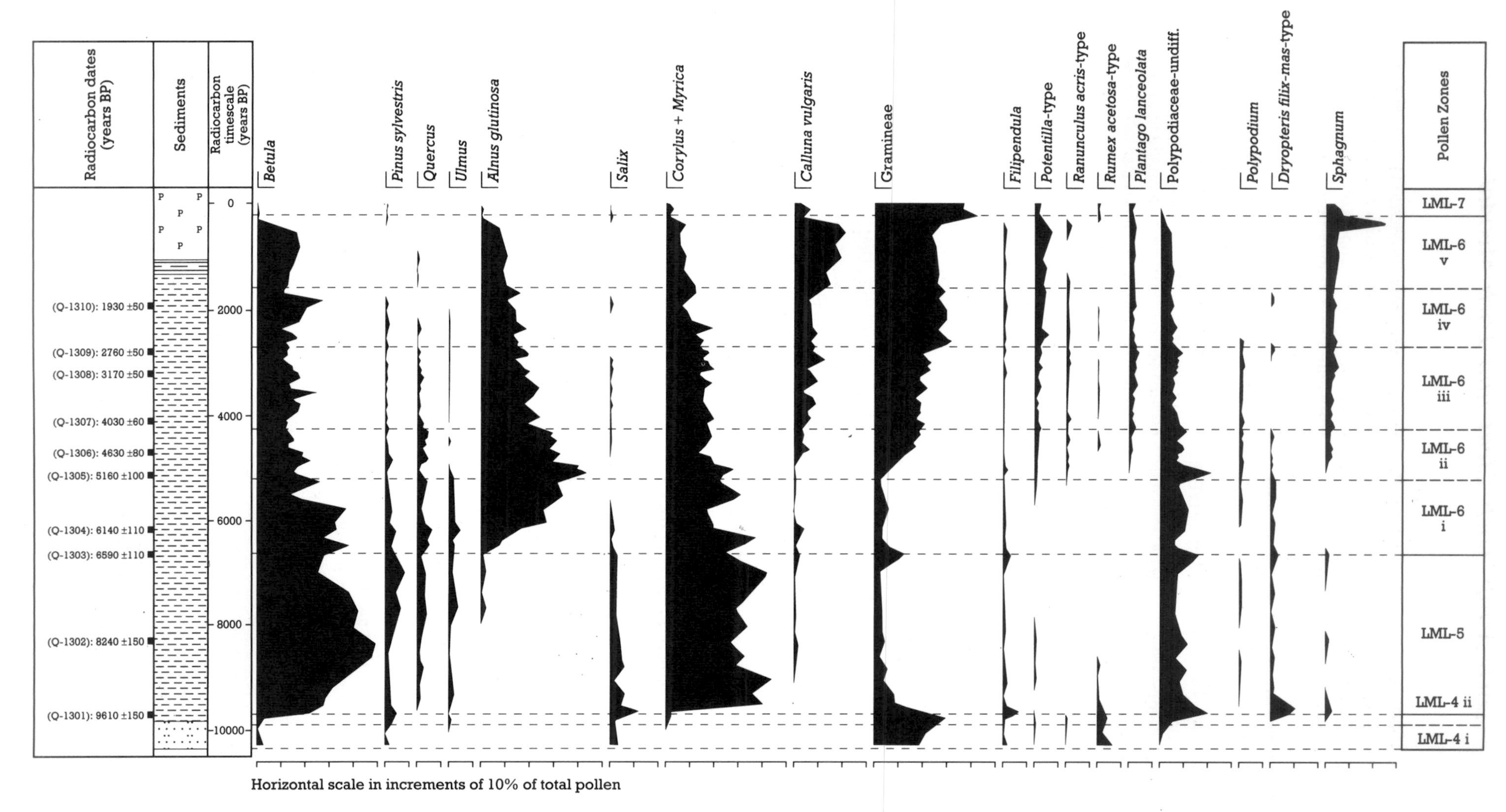

Radiocarbon dates (years BP)
Sediments
Radiocarbon timescale (years BP)
(Q-1310): 1930 ±50
(Q-1309): 2760 ±50
(Q-1308): 3170 ±50
(Q-1307): 4030 ±60
(Q-1306): 4630 ±80
(Q-1305): 5160 ±100
(Q-1304): 6140 ±110
(Q-1303): 6590 ±110
(Q-1302): 8240 ±150
(Q-1301): 9610 ±150
0
2000
4000
6000
8000
10000
Betula
Pinus sylvestris
Quercus
Ulmus
Alnus glutinosa
Salix
Corylus + Myrica
Calluna vulgaris
Gramineae
Filipendula
Potentilla-type
Ranunculus acris-type
Rumex acetosa-type
Plantago lanceolata
Polypodiaceae-undiff.
Polypodium
Dryopteris filix-mas-type
Sphagnum
Pollen Zones
LML-7
LML-6 v
LML-6 iv
LML-6 iii
LML-6 ii
LML-6 i
LML-5
LML-4 ii
LML-4 i
Horizontal scale in increments of 10% of total pollen
Peat
Transition between peat and detritus mud
Detritus mud
Sand and silt

lands. At an inferred age of about 12,200 BP juniper and willow scrub developed. This was replaced by open birch woods with aspen and abundant ferns. Subsequently, the extent of woodland decreased, and *Betula nana* heath expanded, presumably in response to the climatic deterioration associated with the Loch Lomond Stadial.

In the early Holocene (9700 BP) birch and hazel rapidly expanded with some aspen and willow, abundant ferns and tall herbs (Williams, 1977). From about 9000 BP oak and elm may have been present in small amounts within the predominantly birch-and-hazel-dominated landscape. Alder expanded rapidly at about 6500 BP to form mixed birch–hazel–alder woods with some oak, elm, ash, rowan and holly. The oak pollen values are lower than at comparable times on the adjacent mainland, for example the Morar peninsula (Williams, 1977), suggesting that the natural northern limit of predominant oak during the middle Holocene lay near southern Skye. Although oak was certainly present in Sleat, it was never a prominent component of the natural woodland cover. As on the adjacent mainland, *Pinus* was absent in the forests of southern Skye, in contrast to its abundance in parts of Wester Ross (Birks, 1972b; Birks, 1977, 1989) and in eastern Skye (Williams, 1977).

Forest clearance in Sleat began at about 5200 BP. By 4200 BP the landscape was still mainly wooded, but bogs, heaths and acid grasslands became frequent. There was little vegetational change between 2700 BP and 300 BP except for the spread of *Calluna vulgaris* at about 1600 BP. In the last 300 years there has been extensive forest clearance, a large decrease in the extent of heather moor and a massive spread of acid grassland, resulting in the lightly wooded landscape near Loch Meodal today. This widespread forest destruction and spread of grassland may have resulted from the onset of cattle breeding after AD 1650, reaching its peak at about AD 1750 (Williams, 1977). The woodlands surviving in southern Skye are clearly natural relics of the former forests of Sleat.

The Holocene forest history of Loch Meodal has greatest affinities with sites on the adjacent mainland in the 'predominant oak forest with birch' zone of southern and western Scotland (McVean and Ratcliffe, 1962), such as the Morar peninsula and the Loch Sunart area. The main difference between the pollen sequences from the mainland and from Loch Meodal is that oak was rarer in southern Skye than on the mainland.

Loch Meodal is of considerable scientific importance in the reconstruction of Lateglacial and Holocene vegetational history. First, it is situated within the Sleat peninsula of southern Skye and thus lies near the northern limit of McVean and Ratcliffe's (1962) potential 'predominant oak forest with birch' zone (Birks and Williams, 1983). Its detailed pollen stratigraphy shows that the site has been near important distributional limits of two major forest trees (birch and oak) during the last 12,000 years. It also illustrates that very marked vegetational differentiation has existed within Skye and between Skye and the adjacent mainland since the Late Devensian (Birks, 1973; Williams, 1977; Birks and Williams, 1983). Second, southern Skye supports today the largest areas of natural or seminatural woodland on Skye. These birch and birch–hazel woods, often with some oak, elm, ash, alder, holly, rowan and willow, are rich in several internationally rare and biogeographically important, warmth-demanding Atlantic species of ferns, bryophytes and lichens. Many of these species are growing at or near their northernmost known world localities (see Birks, 1973). The Holocene vegetational history of this area, as reconstructed at Loch Meodal (Williams, 1977) is thus of considerable importance in understanding the status of the existing woodlands in Sleat and in elucidating the development of the present ecological landscape of southern Skye. Third, the site contains sediments of Holocene and possible Lateglacial age that have been studied in some palynological detail by Williams (1977) and Birks (1973), respectively. The Lateglacial vegetational history may show that southern Skye was the northernmost known area of tree-birch growth during the Lateglacial Interstadial.

Figure 11.19 Loch Meodal: Holocene relative pollen diagram showing selected taxa as percentages of total pollen (from Birks and Williams, 1983). Note that the data are plotted against a radiocarbon time-scale.

Holocene forest history; an overview

The Holocene forest history of Skye (Williams, 1977; Birks and Williams, 1983) corresponds closely with the present-day distribution of natural or seminatural woodland stands on the

Table 11.1 Generalized comparison of the inferred Holocene vegetational history of the Isle of Skye based on the pollen records from Loch Cleat, Loch Ashik and Loch Meodal (from Birks and Williams, 1983)

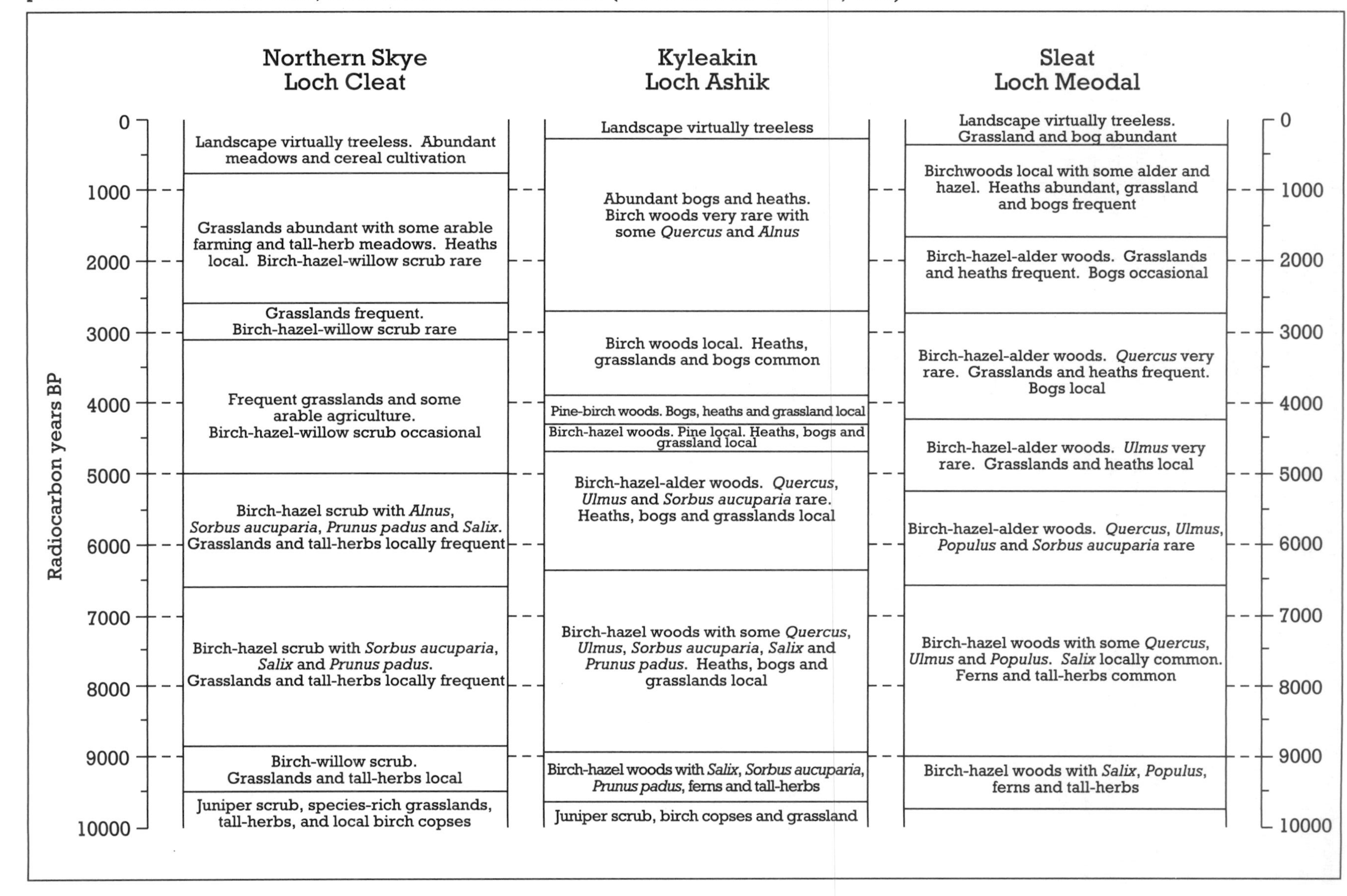

Northern Skye – Loch Cleat

- Landscape virtually treeless. Abundant meadows and cereal cultivation
- Grasslands abundant with some arable farming and tall-herb meadows. Heaths local. Birch-hazel-willow scrub rare
- Grasslands frequent. Birch-hazel-willow scrub rare
- Frequent grasslands and some arable agriculture. Birch-hazel-willow scrub occasional
- Birch-hazel scrub with *Alnus*, *Sorbus aucuparia*, *Prunus padus* and *Salix*. Grasslands and tall-herbs locally frequent
- Birch-hazel scrub with *Sorbus aucuparia*, *Salix* and *Prunus padus*. Grasslands and tall-herbs locally frequent
- Birch-willow scrub. Grasslands and tall-herbs local
- Juniper scrub, species-rich grasslands, tall-herbs, and local birch copses

Kyleakin – Loch Ashik

- Landscape virtually treeless
- Abundant bogs and heaths. Birch woods very rare with some *Quercus* and *Alnus*
- Birch woods local. Heaths, grasslands and bogs common
- Pine-birch woods. Bogs, heaths and grassland local
- Birch-hazel woods. Pine local. Heaths, bogs and grassland local
- Birch-hazel-alder woods. *Quercus*, *Ulmus* and *Sorbus aucuparia* rare. Heaths, bogs and grasslands local
- Birch-hazel woods with some *Quercus*, *Ulmus*, *Sorbus aucuparia*, *Salix* and *Prunus padus*. Heaths, bogs and grasslands local
- Birch-hazel woods with *Salix*, *Sorbus aucuparia*, *Prunus padus*, ferns and tall-herbs
- Juniper scrub, birch copses and grassland

Sleat – Loch Meodal

- Landscape virtually treeless. Grassland and bog abundant
- Birchwoods local with some alder and hazel. Heaths abundant, grassland and bogs frequent
- Birch-hazel-alder woods. Grasslands and heaths frequent. Bogs occasional
- Birch-hazel-alder woods. *Quercus* very rare. Grasslands and heaths frequent. Bogs local
- Birch-hazel-alder woods. *Ulmus* very rare. Grasslands and heaths local
- Birch-hazel-alder woods. *Quercus*, *Ulmus*, *Populus* and *Sorbus aucuparia* rare
- Birch-hazel woods with some *Quercus*, *Ulmus* and *Populus*. *Salix* locally common. Ferns and tall-herbs common
- Birch-hazel woods with *Salix*, *Populus*, ferns and tall-herbs

island (Birks, 1973). Southern Skye has the most woodland today, eastern Skye has woodland confined to slopes that are too steep for blanket-bog development, whereas in northern Skye trees only grow in sheltered, rocky sites. This pattern may reflect climatic differences within Skye, with southern Skye being the mildest and most sheltered part today. The pollen records from Loch Meodal, Loch Cleat and Loch Ashik illustrate the very considerable variation in the Holocene vegetational history and forest composition of Skye (Table 11.1). The vegetational history of southern Skye has its closest affinities with the adjacent mainland of the Morar peninsula. Northern Skye corresponds, in its vegetational history, to sites on the mainland further north such as in northern Wester Ross and West Sutherland (see Loch Sionascaig, Cam Loch and Lochan an Druim). Eastern Skye has affinities with parts of Wester Ross (see Loch Maree) where pine was a major component of the forests for a comparatively short period within the Holocene. Nowhere else in Scotland or elsewhere in north-west Europe can such a range of variation in forest composition and vegetational history be found within such a small area.

Loch Meodal, Loch Ashik and Loch Cleat are three sites of great importance in the reconstruction and understanding of the Holocene forest history of Scotland. Each site has considerable scientific importance. Their importance is even greater, however, when the three sites are considered together. In combination, they provide a unique palynological record of forest history and fine-scale vegetational differentiation. They are of international importance in providing palaeoecological insights into past geographical patterns of forest composition, and into the dynamic nature of Holocene vegetational history.

Conclusion

Together these sites are important for the evidence they provide of the history of vegetation and forest development on Skye during the Holocene (the last 10,000 years), which occupies a critical geographical location in relation to the main woodland zones that occur in Scotland today. The detailed pollen records from each site indicate major regional differences in the vegetational history of the island. Such variations are unique within an area of this size and are of great importance in understanding the geography and dynamics of Holocene vegetation.

Chapter 12

Outer Hebrides

INTRODUCTION

D. G. Sutherland

The Outer Hebrides (Figure 12.1) occupy a critical position for studies of the extent of the Scottish mainland ice-sheet. In addition, their position on the maritime fringe of the country makes them of particular interest for investigation of past vegetation patterns and of palaeoclimate in general. However, despite an initial period of interest in the Quaternary history of the islands during the last century it has only been within the last fifteen years that this fascinating area has begun to be studied in detail.

The first to attempt a synthesis of the glacial history of the Outer Hebrides was J. Geikie (1873, 1877, 1878, 1894). He argued that the whole island chain had been overridden by the ice-sheet originating in the Scottish Highlands flowing from east or south-east to west or north-west. The apparent absence from many of the islands of erratics deriving from this ice movement was explained by Geikie as the result of the lower, dirtier layers of the ice being diverted to the north and south of the islands, whereas the relatively cleaner upper layers of the ice passed across the main area of the islands. Despite brief reports to the contrary (Bryce, in discussion of Geikie, 1873; Campbell, 1873; Milne Home, 1877), the general pattern of ice movement proposed by Geikie became established in the literature of the next 100 years.

Geikie also drew attention to the complex sequence of drift deposits to be found in the north of the islands at Port of Ness and along the north-west Coast of Lewis at Swainbost. These he initially (1874) considered to be the product of a single episode of sedimentation near a glacier margin, but later (1877, 1878) he introduced the concept of two glacial phases separated by an interglacial marine submergence. An early faunal list of the marine fossils contained in these deposits was given by Etheridge (1876).

Subsequent studies until recent years were largely the product of general geological surveys and added only details to the model suggested by Geikie. Thus Jehu and Craig in a series of papers (1923a, 1923b, 1926, 1927, 1934) accepted the dominant role of the Scottish ice-sheet in the glaciation of the islands and provided some detail as to the distribution of exotic erratics, particularly in the southern islands. The existence of erratics on the far-flung islands of Sula Sgeir (Stewart, 1933), North Rona (Dougal, 1928; Stewart, 1932; Gailey, 1959) and the Flannan Isles (Stewart, 1933) appeared to confirm the extension of the Scottish ice-sheet over much of the western continental shelf at maximum glaciation. Whether the ice-sheet reached St Kilda was, however, less certain (Cockburn, 1935; Wager, 1953).

The occurrence of shelly deposits was also reported at Garrabost and Tolsta Head by Dougal (1928). Baden-Powell (1938) examined the fauna of the deposits of the north of Lewis and concurred with the general model of two glaciations separated by an interglacial phase. He considered that at Swainbost an interglacial beach was *in situ*. In a separate publication Baden-Powell and Elton (1937) recorded the occurrence of a raised beach along part of the coast of north-west Lewis. They assigned a Holocene age to the beach, as did McCann (1968) who noted that the beach rested on a wide rock platform from which it was separated, in places, by a till.

In the 1970s a complete revision of the ideas of glaciation of the Outer Hebrides resulted from the work of von Weymarn (1974, 1979), Coward (1977), Flinn (1978b) and Peacock and Ross (1978). These various authors noted that throughout the island chain the last movement of the ice as indicated by ice-moulded landforms, striations and erratic transport was towards the east along the eastern part of the islands. This was in direct opposition to the interpretation of Geikie (1873, 1878) and introduced the concept that the last phase of glaciation of almost the whole of the Outer Hebrides had been by a local ice-cap, and that Scottish mainland ice had only impinged on the extremities of the islands, if at all.

At the time this work was in progress, the complexity of the Pleistocene sequence of the area was also becoming apparent through the discovery of the interstadial sites of Tolsta Head (von Weymarn and Edwards, 1973; Birnie, 1983) and St Kilda (Sutherland *et al.*, 1982, 1984) and the possible interglacial site of Toa Galson (Sutherland and Walker, 1984), the unravelling of the sequence of sediments in northern Lewis (von Weymarn, 1974; Peacock, 1981a, 1984a; Sutherland and Walker, 1984) and the recognition that the raised beach sediments were not Holocene in age, but earlier, and could be used as a stratigraphic marker horizon in subdivision of the glacial deposits (von Weymarn, 1974; Peacock, 1981a, 1984a; Sutherland and Walker, 1984; Selby, 1987).

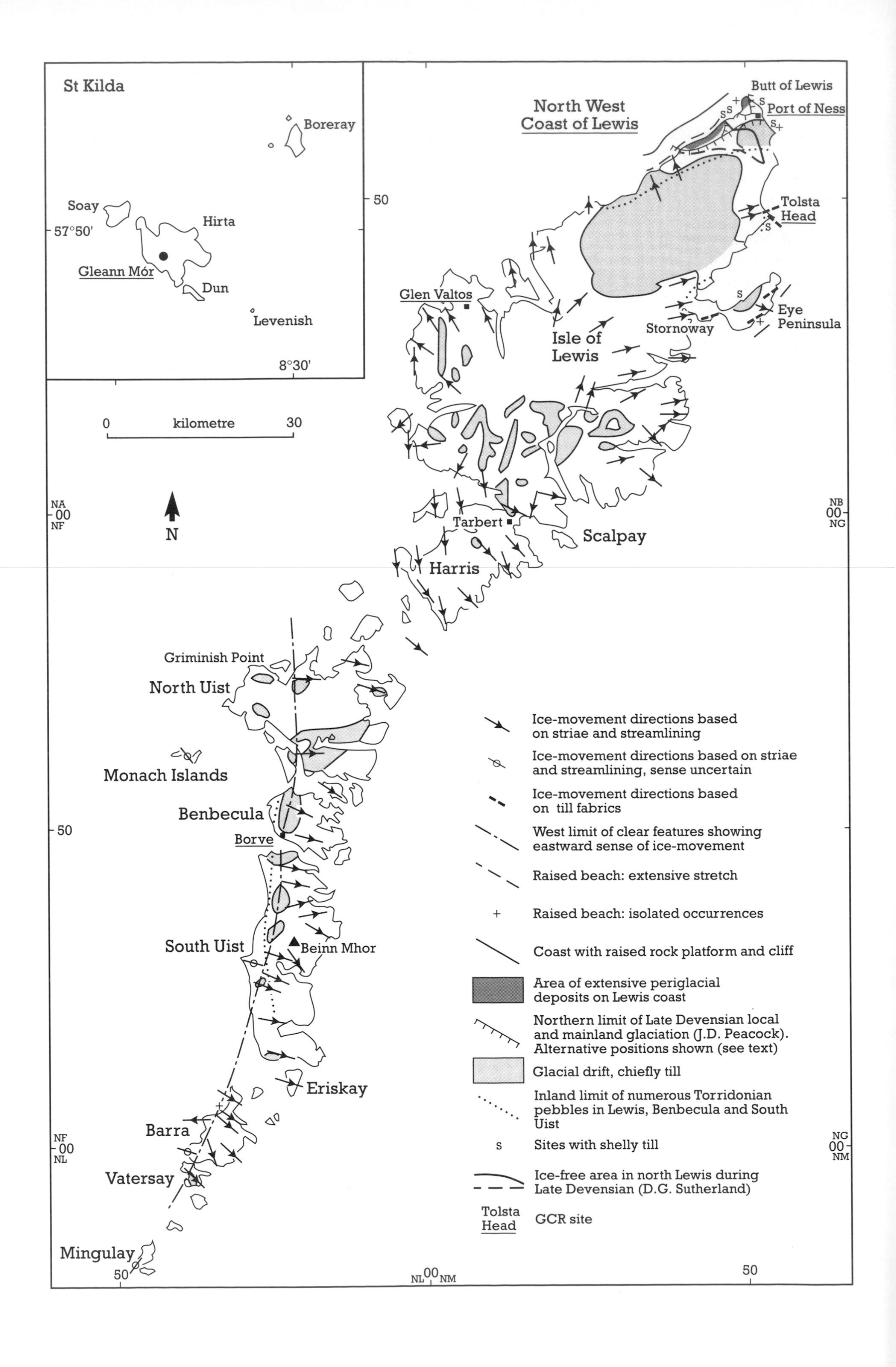
St Kilda
Boreray
Soay
Hirta
57°50'
Gleann Mór
Dun
Levenish
8°30'
0
kilometre
30
N
North West
Coast of Lewis
Butt of Lewis
Port of Ness
Tolsta
Head
Glen Valtos
Isle of
Lewis
Stornoway
Eye
Peninsula
Tarbert
Scalpay
Harris
Griminish Point
North Uist
Monach Islands
Benbecula
Borve
South Uist
Beinn Mhor
Eriskay
Barra
Vatersay
Mingulay
NA
00
NF
NB
00
NG
NF
00
NL
NG
00
NM
NL 00 NM
50
Ice-movement directions based on striae and streamlining
Ice-movement directions based on striae and streamlining, sense uncertain
Ice-movement directions based on till fabrics
West limit of clear features showing eastward sense of ice-movement
Raised beach: extensive stretch
Raised beach: isolated occurrences
Coast with raised rock platform and cliff
Area of extensive periglacial deposits on Lewis coast
Northern limit of Late Devensian local and mainland glaciation (J.D. Peacock). Alternative positions shown (see text)
Glacial drift, chiefly till
Inland limit of numerous Torridonian pebbles in Lewis, Benbecula and South Uist
Sites with shelly till
Ice-free area in north Lewis during Late Devensian (D.G. Sutherland)
Tolsta Head
GCR site

Introduction

The Pleistocene sequence as presently understood in the Outer Hebrides may be summarized as follows. The earliest event known appears to be the erosion of the raised rock platform in northern Lewis. Subsequently there was a glacial event depositing the till on the rock platform, and it is possible that it was at this time that the exotic erratics were emplaced on the outer isles such as North Rona, Sula Sgeir and the Flannan Isles. Traces of similar erratics have been found, reworked in more recent deposits, on St Kilda (Harding *et al.*, 1984; Sutherland *et al.*, 1984), where the emplacement of those erratics has been suggested to pre-date the Devensian. Two sets of moraines have been found on the outer shelf to the north-west and north of Lewis (Stoker and Holmes, 1991) and the glaciation responsible for the earlier of these moraines may have occurred at this time. In north-west Lewis, the Toa Galson peat post-dates this phase of glaciation, and if it is indeed of interglacial origin as suggested by Sutherland and Walker (1984), this too would imply a pre-Devensian age for this glaciation.

Pollen analysis indicates that the Toa Galson peat formed during a period when vegetation changed progressively from a maritime grassland to an acid heath, and the pollen spectra are comparable to those from Sel Ayre, in Shetland (Sutherland and Walker, 1984). They are quite distinct from the spectra from the Tolsta Head and St Kilda interstadial sites, both of these being associated with finite radiocarbon dates implying their attribution to a Middle Devensian interstadial.

The Toa Galson peat is overlain by a periglacial slope deposit representing a period of cold climate, and that is directly overlain by the sediments of the Galson Beach. There is some dispute as to the precise stratigraphic relations of this beach in north Lewis and whether there were one or two periods of beach formation. However, it is generally accepted that in the Galson/Toa Galson area the beach is only overlain by periglacial slope deposits and its upper horizons are periglacially disturbed. To the south, glacial deposits of the last Outer Hebrides ice-cap overlie the beach. Von Weymarn (1974) and Peacock (1981a, 1984a) consider that, to the north, beach sediments overlie the deposits of the last glaciation, whereas Sutherland and Walker (1984) consider them to underlie the glacial sediments, regarding those gravels that occur within or on the glacial deposits as erratics. Radiocarbon and amino acid dating of the included mollusc shells in the glacial sediments of Lewis allowed Sutherland and Walker (1984) to establish the glacial phase post-dating beach formation as Late Devensian in age. The till overlying the Middle Devensian Tolsta Head interstadial deposits also dates from this episode. It is unclear whether the younger set of moraines on the outer shelf north of Lewis is also of this age (Stoker and Holmes, 1991).

During the Late Devensian, therefore, the Outer Hebrides were glaciated by a local ice-cap, with external ice impinging only on the northern tip of Lewis. A small ice-free area existed along the north-west coast of Lewis. On St Kilda a small local glacier existed at this time (Sutherland *et al.*, 1982, 1984). There has been considerable discussion as to the ability of the Outer Hebrides ice-cap to become established on the low ground of the Uists and Benbecula (Flinn, 1980; Peacock, 1980a; Sissons, 1980c). Sissons (1980c) suggested that it was in fact a remnant of a once more extensive Scottish mainland ice-sheet that had become isolated during ice retreat by calving in the deep waters of the Minches to the east. However, this explanation does not allow for the period of beach formation both in the north and the south of the islands (Peacock, 1980a, 1984a; Selby, 1987) prior to expansion of the local ice-cap. Later, Sissons (1983c) proposed that mainland ice crossed the Outer Hebrides during the Early Devensian and that the ice margin subsequently stabilized to the east of the Outer Hebrides during the Late Devensian, allowing the independent development of an Outer Hebrides ice-cap at this time. On present evidence it therefore seems that during at least part of the Late Devensian the climate was sufficiently severe for the ice-cap to become established on low ground.

Features of deglaciation formed during the decay of the ice-cap have only been discussed in general terms (von Weymarn, 1979; Peacock, 1984a). Particular note has been made of the glaciofluvial deposits in the Uig area of west Lewis (Jehu and Craig, 1934; von Weymarn, 1979; Peacock, 1984a), where eskers, kames, kame terraces, glaciolacustrine deltas and meltwater channels, including the remarkable Glen

Figure 12.1 Location map and principal Quaternary features of the Outer Hebrides (from Peacock, 1984a).

Valtos, were produced. It is notable that the direction of meltwater drainage indicated by these features was from west to east, implying thicker ice immediately offshore at this stage of the glaciation.

Within the hills of Lewis and Harris there are abundant moraines, which have been referred to a period or periods of local glaciation (Geikie, 1878; von Weymarn, 1974, 1979; Peacock, 1984a; D. G. Sutherland, unpublished data). More than one phase of glaciation is apparently represented, the last being the Loch Lomond Readvance.

No studies have been published of the Lateglacial vegetation of the Outer Hebrides, and only a limited amount of information is available on the Holocene vegetational history (Lewis, 1907; Erdtman, 1924; Blackburn, 1946; Heslop Harrison and Blackburn, 1946; McVean, 1961; Ritchie, 1966, 1985; Birks and Madsen, 1979; Walker, 1984a; Wilkins, 1984; Angus, 1987; Bohncke, 1988; Whittington and Ritchie, 1988; Bennett *et al.*, 1990; Birks, 1991). Pollen analyses from Little Loch Roag in Lewis (Birks and Madsen, 1979) suggest that the island was essentially treeless throughout the Holocene. However, the recovery of birch, pine and willow macrofossils from the blanket peats of Lewis (Wilkins, 1984), together with pollen records of *Betula, Corylus, Salix, Sorbus* and *Populus* at Callanish (Bohncke, 1988), implies the presence of scattered pockets of woodland in sheltered locations (Bohncke, 1988; Birks, 1991). By way of contrast, a detailed pollen record from Loch Lang in South Uist (Bennett *et al.*, 1990) indicates the presence there of areas of relatively diverse woodland dominated by *Betula* and *Corylus*, but also including *Quercus, Ulmus, Fraxinus exelsior* and *Alnus glutinosa*. On St Kilda, Walker (1984a) found that the middle to late Holocene vegetational changes reflected the broad climatic variations of this period, Man having an apparently negligible impact on the vegetation in this isolated area.

The history of sea-level changes in the outer islands is quite distinct from that in the Inner Hebrides and the greater part of the Scottish mainland by virtue of the relatively minor glacio-isostatic downwarping of the former area. Eustatic rise in sea level only impinged on the present coastal areas of the Outer Hebrides during the middle Holocene. This has resulted in peat cropping out in the present intertidal zone in many areas. The best documented example is at Borve (Ritchie, 1966, 1985), where at least 5 m of sea-level rise has occurred in the last 5000 years. Contemporaneously with this rising sea level, the coastline has evolved through the development of sand dune and machair systems along lengthy sections of the west coasts of the islands (Ritchie, 1966, 1979) and exceptionally on the eastern coasts (Ritchie, 1986; Whittington and Ritchie, 1988).

NORTH-WEST COAST OF LEWIS

J. E. Gordon

Highlights

The landforms and deposits on the coast of north-west Lewis provide important evidence for interpreting the patterns of sea-level change and glaciation in an area located towards the periphery of the main centres of glaciation. Part of the area remained ice-free throughout the Late Devensian.

Introduction

This site comprises a series of sections and landforms along a 16 km stretch of the north-west coast of Lewis, from Cunndal (NB 512655) to Cladach Lag na Greine (NB 387557). These have long provided some of the most important evidence for interpreting the Pleistocene history of the Outer Hebrides (J. Geikie, 1873, 1874, 1877, 1878; Baden-Powell, 1938; McCann, 1968; von Weymarn, 1974, 1979; Peacock, 1984a, 1991; Sutherland and Walker, 1984). The interest comprises a fossil shore platform, pre-Holocene raised beach deposits, peat (?interglacial), till, complex glacigenic sediment sequences and solifluction deposits. Together these features are of the very highest importance for studying glaciation and sea-level change in an area which has a fundamental bearing on the wider understanding of ice-movement patterns, ice sources, ice extent and climatic conditions in north-west Britain during the Pleistocene.

Description

Early descriptions recognized an essentially tripartite sequence in the area of Swainbost (NB 500635), comprising sands, gravels, clays and silts interbedded between two tills, with shells present in each of the layers (J. Geikie, 1873, 1874, 1877, 1878; Baden-Powell, 1938). However, more recent work has shown a much more complex succession of landforms and deposits represented at a number of localities along the coast (Figure 12.2). The principal features in the currently recognized sequence are as follows:

5. Till in the south and north; head and soliflucted till in the central part; multiple drift deposits at Swainbost.
4. Raised beach deposits with a cryoturbated upper horizon.
3. Head.
2. At different locations the shore platform is overlain by till and by peat; the relationship of the till to the peat is undetermined.
1. A raised shore platform and cliffline.

The oldest marine feature clearly recognizable is a low-level (7–10 m OD) raised shore platform and cliffline (Godard, 1965; McCann, 1968; von Weymarn, 1974). It occurs discontinuously in north-west Lewis and on the Eye peninsula (Figure 12.1) but is best seen between Galson (NB 453603) and Dell (NB 472621) where it attains a width of 150 m (McCann, 1968; von Weymarn, 1974). As described by von Weymarn (1974), the principal features of the platform are:

1. it is cut across the bedrock structure;
2. its seaward margin is sometimes covered in modern sand and shingle;
3. there is often a step down to a lower intertidal shore platform;
4. zones of weathered rock occur on its surface;
5. its landward margin is typically drift covered;
6. it is widest in embayments and narrows towards headlands where the backing cliff is best seen (e.g. at the mouth of the Dibadale Burn at NB 470615 – McCann, 1968);
7. abandoned stacks occur on its surface;
8. its distribution is largely confined to the area outside that of local glaciation in Lewis; a few remnants within the latter area are severely ice-modified.

Resting on the shore platform is a complex succession of deposits. In part, following von Weymarn (1974), these can be grouped into four units:

1. Raised beach deposits associated with till. Raised beach deposits form an important stratigraphic marker and were first identified at Galson, where they appear to overlie till (Baden-Powell and Elton, 1937). Baden-Powell and Elton noted that Torridonian sandstone erratics were present in the beach deposits but absent from the till. The occurrence of the beach has since been identified intermittently along the whole coast (Figure 12.2) (McCann, 1968; von Weymarn, 1974; Peacock, 1984a). From the height of the one beach terrace of any extent at Galson, McCann (1968) inferred a maximum sea level of 32–37 ft (9.8–11.3 m) OD. At several localities the upper layers of the beach deposit are cryoturbated. In addition to Galson, the raised beach deposits overlie till at Cladach na Luinge (NB 465611), Toa Dibadale (NB 469615), Cunndal and Swainbost (McCann, 1968; von Weymarn, 1974; Peacock, 1984a). In contrast, in the southern part of the site, to the south of Breivig (NB 414582), a separate till overlies the raised beach deposits.
2. Raised beach deposits associated with head and soliflucted till. Between North Galson (NB 438595) and Breivig, soliflucted till and head overlie the raised beach deposits; in other places the beach and soliflucted deposits are interbedded, for example at South Galson (NB 431591) and Cunndal.
3. At Toa Galson (NB 453603) an organic deposit interbedded with sand occurs above the shore platform and is overlain in turn by head and raised beach gravels (Figures 12.3 and 12.4). The organic deposit yielded radiocarbon dates greater than 39,100 to 47,150 BP (SRR–2365) (Sutherland and Walker, 1984). The pollen spectra (Figure 12.5) indicate development of the vegetation from treeless, open grassland to grassland with acid heath.
4. Complex successions often referred to as multiple drift deposits. These occur in the Swainbost–Dell area and comprise interbedded till, sand, gravel, laminated clay and silt beds. Historically, they were described by J. Geikie (1873, 1874, 1877, 1878) and Baden-Powell (1938), who both recognized a tripartite sequence comprising sand, gravel, clay and silt interbedded between two till layers. Although more recent references acknowledge the complexity of the sequence (von Weymarn, 1974; Peacock, 1984a), there have been no detailed sedimentological

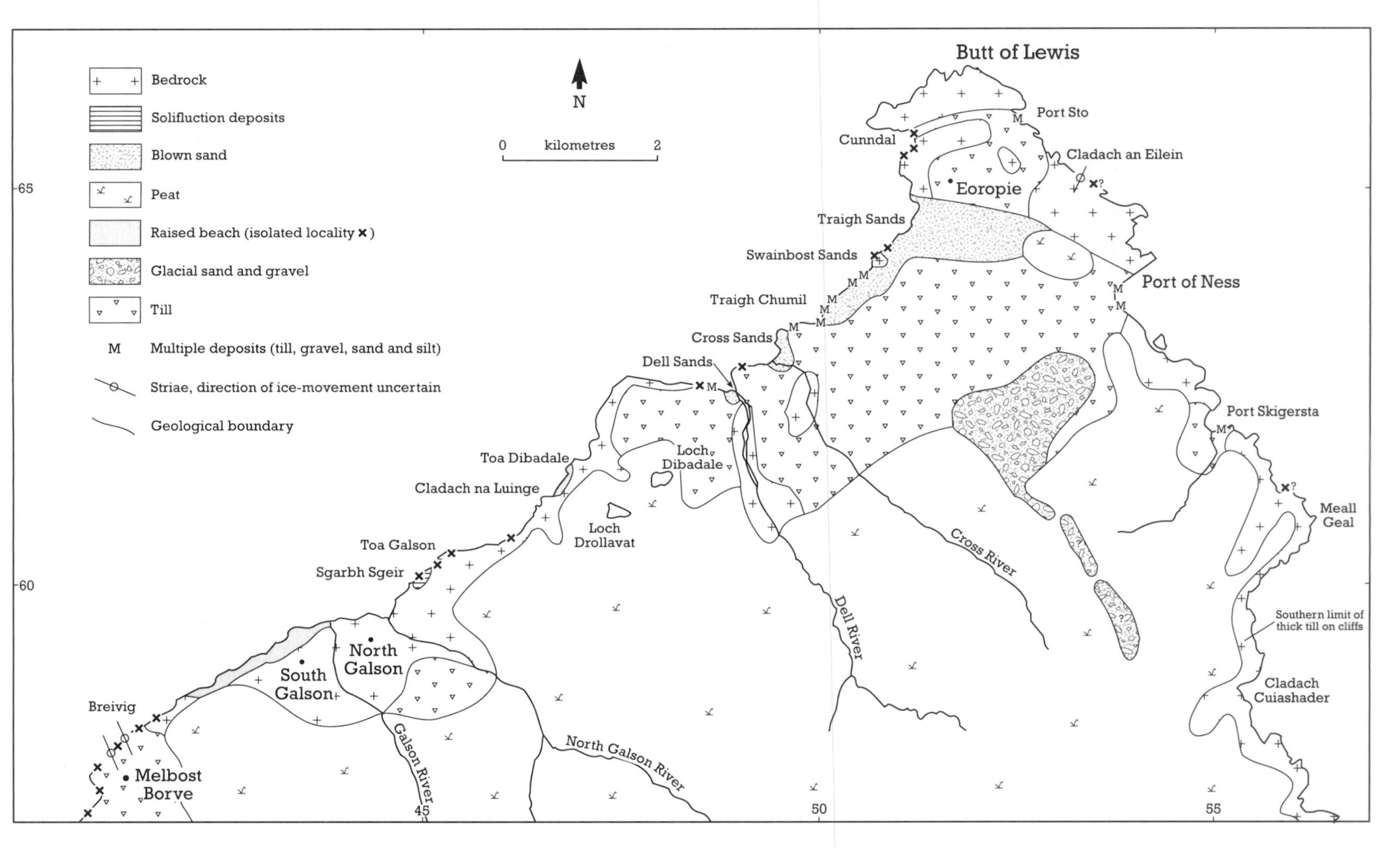

Figure 12.2 Quaternary deposits of north-west Lewis (from Peacock, 1984a).

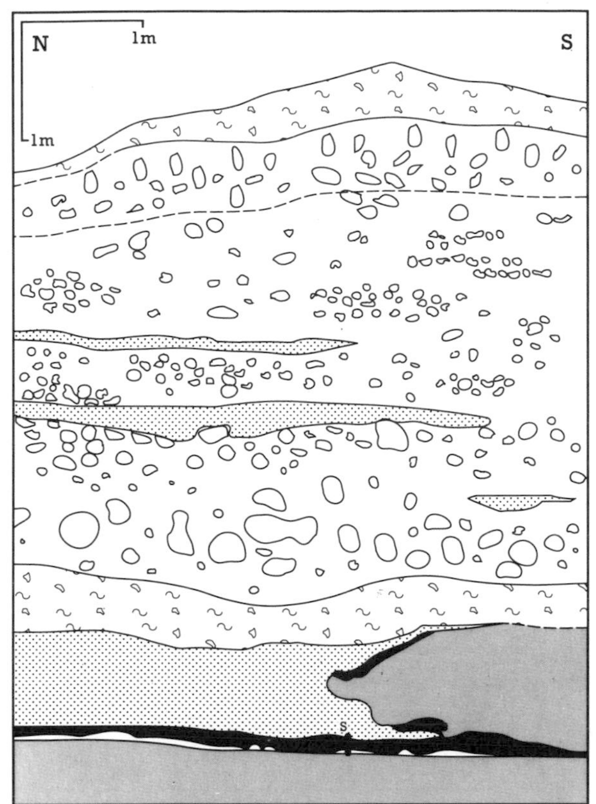

Figure 12.3 Toa Galson: sequence of sediments (from Sutherland and Walker, 1984).

studies. The deposits are glaciotectonically disturbed (J. Geikie, 1877; von Weymarn, 1974; Flinn, 1978b) and are rich in broken shell fragments. A faunal list published by Etheridge (1876) was updated by Baden-Powell (1938). According to Etheridge the faunal assemblage included species of arctic affinity (e.g. *Astarte depressa* (Brown) and *Chlamys islandica* (Müller)) and northern affinity (e.g. *Nuculana pernula* (Müller), *Arctica islandica* (L.), *Saxicava norvegica* (Spengler), *Natica montacuti* (Forbes) and *Fusus gracilis* (Da Costa)); Baden-Powell inferred that only the upper till of the tripartite sequence contained a diagnostic faunal assemblage (e.g. including *Astarte borealis* (Chemnitz), *Mya truncata* (L.), and *Panomya norvegica* (Spengler)), indicating cold-water conditions. Radiocarbon dates on shells from the middle or lower layers of these deposits yielded ages of between 34,470 +720/−660 BP and 39,500 +1270/−1100 BP (inner fractions) (SRR–2368 to SRR–2370) (Sutherland and Walker, 1984).

Amino acid analyses of fragments of *Arctica islandica* from the same beds indicated that the shells were from at least two distinct periods. The older group (6 analyses) had aIle/Ile ratios of 0.114 ± 0.007 and the younger group (9 analyses) had aIle/Ile ratios of 0.079 ± 0.013 (D. G. Sutherland, unpublished data). Ratios from similar analyses on *Arctica islandica* from the last interglacial in the British Isles are 0.15–0.16 (Bowen and Sykes, 1988), indicating that the shells from north-west Lewis are Early and Middle Devensian.

Interpretation

Interpretations of the evidence in north-west Lewis fall into two main groups. The earlier accounts were focused on the drift sequences at Swainbost; later accounts have placed greater emphasis in the wider succession and its spatial variations.

In an early reference to the 'alluvial land' of north Lewis, McCulloch (1819) described superficial accumulations of 'clay and marle, together with a mixture of rolled stones of different kinds'. From their lithological composition he believed they were derived from the gneissic rocks of the mountains but was unable to suggest a mode of derivation.

Initially, from analogies with Greenland glaciers, J. Geikie (1874, p. 215) hinted that sequences such as the multiple drift deposits in north Lewis could have been laid down in a single episode at or near the front of a glacier terminating in the sea (see also J. Geikie, 1877, p. 185 footnote).

Figure 12.4 Section at Toa Galson, north-west Lewis, showing the interglacial peat resting on bedrock and overlain by sand, head and the Galson Beach deposits. The upper part of the beach deposits has been affected by cryoturbation. (Photo: D. G. Sutherland.)

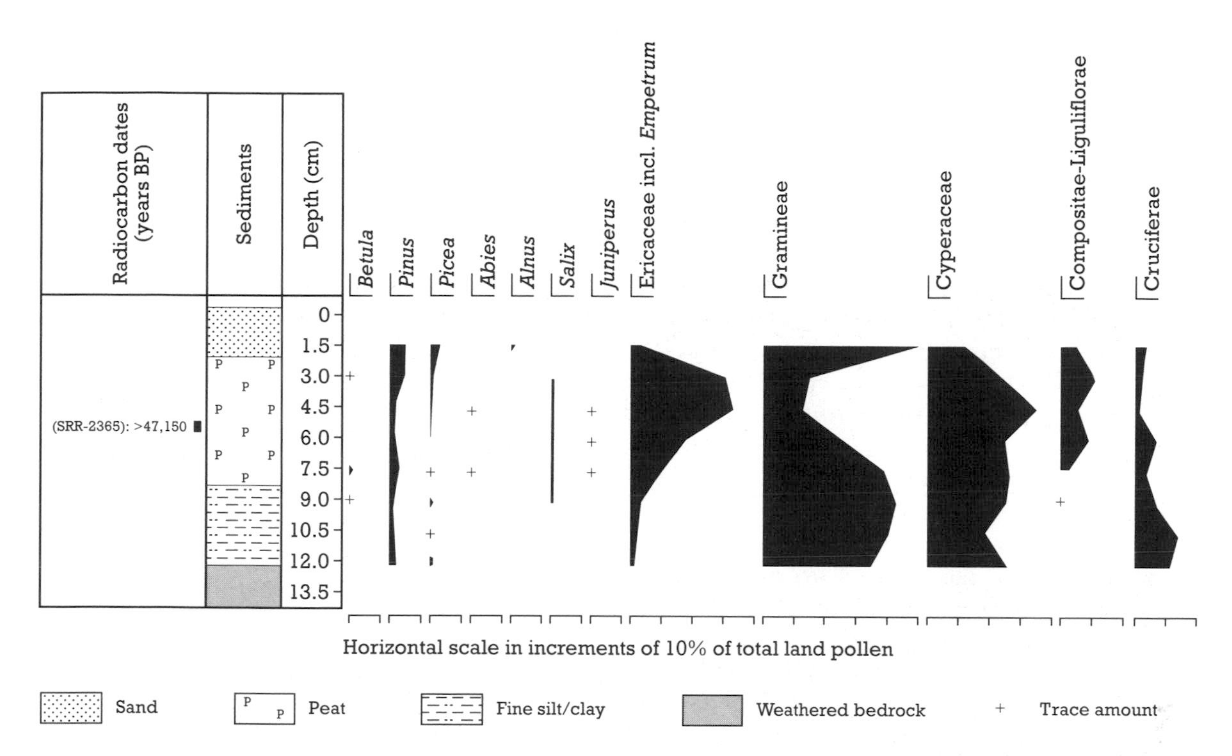

Figure 12.5 Relative pollen diagram for the peat deposit at Toa Galson showing selected taxa as percentages of total land pollen (from Sutherland and Walker, 1984). The location of the sample is shown in Figure 12.3.

Later, however, he interpreted the middle beds of the tripartite sequence as 'interglacial' marine deposits (Geikie, 1878) and postulated a succession of glacial and non-glacial episodes and relative sea-level fluctuations to explain the development of the full sequence over a much longer time-scale (J. Geikie, 1877, 1878).

Geikie (1878) noted the restricted distribution of the shelly till and interbedded deposits. He believed that the latter had once been more extensive but had been removed by the ice which deposited the upper till; the shelly till was confined to the area where basal ice, deflected by the Hebridean land mass, impinged on its northern periphery.

Subsequently, Jehu and Craig (1934) and Baden-Powell (1938) acknowledged Geikie's interpretation of interglacial submergence. Baden-Powell suggested that the submergence was at least 17 m and possibly as much as 60 m, the latter value corresponding with one of the planation surfaces identified by Panzer (1928). Godard (1965) largely concurred, but raised the question as to whether the altitude of the marine deposits might simply reflect older marine material reworked by glacial meltwaters. More recent interpretations, however, are agreed that the multiple drift deposits of north-west Lewis are glacigenic in origin and were formed during a single glacial episode (von Weymarn, 1974; Flinn, 1978b; Peacock, 1984a; Sutherland and Walker, 1984).

Based on what was the first systematic account of the landforms and deposits in north-west Lewis, von Weymarn (1974) inferred a possible sequence of events commencing with formation of the raised shore platform during one or more pre-Devensian interglacials. The platform was overridden by glacier ice from the east and buried by till containing Torridonian sandstone erratics. During the Ipswichian, relative sea level rose by around 7 m and the earlier till was reworked. During the Devensian glacial phases the raised beach gravels were overridden to the south of Borve by a local ice-cap on Lewis, but the coast to the north remained free of ice and subject to intense periglacial activity. The multiple drift deposits probably related to external ice in the north of Lewis during the Late Devensian.

Alternative suggestions were made by Jardine (1977) that the beach deposits could date from the Loch Lomond Stadial rather than the Ipswichian, and by Dawson (1979c) that the solifluction deposits could be of Loch Lomond Stadial age and the beach formed earlier during the Lateglacial. Flinn (1978b) inferred that the multiple drift deposits at Swainbost were deposited prior to the last glacial maximum, when the northern part of Lewis was covered by relatively inactive ice near the margin of the local ice-cap. However, subsequent work by Peacock (1981a, 1984a) and Sutherland and Walker (1984) has substantially confirmed von Weymarn's interpretations and elaborated on the details. Peacock (1981a, 1984a) identified possible limits of the Late Devensian ice in north-west Lewis. The northern limit of the local ice-cap was at Breivig. The limit of the external ice to the north was more difficult to define and he presented two alternative positions. The critical problem concerned the stratigraphic relationship of the raised beach to the multiple drift sequence at Swainbost. Peacock considered that the beach lay above the latter and therefore favoured a relatively small ice-covered area, leaving the coast between Breivig and the Butt of Lewis unglaciated. The alternative explanation that a lobe of ice crossed the coast in the Swainbost area was considered less probable because of the inferred stratigraphic relationship of the raised beach and multiple drift sequence.

Sutherland and Walker (1984) produced a different interpretation of the stratigraphy from that of von Weymarn and Peacock, concluding that the multiple drift sequences on the north-west coast lay stratigraphically above the raised beach. A Late Devensian age for the multiple drift sequence is supported by the radiocarbon dates and amino acid analyses on the shells from the glacial deposits. Therefore the coast between Eoropie and Dell Sands, where the multiple drift sequences occur, was covered by the last ice-sheet, confirming Peacock's alternative hypothesis.

The radiocarbon dates from the peat beneath the raised beach at Toa Galson confirm the antiquity of the deposit. Sutherland and Walker (1984) tentatively considered it to be of interglacial origin although there is no evidence to assign it to a particular stage. Significantly, the pollen spectra differ from those associated with the interstadial sites at Tolsta Head (see below) and St Kilda (Sutherland *et al.*, 1984), both of which produced finite radiocarbon dates.

North-west Lewis is therefore an area of the highest importance for demonstrating a series of glacial, periglacial and marine events of Devensian and possibly pre-Devensian ages. In sequence, the key elements are as follows:

6. Late Devensian features, including till limits, shelly multiple drift sequences and ice-free areas with associated periglacial deposits.
5. Raised beach deposits (pre-Late Devensian).
4. Periglacial deposits.
3. Organic deposits, possibly interglacial (pre-dating the raised beach deposits).
2. Till (pre-dating the raised beach deposits).
1. Raised shore platform of pre-Devensian age.

Key localities within the site are as follows:

1. Cunndal, where till is overlain by a raised beach, which is succeeded by soliflucted till. Cunndal lies outside the Late Devensian ice margin.
2. Traigh Sands (NB 511644) to Peicir (NB 485625), which includes a raised beach and the classic shelly multiple drift sequences now dated as Late Devensian in age.
3. Aird Dell to Toa Galson, which demonstrates the sequence of shore platform, till and raised beach outside the Late Devensian ice limit.
4. Immediately south of Toa Galson organic deposits occur between the shore platform and the raised beach.
5. Farther south, between North Galson and Breivig, the raised beach terrace is best developed and is overlain by soliflucted till and head.
6. At Breivig the Late Devensian ice limit crosses the coast and till overlies the raised beach here and to the south.
7. At Cladach Lag na Greine the raised beach rests on a shore platform and is overlain by till. Locally, where the platform rises above the beach deposit, it is striated, which provides important evidence that the beach was overridden by glacier ice and is overlain by *in situ* lodgement, rather than soliflucted till.

The assemblage of deposits and landforms represented at these localities is unique in Scotland and has significant bearing on the interpretation of the Pleistocene history and chronology of events in north-west Britain, including sea-level change, glacier reconstruction, palaeoclimatology and vegetational history. For example, as Sutherland and Walker (1984) point out, conventional models of the last Scottish ice-sheet require reassessment in the light of the evidence from north-west Lewis. The presence there of an ice-free area indicates a relatively restricted extent of the last ice-sheet in Britain at its north-west margins, which contrasts with some recent reconstructions that placed the margin on, or near, the edge of the continental shelf (Boulton *et al.*, 1977, 1085; Andersen, 1981). It also supports a growing body of evidence (see Orkney, Caithness and North-east Scotland) that the northern margins of the last ice-sheet may have been relatively limited in extent in contrast to the southern margins which reached south into South Wales, central England and the north coast of East Anglia. Sites such as the north-west coast of Lewis therefore provide critical field evidence to constrain and refine mathematical models of the last ice-sheet and the palaeoclimatic inferences derived from them.

North-west Lewis is also unique in Scotland in providing unequivocal evidence for a Late Devensian ice-sheet limit on land and therefore provides exceptional opportunities for comparative studies of pre-Late Devensian and post-Late Devensian rates of weathering and soil development. Further, north-west Lewis is one of only a few sites in Scotland where interstadial or possibly interglacial organic deposits are represented. As such it is of the very highest importance for palaeoecological and palaeoenvironmental studies. Finally, the multiple drift deposits of north-west Lewis are of significant sedimentological interest for interpreting and reconstructing the depositional environments at the margin of the Late Devensian ice-sheet.

Conclusion

The unique assemblage of landforms and deposits on the north-west coast of Lewis has a long history of research and has provided critical information for interpreting the Pleistocene in Scotland. Among the key features of interest are a pre-Devensian shore platform, a pre-Late Devensian raised beach, organic deposits of possible interglacial origin, solifluction deposits and a complex sequence of fossiliferous sediments deposited by the Late Devensian ice-sheet (approximately 18,000 years ago). This area therefore provides a wealth of evidence for interpreting patterns of sea-level change, palaeoenvironmental conditions and ice-sheet history. One of the key findings to emerge from recent work is that part of the area remained ice-free when the Late Devensian glaciation was at its maximum extent (about 18,000 years ago).

PORT OF NESS

J. E. Gordon

Highlights

The sediments exposed in the coastal section at Port of Ness form part of a complex sequence deposited as the Late Devensian ice-sheet melted. They provide important evidence for interpreting the sedimentary environments associated with ice wastage.

Introduction

The site (NB 537636) is a coastal section located immediately to the south of Port of Ness, in north Lewis. It is important for glacial stratigraphy and sedimentology. The exposures show a complex sequence of interbedded shelly tills, sands, silts and gravels which has provided significant evidence for interpreting the Pleistocene succession in the Outer Hebrides for over one hundred years (Geikie, 1873, 1878; Baden-Powell, 1938; von Weymarn, 1974; Peacock, 1981a, 1984a).

Description

In an early account, MacCulloch (1819) referred briefly to the superficial deposits in north Lewis, which he believed were derived from the 'waste of the gneissic mountains'. Later, J. Geikie (1874) presented a general stratigraphy for the same area based on sections in north-west Lewis and at Port of Ness, which essentially comprised a tripartite succession of two till units with interbedded layers of stratified sand, gravel and clay. At Port of Ness the lower till contained shells and lenses of sand incorporated from the sea floor. The stratified sediments above consisted of coarse gravel with shell fragments. The upper till, which also contained shell fragments, was capped by a further layer of stratified gravels and sands.

Baden-Powell (1938) provided additional details of the Port of Ness succession, in particular noting the lateral variability of the deposits. At the southern end the entire section comprised a massive purple-coloured silt with occasional stony layers. Mollusc shell fragments (e.g. of *Chlamys islandica* (Müller), *Astarte borealis* (Chemnitz), *Macoma calcarea* (Gmelin) and *Mya truncata* (L.)) were taken by him to represent a 'cold'-water marine assemblage, and on this basis Baden-Powell correlated the silt deposit ('Glacial Marine Bed') with the upper till of the tripartite sequence at Swainbost on the north-west coast of Lewis (see above). The stone content of the silt increased northwards, merging into boulders and gravel with a silty matrix. At one locality, a hollow in the underlying bedrock showed a layer of till interbedded with sands and gravel. The sands and gravel were correlated with the 'interglacial' marine deposits (the middle bed of the tripartite sequence) identified by Baden-Powell in north-west Lewis.

At Port of Ness, von Weymarn (1974) recorded the following sequence:

5.	sand and silt	*c.* 0.8 m
4.	till	*c.* 5.4 m
3.	sand, silt and gravel	*c.* 1.9 m
2.	till	*c.* 1.8 m
1.	gravel	up to *c.* 1.6 m

Later Peacock (1981a, 1984a) provided further details of the complexity of the sequence and its lateral variability. At the north end of the section, he recorded about 8 m of crudely-bedded, bouldery gravel in which clast imbrication suggested deposition towards the north-west. The gravel is overlain by a brown diamicton, possibly a till. Southwards, the gravel passes laterally into an interbedded sequence, about 10 m thick, of tills or debris-flow deposits and sands and gravels.

Until recently the age of the Port of Ness deposits was unconfirmed. However, the work of Sutherland and Walker (1984), together with unpublished results of amino acid analysis of shells in the glacial deposits of north-west Lewis (see above), suggests that the multiple drift sequence at Port of Ness may be correlated with that on the north-west coast of Lewis and therefore is of Late Devensian age.

Interpretation

J. Geikie (1874) interpreted the stratified deposits at Port of Ness as representing an interglacial marine submergence. This idea was also developed by Baden-Powell (1938). According to von Weymarn (1974), however, the sand, silt and gravel (bed 3 of his succession) could not be correlated with the raised beach gravels on the north-west coast of Lewis, and their characteristics suggested that they might be part of a complex depositional sequence. This suggestion was later supported by Peacock and Ross (1978). Neither von Weymarn nor Peacock and Ross ascribed specific origins to the individual beds or the succession as a whole. Subsequently, however, Peacock (1981a, 1984a) tentatively identified the succession as representing a complex proglacial debris fan from an ice source lying to seaward. Therefore, although the sequence of deposits appears to be established in outline, the site has significant potential for sedimentological study to clarify in detail the lateral variability of the different beds and to provide the basis for a better understanding of their depositional environments.

Port of Ness is important in several respects. It is of historical interest, having been recognized for over one hundred years as a key section providing evidence for the Pleistocene history of northern Lewis, and complementing the interest of the sections on the north-west coast of the island. Formerly it was thought to show both interglacial and glacial marine deposits, but recent studies suggest that the sequence of sediments represents a complex ice-marginal depositional environment. By virtue of its location, the Port of Ness succession also has a direct bearing on interpreting the pattern of the last glaciation and deglaciation in the northern part of the Outer Hebrides. The conventional view, as set out by J. Geikie (1873, 1874, 1877, 1878), that during the last glaciation the whole of the Outer Hebrides was covered by ice moving across the Minch from the Scottish mainland, has now been reassessed and the concept of an extensive local ice-cap firmly established (von Weymarn, 1974, 1979; Flinn, 1978b; Peacock and Ross, 1978; Peacock, 1984a; Sutherland and Walker, 1984). However, a key question, still not fully resolved, is whether ice from the mainland reached and crossed northern Lewis (see Peacock, 1984a; Sutherland and Walker, 1984). Evidence of onshore movement of ice is suggested by the drift deposits in northern Lewis. As first noted by Geikie and later by Baden-Powell (1938) these deposits are distinguished from the drift of the rest of Lewis by their complex stratigraphy, glaciotectonic deformation and their sandstone (ascribed to the Torridonian) erratic and shell content. However, the erratics could possibly have been derived from the Mesozoic sedimentary rocks that underlie the floor of the North Minch close offshore and then transported on land by local Hebridean ice recurving onshore on to north Lewis (Sutherland and Walker, 1984).

Further research on the Port of Ness succession, together with other sites in north Lewis, should help to clarify this issue, which has potentially significant implications for the extent of the Late Devensian ice-sheet, the ice sources and ice-movement patterns and therefore the associated palaeoclimatic conditions in north-west Britain (see also North-west coast of Lewis).

Conclusion

Like the deposits in north-west Lewis, those at Port of Ness have been much studied, featuring in early interpretations of the glacial history of the Outer Hebrides. Recent investigation suggests that they form part of a complex fan of sediments deposited from the margin of an ice-sheet lying to the east. The site has significant research potential for achieving a better understanding both of the pattern of development and melting (deglaciation) of the last ice-sheet in the Outer Hebrides, and of the depositional processes that accompanied ice wastage (about 18,000 years ago).

TOLSTA HEAD

J. E. Gordon and D. G. Sutherland

Highlights

Deposits exposed in the coastal section at Tolsta Head include organic lake detritus formed during a Middle Devensian interstadial around 27,000 years ago. Pollen and diatoms preserved in these sediments provide an exceptionally detailed record of the vegetational and environmental conditions at that time.

Introduction

This site (NB 557468) comprises a cliff-top section located on the south side of Tolsta Head in north Lewis. It shows a sequence of organic interstadial deposits underlying till deposited by Late Devensian ice. Although organic sediments containing marine shells were reported from near Tolsta Head by Dougal (1928), the present site was first described by von Weymarn and Edwards (1973) and further details of both the organic and glacial sediments have subsequently been given by Flinn (1978b), Edwards (1979a) and Birnie (1983).

Description

The section was described by von Weymarn and Edwards (1973), with further details being added by Birnie (1983) (Figure 12.6). At the base, resting on Lewisian gneiss bedrock, is approximately 0.6 m of bedded silts and sands with organic lake detritus. Overlying these is 2.5 m of reddish-brown till notable for a relatively high content of Torridonian sandstone erratics. The top of the organic sediments has been truncated and material from it incorporated as clasts into the base of the till (D. G. Sutherland, unpublished data). The uppermost 1 m of the till has been frost-disturbed.

Palynological, plant macrofossil and diatom studies have been carried out on the organic sediments (von Weymarn and Edwards, 1973; Edwards, 1979a; Birnie, 1983). Von Weymarn and Edwards (1973) reported grass–sedge-dominated pollen spectra (Figure 12.6), with a consistent increase in the percentage of juniper pollen towards the top of the profile. Birnie (1983) provided further details, subdividing the organic deposits into a lower, sandier unit and an upper, more organic unit. The lower unit was considered to have been deposited in more variable flow conditions than the upper one, and it is possible that there may have been a hiatus in deposition between the two. The diatom record indicates that alkaline water conditions were present throughout the upper zone and a succession developed from epipelic and planktonic communities to epiphytic communities. With the notable exception of *Juniperus*, Birnie (1983) confirmed the pollen spectra as reported by von Weymarn and Edwards (1973) and she also recorded *Salix herbacaea* macrofossils. A radiocarbon date of 27,333 ± 240 BP (SRR–87) was obtained from the uppermost 0.15 m of the organic sediments (von Weymarn and Edwards, 1973).

Interpretation

The radiocarbon date confirms that the deposits preserve evidence of a Middle Devensian interstadial. Von Weymarn and Edwards (1973) considered that the pollen spectra indicated a

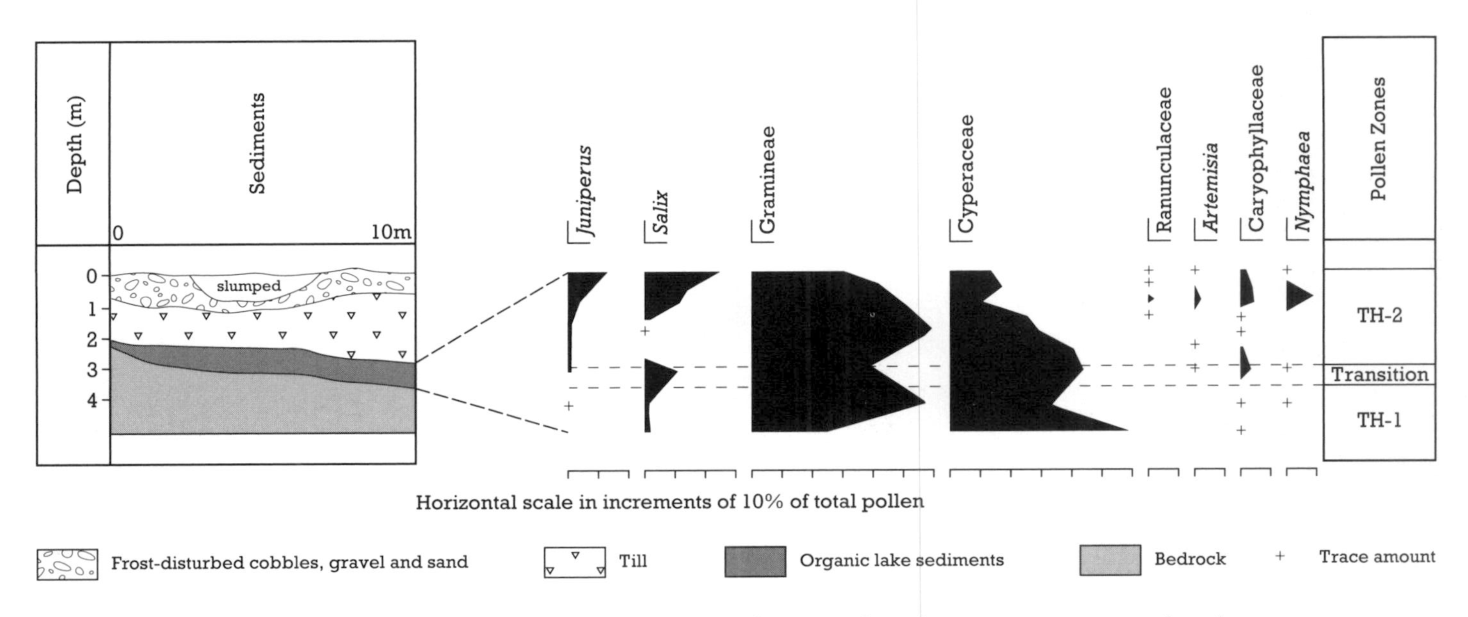

Figure 12.6 Tolsta Head: sediments and relative pollen diagram showing selected taxa as percentages of total pollen (from von Weymarn and Edwards, 1973; Birnie, 1983; Lowe, 1984).

flora not inconsistent with a cool maritime climate. Birnie (1983) considered that the alkaline water, the inwash of minerogenic sediment and the occurrence of both pollen and macrofossils of open-ground herbs suggested soil instability, probably solifluction, throughout the period represented by the organic sediments. However, the degree of severity of the climate is uncertain, as the presence of *Nymphaea* in the uppermost 0.10 m and certain diatoms are compatible with temperatures not necessarily any lower than those of today. The radiocarbon date also indicates that the overlying till is Late Devensian in age, in agreement with the conclusions of Sutherland and Walker (1984) based on radiocarbon dating and amino acid analyses of shells from till in the north of Lewis (see North-west coast of Lewis).

Von Weymarn and Edwards (1973) reported that till fabric measurements showed a predominant N50°W clast orientation and they inferred ice movement from the south-east. However, in the till Flinn (1978b) found fragments of a phyllonite, which crops out at the south-west corner of Tolsta Head, and he concluded that the direction of movement of the ice that deposited the till was towards the north-east. Von Weymarn (1979) suggested that during the Late Devensian, Tolsta Head was close to the junction of the Scottish ice-sheet (ice flow from the south-east) and the Outer Hebrides ice-cap (ice flow from the south-west), which may explain the apparently conflicting evidence (see also Sutherland, 1984a).

There are relatively few sites in Scotland, in addition to Tolsta Head, at which evidence for a Middle Devensian interstadial has been discovered and dated: Bishopbriggs (Rolfe, 1966), Hirta (St Kilda) (Sutherland *et al.*, 1984), Crossbrae (Hall, 1984b), Sourlie (Jardine *et al.*, 1988), Creag nan Uamh (Lawson, 1984), and possibly Teindland (Fitzpatrick, 1965; Edwards *et al.*, 1976). Of these, Tolsta Head is the site which has been studied in most detail and which has provided most information about the environment of that period. Three of the sites (Bishopbriggs, Crossbrae and Sourlie) were temporary exposures and as there is doubt as to whether Teindland does indeed date from the Middle Devensian, Tolsta Head remains the most important interstadial site of this age in Scotland.

Conclusion

Tolsta Head is important for showing one of the few deposits in Scotland that may be dated to a Middle Devensian interstadial (a warmer climatic phase during an otherwise intensely cold glaciation). It has provided the most detailed evidence for environmental conditions during the phase (approximately 27,000 years ago) immediately before the last glacial maximum, indicating a cool maritime climate and unstable soils. Tolsta Head is a key reference site for palaeoenvironmental reconstruction in Scotland.

GLEN VALTOS

D. G. Sutherland

Highlights

Glen Valtos is a notable example of a glacial meltwater channel formed during the melting of the Late Devensian ice-sheet. It provides important evidence for interpreting the pattern of meltwater flow and the ice-sheet configuration.

Introduction

Glen Valtos (NB 060343–NB 084346) is a 2.5 km long meltwater channel located near Uig in west Lewis. In the Outer Hebrides there are few areas in which clear systems of meltwater deposits or landforms are developed. The most impressive of such features occur in south-west Lewis in the neighbourhood of Uig, where the Glen Valtos meltwater channel is particularly prominent. This channel, as well as being impressive in its own right, is significant also as part of a sequence of glaciofluvial features deposited by meltwaters flowing from west to east, at right angles to the direction of ice flow during the last glaciation of the area. Glen Valtos has been noted in several accounts of the glaciation and deglaciation of the Uig area (Jehu and Craig, 1934; Godard, 1965; von Weymarn, 1974, 1979; Peacock, 1984a).

Description

The Glen Valtos channel begins on the watershed overlooking Camas Uig at approximately 40 m OD and continues to the east for over 2.5 km to enter Loch Miavaig (Figure 12.7). At the intake, the channel is small but deepens rapidly against

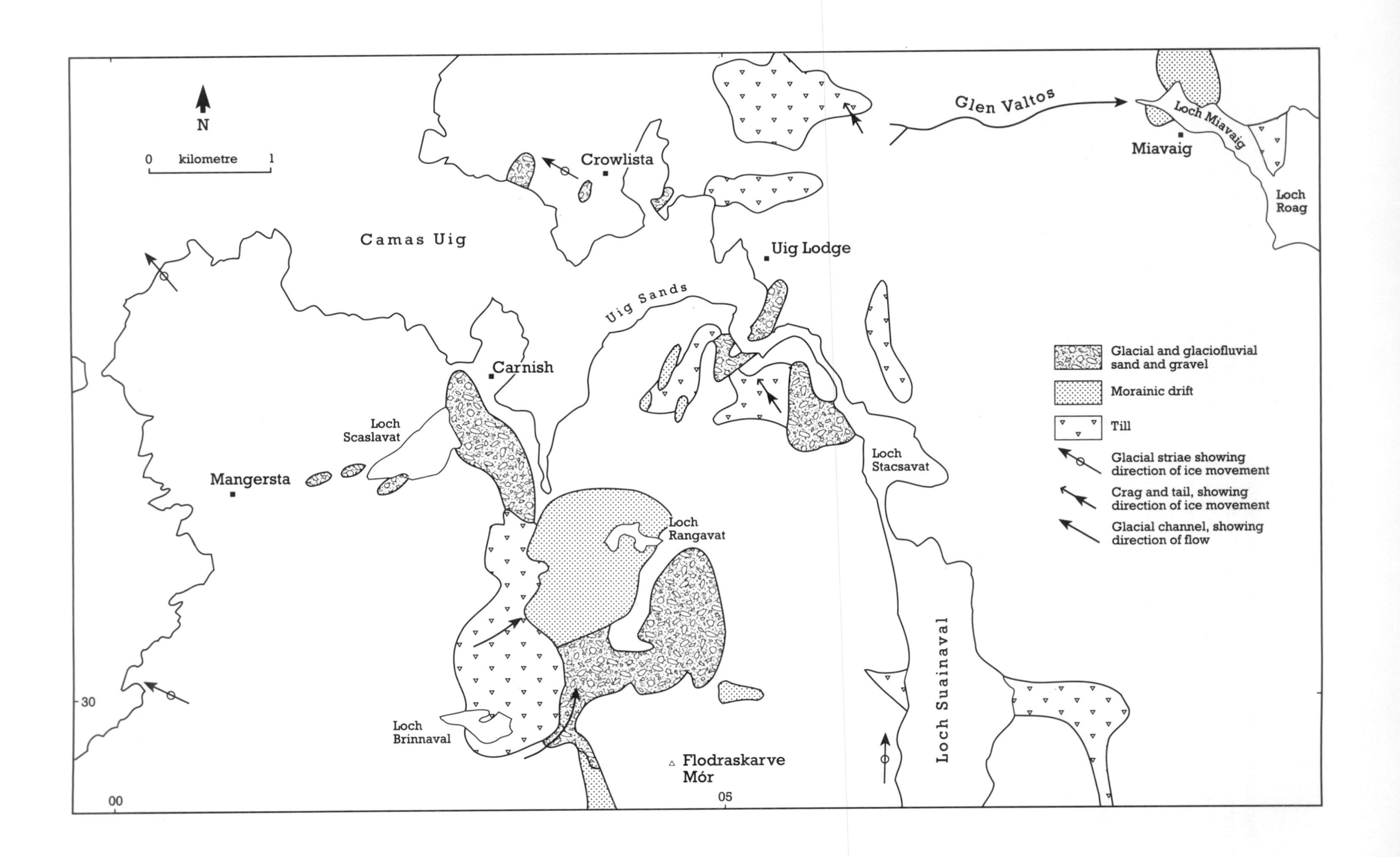
N
0 kilometre 1
Camas Uig
Crowlista
Glen Valtos
Loch Miavaig
Miavaig
Loch Roag
Uig Lodge
Uig Sands
Carnish
Loch Scaslavat
Mangersta
Loch Stacsavat
Loch Rangavat
Loch Brinnaval
Flodraskarve Mór
Loch Suainaval
Glacial and glaciofluvial sand and gravel
Morainic drift
Till
Glacial striae showing direction of ice movement
Crag and tail, showing direction of ice movement
Glacial channel, showing direction of flow
30
00
05

the general slope of the land, and at its eastern end is over 45 m deep (Figure 12.8). Its western portion is sinuous but the eastern, most deeply incised section is linear, suggesting fault control. At its head there are three 'blind' intakes in addition to the main one. Two of these blind ends are continued at the top of the slope by a peat-filled channel system, which extends for approximately 800 m to the south-west where it too has an intake above Camas Uig, at an altitude of about 50 m OD.

There are no glaciofluvial deposits contained within the channel, but at its mouth by Miavaig (Figure 12.7) there is a group of small mounds composed of bouldery, poorly sorted fluvial gravels. In the upper part of the northern channel side (near NB 074346) up to 1 m of grey till is exposed. The implication of these deposits together with the discordant nature of the channel intakes is that the channel was probably formed, or at least initiated, subglacially.

The cliffed bedrock slopes of the channel have weathered to produce angular debris that has formed stratified screes several metres thick, but these deposits have not been investigated.

The various intakes to the channel system occur on bare rocky slopes. However, to the west, around Camas Uig, there are major glaciofluvial accumulations (Figure 12.7) (Peacock, 1984a, figure 3). The principal ones are, first, a major arcuate ridge trending approximately south-west to north-east and terminating by Uig Lodge (NB 055333); second, a large glaciofluvial delta at Carnish (NB 030323) deposited to an altitude of about 53 m OD; third, a large area of dead-ice terrain south of Loch Rangavat (NB 042311); and fourth, a large mound of stratified sand and gravel to the west of Crowlista (NB 041339). All of these deposits relate to meltwater flow from west-south-west to east-north-east.

Interpretation

The glaciofluvial deposits and landforms in the Uig area have been described in a number of publications. Jehu and Craig (1934) considered the Glen Valtos channel to have been eroded by waters overflowing from an ice-dammed lake in the area of Camas Uig, the Carnish delta having been deposited in the same lake. The large arcuate drift mound south-west of Uig Lodge was interpreted by both Godard (1965) and von Weymarn (1974) as an end moraine. The former thought the ice from which it was deposited had occupied Camas Uig, whereas the latter suggested the moraine was related to a glacier emerging northwards from the valley of Loch Suainaval (NB 068290), on the opposite side from Camas Uig. In contrast, Ritchie and Mather (1970) interpreted the ridge as an esker, and Peacock (1984a) argued that it was one of a series of kames, eskers and morainic mounds that were possibly associated with subglacial drainage towards the channel. The topographic situation of the glaciofluvial deposits in the Uig area is such that at the time of their formation, drainage north-westwards to the sea via Camas Uig must have been blocked by ice and it is most probable that it was at this time that Glen Valtos was eroded. Such an interpretation is supported by the only slight difference in altitude between the highest intake to the Glen Valtos channel system (about 50 m OD) and the altitude of the top of the Carnish delta (53 m OD).

Striations, crag-and-tail landforms and ice-moulded bedrock indicate that the last ice movement in the Uig area was to the west of north. Both von Weymarn (1974, 1979) and Peacock (1984a) noted that the west to east direction of flow of the meltwaters draining through the Glen Valtos channel was almost at right angles to that direction of ice flow. Furthermore, the disposition of the features apparently requires ice lying off the west coast of south-west Lewis to have been thicker or to have melted later than ice in the Loch Roag area to the east. This pattern of ice decay is anomalous in terms of the present knowledge of glaciation of the Outer Hebrides and awaits further research to be fully understood.

The Glen Valtos channel is a particularly impressive example of a subglacial meltwater channel. In contrast to the majority of similar channels found elsewhere in Scotland, it has been eroded by waters flowing almost at right angles to the last direction of ice flow in the area. The channel is part of a wider assemblage of glaciofluvial deposits and landforms which were deposited during the same period of ice decay, but the reasons for the apparently anomalous direction of meltwater flow are as yet poorly understood.

Figure 12.7 Landforms and deposits of the Glen Valtos–Uig area (from Peacock, 1984a).

Figure 12.8 Glen Valtos meltwater channel. The channel has the form of a single, narrow gorge. (Photo: D. G. Sutherland.)

They seem to imply thicker ice immediately off the west coast of south-west Lewis than in the area close to the mountains that were one of the sources of the ice.

Conclusion

Glen Valtos is important for glacial geomorphology, in particular for the most impressive meltwater channel in the Outer Hebrides. It forms part of an assemblage of landforms and deposits that together provide important evidence for interpreting the pattern of decay of the last ice-sheet (approximately 18,000–14,000 years ago). In particular, the direction of meltwater flow indicated by Glen Valtos implies the presence of thicker ice off the west coast of Lewis than onshore; this apparent anomaly (thicker ice would be expected nearer the mountain sources inland) remains to be explained in terms of what is known of the configuration of the last ice-sheet.

BORVE

W. Ritchie

Highlights

The coastal sediments in the inter-tidal and sub-tidal areas at Borve comprise a sequence of interbedded sands and organic materials which accumulated in a former freshwater lake. Analysis and radiocarbon dating of these sediments has provided important evidence for interpreting sea-level changes and coastline evolution during the Holocene in the Western Isles.

Introduction

The site (NF 769499) is located on the south coast of North Uist facing the entrance to the South Ford, which separates Benbecula from South Uist. The interest lies in organic deposits preserved in the inter- and sub-tidal zones. Organic deposits, usually described as peat, have been recorded at intertidal locations in the Uists since the 18th century (Macleod, 1794; McRae, 1845; Martin, 1884; Beveridge, 1911; Jehu and Craig, 1927; Elton, 1938). Although the materials at these sites resemble peat, they consist of a variety of compressed organic layers, normally with a high sand content. Some layers contain wood fragments and others reed stems: some layers have a high silt content and a few contain terrestrial gastropods. Investigations in recent years (Ritchie, 1966, 1979, 1985) have demonstrated the importance of these organic horizons in understanding both sea-level change and the accompanying development of the dune and machair landscapes of the western coasts of the Outer Hebrides. The most extensive and best documented site is that at Borve on the south-west coast of Benbecula (Ritchie, 1985).

Description

The deposits occur in a shallow rock basin, now covered by a thin sand beach, between low, irregular rock platforms. At high tide level there is a small shingle storm beach which lies at the base of a sand cliff 1–4 m high. There are no dunes and the coastline is backed by a mature, flat machair plain which stretches inland for up to 2 km. The deposits have been investigated in a series of pits and boreholes and comprise, principally, a succession of interbedded sand and organic layers (Figure 12.9). Some organic layers include wood fragments and reed stems. Pollen analysis and study of macrofossils indicate that the entire site was at one time a freshwater lake which progressively infilled to form a marshy grassland. Later, massive quantities of sand were carried across these deposits to form the existing machair landforms. The complexity of the stratigraphy, however, indicates many changes in local conditions. Throughout the period of accumulation of the organic deposits there was considerable sand-blowing, the episodic nature of which is reflected in some individual sand layers up to 0.2 m in thickness; other beds are mainly organic material with only a few discrete sand particles in the matrix.

Three radiocarbon dates indicate the approximate period of deposition of the organic deposits and phases of sand influx. A fragment of wood from 0.6 m below present mean tide level was dated to 5700 ± 170 BP (I–1543) (Figure 12.9, pit J), and peat-like material, obtained from offshore and 3.7 m below the same datum, gave an age of 5160 ± 45 BP (SRR–1222). These two dates imply that the freshwater lake existed between a time prior to 5700 BP until after 5200 BP. In the machair sands backing the beach, a peaty layer at 2 m above present high water

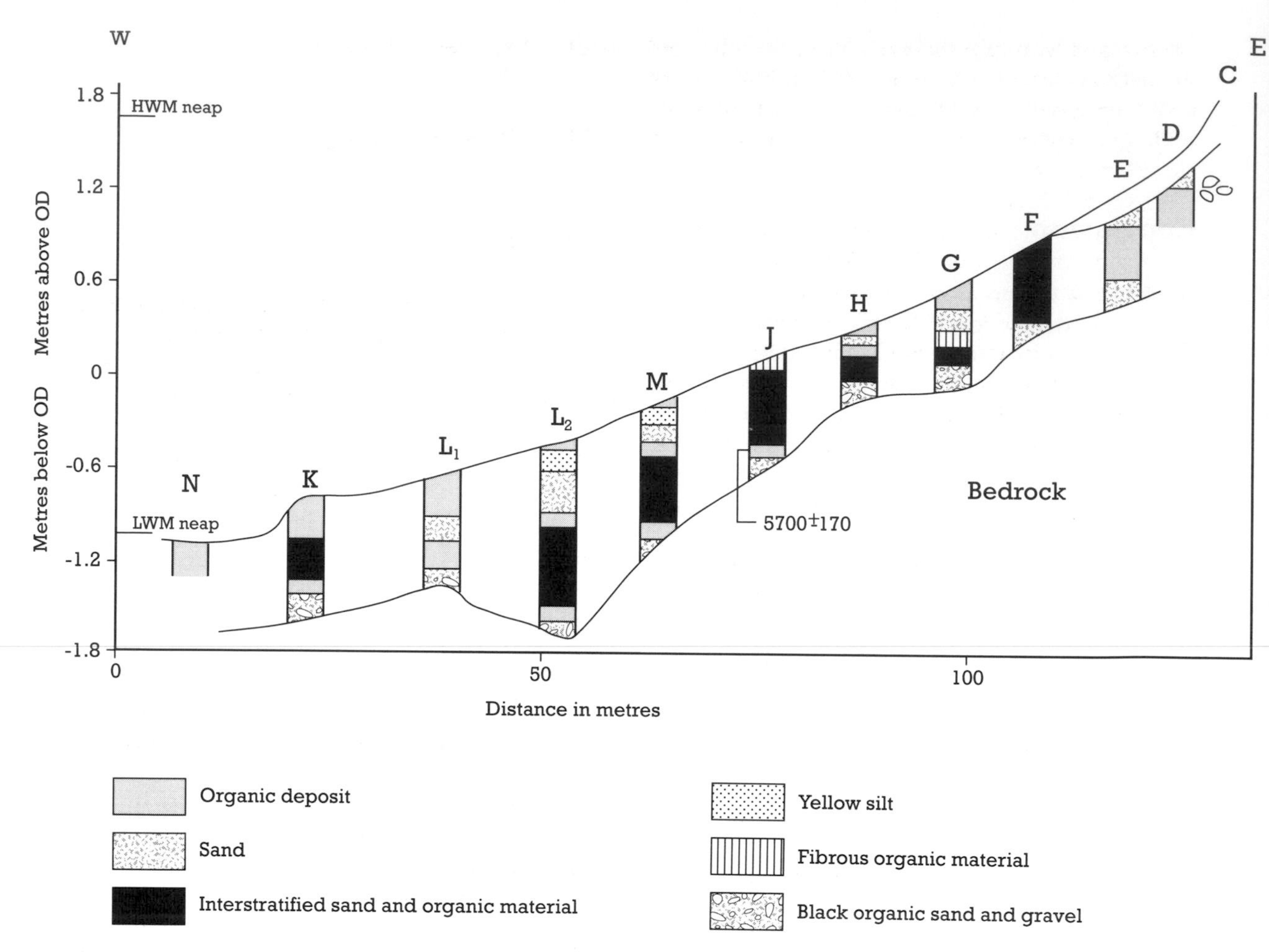

Figure 12.9 Profile across the intertidal deposits at Borve, showing the sediment sequence and its variations (from Ritchie, 1985).

mark has been radiocarbon dated to 3370 ± 60 BP (GU–1096); this layer represents a machair slack similar to those found in most low-lying machair plains today.

Interpretation

The freshwater lake in which the organic remains accumulated is interpreted as being similar to the lochs and marshlands that usually form the landward margins of modern machairs, the water level of which rarely exceeds 2 m above mean sea level (Ritchie, 1985). It seems likely that there has been a sea-level rise of at least 5 m as well as an accompanying landward movement of the shoreline in the last 5200 years. Radiocarbon dates on similar subtidal and intertidal peats of 8330 ± 65 BP (SRR–1223) from Pabbay (Ritchie, 1985) and 8802 ± 70 BP (SRR–396) from Holm in Lewis (von Weymarn, 1974) suggest sea levels at least 3–5 m lower between about 8800 BP and 5200 BP (Ritchie, 1985). The lacustrine origin of these dated organic materials makes it difficult to provide exact values for both the amount and rate of sea-level rise, but all such deposits provide unambiguous evidence of coastline recession.

The radiocarbon date of 3370 ± 60 BP from the machair sands indicates a locally stable land surface at that time, with phases of sand accumulation both before and after. Elsewhere, radiocarbon dates associated with machair coastlines suggest significant sand accumulation after 4366 ± 40 BP (SRR–1225) on Pabbay (Ritchie, 1985), 4550 ± 70 BP (SRR–2988) on Grimsay (Whittington and Ritchie, 1988) and 7810 ± 140 BP (GU–1762) at Claddach More (Balelone) (G. Whittington and W. Ritchie, unpublished data). Thus more evidence is accumulating from a variety of sites to demonstrate a long period of sand drifting both from primary and secondary sources. Unpublished evidence from a variety of

Bronze and Iron Age archaeological sites, located in machair areas, also indicates that there were many episodes of sand drifting which alternate with periods of stability. The extent to which these changes were anthropogenic is open to discussion. Similarly, the question of whether these periods of massive sand movements were synchronous in South Uist, North Uist and the Sound of Harris remains unanswered.

Inter- and sub-tidal organic remains occur at thirteen sites in the Uists (Ritchie, 1985). However, those at Borve are the most extensive and best documented deposits of their type in the Outer Hebrides. They are important for the evidence they provide both for sea-level change in the middle to late Holocene and for the accompanying development of the dune and machair landscapes that occupy 10% of the land area of the Uists. Unlike the coastline of mainland Scotland (see Western Forth Valley, Silver Moss and Philorth Valley) and the Inner Herbrides, sea-level change around the outer isles during the latter part of the Holocene resulted in coastal submergence and landward migration of the shoreline. Borve is one of the few sites in Scotland at which this type of sea-level movement has been dated. The evidence from there and related sites provides a fundamental, if as yet incomplete, stratigraphic and chronological framework that underpins the interpretation of the development and evolution of the beach and machair systems in the Outer Hebrides (see Ritchie, 1966, 1979, 1986; Whittington and Ritchie, 1988, unpublished data).

Conclusion

Borve is important for a sequence of deposits that provides a detailed and dated record of sea-level changes and coastline development in the Outer Hebrides during Holocene times. In particular, the deposits show that the coastline has moved landwards as relative sea level rose by at least 5 m during the last 5200 years. Phases of sand erosion and accumulation occurred both before and after about 3400 years ago. This evidence is important not only for understanding the development of the dune and machair landscapes of the Western Isles, but it also allows valuable comparisons with results from sites elsewhere in Scotland where the pattern has been one of coastal emergence rather than submergence. Borve is therefore a valuable reference site for sea-level studies in Scotland.

GLEANN MÓR, HIRTA

M. J. C. Walker

Highlights

The sediments which infill a topographic basin in Gleann Mór contain a valuable pollen record, supported by radiocarbon dating, of the Lateglacial and Holocene vegetational and environmental changes on St Kilda. This record is particularly significant in view of the location of the site on the extreme Atlantic periphery of the British Isles, where human modifications have been minimal.

Introduction

The site (NF 086997) is a small peat bog located at an altitude of approximately 90 m OD on the lower slopes of Gleann Mór, in the north-west part of the island of Hirta. The islands of St Kilda have long been a focus of ecological interest because of their remote position on the Atlantic fringes of the British Isles. The archipelago, consisting of the islands of Hirta, Dun, Soay and Boreray, lies near the edge of the continental shelf 180 km west of the Scottish mainland and 64 km west-north-west of the westernmost headland of the Outer Hebrides. Although a considerable amount of research has been carried out on the present vegetation (Turrill, 1927; Petch, 1933; Poore and Robertson, 1948; McVean, 1961; Gwynne *et al.*, 1974), until recently only limited palynological investigations (McVean, 1961) had been made and hence the vegetational history of the island group was virtually unknown. Following the discovery of an interstadial polleniferous sand (Sutherland *et al.*, 1984), the publication (Walker, 1984a) of a detailed pollen diagram from Hirta covering much of the Holocene has given important insights into the vegetational development of the islands.

Description

A little over 2 m of sediment, comprising principally sand, organic muds and peat, have accumulated in the peat bog (Figure 12.10).

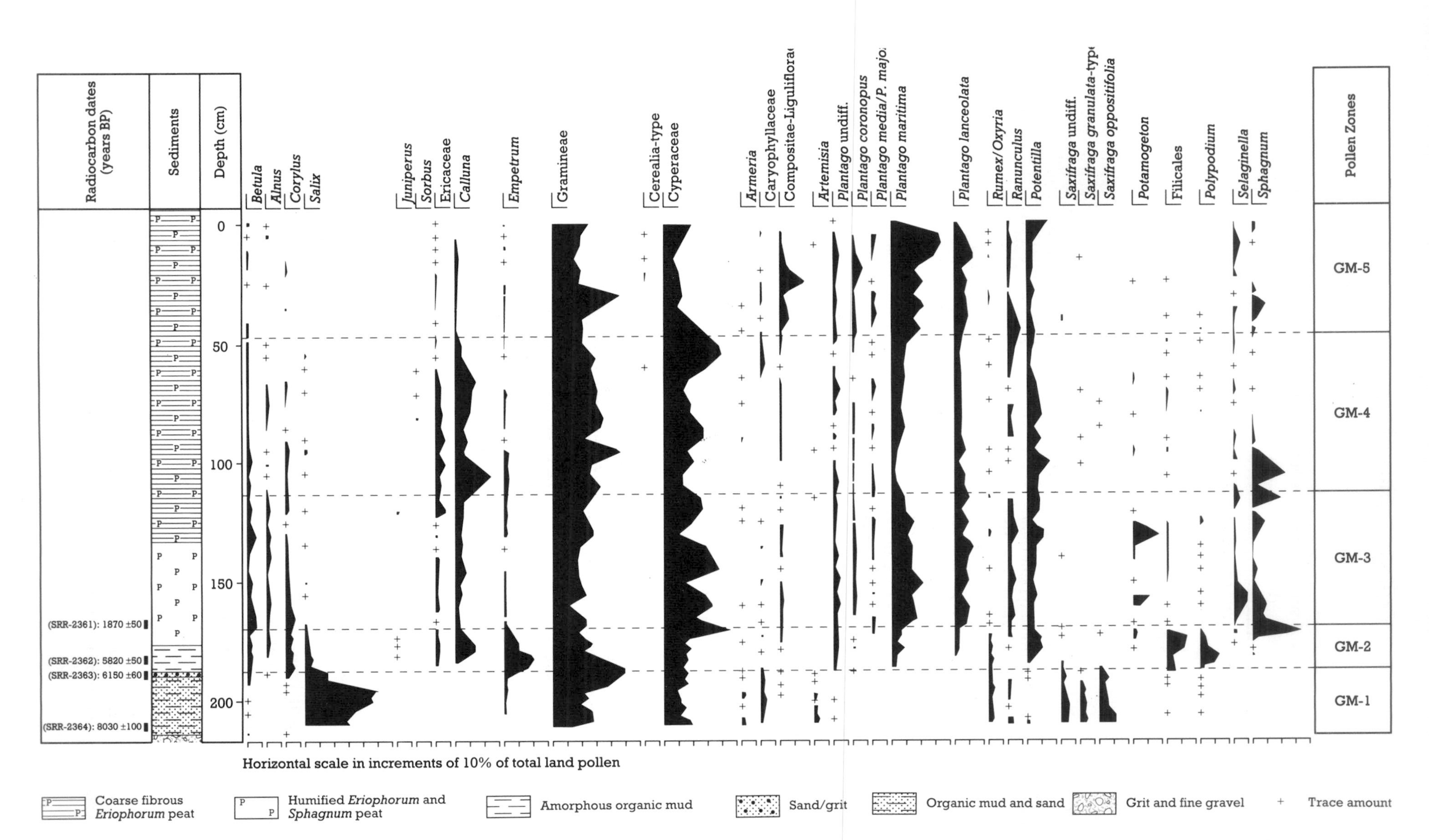

Figure 12.10 Gleann Mór, Hirta: relative pollen diagram showing selected taxa as percentages of total land pollen (from Walker, 1984a).

The lowermost sediments in the profile, comprising beds of grit and fine gravel, and brown organic mud are of probable Loch Lomond Stadial age; the remainder are Holocene deposits. Five pollen assemblage zones have been identified and four radiocarbon dates (SRR–2361 to SRR–2364) have been obtained from the sediments (Figure 12.10).

Interpretation

The earliest pollen zone (GM–1) is dominated by *Salix*, and is characterized by pollen from taxa indicative of bare and disturbed ground such as *Saxifraga oppositifolia*, *Oxyria digyna*, *Artemisia* and species of Caryophyllaceae. An open tundra landscape is indicated and hence the *Salix* pollen indicates the presence of either the least willow (*Salix herbacea*), or such northern or arctic willows as *Salix polaris*, *Salix reticulata* or *Salix glauca*. The basal minerogenic sediments that contain this pollen assemblage are indicative of periglacial conditions, and a Loch Lomond Stadial age (11,000–10,000 BP) is suggested. The radiocarbon date of 8030 ± 100 BP (SRR–2364) obtained from the base of the profile is considered to be in error by around 2000 years as a consequence of groundwater contamination.

The boundary between pollen assemblage zones GM–1 and GM–2 has been dated to 6150 ± 60 BP (SRR–2363) which, if correct, indicates a gap in the sediment record of the Gleann Mór profile spanning over 4000 years during the early and middle Holocene. The pollen spectra in GM–2 reflect a floristically diverse grassland with *Plantago maritima*, *Plantago lanceolata*, *Potentilla*, *Rumex*, Compositae (Liguliflorae) and *Polypodium*, interspersed with heathland communities dominated by *Empetrum*, *Calluna vulgaris* and, probably, *Erica cinerea*. Pollen of *Betula*, *Alnus* and *Corylus* is also present. While it is possible that small numbers of birch, hazel and alder managed to gain a foothold on Hirta, the fact that frequencies for these genera never exceed 6–7% of total land pollen means that the possibility of long-distance transfer from the Scottish mainland cannot be discounted (see Birks and Madsen, 1979). Certainly, there is no unequivocal evidence to support the view (McVean, 1961) that St Kilda once possessed a cover of birch–hazel scrub.

The renewed accumulation of sediment at Gleann Mór a little before 6000 BP is almost certainly a reflection of a regional climatic shift to more oceanic conditions in north-west Europe following the Holocene rise in sea level. The first traces of *Alnus* which are recorded in the Gleann Mór profile around 5800 BP may be a further indication of a general trend towards increasing climatic wetness (Godwin, 1975).

Pollen zone GM–3, the base of which has been dated to 1870 ± 50 BP (SRR–2361), reflects a change from a mixed heath and grassland landscape to a *Plantago*-dominated sward, the principal elements of which were *Plantago maritima* and *P. lanceolata*, along with *P. media*/*P. major* and *P. coronopus*. Many of these species are found in the halophyte swards that have developed on St Kilda at the present day in areas where the sea-spray effect is considerable (Gwynne *et al.*, 1974). The decline in heathland and expansion of the *Plantago* grassland may therefore be indicative of an increase in storm frequency and intensity around St Kilda, with salt spray being blown across large areas of Hirta. Wetter conditions are also indicated by the occurrence of *Potamogeton* and *Littorella* which suggest pools of standing water, by the appearance of *Selaginella* reflecting the expansion of moist and damp habitats, and by the higher counts for *Sphagnum* and Cyperaceae. These records probably reflect the marked deterioration in climate that occurred throughout north-west Europe around 2500 BP (Lamb, 1977, 1982a).

The pollen spectra in zone GM–4, by contrast, appear to indicate a change to drier and less stormy conditions. The decline in *Plantago maritima* frequencies and the increased values for *Empetrum* imply an expansion of heathland at the expense of *Plantago* sward. Some areas of marshy and boggy ground remained but, in general, the extensive maritime communities that had previously been a feature of the Hirta landscape were significantly reduced. No radiocarbon dates were obtained from this level of the profile, and hence dating is speculative, but pollen zone GM–4 may span the first millennium AD, a period of generally more equable climate in the North Atlantic region (Lamb, 1977, 1982a, 1982b).

The uppermost pollen assemblage zone (GM–5) is comparable in a number of respects to GM–3, with the expansion of *Plantago*-dominated plant communities at the expense of heathland. Again, an episode of wetter and more stormy conditions is implied, most probably corresponding with the Little Ice Age, the climatic deterioration experienced throughout north-west Europe and the

North Atlantic region during the second millennium AD.

Anthropogenic effects on the vegetation of Hirta are difficult to establish. The appearance of *Plantago lanceolata* pollen shortly after 5800 BP is a feature that has frequently been associated with human activity (for example, Johansen, 1978), but in the context of St Kilda, the pollen record for this species is more probably a reflection of the growth of ribwort plantain in natural maritime grassland communities. Similarly, although it is possible that some of the fluctuations in the *Calluna* pollen curves may be due to prehistoric grazing activity, there is little independent archaeological evidence, and hence the pollen record is more likely to reflect climatic rather than anthropogenic influences. Only in the uppermost 0.2 m of the profile, where low frequencies of cereal-type pollen are encountered, is there clear evidence of human activity.

The pollen site in Gleann Mór on the island of Hirta is of considerable significance. It provides valuable data on the vegetational history of St Kilda during the Loch Lomond Stadial, and it contains detailed pollen evidence of vegetational changes in this remote area of the British Isles throughout the middle and late Holocene. Of particular importance, however, is the fact that, by contrast with other areas of the British Isles and north-west Europe, human influence on the vegetation of the islands appears to have been minimal and hence the pollen sequence provides a rare proxy record of climatic changes in the North Atlantic province spanning the past 6000 years. Relatively few records of this nature are available because the effects of climatic change are frequently masked in late Holocene pollen records by anthropogenic influences (Birks, 1986).

Conclusion

Gleann Mór is important for its record of vegetation history and environmental change on St Kilda during the Lateglacial and Holocene (approximately the last 11,000 years). The site provides valuable palaeoecological data from a remote and inaccessible part of Britain, and the pollen evidence from Gleann Mór is of particular significance for the insights it allows into the patterns of climatic change during the later part of the Holocene.

Chapter 13

Western Highland Boundary

INTRODUCTION

D. G. Sutherland

The Western Highland Boundary area is taken to extend from the Teith valley, by Callander, west to the Firth of Clyde (Figure 13.1). This area has been of fundamental importance in defining the sequence of events during the closing stages of the Devensian cold phase. Lying adjacent to one of the principal centres of ice dispersal in the south-west Highlands it has been readily invaded by ice, but being outside the Highland zone, the extent of ice erosion has not been as great as in the mountains. Thus for the period since the end of the last ice-sheet glaciation there is an almost complete record of glacial events, sea-level change and environmental change, and these subjects have formed the main themes of research.

In this area there is little evidence of Quaternary events prior to the main Late Devensian ice-sheet. It can reasonably be presumed that the major ice-moulded landforms and overdeepened valleys leading out from the Highlands relate to repeated phases of ice-sheet glaciation, but no dated material has been found that pre-dates the Late Devensian. The last ice-sheet flowed to the south and south-east across this area, depositing a till, termed the Wilderness Till by Rose *et al.* (1988). Subsequently, during ice retreat, relative sea level was high, with marine invasion of the Western Forth Valley, the Firth of Clyde (see Geilston) and the Loch Lomond basin (see South Loch Lomond). Briefly, prior to marine invasion of this last area, an ice-dammed lake formed in the Blane Valley (Rose, 1980e; Rose *et al.*, 1988) as the ice retreated, but still blocked the connection between Loch Lomond and the Firth of Clyde.

At the period of maximum marine invasion, the neck of land between the Loch Lomond basin and the Western Forth Valley was the only land connection between the Highlands and southern Scotland. It is notable that whereas the early phases of retreat of the last ice-sheet were accompanied around the Scottish coast by deposition of sediments (Errol beds) containing high-arctic marine faunas, the final phase of deglaciation as the ice was retreating into the Highlands accorded with a milder marine climate and the deposition of sediments containing boreo-arctic marine faunas (Clyde beds, see Geilston). This is in accord with the radiocarbon dates on marine shells, which suggest that the head of the Firth of Clyde was deglaciated at around 13,000 BP (Sutherland, 1986).

The Lateglacial Interstadial climatic amelioration is also marked in the terrestrial record at such sites as Tynaspirit and Muir Park Reservoir (Donner, 1957; Vasari and Vasari, 1968; Vasari, 1977), where an initially open vegetation typical of disturbed ground was succeeded later in the interstadial by closed vegetation dominated by grasslands, scrub or heath with, possibly, in favoured localities, copses of tree birch. For much of the early to middle interstadial, relative sea level fell rapidly, probably to near or below 5 m above present.

The onset of the severe conditions of the Loch Lomond Stadial is pronounced in both the marine and the terrestrial records. The interstadial vegetation cover was disrupted and open-habitat vegetation again became dominant; in the nearshore waters a restricted fauna with high-arctic affinities replaced the interstadial fauna (Peacock *et al.*, 1978; Peacock, 1981b). The return to severe conditions appears to have been accompanied by increased erosion in the shore zone, and a marked rock-cut shoreline, the Main Rock Platform, was formed (Sissons, 1974d; Gray, 1978a) or at least extensively modified from a pre-existing feature (Browne and McMillan, 1984; Gray and Ivanovich, 1988). The cliff associated with this shoreline can still be clearly observed around the southern shores of Loch Lomond as well as the shores of the Firth of Clyde.

Glaciers once again readvanced into the area down the principal outlet valleys from the Highlands, into Gareloch (see Rhu Point), Loch Lomond (see Croftamie, Aucheneck and Gartness), the Western Forth Valley and the Teith valley (see Mollands). It was in the southern Loch Lomond basin that the readvance nature of this local glaciation was first recognized (Simpson, 1933): hence this has become the most critical area for the definition of the Loch Lomond Readvance. The readvancing ice overrode or reworked the shelly Clyde beds, and radiocarbon dating of the included shells has established conclusively that the glacial event occurred after 11,500 to 11,000 BP. In addition, near Callander in the Teith valley (at NN 63790509), Lateglacial Interstadial lacustrine sediments have been recorded below till deposited by the Loch Lomond Readvance glacier in that valley (Merritt *et al.*, 1990). The interstadial sediments have been radiocarbon dated to 12,750 ± 70 BP (SRR–2317) and are succeeded by 2 m of Loch Lomond

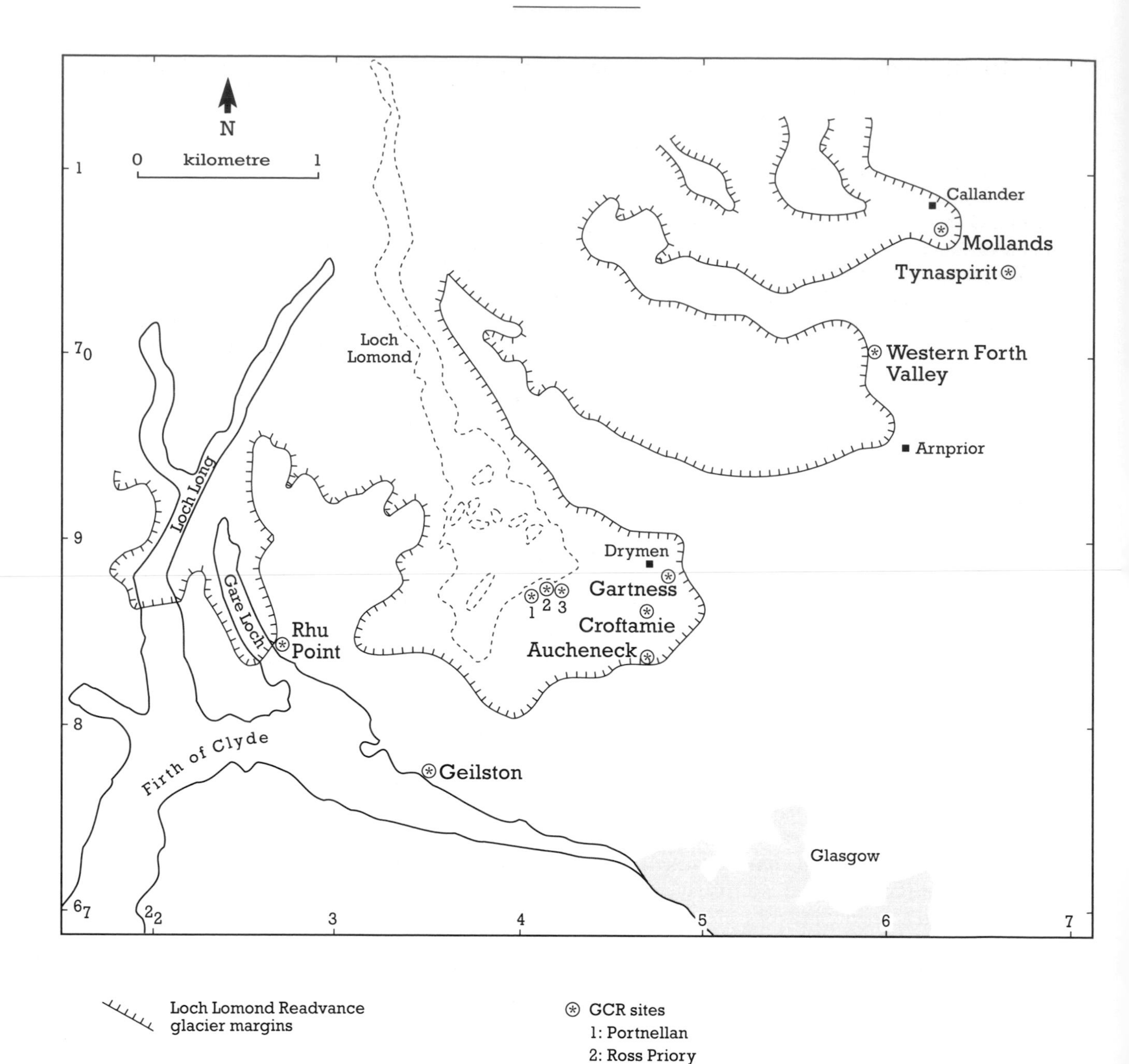

Figure 13.1 Location map of the Western Highland Boundary area.

Stadial lacustrine sediments that were emplaced prior to deposition of the overlying till. An implication of this sequence is that the readvance reached its maximum late in the stadial. More critically, at Croftamie, a layer of plant detritus, with a radiocarbon age of about 10,500 BP, is overlain first, by silts deposited in a lake dammed as the Loch Lomond glacier advanced into the Endrick and Blane valleys, and second, by till deposited directly by the glacier (Rose *et al.*, 1988). This indicates that the Loch Lomond glacier reached its maximum extent after 10,500 BP. This chronology is also supported in the Western Forth Valley, where a marine shoreline, termed the High Buried Shoreline, was formed following a rapid marine transgression at approximately the time of deposition of the Menteith Moraine at the maximum of the readvance. This shoreline has an inferred age of between 10,100 and 10,500 BP (Sissons, 1966, 1976b).

In this area there are major and instructive contrasts in the form of the moraines deposited at or close to the maximum of the readvance. In places they take the form of small, clear end moraines (Aucheneck), whereas elsewhere they are massive push moraines (Western Forth Valley). A further contrast is at Gartness where the Loch Lomond glacier terminated in an ice-dammed lake and constructed a series of large arcuate ridges.

The age and extent of the readvance are also limited by the mutual distribution of basinal sites in which Lateglacial Interstadial sediments are present or where the earliest deposited sediments are of early Holocene age (see Mollands and Tynaspirit). Early deglaciation of the Mollands site relative to the Highland interior is indicated by the relative changes in the pollen assemblages at the base of the sedimentary sequences (Lowe and Walker, 1981).

After retreat of the ice, sea level in the Western Forth Valley fell, with halts to form particular shorelines at approximately 9600 BP and 8600 BP (Sissons, 1966, 1983a; Sissons and Brooks, 1971). The overall regression continued, however, until about 8300 BP when a marked marine transgression began, culminating at approximately 6800 BP with the formation of the Main Postglacial Shoreline (Sissons, 1983a). Subsequently, the sea fell progressively to its present level. Similar, but less well-documented changes of sea level occurred around the Firth of Clyde, with Loch Lomond being freshwater during the early to middle Holocene; later a marine episode started around 6900 BP, continuing to 5500 BP (Dickson *et al.*, 1978). This marine event may in fact have been twofold, with a brief return to freshwater conditions (Stewart, 1987).

The Holocene vegetational history of the area has been investigated at a number of sites (Donner, 1957; Turner, 1965; Vasari and Vasari, 1968; Vasari, 1977; Dickson *et al.*, 1978; Stewart, 1979; Lowe, 1982a; Stewart *et al.*, 1984). Of particular interest is the position of the area between the pine forest zone to the north and the oak forest zone to the south (Stewart *et al.*, 1984).

AUCHENECK
J. E. Gordon

Highlights

The landforms at Aucheneck demonstrate important aspects of the geomorphology of the Loch Lomond Readvance; they include one of the best examples of an end-moraine ridge in the type area for the readvance.

Introduction

The site (NS 478830) is located west of Aucheneck athwart the Finnich Glen and covers an area of *c.* 1 km^2. It is important for an assemblage of landforms associated with the Loch Lomond Readvance glacier in the Loch Lomond basin, the area in which the readvance was first recognized (Simpson, 1933). In particular, it contains one of the finest segments of a till moraine ridge associated with the former Loch Lomond glacier. It is also notable for good sections in the moraine and clear contrasts in landform patterns and soil development 'inside' and 'outside' the ice limit. The only detailed account of the site is by Rose (1981), although the moraine as a whole is discussed in a number of publications (Renwick and Gregory, 1907; Gregory, 1928; Simpson, 1933; Dickson *et al.*, 1978; Rose, 1980d).

Description

The geomorphology of the Aucheneck area is shown in Figure 13.2. The terminal moraine is best developed on the west side of the glen, where it forms a prominent landform, 5–7 m high and about 40 m wide, on the otherwise subdued relief of Cameron Muir. Immediately inside the terminal ridge three further ridge fragments occur, then an area of hummocky drift and meltwater channels. On the east side of Finnich Glen (NS 482828) there are good stream sections in the moraine. These show dark, reddish-brown, sandy till, locally with flow banding (Rose, 1981). Clast fabrics are orientated parallel to the ridge (Rose, 1981). Several subglacial meltwater channels occur on the north side of the ridge, trending towards the north and east. The channels appear to form two groups (Rose, 1981): those aligned west–east were formed by meltwater draining directly from the ice, whereas those aligned north–south may have received part of their discharge from the Carnock catchment.

At NS 483834 one of the channels leads into an esker. Of further interest is the contrast in landform type and development 'outside' and 'inside' the moraine limit (Figure 13.2). On the

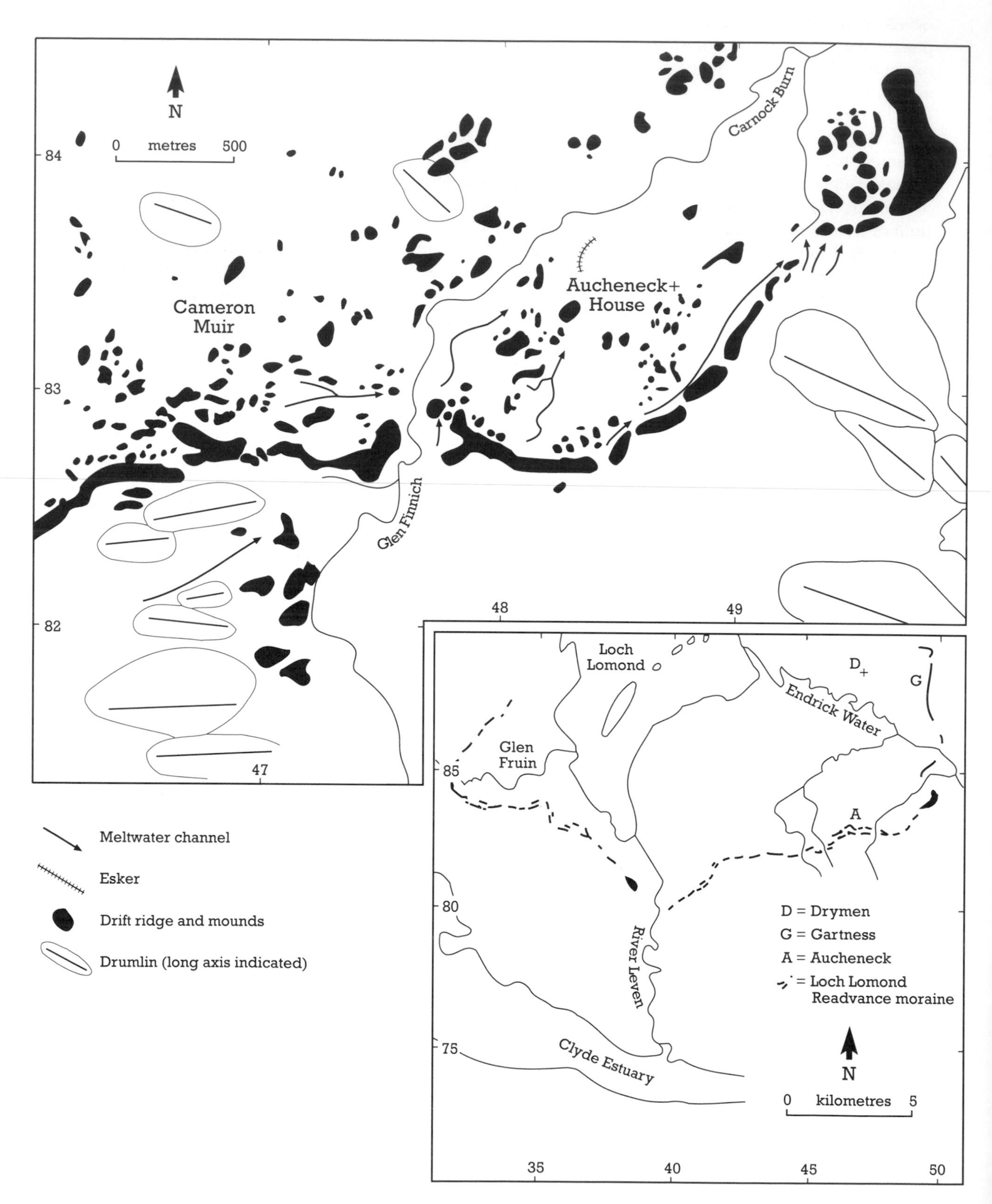

Figure 13.2 Landforms and deposits associated with the Loch Lomond Readvance ice limit at Aucheneck (from Rose, 1981). Inset shows the wider extent of the moraine that marks the ice limit at the southern end of Loch Lomond (from Dickson *et al.*, 1978).

south side, there are low drumlins with ridge crests orientated west–east; on the north side, the drift has a sharper, more irregular, hummocky form, and drumlins (outside the site) have long axes aligned NW–SE. There is also a contrast in the form of the Carnock valley on either side of the ice limit; on the south side, the valley is relatively wide, and bedrock is exposed along the sides; to the north, an earlier valley has been infilled with till, and the present burn occupies a narrow course cut in this material and locally a bedrock gorge. Shells of marine molluscs and Foraminifera have been recorded in the till 'inside' the moraine, for example along the Carnock Burn west of Aucheneck House (NS 487835) (Simpson, 1928, 1933; J. Rose, unpublished data), but have not been found 'outside' the moraine. There is also a contrast in soil development 'inside' and 'outside' the ice limit with higher levels of podsolization and indurated layer development on the 'outer' side (J. Rose, unpublished data).

Interpretation

The Aucheneck moraine is part of a near continuous end-moraine system extending around the south end of Loch Lomond from Conic Hill to Glen Fruin (Renwick and Gregory, 1907; Gregory, 1928; Simpson, 1928, 1929, 1933; Dickson *et al.*, 1978; Rose 1980d). Gregory (1928) considered that the best remnant on the east side was at Aucheneck. Renwick (1895) first recognized that the fragment in Glen Fruin was associated with Loch Lomond basin ice rather than ice coming from the west, as postulated by Bell (1891b, 1893c, 1894, 1896b). Subsequently, from the evidence of the moraine itself and the distinctive shelly till (see Croftamie (below), and Jack, 1875) that occurred inside the line of the moraine, but not outside it, Simpson (1933) inferred a readvance of ice during the wastage of the last ice-sheet, which he called the Loch Lomond Readvance. Charlesworth (1956) identified the moraine as part of his Lateglacial 'Stage M' or 'Moraine glaciation'. From pollen studies Donner (1957) placed the readvance during the climatic deterioration represented in Lateglacial Pollen Zone III of the Jessen–Godwin scheme. Subsequently radiocarbon dating has confirmed that the readvance occurred between *c.* 11,000–10,000 BP (see Gartness and Croftamie), and the period of climatic deterioration has been called the Loch Lomond Stadial (Gray and Lowe, 1977a) after the area where the readvance was first named by Simpson (1933). A summary of the stratigraphic evidence justifying the designation as a type area is given in Rose (1989).

Rose (1981) considered that the moraine at Aucheneck formed by a combination of debris pushed up at the ice margin and also material that slumped off the glacier surface. The ice limit is clearly defined, not only by the moraine, but also by the contrast in landforms 'inside' and 'outside' it. 'Outside', the drumlins have a west–east orientation associated with the last movement of the Late Devensian ice-sheet across the area (Rose, 1981, 1987) and their relatively subdued appearance reflects the effects of gelifluction during the Loch Lomond Stadial. The landforms inside the limit are associated with a north-west to south-east ice movement of the piedmont lobe of the Loch Lomond glacier and have not been modified to the same extent by periglacial mass movement.

Single and multiple end moraines were widely formed during the Loch Lomond Readvance in the Highlands and Southern Uplands of Scotland, the Lake District and North Wales (see Cnoc a'Mhoraire, Tauchers, Gribun, Lochnagar, the Cairngorms, and An Teallach) (Sissons, 1974c, 1979e, 1980a; Gray, 1982a). The particular significance of Aucheneck centres on it being one of the best examples of a till ridge in the type area for the readvance (Simpson, 1933; Jardine, 1981; Rose, 1989) and the clear assemblage of evidence it provides for ice readvance: the end moraine itself and the contrasts in till lithology (shell and Foraminifera content), landforms and soils inside and outside the ice limit. The exposures, additional landform assemblage, including meltwater channels, hummocky drift and an esker, and proximity to dated biostratigraphy make Aucheneck a better representative site than, for example, the equally fine moraine in Glen Fruin. Moreover, the multiple end-moraine sequence at Aucheneck may be significant in illustrating local fluctuations of the ice margin during the stadial. Aucheneck complements the sites at Gartness where Loch Lomond Readvance end moraines were formed subaqueously and Croftamie where the local shelly till of the readvance is well represented.

Conclusion

Aucheneck provides an excellent representative

assemblage of landforms formed by a Loch Lomond Readvance glacier (approximately 11,000–10,000 years ago) in the type area where this readvance was first recognized. In particular, it includes one of the best examples of an end-moraine ridge in the area and is important in demonstrating other characteristic landforms produced by the ice readvance.

CROFTAMIE

J. E. Gordon

Highlights

The sequence of deposits at Croftamie includes plant detritus interbedded between two tills. The organic material has yielded important palaeoenvironmental and dating information, allowing the glacial history of the area during the Late Devensian to be established. The site provides important evidence for the timing of the Loch Lomond Readvance in the type area.

Introduction

The section at Croftamie (NS 473861), which occurs in a cutting along the now disused Forth and Clyde Junction Railway, is important for Late Devensian stratigraphy in the type area for the Loch Lomond Readvance. It shows the stratigraphic relationship between the Wilderness Till (deposited by the main Late Devensian ice-sheet) and the Gartocharn Till (deposited by a piedmont glacier during the Loch Lomond Stadial). Beds of organic detritus (plant remains representative of a dwarf-shrub/grassland association) and laminated lake sediments occur between the two tills. Radiocarbon assay of the organic sediments has allowed a limiting date to be placed on the maximum of the Loch Lomond Readvance in its type area (Rose *et al.*, 1988). This important site also has a long history of research (McFarlane, 1858; Smith, 1858; Jack, 1875; Simpson, 1933; Rose *et al.*, 1988).

Description

At the time of their discovery, the deposits at Croftamie generated considerable interest because of their fossil content. McFarlane (1858) and Smith (1858) described 12 ft (3.7 m) of till overlying 7 ft (2.1 m) of blue clay. Near the bottom of the latter, just above sandstone bedrock, the antler of a reindeer, *Rangifer tarandus* (L.) and a number of marine shells were found. Jack (1875) correlated the till above the clays at Croftamie with the distinctive shelly till unit that he identified in the lower Endrick Valley. He listed the macrofauna found in numerous till exposures in the area, including *Arctica islandica* (L.), *Nicania montagui* (Dillwyn), *Neptunea antiqua* (L.), *Littorina littorea* (L.) and *Balanus* sp. from Croftamie.

The most recent investigations of the Croftamie deposits are by Rose (1981) and Rose *et al.* (1988). The sequence comprises:

5.	Till with marine fauna derived from Clyde beds (Gartocharn Till)	up to 5 m
4.	Rhythmically laminated clays and silts (Blane Valley Silts)	up to 0.4 m
3.	Felted plant detritus	up to 0.08 m
2.	Till, mainly of locally derived bedrock (Wilderness Till)	up to 1.5 m
1.	Old Red Sandstone bedrock	

The Wilderness Till is associated with an ice-sheet moving west–east and represents glaciation during the Late Devensian glacial maximum. The Blane Valley Silts are interpreted as diatactic varves formed in a proglacial lake dammed back in the Blane Valley by the Loch Lomond Readvance. The number of varves indicates that the lake existed for at least 50 years. The overlying Gartocharn Till includes material incorporated from Clyde beds deposited in the Loch Lomond area during the Lateglacial Interstadial. The organic detritus (bed 3) contains a sparse, poorly preserved pollen assemblage (dominantly Caryophyllaceae, *Salix*, Gramineae, Cyperaceae and Cruciferae, but also with *Artemisia*, *Valeriana* and *Selaginella*) typical of an open-habitat, dwarf-shrub/grassland environment, and gave a radiocarbon date of 10,560 ± 160 BP (Q–2673) (Rose *et al.*, 1988). It therefore accumulated during the Loch Lomond Stadial, before the onset of the glacial lake sedimentation.

Interpretation

The significance of the Croftamie deposits in interpreting the Pleistocene sequence in Scotland was recognized in the last century. Together with other fossil remains in the drift of Scotland, the reindeer horn was considered to provide evidence for ameliorating climatic conditions and restricted ice extent at some time during the Pleistocene before a subsequent ice advance (Geikie, 1863a). Jack (1875) concluded that the marine fauna had been alive during one of the milder phases of the glacial period when Loch Lomond was an arm of the sea, although in conditions slightly colder than at present. The shelly till was therefore the product of subsequent glaciation. This explanation of the evidence was also adopted by later workers in the area (Geikie, 1894; Bell, 1895d; Cunningham Craig, 1901), following some discussion as to whether the marine event was a local extension of the sea into Loch Lomond or a great interglacial submergence (J. Geikie, 1874, 1877).

Following detailed fieldwork in the area Simpson (1928, 1929, 1933) developed the ideas of Jack, Geikie and Cunningham Craig, assembling the evidence and formulating carefully the case for a Lateglacial readvance of ice in the Loch Lomond valley following a period of marine occupation. He observed that shelly till only occurred inside a prominent end moraine, which he traced around the southern end of Loch Lomond (see Aucheneck and Gartness). He called the glacial event the Loch Lomond Readvance and noted that it was also represented in the Western Forth Valley (see below).

Charlesworth (1956) subsequently extended the limits of the readvance, incorporating the Loch Lomond and Forth valley landforms and deposits as part of his Lateglacial 'Stage M' or 'Moraine Glaciation'. The name Loch Lomond Readvance, however, was retained by Sissons (1965, 1967a), who revised the ice limits suggested by Charlesworth. More recently these limits have been further established over large areas of the Highlands and Islands and the Southern Uplands (Sissons *et al.*, 1973; Sissons, 1979e, 1983b).

The detailed work of Rose (1981) and Rose *et al.* (1988) has allowed a clearer interpretation of the full sequence of events at Croftamie: (1) deposition of till by the Late Devensian ice-sheet (Wilderness Till); (2) deposition of marine clayey silts (Clyde beds) (elsewhere in the area); (3) formation of a proglacial lake by the advancing Loch Lomond Readvance (Blane Valley Silts); (4) deposition of the shelly till (Gartocharn Till), in part derived from Clyde beds sediments; and (5) maintenance of the proglacial lake which drained into the Forth valley (Blane Valley Silts). The radiocarbon date from the organic sediments implies that the Loch Lomond glacier reached its maximum extent after 10,560 ± 160 BP, in agreement with the evidence from the Vale of Leven that glaciomarine sedimentation continued until after 10,350 ± 125 BP (SRR–1529) (Browne and Graham, 1981). This also accords with the inference that the Loch Lomond Stadial glacier in the Western Forth valley reached its maximum extent between 10,500 BP and 10,100 BP (Sissons, 1983a; Sutherland, 1984a), and that the Creran glacier reached its maximum after 10,500 BP and perhaps as late as 10,000 BP (see South Shian and Balure of Shian) (Peacock *et al.*, 1989). However, these dates are significantly at variance with dates previously obtained from elsewhere that have been used to infer the time of the maximum glacier extent (see Sutherland, 1986; Rose *et al.*, 1988): they are at least 400 years younger than the other dates from sediments overridden by Loch Lomond Stadial glaciers, and they also overlap in age ranges with dates from lacustrine sediments that were deposited immediately on retreat of the stadial glaciers (Rose *et al.*, 1988) (see also Kingshouse, Loch an t-Suidhe, Mollands, Tynaspirit, Rhu Point and Loch Cleat). Rose *et al.* (1988) argued that this variance could relate to three factors. First, many of the dates are from marine shells, but environmental and sedimentation factors tend to favour a higher probability of sampling shells from the middle part of the interstadial, rather than the later part (see also Sutherland, 1986). Second, there may be errors in the dates on basal organic lake sediments caused by mineral carbon and hard-water contamination (see Sutherland, 1980). Third, because of variations in individual glacier dynamics and glacier response to climate, the fluctuations of the glaciers are likely to have been diachronous. Since the material dated at Croftamie is plant detritus, the date obtained should be less susceptible to mineral carbon and hard water errors than dates obtained from organic lake sediments (gyttja); Rose *et al.* (1988) suggested that dates on the latter material may be slightly too old. The evidence from Croftamie and Inverleven therefore indicates relatively later deglaciation of Loch Lomond Stadial glaciers than suggested elsewhere (Lowe, 1978; Lowe and Walker, 1980, 1984; Walker and Lowe, 1980, 1982; Dawson *et al.*, 1987a) (but see Kingshouse). A further complication in interpreting the different radiocarbon dates arises from the variations in atmospheric radiocarbon production known to have occurred during the Lateglacial (Ammann and

Lotter, 1989; Zbinden *et al.*, 1989; Bard *et al.*, 1990), but this may have affected all the dates.

Croftamie is a site of highest importance in several respects. It provides a key sequence of Lateglacial deposits in the type area for the Loch Lomond Readvance. In stratigraphic terms, it unequivocably demonstrates the superposition of till deposited during the Loch Lomond Stadial on till deposited by the main Late Devensian ice-sheet. It is the only site at which plant material has been used to date the maximum of the Loch Lomond Readvance. Moreover, the results of this dating, corroborated by additional dates from Inverleven and South Shian and Balure of Shian, provide significant evidence that the maximum extent of the Loch Lomond Readvance glaciers may have occurred after 10,500 BP. The interest at Croftamie complements that at Aucheneck, Gartness and South Loch Lomond (Portnellan, Ross Priory and Claddochside), and together these sites provide a comprehensive demonstration of the Lateglacial stratigraphy and landforms, both glacial and marine, in the area where the Loch Lomond Readvance was first recognized.

Conclusion

Croftamie is important for interpreting the glacial history in the type area for the Loch Lomond Readvance. The site has a long history of research and is particularly significant in showing organic deposits interbedded between tills deposited by the Late Devensian ice-sheet (approximately 18,000 years ago) and a Loch Lomond Stadial glacier (approximately 10,500 years ago). The organic deposits represent a warmer phase between the two glacial episodes, and the pollen they contain provides details of environmental conditions during that period. The organic deposits also provide an important means of dating the readvance. Croftamie is therefore a key reference site for establishing particular details of the nature and timing of the Loch Lomond Readvance in its type area.

GARTNESS
J. E. Gordon

Highlights

The interest of Gartness includes an assemblage of glacial, glaciolacustrine and marine landforms and deposits. These provide evidence for the glacial history of the Western Highland Boundary area and the associated landscape changes during the Late Devensian. Of particular note is a series of end-moraine ridges formed in a glacial lake.

Introduction

The interests at Gartness extend over an area of *c.* 10 km^2 (centred on NS 495875) in the Endrick Valley to the east of Drymen. They include an assemblage of Late Devensian sediments and landforms located in the type area for the Loch Lomond Readvance. The sequence of sediments includes evidence for glaciation by the Late Devensian ice-sheet, proglacial lake formation, marine transgression, glaciation during the Loch Lomond Stadial and further proglacial lake development (Rose, 1980e, 1981). The key landforms are a series of Loch Lomond Readvance end-moraine ridges formed partly subaqueously, in a proglacial lake, and partly subaerially. The area has a long history of investigation (Jamieson, 1865, 1905; Jack, 1875; J. Geikie, 1877, 1894; Clough *et al.*, 1925; Gregory, 1928; Simpson, 1933; Sissons, 1967a; Browne and McMillan, 1989), but the most detailed modern accounts are by Rose (1980e, 1981); it has also been described by Price (1983).

Description

Early descriptions of the area (Jamieson, 1865; Jack, 1875; J. Geikie, 1877) drew attention to the occurrence of extensive deposits (about 13 km^2) of sand and gravel, sometimes interbedded with clay. These deposits, which formed a series of rounded hills resembling 'drumlins' (Jack, 1875), contained marine shells (e.g. *Arctica islandica* (L.), *Mya truncata* (L.), *Chlamys islandica* (Müller) and *Boreotrophon clathratus* (Ström)) (Jamieson, 1865; Jack, 1875) similar in character of assemblage to that in the Clyde beds (Jamieson, 1865). According to Jack (1875), the sands and gravels were stratigraphically above the distinctive shelly till of the area (see Croftamie) and were locally overlain by till.

Additional details were given by Simpson (1933), and he included the Gartness deposits as part of the end moraine system of the Loch Lomond Readvance glacier that occupied the

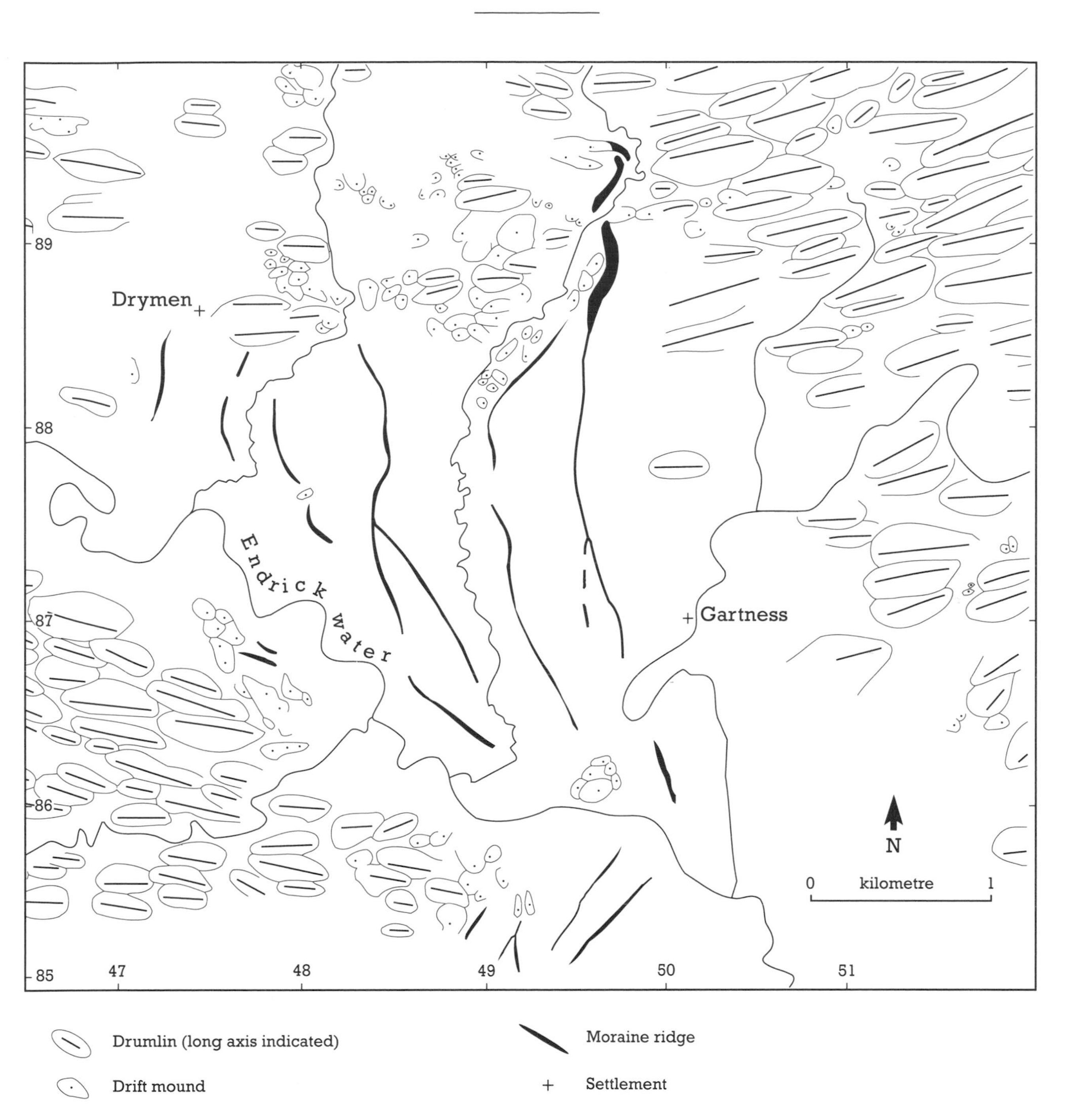

Figure 13.3 Landforms and deposits associated with the Loch Lomond Readvance ice limit at Gartness (from Rose, 1980e, 1981).

Loch Lomond basin.

In a comprehensive investigation, Rose (1980e, 1981) has described and mapped in detail the landforms and sediments of the Gartness area. The principal landform is an end-moraine ridge running approximately north–south from about NS 495895 to NS 501860 (Figure 13.3). The northern part of the moraine ridge comprises sand and gravel near Drumhead (NS 495882), and till where it curves round to the west upslope (NS 495895). The southern part (between NS 495882 and NS 501860) comprises deformed silts resting on a succession of earlier deposits. On the ice-distal (east) side these deposits are largely undisturbed; on the ice-proximal (west) side they are glaciotectonically deformed. These deformations have also been noted by Browne and McMillan (1989) but have not been described in detail as sections are poor due to landslipping. The deposits are best exemplified in sections in and through the moraine at NS 498864 and NS 500859; they include the following succession:

5. Blane Valley Silts
4. Gartocharn Till
3. Clyde beds
2. Gartness Silts
1. Wilderness Till

A similar sequence was confirmed in a British Geological Survey borehole located on the end moraine (Browne and McMillan, 1989). Bed 1 is typical of the stiff, reddish-brown till deposited by the Late Devensian ice-sheet, which moved towards the north-east across the area and formed the extensive drumlins in this part of the Midland Valley, of which there are good examples to the east of the moraine ridge (Figure 13.3). This till is overlain by a lag gravel and locally by well-sorted sands, which are succeeded by laminated clays and silts with particle size range and sorting typical of diatactic varves (Sauramo, 1923) (Gartness Silts – bed 2). Over 100 couplets have been counted (Rose, 1981). Clyde beds (bed 3) (see South Loch Lomond and Geilston) overlie the laminated sediments and comprise pink and grey silt laminations with a marine microfauna and occasional shells of marine molluscs. On the ice-proximal side of the moraine the Clyde beds are glaciotectonically disturbed and mixed with glaciofluvial sediments. Rose interpreted these deposits as a deformation facies of the Gartocharn Till (bed 4). The Blane Valley Silts (bed 5) consist of pink and brown silt laminae but become coarser and sheared towards the ice-proximal side of the moraine ridge where they merge into sand and gravel, for example at NS 500859. The silts occur extensively in the Endrick and Blane valleys below an altitude of 65 m OD. To the west of the end-moraine ridge there is a series of five further moraine ridges (Figure 13.3), partly buried by Blane Valley Silts. The ridge at Drumbeg (NS 483880), near Drymen, is associated with a large accumulation of deltaic sands and gravels seen exposed in quarry sections (NS 484882) (Browne and McMillan, 1989).

Shells derived from the Clyde beds and reworked into glaciofluvial gravels exposed in the quarry at Drumbeg gave a radiocarbon date of 11,700 ± 170 BP (I–2235) (Sissons, 1967b) and plant detritus, below Blane Valley Silts and Gartocharn Till at Croftamie, has been dated at 10,560 ± 160 BP (Q–2673) (Rose *et al.*, 1988); this confirms readvance of the ice during the climatic deterioration of the Loch Lomond Stadial as previously suggested by Donner (1957).

Interpretation

Early interpretations focused on the marine aspects of the deposits at Gartness. Jamieson (1865) correlated the shelly sands and gravels with the marine clays of Aberdeenshire and assigned them to a period of submergence at the end of the main period of land ice.

Jack (1875) believed that the latter could not have survived glaciation and referred them to a great submergence during the later phases of the glacial period after the shelly till was deposited. This till was derived from marine sediments laid down during a limited interglacial submergence and represented by the shelly clay under the till at Croftamie.

J. Geikie (1877), on the other hand, considered the stratified deposits (beds 2 and 3) at Gartness to be interglacial and the till on top to be that of the last ice-sheet and not iceberg transported, as Jack had postulated. However, in the third edition of his book, Geikie (1894) argued that since all the shells were worn and could have been derived from the underlying boulder clay, the sands and gravels were probably formed in a proglacial lake, the material being washed in from the receding glacier. Clough *et al.* (1925) also followed this view, postulating a lake dammed by ice both in the Loch Lomond and Blane valleys and with an outlet to the north of Balfron station. Gregory (1928), however, still adhered to the submergence theory of Jack.

Simpson (1928) initially adopted the hypothesis of Geikie, but following further investigation of the area, reconsidered his views since there was no evidence elsewhere to substantiate the required submergence up to 65 m OD and, moreover, the whole aspect of the deposits was typically morainic (Simpson, 1933). They comprised englacial debris laid down as a moraine ridge in front of a receding glacier, partly in water and partly on land. As Renwick and Gregory (1907) had earlier done, Simpson (1928, 1929, 1933) traced the moraine almost continuously around the south end of Loch Lomond and linked it with the well-known ridge in Glen Fruin (Bell, 1891b, 1893c, 1894, 1896b; Renwick, 1895). He recognized that it represented a readvance of ice (see Aucheneck), which he called the Loch Lomond Readvance, following a period of ice-sheet recession when the sea penetrated into Loch Lomond and shelly marine sediments (Clyde beds) were deposited. The Loch Lomond moraine ridge was also contemporaneous with a similar moraine ridge in the Western Forth Valley.

Charlesworth (1956) traced the Loch Lomond moraine ridge and mapped it as part of a Lateglacial readvance which he called the 'Moraine Glaciation' or 'Stage M'. Sissons (1967a) described the damming of lakes in the Endrick and Blane valleys by ice advancing down Loch Lomond during the Loch Lomond Stadial, thick silt and clay deposits in the Blane Valley possibly representing the bottom sediments. Sissons also suggested that in the area between Killearn and Drymen considerable deposits of sands and clays were laid down in an area of stagnating ice up to a level controlled by a meltwater channel spillway to the Forth valley (see also Price, 1983).

Rose (1980e, 1981) interpreted the fining-upwards sequence of gravels, sands and Gartness Silts (bed 2) as freshwater lake sediments deposited in a proglacial lake as the Late Devensian ice-sheet, which deposited the Wilderness Till (bed 1), retreated westwards. The Clyde beds (bed 3) indicate extension of the sea into the area during the Lateglacial Interstadial. The characteristics of the Gartocharn Till (bed 4) reflect the incorporation of Clyde beds into the moraine ridge by the glacier that crossed the area from the north-west during the Loch Lomond Stadial. The Blane Valley Silts represent proglacial lake sediments deposited in an ice-dammed lake in the Blane and Endrick valleys (cf. Price, 1983), occupying this locality both before the moraine ridge was being formed and after the ice had receded (see also Croftamie). The level of the lake and the upper level of sedimentation were controlled by a meltwater channel at 65 m OD across the lowest col on the Endrick–Blane watershed at Ballat (NS 528907). The water drained northwards and into a contemporaneous glacier in the Forth valley. The southern part of the moraine ridge was formed subaqueously by the deposition and subsequent deformation of lake sediments at the ice margin. Where the ice grounded above the lake level, the moraine ridge was formed of sand and gravel or till. As the glacier retreated from its maximum extent, five further moraine ridges were formed, also partly subaqueously, and deltaic sediments were deposited in the proglacial lake. Rose (1980e, 1981) considered that each ridge represented an oscillation or standstill of the ice margin in an environment of rapid sedimentation of large volumes of unconsolidated material.

The significance of Gartness is that it demonstrates an exceptional assemblage of landforms and sediments in the type area for the Loch Lomond Stadial. It provides clear geomorphological and stratigraphic evidence for the Loch Lomond Readvance in the form of a sequence of deposits including non-glacial marine sediments interbedded between till deposited by the Late Devensian ice-sheet and till formed during the Loch Lomond Stadial. It is of considerable sedimentological interest, demonstrating a succession of glacial, lacustrine and marine events, and of considerable geomorphological interest for a series of sublacustrine moraines and associated proglacial lake deposits. The site has significant research potential for detailed studies of the lake sediments, the glaciotectonic deformations and the processes of sublacustrine moraine formation. Stratigraphically, the site complements the interest at Croftamie, including Gartness gravels, sands, silts and laminated clays; geomorphologically, that at Aucheneck, demonstrating moraine ridges formed in a glacial lake and glaciotectonic deformation.

The sequence of sediments at Gartness compares with that at other sites in showing deformation of Clyde beds by a glacier during the Loch Lomond Stadial; for example at Rhu Point (Rose, 1980c), Western Forth Valley, Kinlochspelve (Gray and Brooks, 1972) and South Shian and Balure of Shian (Peacock, 1971b). The deposits at Gartness also represent a lowland example of sedimentation in one of a number of ice-dammed lakes that formed during the Loch Lomond Readvance (see Glen Roy and the Parallel Roads of Lochaber, Coire Dho and Achnasheen). However, apart from Achnasheen (Benn, 1989a) and parts of Glen Roy (Peacock, 1986), there have been no detailed sedimentological studies. In addition to Gartness, sublacustrine moraine ridges of Loch Lomond Stadial age are also reported from Coire Dho (Sissons, 1977b), Glen Spean (Sissons, 1979c) (see Glen Roy and the Parallel Roads of Lochaber) and Achnasheen (Sissons, 1982a). Together these sites provide a range of sedimentary environments that merit further detailed, comparative study of sedimentation in glacial lakes.

Conclusion

The landforms and deposits at Gartness demonstrate the sequence of landscape changes that occurred in the Loch Lomond area during the Late Devensian. The sediments provide evidence for ice-sheet glaciation (approximately 18,000 years ago), the formation of a proglacial lake, invasion by the sea, then renewed glaciation during the Loch Lomond Stadial (approximately 11,000–10,000 years ago) with contemporary glacial lake formation. The principal landforms are a series of end-moraine ridges that formed in the later ice-dammed lake. Gartness is important for interpreting key facets of the geomorphological changes that occurred during the Late Devensian in the type area for the Loch Lomond Stadial and the Loch Lomond Readvance.

SOUTH LOCH LOMOND: PORTNELLAN, ROSS PRIORY AND CLADDOCHSIDE

J. E. Gordon

Highlights

The assemblage of landforms and deposits on the south shore of Loch Lomond provides important evidence for the geomorphological changes that occurred at the end of the Late Devensian in the type area for the Loch Lomond Stadial and its associated glacier readvance. These include the formation of marine shorelines, fossiliferous marine deposits and glaciation of the Loch Lomond Readvance.

Introduction

Three localities, at Portnellan (NS 404873), Ross Priory (NS 413876) and Claddochside (NS 427878) on the south shore of Loch Lomond *c.* 6 km north-east of Balloch, illustrate important aspects of the Lateglacial and Holocene history of the Loch Lomond area, the type area for the Loch Lomond Readvance. They provide important stratigraphic and geomorphological evidence, described by Rose (1980f), particularly for the sequence of marine and glacial events, and show that the Main Rock Platform of western Scotland was significantly developed before the time of the maximum extent of the readvance. The sediments from the floor of the southern basin of the loch also provide a record of Holocene marine incursions, vegetational history and palaeomagnetism (Dickson *et al.*, 1978; Thompson and Morton, 1979; Turner and Thompson, 1979; Stewart, 1987).

Description

The three localities on the shore of the loch have been described by Rose (1980f). Together they show a succession of marine and glacial features. One of the most distinctive is a shore platform developed extensively along the southern shore. At Ross Priory it extends for about 0.5 km offshore to about 1 m below the loch level; it is also well displayed at Portnellan and Claddochside. In places it is associated with an impressive backing cliff up to about 15 m high, for example at Portnellan. At Claddochside, the cliff is cut in Old Red Sandstone bedrock. The junction of the cliff and platform is at about 12 m OD. At Portnellan, the platform and cliff are covered by till deposited during the Loch Lomond Readvance. This is the Gartocharn shelly till of the area, partly derived from reworked Clyde beds (see Croftamie) (Jack, 1875). Gartocharn Till also buries Clyde beds in foreshore sections at Claddochside, and a similar succession has been recorded on Inchlonaig (NS 380935) to the north (Cunningham Craig, 1901).

The sediments laid down in the Loch Lomond basin, immediately following retreat of the Loch Lomond Readvance, have been described from a borehole in the lower Endrick Valley (NS 44838829), *c.* 3 km north-east of Claddochside (Elliot, 1984; Browne and McMillan, 1989). In that borehole *c.* 7 m of finely-bedded silty clay with occasional laminae of silt or sand overlay the Gartocharn Till (see Croftamie). The sediments were characterized by a marine microflora and microfauna (dinoflagellate cysts and Foraminifera) and occasional shell fragments of marine molluscs and crustacea. Browne and McMillan (1989) noted the possibility that the fossils may have been derived from the underlying shelly sediments but considered that it was more likely that the fauna was contemporaneous with the enclosing sediment. These sediments were overlain in the borehole by over 13.5 m of lacustrine and fluvial deposits of presumed Holocene age.

A middle Holocene marine invasion of the Loch Lomond basin is indicated by raised shorelines around the southern part of the loch at about 13 m and 12 m OD. The latter is the most conspicuous, its width in part reflecting exhumation of the Lateglacial shore platform (Dickson *et al.*, 1978; Rose, 1980f). It is best seen at Portnellan and Ross Priory. Rose (1980f) traced these shorelines through the Vale of Leven to the Clyde estuary, thereby confirming their marine origin (see Dickson *et al.*, 1978, figure 1). Shells in deposits related to the transgression have been recorded near Luss (Robertson, 1868; Brady *et al.*, 1874). A further shoreline at 9 m OD was considered by Rose (1980f) to be lacustrine in origin, related to present lake level.

The middle Holocene marine transgression is also represented in sediment cores from the southern basin of Loch Lomond (Dickson *et al.*, 1978). The marine deposits, interbedded with freshwater sediments, are distinguished by low remnant magnetic susceptibility and intensity and by the presence of marine plankton and absence of freshwater plants. Radiocarbon dates on organic material from one core place the transgression between 6900 BP and 5450 BP. More detailed results were presented by Stewart (1987). He showed that there had been two phases to the marine transgression, the earlier one lasting from approximately 7200 BP until about 6400 BP and the later one, following a brief non-marine period, terminated at approximately 5500 BP.

Interpretation

The three localities show a sequence of events (Rose, 1980f) beginning with a Lateglacial marine transgression and deposition of Clyde beds (see Geilston), seen at Claddochside. The existence of such marine deposits in the Loch Lomond basin has long been recognized principally on account of their fossiliferous nature (Adamson, 1823; Brady *et al.*, 1874; Jack, 1875; Cunningham Craig, 1901). Deposition of the Clyde beds was followed by the formation of the shore platform. Rose (1980f) has correlated this platform with the Main Rock Platform of western Scotland (see Isle of Lismore) (Gray, 1974a, 1978a; Sissons, 1974d). Subsequently, during the Loch Lomond Stadial, a glacier advanced across the area depositing till over both the shore platform and the Clyde beds. Together with Gare Loch (Rose, 1980b) and Loch Spelve (Gray, 1974a), Portnellan is one of only a few localities providing important evidence that the Main Rock Platform was predominantly formed before the Loch Lomond Readvance reached its maximum extent. Also, the burial by till of the Clyde beds, which postdate *c.* 13,000 BP (see Geilston), provides clear stratigraphic evidence for the Loch Lomond Readvance (Cunningham Craig, 1901; Simpson, 1933).

The evidence from the borehole in the lower Endrick Valley for a marine deposit immediately overlying the Gartocharn Till deposited during the Loch Lomond Readvance, implies that the sea entered the Loch Lomond basin towards the end of the Loch Lomond Stadial (Browne and McMillan, 1989). This conclusion is dependent upon the marine fossils contained in the sediment being *in situ* and not derived from the underlying shelly deposits. Comparison may be made, however, with the Western Forth Valley, where Sissons (1966, 1976b) has demonstrated that a marine transgression, culminating between 10,500 BP and 10,100 BP, was approximately contemporaneous with the maximum of the readvance.

During the Holocene, the radiocarbon dates indicate that entry of the sea into Loch Lomond was some 1100 years later than the start of the Main Postglacial Transgression in the Western Forth Valley (Sissons and Brooks, 1971), a delay which reflects the time required for the sea to surmount the outwash fan barrier formed by the Loch Lomond Readvance across the Vale of Leven at Alexandria, between Loch Lomond and the Clyde estuary (Dickson *et al.*, 1978). This early marine phase also coincides with the maximum of the transgression (6800–6650 BP) in the Forth valley (Sissons, 1983a).

The Holocene palaeomagnetic record from Loch Lomond shows similarities to those from Lake Windermere and Lough Neagh, but is particularly significant in providing much finer detail because of a higher rate of sedimentation (Dickson *et al.*, 1978). It provides the most precise and detailed geomagnetic record so far obtained for the last 7000 years (Turner and Thompson, 1979).

Portnellan, Ross Priory and Claddochside complement the interest at Croftamie, Gartness and Aucheneck by illustrating the sequence of Lateglacial and early and middle Holocene marine episodes and their relationships to the glacial deposits of the Loch Lomond Readvance in the type area for the readvance and the Loch Lomond Stadial. The lake-bed sediments are also of considerable importance for the detailed palaeomagnetic record they preserve and for the succession showing Holocene marine transgression deposits interbedded with freshwater lake deposits.

Conclusion

The landforms and sediments at this site are important for the evidence they provide for the sequence of glacial and marine events at the end of the Devensian. In particular, they show a shore platform and fossiliferous Lateglacial marine deposits that formed before the Loch Lomond Stadial (approximately 11,000–10,000 years ago). During the latter phase, a shelly till was deposited by a glacier readvance (Loch Lomond Readvance). Raised shorelines and sediments from the floor of the loch also indicate two episodes during the Holocene (the last 10,000 years) when the sea encroached into Loch Lomond. This sequence is important because of its location in the type area for the Loch Lomond Stadial.

GEILSTON

D. G. Sutherland

Highlights

The deposits exposed in stream sections at Geilston include a sequence of fossiliferous marine sediments, the Clyde beds, which provide important evidence for marine palaeoenvironmental conditions during the latter part of the Late Devensian. Geilston is one of the few sites with good exposures which have been studied in detail.

Introduction

The site (NS 341777) comprises a series of stream sections along a 0.4 km reach of the Geilston Burn at Cardross. It is one of the few localities where *in situ*, fossiliferous Clyde beds sediments can be observed in sections. Radiocarbon dates from the site also provide a limiting date on the time of deglaciation of the area, and the sediments record the changing environment from glacial to glaciomarine to marine as the last ice-sheet decayed and the area became ice-free (Rose, 1980a).

It has long been known that certain of the silts, sands and clays exposed at low altitudes (generally below 35 m OD) and in many foreshore areas around the head of the Firth of Clyde contain marine fossil faunal assemblages that are indicative of a climate colder than that of the present (Smith, 1838; Jamieson, 1865; Crosskey and Robertson, 1867–1875; Brady *et al.*, 1874). Furthermore, it was also realized by these researchers that the changes in the fossil faunas reflected changes in the level of the sea (Robertson, 1883) as well as changing climates. In recent years, radiocarbon dating and quantitative analysis of the faunal assemblages have clarified the age of the sediments and the changes in climate and sea level that accompanied their deposition (Peacock *et al.*, 1977, 1978; Peacock, 1981b, 1983a, 1989b). The informal term 'Clyde beds' was proposed by Peacock (1975c) for those marine sediments deposited around the Scottish coast (but known principally in the area of the Firth of Clyde) subsequent to the deposition of the arctic 'Errol beds' (see Inchcoonans and Gallowflat), but prior to the establishment of a marine climate similar to that of today in the early Holocene. The Clyde beds thus cover the period from approximately 13,000 BP to around 10,000 BP.

Description

Rose (1980a) identified the following stratigraphic succession in the Geilston deposits (Figure 13.4):

5. Beach sands and gravels unconformably overlying bed 4
4. Clayey silt } Clyde beds
3. Laminated silts } Clyde beds
2. Glaciomarine till
1. Lodgement till

At the base the stiff, reddish-brown lodgement till (bed 1) has a relatively high silt–clay content and a dominant NW–SE fabric. The overlying glaciomarine till (bed 2) is similar in colour and lithology to the lodgement till but has a much lower silt–clay content. It contains a marine microfauna. Resting on the glaciomarine till is a sequence of colour laminated (pale olive-brown to brown) silts (bed 3) of which 26 couplets were reported by Rose (1980a). The individual laminae fine upwards and the colour change relates not to differences in particle size, but to the relative proportions of materials derived from Dalradian (pale laminae) or Old Red Sandstone bedrock. Foraminiferal remains increase in abundance through the lower half of the bed, remaining constant in frequency in the sediments

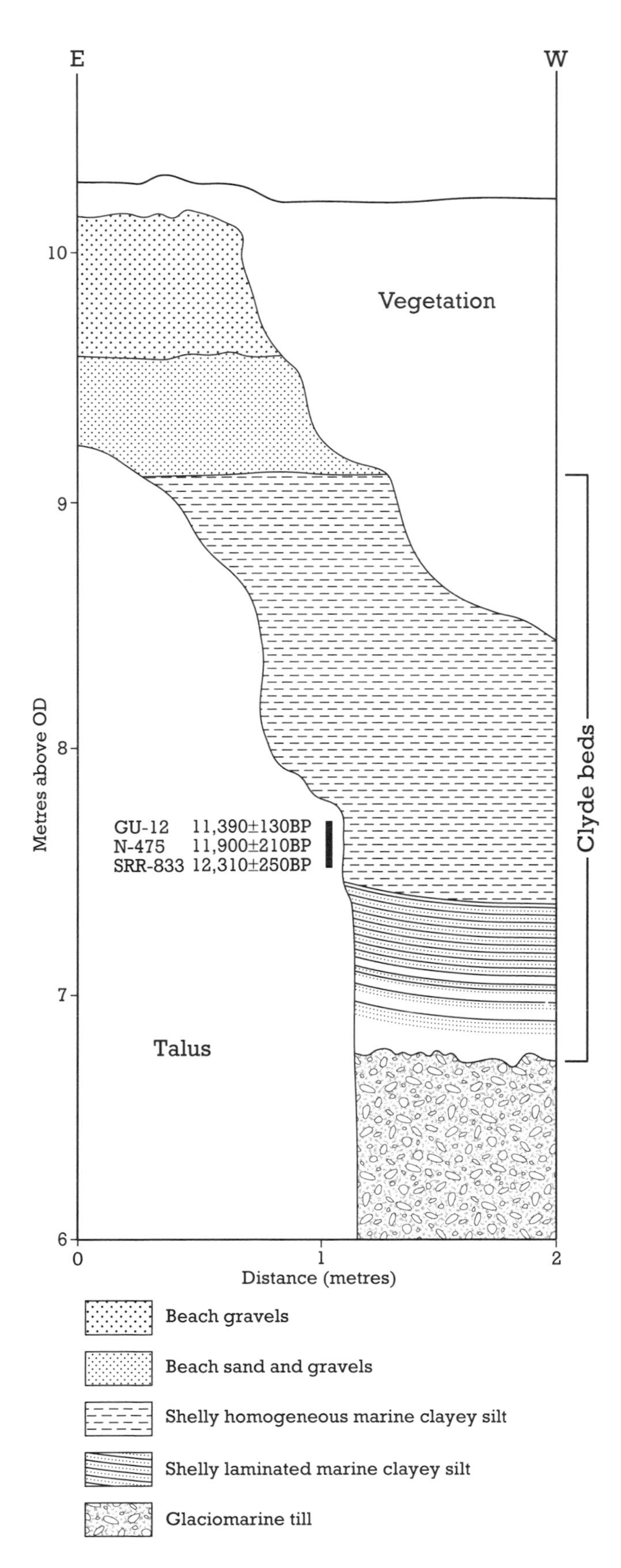

Figure 13.4 Geilston: sequence of sediments (from Rose, 1980a).

above this level. Mollusc shells (*Yoldiella lenticula* (Müller)) occur in the upper part of the laminated silts. The upper laminae merge into an overlying homogeneous clayey silt (bed 4) in which shells are abundant, in particular *Arctica islandica* (L.). The clayey silts are truncated by a marked unconformity, above which are horizontally bedded sands and gravels. The clasts are generally subrounded and become progressively more disc-like or blade-shaped upwards.

Three samples of *Arctica islandica* from close to the base of the homogeneous clayey silt have been radiocarbon dated and have given ages (Sutherland, 1986) of 11,390 ± 130 BP (GU–12) (Baxter *et al.*, 1969); 11,900 ± 210 BP (N–475) (Yamasaki *et al.*, 1969); and 12,310 ± 250 BP (SRR–833) (Harkness and Wilson, 1979).

Interpretation

The sediments were interpreted by Rose (1980a) in the following manner. The basal lodgement till was deposited by the Late Devensian ice-sheet, which flowed from the south-west Highlands towards the east along the Clyde valley. The long axes of drumlins in the area accord with this direction of ice movement. During deglaciation, the tidewater ice front retreated along the Clyde estuary, allowing deposition of the glaciomarine till beneath the floating ice margin. With further ice retreat the laminated sediments were deposited, the upward thinning of the laminae reflecting the increasing distance of the ice margin from the site. Each lamina is considered to represent the suspension fraction of subaqueous sediment plumes discharged from the ice margin. The homogeneous clayey silt with abundant macrofauna was deposited after the area was completely deglaciated, and when clearer water allowed *Arctica islandica* to become established (Peacock, 1981b).

The radiocarbon dates confirm the silts to have been deposited during the Lateglacial Interstadial but the considerable range in age is either due to mixing of the sediments by bottom current activity or very low sedimentation rates (Peacock *et al.*, 1978; Peacock, 1981b; Sutherland, 1986). The oldest radiocarbon date also places a minimum age on deglaciation of this area although consideration of dates from a variety of sites in the Clyde estuary suggests deglaciation had occurred by 12,600 BP (Sutherland, 1986).

The unconformity which truncates the Clyde beds is shown by Rose (1980a) to correlate with the period of erosion responsible for the cutting of the Main Rock Platform during the Loch Lomond Stadial. Both in accord with that age for the unconformity and their altitude, the sands and gravels capping the section are considered to be part of the Holocene raised beach deposits which occur widely around the coasts of the head of the Firth of Clyde.

Despite the wide distribution of Clyde beds sediments around the coast of Scotland and, in particular, around the Firth of Clyde, there are very few localities where they can be examined in section. Geilston is one such locality and one of the few where detailed sedimentological work has been carried out and the stratigraphic sequence and origin of the sediments is well established. Furthermore, radiocarbon dating here has placed the deposits in a clear geochronometric framework. The Clyde beds are a very important sedimentary unit, not only for historical reasons related to the evolution of ideas on climatic change and the ice age, but also for the evidence they contain of climatic and sea-level changes during the Lateglacial. Such evidence is an important counterpoint to the terrestrially-derived models of climatic change during the same time interval (cf. Peacock, 1989b; Peacock and Harkness, 1990). As a representative of the Clyde beds, Geilston is hence a particularly important site.

Conclusion

Geilston is a key reference site for the Clyde beds, a sequence of fossiliferous glaciomarine and marine sediments that formed in the period during and following the wastage of the last ice-sheet (approximately 13,000–11,000 years ago). These deposits provide important evidence for changing conditions in the marine environment during the phase at the end of Devensian times, the Lateglacial. Geilston is one of the few sites where there are both good sections available and the sediments have been studied and dated.

RHU POINT

J. E. Gordon

Highlights

Deposits exposed in the coastal section at Rhu

Point provide important evidence for interpreting the glacial history of the Western Highland Boundary area and the associated changes in sea-level during the Late Devensian and early Holocene. The evidence from Rhu Point allows the nature and timing of the Loch Lomond Readvance to be established.

Introduction

The site (NS 264841) is a coastal section on the north side of Rhu Point, cut into the terminal moraine ridge of the Loch Lomond Readvance glacier that occupied the Gare Loch basin. The deposits are of long-standing importance for demonstrating the succession of marine and glacial episodes in the coastal areas of west-central Scotland during the Lateglacial and early Holocene (Maclaren, 1845, 1846; Anderson, 1896; McCallien, 1937b; Anderson, 1949; Rose, 1980c). Excellent glaciotectonic deformation structures formed by the advancing glacier are of additional sedimentological interest. Part of the sequence is now concealed behind sea defences.

Description

Rose (1980c) provided a detailed description of the deposits at Rhu Point (Figure 13.5) noting the following succession:

3. Raised beach gravels
2. Glaciofluvial gravels (Rhu Gravels), with included blocks of till
1. Marine, clayey silt (Clyde beds)

At the base of the section, the typical Clyde beds of the region (see Geilston), comprise homogeneous clayey silts and clayey silts with size-graded laminations (bed 1). These deposits occur in both undisturbed and glacially disturbed states (Figure 13.5). The Rhu Gravels (bed 2) are a suite of glaciofluvial sands and gravels often with current bedding and flow structures. In places they are glaciotectonically folded and sheared and sometimes contain inclusions of Clyde beds and shelly till derived from the Clyde beds. The till occurs as detached blocks interdigitated with the gravels. Radiocarbon dating of a sample of shells from the till gave an unadjusted radiocarbon age of 11,520 ± 250 BP (HAR–931) (Otlet and Walker, 1979). Adjusting the radiocarbon date for the apparent age of seawater (Harkness, 1983), gives a best estimate for the age of the dated shells of *c.* 11,100 BP. The molluscan shells indicate a fauna of interstadial aspect, but also include fragments of *Portlandia arctica* (Gray) which almost certainly date from the Loch Lomond Stadial (J. D. Peacock, unpublished data). The Rhu Gravels are truncated by an erosion surface and overlain unconformably by raised beach gravels associated with successive shorelines at 14 m, 10 m and 8 m OD.

Interpretation

The terminal moraine at Rhu Point was one of the first such features in Scotland to be comprehensively investigated and explained in terms of the presence of former glaciers. In two remarkably perceptive papers for their time, not long after the introduction of the glacial theory to Scotland (Maclaren, 1840; Agassiz, 1841b) and before this theory was widely accepted, Charles Maclaren convincingly described the evidence for the former existence of glacier ice in the Gare Loch valley (Maclaren, 1845, 1846). He inferred that the moraine at Rhu Point was probably the last of a number of such features to be formed before the ice finally disappeared. Maclaren (1846) also recognized lateral moraine ridges on the east side of the valley and described the terminal moraine at Rhu as consisting of clay overlain by sand and gravel. He noted that it had been trimmed by the sea and also reported a sequence of raised shorelines along the loch side. 'In short, marks of the ancient existence of a glacier in the valley are numerous and remarkably complete' (Maclaren, 1849, p. 165). Maclaren's papers represent an important development at a time when the glacial theory was still in its infancy. Although moraines had been identified elsewhere, notably by Agassiz and Buckland during their tour of Scotland (Agassiz, 1841b, 1842; Buckland, 1841a; Davies, 1968b), Maclaren's contribution was one of the first detailed local studies of this type of evidence, together with that of Forbes (1846) in the Cuillin Hills. Maclaren's work in the Gare Loch area was also significant in identifying geomorphological links between glaciation and raised shorelines, although it was Jamieson (1865) who later developed formally the concept of glacio-isostasy. Maclaren (1842a) was also first to introduce the idea of glacio-eustasy.

In a subsequent account of the Gare Loch area, Anderson (1896) largely followed Maclaren's

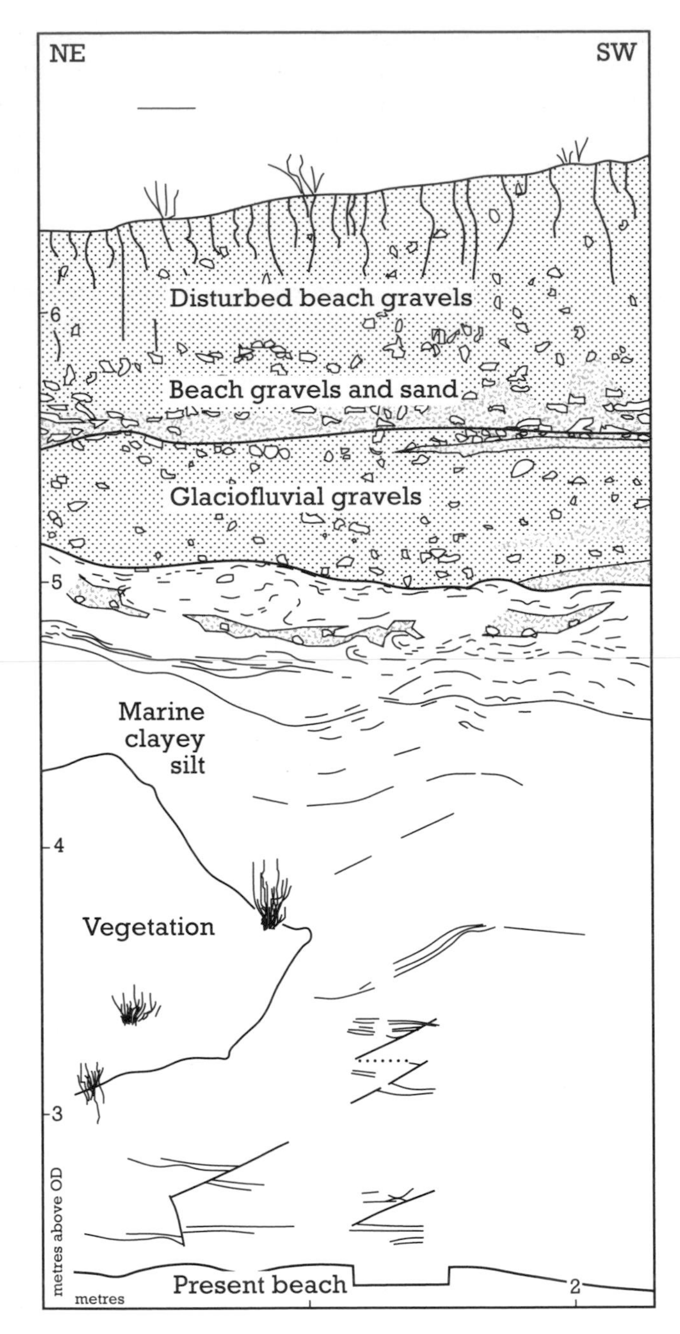

Figure 13.5 Rhu Point: section showing glacially deformed Clyde beds, Rhu Gravels and Holocene raised beach deposits (from Rose, 1980c).

interpretation of the Rhu Point deposits. Later, McCallien (1937) described in detail sections on the north side of the promontory at Rhu, noting, in particular, raised beach sediments unconformably overlying deformed 'morainic' material comprising sand and gravel which in turn rested on finely laminated clays. He interpreted the sediments as representing a readvance of ice which pushed up marine clays from the sea floor and dumped them along with morainic sands and gravels. Subsequently the deposits were trimmed by the sea. McCallien suggested that the Rhu moraine was a product of the Loch Lomond Readvance postulated by Simpson (1933) from evidence in the Loch Lomond and Western Forth valleys.

Anderson (1949) described the same sequence of deposits as McCallien, but also reported the presence of molluscan shell fragments, ostracods, Foraminifera, plant seeds and beetle fragments in the marine clay, which he stated were identical with those occurring widely elsewhere in the Clyde estuary (Clyde beds). He also traced the lateral continuation of the moraine ridge on both sides of Gare Loch and observed that the '100 foot' raised beach terminated at the moraine ridge, whereas the '25 foot' beach continued along the side of the loch inside the moraine. Anderson therefore concluded that the moraine ridge marked the limit of a late valley glacier readvance which, from its relationship with the raised shorelines, was contemporaneous with the Loch Lomond Readvance.

Charlesworth (1956) correlated the Rhu moraine ridge with his 'Stage M' readvance in the Highlands, which included Simpson's Loch Lomond Readvance moraines. Later, Sissons (1967a) mapped it as part of the Loch Lomond Readvance.

From the evidence at Rhu Point and in the adjacent area, Rose (1980c) interpreted the following sequence of events. As the Late Devensian ice-sheet wasted, the sea flooded into Gare Loch to a marine limit at 24 m OD. Deposition of the Clyde beds began and continued until at least 11,100 BP. A glacier then advanced along the Gare Loch valley forming a moraine ridge at Rhu Point, which consists primarily of glaciotectonized Clyde beds and sand and gravel deposited by meltwater streams flowing parallel to the ice front and towards the valley centre. The radiocarbon date on shells in the shelly till confirms that the moraine was formed during the Loch Lomond Stadial. The deformation structures in the earlier deposits indicate an oscillating ice front. Subaerial meltwater flow is suggested by the sedimentary characteristics of the sands and gravels, and this implies that sea level must have stood at or below 2.3 m OD, the lowest elevation at which these deposits occur. Subsequently, relative sea level rose during the Holocene transgression to 14 m OD, then fell to its present level via intermediate shorelines at 10 and 8 m OD.

The sequence of deposits at Rhu Point provides a key stratigraphic record of Lateglacial and early Holocene environmental and geomorphological changes in western Scotland. It is particularly important in integrating both terrestrial and marine evidence in a single radiocarbon-dated succession. It demonstrates the period of incursion of 'sub-arctic' seas, represented by the Clyde beds, following the recession of outlet glaciers from the main Late Devensian ice-sheet. Clear evidence for a subsequent readvance of ice is provided by the deformation and incorporation of Clyde beds into an end moraine ridge and the superimposition of glaciofluvial deposits with till inclusions. Radiocarbon assay constrains the date of this ice readvance to a period after the later part of the Lateglacial Interstadial and provides a maximum date for the local maximum extent of the Loch Lomond Readvance in Gare Loch. The two lower shorelines demonstrate temporary stillstands in the fall of relative sea level to its present position.

Broadly similar sequences, but without the Holocene beach sediments, occur at South Shian and Balure of Shian (see above), where glacially disturbed marine sediments (Clyde beds) are overlain by outwash from a Loch Lomond Stadial glacier in the valley of Loch Creran; at Gartness (see above), where deformed marine sediments are overlain by Loch Lomond Readvance till, and in the Western Forth Valley (see below) and at Loch Spelve (Gray and Brooks, 1972), where marine sediments are incorporated into Loch Lomond Readvance moraines. The significance of Rhu Point is, first, as a key reference site demonstrating the sequence of marine and glacial events during the Lateglacial and early Holocene in west-central Scotland. Second, Rhu Point is a site of considerable historical interest as one of the first localities in Scotland where glacier theory was applied in a detailed study to explain surface landforms and deposits. Finally, Rhu Point is notable for the glaciotectonic deformation structures in both the marine and glaciofluvial deposits; these have not been investigated in detail but have significant potential for research.

Conclusion

The deposits at Rhu Point have a long history of research and have provided important information for interpreting the sequence of Late Devensian glacial events in western Scotland. They show clear evidence for glaciation during the Loch Lomond Stadial (approximately 11,000–10,000 years ago) after a phase of marine sedimentation following the wastage of the main Late Devensian ice-sheet (approximately 13,000 years ago). Rhu Point forms part of the network of sites that demonstrate geomorphological processes associated with the Loch Lomond glacier readvance and provide evidence for its timing.

WESTERN FORTH VALLEY

D. E. Smith

Highlights

The interest of this area includes an outstanding assemblage of moraine ridges, outwash sediments and a sequence of buried estuarine deposits and peats. These landforms and deposits provide an exceptionally detailed record of the glacial history and sequence of sea-level changes in central Scotland at the end of the Late Devensian and during the early and middle Holocene.

Introduction

At the head of the Forth valley, around the Lake of Menteith (Figure 13.6), a belt of moraine ridges, glaciofluvial deposits and sequences of marine deposits record the advance and decay of a Loch Lomond Readvance glacier, together with related and subsequent changes in relative sea level. This assemblage of landforms and deposits has been studied for over 100 years (Jamieson, 1865; Simpson, 1933; D. E. Smith, 1965, 1968; Newey, 1966; Sissons, 1966, 1972b, 1976e; Sissons and Smith, 1965b; Sissons *et al.*, 1965; Sissons and Brooks, 1971; Brooks, 1972; Gray and Brooks, 1972; Kemp, 1971, 1976). It represents probably the most detailed evidence for readvance, deglaciation and relative sea-level change during the Late Devensian and early Holocene in Scotland. The extensive early literature on the area has been reviewed by Smith (1965).

The main features of this area are particularly well-illustrated at four key sites: these are identified on the accompanying generalized geomorphological map (Figure 13.6) as site I (Inchie), site II (Easter Garden), site III (West Moss-side) and site IV (Easter Poldar). In the account that follows, the landforms and deposits attributable to glacial advance, retreat and relative sea-level change are summarized before more detailed descriptions and interpretations are given for the key sites.

Description

The Forth valley is a broad lowland running some 40 km eastwards from the edge of the Highlands to the head of the Firth of Forth. The surrounding landscape reflects the varied geology of the area, but the lowland is distinguished by extensive areas of raised estuarine deposits, which occur on both sides of the River Forth throughout its length. These deposits consist of a grey silty clay with occasional lenses of sand, and are known locally as carse clay. They form a remarkably uniform surface, in which local changes of level of more than 1 m are rare. These carselands extend up to 3 km either side of the Forth, and occupy an area in excess of 50 km. A sharp break of slope occurs where they meet the sides of the lowland, which are for the most part mantled in glacial and glaciofluvial deposits.

West of the village of Kippen (NS 650948), the head of the lowland is occupied by large numbers of glacial and glaciofluvial features, and the carselands are restricted to narrow areas along the Forth and Goodie Water. It is these landforms and deposits which represent the key interest.

Glacial events

The glacial and glaciofluvial landforms and deposits of the Western Forth Valley have long attracted attention, with Jamieson (1865) apparently the first to have recognized an ice limit. In 1933, Simpson maintained that subsequent to general ice-sheet glaciation, ice had readvanced eastwards into the area, citing as evidence for this a section near Inchie where grey clay with marine shells was overlain by sands and gravels (see below). He called this event the Loch Lomond Readvance. In 1956, Charlesworth supported the view that an important ice limit could be identified here, correlating it with his 'Moraine Glaciation'.

Since 1962, more systematic studies of the glacial and glaciofluvial landforms and deposits of the area have been undertaken (for example, Sissons *et al.*, 1965; Smith, 1965). These studies confirm the evidence of a readvance moraine, and describe the landforms and deposits of the area in some detail. The frontal margin of a major ice limit can be traced in a broad belt of moraine ridges which forms an arc across the valley between Port of Menteith (NN 584012) and Buchlyvie (NS 575937). The northern and southern limbs of this belt (A and B, Figure 13.6) are elongate areas each composed of a large number of small ridges, most of which trend across the valley. These areas are dissected by deep channels leading eastward, and in some of these channels there are terraces composed of coarse

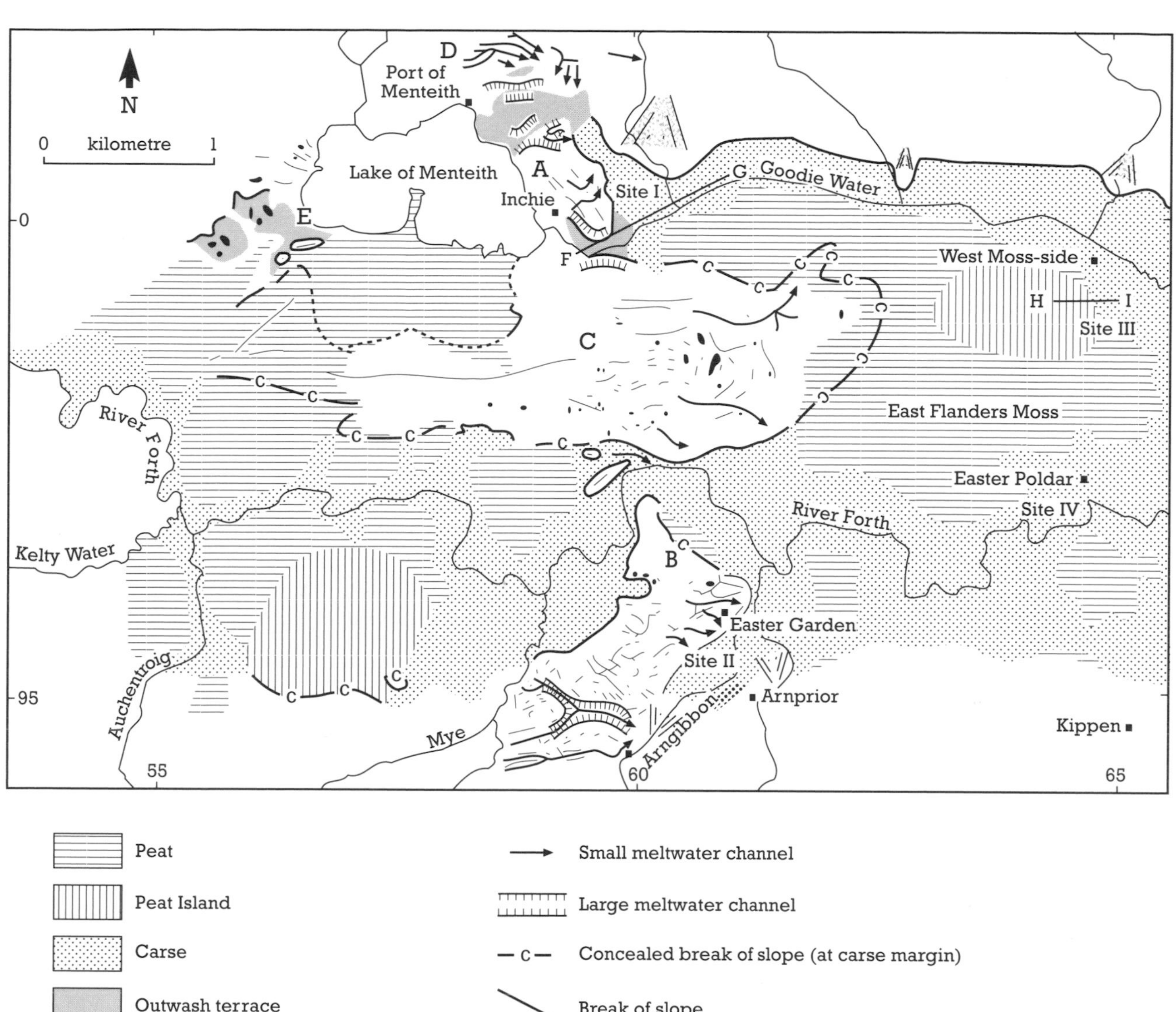

Figure 13.6 Geomorphology of the Western Forth Valley.

sands and gravels. The two areas of ridges are thought to be push moraines; the channels, routeways for proglacial meltwaters (Smith, 1965).

The central part of the belt (C, Figure 13.6) does not contain such clear evidence. It consists of a broad ridge running parallel to the axis of the valley, with a number of smaller ridges on its surface. The area extends about 1 km east of the ridged areas to the north and south. Exposures show the main ridge to be composed of bedrock (Smith, 1965; Laxton, 1984), whereas the smaller ridges are apparently formed of sands and gravels. Many of the smaller ridges are orientated parallel to the valley axis and to the axis of the main ridge, rather than across it; only a few of them continue the trends of the northern and southern limbs of the belt. Through this area thread a number of sinuous meltwater channels. It is possible that some of the ridges may have been crevasse fillings, and some channel systems may have operated proglacially. Smith (1965) argued that, in view of the fact that this central area extends beyond the ridged areas to the north and south, some of the deposits in it may have been formed earlier than in the latter two areas, perhaps during earlier ice-sheet deglaciation. It seems that the contrast between the central part of the belt and the areas to the north and south reflects the bedrock topography over which the glacier advanced.

The recession of ice from its limit is recorded in a number of glaciofluvial landforms lying west of the moraine belt. Thus, on the hillslopes above Port of Menteith, 'staircases' of kame terraces with associated marginal meltwater channels and, below these, subglacial chutes (D, Figure 13.6) record the downwasting of the ice mass which had formed the moraine belt; west of the Lake of Menteith, outwash terraces (E, Figure 13.6) probably record the further retreat of the ice mass. Kettle holes in the area west of the moraine belt indicate the wastage of ice detached from the formerly more extensive glacier.

In 1957, Donner concluded from pollen analysis of sites in the general area that the ice limit here belonged to Pollen Zone III of the Jessen–Godwin system of pollen zonation, or some time between 10,800 and 10,300 BP. This inferred date has subsequently been supported by radiocarbon dating of shells from Simpson's section (Sissons, 1967b) (see below), and the limit (the Menteith Moraine) is correlated with the Loch Lomond Readvance elsewhere in Scotland (Sissons *et al.*, 1973; Sissons, 1974c).

Relative sea-level changes

The study of changes in relative sea level in the area both during and following the Loch Lomond Readvance, has been an increasing focus of interest in recent years. It has long been known that the carse clay is an estuarine deposit, from the many faunal remains found and recorded during a period of over 300 years (Smith, 1965). However, detailed stratigraphical investigations since the mid–1960s (for example, Sissons and Smith, 1965b; Sissons, 1966, 1972b; Smith, 1968; Sissons and Brooks, 1971; Kemp, 1976; D. E. Smith *et al.*, 1978; M. Robinson, unpublished data) have identified a detailed sequence of relative sea-level changes in the area, in which the carse clays are only one element.

Relative sea levels prior to the Loch Lomond Readvance are not well known. Simpson (1933) believed that the altitude of the top of the shelly clay at Inchie, 65 ft (19.8 m) OD, reflected contemporary relative sea level, but this is now not thought likely in view of the possibility that the sediments may have been disturbed by the readvancing ice. Francis *et al.* (1970) and Laxton (1984) have suggested that certain terraces underlain by thin beds of stratified sands on the valley slopes to the east of the Menteith Moraine were formed during ice-sheet retreat, when sea level was between 18 m and 40 m OD. This interpretation is in contrast to that of Sissons and Smith (1965a) who implied that by the time the ice-sheet had retreated to the Western Forth Valley, relative sea level had fallen to a low level.

During the readvance, relative sea level is thought to have been low. The outwash terraces which lead from the channels in the moraine belt descend to at least 8.8 m OD (Sissons, 1966), and the large outwash fan related to the contemporaneous Teith glacier descends to 6 m OD (Smith *et al.*, 1978; Laxton and Ross, 1983). Subsequently, however, relative sea level rose and three marine terraces, now buried beneath later deposits, were formed in the area of the moraine belt (Sissons, 1966). The highest of these reaches 11.9 m to 12.2 m OD at its inland margin, and has been called the High Buried Beach. It has not been found within the morainic arc, and is considered to have formed while ice stood at the moraine. On this basis, a date of between 10,500 and 10,100 BP has been suggested by Sissons (1966, 1983a) for this shoreline.

The lower terraces (the Main and Low Buried Beaches) are found within the morainic arc and were formed after the ice had withdrawn from the vicinity. They reach, respectively, 11.5 m and 8.0 m OD. Overlying these two lower terraces is a buried peat; detailed study shows that it began accumulating as the sea withdrew from the terraces (Newey, 1966). Radiocarbon dates and pollen analyses from the base of the peat give ages of around 9600 BP for the Main Buried Beach and around 8700 BP for the Low Buried Beach (Sissons, 1966, 1983a; Sissons and Brooks, 1971). From the form of the three buried beaches, Sissons concluded that each had resulted from a brief transgression.

After falling to an unknown level, reached at possibly 8500 BP, relative sea level again rose, and the carse clays began to accumulate in the now expanded Forth estuary. This transgression is known as the Main Postglacial Transgression (Sissons, 1974c). As the sea rose, silts and clays were deposited on the peat on the surface of the buried beaches. In two areas, however, peat continued to accumulate faster than the rise in relative sea level. These areas lie on either side of the morainic arc, at East and West Flanders Moss (Figure 13.6). There, islands of peat formed (Sissons and Smith, 1965b) (see below). During this period, a North Sea storm surge (Smith *et al.*, 1985a; Haggart, 1988b) or more probably a tsunami (Dawson *et al.*, 1988; Long *et al.*, 1989a) penetrated the Western Forth Valley and a layer of fine sand accumulated up to at least 11.2 m OD in the estuarine sediments and surrounding peat at approximately 7000 BP (Sissons and Smith, 1965b; Smith *et al.*, 1985a). As the Main Postglacial Transgression continued, carse clays were deposited up to about 15 m OD (D. E. Smith, 1968). The consistent altitude of the inner margin of the carse clays in this area is taken to mark a shoreline, the Main Postglacial Shoreline (Sissons, 1974c). Subsequently, relative sea level fell and peat began to accumulate on the carse clay surface, and the peat islands expanded over the adjacent clays. A radiocarbon date from a site within the morainic arc and south of the Forth gave the age of peat immediately beneath the carse clay at 7480 ± 125 BP, and another in the same area for peat resting upon the carse surface gave 6490 ± 125 BP (Sissons and Brooks, 1971), thus providing a range for the age of culmination of the Main Postglacial Transgression and of the Main Postglacial Shoreline in the area. Sissons (1983a) has indicated that the transgression culminated at about 6800 BP in the Western Forth Valley, an interpretation supported by more recent radiocarbon dating (Cullingford *et al.*, 1991).

In a very detailed study of their altitudes, Sissons (1972b) demonstrated that both the Main Postglacial Shoreline and the shoreline of the Main Buried Beach in the area east of the moraine had been faulted or warped at two locations. Since between these points the gradients of the shorelines resulting from isostatic uplift were the same, it was concluded that during the period between the formation of the two shorelines (around 3000 years), there had been no differential uplift in the area except at the dislocations. This demonstration of dislocation and non-uniform uplift has since been corroborated by studies elsewhere in Scotland, and may indicate that raised shoreline altitudes in isostatically affected areas may be more complex than had been previously thought.

A curve of relative sea-level change for the area of the morainic arc, for the period from 10,500 to 4000 BP, together with a curve of land uplift for the last 10,000 years for the same area has been published by Sissons and Brooks (1971).

Inchie

The site (Site I, Figure 13.6) stretches from Port of Menteith in the north to Inchie Farm (NN 592000) in the south. It is bordered by the Lake of Menteith to the west and extends along the Goodie Water to the east. This area contains arguably the most distinctive part of the morainic arc and includes a transect along the Goodie Water where Loch Lomond Readvance outwash descends beneath buried beach and later Holocene deposits.

The northern limb of the morainic arc runs from north to south through this area, as an elongate belt consisting of many small ridges, each aligned across the main valley. These small ridges are rarely more than 5 m high, although the moraine reaches over 30 m above the adjacent lake. The surface of the moraine is furrowed in places by sinuous meltwater channels, leading eastwards, and is breached at three places by very deep channels with coarse gravels on their floors. The few exposures in the area indicate that the composition of the ridges is complex, with shelly marine clay, till and sands and gravels having been seen. It is likely that the area was formed by an advancing glacier pushing a variety of deposits before it. The marked asymmetry of the moraine ridge as a whole is in agreement with this interpretation, the steeper slope facing up-valley.

The section where Simpson (1933) reported evidence for a readvance was near Inchie Farm, probably close to NN 5920000. He found 10 ft (3 m) of grey clay with fragments of *Mytilus edulis* (L.) overlain by 30 ft (9.1 m) of sand and gravel. In 1967, Sissons identified a similar sequence near that location and on the lake shore, and obtained a radiocarbon date of 11,800 ± 170 BP (I–2234) for a specimen of *Mytilus edulis* (L.) (Sissons, 1967b), concluding that the readvance had occurred during the climatic deterioration at the end of the Lateglacial, the Loch Lomond Stadial. The fauna from the grey clay in Sisson's section (S. M. Smith, *in* Gray and Brooks, 1972) is as follows:

Littorina littorea (L.)
Littorina obtusata (L.)
Buccinum undatum L.

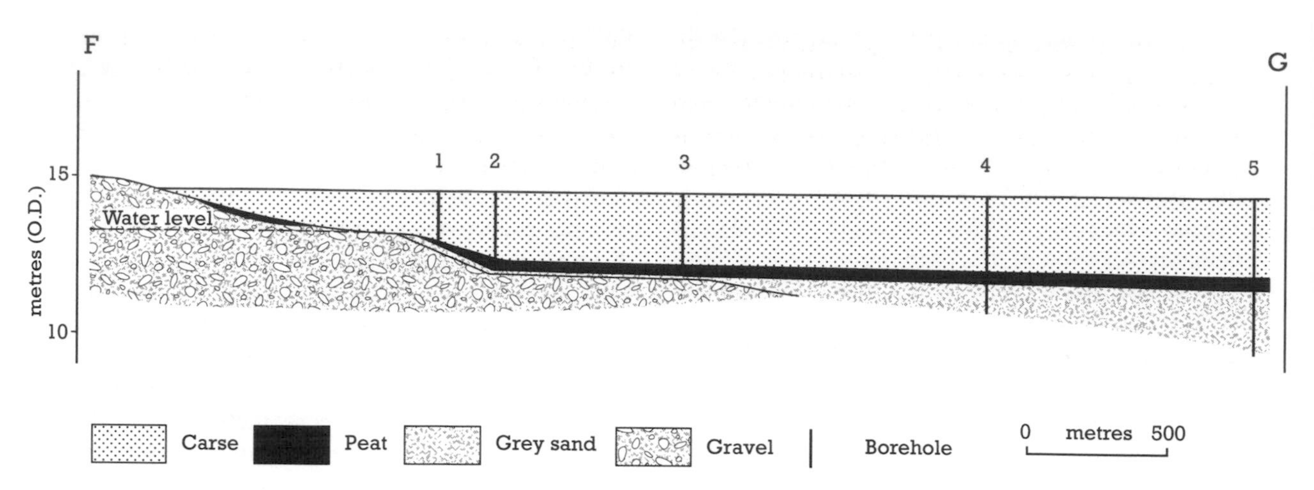

Figure 13.7 Western Forth Valley: section along the Goodie Water (from Sissons *et al.*, 1965). See Figure 13.6 for location of section.

Mytilus edulis (L.)
Nuculana pernula (Müller)
Chlamys sp.
Macoma sp.

Gray and Brooks (1972) also concluded that pollen from the grey clay indicated a Lateglacial Interstadial age for the deposit, and a largely open habitat locally was suggested by the low percentages of arboreal pollen. Taken with the stratigraphical evidence, these observations further support the correlation with the Loch Lomond Readvance.

The deep channels through the site carried outwash deposits eastward, as is indicated from the coarse sands and gravels on their floors. The largest channel, in the south of the site, contains a wide outwash terrace, part of which occurs on the proximal side of the moraine, indicating that it was deposited as ice began to waste back. This terrace, near Inchie Farm, slopes eastwards from 23.5 m OD. Beyond the moraine it descends beneath the carse clays, but its descent can be followed in the banks of the Goodie Water, where it passes beneath peat and fine-grained grey sand. The sequence was traced by Sissons *et al.* (1965) in exposures and boreholes and is shown in Figure 13.7. The outwash descends to at least 9 m OD and it was concluded that relative sea level at the time of deposition lay below that altitude. The fine grey sand above has a level surface at 11 m OD and was interpreted as forming part of the Main Buried Beach, which thus appears to have accumulated following a marine transgression across the outwash. The peat above the sand demonstrates a marine regression, and the estuarine (carse) clay which overlies this peat, and reaches 14.7 m OD here, was laid down during the Main Postglacial Transgression.

Although the floors of the channels at the northern end of the moraine in this area descend to below 20 m OD there is no indication that the carse clays were deposited in them, and therefore no indication that the sea at the culmination of the Main Postglacial Transgression entered the Lake of Menteith.

Thus the landforms and deposits at Inchie demonstrate glacial advance, retreat and subsequent relative sea-level changes. Though the sequence is known in detail, it is interesting that so far no evidence of the High Buried Beach has been found in this area. It may be that the beach is confined to the sides of the depression through which the Goodie Water flows, in a similar manner to the situation at Easter Garden, discussed below.

Easter Garden

The site (Site II, Figure 13.6) lies on the left bank of the Garden Burn near Arnprior (NS 612948), south of the River Forth. It includes the ice-distal (outer) portion of the morainic arc, together with carselands south of the Forth and along the Arngibbon and Garden Burns. The area was studied in detail by Sissons (1966). It contains some fine morainic topography, but is particularly notable for the sub-carse deposits, which include outwash gravels, together with High, Main, and Low Buried Beaches.

The southern limb of the morainic arc consists

of a large number of small ridges intersected by shallow meltwater channels. It is kettled in places and appears to be largely composed of sands and gravels. To the south of the site, the moraine is dissected by a large meltwater channel system, in which the main channel exceeds 20 m in depth. This was probably excavated largely by proglacial meltwaters.

The carseland stratigraphy in this area (Sissons, 1966) reveals a number of interesting features. South of Easter Garden (NS 607957), the carselands at the foot of the moraine are underlain by outwash gravels, which come from the mouth of the meltwater channel to the south. The surface of these gravels is slightly irregular, but falls north-eastward down to 11 m OD. Between Easter Garden and the moraine, and in a small area south of Easter Garden, the outwash is overlain by a deposit of silty sand which is pinkish or pale brown in colour except at its surface, where it is grey. In places, this deposit is laminated and its surface is level at about 12 m OD. It is part of the High Buried Beach (Sissons, 1966) and this location is the farthest up-valley it has been found.

South of Easter Garden and north of the Arngibbon Burn, the outwash surface and High Buried Beach have been dissected by a channel up to 200 m broad and 3 m deep, which appears to originate at the mouth of the meltwater channel system referred to above. This channel is largely floored with deposits of a grey silty fine-grained sand with occasional bands of silt or clay, lying at around 11 m OD. Where the channel ends north-east of Easter Garden, these sediments form a wide surface lying at 10.7 m OD. Sissons identified this as the Main Buried Beach. North of the abandoned railway line, a sharp break of slope occurs in the Main Buried Beach deposits and a lower area of grey silty clay, becoming more sandy with depth, ensues, lying at 6.4 m OD. This has been interpreted as the Low Buried Beach.

The buried surfaces of the outwash, and the High, Main and Low Buried Beaches are covered in peat, which becomes thicker over the lower surfaces. This peat is, in turn, overlain by carse clays, which reach 15 m OD in this area and belong to the Main Postglacial Shoreline. In some areas nearby the carse clays are covered in peat.

At the mouth of the large meltwater channel system mentioned above, the Arngibbon Burn has deposited an alluvial fan, which extends into the southern part of the site (Sissons *et al.*, 1965). Boreholes through the fan show that it overlies carse clay, whereas boreholes nearby in the floor of the main meltwater channel show peat, overlying lacustrine clay, which overlies estuarine (carse) clay. The alluvial fan was deposited across the mouth of the meltwater channel, and it seems that here the alluvial fan, built out across the carse clay, temporarily dammed a lake in the mouth of the meltwater channel system.

The sequence of events recorded in this area is therefore as follows. First, a readvance of ice formed a large moraine belt, and meltwaters discharged proglacially across it. At the southern end of the moraine a large proglacial meltwater system deposited an outwash fan in a north-easterly direction. The lowest elevation of the outwash, 8 m OD, indicates that relative sea level could not have been higher than this at the time. Then, relative sea level rose, and the High Buried Beach was deposited on part of the outwash surface. The ice must have been at the moraine during this period, because a large channel, emanating from the mouth of the meltwater channel system, was subsequently cut in both the outwash and the High Buried Beach. The size and position of this channel suggests that it was probably cut by proglacial meltwaters. Subsequently, relative sea level fell and deposits of the Main Buried Beach were laid down across lower parts of the outwash and in the channel. A further fall in relative sea level was followed by deposition of the Low Buried Beach. Eventually, relative sea level fell further and peat accumulated on the surfaces revealed. Later, the Main Postglacial Transgression led to the deposition of the carse clays. Subsequently, relative sea level fell again, and as peat began to accumulate on the carse clay surface, a small fan was built up by the Arngibbon Burn at the mouth of the large meltwater channel system, trapping a small lake in the lower part of the main channel. The lake eventually filled with clay, and peat also accumulated on these sediments.

West Moss-side

South of the Farm of West Moss-side (NS 648996), an area of East Flanders Moss (Site III, Figure 13.6) has provided a focus of interest for studies of Holocene relative sea-level changes. It is in this area that peat continued to accumulate during the Main Postglacial Transgression, form-

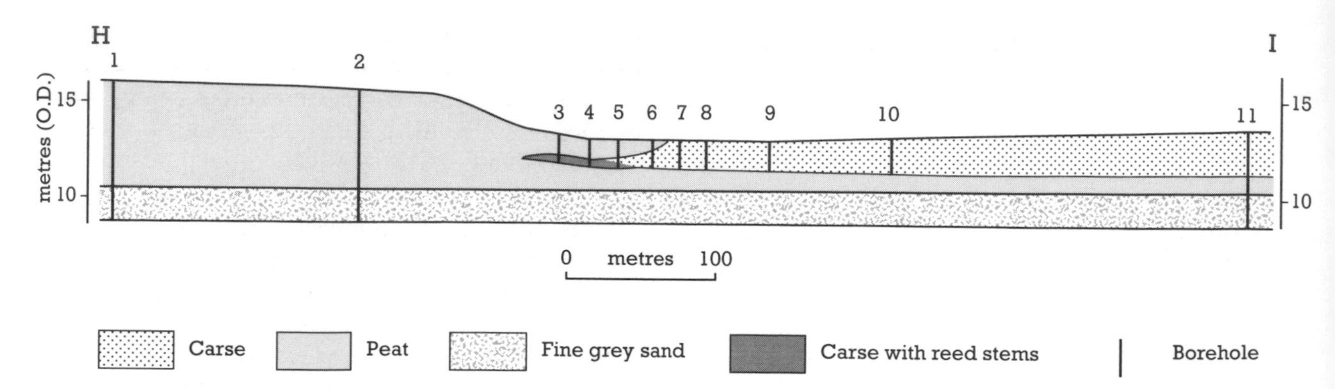

Figure 13.8 Western Forth Valley: section at West Moss-side, East Flanders Moss (from Sissons and Smith, 1965b). See Figure 13.6 for location of section.

ing an island in the then expanded estuary of the Forth. The peat stratigraphy of this area has been examined in detail by Smith (1965) and by Sissons and Smith (1965b).

East Flanders Moss is a remnant of a much more extensive peat bog that once covered the carselands of the Forth Valley. Clearance of the peat in the Forth Valley began in the eighteenth century (Tait, 1794) and ended in 1865 (Cadell, 1913). The termination of most peat clearances appears to have been for economic or social reasons, but near West Flanders Moss there is evidence that as the clearance progressed, the farmers encountered an increasingly boggy carse surface, which declined in altitude sharply.

In 1950, a survey of East Flanders Moss was undertaken by the Department of Agriculture for Scotland (Peat Section). From a number of boreholes they found that the peat was widely underlain by the level surface of the carseland, but that near West Moss-side a depression of over 4 m lay beneath the peat. Durno (1956) undertook pollen analysis of a core from this depression, finding that the basal peat began to accumulate in Zone V of the Jessen–Godwin system of pollen zonation, significantly earlier than peat elsewhere on the carseland surface. He concluded that the marine transgression in which the estuarine (carse) clay had accumulated had ended east of East Flanders Moss.

Smith (1965) and Sissons and Smith (1965b) later examined the stratigraphy in detail (Figure 13.8). They found that beneath the peat the carse clay formed a wedge at the edge of the depression. They identified a similar sequence at West Flanders Moss, west of the morainic arc. They also found that the peat at the base of the depression in East Flanders Moss continued beneath the carse clay, and that it rested upon a remarkably level surface lying at *c.* 10.0 m OD. They concluded that the surface upon which the peat had commenced accumulating was that of a buried beach (later identified as the Main Buried Beach by Kemp (1976)), and that this peat had managed to continue accumulating in a small area throughout the Main Postglacial Transgression. After the sea subsequently withdrew, the peat island expanded across the adjacent carse surface, at an altitude of approximately 12 m OD.

In West Flanders Moss, a thin (about 0.03 m) layer of grey, micaceous, silty fine sand has been identified within the carse and continuing into the peat, beneath the carse clay 'wedge'. This is thought to have been deposited during a North Sea storm surge (Smith *et al.*, 1985a; Haggart, 1988b) or more probably a tsunami event (Dawson *et al.*, 1988; Long *et al.*, 1989a). Recent work suggests that this layer may also be present beneath East Flanders Moss.

Easter Poldar

The detailed and often complex sub-carse stratigraphy in the Western Forth Valley can normally only be examined through detailed coring. However, at a number of sites along the Forth and Goodie Water, some elements of the stratigraphy can be examined in exposures. The best individual sites are along the river bank from Faraway (NS 615964) to Easter Poldar (NS 647972).

Near the Farm of Easter Poldar, on the left bank of the River Forth, an exposure in the carse clay and the underlying deposits was first recorded in 1964 (Smith, 1965). The sequence of deposits is as follows:

	Thickness (m)	Altitude (m) OD (surface = 12.56)
5. Brown silty clay, with reed stems	4.31	8.25
4. Blue-grey silty clay, with reed stems	0.15	8.10
3. Woody peat	0.24	7.86
2. *Sphagnum* peat	0.09	7.77
1. Fine-grained blue-grey silty sand with reed stems, especially in the upper part (base not seen), to	0.06	7.71

In 1966, Sissons identified the surface of the fine-grained, blue-grey, silty sand (bed 1) as part of the Low Buried Beach. The presence of increasing numbers of reeds in its upper part suggests that the overlying peat may have started to accumulate shortly after the sea withdrew from the surface. The peat (beds 2 and 3) is much compressed, but the transition from *Sphagnum* peat (bed 2) to a more woody peat (bed 3) indicates that conditions became less moist. The sediments above the peat are estuarine carse clay (beds 4 and 5), the brown coloration in the upper part (bed 5) probably being due to oxidation from the face of the exposure.

Interpretation

The assemblage of landforms and the sequences of deposits in the Western Forth Valley are among the most detailed for the Late Devensian and early Holocene in Scotland. The glacial and glaciofluvial landforms represent arguably the finest evidence for a readvance ice limit and subsequent ice decay anywhere in the British Isles, and the detail of the record of relative sea-level change is unsurpassed in Scotland.

At a number of other sites (Loch Etive, South Shian and Balure of Shian, and Rhu Point) it can be demonstrated that the Loch Lomond Readvance glaciers discharged into the sea or constructed outwash plains graded to sea level. However, it is only in the Western Forth Valley that a specific sequence of sea-level changes can be established at, or close to the time of the maximum of the readvance.

Early to middle Holocene sea-level changes have been studied in greater detail in the Western Forth Valley than in any other locality in Scotland. Both the sedimentary sequences and the landforms arising from sea-level change have been analysed, unlike in most areas where one or the other approach to sea-level change studies has been adopted. In consequence, fundamental information on the nature of isostatic uplift has been established in this area, notably the role of fault-reactivation and block movement. This last aspect has had important consequences for the study of neotectonic activity in Scotland.

The geomorphology of the morainic system has yet to be fully mapped, but it is evident from the details so far revealed, especially from exposures, that the moraine sediments will greatly repay further study. In this regard, two sites where the moraine is displayed (Inchie and Easter Garden) will be crucial. At all the key sites discussed here, the carse and sub-carse deposits have yielded a remarkably detailed story, but further work is needed to reveal the extent of tsumani or storm surge deposits, the progress of marine transgression and regression, and in particular further evidence for neotectonic activity. It is evident that the Western Forth Valley will play an important part in future scientific enquiry into both glacial events and relative sea-level changes in Scotland.

Conclusion

The Western Forth Valley is a critical reference area for studies of the Loch Lomond Readvance, and particularly for the sequence of sea-level changes that occurred around and following the maximum of the ice readvance (approximately 10,500 years ago). The key interest includes a complex of moraine ridges and outwash sediments (deposited in front of the glacier by meltwater rivers) and a sequence of buried estuarine deposits. The latter provide the most detailed record of sea-level changes in Scotland at the time of the readvance and later during early to middle Holocene times (approximately 10,500–6000 years ago). The area has a long history of research, and its landforms and sediments have been studied in great detail. In addition, the results of the work provide a standard framework of sea-level changes for comparison with other areas. Furthermore, the Western Forth Valley has allowed important insights into the processes that accompany isostatic recovery, that is, upward movement of the Earth's crust following ice melting.

MOLLANDS

J. J. Lowe

Highlights

The sediments that infill the floor of a kettle hole at Mollands provide an important record, supported by radiocarbon dating, of the Holocene vegetational history of this important ecological area located at the boundary between the Highlands and the Central Lowlands. Together with Tynaspirit, Mollands also provides significant evidence for establishing a glacial chronology for the area.

Introduction

The site at Mollands (NN 628068) is an infilled kettle hole located *c.* 1 km south of Bridgend, Callander. It is important for pollen stratigraphy at the Lateglacial–Holocene boundary and during the Holocene. It has also been significant in establishing a glacial chronology for the Callander area. The results obtained from Mollands (Lowe, 1977, 1978, 1982a, 1982b; Lowe and Walker, 1977) complement those from Tynaspirit (see below), and together these two sites provide one of the strongest lines of support for recently proposed schemes of the Late Devensian glacial chronology of Scotland (see Gray and Lowe, 1977b; Price, 1983; Lowe and Walker, 1984; Sutherland, 1984a).

Description

The limit of the last glacier advance in the Teith valley is marked by a well-defined terminal moraine (the Callander Moraine) which can be observed on both sides of the river just downstream from Callander (Figure 13.9). Upstream of the moraine some sharply delimited eskers occur, as for example around Clash and Gart (NN 636068 to NN 644064) and on both banks of the River Teith downstream of the site of Callander Castle (NN 632075 to NN 635073 and NN 632073 to Clash). A series of kame terraces has also been mapped in the area between the Callander Moraine and the town of Callander (Thompson, 1972; Merritt and Laxton, 1982). At Mollands, on the ice-proximal side of the moraine ridge (Figure 13.9), lake sediments and peats have filled a depression which exceeds 8 m in depth from the ground surface and which is bounded on its eastern and south-eastern edges by kame terraces. The site is therefore interpreted as a kettle hole, formed by melting of a block of buried ice that persisted until after the formation of the kame terraces. The sediment succession at Mollands should therefore provide a minimum age for the retreat of the Loch Lomond Readvance ice in the vicinity of Callander (see Sissons *et al.*, 1973; Lowe and Walker, 1977 for details of methodology).

The basal sediments are very variable, with a number of thinly bedded units, including rhythmites (Figure 13.10). Collectively, the basal sediments comprise a fining-upwards sequence from compact sands at the base (8.32 m to 7.50 m) through finely laminated silts, sands and clays (7.50 m to 7.30 m) to laminated organic and inorganic muds (see Lowe, 1978). The sediments proved to be non-polleniferous below 7.32 m, but a detailed pollen stratigraphy was based on samples obtained from the remainder of the succession. Three radiocarbon dates (HV–5645 to HV–5647) were obtained from the profile (Figure 13.10).

Interpretation

The vegetational succession (Figure 13.10) that can be inferred is one of successive plant colonization of the slopes around the site from initially bare, stoney surfaces to a woodland cover. The lowermost pollen zone (Mo a I) is dominated by pollen of grasses, sedges, sorrel, the dandelion group (*Taraxacum*) and meadow rues (*Thalictrum*), with spores of the clubmosses (*Selaginella* and *Lycopodium*) also well represented. These species thrive on bare, gravelly or stoney surfaces free of competition from shrub and woodland associations. The zones Mo a II to Mo c record successive invasions of the district,

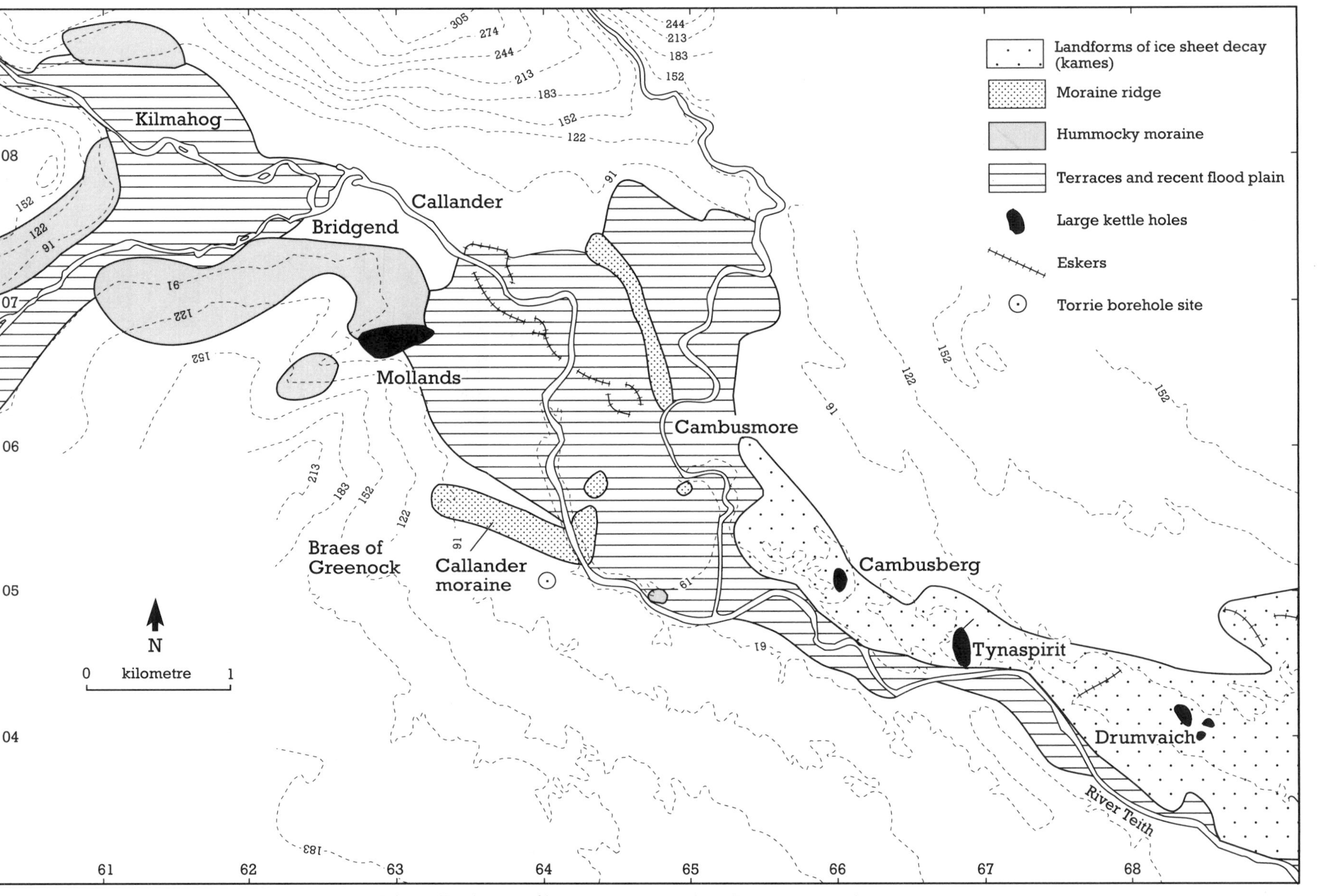

Figure 13.9 Glacial and glaciofluvial landforms of the Callander area (from Lowe, 1978, after Thompson, 1972).

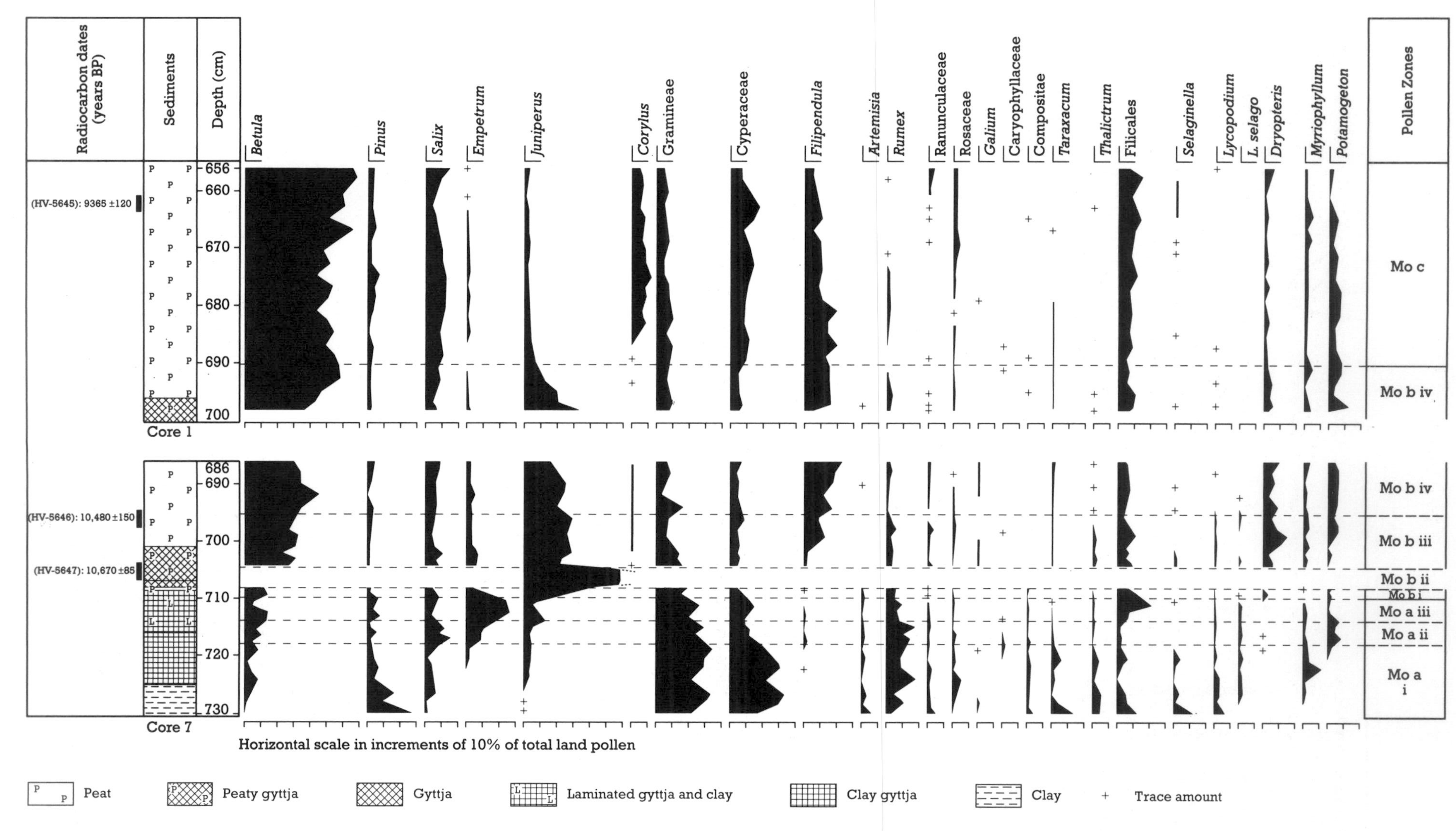

Figure 13.10 Mollands: Lateglacial and early Holocene relative pollen diagram showing selected taxa as percentages of total land pollen (from Lowe, 1978).

first by heath (crowberry and dwarf willows), followed by juniper, then birch and eventually hazel. This sequence is characteristic of the early Holocene at Scottish sites.

The dominant juniper phase at the site has been dated to 10,670 ± 85 BP (Hv–5647), a date which compares well with that of 10,420 ± 160 BP (Hv–4985) obtained for the immigration of juniper at Tynaspirit. Juniper remained a dominant species until shaded out by birch, and hazel colonization appears to have been well under way in southern Perthshire by 9365 ± 120 BP (Hv–5645), an age estimate that compares favourably with that of 9260 ± 100 BP (Hv–4984) obtained from a comparable biostratigraphic level at Tynaspirit.

Figure 13.11 presents the results of pollen analysis of the main Holocene organic sequence at Mollands (excluding the uppermost 0.4 m of disturbed sediments). The diagram indicates that an early Holocene phase of birch–hazel woodland (zones Mo b to Mo e) gave way to a mixed deciduous woodland, where alder, elm and oak were important constituents. The relatively high representation of pine at the site (zones Mo f to Mo g) suggests that the middle Holocene woodlands of the Teith valley were transitional between the deciduous forests of the central lowlands of Scotland and the dominant pine woodlands of the Highlands. An 'elm decline' occurs at the Mo g/Mo h boundary, suggesting an age for this part of the record of about 5000 BP (Hibbert *et al*., 1971; Smith and Pilcher, 1973; Godwin, 1975). Zone Mo h is characterized by the virtual disappearance of pine, by reductions in the percentages of oak and tree pollen generally, and by renewed growth of birch and hazel. These vegetational changes are likely to reflect anthropogenically induced, selective forest clearance in the Teith valley from about 5000 BP onwards (Lowe, 1982a).

The principal points to emerge from the studies completed at Mollands so far, excluding the vegetational history which has been summarized above, relate to the time and sequence of deglaciation at the end of the Loch Lomond Readvance and to aspects of Holocene vegetational history. Taken together, the evidence from Mollands, Tynaspirit and Torrie (NN 638051) (Merritt *et al*., 1990) confirms that the ice limit at Callander is that of a Loch Lomond Readvance glacier. The date on the basal sediments at Mollands of 10,670 ± 85 BP (Hv–5647) suggested a much earlier withdrawal of the readvance ice in parts of Scotland than was generally accepted at the time of publication (Lowe, 1977; Lowe and Walker, 1977). Most authorities generally accept an age of around 10,000 to 10,250 BP for the end of the Loch Lomond Stadial and accordingly for deglaciation. The data from Mollands contrast with this interpretation. There is much debate about the reliability of radiocarbon dates obtained from organic lake sediments that have accumulated in newly deglaciated terrain (see Lowe and Walker, 1980; Sutherland, 1980), but the following points lend credence to the conclusion that deglaciation may have occurred in parts of Scotland as early as about 10,700 BP. First, two radiocarbon dates are available from the base of the sequence at Mollands, which are internally consistent and have age ranges older than 10,250 BP. Second, those dates are consistent with the earliest post-stadial date from Tynaspirit, which suggests major climatic improvements some time prior to 10,420 ± 160 BP (Hv–4985). Third, a number of dates have now been obtained from several other sites in Scotland which may indicate that deglaciation was probably under way before 10,250 BP (Lowe and Walker, 1980; Walker and Lowe, 1980, 1982). However, results from Inverleven (Browne and Graham, 1981), Croftamie (Rose *et al*., 1988) and South Shian and Balure of Shian (Peacock *et al*., 1989) support the earlier model of relatively late deglaciation and that marine conditions were arctic until after 10,350 BP (Peacock *et al*., 1978).

The interpretation of radiocarbon dates obtained from samples of last glacial–interglacial transition age (*c*. 14,000–10,000 BP) is, however, fraught with difficulty. Consistency in age measurements is no longer taken to be an indication of reliability, for there appear to have been short-term, temporal variations in radiocarbon activity (Ammann and Lotter, 1989), and it has been argued that the radiocarbon time-scale deviated significantly from the calendar time-scale over the period concerned (Bard *et al*., 1990). The various problems affecting the dating of Lateglacial and early Holocene deposits are reviewed by Lowe (1991) and Pilcher (1991). In view of these problems, the precise timing of events and the conflicts in interpretation of chronology cannot, at present, be satisfactorily resolved.

The Loch Lomond Readvance ice masses are generally thought to have decayed rapidly, within the space of a few hundred years. This means that it is difficult to use conventional radiocarbon

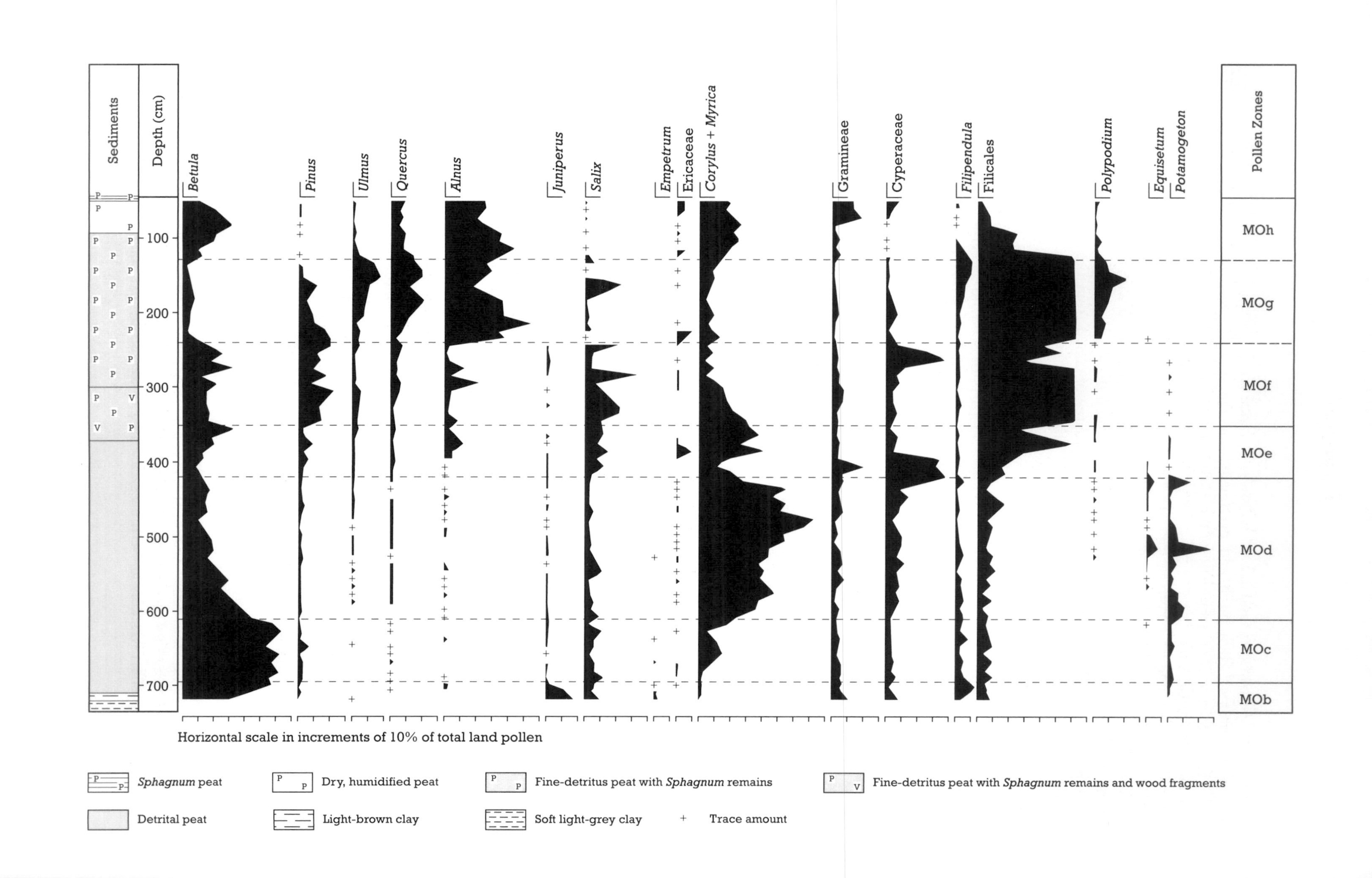

Sediments
Depth (cm)
100
200
300
400
500
600
700
Betula
Pinus
Ulmus
Quercus
Alnus
Juniperus
Salix
Empetrum
Ericaceae
Corylus + Myrica
Gramineae
Cyperaceae
Filipendula
Filicales
Polypodium
Equisetum
Potamogeton
Pollen Zones
MOh
MOg
MOf
MOe
MOd
MOc
MOb
Horizontal scale in increments of 10% of total land pollen
Sphagnum peat
Dry, humidified peat
Fine-detritus peat with Sphagnum remains
Fine-detritus peat with Sphagnum remains and wood fragments
Detrital peat
Light-brown clay
Soft light-grey clay
Trace amount

dates to assess the pattern and timing of deglaciation in Scotland owing mainly to a lack of resolution in the method (standard error ranges of Lateglacial/early Holocene dates are commonly of the order of 200 to 300 years) (see also Price, 1983) and to the methodological problems mentioned above. Recently, however, it has been proposed that patterns of ice retreat can be inferred from contrasts between the pollen assemblages recorded in the basal sediments from depressions that lie within the Loch Lomond Readvance limits and which have received sediment from the time of deglaciation (Lowe and Walker, 1981; Tipping, 1988). First proposed by Pennington (1978) for sites in the English Lake District, the conclusions are based upon the recognition of a full stadial–early Holocene pollen-stratigraphic sequence at sites where deglaciation occurred early, whereas curtailed sequences (the earlier pollen assemblages are not represented) characterize sites in areas of delayed deglaciation. The Mollands site has proved crucial in this developing argument.

Using this approach, Lowe and Walker (1981) have concluded that climatic amelioration, which promoted widespread glacier retreat, resulted in an immediate response near the termini of long valley glaciers (such as at Mollands), but ice-melt was delayed in areas where the ice was thicker or closer to source catchments (for example, parts of Rannoch Moor – see Kingshouse). On the basis of a comparison of pollen spectra from basal samples at three sites in Glen More (Isle of Mull), Walker and Lowe (1985) have also proposed that a pattern of valley retreat of Loch Lomond Readvance ice can be deciphered. A more recent study of four sites in the Varragill–Sligachan valleys of Skye, which lay within the Loch Lomond Readvance limits, also indicates progressive ice-front retreat (Walker *et al.*, 1988; Lowe and Walker, 1991; Benn *et al.*, 1992). Tipping (1988), however, has sounded a note of caution about the general applicability of this approach, finding that sites along Loch Awe did not show progressive changes in the pollen sequences in the basal sediments from the sites he examined.

This methodology therefore offers the prospect of determining general patterns of regional ice decay in Scotland, and may enable the distinction of areas characterized by progressive ice-margin retreat from those characterized by widespread downwasting of ice. The Mollands site contains the most detailed and fullest Late Devensian–Holocene pollen record so far reported from Britain. It has the added interest that radiocarbon dates obtained from the basal sediments support the contention that ice retreat may have occurred early at some valley-glacier termini as an immediate response to changes in glacier budgets.

The site offers one of the best resolutions of Holocene vegetational development reported from the south-east Grampians and adjacent lowlands, in particular in relation to the early Holocene part of the sequence. The biostratigraphic boundaries are distinct throughout, including the late Holocene 'elm decline'. The site appears to be located in what was an important ecotonal transition between the dominant mixed deciduous woodlands of the central lowlands of Scotland and the pine woods that dominated the Highlands (McVean and Ratcliffe, 1962). It is therefore an important reference site for the succession of Holocene pollen zones in the Western Highland Boundary area.

Conclusion

Mollands is important in two main respects. First, the pollen grains preserved in the sediments provide a valuable and detailed record of vegetational history during the Holocene (last 10,000 years) in the West Highland Boundary area, an important ecological zone of transition between the Highlands and Central Lowlands. Second, Mollands is important for establishing the timing of Loch Lomond Readvance deglaciation and, together with Tynaspirit, for establishing a chronological sequence of ice advance and retreat in the Callander area during the latter part of the last ice age.

Figure 13.11 Mollands: main Holocene relative pollen diagram showing selected taxa as percentages of total land pollen (from Lowe, 1982a).

TYNASPIRIT

J. J. Lowe

Highlights

The infilled kettle hole at Tynaspirit contains detailed sedimentary and pollen records, supported by radiocarbon dating, of the vegetational history and environmental changes that occurred during the Lateglacial and early Holocene. In conjunction with evidence from nearby Mollands, Tynaspirit is important for establishing the glacial chronology at the end of the Late Devensian.

Introduction

The site at Tynaspirit (NN 666047) is an infilled kettle hole located on the north side of the A84, about 0.5 km south-east of Cambusbeg Farm. The identification of the limits of the Loch Lomond Readvance has been based upon geomorphological evidence, but the chronology of the glacial event has in many areas rested largely upon relative dating by pollen stratigraphy, supported at some sites by radiocarbon dating. The basis of this approach has been outlined by Sissons *et al.* (1973). The site of Tynaspirit not only illustrates the methodology employed, but together with the evidence from the neighbouring site of Mollands (see above) provides a critical test of its application (Lowe, 1977, 1978; Lowe and Walker, 1977). Together these sites offer one of the strongest lines of support for dating the readvance and they are unique in their close geographical proximity to the local geomorphological evidence. Palynological data for later parts of the Holocene have also been reported from this site (Lowe, 1982a, 1982b).

Description

The drift geology of the area around Callander in Perthshire has been evaluated most recently by Thompson (1972) and Merritt and Laxton (1982). In summary, there are two major suites of glacial/fluvial landforms in the Teith Valley that are separated by a clear arcuate terminal moraine (the Callander Moraine, Figure 13.9), which is best observed from the A84 looking north towards Drumdhu Wood (NN 644074). To the south-east of this moraine, from the quarries near Cambusbeg to those at Easter Coillechat (NN 688038), there is an extensive spread of sand and gravel mounds, mapped by Thompson as a series of kames and kame terraces with occasional eskers. To the north-west of the moraine, as far as Kilmahog (NN 611083), smaller-scale terraces, eskers and moraines have been mapped. Those deposits lying 'outside' (to the south-east of) the terminal moraine have been dated to the time of decay of the Late Devensian ice-sheet, whereas the Callander Moraine marks the maximum position of the Loch Lomond Readvance glacier (Merritt *et al.*, 1990), which originated in the higher ground to the north and west, the ice moving down the valleys currently occupied by Lochs Lubnaig, Katrine and Venacher. The landforms lying 'within' the moraine were thus attributed to melting of this younger ice mass (see also Smith *et al.*, 1978).

The boggy, peat-filled depression at Tynaspirit occupies a kettle hole within the suite of sand and gravel mounds considered to date from the melting of the Late Devensian ice-sheet. A small lochan formerly occupied the depression, and a suite of lake sediments and peats has built up to the present ground level. The site was chosen for detailed study since it provided an opportunity to test the suggested chronology of events in the Teith valley. However, the results of stratigraphic investigations undertaken at this site also have much wider significance.

Coring at the site revealed several minor basins within the boggy area, but detailed investigations concentrated on two of these: the deepest basin (T1 – 6.65 m depth), and a shallower one (T2 – 4.45 m depth) which contained richer organic sediments more suitable for radiocarbon dating. Both basins contain a 'tripartite' sediment sequence at their base (Figure 13.12). For example, at T2 (Figure 13.13) the lowermost 0.33 m comprises (1) a basal set of inorganic sediments (beds 1 to 3), (2) an organic-rich layer (beds 4 to 6) and (3) an upper bed of inorganic sediments (bed 7). Experience has shown that this 'tripartite' sequence is typical of successions spanning the Lateglacial period (13,000 to 10,000 BP), an assertion which can be readily confirmed by pollen stratigraphy (see Lowe and Walker, 1977, 1984).

At Tynaspirit the relative age of the sediment succession was established through pollen analysis and confirmed by radiocarbon dating (Lowe, 1977, 1978; Lowe and Walker, 1977); five radiocarbon dates (Hv–4984, Hv–4985 and Hv–4987 to Hv–4989) were obtained (Figure 13.13). Seven local pollen assemblage zones have been defined (T2a to T2g) (Figure 13.13), but the sequence can be simplified as follows.

Interpretation

The earliest pollen zone (T2a) is dominated by pollen of herbaceous taxa, such as *Rumex*,

Figure 13.12 Core from the basal sediments at Tynaspirit. From the left, the sequence comprises Late Devensian minerogenic sediments, organic Lateglacial Interstadial sediments, Loch Lomond Stadial silts and clays, and early Holocene organic lake muds. (Photo: M. J. C. Walker.)

grasses, sedges and *Thalictrum*, and of dwarf shrubs. The assemblage reflects an open, treeless landscape varying from stoney, bare soils on steep or exposed sites to a luxuriant heath associated with rich herb vegetation in more favoured localities. The end of this phase is dated to 12,750 ± 120 BP (Hv–4989) confirming that the sediments began to accumulate at Tynaspirit immediately or shortly after wastage of the Late Devensian ice-sheet in the Teith valley.

The next two pollen zones (T2b and T2c) record the increasing importance of birch and juniper in the vicinity of the site. The data are interpreted as indicating the gradual spread of birch and juniper copses across the Teith valley, with perhaps more extensive birch woodland in sheltered valley-bottom sites. At the same time it is likely that the ground cover of dwarf shrubs and herb vegetation continued to fill across unwooded sections of the valley, and that organic detritus built up within the developing soils. It appears that a major expansion of birch and juniper occurred in the area at 12,395 ± 195 BP (Hv–4988), but that these communities were severely disturbed and all but disappeared from the area by 11,385 ± 290 BP (Hv–4987). Overall, the pollen data imply a relatively mild climatic episode from perhaps as early as 12,750 to about 11,400 years ago.

Zone T2d is characterized by a return to minerogenic sedimentation, a dramatic reduction in birch percentages, an initial reduction in juniper and a return to high percentages of dwarf-shrub and herbaceous taxa. Those taxa common in the basal assemblage (zone T2a) (for example, *Rumex*, Compositae, grasses and sedges) are also well represented in zone T2d. The term 'revertance' is used to denote a return to the open, treeless conditions inferred for zone T2a. However, the later phase (zone T2d) records much higher representations of taxa indicative of bare, disturbed stoney surfaces, such as the clubmosses (*Selaginella* and *Lycopodium*) and the wormwoods or mugwort (*Artemisia*). The evidence suggests a harsh climatic environment, dated to between 11,385 ± 290 and 10,420 ± 160 BP (Hv–4985).

These conclusions are supported by the results of more recent investigations in the area by Merritt *et al.* (1990). Organic silts discovered

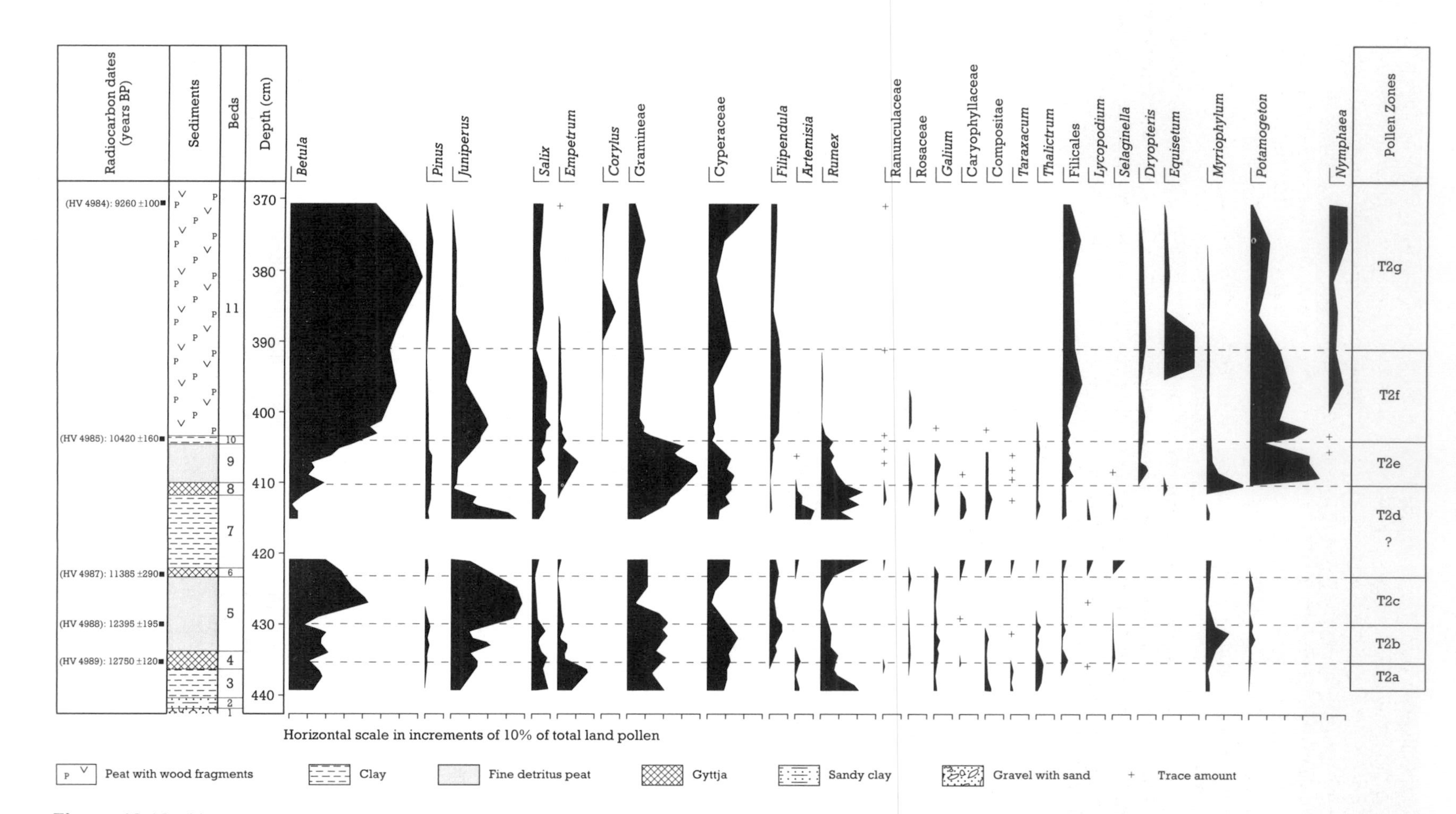

Figure 13.13 Tynaspirit: relative pollen diagram showing selected taxa as percentages of total land pollen (from Lowe, 1978).

beneath till at Callander have been assigned to the Lateglacial Interstadial on the basis of radiocarbon dating, pollen stratigraphy and their arthropod fauna. A radiocarbon date of 12,750 ± 70 BP (SRR–2317) from basal sediments matches almost exactly the date from the base of the Tynaspirit organic sediments, and the till overlying the dated organic silts is ascribed to the Loch Lomond Readvance.

Zones T2e to T2g record the development of woodland within the Teith valley. A progressive succession of (from base) *Empetrum*, juniper and birch, followed by the immigration of *Corylus* is a sequence recognized at numerous sites in Britain, one which characterizes the start of the Holocene Stage. From the dominance in the pollen assemblages of birch and juniper in particular, and from the highly organic nature of the lake sediments (beds 8 to 11), it is concluded that a woodland cover clothed much of the lower lying area of the Teith valley, and that correspondingly organic-rich soils had developed by about 9,260 ± 100 BP (Hv–4984).

The simplest explanation of the Tynaspirit succession, and the most important conclusions based upon the reported data, are as follows.

The Late Devensian ice-sheet must have disappeared from the lower parts of the Teith valley some time prior to about 12,800 BP. Several dates from other sites in Scotland (Sissons, 1976b; Gray and Lowe, 1977b; Sutherland, 1984a) indicate that around 13,000 BP or shortly after, most of the British Isles was ice-free.

Relatively stable and probably mild climatic conditions persisted for almost 2000 years. Whether or not ice disappeared entirely from Scotland during this period remains a contentious point (Sutherland, 1984a), one that is difficult to resolve on the basis of available evidence. Certainly the evidence from Tynaspirit would seem to imply substantial retreat of ice from the Callander area.

The data from Tynaspirit support Thompson's view (1972) that any glacier readvance into this area during the Loch Lomond Stadial did not extend beyond the vicinity of Callander and that an earlier suggestion by Francis *et al.* (1970) that the readvance extended to Drumvaich (*c.* 1 km east-south-east of Tynaspirit) is incorrect.

The 10,420 ± 160 BP date from Tynaspirit suggests that the harsh climatic conditions that promoted the Loch Lomond Readvance had substantially improved by some time earlier than 10,400 years ago, since the dated sample is not from the earliest post-stadial sediments at the site. This date, together with the evidence from the neighbouring site of Mollands, led to a revision of the estimated age of deglaciation in parts of Scotland at the close of the Loch Lomond Stadial (see Gray and Lowe, 1977b; Lowe, 1978; Lowe and Walker, 1980). However, recent studies at Croftamie (Rose *et al.*, 1988) and South Shian and Balure of Shian (Peacock *et al.*, 1989) have indicated that Loch Lomond Readvance glaciers at those localities attained their maximum extents relatively late in the Stadial, after 10,500 BP.

The close association of (1) Late Devensian ice-sheet landforms of deglaciation, (2) Loch Lomond Readvance landforms, including a well-defined ice limit, and (3) two sites (Tynaspirit and Mollands) providing relative and absolute dating controls on the ages of those landforms, is rare in the British Isles. The internally consistent radiocarbon dates and very close association of the field evidence collectively provide one of the strongest lines of support presently available for the Late Devensian chronology outlined earlier in this report. The area around Callander is therefore outstanding for exemplifying the field relations of the geomorphological evidence together with the pollen stratigraphy and radiocarbon dating evidence for the Lateglacial in Scotland. Tynaspirit is therefore a site of major importance in the Quaternary geology of the British Isles.

Tynaspirit, together with Mollands, is also significant in illustrating one of the two methodological approaches used in dating the Loch Lomond Readvance. In many areas the chronology of the event has been established through pollen stratigraphy and sometimes radiocarbon dating of organic lake sediments at sites inside and outside the inferred ice limits (Donner, 1957; Sissons *et al.*, 1973; Walker, 1975a; Walker *et al.*, 1988; Tipping, 1988). The other approach has involved radiocarbon dating of marine shells and organic deposits incorporated or overridden by Loch Lomond Readvance glaciers (see South Shian and Balure of Shian, Rhu Point, Croftamie, and the Western Forth Valley). These two approaches have tended to produce contrasting conclusions, the time of the maximum of the readvance inferred from the first approach often being apparently earlier than that inferred from the second approach. It remains to be established whether this contrast is due to methodological problems (see Mollands) or whether it does indeed reflect diachroneity in the response of the various glaciers to rapid climatic change at the

Late Devensian–Holocene boundary.

Conclusion

Evidence from Tynaspirit in conjunction with that from Mollands, is important for establishing the pattern and timing of glaciation and environmental change during the Lateglacial, the period between about 13,000 and 10,000 years ago. The sediment and pollen data from the site, supported by radiocarbon dating, provide a full record of Lateglacial environmental conditions from the time of wastage of the main Late Devensian ice-sheet (approximately 13,000 years ago), indicating that Loch Lomond Readvance glaciers did not extend as far as the site. In contrast, Mollands contains only a Holocene pollen record, indicating that the ice limit which lies between it and Tynaspirit was formed during the Loch Lomond Readvance (approximately 11,000–10,000 years ago). Together these two sites provide a particularly good illustration of this type of approach used to date Loch Lomond Readvance deposits.

Chapter 14

Eastern Highland Boundary

INTRODUCTION

D. G. Sutherland

The Eastern Highland Boundary region extends from west of Perth and the Tay Valley through Strathmore to Aberdeen (Figure 14.1). It incorporates the coastline from south of Montrose to Aberdeen. Glaciation of the area has been dominantly by ice originating in the western Highlands, with relatively little contribution from ice sources in the south-eastern Grampians. The eastern parts of the region are therefore located towards the periphery of the intensely glaciated zone and contain evidence of Pleistocene events pre-dating the Late Devensian; to the west the increasing intensity of glaciation has resulted in only deposits relating to the last phase of ice-sheet glaciation (the Late Devensian) being present. The region is important, too, for the development of vegetation during both the Lateglacial Interstadial and the Holocene as it is adjacent to the Highland Boundary Fault, one of the major topographical, geological and ecotonal boundaries in Scotland.

The eastern seaboard of the region contains a number of pre-Late Devensian landforms and deposits. The coast is flanked by two glaciated rock platforms, a higher one at around 23 m OD (Bremner, 1925b) and a lower one only slightly above present sea level (Synge, 1956). This latter platform is a clear feature at Milton Ness. The ages of the platforms are unknown. Evidence for glaciation prior to the Late Devensian is also preserved in the basal deposits at Nigg Bay (Bremner, 1934a; Synge, 1956, 1963; Chester, 1975) where the presence of erratics with a Norwegian origin points to an earlier glacial event with ice possibly moving onshore. This latter idea has also been suggested to explain shelly clays below till at Burn of Benholm and other neighbouring localities (Campbell, 1934).

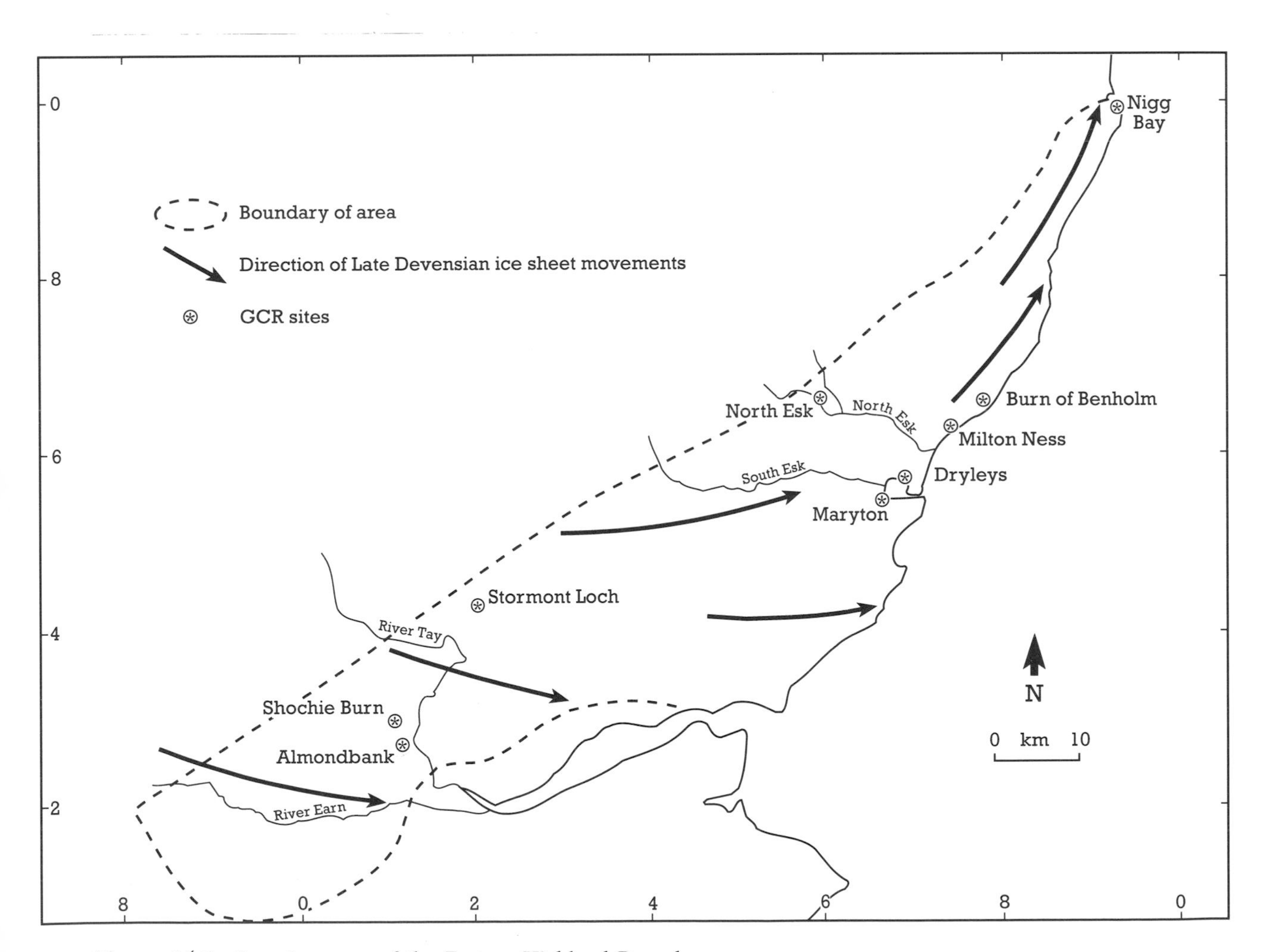

Figure 14.1 Location map of the Eastern Highland Boundary area.

An alternative explanation for these shelly deposits was advanced by Sutherland (1981a), who suggested that they may have formed from *in situ* marine clays deposited during the build-up of the last ice-sheet during the Early Devensian. More information is awaited on these deposits before their true nature is clear. Also at Burn of Benholm, lenses of peat have been found interbedded with the red till that overlies the shelly clays. Pollen analysis initially led Donner (1960) to hypothesize a Lateglacial age for the peat, and consequently that the enclosing till was soliflucted. However, radiocarbon dating indicated the age of the peat to be greater than 42,000 BP and Donner (1979) revised his interpretation, suggesting that the peat accumulated during a Middle or Early Devensian interstadial.

The great majority of the glacial deposits and landforms of the region have been interpreted as the product of the last ice-sheet. Armstrong *et al.* (1985) have provided evidence of two distinct phases of ice movement, an earlier period in which ice was directed towards the south-east and a later period when ice flowed towards the east-north-east. Sutherland (1984a) suggested that this reorientation of ice flow on the eastern side of Scotland was a result of the expansion of the Southern Uplands ice as the last glaciation progressed (see Chapters 15, 16 and 17). The east-north-east flow of ice originating in the south-west Highlands extended throughout the region and beyond the coast. It was sufficiently powerful to flow into the mouths of the Highland valleys such as Glen Clova and Glen Esk (Bremner, 1934b, 1936; Synge, 1956) and it left behind a characteristic red till. This till is well exposed at both Burn of Benholm and Nigg Bay and at the latter locality it overlies a grey till deposited by ice moving from the eastern Grampians, again indicating the relative timing and strengths of the two ice flows (cf. Chapter 8).

Deglaciation took place broadly from north-east to south-west, although large areas of 'dead' ice became isolated in the low ground of Strathmore and in the adjacent Highland valleys, giving rise to extensive areas of kames, kettle holes and eskers (Rice, 1959; Rose and McLellan, 1967; Paterson, 1974; Ellis, 1975; Insch, 1976; Armstrong *et al.*, 1985; Auton *et al.*, 1988; Auton *et al.*, 1990). Meltwater channels, both ice-directed and topographically controlled, occur extensively along the Highland margins (Bremner, 1925b, 1934b; Watson, 1945; Synge, 1956; Sissons, 1961a; Rose and McLellan, 1967; Ellis, 1975; Insch, 1976; Auton *et al.*, 1988; Auton *et al.*, 1990). The earliest area deglaciated was the eastern coast where a series of raised shorelines (Cullingford and Smith, 1980) was formed while ice decayed in the adjacent valleys. These shorelines are well exposed at Milton Ness and at Dryleys in the Montrose basin where their relationship to the fossiliferous Errol beds (Late Devensian estuarine sediments) has also been established. These shorelines are likely to have been formed between 17,000 BP and 14,000 BP (Sutherland, 1984a).

For some time after the deglaciation of the eastern coast the Tay estuary remained blocked by ice, and meltwater drainage was directed through the 'dead' ice in Strathmore. Eventually the ice in the Tay–Earn area receded and the sea flooded as far west as Crieff (Browne, 1980) and into the lower Tay valley to the north of Perth (Armstrong *et al.*, 1985). This is thought to have occurred at about or slightly before 13,000 BP. The meltwater from decaying ice in the Highland valleys produced large outwash fans at the mouths of those valleys, extending considerable distances into the lowland area; for example, along the River North Esk (Sissons, 1967a; Paterson, 1974; Insch, 1976; Maizels, 1983a). The earliest interpretation of the sequences where these outwash deposits were found to overlie laminated estuarine sediments (Errol beds) was that they were the product of a major readvance of the ice-sheet, the Perth Readvance (Simpson, 1933; Sissons, 1963a, 1964). The classic section in such a sequence is at Almondbank. However, Paterson (1974) has demonstrated that it is not necessary to invoke a readvance to explain the stratigraphic sequence in this area. The only section that may indicate at least a local readvance of the retreating ice-sheet is at Shochie Burn.

Kettle holes that developed in the outwash deposits and 'dead'-ice terrain have accumulated sediment since the time of deglaciation, and pollen analysis of these sediments has revealed the nature of the vegetational succession during the Lateglacial Interstadial once the region was ice-free and the climate had ameliorated. One particularly interesting sequence is that at Stormont Loch where, following an initial succession of open habitat taxa followed by birch, juniper and *Empetrum* heath, there was a notable return to more open vegetation suggestive of disturbed soils (Caseldine, 1980a). This distinct 'revertance' phase, which is not found in all Lateglacial pollen diagrams (Walker, 1984b; Tipping, 1991b), has been correlated with the Older Dryas 'chronozone' of the Scandinavian sequence. However, there is no radiocarbon evidence from Stormont Loch to support this correlation. Subsequent to this period, woody species again expanded, giving rise to a juniper-dominated heath with scattered copses of tree birches, the more typical vegetation of the Lateglacial Interstadial in this part of Scotland.

During the Loch Lomond Stadial there was a return to unstable soils, with a marked reduction in woody species and an increase in open-habitat species. There is an interesting trend in the values for pollen of *Artemisia* species during the stadial, from low values at the beginning to higher values in the latter part (Caseldine, 1980a). This has been interpreted as the result of a change from higher to lower precipitation as the stadial progressed.

Sea level was relatively low along the coasts of the region during the latter part of the Lateglacial and the early Holocene, and in the estuarine areas peat deposits accumulated on the poorly drained surfaces of the Lateglacial sediments. During the Main Postglacial Transgression these estuarine areas were once again invaded by the sea and the peat deposits were buried by further silts, sands and clays. Such sediment sequences are well exposed around the Montrose basin, for example at Maryton and Dryleys. The progress of the transgression has been studied using pollen and diatom analyses and radiocarbon dating (Smith *et al.*, 1980), and the raised shorelines formed at the maximum of the transgression and subsequently have been mapped and surveyed by Smith and Cullingford (1985). Of particular note in the sediment sequence is a distinctive sand layer, exposed at Maryton, that was deposited slightly prior to the maximum of the transgression. A similar sand layer, dated to about 7000–6800 BP, has been found at a number of sites along the east coast of Scotland and it has been interpreted as the product of a major storm surge in the North Sea (Smith *et al.*, 1985a) or a tsunami resulting from a large submarine slide (Dawson *et al.*, 1988).

The Holocene forest history is of interest because this region lies close to the border of the mixed deciduous forest and the pine forest zones (Birks, 1977). Following an early Holocene succession of juniper scrub to birch–hazel woodland, mixed deciduous forest with oak, birch and elm became established throughout most of the region, although pine may have occurred during the middle Holocene in edaphically favourable locations. Man's impact on the forests was initially recorded by the decline in elm at around 5000 BP.

NIGG BAY

J. E. Gordon

Highlights

The deposits exposed in the coastal section at Nigg Bay comprise a sequence of tills and glaciofluvial sediments. These provide important evidence for establishing the sequence and patterns of glaciation, including the interactions of different ice masses, in eastern Scotland.

Introduction

The site at Nigg Bay (NJ 965046) is a section on the coast immediately to the south of Aberdeen. It shows a succession of till and glaciofluvial deposits and has long been regarded as a classic locality for glacial stratigraphy. The deposits at Nigg Bay have been discussed in the literature for almost a century, with debate focused on the number of separate ice advances and readvances that may be represented. The deposits have been described by Jamieson (1882b), Bremner (1928), Simpson (1948), Synge (1963), Chester (1975), McLean (1977) and Munro (1986), and also discussed in more general reviews by Charlesworth (1956), Synge (1956), Clapperton and Sugden (1972, 1977) Murdoch (1977) and Hall and Connell (1991).

Description

The section occurs on the south side of Nigg Bay in a suite of deposits infilling a former valley of the River Dee (Simpson, 1948; Law, 1962; Munro, 1986). Borehole evidence on the shore and seismic evidence from the bay show this channel to descend to below −40 m OD and extend at least 1 km offshore. Six distinctive beds have been recognized in the exposure (Jamieson, 1882b; Bremner, 1928; Simpson, 1948; Synge, 1963; Chester, 1975):

6.	Head	0.5–1.0 m
5.	Gravels	1–3 m
4.	Red sands with laminated silts and clays	1–2 m
3.	Red till	2–3 m
2.	Grey till	10 m
1.	Sands and gravels	3–6 m

The horizontally bedded sands and gravels (bed 1) are seen only at the southern end of the section and rest on a weathered granite–gneiss. Jamieson (1882b) described their thickness as 10–20 ft (3–6 m), but only part of this bed has been exposed in recent years. Diagnostic erratics include ultrabasic material which can be traced to the Belhelvie igneous complex 10 km to the north-north-west, larvikite and rhomb porphyry (see Bremner, 1922) inferred to be of Scandinavian origin, and andalusite schist (Read *et al.*, 1923; Bremner, 1928, 1934a, 1939b; Synge, 1963). Armoured till balls have also been found incorporated in the gravels (E. R. Connell, unpublished data). A 5 m thick layer of sands and gravels, underlying a till, was encountered at the base of a borehole at Nigg Bay, and this may correlate with the basal sands and gravels of the cliff section (Munro, 1986).

Unconformably overlying the lower sands and gravels is a tough, compact, grey till up to 10 m thick (bed 2). It contains local erratics predominantly of granite and gneiss and in its lower part the clasts have a south-south-west to north-north-east preferred orientation (Synge, 1963). This unit correlates with a widely occurring local till in the Aberdeen area (McLean, 1977; Murdoch, 1977). There is a transition upwards from the grey till to the red till (bed 3). Elsewhere, in pipeline trench exposures near Aberdeen, these two tills have been seen interbedded (Chester, 1974, 1975; Murdoch, 1975, 1977; Munro, 1986). Apart from its colour the red till is distinguished by the presence of Old Red Sandstone conglomerate and volcanic erratics and by a more northerly preferred stone orientation than the grey till (Jamieson, 1882b, 1906; Synge, 1963). The source of this material was traditionally held to be the Old Red Sandstone rocks of Strathmore and the volcanic rocks between Lunan Bay and Montrose (Jamieson, 1906; Munro, 1986). However, heavy-mineral studies by Glentworth *et al.* (1964) point to the floor of the North Sea as an additional source, a possibility in fact admitted by Jamieson (1882b). The red till forms part of a suite of red deposits extending along the north-east coast of Scotland from south of Stonehaven to north of Peterhead (Jamieson, 1882b; Bremner, 1916b; Synge, 1956; McLean, 1977; Murdoch, 1975, 1977; Hall, 1984b; Auton and Crofts, 1986; Munro, 1986).

The sands (bed 4) above are reddish brown in colour and contain thin bands of laminated red clays. Southwards these sands merge into and are succeeded by gravels (bed 5) containing the same erratics as the red till. There is no sharp junction between the sands and gravels, implying that they are probably contemporaneous (Simpson, 1948). The gravels are coarse and poorly bedded with lenticles of sand, and in their upper part (bed 6) are highly shattered and cracked (Jamieson, 1882b), typical of a head deposit. Simpson (1948) suggested that sorted coarse and fine fractions in this deposit represented stone stripes and polygons seen in section. The gravels have a hummocky surface expression (termed 'morainic' by Simpson (1948)), and continue inland and to the south of Nigg Bay where they are associated with kettle holes in the Loirston area (NJ 940010).

Interpretation

The erratic content of the lower sands and gravels led most workers to postulate early glaciation from the north or north-west contemporaneous with the presence of Scandinavian ice in the North Sea Basin. Bremner (1928) suggested that the lower sands and gravels consisted in part of deposits from his first glaciation from the north that were subsequently reworked by meltwater. Simpson (1948) also thought that they were derived from the earliest glacial deposits of the area, which might be associated with Bremner's first ice-sheet (Bremner, 1928, 1934a, 1938). From the distribution of erratics in north-east Scotland, Bremner inferred ice movement during this glaciation from north-west to south-east across this area; for example, near Aberdeen he recorded boulders thought to be from the northern Highlands and the Elgin area, and, near Inverbervie, boulders from Huntly. At the same time Scandinavian ice in the North Sea Basin blocked the escape of Scottish ice to the east and, as the former approached from the north-east, it impinged on the coast at Aberdeen and in Kincardineshire. It was possible, he argued, that the shelly indigo-coloured till in the Ellon area (see Bellscamphie) reported by Jamieson (1906) and thought by him to have been transported from the north during the first glaciation was in fact related to the Scandinavian ice moving onshore (Bremner, 1928, 1934a). Clapperton and Sugden (1977), however, have advocated the need for caution in interpreting igneous and metamorphic 'erratics' since more recent work on bedrock geology in north-east Scotland has shown greater diversity of rock types and more widespread distribution of the sources of indicator types. Moreover, in the light of evidence from the North Sea (Stoker *et al.*, 1985), Hall and Connell (1991) considered it unlikely that Scandinavian ice ever reached the coast of Scotland, and that the erratics of presumed Scandinavian origin were ice-rafted and subsequently entrained by Scottish ice.

Synge (1956, 1963) also proposed early glaciation from the north-east (Scandinavian glaciation) followed by an expansion of Scottish ice (Greater Highland Glaciation) moving south-south-east along the Aberdeenshire coastal area, but he thought the lower sands and gravels at Nigg Bay were beach deposits derived from the earliest till rather than glaciofluvial deposits, as implied by Bremner (1928) and stated by Read *et al.* (1923). Possibly the armoured till balls in the lower sands and gravels at Nigg Bay are the reworked remnants of an early till. Further evidence for an early north-west to south-east ice movement was provided by Murdoch (1975, 1977) from clast-fabric studies of an argillaceous lodgement till found locally in the Aberdeen area but not seen at Nigg Bay. This till was considered (Murdoch, 1975; Munro, 1986) to be of Late Devensian age, but Sutherland (1984a), on the basis of evidence for ice-free conditions to the north-west of Aberdeen (Hall, 1984b), suggested that the north-west to south-east glaciation that deposited this till must have pre-dated the Late Devensian.

There is general agreement among all workers that the red and grey tills were broadly contemporaneous and reflect the interaction of two distinctive ice masses (Jamieson, 1882b, 1906; Bremner, 1928, 1934a, 1938; Simpson, 1948, 1955; Synge, 1956, 1963; Clapperton and Sugden, 1972, 1977; Murdoch, 1975, 1977; McLean, 1977; Auton and Crofts, 1986; Munro 1986; Hall and Connell, 1991). The grey till with its local erratics was deposited by ice moving down the Dee Valley; the red till, with its Old Red Sandstone erratics, was deposited by Strathmore ice moving north-east along the coast. Both Jamieson and Bremner related the two tills to their respective second ice-sheets and suggested that the local Dee ice had receded before the red till was deposited. Most workers now agree, however, that the two ice streams coalesced, although the initial influx of ice from the west weakened and was succeeded by the Strathmore ice (Hall and Connell, 1991). Synge interpreted the deposits as being the product of his Moray Firth–Strathmore Glaciation of Devensian age; Simpson, Clapperton and Sugden, McLean and Murdoch assigned them to the Late Devensian ice-sheet.

The presence of Scandinavian ice in the North Sea Basin has often been cited as the reason for deflection of the Strathmore ice along the east coast of Scotland. Recent studies of offshore sediments, however, indicate that the Scottish and Scandinavian ice-sheets did not meet or coalesce during the Late Devensian (Jansen, 1976; Cameron *et al.*, 1987; Sejrup *et al.*, 1987; Hall and Bent, 1990). Alternative explanations for the flow pattern must therefore be sought. Following the suggestion of Hall and Bent (1990) for the Moray Firth area, one possibility is that a change in flow pattern developed where ice from sources in the south-west Highlands emerged from valleys with rigid rock beds on to the potentially mobile beds formed by the thick sediments of the lowlands and nearshore coastal zone. That such changes can theoretically occur has been demonstrated for the Late Weichselian Laurentide ice-sheet (Fisher *et al.*, 1985).

Influenced by the presence of red clay deposits at Tullos and to the north of Aberdeen, Jamieson envisaged a marine or lacustrine depositional environment for the Nigg Bay tills. However, their undoubted glacial nature was established by later investigators. Recent ideas suggest they may have been formed by melt-out and flow processes (McLean, 1977; Murdoch, 1977).

The origin of the upper gravels (bed 5) has been more contentious. Agassiz (see Jamieson, 1874, p. 323) suggested that the gravel mounds were moraines, a view endorsed by Buckland (1841a). According to Jamieson (Jamieson, 1865, 1874, 1882b, 1906) they were the ice-marginal moraines of his third and last glaciation during which ice advanced down the Dee Valley and extended just offshore, forming an 8 km wide ice-front between Nigg Bay and Belhelvie to the north of Aberdeen where another suite of gravel mounds occurs. He also argued that there had been a longer and possibly warmer time interval between his second and third ice-sheets than between the first two. Bremner (1928, 1934a, 1938) proposed a similar interpretation, but placed the northern limit of the ice near the Ythan. He also related the gravels to his third ice-sheet, the existence of which he had established elsewhere largely on the basis of supposedly marginal meltwater channels.

Charlesworth (1956) included the upper morainic gravels as part of his Highland Glaciation, which he recognized as a separate readvance after the last ice maximum, with a local ice limit near Belhelvie. Synge (1956, 1963), too, thought the gravels were ice-marginal deposits, part of what he called a 'kame moraine' marking the limit of a separate glacial event of pre-Loch Lomond Stadial age, the Aberdeen Readvance. Synge correlated this glacial event with the Perth Readvance (Simpson, 1933), but Sissons (1965, 1967a) later linked it with features in the south of Scotland to mark another event known as the Aberdeen–Lammermuir Readvance. However, subsequent re-examination of the evidence has failed to substantiate the existence of these readvances (Sissons, 1974c).

In important contributions, Simpson (1948, 1955) was first to perceive that the upper 'morainic' gravels did not represent a separate ice readvance but were contemporaneous with the last ice-sheet to cover the area. The gravels contained the same erratics as the red till and were not derived by ice readvance from the west. Moreover, he suggested that the gravels were not moraines, but were glaciofluvial kames which formed a continuous suite of deposits extending from south of Nigg Bay to the River Ythan. In places the kames were overlain by till and he proposed that many were of englacial origin. Subsequent work by Clapperton and Sugden (1972, 1975, 1977) and Murdoch (1975, 1977) has substantiated Simpson's interpretation. These authors argued that the continuity of glaciofluvial landforms across the supposed limit of the Aberdeen Readvance reflected the integrated subglacial hydrological system of a single ice-sheet. Moreover, the continuity of the landforms and the undisturbed nature of the red and grey tills implied contemporaneous deposition and ice decay during a single glaciation. Furthermore, ice-wedge casts occur both within and outside the supposed Aberdeen Readvance ice limit (Sissons, 1974c; Gemmell and Ralston, 1984), and therefore their distribution cannot be used to support the readvance concept, as was suggested by Synge (1963).

The Nigg Bay section is a key reference locality for the interpretation of the glacial history of the eastern Highland boundary area. At least two distinct glaciations can be inferred from the sequence of deposits: a pre-Devensian glaciation represented by the lower sands and gravels, and a Late Devensian glaciation by the overlying succession of tills, sands and gravels. The succession of Late Devensian deposits illustrates particularly well the complexity of depositional environments that may be associated with a single glacial episode. Nigg Bay also has significant potential for further research, for example on the sedimentology of the deposits. In addition, further study of Nigg Bay and other multiple till sections will provide useful field evidence of changing ice-sheet flow patterns that can be used to test and refine theoretical models of ice-sheet growth. For example, Nigg Bay clearly illustrates that early expansion of inland ice from the mountains and valleys to the west was succeeded by a dominant flow of ice from sources in the south-west Highlands.

Conclusion

The deposits at Nigg Bay are important for interpreting the glacial history of the Eastern Highland Boundary area. They have a long history of research and illustrate the interaction of ice masses from different source areas. Although it has been proposed that the deposits were formed by different ice-sheets in a series of separate glaciations, current interpretations ascribe most of the sequence to the Late Devensian glaciation (approximately 18,000 years ago).

BURN OF BENHOLM

J. E. Gordon

Highlights

Stream sections at Burn of Benholm have revealed a sequence of deposits including a pre-Late Devensian shelly clay with marine molluscs and an interstadial peat of Early or Middle Devensian origin. These deposits provide important evidence for interpreting the Quaternary history of eastern Scotland.

Introduction

This site (NO 795691) comprises a series of stream sections at *c.* 45 m OD along the Burn of Benholm, 14 km north of Montrose. It is notable for a dark shelly deposit interpreted as a till, and for peat lenses incorporated near the base of an overlying red till. These deposits have played an important role in reconstructing the glacial history and environmental changes in the area (Campbell, 1934; Donner, 1960, 1979). More recently, Sutherland (1981a) has proposed an alternative explanation that the shelly sediment is an *in situ* marine deposit. He described its possible wider correlations with other sites in Scotland and how it might relate to a model of glacio-isostatic sea-level changes associated with the inception of the last ice-sheet during the Early Devensian.

Description

The deposits at Burn of Benholm were first described by Campbell (1934). He recorded the following sequence in several sections:

4.	Coarse gravel	0.3–0.8 m
3.	Red till derived from the Old Red Sandstone rocks of Strathmore and typical of the tills of this part of eastern Scotland	0.3–1.2 m
2.	Greyish-black till containing abundant 'arctic' shells, the commonest being *Arctica islandica* (L.), and relatively few clasts, but of lithologies quite distinct from those in the overlying red till (Lower Old Red Sandstone, basalt, gneiss, troctolite, limestone (including chalk), shale, flint and jet)	0.6–1.8 m
1.	Andesite bedrock	

In the sections described by Campbell the shelly till (bed 2), which extended almost continuously for a distance of over 550 m, was locally underlain by sands and gravels (0.2–0.3 m thick) and, in places, it was mixed with the red till (bed 3). At one locality, finely laminated silts and a band of peat (total thickness 0.3 m) were incorporated at the base of the red till (bed 3).

The present exposures clearly demonstrate (1) red till overlying grey, shelly clay with a low stone content, as described by Campbell (1934); (2) in some sections a sharp, undulating contact between the grey clay and the red till, varying from subhorizontal to steeply dipping; (3) in other sections, a zone of mixing up to 0.4 m thick of red till and grey clay; (4) deformation of the upper surface of the grey clay and interfingering with the red till; (5) stringers and bands of grey clay incorporated into the red till; (6) a layer of reddish sand (0.01–0.02 m thick) 0.03–0.04 m below the contact of the red till and the grey clay; (7) deformation of the grey clay and underlying bands of sand and gravel against a bedrock knoll. Hall and Connell (1991) have reported that the grey clay contains reworked Upper Cretaceous and Tertiary dinoflagellate cysts which they considered were derived from the North Sea Basin to the south-east.

Bremner (1943a) recorded that analysis of the peat (base of bed 3) by I. M. Robertson had revealed pollen of oak, pine, alder and elm. Donner (1960) noted that the peat, which is no longer exposed, occurred in thin lenses. The pollen content, however, was dominated by non-arboreal types, notably Gramineae and Cyperaceae, representing herb communities (Donner, 1979). Radiocarbon dating of the peat gave an age of >42,000 BP (Hel–1098) (Donner, 1979).

Interpretation

The dark shelly deposit at Burn of Benholm is one of several such occurrences in Kincardineshire described by Campbell (1934), but it is the only one presently exposed. It is also the only locality where peat deposits have been recorded in the sequence. Campbell (1934) concluded that the shelly deposit was a till emplaced by ice moving across the floor of the North Sea when Scottish and Scandinavian ice-sheets coalesced (but see Nigg Bay). The peat represented interglacial conditions, being formed after retreat of the North Sea ice and before the advance of the Strathmore ice-sheet.

Donner (1960) reinterpreted the sequence inferring that the red till was not *in situ*, but had been transported by solifluction or a small landslide during the Loch Lomond Stadial. In support he cited the results of pollen analysis on the peat which showed similarities with Lateglacial Interstadial deposits elsewhere in Scotland, and he concluded that the peat had also been displaced by solifluction during the stadial. Subsequently, however, from the radiocarbon date and a re-evaluation of the pollen data, Donner (1979) revised his conclusions. He considered that the peat represented the remains of an organic deposit formed in a tundra environment of predominantly grassland communities, probably during the Early or Middle Devensian. It was then incorporated into the base of the red till during the Late Devensian glaciation in Strathmore. As noted by Edwards and Connell (1981), the discrepancy between the pollen records of Bremner (1943a) and Donner (1960, 1979) either casts doubts on the earlier identifications or suggests that peat of more than one age or environment was present.

Sutherland (1981a) presented a radically different interpretation of the shelly deposit at Burn of Benholm. He inferred that it was an *in situ* marine bed on the basis that it formed part of a suite of high-level marine shell beds buried by till and that its altitude conformed with a model that involved a marine transgression associated with depression of the Earth's crust as the last ice-sheet began to accumulate. Sutherland argued that this occurred during the Early Devensian rather than during the Late Devensian, as in more conventional interpretations (Sissons, 1976b).

The Burn of Benholm deposits therefore have a significant bearing on the interpretation of the Late Pleistocene history of eastern Scotland. In the conventional view the shelly deposit is a till and provides support for the onshore movement of ice, possibly deflected by the presence of Scandinavian ice in the North Sea during a glacial episode pre-dating the last ice-sheet. Although the presence of Scandinavian ice has also been held responsible for the movement of Strathmore ice north-eastwards along the east coast during the last glaciation (see Nigg Bay), recent evidence from the central North Sea (Sutherland, 1984a; Stoker *et al.*, 1985; Sejrup *et al.*, 1987) suggests this to have been unlikely. Hall and Connell (1991) have maintained the interpretation of the shelly deposit as a till, and proposed deposition by ice flowing from the east or south-east. This, they suggested, may have occurred during the 'Wolstonian' glaciation.

In the model of Sutherland (1981a), the shelly deposit at Burn of Benholm forms part of a network of *in situ*, high-level shell beds in Scotland (see Afton Lodge, Clava and Tangy Glen). Together these were considered to reflect ice-sheet growth and high relative sea level during the Early Devensian. However, correlations with other high-level shell beds are, at present, conjectural; before either interpretation can be tested, several key questions remain to be answered. These concern first, the origin of the shelly deposit and whether it is a till, an *in situ* marine deposit, or possibly a glaciomarine deposit or an ice-rafted marine deposit; second, the age of the shelly deposit; and third, whether the shelly deposit and the overlying red till are broadly contemporaneous or separated by a significant time interval, possibly represented by the deposition of the peat. The deposits therefore have significant potential for detailed sedimentary and faunal studies. In addition, amino acid analyses of the shells should help to clarify the age of the shelly deposit and provide a firmer basis for any wider correlations. Reported results indicate isoleucine ratios of 0.36 from *Arctica* shells (Bowen, 1991). Therefore, according to Bowen (1991), if the shelly clay is glaciomarine in origin, then it dates from the time of deglaciation of the Anglian ice-sheet (Oxygen Isotope Stage 12). However, if the deposit is a till, then it is younger than Stage 11.

The peat at Burn of Benholm, although no longer exposed, is also significant in having provided a rare pollen record of interstadial environmental conditions during Early or Middle Devensian times. Middle Devensian pollen profiles are also recorded from Tolsta Head, Crossbrae, St Kilda, and possibly Teindland (see Lowe, 1984), but as yet the record of this time period in Scotland is highly fragmentary. The available radiocarbon dates suggest that the Burn of Benholm peat represents an earlier phase than the organic deposits at the other sites.

Conclusion

Burn of Benholm is notable for a sequence of deposits that provide important evidence for interpreting the glacial history of eastern Scotland. Of particular interest is a clay deposit, the origin and age of which are controversial. It contains shells of marine molluscs and may be an *in situ* marine deposit representing a high sea-level stand pre-dating the Late Devensian, possibly as old as 450,000 years, or it may be a deposit transported by ice from offshore, that is by the last or earlier ice-sheets. In either case it is a deposit of great interest for Quaternary studies. Also of note is an interstadial peat bed formerly exposed at the site. This provided a rare pollen record of environmental conditions during Early or Mid-Devensian times.

ALMONDBANK

J. E. Gordon and D. G. Sutherland

Highlights

The sequence of deposits exposed in the river bank section at Almondbank has provided important evidence for interpreting the pattern of deglaciation of the Late Devensian ice-sheet. Historically, this evidence was used to support a major readvance of the ice, but current understanding indicates progressive ice decay and deposition of the sediments in a prograding marine delta.

Introduction

This site (NO 084262) comprises a section on the north bank of the River Almond, located 1.75 km west of its junction with the River Tay and *c.* 1 km west of the outskirts of Perth. The sequence of deposits has provided important evidence for interpreting the pattern of decay of the Late Devensian ice-sheet. In particular, it was first described and used by Simpson (1933) as evidence for a readvance, the Perth Readvance, which interrupted wastage of the ice-sheet. The concept of this readvance was supported by Sissons (1963a, 1964, 1967a) as well as by Cullingford (1972), but the investigations of Paterson (1974), Browne (1980), Paterson *et al.* (1981) and Armstrong *et al.* (1985) have led to a rejection of the concept. The section at Almondbank, however, retains its importance in the interpretation of the deglaciation of this part of eastern Scotland.

Description

The sediments exposed in the Almondbank section comprise the following sequence, described by Simpson (1933) and Paterson (1974):

3.	Sands and gravels	4.6 m
2.	Laminated silts and clays	7.2 m
1.	Red-brown till	at least 3.7 m

Although Simpson (1933) described bed 3 as 'morainic deposits', he recognized that the deposits were outwash, a view since confirmed by Sissons (1963a) and Paterson (1974). The sands and gravels (bed 3) occur as terraces on both sides of the River Almond and contain large kettle holes immediately to the north of the section (Figure 14.2).

Paterson (1974) reported that the laminated sediments contained the mollusc *Portlandia arctica* (Gray), Foraminifera and ostracods. Further details were provided by Browne (1980) who reported a restricted microfauna with the Foraminifera mainly being *Elphidium clavatum* (Cushman). Browne also established that the laminated silts and clays extended as far west as Crieff and that they had not been disturbed subsequent to deposition.

Interpretation

Simpson (1933) considered that the red-brown till (bed 1) was deposited by the last ice-sheet to cover the area and that the laminated silts and clays (bed 2) were laid down immediately upon retreat of that ice-sheet. Although he had found no fossils in the immediate vicinity, Simpson argued that the silts and clays correlated with the extensive fossiliferous marine clays (Errol beds, Peacock, 1975c) farther east in the Tay estuary (Jamieson, 1865; Brown, 1867; Davidson, 1932). They could be traced as far west as Templemill (NN 875187), near Crieff, and Simpson inferred ice retreat to at least this locality during their deposition. In addition, Simpson interpreted the rhythmic bedding of these sediments as evidence of annual deposition (that is, they were varves) and estimated that the whole 12.2 m thickness at Almondbank took 640 years to accumulate. It is of note that measurements on a similar sequence of laminated sediments at Dunning in the Earn valley were matched by De Geer (1935) with a Swedish varve sequence from which he inferred a date of deglaciation for that area of 13,013 to 13,071 BP.

The sands and gravels (bed 3) that overlie the laminated sediments are in a similar stratigraphic position to other, frequently poorly sorted, gravels which Simpson mapped in the Earn valley as far west as Crieff. Simpson considered that these were the product of a readvance of the ice-sheet. This conclusion was apparently confirmed by the occurrence of kettle holes in the sands and gravels, which implied the presence of ice at the

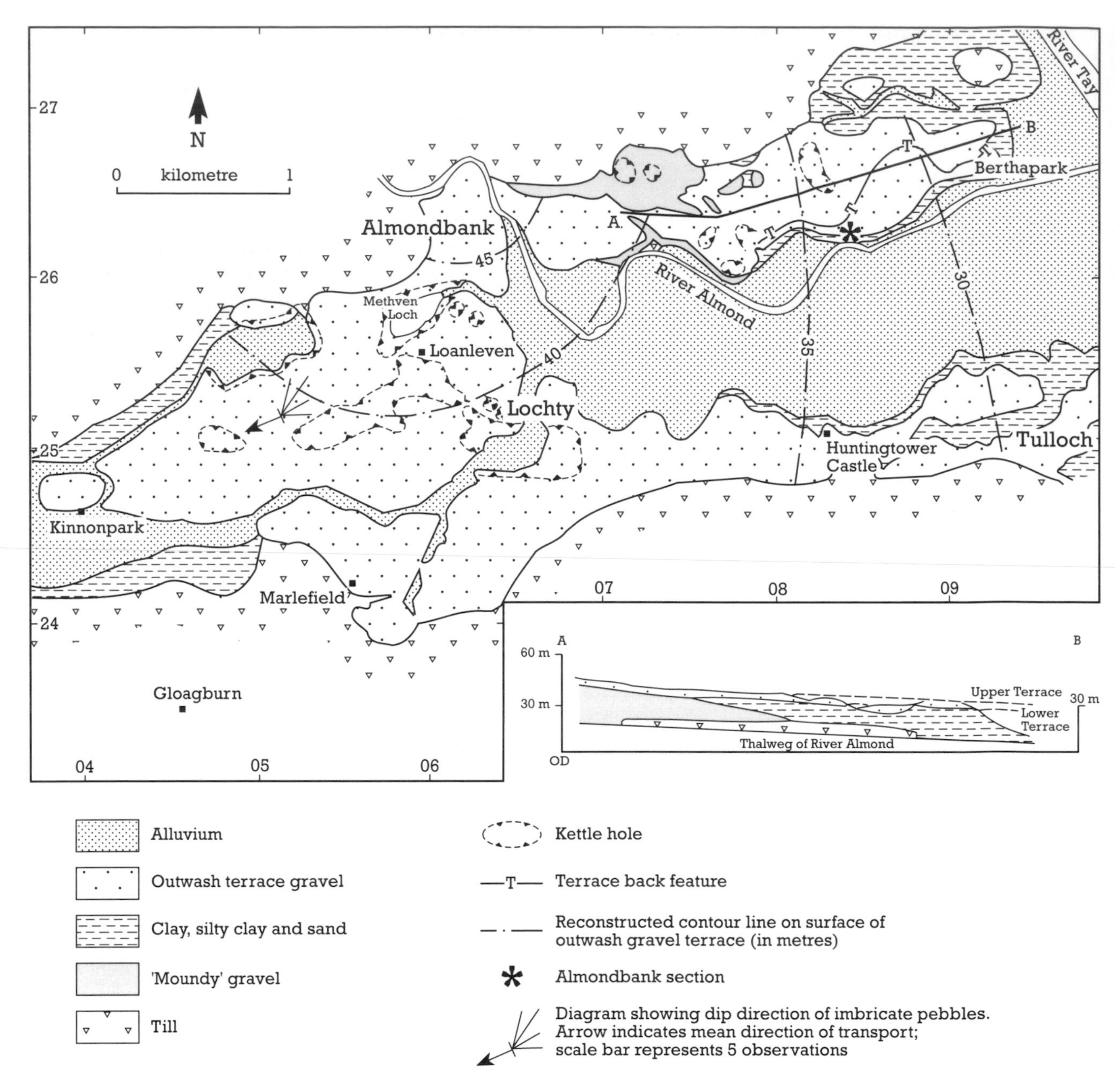

Figure 14.2 Map and section of Late glacial deposits in the Almond Valley (from Paterson, 1974).

time of their deposition.

This interpretation of the Almondbank sequence was accepted by Sissons (1963a), although he suggested a modification to Simpson's readvance limit, placing it approximately 5 km north-west of Perth, where outwash graded into ice-contact glaciofluvial deposits. This latter limit was accepted by Cullingford (1972) who correlated the maximum of the readvance with a pronounced raised shoreline (the Main Perth Shoreline) which had been traced widely along the coasts of south-east Scotland; this shoreline also apparently correlated with a corresponding ice margin in the Forth Valley (Sissons and Smith, 1965a; Sissons *et al.*, 1966; Cullingford, 1977).

Subsequent work by the Geological Survey in the Earn–Tay area (summarized in Paterson *et al.*, 1981; Armstrong *et al.*, 1985) has confirmed Simpson's (1933) stratigraphic sequence, but effectively demonstrated that it is not necessary to invoke a readvance to explain it. Paterson (1974) found evidence that could be interpreted as indicating a readvance at only two localities; Almondbank, where the kettled outwash overlay laminated sediments, and Shochie Burn (see below), where two tills were separated by a layer of sand and gravel. However, Paterson argued that the general absence in the area of an overlying second till together with the lack of disturbance of the laminated sediments precluded a readvance. The contained fossils confirmed the marine origin of the silts and clays, but Paterson argued that, if they were varves, then their period of formation would have been considerably greater than that estimated by Simpson since their thickness was proven to be much greater in boreholes down-valley from the Almondbank section. Alternatively, Paterson contended that these sediments were not varves, but that they accumulated by repeated discharge of material, possibly several times annually, in a marine delta advancing down the lower Almond valley where there were still areas of stagnant ice (Figure 14.2). With the final decay of the 'dead' ice, outwash containing kettle holes was left on top of the deltaic sediments.

According to Browne (1980), the altitudinal distribution of the laminated sediments implies that the sea at the time of the formation of the Main Perth Shoreline had penetrated much farther to the west than Cullingford (1972, 1977) had suggested and that hence the Main Perth Shoreline was not associated with an ice margin immediately north-west of Perth. The overall picture to emerge from the work of the Geological Survey was one of continuous ice retreat with deltaic sedimentation by the major rivers such as the Tay, Earn and Almond into a marine embayment stretching as far west as Crieff and occupying, at its maximum extent, the Methven depression between there and Almondbank.

The Almondbank section is one of considerable historical significance in the development of understanding of the glacial history of Scotland. The original interpretation of the sequence as resulting from a major readvance of the last ice-sheet at approximately 13,000 BP has now been rejected in favour of an origin as the product of deltaic progradation following earlier uninterrupted deglaciation of the area. The sediments retain their interest, however, as one of the few accessible exposures of the marine laminated deposits (equivalent to the Errol beds) in east-central Scotland.

Conclusion

The sequence of deposits at Almondbank is important for interpreting the mode of deglaciation of the Late Devensian ice-sheet (approximately 14,000–13,000 years ago). Historically, it was the key reference site used to substantiate a major ice readvance, the Perth Readvance. Although the deposits have now been reinterpreted in terms of a delta that built seawards during uninterrupted decay of the ice-sheet, the site retains both its historical significance and its high value for sedimentological studies.

SHOCHIE BURN

J. E. Gordon

Highlights

The sequence of deposits exposed in the stream section at Shochie Burn shows glaciotectonically disturbed sands and gravels overlain by till. It shows that a local readvance of ice interrupted the decay of the Late Devensian ice-sheet.

Introduction

The site (NO 071292) is a stream section located on the south bank of the Shochie Burn near Moneydie, 5 km north of Perth. The deposits comprise a succession of two tills separated by sands and gravels, which are important in interpreting the pattern of recession of the Late Devensian ice-sheet margin. The site occurs within the area of the formerly recognized Perth Readvance, and although the succession represents a readvance of the ice margin, this was probably only a minor oscillation. The only description of the site is by Paterson (1974).

Description

The section at Shochie Burn was described by Paterson (1974). It shows a sequence of:

4. Coarse gravel
3. Reddish-brown, sandy till
2. Silt and sand with clay laminae and gravel lenses
1. Reddish-brown, clayey till

The geometry of the deposits is illustrated in Figure 14.3. The lower till (bed 1) is separated from the overlying silt and fine sand (bed 2) by what appears to be a shear plane (Paterson, 1974). The sediments of bed 2 are compacted and deformed: clay laminae are folded, streaked out and displaced by many small faults. The lamination dips generally at 15–20° to the west, but towards the western end of the section it is vertical where the deposit abuts a mass of reddish-brown, clayey till. Overlying bed 2 is a reddish-brown, sandy till (bed 3) which has the form of a wedge. The uppermost deposit (bed 4) is a coarse gravel 4–5 m thick.

Interpretation

Although historical arguments that tripartite sequences, comprising sand and gravel between two tills, imply ice readvance have been shown to be unfounded by work in modern glacier environments (Boulton, 1972b), Paterson (1974) considered that additional evidence from the Shochie Burn deposits supported a minor readvance or brief surge of the Late Devensian ice-sheet. The compacted nature of bed 2, interpreted as fluvial in origin, and the deformation of the sediments, suggested that it had been overridden and glaciotectonized by the ice that deposited the upper till (bed 3); at the same time blocks of till detached from bed 1 were emplaced into the silts and fine sands of bed 2. The gravels of bed 4 form part of an extensive deposit, probably an outwash fan produced as the ice subsequently retreated westwards.

Shochie Burn lies within the limits of the formerly hypothesized Perth Readvance (Simpson, 1933; Sissons, 1963a, 1964). However, the evidence for this event has been reinterpreted, and the readvance is no longer recognized (Paterson, 1974; Sissons, 1974c). The succession at Shochie Burn, the only locality in the Perth area where two tills are known to occur (Paterson, 1974), therefore provides a valuable record of a localized oscillation of the Late Devensian ice-sheet margin and the accompanying sedimentary processes. In particular, the glaciotectonic features are potentially of considerable interest although they have not been studied in any detail.

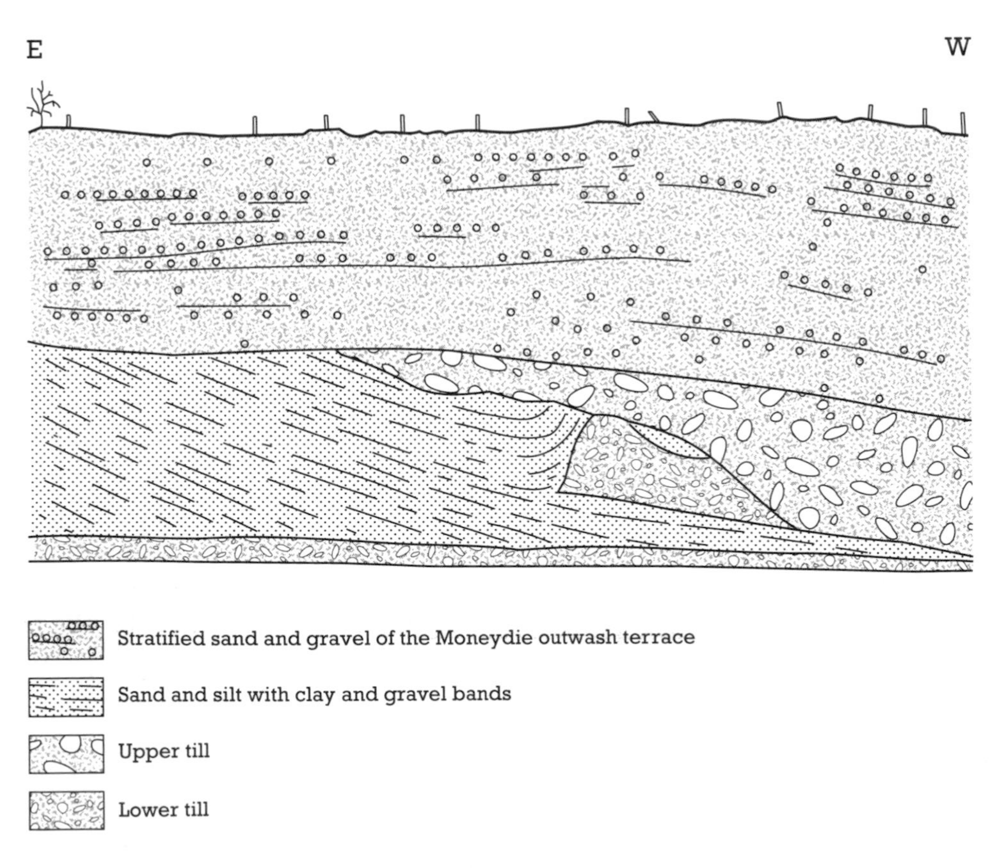

Figure 14.3 Sketch section of glacial deposits at Shochie Burn (unscaled) (from Paterson, 1974).

Conclusion

The deposits at Shochie Burn provide evidence for interpreting the pattern of decay of the last ice-sheet during Late Devensian times (approximately 14,000–13,000 years ago). They show that there was a local readvance of the icefront which overrode and disturbed previously deposited sands and gravels. Although there is no evidence for a widespread readvance, the site is important in demonstrating aspects of the complexity of depositional environments at an ice margin.

DRYLEYS

D. E. Smith

Highlights

The sequence of estuarine sediments in the former clay pit at Dryleys has yielded an important fossil fauna which, in conjunction with geomorphological and sedimentary evidence from the adjacent area, provides a detailed picture of sea-level changes and conditions in the marine environment during the Lateglacial.

Introduction

The Dryleys site (NO 709604) is a former claypit in Late Devensian and Holocene estuarine sediments near Dryleys Farm, north of Montrose. During the latter part of the 19th century, the pit yielded the shells of a largely marine, cold-climate fauna, collected by J. C. Howden. A partial list of the fauna is contained in his account (Howden, 1868) of the sequence of glacial events and relative sea-level changes in the Montrose area. Recent studies by Smith *et al.* (1977), Cullingford and Smith (1980) and Smith and Cullingford (1985) have further contributed to knowledge of Late Devensian and Holocene relative sea-level change in the area, and they enable the Dryleys fauna and deposits to be placed in a more detailed palaeoenvironmental context.

Description

The hillslopes on the northern side of the Montrose Basin are marked by a large number of terraces between Langleypark and Hillside (Figure 14.4). These terraces have been mapped and surveyed by Cullingford and Smith (1980) and Smith and Cullingford (1985). Above 25 m OD, the terraces are glaciofluvial in origin, declining rapidly in altitude eastwards and, from available exposures, are largely composed of poorly sorted sands and gravels. Below 25 m OD, the terraces are marine in origin. Between 10 m and 25 m OD, they are composed of generally fine-grained deposits, largely fine sands and silts, becoming increasingly clayey with depth. These terraces decline only gently eastwards, and at least one appears as the continuation of a higher glaciofluvial terrace, where glacial outwash reached standing water. The marine terraces are, however, fragmentary, being deeply dissected by gullies, most of which formed during the Late Devensian and are now dry. The lower reaches of these gullies are infilled by grey silty clay, reaching a maximum altitude of 6 m to 7 m OD. These sediments underlie a large flat area along the Tayock Burn, reaching the edge of the Montrose Basin. The grey silty clay is similar to the estuarine (carse) clay found in similar areas of central Scotland, and is considered to be the local equivalent of the carse clay. The surface which it forms, here as elsewhere, is distinguished by its uniformity and consistent altitude.

The Dryleys claypit lies on the eastern side of a gully at the eastern margin of this area (Figure 14.4). On the slopes around the gully, terraces formed at approximately 23 m, 19 m, 18 m, 15 m and 13 m OD as relative sea level fell. The claypit is excavated into the 13 m surface which is the main surface in the area, and several exposures, one over 4 m high, occur in the southern face of the pit. This shows laminated sandy silts becoming coarser with depth. These marine sediments are older than the carse clay that occupies the floor of the gully here, although it is not exposed.

During the 19th century, J. C. Howden made a collection of fossil remains from excavations in the Montrose area. This collection was given to the Montrose Museum, where it is still held (1990). Much of the collection was obtained from the carse clay, but from the older terrace deposits above the carselands Howden also obtained a largely arctic fauna, most of which appears to have come from claypits at Dryleys and Puggieston (Figure 14.4). This fauna is listed in Howden's (1868) paper, differing slightly from the museum collection in Table 14.1.

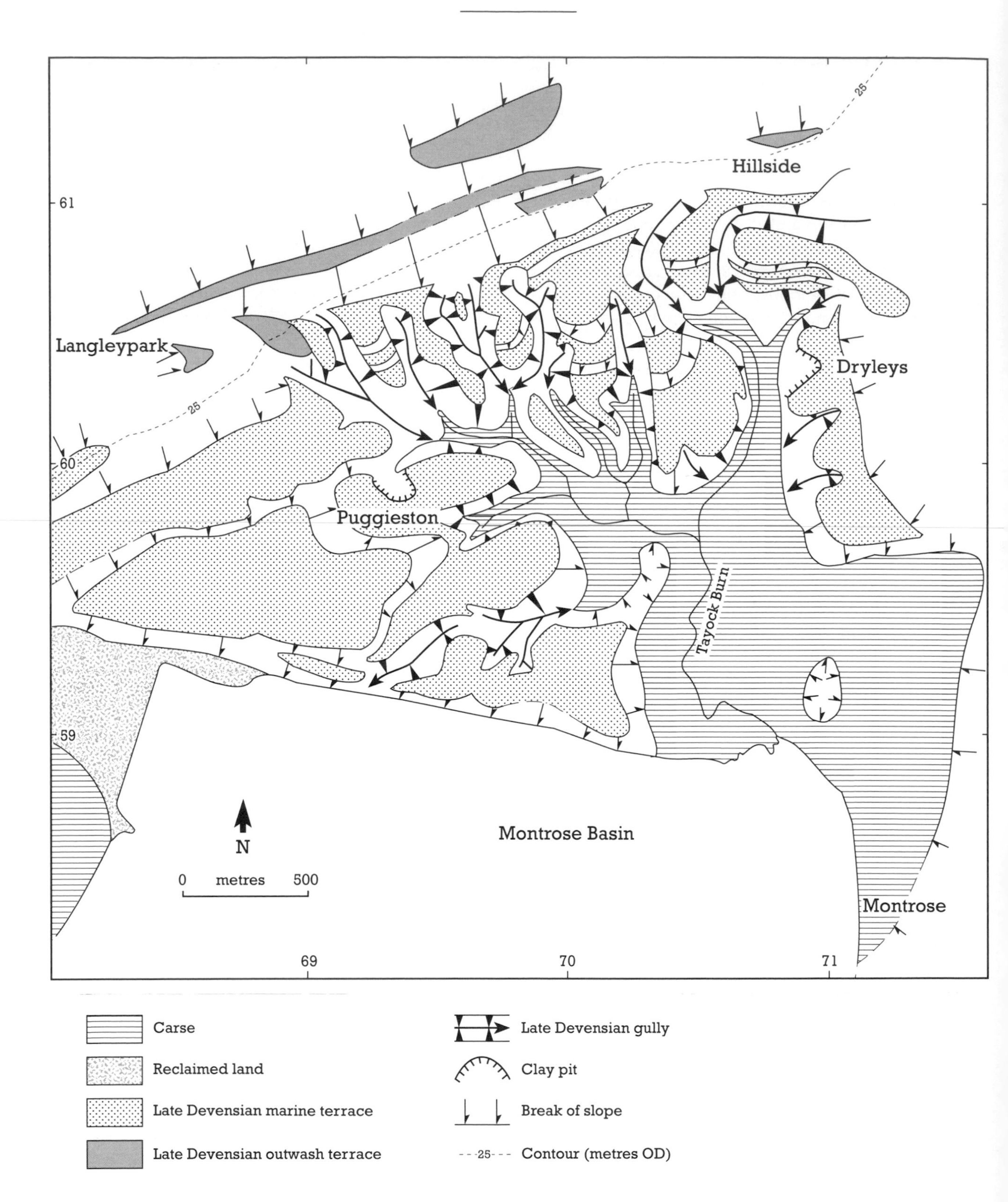

Figure 14.4 Lateglacial and Holocene raised marine deposits in the Dryleys area.

The following details are also provided by J. D. Peacock (unpublished data). The preservation of the specimens of *Arctica islandica* is good, ranging from slightly weathered to fresh and unweathered, whereas that of the other fossils is poor. This contrast suggests that the *Arctica*

Table 14.1 Faunal remains (collected by J. C. Howden) in the Montrose area and attributed to a cold-climate environment

Museum specimen	Location	Modern name
Cyprina islandica	Dryleys	*Arctica islandica* (L.)
Leda arctica	Not given	*Portlandia arctica* (Gray)
Nucula tenuis	Not given	*Nuculoma tenuis* (Montagu)
Pecten greenlandicus	Dryleys	*Arctinula greenlandica* (Sowerby)
Saxicava sulcata	Dryleys	*Hiatella arctica* (L.)
Yoldia arctica	Dryleys	*Portlandia arctica* (Gray)
Somateria sp.	Puggieston	
Cythere sp.	Not given	
Ophiolepis gracilis	Dryleys	
Phoca vitellinus	Balwyllo	*Phoca vitulina* L.

An investigation of the Howden Collection in the Montrose Museum showed the following molluscan fauna to be present (J.D. Peacock, unpublished data).

Dryleys: *Arctinula greenlandica* (Sowerby), *Hiatella arctica* (L.), *Mya truncata* (L.) *Nuculoma belloti* (Adams), *Portlandia arctica* (Gray) and *Yoldiella* cf. *lenticula* (Müller).

Puggieston: *Arctica islandica* (L.), *Hiatella arctica* (L.), *Nuculoma* cf. *belloti* (Adams), *Yoldiella solidula* Warén and *Y. lenticula* (Müller).

Ballwyllo: *Arctinula greenlandica* (Sowerby).

Unplaced specimens: *Arctica islandica* (L.), *Arctinula greenlandica* (Sowerby), *Hiatella arctica* (L.) and *Portlandia arctica* (Gray).

Table 14.2 Radiocarbon dates from faunal remains in the Montrose area

Sample	Location	Age (14C years BP)	Laboratory number
[1]*Somateria mollissima*	Puggieston	10,610 ± 220	Birm–660
[1]*Somateria* sp. or *Melanitta* sp.	Puggieston	11,110 ± 210	Birm–661
[2]*Arctica islandica*	Dryleys		
Outer fraction		3830 ± 140	Birm–737(1)
First middle fraction		4180 ± 120	Birm–737(2)
Second middle fraction		4170 ± 160	Birm–737(3)
Inner fraction		4020 ± 200	Birm–737(4)

[1]Smith *et al.* (1977); Williams and Johnson (1976).
[2]Smith (1986).

specimens are an intrusion into the high-arctic, fully marine fauna, a conclusion that accords with radiocarbon dating (see above). The specimens of *Yoldiella solidula* and *Y. lenticula* from Puggieston presumably came (as *Leda pygmaea*) from the tympanic bones of the seal *Phoca vitulina* L. (see Howden, 1868, p. 141). The list of ostracods from Dryleys given by Brady *et al.* (1874) supports the high-arctic affinities of the fauna.

The principal difference between the faunal list contained in the 1868 paper and the specimens preserved in the museum is the presence of duck bones in the museum, identified as those of Eider (*Somateria* sp.). These bones, collected in 1891 from the claypit at Puggieston, were re-examined by D. Bramwell in 1976 and are believed to belong to two individuals, identified as Eider (*Somateria mollissima* (L.)) and either Eider, or Scoter (*Melanitta* sp.) (Smith *et al.*, 1977).

Smith *et al.* (1977) sought to determine the age and environment of the fauna. They examined pollen from the clay in which the Eider bones were embedded, finding evidence of an open habitat, with Gramineae, Cyperaceae, *Empetrum*, *Artemisia* and Rosaceae pollen grains present. Table 14.2 gives the radiocarbon dates from the duck bones and a specimen of *Arctica islandica*.

Smith and Cullingford (1985) identified a buried peat beneath the grey silty clay, and, in addition, found a layer of grey, micaceous, silty fine sand within the deposits in several of the gullies, including that at Dryleys. From peat above and below this layer, in the nearby gully at Puggieston, they obtained radiocarbon dates (Table 14.3).

Table 14.3 Radiocarbon dates on a possible storm surge layer at Puggieston, after Smith and Cullingford (1985)

Sample	Date (14C years BP)	Laboratory number
0.02 m thick slice of peat above layer	6850 ± 75	SRR–2119
0.02 m thick slice of peat below layer	7120 ± 75	SRR–2120

Interpretation

The morphology of the area surrounding Dryleys indicates that the pit was excavated in deposits belonging to one of a series of eight Late Devensian shorelines in the region, formed as ice withdrew from the Montrose Basin area. The five terraces surrounding the gully at Dryleys are correlated with shorelines DS3 to DS7 of Cullingford and Smith (1980), with the pit having been excavated in the deposits of DS7. Cullingford and Smith identified the nearby deposit at Puggieston as belonging to the same sequence, but related to the lowest shoreline, DS8, one below that at Dryleys. The largely cold-climate faunal remains recovered from the pit during the 19th century are in accord with this interpretation, as is the limited pollen evidence from the deposit. The radiocarbon dates obtained from faunal remains from Dryleys and Puggieston are of limited value. The dates for the *Arctica islandica* shell specimen, although internally consistent, are far too young for a Late Devensian deposit, and since this bivalve has a wide environmental range, the possibility of incorrect labelling in the museum collection must be considered. The dates for the duck bones are older, but apparently also too young, and museum conservation practices may have produced contamination. Thus, in view of the uncertainty of the radiocarbon evidence all that can be said at present is that the deposits are of Late Devensian age.

The gully system at Dryleys is largely dry, and in view of the extensive Holocene estuarine deposits in the floor of the gullies, it seems likely that, by analogy with similar features elsewhere in Scotland (see Sissons *et al.*, 1965), the gully system was largely excavated under periglacial conditions during the Late Devensian.

Smith and Cullingford (1985) confirmed that the grey silty clay infilling the floors of the gullies was a Holocene estuarine deposit accumulated during the Main Postglacial Transgression, and that the surface on the deposit was formed at the maximum of that event in the area, correlating with the Main Postglacial Shoreline. A layer of grey, micaceous, silty fine sand was formed in the course of the transgression and then buried by later deposits. This layer has been identified at a number of estuarine sites in eastern Scotland (see Maryton, Silver Moss and Western Forth Valley), and must have been a widespread event (Smith *et al.*, 1985a). It has been interpreted as a storm surge event (Smith and Cullingford, 1985; Smith *et al.*, 1985a; Haggart, 1988b), or more probably as a tsunami associated with a submarine slide on the Norwegian continental slope (Dawson *et al.*, 1988; Long *et al.*, 1989a).

The Dryleys claypit is of great importance for studies of the Late Devensian environment in Scotland. Excavated into a terrace which forms part of a suite of Late Devensian marine shorelines in the area, it yielded during the 19th century a largely cold-climate fauna which is still preserved and available for study; present-day exposures provide the potential for further sedimentological and palaeoecological investigation. In addition, the site lies in an area rich in Late Devensian and Holocene landforms and deposits. The evidence it contains can thus be set in a wider context, in which the relationships between morphology and stratigraphy are clearly demonstrated. There are few other Late Devensian sites in eastern Scotland where such an extensive and varied faunal record has been identified.

Conclusion

Dryleys is important for studies of Late Devensian sea levels in eastern Scotland. The deposits form part of a series of marine terraces that formed during and following the deglaciation of the area (possibly between 16,000 and 14,000 years ago) and were first studied last century. They have yielded a range of fossils which have been used to interpret a cold-climate marine environment at their time of deposition. Adjacent deposits include later Holocene marine sediments. Dryleys is a valuable reference site in an area of long-standing interest for sea-level studies.

MARYTON

J. E. Gordon

Highlights

Deposits in the coastal section at Maryton include a sequence of estuarine sediments and buried peat. These deposits provide a record of the changes that occurred in sea level and coastal environmental conditions during the Holocene. The site is particularly notable for the evidence it displays for a major coastal flood in eastern Scotland during the middle Holocene.

Introduction

The site at Maryton (NO 683565) is a cliff section in raised estuarine deposits on the west side of the Montrose Basin. It is important for demonstrating Holocene sea-level fluctuations in eastern Scotland and provides one of the best exposures showing the stratigraphic relationships of the estuarine (carse) deposits of the Main Postglacial Shoreline. It is also the only site known where a possible storm-surge or tsunami deposit dating to slightly prior to the maximum of the Main Postglacial Transgression is currently exposed. The site has been described by Smith *et al.* (1980) and discussed in a wider context by Smith *et al.* (1985a); it has also been illustrated in Smith *et al.* (1985a), Long *et al.* (1989b) and Smith and Dawson (1990).

Description

The superficial deposits of the Montrose area were first described in detail by Howden (1868) and later summarized by him (Howden, 1886) in the following general succession:

6. Blown sand
5. Estuarine (carse) clay and sand
4. Peat
3. Glaciofluvial sands and gravels
2. Laminated marine clay
1. Till

The basal unit of till was overlain by laminated marine clay extending up the South Esk Valley to an altitude of 40 ft (12.2 m). Although the contact between the two was nowhere exposed, Howden was in no doubt of their stratigraphic sequence. The clay (bed 2) contained abundant fossil remains of arctic molluscs which Howden (1868) listed (see Dryleys). Jamieson (1865) gave additional details, and Brady *et al.* (1874) listed the ostracods.

Howden described sands and gravels (bed 3) overlying the marine clay and the till, which extended as far inland as the mouths of the Highland glens. These gravels were part of the sequence first described by Buckland (1841a), and in mineralogical composition were identical to the moraines of the Highland glens. Howden suggested that the sands and gravels were formed by meltwaters at the end of the great ice-sheet glaciation and subsequently while valley glaciers remained in the glens. Cullingford and Smith (1980) have mapped these deposits to the west of the Montrose Basin and noted that four separate outwash terraces can be related to distinct shorelines at altitudes of between 15 m and 25 m OD around the Montrose Basin.

In a number of stream courses tributary to the South Esk, Howden observed peat (bed 4) resting on the marine clays and extending on to the gravels. At one locality the peat was overlain by carse clays. Generally, however, the latter deposits rested on the lower marine clay and reached an altitude of 15 ft (4.6 m). Shells similar to those found in the present-day estuary were locally abundant in the carse clay. Details were given by Jamieson (1865) and Howden (1868). More recently, the pattern of Holocene sea-level changes in the Montrose area has been investigated in detail by Crofts (1971, 1972, 1974) and Smith and Cullingford (1985). In particular, Maryton has been described in detail by Smith *et al.*

(1980). A sequence of marine and non-marine organic deposits infills a number of gullies dissecting Late Devensian marine and glaciofluvial terraces at the western end of the Montrose Basin (Cullingford and Smith, 1980; Smith *et al.*, 1980). A section in a terrace bluff at Maryton shows (Figure 14.5):

Figure 14.5 Section at Maryton. Laminated Late Devensian marine clays are overlain by peat then a layer of grey, micaceous, silty, fine sand interpreted as a tsunami deposit. Above the latter is silty peat then carse clay. (Photo: D. E. Smith.)

5.	Grey silty clay (carse)	0.85 m
4.	Peat	0.10 m
3.	Grey, micaceous, silty fine sand	0.18 m
2.	Peat	0.15 m
1.	Laminated, pink silty clay	0.20 m

The sequence in the section is broadly similar to that revealed in boreholes in the Fullerton area 0.5 km to the west (Smith *et al.*, 1980). Samples for pollen analysis were taken from the Maryton section and two coring sites at Fullerton by Smith *et al.* (1980). They recognized three pollen assemblage zones at Maryton (Figure 14.6). Zone M–1 includes poorly preserved pollen from most of the laminated, pink silty clay (bed 1). Although the pollen frequencies are low and the range of taxa restricted, open habitat conditions are indicated. Zone M–2 includes pollen from the horizons between the upper layers of the pink silty clay (bed 1) and the lowest part of the carse clay (bed 5). Contrasts in pollen frequency and assemblage suggest a break in deposition between zones M–1 and M–2. The main characteristics of zone M–2 are the consistent representation of *Betula* and *Pinus sylvestris*, variable amounts of *Corylus*-type and *Salix*-type pollen and the presence of *Ulmus* and *Quercus*. Several taxa also suggest the proximity of damp conditions (Gramineae, Cyperaceae, Chenopodiaceae, Nymphaceae and *Filipendula*). Zone M–3 includes the greater part of the carse clay. The main representatives of M–3 are *Pinus sylvestris*, *Corylus* type, *Betula*, *Ulmus* and *Quercus*. Other taxa (Nymphaceae and Cyperaceae) again indicate local damp conditions.

Smith *et al.* (1980) correlated pollen zone M–1 with an equivalent zone from a similar, basal, laminated clay layer in one of the Fullerton boreholes. It was thought to reflect an open-habitat, Late Devensian environment. Zones M–2 and M–3 also have equivalents with similar tree and herbaceous pollen in both the Fullerton boreholes. These assemblages indicate a boreal environment, corresponding with Holocene 'chronozone' Fl I of West (1970). An additional zone, corresponding with 'chronozone' Fl II, is represented in one of the Fullerton profiles.

Smith *et al.* (1980) obtained radiocarbon dates for several horizons at Maryton and Fullerton (Table 14.4).

Interpretation

Sequences similar to those described by Howden (1868, 1886) from the Montrose area have long been recognized elsewhere in the Tay, Forth and Ythan estuaries. Together with the evidence from these other sites, the Montrose deposits have formed an important part of the wider interpretation of Lateglacial and Holocene sea-level changes in eastern Scotland (Jamieson, 1865; J. Geikie, 1874; Wright, 1937; Sissons, 1967a; Cullingford *et al.*, 1986, 1991). A full sequence of Lateglacial and Holocene deposits is nowhere exposed, but the section at Maryton demonstrates a key part of the succession spanning the period of the

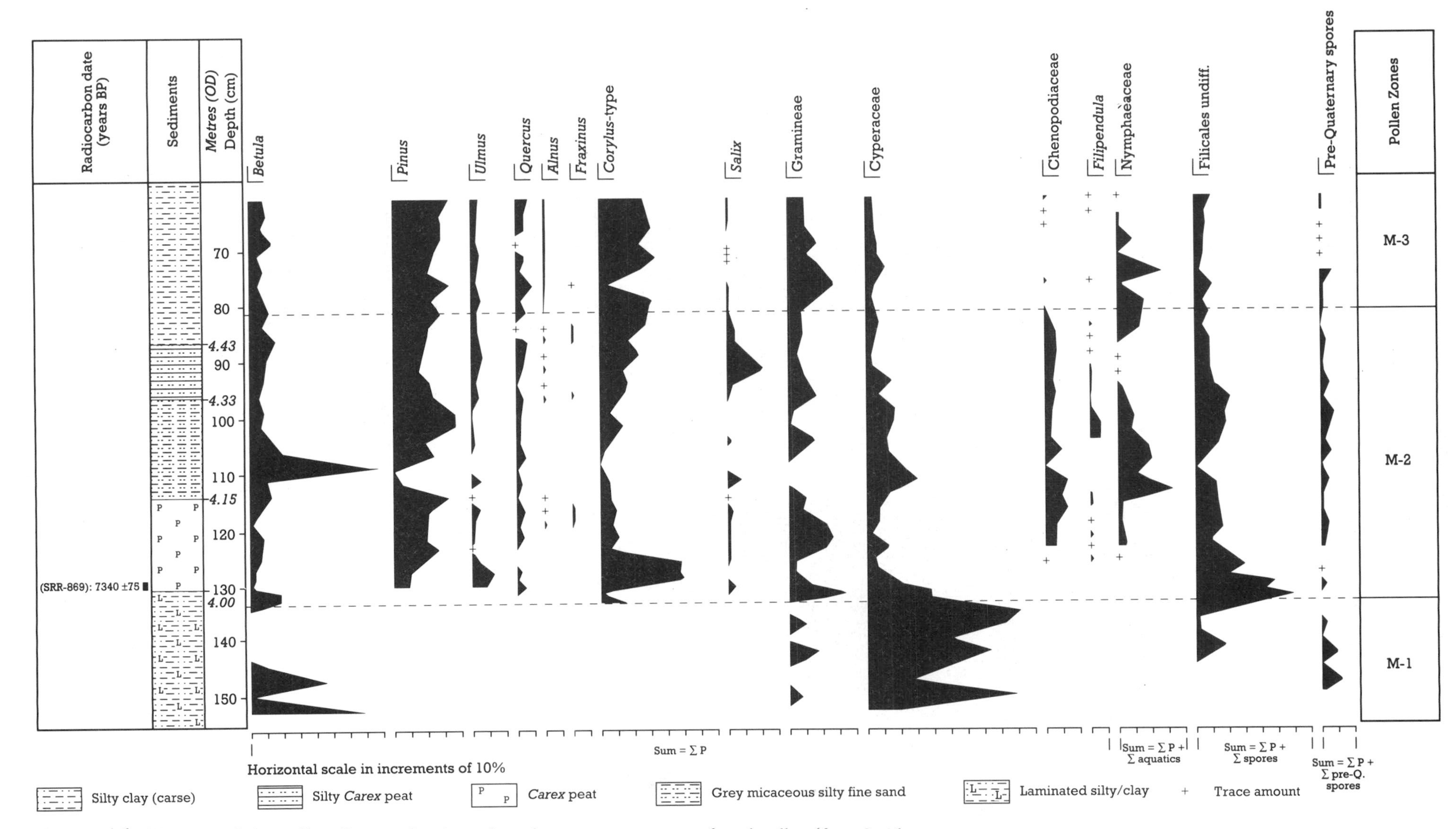

Figure 14.6 Maryton: relative pollen diagram showing selected taxa as percentages of total pollen (from Smith *et al.*, 1980).

Table 14.4 Radiocarbon dates on sand layer in peat at Maryton and Fullerton (Smith *et al.*, 1980)

Sample	Location	Date (14C years BP)	Laboratory number
0.02 m thick band of peat above basal laminated silty clay (bed 1)	Maryton	7340 ± 75 BP	SRR–869
0.02 m thick band of peat below sand layer	Fullerton	7140 ± 120 BP	Birm–823
0.02 m thick band of peat above sand layer	Fullerton	6880 ± 110 BP	Birm–867
0.02 m thick band from top of peat below the carse	Fullerton	7086 ± 50 BP	SRR–1149
0.02 m thick band from base of peat above the carse	Fullerton	6704 ± 55 BP	SRR–1148

Although the inversion of SRR–1149 and Birm–867 could reflect reworking or contamination, Smith *et al.* (1980) point out that the two dates are not statistically different at the 95% confidence level, and therefore suggest a mean age of 6983 ± 60 BP for peat above the sand layer.

transgression associated with the Main Postglacial Shoreline; the Lateglacial part of the succession is represented at Dryleys (see above).

In the Maryton sequence the basal, laminated, silty clay is interpreted as a Late Devensian estuarine deposit (see Dryleys above) (Smith *et al.*, 1980). There is then a break in the sequence of deposits until peat began to form during the Holocene about 7340 BP. About 7140 BP, the peat accumulation was interrupted by deposition of the grey, micaceous, silty sand layer, before resuming until about 6980 BP, when it was terminated by deposition of the carse.

On geomorphological and stratigraphic grounds Smith *et al.* (1980) correlated the carse surface in the Maryton area with the Main Postglacial Shoreline of eastern Scotland (see Sissons *et al.*, 1966). The radiocarbon dates show that the Main Postglacial Transgression associated with this shoreline was in progress at 7140 BP and possibly culminated between about 6983 and 6704 BP. In the context of the age of the Main Postglacial Shoreline, these dates are older than those from stratigraphically similar sites in the Tay estuary and eastern Fife (Chisholm, 1971; Cullingford *et al.*, 1980; Morrison *et al.*, 1981; Smith *et al.*, 1985b), but similar to the dates obtained in the Western Forth Valley (Sissons and Brooks, 1971).

The grey, micaceous, silty sand layer has recently attracted particular attention since it appears to represent a high-magnitude marine event, and Maryton has become a reference site for demonstrating its stratigraphic position and sedimentary characteristics. It has also been identified in boreholes at Fullerton, Dryleys and other sites around the margins of the Montrose Basin. At Puggieston (see Dryleys), the top and base of peats below and above the sand have been dated respectively to 7120 ± 75 BP (SRR–2120) and 6850 ± 75 BP (SRR–2119), dates that are indistinguishable from the similarly placed samples at Fullerton. A similar sand layer occupying the same stratigraphic position and giving approximately the same radiocarbon age has now been detected at a number of sites on the east coast of Scotland ranging from the Beauly Firth in the north to the Carse of Stirling in the south (see Western Forth Valley, Silver Moss and Barnyards). It has been argued that this unique deposit was possibly the result of a major storm surge in the North Sea Basin around 7000 BP (Smith *et al.*, 1985a; Haggart, 1988b) or, more probably, a tsunami associated with a major submarine slide on the Norwegian continental slope (Dawson *et al.*, 1988; Long *et al.*, 1989a, 1989b).

The section at Maryton therefore provides important stratigraphic evidence for Holocene sea-level changes in the Montrose Basin area. It is significant in several respects. First, as one of the best sections in eastern Scotland illustrating the Main Postglacial Transgression, Maryton is an integral member of a network of sites in this area (see Western Forth Valley, Pitlowie, Silver Moss and Philorth Valley) which together have important research potential for establishing the diachroneity of the culmination of the Main Postglacial Transgression and the formation of the Main Postglacial Shoreline and therefore have potential for elucidating the wider regional patterns of glacio-isostatic and eustatic changes. Second, since Maryton is the only location where the tsunami deposit is exposed, it is critically important in further studies of this event. Third, in a historical context, Maryton is also significant for its location in one of a number of key reference areas recognized for over 100 years as illustrating the pattern of Lateglacial and Holocene sea-level change.

Conclusion

Maryton forms part of a network of sites that is important for establishing the pattern of sea-level variations in eastern Scotland during the Holocene (the last 10,000 years). It is particularly notable for one of the best exposures in the estuarine deposits of the Main Postglacial Shoreline (an extensive raised shoreline that formed approximately 6800 years ago), and also the only available exposure of the deposits formed by a major flood that affected coastal areas around 7000 years ago.

MILTON NESS

D. E. Smith

Highlights

The assemblage of raised beaches and a shore platform at Milton Ness provides important geomorphological and sedimentary evidence for Quaternary sea-level changes in eastern Scotland. This evidence, from an exposed headland location, allows valuable comparisons with nearby estuarine sites.

Introduction

Milton Ness (NO 770649) lies 8 km north of Montrose and is important for a series of raised shorelines and associated deposits. The headland displays evidence of both Late Devensian and Holocene relative sea levels, together with an extensive intertidal rock platform, the age of which is uncertain. The site was first described by Campbell (1935), and subsequently by Cullingford and Smith (1980), Smith and Cullingford (1985) and Smith (1986).

Description

The bedrock at Milton Ness is composed of resistant Upper Old Red Sandstone sediments (Hickling, 1908; Campbell, 1913) overlain by Quaternary deposits largely consisting of till, with some sands and gravels at the surface. The oldest Quaternary feature on Milton Ness is undoubtedly the extensive intertidal rock platform, particularly well developed on the northern side of the headland (Figure 14.7). This platform passes beneath both the till and sands and gravels. Dawson (1980a, in Smith, 1986) has remarked on its extent and correlated it with the Low Rock Platform elsewhere in Scotland. It is well developed along this stretch of coast and northwards to Inverbervie (Myers, 1872; Campbell, 1935) but whether it was formed in part by periglacial processes during glacial episodes, or by temperate marine erosional processes during interglacials, is unknown.

In his paper on the Mearns coastline, Campbell (1935) noted that three raised shorelines were well developed at Milton Ness, reflecting former changes in relative sea level. More recently, Cullingford and Smith (1980) and Smith and Cullingford (1985) have identified a sequence of both Late Devensian and Holocene shorelines there, each shoreline marked by a terrace formed of sands and gravels, often with shell fragments. The upper two terraces are thought to be of Late Devensian age (Cullingford and Smith, 1980). They are associated with shorelines at 20.1–20.7 m OD and 21.1–23.7 m OD. The highest terrace is almost 0.5 km long. An exposure in the middle terrace on the south side of the headland shows 5 m of bedded gravel and sand with shell fragments. The highest terrace is correlated with the shoreline DS1 of Cullingford and Smith (1980), which slopes eastward across the general area of Fife, Angus and Kincardine at 0.85 m km^{-1}, whereas the middle terrace is correlated with shoreline DS2, also sloping eastward, at 0.73 m km^{-1}. These shorelines are the highest in the sequence identified by Cullingford and Smith, who maintain that they were formed very early in deglaciation (prior to 14,000 BP), when ice still occupied much of the surrounding area. The lowest terrace was found to be of Holocene age by Smith and Cullingford (1985). It forms a shoreline at 5.2–5.5 m OD, which they have suggested may correlate with the Main Postglacial Shoreline in the Montrose Basin carselands.

On the south side of Milton Ness, the present cliffline lies at right angles to the shoreline and there are extensive exposures which reveal the composition and internal structure of the middle and lowest raised beach terraces. Although these

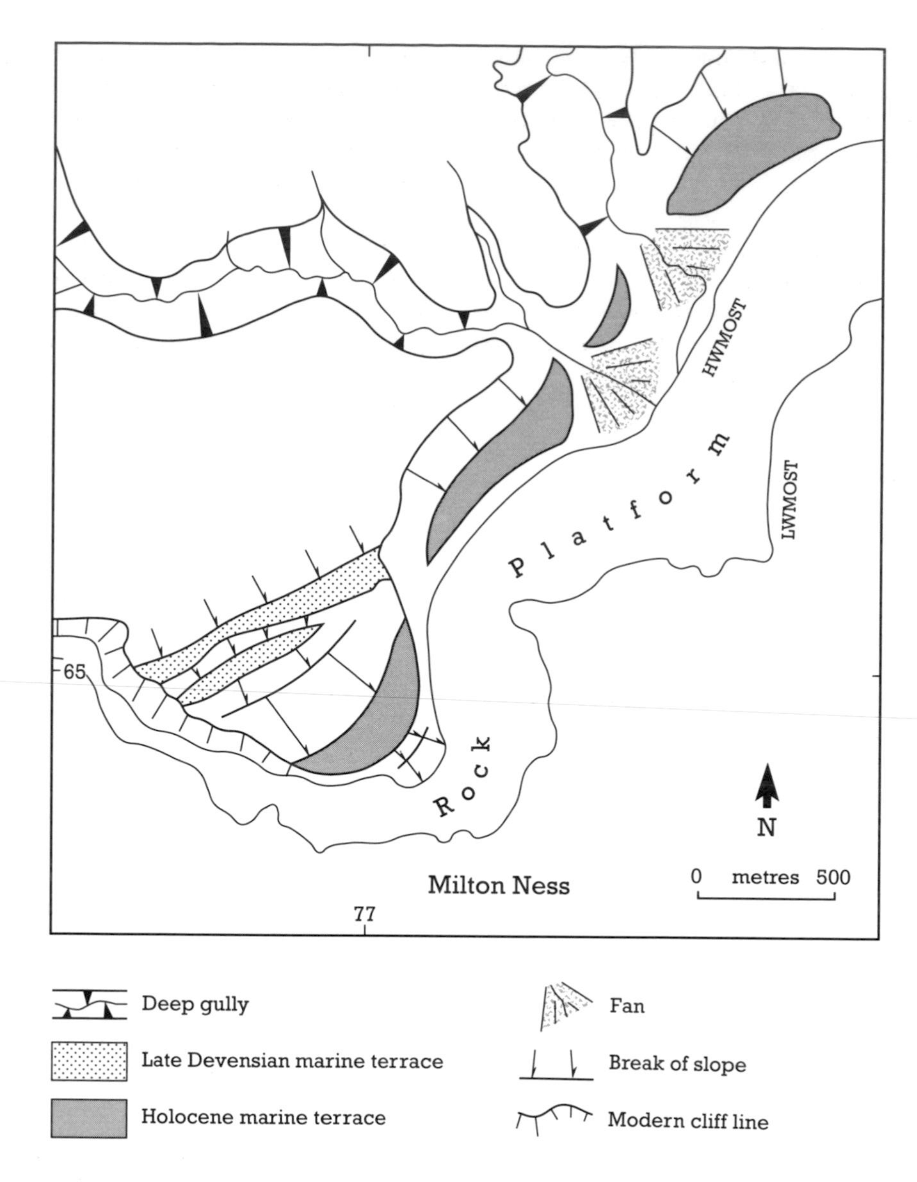

Figure 14.7 Raised shorelines and intertidal shore platform at Milton Ness.

have not been studied in detail, they provide a potential opportunity to relate raised beach morphology and sediments.

Interpretation

Milton Ness demonstrates an assemblage of several important features associated with relative sea-level changes in eastern Scotland. The interest includes first, an intertidal shore platform of uncertain age but pre-dating, at least in part, the last glaciation; second, two Lateglacial raised beaches, which are part of the earliest shorelines to form in the general area following deglaciation; and third, a prominent Holocene raised beach terrace possibly related to the Main Postglacial Shoreline. Each feature is clearly identifiable and extensive exposures illustrate the structure and composition of the raised beach terraces. Elsewhere in eastern Scotland, comparable sequences of raised beaches occur in East Fife (Cullingford and Smith, 1966), and between Dundee and Arbroath and north of Johnshaven (Cullingford and Smith, 1980), but they lack comparable exposures to those at Milton Ness. This rare combination of both morphological and sedimentary evidence is important in the study of coastal evolution in eastern Scotland. The morphological and sedimentary evidence from the exposed headland at Milton Ness complements the stratigraphic evidence for sea-level change represented nearby at Maryton and Dryleys in the more sheltered estuary of the South Esk. Milton Ness, together with Dunbar and Kincraig Point, provides good examples of the range of coastal landforms and sediments developed on the exposed (as opposed to estuarine) coasts of the east of Scotland. They can be compared directly with the similarly exposed coastal sites in the west (Glenacardoch Point, Northern Islay and West Coast of Jura) to illustrate the differing histories of coastal evolution during the Late Quaternary.

Conclusion

Milton Ness is important for the study of coastal changes in eastern Scotland during pre-Late Devensian times and during the Late Devensian and Holocene (approximately the last 16,000 years). The features of interest include a shore platform and raised beaches with good exposures in the deposits. The particular value of Milton Ness lies in this combination of geomorphological and sedimentary evidence and its location on an exposed headland, which contrasts with sedimentary evidence in the estuarine situation of Dryleys and Maryton.

NORTH ESK AND WEST WATER GLACIOFLUVIAL LANDFORMS

J. E. Gordon and L. J. McEwen

Highlights

The assemblage of outwash and river terraces at this site illustrates the range of geomorphological processes that accompanied the decay of the Late Devensian ice-sheet in the Eastern Highland Boundary area. The site is particularly noted for its palaeochannels which have allowed changing discharge patterns to be reconstructed since the time of deglaciation.

Introduction

This site comprises two areas in Strathmore located at the Highland edge near Edzell. The larger ($c.2.5\ km^2$) lies to the west of the village between NO 565695 and NO 597679; the smaller ($c.\ 0.6\ km^2$) to the south-east between NO 614686 and NO 620673. Together these areas are important for an assemblage of glaciofluvial landforms and deposits. These comprise an extensive spread of outwash (palaeosandur) deposits built out eastwards across Strathmore during the wastage of Late Devensian glaciers in the adjacent glens of the West Water and River North Esk (Synge, 1956; Sissons, 1967a; Maizels, 1976). In the former case the sandur deposits are associated with an ice-marginal position marked by a distinctive area of kame and kettle topography and meltwater channels.

The palaeosandur deposits associated with the North Esk and its tributary, the West Water, extend for 10 km downstream from the Highland Boundary Fault zone. They provide an excellent example of Late Devensian outwash deposits which have been dissected to form four main terrace systems. The terraces display systems of palaeochannels which have been mapped in detail by Maizels (1976, 1983a, 1983b), and used in palaeohydrological reconstructions and modelling (Maizels, 1983a, 1983b, 1983c, 1986; Maizels and Aitken, 1991).

Description

In the valley of the West Water, 3 km west of Edzell, an area of hummocky kame and kettle topography and ice-contact slopes marks the position of a former glacier margin. A section at NO 567688 shows that, at least in part, the landforms are developed in sand and gravel. Although formerly ascribed to readvances of the last ice-sheet (Synge, 1956; Sissons, 1967a), these deposits probably reflect either a local halt, or change in glacier dynamics and sedimentation as the Late Devensian ice wasted back from Strathmore into the confined Angus glens, probably between 14,000 and 13,000 BP (Sutherland, 1984a). The former ice margin is not marked by an end moraine in the classic sense, but rather by a landform assemblage typical of modern glacier environments dominated by glaciofluvial activity and the formation of kame and kettle topography (Price, 1973; Boulton and Paul, 1976).

Meltwater channels are associated with the ice-front deposits and they also occur in a proglacial location, for example, at Edzell Castle (NO 585692). Outwash terraces lead away from the former ice-front and extend out across Strathmore, and also from adjacent glens (Buckland, 1841a; Lyell, 1841a; Howden, 1868; Bremner, 1939a; Synge, 1956; Sissons, 1967a, 1976b;

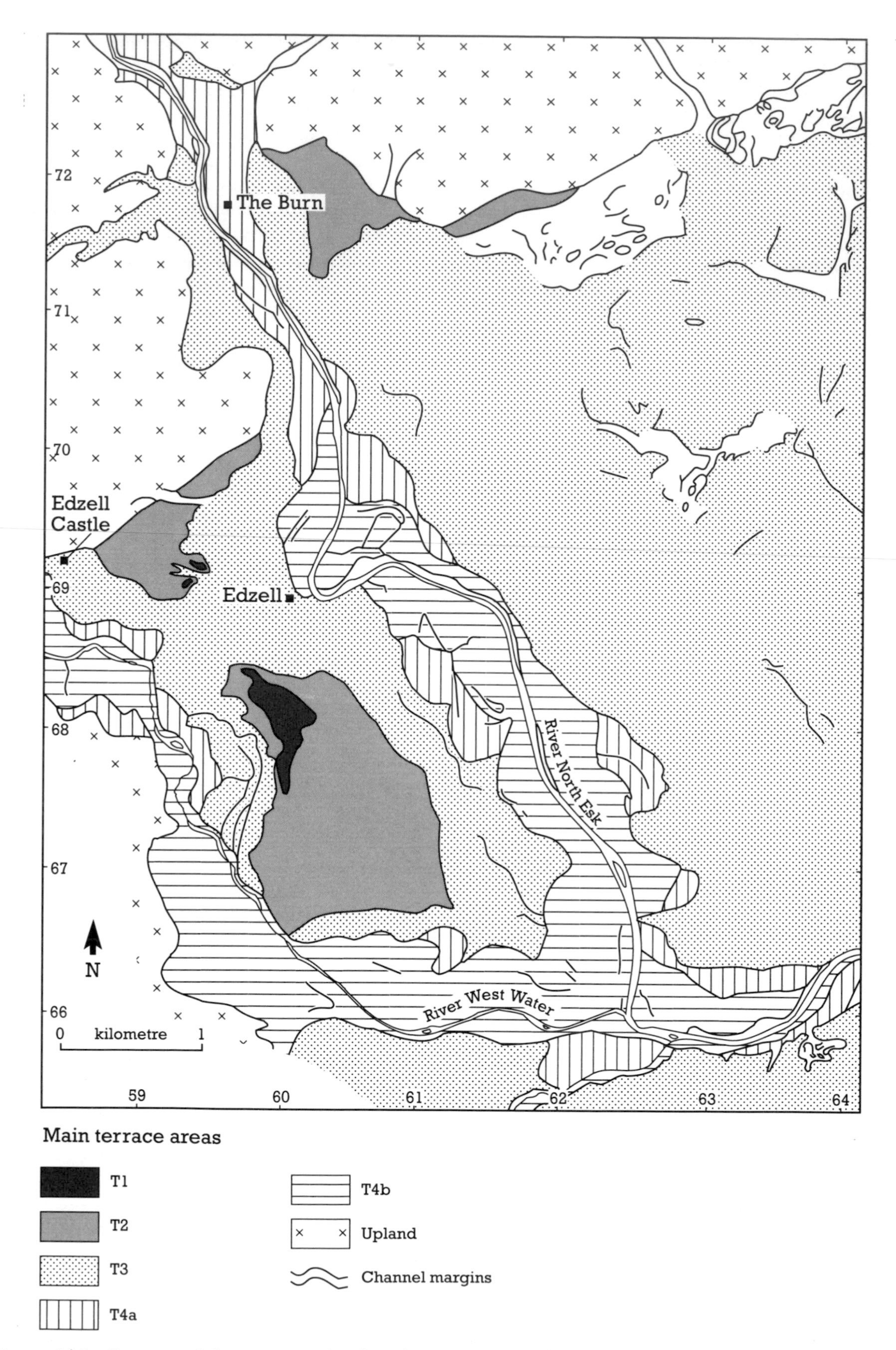

Figure 14.8 Terraces of the River North Esk and West Water in the Edzell area (from Maizels, 1983a).

Crofts, 1974). In Glen Esk, outwash terraces extend from north-west of The Burn (Sissons, 1967a). As deglaciation progressed, the outwash deposits and stream channels would have continued to adjust to changes in water discharge and sediment supply. During and following deglaciation, the area also underwent isostatic uplift, with consequent changes in base level.

The sandur deposits are up to 6 m in depth and are characterized as 'massive, coarse, poorly-sorted, imbricated gravels and cobbles, with isolated lenses of cross-bedded and plane-bedded coarse and medium sands, characteristic of Miall's (1978) 'Gm' gravel lithofacies type, and similar to Scott outwash sediments (facies assemblage GII of Rust, 1978) comprising over 90% gravel content' (Maizels, 1983b, p. 256). The sedimentary characteristics of the sediments indicate deposition in an aggrading, proglacial, braided river environment (Maizels, 1983a).

The four main terraces, associated with both the North Esk and the West Water, have been mapped by Maizels (1983a, 1983b) (Figure 14.8). The upper two terraces (T1 and T2) are evident only as isolated fragments; the lower two (T3 and T4) are much more extensive (Figure 14.8). Study has focused on the nature, direction and magnitude of change within this terrace sequence (Maizels, 1983a, 1983b, 1983c, 1986; Maizels and Aitken, 1991). For example, large-scale changes in channel pattern and morphology have been identified between terrace fragments and attributed to a decline in the amounts of meltwater discharge and sediment supplied during and after deglaciation. The resulting palaeoforms thus reflect channel adjustment from a proglacial environment to present-day fluvial controls. They also demonstrate a south-eastward migration of the North Esk/West Water confluence by 2.8 km; a shift that was clearly accompanied by periods of aggradation and incision.

As well as macro-scale change, more intricate localized channel incision and minor terrace fragments have been mapped. Three different types of palaeochannel have been identified, each type associated with different rates of discharge and sediment availability (Maizels, 1983a). These include complex braided systems (Type A), deeper wider channels (Type B) and deeply incised and relatively localized meander scars (Type C). In terms of inter-terrace variation in palaeochannel type, the two upper terraces have been identified as having more braided (Type A) and periglacial (Type B) channel systems, whereas the lower terraces have more sinuous (Type C) channels, although all terraces exhibit complex braided palaeochannels to a certain degree. The lower terraces (T3 and T4) are characterized by multiple channel networks, longitudinal bars, high width-to-depth ratios and low sinuosities. The lower terrace surfaces are also locally incised by relatively well-defined, deep, low-gradient, sinuous channels and many of the adjacent terrace bluffs possess major meander scars. The number of sinuous channels increases from the highest terrace (T1) to the lowest (T4), whereas the degree of braiding declines. Mean width-to-depth ratios decline from about 108 to about 40 between T1 and T4 (Maizels, 1983a).

Interpretation

Maizels (1983a, 1983b) concluded that the deep, sinuous channels (Type C) were responsible for terrace formation and that each phase of terrace formation involved a change from straight, multiple channels to single-thread channels in response to threshold changes in meltwater and sediment discharges, which could relate to glacier fluctuations or episodic flood events. Base-level variations appear to have had only a minor effect on the channel adjustments.

As well as inter-terrace variation in palaeochannel type, both lateral and downstream intra-terrace changes have been identified. It is important to note, however, that many of the palaeochannel features are of low amplitude; the palaeochannel patterns are better viewed from the air, and are especially highlighted on infra-red aerial photographs. This site, with its extensive suite of palaeochannels in the terrace surfaces, thus provides a marked contrast to the well-defined suite of terraces with steep 'risers' along the River Findhorn (see above).

The present channel discharges are small compared with estimated palaeodischarges, and Maizels (1983a, 1983b, 1986; Maizels and Aitken, 1991) estimated order-of-magnitude velocities and discharges for particular palaeochannels, using empirical formulae (cf. Church, 1978; Ryder and Church, 1986). Peak flows calculated for the terrace sequence decreased from a maximum of *c.* 18,000 m^3s^{-1} on the highest terrace to *c.* 1,300 m^3s^{-1} on the lowest. The decline in peak discharge to the present day value of *c.* 330 m^3s^{-1} is thus as much as fifty times. At best, these values provide only order-of-magnitude estimates (cf. Maizels and Aitken, 1991)

The River North Esk and West Water site is significant in several respects. First, it provides a good illustration of a suite of glaciofluvial landforms which are characteristic of the Highland margin in Strathmore, where extensive spreads of outwash have built out from the mouths of the Highland glens (Sissons, 1967a, 1976b; Paterson, 1974; Insch, 1976). Second, it provides a good illustration of a former glacier-margin landform assemblage dominated by glaciofluvial activity and the close association of ice-contact and proglacial meltwater features. In this respect the site bears comparison with, for example, Muir of Dinnet, Glen Feshie and Almondbank, and with the Loch Lomond Readvance features at Moss of Achnacree and the Western Forth Valley. Third, the River North Esk and West Water site demonstrates the effects of topographic controls on deglaciation. As the Late Devensian ice receded from the open, piedmont area of Strathmore, the tributary glacier fronts in the narrow Highland glens may have become temporarily stabilized in the lower reaches of the valleys as their ablation areas changed configuration. Fourth, the site is significant for the development of palaeochannels on the terrace surfaces. These features are among the best preserved of their type studied in Scotland and have allowed significant insights into the controls and thresholds governing channel change during and since deglaciation, particularly in relation to discharge and sediment supply. It is an important research site for assessing the extent of adjustments within the fluvial system since the Late Devensian in a lowland area with upland headwaters. Although Late Devensian palaeochannels are known to exist on terrace surfaces at other locations in Scotland; for example, along the River Dee (Maizels, 1985) and River Don (Maizels and Aitken, 1991), in Glen Feshie (Robertson-Rintoul, 1986a) and in the area south of Fraserburgh (British Geological Survey 1:50,000 Sheet 97) these are generally less extensive and have not been studied in comparable detail to the North Esk and West Water features.

Conclusion

This site is important for understanding the geomorphological changes that occurred in the landscape during and following deglaciation of the Late Devensian ice-sheet (approximately 14,000–13,000 years ago) when large volumes of meltwaters were released from the decaying ice. It shows an excellent assemblage of landforms characteristic of the eastern Highland boundary area. These include outwash and river terraces that were formed as the ice melted and wasted back into the Highland glens. The higher terraces contain kettle holes indicating the former presence of the glacier, whereas the lower terraces are most notable for particularly good examples of fossil river channels. The latter provide valuable evidence for reconstructing the changes that occurred in river characteristics and behaviour during and following the period of ice melting.

STORMONT LOCH

C. J. Caseldine

Highlights

The sediments which infill the floor of Stormont Loch provide an important pollen record, supported by radiocarbon dating, of the vegetational history and environmental changes in eastern Scotland during the Lateglacial.

Introduction

Stormont Loch (NO 190423) is a partially infilled, shallow (less than 3 m) lake which occupies part of a large kettle hole complex in the outwash fan that spreads across Strathmore from the Ericht valley, draining both Glenshee and Strathardle (Paterson, 1974; Insch, 1976). The site lies at an altitude of 61 m OD within freely drained iron podsols (Corby Association; Laing, 1976), where they closely abut brown earths (Balrownie Association) which comprise the fertile, till-derived soils of much of Strathmore. Stormont Loch is one of the few lowland sites in Strathmore which retains a complete Lateglacial and Holocene sedimentary sequence, having escaped drainage and marl extraction during the 18th and 19th centuries, which either removed or disturbed deposits on a wide scale throughout the agricultural lowlands of eastern Scotland. Limited stratigraphical investigations and detailed pollen analysis of one part of the loch basin have been undertaken by Caseldine (1980a, 1980b).

Description

Borings in the western fringe of the basin have shown relatively rapid thickening of the sediment infill to over 6 m within 50 m of the edge; beyond this, surface water conditions prevented further boring. The Lateglacial sediments comprise a straightforward sequence of a basal, dark grey clay which becomes increasingly coarse with depth, followed by a brown organic mud incorporating occasional lenses of both gyttja and sedge peat, overlain in turn by a very fine grey clay (Figure 14.9). Overlying the Lateglacial deposits there are over 4 m of Holocene sediments, largely telmatic peats and dystrophic gel muds.

Pollen diagrams have been prepared for the Lateglacial and Holocene sediments from the basin by Caseldine (1980a, 1980b). Of particular interest is the record for the Lateglacial derived from two parallel cores and comprising both relative pollen (Figure 14.9) and pollen concentration diagrams supplemented by four radiocarbon dates (SRR–1732 to SRR–1735).

Interpretation

Within the lowest clay there is a Gramineae–*Rumex* pollen assemblage typical of the earliest phase of the Lateglacial Interstadial throughout Scotland (Gray and Lowe, 1977b). The later interstadial record that occurs within the mud varies from the sequence more commonly found in eastern Scotland (Lowe and Walker, 1977), in that it has two clear peaks for *Juniperus* separated by an assemblage having increased *Rumex* and Cyperaceae. This pattern appears both in the relative and concentration diagrams and is interpreted as representing a brief climatic deterioration within the interstadial. This has been tentatively correlated with the Older Dryas climatic oscillation which is found in north-west Scotland (Pennington, 1977b) and in other parts of north-west Europe, but a radiocarbon date from this level of 13,820 +670/−580 BP (SRR–1735), is thought to be affected by older carbon and hence inaccurate. Sites demonstrating such an oscillation are rare in eastern Scotland; the only other clear sequence is found at Corrydon in Glenshee (Walker, 1977).

The change to the Loch Lomond Stadial is demonstrated by the lithological change from organic mud to grey clay and in the pollen record by the virtual disappearance of *Juniperus* and the presence of a range of herb taxa characteristic of the cold period. This occurred after 11,510 ± 140 BP (SRR–1734). At Stormont Loch there is a very full and detailed representation of the local stadial vegetation cover, which suggests a delayed expansion of *Artemisia* after an initial phase in which *Rumex* and Gramineae were dominant. This is assumed to reflect an initial period of high precipitation followed by a very much drier, but still very cold phase, a pattern similar to that found further north in the Strathspey by MacPherson (1980) and which lends support to the stadial palaeoclimatic interpretations of Sissons (1979d). There is evidence for increasing warmth at 10,150 ± 110 BP (SRR–1733) in an expansion of aquatic taxa and a decrease in *Artemisia*, but local development of *Juniperus* and *Empetrum* is dated rather late at 9700 ± 90 BP (SRR–1732).

The Holocene sequence conforms with the expected pattern for eastern Scotland (Birks, 1977), showing early dominance of *Betula* and *Corylus*, and the eventual development of *Quercus*, *Ulmus* and *Alnus*. Strathmore lies well to the south of the extensive pine-dominated woodland of northern Scotland and during the middle Holocene exhibited a mixed-oak woodland with enhanced frequences of birch, which itself was the major woodland element on the higher ground immediately north of the Highland Boundary Fault. An elm decline is sharply delimited in the pollen record and is associated with the first appearance of *Plantago lanceolata* and the expansion of other open-ground indicators. The more recent record shows further woodland clearance and the eventual local expansion of *Calluna* as the soils immediately around the site became heavily podsolized.

Stormont Loch is of national importance in that it provides a complete picture of vegetational change over the last approximately 13,000 years from an inland lowland area close to the main centres of Lateglacial ice development, but located away from the direct influence of the ice (for example, compare with Tynaspirit and Loch Etteridge). It also allows potentially valuable comparisons with the many upland sites that have been studied, which may help clarify environmental (for instance, altitudinal) influences in the Lateglacial climatic record; as yet there is insufficient evidence to evaluate fully such influences (see, for example, Tipping, 1991b). In its detailed and generally atypical Lateglacial pollen record, Stormont Loch is of wider importance in con-

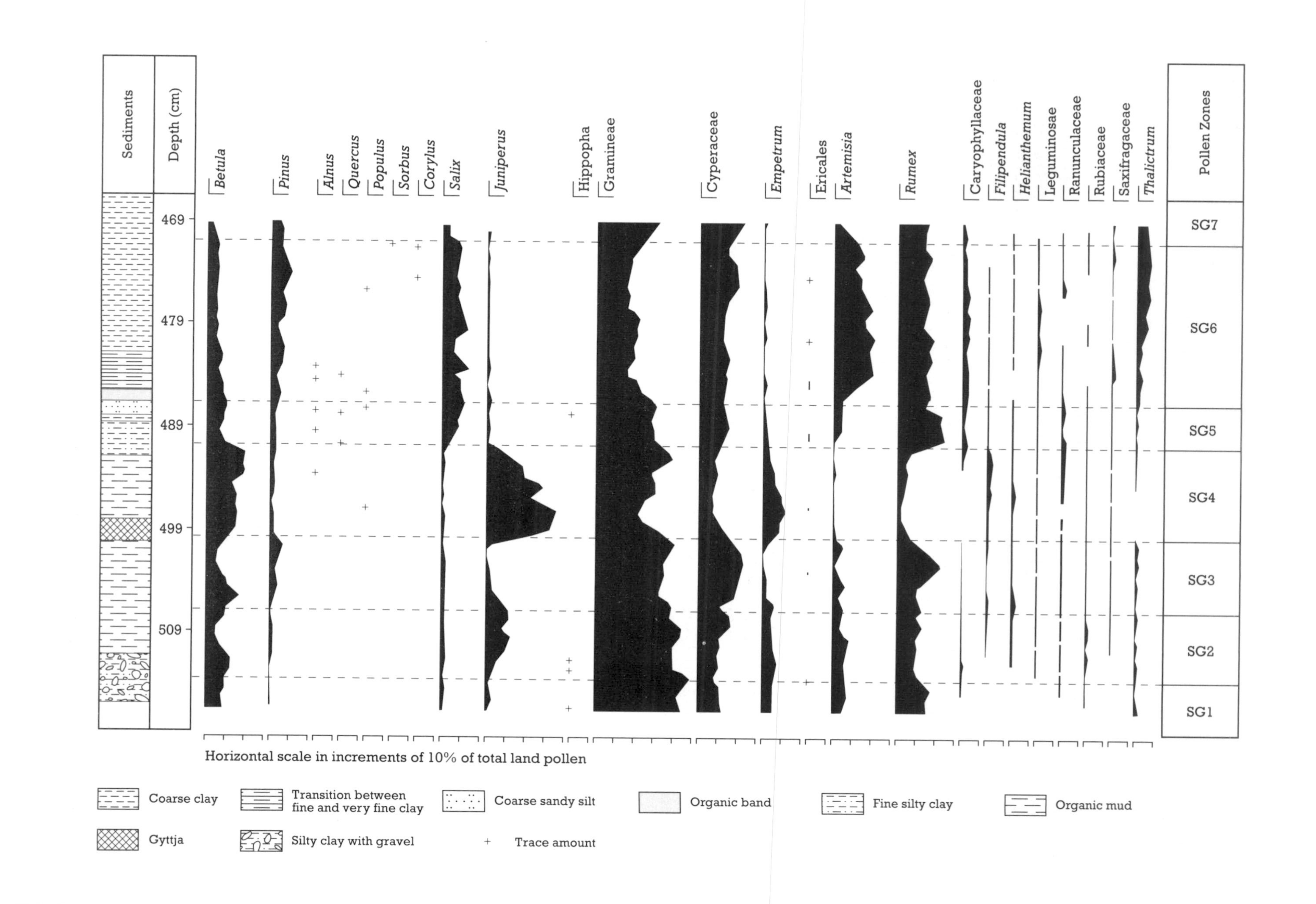

Sediments
Depth (cm)
469
479
489
499
509
Betula
Pinus
Alnus
Quercus
Populus
Sorbus
Corylus
Salix
Juniperus
Hippopha
Gramineae
Cyperaceae
Empetrum
Ericales
Artemisia
Rumex
Caryophyllaceae
Filipendula
Helianthemum
Leguminosae
Ranunculaceae
Rubiaceae
Saxifragaceae
Thalictrum
Pollen Zones
SG7
SG6
SG5
SG4
SG3
SG2
SG1
Horizontal scale in increments of 10% of total land pollen
Coarse clay
Transition between fine and very fine clay
Coarse sandy silt
Organic band
Fine silty clay
Organic mud
Gyttja
Silty clay with gravel
Trace amount

tributing to a better understanding of the complexity and character of Lateglacial climatic change. By demonstrating consistent vegetational changes in duplicate cores, particularly within the interstadial part of the record, but also within the stadial, the results confirm Stormont Loch as an important reference site not only for eastern Scotland, but also for making the comparisons necessary to establish the wider patterns of Lateglacial climatic change (see Tipping, 1991b). Further, comparisons of the pollen record with the coleopteran record (Atkinson *et al.*, 1987) are also essential and in this respect Stormont Loch is significant in showing early climatic deterioration during the Lateglacial, a feature matched in the coleopteran record established at other sites (Atkinson *et al.*, 1987). Stormont Loch is also notable for its full and clear stadial record, showing increasing aridity as the stadial progressed, an aspect now attracting attention in explaining the pattern of glacier changes.

Further, Stormont Loch has great potential for the study of Holocene vegetational history in a dominantly agricultural landscape whose past land-use history is but poorly understood. This potential lies in the probable thick depth of sediments in the main part of the loch, affording a high level of analytical resolution, and in the relatively long period of anthropogenic influences identified in the preliminary Holocene pollen diagram.

Conclusion

Stormont Loch is an important reference site for studies of the environmental history of eastern Scotland during the Lateglacial and Holocene (approximately the last 13,000 years). It is particularly notable for the detail of its Lateglacial pollen record, which has allowed revealing insights into the environmental changes that occurred; for example, it is one of only a few sites in eastern Scotland to show that vegetational development was interrupted by a short, but separate climatic deterioration before the Loch Lomond Stadial (about 11,000–10,000 years ago). By virtue of its geographical location, Stormont Loch is also significant for the potential comparisons it allows between the different vegetational histories at sites in the uplands and lowlands.

Figure 14.9 Stormont Loch: relative pollen diagram for core SGI showing selected taxa as percentages of total land pollen (from Caseldine, 1980a).

Chapter 15

Fife and lower Tay

INTRODUCTION

D. G. Sutherland

This area includes the peninsula of Fife between the Firth of Tay and Firth of Forth, the lower Tay valley and Strathearn (Figure 15.1). The low-lying coastal areas and the lower parts of the valleys contain extensive accumulations of Late-glacial and Holocene marine deposits, the investigation of which has been the principal research theme in this area. The Fife and lower Tay area contains evidence for only one period of glaciation, that of the Late Devensian ice-sheet. It can reasonably be inferred, however, that the area was glaciated on more than one occasion and that the ice-moulded nature of most of the hills is the cumulative result of successive glaciations rather than solely the product of the last ice-sheet. Good evidence for glaciation of this region during the early Middle Pleistocene has been provided from the immediate offshore zone by Stoker and Bent (1985), and subsequent ice-sheet glaciation on at least three occasions may be inferred from evidence in neighbouring regions (Bowen *et al.*, 1986). The only feature that is known from the region to pre-date the last ice-sheet is the rock platform at, or close to, present sea level which can be followed around much of the coast of eastern Fife. This platform and its associated cliffline are overlain in places by glacial deposits and they have been presumed to be interglacial in origin (Sissons, 1967a).

Ice-moulded landforms and striated rock surfaces indicate that during the last ice-sheet glaciation, ice from the western Highlands moved into the region from the north-west (Geikie, 1900, 1902; Forsyth and Chisholm, 1977; Armstrong *et al.*, 1985). As the glaciation proceeded, the western part of the region continued to be affected by ice flowing from that direction, but to the east there was latterly a change to an easterly or even north-easterly movement. The transport of erratics and the general colour and composition of the till are in accord with these ice movements.

The most notable glacial deposits and landforms were produced during the period of ice-sheet retreat when extensive areas of sand and gravel were laid down and there was a widespread marine invasion of the lower ground around the coasts. Major accumulations of glaciofluvial sediments were deposited in the Wormit Gap, the northern Howe of Fife, near Barry in Angus and west of Loch Leven, these being areas that were topographically suitable for the isolation of 'dead'-ice masses and the concentration of meltwater drainage (Rice, 1961, 1962; Chisholm, 1966; Cullingford, 1972; Browne, 1977; Paterson, 1977; Armstrong *et al.*, 1975, 1985).

Certain of these areas of 'dead' ice terminated in the sea which during deglaciation attained altitudes of between 30 m and 40 m OD around the coasts. The marine deposits from this period are typically red, laminated clays, the Errol beds (see Inchcoonans and Gallowflat) that contain a high-arctic faunal assemblage (Brown, 1867; Geikie, 1902; Davidson, 1932; Peacock, 1975c; Paterson *et al.*, 1981). These are found around all the coasts of the region and also within the Howe of Fife into which the sea penetrated at the time of deglaciation. The surface morphology of the marine sediments consists of a series of distinct terraces, which have been mapped and levelled by Cullingford (1972, 1977), Cullingford and Smith (1966, 1980) and Sissons and Smith (1965a). These studies have demonstrated that a succession of easterly sloping shorelines was formed progressively as the ice retreated to the west, each shoreline having a lower gradient than its predecessor. The 'staircase' of shorelines around Kincraig Point (Geikie, 1902; Smith, 1965) probably formed at this time.

The shoreline sequence and the associated marine and estuarine clays are as yet undated. However, the progressive change in gradient of the shorelines (a consequence of isostatic uplift) allows approximate ages to be extrapolated from the known ages of younger and lower-gradient shorelines (Andrews and Dugdale, 1970). Such a calculation suggests that eastern Fife was deglaciated prior to 15,500 BP and the remainder of the region became ice-free during the ensuing 2000 years (Sutherland, 1991a).

Only a limited number of sites (as at Creich Castle (Cundill and Whittington, 1983) and Black Loch (Whittington *et al.*, 1991a) have been investigated for Lateglacial pollen in order to provide evidence of terrestrial environmental change following deglaciation. The marine record makes it clear that deglaciation occurred when the climate was still very cold, but no pollen evidence has been reported which accords with such conditions. This implies that either the sites investigated to date were locations of 'dead'-ice masses that did not melt until the climate ameliorated or that there was insufficient vegetation to provide enough identifiable contem-

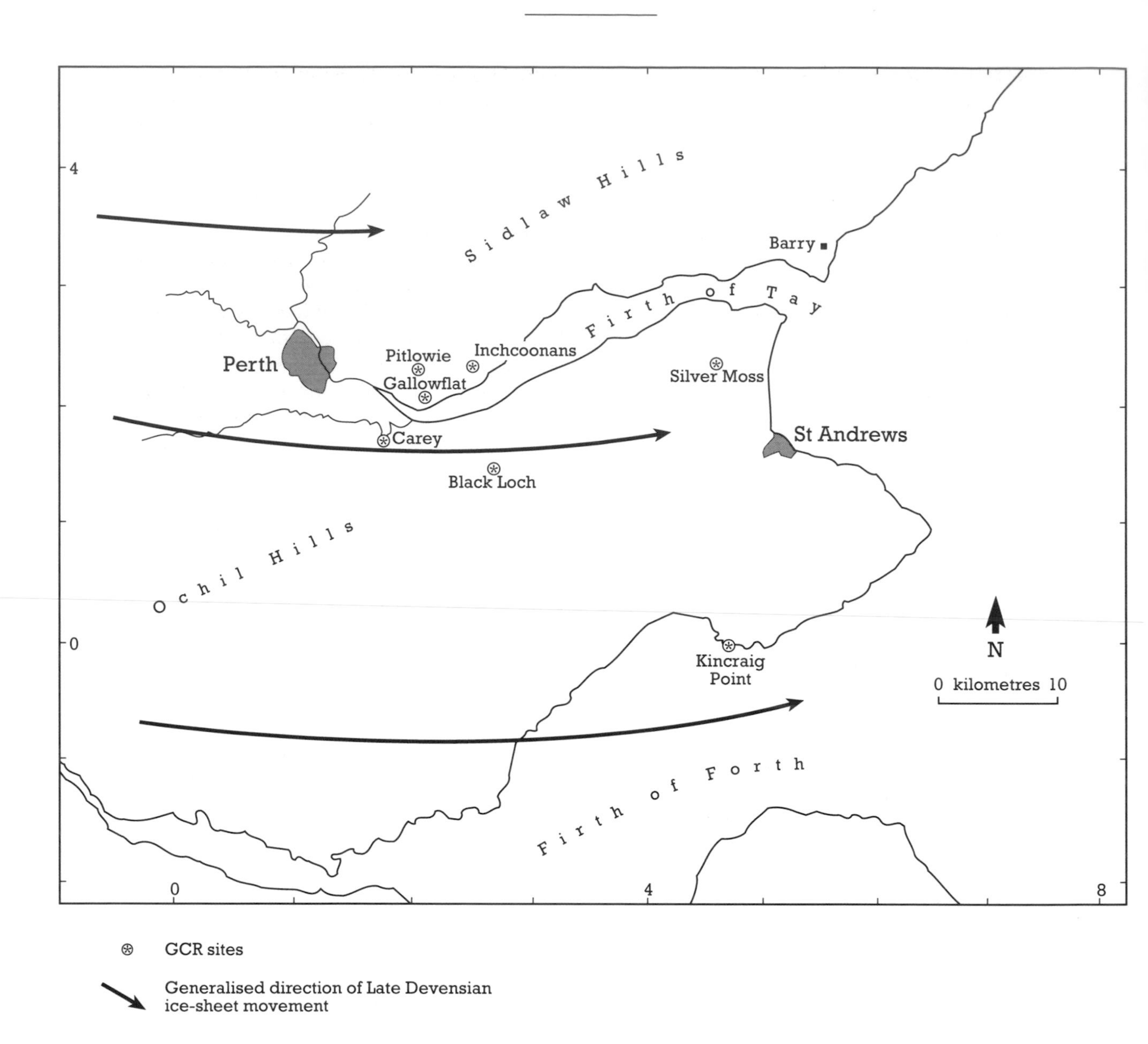

Figure 15.1 Location map of the Fife and the Lower Tay area.

poraneous pollen, given the apparently high background of derived or long-distance transported pollen grains (Cundill and Whittington, 1983).

During the Lateglacial Interstadial an open grass–herb vegetation was dominant, with a lower representation of woody taxa, such as birch, than in the neighbouring regions.

Sea level fell during the early part of the interstadial, and throughout the region in the latter part of the Lateglacial it was below the level subsequently attained during the Holocene. There has been some disagreement as to the exact course of sea-level change during the Lateglacial and, in particular, over the age of a widespread marine erosional episode, evidence for which is found at or below present sea level (Sissons and Rhind, 1970; Browne and Jarvis, 1983). Sissons (1969, 1974d, 1976a) first identified this period of Lateglacial marine erosion (the landward margin of which he termed the Main Lateglacial Shoreline) on the southern side of the Forth valley and Firth of Forth, and Cullingford (1972) suggested correlation with gravel horizons in the Tay and Earn valleys. It was argued by Sissons (1974d, 1976a) that the erosion of this feature occurred during the latter part of the Lateglacial and was promoted by the severe climate of this period. The Main Lateglacial Shoreline is tilted to the south-east at a gradient of $0.17\ \mathrm{m\,km^{-1}}$.

Paterson *et al.* (1981) and Armstrong *et al.* (1985) have discussed erosional features in the Tay and Earn area which they have correlated with the Main Lateglacial Shoreline. However, they suggested that these features formed during the early part of the Lateglacial when sea level was falling from the Main Perth Shoreline. They further suggested that during the latter part of the Lateglacial, sea level was particularly low resulting in the erosion of deep channels along the Tay estuary (see also Buller and McManus, 1971; McManus, 1972). It may be noted with respect to the correlations proposed by Paterson *et al.* (1981) that they provide no mechanism for an erosional event in the early Lateglacial in areas that are otherwise characterized by fine-grained sedimentation throughout the Lateglacial and Holocene. Browne and Jarvis (1983) have reported marine erosional features in St Andrews Bay. They correlated a surface cut across glacial and glaciomarine (Errol beds) sediments with the erosional surface identified by Sissons (1969, 1976a), but suggested that a bedrock surface at approximately the same altitude could have been inherited from an early phase of marine erosion prior to the Devensian. A further erosional surface cut across glacial and glaciomarine sediments at Buddon Ness, but overlain by Holocene deposits, has been described by Paterson (1981).

Sea-level changes during the early and middle Holocene are more clearly understood and are particularly well documented in this region. The sites at Carey (Cullingford *et al.*, 1980) and Silver Moss (Chisholm, 1971; Morrison *et al.*, 1981) provide details of the early Holocene changes, including the Main Postglacial Transgression, and the site at Pitlowie (Smith *et al.*, 1985b) has been the focus of the most detailed study in Scotland to date of the minor changes in sea level at the time of the maximum of that transgression, when the Main Postglacial Shoreline was formed. These various sites show that at the beginning of the Holocene, sea level was about 3 m OD in the lower Earn valley but that it progressively fell, reaching a low, probably below present sea level, at around 8000 BP. Subsequently there was a rapid rise in sea level culminating at around 6100 BP at Pitlowie, and at prior to 5900 BP at Silver Moss. The Silver Moss site also contains evidence of a brief marine invasion of the coastal zone at around 7000 BP (Morrison *et al.*, 1981). This event (described above – see Maryton) has been observed at sites throughout the east coast of Scotland (Smith *et al.*, 1985a; Dawson *et al.*, 1988; Haggart, 1988b; Long *et al.*, 1989a).

The Main Postglacial Shoreline formed at the maximum of the transgression has been shown to be isostatically tilted towards the south-east at a gradient of approximately 0.08 m km^{-1} (Sissons, 1983a). During the subsequent fall of sea level to the present level, lower shorelines were formed (Cullingford, 1972) but these are not well dated.

Holocene terrestrial environmental change has been studied in considerable detail at Black Loch. This site provides evidence of early and middle Holocene forest expansion and development, but it is particularly notable for the detail of the late Holocene changes in vegetation consequent upon Man's impact from the Neolithic onwards.

INCHCOONANS AND GALLOWFLAT

D. G. Sutherland

Highlights

Inchcoonans and Gallowflat are important reference sites for the Errol beds, a sequence of fossiliferous estuarine sediments deposited largely in eastern Scotland as the Late Devensian ice-sheet melted. They provide important evidence for the high-arctic nature of the marine environment during the early part of the Lateglacial.

Introduction

The sites at Inchcoonans (NO 242233) and Gallowflat (NO 211202) are located, respectively, 1 km north-west and 4.5 km south-west of Errol, between Perth and Dundee. Both occur in an area of fossiliferous Late Devensian raised estuarine deposits (Errol beds). Inchcoonans comprises a small area of undisturbed deposit adjacent to a former claypit (now infilled) which provided the type sequence for these deposits; Gallowflat is a working claypit, where the sediments are exposed.

A broad twofold subdivision of the marine and estuarine sands, silts and clays that were laid down around the Scottish coasts during Late Devensian ice-sheet retreat and in the Lateglacial period has long been recognized; between those deposits containing a restricted high-arctic fauna and those containing a much more diverse boreal to arctic fauna (Jamieson, 1865; Brady *et al.*, 1874; Robertson, 1875). This subdivision has been noted to correspond, in general, to the geographical distribution of the deposits, those on the east coast being predominantly high arctic in character and those on the west being predominantly boreal to arctic (Robertson, 1875; Anderson, 1948; Sissons, 1965). Peacock (1975c) proposed the informal terms Errol beds and Clyde beds to apply, respectively, to the high-arctic and the boreal-to-arctic deposits, this terminology reflecting the locations where the different deposits had been first described in detail. The Inchcoonans and Gallowflat area, by Errol to the north of the Tay estuary, is therefore the principal reference area of the high-arctic deposits.

At the western end of the Carse of Gowrie on the north of the Tay estuary the surface of an area of higher ground is mantled by Late Devensian estuarine clays. The clays extend to altitudes of approximately 30 m OD and their surface forms a series of terraces, the lowest of which is below 12 m OD. The western, southern and eastern sides of this higher ground are flanked by Holocene estuarine deposits. A number of claypits excavated in the upper deposits resulted in the discovery during the last century of marine fossils of species indicative of very cold conditions when the clays were deposited (Jamieson, 1865; Brown, 1867; Brady *et al.*, 1874). The most detailed studies were carried out in the Inchcoonans claypit (now infilled) by Davidson (1932), whose work has in recent years been verified and amplified by Paterson *et al.* (1981) and Graham and Gregory (1981). The deposits at Gallowflat have been described by McManus (1972), MacGregor (1973) and Duck (1990). In addition, the raised shorelines contemporaneous with deposition of the marine clays have been studied in detail by Cullingford (1972, 1977). A summary of the current understanding of these deposits is given in Armstrong *et al.* (1985).

Description

The first systematic description of the stratigraphy of the Inchcoonans claypit was given by Davidson (1932). He identified three principal units:

3.	Yellowish-brown, sandy clay	2.5–3.0 m
2.	Fine, blue clay, coarsening upwards	1.5–2.1 m
1.	Fine, red clay	over 1.2 m

Pebbles, cobbles and even boulders were scattered throughout the deposits, the surface of which was at about 12 m OD. More recent investigations have added detail to this outline, and Paterson *et al.* (1981) determined the following sequence (Figure 15.2):

5.	Yellowish-grey sandy clay	
	erosion surface	1.0–1.75 m
4.	Yellowish-grey silty clay	1.25–2.9 m
3.	Brownish-grey clay	0.9–1.5 m
2.	Reddish brown clay	1.3 m
1.	Sand and gravel	

The main difference from Davidson's section was the recognition of an erosion surface formed during the deposition of the upper yellowish-grey sediments. In a neighbouring borehole (Figure 15.2) clayey gravels were encountered within beds 3 and 4. These were considered to be due to slumping from a mound of submerged glaciofluvial sand and gravel (Paterson *et al.*, 1981) or deposited from an iceberg (Armstrong *et al.*, 1985). The change in colour during the period of deposition of the clays was thought to relate to the retreat of the ice-sheet towards the west: the reddish clays at the base reflected derivation from local Old Red Sandstone, whereas the upper yellowish clays received their colour from material principally derived from Highland rock types.

The fauna described by Davidson (1932; Graham and Gregory, 1981) came principally from beds 3 and 4. The molluscan fauna consisted of the gastropods *Buccinum groenlandicum* (Chemnitz) and *Lunatia pallida*? (Broderip and Sowerby) and the bivalves *Astarte borealis* (Schumacher), *Hiatella arctica* (L.), *Macoma calcarea* (Chemnitz), *Musculus laevigatus* (Gray), *Musculus niger* (Gray), *Palliolum groenlandicum* (Sowerby) (=*Arctinula greenlandica*), *Portlandia arctica* (Gray), and *Thracia* cf. *septentrionalis* (Jeffreys). Although some of these species have wide geographical ranges, others, such as *Portlandia arctica* and *Palliolum groenlandicum* are strongly indicative of high-arctic conditions. Certain of the ostracods recovered are similarly indicative, for example, *Krithe glacialis* (Brady, Crosskey and Robertson), *Rabimilis mirabilis* (Brady) and *Cytheropteron montrosiense* (Brady, Crosskey and Robertson). Other macrofossils recovered include bones of the common seal, *Phoca vitulina* L.

The reinvestigation of the site allowed a correlation to be established between the stratigraphy and the microfaunal distribution. The lowest marine deposits (bed 2) are characterized

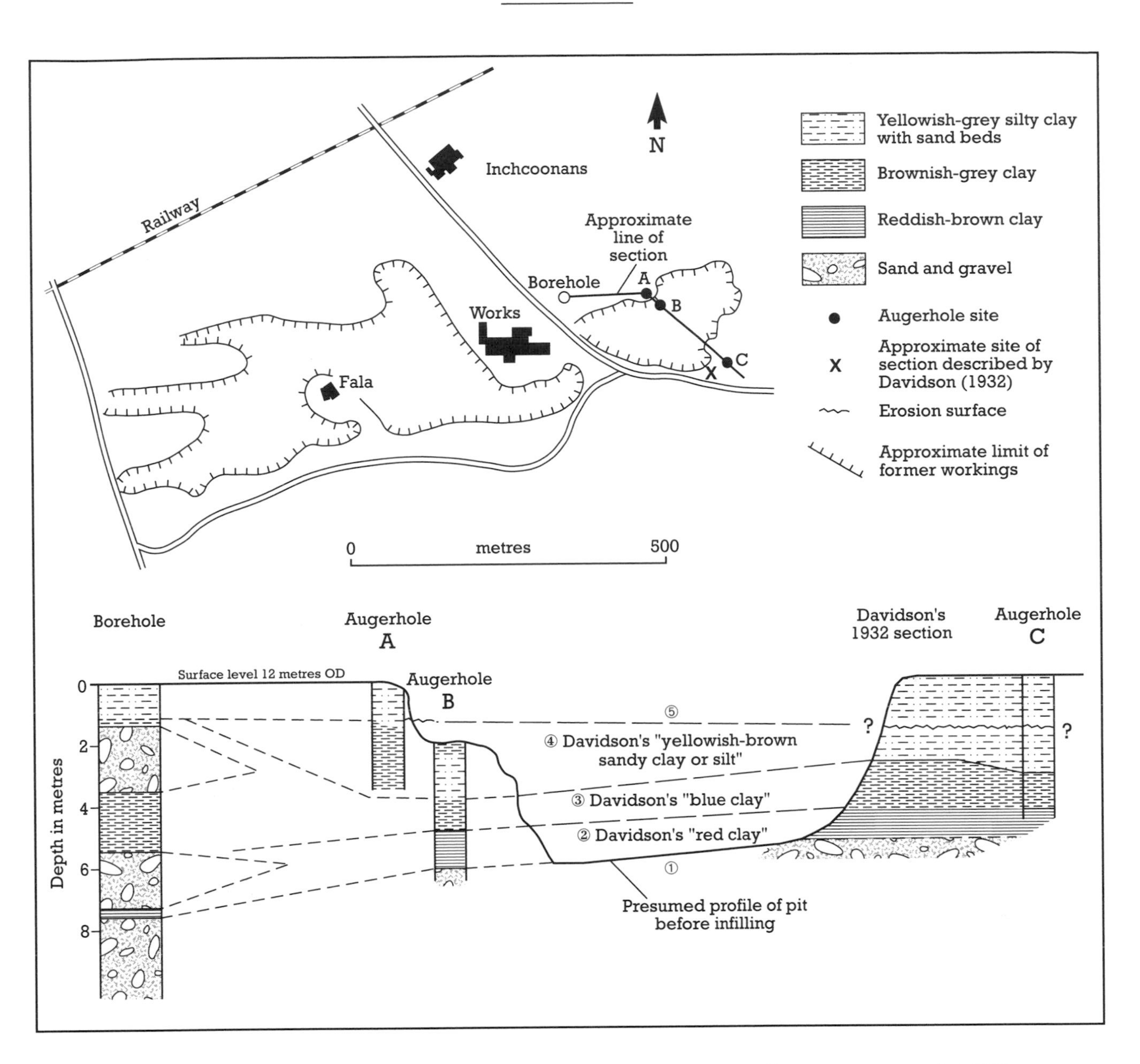

Figure 15.2 Inferred stratigraphy of the Errol beds at Inchcoonans claypit (from Paterson *et al.*, 1981).

by the above-mentioned ostracods and in the foraminiferal assemblage, *Elphidium clavatum* (Cushman) predominates over *Elphidium bartletti* (Cushman). In bed 3, *Elphidium bartletti* attains dominance and there is a marked reduction in the occurrence of the ostracods *Rabimilis mirabilis* and *Cytheropteron arcuatum* (Brady, Crosskey and Robertson). *Krithe glacialis* disappears in bed 4, and *Elphidium clavatum* regains dominance in the foraminiferal assemblage. Bed 5 was barren of both micro- and macrofauna. Paterson *et al.* (1981) have suggested that it is not part of the Errol beds but should be correlated with the Powgavie Clays, a later deposit lacking the high-arctic indicator species: these clays were intersected in boreholes in the Carse of Gowrie to the east of Errol.

The variation in the microfauna appears mainly to reflect variations in salinity, bed 3 with the dominance of *Elphidium bartletti* being indicative of more fully marine conditions than either beds 2 or 4. The reduced salinity of bed 2 may be due to meltwater influx from the retreating icesheet, whereas bed 4 may have been deposited in shallower water as sea level fell consequent upon isostatic uplift.

At the Gallowflat claypit (surface altitude about 25 m OD), deposits similar to those at Inchcoonans are revealed. McManus (1972) described the sedimentary characteristics of the deposits, noting that they comprised laminated silty clay with thin sand layers. The lower part of the succession shows rhythmic bedding in silty clays or clayey silts, whereas the upper part comprises fine and medium sands. Pebbles and boulders up to 1.3 m in size occur as drop stones in the succession, and calcareous concretions are also present (Duck, 1990). The erratic material includes dolerite, metamorphic rocks and Old Red Sandstone sediments, sometimes striated (MacGregor, 1973; Duck, 1990). The only macrofossils recovered were *Portlandia* spp., and the microfauna consisted of the ostracod *Cythere montrosiense* (Brady, Crosskey and Robertson) and the Foraminifera *Elphidium clavatum* and *Cassidulina obtusa* (Williamson), *E. clavatum* being dominant (Paterson *et al.*, 1981).

Interpretation

By analogy with modern polar environments, McManus (1972) considered that the Errol beds were deposited in association with seasonal pack-ice and icebergs in water depths of up to 100 m. Analysis of the calcareous concretions led Duck (1990) to support McManus's suggestion that the clays were deposited from flocculated suspensions in a strongly stratified water body and in the absence of significant currents. As the water shallowed and became more mixed, sedimentation of coarser particles occurred.

The earliest of the Errol beds were deposited when relative sea level was at least 28 m OD in this area. During the subsequent fall in sea level particularly pronounced terraces were formed at 24–25 m OD, and these have been correlated with the Main Perth Shoreline (Cullingford, 1972, 1977). This shoreline was formed when ice lay some distance to the west of the present area (see Almondbank) and, it has been argued (Paterson *et al.*, 1981), at approximately the time of cessation of deposition of the Errol beds and the start of deposition of the Powgavie Clays, with their fauna indicative of a milder climate. The shoreline has been traced widely along the coasts of east-central Scotland (Sissons *et al.*, 1966; Smith *et al.*, 1969) and has a marked tilt to the south-east of 0.43 m km^{-1} resulting from isostatic uplift subsequent to the formation of the shoreline. The lower terraces in the Inchcoonans and Gallowflat area have been correlated with shorelines formed as sea level continued to fall. These terraces have successively lower tilts, reflecting the decrease in isostatic uplift during the period of their formation (Cullingford, 1972, 1977).

The Errol beds have not been dated directly. Their base is clearly diachronous as they were laid down in front of a retreating ice-sheet. On the basis of shoreline gradient calculations (Andrews and Dugdale, 1970) the start of deposition of the Errol beds can be placed at as early as 17,000 BP (Sutherland, 1984a), and if deposition ceased at the time of change from arctic to more boreal conditions as the oceanic polar front retreated north of the Scottish coast (Ruddiman and McIntyre, 1973, 1981b; Peacock, 1981b, 1989b), then this may be placed at approximately 13,500 to 13,000 BP.

The Errol beds are an important element in the Late Devensian stratigraphy of Scotland. They are typified by a high-arctic fauna and their wide distribution along the east coast of Scotland indicates that the majority of the last ice-sheet had melted prior to the climate amelioration at the opening of the Lateglacial Interstadial at around 13,000 BP. They are the equivalents, now on land, of the St Abbs Formation of the North Sea Basin (Stoker *et al.*, 1985). There are apparently few deposits on the west coast that may be correlated with the Errol beds, with two exceptions possibly at Stranraer (Brady *et al.*, 1874) and in the North Minch (Gregory, 1980; Graham *et al.*, 1990), and their absence from much of the west coast has been attributed to these areas being covered by ice during the period of their deposition (for example, Sissons, 1965; Peacock, 1975c). This hypothesis, however, awaits full substantiation (Sutherland, 1984a). Inchcoonans and Gallowflat constitute the principal reference area for the Errol beds, where the most abundant macro- and microfauna has been recovered and where the deposits have been examined in most detail in recent years.

The former pit at Inchcoonans, which yielded the most abundant faunas is now infilled, but the deposits can still be examined in sections at Gallowflat claypit.

Conclusion

Inchcoonans and Gallowflat are reference localities for a sequence of fossiliferous estuarine deposits (Errol beds) restricted almost entirely to eastern

Scotland and formed during the melting of the last ice-sheet, about 17,000–13,000 years ago. These sediments and the fossil fauna (marine mollusc shells) they contain, provide important evidence for marine environmental conditions at this time. In particular, they indicate that the estuarine waters were high-arctic in character.

CAREY

R. A. Cullingford

Highlights

The deposits exposed in the river-bank section at Carey include a sequence of estuarine sediments and buried peat. They provide important evidence for changes in sea level and coastal environmental conditions in eastern Scotland during the Holocene. In particular, they allow a rare opportunity to study and date the Main Buried Shoreline.

Introduction

The site at Carey (NO 173171) is an exposure on the south bank of the River Earn, 8 km south-east of Perth. It is important for the study of early and middle Holocene relative sea-level changes and associated environmental changes in an area affected by glacio-isostatic recovery. The stratigraphy of the estuarine sediments revealed in this and similar exposures nearby excited early scientific interest, the first clear descriptions being those of Taylor (1792) and Duncan (1794). The nature and scientific significance of the evidence revealed at Carey were discussed in a wider context by Buist (1841), Jamieson (1865) and Geikie (1881, 1894). The Carey site has figured prominently in morphological, stratigraphic and palaeobotanical studies by Cullingford (1972) and Cullingford *et al.* (1980), and is one of the few locations where the age of the Main Buried Shoreline has been determined by radiocarbon dating.

Description

Although in its broader Scots usage the term 'carse' denotes an alluvial flat or river floodplain, it has also long been applied more specifically to clay flat-lands composed of estuarine deposits. These carselands extend with impressive flatness over an area of about 18 km^2 in Lower Strathearn. They are backed by the Main Postglacial Shoreline, the local altitude of which is 9.8–10.2 m OD, and which has a radiocarbon age (determined elsewhere in eastern Scotland) in the range 6800–5700 BP (Cullingford *et al.*, 1991). The sinuous tidal portion of the River Earn crosses the carselands in a 6 m deep trench, in the cliffed walls of which are exposed the three main elements of the carseland stratigraphy (Cullingford *et al.*, 1980, 1989a):

4.	Estuarine silty clay and clayey silt (carse)	>6.0 m
3.	Terrestrial peat	0.59 m
2.	Micaceous fine to medium sand (estuarine buried beach deposits)	0.9 m
1.	Coarse sand with fine gravel	?

At the base of the succession is an unknown thickness of coarse sand with fine gravel (bed 1), of presumed fluvial origin, which can only be inspected at low tide. The sandy estuarine deposits (bed 2) consist of bedded, micaceous, fine to medium sand, coarser than the silty fine sand that typifies the buried estuarine materials in the area. The surface of these deposits has been shown to occur as a series of terraces, separated by bluffs, forming a series of buried beaches or estuarine flats. The base of each bluff represents a buried raised shoreline. The surface of the sands at Carey lies at an altitude of 3.2 m OD and the site is located close to a buried shoreline at that altitude. No faunal remains have been found in these deposits at Carey or elsewhere in Lower Strathearn.

The sub-carse peat (bed 3) consists of sedges and grasses, with occasional mosses and abundant woody remains. Macroscopic remains of non-arboreal plants include stems, leaves, roots, and seeds of various marsh and heath plants, including *Carex*, *Sarothamnus* and *Equisetum*. Macroscopic remains of arboreal plants include bark, leaves, fruit, roots, branches, and trunks of *Alnus*, *Corylus*, *Betula*, *Pinus*, *Salix* and *Quercus*, the first three being of most common occurrence, and the last rarest. The peat is highly compressed, as shown by its toughness and by the oval cross-sections of flat-lying branches and twigs, and it 'readily splits into laminae, on the surface of which many small seeds . . . appear, together with occasional wing cases of beetles' (Geikie, 1894, p. 292). The transition with the overlying carse

Table 15.1 Radiocarbon dates on the buried peat layer at Carey

Sample	Altitude OD (m)	Date (^{14}C years BP)	Laboratory number
Bottom 0.01 m of peat	3.19	9640 ± 140	I–2796
Bottom 0.04 m of peat	3.19	9524 ± 67	SRR–72
Top 0.01 m of peat	3.78	7605 ± 180	NPL–127
Top 0.04 m of peat	3.78	7778 ± 55	SRR–71

deposits is gradual, the top few centimetres of peat being silty, and the basal 0.3–0.5 m of silts and clays being heavily charged with both horizontal black streaks of vegetal material and plant stems passing vertically upwards from the peat. The base of the peat is more sharply defined, but roots penetrate from it into the sands below. The thin (0.005–0.05 m) iron pan between the peat and underlying sands is a highly localized consequence of concentrated groundwater flow in the immediate vicinity of the river cliff, for more than 250 boreholes sunk throughout the Lower Strathearn carselands show it to be generally absent. Cullingford *et al.* (1980) obtained radiocarbon dates from the top and bottom of the peat bed (Table 15.1).

The estuarine clays (bed 4) are bluish-grey when freshly dug, but quickly turn brown on exposure to the air. There is, in general, no distinct stratification, and stones are absent except occasionally near the rising ground at the edge of the carselands. There are occasional thin layers and lenses of sand. Plant remains, chiefly of reeds, are common throughout the deposit and are often fetid. No shells of estuarine molluscs have been recorded in the carse deposits of Lower Strathearn, though they occur in the Carse of Gowrie, where a sparse estuarine microfauna has also been identified (Paterson *et al.*, 1981).

Detailed pollen and diatom analyses have been carried out at Carey by P. Gotts (unpublished data). The pollen record established by analysis of samples at 0.01–0.02 m intervals throughout the peat and the upper 0.07 m and lower 0.06 m of underlying and overlying estuarine sediments predominantly reflects local vegetation changes, but does also have regional indicators. High values for Gramineae occur in the lower peat coincident with the macroscopic remains of *Phragmites*, accompanied by even higher values for Cyperaceae, thus representing a typical reed-swamp environment. Grass pollen is dominant throughout most of the peat, with very few other herb taxa represented. At the lower transition there are continuous curves for Cruciferae and *Filipendula* and occasional grains of aquatics, such as *Alisma*, *Lemna* and *Equisetum*, but very little unequivocal evidence of vegetation succession from saltwater to freshwater environments. At the upper transition to the carse silts and clays there is more evidence for the rising water table with a peak for *Typha latifolia*, but overall there is the same lack of variety in taxa indicative of changing hydroseral communities. At the peat/sand transition, dated to 9640 ± 140 BP (I–2796), there is a peak in *Betula*, with continuous curves for *Juniperus* and *Salix*. This slightly late date for the presence of juniper suggests that in the birch woodland which characterized eastern Scotland in the early Holocene, juniper was able to survive in marginal environments, as at Carey. With the immigration of hazel, demonstrated in almost all pollen diagrams in Scotland as a significant rise in *Corylus/Myrica* pollen, juniper disappears. This is dated to 8740 ± 55 BP (SRR–1392), which agrees well with other dates for this event (Huntley and Birks, 1983). Locally, close to the peat, hazel would have taken advantage of the marginal drier environment, with the occasional presence of birch and *Salix* on the slightly wetter margins. Elements of mixed oak forest are only represented to any extent in the peat/silt–clay transition and in the carse clay itself. It seems likely that oak woodland would have been present in Strathearn before 7600 BP, and the pollen record probably represents the local exclusion of pollen of mixed oak forest provenance by the surrounding hazel. With the change in sedimentary environment this pollen was then brought to Carey in the estuarine clays.

The diatom assemblage of the basal 0.12 m of carse clays and silts is dominated by several species of *Fragilaria*, averaging over 70% of the total count (1000 valves). Diatoms are abundant only within and above the peat/silt–clay transition, and the micaceous sand deposits contain only small numbers, which include both marine and brackish-water taxa.

Interpretation

It may be inferred from the sedimentary and palaeobotanical information above that the abandonment of the buried beach deposits by the sea was followed without a significant break by the colonization of their surface by terrestrial vegetation, which eventually took the form of a peat bog. The top of the peat in turn represents a former land surface that was gradually inundated by rising estuarine waters in which the silt and clay deposits accumulated.

The base of the peat (altitude 3.2 m OD) at two locations along the Carey exposure has been radiocarbon dated at 9640 ± 140 BP (I–2796) and 9524 ± 67 BP (SRR–72) (Cullingford *et al.*, 1980), giving the approximate date of initiation of peat growth following the withdrawal of the sea. As the Carey exposure is located close to the buried shoreline (altitude 3.2 m OD), the radiocarbon dates must relate closely to the abandonment of the latter. The buried shoreline is believed to correlate with the Main Buried Shoreline of the Forth Valley, which has been similarly dated at about 9600 BP (Sissons, 1983a). The top of the peat at Carey (present altitude 3.8 m OD) gave radiocarbon ages of 7605 ± 180 BP (NPL–127) and 7778 ± 55 BP (SRR–71), dating the onset of peat burial beneath the silt and clay. Use of peat-top dates in constructing a relative sea-level curve requires account to be taken of the peat compaction that accompanied and followed burial by the carse deposits, and a method for estimating compaction was employed at Carey and other sites in Lower Strathearn (Cullingford *et al.*, 1980).

A recently published account of a site nearby at Wester Rhynd (Cullingford *et al.*, 1989a) has further enhanced the importance of Lower Strathearn for sea-level studies in eastern Scotland. The results have suggested 8765 ± 75 BP (GU–1250) as an approximate date for the abandonment of the local equivalent of the Low Buried Beach of the Forth Valley, and have also demonstrated the presence of two brief marine incursions shortly after 8565 + 85 BP (GU–1518) and between 8485 ± 80 BP (GU–1517) and 8510 ± 85 BP (GU–1516), which are consistent with storm surge or tsunami events (Cullingford *et al.*, 1989a). The latter are earlier than the widely recognized event at about 7000 BP in eastern Scotland (Smith *et al.*, 1985a; Dawson *et al.*, 1988; Haggart, 1988b; Long *et al.*, 1989a) and raise the possibility that the record of high-magnitude events is more detailed than previously recognized. Dawson *et al.* (1989) suggested that the 7000 BP event might be represented by a layer of fine silty sand at the base of the carse deposits at Wester Rhynd, but Cullingford *et al.* (1989b) argued that this layer was of very limited extent and that the 7000 BP event might be represented by a sand layer higher in the carse deposits in the Carse of Gowrie on the north side of the Tay Estuary. Further work in Lower Strathearn and the Carse of Gowrie should help to clarify the sequence and depositional environments of these high-magnitude events.

Lower Strathearn is important for studies of relative sea-level change in eastern Scotland, and evidence from the area, including Carey, has allowed the construction of relative sea-level and uplift curves for the Lateglacial and early Holocene (Cullingford *et al.*, 1980; Paterson *et al.*, 1981).

A Lateglacial and early Holocene phase of generally falling relative sea level was punctuated by stillstands and/or transgressive episodes resulting in the formation of now buried estuarine flats (buried beaches) in descending order of age and altitude, the abandonment of the flats being followed by the growth of vegetation, including peat. Later, relative sea level rose again, causing the progressive burial of successive peat-covered flats by the carse deposits, and culminating in the formation of the extensive carseland surface visible today (see also Western Forth Valley). The Carey exposure has afforded vital evidence in dating and elucidating the nature of the environmental changes that accompanied these relative sea-level changes, and is still the only known site in eastern Scotland where exposure of the peat-covered Main Buried Beach deposits occurs close to the former shoreline, allowing accurate dating of the latter. With its thick sub-carse peat and clearly displayed sequence, the Carey site is of great value for demonstration purposes, and it is likely that, with improved analytical techniques in the future, it will have an important research role to play in the further study of Holocene relative sea-level changes.

Conclusion

The sediments at Carey provide important evidence for interpreting the sequence of sea-level changes that occurred during Holocene times (the last 10,000 years) in eastern Scotland. In particular, it allows a rare opportunity for dating the early Holocene Main Buried Shoreline (about 9600 years ago). In addition, subsequent changes in the coastal environment during the middle Holocene have been revealed by detailed pollen, diatom and sediment analyses. Carey is therefore an integral component of the network of reference sites for sea-level history.

SILVER MOSS

D. E. Smith

Highlights

The sub-surface deposits at Silver Moss include a sequence of estuarine and buried peat sediments which provide a detailed and dated record of sea-level and coastal changes during the Holocene. They are particularly significant for studying the Main Postglacial Transgression and a major coastal flood which occurred in eastern Scotland during the middle Holocene.

Introduction

Silver Moss (NO 450233) is a small peat bog in St Michael's Wood, 8 km north of St Andrews, in Fife. It occupies a gully which once formed a narrow embayment in the coastline when relative sea level stood at the Main Postglacial Shoreline in that area. Silver Moss contains a sequence of deposits which record the culmination of the Main Postglacial Transgression, together with a unique coastal flood, possibly a storm surge or tsunami. The site was first investigated by Chisholm (1971), and later studied by Morrison *et al.* (1981).

Description

In East Fife, the lower ends of the valleys of the Motray Water, Moonzie Burn and River Eden are occupied by extensive areas of raised estuarine sediments, which extend northwards towards the Tay estuary and southwards towards St Andrews, and which continue beneath coastal sand dunes to the east. These sediments, the local equivalent of the carse of central Scotland, consist of grey silty clay or clayey silt with lenses of sand. They underlie a remarkably flat surface which contrasts sharply with the rising ground inland. From the fossil content of these sediments (Chisholm, 1971) it is evident that they accumulated in a marine/estuarine environment. The break of slope at their inland limit is taken to mark a shoreline, which lies between 7 m and 9 m OD (Cullingford, 1972).

Towards the northern region of this carse area, near Craigie (NO 453243), the rising ground inland consists of Late Devensian raised beaches which are dissected by a number of gullies leading eastwards. These gullies reach the edge of the carse, and in the lower ends of some of them small peat bogs occur. The largest of these is Silver Moss, which lies in a gully draining through St Michael's Wood. From boreholes made near the mouth of the gully, Chisholm (1971) showed that the raised estuarine (carse) sediments to the east of the gully mouth continued into it to form a layer within the peat (Figure 15.3). He noted that east of the gully mouth the carse sediments include a sand layer, which appears at the base of the carse within the gully. He also identified a sand bar at the mouth of the gully (Figure 15.3), which appears to have formed contemporaneously with the carse deposit when the gully was an arm of the sea. Chisholm obtained radiocarbon dates on two 6 in. (0.15 m) thick samples of peat: the upper sample (above the clays and silts) gave 5830 ± 110 BP (St–3062) and the lower sample (below the sand at the base of the clays and silts), 7605 ± 130 BP (St–3063). He concluded that the clastic sediments in Silver Moss had accumulated between those dates, but did not exclude the possibility that the clays and silts east of the gully mouth and sand bar may have continued to accumulate for some time after the younger date.

Morrison *et al.* (1981) later undertook further stratigraphical investigations of Silver Moss. They were able to trace the inland limit of the clays and silts, which form a tapering wedge within the peat (Figure 15.3). They also proved the extent of the sand, which they described as a grey, micaceous, silty fine sand. They found that it occurs within the clays and silts east of the gully but forms a separate, tapering wedge below the clays and silts within Silver Moss (Figure 15.3). Within the peat the sand extends over 250 m farther up the gully than does the wedge of clays and silts, eventually reaching a higher elevation than the latter (Figure 15.3).

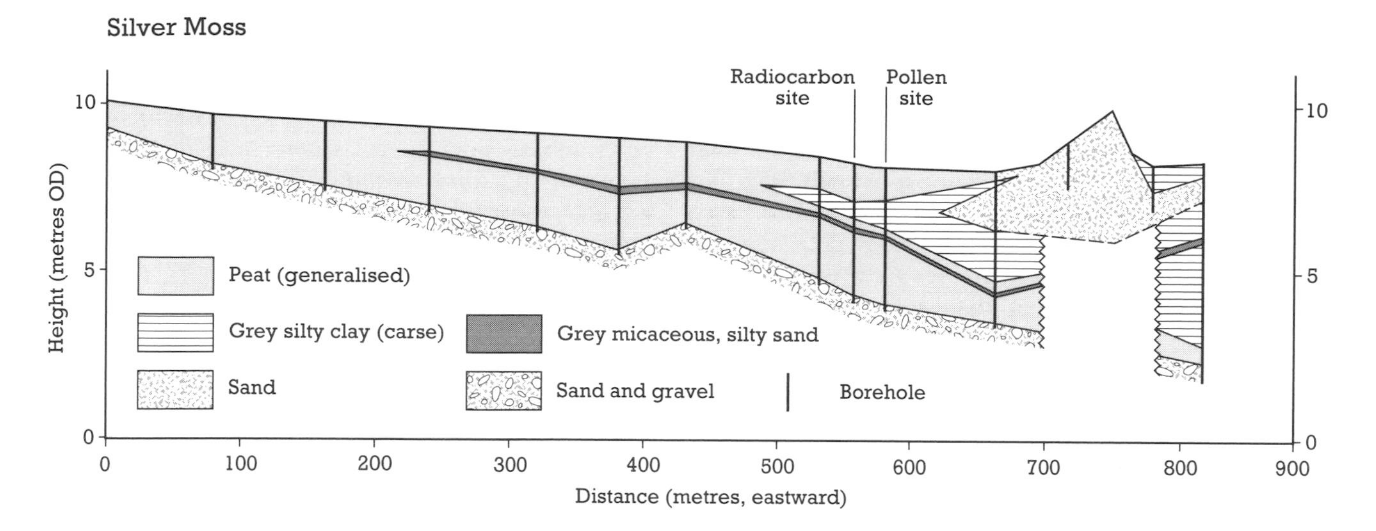

Figure 15.3 Silver Moss: section along the gully showing the sequence of sediments (from Morrison *et al.*, 1981).

Morrison *et al.* carried out pollen analysis through the sequence of deposits and also obtained further radiocarbon dates. From the pollen evidence, they found that peat had begun to accumulate in the early Holocene. The pollen sequences identified (Figure 15.4) were found to be similar to those of other Holocene coastal sites in eastern Scotland (Smith *et al.*, 1982, 1983). The gradual changes in the pollen record suggest that no breaks in sedimentation are present. High values of *Pinus* and *Quercus* pollen coincide with the wedge of clays and silts (Figure 15.4). Such increases, however, are common in marine sediments (Traverse and Ginsberg, 1966), and the presence of Chenopodiaceae in this layer is characteristic of a marine environment. However, Morrison *et al.* could find no changes in the pollen spectrum based on samples from the grey, micaceous, silty fine sand layer. The radiocarbon dates obtained by Morrison *et al.* (1981) are listed in Table 15.2.

Interpretation

The gully in which Silver Moss lies had probably been formed before the early Holocene, when the peat began to accumulate. It seems possible that the gully may have been cut during the Late Devensian under periglacial conditions, in a similar manner to other gullies in raised marine deposits in central Scotland (Sissons *et al.*, 1965). As the peat of Silver Moss accumulated, the lower end of the gully became inundated by a marine transgression, and a sand bar began to form at its mouth. During this time, the prominent layer of grey, micaceous, silty fine sand was deposited.

The radiocarbon dates obtained by Morrison *et*

Table 15.2 Radiocarbon dates at Silver Moss (from Morrison *et al.*, 1981)

Sample	Altitude (m) OD (surface = 8.30 m)	Date (^{14}C years BP)	Laboratory number
Bottom 0.01 m of surface peat	7.23–7.24	5890 ± 5	SRR–1331
Top 0.01 m of peat beneath grey silty clay ('carse')	6.75–6.76	7310 ± 100	SRR–1332
Bottom 0.01 m of peat above grey, micaceous, silty fine sand	6.38–6.39	7050 ± 100	SRR–1333
Top 0.01 m of peat below grey, micaceous, silty fine sand	6.19–6.20	7555 ± 110	SRR–1334

al. place the age of the sand at between 7555 ± 110 BP and 7050 ± 110 BP. Chisholm (1971), who did not identify the sand as separate from the clays and silts, nevertheless obtained a radiocarbon date of 7605 ± 130 BP for the base of the layer (and his figure 6 actually identifies organic material between the sand and the clays and silts above). It is likely, however, that the layer was deposited in the gully over a much shorter period than the radiocarbon evidence implies. Chisholm remarked on evidence of erosion of the peat at the base of the sand (which would therefore make the basal date at best a maximum age estimate); the dates of Morrison *et al.* are reversed in the middle of the sequence (see Table 15.2), implying an older date for the base of the peat above the layer. Elsewhere, in eastern Scotland, a similar bed of sand in a similar stratigraphic position has been found at a number of sites (see Western Forth Valley, Maryton, Dryleys and Barnyards) (Smith *et al.*, 1985a). At each of these other sites the bed appears to have accumulated rapidly, probably over a period of much less than 100 years. The radiocarbon ages at these sites range between 6900 and 7200 BP. Given the radiocarbon dates at Silver Moss, it seems likely that the sand at this location correlates with these other sand beds. It was originally thought that they are the deposits of a major storm surge in the North Sea which was particularly effective for having occurred at a time of rapidly rising relative sea level (Smith *et al.*, 1985a; Haggart, 1988b), but a recent interpretation suggests that they relate to a tsunami associated with a major submarine slide on the Norwegian continental slope (Dawson *et al.*, 1988; Long *et al.*, 1989a). It is interesting to note that a similar sand layer has been found quite close to Silver Moss, near Craigie (Haggart, 1978), though this has not been dated.

The clays and silts (carse) continued to accumulate after the sand was deposited but, as relative sea level fell, peat began to form on the surface. The radiocarbon dates obtained by Morrison *et al.* below and above the wedge of clays and silts in the gully probably embrace the culmination of this marine transgression. The date for the base of the layer may well be younger than 7310 ± 100 BP (SRR–1332) in view of the age reversal referred to above; the age for the top of the layer, at 5890 ± 95 BP (SRR–1331), agrees well with Chisholm's date of 5830 ± 110 BP (St–3062) at the same horizon. These dates indicate that the event involved was the Main Postglacial Transgression. Morrison *et al.* maintain that the date for the end of the deposition of the clays and silts probably applies to the wider local area, despite Chisholm's reservations, since there is very little difference between the local altitude of the carse surface beyond the gully (up to 8.3 m OD) and the altitude reached by the wedge of the clays and silts within the gully (7.9 m OD).

The sequence of deposits in Silver Moss contains an excellent record of relative sea level change during much of the Holocene. The sheltered nature of the gully ensured that the sedimentary record was relatively undisturbed, and that the evidence of two significant events was preserved.

The deposits which record the culmination of the Main Postglacial Transgression at this site will repay further study; notably, additional dating will contribute towards a better understanding of the wider pattern of diachroneity of the Main Postglacial Shoreline in Scotland. The grey, micaceous, silty fine sand layer is remarkably extensive in the moss, and provides an opportunity for reconstruction of the details of the event which led to its deposition. The layer can be identified so widely in the moss that its altitude and inland limit can be studied in great detail, and will provide evidence for the 'run-up' of the wave from which it was deposited. This should, in turn, provide evidence for the magnitude of the tsumani waves as anticipated in Bugge (1983) and discussed in Long *et al.* (1989a).

Conclusion

The deposits at Silver Moss provide a detailed record of sea-level changes during the Holocene (the last 10,000 years). In particular, they are significant for studies of the Main Postglacial Transgression (an encroachment of the sea on to the land that occurred around 6000 years ago as ice age glaciers in North America and Scandinavia melted away releasing large volumes of meltwater into the oceans) and a major coastal flood that occurred about 7000 years ago. Silver Moss forms

Figure 15.4 Silver Moss: relative pollen diagram showing selected taxa as percentages of total pollen and spores (from Morrison *et al.*, 1981).

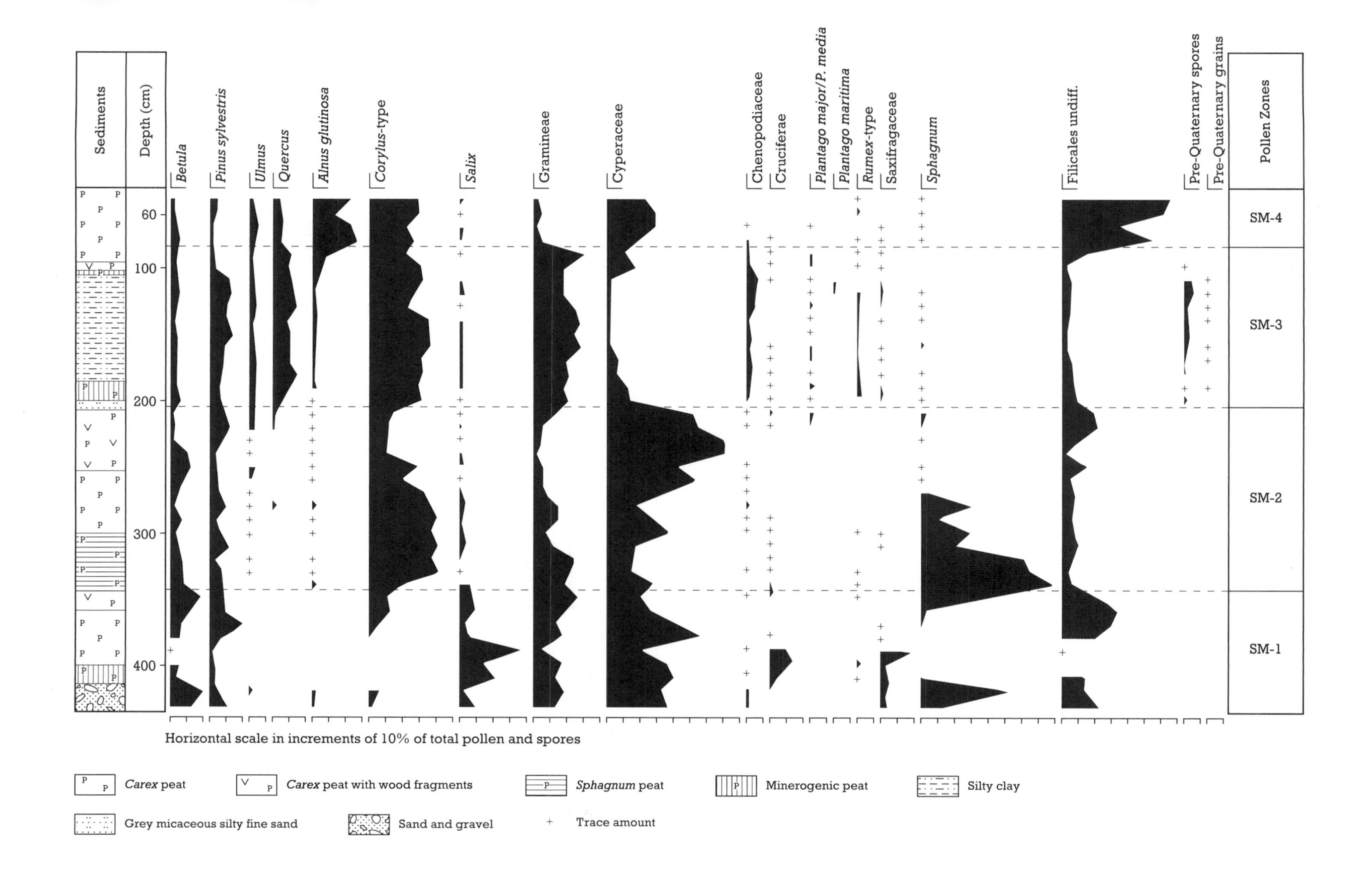
Sediments
Depth (cm)
60
100
200
300
400
Betula
Pinus sylvestris
Ulmus
Quercus
Alnus glutinosa
Corylus-type
Salix
Gramineae
Cyperaceae
Chenopodiaceae
Cruciferae
Plantago major/P. media
Plantago maritima
Rumex-type
Saxifragaceae
Sphagnum
Filicales undiff.
Pre-Quaternary spores
Pre-Quaternary grains
Pollen Zones
SM-4
SM-3
SM-2
SM-1
Horizontal scale in increments of 10% of total pollen and spores
Carex peat
Carex peat with wood fragments
Sphagnum peat
Minerogenic peat
Silty clay
Grey micaceous silty fine sand
Sand and gravel
Trace amount

part of a network of sites for establishing the wider extent and timing of these events.

PITLOWIE
D. E. Smith

Highlights

The sub-surface deposits in a series of gullies at Pitlowie comprise a sequence of estuarine sediments and buried peat. These have been intensively studied and dated, allowing a detailed reconstruction of the progress of the Main Postglacial Transgression close to its maximum and the subsequent fall in relative sea level.

Introduction

Near Pitlowie, 8 km east of Perth on the north side of the Tay estuary, Late Devensian raised marine deposits are dissected by a small gully system (NO 205230), now largely dry. The system forms part of a larger system of gullies originally studied by Sissons *et al.* (1965) and Morrison *et al.* (1981). These studies show that the gullies were largely formed during the Late Devensian, and that their lower ends contain Holocene sediments which record the Main Postglacial Transgression in the area. The Pitlowie gullies have been examined in detail by Smith *et al.* (1985b), who demonstrate that the final stages of the Main Postglacial Transgression and the beginning of the subsequent regression in the area are recorded in detail in the sediments.

Description

On the northern side of the Firth of Tay, Late Devensian raised marine deposits form two well-defined surfaces extending for several kilometres on either side of the village of Glencarse (NO 197217). Both surfaces slope eastwards: the upper one from 20.1 m OD to 19.4 m OD, and the lower one sloping eastwards from 16.1 m OD to 14.6 m OD (Cullingford, 1972, 1977). These surfaces overlie fine to medium-grained sands, in which excellent fossil cryoturbation structures have been observed (Smith *et al.*, 1985b). East of Glencarse, the features are very extensively dissected by a system of gullies. The system consists of a main gully, drained by the Pow of Glencarse, fed by a number of tributary gullies, most of which are dry (Figure 15.5). Many of the gullies contain small peat bogs. The Pow of Glencarse runs south-westward to join the Tay estuary, which in this area is surrounded by raised estuarine sediments forming the remarkably flat surfaces of the Tay carselands. The estuarine clays and silts extend into the gully system along the Pow of Glencarse, and occur in the mouths of most of the smaller gullies. The Pitlowie gullies lie at the head of the system.

Sissons *et al.* (1965) concluded that the larger gully system had been initiated during the Late Devensian following exposure of marine deposits when relative sea level fell. They showed that in addition to peat being present in the floors of the gullies, it also lay beneath the estuarine clays and silts there, and they therefore concluded that the gullies had largely ceased to form by the time peat accumulation started. Since it was likely that the peat had begun to accumulate early in the Holocene, they concluded that the gullies formed during the Late Devensian. They maintained that although the processes were not known for certain, it seemed possible that the gullies had been formed under periglacial conditions with a high surface runoff.

Morrison *et al.* (1981) examined the sediments within the larger gully system. They showed that the estuarine clays and silts were extensively underlain by peat, and that in the tributary gullies they formed a wedge within small peat bogs on the gully floors (Figure 15.6). Pollen analysis through the sequence of basal peat, grey silty clay (carse), and surface peat in the main Pitlowie gully near Hole of Clien farm (NO 204234), showed that the peat had begun to accumulate in the early Holocene, and that the episode of silty clay sedimentation occurred in the middle Holocene. The pollen record disclosed evidence that the silty clays were indeed marine, with Chenopodiaceae pollen associated with them as well as the high values of *Pinus* and *Quercus*, pollen characteristic of selective preservation in a marine environment (Traverse and Ginsberg, 1966). Morrison *et al.* (1981) obtained radiocarbon dates from samples of peat at the upper and lower contacts with the silty clays of the Hole of Clien gully and in nearby Glencarse gully. At Hole of Clien, 0.02 m thick samples gave, respectively, 6170 ± 90 BP (SRR–1510) and 7500 ± 90 BP (SRR–1511) for the upper and

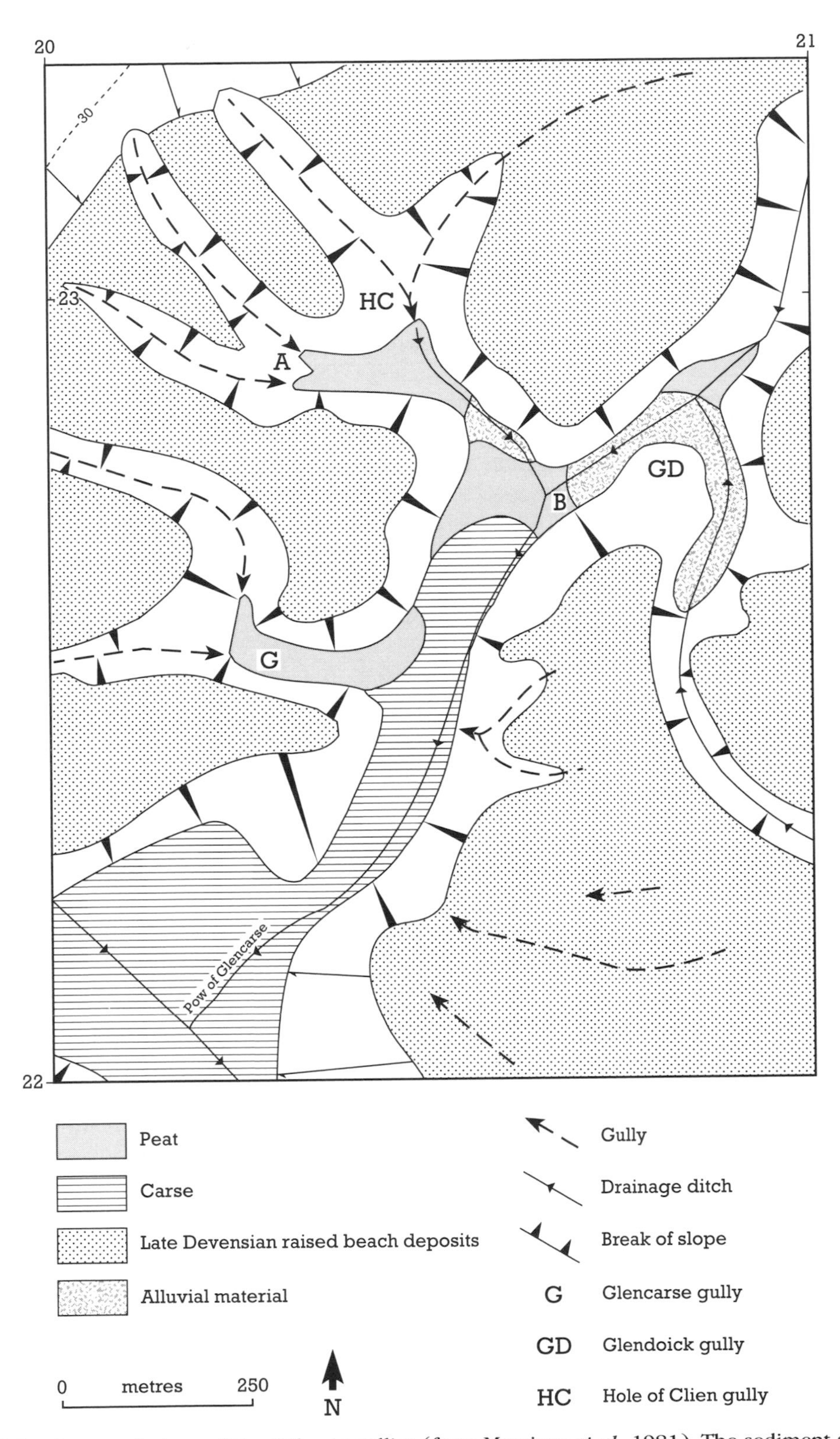

Figure 15.5 Geomorphology of the Pitlowie gullies (from Morrison *et al.*, 1981). The sediment sequence along the line AB is shown in Figure 15.6.

lower contacts. At Glencarse, 0.01 m thick samples gave, respectively, 6083 ± 40 BP (SRR–1151) and 6679 ± 40 BP (SRR–1150) for the upper and lower contacts. Morrison *et al.* concluded that the Main Postglacial Trangression had culminated in the area between 6679 ±

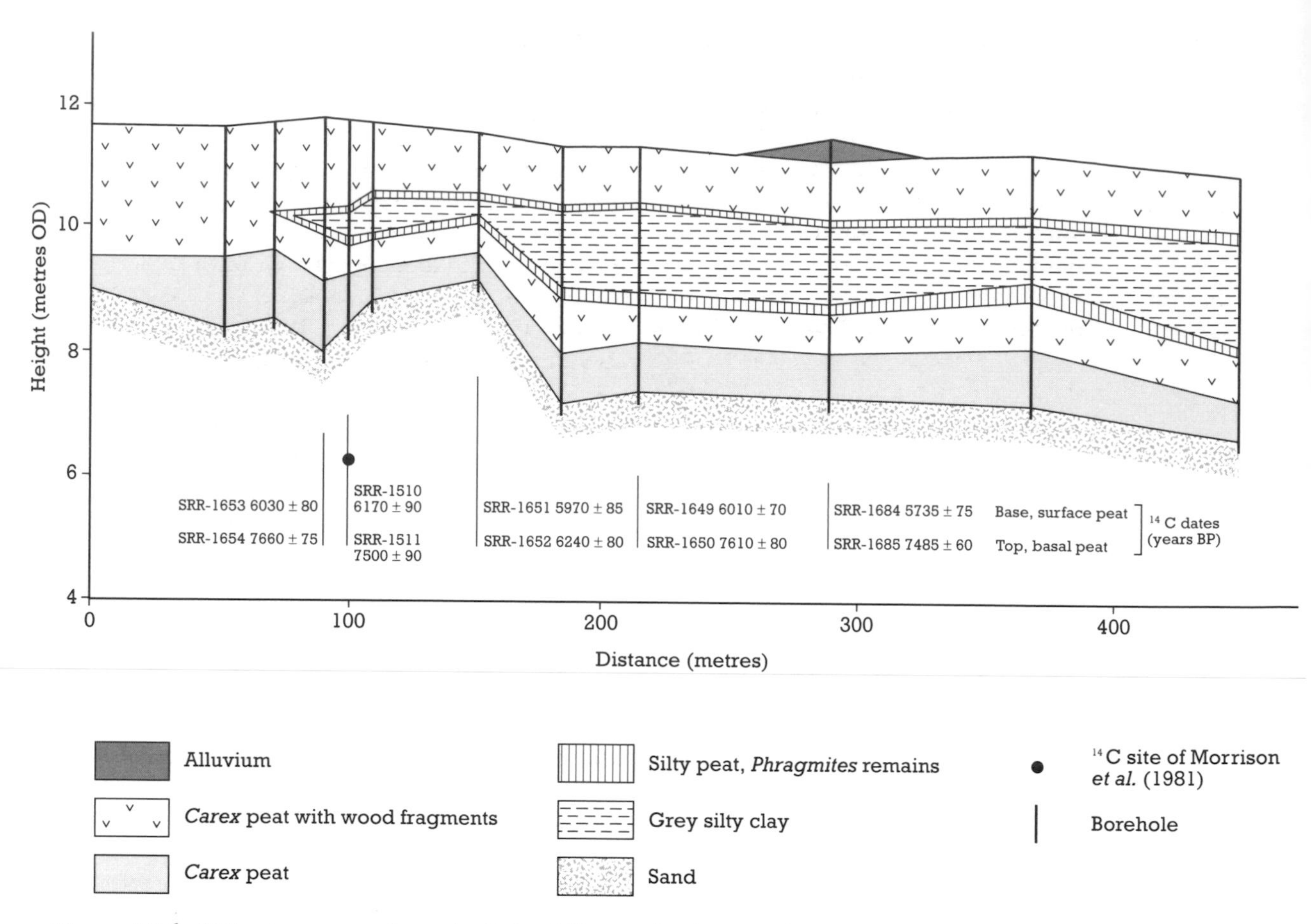

Figure 15.6 Pitlowie: section along the Hole of Clien gully showing the sequence of sediments and radiocarbon dates (from Smith *et al.*, 1985b).

40 BP (the younger of the two dates for the lower contact) and 6100 ± 35 BP (the weighted mean of the statistically indistinguishable dates on the base of the surface peat at the two locations).

Smith *et al.* (1985b) confined their detailed study to the Holocene sediments in the Pitlowie gullies. They determined the stratigraphy closely, and in the Hole of Clien gully studied the geochemistry of the sediments in an effort to determine whether or not deposition had been continuous. In addition they obtained eight further radiocarbon dates at the peat/silty clay interface along the wedge of the carse sediments. The dates and the detailed stratigraphy they obtained along the Hole of Clien gully are shown in Figure 15.6. Their studies of the geochemistry of the silty clay showed no major changes in organic carbon, $\delta^{13}C_{PDB}$, aluminium or magnesium, and they concluded that no hiatus in deposition within the silty clays was indicated.

Interpretation

Smith *et al.* (1985b) concluded that since no hiatus was indicated within the silty clays (carse), and since the earlier pollen work of Morrison *et al.* (1981) demonstrated no hiatus in the pollen record through the Holocene deposits, it was likely that a continuous depositional sequence obtained in the Hole of Clien gully. The radiocarbon dates indicated an initially rapid invasion of the gully by the Main Postglacial Transgression at around 7600 BP; culmination of the transgression possibly between 6240 ± 80 BP (SRR–1652) and 6170 ± 90 BP (SRR–1510), and regression from the mouth of the gully by 5735 ± 75 BP (SRR–1684). The sea had thus occupied that part of the gully where the silty clays occur for nearly 2000 years, yet sedimentation, though apparently continuous, was relatively slight. Little sediment was evidently derived from the land, which emphasizes the lack of gully development during that time.

The Pitlowie gully system was developed during the Late Devensian, possibly under periglacial conditions. Cut into sands and fine gravels which frequently display fossil periglacial structures in section, the gully system is one of the best examples of its type in eastern Scotland. The sediments which lie in the gullies record an apparently complete history of vegetational and environmental change in the area during the Holocene. The progress of the Main Postglacial Transgression and subsequent regression are recorded in detail over a period of around 2000 years in the Hole of Clien gully. Other aspects of the Main Postglacial Transgression are recorded at Silver Moss, Maryton and Western Forth Valley on the east coast and Newbie, Redkirk Point and Dundonald Burn on the west coast.

It is probably the sheltered nature of this gully system which has enabled the Holocene marine sediments to be preserved in such detail and with such apparent continuity. It is the fine and detailed preservation of the Holocene sequence which distinquishes this site and will make it a focal point for detailed studies of relative sea-level change in Scotland in the future.

Conclusion

The sediments at Pitlowie are important for establishing the history of changing sea levels in eastern Scotland during the Holocene (the last 10,000 years). They have been studied in considerable detail and provide a record of changes in the coastal environment during the middle Holocene. In particular, they allow a detailed reconstruction of the Main Postglacial Transgression (see Silver Moss above) and the subsequent fall in relative sea level. Pitlowie is therefore a valuable reference site for studies of this event.

KINCRAIG POINT

J. E. Gordon

Highlights

The Lateglacial and Holocene raised shorelines at Kincraig Point are notable as erosional features formed in bedrock. They form a striking sequence of landforms that contrasts with the sedimentary evidence for coastal changes recorded elsewhere in the estuaries of eastern Scotland.

Introduction

Kincraig Point (NT 465998) is a headland on the south coast of Fife, 2.5 km west of Elie. It is notable for its geomorphology, demonstrating a series of raised shorelines (shore platforms) cut into the western flank of the headland. The interest and striking appearence of those features has long been noted (Wood, 1887; Geikie, 1901; Geikie, 1902; MacGregor, 1973) and they have also featured frequently in book illustrations (Geikie, 1902, figure 66; MacGregor, 1973, figure 18; Forsyth and Chisholm, 1977, plate 1 (frontispiece); Price, 1983, plate 4.1B). The position of the shorelines in the wider regional pattern of Lateglacial sea-level changes in eastern Scotland has been assessed by Cullingford and Smith (1966).

Description

Although frequently described as beaches (Wood, 1887; Geikie, 1902; MacGregor, 1973) the raised shorelines at Kincraig Point are in fact erosional features (Geikie, 1901). They consist of four raised rock benches cut into the volcanic agglomerate of the headland at approximately 4 m, 11 m, 22 m and 24 m OD and are veneered with sand and shells (Cullingford and Smith, 1966). Also present is an intertidal shore platform. The raised shorelines at Kincraig Point are particularly prominent when viewed from the west (see illustrations in Geikie, 1902; MacGregor, 1973; Forsyth and Chisholm, 1977; Price, 1983).

Interpretation

Cullingford and Smith (1966) regarded the upper three raised shorelines as Lateglacial in age, the lowest as Holocene. The two highest shorelines form part of the sequence of early Lateglacial shorelines in East Fife but were excluded by Cullingford and Smith (1966) from their detailed height analyses for the area because of uncertainty regarding the relationships of shore platforms to relative sea-level heights at the time of their formation.

Andrews and Dugdale (1970) originally calculated the ages of the East Fife shorelines to be between 18,250 and 15,100 BP, but more recent information suggests that they formed between *c.*

16,000 BP and 14,000 BP (Sutherland, 1991a). This revised estimate is more consistent with evidence that the last ice-sheet was at its maximum extent *c.* 18,000–17,000 BP (cf. Sissons, 1976b; Cullingford and Smith, 1980).

The third highest shoreline was interpreted as part of the Main Perth Shoreline. The latter is one of the most prominent raised shorelines in eastern Scotland, sloping E17°S at 0.43 m km^{-1}, and was formed after the East Fife shorelines at a time when the last ice-sheet had retreated west up the Forth valley (Sissons and Smith, 1965a; Smith *et al.*, 1969; Sissons, 1976b; Cullingford, 1977). The intertidal platform may be part of the platform that is extensively developed in eastern Scotland (see Milton Ness and Dunbar), and which Dawson (1980a) has correlated with the Low Rock Platform of western Scotland (see Northern Islay and the West coast of Jura).

Examples of raised shorelines are widespread in eastern Scotland, but Kincraig Point provides a particularly clear example of a suite of such features which has a very striking 'staircase' appearance when seen in profile. As erosional features they are exceptional among Lateglacial and Holocene raised shorelines in eastern Scotland (Sissons, 1967a), which typically comprise beaches or estuarine terraces. The prominence of the features at Kincraig Point probably relates to the relatively weak nature of the volcanic agglomerate into which they have been cut. As clear geomorphological features developed on an exposed headland (see also Milton Ness), the shorelines at Kincraig Point complement the interest of other sites in eastern Scotland (see Western Forth Valley, Silver Moss, Dryleys and Maryton) where the emphasis is on sedimentary and geochronological evidence for relative sea-level change in estuarine environments.

Conclusion

Kincraig Point is important for Quaternary coastal geomorphology. It is a good example of a sequence of raised shorelines eroded in bedrock following the retreat of the Late Devensian ice-sheet between about 16,000 and 13,000 years ago: each indicating a different sea level. The shorelines have a striking landscape appearance and complement the interest of sites selected for the sedimentary evidence that they provide for sea-level changes in more estuarine situations.

BLACK LOCH

G. Whittington, K. J. Edwards and P. R. Cundill

Highlights

The sediments which infill the topographic basin at Black Loch have provided a great wealth of information on the palaeoecological and palaeoenvironmental changes in eastern Scotland during the Lateglacial and Holocene. Variations with depth in pollen content and type in the sediments have been studied in considerable detail, and extensive use has been made of radiocarbon dating, so that Black Loch is an indispensable reference site for future work in this field.

Introduction

Black Loch (NO 261150) lies in the Ochil Hills of northern Fife at an altitude of 90 m OD. It has a relatively small surface area (0.015 km^2) and a maximum depth of approximately 3 m. The sediments of Black Loch have been accumulating since Late Devensian times and their pollen and spore records have proved to be of considerable interest for the vegetational history of eastern Scotland (Whittington *et al.*, 1990, 1991a). No site comparable to this has been investigated in Fife (although the detailed study at Pickletillem (Whittington *et al.*, 1991b) is worthy of note), but of greater importance than this purely regional consideration is the fact that eastern Scotland in general is poorly served by sites which can contribute to an understanding of the vegetational history on the national scale. Thus the sediments of Black Loch provide a potential link between the vegetational histories described from the areas to the north in Aberdeenshire (such as at Loch Davan and Braeroddach Loch; Edwards, 1978), the north-west in Perthshire (Stormont Loch; Caseldine, 1980a) and those to the south-east (for example, Newey, 1965b; Hibbert and Switsur, 1976; Mannion, 1978a).

Description

The sediments at Black Loch comprise a succession of clays and detritus muds (Figure 15.7). A representative relative pollen diagram showing selected pollen and spore taxa (Figure 15.7) is based on data from one of four cores, and which was 7.0 m in length taken in 3.0 m of water. There are 14 radiocarbon dates (SRR–2613 to SRR–2626) for the profile, which features both Lateglacial and Holocene age deposits. In addition there are a further eleven radiocarbon dates (UB–2290 to UB–2300) from a second core. The vegetational history is based on the information available from all four cores.

Interpretation

The basal sediments (clay) of Black Loch are not polleniferous, but at some time before 12,670 ± 150 BP (SRR–2626; determination on organic clay) pollen and spore taxa indicate the presence of a cold-climate vegetation pattern dominated by Gramineae and *Salix* with contributions from Cyperaceae, Caryophyllaceae, *Artemisia*, *Selaginella selaginoides*, and *Lycopodium clavatum*. Present also is the pollen of *Koenigia islandica*, a taxon often found in the Lateglacial deposits of sites in the west of Scotland; its presence here shows that it was probably common to the whole of Scotland at this time.

The widespread effect of the lowering of temperatures during the Loch Lomond Stadial is confirmed by the pollen taxa for this period at Black Loch (pollen zone BL II b), and the features generally described for Stormont Loch (Caseldine, 1980a) and at Pickletillem (Whittington *et al.*, 1991b) are reproduced. Conditions of extreme soil instability affected the vegetation and those, together with the restricted growing season, led to a ground cover which principally included *Empetrum nigrum*, *Juniperus communis*, Compositae, *Artemisia*, Caryophyllaceae, Cruciferae, *Rumex* and *Thalictrum*.

Following the amelioration of climatic conditions at the end of the Loch Lomond Stadial, *Betula* pollen totals increased at Black Loch, indicating the presence of birch woodland. The isochrone map of Birks (1989) suggests that this vegetation stage dates to around 10,000 BP in Fife and the Black Loch pollen diagrams add support to this. By approximately 9000 BP the immigrating *Corylus avellana* had achieved a dramatic rise (pollen zone BL II c). This timing is later than that of *c.* 9350 BP for Pickletillem and lies between the date of 9300 BP for southern Scotland (Hibbert and Switsur, 1976) and that of 8700 BP for the central Grampians (Birks and Mathewes, 1978). Other extrapolated dates for Fife are closer to the central Grampians date (for example, 8640 BP at Creich, Cundill and Whittington, 1983; 8690 BP for Loch Rossie in the Howe of Fife, P. R. Cundill, unpublished data).

As elsewhere in eastern Scotland the domination of the vegetation by arboreal taxa reached an important stage with the arrival of *Ulmus* and *Quercus*. At Black Loch, *Ulmus* appears to have expanded at the same time as *Corylus*, whereas *Quercus* made a later entry. The pollen spectra indicate that a mixed deciduous woodland of *Ulmus* and *Quercus*, along with *Betula* and *Corylus*, existed around Black Loch from approximately 8500 BP; isochrone maps (Birks, 1989) had suggested that such a tree cover did not come into being until after 8000 BP.

Relative *Pinus* pollen values at Black Loch were low throughout the Holocene but increased to about 8% after the *Ulmus* and *Quercus* rise. Bennett (1984) has argued that where *Pinus* values are below 20% of total pollen, pine trees did not grow locally, and where deciduous woodland dominated by *Quercus* and *Corylus* occupied an area, *Pinus* could not compete successfully. Thus the presence of *Pinus* pollen at Black Loch at this time would have to be attributed to wind transport, but a major problem exists in finding a source which could have provided such a consistent and large volume of pollen. The pine expansion at Black Loch is dated to before 8000 BP, which is earlier than dates given by O'Sullivan (1976) and Huntley and Birks (1983) for the major development of pine in upland areas to the north and south of Fife. Thus it appears that a more local source needs to be found and one might be provided by the stretch of glaciofluvial sands in the Howe of Fife, which lies several kilometres to the south of Black Loch. This would parallel the situation predicated by Birks (1972a) for south-west Scotland.

The beginning of the middle Holocene period at Black Loch is associated with a marked change in the woodland composition (pollen zone BL II d). *Alnus* becomes a major component and it seems to have taken about 400 radiocarbon years to achieve its main expansion. Again, the importance of Black Loch as a source of evidence for an understanding of the vegetational history of

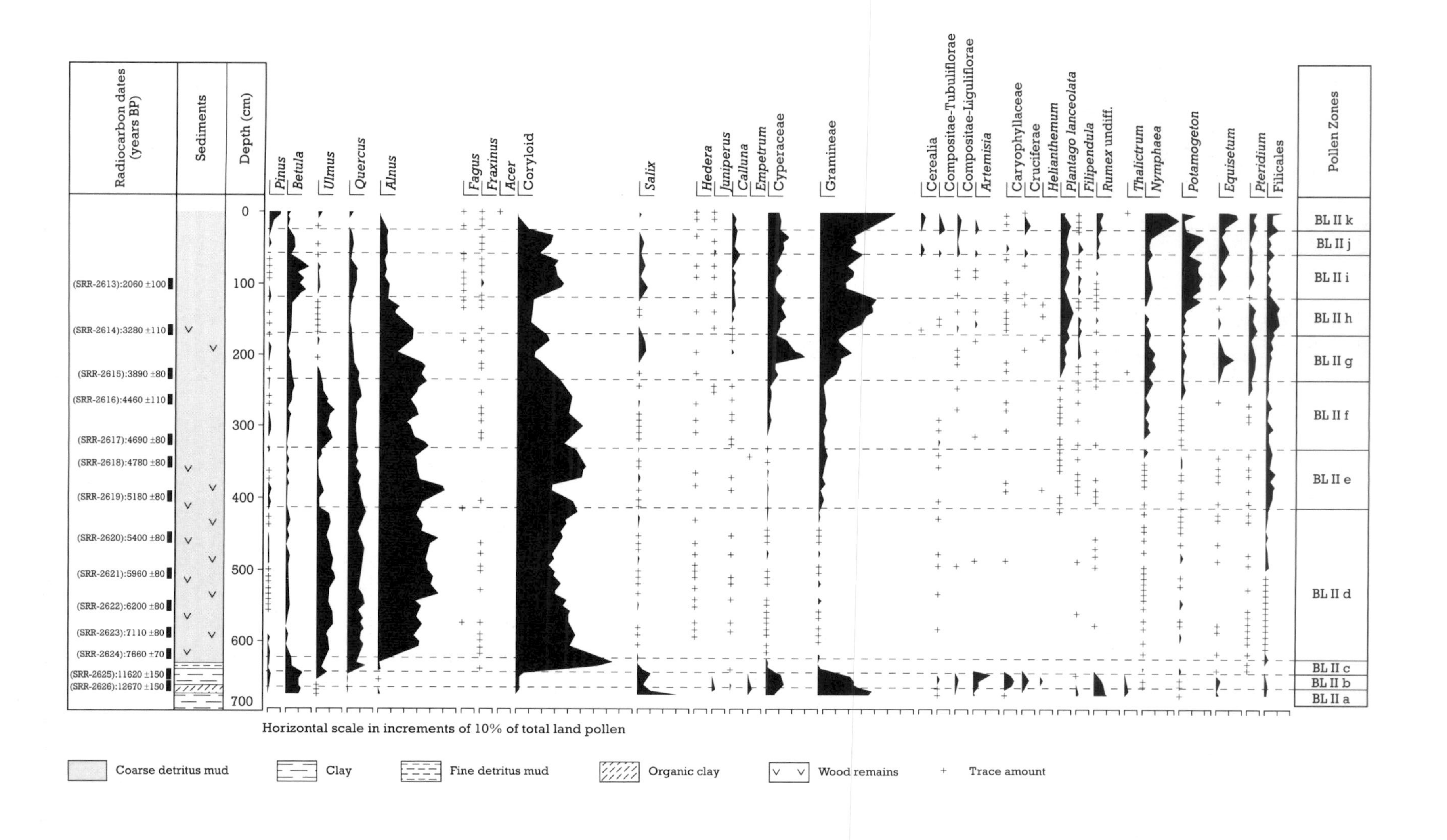

Radiocarbon dates (years BP)
Sediments
Depth (cm)
(SRR-2613):2060 ±100
(SRR-2614):3280 ±110
(SRR-2615):3890 ±80
(SRR-2616):4460 ±110
(SRR-2617):4690 ±80
(SRR-2618):4780 ±80
(SRR-2619):5180 ±80
(SRR-2620):5400 ±80
(SRR-2621):5960 ±80
(SRR-2622):6200 ±80
(SRR-2623):7110 ±80
(SRR-2624):7660 ±70
(SRR-2625):11620 ±150
(SRR-2626):12670 ±150
0
100
200
300
400
500
600
700
Pinus
Betula
Ulmus
Quercus
Alnus
Fagus
Fraxinus
Acer
Coryloid
Salix
Hedera
Juniperus
Calluna
Empetrum
Cyperaceae
Gramineae
Cerealia
Compositae-Tubuliflorae
Compositae-Liguliflorae
Artemisia
Caryophyllaceae
Cruciferae
Helianthemum
Plantago lanceolata
Filipendula
Rumex undiff.
Thalictrum
Nymphaea
Potamogeton
Equisetum
Pteridium
Filicales
Pollen Zones
BL II k
BL II j
BL II i
BL II h
BL II g
BL II f
BL II e
BL II d
BL II c
BL II b
BL II a
Horizontal scale in increments of 10% of total land pollen
Coarse detritus mud
Clay
Fine detritus mud
Organic clay
Wood remains
Trace amount

Scotland is underlined because the main *Alnus* rise is dated here to near 7300 BP, a date which is earlier than most others from Scotland, for example, around 6800 BP for southern Scotland (Hibbert and Switsur, 1976) and 6200 BP for western Scotland (for example, Birks, 1972a; Pennington *et al.*, 1972; Williams, 1977). The relative isolation of Fife flanked by the Firths of Forth and Tay, and the attendant widespread development of estuarine conditions, could well have encouraged an earlier migration of *Alnus* into eastern Scotland (cf. Smith, 1984) but this is difficult to reconcile with a date of *c.* 6500 BP for the main alder rise at the near coastal site of Pickletillem. Interpreting the spread of alder in the British Isles is, however, a notoriously difficult process (Chambers and Elliott, 1989; Bennett and Birks, 1990).

Signs of anthropogenic impact on the middle Holocene vegetation (Edwards and Ralston, 1985) have not been discovered at Black Loch. This was the period of densest forest cover at the site, with samples yielding non-arboreal pollen values frequently below 5% of total pollen.

Around 5180 ± 80 BP (SRR–2619) a decline occurred in *Ulmus* pollen, accompanied by the appearance of *Plantago lanceolata* and increased values for Gramineae. The dating of these events is in accordance with those to the south (for example, 5390 BP at Din Moss, Hibbert and Switsur, 1976) and to the north (for example, 5295 BP and 5105 BP, respectively at Braerod-dach Loch and Loch Davan; Edwards, 1978). It is probable that human interference with the vegetation occurred at this time, a suggestion supported by the first appearance of cereal-type taxa, but natural explanations (disease, climate change) are not disproved. There is no evidence for a widespread environmental disturbance at this elm decline; the chemical and particle size analysis of the loch sediments also indicate this (Whittington *et al.*, 1991c).

Above the level of this elm decline, the pollen values for *Ulmus* and *Quercus* are reduced and Gramineae and *Plantago lanceolata* values rise (pollen zone BL II e). Taken together with an increased sediment accumulation rate (from 0.125 cm a^{-1} to a profile maximum of 0.333 cm a^{-1}) and a sharp increase in clastic inputs, there would appear to have been a period of considerable environmental disturbance around the Loch. Subsequently *Ulmus* pollen totals recovered to pre-elm decline values (pollen zone BL II f) and are matched by a decrease in sedimentation rates. This apparent return to vegetation conditions of the period before 5180 BP was brought to an end between 4460 ± 110 BP (SRR–2616) and 3890 ± 80 BP (SRR–2615), when a second major *Ulmus* decline occurred. This is accompanied in the samples by a fall in *Quercus* values, increased frequencies of herbaceous taxa (especially Gramineae, Cyperaceae, *Plantago lanceolata* and the reappearance of cereal-type pollen) and a rise in values for the spores of *Pteridium aquilinum* and *Equisetum*. (In core BL II only, a partial recovery of elm, followed by a minor decline, takes place at the 5070 BP and 4090 BP levels respectively; this event, intermediate between the two major *Ulmus* declines, is discussed fully elsewhere – Whittington *et al.*, 1991c).

Black Loch's importance as a site for demonstrating the vegetational history of this part of the Holocene is intimately bound up with the events associated with the apparent behaviour of *Ulmus*. Following the first decline of elm the area experienced vegetational disturbance, although it is unlikely, if solely of anthropogenic origin, that it was a discrete event, but rather an amalgam of different periods of human activity. The recovery of *Ulmus* pollen levels at 4690 ± 80 BP (SRR–2617) presumably indicates a reduction in cultural activity (or a cessation of natural limiting factors) which lasted until the second major *Ulmus* decline. Those events occurred at other sites in Britain and western Europe although their marked collective nature at Black Loch, as at the Welsh site of Waun-Fignen-Felen (Smith and Cloutman, 1988), puts the site in the forefront of the continuing controversy and discussion regarding the cause (or perhaps more accurately the causes) of the repeated sudden demise of *Ulmus* (Janssen and Ten Hove, 1971; Tolonen, 1980; van Zeist and van der Spoel-Walvius, 1980; Sturlurdottir and Turner, 1985; Aaby, 1986; Hirons and Edwards, 1986; Perry and Moore, 1987).

After the second *Ulmus* decline the main vegetational characteristics at Black Loch were clearly related to an intensification of human activity (pollen zones BL II g and h). Pollen of

Figure 15.7 Black Loch: relative pollen diagram at coring site BL II showing selected taxa as percentages of total land pollen (from Whittington *et al.*, 1990). *Cannabis sativa* is not represented in this diagram but in the other cores occurs in zones equivalent to BL II j.

arboreal taxa decline continuously and overall, and are replaced reciprocally by expansions, in particular, of Gramineae, Cyperaceae, *Plantago lanceolata*, and *Rumex*. The presence of the light-demanding *Fraxinus excelsior* denotes the existence of open ground and the identification of *Hordeum*-type cereal pollen is an indication of the reasons for the clearance of the woodland. Such events are usually recorded in pollen records from eastern Scotland, but the fine resolution possible due to the great depth of sediment (7.52 m after 3750 BP in the lake-centre core I) has allowed the identification of a phase of interruption in the progressive destruction of woodland. During the period that appears to correlate with the Roman incursion into Scotland (pollen zone BL II i) there was a recovery (reflected in the relative and concentration pollen levels) of *Quercus*, *Ulmus* and Coryloid, with a concomitant decline in *Calluna vulgaris*, Gramineae, *Plantago lanceolata* and *Pteridium aquilinum*. This evidence tends to confirm a similar event recorded in Loch Davan and Braeroddach Loch in Aberdeenshire (Edwards, 1978). Thereafter by 1429 BP, woodland was once more in retreat, agriculture was re-established, and the pollen spectra reveal a continuous curve for cereal-type pollen, not only of *Hordeum* type, but also of *Avena/Triticum* type. Chemical analyses of the sediments show increasing erosional activity, suggesting that the level of arable agriculture was becoming more intense.

By 1000 BP the vegetational history at Black Loch enters its final stage, but one which shows two phases (pollen zones BL II j and k). The first, approximately 1000–400 BP, continues the woodland shrinkage, apart from *Alnus* and *Salix* which survived in the carr at the edge of the loch. The pollen spectra record the strongest representation, in both relative and concentration terms, of Compositae, Cruciferae, *Plantago lanceolata*, *Artemisia*, Umbelliferae and *Calluna vulgaris*.

A feature which attests to the importance of Black Loch in tracing vegetational change in Fife in particular, and in eastern Scotland in general, occurred near the beginning of this period. There is a sudden, strong (up to 12% TLP) and maintained presence of Cannabinaceae pollen. Such pollen can be derived from *Humulus lupulus*, which occurs naturally in fenland, or from *Cannabis sativa* which appears to have been introduced into Britain by the Romans. A more secure method now exists (Whittington and Gordon, 1987) for the separation of these two taxa, and at Black Loch it is the latter which provides the majority (up to 70% at some levels) of the pollen. From around 1000 to 825 BP the growing of hemp, presumably for its useful fibres (Whittington and Edwards, 1990), must have added a distinctive aspect to the vegetation of the Black Loch area (Edwards and Whittington, 1990), providing the introduction of an alien vegetation component, to which might also be added *Juglans regia* (walnut). The growth of *Cannabis sativa* may have received encouragement from the Anglo-Norman penetration of Fife, a period which also witnessed the founding of a Tironesian abbey near Black Loch. The innovatory farming techniques of the monastic community could well have led to the intensification of arable practice, which is revealed by the strong presence in the pollen spectra of cereal-type pollen, whereas the pastoral activities probably brought about the appearance of the pollen of *Vicia cracca* and *Trifolium* types. The increase in pollen influx rates and the sediment chemistry evidence indicate that throughout this period there must have been a continuing removal of the naturally occurring vegetation species, and their replacement by cultigens and ruderals following increased ploughing activity.

In early modern and succeeding times, the agricultural modifications were maintained but appear to intensify. Again the fine resolution obtainable at Black Loch reveals quite clearly the marked impact on the vegetation engendered by changed attitudes to land management. The most striking feature lies in the re-establishment of arboreal species. The creation of coniferous plantations is marked by the sudden resurgence of *Pinus* pollen, to be followed by the appearance of such introduced species as *Abies*, *Larix* and *Picea*. The desire to beautify the landscape led to the planting of *Fagus* and encouraged a greater representation of *Ulmus* and *Fraxinus*.

In terms of unravelling the vegetational history of Scotland, the pollen spectra established from analysis of cores from Black Loch help to fill a large gap. Not only are pollen records from the Late Devensian to the present day infrequent in eastern Scotland, but the investigations undertaken at Black Loch have a base of multiple coring, frequent radiocarbon dating and sedimentological analysis; such a methodology enables corroborative checks to be made upon the representativeness of each core studied, while the differences between them provide important indications of taphonomic and spatial variability.

The findings have confirmed in some instances the results of investigation at other sites (see Din Moss, Stormont Loch and other sites mentioned above). In others they have shown that the isopollen and isochrone maps for Scotland can now be further refined; a mixed deciduous woodland was present locally for at least 500 radiocarbon years before predicted, a local source for *Pinus* is suggested, and *Alnus* expanded some 500 radiocarbon years earlier than suggested by the mapping exercises. Above all, the findings reveal new aspects of the change and development of the vegetational history in Fife, in particular, and eastern Scotland in general. Of considerable interest in this connection is the existence of a multiple elm decline, which continues to fuel the debate surrounding the frequently marked, but solitary, fall in *Ulmus* pollen values at *c.* 5100 BP, and the expansion of *Cannabis* pollen at *c.* 1000 BP, which denotes the probable cultivation of hemp within the local agricultural economy.

Conclusion

Black Loch is a key reference site, providing a record of the vegetational history in eastern Scotland from the time of melting and shrinkage (deglaciation) of the last ice-sheet (around 13,000 years ago) up to the present. Its pollen and sediments have been studied in great detail and have provided a wealth of palaeoenvironmental and palaeoecological information, supported by extensive use of radiocarbon dating. Black Loch is also an integral member of the network of sites for establishing the pattern of variations in vegetation history in eastern Scotland.

Chapter 16

Western Central Lowlands

INTRODUCTION

D. G. Sutherland and J. E. Gordon

The commonly termed Central Lowlands of Scotland (Figure 16.1) do in fact contain a number of hill groups reaching over 600 m OD. The term, therefore, is somewhat inexact but it is useful in highlighting the contrast between the Midland Valley and the mountainous areas to both the north and south. As both these mountain areas were major sources of ice during the Pleistocene, and as there is no evidence for a significant build up of ice within the central belt itself, the history of glaciation of that area is indeed that of a lowland that has been repeatedly invaded by external ice. At times the ice originating in the south-west Highlands has been dominant, whilst at other times Southern Uplands ice has expanded to the north over ground once glaciated from the opposite direction. This led J. Geikie (1877) to coin the phrase, by analogy with social history, of the 'debatable ground' for the southern central belt where the two ice masses held alternate dominance. The Quaternary history of the western Central Lowlands has been reviewed in part recently by Price (1980) and Jardine (1986).

The western Central Lowlands have long been known for mammalian fossil occurrences, including mammoth, reindeer and woolly rhinoceros, that pre-date the last ice-sheet glaciation. These derive from two distinct areas, the Ayrshire Lowlands, as at Kilmaurs (Bryce, 1865b; Young and Craig, 1869), Dreghorn (Craig, 1888) and, most recently, Sourlie (Jardine and Dickson, 1987; Jardine *et al.*, 1988), and from the lower Clyde Valley, as at Bishopbriggs (Bryce, 1859; Rolfe, 1966), Baillieston (Kirsop, 1882), Mount Florida (Macgregor and Ritchie, 1940), Chapelhall (Bryce, 1859) and Carluke (Smith, 1871). In all the locations the fossils are described as occurring within or below deposits of the last ice-sheet, and radiocarbon dates are available from three localities. Samples from Bishopbriggs (Rolfe, 1966) and Sourlie (Jardine and Dickson, 1987; Jardine *et al.*, 1988) suggest ages of about 27,000–30,000 BP for the fauna, but the two museum samples dated from Kilmaurs gave contradictory ages of 13,700 + 1700 − 1300 BP (GX–0634) (Sissons, 1967b) and >40,000 BP (Birm–93) (Shotton *et al.*, 1970).

Two other types of deposit pre-dating the last glaciation have been reported from the same general areas: terrestrial organic sediments, and sands, silts and clays containing marine shells. These have not, in general, been studied in detail, although the organic sediments at Sourlie contain plant macro- and micro-fossils indicative of a treeless vegetation (Jardine and Dickson, 1987; Jardine *et al.*, 1988). The marine deposits are best exposed at Afton Lodge. As with other high-level shell beds in Scotland, it has been disputed whether they are *in situ* (for example, Holden, 1977a) or have been transported as large glacial erratics (Eyles *et al.*, 1949). Sutherland (1981a) has suggested that they are indeed *in situ* and result from a marine incursion during the period of build-up of the last ice-sheet, which he further suggested occurred during the Early Devensian. He also considered that the shelly deposit, which apparently overlay the stratum containing the terrestrial fossils at Kilmaurs (Bryce, 1865b), confirmed that such a marine transgression had occurred. Similar deposits may occur in the south of Arran (Bryce, 1865a; Tyrrell, 1928; Sutherland, 1981a). The dating and sedimentology of all these deposits, however, is uncertain and awaits new evidence.

The last expansion of ice into the western Central Lowlands occurred during the Late Devensian. The initial advance was from the south-west Highlands, and as this ice advanced into the lower Clyde Valley, it dammed the river, producing a sequence of glacial sediments overlying terrestrial and lacustrine deposits (Sissons, 1964; Price, 1975). Throughout the lower Clyde Valley area there are also numerous buried channels, possibly the result of subglacial meltwater erosion (Clough *et al.*, 1916, 1920; Sissons, 1967a; Jardine, 1977; Menzies, 1981). The advancing ice overrode these channels and buried them under glacial deposits. The most spectacular results of this process are seen at the Falls of Clyde, for here, on deglaciation, the river did not regain its old buried valley (Ross, 1927) but cut a new one, with the consequent formation of a sequence of waterfalls.

In the Glasgow area there has long been recognized two till units, red and grey in colour. In places the red till (containing Old Red Sandstone lithologies) has been observed to overlie the grey till (containing Carboniferous lithologies) and they are sometimes seen to be separated by, or overlie, sands and gravels (Bennie, 1868; Clough *et al.*, 1925; Jardine, 1973; Browne and McMillan, 1989). Although it has been proposed that the till units might represent

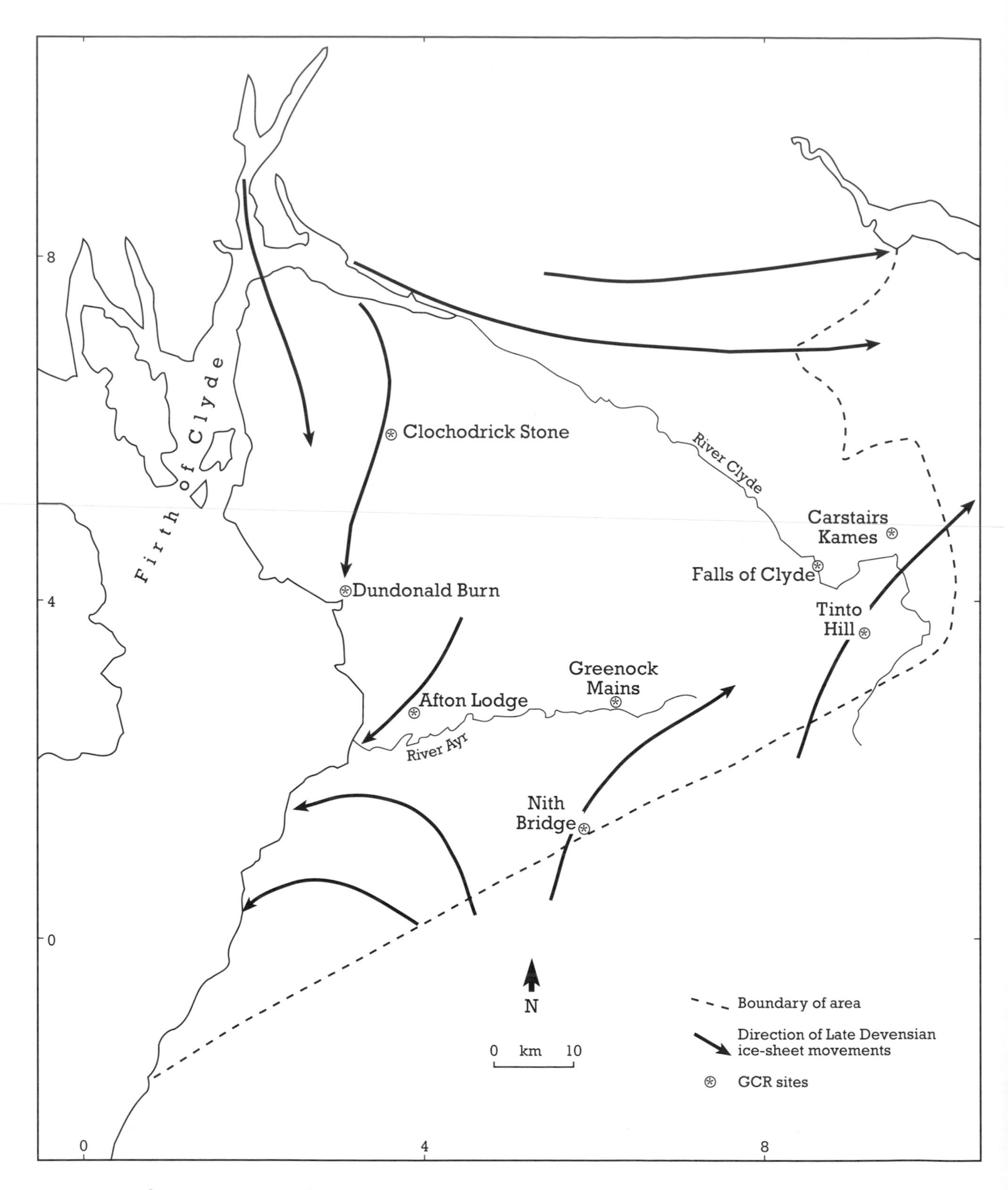

Figure 16.1 Location map of the western Central Lowlands.

separate glacial periods (Sissons, 1967a; Jardine, 1968), more recent work (Menzies, 1976, 1981; Abd-Alla, 1988) suggests that the contrasts in till colour and composition simply reflect the distribution of different bedrock lithologies eroded by the ice. Locally, the till contains marine fossils

(Wright, 1896; Browne and McMillan, 1989). However, an earlier till has been recognized by Browne and McMillan (1989). This till is stratigraphically below the till units described above, has not proved to be fossiliferous and may have been subject to a period of subaerial weathering (Browne and McMillan, 1989). The glacial deposits in the Glasgow area are extensively drumlinized (Elder *et al.*, 1935; Rose and Letzer, 1975, 1977; Menzies, 1981; Rose, 1987). The orientations of the drumlins clearly indicate how the ice flowlines diverged as the ice moved out from the Highlands and across the Midland Valley (Sissons, 1967a).

The Highland ice transported a characteristic sequence of erratics, a fine example of which is Clochodrick Stone. These erratics can be traced into the northern margins of the Southern Uplands (Sutherland, 1984a), indicating that the expansion of Highland ice occurred at a relatively early stage of the last glaciation, for subsequently the Southern Uplands ice became more dominant, advancing northwards into the southern Central Lowlands and deflecting the Highland ice to both east and west. The stratigraphic consequences of this differential development of the two ice centres is a bipartite till sequence throughout much of the southern Central Lowlands, in which a basal till derived from Highland ice is overlain by a till carrying Southern Uplands erratics. Such a sequence is well illustrated at Nith Bridge (Holden, 1977a, 1977c).

On deglaciation very extensive sequences of meltwater channels, eskers and kames were formed (MacLellan, 1969; Cameron *et al.*, 1977). From slightly to the west of the Ayrshire–Clyde watershed, these glaciofluvial features are oriented to the east or north-east and drainage was to the North Sea (Sutherland, 1984a, 1991a). The most outstanding examples of the glaciofluvial deposits formed at this time are the Carstairs Kames (Goodlet, 1964; Sissons, 1967a; MacLellan, 1969), but they are only part of a wider integrated glacial drainage system that operated during the greater part of the deglaciation of central Scotland. Only after the lower Clyde Valley had become free of ice could meltwaters abandon their easterly routes and flow towards the north-west (Browne and McMillan, 1989).

In Ayrshire, west of the Clyde watershed, there is a relative paucity of meltwater phenomena (Holden, 1977a). However, in the upper basin of the River Ayr, there is a particularly interesting sequence of deposits in which an upper till unit, as seen at Greenock Mains, has been interpreted as resulting from a late readvance of the ice-sheet (Holden, 1977a, 1977d). If this is indeed correct, then it implies that active ice occupied the Firth of Clyde after much of the Central Lowlands had been deglaciated. A westwards flow of ice across Kintyre from Arran and carrying Ailsa Craig microgranite to the north coast of Ireland may have occurred at this time and been the counterpart of the readvance in Ayrshire. For reasons of climatic amelioration and the assistance that the deeper waters of the inner Firth of Clyde would have given to ice wastage by calving, this late-stage ice mass may have collapsed rapidly (Sutherland, 1984a). Radiocarbon dating of marine shells around the head of the Firth of Clyde has indicated that these events took place prior to 13,000 BP (Browne *et al.*, 1977; Sutherland, 1986).

During the Loch Lomond Stadial small glaciers formed on Arran (Gemmell, 1973; D. G. Sutherland, unpublished data), and periglacial activity was intense. Ice wedges formed near sea level (Rose, 1975), and radiocarbon dating of buried organic sediments on the flanks of Late Devensian drumlins indicates significant slope instability and solifluction at this time (Dickson *et al.*, 1976). The summits and slopes of the hills in the south of the area are extensively covered in frost-weathered debris and solifluction deposits (Galloway, 1961a; Tivy, 1962; Ragg and Bibby, 1966). On Tinto Hill, where the vegetation has been stripped back, stone stripes are actively forming today.

On deglaciation of the Ayrshire coast the marine limit was at about 25 m OD (Jardine, 1971), or as much as 28 m OD (Boyd, 1986a), with the sea penetrating inland along the lower part of the river valleys, such as that of the Irvine (Boyd, 1986a). Around Arran, the marine limit was also formed on deglaciation, the sea reaching to around 27 m OD in the south of the island and over 30 m OD in the north (Gemmell, 1973; D. G. Sutherland, unpublished data). There has been debate as to whether the sea entered the Clyde estuary area at this time via the Lochwinnoch Gap, while the lower estuary was still occupied by ice (Peacock, 1971c; Price, 1975; Price *et al.*, 1980). The situation remains unclear, not least because of the difficulties in defining the marine limit in the Glasgow area (Sutherland, 1984a), and although rockhead altitude in the Lochwinnoch Gap is sufficiently low for the sea to have passed along its length (Ward, 1977; Institute of

Geological Sciences, 1982), such an event still awaits demonstration by the identification of suitably placed marine sediments.

The western Central Lowlands have been the focus of much research on Quaternary deposits during the last 150 years but, curiously, relatively little has been published in recent times on the Lateglacial and Holocene vegetational history of the area. An exception is the work of Boyd (1982, 1986c, 1988; Boyd and Dickson, 1986) who has studied Holocene environmental change in the Irvine area. On Arran, vegetational change during the Holocene has been studied at a number of sites (Robinson, 1981, 1983; Boyd and Dickson, 1987; Afleck *et al*., 1988; Edwards and McIntosh, 1988; Robinson and Dickson, 1988) and, in Ayrshire, the history of forest clearance has been investigated at Bloak Moss (Turner, 1965, 1970).

In the Irvine area, as well as more widely along the Ayrshire coast, there is well-preserved evidence for the sea-level changes that occurred during the Holocene (see Dundonald Burn). Following a period of relatively low sea level in the early Holocene, the Main Postglacial Transgression resulted in marine invasion of the coastal zone and the burial of terrestrial peats by marine sediments (Jardine, 1964, 1971; Boyd, 1988). At and following the maximum of the transgression, estuarine sediments were laid down in sheltered embayments; on the more open coasts extensive sequences of shingle ridges were deposited and large sand-dune systems developed on their surfaces, as near Irvine.

AFTON LODGE

J. E. Gordon

Highlights

The sediments exposed in stream sections at Afton Lodge include a high-level shelly clay with a fossil marine fauna. They form part of a suite of such deposits in Scotland which have featured prominently in studies of glacial history and sea-level changes.

Introduction

The site at Afton Lodge (NS 417259) comprises a series of exposures along the Ladykirk Burn, 8 km north-east of Ayr, at an altitude of about 85 m OD. It is important for representing the high-level marine deposits of west-central Scotland. Their origin, like that of similar deposits at Clava and Kintyre, has been the subject of considerable debate. Originally, the deposits were interpreted as *in situ* marine sediments, and later as ice-rafted blocks of frozen sea-floor sediments. Holden (1977a) reopened the debate, arguing that the deposits represent a pre-Late Devensian sea-level stand between 115 m to 150 m OD. Sutherland (1981a) supported Holden's interpretation and further suggested that the high-level marine deposits found around Scotland were the product of isostatic downwarping in front of an expanding ice-sheet, possibly during the Early Devensian. The deposits at Afton Lodge were first described by Eyles (1922) and subsequently by Eyles *et al.* (1949), Holden (1977a, 1977b) and Abd-Alla (1988).

Description

Eyles (1922) and Eyles *et al.* (1949) recorded the following sequence at Afton Lodge:

3.	Red till	up to 4.9 m
2.	Laminated, greyish-green, variegated clay	1.5 m
1.	Stiff, blue to black, stoneless clay with shells	0.6 m

The shelly clay extended for 300 m at an altitude of between 76 m and 91 m OD and yielded shells of nine species of mollusc (Table 16.1).

In his detailed study of the glaciation of central Ayrshire, Holden (1977a, 1977b) re-examined the Afton Lodge section and noted the following sequence of deposits:

3.	Sand and gravel comprising angular clasts up to 0.3 m in a red-brown, coarse, sandy, open-textured matrix	3 m
2.	Coarse gravel, dipping east-south-east at 5°	0.3 m
1.	Stiff, black, shelly, silty clay containing bivalves and Foraminifera	5 m

He considered that bed 3 was not till as suggested by Eyles *et al.* (1949), but rather a slope-wash deposit. The typical till of the area did not overlie the shelly clay, but is its lateral equivalent and could be traced upstream from it. Holden described a sharp, vertical contact between these two units, but in fact they are

Table 16.1 List of mollusc shells recorded at Afton Lodge by Eyles (1922) and Eyles *et al.* (1949). (Modern names are from J. D. Peacock, unpublished data.)

Specimen	Modern name
Astarte compressa Mont.	*Tridonta montagui* (Dillwyn)
A. sulcata da Costa	*Astarte sulcata* (da Costa)
Cyprina islandica L.	*Arctica islandica* (L.)
Pecten islandicus Chemnitz	*Chlamys islandica* (Müller)
Leda pernula Müller	*Nuculana pernula* (Müller)
Hydrobia ulvae Pennant	
Natica affinis Gmelin	*Tectonatica clausa* (Broderip and Sowerby)
Drillia turricula Sowerby	*Oenopota turricula* (Montagu)
Trophon clathratus L.	*Boreotrophon clathratus* (L.)

interdigitated. Abd-Alla (1988) considered that the shelly clay was a marine deposit that could be distinguished on the basis of grain size distribution and geochemistry from the shelly till at locations elsewhere in Ayrshire.

Holden (1977a, 1977b) described stratification in the upper part of bed 1, consisting of alternate layers of coarse sand and silty clay; below, the deposit was a homogeneous silty clay. Shells of molluscs in the clay were nearly perfectly preserved. Small ones occurred whole, larger ones as fragments. Foraminifera and ostracods identified by M. Kean were listed by Holden (Table 16.2). As quoted by Holden, Kean reported that the assemblage was one which could be found in the Firth of Clyde today between 15 m and 50 m depth.

Table 16.2 Fauna recovered from the Afton Lodge marine clay listed in Holden (1977a)

Mollusca
Arctica islandica (L.)
Astarte elliptica (Brown)
Foraminifera
Ammonia beccarri (L.)
Quinqueloculina seminulum (L.)
Elphidium excavatum (Terquem)
Elphidium articulatum (d'Orbigny)
Elphidium clavatum (Cushman)
Elphidium sp.
Fissurina cf. *lucida* (Williamson)
Ostracoda
Cyprideis torosa Jones
Cytheropteron latissimum (Norman)

Interpretation

High-level marine deposits with shells in west-central Scotland were first described by Smith (1850b) in clays beneath till at 510 ft (155 m OD) at Chapelhall, near Airdrie. Crosskey (1865), however, believed that the shelly clay was in fact on top of the till and therefore conformed with the position of similar deposits throughout western Scotland (Clyde beds). Geikie essentially followed Smith's interpretation of the deposits and, indeed, until the discovery of a similar shell bed at Clava, the Chapelhall site became a keystone in the marine submergence hypothesis of the 19th century, representing the minimum level of the transgression (J. Geikie, 1874, 1877). However, reinvestigation of the site by a British Association Committee failed to reveal any evidence of a shelly clay (Horne *et al.*, 1895).

Despite the setback at Chapelhall and doubts over Clava (Horne *et al.*, 1894 minority report; Bell, 1895a, 1895b, 1897a), the submergence hypothesis still persisted (Reade, 1896; Smith, 1896a, 1898; Gregory, 1927), in part on the strength of the presence of marine shells in the till of west-central Scotland. Numerous instances of these were reported by Smith (1862, 1896c, 1898, 1901), Geikie *et al.* (1869), Eyles (1922), Richey *et al.* (1930) and Eyles *et al.* (1949). Generally the shells are scattered throughout the till, and they are clearly glacially derived (Eyles, 1922; Richey *et al.*, 1930; Eyles *et al.*, 1949) despite the contrary views of Smith (1898) and Gregory (1926). Locally, however, the shells occur in denser concentrations associated with intact masses of blue-grey, stoneless clay; for example at Afton Lodge where the best exposures now exist, Tarshaw and Catrine (Eyles *et al.*, 1949).

Eyles *et al.* (1949) argued that the shelly clays were sea-floor sediments transported onshore in a frozen state by ice and deposited as erratics. They considered that the clays were too isolated in occurrence to be *in situ*, pre-glacial or interglacial marine deposits.

The hypothesis of ice moving onshore in Ayrshire from the Firth of Clyde had earlier been suggested by Bell (1871), Craig (1873) and Smith (1891). From the distribution of erratics, drumlin orientations and the very existence of the shelly till, Eyles *et al.* (1949) also concurred that the ice movement during the glacial maximum in central Ayrshire was from the west. During a later phase, however, this trend was reversed. In north Ayrshire the same patterns had been established by Richey *et al.* (1930) and the shelly till there explained in a similar manner. Further support for the movement of ice eastwards was later provided by McLellan (1969) working in Lanarkshire.

Holden argued that bed 1 was a marine clay and was *in situ*. His key evidence was that the ice movement in the area was offshore. In his study of the glacial evidence in central Ayrshire he found no support for any ice movement from the west. The northern part of central Ayrshire was glaciated by Highland ice moving south, then bifurcating to the east and west as it encountered ice from the Southern Uplands (see also Goodlet, 1970). The topographic location of Afton Lodge, in the lee of a ridge to ice moving from the north-east, was admirably suited to the preservation of a marine deposit. Holden's other arguments for the deposit being *in situ* are not altogether convincing. His doubts regarding the feasibility of ice transporting large masses of unconsolidated sediment with the bedding preserved are not supported by reports of large-scale block inclusions or till rafts elsewhere (Moran, 1971; Dreimanis, 1976; Aber, 1985; see also Clava). Holden also argued that the marine clays at Clava and Kintyre were remarkably similar to those at Afton Lodge and proposed that they too were *in situ*. Since there was no evidence of a Lateglacial or Holocene submergence of the required magnitude, and the fauna at Afton Lodge were warm temperate, he concluded that the Afton Lodge and Kintyre deposits represented a pre-Late Devensian sea-level stand of between 115 m and 150 m OD.

The general conclusions reached by Holden (1977a) were supported by Sutherland (1981a) in a wider study of the high-level shell beds in Scotland. Sutherland argued that these deposits were *in situ* and demonstrated that, with the possible exception of Clava, isostatic depression in front of an expanding Scottish ice-sheet could have been sufficiently great to explain the altitudes at which the marine clays occurred. The faunas associated with the shell beds were indicative of the North Atlantic Drift reaching the Scottish coasts at the time of the formation of the deposits, and Sutherland suggested that this was compatible with the evidence of Ruddiman *et al.* (1980) that a relatively mild oceanic climate in the North Atlantic accompanied the build up of ice-sheets in the Northern Hemisphere. The explanation offered by Sutherland apparently entailed an Early Devensian expansion of the Scottish ice. However, to date, no unambiguous evidence has been discovered for an Early Devensian glaciation of Scotland (Bowen *et al.*, 1986; see also chapter 5). Furthermore, amino acid analyses of shells from the high-level marine deposits suggest that they may not all have formed contemporaneously (see Tangy Glen and Clava; D. G. Sutherland, unpublished data). However, the general mechanism for the formation of the high-level shell beds proposed by Sutherland (1981a) may be correct even if the chronology is in error.

Afton Lodge is therefore an important site representing the high-level marine clays of west-central Scotland. There is a continuing debate as to whether these sediments are indeed *in situ* and represent a marine transgression pre-dating the Late Devensian ice-sheet glaciation or whether they are very large ice-transported erratics (see Clava, Tangy Glen and Burn of Benholm). On either interpretation the sediments preserve a marine fauna indicative of the climate at the time of deposition. Amino acid analysis of the contained shells is likely to help resolve the outstanding question as to the age of the marine event represented by the clays.

Conclusion

Afton Lodge is notable for a high-level deposit of marine clay containing shells of marine molluscs. It is one of several such deposits in Scotland, which are of critical importance for studies of Quaternary history. It has been questioned whether these sediments, with their marine fossils, are evidence of a former high sea level or of the transport of marine sediments by a former ice-sheet on to the land. Although the precise age, origin and correlations of the deposit at Afton Lodge have yet to be firmly established, its importance for research, as part of the network of high-level shell beds, is unquestioned.

NITH BRIDGE

D. G. Sutherland

Highlights

The river-bank section at Nith Bridge demonstrates a multiple till sequence. This shows that during the main Late Devensian glaciation, the area was successively crossed by ice from sources in the Highlands and Southern Uplands.

Introduction

This site (NS 594141) comprises a section on the south-west bank of the River Nith at Nith Bridge, 6 km south of Cumnock. It shows a sequence of tills and glaciofluvial deposits that is important in illustrating ice-movement patterns in south-central Scotland and, in particular, the interaction of two ice masses with their respective sources in the western Highlands and the Southern Uplands. The intercalation of tills deposited by these two ice masses has long been noted (Geikie, 1863a). It is now generally thought that the tills were deposited by the Late Devensian ice-sheet and that the relative strengths of the two ice masses reflects variations in the climate during the progress of the last glaciation (Bowen *et al.*, 1986). The Nith Bridge section has been investigated by Holden (1977a, 1977c; see also Holden and Jardine, 1980).

Description

The sediments exposed on the south-west bank of the River Nith at Nith Bridge have been described by Holden (1977a, 1977c; Holden and Jardine, 1980). He described the following sequence (Figure 16.2):

5. Till, grey, coarse-grained and gravelly — up to 3.5 m
4. Gravel — <0.5 m
3. Till, brown, less compact and more clayey than bed 1 and containing shell fragments — 2 m

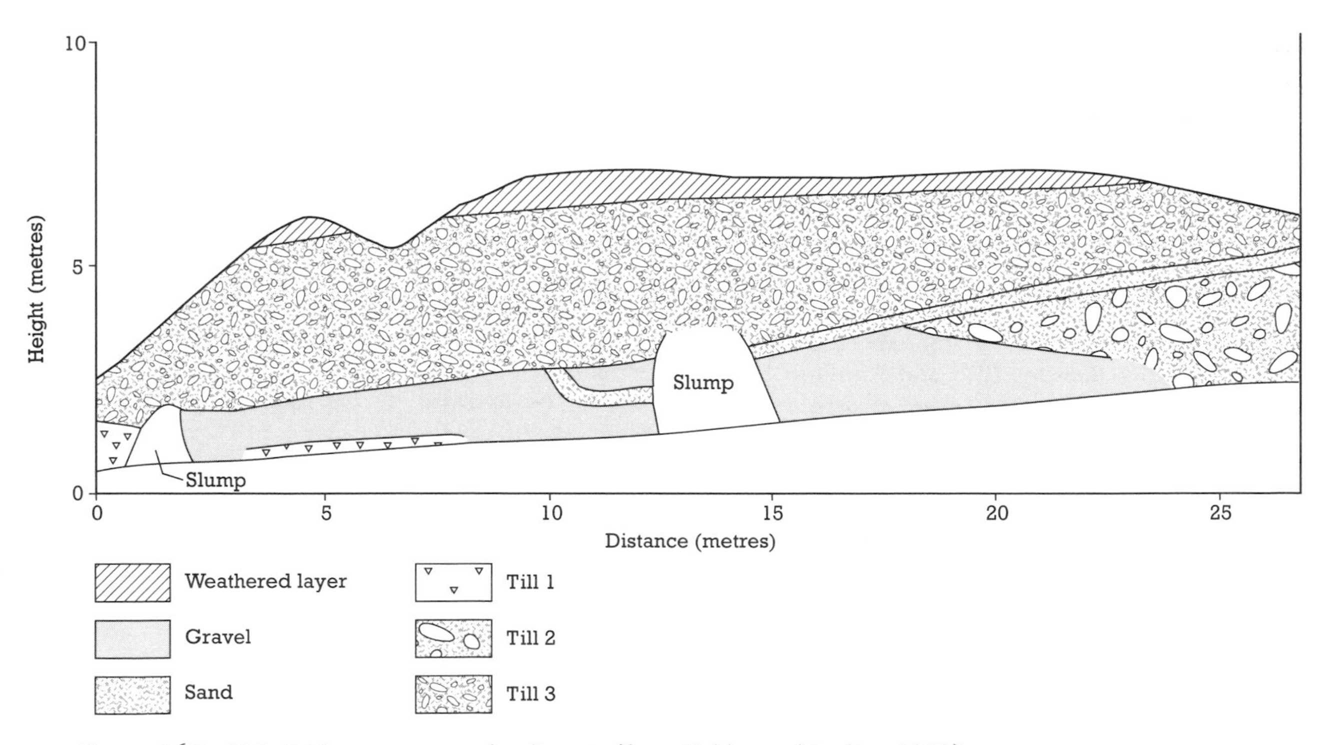

Figure 16.2 Nith Bridge: sequence of sediments (from Holden and Jardine, 1980).

2. Gravels and minor sand horizons *c.* 2 m
1. Till, purple, stiff and sandy at least 1.5 m

Bed 1 occurs immediately above river level. Beds 2 and 3 are truncated and unconformably overlain by bed 4. However, when examined in the field, bed 4 appears to be a line of stones in the till.

Interpretation

Holden investigated the particle size, lithology and clast fabric of the tills. He concluded that the two lower tills had a similar provenance, both being deposited by ice originating in the Highlands and probably flowing across central Ayrshire from the Firth of Clyde area (Holden, 1977c). This view of ice movement is in accord with that of Bell (1871), Craig (1873), Smith (1891), Richey *et al.* (1930), Eyles *et al.* (1949) and McLellan (1967a, 1969) based on those authors' investigations of other sites in Ayrshire and Lanarkshire.

The uppermost till, in contrast, has a southern provenance, as indicated by its erratic content and clast fabrics (Holden, 1977a). This difference demonstrated the former existence of two separate ice masses in central Ayrshire, long established by Geikie (1863a, 1901), Geikie *et al.* (1871) and Geikie (1894). From the stratigraphic relations of the tills at Nith Bridge and other sites, Holden (1977a, 1977c) concluded that there had been at least two distinctive phases of ice movement in central Ayrshire, with an initial advance of Highland ice into the area being succeeded by Southern Uplands ice. Other than locally, there was no evidence for breaks in the deposition of the deposits relating to the separate ice movements. It was therefore concluded that the ice movements related to differences in pressure exerted by the Highland and Southern Uplands ice masses during the progress of the last ice-sheet glaciation.

The stratigraphic evidence at Nith Bridge conforms with other similar evidence across the southern Central Lowlands of Scotland (see Hewan Bank) in demonstrating a significant, regional change in the relative strengths of the ice masses emanating from the Southern Uplands and the western Highlands during the Late Devensian ice-sheet glaciation. This relationship is important in understanding the ice dynamics and the palaeoclimate of Scotland during that glaciation, since it clearly demonstrates a significant shift in the relative importance of the different ice accumulation areas. Such a shift was presumably climatically driven (Bowen *et al.*, 1986) and merits further investigation through modelling of palaeoclimate and the dynamic behaviour of the last ice-sheet.

Conclusion

The deposits at Nith Bridge are important for interpreting the glacial history of the western Central Lowlands. They demonstrate that during the Late Devensian glaciation (about 18,000 years ago), ice from the Highlands first crossed the area and was then replaced by ice coming from the Southern Uplands. Nith Bridge is a valuable reference site for the glacial sequence in this area and for studying the interactions between ice masses from different sources.

GREENOCK MAINS

J. E. Gordon

Highlights

The sequence of glacial and glaciofluvial deposits exposed in stream sections at Greenock Mains is significant for interpreting the movement patterns of the Late Devensian ice-sheet.

Introduction

The site at Greenock Mains (NS 635277) comprises a series of exposures on the Greenock Water, 8 km west of Sorn, which show a layer of sand and gravel interbedded between two tills. This sequence is important for interpreting the glacial history of Ayrshire. It has been described by Smith (1898), and by Holden (1977a, 1977d), who considered that it represented a local readvance of the Late Devensian ice-sheet.

Description

In his classic paper on the glacial deposits of Ayrshire, Smith (1898) first described the sections at Greenock Mains, noting 47 ft (14.3 m) of upper boulder clay, stratified in part, overlying

32 ft (11.1 m) of sand, gravel and lower boulder clay. He recorded that the upper boulder clay was shelly in its upper part.

More recent descriptions of the site have been given by Holden (1977a, 1977d) and Abd-Alla (1988). The basic sequence comprises:

3.	Red-brown, sandy till with clasts of sandstone, shale, siltstone and occasional igneous rocks	up to 6 m
2.	Sand and gravel	up to 20 m
1.	Chocolate-brown, sandy till with shells and clasts of sandstone, shale, igneous and metamorphic rocks; locally with dark grey, silty-clay matrix	up to at least 7 m

Abd-Alla (1988) found that the deposits in bed 1 differed significantly between the two sections that he studied in terms of their grain-size distributions and geochemistry, and to a lesser extent in the relative proportions of the clay minerals present. The deposits in one section closely resembled the shelly till present at other localities in Ayrshire. They also showed an increase in clay and silt content with depth and were weathered in their upper part to a depth of up to 5 m. At the other section, the deposits were comparable in their properties to the shelly clay at Afton Lodge (see above).

In the sand and gravel (bed 2) above the shelly till (bed 1) Holden recognized channel forms, arched-bedding, poor sorting of materials and the presence of large cobbles, all suggestive of a high-energy fluvial environment of deposition typical of a sandur. Palaeocurrent analysis indicated water flow from the north-east.

Holden recorded no shells in the upper till, but apart from this and colour, both tills were similar in particle size composition and their relatively high percentage of clasts of local origin and small percentage of clasts of Highland origin. Abd-Alla (1988) noted that the upper till was weathered to a depth of up to 3 m.

Interpretation

From the clast fabric, Holden concluded that the lower till (bed 1) was deposited by ice originating in the Highlands and moving locally from west to east across the Greenock Mains area after it had bifurcated in central Ayrshire in the presence of Southern Uplands ice. It is part of the distinctive shelly till unit of central and southern Ayrshire (Geikie *et al.*, 1869; Smith, 1898; Richey *et al.*, 1930; Eyles *et al.*, 1949) (see also Afton Lodge) and is one of a number of exposures in the Ayr Valley and near Muirkirk referred to by Geikie (1863a), Smith (1901) and Holden (1977a). Abd-Alla (1988) argued that at one section at Greenock Mains, bed 1 was an *in situ* marine deposit on the grounds of the similarity of its properties to those of the shelly clay at Afton Lodge, for which he accepted an *in situ* marine origin. However, as at Afton Lodge, such a contention requires further investigation, together with full evaluation of the alternative hypothesis of ice transport of a large block of sediment (see also Clava).

From a study of field relations Holden inferred that the overlying sands and gravels were part of the same series of deposits as the extensive surface sands and gravels in the Ayr valley. The latter were interpreted by Geikie (1894) and Charlesworth (1926b) as proglacial lake sediments, but Holden considered them to be ice-contact deposits. From the weathering indicated by the clay minerals and the downwashing of fines in the shelly till of bed 1, Abd-Alla suggested that at least a short interval occurred before the deposition of the overlying sand and gravel; such a suggestion is compatible with observations in modern glacier forefields (Boulton and Dent, 1974).

Holden found that the upper till above the sands and gravels was confined to an area of about 8 km^2 and considered it to represent a local readvance of the Highland ice that last covered the area. He apparently did not investigate the possibility of whether it might be a flow till or part of a single, complex sequence of deposits similar to those recorded in modern glacier environments (Boulton, 1972b; Boulton and Paul, 1976; Paul, 1983). Such a possibility merits further investigation, particularly since many other tripartite sequences in Britain have now been reinterpreted in such terms (see Hewan Bank; Martin, 1981; Eyles *et al.*, 1982).

Sutherland (1984a) has indicated that if the till was indeed the product of a readvance of Highland ice then this implies that active ice continued to occupy the Firth of Clyde and Ayrshire lowlands after the eastern Central Lowlands had been deglaciated.

Greenock Mains is an important reference site for the glacial stratigraphy of Ayrshire, representing the classic shelly till and glaciofluvial sediments of the area. It is also notable for the presence of an upper till which may represent a local readvance of the Late Devensian ice-sheet or may be part of a single, complex sequence of deposits. The upper till is distinct from other tills in a similar stratigraphic position farther south in Ayrshire (such as at Nith Bridge) in that it was deposited by ice flowing outwards from the Firth of Clyde and not from the Southern Uplands.

Conclusion

The ice-deposited sediments at Greenock Mains are important for interpreting the glacial history of Ayrshire. They include two tills derived from the last (Late Devensian) ice-sheet: a lower shelly till characteristic of this area and an upper till which may represent a local readvance of the ice. Greenock Mains forms part of a network of reference sites for reconstructing the pattern of movement and retreat of the last ice-sheet (approximately 18,000–13,000 years ago).

CARSTAIRS KAMES

J. E. Gordon

Highlights

Carstairs Kames is one of the best examples of an esker system in Britain. The landforms and deposits provide important morphological and sedimentological evidence for interpreting the processes of glacial drainage development.

Introduction

Carstairs Kames is one of the most famous geomorphological sites in Britain for an assemblage of glaciofluvial landforms. Although the features, which extend over a distance of *c.* 5.5 km (between NS 937467 and NS 981497), 8 km east of Lanark, have been described in the scientific literature for almost 150 years, their mode of origin is still a source of debate. The main explanations are that they are either subglacial eskers or a form of ice-marginal deposit. The site has been described by Chambers (1848), A. Geikie (1863a, 1874, 1901), Milne Home (1871, 1881c), Dougall (1868), Jamieson (1874), Ramsay (1878), Smith (1901), Gregory (1912, 1913, 1915a, 1915b, 1915c, 1926), Charlesworth (1926b), Macgregor (1927), Sissons (1961c, 1967a), Goodlet (1964), McLellan (1967a, 1967b, 1969), Boulton (1972b), Jardine and Dickson (1980), Laxton and Nickless (1980) and Jenkins (1991).

Description

The Carstairs Kames consist largely of sand and gravel and comprise a series of anastomosing, subparallel ridges and mounds interspersed with kettle holes (Figures 16.3 and 16.4) (A. Geikie, 1874; Gregory, 1913, 1915a; Goodlet, 1964). They formerly extended over a distance of about 7 km, running west-south-west to east-north-east from Newmill (NS 920455) to Woodend Moss (NS 980495) but have been extensively quarried, and the GCR site is restricted to the morphologically most impressive remnants north-east of Carstairs village where the ridges attain heights of 25 m above the adjacent peat bogs. On its northern side, the ridge complex presents a relatively steep face to the flat peat bogs, but on the opposite side glaciofluvial deposits continue southwards in a zone of lower, more subdued mounds. Over a wider area, the Carstairs Kames form part of an extensive belt of glaciofluvial deposits extending south-west to north-east along the valley of the Douglas Water to the Carstairs area (Sissons, 1967a; Goodlet, 1970; Cameron *et al.*, 1977) and north-east from there towards the Edinburgh area (Sutherland, 1984a, 1991a; Jenkins, 1991). Many workers have mentioned in passing the composition of the kames. More detailed accounts have been given by Gregory (1913, 1915a) and particularly Goodlet (1964), McLellan (1969) and Laxton and Nickless (1980). Gregory described bedded gravels with layers of coarse pebbly sand resting upon boulder clay. Gravel was most prevalent and coarsest on the north side of the kames. Goodlet, who monitored the changing faces in working pits over a period of several years, established a threefold stratigraphic succession for the Carstairs area:

3. Later Beds: peat
sand
gravel
2. Middle Beds: Carstairs Station Sands
1. Earlier Beds: Upper Gravels
Main Sands
Lower Gravels
Boulder Drift

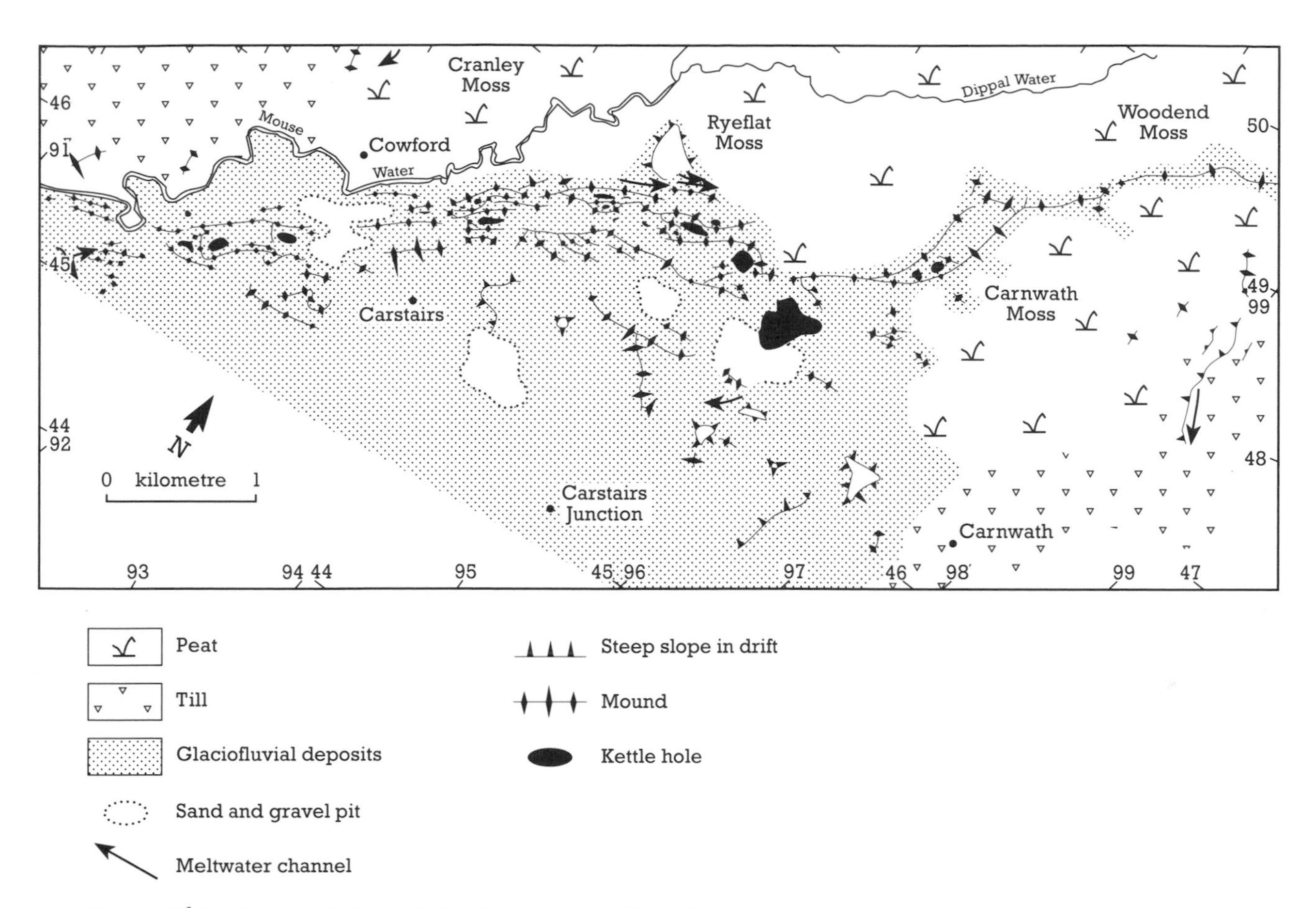

Figure 16.3 Geomorphology of the Carstairs area (from McLellan, 1969).

The kames consisted of the lowest member (1), which he sub-divided into the four units listed. The Boulder Drift had an extremely variable matrix ranging from fine to coarse, clayey sand, with embedded boulders up to 1–2 m in size. The Lower Gravels consisted of clasts 0.07–0.1 m in size and larger stones in a sandy matrix. Sometimes poorly defined bedding was present. The Main Sands, a unit of relatively pure sand with a few bands of gravel, were up to 12–15 m thick and showed clear current bedding with palaeocurrents directed to the east and north-east. They were interdigitated with the poorly bedded Upper Gravels, or separated from them by a thin clay layer. Goodlet also described extensive folding and faulting in the deposits. McLellan (1967a, 1967b, 1969) later recorded decreasing grain size of materials and an increasing proportion of sand in an easterly direction along the kames.

Laxton and Nickless (1980) investigated the sediments of the kames both in sections and from boreholes. They concluded that although Goodlet's stratigraphic framework was valid for the pits which he examined, it was not substantiated over a wider area where their own results suggested a complex variation of facies through time. The ridges largely comprise what they term 'glacial sands and gravels', a poorly sorted cobble and boulder gravel with a clayey, sandy matrix. The ridges are underlain by well-bedded glaciofluvial sand and gravel rapidly grading laterally into glaciolacustrine deposits, noted earlier by Gregory (1915a). In some of the boreholes, till occurs below the glaciofluvial deposits. In the north-east part of the site the mounds consist of glaciolacustrine deposits – laminated clays, silts and fine sand.

Jenkins (1991) distinguished two types of landforms on the basis of morphology and sedimentary characteristics. The first type are the esker-like ridges composed of coarse boulder gravel in a matrix of poorly sorted finer gravel, sand and silt. These deposits are sometimes covered in finer gravel and draped by laminated, fine-to-medium sand, in places with trough cross-bedding. The second type are elongate mounds, up to 20 m high and several hundreds of metres

Figure 16.4 Carstairs Kames showing the interlinked form of the ridges and mounds, with intervening kettle holes. (© British Crown copyright 1992/MOD reproduced with the permission of the Controller of Her Britannic Majesty's Stationery Office.)

long. They mainly comprise medium–coarse sand with trough cross-bedding and channel scours. The mounds are often capped by massive or trough cross-bedded gravel up to 3 m thick and are interspersed with kettle holes.

The lithological composition of the kames is predominantly of materials from sources to the south and south-west, with only a very few rocks of Highland origin present (Gregory, 1912, 1915c; Goodlet, 1964; McLellan, 1969). This agrees with the general pattern established for the drift over a wider area on the northern side of the Southern Uplands (Stark, 1902; Gregory, 1915c; McCall and Goodlet, 1952).

Interpretation

In view of their striking landscape form the Carstairs Kames have long attracted the attention of geomorphologists and geologists. They have been referred to in many textbooks and papers, including the several geological guides to the Glasgow region (McCallien, 1938; Bassett, 1958; Jardine, 1980a). Considerable debate has centred on the origin of the features (Sissons, 1976b) and their implications for the pattern of glacial events in the area. In an early reference, Chambers (1848, p. 213) noted a range of sandhills at Carstairs which he believed were the remains of former marine plains. Later, Geikie (1863a) lyrically described their form and composition. He acknowledged that their origin was probably complex and suggested water currents in association with ice as a possible mechanism, probably operating also with marine processes and drift ice. Milne Home (1871, 1881c) advocated a marine origin for the kames. He interpreted stratified sand and gravel layers overlying boulder clay as submarine banks or deposits formed by the action of marine currents (Milne Home, 1881c). The ridges were formed by water currents. Dougall (1868), too, favoured marine processes and referred to the 'beach' at Carstairs. Jamieson (1874), however, dismissed the marine hypothesis and proposed that eskers and kames such as those seen at Carstairs were formed by meltwaters along the margins of the later glaciers which covered the area.

A. Geikie (1874) provided a systematic description of the morphology and sediments of the kames, noting in particular their variable composition, the stratification and dip of the sands and gravels and the occasionally contorted nature of the beds. He concluded that the kames had not been formed through erosion of a pre-existing deposit by rain or rivers, but he did not speculate on their origin.

Ramsay (1878, p. 386) referred to the 'beautiful' examples of kames at Carstairs and other localities in Scotland, including them in the assemblage of landforms and deposits of the glacial epoch. Somewhat surprisingly, he later described kames or eskers as marine gravelly mounds (p. 430).

Geikie (1901) described the characteristics of the Carstairs Kames, noting that kames and eskers in general were 'a fruitful source of wonder and legend to the people'. After paying due respect to several mythological theories, he stated that no satisfactory explanation so far accounted for them. They were superficial deposits overlying the till and formed during the closing stages of the glacial period. Lacking the characteristics of moraines in the normal sense, they seemed to be associated with meltwater. Smith (1901), however, was rather more specific, suggesting that the Carstairs Kames and related features were the product of fluvial and subaerial denudation of a large delta of drift.

In a series of papers on kames, Gregory (1912, 1913, 1915a, 1915c, 1926) described and discussed many of the characteristics of the Carstairs features. He considered them to be the most famous and typical representative of the Scottish kames (Gregory, 1913). They were glaciofluvial deposits derived mainly from the south and laid down as a marginal formation along the front of a receding glacier. Although they shared similarities in form with eskers, they were not eskers in the true sense, that is deposited on the beds of glacial rivers, since they lacked the seasonal bedding characteristic of these features seen in Sweden and Ireland. Moreover, unlike typical eskers they were aligned across the inferred former direction of glacial drainage in the area.

Macgregor (1927) proposed another mode of origin for the kames. He believed that they were either deposited against a remnant of stagnant ice in the low ground to the north of Carstairs by meltwaters issuing from the glaciers of the Clyde and Douglas valleys, or else they were deposited in a narrow, ice-walled channel between stagnant ice to the north and still-active ice to the south in the Southern Uplands.

Despite the predominantly southern provenance of the constituents, Charlesworth (1926b) argued in the face of previous opinion that the kames were formed by northern ice originating in the Highlands. In his view they were ice-marginal features associated with drainage off a wasting ice-sheet receding from the maximum of the Lammermuir–Stranraer glacial 'stage', supposedly a readvance of the last ice-sheet. In support he pointed to their apparent continuity with deposits both to the north-east and south-west, which he argued were associated with the northern ice of the readvance. He also argued that the steep north face of the kames represented an ice-contact slope (see Charlesworth, 1957, plate 15B). Several factors could explain the high content of southern erratics: they could have been deposits of southern ice reworked and incorporated during the readvance from the north or during fluctuations in the zone of convergence of northern and southern ice at the time of the ice maximum, or even brought to the area by river transport from the south.

Subsequently, Linton (1933) supported Charlesworth's views. However, from a reassessment of the field evidence, Sissons (1961c) showed that in its central and eastern areas the supposed moraine almost everywhere consisted of glaciofluvial 'dead'-ice deposits. Furthermore, over much of the area the last ice came from the Southern Uplands, not the Highlands as Charlesworth had suggested. The Carstairs Kames and their predominantly southern lithologies were associated with this southern ice, which probably advanced over some older Highland till, as happened farther east (Eckford, 1952; McCall and Goodlet, 1952). On morphological grounds Sissons argued that the kames were typical of subglacially formed eskers, and the whole system related to a complex meltwater drainage network under the Southern Uplands ice. The trend of this drainage system accorded with the direction of movement of the last ice-sheet (Sissons, 1967a) and was associated with the supposed Perth Readvance (Sissons, 1963a, 1964).

From his detailed studies of the sediments and structures in the ridges, Goodlet (1964) concluded that the Carstairs Kames were part of a complex terminal moraine formed during a halt-stage in the retreat of Southern Uplands ice, which had readvanced northwards over a wide area (see McCall and Goodlet, 1952) after Highland ice had withdrawn. The Boulder Drift ridges were deposited partly as ice-cored moraines, and as the ice receded, the Lower Gravels and Main Sands were superimposed as outwash deposits. During the later melting of the buried ice, these sediments were modified by erosion and subsidence. Goodlet argued that the deposits were laid down transverse to the last ice movement which, together with their internal composition, made it improbable that they were eskers as Sissons had suggested.

McLellan (1967a, 1967b, 1969) interpreted the available exposures at Carstairs to indicate that the deposits were entirely water-laid, and in his conclusions he fully supported Sissons' view that the kames were part of a subglacial esker system. In McLellan's opinion the Boulder Drift material simply represented coarse proximal sediments which became finer in a downstream direction. However, Boulton (1972b) has shown that there is support from modern glacial environments for the type of interpretation proposed by Goodlet. Boulton argued that the morphology, sediments and internal structure of the kames were unlike those of any known subglacial features, but were identical to those of many supraglacial fluvial deposits accumulated in association with ice-cored moraines. If the analogy is valid, the Carstairs Kames are neither moraines nor eskers, but glaciofluvial deposits originally laid down between ice-cored moraine ridges which blocked and controlled the supraglacial and proglacial drainage. Upon melting of the buried ice in the moraine ridges, inversion of relief gave topographic prominence to the glaciofluvial sediments. Boulton also suggested that many other Pleistocene stratified ice-contact deposits, such as the Bar Hill–Wrexham moraine (Boulton and Worsley, 1965; Yates and Moseley, 1967) and the Escrick moraine (Gaunt, 1970), may have formed in a similar fashion.

Laxton and Nickless (1980) emphasized the complex variation of facies within the whole suite of deposits in the Carstairs area in which they distinguished glacial sand and gravel, glaciofluvial sand and gravel and glaciolacustrine deposits. They envisaged an integrated assemblage of subglacial and proglacial depositional environments in which subglacial streams formed the ridges, then discharged from below the ice. The ridges showed strong similarities with the Guelph Esker in Ontario (Saunderson, 1977), which comprises poorly sorted and bedded material, becoming more graded distally. As suggested by Saunderson (1977) such deposits could have been laid down under sliding bed conditions during high-velocity flow in subglacial tunnels. The underlying glaciofluvial deposits could reflect earlier subglacial deposition under lower velocity in less physically constrained conditions. Laxton and Nickless (1980) therefore supported the esker hypothesis for the origin of the Carstairs Kames. However, they did not consider the origin of the folding and faulting recorded by Goodlet (1964).

More recently, Jenkins (1991) has developed these ideas, drawing analogies with the formation of meltwater drainage systems in modern glacier environments. From the geomorphology and sediments he inferred that the overall environment of deposition at Carstairs was one of englacial or supraglacial streams 'feeding' a subaerial fan overlying buried ice. The ridges of boulder gravel were formed by high-energy flows in englacial tunnels or supraglacial channels near to the ice margin. The sandy mounds were formed by lower-energy, braided streams that fanned out across a surface of buried ice. The laminated finer sand was probably deposited by shallow sheet-floods, whereas the capping gravels represented high-energy floods. Subsequent melting of the buried ice, indicated by the presence of kettle holes and faulting in the sediments, resulted in topographic inversion, with the areas of greatest sediment thickness in the channels now forming upstanding ridges and mounds (cf. Boulton, 1972b).

Jenkins (1991) also considered the deposits at Carstairs in their wider regional context, particularly in relation to the style of deglaciation. He showed that the Carstairs features form part of a sequence of ice-marginal landform and sediment associations that together indicate continuous recession of the last ice-sheet on the south side of the Pentland Hills from the southern outskirts of Edinburgh towards the south-west, accompanied in places by the development of zones of partly buried, stagnant ice. Such a pattern accords with the interpretation of Sutherland (1984a).

In terms of their size, extent and morphology, the Carstairs Kames are one of the most striking assemblages of glaciofluvial landforms in Britain. In terms of their braided morphology, the Carstairs Kames resemble most closely the Kildrummie Kames near Nairn (see above), but differ from a number of large single-ridge features that are typical subglacial eskers, for example at Bedshiel (Stevenson, 1868; McGregor, 1974), Torvean (see above) and Littlemill (see above). Although they are frequently acknowledged to be classic examples of kames or eskers, the precise origin of the Carstairs Kames has been disputed among geomorphologists and geologists for almost 150 years. The most recent studies suggest that they are either subglacial eskers or eskers formed in an englacial or supraglacial position in association with proglacial outwash fans that later became topographically inverted. A key feature to emerge from these studies is that the ridges do not exist in isolation, but are part of a complex and integrated suite of glacial, glaciofluvial and glaciolacustrine deposits in the Carstairs area. To approach a fuller understanding of their formation, they need to be viewed in this wider context. The site clearly has further important research potential in the field of glacial sedimentology and its application to the interpretation and reconstruction of Pleistocene glacial environments and hydrological processes (cf. Kildrummie Kames). Carstairs is a particularly appropriate site for such work because of the existing body of information on the sediments.

Conclusion

Carstairs Kames is a classic site, long renowned for its glacial landforms. These comprise a series of esker ridges, kames and kettle holes, one of the best of such assemblages in Britain formed by the meltwater rivers of the last ice-sheet as it decayed approximately 14,000–13,000 years ago. The site has a long history of research and has featured in many publications. Although the precise origin of the landforms has been much debated, the site is unquestionably of the highest value for studies of meltwater drainage development, processes of meltwater sedimentation and patterns of glaciofluvial landscape development.

CLOCHODRICK STONE

J. E. Gordon

Highlights

Clochodrick Stone is a notable example of a glacial erratic boulder. Historically, such features were among the first considered to require systematic documentation and conservation.

Introduction

The Clochodrick Stone (NS 374613) is a particularly good example of a lowland glacial erratic boulder. It is located 3 km north-east of Lochwinnoch and has been described by Milne Home (1872b).

Description

The large glacial erratic known as the Clochodrick Stone (Clach a'Druidh) (Figure 16.5) is located at an altitude of about 100 m above sea level. It measures 6.7 m in length, 6.1 m in width, 20.6 m in circumference and stands up to 4.0 m above ground level. The rock of which it is composed is a trachytic porphyritic olivine basalt and it is crossed by a series of hematite veins. The boulder rests on lavas of slightly different composition, and it was recorded in the First Report of the Boulder Committee (Milne Home, 1872b) that bedrock of the same type as the boulder occurs in the hills two or three miles to the west and north.

Interpretation

The Clochodrick Stone is a particularly good example of a large erratic boulder. Although erratics and erratic trains are relatively widespread

Figure 16.5 Clochodrick Stone. (Photo: J. E. Gordon.)

in Scotland (see for example, Bell, 1874; Milne Home, 1884; Cumming and Bate, 1933; Sissons, 1967a; Shakesby, 1978; Sutherland, 1984a), the Clochodrick Stone is particularly striking in terms of its size and lowland setting. It was probably transported to its present position by ice from the south-west Highlands moving across the Clyde estuary. This ice moved across and around the Renfrewshire hills towards the south-east (Price, 1975; Paterson *et al.*, 1990).

The Clochodrick Stone is also of more general historical interest in the field of earth-science conservation. It is representative of a suite of features that were among the first to be considered worthy of protection. In 1871 the Royal Society of Edinburgh established the Boulder Committee under the direction of D. Milne Home to identify all the glacial erratics in Scotland that appeared remarkable in terms of size and superficial markings and to recommend measures for their conservation (Milne Home, 1872a, 1872b). This exercise, to some extent a forerunner of the Geological Conservation Review, represents a far-sighted attempt to recognize geomorphological features under threat and to address the need for site survey, assessment and protection. Unfortunately, and unlike the GCR, there was no contemporary legislative framework to underpin the work of the Committee and positive action was to be confined to persuading landowners not to destroy those boulders that merited preservation for further study (Milne Home, 1872a, 1872b). In all, the Committee produced ten reports, the tenth and final one providing a county by county compendium of the boulders listed in the earlier reports (Milne Home, 1884).

Conclusion

Clochodrick Stone is a representative example of a large ice-transported (erratic) boulder. Such boulders were among the first geological features recognized to require systematic survey for conservation during the 19th century; they provide graphic evidence of former ice-sheet movements, in this case from the north-west.

FALLS OF CLYDE

L. J. McEwen and A. Werritty

Highlights

This site is selected for an excellent example of the glacial diversion of drainage. The present route of the River Clyde occupies a bedrock gorge which was cut following the infilling of its former course by glacial deposits.

Introduction

The Falls of Clyde (NS 885406 to NS 882421), 3 km south of Lanark, provide a notable example of glacial disruption of drainage. The Falls of Clyde serve as the local base level for the upper River Clyde, effectively isolating the upper part of the river from its lower reaches. This is the result of glacial deposits blocking the original channel of the Clyde, forcing the river to cut a new bedrock channel. The origin of the falls has been considered by Ross (1927), George (1958), Linton (1963), Sissons (1976b) and Whittow (1977).

Description

The Falls of Clyde are located to the south of Lanark at a point where the River Clyde abruptly changes in both character and direction. Whereas 3 km upstream it is an alluvial, meandering channel flowing south-west through a wide valley, 5 km south-east of Lanark it becomes a relatively straight, narrow, rock-controlled channel flowing to the north-west. Within the designated site the river drops 55 m within 1.8 km. In contrast, the long profile of the Clyde upstream of the site is graded to a base-level at about 183 m OD and requires 32 km to register a comparable descent (George, 1958). The gorge itself, which contains the site, is 7 km long overall and locally up to 50 m deep. The gorge is incised into a pre-glacial surface cut across gently-dipping greywackes of the Lower Old Red Sandstone (Whittow, 1977).

The site consists of two major waterfalls, Bonnington Linn and Cora Linn, separated from each other by a slot gorge. Cora Linn, the larger of the two falls (27 m high), comprises a series of cascades over benches formed from the near horizontally bedded and more resistant sandstone units within the greywackes. The angle made by the top of the falls is oblique to the flow of the main channel indicating that the falls have retreated asymmetrically upstream leaving an enlarged section immediately downstream which now forms the plunge pool. Thus Cora Linn provides a very good example of a waterfall whose configuration is controlled by the detailed stratigraphy and relative resistance of the underlying bedrock. The upper of the two falls, Bonnington Linn, is wider than Cora Linn but not so high. It consists of a single cascade segmented into three parts with large rocky 'islands' separating the individual units. As with the lower falls, the angle is oblique to the main flow of the river.

The shales within the Lower Old Red Sandstone (which dip downstream very gently) provide the risers of the 'staircase' into which the falls are incised. At low flows, it is clear at Cora Linn that there is minimal development of an inner channel within each riser, and very little bedload is at present being transported through the whole rock-controlled section. As a result, the edges of the more massive sandstone units exposed in the bed of the falls have undergone minimal abrasion and rounding.

Between the lower falls at Cora Linn and the upper falls at Bonnington Linn the river descends steeply in a series of rapids over bedrock steps masked by occasional bouldery deposits which, because of their lithology and minimal rounding, are clearly recent and local in origin. The resulting 'step–pool system' is controlled in terms of its detailed morphology (height of 'steps' and dimensions of 'pools') by the spacing of the local joint systems and variation in the relative resistance of the constituent strata. This 1 km long gorge separating Bonnington Linn and Cora Linn is relatively straight, has near vertical sidewalls, 25 m high, and displays a well-developed set of rapids over a very bouldery bed. The local sandstone here is virtually flat-bedded, permitting only limited development of potholes. However, some have developed at the margins of the gorge in response to abrasion and selective exploitation of joint planes.

The current flow over the falls is regulated to some extent by extraction of water at Bonnington Linn for hydro-electric power. Under normal flow conditions this represents only a small proportion of the total.

Interpretation

The explanation for this dramatic change in river level and river character has been attributed to a number of causes. George (1958) described the gorge as an outstanding example of rejuvenation related to a Tertiary lowering of sea levels. Linton (1963), on the other hand, explained the gorge as a product of lowered base-level, where the river descended into an 'ice-cut trough' that Sissons (1976b) subsequently claimed was scoured by a Highlands ice stream which flowed up this part of the Clyde Valley. A third, but less plausible, explanation is that offered by Whittow (1977) who argued that the gorge was the result of rejuvenation caused by tectonic uplift. However, construction of the hydro-electric power station at Bonnington revealed a buried former channel of the Clyde to the east of the present river course (Ross, 1927), and McLellan (1969) mapped the full extent of an area of glacial deposits blocking this channel. Upstream of these deposits a former lake existed in which abundant silts, sands and clays were deposited (Laxton and Nickless, 1980). This former lake basin explains the low river gradient upstream of the falls. The latter were cut upon deglaciation, the Clyde eroding a new channel in bedrock before regaining its original course at the mouth of the gorge.

Disruption of pre-existing river courses has occurred in the Highlands due to glacial erosion of cols and valley heads and the production of glacial breaches (Linton, 1949a, 1951a, 1963) (see also the Cairngorms). The Falls of Clyde, however, provide the most dramatic example in Scotland of disruption of a major river course resulting from glacial deposition in the original river channel and therefore differs from other sites such as Corrieshalloch Gorge (see above). The individual landforms demonstrate specific geological controls on the form and configuration of the waterfalls, rapids and slot gorge which collectively comprise the site.

Conclusion

The Falls of Clyde provide a particularly striking example of the effects of glacial disruption of drainage. The original channel of the Clyde was infilled with glacial deposits, forcing the river to adopt a new course. The latter takes the form of a bedrock gorge and is distinguished by the presence of two waterfalls. The site is important in illustrating some of the indirect, but nevertheless significant, effects of glaciation on the landscape.

DUNDONALD BURN

D. G. Sutherland

Highlights

The sediments exposed in the stream section at Dundonald Burn include a sequence of estuarine, littoral, aeolian and buried peat deposits. These provide important sedimentary, pollen and marine fossil evidence for changes in sea level and coastal environmental conditions during the Holocene.

Introduction

The site at Dundonald Burn (NS 337372) comprises a stream section, located 2 km south-east of Irvine. Along the Ayrshire coast, from south of Ayr to north of Ardrossan, major coastal embayments have existed at two distinct periods. During the Lateglacial, sea level was initially at 26–28 m OD (Jardine, 1971; Boyd, 1986a) and in areas such as the valley of the River Irvine the sea penetrated inland for over 10 km. Sea level fell from this altitude during the Lateglacial and early Holocene only to flood the lowlands again during the middle Holocene, attaining a maximum altitude of approximately 12 m OD (Jardine, 1971; Boyd, 1982, 1986b). Sedimentary sequences related to the period of low sea level during the early Holocene and the subsequent Main Postglacial Transgression have been studied along the Dundonald Burn close to its confluence with the River Irvine and at the nearby 'Great Bend' (NS 324372) on the River Irvine (Crosskey, 1864; Smith, 1896b; Jardine, 1971; Jardine and Morrison, 1980; Akpan and Farrow, 1984; Boyd, 1986b, 1988).

Description

The stratigraphic sequence exposed by the Dundonald Burn has been most recently recorded by Boyd (1988) as follows:

5.	Orange sand with occasional organic detritus	?
4.	Peat with occasional silt and fine gravel	0.04 m

3. Bedded organic detritus containing some sand 0.15 m
2. Dark, bedded organic detritus 0.14 m
1. Finely, horizontally bedded clay grading upwards into organic detritus 0.03 m

The basal clays (bed 1) may be equivalent to the grey sands exposed at the base of the Great Bend section (Smith, 1896b; Boyd, 1986b), since Smith (1896b) reported that the sands became peaty towards the eastern end of the section. *Pholas* borings and shells occur *in situ* in these grey sands. At the Great Bend, these sands are unconformably overlain by sands and gravels (Boyd, 1986b), the lateral equivalents of bed 5 at Dundonald Burn. At the Great Bend, a 'basal gravel' layer has been recognized as being distinct from the remainder of these sands and gravels. This 'basal gravel' rests on the grey sands.

The sands and gravels are fossiliferous, containing abundant shells of marine molluscs as well as various types of algae (Smith, 1896b; Jardine and Morrison, 1980; Akpan and Farrow, 1984; Boyd, 1986b). From this area also a number of whale (*Balaena glacialis* (Müller)) bones have been recovered (Crosskey, 1864; Smith, 1896b; Jardine and Morrison, 1980). The marine sands and gravels are up to 4.5 m thick and comprise a series of ridges with an amplitude between 0.5 m and 2.0 m (Jardine, 1971; Jardine and Morrison, 1980). Wind-deposited sands with interstratified peat lenses rest upon these marine deposits and extend as much as 2 km inland. Abundant Mesolithic and younger artifacts have been found among the sand dunes (Jardine and Morrison, 1980).

Analysis of the contained fauna has allowed inferences to be made about the conditions of deposition. The occurrence of *Pholas* shells in the grey sands at the Great Bend implies that these were exposed in the intertidal zone following deposition. The overlying 'basal gravel', however, contains faunal elements indicative of deposition in water depths of around 10 m (Akpan and Farrow, 1984; Boyd, 1986b), whereas the upper marine sands and gravels were laid down in sublittoral water (Akpan and Farrow, 1984; Boyd, 1986b).

Further information as to the chronology of events and the local environment during the early Holocene derives from pollen analysis and radiocarbon dating of the Dundonald Burn organic deposits (beds 2, 3 and 4). Boyd (1988) recognized five local pollen assemblage zones in these organic deposits. The basal *Salix–Filipendula*–Filicales zone indicates a period of open vegetation, and was considered to be of Loch Lomond Stadial or very early Holocene age. The next pollen zone is characterized by pollen of taxa which indicate the expansion of sedges and birch and thereafter a pronounced expansion of coryloid pollen, these vegetational changes being typical of the early Holocene in central Scotland. Following the *Corylus* rise there was a period of locally dense *Salix*-dominated woodland. The final pollen zone was defined on the basis of two samples from an isolated peat fragment in the overlying sands (bed 5), and is characterized by *Quercus*, *Ulmus* and *Alnus* pollen indicating immigration of mixed boreal forest into the area prior to marine inudation and erosion of the top of the organic horizon. By comparison with other dated pollen diagrams, the *Corylus* rise may be placed at around 9300 BP and the arrival of alder at approximately 7000 BP (Boyd, 1988).

Four separate samples have been radiocarbon dated from the organic beds. The basal 0.02 m gave an age of 9780 ± 90 BP (SRR–382) and the top 0.02 m 8070 ± 70 BP (SRR–381) (Harkness and Wilson, 1979), both of these dates being in agreement with the relative dating based on pollen analysis. Two further samples from within the organic beds, although not so critically placed, are in accord with the other radiocarbon dates: 8950 ± 90 BP (GU–373) (Ergin *et al.*, 1972) from the top 0.05 m of the organic horizon and 9530 ± 150 BP and 9620 ± 150 BP (Q–642) (Godwin and Willis, 1962) from two assays on wood from within the organic horizon.

Two other dates are of relevance. A thin peat layer resting on littoral sands and gravels at 10.4 m OD and overlain by blown sand a few hundred metres to the south-west of the Dundonald Burn exposure has been dated to 3944 ± 190 BP (Birm–221) (Shotton and Williams, 1971). A biserially barbed point, manufactured from a red deer antler and probably contemporaneous with Mesolithic occupation of the area, was recovered from the bed of the River Irvine approximately 1 km from Dundonald Burn (Lacaille, 1954; Jardine and Morrison, 1980) and this has been dated to 5840 ± 80 BP (OxA–1947) (Bonsall and Smith, 1990).

Interpretation

Based on the available information, the following

inferences may be made as to relative sea-level change during the early to middle Holocene along this part of the Ayrshire Coast. During the early Holocene, sea level was low, the intertidal zone occurring at around 2 m OD (the *Pholas* bed). Peat accumulated on low ground inland of the coast, as at Dundonald Burn. Thereafter sea level started to rise, this rise continuing until after 8000 BP and probably after 7000 BP. During this transgression, the sands overlying the peat at Dundonald Burn were deposited. At the time of the maximum of the transgression, the basal gravel at the Great Bend was deposited in a water depth of about 10 m, and sand and gravel ridges were built up to an altitude of 12 m OD. During the subsequent regression towards present sea level, the series of sand and gravel ridges were formed in the littoral zone, and aeolian sands accumulated on their surface.

An organic bed in a stratigraphically similar position to that at Dundonald Burn has been reported near Troon, and radiocarbon dates on the top and base of that deposit support the concept of a low early Holocene sea level. The dates are 8015 ± 120 BP (IGS-C14/149) for the top and 9090 ± 320 BP (IGS-C14/150) for the base (Welin *et al.*, 1975).

The Dundonald Burn area is the only location on the North Ayrshire coast at which the Holocene coastal sediments have been studied in detail. The information obtained has revealed complex changes in both the terrestrial and marine environment, particularly during the early to middle Holocene. These changes have been in response to the climatic amelioration at the onset of the Holocene, but most especially to the variations in sea level and the corresponding migration of the shoreline.

A number of other localities in west-central Scotland, such as Linwood Moss (Jardine, 1971) and Girvan (Jardine, 1962, 1963, 1971) have broadly similar records of environmental change but none of these has provided the range of detail comparable to that in the evidence from the Dundonald Burn area.

Conclusion

The sediments at Dundonald Burn provide valuable evidence for interpreting the sea-level history of the western Central Lowlands. Changes in the coastal environment during early and middle Holocene times (approximately 10,000–6,000 years ago), culminating in the advance of the sea known as the Main Postglacial Transgression (see Silver Moss above), have been revealed by detailed analyses of the sediments and the fauna and pollen that they contain. Dundonald Burn is an integral component in the network of sites for demonstrating Holocene sea-level change.

TINTO HILL

C. K. Ballantyne

Highlights

Tinto Hill is important for studies of periglacial processes, illustrating the best examples of active stone stripes in Scotland.

Introduction

Tinto Hill (NS 953343) is a broad, rounded hill, elongated in an east–west direction. It is composed of felsite intruded into Old Red Sandstone and, rising to 707 m OD, dominates the middle Clyde Valley. Its lower slopes are covered by drift, and along the northern flank, generally at an altitude of less 300 m OD, there is an impressive sequence of meltwater channels (Sissons, 1961b). However, Tinto Hill is most important for periglacial geomorphology, demonstrating an assemblage of active stone stripes (Miller *et al.*, 1954; Galloway, 1958; Ballantyne, 1981, 1987a).

Description

The upper parts of Tinto Hill are apparently bare of glacial deposits and, where the vegetation cover is broken, are covered by frost-shattered debris consisting of angular felsite clasts set in a peaty, sandy matrix. On vegetation-free areas this debris is being arranged at the surface into outstanding examples of active stone stripes (Figure 16.6), the first and possibly finest examples of their kind reported in Scotland (Miller *et al.*, 1954; Galloway, 1958).

Miller *et al.* (1954) reported patterned ground occurring in three main areas: to the south of the summit of Tinto, south-east of there in the gully known as the Dimple, and on the northern side of the hill in Maurice's Cleugh. On the south side of the hill the features were found down to an

Figure 16.6 Stone stripes are particularly well developed on Tinto Hill. (Photo: J. E. Gordon.)

altitude of 580 m OD and on the north they were reported as low as 400 m OD. At present, the best-developed area of stripes lies south of the summit where they descend to 570 m OD.

On vegetation-free debris slopes, with a gradient of approximately 20°, and where the coarse fraction consists of material generally less than 0.15 m in length, the regolith is being frost-sorted into well-developed 'gutters' between lines of slightly updomed fine-grained material. The stripes are aligned directly downslope and the long axes of pebbles tend to be oriented downslope. Both coarse and fine stripes range in width from about 0.1 m to 0.3 m, and sorting extends to a depth of between 0.08 m and 0.15 m. Below the sorted surface layer, coarse and fine material is intermingled amongst black peaty humus.

In the Dimple, Miller *et al.* (1954) also found a form of sorted circle that they termed 'inverted garlands'. These consisted of convex-upslope, arcuate arrangements of flat or flat-lying clasts, 0.15–0.3 m across. The long axes of the clasts were arranged along the arc with the clasts lying on edge. Excavations to a depth of 0.5–0.6 m revealed a preponderance of similar clasts to those at the surface but with little development of the matrix found under the areas of stripe development.

Miller *et al.* (1954) reported that stripes had started to reform on artificially disturbed ground after two winters. Ballantyne (1981, 1987a) found that a single winter was sufficient for perfect regeneration of stripes on ground dug over to a depth of 0.3 m, and recorded downslope movement of surface clasts averaging 0.25 m (maximum 0.63 m) over a six month period covering the winter of 1977–78. Such experiments prove conclusively the present-day activity of these features. By comparing different editions of Ordnance Survey 1:10,560 maps, Miller *et al.* (1954) observed that the extent of bare ground on the surface of Tinto Hill had increased greatly since 1862. The present extent of development of the patterned ground seems likely therefore to be due to increased grazing pressure on the hill over the last 100 years (see Ballantyne, 1991a).

Interpretation

The processes responsible for the formation of the stripes are not fully understood. Frost heave resulting from the development of needle ice is important, but Miller *et al.* (1954) considered that the stripes also played a role in the drainage of the hill, noting that there was no gullying of the areas where stripes were developed. Ballantyne (1987a) also suggested a possible role for running water, speculating that the development of rill networks on a bare ground surface may have been an initial condition for the development of lateral sorting. He also suggested that surface wash may have been in part responsible for the high rates of movement he observed for surface clasts. To explain the origin of similar features in the Lake District, Warburton (1987) invoked density-driven convection of soil water, upfreezing of clasts, and downslope movement of debris resulting from creep and rillwash.

Scottish mountains carry a wide range of relict and active periglacial features (Ballantyne, 1984, 1987a). In southern Scotland, the summits and slopes of the main hill groups are mantled by frost-weathered detritus that has been extensively soliflucted (Galloway, 1961a; Tivy 1962; Ragg and Bibby, 1966). On Tinto Hill, where the vegetation has been stripped clear, the surface of this material displays the most outstanding development hitherto reported of active stone stripes in Scotland. Although active stone stripes are known from a number of mountains in the Highlands and Islands (see the Cairngorms and Western Hills of Rum; Godard, 1959; Ballantyne, 1987a; Carter *et al.*, 1987), the examples on Tinto Hill are exceptional for their size, clarity and degree of development. Tinto Hill is also the first locality in Scotland at which such features were described in detail.

Conclusion

Tinto Hill is an important component of the network of sites for periglacial geomorphology, that is for features formed under extremely cold, but non-glacial conditions. The results are evident in intense weathering of bedrock by frost action and other related processes and in frost disturbance of the soil, producing surface arrangements of stones and finer material in the form of stripes and circles, known as patterned ground. Tinto Hill is particularly notable for active-process studies of the formation of stone stripes.

Chapter 17

Lothians and Borders

INTRODUCTION

D. G. Sutherland

This area includes the lowlands of the Lothians along the southern shore of the Firth of Forth and the coastal area south to Berwick (Figure 17.1). Inland it extends west to the A74 across the Southern Uplands and south to the Tweed Valley. The area of highest ground, rising to 840 m OD at Broad Law, lies between Peebles and Moffat, and other hill groups (Lammermuir Hills, Moorfoot Hills) have summits up to 500–700 m OD. The Lothians area was one of the key areas in Scotland for the elaboration of many of the concepts related to ice-sheet glaciation and glacial sediments and landforms. In this area many of the early observations on both large- and small-scale features of ice moulding were made and the relationships between clast orientation, striations on rock surfaces and ice-flow direction were noted.

The earliest interpretation of striations in the region as being the product of ice flow was that of Agassiz (1841b) for a rock surface on Blackford Hill, which is today commemorated by a plaque at Agassiz Rock. Although the concept of multiple glaciation was also advanced at an early stage in this area (Croll, 1870b; J. Geikie, 1877, 1894) and sequences containing distinct glacial sedimentary units were described (see below), all the presently known glacial deposits are now attributed to the advance and retreat of the last ice-sheet and the subsequent Loch Lomond Readvance. However, marine erosional features are known that pre-date at least the last glaciation and these are well displayed at Dunbar. There, an ice-moulded high rock platform occurs at approximately 23 m OD (Sissons, 1967a), and a platform at about present sea level can also be seen to pass under glacial sediments along the East Lothian coast.

The earliest phase of the last glaciation was an advance into the Lothians and northern Southern Uplands by ice originating in the south-west Highlands. This ice deposited a basal till, with a characteristic erratic assemblage, found widely in the Midlothian basin (Kirby, 1968) as well as across the Moorfoot Hills (Aitken *et al.*, 1984). The direction of movement was generally from west to east in the north-west of the region, with a south-easterly component near the east coast. Subsequently, Southern Uplands ice expanded to exclude the Highland ice from virtually the whole of the region, with the exception of the coastal fringe of the Firth of Forth. As elsewhere along the southern Central Lowlands (Chapter 16) this second phase of glaciation resulted in the deposition of a till containing Southern Uplands erratics on top of the earlier till (see Hewan Bank and Keith Water). In the Lothians this ice moved in a north-easterly direction carrying erratics of Tinto Hill felsite to the outskirts of Edinburgh and the Midlothian basin (McCall and Goodlet, 1952; Mitchell and Mykura, 1962). In the Borders area the only evidence of glaciation is that of Southern Uplands ice which flowed down the Tweed Valley, the direction of flow being demonstrated by transport of erratics (Kerr, 1978) and by a major drumlin field (Sissons, 1976b). This ice was dominantly nurtured in the hills of the eastern Southern Uplands, but a small ice-cap may have developed on the Cheviot Hills (Clapperton, 1970).

The retreat phase of the ice-sheet resulted in the production of spectacular sequences of meltwater channels and glaciofluvial sediments. These are best developed and have been most intensively studied in the Lothians and northern hills of the Southern Uplands (Sissons, 1958a, 1958b, 1960, 1961a, 1963b; Price, 1960, 1963a; Kirby, 1969c; McAdam and Tulloch, 1985; Davies *et al.*, 1986). Particularly outstanding meltwater channels occur along the northern face of the Lammermuir and Moorfoot Hills, as at Rammer Cleugh and by Carlops in the Midlothian basin. These channels were formed as the Southern Uplands ice retreated towards the south-west. In the area between the retreating Southern Uplands ice and the Highlands ice, which still occupied the Firth of Forth, a sequence of outwash terraces was deposited (Kirby, 1969c). A third till unit (the Roslin Till) in the Midlothian basin was interpreted by Kirby (1968, 1969b) as having been deposited on top of these outwash gravels during a readvance of the Highland ice, but Martin (1981) disputed this interpretation. Instead, he suggested that the Roslin Till was not a single lodgement till but the result of deposition of debris flows during the general deglaciation sequence.

Few details are available on shorelines formed during ice retreat and it is not known whether there are shorelines that correlate with those identified in eastern Fife by Cullingford and Smith (1966) (Chapter 15). However, the proximity of the two areas suggests that deglaciation of the East Lothian coastline probably occurred at some

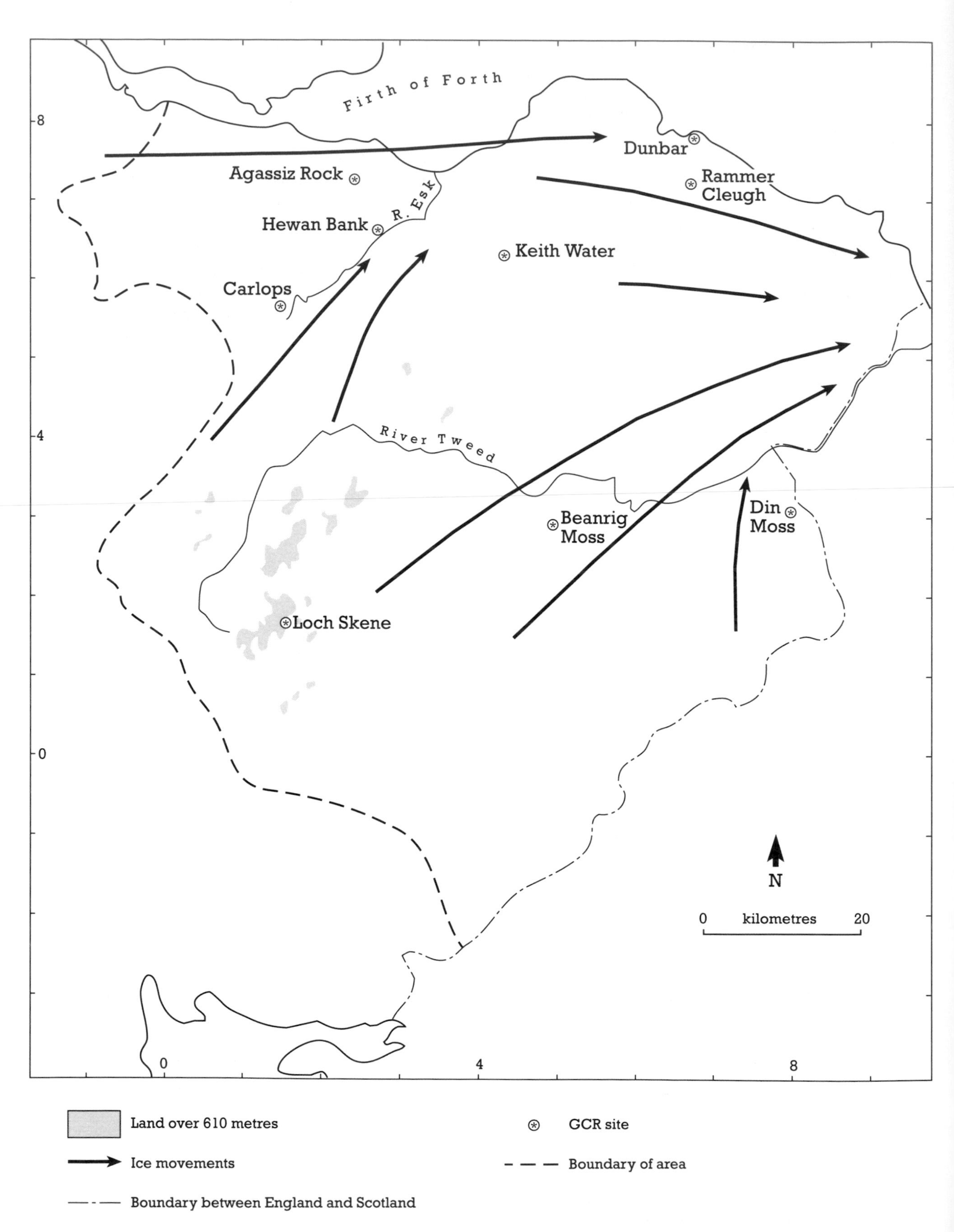

Figure 17.1 Location map of the Lothians and Borders area.

time between 17,000 and 14,000 BP. The marine limit, formed at the time of deglaciation, rises westwards along the coast from around 14 m OD (Sissons *et al.*, 1966) or 22 m OD (Davies *et al.*, 1986) near Dunbar to over 34 m west of Edinburgh and up to 38 m at Stirling (Sissons *et al.*, 1966). Deglaciation of the coast west of Edinburgh probably occurred between 14,000 and 13,000 BP, and a particularly pronounced shoreline was formed when the ice-front was in the Stirling/Larbert area. This shoreline, termed the Main Perth Shoreline (Sissons and Smith, 1965a), slopes towards the south-east at a gradient of about 0.43 m km^{-1} (Smith *et al.*, 1969), declining in altitude to about 5 m OD near Dunbar. As the ice retreated up the Firth of Forth, poorly-sorted glaciomarine deposits were laid down close to the ice-front (Sissons and Rhind, 1970); these were succeeded at greater distance from the ice by laminated silts and clays containing a restricted arctic marine fauna (Browne *et al.*, 1984).

A number of sites have been investigated that provide information on the evolution of the vegetation during the Lateglacial (Newey, 1965a, 1970; Mannion, 1978a; Webb and Moore, 1982; Alexander, 1985). Particularly interesting among these sites is Beanrig Moss (Webb and Moore, 1982) because of its detailed macrofossil record. The earliest vegetational communities following deglaciation were of pioneer plants. No radiocarbon dates are available from the earliest phase of vegetational development, although it is reasonable to infer that some areas were deglaciated well before the opening of the Lateglacial Interstadial (at around 13,000 BP). It was probably at this time of cold climate that the fossil ice-wedge polygons identified in Berwickshire (Greig, 1981) were formed.

During the interstadial itself a mosaic of different vegetational communities developed, Webb and Moore (1982) identifying juniper scrub with sparse tree birches, well-drained grasslands, damp, tall-herb meadows, dwarf-shrub heaths and open-ground communities as the principal elements. As elsewhere in Scotland, certain sites suggest a single uninterrupted phase of vegetational development during the interstadial, whereas others show a period of vegetational 'revertence' during the first half of the interstadial, (Alexander, 1985). This is typically reflected in a reduction in the proportion of woody plants, as represented in the pollen record. No radiocarbon dates closely bracket this brief phase and its status remains uncertain.

The Loch Lomond Stadial is registered unequivocally across the region as a period of severe climate during which small valley and corrie glaciers developed in the higher hills of the Southern Uplands, as around Loch Skene. The vegetation was composed of bare-ground communities, even at low altitudes, and a subdivision of the stadial sediments can be made into an early phase with relatively low values of *Artemisia* pollen and a later phase with much increased values of *Artemisia*. This may reflect an earlier more moist and a later drier climate. A similar inference may be drawn from an assemblage of fossil beetles from Corstorphine in Edinburgh (Coope, 1968). Along the coast marine erosion was apparently particularly effective during the stadial and a pronounced erosional surface and accompanying shoreline, the Main Lateglacial Shoreline, was produced (Sissons, 1969, 1976a; Browne *et al.*, 1984; Browne, 1987). This shoreline is isostatically tilted at approximately 0.17 m km^{-1} towards the south-east, descending from close to present sea level near Grangemouth to well below it along the Berwickshire coast (Eden *et al.*, 1969; Sissons, 1976a), and rivers such as the Tweed (Rhind, 1972) may have eroded deep valleys in their lower reaches at this period. Fluvial and slope activity were also enhanced inland, as is indicated by the contemporaneous deposition of a large alluvial fan in the former Corstorphine Loch (Bennie, 1894; Tait, 1934; Newey, 1970), and solifluction deposits in the former Holyrood Loch (Sissons, 1967a, 1971).

The early Holocene witnessed a rapid progression of vegetational communities from a period of juniper scrub dominance at around 10,000 BP through a period of birch and hazel woodland to the development of mixed deciduous forest of mainly oak and birch throughout most of the region by 8500 BP. A detailed and well-dated record of these changes is available from Din Moss (Hibbert and Switsur, 1976). Just before 5000 BP a reduction in elm pollen marks the start of human interference, with the eventual expansion of heath, grassland and bog communities as the forest was cleared. A trend in climate to more moist and cooler conditions over the same period may also have contributed to this change in the vegetation. A detailed record of Holocene environmental changes based on pollen and diatoms is also available from Linton Loch (Mannion, 1978a, 1978b, 1978c, 1978d, 1981a, 1981b, 1982).

Sea level was relatively low during the early Holocene but marine transgression was under way by about 7500 BP (Robinson, 1982). The transgression culminated in the formation of the Main Postglacial Shoreline some time between the above date and 5500 BP, at which time extensive shell beds were deposited along part of the East Lothian coast (Smith, 1971; McAdam and Tulloch, 1985). After the formation of the Main Postglacial Shoreline a series of lower shorelines was formed. These have not been accurately dated.

THE GLACIATION OF THE EDINBURGH AND LOTHIANS AREA

J. E. Gordon

The Quaternary landforms and deposits of this area have a long record of scientific study. The earlier 19th century work is dominated by accounts of the superficial deposits and bedrock markings which were explained in terms of diluvial or marine processes. Nevertheless, many of the original field observations and reports stand the test of time, and they remain valuable and pertinent contributions to the literature, most notably those of Maclaren (1828) and Milne Home (1840). In the years following the visit of Agassiz to Scotland in 1840 (see Agassiz Rock), the glacial theory gradually became established.

In an early keynote paper Milne Home (1840) described in some detail the superficial deposits of the Edinburgh and Lothians area, erecting an eight-unit stratigraphic succession. Boulder clay, sometimes resting on a layer of sand and gravel, was overlain by a sequence of sands, gravels and clays. Although he explained the full sequence in terms of marine inundation, his carefully set out field observations represent a landmark in Scottish Quaternary stratigraphy. Later, Nicol (1844) and Chambers (1853) quoted extensively from Milne Home's succession, Chambers adding further observations of his own in support. In the first editions of the Geological Survey Memoirs for the Edinburgh and East Lothian areas, Geikie (in Howell and Geikie, 1861) and Young (in Howell *et al.*, 1866) presented broadly similar successions to those of Milne Home: boulder clay overlain by sands and gravels and brick clays, then succeeded by raised beach deposits.

Fleming (1847, 1859) described essentially the same succession as Milne Home, but classified the deposits using different terminology. At one locality he noted an upper boulder clay resting on sand (Fleming, 1859, p. 75).

The existence of more than one till in the area was possibly first recognized by Maclaren (1828; see also Maclaren, 1838, 1866). He described the following sequence of deposits in sections along the Dalkeith railway:

3. Bedded sands.
2. Upper boulder clay; redder, looser texture and with fewer and more angular stones than the lower boulder clay.
1. Lower boulder clay; stiff, blackish and bluish clay interspersed with boulders and stones.

This sequence is similar to that at Hewan Bank (see below) and has recently been widely recognized throughout the Edinburgh area in the regional stratigraphic scheme of Kirby (1966, 1968, 1969b). Maclaren's valuable early observations, however, have largely been neglected by subsequent workers.

Maclaren (1828, 1838) originally interpreted the deposits as the product of ocean currents or a succession of great waves in the style of Hall (1815). Later, as a confirmed glacialist, he proposed that the lower boulder clay was formed as a glacial deposit, but, surprisingly, suggested that the upper one was formed by icebergs and ocean currents. However, in the preface to the second edition of his book (Maclaren, 1866) he does note that his interpretations of the surface deposits of the area were not fully recast with his revised views on glaciation. Significantly, Maclaren (1866) inferred that the two compositionally different boulder clays were formed in one uninterrupted depositional event.

J. Geikie (1877, p. 72) first introduced the idea of the interaction of more than one ice mass in the southern part of the Midland Valley. He suggested the presence of ice from separate sources in the Highlands and Southern Uplands in what he termed 'the debatable ground'. As part of his evidence he cited the intermingling of stones of both northern and southern origins in the till of the Esk Valley in Midlothian and noted the occurrence of Highland erratics as far south as Tynehead. Subsequently, Somervail (1879) inferred two directions of erratic transport in the Pentland Hills: from south-west to north-east and from north-west to south-east. This idea of shifts in ice-movement direction later assumed considerable importance in 20th century investigations of till clast fabric and till lithology patterns.

The glaciation of the Edinburgh and Lothians area

In the second edition of the Geological Survey Memoir, Peach *et al.* (1910a) envisaged a composite ice-sheet from sources in the Highlands and Southern Uplands moving eastwards across the Edinburgh area. They emphasized that variations in the character of the till were related to variations in the underlying bedrock. However, they made no references to multiple till sequences nor the previous accounts of them.

Peach *et al.* (1910a) considered that the thick sequences of sand and gravel above the till in the Midlothian basin were deposited when the southern ice had receded, but drainage was still obstructed by ice to the north. Significantly, they noted that the gravels largely comprised greywacke pebbles derived from the Silurian rocks to the south.

Campbell (1951) identified three separate boulder clays in the Esk Valley in the Pentland Hills. The lithological character of the deposits suggested that the lowest one was associated with ice moving east-north-east, the middle one with local ice in the Pentland Hills and the uppermost one with ice moving east-south-east. Eckford (1952), however, considered that the yellow colour of the middle till could reflect weathering, whereas the other two could be explained by changes in direction of flow of the dominant ice mass. Eckford in fact described two tills from the southern part of the Pentland Hills, but suggested that both were associated with the same ice-sheet, the lower one being a basal lodgement till, the upper one being formed by ablation of englacial and supraglacial debris. On lithological grounds Eckford considered that the last ice invasion of the area was from the Southern Uplands and was represented by extensive sand and gravel deposits overlying the Highland drift.

McCall and Goodlet (1952) studied indicator stones (erratics indicative of a particular lithological source), especially felsites, in various Midlothian deposits and concluded that the lower boulder clay of the area (which they did not differentiate) was the product of Highland ice.

Tulloch and Walton (1958) reached a similar conclusion to Peach *et al.* (1910a) that the composition of the till in the area of the Midlothian Coalfield closely reflected the character of the local bedrock. This view was again echoed by Mitchell and Mykura (1962). The latter summarized the glacial sequence in the Edinburgh area as:

3. Southern Uplands Readvance Boulder Clay and Gravel.
2. 'Middle' Sands and Gravels.
1. Basal Boulder Clay.

The available evidence suggested that a single ice-sheet flowed eastwards across the area. From erratics of distinctive lithology they inferred that most of this ice originated in the western Highlands but coalesced in the southern part of the Edinburgh area with ice from the Southern Uplands. Like Eckford (1952), they found no definite evidence that the two tills were deposited by separate ice-sheets, or even different phases of a single ice-sheet. Significantly, also, they found no evidence of three tills as recorded by Campbell (1951). During the retreat of the Highland ice, the extensive suite of 'Middle' sands and gravels was deposited.

Charlesworth (1926b) introduced the idea of a Lammermuir–Stranraer 'moraine' formed during the retreat phase of the last ice-sheet. In its eastern part, he correlated the 'moraine' with the extensive deposits of sand and gravel along the northern flanks of the Lammermuir and Moorfoot Hills. The 'moraine' was formed by a readvance of Highland ice from the north. Sissons (1961c), however, showed that the sand and gravel deposits were not ice-marginal moraines but were rather 'dead'-ice features formed diachronously. Moreover, several lines of evidence in the area to the west strongly suggested that the last ice movement was from the south and south-west, not from the north (McCall and Goodlet, 1952; Bailey and Eckford, 1956; Sissons, 1958b, 1961c).

Kirby (1966, 1968, 1969a, 1969b, 1969c) carried out a detailed investigation of the glacial deposits in the Lothians area. On the basis of stratigraphy, fabric analyses and stone counts he recognized a sequence of:

5. Roslin Till
4. Sand
3. Intermediate Till
2. Sand
1. Basal Till

The Basal Till occurs widely above bedrock throughout the area, from sea level up to about 250 m in the Esk basin. The Intermediate Till is best developed at the southern end of the basin on higher ground, but appears to be absent as a separate unit near the coast. At a number of localities the two tills are superimposed, and the

Intermediate Till occupies an analogous position to the reddish-brown till in the Pentland Hills, recorded by Eckford (1952) and Mitchell and Mykura (1962). Indeed the stratigraphy described by Kirby is basically that noted first by Maclaren in 1828. Kirby also recorded a similar stratigraphy for East Lothian (see Keith Water).

From its lithology and fabric, Kirby inferred that the Basal Till was deposited by ice moving from the west, as noted by many of the early workers in the area (see above) and also more recently by Burke (1968); the Intermediate Till was deposited by ice from the south. Since both tills graded up into their own meltwater deposits, Kirby envisaged two temporally separate stages of ice movement, with ice from Highland sources being succeeded by ice from the Southern Uplands. Locally, however, he also identified complex sequences of till and glaciofluvial sediments formed during a single phase of glaciation (Kirby, 1969c) (see also Young, 1966).

Several authors have referred to washed tills in the Lothians area, usually comprising a sandy clay overlying the more typical tills (for example, Geikie, 1863a; Burke, 1968, Kirby, 1968). Kirby considered them to be subglacial deposits; Burke, the result of washing in the immediate post-glacial period.

In addition to the Basal and Intermediate tills of Kirby, a third and later till unit has been identified in the Midlothian area. Peach *et al.* (1910a) first referred to a reddish-brown boulder clay overlying sands and gravels in the Eskbank and Newton Grange areas, but did not attach any particular significance to it. Later, Anderson (1940) described between 3 and 10 ft (0.9–3.0 m) of what he called 'Upper Boulder Clay' overlying considerable thicknesses of sand and gravel in a number of sandpits around Roslin. Anderson also recorded clay-filled fissures in the sands and gravels, interpreting them as frost cracks. Additional examples of the latter were subsequently reported by Common and Galloway (1958). Anderson considered the Upper Boulder Clay to represent a readvance of ice. He did not specify Highland ice as Mitchell and Mykura (1962) stated, although it might be a logical inference from the indicator stones that he recorded.

Carruthers (1941, 1942), however, considered that the frost wedges described by Anderson were post-glacial and that the till–sand contact was not typical of readvancing ice, but rather was the product of subglacial meltout during a single event (Carruthers, 1939). Anderson (1941, 1942) countered by arguing that there was no trace of the fissures in the boulder clay.

Contrary to Anderson's results, McCall and Goodlet (1952) found no rocks of Highland origin in what they now called the 'Roslin Upper Boulder Clay', but there was abundant felsite which they referred to the Tinto Hill outcrop. They therefore concluded that the boulder clay was laid down during a readvance of Southern Uplands ice after the retreat of the main Highland ice which deposited the lower till of the area.

Kirby (1966, 1968) proposed the term 'Roslin Till' for this third till after the area where it was best exposed. He noted that it was only identified there because it overlay thick glaciofluvial deposits (cf. Tulloch and Walton, 1958). Elsewhere it was visually indistinguishable from the underlying till. However, Kirby put forward a variety of evidence to show that the Roslin Till represented a distinctive readvance of ice. His main arguments included the presence of frost wedges and deformation in the top of the sands below the Roslin Till and variations in lithological content between the Roslin Till and the two earlier tills. Mechanical analysis and fabrics excluded the possibility of the Roslin Till being an ablation moraine (Kirby, 1969b, 1969c). The fabric also suggested that it was not a solifluction deposit and, although its orientation was similar to that of the underlying intermediate till, the lithologies of the two units were quite distinctive. On the basis of the fabrics and stone counts he inferred that after a phase of glaciation by Highland ice, ice from the Southern Uplands entered the area from the south. From geomorphological evidence (Kirby, 1969c), he suggested that the ice then bifurcated near the present watershed at Kingside. One lobe retreated southwards (Sissons, 1958b, 1963b). The northern lobe receded northwards downslope into Midlothian and, during a subsequent readvance of this lobe, the Roslin Till was deposited on top of some of the earlier deglaciation features.

Subsequently, Martin (1981) has reassessed the status of the Roslin Till and presented a revised interpretation of the glacial stratigraphy. From sedimentological studies and comparisons with contemporary glacial environments, he concluded that there is no basis for regarding the Roslin Till as a separate stratigraphic unit. The sequence of glacial deposits in Midlothian shows considerable lateral and vertical variation and is clearly analogous to that seen along present-day glacier margins, for example in Iceland, where outwash, flow till and subaerial fan depositional environments are all closely related. Thus the till overlying the sands and gravels of the Roslin area has been interpreted by Martin as part of a complex sequence of diachronous debris flows draped over sediments deposited at receding ice margins. Martin considered features such as the till-filled cracks in the sand and gravel to reflect loading pressures rather than ice wedging. Martin also questioned whether the two lower tills are temporally discrete, and suggested rather that they represent a single complex meltout sequence such as described by Young (1966) and Kirby (1969a). He argued that the clast fabrics were not sufficient evidence to assign the tills to separate glacial episodes. Variations in flow conditions in a single glacier could produce clast fabrics both normal and parallel to the ice flow. Also, the observed lateral variations in clast composition could be explained by variations in solid geology, whereas the vertical variations could reflect the reverse order of lithologies traversed by the glacier (cf. Boulton, 1970).

Martin therefore proposed a single Late Devensian ice-sheet derived from the west, but with a flow component from the south to account for the transport of the Southern Upland erratics into the Lothian basin. This ice-sheet deposited a heterogeneous lodgement till, possibly with a melt-out component, with sand and gravel deposits formed as diachronously off-lapping wedges during the recession and marginal stagnation of the ice; the latter deposits were then buried by debris flows. Such a pattern accords with observed associations of depositional environments at modern glacier margins (Boulton, 1972b; Lawson, 1979). These observations demonstrate that the tripartite sequences comprising sands and gravels between tills, conventionally interpreted in terms of multiple glaciation, can quite normally relate to a single phase of ice retreat (Boulton, 1972b). Similarly, reappraisal of lodgement till complexes (Eyles *et al.*, 1982) has shown that observed variations in their sedimentary characteristics can be satisfactorily explained in terms of a single glacial episode. For example, till deposition may be interrupted by phases of erosion or deposition of subglacial meltwater sediments, and lateral migration of basal flow-lines may produce the unconformable superimposition of lodgement till units derived from different source areas.

AGASSIZ ROCK

J. E. Gordon

Highlights

The striated rock surface at Agassiz Rock was first recognized by Louis Agassiz in 1840 to have been eroded by glacier ice. The site is historically significant for its part in the development of the glacial theory in Scotland.

Introduction

Agassiz Rock (NT 254702) is located on the south side of Blackford Hill in Edinburgh. It is principally of historical interest as a striated rock surface that was associated with the early development of glacial theory in Scotland. In addition, Agassiz Rock represents an important landmark in geological conservation, being one of the earliest Quaternary sites recognized as requiring safeguarding. The site has been referred to in a number of papers (Rhind, 1836; Milne Home, 1840, 1846, 1847a; Buckland, 1841a; Maclaren, 1841, 1842a; Fleming, 1859; Panton, 1873; Brown, 1874; Peach *et al.*, 1910; Mitchell and Mykura, 1962). It was included in field excursions of the 1948 International Geological Congress in Britain (International Geological Congress, 1948) and, in addition, it features in the itineraries recommended by Geikie (1901), Campbell (1951) and Waterston (1960).

Description

Agassiz Rock is located on the south side of Blackford Hill where an andesite cliff has been undercut to form a shallow cave, the rock surfaces of which are grooved and striated like the overhanging cliff (Figure 17.2). Early descriptions of the site include those of Rhind (1836) and Milne Home (1840). The former explained the grooving by molten rock falling on a bed of sand and retaining the moulded impression of its surface; the latter in terms of marine submergence. However, it was on 27 October 1840 that the site attained its fame when it was visited by Louis Agassiz. A few weeks earlier, Agassiz had delivered a paper at a British Association meeting in Glasgow in which he argued that all the northern parts of Europe, Asia and America were

Figure 17.2 Part of the smoothed and grooved rock surface at Agassiz Rock, Edinburgh, which has been attributed to glacial abrasion. The form of the rock surface bears a strong resemblance to glacially abraded surfaces elsewhere in Scotland and in modern glacial environments (Photo: J. E. Gordon.)

formerly covered with a mass of ice (Agassiz, 1841a). Although it is not recorded in the abstract of his paper, Agassiz apparently alluded to the former existence of glaciers in Scotland (Anon, 1840; Maclaren, 1840). After the meeting he departed on a tour of Scotland to investigate the field evidence (Davies, 1968a, 1968b). In the course of this journey, accompanied by William Buckland, he found many striking and convincing traces of former glaciers. When he visited Edinburgh he was taken on a tour to search for glacier markings on the south side of the city by a group of Edinburgh geologists, including Charles Maclaren, then editor of *The Scotsman*. Agassiz was doubtful about some of the features initially shown to him, but on seeing the cave at Blackford Hill is reputed to have exclaimed 'That is the work of the ice' (Maclaren, 1841, 1842a; Cox and Nicol, 1869).

The striations at Agassiz Rock form part of a local assemblage of features that indicate ice moving eastwards across the area (Figure 17.3). Blackford Hill itself is a crag and tail, 1.5 km long, with deep erosional grooves on both its north and south sides, comparable to those around Edinburgh Castle Rock (Sissons, 1971). A smaller superimposed crag and tail occurs at Corbie's Craig south of the hill top, and to the east clast fabric measurements in the main drift tail also conform with ice flowing to the east (Kirby, 1969b).

In the Edinburgh area, a prominent theme in many of the earlier 19th century accounts is the recognition of the overall easterly movement of the agent responsible for the superficial deposits and bedrock striations (Figure 17.4). Typical lines of evidence included the disposition of crag-and-tail forms (Hall, 1815; Maclaren, 1828, 1866); the transport of erratics (Milne Home, 1840, 1871, 1874a, 1874b; Nicol, 1848; Fleming, 1859; Campbell and Anderson, 1909); deformation and overfolding of strata to the east (Milne Home, 1840, 1871; Fleming, 1859; Brown, 1874); bed-rock striations and moulding (Imrie, 1812; Hall, 1815; Maclaren, 1828, 1842b,, 1866; Milne Home, 1840; Fleming, 1847, 1859; Chambers, 1853; Miller, 1864; Henderson, 1872; Richardson, 1877a, 1877b; Goodchild, 1896); and striations on stones in till (Milne Home, 1840; Maclaren, 1849; 1866; Miller, 1864, 1884; Henderson,

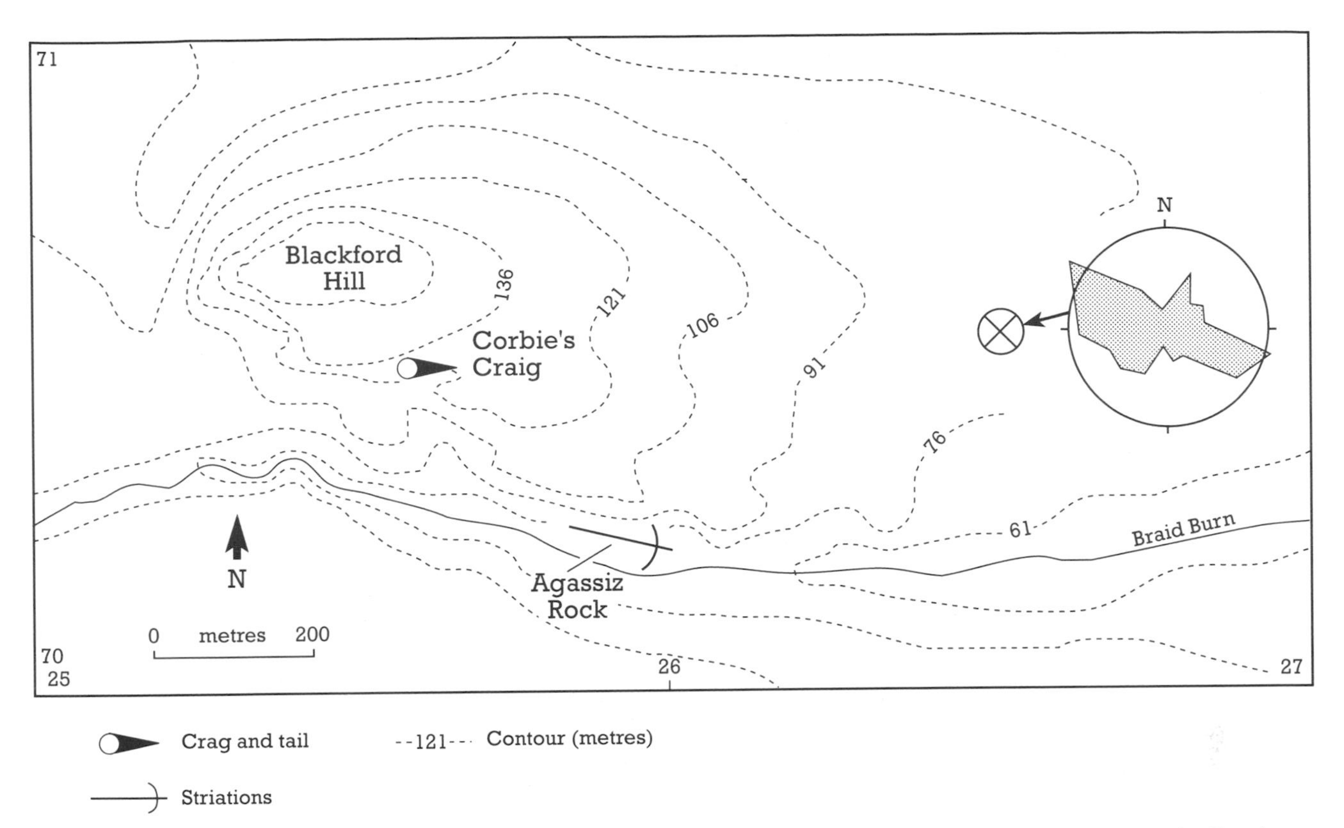

Figure 17.3 Blackford Hill crag and tail, showing a till clast fabric in the tail, the Corbie's Craig crag and tail and the direction of ice movement inferred from the striations at Agassiz Rock (from Kirby, 1969b).

1874). Miller (1884) produced the first map showing the pattern of striations on bedrock and till clasts. Further instances of striations and movements of erratics are given by Peach *et al.* (1910a), and Peach (1909) described the classic Lennoxtown essexite boulder train (see also Shakesby 1978, 1979, 1981). More recently Burke (1968, 1969) has quantified some of the evidence for these ice-movement trends, and Sissons (1971) has described the strong imprint of glacial erosion in central Edinburgh.

Interpretation

Although Agassiz Rock does not bear the distinction of being the first site in Scotland to have been recognized as the product of land ice, it was nevertheless of considerable significance (Buckland, 1841a; Davies, 1968a, 1968b), since the striations under the overhang could not have been produced by marine-drifted icebergs, the hypothesis of many contemporary geologists to explain such phenomena. Nor could they have been formed by debris-laden catastrophic deluges or floods as suggested by Hall (1815) in order to explain striations on nearby Corstorphine Hill, because of their close parallel arrangement over short distances.

Nevertheless, Milne Home (1846, 1847a) and Fleming (1859) were not convinced of the glacial origin of the striations at Blackford Hill and other localities around Edinburgh. The former persisted with the diluvial hypothesis, and the latter explained them as a local phenomenon associated with the Braid Burn. Geikie (1863a), however, in his important exposition on the evidence for former glaciers in Scotland clearly established that striations, including those at Blackford Hill, were the product of land ice. Brown (1874) also believed the striations to be glacial but considered that a large landslip had brought the striated rock to its present position.

Subsquent references in the literature to Agassiz Rock (Panton, 1873; Peach *et al.*, 1910a; Mitchell and Mykura, 1962) acknowledge the historical significance of the site, although critics have suggested that some of the striations may in fact be tectonic slickensides (Mitchell and Mykura, 1962).

Agassiz Rock is a site of considerable historical interest as one of the classic localities that played a significant part in the development of glacial

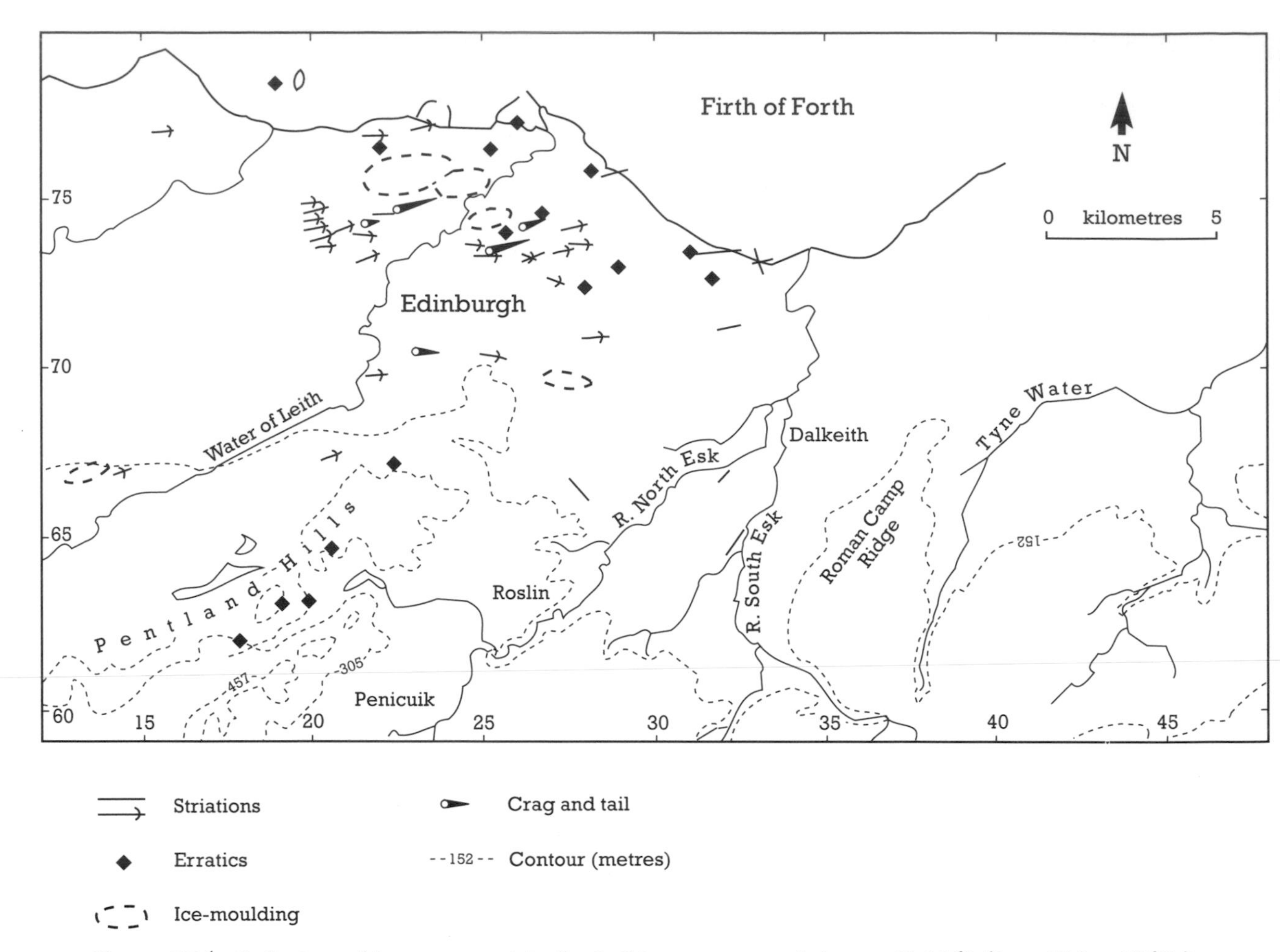

Figure 17.4 Indicators of ice movement in the Lothians area recorded up until 1863 (from Kirby, 1969b).

theory in Scotland. Its striated rock surface was among the first of such features to be recognized as the product of glacier ice by Louis Agassiz in 1840. It is also significant in another historical context, being one of the first geological sites recognized to require practical measures to ensure the preservation of its interest. In 1908 the Council of Edinburgh Geological Society successfully negotiated with Edinburgh Town Council to place a railing around the site and erect a memorial tablet (Watson, 1934). The railing and tablet are now dismantled, but it is planned to restore the plaque to mark the significance of the site.

Conclusion

Agassiz Rock is a site of considerable historical importance. Its significance stems from its association with Louis Agassiz, one of the principal figures in the introduction of the glacial theory in Scotland. The striated rock surface was unequivocally attributed by Agassiz to the effects of the passage of glacier ice. Agassiz Rock was also one of the first geological sites in Scotland to be conserved.

HEWAN BANK

J. E. Gordon

Highlights

The deposits in the section at Hewan Bank include two superimposed tills. These provide sedimentary evidence for the sequence and pattern of ice flow in the Lothians area during the Late Devensian and show successive ice movements from sources in the Highlands and Southern Uplands.

Introduction

The Hewan Bank (NT 285647) site comprises a section on the west bank of the River North Esk near Polton, 10 km south of the centre of Edinburgh. The deposits consist of a sequence of till and glaciofluvial sediments. Hewan Bank is the best natural exposure currently available showing facies of the type which have been used in the past to erect a regional glacial stratigraphy for the Edinburgh and Lothians area. Details of the sediments and stratigraphy are given by Martin (1981).

Description

Several exposures in the bank above an extensive area of landslips show the following sequence of deposits:

4. Gravel
3. Sand
2. Reddish-brown till
1. Blue-grey till

Details of these sediments are given by Martin (1981); the site corresponds with his Polton site. Martin described bed 1 as a massive, fissile diamicton. The upper diamicton (bed 2) is weathered with larger, more widely dispersed clasts; primary fissility cannot be distinguished. The contact with bed 3 is sharp and irregular, although lenses of till and clasts occur in the overlying sand. Deposits of sand, gravel, silt and clay are exposed above, and Martin (1981) recorded several facies variations including sands with load casts, stratified diamictons and stratified, reworked diamictons. Various sedimentary structures, notably sand pillars, are also present, as are small faults.

Interpretation

In terms of the conventional stratigraphies, the tills (beds 1 and 2) at Hewan Bank correspond with the Lower Boulder Clay of Mitchell and Mykura (1962) and the Basal and Intermediate Tills, respectively, of Kirby (1968); the overlying sands and gravels (beds 3 and 4), with the 'Middle' Sands and Gravels of Mitchell and Mykura (1962) and the outwash deposits associated with Kirby's Intermediate Till. The so-called Roslin Till or Southern Uplands Boulder Clay is not represented at Hewan Bank but can be seen in a number of working sandpits nearby (for example at NT 259626 and NT 297667).

According to Martin (1981), however, the sequence of deposits at Hewan Bank can be interpreted in terms of a single glacial episode during the Late Devensian. Bed 1 is a lodgement till; bed 2 is possibly transitional between a lodgement till and a melt-out till. The overlying sands and gravels are ice-marginal deposits, and the stratified diamictons are debris-flow deposits.

Deposits similar to those at Hewan Bank have been widely recognized around Edinburgh and in the Lothians for over 170 years (see above) and have provided the basis for interpreting the sequence of glacial episodes in one of the historically core areas for Quaternary studies in Scotland. There is no long history of investigations at Hewan Bank, but the site provides one of the best natural exposures demonstrating examples of the main sedimentary units recognized in the various glacial successions identified for the area. Historically, a sequence of the type at Hewan Bank was first described in Midlothian by Maclaren (1828) and is known to occur at a number of other localities (Kirby, 1968). However, apart from Hewan Bank, few sections are currently exposed.

In addition to its stratigraphic interest, Hewan Bank is also of glaciological and sedimentological note for the multiple-unit depositional sequence that appears to reflect the convergence and interaction of ice flow-lines from separate source areas during the Late Devensian glaciation. In this respect close analogies exist between Hewan Bank and several other sites, for example at Den Wick, Baile an t-Sratha, Nigg Bay, Boyne Quarry and Nith Bridge. Together these sites have important research potential for interpreting processes and patterns of debris entrainment and sedimentation beneath former ice-sheets where the interaction of ice masses from different sources has produced distinctive sedimentary units. Such sites will also provide the field evidence to underpin mathematical modelling and reconstruction of ice-sheet dynamics and the controls that determine changes in ice-sheet flow patterns.

Conclusion

The deposits at Hewan Bank are important for interpreting the glacial history of the Lothians area. They show that during the Late Devensian glaciation (around 18,000 years ago) the area was first covered by Highland ice issuing from the west, then by ice from sources in the Southern Uplands. There has been much discussion about whether these ice movements represent separate glacial episodes, but current interpretations favour a single ice-sheet in which the direction of flow shifted. Hewan Bank is a valuable reference site not only for establishing the glacial sequence in this area, but also for studying the interaction between ice masses from different sources.

KEITH WATER
J. E. Gordon

Highlights

Stream sections along the Keith Water reveal sequences of tills and glaciofluvial sediments. These deposits are important both for interpreting the sequence and patterns of movement of the Late Devensian ice-sheet, and for demonstrating the complex sedimentary environments associated with the melting of the ice which included flow components from separate sources in the Highlands and Southern Uplands.

Introduction

The site comprises a series of stream and gravel pit exposures along the valley of the Keith Water (NT 440621 to NT 452639), a tributary of the River Tyne, located about 20 km south-east of the centre of Edinburgh. On the north side of the Keith Water and in the area between the Keith Water and the Humbie Water, the landscape is underlain by thick drift deposits heavily dissected by former meltwater streams and the present rivers. The deposits comprise thick sequences of till and glaciofluvial materials which appear at the surface as flat-topped areas or mounds. The hummocky glaciofluvial deposits form part of an extensive suite extending along the northern flanks of the Lammermuir and Moorfoot Hills (Young, in Howell *et al.*, 1866; J. Geikie, 1877; Kendall and Bailey, 1908; Charlesworth, 1926b; Sissons, 1958a; McAdam, 1978; McAdam and Tulloch, 1985).

A number of the sections along the Keith Water are of longstanding interest and have sometimes been taken to represent a regional glacial stratigraphic standard for the Edinburgh–Lothians area, for which up to three separate glacial episodes have been recognized (see above). An alternative view, however, holds that the considerable lateral and vertical variability in the stratigraphy is a function of purely local ice-margin conditions associated with a single ice-sheet (Young, 1966, 1969; Martin, 1981). The deposits at Keith Water have been described by Kendall and Bailey (1908), Kirby (1966, 1968, 1969b), Young (1966, 1969), Ragg and Futty (1967) and Martin (1981).

Description

Six main sections and a number of smaller exposures have been described from the valley of the Keith Water. These show multiple, interbedded sequences of till and glaciofluvial deposits. Section 1, at NT 440621, was described by Young (1966) as showing the following beds:

3.	Reddish-brown till containing Carboniferous sandstone and greywacke erratics	1.2–1.4 m
2.	Current-bedded sands with inclusions of reddish-brown till at the top	at least 4.6 m
1.	Dark-grey till containing coal fragments and Carboniferous sandstone, greywacke and limestone erratics	at least 3.1 m

In 1980 only beds 1 and 2 were exposed.

Section 2, at NT 438623, has been the most intensively studied exposure, particularly through the work of Young (1966, 1969) who examined the stratigraphy, till clast-fabrics, particle-size distributions, heavy mineral assemblages, pH and soluble carbonates. The sequence comprises:

6.	Dark brown sands	0.9–1.1 m
5.	Dark reddish-brown till containing coal fragments, Carboniferous sandstone and greywacke erratics	1.2–1.5 m
4.	Pebbly sand	0.6 m
3.	Dark-brown till containing coal fragments, Carboniferous sandstone, greywacke and tuff erratics	at least 3.7 m
2.	Current-bedded sands with bands of coal fragments	0.9 m

1.	Dark-grey till containing coal, Carboniferous sandstone, greywacke and limestone erratics	at least 3.0 m

Young found that variations in clast fabrics were often greater within the same till units than between different ones, and that the dip analyses were inconclusive in demonstrating the ice-movement direction. Particle size, mineralogy and stone-count studies showed no significant differences between the till units. Greywacke and Carboniferous sandstone clasts were the dominant constituents of the deposit; limestone clasts occurred only in the bottom part of the middle till and in the lowest till. No identifiable Highland rocks were found.

Section 3, at NT 438631, is the notable Red Scar exposure first described by Kendall and Bailey (1908). They described the following sequence:

4.	Sand	16.1 m
3.	Boulder clay	4.0 m
2.	Sand	4.6 m
1.	Boulder clay	up to 11.3 m

Gravel layers interbedded with the sands contained a noticeable proportion of greywacke pebbles, whereas the larger blocks in the tills comprised sandstone.

The succession of deposits was confirmed by Kirby (1966, 1968) and Young (1966). Young described the lower till as very dark grey and containing Carboniferous sandstone, greywacke and limestone erratics and coal fragments; the upper till as reddish brown and containing coal fragments and Carboniferous sandstone and greywacke erratics. Kirby noted that the lower till merged, and was interbedded with, the sand above; inorganic laminated clays occurred near the top of the lower sand layer; a sharp junction existed between the lower and upper till; and there was a transitional change from the upper till to the upper sands.

Section 4, at NT 449637, was first described by Kendall and Bailey (1908) who recorded:

3.	Boulder clay	1.8 m
2.	Sand	19.8 m
1.	Boulder clay	10.1 m

In more detail, Kirby (1966, 1968) described:

6.	Red till	2.1 m
5.	Fine- and medium-grained bedded sands	*c.* 20 m
4.	Red till	0.9 m
3.	Sand with red till inclusions	1.8 m
2.	Silty till	0.9 m
1.	Dark till grading up into bed 2	5.2 m

In 1980 the section was completely vegetated over. In the same area, in sections also no longer exposed, Young (1966) had recorded sands and gravels variously overlying dark grey till containing coal fragments and Carboniferous sandstone and greywacke erratics, and reddish-brown till containing similar erratics overlying current-bedded sands.

Between sections 3 and 4 both Kirby (1966) and Young (1966) recorded several small exposures showing either sands and gravels overlying dark grey till or reddish brown-till overlying sands.

Section 5 is at Keith Marischal sandpit (NT 450640) (Kirby, 1966; Young, 1966; Ragg and Futty, 1967). As described by Young in greatest detail it shows:

4.	Dark, reddish-brown till containing Carboniferous sandstone, Old Red Sandstone, coal and greywacke erratics	1.2 m
3.	Current-bedded, dark reddish-brown sands	0.08–0.13 m
2.	Dark, reddish-brown till containing coal fragments, greywackes, Old Red Sandstone and Carboniferous sandstone erratics	0.05–0.10 m
1.	Current-bedded sands	at least 10.7 m

Section 6 is a sandpit, at NT 453637. Here Kirby (1966) described thick, horizontally bedded sands overlain by beds of till of variable thickness interlayered with sands and clays, capped by a horizon of red till.

Interpretation

The first significant synthesis of the glacial stratigraphy of East Lothian was that of Young (in Howell *et al.*, 1866) who established a basic succession of till overlain by sands and gravels. He noted variations in the colour of the till according to the local bedrock and also referred to a sandier till near the coast which merged with the main till of the area.

Following J. Geikie (1877), Kendall and Bailey (1908) confirmed the presence of Highland erratics in East Lothian and distinguished on lithological grounds the till containing Midland Valley material and the overlying sands and gravels dominated by Silurian greywackes. They explained the sequence of deposits in the Keith Water area in terms of an oscillating ice margin. Following deposition of the lower till the ice retreated northwards, and great accumulations of sand and gravel were laid down either in temporary ice-dammed lakes or as gravel spreads extending between the ice and the Lammermuir Hills. Debris washed in from the hills to the south explained the high content of local lithologies. The upper till was subsequently deposited during a forward oscillation of the ice margin. Kendall and Bailey's observations were substantially incorporated into the second edition of the Geological Survey Memoir for East Lothian (Clough *et al.*, 1910). Sissons (1958a), however, disputed Kendall and Bailey's interpretation of oscillatory retreat, and suggested instead that the balance of evidence favoured widespread thinning and stagnation of the ice.

Kirby (1966, 1968) correlated the beds in the different sections along the Keith Water. Using clast fabrics and stone counts as evidence, he inferred that the stratigraphy in the Keith Water area was similar to that in the Esk basin (see Hewan Bank). The two tills in sections such as that at Red Scar corresponded to the Basal and Intermediate tills of his regional stratigraphic scheme. The Intermediate Till was distinguished by a higher percentage of greywackes of southern origin and clast fabrics indicating ice moving to the north, whereas the Basal Till consisted largely of material of western derivation and had clast fabrics orientated to the east. Although the topmost till at sections 5 and 6 corresponded in its stratigraphic position with the Roslin Till (see Hewan Bank), it could be distinguished from the latter on the basis of its fabric and lithology (Kirby, 1968, 1969b). Its characteristics suggested that it was an ablation till associated with the decay of the Southern Uplands ice that produced the Intermediate Till.

Young (1966) studied the stratigraphy of a large number of individual sites in the Upper Tyne area. He stressed the appearance of 'rapid and radical changes' in the succession over very short distances and concluded that 'it would be impossible to construct an isopachyte map of any clarity for any one strata from the results obtained' (p. 15). However, a broad pattern emerged of a dark grey till underlying much of the area, with an overlying reddish-brown till restricted largely to the Keith Water area. From his detailed studies of section 2, Young concluded that all the deposits were laid down by the same ice flowing northwards into the Midlothian basin from the Southern Uplands, subsequently being directed south-eastwards into the Upper Tyne area. In contrast to Kirby, Young found no significant differences in lithology or clast fabric among the tills, including those of section 2 which he studied in great detail.

More recently, Martin (1981) has supported Young's interpretation. From detailed sedimentological studies and comparisons with contemporary glacial environments he concluded that the Keith Valley deposits are best interpreted in terms of a basal lodgement till succeeded by an interbedded sequence of flow tills and delta-fan deposits formed at the receding or thinning ice margin of the Late Devensian ice-sheet.

The Keith Water deposits are of considerable interest from a historical viewpoint because of their role in the development of a regional stratigraphy (see also Hewan Bank). However, more recent studies have emphasized the local variability of the deposits and their spatially restricted distribution. Taken together, therefore, the various sections provide a valuable record of the complex depositional environments associated with Pleistocene ice-sheets. They provide significant evidence for interpreting the patterns and processes of sedimentation beneath and at the margins of former ice-sheets, and thus may allow analogies with modern glacier sedimentary systems (for example see Boulton, 1972b; Lawson, 1979).

Conclusion

The sequences of deposits at Keith Water are important for studying the glacial history of the Lothians area and processes of glacial sedimentation. They were formed by ice from sources both to the west and south during the Late Devensian glaciation (around 18,000 years ago) and are particularly significant in illustrating the complex depositional patterns arising from the interactions of these sediment-carrying ice masses and their subsequent melting. The site is therefore valuable both as a reference locality for the Lothians area and for studies of glacial sedimentary environments.

CARLOPS

J. E. Gordon and D. G. Sutherland

Highlights

The landforms at Carlops comprise an outstanding assemblage of subglacial meltwater channels. These are particularly well developed and are noted for their anastomosing pattern.

Introduction

One of the most outstanding examples of meltwater channels in Scotland is located near Carlops, 21 km south-west of Edinburgh. The principal features extend over a distance of 3 km (from *c*. NT 140538 to NT 160557). The Carlops channels are part of an extensive glacial drainage system running south-west to north-east from the Clyde Valley to the Firth of Forth (see Sissons, 1967a, figure 47; Price, 1973, figure 43; Sutherland, 1984a, figure 9) in which meltwater flow was concentrated through the gap between the Pentland Hills and Moorfoot Hills during the wastage of the Late Devensian ice-sheet. The most detailed description of the channels at Carlops is that of Sissons (1963b) although the features were earlier mentioned by Milne Home (1840), Maclaren (1866), Day (1923), Charlesworth (1926b) and Eckford (1952).

Description

The channels to the south-west of Carlops (Figure 17.5) are cut in bedrock and range in depth from 1 m to over 20 m. In plan, they form an anastomosing pattern with a number of isolated rock 'islands' occurring within the larger channels (for example, Windy Gowl) (Figure 17.6). In profile, certain of the channels have 'up and down' forms (i.e. the floor of the channel first rises and then falls in the former direction of the water movement), and there are numerous discordant junctions as well as channels that trend obliquely across slopes. Several of the channels cut across the present drainage divides. Sissons (1963b) noted that certain of the channels were aligned along faults, but not all faults are followed by channels and many channels are not apparently fault-guided, so the principal controls on channel location may be presumed to be topographical and glaciological. To the north-east of Carlops, the channels cut into an area of glaciofluvial deposition and join the major meandering channel, over 30 m deep, which is now occupied by the River North Esk. In this area, Sissons (1963b) was able to demonstrate that there is an ancient channel, which pre-dates the last glaciation, below the drift deposits and which largely controls the direction of the subsequent drainage.

Interpretation

The general alignment of the channels is in the SW–NE direction of flow of the last ice to cover the Carlops area, this direction being indicated by both striations and ice-moulded drift and bedrock ridges along the foot of the Pentland Hills. Charlesworth (1926b) considered that this ice originated in the Highlands and had recurved around the south of the Pentland Hills, but the erratic content of the local drift, particularly the presence of Tinto felsite (McCall and Goodlet, 1952), clearly indicates that the last ice originated in the Southern Uplands (see Sutherland, 1984a, figure 6). It is notable that the Carlops channels occur on the highest part of the meltwater drainage system between the River Clyde and the Firth of Forth and that, to both the south-west and the north-east, the drainage system is dominantly represented by glaciofluvial depositional features (see Carstairs Kames). Such a spatial pattern accords broadly with that predicted in the model of Sugden and John (1976) for meltwater flow across irregular topography under active, warm-based ice. The model, based on Shreve's (1972) analysis of the variations in the pressure gradient that drives subglacial meltwater flow, predicts erosion across the crests and immediately on the lee sides of divides, and deposition on intervening lower ground.

Charlesworth (1926b) interpreted the major Windy Gowl channel at Carlops as ice marginal in origin; Eckford (1952) thought it to be a lake overflow, with the reversed slope on part of it being formed by post-glacial stream erosion. Sissons (1963b), however, showed these interpretations to be invalid. He argued that the forms, positions and relationships of the channels and glaciofluvial deposits were indicative of a subglacial origin through superimposition of an englacial stream system on to the underlying topography, an interpretation supported by Price

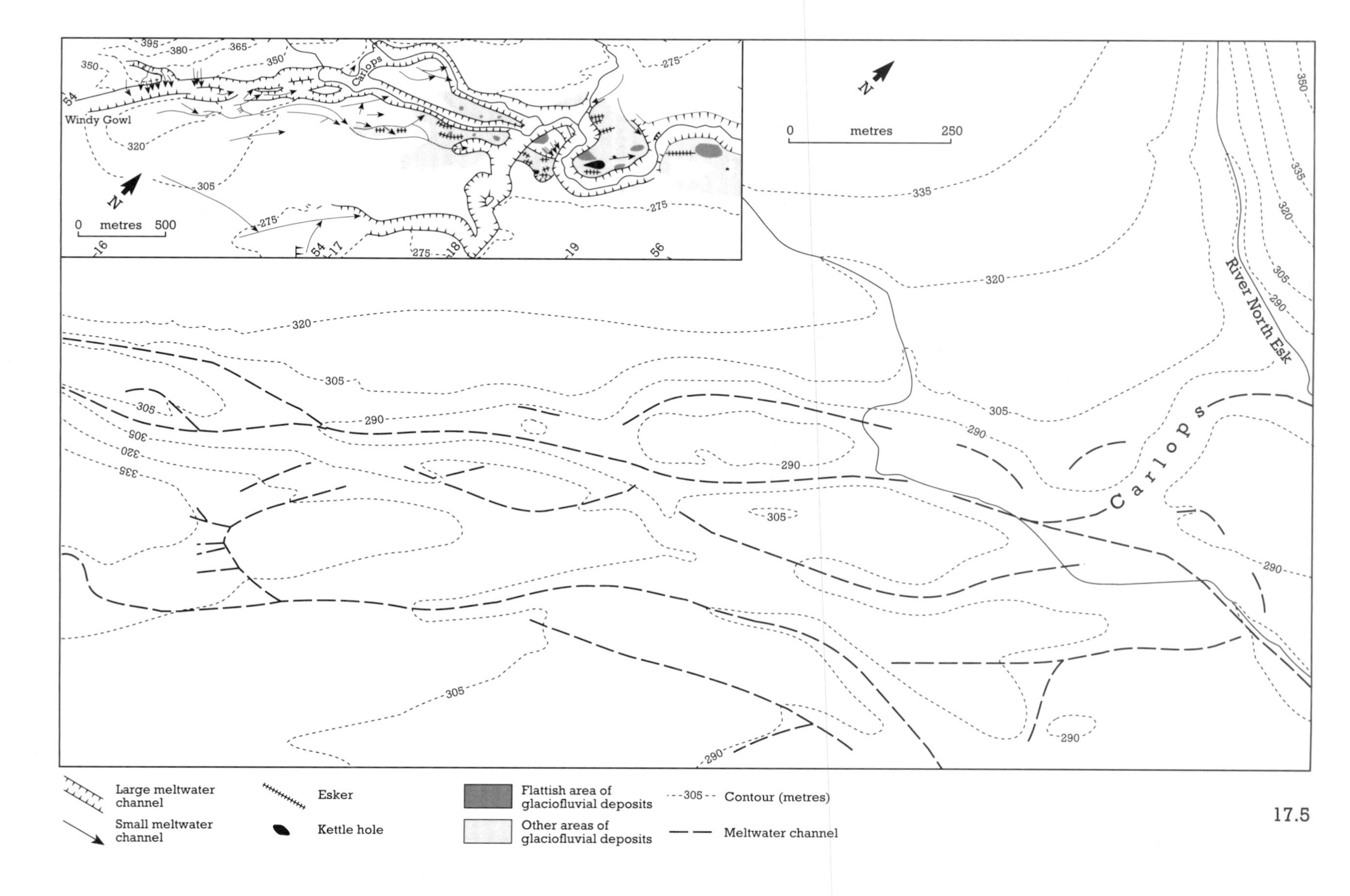
Windy Gowl
Carlops
River North Esk
N
0 metres 500
0 metres 250
395
380
365
350
335
320
305
290
275
54
16
17
18
19
56
Large meltwater channel
Small meltwater channel
Esker
Kettle hole
Flattish area of glaciofluvial deposits
Other areas of glaciofluvial deposits
305 Contour (metres)
Meltwater channel

17.5

Figure 17.6 The main meltwater channel at Carlops showing the anastomosing form of the channel system and the isolated interfluve 'islands' between constituent channels. (Photo: J. E. Gordon.)

(1973). In the overall evolution of the system, Sissons envisaged a progressive change from subglacial to open ice-walled to proglacial drainage as the ice-sheet downwasted and the Carlops area became ice-free.

The features at Carlops are classic landforms which demonstrate strikingly the morphology of subglacial meltwater channels. They illustrate very clearly the anastomosing pattern of subglacial meltwater flow and the hydraulic gradient reflected in the 'up and down' channel profiles. The Carlops channels are representative of the ice-directed type of meltwater channel (cf. Sugden and John, 1976) and in this respect are similar to certain of the channels in the Cairngorms, at Muir of Dinnet and at Rammer Cleugh. In contrast to these other areas, they demonstrate particularly well an anastomosing subglacial meltwater flow pattern. Further, in their overall pattern, the Carlops channels differ from the channel systems in the Cairngorms and at Muir of Dinnet, where they formed in submarginal positions in association with a progressively downwasting ice-sheet. The Carlops channels are also part of a much more extensive system of meltwater landforms formed during the retreat of the Late Devensian ice-sheet and hence are important in a regional context for reconstructing the pattern of glacial meltwater drainage.

Figure 17.5 The meltwater channel system at Carlops. The detailed pattern of channels immediately to the south-west of the village (main diagram) forms part of a more extensive glacial drainage system represented by channels and glaciofluvial deposits (inset) (from Sissons, 1963b).

Conclusion

Carlops is important for glacial geomorphology. It is an outstanding example of a subglacial meltwater channel system that formed beneath the last ice-sheet. It is particularly noted for the fine development of its individual channels, which include a variety of anastomosing forms. The Carlops channels form part of a wider regional pattern of glacial meltwater drainage that developed during the melting of the last ice-sheet (around 14,000 years ago).

RAMMER CLEUGH
J. E. Gordon

Highlights

The assemblage of glacial meltwater channels and deposits at Rammer Cleugh provides a particularly clear illustration of the development and evolution of a glacial drainage system during the wastage of the Late Devensian ice-sheet.

Introduction

The Rammer Cleugh site, 8 km south-west of Dunbar, covers an area of 5 km^2 between Stoneypath (NT 616711) and Hartside (NT 653721). It is important for an assemblage of glacial meltwater drainage channels, including the large and spectacular Rammer Cleugh, probably formed subglacially, and a series of subglacial chutes, ice-marginal benches and small marginal and submarginal channels. Glaciofluvial deposits, including kame terraces and an esker, also form part of the landform assemblage. Rammer Cleugh has been described by Young (in Howell *et al.*, 1866), Kendall and Bailey (1908) and Sissons (1958a, 1961a, 1975c), the last of whom published detailed geomorphological maps of the area (Sissons, 1958a, figure 2; Sissons, 1961a, figure 13).

Description

The meltwater channels at Rammer Cleugh comprise three principal sets of features (Sissons, 1958a): the large channel of Rammer Cleugh itself, channels running steeply downslope and a series of channels and benches with a much gentler gradient. The main feature, called Rammer Cleugh (see plate I of Kendall and Bailey, 1908, and plate IX of Wright, 1937), occupies the col between Deuchrie Dod and Lothian Edge. It comprises a 60 m deep glacial drainage channel extending some 5 km from Stoneypath to east of Pathhead. It has an 'up and down' long profile and at its western end near Stoneypath is cut on its southern side by numerous deep gullies and channels joining it at right angles (Figure 17.7). Above the main channel of Rammer Cleugh, particularly on the south side, several smaller channels and benches cut in bedrock run parallel, or at a low angle to the hillside contours. A number of these features turn steeply downslope at their eastern ends. To the west and north of Hartside Farm a complex assemblage of rock benches and anastomosing, rock-walled channels (occasionally with 'up and down' long profiles – cf. Sissons, 1975c) occurs in a broad embayment in the hills (Figure 17.7).

Meltwater deposits occur in two main areas within the site. To the east of Deuchrie, a series of aligned gravel mounds up to 12 m high form an esker on the floor of Rammer Cleugh. Farther east in the area between Hartside and Pathhead (Figure 17.7), kame terraces occur both to the north and south of Rammer Cleugh, locally interspersed with kame mounds. Further examples of kames occur north of Stoneypath at the west end of Rammer Cleugh.

In a wider context, Rammer Cleugh and its associated features are part of an extensive sequence of meltwater drainage phenomena along the northern flanks of the Moorfoot and Lammermuir Hills (Sissons, 1958a, 1967a; McAdam and Tulloch, 1985; Davies *et al.*, 1986).

Interpretation

Young (in Howell *et al.*, 1866) provided the first account of Rammer Cleugh and other dry valleys on the northern flanks of the Lammermuir Hills. He noted that their consistent north-east orientation was discordant with the present drainage system and that they shared similarities with dry channels in Berwickshire described earlier by Geikie (1863b). Although further investigation was necessary to explain their origin, Young discounted present stream processes and direct glacial erosion. Geikie (1894) later identified deposits associated with the channels as glaciofluvial in origin.

Kendall and Bailey (1908) mapped the large drainage channels in their study of the deglaciation of East Lothian. They explained them as the product of a retreating ice-sheet, applying the ideas developed by Kendall (1902) to account for similar features in Yorkshire. In their view, the channels were formed by meltwater torrents linking marginal ice-dammed lakes along the hillsides. Kendall and Bailey inferred oscillations in the receding ice-front from several lines of evidence and explained the mounds at the western end of Rammer Cleugh as the result of a minor readvance. In the second edition of the

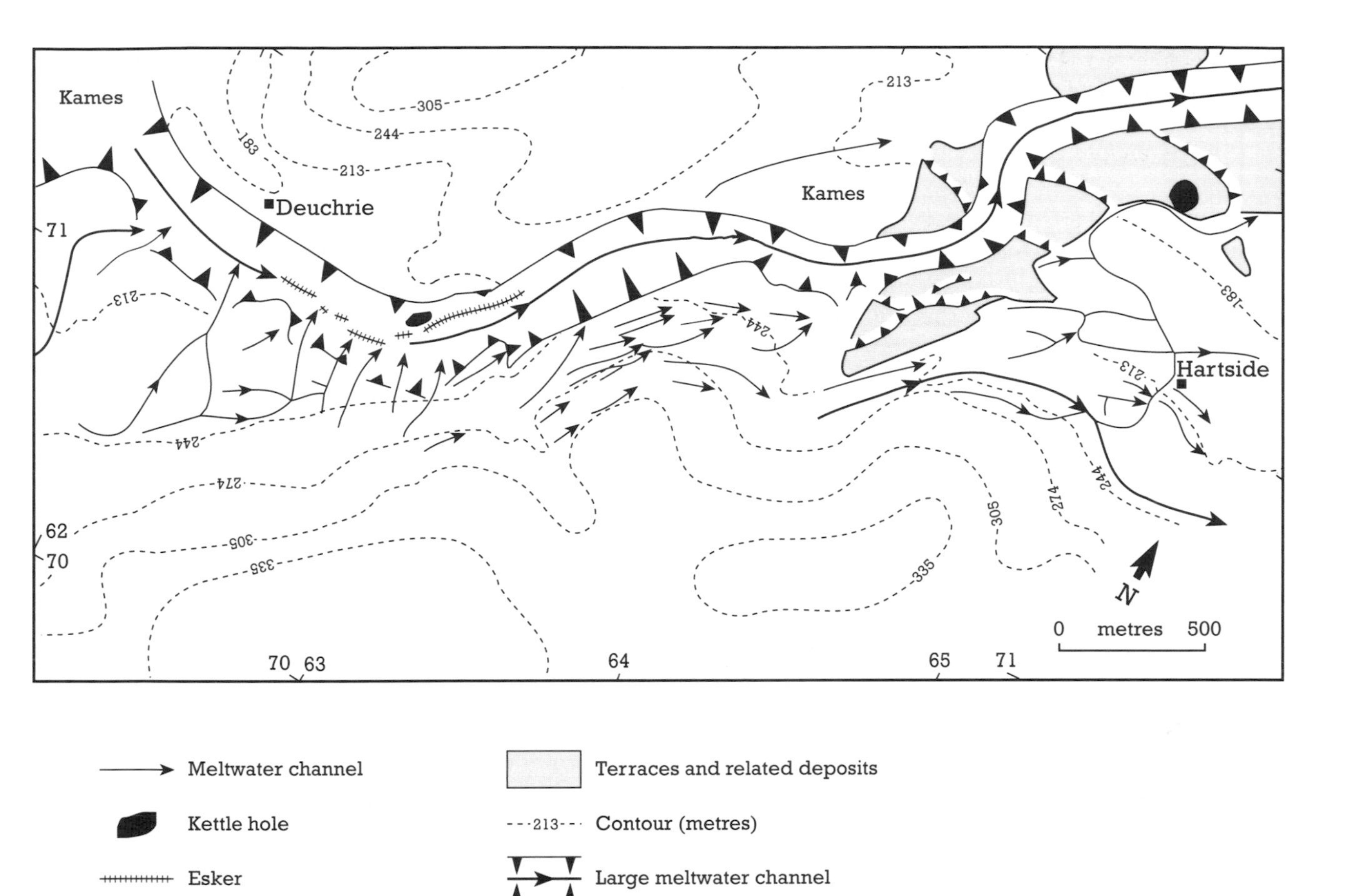

Figure 17.7 Geomorphology of the meltwater channel system at Rammer Cleugh (from Sissons, 1958a).

Geological Survey Memoir on the geology of East Lothian Clough *et al.* (1910) drew heavily on Kendall and Bailey's work.

Charlesworth (1926b) related the meltwater channels and deposits along the northern flanks of the Lammermuir Hills to his Lammermuir–Stranraer Readvance.

Based on his detailed mapping, Sissons (1958a) reinterpreted the Rammer Cleugh features and placed them in the context of the wider pattern of deglaciation in East Lothain. Many of the meltwater channels and benches running across the slopes at low angles to the contours were largely formed by ice-marginal streams, and not by glacial lake overspill as suggested by Kendall and Bailey (1908). Some of the channels, for example those running steeply downslope on the south side of Rammer Cleugh near Stoneypath, were subglacial chutes similar to features described from Scandinavia by Mannerfelt (1945, 1949). In several cases, the marginal channels terminated abruptly at the chutes, indicating that the meltwaters turned abruptly to flow beneath the ice margin.

Subsequently, in an important contribution, Sissons (1960, 1961a) offered further detailed criticisms of the haphazard application of the lake overspill hypothesis (Kendall, 1902; Kendall and Bailey, 1908), and he elaborated on alternative processes of meltwater channel formation that agreed better with the field evidence. According to Sissons (1961a) the main Rammer Cleugh channel was probably a subglacial feature, its 'up and down' long profile being cut by meltwaters under hydrostatic pressure at a stage before the subglacial chutes were formed on its flanks and the esker on its floor. Subsequent work elsewhere suggests that Rammer Cleugh and the other channels with 'up and down' long profiles may have been formed by superimposition of submarginal, englacial meltwaters on to the underlying topography, in the manner outlined for similar channels in southern Scotland and northern England (Price, 1960, 1963a; Clapperton, 1968, 1971b).

From his detailed mapping of benches, channels and kame terraces, including those of the Rammer Cleugh area, Sissons (1958a) reconstructed a series of former ice-margin positions and inferred extensive thinning and marginal stagnation of the last ice-sheet in East Lothian. According to Sissons, the higher slopes became ice-free before the lower ground. Significantly, he found no evidence for an oscillating ice front as suggested by Kendall and Bailey.

The Rammer Cleugh site is important in demonstrating an interlinked assemblage of subglacial, submarginal and marginal meltwater phenomena, including drainage channels and kame terraces associated with a thinning and downwasting ice margin. Although meltwater phenomena are widely known throughout Scotland and northern England (Sissons, 1958a, 1958b, 1960, 1961a, 1961b, 1963b, 1974b; Price, 1960, 1963a; Clapperton, 1968, 1971b; Sugden, 1970; Clapperton and Sugden, 1972; Young, 1975a, 1975b, 1977b, 1978, 1980), the majority of features appear to have formed subglacially or submarginally and there are few accounts of ice-marginal phenomena as well developed as those of the Rammer Cleugh area. As far as individual examples of subglacial channels are concerned, arguably comparable or better examples occur elsewhere; for example at Carlops (Sissons, 1963b). Subglacial chutes are also often recorded, notably by Price (1960) near Tweedsmuir and by Sissons (1961a) near Fettercairn. However, it is the total assemblage of different types of feature and the particular component of marginal channels which distinguishes Rammer Cleugh.

The Rammer Cleugh area is also significant in a historical context. It forms an integral part of the landform assemblage on the northern flanks of the Lammermuir Hills. This area was one of the first to be interpreted by Sissons (1958a) in terms of a downwasting model of ice-sheet deglaciation, in contrast to earlier interpretations that involved active ice-margin recession. The downwasting model has formed the basis of most subsequent studies of the pattern of deglaciation of the last ice-sheet. However, as a universal model, like those before it, it is probably an oversimplification. Deglaciation probably proceeded by a combination of downwasting and frontal recession, with considerable local variations controlled by factors such as ice dynamics, topography, debris content, and, where appropriate, calving.

Conclusion

Rammer Cleugh is notable for an assemblage of landforms produced by meltwater rivers flowing from an ice-sheet. It is especially important for the range of glacial meltwater channel types present, together with associated depositional features that include kame terraces and an esker ridge. These landforms illustrate the development of a glacial drainage system which involved the meltwaters flowing at, beneath and near the margin of the last ice-sheet (around 14,000 years ago).

LOCH SKENE

J. E. Gordon

Highlights

Loch Skene demonstrates an excellent assemblage of moraines formed by a Loch Lomond Readvance glacier in the Southern Uplands. These landforms provide evidence for different processes of glacier deposition and illustrate clearly the pattern of ice wastage. The evidence from Loch Skene is also important for wider palaeoclimatic reconstructions for the Loch Lomond Stadial.

Introduction

The Loch Skene site (NT 168162) covers an area of *c.* 5.75 km^2 on the north-west side of Moffat Dale, a classic, fault-guided, glacial trough in the Southern Uplands. It is important for an assemblage of glacial landforms, including fine examples of a corrie and hanging valley which were occupied by part of the largest glacier system in the Southern Uplands during the Loch Lomond Stadial. The locality is therefore significant for glacier and palaeoclimatic reconstructions, and good examples of end and hummocky moraines also provide a valuable means of interpreting the recession of the glacier. The Loch Skene landforms have a long history of research (Chambers, 1855a, 1855b; Geikie, 1863a, 1901; Young, 1864; Brown, 1868; Eckford and Manson, 1927; Price, 1963b, 1983; Sissons, 1967a; May, 1981), and aspects of vegetational history recorded in the peat deposits on the corrie floor have also been examined (Lewis, 1905; Erdtman, 1928).

Description

The hills around Loch Skene rise to an altitude of 822 m OD at White Coomb. The steep headwalls of a corrie enclose the valley at the head of the loch and a second, shallower corrie, drained by the Midlaw Burn, forms a tributary basin to the south-west. The glacial deposits at Loch Skene have long been recognized as being significant. Chambers (1855a, 1855b) briefly noted the presence of a moraine-dammed lake associated with local glaciers. Later Geikie (1863a) and Young (1864) described the area in more detail, the latter including a geomorphological map compiled by Geikie. Both Geikie and Young recorded clear end, lateral and hummocky moraines associated with glaciers flowing down from an ice-field on the adjacent plateau slopes. They noted the arcuate alignments of many of the moraines across the valley at the southern end of Loch Skene and in the valley of the Midlaw Burn. In the intervening area the hummocky moraine was irregular in its form. A particularly prominent lateral moraine called 'The Causey' marked the position where the ice flow diverged into the head of the Winterhope Valley.

Subsequent accounts of the Loch Skene area appeared in Brown (1868), Geikie (1901), Eckford and Manson (1927), Price (1963b, 1983) and Sissons (1967a). The ground has also been re-mapped by May (1981) who confirmed the observations of Geikie and Young. May (1981) provided a detailed map of the area showing the form and distribution of the moraines (Figure 17.8). He also described their main characteristics, elaborating on the earlier accounts. Sections in the ridges along the Tail Burn revealed locally derived till comprising angular and subangular clasts in a gravelly matrix.

In a brief reference, Sissons (1977a) recorded the presence of fluting on the hummocky moraine at Loch Skene, implying that active ice overrode 'dead'-ice topography.

As part of a wider study of the Holocene vegetational history of Scotland, Lewis (1905) examined the peat deposits in the Tweedsmuir area, including Loch Skene, and identified layers of birch tree remains and two 'arctic beds', layers in the peat where the plant macrofossils had predominantly northern affinities. However, from pollen analysis and a re-examination of the peat deposits on the north-east side of Loch Skene, Erdtman (1928) questioned some of Lewis' interpretations, particularly the occurrence of a second 'arctic bed' (see Loch Dungeon).

Interpretation

As first recognized by Chambers (1855a, 1855b) and followed by later authors (for example, Geikie, 1894), the Loch Skene moraines relate to an episode of local glaciation following the last ice-sheet and now recognized to have occurred during the Loch Lomond Stadial (Sissons, 1967a; May, 1981; Price, 1983). Similar moraines occur in adjacent valleys (Young, 1864; Price, 1963b, 1983; Sissons, 1967a; May, 1981;) indicating the extent of the former glaciers associated with the ice-field that developed in the White Coomb area. The moraines at Loch Skene mark successive stages in ice wastage as the glacier retreated back into the corries at the head of Loch Skene and the Midlaw Burn.

Loch Skene is important in several respects. First, in a historical context the work of Geikie and Young provides a good example of early, large-scale geomorphological mapping and description. Their basic field observations demonstrated the value of this type of approach and provided a good contemporary record which has stood the test of time.

Second, Loch Skene provides an ideal area for the study of the formation of hummocky moraine, the origin of which is controversial (see Coire a'Cheud-chnoic). Three main hypotheses exist: it is a form of stagnant-ice topography formed by rapid wastage of ice with a thick cover of supraglacial debris (Sissons, 1967a; Thompson, 1972); it is a product of controlled or uncontrolled deposition by actively retreating glaciers (Eyles, 1983; Day, 1983; Horsfield, 1983; Benn, 1990, 1991; Bennett, 1990, 1991; Bennett and Glasser, 1991); or it is a subglacial deposit formed by deformation of pre-existing till (Hodgson, 1982, 1987; Ballantyne, 1989a; Benn, 1991). All three types probably exist (see the Cuillin). Loch Skene provides a particularly good opportunity to investigate the genesis of different forms of hummocky moraine and their relationships since features of all three models exist in the area: chaotic assemblages of mounds, arcuate alignments of mounds and ridges which have the form of recessional moraines, and overridden and fluted mounds. The results of such work and comparative studies with other sites (see the Cuillin, Coire a'Cheud-chnoic and the Cairngorms) will facilitate the recognition of styles of deglaciation from the landform assemblages and sediment facies (see Eyles, 1979, 1983; Sharp, 1985; Evans, 1989; Benn, 1990, 1991; Benn *et al.*, 1992;

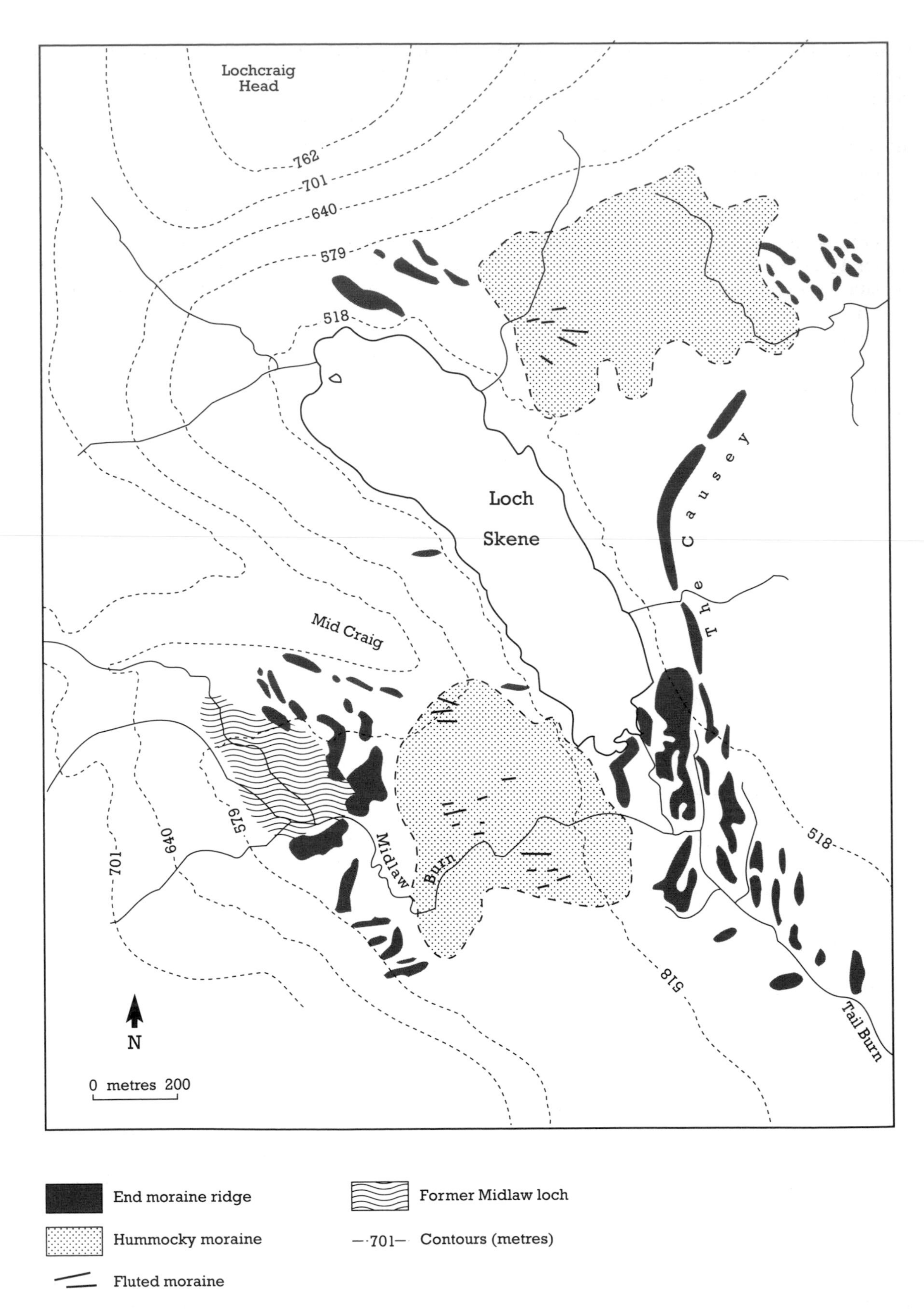

Figure 17.8 Loch Lomond Readvance moraines at Loch Skene (from May, 1981).

Bennett, 1991). In some cases they will also provide field evidence for the characteristics of deformable glacier beds and information on sediment transfer patterns during the Loch Lomond Readvance (see Coire a'Cheud-chnoic).

Third, Loch Skene is important for palaeoclimatic reconstructions. The end moraines provide a clear geomorphological record of the pattern of glacier wastage during the Loch Lomond Stadial. At present little is known of climatic variations during the stadial, but comparative studies of the moraines at Loch Skene and other sites (e.g. Tauchers, the Cairngorms, Lochnagar, Cnoc a'Mhoraire and the Cuillin) should provide important information on glacier–climate relationships and the effects of other variables, such as topography, in controlling the mass balance and fluctuations of Loch Lomond Readvance glaciers.

Fourth, glacier development was of very restricted extent in the Southern Uplands during the Loch Lomond Stadial (Sissons, 1979d, 1980b; Cornish, 1981; Price, 1983), and the Loch Skene glacier is therefore significant in forming part of the largest ice mass in the area. This ice mass provides an important geographical link between those in the Highlands, Lake District and Wales (Sissons, 1979d, 1979e) and is a significant element in interpreting the wider national pattern of Loch Lomond Stadial glaciers and underlying climatic conditions (Sissons, 1979d, 1979e). Significant contrasts in glacier development between these areas are explicable in terms of regional precipitation differences associated with variations in the position of the Polar Front and Atlantic depression tracks (Sissons, 1979d, 1979e, 1980b). Such variations can also account for the marked contrast in the degree of glacierization during the Loch Lomond Stadial compared with earlier in the Late Devensian when the Southern Uplands formed a major centre of ice-sheet accumulation (Sissons, 1979d; Sutherland, 1984a). The evidence from Loch Skene therefore contributes significantly towards the establishment of wider palaeoclimatic reconstructions.

Conclusion

Loch Skene is important for an assemblage of landforms in the Southern Uplands representing the resurgence of glacier ice which occurred during the Loch Lomond Stadial, approximately 11,000–10,000 years ago. Principally, these landforms comprise an excellent range of moraine types which illustrate clearly the different processes of glacier deposition. The detailed form of the moraines also shows the pattern of ice decay in the corrie, which is significant for interpreting the glacier behaviour and its possible climatic controls. As part of a wider network of sites representing the pattern of glacier growth and retreat during the Loch Lomond Stadial, Loch Skene also contributes important geomorphological evidence for palaeoclimatic reconstructions.

BEANRIG MOSS

P. D. Moore

Highlights

The pollen and plant macro-fossils preserved in the sediments which infill the topographic depression at Beanrig Moss provide a detailed record of vegetational history and environmental change in south-east Scotland during the Lateglacial. The plant macro-fossils, in particular, have yielded a great wealth of palaeoecological information.

Introduction

Beanrig Moss (NT 517293) is part of the Whitlaw Mosses, a series of peat bogs located 5 km east of Selkirk and 6 km south-west of Melrose. It is about 200 m long and occurs at an altitude of about 240 m OD. The sediments of Beanrig Moss are of considerable palaeoecological value in that their rich fossil content provides evidence for a continental flora in eastern Scotland during Late Devensian times, several members of which are now rare or extinct in the British Isles. The site and its deposits have been described by Daniels (1972) and Webb and Moore (1982).

Description

The drift-covered Silurian shales of the Melrose area of the Borders have given rise to alternating ridges and hollows depending upon their susceptibility to glacial erosion (Ragg, 1960). Within the hollows is developed a series of valley fens, the Whitlaw Mosses, of which the most thoroughly studied is Beanrig Moss (Daniels, 1972; Webb and Moore, 1982). Calcareous groundwater during the Late Devensian led to the development of marl and clay deposits in these fens, which form the main feature of sedimentary interest. Similar sediments have been described at Whitrig Bog (Mitchell, 1948; Connolly, 1957) some 10 km to the north-east, and at Blackpool and Murder Mosses within the Whitlaw complex.

The deposits at Beanrig Moss comprise grey and pink, banded clays containing angular shale fragments. The succession at the deepest part of the site is as follows:

5.	Detritus (organic) muds and swamp peats interrupted at a depth of about 1 m below the surface by a clear depositional hiatus, followed by fresh, unconsolidated swamp peat	1.55 m
4.	Grey clay with angular rock fragments and with abundant bryophyte remains	0.50 m
3.	Coarse detritus muds	0.38 m
2.	Marls, muds and clays	0.60 m
1.	Grey and pink, banded clays containing angular shale fragments	0.97 m

The Late Devensian sediments of bed 2 have their greatest depth (about 3 m) in the centre of the basin. The hiatus in bed 5 can be traced across the entire basin and has evidently resulted from the harvesting of peat from the basin in the past. The current vegetation at the site is thus the consequence of secondary reinvasion of abandoned peat cuttings.

The sediments at Beanrig Moss contain an abundance of plant microfossils and macrofossils which provide material for a detailed reconstruction of the vegetation of the Lateglacial Interstadial and Loch Lomond Stadial (Figure 17.9). Evidence from the fossil plant materials within the sediments indicates that they represent the full Lateglacial sequence. The abundance of marl at the site, however, has so far precluded the use of radiocarbon dating techniques.

Interpretation

The basal clays (bed 1) do not contain a sufficient density of pollen for analysis, but plant macrofossils and other microfossils have been extracted (upper part of bed 2) and identified (Webb and Moore, 1982). These included ostracods, *Daphnia*, Chironomidae, and *Nitella* and *Tolypella* oospores. Terrestrial plants included *Artemisia* (a capitulum of *A.* sect. *dracunculus*), *Salix* and *Papaver* sect. *scapiflora*, reflecting open, cold tundra conditions.

Warmer conditions are indicated in the succeeding sediments (bed 3) by an increasing abundance of *Potamogeton* fruit stones, particularly within the marls, but terrestrial plant macrofossils, including *Dryas octopetala*, *Vaccinium vitis-idaea* and many mosses (for example, *Rhytidium rugosum*) show that dwarf-shrub vegetation persisted. Pollen density increases within these sediments and has provided evidence of local fens and willow thickets, together with both tree and dwarf birch (pollen assemblage zone BRM-a). Open ground was clearly still present in the early part of the interstadial, however, as is demonstrated both by flowering plants (such as *Saxifraga oppositifolia*, *Artemisia* cf. *norvegica*, *Minuartia rubella*), and mosses (for example, *Tortella fragilis*). Basic short-turf communities were also present forming a 'continental grass heath' (for example, *Medicago falcata*, *Astragalus danicus*).

A second pollen assemblage zone is distinguishable within the interstadial (BRM-b) (upper part of bed 3), differing from the lower zone (BRM-a) mainly in its greater abundance of *Juniperus* and *Filipendula*, and lower proportions of tree birch and willow. *Helianthemum* is extremely abundant in this zone, showing the persistence, perhaps extension, of basic grassland, and the continental grassland species *Gypsophila fastigiata/repens* and *Hedysarum* cf. *hedysaroides* also occur, together with *Astragalus alpinus* and *Artemisia* cf. *norvegica*. These changes of vegetation indicate a climatic cooling, and the grassland species show very distinct continental affinities, supporting an argument for a continental climate at the time.

The Loch Lomond Stadial is marked lithologically by clays (bed 4), and corresponds with a distinct pollen assemblage zone (BRM-c), the most marked features of which are high proportions of *Artemisia*, *Oxyria*-type and *Thalictrum* pollen. The decrease in warmth-demanding taxa, such as *Juniperus* and tree birches, confirms the onset of colder conditions. Of the *Artemisia* pollen, the most abundant type was *A.* cf. *norvegica*, indicating a 'fell-field' habitat, and also containing such taxa as *Papaver* sect. *scapiflora* and *Polytrichum alpinum*. Both snow patch (indicated by *Salix herbacea* and *Polytrichum norvegicum*) and wind-exposed areas (indicated by *Minuartia rubella*) were evidently present at this time.

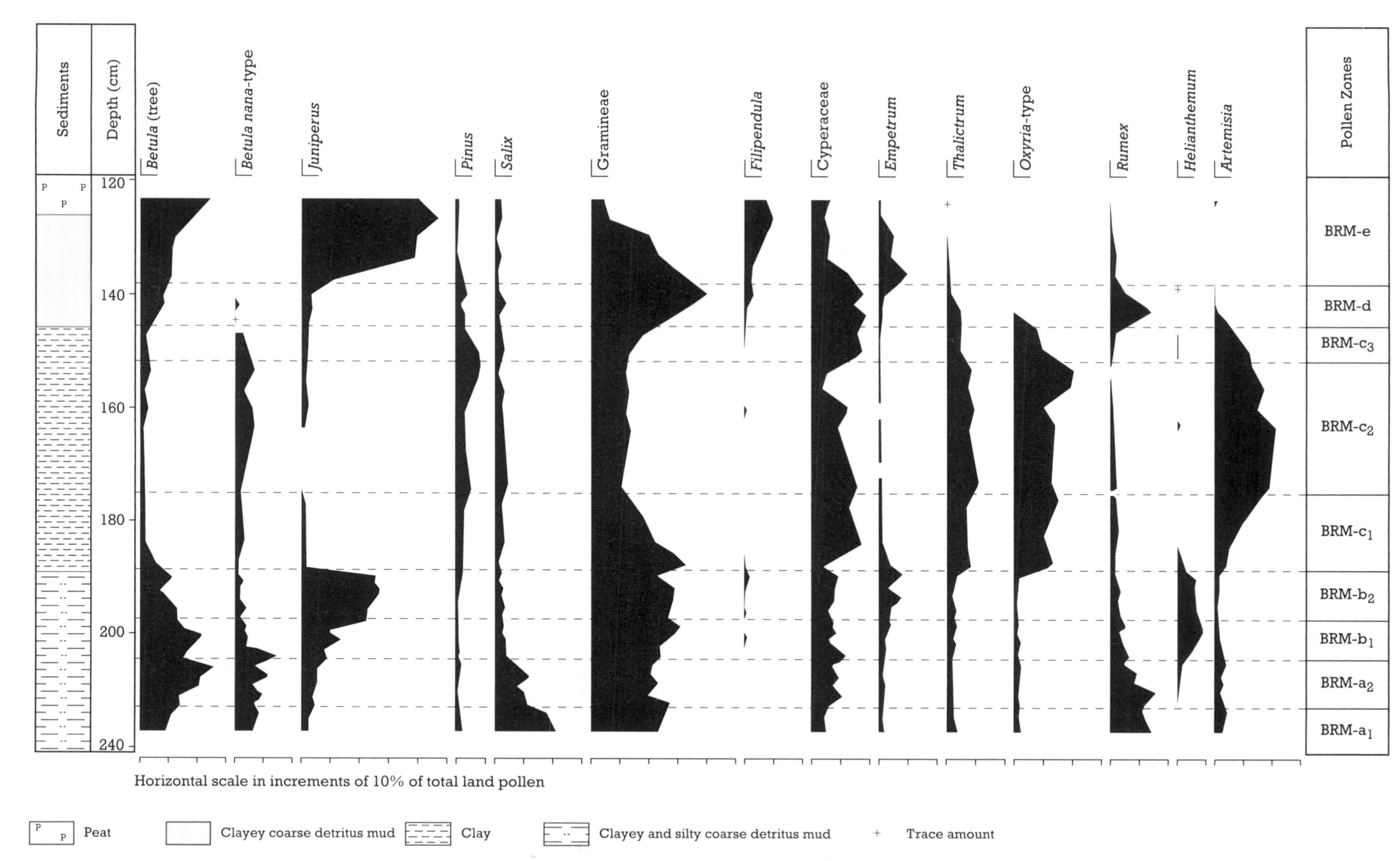

Figure 17.9 Beanrig Moss: relative pollen diagram showing selected taxa as percentages of total land pollen (from Webb and Moore, 1982).

Subsequent detritus muds (bed 5), up to the level of the peat cutting, correspond with pollen assemblage zones BRM-d and BRM-e, in which the warmth-demanding taxa of the Holocene increase in abundance. Gramineae and *Rumex acetosa* dominate BRM-d, followed by tree birch and *Juniperus* in BRM-e.

The most important features of the Beanrig Moss sediments are first, the floristic richness of the Lateglacial fossil material and second, the phytogeographical and climatic implications of the plant fossil assemblages.

Beanrig Moss is one of very few sites in Scotland where abundant Lateglacial plant macrofossils have been recovered. Much of the reconstruction of Lateglacial terrestrial environments in Scotland has been based on analysis of pollen grains which cannot, in general, be identified to species level, thus limiting the environmental inferences that can be made. Plant macrofossils, however, can normally be identified to species level and hence provide pollen analysis with complementary and more precise evidence as to the nature of the local environments and how these have changed through time. The sites of Beanrig Moss (lowland, moderate altitude), Morrone (highland, high altitude) and Abernethy Forest (highland, moderate altitude) therefore not only have particular value because of their detailed plant macrofossil records, but also have a wider significance in reconstruction of the regional variation in vegetational communities during the Lateglacial.

At Beanrig Moss there are records of many plant species that are now rare or extinct in Scotland and disjunct in their current distributions. The fossils provide the means of reconstructing former distribution patterns, many of which (such as *Artemisia norvegica*, *A.* cf. *campestris*, *Gypsophila fastigiata/repens* and *Hedysarum* cf. *hedysaroides*) show phytogeographical links with Scandinavia and other parts of northern continental Europe. The 'steppic' element in the flora of the Lateglacial Interstadial at Beanrig Moss suggests a more continental climate in the eastern part of southern Scotland at that time than has been proposed for any other British Late Devensian site.

Conclusion

Beanrig Moss is an important reference site for elucidating the vegetational history of south-east Scotland during the Lateglacial phase in Late Devensian times (about 13,000–10,000 years ago), based on the records of pollen and larger plant remains contained in the sediments. It is particularly significant for the wealth of larger plant remains present, and these have allowed a more detailed reconstruction of Lateglacial environmental conditions than is normally possible from pollen alone. The records show the development of a distinctively more continental climate than indicated from other sites in Britain.

DIN MOSS

H. J. B. Birks

Highlights

The sediments from the raised bog at Din Moss provide a detailed pollen record, supported by radiocarbon dating, of vegetational history and environmental change in south-east Scotland during the Holocene. This record is particularly important in showing the timings of arrival and expansion of different woodland types in the area.

Introduction

Din Moss (NT 805314) is located 8 km south-west of Coldstream, at an elevation of 170 m OD, near the Scotland–England border. The site is a raised bog developed at the south-west corner of Hoselaw Loch. It is important for its Holocene pollen record, in particular for reconstructing the vegetational dynamics of the Holocene. It provides one of the few well-dated records from south-east Scotland (Hibbert and Switsur, 1976).

Description

At Din Moss there is a total thickness of 4.6 m of sediment consisting of 0.6 m of fine detritus muds, overlain by 1.4 m of fen peat and 2.6 m of *Sphagnum–Eriophorum* peat (Figure 17.10). A series of 18 radiocarbon dates was obtained

Figure 17.10 Din Moss: relative pollen diagram showing selected taxa as percentages of the pollen sums indicated (from Hibbert and Switsur, 1976).

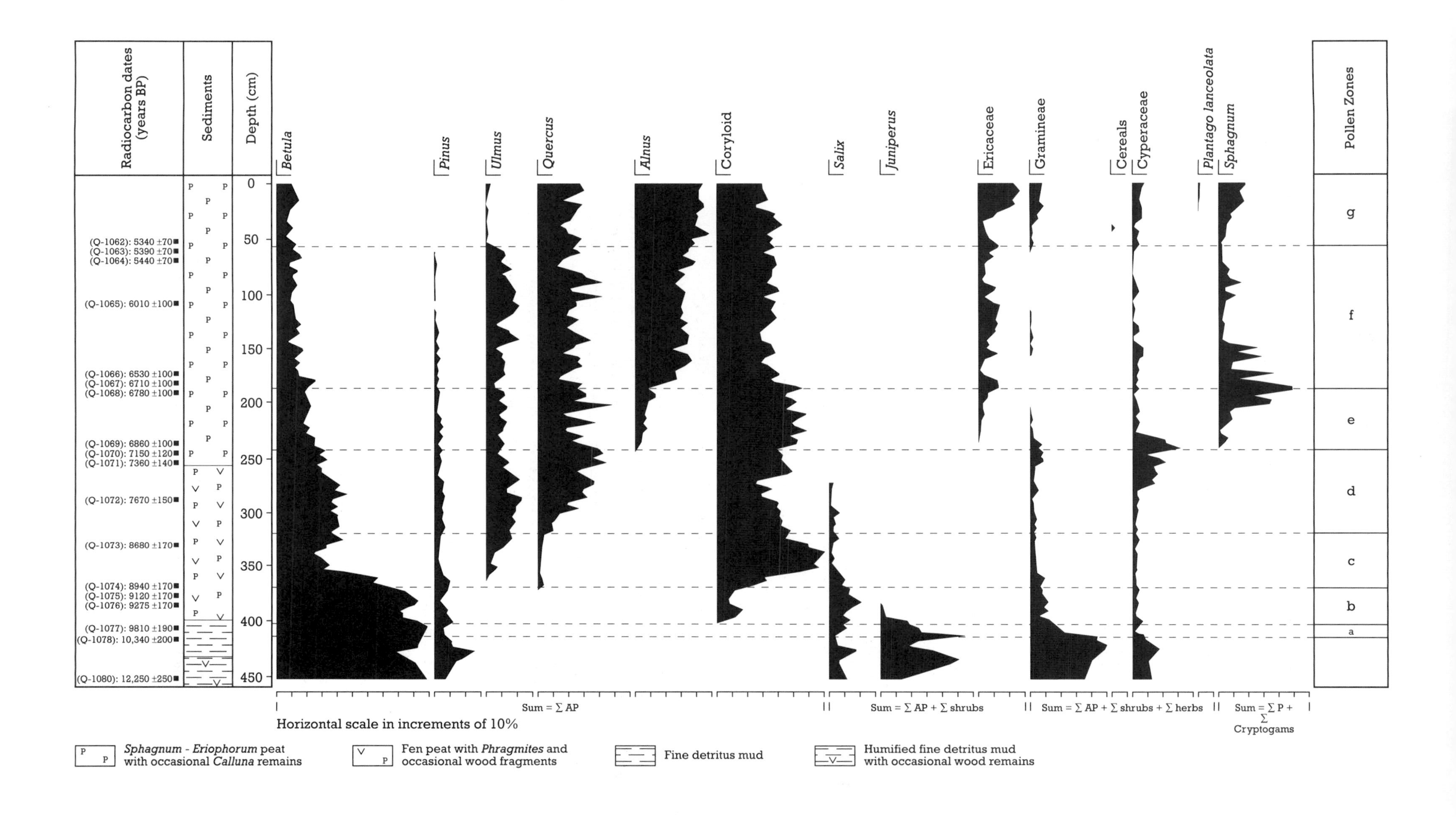

Radiocarbon dates (years BP)
Sediments
Depth (cm)
Betula
Pinus
Ulmus
Quercus
Alnus
Coryloid
Salix
Juniperus
Ericaceae
Gramineae
Cereals
Cyperaceae
Plantago lanceolata
Sphagnum
Pollen Zones
(Q-1062): 5340 ±70
(Q-1063): 5390 ±70
(Q-1064): 5440 ±70
(Q-1065): 6010 ±100
(Q-1066): 6530 ±100
(Q-1067): 6710 ±100
(Q-1068): 6780 ±100
(Q-1069): 6860 ±100
(Q-1070): 7150 ±120
(Q-1071): 7360 ±140
(Q-1072): 7670 ±150
(Q-1073): 8680 ±170
(Q-1074): 8940 ±170
(Q-1075): 9120 ±170
(Q-1076): 9275 ±170
(Q-1077): 9810 ±190
(Q-1078): 10,340 ±200
(Q-1080): 12,250 ±250
0
50
100
150
200
250
300
350
400
450
g
f
e
d
c
b
a
Sum = Σ AP
Sum = Σ AP + Σ shrubs
Sum = Σ AP + Σ shrubs + Σ herbs
Sum = Σ P + Σ Cryptogams
Horizontal scale in increments of 10%
Sphagnum - Eriophorum peat with occasional Calluna remains
Fen peat with Phragmites and occasional wood fragments
Fine detritus mud
Humified fine detritus mud with occasional wood remains

from the profile (Hibbert and Switsur, 1976) (Figure 17.10).

The base of the sequence has a date of 12,250 BP (Q–1080), indicating a Lateglacial age. This part of the profile has not, however, been investigated in detail palynologically. The Holocene part has been studied in considerable detail. Radiocarbon dates, often two or three for each horizon, are available for the first appearance and expansion of *Corylus*, *Quercus* and *Alnus*, and the decline of *Ulmus*. Several other dates are also available, permitting the reconstruction of a reliable age–depth curve from which the age of other changes in pollen frequences can be interpolated with confidence. The profile only extends to about 4500 BP, possibly because of recent peat cutting.

Interpretation

The sequence indicates that during the closing parts of the Lateglacial, *Juniperus* and *Betula* quickly expanded. In contrast to other sites in eastern Scotland and north-east England, *Corylus* expanded as early as 9800 BP (Birks, 1989). *Ulmus* arrived and spread at about 8800 BP, as it did between 8500 and 9000 BP in much of northern England and southern Scotland (Birks, 1989). *Quercus*, although possibly present in low amounts from about 9100 BP, did not expand until about 8400 BP. As elsewhere in southern Scotland (except Galloway) (Birks, 1989), pine does not appear to have been important around Din Moss at any time in the Holocene. *Alnus* was present from 7000 BP but did not expand until after 6500 BP (Bennett and Birks, 1990). There is a well-marked *Ulmus* pollen decline, dated to 5400 BP, associated with the first occurrences of cereal-type and *Plantago lanceolata* pollen.

The dating of major pollen-assemblage changes in the Holocene and the study of the temporal and spatial patterns of important palaeobotanical events within Britain are critical aspects of Holocene palaeoecological research. Such studies permit the reconstruction of the directions, routes, rates and timings of the spread of forest trees (Birks, 1989), the spatial patterns of pollen assemblages at selected points in time (Birks *et al.*, 1975), and the spatial patterns in past vegetation (Bennett, 1989). They also permit the testing of the geochronometric validity of pollen-zone boundaries (Smith and Pilcher, 1973). Such studies require well-dated, detailed, pollen sequences from carefully selected regional sites in different parts of Britain. Radiocarbon dates are generally more reliable from ombrotrophic raised-bog peat than from lake sediments, with their associated problems of hard-water errors, inwashing and redeposition. Din Moss is one such bog site and is the only such site studied in detail and well-dated in the south-east lowland fringe of Scotland and northernmost north-east England (Hibbert and Switsur, 1976).

In the context of Holocene vegetational history Din Moss is important in providing a well-dated pollen sequence from this part of Scotland. The pollen record demonstrates that *Pinus* was never an important part of the forest vegetation: indeed pine was probably absent from this part of Scotland (Birks, 1977, 1989). This has important implications for our understanding of the history of the endemic *Pinus sylvestris* var. *scotica* of the Highland pine forests. The timings of arrival and expansion of *Betula*, *Ulmus*, *Quercus* and *Alnus* at Din Moss are in close agreement with dates from other sites in northern England (Birks, 1989). The nearest sites with radiocarbon dates for these events in southern Scotland are all in the west (Birks, 1989). At a British scale, features of particular interest at Din Moss are the early expansion of *Betula* (about 10,300 BP), the early expansion of *Corylus* (about 9800 BP), the long and protracted expansion of *Quercus* (from about 9100 to 8400 BP), and the comparatively late expansion of *Alnus* (about 6400 BP).

In the light of all these features, Din Moss is an important Holocene site both within Scotland and within Britain as a whole.

Conclusion

Din Moss is an important reference site for the vegetational history of south-east Scotland during the Holocene (last 10,000 years). In particular, analysis of the pollen contained in the sediments, supported by numerous radiocarbon dates, has provided a detailed record of the spread of successive woodland types. In view of its detailed and well-dated pollen record, the only such one available for this area, Din Moss forms an important part of the network of sites representing the wider regional patterns of vegetational history and environmental change in Scotland.

DUNBAR

J. E. Gordon

Highlights

The coastal landforms at Dunbar are notable for a series of shore platforms, including features which pre-date the Late Devensian. These landforms are representative of erosional coastal features found along the east coast of Scotland and are important for interpreting former sea-level changes and processes of rock coast development.

Introduction

This site (NT 661788) is a 1.9 km long stretch of coast in the immediate vicinity of Dunbar. Within its limits are preserved four distinct shore platforms, which range in altitude from 25 m above, to 11 m below present sea level (Rhind, 1965; Sissons, 1967a, 1976b; Hall, 1989a).

Description

The highest (A) of the four shore platforms known to be present at Dunbar is one of a number of fragments which occur at 18 m to 25 m OD between North Berwick and Berwick in south-east Scotland (Rhind, 1965). That this platform pre-dates the last ice-sheet may be inferred from the preservation of the drift tail of a crag and tail on its surface at Dunbar (see Sissons, 1967a, figure 83). A second platform (B) occurs in the intertidal zone, and for about 1 km west of the Castle ruins it is backed by a 20 m high cliff cut in volcanic tuffs and sandstones (Clough *et al.*, 1910; Francis, 1975). Several stacks protrude above the platform surface, including the Dove Rock, a relatively more resistant volcanic plug, and shallow caves are cut into the backing cliffs. Present-day beach deposits, mainly shingle and some coarser debris, occur at the heads of embayments in the cliffline. The platform attains its greatest width, about 350 m, west of Long Craigs, where it clearly truncates the underlying sediments and agglomerates (Clough *et al.*, 1910; Francis, 1975) (Figure 17.11).

To the west, the backing cliff of the intertidal

Figure 17.11 Intertidal shore platform at Dunbar, which has been planed across a series of Devonian–Carboniferous and Carboniferous sediments and agglomerates. (Photo: J. E. Gordon.)

platform is degraded and is fronted by Holocene raised beach deposits resting on a third platform (C) at an intermediate level and separated from the lower intertidal one by a rock step 1 m to 2 m high. These relationships are best seen in section at NT 66337899. A further platform (D) occurs offshore at about −11 m OD.

Interpretation

The highest platform (A) clearly pre-dates the Late Devensian glacation, although its age is unknown. Similarly the age or ages of the next two lower platforms (B and C) are unknown, although from evidence elsewhere on this part of the coast they pre-date the last ice-sheet (Hall, 1989a). Sissons (1976b) suggested the possibility that they were originally a single feature, the step between them having been cut by recent marine erosion since the sea attained its present level. Hall (1989a), however, has argued that the presence of till at the rear of the intertidal platform (B) at two sites nearby precludes significant erosion and retreat of the backing cliff during the Holocene. According to Hall (1989a) the two platforms (B and C) existed as separate features prior to the Late Devensian, and Holocene marine erosion has been confined to stripping a till cover. These platforms may form part of the intertidal platform that is developed extensively elsewhere in eastern Scotland (see Milton Ness and Kincraig Point). Dawson (1980a) has correlated this platform with the Low Rock Platform of western Scotland (see Northern Islay and West coast of Jura).

The offshore platform (D) has been correlated with a buried gravel layer and platform in the Firth of Forth and a submerged platform near Burnmouth to define the Main Lateglacial Shoreline in south-east Scotland (Sissons, 1976a, 1976b). Sissons (1974d) also correlated the Main Rock Platform of western Scotland (see Isle of Lismore) with this shoreline, suggesting that they were formed during the severe climatic conditions of the Loch Lomond Stadial.

Dunbar is important for demonstrating the geomorphology of shore platforms formed during the Pleistocene. Such platforms occur at a variety of altitudes along the east coast of Scotland (Walton, 1959; Rhind, 1965; Sissons, 1967a; Crofts, 1975; Browne and Jarvis, 1983; Stoker and Graham, 1985; Hall, 1989a) (see also Kincraig Point and Milton Ness) but are particularly well-preserved in the vicinity of Dunbar. Here also, the relationship of the uppermost platform (A) to glaciation is indicated by the superimposition of a drift tail on the platform, and two lower platforms (B and C) are also inferred to pre-date the last glaciation. The site therefore emphasizes the importance of inherited features in the coastal geomorphology of eastern Scotland (cf. Walton, 1959; Sissons, 1967a; Hall, 1989a).

Conclusion

Dunbar forms part of the site network demonstrating Quaternary coastal geomorphology and sea-level change. In particular, it is notable for an excellent series of rock platforms of different ages, including examples that pre-date the last ice age (i.e. formed before 26,000 years ago). Dunbar is one of the best sites in eastern Scotland illustrating the development of multiple shore platforms and it also highlights the contribution of older elements to the form of the present coastal landscape.

Chapter 18

South-west Scotland

INTRODUCTION

D. G. Sutherland

South-west Scotland comprises that part of the Southern Uplands west of the A74 and the coastal lowlands along the Solway Firth (Figure 18.1). Little is known of the Quaternary history of the area prior to the last, Devensian ice-sheet glaciation, although, as in the rest of Scotland multiple glaciation may be inferred from the evidence from neighbouring regions, as well as the presence of landforms of glacial erosion that have developed over a long period.

The mountains in the Loch Doon area acted as one of the major centres of the last ice-sheet, as is indicated by the transport of various erratics to the north, west and south as well as the distribution of ice-moulded landforms and striated bedrock surfaces (Geikie, 1894; Charlesworth, 1926a; Cornish, 1982, 1983). The ice-sheet centred in the Galloway hills appears to have built up more slowly than that in the south-west Highlands, for the Highland ice initially impinged on the Southern Uplands, transporting erratics into the northern valleys of the hill mass as well as into the western coastal areas in Ayrshire and the Rhins of Galloway. The subsequent expansion of the Southern Uplands ice deflected the Highland ice to the east and west, producing typical till sequences in which lower, frequently shelly, tills with Highland or Ayrshire erratics are overlain by tills with Southern Uplands erratics. This classic sequence of the 'debatable ground' (J. Geikie, 1877) is shown at Nith Bridge (Chapter 15) and Port Logan.

On the southern and western side of the ice shed the pattern of ice flow is recorded in one of the largest drumlin fields in Scotland, covering much of lowland Kirkcudbrightshire as well as the Machars and Rhins of Wigtownshire (Cutler, 1978; Kerr, 1982); drumlins also occur in the eastern Solway lowlands (Hollingworth, 1931). Cornish (1979) has demonstrated that towards the ice-shed zone the drumlins merge into rogen moraine, the only instance of this type of moraine documented in Scotland. Charlesworth (1926a) noted a change over a short distance in the direction of the long axes of the drumlins in the Machars of Wigtownshire, which he interpreted as being due to a readvance of the ice, correlated with his Lammermuir–Stranraer Readvance. However, such changes in direction of drumlin long axes are not infrequent (Kerr, 1982; D. G. Sutherland, unpublished data) and appear to relate to topographical influence on ice flow rather than readvancing ice.

Many of the valleys on the southern side of the uplands contain massive spreads of glaciofluvial sands and gravels (Stone, 1959; Sissons, 1967a), with kames, kettle holes, kame terraces, eskers and meltwater channels forming complex and extensive areas of 'dead'-ice topography. The deposits around Stranraer where such features are fronted by a major outwash terrace merging into raised beaches (Sutherland, unpublished data) were considered by Charlesworth (1926b) to have been formed at the western end of his Lammermuir–Stranraer Readvance kame-moraine, but there is no evidence for such a readvance (e.g. Cutler, 1979), as has also been pointed out by Sissons (1961c) for the central and eastern parts of this putative limit.

Along the eastern Solway Firth coast the features of ice decay cannot be related to raised shorelines (Jardine, 1977, 1982; cf. Eyles and McCabe, 1989). In western Wigtownshire, however, such shorelines occur at altitudes of at least 20 m OD implying greater isostatic depression and/or later deglaciation in this latter area. In the Rhins of Wigtownshire there are possibly the only known examples of Errol beds on-shore in the west of Scotland (Brady *et al.*, 1874; Peacock, 1975c), these being deposited shortly after the deglaciation of this area. These beds, with their arctic fauna, suggest that deglaciation occurred prior to the oceanic polar front moving to the north of the British Isles, that is some time prior to about 13,000 BP.

There are no radiocarbon dates, however, that closely limit the time of deglaciation. A number of sites containing sediments deposited during the Lateglacial Interstadial have been investigated for both their pollen and beetle remains which have provided limiting relative ages on deglaciation as well as information about the progress of environmental change during the interstadial (Bishop, 1963; Moar, 1969b; Bishop and Coope, 1977; Jones, 1987). The sites with the earliest dated sediments are those of Roberthill and Redkirk Point in Dumfrieshire. At these localities coleopteran remains, from around 13,000 BP, indicate that the climate was almost as mild as at present although, due to low rates of migration, the vegetation was dominated by grasses and open-habitat taxa (Moar, 1963, 1969b; Bishop and Coope, 1977). Subsequently, during the interstadial, temperatures are inferred on the

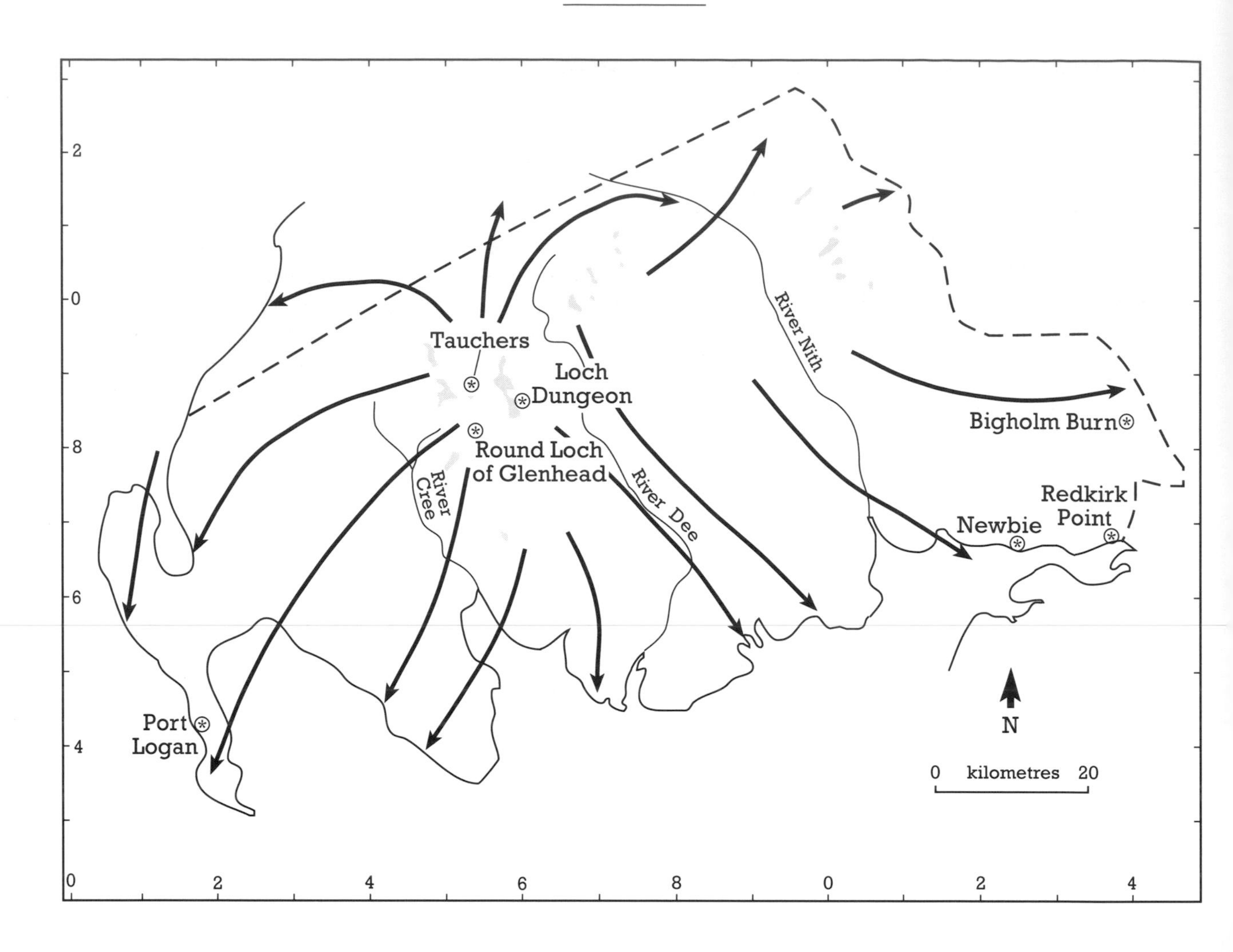

Figure 18.1 Location map of south-west Scotland and generalized directions of ice movement.

basis of the coleopteran faunas to have declined by 2–3°C, although there was during this period an increase in vegetation cover with the occurrence of stands of tree birch, willow and juniper. The freshwater and terrestrial deposits of Lateglacial age at Redkirk Point occur in the present intertidal area, indicating sea level to have been below that of the present throughout that period.

At the end of the Lateglacial Interstadial, at about 11,000 BP, there was a severe climatic deterioration and small glaciers became re-established in the Galloway hills (Cornish, 1981), depositing particularly clear end moraines at the Tauchers and Loch Dungeon. At lower altitudes there was slumping and solifluction of glacial deposits, as demonstrated at Bigholm Burn near Langholm (Bishop and Coope, 1977). This is one of the few lowland sites in Scotland where Lateglacial organic deposits have been found disrupted by soliflucted debris, allowing dating of the period of solifluction and analysis of the contemporaneous environmental conditions. Summer temperatures at that time were at least 6°C colder than at present (Bishop and Coope, 1977), which gave rise to a sparse, tundra-type vegetation.

The Holocene vegetational history of both the uplands and coastal lowlands has been investigated by Nichols (1967), Moar (1969b), and Birks (1969, 1972a, 1975). Mean summer temperatures rose very rapidly at the beginning of the Holocene and plant migration resulted in an early phase of juniper and herb dominance at around 10,000 BP. Birch and hazel expanded rapidly after this, and woodlands with those two species dominating became widely established by 9500 BP. Mixed deciduous woodlands with the addition of oak and elm developed in the lowlands by 8500 BP; pine appeared in the area by 8000 BP but was never a dominant species, attaining its widest extent in the Galloway hills at between 7500 and 6800 BP. The treeline at the time of maximum expansion of trees was at least at 600 m OD.

As demonstrated at both the Loch Dungeon and Round Loch of Glenhead sites, blanket peat began to expand in the middle Holocene, and pine was reduced significantly as an element in the forest composition. Man's initial impact on the forests is manifest in the decline of elm at slightly prior to 5000 BP, but the diatom and radiocarbon analyses carried out at the Round Loch of Glenhead are particularly important for the temporal framework they provide for the much more recent human impact related to acidification. This can be demonstrated to have increased markedly in this area in very recent times (Jones, 1987; Jones *et al.*, 1989).

During the early Holocene, sea level was several metres, at least, below its present level (Jardine, 1975, 1980b) and peat accumulated across the low-gradient coastal fringe. With the onset of the Main Postglacial Transgression these peats were successively transgressed, and radiocarbon dating at sites such as Redkirk Point and Newbie (Jardine, 1971, 1975, 1980b) allow the progress of the rising sea level to be followed. At its maximum, major marine embayments occurred at the head of the Solway Firth, at the mouth of the Nith valley and at the head of Wigtown Bay. Subsequent regression of the sea resulted in the renewed accumulation of peat on the surface of the estuarine deposits. Curiously, this sequence of events, although taking place between 8000 and 3000 years ago has been incorporated into folklore in the form of a witch's curse in which Lochar Moss by Dumfries was 'once a moss and then a sea' and became 'again a moss and aye will be' (Wood, 1975).

PORT LOGAN

D. G. Sutherland

Highlights

The coastal section at Port Logan displays a sequence of glacial deposits resting on a shore platform. These provide sedimentary evidence for the pattern of Late Devensian ice-sheet flow in south-west Scotland and show successive ice movements from sources in the Highlands and Southern Uplands.

Introduction

Port Logan (NX 092402) is a coastal section 20 km south of Stranraer in the Rhins of Galloway. This area was among the first to be surveyed by the Geological Survey of Scotland (Irvine, 1872) and it was noted that the glacial deposits contained a considerable variety of erratics. Particular attention was given to the occurrence of marine shells associated with the glacial deposits, which had previously been reported in that area by Moore (1850). It was noted that shells occurred in both laminated clays as well as in the underlying till (Irvine, 1872), and the cliffs immediately to the south of Port Logan were reported as a good example of the latter type of deposit.

Description

The Mull of Galloway was last glaciated by ice moving from the north-east, as is indicated by the long axes of the many drumlins on the peninsula (Kerr, 1982). Thick drift sections are present along the coast where these drumlins have been truncated by marine erosion. A lower, calcareous, shelly till associated with an earlier phase of glaciation from the north is also recorded (Kerr, 1982). In the cliffs south of Port Logan, a brown clayey till, with only occasional clasts near the base, but becoming more stoney higher up the section, rests directly on bedrock. Fragments of shells occur within the till (Brady *et al.*, 1874; D. G. Sutherland, unpublished data) or within a sandy lens in the till (Irvine, 1872). Approximately 2–3 m of the deposit are presently exposed, with the greater part of the 20 m of drift being vegetated. Near the top of the deposits approximately 1 m of sand and gravel with interbedded massive red clays overlie the till (D. G. Sutherland, unpublished data).

Only a limited fauna has been reported from Port Logan. Brady *et al.* (1874) recorded the molluscs *Astarte sulcata* (da Costa), *Nuculana pernula* (Müller) and *Yoldiella lenticulata* (Müller) and the ostracod *Cytheridea punctillata* Brady; Irvine (1872) recorded *Astarte sulcata* (da Costa), *Macoma baltica* (L.), *Nuculana* cf. *pernula* (Müller) and *Hiatella arctica* (L.).

The bedrock surface on which the shelly till rests is part of a shore platform which can be traced around the coast of the Mull of Galloway at approximately 10 m above present sea level. It clearly pre-dates the last glaciation of the area and may be equivalent to the rock platforms described by Stephens (1957) on the opposite side of the North Channel and by Gray (1978a) on the Mull of Kintyre (see Glenacardoch Point).

Interpretation

The shelly till of the Rhins of Galloway may relate to an early phase of the last (Late Devensian) ice-sheet glaciation when Highland ice extended across this area from the north, transporting, *inter alia*, erratics of Ailsa Craig microgranite and Arran granite on to the peninsula (Charlesworth, 1926a; Kerr, 1982; Sutherland, 1984a). Subsequently, Southern Uplands ice expanded, and a north-east to south-west ice flow developed across the area, moulding the drumlin landscape and carrying erratics of Loch Doon granite. This sequence of events is thus similar to that in the southern Central Lowlands (see Nith Bridge and Hewan Bank) where expansion of the Southern Uplands ice displaced Highland ice as the last glaciation progressed. Port Logan is thus important in a regional context, providing evidence of the relative pressures exerted by the two principal ice masses in Scotland during the last glaciation. This field evidence will help provide the necessary constraints for models of ice-sheet dynamics and palaeoclimatic reconstruction.

Additional interest in the site is provided by the glaciated rock platform, as there is currently considerable debate as to the age and origin of rock platforms along the west coast of Scotland (Sissons, 1981a, 1982b; Dawson, 1984; Sutherland, 1984a).

Conclusion

Port Logan is a representative site for the glacial sequence in south-west Scotland and forms part of the site network showing the major regional variations in the flow patterns of the last ice-sheet (around 18,000 years ago). In particular, the deposits reveal successive movements of ice across the area from the Highlands and Southern Uplands during the Late Devensian ice age.

TAUCHERS

D. G. Sutherland

Highlights

Tauchers displays an excellent example of a double end moraine formed by a corrie glacier of the Loch Lomond Readvance in the Southern Uplands. The site demonstrates clearly the importance of topographic factors in influencing glacier location and development.

Introduction

The Loch Doon granitic intrusion forms a large basin in the western Southern Uplands, the surrounding hills being composed of contact-metamorphosed greywackes and shales. In the centre of the basin is a prominent north–south ridge which culminates in the hill of Mullwharchar (692 m OD). The corrie-like embayment on the north-east flank of this hill is known as the Tauchers (NX 462876). The Southern Uplands was a major centre of ice dispersal during the Late Devensian ice-sheet glaciation (Sissons, 1967a; Sutherland, 1984a), but during the Loch Lomond Readvance only minor valley and corrie glaciers developed (Sissons, 1979d) in contrast to the much more extensive glaciation of the western Highlands at that time. In the western Southern Uplands, Cornish (1979, 1981) mapped 11 glaciers considered to be part of the Loch Lomond Readvance. Of these, the geomorphological evidence defining the former glaciers is most clear and most complete for the Tauchers glacier and has been described by Cornish (1979, 1981).

Description

Cornish (1979, 1981) has described the evidence that delimits a former glacier that flowed eastwards out of the Tauchers embayment. The terminus of the glacier is defined by a double end moraine complex (Figure 18.2). The outer moraine is up to 200 m wide and its outer edge has been eroded by the Gala Lane, showing it to be composed of an olive-brown, bouldery till with a sandy matrix. At its southern end the moraine has a maximum height of 10–12 m and comprises a series of ridges and depressions with an amplitude of 2–3 m. The inner end moraine is 3–5 m high and composed of olive-brown sandy till. It is somewhat sinuous in detail but broadly forms an arc parallel with the outer moraine.

On its north side, twin lateral moraines extend upslope from the ice-proximal and ice-distal slopes of the outer end moraine, terminating at 290 m and

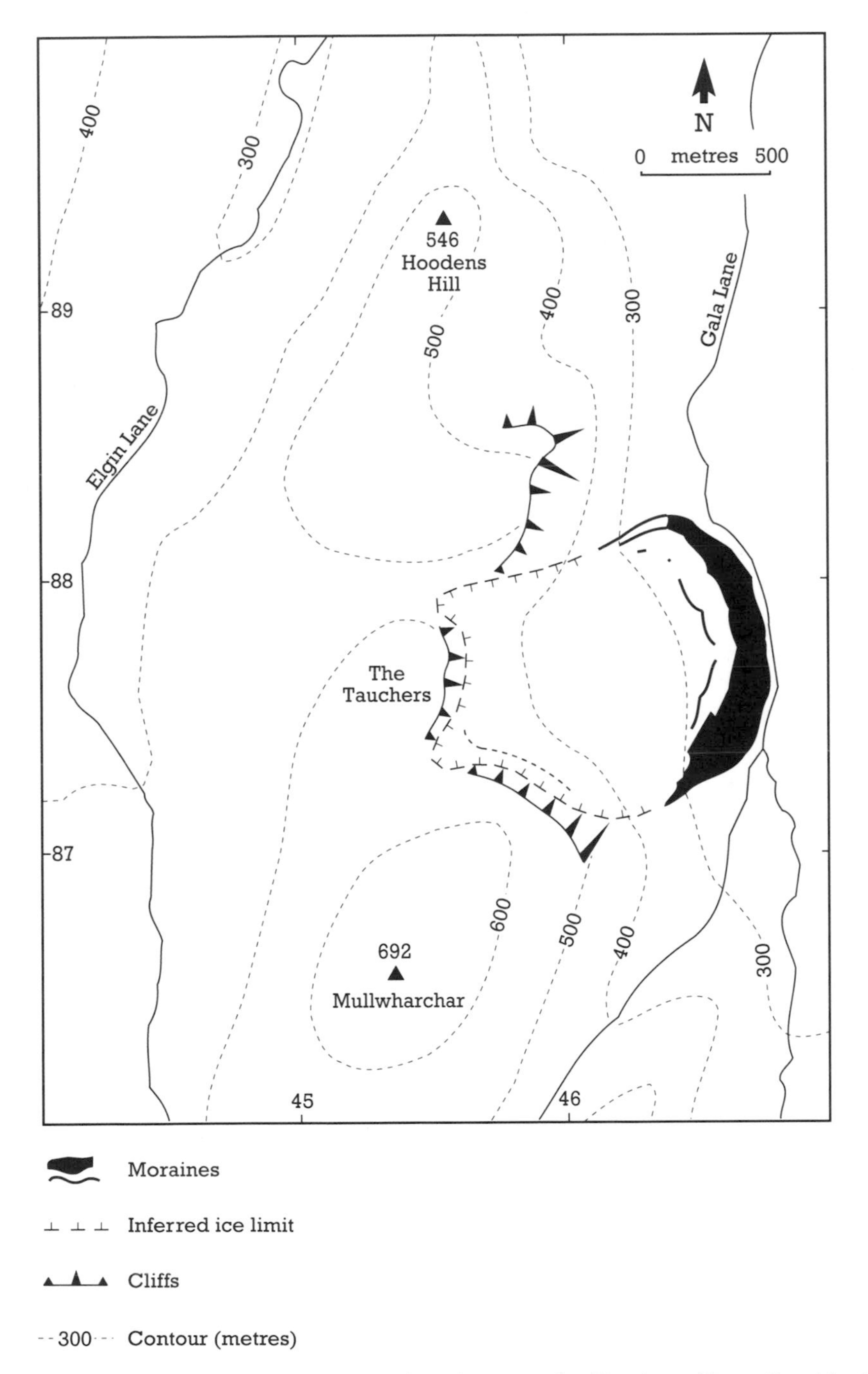

Figure 18.2 Loch Lomond Readvance moraines and ice limits at the Tauchers (from Cornish, 1981).

335 m OD respectively. These lateral moraines are composed of granite boulders up to 5 m in diameter. The northern margin of the inner moraine is also continued upslope to an altitude of 335 m OD by a lateral moraine composed of boulders.

The ground 'outside' the moraines of the former glacier is littered with large granite boulders, yet 'inside' the moraines such boulders are almost absent, those present apparently having rolled down from the slopes above. The implication is that the granite boulders were produced by periglacial activity before or during the period of local glaciation and that the glacier incorporated those within the area of its development into its lateral and end moraines.

Interpretation

It has long been recognized that following ice-sheet decay local glaciers developed in the western Southern Uplands (Geikie, 1863a; Jolly, 1868). Charlesworth (1926a) noted the Tauchers moraine as well as others in the area, placing them in his 'Corrie Moraine' stage of ice-sheet retreat. No direct dating of the glacial event responsible for the formation of the moraines is available, although they have been most recently accepted as part of the Loch Lomond Readvance (Sissons, 1979d; Cornish, 1981; Sutherland, 1984a). Pollen analyses of two cores from within the limits of the former glaciers, at Tauchers (Moar, 1969b) and Loch Dungeon (Birks, 1972a), revealed only Holocene sediments. This is in conformity with their presumed Loch Lomond Stadial age, although it is not conclusive evidence (Cornish, 1981).

Cornish (1981) calculated that the Tauchers glacier was 0.91 km^2 in area and that its equilibrium line altitude was approximately 330 m OD. This latter figure conforms with the altitude attained by the lateral moraines (cf. Andersen, 1968). The equilibrium line altitude for the Tauchers glacier was the lowest of the presumed Loch Lomond Readvance glaciers identified by Cornish, apparently in response to large amounts of snow being blown on to the glacier from the extensive slopes to the south-west.

The Tauchers landforms are an outstanding example of a terminal moraine complex of a corrie glacier. The moraine, representing one of only a small number in the western Southern Uplands, is presumed to have been formed during the Loch Lomond Readvance, and is hence an integral part of the glacial history of that region. The evidence for only minor glaciers existing in the region during the Loch Lomond Stadial (see also Loch Skene) is an important part of the national pattern of glaciation at that time, which is fundamental to the understanding of the Loch Lomond Stadial environment and climate (cf. Sissons, 1979d). The Tauchers site illustrates particularly well the relationships between topography (snow-blowing area) and glacier growth in a marginal situation for glacier development.

Conclusion

Tauchers is important for studies of the Loch Lomond Readvance, the last resurgence of glaciers in Scotland (approximately 11,000–10,000 years ago). It is a representative site for corrie moraines of the readvance in the Southern Uplands and includes an excellent example of a double end-moraine system. The site also shows particularly well the interaction of local topography and climate in affecting glacier development, and the Tauchers is an integral part of the network of sites providing geomorphological evidence for the Loch Lomond Readvance.

BIGHOLM BURN

J. E. Gordon

Highlights

The stream section at Bigholm Burn demonstrates a sequence of glacial, lacustrine and fluvial deposits, with interbedded organic sediments including peat. These sediments, together with the pollen and coleopteran records from the organic horizons, have provided important evidence for palaeoclimatic conditions and environmental change during the Lateglacial.

Introduction

This site (NY 316812) comprises a stream section located on the east bank of Bigholm Burn, 6 km south-west of Langholm. It is important in providing sedimentary and biostratigraphic evidence for the Lateglacial and early Holocene environmental history of south-west Scotland. The key interest includes assemblages of fossil

Coleoptera, which, together with those from Redkirk Point and other sites in the region, reveal a significantly different pattern of Lateglacial climatic change to that suggested by pollen evidence: the mildest climate occurring about 13,000 BP, followed by a progressive deterioration throughout the Lateglacial Interstadial, which led into the severe conditions of the Loch Lomond Stadial. The sequence of sediments at Bigholm Burn and their lateral variations have been described by Moar (1964, 1969b) and Bishop and Coope (1977).

Description

The sequence of sediments is as follows (Moar, 1969b; Bishop and Coope, 1977):

7.	Peat, comprising sedge peat passing up into woody peat and blanket bog peat	up to 2 m
6.	Brown organic mud uncomformably overlying bed 5	up to *c.* 0.5 m
5.	Lenses of dark-grey silt	0.15 m
4.	Poorly bedded, subangular gravels dipping north-west, locally incorporating blocks of peat and clay	up to 3 m
3.	Folded sedge peat	0.32 m
2.	Blue-grey clay	*c.* 0.2 m
1.	Grey, pebbly and clayey till largely derived from local Silurian rocks	1.5 m exposed

This sequence of sediments is calibrated by seven radiocarbon dates (Godwin and Willis, 1964; Godwin *et al.*, 1965): 11,820 ± 180 BP and 11,580 ± 180 BP (Q–694) from the peat below the gravel (bed 3); 10,820 ± 170 BP (Q–695) from a block of peat incorporated in the gravel (bed 4); 9590 ± 170 BP and 9470 ± 170 BP (Q–697) from the organic mud in bed 6; 8650 ± 165 BP (Q–699) from the transition between sedge and woody peat in bed 7; 7735 ± 155 BP(Q–700) from the base of the woody peat in bed 7; 7640 ± 160 BP (Q–701) from the base of the woody peat in bed 7 above the previous sample, and 5475 ± 120 BP (Q–702) in the blanket bog above in bed 7.

Interpretation

Bishop (Bishop and Coope, 1977) interpreted the succession as beginning with the accumulation of fine-grained sediments (bed 2) and then sedge peat (bed 3) in pools in the till surface. The coarse angular gravel above (bed 4) represented solifluction deposits formed during the Loch Lomond Stadial. The weight of this gravel compressed the peat and caused it to buckle, and in places to rupture, so that blocks became incorporated into the gravels. The radiocarbon date of 10,820 ± 170 BP on one such block may indicate the end of the period of peat formation, but could be too young owing to enrichment by percolation of humic acid from the peat above. Small ponds appeared on the gravel surface towards the end of the stadial, and organic silts (bed 5) accumulated. These were subsequently eroded by lateral migration of the stream that deposited the alluvial mud (bed 6). Again the radiocarbon age of the organic mud is probably too young, since the coleopteran remains indicate the presence of species that imply stadial conditions (see below). Peat (represented by bed 7) finally became established on the alluvial surface during the Holocene.

Recent observations (A. M. Hall and C. M. White, unpublished data) have added further details to the stratigraphy. Bed 1 is poorly exposed at the base of the section and is much better represented in stream sections *c.* 1 km downstream (NY 322815 and 324817), where up to 2 m of grey till, with Silurian pebbles, rests on a similar thickness of red-brown till, with abundant clasts of red sandstone. Bed 2 shows fine horizontal lamination and passes downstream into laminated clays and silts, with drop stones, and interbedded sand and gravel bands with a total exposed thickness of *c.* 4 m (at NY 324817). These deposits suggest that during ice retreat from the area, a small lake was ponded within the Bigholm valley and partially infilled by glaciolacustrine deposits. Bed 4 includes locally imbricate gravel horizons and cross- and flat-bedded sand lenses, indicating that it is a braided stream deposit. The occasional presence of sand clasts within the gravels demonstrates transport of frozen sand blocks. The silts of bed 5 may represent fine-grained overbank deposition.

Moar (1964, 1969b) investigated the pollen stratigraphy at Bigholm Burn as part of a wider study of Late Devensian and Holocene vegetational history in south-west Scotland. Pollen grains in the basal peat (bed 3) and the peat included in the gravel (bed 4) are representative of an open-vegetation assemblage (including Cyperaceae, Gramineae, Compositae, *Filipendula* and *Empetrum*) and compatible with a Lateglacial Interstadial age. Pollen grains in the silts (bed 5) above the gravels are from dominantly herbaceous plants of types associated with open, unstable habitats (for example, Cyperaceae, Gramineae, Compositae, Cruciferae, *Epilobium*, *Saxifraga oppositifolia* type, *Thalictrum*, *Artemisia* and *Koenigia*) and suggest that the deposit originated during the Loch Lomond Stadial. In the transition to the milder climate of the Holocene, pollen zone FI (lower part of bed 6) shows a rise in *Betula*, *Juniperus* and *Salix*, high frequencies of non-arboreal pollen and a remarkable development of aquatic types. *Betula* is dominant in zone FII (upper part of bed 6) and in the early part of zone FIII (bed 7) before being replaced by *Pinus*, *Ulmus*, *Quercus* and *Alnus*. Radiocarbon dates from the Holocene sediments at Bigholm Burn are comparatively younger than those for equivalent pollen zones at Scaleby Moss 21 km to the south-east (Godwin *et al.*, 1957) and may reflect rootlet penetration from above or contamination by humic acid percolation (Bishop and Coope, 1977).

The overall regional pattern of the vegetational history of south-west Scotland to emerge from Moar's work was one of open, treeless vegetation during the Lateglacial. The regional Lateglacial pollen zone I was dominated by Gramineae, Cyperaceae and *Rumex* together with dwarf shrubs. In zone II the vegetation became more stable and denser. Herbs remained dominant, and shrubs increased. Zone III was characterized by a sharp increase of open, unstable habitats. Open vegetation persisted at the start of the Holocene before the slow spread of birch woodland into the area and its subsequent dominance until partly replaced by mixed oak forest in Holocene zone IV.

Coope (Bishop and Coope, 1977) investigated the assemblages of fossil Coleoptera in the organic layers at Bigholm Burn and other sites in the region. The assemblage from the lower peat (bed 3) at Bigholm Burn includes *Diacheila arctica* Gyll., *Agonum consimile* Gyll., *Hydroporous arcticus* Th., *H. longicornis* Sharp, *Ilybius angustior* Gyll., *Helophorous aquaticus* L., *H. flavipes* F., *Pycnoglypta lurida* Gyll., *Olophrum assimile* Payk., *O. fuscum* Gr. and *Stenus* spp. It closely resembles the assemblage from a broadly contemporaneous peat dated at 11,205 ± 177 BP (Birm–41) at Redkirk Point (see below), although there is some indication from a slightly higher representation of phytophagous species that the vegetation was more diverse at Bigholm Burn. The assemblage at Bigholm Burn has a definite northern chatacter, but the presence of two relatively southern species (*Eubrychius velutus* Beck. and *Gymnetron beccabungae* L.) suggests slightly less severe conditions than at Redkirk Point, with an average July temperature of about 12°C. A second sample from the organic silt (bed 5) yielded an impoverished fauna dominated by *Olophrum boreale* Payk., suggesting little or no plant cover and average July temperatures well below 10°C at the time of the Loch Lomond Stadial. This sample from Bigholm Burn demonstrates further climatic deterioration after deposition of the Redkirk Point sample dated at 10,898 ± 127 BP (Birm–40) and is the only one of stadial age in Scotland so far investigated for fossil Coleoptera.

The combined evidence from the assemblages of fossil Coleoptera at Bigholm Burn, Redkirk Point and other sites discussed by Bishop and Coope (1977), each covering a slightly different time period, suggests a pattern of Lateglacial climatic change in which temperatures rose to as warm as present by about 13,000 BP (or possibly later – cf. Atkinson *et al.*, 1987). Temperatures then fell sharply in two stages between about 12,500–12,000 BP and 11,000–10,500 BP, culminating in the Loch Lomond Stadial. The drop in July temperatures inferred from the coleopteran assemblages was as much as 6–7°C. At the end of the stadial a temperature rise of similar magnitude occurred within 700 years. This interpretation differs from the traditional view based on pollen analysis (see above) that the Lateglacial thermal maximum occurred during pollen zone II of the Jessen–Godwin Scheme, usually equated with the Allerød Interstadial between about 11,800 and 11,000 BP. The evidence from the Coleoptera, however, indicates that the thermal maximum was already past by this time and that temperatures were decreasing. Such a pattern is substantiated by similar studies elsewhere in Britain (see Coope, 1977; Atkinson *et al.*, 1987). It appears that vegetation recolonization at the end of the Late Devensian ice-sheet glaciation lagged behind the climatic changes. In contrast, the more mobile beetles responded more rapidly to the changing climate, and hence fossil coleoptera are more sensitive indicators of past climate change than are pollen assemblages (cf. Coope and Brophy, 1972;

Coope, 1975, 1981; Coope and Joachim, 1980).

Bigholm Burn is an important locality for interpreting the Lateglacial and early Holocene environmental history of south-west Scotland. The sedimentary and biostratigraphic evidence complements that at Redkirk Point, particularly the fossil coleopteran assemblages. The deposits show:

1. the transition from cold conditions, accompanied by glacial and glaciolacustrine sedimentation, at the end of the Late Devensian glaciation to the establishment of temperate conditions during the Lateglacial Interstadial;
2. subsequent climatic deterioration culminating in the Loch Lomond Stadial, with accelerated mass movement, frost weathering and associated cold-climate fluvial activity;
3. climatic amelioration at the start of the Holocene.

Conclusion

Bigholm Burn is important for studies of environmental history during the final phase of the Late Devensian, the Lateglacial (approximately 13,000–10,000 years ago). Detailed study of the sequence of deposits, including analysis of the pollen and beetle remains they contain, has provided a vital record of Lateglacial environmental conditions and geomorphological changes in south-west Scotland. The evidence from the fossil beetles, in conjunction with that from Redkirk Point, is particularly significant in showing rapid climatic warming early in the Lateglacial, followed by a considerable climatic deterioration in two steps. Bigholm Burn forms a key part of the network of sites for establishing Lateglacial environmental conditions.

REDKIRK POINT

J. E. Gordon

Highlights

The coastal section at Redkirk Point displays a sequence of estuarine deposits and buried peat. Analysis of these deposits, and the pollen and coleopteran remains they contain, has provided detailed information about palaeoclimatic conditions, environmental change and coastline development during the Lateglacial and early Holocene.

Introduction

Redkirk Point (NY 301652) is located on the coast of the Solway Firth, 11 km east of Annan. It shows a sequence of interbedded organic and marine sediments exposed on the foreshore and in the backing cliff. These deposits are important for interpreting the patterns of Lateglacial and early Holocene environmental history and sea-level change in south-west Scotland. The sediments exposed at Redkirk Point have been described by Jardine (1964, 1971, 1975, 1980b) and in greatest detail by Bishop (Bishop and Coope, 1977).

Description

Bishop and Coope (1977) recorded the following sequence partly infilling a shallow channel cut in the New Red Sandstone bedrock (see also Figure 18.3) (see also Jardine, 1980b):

Bed	Description	Thickness
9.	Disturbed ground and soil profile	0.15 m
8.	Sandy silts and alluvium	0.40 m
7.	Grey clays and silts (carse clays) with thin peat layer	3.00 m
6.	Grey clays and fine sands with several discontinuous peat lenses and disturbed bedding	3.0 m
5.	Highly compacted and disturbed woody peat with *in situ* tree stumps	0.15 m
4.	Grey silts and fine sands with disturbed bedding	1.60 m
3.	Highly compacted and disturbed peat, with local peat lens below	0.25 m
2.	Carbonaceous silts and fine sands	0.30 m
1.	Red, sandy and pebbly till	1.50 m

The peat layers are typically deformed and buckled under the weight of the overlying clays (carse). Five radiocarbon dates have been obtained on material from the organic layers (Bishop and Coope, 1977): 12,290 ± 250 BP (Q–816) from the peat lens below bed 3 (Godwin *et al.*, 1965); 11,205 ± 177 BP (Birm–41) from the bottom 0.06 m of bed 3 (Shotton *et al.*, 1968); 10,898 ± 127 BP (Birm–40) from the top 0.03 m of bed 3 (Shotton *et al.*, 1968); 10,300 ± 185 BP (Q–815) from wood (*Populus*) from the eroded top of bed 3 (Godwin *et al.*, 1965), and 8135 ± 150 BP (Q–637) from the outer rings of an *in situ* tree stump in bed 5 (Godwin and Willis, 1962).

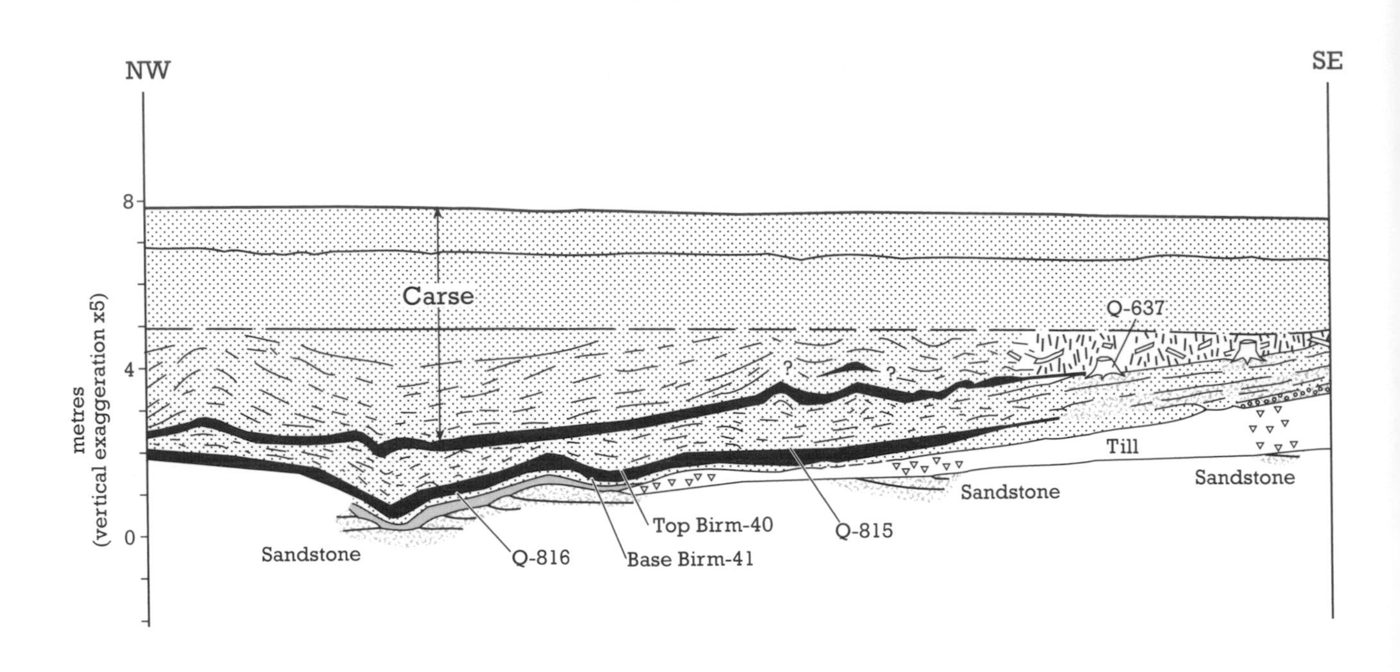

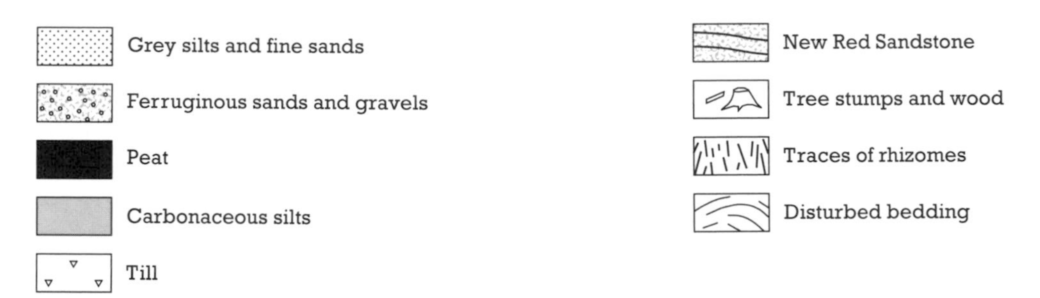

Figure 18.3 Redkirk Point: sediment succession (from Bishop and Coope, 1977).

Interpretation

Bishop (Bishop and Coope, 1977) interpreted the sequence of events beginning with a period of low sea level following glaciation, during which a shallow channel was eroded in the till and sandstone. Sedimentation of the carbonaceous silts and sands (bed 2) under fluvial conditions began before 12,000 BP. The pollen of the peat lens below bed 3 indicates a typical Lateglacial floral assemblage with abundant herbaceous types (including *Selaginella*, *Empetrum*, *Thalictrum* and cf. *Saussurea alpina*) and limited arboreal types (*Betula*, *Pinus* and *Salix*) (Godwin and Switsur, 1966). The peat of bed 3 probably attained a thickness of about 1.25 m before compression. Its development was curtailed by freshwater ponding associated with a rising sea level around 10,300 BP, followed by deposition of the grey silts and fine sands of bed 4. Thereafter, a relative fall in sea level allowed re-establishment of vegetation and development of a tree cover (represented in the deposits of bed 5). A subsequent rise in relative sea level after about 8100 BP associated with the Main Postglacial Transgression (Jardine, 1964, 1971, 1975, 1980b) was accompanied by deposition of over 6 m of carse clays (beds 6, 7 and 8).

The peat beds in the sequence at Redkirk Point preserve a valuable record of changing environmental conditions. The pollen record has not been investigated in detail, although the regional pattern of the Late Devensian and Holocene vegetational succession has been established from studies at a number of other lowland sites in south-west Scotland (Moar, 1964, 1969b; Nichols, 1967). The key palaeoecological evidence at Redkirk Point lies in the remains of Coleoptera preserved in the sediments. These, together with the fossil assemblages from Bigholm Burn (see above) and other sites, each covering a slighty different time period, provide critical information on Lateglacial climatic conditions (Bishop and Coope, 1977).

The assemblage of fossil Coleoptera in bed 2 at Redkirk Point, below the peat dated at 12,290 ± 250 BP, is characterized by a relative abundance of species of running water (e.g. Elminthidae and *Hydraena gracilis* Germ.) and stream bank (e.g. *Bembidion schueppeli* Dej. and *Hypnoidus riparius* F.) habitats. The presence of species with both northern (e.g. *Bembidion schueppeli* Dej. and *Arpedium brachypterum* Gr.) and southern (e.g. *Bembidion gilvipes* Sturm, *Cymindis angularis* Gyll. and *Esolus parallelepipedus* Müll.) distributions today suggests a temperate climate similar to that of south-west Scotland at present, reflecting moderately oceanic conditions and a mean July temperature of about 15°C. Sparse vegetation cover is also indicated by the rarity of phytophagous species, and the overall environmental conditions are similar to those deduced by Coope (Bishop and Coope, 1977) from the assemblage in a peat layer at Roberthill (NY 110797) dated at 12,940 ± 250 BP (Q–643) (Bishop, 1963). The remains from the horizon in bed 3, dated at 11,205 ± 177 BP, at Redkirk Point are of species which have overall northern affinities (e.g. *Diacheila arctica* Gyll., *Elaphrus lapponicus* Gyll., *Patrobus septentrionis* Dej., *Amara torrida* Ill., *Agonum consimile* Gyll., *Hydroprous tartaricus* Lec., *Ilybius anqustior* Gyll., *Olophrum boreale* Payk., *Acidota quadrata* Zett., *Boreaphilus henningianus* Sahlb. and *Otiorhynchus nodosus* Müll.), and eastern affinities (e.g. *Chlaenius costulatus* Mtsch. and *Bembidion transparens* Gebl.), reflecting a marked contrast in environmental conditions compared with those indicated by the sample from bed 2. Climatic deterioration was accompanied by increased continentality, a fall in average July temperatures to about 12°C and widespread development of acid bog and wetlands. Similar conditions at this time are also implied by the Coleoptera in a bed, dated at 11,580 ± 180 BP to 11,820 ± 180 BP (Q–694,) at Bigholm Burn (see above). Further climatic deterioration is indicated by increased numbers of the northern species in the horizon in bed 3 dated at 10,898 ± 127 BP. The greater abundance of species such as *Pycnoglypta lurida* Gyll., *Olophrum fuscum* Gr., *Arpedium brachypterum* Gr. and *Boreaphilus henningianus* Sahlb. is indicative of increasingly more open tree cover. However, the absence of species characteristic of alpine and tundra environments suggests average July temperatures of about 10°C.

The organic deposits at Redkirk Point from which the beetle assemblages have been studied span the period from prior to 12,290 BP to about 10,890 BP, covering much of the Lateglacial Interstadial and the beginning of the Loch Lomond Stadial. Redkirk Point is one of only a few sites in Scotland where Lateglacial environmental conditions have been interpreted from beetle remains (see Bigholm Burn). The significance of the coleopteran evidence from these sites is that it points to a pattern of Lateglacial climatic conditions quite distinct from that suggested by pollen assemblages for the same period (Bishop and Coope, 1977). The beetles indicate that climatic amelioration occurred early in the interstadial, with temperatures as warm as those of the present day attained by about 13,000 BP (or possibly later – cf. Atkinson *et al.*, 1987). Subsequently, climatic deterioration began about 12,500 BP and intensified between 11,000 BP and 10,500 BP during the Loch Lomond Stadial. Conversely, the pollen record shows open habitat, treeless conditions prior to 13,000 BP followed by the main expansion of vegetation between 11,800 BP and 11,000 BP. The pattern of Lateglacial climatic change inferred from the evidence at Redkirk Point is similar to that established from beetle evidence elsewhere in Britain (Coope, 1977; Atkinson *et al.*, 1987) and reflects the great sensitivity and response rate of the beetle populations to changing environmental conditions (cf. Coope and Brophy, 1972; Coope, 1975, 1981; Coope and Joachim, 1980).

Redkirk Point is also a key site providing stratigraphic and geochronometric evidence for the pattern of Lateglacial and early Holocene sea-level change in south-west Scotland, complementing the interest at Newbie. This evidence demonstrates that from the time of deglaciation (prior to 13,000 BP) and throughout almost the whole of the Lateglacial, sea level was below that of the present day. The first evidence of marine influence apparently occurs at the end of the Lateglacial and the start of the Holocene (bed 4), which invites comparison with the marine transgression that occurred at this time in the Western Forth Valley (see above). Subsequently, sea level fell and peat (bed 5) accumulated. During the Main Postglacial Transgression grey silts and clays (beds 6 and 7) were deposited on top of the peat.

These last events have been studied in greater detail at Newbie (see below).

Conclusion

The deposits at Redkirk Point provide important evidence for changes in sea level and coastal environmental conditions in south-west Scotland during the phase which closed Devensian times (the Lateglacial) and the succeeding and warmer early Holocene (between approximately 13,000 and 8000 years ago). This evidence is derived from detailed analysis of the sediments and the pollen and beetle remains they contain, and is supported by radiocarbon dating. The length and detail of the record, and in particular the combined evidence from the pollen and beetles, make Redkirk Point a key reference site for studies of Lateglacial environmental history in south-west Scotland and an integral component in the national site network.

NEWBIE

J. E. Gordon

Highlights

The coastal section at Newbie shows a sequence of interbedded estuarine and organic deposits, including peat. These deposits provide important evidence for interpreting sea-level changes on the Solway coast during the Holocene.

Introduction

The site at Newbie (NY 165651) is a section on the coast of the Solway Firth, 2.5 km south-west of Annan. It shows a sequence of interbedded marine and organic sediments that provides stratigraphic and geochronometric evidence for Holocene sea-level change in the Solway Firth area, notably the Main Postglacial Transgression. The deposits at Newbie have been described in several papers by Jardine (1964, 1971, 1975, 1980b) from section and borehole evidence.

Description

At Newbie Cottages, marine deposits infill the remnants of several small kettle holes in an area of Late Devensian glaciofluvial deposits (Jardine, 1964, 1971, 1975, 1980b). Jardine (1980b) described several sections in detail, but the full generalized sequence is as follows (see also Figure 18.4):

5.	Blown sands with interbedded layers of sand containing fragments of charcoal and organic matter	up to 3.5 m
4.	Peat	up to 1 m
3.	Sands, silts or clayey silts (carse)	up to >4.5 m
2.	Organic detritus	up to 0.3 m
1.	Glaciofluvial sands and gravels displaying at the top, podsolic A_2 and B_2 soil horizons	

The carse deposits (bed 3) comprise two units. The lower one consists of medium and fine-grained sand, locally laminated, and the upper, of silt and clay. The lower sediments also have a relatively higher content of microfaunal remains, and plant debris is present in the upper sediments. Jardine (1967, 1975, 1980b) and Jardine and Morrison (1976) have discussed the sediment characteristics and environment of accumulation of carse deposits both in general terms and in relation to Newbie.

Radiocarbon assays on material from the upper part of the lower organic layer (bed 2) gave the following dates (Jardine, 1975): 7254 ± 101 BP (GU–64) (Baxter *et al.*, 1969), 7540 ± 150 BP (Birm–222) (Shotton and Williams, 1971) and 7400 ± 150 BP (Birm–325) (Shotton and Williams, 1973). Dates obtained from the lower part of the upper organic layer (bed 4) were (Jardine, 1975): 5630 ± 116 BP (Birm–220) (Shotton and Williams, 1971) and 4290 ± 100 BP (I–5070) (Buckley and Willis, 1972). Of the dates on bed 4, the latter may be less reliable than the former as the sample was not pre-treated for humic acid extraction. Charcoal from the wind-blown sands (bed 5) was dated at 3480 ± 110 BP (Birm–218) (Shotton and Williams, 1971).

A borehole located a few hundred metres inland from the coast section at Newbie Cottages penetrated 0.01 m of organic material dated at 7812 ± 130 BP (GU–375) (Ergin *et al.*, 1972) within the sequence of carse sediments (Jardine, 1975). Sand below this organic layer contained frequent Foraminifera tests and a few small fragments of mollusc shells and echinoid spines; the sand above contained occasional fragments of

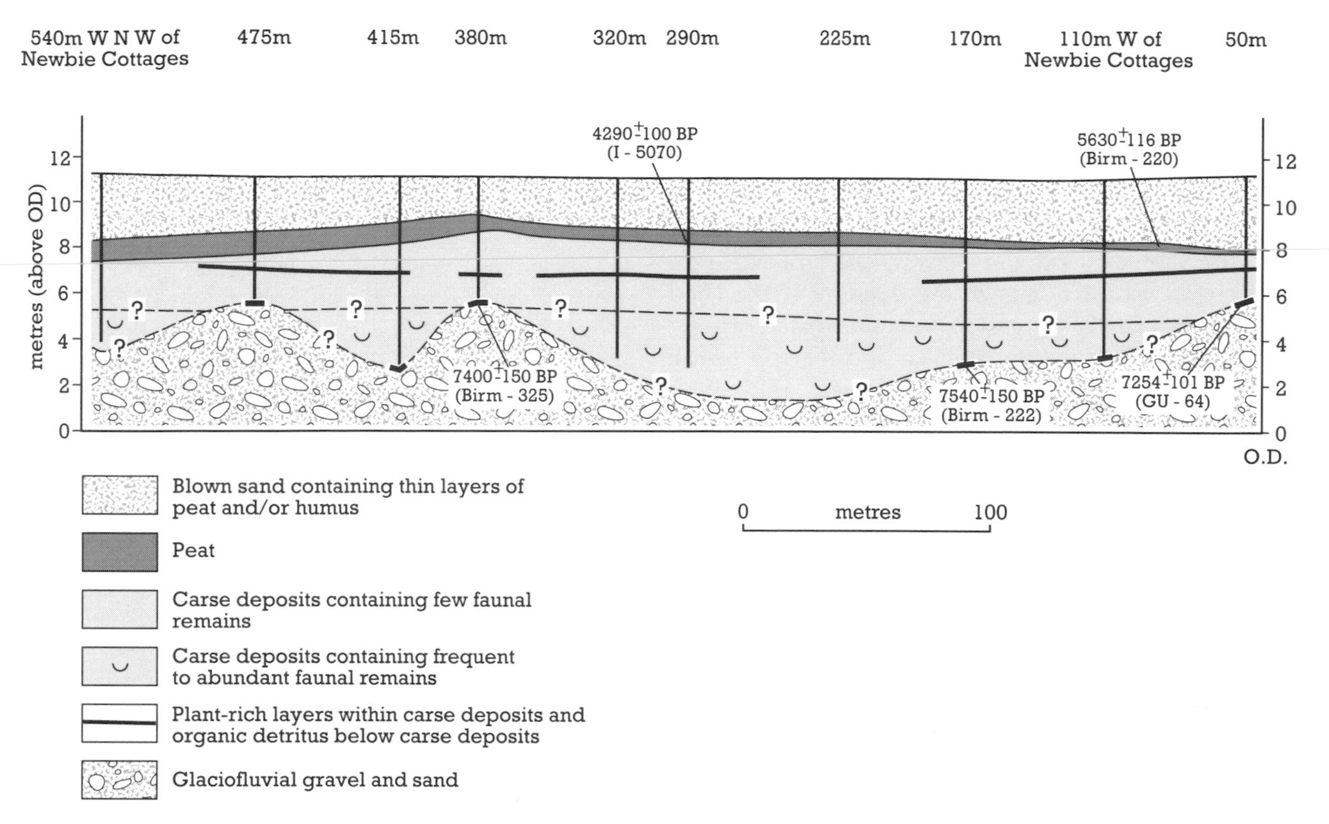

Figure 18.4 Newbie: sediment succession (from Jardine, 1980b).

echinoid spines and sponge spicules.

The biostratigraphy of the organic beds at Newbie has not been investigated. However, Nichols (1967) has studied the pollen stratigraphy of comparable peat layers above and below Main Postglacial Transgression sediments in a borehole at Lochar Moss, and Moar (1964, 1969b) has established the Late Devensian and Holocene vegetational successions at a number of sites in the area.

Interpretation

Jardine (1975, 1980b) interpreted the Newbie deposits to represent a locally diachronous marine transgression controlled by the form of the topography. The lower area where the borehole was located was inundated by the sea before 7800 BP. This was followed by a brief interruption in marine sedimentation at around 7800 BP, then a further transgressive overlap. The kettle holes at Newbie Cottages were not breached by the sea until about 7450 BP. The marine transgression then culminated at, or some time prior to, 5600 BP.

The position of the local succession at Newbie in the overall chronology and pattern of Holocene marine transgression and regression in the Solway area was discussed by Jardine (1964, 1971, 1975, 1980b). Depending on the altitude of the individual sites, the date of the first evidence for the transgression varies from place to place along the north coast of the Solway, ranging from 9400 BP to 7200 BP. The earliest occurrence is recorded in a borehole at Carsethorn (NX 988594), south of Dumfries (−1.05 m OD: 9400 BP). In the eastern Solway the transgression was first registered at Redkirk Point (2.90 m OD: 8100 BP) and then at Newbie (2.95 m OD to 5.80 m OD: between 7500 BP and 7200 BP). The Newbie dates are broadly similar to that for the transgression into the northern part of the Lochar 'gulf' (7400 BP). In the western Solway at the head of Wigtown Bay the transgression started prior to 7900 BP.

Jardine (1975, 1980b) inferred from the radiocarbon dates on the peat overlying the grey silts and clays that the culmination of the transgression was also diachronous. He suggested that the Lochar 'gulf' was abandoned by 6600 BP, but that regression did not start at Newbie until 5600 BP and at the head of Wigtown Bay until 4700 BP. This interpretation has been disputed by Sissons (1983a), Sutherland (1984a) and Haggart (1988a, 1989) who have indicated that the evidence available does not conflict with a synchronous maximum to the Main Postglacial Transgression along the north coast of the Solway Firth.

The importance of Newbie is that it is one of the few sites on the Solway coast providing a clear exposure through the deposits of the Main Postglacial Transgression, including dated organic deposits at both the bottom and the top of the sequence. It complements the site at Redkirk Point (see above), where the close of the transgression is less clear. Newbie is also important in illustrating the effects of local topographic controls on marine sedimentation during a transgressive episode.

The coastal lowlands around the Solway Firth provide some of the most extensive areas of estuarine sedimentation on the west coast of Scotland. During the Holocene these areas of quiet-water sedimentation provided excellent environments for the deposition and preservation of sedimentary sequences related to sea-level change. They contrast with the greater part of the west coast, where higher energy environments were typical, resulting in the reworking and destruction of much of the sedimentary evidence. The Solway Firth estuarine sequences are more akin to those found around the large estuaries on the east coast and comparison between these various locations is instructive. For example, it is notable that no equivalent has been found in the Solway of the grey, micaceous, silty, sand layer deposited widely along the east coast at around 7000 BP (see Western Forth Valley, Silver Moss and Maryton). This suggests that the event that led to deposition of this sand was confined to the North Sea Basin, thus lending support to its interpretation as a storm surge or tsunami (Smith *et al*., 1985a; Dawson *et al*., 1988). The date of the maximum of the Main Postglacial Transgression may also provide an instructive contrast between the two regions. Most east coast sites suggest that the transgression reached its maximum increasingly late with increasing distance from the western Highlands ice centre. However, in the Solway Firth area, the transgression may have reached its maximum relatively early compared with its distance from the western Highlands, thus suggesting a significant role for the Southern Uplands ice centre in glacio-isostatic depression and recovery.

Conclusion

The deposits at Newbie provide a detailed record of sea-level changes in south-west Scotland during the Holocene (the last 10,000 years). In particular, they are significant for studies of the Main Postglacial Transgression (see Silver Moss above) and its timing. Newbie is an important reference site for this event on the Solway Firth coast, where it reached its maximum approximately 6500 years ago and forms part of the network of sites for establishing the regional variations in the pattern of sea-level change during Holocene times.

LOCH DUNGEON

H. J. B. Birks

Highlights

Pollen and plant macrofossils from the sediments of Loch Dungeon and the adjacent blanket peat provide a detailed record, supported by radiocarbon dating, of vegetational history and environmental change in the uplands of south-west Scotland during the Holocene.

Introduction

Loch Dungeon (NX 525846) lies within a complex corrie at an altitude of 305 m OD on the east side of the Rhinns of Kells. The pollen and plant macrofossils contained in the lake sediments and in the adjacent area of blanket peat provide a detailed record of the local and regional Holocene vegetational history of the Galloway hills of south-west Scotland. Loch Dungeon is the site at which many of the Galloway regional pollen assemblage zones were delimited and defined by Birks (1972a). Fossil pine stumps occur locally in the blanket peats of the Galloway hills, forming the southernmost area of pine stumps in Scotland (Birks, 1975). Loch Dungeon is one of the few areas in the British Isles where pollen diagrams have been constructed from both a lake and nearby bog profiles.

Figure 18.5 Geomorphology of the Loch Dungeon area (from Cornish, 1981).

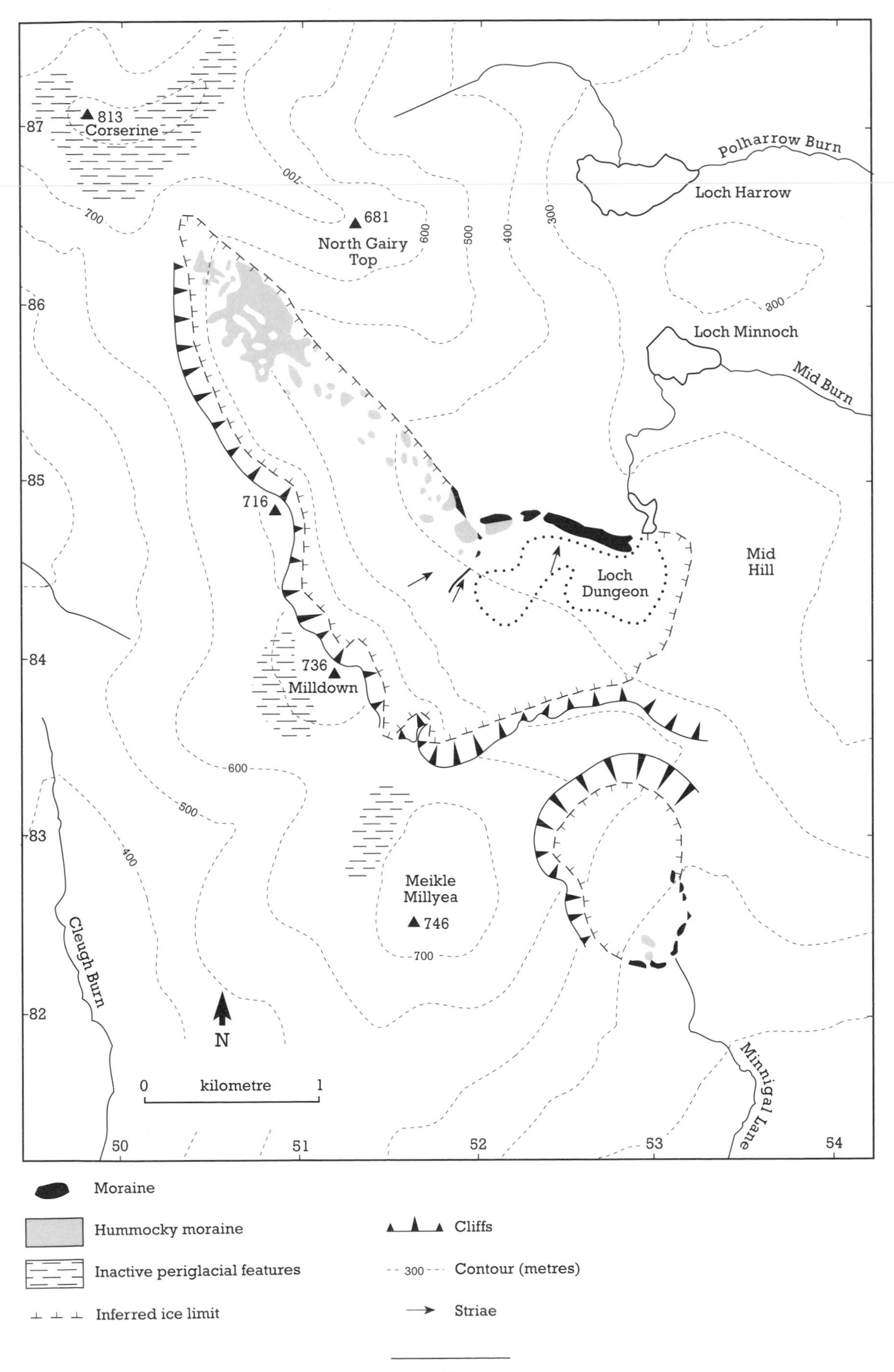
813
Corserine
Polharrow Burn
Loch Harrow
681
North Gairy Top
Loch Minnoch
Mid Burn
716
Loch Dungeon
Mid Hill
736
Milldown
Meikle Millyea
746
Cleugh Burn
N
0 kilometre 1
Minnigal Lane
Moraine
Hummocky moraine
Inactive periglacial features
Inferred ice limit
Cliffs
Contour (metres)
Striae

Description

Precipitous cliffs of Silurian shales, greywackes and grits of Meikle Millyea rise from the south-east shore of Loch Dungeon, and a subsidiary corrie on the south-east flank of Corserine opens out on the north-west shore. A Loch Lomond Readvance glacier emerged from the corries and its terminus is marked by a large end moraine to the west of the loch (Figure 18.5) and by a shallow area within the loch itself (Cornish, 1981). The bottom of the loch is very rocky, with only about 1 m of organic sediment. In the outer, eastern basin, fine-grained organic sediments up to 4 m in thickness occur.

The palaeoenvironmental record of the Southern Uplands peats was first investigated by Lewis (1905) and later by Samuelsson (1910). The Loch Dungeon peat sequence is from the area of blanket peat on the flat top of the spur running from Meikle Millyea to Mid Hill at an altitude of 396 m OD. The peat is presently severely hagged and subject to sheet erosion. A sequence 2.65 m thick was analysed by Birks (1975). The stratigraphy consists of humified *Eriophorum–Calluna* peat overlying a black humus soil rich in carbonized fragments around or just above pine stumps at a depth of 1.45 m. The basal 1.2 m consist of *Phragmites–Carex* wood peat. The pine stump layer is radiocarbon dated to 7165 ± 180 BP (Q–876) and the humus layer to 6787 ± 200 BP (Q–877).

Interpretation

The pollen sequence from Loch Dungeon extends from the Devensian/Holocene boundary to the last 100 years. It is divided into six local pollen zones, which are correlated with five regional (Galloway) pollen assemblage zones (Figure 18.6A) (Birks, 1972a). Three of these regional zones are best represented at Loch Dungeon. The pollen records the colonization of open, pioneer species-rich, herbaceous assemblages following the melting of local corrie glacier ice at the end of the Loch Lomond Stadial at about 10,000 BP. These were rapidly replaced by fern-rich juniper and birch scrub with abundant tall herbs during the transition to the milder climate of the Holocene. In the early Holocene, *Betula* and *Corylus avellana* expanded to form extensive woods with willow, aspen and rowan. At about 8500 BP, *Ulmus* and *Quercus* expanded and at about 7400 BP, *Alnus* expanded. By middle Holocene times the area must have been extensively covered by forest consisting primarily of elm, oak, alder, birch and hazel. The treeline of birch, aspen and rowan probably reached to the summits of most of the hills. Within Galloway, *Pinus* appears to have been restricted to upland areas, free from competition with other trees, and to dried peat surfaces, as at Loch Dungeon. Pine arrived in Galloway at about 7500 BP, possibly from northern Ireland rather than from the Scottish Highlands or from northern England (Birks, 1989). It was largely eliminated by increased wetness and bog growth at about 7000 BP and became rare in the area, becoming extinct at about 5000 BP (Birks, 1975). There was a well-marked decline in elm pollen values at 5000 BP, associated with the expansion of grasses, sedges and *Calluna vulgaris* following forest clearance in the Neolithic, Bronze Age, Iron Age, and later. Soil erosion and the inwashing of terrestrial material appears to have begun at about 5000 BP. Deforestation was virtually complete by the last century. Changes within the loch from a clear-water oligotrophic loch to a humus-coloured dystrophic loch occurred in the last 200 years owing to the inwashing of humified blanket peat following the onset of widespread peat erosion within the loch's catchment (see Round Loch of Glenhead).

The adjacent blanket peat profile (Figure 18.6B) (Birks, 1975) extends from the early Holocene to about 4000 BP and it records the local vegetational development. Initially base-rich, sedge-dominated communities with *Phragmites communis* and *Salix* were present. With peat accumulation and infilling of waterlogged hollows, herb-rich fen communities developed at about 8000 BP, and their litter formed the black humus layer. The cause of death of the pines is uncertain. It is possible that increased wetness due to a small climatic change may have eliminated pine locally. Alternatively the mire may have become completely ombrotrophic and dominated by *Calluna vulgaris* and *Eriophorum*

Figure 18.6A Loch Dungeon: relative pollen diagram from the loch showing selected taxa as percentages of the pollen sums indicated (from Birks, 1972a).

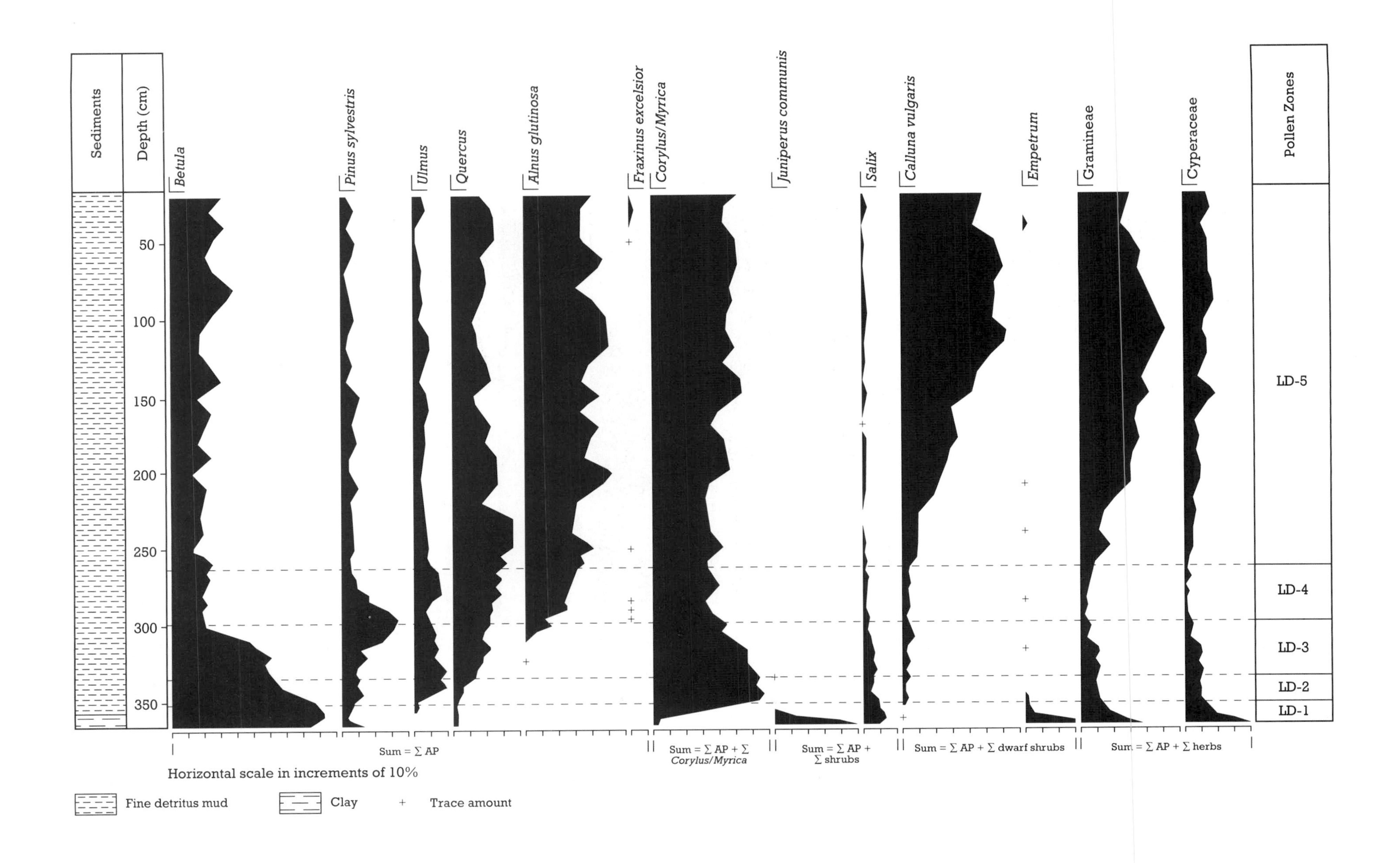

Sediments
Depth (cm)
50
100
150
200
250
300
350
Betula
Pinus sylvestris
Ulmus
Quercus
Alnus glutinosa
Fraxinus excelsior
Corylus/Myrica
Juniperus communis
Salix
Calluna vulgaris
Empetrum
Gramineae
Cyperaceae
Pollen Zones
LD-5
LD-4
LD-3
LD-2
LD-1
Sum = Σ AP
Sum = Σ AP + Σ Corylus/Myrica
Sum = Σ AP + Σ shrubs
Sum = Σ AP + Σ dwarf shrubs
Sum = Σ AP + Σ herbs
Horizontal scale in increments of 10%
Fine detritus mud
Clay
+ Trace amount

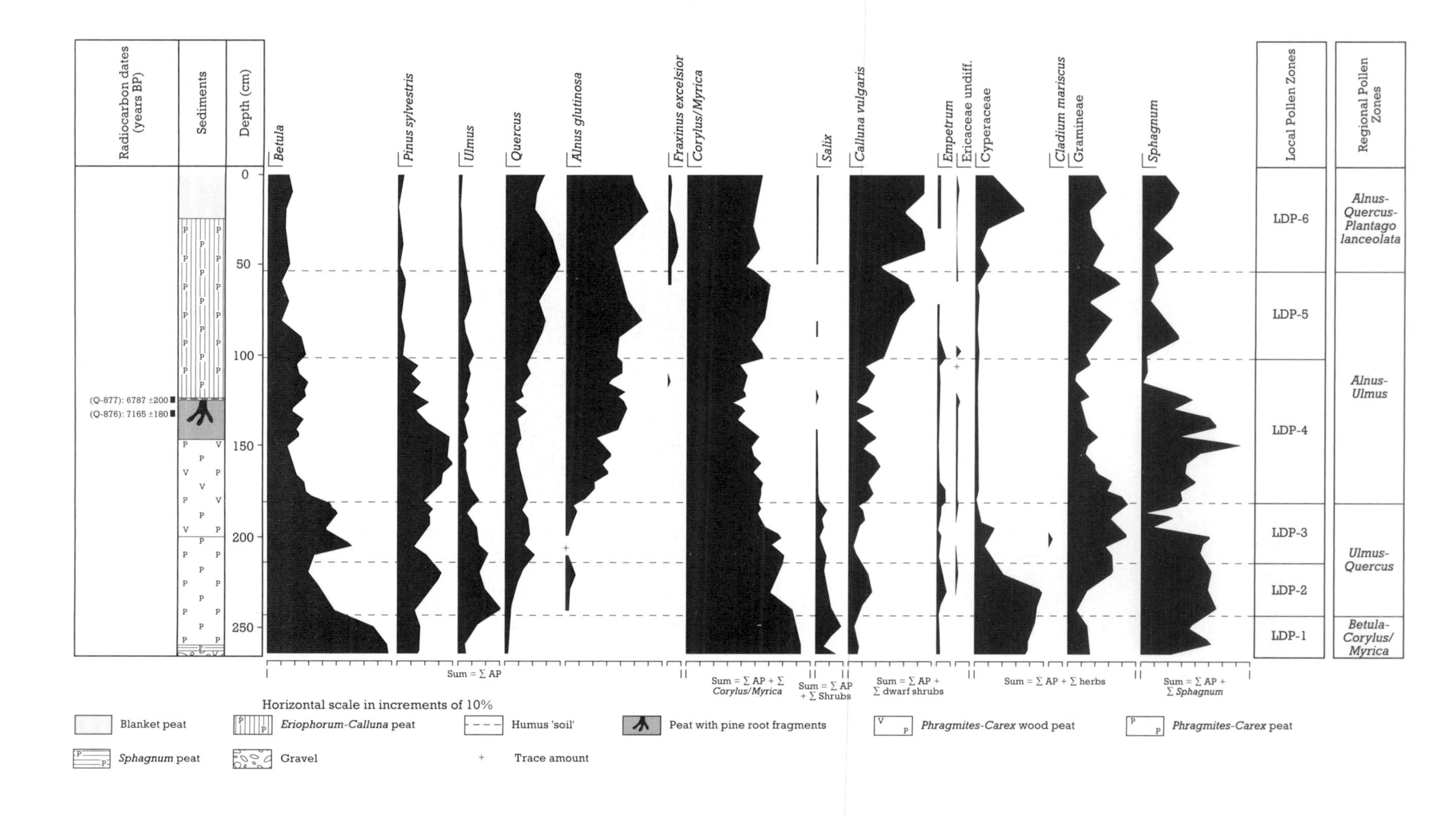

Radiocarbon dates (years BP)
Sediments
Depth (cm)
(Q-877): 6787 ±200
(Q-876): 7165 ±180
0
50
100
150
200
250
Betula
Pinus sylvestris
Ulmus
Quercus
Alnus glutinosa
Fraxinus excelsior
Corylus/Myrica
Salix
Calluna vulgaris
Empetrum
Ericaceae undiff.
Cyperaceae
Cladium mariscus
Gramineae
Sphagnum
Local Pollen Zones
LDP-6
LDP-5
LDP-4
LDP-3
LDP-2
LDP-1
Regional Pollen Zones
Alnus-Quercus-Plantago lanceolata
Alnus-Ulmus
Ulmus-Quercus
Betula-Corylus/Myrica
Sum = Σ AP
Sum = Σ AP + Σ Corylus/Myrica
Sum = Σ AP + Σ Shrubs
Sum = Σ AP + Σ dwarf shrubs
Sum = Σ AP + Σ herbs
Sum = Σ AP + Σ Sphagnum
Horizontal scale in increments of 10%
Blanket peat
Eriophorum-Calluna peat
Humus 'soil'
Peat with pine root fragments
Phragmites-Carex wood peat
Phragmites-Carex peat
Sphagnum peat
Gravel
Trace amount

vaginatum, with *Calluna vulgaris* developing at about 6800 BP.

Loch Dungeon and its associated peat profile are of considerable importance because of their wealth of palaeobotanical and palaeoecological data. They provide one of the most detailed and stratigraphically complete pollen profiles from south-west Scotland, and are thus important in the reconstruction and understanding of the vegetational history of this part of Scotland. It is an area whose vegetational history has affinities with the western Scottish Highlands (Birks, 1977), the English Lake District (Birks, 1972a), and northern Ireland (Birks, 1989). Loch Dungeon is also important because it is one of the few areas in the British Isles where pollen sequences are available from a lake and a nearby bog. Insights into both regional and local vegetational history are thus possible. Comparison of the two pollen sequences highlights the advantages and disadvantages of the two types of sites as repositories of Holocene vegetational history (Birks, 1972a). Loch Dungeon, with its adjacent blanket peat is thus a site of very considerable interest to Quaternary palaeoecologists. It also provides an important contrast with the nearby Round Loch of Glenhead (see below) situated on the Craignaw granite (Jones *et al.*, 1989), particularly in terms of recent lake changes.

Conclusion

Loch Dungeon is an important reference site for studies of vegetational history in south-west Scotland during the Holocene (the last 10,000 years). It is particularly notable for the length and detail of its record, and the combination of evidence available from the sediments of both the loch and the adjacent blanket peat. The pollen record indicates past vegetational changes in the area, with the development of successive woodland types, including the expansion and decline of pine. Loch Dungeon is an important component of the network of sites showing the major regional variations in Holocene vegetational history.

Figure 18.6B Loch Dungeon: relative pollen diagram from adjacent peat profile showing selected taxa as percentages of the pollen sums indicated (from Birks, 1975).

ROUND LOCH OF GLENHEAD

V. J. Jones and A. C. Stevenson

Highlights

Pollen and diatoms preserved in the sediments of the Round Loch of Glenhead provide a valuable record of vegetational history and environmental change during the Holocene. The diatom record, in particular, allows the impact of lake acidification during the industrial period to be placed in the context of longer-term trends during the Holocene.

Introduction

The Round Loch of Glenhead (NX 450804) is situated on the Loch Doon granite mass, some 5 km north-north-west of Loch Dee, at an altitude of 295 m OD. To the north and north-east of the loch steep cliffs rise up to Craiglee (531 m OD). The land to the south and south-west of the loch is less steep and the twin loch, the Long Loch of Glenhead, is separated by gently undulating land to the north-west.

The sediments in the Round Loch of Glenhead provide not only a detailed radiocarbon-dated history of Holocene vegetation for the Galloway hills in south-west Scotland, but also demonstrate the relationship between catchment vegetation change and lake water quality. A detailed study of the Round Loch of Glenhead and its catchment has been made by Jones (1987) and Jones *et al.* (1986, 1989).

Description

The majority of the catchment (area 0.95 km^2) is covered by a thin deposit of blanket peat. Peats, over 5 m deep in places, occur in small valley depressions. Two transects across the depressions were selected for a radiocarbon and palaeoecological study into peatland initiation and development (Jones, 1987). As in the Loch Dungeon study, analysis of pollen from peat profiles enables a picture of local vegetational change to be contrasted with regional changes obtained from the lake sediment. At present, peat hags are found in the catchment.

A core (RLGH3) was obtained from the deepest part of the loch (13 m). A Lateglacial sequence of two clay layers separated by an organic mud corresponding to the Lateglacial Interstadial occurs below a depth of 2.27 m (Figure 18.7). The Holocene sediments above are relatively uniform, with loss-on-ignition values of 20–30%. However, there was a change in the sediment type above 0.47 m in the core to a blackish, fine detritus mud with higher loss-on-ignition values (40–50%). From the core, a series of 20 radiocarbon dates has been obtained (Figure 18.7) which shows a conformable sequence from 9280 ± 80 BP (SRR–2821) to 3970 ± 70 BP (SRR–2815). After the last date, erosion of organic material from the catchment occurred, resulting in reworking and ages older than expected. This problem was intensified at about AD 1600 with the onset of a major phase of peat erosion in the catchment (Stevenson *et al.*, 1990).

From the peat profiles, ten radiocarbon dates of the basal peats show that peat accumulation began in wet hollows early in the Holocene; for example a date of 9390 ± 60 BP (SRR–2865) was found at one site. On the better drained slopes peat accumulation began later, for example at 5450 ± 40 BP (SRR–2871), and it is probable that the shallow blanket peats which cover the majority of the catchment began to form at that time.

Interpretation

For the loch, six local pollen assemblage zones were defined (Figure 18.7) which can be correlated with the five regional zones of Birks (1972a). Early in the Holocene an open juniper-dominated community colonized the mineral soils, and birch and hazel invaded by about 9000 BP (Figure 18.7). Oak, elm and pine became much more important by about 8600 BP when it is likely that birch became restricted to the more acidic, wetter soils in the catchment. The importance of pine was short lived with values reaching a peak at about 7350 BP, a pattern characteristic of many pollen diagrams derived from sites in south-west Scotland. From 7650 BP there was a rise in *Alnus* pollen which remained important, along with *Quercus* and *Ulmus*, to about 5400 BP. From 5400 to 4200 BP there is evidence of progressive podsolization and peat expansion, although an elm, oak and alder woodland still persisted, with open areas of peaty soils and some *Calluna* as an understorey. There is also evidence of anthropogenic effects, as disturbance indicators such as *Plantago lanceolata* and *Rumex* become more important. The elm decline is not clearly distinguished in this core, although *Ulmus* values fell at about 4200 BP.

In sediments deposited after 4200 BP the total tree pollen falls suddenly, marking the rapid disappearance of the forest cover from the area. The forest was replaced by a blanket mire community probably dominated by *Calluna*, *Molinia*, *Eriophorum* and *Trichophorum*. This change was associated with the erosion of catchment soils, possibly caused by anthropogenic activity. Peatland communities continued to dominate the vegetation of the area until the present day.

Pollen analysis on the peat in 10 cores and a 5.12 m profile (S18) was used to define 7 local pollen assemblage zones (Figure 18.8) (Jones, 1987). Organic sedimentation began early in the Holocene in wet hollows where nutrient-rich water existed. Aquatic taxa, such as *Nymphaea*, *Isoetes*, *Myriophyllum* and *Typha*, are found together with indicators of an open *Betula–Juniper–Salix* scrub (Figure 18.8). Hazel invaded the surrounding area shortly after 9400 BP. By about 8800 BP open-water conditions ceased to exist and there was an expansion of elm and oak woodland into which pine and alder later invaded. Initially, the rate of peat growth was slow, but as conditions became more ombrotrophic between 4700 and 3500 BP more rapid accumulation occurred. After about 4000 BP there was a major expansion of *Calluna*, and *Narthecium* pollen appeared in substantial amounts. Values of Cyperaceae and *Sphagnum* also increased as peatland growth swamped the woodland. The peatland community formed at this stage continued to be important until the present day.

Figure 18.7 Round Loch of Glenhead: relative pollen diagram showing selected taxa from core RLGH3 as percentages of total land pollen. Values of herbs except Gramineae and Cyperaceae are expanded ×10. Peatland indicators are based on the sum of *Calluna*, *Potentilla*, *Sphagnum* and Cyperaceae (from Jones, 1987).

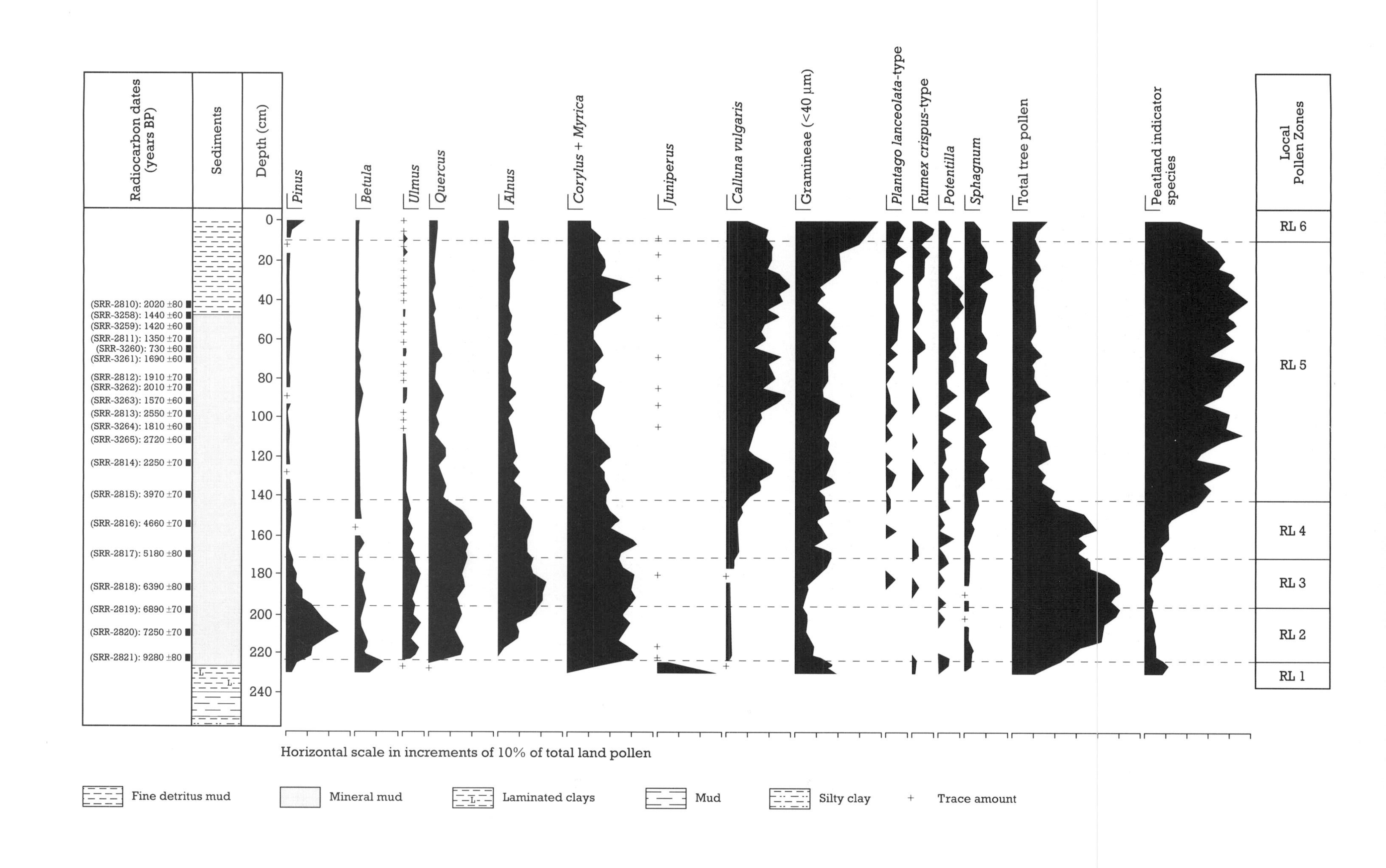

Radiocarbon dates (years BP)
Sediments
Depth (cm)
(SRR-2810): 2020 ±80
(SRR-3258): 1440 ±60
(SRR-3259): 1420 ±60
(SRR-2811): 1350 ±70
(SRR-3260): 730 ±60
(SRR-3261): 1690 ±60
(SRR-2812): 1910 ±70
(SRR-3262): 2010 ±70
(SRR-3263): 1570 ±60
(SRR-2813): 2550 ±70
(SRR-3264): 1810 ±60
(SRR-3265): 2720 ±60
(SRR-2814): 2250 ±70
(SRR-2815): 3970 ±70
(SRR-2816): 4660 ±70
(SRR-2817): 5180 ±80
(SRR-2818): 6390 ±80
(SRR-2819): 6890 ±70
(SRR-2820): 7250 ±70
(SRR-2821): 9280 ±80
0
20
40
60
80
100
120
140
160
180
200
220
240
Pinus
Betula
Ulmus
Quercus
Alnus
Corylus + Myrica
Juniperus
Calluna vulgaris
Gramineae (<40 μm)
Plantago lanceolata-type
Rumex crispus-type
Potentilla
Sphagnum
Total tree pollen
Peatland indicator species
Local Pollen Zones
RL 6
RL 5
RL 4
RL 3
RL 2
RL 1
Horizontal scale in increments of 10% of total land pollen
Fine detritus mud
Mineral mud
Laminated clays
Mud
Silty clay
+ Trace amount

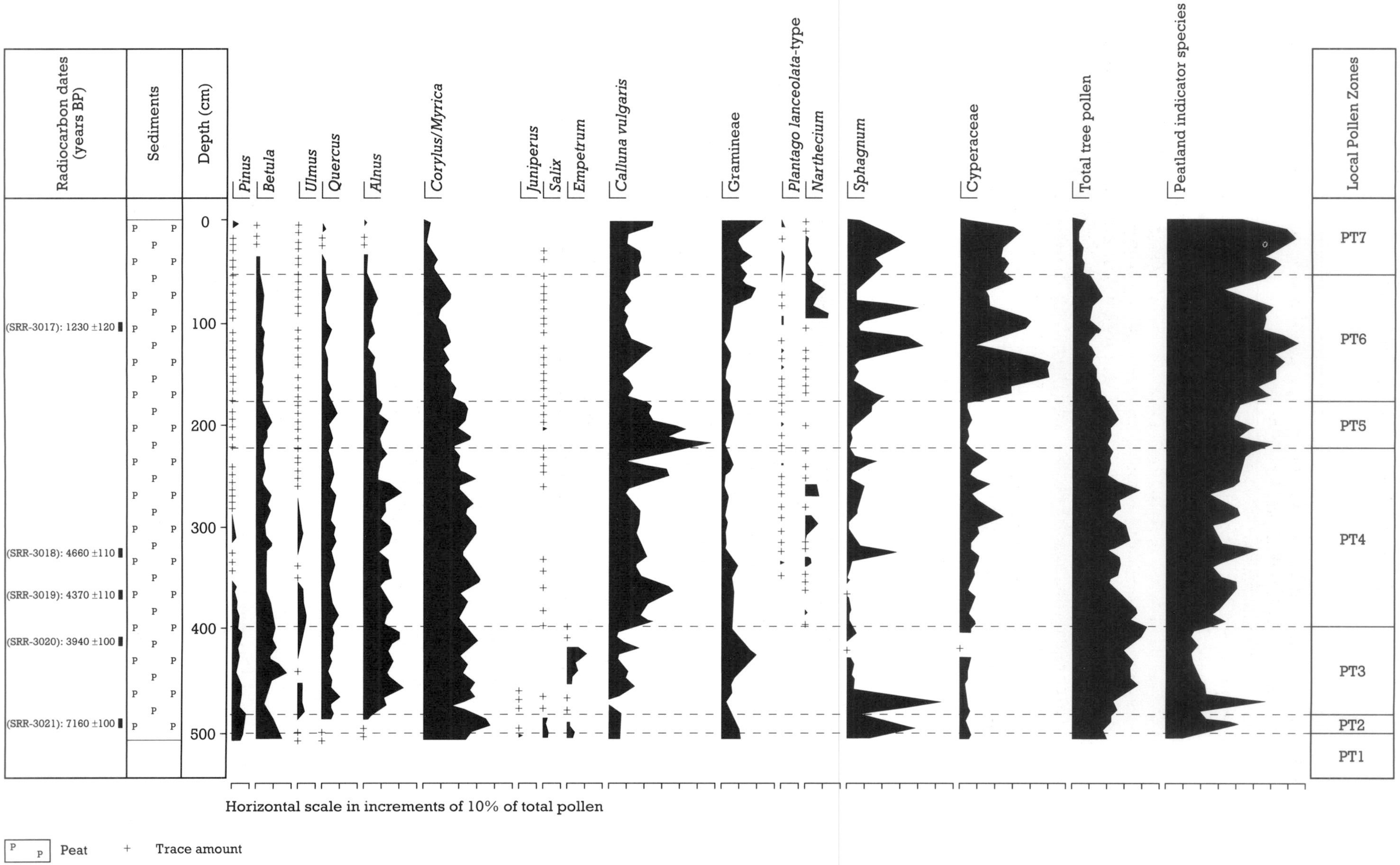

Radiocarbon dates (years BP)

(SRR-3017): 1230 ±120

(SRR-3018): 4660 ±110

(SRR-3019): 4370 ±110

(SRR-3020): 3940 ±100

(SRR-3021): 7160 ±100

Sediments

Depth (cm)

0

100

200

300

400

500

Pinus

Betula

Ulmus

Quercus

Alnus

Corylus/Myrica

Juniperus

Salix

Empetrum

Calluna vulgaris

Gramineae

Plantago lanceolata-type

Narthecium

Sphagnum

Cyperaceae

Total tree pollen

Peatland indicator species

Local Pollen Zones

PT7

PT6

PT5

PT4

PT3

PT2

PT1

Horizontal scale in increments of 10% of total pollen

P Peat

\+ Trace amount

A pH curve was constructed based on the analysis of diatoms in the lake sediments. The lake was acid in the Lateglacial (pH 5.3–5.7) and the early Holocene (pH 5.3–5.9) and remained at a pH of 5.3–5.7 until about 4000 BP. This is unusual since at most sites in the British Isles an early alkaline flora was found to be replaced by more acid-loving species as soil acidification took place. Because the lake is on granite bedrock with little till, the lake and catchment initially had a low buffering capacity, and soils formed in the catchment were probably low in cations and acidified rapidly. The initial development and the subsequent expansion of peats in the catchment also had no apparent acidifying effect on the lake, the pH of which remained between 5.3 and 5.8 (Jones *et al.*, 1986).

Modern changes in the pH of the Round Loch of Glenhead are quite clear. From about AD 1840, pH started to fall with a major decline from AD 1900 to the present-day pH of 4.7 (Battarbee and Flower, 1985). Both the diatom assemblages and the range of reconstructed pH values found in the recent period are unique in the history of the lake. The recent acidification is associated with heavy-metal and soot-particle evidence for the deposition of atmospheric pollutants, and the lake lies in an area known to be affected by acid deposition (Battarbee *et al.*, 1989; Jones *et al.*, 1990). As a naturally acid lake the Round Loch of Glenhead was unable to counter the effects of industrial emissions and acidified rapidly.

The record of Holocene vegetational development at the Round Loch of Glenhead constitutes an important contribution to the regional vegetation history of the Galloway hills. Pollen profiles from the loch and catchment peats complement those obtained by Birks (1972a) from other sites in the area (see Loch Dungeon), and correlations with regional pollen assemblage zones have been made.

The Round Loch of Glenhead is also of great importance in the evaluation of the role of vegetation and soil development processes in lake acidification. The history of the site involved forest and soil development in the first half of the Holocene, widespread paludification in the second half, and recent peat erosion. Although Holocene diatoms have been studied elsewhere in Britain (e.g. Round, 1957, 1961; Haworth, 1969), previous work has not involved reconstruction of the pH of lake waters using recently developed statistical methods. In contrast, the pH of the Round Loch of Glenhead has been reconstructed for the whole of the Holocene using diatom analysis, and the site is unique in this respect. The recent acidification (from about AD 1840) associated with industrial emissions can therefore be placed in a long-term context. Significantly, no trend of increasing acidity through the Holocene has been found. This is in contrast to other sites where diatoms show evidence of a slow natural fall in pH in response to soil leaching and acidification (Round, 1957, 1961; Renberg and Hellberg, 1982). The Round Loch of Glenhead is unusual in that consistently acid conditions have prevailed throughout its entire history. Thus, despite a clear change from mineral to acid organic soils in the catchment in the middle Holocene, feedback mechanisms have operated to maintain a loch with pH stable at 5 and above for most of the Holocene.

Figure 18.8 Round Loch of Glenhead: relative pollen diagram showing selected taxa from core S18, from a peat deposit adjacent to the loch, as percentages of total pollen. Values of the herbs, Gramineae (40–49 μm), Liguliflorae, *Artemisia*, Cruciferae and *Potentilla* are expanded ×10. Peatland indicators are based on the sum of *Calluna*, *Potentilla*, *Sphagnum* and Cyperaceae (from Jones, 1987).

Conclusion

Round Loch of Glenhead is important for studies of environmental change during the Holocene (the last 10,000 years). The sediments not only provide a detailed record of vegetational history, but also allow an assessment of the contribution of vegetational and soil changes to loch acidification. It has been possible from analysis of diatoms to reconstruct the pH of the loch for the whole of the Holocene, therefore enabling the environmental impact of recent industrial pollution to be placed in a wider context. Round Loch of Glenhead is therefore a key reference locality in the network of sites for Holocene environmental change.

References

Aaby, B. (1986) Trees as anthropogenic indicators in regional pollen diagrams. In *Anthropogenic Indicators in Pollen Diagrams* (ed. K-E. Behre). A.A. Balkema, Rotterdam, pp. 73–93.

Abd-Alla, M.A.A. (1988) Mineralogical and geochemical studies of tills in south-western Scotland. Unpublished PhD thesis, University of Glasgow.

Aber, J.S. (1985) The character of glaciotectonism. *Geologie en Mijnbouw*, **64**, 389–95.

Aber, J.S. (1989) Spectrum of constructional glaciotectonic landforms. In *Genetic Classification of Glacigenic Deposits* (eds R.P. Goldthwait and C.L. Matsch). A.A. Balkema, Rotterdam, pp. 281–92.

Adamson, J. (1823) Notice of marine deposits on the margin of Loch Lomond. *Memoir of the Wernerian Natural History Society of Edinburgh*, **4**, 334–38.

Affleck, T.L., Edwards. K.J. and Clarke, A. (1988) Archaeological and palynological studies at the Mesolithic pitchstone and flint site of Auchareoch, Isle of Arran. *Proceedings of the Society of Antiquaries of Scotland*, **118**, 37–59.

Agassiz, L. (1841a) On glaciers and boulders in Switzerland. *Report of the British Association for 1840*, 113–4.

Agassiz, L. (1841b) On glaciers, and the evidence of their having once existed in Scotland, Ireland, and England. *Proceedings of the Geological Society of London*, **3**, 327–32. (Also in *Philosophical Magazine*, **18**, 1841, 569–74).

Agassiz, L. (1842) The glacial theory and its recent progress. *Edinburgh New Philosophical Journal*, **33**, 217–83.

Aguirre, E. and Pasini, G. (1985) The Pliocene–Pleistocene boundary. *Episodes*, **8**, 116–20.

Aitken, A.M., Lovell, J.H., Shaw, A.J. and Thomas, C.W. (1984) The sand and gravel resources of the country around Dalkeith and Temple, Lothian Region. Description of 1:25,000 sheets NT25 and 35, and NT26 and 36. *Mineral Assessment Report, British Geological Survey*, No. 140, 170 pp.

Aitken, J.F. (1990) Glaciolacustrine deposits in Glen Nochty, Grampian Region, Scotland, UK. *Quaternary Newsletter*, **60**, 13–20.

Aitken, J.F. (1991) Sedimentology and palaeoenvironmental significance of Late Devensian to mid Holocene deposits in the Don Valley, north-east Scotland. Unpublished PhD thesis, University of Aberdeen.

Aitken, T. (1880) The formation of Tomnahurich and Strathnairn. *Transactions of Inverness Scientific Society and Field Club*, **1**, 266–9.

Akpan, E.B. and Farrow, G.E. (1984) Depth of deposition of early Holocene raised sediments at Irvine deduced from algal borings in mollusc shells. *Scottish Journal of Geology*, **20**, 237–47.

Alexander, A.J. (1985) Palynological, stratigraphic and chemical analyses of sediments in the Lothians with particular reference to the Lateglacial. Unpublished PhD thesis, University of Edinburgh.

Alexander, H. (1928) *The Cairngorms*. Scottish Mountaineering Club, Edinburgh, 218 pp.

Alley, R.B., Blankenship, D.D., Bentley, C.R. and Rooney, S.T. (1986). Deformation of till beneath ice stream B, West Antarctica. *Nature*, **322**, 57–9.

References

Ambrose, K. and Brewster, J. (1982) A reinterpretation of parts of the 400 ft bench of south-east Warwickshire. *Quaternary Newsletter*, **36**, 21–4.

Ammann, B. and Lotter, A.F. (1989) Late-Glacial radiocarbon and palynostratigraphy on the Swiss plateau. *Boreas*, **18**, 109–26.

Andersen, B.G. (1968) Glacial geology of western Troms, North Norway. *Norges Geologiske Undersøkelse*, **256**, 160 pp.

Andersen, B.G. (1981) Late Weichselian ice sheets in Eurasia and Greenland. In *The Last Great Ice Sheets* (eds G.H. Denton and T.J. Hughes). John Wiley, New York, pp. 1–65.

Andersen, S.Th. (1966) Interglacial vegetational succession and lake development in Denmark. *Palaeobotanist*, **15**, 117–27.

Andersen, S.Th. (1969) Interglacial vegetation and soil development. *Meddelelser fra Dansk Geologisk Förening*, **19**, 90–102.

Anderson, F.W. and Dunham, K.C. (1966) *The Geology of Northern Skye*. Memoirs of the Geological Survey of Scotland. HMSO, Edinburgh, 216 pp.

Anderson, J. (1896) Evidences of the most recent glaciers in the Firth of Clyde district. *Transactions of the Geological Society of Glasgow*, **10**, 198–209.

Anderson, J.B., Brake, C., Domack, E., Myers, N. and Wright, R. (1983) Development of a polar glacial-marine sedimentation model from Antarctic Quaternary deposits and glaciological information. In *Glacial-marine Sedimentation* (ed. B.P. Molnia). Plenum Press, New York, 233–64.

Anderson, J.G.C. (1940) Glacial drifts near Roslin, Midlothian. *Geological Magazine*, **77**, 470–3.

Anderson, J.G.C. (1941) Glacial drifts. *Geological Magazine*, **78**, 470–1.

Anderson, J.G.C. (1942) Glacial drifts. *Geological Magazine*, **79**, 202.

Anderson, J.G.C. (1948) The fauna of the '100 feet beach' clays. *Transactions of the Edinburgh Geological Society*, **14**, 220–9.

Anderson, J.G.C. (1949) The Gareloch readvance moraine. *Geological Magazine*, **86**, 239–44.

Andrews, I.J., Long, D., Richards, P.C., Thomson, A.R., Brown, S., Chesher, J.A. and McCormac, M. (1990). *United Kingdom Offshore Regional Report: the Geology of the Moray Firth*. British Geological Survey, HMSO, London, 96 pp.

Andrews, J.T. (1963a) Cross-valley moraines of the Rimrock and Isortoq River valleys, Baffin Island, N.W.T. – a descriptive analysis. *Geographical Bulletin*, **19**, 49–77.

Andrews, J.T. (1963b) The cross-valley moraines of north-central Baffin Island: a quantitative analysis. *Geographical Bulletin*, **20**, 82–129.

Andrews, J.T. (1975) *Glacial Systems. An Approach to Glaciers and their Environments*. Duxbury Press, North Scituate, 191 pp.

Andrews, J.T. and Dugdale, R.E. (1970) Age prediction of glacio-isostatic strandlines based on their gradients. *Geological Society of America Bulletin*, **81**, 3769–71.

Andrews, J.T. and Smithson, B.B. (1966) Till fabrics of the cross-valley moraines of north-central Baffin Island, Northwest Territories, Canada. *Bulletin of the Geological Society of America*, **77**, 271–90.

Andrews, J.T., Clark, P. and Stravers, J.A. (1985) The patterns of glacial erosion across the eastern Canadian Arctic. In *Quaternary Environments. Eastern Canadian Arctic, Baffin Bay and Western Greenland* (ed. J.T. Andrews). Allen and Unwin, London, pp. 69–92.

Andrews, M.V., Beck, R.B., Gilbertson, D.D. and Switsur, V.R. (1987) Palaeobotanical investigations at An t-Aoradh (Oronsay), the Strand and Loch Cholla (Colonsay). In *Excavations on Oronsay: Prehistoric Human Ecology on a Small Island* (ed. P. Mellars). Edinburgh University Press, Edinburgh, pp. 57–71.

Angus, S. (1987) Lost woodlands of the Western Isles. *Hebridean Naturalist*, **9**, 24–30.

Anon (1840) Tenth Meeting of the British Association for the Advancement of Science. Section C. Geology and physical geography. *The Athenaeum*, No. **677**, 823–4.

Armstrong, M. and Paterson, I.B. (1985) Some recent discoveries of ice-wedge cast networks in north-east Scotland – a comment. *Scottish Journal of Geology*, **21**, 107–8.

Armstrong, M., Paterson, I.B. and Browne, M.A.E. (1975) Late-glacial ice limits are raised shorelines in east central Scotland. In *Quaternary Studies in North East Scotland* (ed. A.M.D. Gemmell). Department of Geography, University of Aberdeen, Aberdeen, pp. 39–44.

Armstrong, M., Paterson, I.B. and Browne, M.A.E. (1985) *Geology of the Perth and Dundee District*. 1:50,000 geological sheets 48W, 48E, 49. Memoirs of the British Geological Survey (Scotland), London, HMSO, 108 pp.

Atkinson, K.M. and Haworth, E.Y. (1990) Devoke

Water and Loch Sionascaig: recent environmental changes and the post-glacial overview. *Philosophical Transactions of the Royal Society of London*, **B327**, 349–55.

Atkinson, T.C., Briffa, K.R. and Coope, G.R. (1987) Seasonal temperatures in Britain during the last 22,000 years, reconstructed using beetle remains. *Nature*, **325**, 587–93.

Atkinson, T.C., Lawson, T.J., Smart, P.L., Harmon, R.S. and Jess, J.W. (1986) New data on speleothem deposition and palaeoclimate in Britain over the last forty thousand years. *Journal of Quaternary Science*, **1**, 67–72.

Auton, C.A. (1990a) Early models for the late-Devensian glaciation and deglaciation of the upland areas. In *Beauly to Nairn: Field Guide* (eds C.A. Auton, C.R. Firth and J.W. Merritt). Quaternary Research Association, Cambridge, pp. 3–5.

Auton, C.A. (1990b) The middle Findhorn Valley. In *Beauly to Nairn: Field Guide* (eds C.A. Auton, C.R. Firth and J.W. Merritt). Quaternary Research Association, Cambridge, pp. 74–96.

Auton, C.A. and Crofts, R.G. (1986) The sand and gravel resources of the country around Aberdeen, Grampian Region. Description of 1:25,000 resource sheets NJ71, 80, 81 and 91 with parts of NJ61, 90 and 92 and with parts of NO89 and 99. *Mineral Assessment Report of the British Geological Survey*, No. 146, 46 pp. + appendices.

Auton, C.A., Merritt, J.W. and Ross, D.L. (1988) The sand and gravel resources of the country around Inverurie and Dunecht, and between Banchory and Stonehaven, Grampian Region. Description of 1:25,000 sheets NJ70, 72 and NO79, and parts of NO88, 89 and 99. *Mineral Assessment Report of the British Geological Survey,* No. 148; *British Geological Survey Technical Report*, No. WF/88/1. Part 1: Report, 89 pp.

Auton, C.A., Thomas, C.W. and Merritt, J.W. (1990) The sand and gravel resources of the country around Strachan and between Auchenblae and Catterline, Grampian Region. Description of parts of 1:25,000 sheets NO68, 69, 77, 78, 87 and 88. *Mineral Assessment Report of the British Geological Survey,* No. 149; *British Geological Survey Technical Report*, No. WF/90/7. Part 1: Report, 75 pp.

Babbage, C. (1868) Observations on the Parallel Roads of Glen Roy. *Quarterly Journal of the Geological Society of London*, **24**, 273–7.

Baden-Powell, D.F.W. (1938) On the glacial and interglacial marine beds of north Lewis. *Geological Magazine*, **75**, 395–409.

Baden-Powell, D.F.W. and Elton, C. (1937) On the relation between a raised beach and an Iron Age midden on the Island of Lewis, Outer Hebrides. *Proceedings of the Society of Antiquaries in Scotland*, **71**, 347–65.

Bailey, E.B. and Anderson, E.M. (1925) *The Geology of Staffa, Iona, and Western Mull.* (Description of Sheet 43 of the Geological Map). Memoirs of the Geological Survey of Scotland. HMSO, Edinburgh, 107 pp.

Bailey, E.B. and Eckford, R.J.A. (1956) Eddleston gravel–moraine. *Transactions of the Edinburgh Geological Society*, **16**, 254–61.

Bailey, E.B. and Maufe, H.B. (1916) *The Geology of Ben Nevis and Glencoe, and the Surrounding Country*. Explanation of Sheet 53. Memoirs of the Geological Survey of Scotland. HMSO, Edinburgh, 247 pp.

Bailey, E.B., Clough, C.T., Wright, W.B., Richey, J.E. and Wilson, G.V. (1924) *Tertiary and Post-Tertiary Geology of Mull, Loch Aline and Oban*. Description of parts of Sheets 43, 44, 51, and 52 of Geological Map. Memoirs of the Geological Survey of Scotland. HMSO, Edinburgh, 445 pp.

Baird, P.D. and Lewis, W.V. (1957) The Cairngorm floods, 1956: summer solifluction and distributory formation. *Scottish Geographical Magazine*, **73**, 91–100.

Ball, D.F. and Goodier, R. (1974) Ronas Hill, Shetland: a preliminary account of its ground pattern features resulting from the action of frost and wind. In *The Natural Environment of Shetland* (ed. R. Goodier). Nature Conservancy Council, Edinburgh, 89–106.

Ballantyne, C.K. (1979) A sequence of Lateglacial ice-dammed lakes in east Argyll. *Scottish Journal of Geology*, **15**, 153–60.

Ballantyne, C.K. (1981) Periglacial landforms and environments on mountains in the northern Highlands of Scotland. Unpublished PhD thesis, University of Edinburgh.

Ballantyne, C.K. (1984) The Late Devensian periglaciation of upland Scotland. *Quaternary Science Reviews*, **3**, 311–43.

Ballantyne, C.K. (1985) Nivation landforms and snowpatch erosion on two massifs in the northern Highlands of Scotland. *Scottish Geographical Magazine*, **101**, 40–9.

Ballantyne, C.K. (1986a) Protalus rampart development and the limits of former glaciers in the vicinity of Baosbheinn, Wester Ross. *Scottish Journal of Geology*, **22**, 13–25.

Ballantyne, C.K. (1986b) Non-sorted patterned ground on mountains in the northern Highlands of Scotland. *Biuletyn Peryglacjalny*, **30**, 15–34.

Ballantyne, C.K. (1986c) Late Flandrian solifluction on the Fannich Mountains, Ross-shire. *Scottish Journal of Geology*, **22**, 395–406.

Ballantyne, C.K. (1986d) Landslides and slope failures in Scotland: a review. *Scottish Geographical Magazine*, **102**, 134–50.

Ballantyne, C.K. (1987a) The present-day periglaciation of upland Britain. In *Periglacial Processes and Landforms in Britain and Ireland* (ed. J. Boardman). Cambridge University Press, Cambridge, pp. 113–27.

Ballantyne, C.K. (1987b) An Teallach. In *Wester Ross Field Guide* (eds C.K. Ballantyne and D.G. Sutherland). Quaternary Research Association, Cambridge, pp. 72–92.

Ballantyne, C.K. (1987c) The Beinn Alligin 'rock glacier'. In *Wester Ross Field Guide* (eds C.K. Ballantyne and D.G Sutherland). Quaternary Research Association, Cambridge, pp. 134–7.

Ballantyne, C.K. (1987d) The Baosbheinn protalus rampart. In *Wester Ross Field Guide* (eds C.K. Ballantyne and D.G. Sutherland). Quaternary Research Association, Cambridge, pp. 167–71.

Ballantyne, C.K. (1987e) The solifluction features of the Fannich Mountains. In *Wester Ross Field Guide* (eds C.K. Ballantyne and D.G. Sutherland). Quaternary Research Association, Cambridge, pp. 171–2.

Ballantyne, C.K. (1988) Ice-sheet moraines in southern Skye. *Scottish Journal of Geology*, **24**, 301–4.

Ballantyne, C.K. (1989a) The Loch Lomond Readvance on the Isle of Skye, Scotland; glacier reconstruction and palaeoclimatic implications. *Journal of Quaternary Science*, **4**, 95–108.

Ballantyne, C.K. (1989b) Avalanche impact landforms on Ben Nevis, Scotland. *Scottish Geographical Magazine*, **105**, 38–42.

Ballantyne, C.K. (1990) The late Quaternary glacial history of the Trotternish escarpment, Isle of Skye, Scotland, and its implications for ice-sheet reconstruction. *Proceedings of the Geologists' Association*, **101**, 171–86.

Ballantyne, C.K. (1991a) Late Holocene erosion in upland Britain: climatic deterioration or human influence? *The Holocene*, **1**, 81–5.

Ballantyne, C.K. (1991b) Periglacial features on the mountains of Skye. In *The Quaternary of the Isle of Skye: Field Guide* (eds C.K. Ballantyne, D.I. Benn, J.J. Lowe and M.J.C. Walker). Quaternary Research Association, Cambridge, pp. 68–81.

Ballantyne, C.K. (1991c) Holocene geomorphic activity in the Scottish Highlands. *Scottish Geographical Magazine*, **107**, 84–98.

Ballantyne, C.K. and Benn, D.I. (1991) The glacial history of the Isle of Skye. In *The Quaternary of the Isle of Skye: Field Guide* (eds C.K. Ballantyne, D.I. Benn, J.J. Lowe and M.J.C. Walker). Quaternary Research Association, Cambridge, pp. 11–34.

Ballantyne, C.K. and Eckford, J.D. (1984) Characteristics and evolution of two relict talus slopes in Scotland. *Scottish Geographical Magazine*, **100**, 20–33.

Ballantyne, C.K. and Kirkbride, M.P. (1986) The characteristics and significance of some Lateglacial protalus ramparts in upland Britain. *Earth Surface Processes and Landforms*, **11**, 659–71.

Ballantyne, C.K. and Kirkbride, M.P. (1987) Rockfall activity in upland Britain during the Loch Lomond Stadial. *Geographical Journal*, **153**, 86–92.

Ballantyne, C.K. and Wain-Hobson, T. (1980) The Loch Lomond Advance on the Island of Rhum. *Scottish Journal of Geology*, **16**, 1–10.

Ballantyne, C.K. and Whittington, G. (1987) Niveo-aeolian sand deposits on An Teallach, Wester Ross, Scotland. *Transactions of the Royal Society of Edinburgh: Earth Sciences*, **78**, 51–63.

Ballantyne, C.K., Sutherland, D.G. and Reed, W.J. (1987) Introduction. In *Wester Ross Field Guide* (eds C.K. Ballantyne and D.G. Sutherland). Quaternary Research Association, Cambridge, pp. 1–63.

Banham, P.H. (1975) Glaciotectonic structures: a general discussion with particular reference to the Contorted Drift of Norfolk. In *Ice Ages: Ancient and Modern* (eds A.E. Wright and F. Moseley). Seel House Press, Liverpool, pp. 69–94.

Bard, E., Arnold, M., Maurice, P., Duprat, J., Moyes, J. and Duplessy, J-C. (1987) Retreat velocity of the North Atlantic polar front during the last deglaciation determined by ^{14}C accelerator mass spectrometry. *Nature*, **328**, 791–4.

Bard, E., Hamelin, B., Fairbanks, R.G. and Zindler,

References

A. (1990) Calibration of the ^{14}C timescale over the past 30,000 years using mass spectrometric U–Th ages from Barbados. *Nature*, **345**, 405–10.

Barnett, D.M. and Holdsworth, G. (1974) Origin, morphology and chronology of sublacustrine moraines, Generator Lake, Baffin Island, Northwest Territories, Canada. *Canadian Journal of Earth Sciences*, **11**, 380–408.

Barnola, J.M., Raynaud, D., Korotkevich, Y.S. and Lorius, C. (1987) Vostok ice core provides 160,000-year record of atmospheric CO_2. *Nature*, **329**, 405–14.

Barrett, P.H. (1973) Darwin's 'gigantic blunder'. *Journal of Geological Education*, **21**, 19–28.

Barrow, G., Grant Wilson, J.S. and Cunningham Craig, B.A. (1905). *The Geology of the Country round Blair Atholl, Pitlochry and Aberfeldy*. Explanation of Sheet 55. Memoirs of the Geological Survey of Scotland. HMSO, Glasgow, 161 pp.

Barrow, G., Cunningham Craig, E.H. and Hinxman, L.W. (1912) *The Geology of the Districts of Braemar, Ballater and Glen Clova*. Explanation of Sheet 65. Memoirs of the Geological Survey of Scotland. HMSO, Edinburgh, 138 pp.

Barrow, G., Hinxman, L.W. and Cunningham Craig, E.H. (1913) *The Geology of Upper Strathspey, Gaick and the Forest of Atholl.* Explanation of Sheet 64. Memoirs of the Geological Survey of Scotland. HMSO, Edinburgh, 116 pp.

Barsch, D. (1969) Studien und Messungen an Blockgletschern in Macun, Unterengadin. *Zeitschrift für Geomorphologie*, Supplementband **8**, 11–30.

Barsch, D. (1988) Rock glaciers. In *Advances in Periglacial Geomorphology* (ed. M.J. Clark). John Wiley, Chichester, 69–90.

Basham, J.R. (1974) Mineralogical changes associated with deep weathering of gabbro in Aberdeenshire. *Clay Minerals*, **10**, 189–202.

Bassett, D.A. (1958) *Geological Excursion Guide to the Glasgow District.* Geological Society of Glasgow, Glasgow, 104 pp.

Battarbee, R. and Flower, R.J. (1985) Palaeoecological evidence for the timing and causes of lake acidification in Galloway, south-west Scotland. *Working Paper No. 8*. Palaeoecology Research Unit, Department of Geography, University College London, 79 pp.

Battarbee, R.W., Stevenson, A.C., Rippey, B., Fletcher, C., Natkanski, J., Wik, M. and Flower, R.J. (1989) Causes of lake acidification in Galloway, south-west Scotland: a palaeoecological evaluation of the relative roles of atmospheric contamination and catchment change for two acidified sites with non-afforested catchments. *Journal of Ecology*, **77**, 651–72.

Battiau-Queney, Y. (1984) The pre-glacial evolution of Wales. *Earth Surface Processes and Landforms*, **9**, 229–52.

Baxter, M.S., Ergin, M. and Walton, A. (1969) Glasgow University radiocarbon measurements I. *Radiocarbon*, **11**, 43–52.

Bayfield, N.G. (1984) The dynamics of heather (*Calluna vulgaris*) stripes in the Cairngorm Mountains, Scotland. *Journal of Ecology*, **72**, 515–27.

Beaumont, P., Turner, J. and Ward, P.F. (1969) An Ipswichian peat raft in glacial till at Hutton Henry, Co. Durham. *New Phytologist*, **68**, 797–805.

Bell, D. (1871) On the aspects of Clydesdale during the glacial period. *Transactions of the Geological Society of Glasgow*, **4**, 63–9.

Bell, D. (1874) Notes on the glaciation of the west of Scotland, with reference to some recently observed instances of cross-striation. *Transactions of the Geological Society of Glasgow*, **4**, 300–10.

Bell, D. (1891a) Phenomena of the glacial epoch: II. The 'great submergence'. *Transactions of the Geological Society of Glasgow*, **9**, 100–38.

Bell, D. (1891b) On a glacial mound in Glen Fruin, Dumbartonshire. *Geological Magazine*, Decade 3, **8**, 415–8.

Bell, D. (1893a) On the alleged proofs of submergence in Scotland during the glacial epoch. *Report of the British Association for 1892*, 713–4.

Bell, D. (1893b) On the alleged proofs of submergence in Scotland during the glacial epoch – I. Chapelhall, near Airdrie. *Transactions of the Geological Society of Glasgow*, **9**, 321–44.

Bell, D. (1893c) On a glacial mound in Glen Fruin, Dumbartonshire. *Transactions of the Geological Society of Glasgow*, **9**, 345–54.

Bell, D. (1894) On the glaciation of the west of Scotland. *Proceedings of the Philosophical Society of Glasgow*, **25**, 118–36.

Bell, D. (1895a) On the alleged proofs of submergence in Scotland during the glacial epoch – II. Clava and other northern localities. *Transactions of the Geological Society of Glasgow*, **10**, 105–20.

Bell, D. (1895b) The shelly clays and gravels of Aberdeenshire considered in relation to the question of submergence. *Quarterly Journal of the Geological Society*, **51**, 472–9.

Bell, D. (1895c) Reply to Professor Hull on the glacial deposits of Aberdeenshire. *Geological Magazine*, Decade 4, **2**, 524–7.

Bell, D. (1895d) Notes on 'The Great Ice Age' in relation to the question of submergence. *Geological Magazine*, Decade 4, **2**, 321–6; 348–55 and 402–5.

Bell, D. (1896a) The Ayrshire 'shell beds'. *Geological Magazine*, Decade 4, **3**, 335–6.

Bell, D. (1896b) Additional notes on Glen Fruin. *Transactions of the Geological Society of Glasgow*, **10**, 380–1.

Bell, D. (1897a) The 'great submergence' again Clava, etc, Part II. *Geological Magazine*, Decade 4, **4**, 63–8.

Bell, D. (1897b) The high-level shelly clays and Mr Mellard Reade. *Geological Magazine*, Decade 4, **4**, 189–90.

Benn, D.I. (1989a) Controls on sedimentation in a late Devensian ice-dammed lake, Achnasheen, Scotland. *Boreas*, **18**, 31–42.

Benn, D.I. (1989b) Debris transport by Loch Lomond Readvance glaciers in northern Scotland: basin form and within-valley asymmetry of lateral moraines. *Journal of Quaternary Science*, **4**, 243–54.

Benn, D.I. (1990) Scottish Lateglacial moraines: debris supply, genesis and significance. Unpublished PhD thesis, University of St Andrews.

Benn, D.I. (1991) Glacial landforms and sediments on Skye. In *The Quaternary of the Isle of Skye: Field Guide* (eds C.K. Ballantyne, D.I. Benn, J.J. Lowe and M.J.C. Walker). Quaternary Research Association, Cambridge, pp. 35–67.

Benn, D.I. (1992) Scottish landform examples – 5. The Achnasheen terraces. *Scottish Geographical Magazine*, **108**, 128–31.

Benn, D.I. and Dawson, A.G. (1987) A Devensian glaciomarine sequence in western Islay, Inner Hebrides. *Scottish Journal of Geology*, **23**, 175–87.

Benn, D.I., Lowe, J.J. and Walker, M.J.C. (1992) Glacier response to climatic change during the Loch Lomond Stadial and early Flandrian: geomorphological and palynological evidence from the Isle of Skye, Scotland. *Journal of Quaternary Science*, **7**, 125–44.

Bennett, K.D. (1984) The post-glacial history of *Pinus sylvestris* in the British Isles. *Quaternary Science Reviews*, **3**, 133–55.

Bennett, K.D. (1989) A provisional map of forest types for the British Isles 5000 years ago. *Journal of Quaternary Science*, **4**, 141–4.

Bennett, K.D. and Birks, H.J.B. (1990) Postglacial history of alder (*Alnus glutinosa* (L.) Gaertn) in the British Isles. *Journal of Quaternary Science*, **5**, 123–33.

Bennett, K.D., Fossitt, J.A., Sharp, M. and Switsur, V.R. (1990) Holocene vegetational and environmental history at Loch Lang, South Uist, Western Isles, Scotland. *New Phytologist*, **114**, 281–98.

Bennett, K.D., Boreham, S., Sharp, M.J. and Switsur, V.R. (1992) Holocene history of environment, vegetation and human settlement on Catta Ness, Lunnasting, Shetland. *Journal of Ecology*, **80**, 241–73.

Bennett, M.R. (1990) The deglaciation of Glen Croulin, Knoydart. *Scottish Journal of Geology*, **26**, 41–6.

Bennett, M.R. (1991) Scottish 'hummocky moraine': its implications for the deglaciation of the north west Highlands during the Younger Dryas or Loch Lomond Stadial. Unpublished PhD thesis, University of Edinburgh.

Bennett, M.R. and Glasser, N.F. (1991) The glacial landforms of Glen Geusachan, Cairngorms: a reintepretation. *Scottish Geographical Magazine*, **107**, 116–23.

Bennie, J. (1868) On the surface geology of the district round Glasgow, as indicated by the journals of certain bores. *Transactions of the Geological Society of Glasgow*, **3**, 133–48.

Bennie, J. (1894) Arctic plant beds in the old lake deposits of Scotland. *Annals of Scottish Natural History*, **9**, 46–52.

Bent, A.J.A. (1986) Aspects of Pleistocene glaciomarine sequences in the North Sea. Unpublished PhD thesis, University of Edinbugh.

Berger, A. (1988) Milankovitch and climate. *Reviews of Geophysics*, **26**, 624–57.

Beveridge, E. (1911) *North Uist: Its Archaeology and Topography, with Notes upon the Early History of the Outer Hebrides*. William Brown, Edinburgh, 348 pp.

Bindschadler, R. (1983) The importance of pressurised subglacial water in separation and sliding at the glacier bed. *Journal of Glaciology*, **29**, 3–19.

Binns, P.E., Harland, R. and Hughes, M.J. (1974) Glacial and post-glacial sedimentation in the Sea of the Hebrides. *Nature*, **248**, 751–4.

References

Birks, H.H. (1969) Studies in the vegetational history of Scotland. Unpublished PhD thesis, University of Cambridge.

Birks, H.H. (1970) Studies in the vegetational history of Scotland I. A pollen diagram from Abernethy Forest, Inverness-shire. *Journal of Ecology*, **58**, 827–46.

Birks, H.H. (1972a) Studies in the vegetational history of Scotland II. Two pollen diagrams from the Galloway Hills, Kirkcudbrightshire. *Journal of Ecology*, **60**, 182–217.

Birks, H.H. (1972b) Studies in the vegetational history of Scotland III. A radiocarbon dated pollen diagram from Loch Maree, Ross and Cromarty. *New Phytologist*, **71**, 731–54.

Birks, H.H. (1975) Studies in the vegetational history of Scotland IV. Pine stumps in Scottish blanket peats. *Philosophical Transactions of the Royal Society of London*, **B270**, 181–226.

Birks, H.H. (1984) Late-Quaternary pollen and plant macrofossil stratigraphy at Lochan an Druim, north-west Scotland. In *Lake Sediments and Environmental History* (eds E.Y. Haworth and J.W.G. Lund). Leicester University Press, Leicester, pp. 377–405.

Birks, H.H. and Mathewes, R.W. (1978) Studies in the vegetational history of Scotland. V. Late Devensian and early Flandrian pollen and macrofossil stratigraphy at Abernethy Forest, Inverness-shire. *New Phytologist*, **80**, 455–84.

Birks, H.J.B. (1973) *Past and Present Vegetation of the Isle of Skye. A Palaeoecological Study*. Cambridge University Press, Cambridge, 415 pp.

Birks, H.J.B. (1977) The Flandrian forest history of Scotland: a preliminary synthesis. In *British Quaternary Studies: Recent Advances* (ed. F.W. Shotton). Clarendon Press, Oxford, pp. 119–35.

Birks, H.J.B. (1980) *Quaternary Vegetational History of West Scotland*. 5th International Palynological Conference, Guidebook for Excursion C8. The Botany School, University of Cambridge, Cambridge, 70 pp.

Birks, H.J.B. (1986) Late Quaternary biotic changes in terrestrial and lacustrine environments, with particular reference to north-west Europe. In *Handbook of Holocene Palaeoecology and Palaeohydrology* (ed. B.E. Berglund). John Wiley, Chichester, pp. 1–65.

Birks, H.J.B. (1987) The vegetational context of mesolithic occupation on Oronsay and adjacent areas. In *Excavations on Oronsay: Prehistoric Human Ecology on a Small Island* (ed. P.A. Mellars). Edinburgh University Press, Edinburgh, pp. 71–7.

Birks, H.J.B. (1988) Long-term ecological change in the British uplands. In *Ecological Change in the Uplands* (eds M.B. Usher and D.B.A. Thompson). Special Publication of the British Ecological Society, No. 7. Blackwell Scientific Publications, Oxford, 37–56.

Birks, H.J.B. (1989) Holocene isochrone maps and patterns of tree-spreading in the British Isles. *Journal of Biogeography*, **16**, 503–40.

Birks, H.J.B. (1991) Floristic and vegetational history of the Outer Hebrides. In *Flora of the Outer Hebrides* (eds R.J. Pankhurst and J.M. Mullin). Natural History Museum Publications, London, pp. 32–7.

Birks, H.J.B. and Birks, H.H. (1980) *Quaternary Palaeoecology*. Edward Arnold, London, 289 pp.

Birks, H.J.B. and Madsen, B.J. (1979) Flandrian vegetational history of Little Loch Roag, Isle of Lewis, Scotland. *Journal of Ecology*, **67**, 825–42.

Birks, H.J.B. and Peglar, S.M. (1979) Interglacial pollen spectra from Sel Ayre, Shetland. *New Phytologist*, **83**, 559–75.

Birks, H.J.B. and Ransom, M.E. (1969) An interglacial peat at Fugla Ness, Shetland. *New Phytologist*, **68**, 777–96.

Birks, H.J.B. and Williams, W. (1983) Late Quaternary vegetational history of the Inner Hebrides. *Proceedings of the Royal Society of Edinburgh*, **83B**, 269–92.

Birks, H.J.B., Deacon, J. and Peglar, S. (1975) Pollen maps for the British Isles 5000 years ago. *Proceedings of the Royal Society of London*, **B189**, 87–105.

Birnie, J. (1981) Environmental changes in Shetland since the end of the last glaciation. Unpublished PhD thesis, University of Aberdeen.

Birnie, J. (1983) Tolsta Head: further investigations of the interstadial deposit. *Quaternary Newsletter*, **41**, 18–25.

Birnie, J. (1984) Trees and shrubs in the Shetland Islands: evidence for a postglacial climatic optimum. In *Climatic Changes on a Yearly to Millenial Basis* (eds N.A. Morner and W. Karlén). D. Reidel, Dordrecht, pp. 155–61.

Birse, E.L. (1971) *Assessment of Climatic Conditions in Scotland. 3. The Bioclimatic Subregions*. Soil Survey of Scotland. Macaulay Institute for Soil Research, Aberdeen, 12 pp.

Birse, E.L. (1974) Bioclimatic characteristics of

Shetland. In *The Natural Environment of Shetland* (ed. R. Goodier). Nature Conservancy Council, Edinburgh, pp. 24–33.

Birse, E.L. (1980) Suggested amendments to the world soil classification to accommodate Scottish mountain and aeolian soils. *Journal of Soil Science*, **31**, 117–24.

Bishop, W.W. (1963) Lateglacial deposits near Lockerbie, Dumfriesshire. *Transactions of the Dumfriesshire and Galloway Natural History and Antiquarian Society*, **40**, 117–32.

Bishop, W.W. and Coope, G.R. (1977) Stratigraphical and faunal evidence for Lateglacial and early Flandrian environments in south-west Scotland. In *Studies in the Scottish Lateglacial Environment* (eds J.M. Gray and J.J. Lowe). Pergamon Press, Oxford, pp. 61–88.

Blackburn, K.B. (1946) On a peat from the island of Barra, Outer Hebrides. Data for the study of post-glacial history. X. *New Phytologist*, **45**, 44–9.

Bloodworth, A.J. (1990) Clay mineralogy of Quaternary sediments from the Dalcharn interglacial site. In *Beauly to Nairn: Field Guide* (eds C.A. Auton, C.R. Firth and J.W. Merritt). Quaternary Research Association, Cambridge, pp. 60–61.

Bohncke, S.J.P. (1988) Vegetation and habitation history of the Callanish area, Isle of Lewis, Scotland. In *The Cultural Landscape – Past, Present and Future* (eds H.H. Birks, H.J.B. Birks, P.E. Kaland and D. Moe). Cambridge University Press, Cambridge, pp. 445–61.

Bonney, T.G. (1871) On a cirque in the syenite hills of Skye. *Geological Magazine*, **8**, 535–40.

Bonsall, C. and Smith, C. (1990) Bone and antler technology in the British Late Upper Palaeolithic and Mesolithic: the impact of accelerator dating. In *Contributions to the Mesolithic in Europe: Papers Presented at the Fourth International Symposium, 'The Mesolithic in Europe', Leuven 1990* (eds P.M. Vermeersch and P. Van Peer). Leuven University Press, Leuven, pp. 359–68.

Boulton, G.S. (1968) Flow tills and related deposits on some Vest-spitsbergen glaciers. *Journal of Glaciology*, **7**, 391–412.

Boulton, G.S. (1970) On the origin and transport of englacial debris in Svalbard glaciers. *Journal of Glaciology*, **9**, 213–29.

Boulton, G.S. (1972a) The role of thermal regime in glacial sedimentation. *Institute of British Geographers Special Publication*, **4**, 1–19.

Boulton, G.S. (1972b) Modern Arctic glaciers as depositional models for former ice sheets. *Quarterly Journal of the Geological Society of London*, **128**, 361–93.

Boulton, G.S. (1974) Processes and patterns of glacial erosion. In *Glacial Geomorphology* (ed. D.R. Coates). State University of New York, Binghampton, pp. 41–87.

Boulton, G.S. (1975) Processes and patterns of subglacial sedimentation: a theoretical approach. In *Ice Ages: Ancient and Modern* (eds A.E. Wright and F. Moseley). Seel House Press, Liverpool, pp. 7–42.

Boulton, G.S. and Dent, D.L. (1974) The nature and rates of post-depositional changes in recently deposited till from south-east Iceland. *Geografiska Annaler,* **56A**, 121–34.

Boulton, G.S. and Hindmarsh, R. (1987) Sediment deformation beneath glaciers: rheology and geological consequences. *Journal of Geophysical Research*, **92**, 9059–82.

Boulton, G.S. and Jones, A.S. (1979) Stability of temperate ice caps and ice sheets on beds of deformable sediment. *Journal of Glaciology*, **24**, 29–42.

Boulton, G.S. and Paul, M.A. (1976) The influence of genetic processes on some geotechnical properties of glacial tills. *Quarterly Journal of Engineering Geology*, **9**, 159–94.

Boulton, G.S. and Worsley, P. (1965) Late Weichselian glaciation in the Cheshire–Shropshire basin. *Nature*, **207**, 704–6.

Boulton, G.S., Jones, A.S., Clayton, K.M. and Kenning, M.J. (1977) A British ice-sheet model and patterns of glacial erosion and deposition in Britain. In *British Quaternary Studies. Recent Advances* (ed. F.W. Shotton). Clarendon Press, Oxford, pp. 231–46.

Boulton, G.S., Chroston, P.N. and Jarvis, J. (1981) A marine seismic study of late Quaternary sedimentation and inferred glacier fluctuations along western Inverness-shire, Scotland. *Boreas*, **10**, 39–51.

Boulton, G.S., Smith, G.D., Jones, A.S. and Newsome, J. (1985) Glacial geology and glaciology of the last mid-latitude ice sheets. *Journal of the Geological Society*, **142**, 447–74.

Boulton, G.S., Peacock, J.D. and Sutherland, D.G. (1991) Quaternary. In *Geology of Scotland*, 3rd edn (ed. G.Y. Craig). Geological Society, London, pp. 503–43.

Bourcart, J. (1938) La marge continentale. Essai sur les régressions et transgressions marines.

Bulletin de la Société Géologique de France, Ser. 5, **8**, 393–474.

Bowen, D.Q. (1978) *Quaternary Geology. A Stratigraphic Framework for Multidisciplinary Work*. Pergammon Press, Oxford, 221 pp.

Bowen, D.Q. (1989) The last interglacial–glacial cycle in the British Isles. *Quaternary International*, **3/4**, 41–7.

Bowen, D.Q. (1991) Time and space in the glacial sediment systems of the British Isles. In *Glacial Deposits in Great Britain and Ireland* (eds J. Ehlers, P.L. Gibbard and J. Rose). A.A. Balkema, Rotterdam, pp. 3–11.

Bowen, D.Q. and Sykes, G.A. (1988) Correlation of marine events and glaciations on the northeast Atlantic margin. *Philosophical Transactions of the Royal Society of London*, **B318**, 619–35.

Bowen, D.Q., Rose, J., McCabe, A.M. and Sutherland, D.G. (1986) Correlation of Quaternary glaciations in England, Ireland, Scotland and Wales. *Quaternary Science Reviews*, **5**, 299–340.

Boyd, W.E. (1982) The stratigraphy and chronology of Late Quaternary raised coastal deposits in Renfrewshire and Ayrshire, western Scotland. Unpublished PhD thesis, University of Glasgow.

Boyd, W.E. (1986a) Late Devensian shoreline position in north Ayrshire. *Scottish Journal of Geology*, **22**, 412–6.

Boyd, W.E. (1986b) Fossil *Lithothamnium* (calcareous algae) rhodoliths from Late Quaternary raised coastal sediments, Irvine, Ayrshire. *Scottish Journal of Geology*, **22**, 165–77.

Boyd, W.E. (1986c) Vegetation history at Linwood Moss, Renfrewshire, central Scotland. *Journal of Biogeography*, **13**, 207–23.

Boyd, W.E. (1988) Early Flandrian vegetational development on the coastal plain of north Ayrshire, Scotland: evidence from multiple pollen profiles. *Journal of Biogeography*, **15**, 325–37.

Boyd, W.E. and Dickson, J.H. (1986) Patterns in the geographical distribution of the early Flandrian *Corylus* rise in south-west Scotland. *New Phytologist*, **102**, 615–23.

Boyd, W.E. and Dickson, J.H. (1987) A post-glacial pollen sequence from Loch a' Mhuillinn, North Arran: a record of vegetation history with special reference to the history of endemic *Sorbus* species. *New Phytologist*, **107**, 221–44.

Boyer, S.J. and Pheasant, D.R. (1974) Delimitation of weathering zones in the fjord area of eastern Baffin Island, Canada. *Bulletin of the Geological Society of America*, **85**, 805–10.

Bradley, R.S. (1985) *Quaternary Palaeoclimatology: Methods of Palaeoclimatic reconstruction*. Allen and Unwin, Boston, 472 pp.

Brady, G.S., Crosskey, H.W. and Robertson, D. (1874) *A Monograph of the Post-Tertiary Entomostraca of Scotland*. Palaeontographical Society Monograph, London, 229 pp.

Bramwell, D. (1977) Archaeology and palaeontology. In *Limestones and Caves of the Peak District* (ed. T.D. Ford). Geo Abstracts, Norwich, pp. 263–91.

Brazier, V. (1987) Late Quaternary alluvial fans, debris cones and talus cones in the Grampian Highlands, Scotland. Unpublished PhD thesis, University of St Andrews.

Brazier, V. and Ballantyne, C.K. (1989) Late Holocene debris cone evolution in Glen Feshie, western Cairngorm Mountains Scotland. *Transactions of the Royal Society of Edinburgh: Earth Sciences*, **80**, 17–24.

Brazier, V., Whittington, G. and Ballantyne, C.K. (1988). Holocene debris cone formation in Glen Etive, Western Grampian Highlands, Scotland. *Earth Surface Processes and Landforms*, **13**, 525–31.

Bremner, A. (1912) *The physical geology of the Dee Valley*. Aberdeen Natural History and Antiquarian Society. Survey of the Natural History and Antiquities of the Valley of the Dee, Vol. 1, Part 2. The University Press, Aberdeen, 89 pp.

Bremner, A. (1915) The capture of the Geldie by the Feshie. *Scottish Geographical Magazine*, **31**, 589–96.

Bremner, A. (1916a) The Vat near Loch Kinord, Aberdeenshire. Is it a giant's kettle (moulin pot-hole) or a stream pot-hole. *Transactions of the Edinburgh Geological Society*, **10**, 326–33.

Bremner, A. (1916b) Problems in the glacial geology of northeast Scotland and some fresh facts bearing on them. *Transactions of the Edinburgh Geological Society*, **10**, 334–47.

Bremner, A. (1919) A geographical study of the high plateau of the south-eastern Highlands. *Scottish Geographical Magazine*, **35**, 331–51.

Bremner, A. (1920) Limits of valley glaciation in the basin of the Dee. *Transactions of the Edinburgh Geological Society*, **11**, 61–8.

Bremner, A. (1921) The physical geology of the Don Basin. *Aberdeen University Studies*, **83**, 129 pp.

Bremner, A. (1922) Deeside as a field for the study of geology. *The Deeside Field*, **1**, 33–6.

Bremner, A. (1925a) The Vat. *The Deeside Field*, **2**, 40–4.

Bremner, A. (1925b) The glacial geology of the Stonehaven district. *Transactions of the Edinburgh Geological Society*, **11**, 25–41.

Bremner, A. (1928) Further problems in the glacial geology of north-eastern Scotland. *Transactions of the Edinburgh Geological Society*, **12**, 147–64.

Bremner, A. (1929) The glaciation of the Cairngorms. *The Deeside Field*, **4**, 29–37.

Bremner, A. (1931) The valley glaciation in the district round Dinnet, Cambus o'May, and Ballater. *The Deeside Field*, **5**, 15–24.

Bremner, A. (1934a) The glaciation of Moray and ice movements in the north of Scotland. *Transactions of the Edinburgh Geological Society*, **13**, 17–56.

Bremner, A. (1934b) Meltwater drainage channels and other glacial phenomena of the Highland border belt from Cortachy to the Bervie Water. *Transactions of the Edinburgh Geological Society*, **13**, 174–5.

Bremner, A. (1936) The glaciation of Glenesk. *Transactions of the Edinburgh Geological Society*, **13**, 378–82.

Bremner, A. (1938) The glacial epoch in north-east Scotland. *The Deeside Field*, **8**, 64–8.

Bremner, A. (1939a) The Late-glacial geology of the Tay basin from Pass of Birnam to Grandtully and Pitlochry. *Transactions of the Edinburgh Geological Society*, **13**, 473–4.

Bremner, A. (1939b) Notes on the glacial geology of East Aberdeenshire. *Transactions of the Edinburgh Geological Society*, **13**, 474–5.

Bremner, A. (1939c) The River Findhorn. *Scottish Geographical Magazine*, **55**, 65–85.

Bremner, A. (1942) The origin of the Scottish river system. Parts I, II and III. *Scottish Geographical Magazine*, **58**, 15–20; 54–59; 99–103.

Bremner, A. (1943a) The glacial epoch in the north-east. In *The Book of Buchan* (Jubilee volume) (ed. J.F. Tocher). Aberdeen University Press, Aberdeen, pp. 10–30.

Bremner, A. (1943b) The later history of the Tilt and Geldie drainage. *Scottish Geographical Magazine*, **59**, 92–7.

Bretz, J.H. (1935) Physiographic studies in East Greenland. In *The Fjord Region of East Greenland* (ed. L.A. Boyd). *American Geographical Society, Special Publication*, **18**, 161–266.

Brewster, D. (1829) *The New Edinburgh Encyclopaedia* (American Edition). Joseph Parker, Philadelphia.

Brickenden, L. (1851) On the occurrence of the boulder clay in the limestone quarry, Linksfield, Elgin, N.B. *Quarterly Journal of the Geological Society*, **7**, 289–92.

Bridge, M.C., Haggart, B.A. and Lowe, J.J. (1990) The history and palaeoclimatic significance of subfossil remains of *Pinus sylvestris* in blanket peats from Scotland. *Journal of Ecology*, **78**, 77–99.

Broccoli, A.-J. and Manabe, S. (1987) The influence of continental ice, atmospheric CO_2, and land albedo on the climate of the last glacial maximum. *Climate Dynamics*, **1**, 87–99.

Broecker, W.S. and Denton, G.H. (1989) The role of ocean-atmosphere reorganizations in glacial cycles. *Geochimica et Cosmochimica Acta*, **53**, 2465–501.

Broecker, W.S. and Denton, G.H. (1990) What drives glacial cycles? *Scientific American*, **262(1)**, 43–50.

Brooks, C.L. (1972) Pollen analysis and the Main Buried Beach in the western part of the Forth Valley. *Transactions of the Institute of British Geographers*, **55**, 161–70.

Brown, D.J. (1868) On the glaciation of Loch Skene and surrounding districts; being a journey across the hills from Moffat to Tweedsmuir. *Transactions of the Edinburgh Geological Society*, **1**, 81–5.

Brown, D.J. (1874) On some of the glacial phenomena of the neighbourhood of Edinburgh, as observed in the Pentlands, Blackford Hill, Bruntsfield Links, and Tynecastle Sandpit. *Transactions of the Edinburgh Geological Society*, **2**, 351–7.

Brown, T. (1867) On the arctic shell clay of Elie and Errol, viewed in connection with our other glacial and more recent deposits. *Transactions of the Royal Society of Edinburgh*, **24**, 617–33.

Brown, T. (1875) On the Parallel Roads of Glen Roy. *Proceedings of the Royal Society of Edinburgh*, **8**, 339–42.

Browne, M.A.E. (1977) Sand and gravel resources of the Fife Region. *Report of the Institute of Geological Sciences*, **77/5**, 14 pp.

Browne, M.A.E. (1980) Late-Devensian marine

limits and the pattern of deglaciation of the Strathearn area, Tayside. *Scottish Journal of Geology*, **16**, 221–30.

Browne, M.A.E. (1987) The physical geography and geology of the estuary and Firth of Forth, Scotland. *Proceedings of the Royal Society of Edinburgh*, **93B**, 235–44.

Browne, M.A.E. and Graham, D.K. (1981) Glaciomarine deposits of the Loch Lomond stade glacier in the Vale of Leven between Dumbarton and Balloch, west-central Scotland. *Quaternary Newsletter*, **34**, 1–7.

Browne, M.A.E. and Jarvis, J. (1983) Late-Devensian marine erosion in St Andrews Bay, east-central Scotland. *Quaternary Newsletter*, **41**, 11–7.

Browne, M.A.E. and McMillan, A.A. (1984) Shoreline inheritance and coastal history in the Firth of Clyde. *Scottish Journal of Geology*, **20**, 119–20.

Browne, M.A.E. and McMillan, A.A. (1989) Quaternary geology of the Clyde Valley. *British Geological Survey Research Report* SA/89/1, Onshore Geology Series, HMSO, London, 63 pp.

Browne, M.A.E., Harkness, D.D., Peacock, J.D. and Ward, R.G. (1977) The date of deglaciation of the Paisley–Renfrew area. *Scottish Journal of Geology*, **13**, 301–3.

Browne, M.A.E., McMillan, A.A. and Hall, I.H.S. (1983) Blocks of marine clay in till near Helensburgh, Strathclyde. *Scottish Journal of Geology*, **19**, 321–5.

Browne, M.A.E., Graham, D.K. and Gregory, D.M. (1984) Quaternary estuarine deposits in the Grangemouth area, Scotland. *Report of the British Geological Survey*, **16/3**, 14 pp.

Brunsden, D. (1964) The origin of decomposed granite on Dartmoor. In *Dartmoor Essays* (ed. I.G. Simmons). The Devonshire Association for the Advancement of Science, Literature and Art, Torquay, pp. 97–116.

Bryce, J. (1855) On the Parallel Roads of Lochaber. *Proceedings of the Philosophical Society of Glasgow*, **3**, 99–109.

Bryce, J. (1859) *Geology of Clydesdale and Arran*. Griffin, London, pp. 199.

Bryce, J. (1865a) On the order of succession in the drift-beds of the Island of Arran. *Quarterly Journal of the Geological Society of London*, **21**, 204–13.

Bryce, J. (1865b) On the occurrence of beds in the west of Scotland beneath the boulder-clay. *Quarterly Journal of the Geological Society of London*, **21**, 213–8.

Bryden, J. (1845) United parishes of Sandsting and Aithsting. Presbytery of Lerwick, Synod of Shetland. *The New Statistical Account of Scotland*, Vol. 15. William Blackwood and Sons, Edinburgh and London, pp. 97–144.

Buck, S.G. (1978) The sedimentology of a coarse bed-load braided river, the R. Feshie, Inverness-shire, Scotland. Unpublished MSc thesis, University of Reading.

Buckland, W. (1841a) On the evidences of glaciers in Scotland and the north of England. *Proceedings of the Geological Society of London*, **3**, 332–7.

Buckland, W. (1841b) Address delivered on the anniversary, February 19th, by the Rev. Professor Buckland, D.D., P.G.S. *Proceedings of the Geological Society of London*, **3**, 468–540.

Buckley, J. and Willis, E.H. (1972) Isotopes' radiocarbon measurements IX. *Radiocarbon*, **14**, 114–39.

Buist, G. (1841) Outline of the geology of the south-east district of Perthshire. *Transactions of the Royal Highland Agricultural Society of Scotland*, **13**, 17–49.

Bugge, T. (1983) Submarine slides on the Norwegian continental margin, with special emphasis on the Storegga area. *Institutt for Kontinentalsokkelundersokkelser (Trondheim)*, Publication no. **110**, 152 pp.

Buller, A.T. and McManus, J. (1971) Channel stability in the Tay estuary, control by bedrock and unconsolidated postglacial sediments. *Engineering Geology*, **5**, 227–37.

Burke, M.J. (1968) Some stone-orientation results from the Forth Valley. *Scottish Journal of Geology*, **5**, 286–92.

Burke, M.J. (1969) The Forth Valley: an ice-moulded lowland. *Transactions of the Institute of British Geographers*, **48**, 51–9.

Busby, J. (1802) Minutes of observations drawn up in the course of a mineralogical survey of the county of Caithness in 1802. Edinburgh. (Cited by Crampton *et al.*, 1914, p. 118.)

Cadell, H.M. (1913) *The Story of the Forth*. James Maclehose and Sons, Glasgow, 299 pp.

Caine, N. (1967) The tors of Ben Lomond, Tasmania. *Zeitschrift für Geomorphologie*, **11**, 418–29.

Callender, J.G., Cree, J.E. and Ritchie, J. (1927) Preliminary report on caves containing Palaeolithic relics, near Inchnadamph, Sutherland.

Proceedings of the Society of Antiquaries of Scotland, **61**, 169–72.

Cambridge, P.G. (1982) A note on the supposed Crag shells from the Kippet Hills, Aberdeenshire. *Bulletin of the Geological Society of Norfolk*, **32**, 37–8.

Cameron, D. (1882a) The glaciers of lower Strathnairn. *Transactions of the Edinburgh Geological Society*, **4**, 160–3.

Cameron, D. (1882b) Notes on the submergence of Scotland. *Transactions of the Edinburgh Geological Society*, **4**, 263–8.

Cameron, I.B., Forsyth, I.H., Hall, I.H.S. and Peacock, J.D. (1977) Sand and gravel resources of the Strathclyde Region of Scotland. *Report of the Institute of Geological Sciences*, **77/8**, 51 pp.

Cameron, T.D.J., Stoker, M.S. and Long, D. (1987) The history of Quaternary sedimentation in the UK sector of the North Sea. *Journal of the Geological Society*, **144**, 43–58.

Campbell, A.C. and Anderson, E.M. (1909) Notes on a transported mass of igneous rock at Comiston sand-pit, near Edinburgh. *Transactions of the Edinburgh Geological Society*, **9**, 219–24.

Campbell, J.F. (1865) *Frost and Fire. Natural Engines, Tool-marks and Chips with Sketches Taken at Home and Abroad by a Traveller*, Vol. 2. Edmonston and Douglas, Edinburgh, 519 pp.

Campbell, J.F. (1873) Notes on the glacial phenomena of the Hebrides. *Quarterly Journal of the Geological Society*, **29**, 545–8.

Campbell, J.F. (1877) *Glen Roy. The Parallel Roads of Lochaber*. Privately printed, London, 28 pp.

Campbell, R. (1913) The geology of south-eastern Kincardineshire. *Transactions of the Royal Society of Edinburgh*, 48, 923–60.

Campbell, R. (1934) On the occurrence of shelly boulder clay and interglacial deposits in Kincardineshire. *Transactions of the Edinburgh Geological Society*, **13**, 176–83.

Campbell, R. (1935) The seaboard of the Mearns: a geological sketch. *The Deeside Field*, **7**, 86–90.

Campbell, R. (1951) Geology. In *Scientific Survey of South-eastern Scotland*. British Association, Edinburgh, 184–99.

Campbell, S. and Bowen, D.Q. (1989) *Quaternary of Wales. Geological Conservation Review*. Nature Conservancy Council, Peterborough, 237 pp.

Carling, P.A. (1987) A terminal debris flow lobe in the northern Pennines, United Kingdom. *Transactions of the Royal Society of Edinburgh: Earth Sciences*, **78**, 169–76.

Carlsson, A. and Olsson, T. (1982) High rock stresses as a consequence of glaciation. *Nature*, **298**, 739–42.

Carruthers, R.G. (1911) On the occurrence of a Cretaceous Boulder of unusual size, at Leavad, in Caithness. *Summary of Progress of the Geological Survey of Great Britain for 1910*. HMSO, London, 80–4.

Carruthers, R.G. (1939) On northern glacial drifts: some pecularities and their significance. *Quarterly Journal of the Geological Society*, **95**, 299–333.

Carruthers, R.G. (1941) Glacial drifts. *Geological Magazine*, **78**, 317–8.

Carruthers, R.G. (1942) Glacial drifts. *Geological Magazine*, **79**, 153–4.

Carter, S.P., Proctor, J. and Slingsby, D.R. (1987) Soil and vegetation of the Keen of Hamar serpentine, Shetland. *Journal of Ecology*, **75**, 21–41.

Caseldine, C.J. (1980a) A Lateglacial site at Stormont Loch, near Blairgowrie, eastern Scotland. In *Studies in the Lateglacial of North-west Europe* (eds J.J. Lowe, J.M. Gray and J.E. Robinson). Pergamon Press, Oxford, pp. 69–88.

Caseldine, C.J. (1980b) Aspects of the vegetation history of south-east Perthshire. Unpublished PhD thesis, University of St Andrews.

Caseldine, C.J. and Edwards, K.J. (1982) Interstadial and last interglacial deposits covered by till in Scotland: comments and new evidence. *Boreas*, **11**, 119–22.

Caseldine, C.J. and Whittington, G. (1976) Pollen analysis of material from the Stones of Stenness, Orkney. In *The Stones of Stenness, Orkney* (ed. J.W.G. Ritchie). *Proceedings of the Society of Antiquaries of Scotland*, **107**, 37–40.

Chambers, F.M. and Elliott, L. (1989) Spread and expansion of *Alnus* Mill. in the British Isles: timing, agencies and possible vectors. *Journal of Biogeography*, **16**, 541–50.

Chambers, R. (1848) *Ancient Sea-margins, as Memorials of Changes in the Relative Level of Sea and Land*. W. and R. Chambers, Edinburgh, 338 pp.

Chambers, R. (1853) On glacial phenomena in Scotland and parts of England. *Edinburgh New Philosophical Journal*, **54**, 229–81.

References

Chambers, R. (1855a) Further observation on glacial phenomena in Scotland and the north of England. *Edinburgh New Philosophical Journal*, NS **1**, 97–103.

Chambers, R. (1855b) On glacial phenomena in Peebles and Selkirk Shires. *Edinburgh New Philosophical Journal*, NS **2**, 184.

Chambers, R. (1857) Geological notes on Banffshire. *Proceedings of the Royal Society of Banffshire*, **3**, 332–3.

Chapelhowe, R. (1965) On glaciation in North Roe, Shetland. *Geographical Journal*, **131**, 60–71.

Chapman, C.A. and Rioux, R.L. (1958) Statistical study of topography, sheeting and jointing in granite, Acadia National Park, Maine. *American Journal of Science*, **256**, 111–27.

Charlesworth, J.K. (1926a) The glacial geology of the Southern Uplands of Scotland, west of Annandale and Upper Clydesdale. *Transactions of the Royal Society of Edinburgh*, **55**, 1–23.

Charlesworth, J.K. (1926b) The readvance, marginal kame-moraine of the south of Scotland, and some later stages of retreat. *Transactions of the Royal Society of Edinburgh*, **55**, 25–50.

Charlesworth, J.K. (1956) The late-glacial history of the Highlands and Islands of Scotland. *Transactions of the Royal Society of Edinburgh*, **62**, 769–928.

Charlesworth, J.K. (1957) *The Quaternary Era*, 2 Vols. Edward Arnold, London, 591 pp and 1700 pp.

Charman, D.J. (1990) Origins and development of blanket mire in the flow country, northern Scotland, with special reference to patterned fens. Unpublished PhD thesis, University of Southampton.

Charman, D.J. (1992) Blanket mire formation at the Cross Lochs, Sutherland, northern Scotland. *Boreas*, **21**, 53–72.

Chattopadhyay, G.P. (1982) Periglacial geomorphology of parts of the Grampian Highlands of Scotland. Unpublished PhD thesis, University of Edinburgh.

Chattopadhyay, G.P. (1984) A fossil valley-wall rock glacier in the Cairngorm mountains. *Scottish Journal of Geology*, **20**, 121–5.

Chattopadhyay, G.P. (1986) Ploughing block movement on the Drumochter Hills in the Grampian Highlands, Scotland. *Biuletyn Peryglacjalny*, **30**, 57–60.

Chester, D.K. (1974) Description of the section exposed by the BP oil pipeline trench between Stonehaven and the River Don. Unpublished Report, Department of Geography University of Aberdeen, 7 pp.

Chester, D.K. (1975) Bay of Nigg section. *Aberdeen Field Excursion Guide*. Quaternary Research Association, Aberdeen, pp. 2–4.

Chisholm, J.I. (1966) An association of raised beaches with glacial deposits near Leuchars, Fife. *Bulletin of the Geological Survey of Great Britain*, **24**, 163–74.

Chisholm, J.I. (1971) The stratigraphy of the post-glacial marine transgression in NE Fife. *Bulletin of the Geological Survey of Great Britain*, **37**, 91–107.

Chorley, R.J., Schumm, S.C. and Sugden, D.E. (1984) *Geomorphology*. Methuen, London, 605 pp.

Christie, J. (1830) Blackpotts Clay, near Banff. *Edinburgh New Philosophical Journal*, **9**, 382.

Christie, J. (1831) On the occurrence of chalk-flints in Banffshire. *Edinburgh New Philosophical Journal*, **10**, 163–4.

Church, M. (1978) Palaeohydrological reconstructions from a Holocene valley fill. In *Fluvial Sedimentology* (ed. A.D. Miall). *Memoir of the Canadian Society of Petroleum Geologists, Calgary*, **5**, 743–72.

Church, M. and Ryder, J.M. (1972) Paraglacial sedimentation: a consideration of fluvial processes conditioned by glaciation. *Bulletin of the Geological Society of America*, **83**, 3059–72.

Clapperton, C.M. (1968) Channels formed by the superimposition of glacial meltwater streams with special reference to the east Cheviot Hills, north-east England. *Geografiska Annaler*, **50A**, 207–20.

Clapperton, C.M. (1970) The evidence for a Cheviot ice cap. *Transactions of the Institute of British Geographers*, **50**, 115–27.

Clapperton, C.M. (1971a) The pattern of deglaciation in part of north Northumberland. *Transactions of the Institute of British Geographers*, **53**, 67–78.

Clapperton, C.M. (1971b) The location and origin of glacial meltwater phenomena in the eastern Cheviot Hills. *Proceedings of the Yorkshire Geological Society*, **38**, 361–80.

Clapperton, C.M. (ed). (1977) *The Northern Highlands of Scotland*. INQUA X Congress.

Guidebook for Excursions A10 and C10. Geo Abstracts, Norwich, 44 pp.

Clapperton, C.M. (1986) Glacial geomorphology of north-east Lochnagar. In *Essays for Professor R.E.H. Mellor* (ed. W. Ritchie, J.C. Stone and A.S. Mather). Department of Geography, University of Aberdeen, Aberdeen, pp. 390–6.

Clapperton, C.M. and Crofts, R.S. (1969) Physiography and terrain analysis. In *Royal Grampian Country* (ed. K. Walton). Department of Geography, University of Aberdeen, Aberdeen, pp. 1–8.

Clapperton, C.M. and Sugden, D.E. (1972) The Aberdeen and Dinnet glacial limits reconsidered. In *North-east Scotland Geographical Essays* (ed. C.M. Clapperton). Department of Geography, University of Aberdeen, Aberdeen, pp. 5–11.

Clapperton, C.M. and Sugden, D.E. (1975) The glaciation of Buchan – a reappraisal. In *Quaternary Studies in North East Scotland* (ed. A.M.D. Gemmell). Department of Geography, University of Aberdeen, Aberdeen, pp. 19–22.

Clapperton, C.M. and Sugden, D.E. (1977) The Late Devensian glaciation of north-east Scotland. In *Studies in the Scottish Lateglacial Environment* (ed. J.M. Gray and J.J. Lowe). Pergamon Press, Oxford, pp. 1–13.

Clapperton, C.M., Gunson, A.R. and Sugden, D.E. (1975) Loch Lomond Readvance in the eastern Cairngorms. *Nature*, **253**, 710–12.

Clark, R. (1962) Some landforms and processes of Southern Rhum. Unpublished report. Nature Conservancy Council, Edinburgh.

Clark, R. (1970) Aspects of glaciation in Northumberland. *Proceedings of the Cumberland Geological Society*, **2**, 133–56.

Clark, R. (1971) Periglacial landforms and landscapes in Northumberland. *Proceedings of the Cumberland Geological Society*, **3**, 5–20.

Clayton, K.M. (1965) Glacial erosion in the Finger Lakes region (New York State, U.S.A.). *Zeitschrift für Geomorphologie*, **9**, 50–62.

Clayton, K.M. (1974) Zones of glacial erosion. *Institute of British Geographers Special Publication*, **7**, 163–76.

Cleghorn, J. (1850) On the till near Wick in Caithness. *Quarterly Journal of the Geological Society of London*, **6**, 385–6.

Cleghorn, J. (1851) On the till of Caithness. *Quarterly Journal of the Geological Society of London*, **7**, 200–1.

Clement, P. (1984) The drainage of a marginal ice-dammed lake at Nordbogletscher, Johan Dahl Land, south Greenland. *Arctic and Alpine Research*, **16**, 209–16.

Clough, C.T. and Harker, A. (1904) *The Geology of West-central Skye, with Soay*. Explanation of Sheet 70. Memoirs of the Geological Survey of Scotland. HMSO, Glasgow, 59 pp.

Clough, C.T., Barrow, G., Crampton, C.B., Maufe, H.R., Bailey, E.B. and Anderson, E.M. (1910) *The Geology of East Lothian, including parts of the Counties of Edinburgh and Berwick*. Explanation of Sheet 33, with parts of 34 and 31. 2nd edn. Memoirs of the Geological Survey of Scotland. HMSO, Edinburgh, 226 pp.

Clough, C.T., Hinxman, L.W., Wright, W.B., Anderson, E.M. and Carruthers, R.G. (1916) *The Economic Geology of the Central Coalfield of Scotland*. Area V. Memoirs of the Geological Survey of Scotland. HMSO, Edinburgh, 146 pp.

Clough, C.T., Wilson, J.S.G., Anderson, E.M. and Macgregor, M. (1920) *The Economic Geology of the Central Coalfield of Scotland*. Area VII. Memoirs of the Geological Survey of Scotland. HMSO, Edinburgh, 144 pp.

Clough, C.T., Hinxman, L.W., Wilson, J.S.G., Crampton, C.B., Wright, W.B., Bailey, E.B., Anderson, E.M. and Carruthers, R.G. (1925) *The Geology of the Glasgow District*. Glasgow District Map, including parts of Sheets 30, 31, 22 and 23. (Revised edition by M. Macgregor, C.H. Dinham, E.B. Bailey and E.M. Anderson). Memoirs of the Geological Survey of Scotland, HMSO, Edinburgh, 299 pp.

Cockburn, A.M. (1935) The geology of St Kilda. *Transactions of the Royal Society of Edinburgh*, **58**, 511–47.

Colhoun, E.A. (1981) A protalus rampart from the western Mourne Mountains, Northern Ireland. *Irish Geography*, **14**, 85–90.

Comber, D.P.M. (1991) Holocene coastal evolution of Culbin Sands and Forest. In *Late Quaternary Coastal Evolution in the Inner Moray Firth. Field Guide* (eds C.R. Firth and B.A. Haggart). West London Press, London, pp. 23–9.

Common, R. (1954a) A report on the Lochaber, Appin and Benderloch floods, May, 1953. *Scottish Geographical Magazine*, **70**, 6–20.

Common, R. (1954b) The geomorphology of the east Cheviot area. *Scottish Geographical Magazine*, **70**, 124–38.

Common, R. and Galloway, R.W. (1958) Ice wedges in Midlothian: a note. *Scottish Geographical Magazine*, **74**, 44–6.

Conacher, H.R.J. (1932) A group of 'kettle-holes' near Connel Ferry. *Transactions of the Geological Society of Glasgow*, **19**, 152–7.

Connell, E.R. (1984a) Kirkhill. Stratigraphy. B. Deposits between the lower and upper palaeosols. In *Buchan Field Guide* (ed. A.M. Hall). Quaternary Research Association, Cambridge, pp. 62–4.

Connell, E.R. (1984b) Kirkhill. Chronology. In *Buchan Field Guide* (ed. A.M. Hall). Quaternary Research Association, Cambridge, pp. 78–9.

Connell, E.R. and Hall, A.M. (1984a) Kirkhill Quarry: correlation. In *Buchan Field Guide* (ed. A.M. Hall). Quaternary Research Association, Cambridge, pp. 80.

Connell, E.R. and Hall, A.M. (1984b) Boyne Bay Quarry. In *Buchan Field Guide* (ed. A.M. Hall). Quaternary Research Association, Cambridge, pp. 97–101.

Connell, E.R. and Hall, A.M. (1987) The periglacial stratigraphy of Buchan. In *Periglacial Processes and Landforms in Britain and Ireland* (ed. J. Boardman). Cambridge University Press, Cambridge, pp. 277–85.

Connell, E.R. and Romans, J.C.C. (1984) Kirkhill. Palaeosols. In *Buchan Field Guide* (ed. A.M. Hall). Quaternary Research Association, Cambridge, pp. 70–6.

Connell, E.R., Edwards, K.J. and Hall, A.M. (1982) Evidence for two pre-Flandrian palaeosols in Buchan, Scotland. *Nature*, **297**, 570–2.

Connell, E.R., Hall, A.M., Shaw, D. and Riley, L.A. (1985) Palynology and significance of radiocarbon dated organic materials from Cruden Bay Brick Pit, Grampian Region, Scotland. *Quaternary Newsletter*, **47**, 19–25.

Connolly, A.P. (1957) The occurrence of seeds of *Papaver* sect. *Scapiflora* in a Scottish Late Glacial site. *Veröffentlichungen Geobotanischen Instituts, Zürich*, **34**, 27–9.

Coope, G.R. (1959) A Late Pleistocene insect fauna from Chelford, Cheshire. *Proceedings of the Royal Society of London*, **B151**, 70–86.

Coope, G.R. (1968) Fossil beetles collected by James Bennie from Late Glacial silts at Corstorphine, Edinburgh. *Scottish Journal of Geology*, **4**, 339–48.

Coope, G.R. (1975) Climatic fluctuations in northwest Europe since the last interglacial, indicated by fossil assemblages of coleoptera. In *Ice Ages: Ancient and Modern* (eds A.E. Wright and F. Moseley). Seel House Press, Liverpool, pp. 153–68.

Coope, G.R. (1977) Fossil coleopteran assemblages as sensitive indicators of climatic changes during the Devensian (Last) cold stage. *Philosophical Transactions of the Royal Society of London*, **B280**, 313–40.

Coope, G.R. (1981) Episodes of local extinction of insect species during the Quaternary as indicators of climatic change. In *The Quaternary in Britain* (eds J. Neale and J. Flenley). Pergamon Press, Oxford, pp. 216–21.

Coope, G.R. and Brophy, J.A. (1972) Lateglacial environmental changes indicated by a coleopteran succession from north Wales. *Boreas*, **1**, 97–142.

Coope, G.R. and Joachim, M.J. (1980) Lateglacial environmental changes interpreted from fossil coleoptera from St Bees, Cumbria, NW England. In *Studies in the Lateglacial of North-west Europe* (eds J.J. Lowe, J.M. Gray and J.E. Robinson). Pergamon Press, Oxford, pp. 55–68.

Coque-Delhuille, B. and Veyret, Y. (1988) Recherches géomorphologiques préliminaires dans l'archipel des Shetland (G.B.). *Hommes et Terres du Nord*, **3**, 137–54.

Cornish, R. (1979) Glacial Geomorphology of the west-central Southern Uplands of Scotland, with particular reference to 'rogen moraines'. Unpublished PhD thesis, University of Edinburgh.

Cornish, R. (1981) Glaciers of the Loch Lomond Stadial in the western Southern Uplands of Scotland. *Proceedings of the Geologists' Association*, **92**, 105–14.

Cornish, R. (1982) Glacier flow at a former ice-divide in SW Scotland. *Transactions of the Royal Society of Edinburgh: Earth Sciences*, **73**, 31–41.

Cornish, R. (1983) Glacial erosion in an ice-divide zone. *Nature*, **301**, 413–5.

Coward, M.P. (1977) Anomalous glacial erratics in the southern part of the Outer Hebrides. *Scottish Journal of Geology*, **13**, 185–8.

Cox, R. and Nicol, J. (1869) *Select Writings, Political, Scientific, Topographical and Miscellaneous, of the Late Charles Maclaren, FRSE*, 2 Vols. Edmonston and Douglas, Edinburgh.

Craig, A.J. (1978) Pollen percentage and influx analyses in south-east Ireland: a contribution to the ecological history of the Late-glacial period. *Journal of Ecology*, **66**, 297–324.

Craig, R. (1873) On the glacial deposits of north

Ayrshire and Renfrewshire. *Transactions of the Geological Society of Glasgow*, **4**, 138–64.

Craig, R. (1888) On the post-Pliocene beds of the Irvine valley, Kilmaurs and Dreghorn districts. *Transactions of the Geological Society of Glasgow*, **8**, 214–26.

Crampton, C.B. (1911) *The Vegetation of Caithness considered in relation to the Geology*. Published under the auspices of the Committee for the Survey and Study of British Vegetation, Edinburgh, 132 pp.

Crampton, C.B. and Carruthers, R.G. (1914) *The Geology of Caithness*. Sheets 110 and 116, with parts of 109, 115 and 117. Memoirs of the Geological Survey of Scotland, HMSO, Edinburgh, 194 pp.

Cranwell, P.A. (1977) Organic geochemistry of Cam Loch (Sutherland) sediments. *Chemical Geology*, **20**, 205–21.

Cree, J.E. (1927) Palaeolithic Man in Scotland. *Antiquity*, **1**, 169–72.

Crofts, R.S. (1971) Coastal process and evolution around St Cyrus, Angus–Kincardine. Unpublished MLitt thesis, University of Aberdeen.

Crofts, R.S. (1972) Coastal sediments and process around St Cyrus. In *North East Scotland Geographical Essays* (ed. C.M. Clapperton). Department of Geography, University of Aberdeen, Aberdeen, pp. 15–9.

Crofts, R.S. (1974) A method to determine shingle supply to the coast. *Transactions of the Institute of British Geographers*, **62**, 115–27.

Crofts, R.S. (1975) Sea bed topography off northeast Scotland. *Scottish Geographical Magazine*, **91**, 52–64.

Croll, J. (1867) On the excentricity of the Earth's orbit, and its physical relations to the glacial epoch. *Philosophical Magazine*, **33**, 119–31.

Croll, J. (1870a) The boulder-clay of Caithness a product of land-ice. *Geological Magazine*, **7**, 209–14 and 271–8.

Croll, J. (1870b) On two river channels buried under drift, belonging to a period when the land stood several hundred feet higher than at present. *Transactions of the Edinburgh Geological Society*, **1**, 330–45.

Croll, J. (1875) *Climate and Time in their Geological Relations. A Theory of Secular Changes of the Earth's Climate*. Daldy, Isbister and Co., London, 577 pp.

Croll, J. (1885) *Climate and Cosmology*. Adam and Charles Black, Edinburgh, 327 pp.

Crosskey, H.W. (1864) On the recent discovery of the remains of a cetacean in the banks of the River Irvine. *Proceedings of the Philosophical Society of Glasgow*, **5**, 243–6.

Crosskey, H.W. (1865) On the *Tellina calcarea* bed at Chappel Hall, near Airdrie. *Quarterly Journal of the Geological Society*, **21**, 219–21.

Crosskey, H.W. (1887) Dr Crosskey on the Clava shell bed. *Transactions of the Inverness Scientific Society and Field Club*, **3**, 277–85.

Crosskey, H.W. and Robertson, D. (1867) The post-Tertiary fossiliferous beds of Scotland. *Transactions of the Geological Society of Glasgow*, **2**, 267–82.

Crosskey, H.W. and Robertson, D. (1868a) The post-Tertiary fossiliferous beds of Scotland. II. Cumbrae College. *Transactions of the Geological Society of Glasgow*, **3**, 113–8.

Crosskey, H.W. and Robertson, D. (1868b) The post-Tertiary fossiliferous beds of Scotland. III. Loch Gilp. *Transactions of the Geological Society of Glasgow*, **3**, 118–25.

Crosskey, H.W. and Robertson, D. (1868c) The post-Tertiary fossiliferous beds of Scotland. IV. Boulder Clay – Caithness. *Transactions of the Geological Society of Glasgow*, **3**, 125–7.

Crosskey, H.W. and Robertson, D. (1868d) The post-Tertiary fossiliferous beds of Scotland. V. Lucknow Pit, Ardeer Iron Works, Ayrshire. *Transactions of the Geological Society of Glasgow*, **3**, 127–9

Crosskey, H.W. and Robertson, D. (1869a) The post-Tertiary fossiliferous beds of Scotland. VI. East Tarbert – Loch Fyne. *Transactions of the Geological Society of Glasgow*, **3**, 321–4.

Crosskey, H.W. and Robertson, D. (1869b) The post-Tertiary fossiliferous beds of Scotland. VII. West Tarbert. *Transactions of the Geological Society of Glasgow*, **3**, 324–7.

Crosskey, H.W. and Robertson, D. (1869c) The post-Tertiary fossiliferous beds of Scotland. VIII. Crinan. *Transactions of the Geological Society of Glasgow*, **3**, 327–8.

Crosskey, H.W. and Robertson, D. (1869d) The post-Tertiary fossiliferous beds of Scotland. IX. Duntroon. *Transactions of the Geological Society of Glasgow*, **3**, 328–31.

Crosskey, H.W. and Robertson, D. (1869e) The post-Tertiary fossiliferous beds of Scotland. X. Old Mains, Renfrew. *Transactions of the Geological Society of Glasgow*, **3**, 331–4.

Crosskey, H.W. and Robertson, D. (1869f) The post-Tertiary fossiliferous beds of Scotland. XI. Paisley. *Transactions of the Geological Society of Glasgow*, **3**, 334–41

Crosskey, H.W. and Robertson, D. (1871) The post-Tertiary fossiliferous beds of Scotland. XII. Garvel Park New Dock, Greenock. *Transactions of the Geological Society of Glasgow*, **4**, 32–45.

Crosskey, H.W. and Robertson, D. (1873a) The post-Tertiary fossiliferous beds of Scotland. XIII. Kilchattan Tile-Works, Bute. *Transactions of the Geological Society of Glasgow*, **4**, 128–33.

Crosskey, H.W. and Robertson, D. (1873b) The post-Tertiary fossiliferous beds of Scotland. XIV. Tangy Glen, near Campbeltown. *Transactions of the Geological Society of Glasgow*, **4**, 134–7.

Crosskey, H.W. and Robertson, D. (1874a) The post-Tertiary fossiliferous beds of Scotland. XV. Jordanhill Brick Works. *Transactions of the Geological Society of Glasgow*, **4**, 241–5.

Crosskey, H.W. and Robertson, D. (1874b) The post-Tertiary fossiliferous beds of Scotland. XVI. Stobcross. *Transactions of the Geological Society of Glasgow*, **4**, 245–51.

Crosskey, H.W. and Robertson, D. (1874c) The post-Tertiary fossiliferous beds of Scotland. XVII. Fairfield, near Govan. *Transactions of the Geological Society of Glasgow*, **4**, 251–2.

Crosskey, H.W. and Robertson, D. (1874d) The post-Tertiary fossiliferous beds of Scotland. XVIII. Paisley Canal. *Transactions of the Geological Society of Glasgow*, **4**, 252–4.

Crosskey, H.W. and Robertson, D. (1874e) The post-Tertiary fossiliferous beds of Scotland. XIX. Dipple Tile-Works. *Transactions of the Geological Society of Glasgow*, **4**, 255–6.

Crosskey, H.W. and Robertson, D. (1875) The post-Tertiary fossiliferous beds of Scotland. XX. Kyles of Bute. *Transactions of the Geological Society of Glasgow*, **5**, 29–35.

Cruden, D.M. and Hungr, O. (1986) The debris of the Frank Slide and theories of rockslide-avalanche mobility. *Canadian Journal of Earth Sciences*, **23**, 425–32.

Cullingford, R.A. (1972) Lateglacial and postglacial shoreline displacement in the Earn–Tay Area and Eastern Fife. Unpublished PhD thesis, University of Edinburgh.

Cullingford, R.A. (1977) Lateglacial raised shorelines and deglaciation in the Earn–Tay area. In *Studies in the Scottish Lateglacial Environment* (eds J.M. Gray and J.J. Lowe). Pergamon Press, Oxford, pp. 15–32.

Cullingford, R.A. and Smith, D.E. (1966) Lateglacial shorelines in eastern Fife. *Transactions of the Institute of British Geographers*, **39**, 31–51.

Cullingford, R.A. and Smith, D.E. (1980) Late Devensian raised shorelines in Angus and Kincardineshire, Scotland. *Boreas*, **9**, 21–38.

Cullingford, R.A., Caseldine, C.J. and Gotts, P.E. (1980) Early Flandrian land and sea-level changes in lower Strathearn. *Nature*, **184**, 159–61.

Cullingford, R.A. Firth, C.R. and Smith, D.E. (1986) *Relative Sea Level Changes in Eastern Scotland from the Loch Lomond Stadial to the Present: A Summary of Present Knowledge*. Review Paper, IGCP Project 200 Field Meeting, Montrose, May 1986, 29 pp.

Cullingford, R.A., Caseldine, C.J. and Gotts, P.E. (1989a) Evidence of early Flandrian tidal surges in lower Strathearn, Scotland. *Journal of Quaternary Science*, **4**, 51–60.

Cullingford, R.A., Caseldine, C.J. and Gotts, P.E. (1989b) Reply to A.G. Dawson, D.E. Smith and D. Long. *Journal of Quaternary Science*, **4**, 273–4.

Cullingford, R.A. Smith, D.E. and Firth, C.R. (1991) The altitude and age of the Main Postglacial Shoreline in eastern Scotland. *Quaternary International*, **9**, 39–52.

Cumming, G.A. and Bate, P.A. (1933) The Lower Cretaceous erratics of the Fraserburgh district, Aberdeenshire. *Geological Magazine*, **70**, 397–413.

Cundill, P. and Whittington, G. (1983) Anomalous arboreal pollen assemblages in Late Devensian and Early Flandrian deposits at Creich Castle, Fife, Scotland. *Boreas*, **12**, 297–311.

Cunningham Craig, E.H. (1901) Dumbartonshire. In *Summary of Progress of the Geological Survey of the United Kingdom for 1900*. Memoirs of the Geological Survey, HMSO, London, 138–9.

Cunningham Craig, E.H., Wright, W.B. and Bailey, E.B. (1911) *The Geology of Colonsay and Oronsay, with part of the Ross of Mull*. Explanation of Sheet 35, with part of 27. Memoirs of the Geological Survey of Scotland. HMSO, Edinburgh, 109 pp.

Curtis, L.F., Courtney, F.M. and Trudgill, S.T. (1976) *Soils in the British Isles*. Longman, London, 364 pp.

Cutler, H.D. (1978) The glaciation and the drumlins of the moors and machers of Galloway, south-west Scotland. Unpublished PhD thesis, University of Liverpool.

Dahl, R. (1965) Plastically sculptured detail forms

on rock surfaces in northern Nordland, Norway. *Geografiska Annaler*, **47A**, 83–140.

Dahl, R. (1966) Block fields, weathering pits and tor-like forms in the Narvik mountains, Nordland, Norway. *Geografiska Annaler*, **48A**, 55–85.

Dakyns, J.R. (1879) The Parallel Roads of Glen Roy. *Geological Magazine*, **6**, 529–31.

Dale, M.L. (1981) Rock walls in glacier source areas in part of the Highlands of Scotland. Unpublished PhD thesis, University of Edinburgh.

Daniels, R.E. (1972) A preliminary survey of Beanrig Moss, a fen in south Scotland. *Transactions of the Botanical Society of Edinburgh*, **42**, 507–16.

Dansgaard, W. (1964) Stable isotopes in precipitation. *Tellus*, **16**, 436–68.

Dansgaard, W., White, J.W.C. and Johnsen, S.J. (1989) The abrupt termination of the Younger Dryas climate event. *Nature*, **339**, 532–4.

Dardis, G.F. (1985) Genesis of Late Pleistocene cross-valley moraine ridges, south-central Ulster, Northern Ireland. *Earth Surface Processes and Landforms*, **10**, 483–95.

Darwin, C. (1839) Observations on the Parallel Roads of Glen Roy and of other parts of Lochaber in Scotland, with an attempt to prove that they are of marine origin. *Philosophical Transactions of the Royal Society of London*, **129**, 39–81.

Darwin, F. (1887) *The life and letters of Charles Darwin*. Vol. 1. John Murray, London, 395 pp.

Davenport, C.A. and Ringrose, P.S. (1985) Fault activity and palaeoseismicity during Quaternary time in Scotland – preliminary studies. In *Earthquake Engineering in Britain*. Thomas Telford, London, pp. 143–55.

Davenport, C.A. and Ringrose, P.S. (1987) Deformation of Scottish Quaternary sediment sequences by strong earthquake motions. In *Deformation of Sediments and Sedimentary Rocks* (eds M.E. Jones and R.M.F. Preston). Geological Society Special Publication No. 29, 299–314.

Davenport, C.A., Ringrose, P.S., Becker, A., Hancock, P. and Fenton, C. (1989) Geological investigations of late and post glacial earthquake activity in Scotland. In *Earthquakes at North Atlantic Passive Margins: Neotectonics and Postglacial Rebound* (eds S. Gregersen and P. Basham). Kluwer Academic Publishers, Dordrecht, pp. 175–94.

Davidson, C.F. (1932) The arctic clay of Errol, Perthshire. *Transactions of the Perth Society for Natural Science*, **9**, 55–68.

Davidson, D.A., Jones, R.L. and Renfrew, C. (1976) Palaeoenvironmental reconstruction and evaluation: a case study from Orkney. *Transactions of the Institute of British Geographers*, NS, **1**, 346–61.

Davies, A., McAdam, A.D. and Cameron, I.B. (1986) *Geology of the Dunbar District. 1:50,000 Sheet 33E and part of Sheet 41*. Memoirs of the British Geological Survey (Scotland), HMSO, London, 69 pp.

Davies, G.L. (1968a) *The Earth in Decay. A History of British Geomorphology 1578–1878*. MacDonald Technical and Scientific, London, 390 pp.

Davies, G.L. (1968b) The tour of the British Isles made by Louis Agassiz in 1840. *Annals of Science*, **24**, 131–46.

Davison, R.W. and Davison, S.K. (1987) Characteristics of two full-depth slab avalanches on Meall Uaine, Glen Shee, Scotland. *Journal of Glaciology*, **33**, 51–4.

Dawson, A.G. (1977) A fossil lobate rock glacier in Jura. *Scottish Journal of Geology*, **13**, 37–42.

Dawson, A.G. (1979a) Raised shorelines of Jura, Scarba and NE Islay. Unpublished PhD thesis, University of Edinburgh.

Dawson, A.G. (1979b) A Devensian medial moraine in Jura. *Scottish Journal of Geology*, **15**, 43–8.

Dawson, A.G. (1979c) Former sea-level changes in the Scottish Hebrides. *Hebridean Naturalist*, **3**, 16–22.

Dawson, A.G. (1980a) The Low Rock Platform in western Scotland. *Proceedings of the Geologists' Association*, **91**, 339–44.

Dawson, A.G. (1980b) Shore erosion by frost: an example from the Scottish Lateglacial. In *Studies in the Lateglacial of North-West Europe* (eds J.J. Lowe, J.M. Gray and J.E. Robinson). Pergamon Press, Oxford, pp. 45–53.

Dawson, A.G. (1982) Lateglacial sea-level changes and ice-limits in Islay, Jura and Scarba, Scottish Inner Hebrides. *Scottish Journal of Geology*, **18**, 253–65.

Dawson, A.G. (1983a) *Islay and Jura, Scottish Hebrides: Field Guide*. Quaternary Research Association, Cambridge, 31pp.

Dawson, A.G. (1983b) Report on a Quaternary Research Association Field Excursion to Islay and Jura, Scottish Inner Hebrides, 24–28 May 1983. *Quaternary Newsletter*, **41**, 26–34.

Dawson, A.G. (1983c) Glacier-dammed lake investigations in the Hullet lake area, South Greenland. *Meddelelser om Grønland, Geoscience*, **11**, 22 pp.

Dawson, A.G. (1984) Quaternary sea-level changes in western Scotland. *Quaternary Science Reviews*, **3**, 345–68.

Dawson, A.G. (1988a) The Main Rock Platform (Main Lateglacial Shoreline) in Ardnamurchan and Moidart, western Scotland. *Scottish Journal of Geology*, **24**, 163–74.

Dawson, A.G. (1988b) Western Jura. In *Field Excursion and Symposium on Late Quaternary Sea Levels and Crustal Deformation* (eds A.G. Dawson, D.G. Sutherland and D.E. Smith). INQUA Subcommission on Shorelines of NW Europe. Coventry Polytechnic, Coventry, pp. 44–67.

Dawson, A.G. (1989) Distribution and development of the Main Rock Platform, western Scotland: reply. *Scottish Journal of Geology*, **25**, 233–8.

Dawson, A.G. (1991) Scottish landform examples – 3. The raised shorelines of northern Islay and western Jura. *Scottish Geographical Magazine*, **107**, 207–12.

Dawson, A.G. (1992) *Ice Age Earth. Late Quaternary Geology and Climate*. Routledge, London, 293 pp.

Dawson, A.G., Smith, D.E. and Long, D. (1990) Evidence for a tsunami from a Mesolithic site in Inverness, Scotland. *Journal of Archaeological Science*, **17**, 509–12.

Dawson, A.G., Lowe, J.J. and Walker, M.J.C. (1987a) The nature and age of the debris accumulation at Gribun, western Mull, Inner Hebrides. *Scottish Journal of Geology*, **23**, 149–62.

Dawson, A.G., Matthews, J.A. and Shakesby, R.A. (1987b) Rock platform erosion on periglacial shores: a modern analogue for Pleistocene rock platforms in Britain. In *Periglacial Processes and Landforms in Britain and Ireland* (ed. J. Boardman). Cambridge University Press, Cambridge, pp. 173–82.

Dawson, A.G., Long, D. and Smith, D.E. (1988) The Storegga Slides: evidence from eastern Scotland of a possible tsunami. *Marine Geology*, **82**, 271–6.

Dawson, A.G., Smith, D.E. and Long, D. (1989) Early Flandrian tidal surges in the Tay estuary, Scotland. *Journal of Quaternary Science*, **4**, 273.

Day, T.C. (1923) Note on the dry valley of Windy Gowl, Carlops. *Transactions of the Edinburgh Geological Society*, **11**, 266.

Day, T.E. (1983) The remanent magnetism of till and other glacial sediments. Unpublished PhD thesis, University of East Anglia.

Debenham, F. (1919) A new mode of transportation by ice: the raised marine muds of South Victoria Land (Antarctica). *Quarterly Journal of the Geological Society*, **75**, 51–76.

De Geer, G. (1935) Dating of late-glacial varves in Scotland. *Proceedings of the Royal Society of Edinburgh*, **55**, 23–6.

Demek, J. (1964) Castle koppies and tors in the Bohemian Highland (Czechoslovakia). *Biuletyn Peryglacjalny*, **14**, 195–216.

Denton, G.H. and Hughes, T.J. (ed.) (1981) *The Last Great Ice Sheets*. John Wiley, New York and Chichester, 484 pp.

Derbyshire, E. (1972) Tors, rock weathering and climate in southern Victoria Land, Antactica. *Institute of British Geographers Special Publication*, **4**, 93–105.

Dick, T.D. (1823) On the Parallel Roads of Lochaber. *Transactions of the Royal Society of Edinburgh*, **9**, 1–64.

Dickson, J.H., Jardine, W.G. and Price, R.J. (1976) Three Late-Devensian sites in west-central Scotland. *Nature*, **262**, 43–4.

Dickson, J.H., Stewart, D.A., Thompson, R., Turner, G., Baxter, M.S., Drndarsky, N.D. and Rose, J. (1978) Palynology, palaeomagnetism and radiometric dating of Flandrian marine and freshwater sediments of Loch Lomond. *Nature*, **274**, 548–53.

Donner, J.J. (1955) The geology and vegetation of Late-glacial retreat stages in Scotland. Unpublished PhD thesis, University of Cambridge.

Donner, J.J. (1957) The geology and vegetation of Late-glacial retreat stages in Scotland. *Transactions of the Royal Society of Edinburgh*, **63**, 221–64.

Donner, J.J. (1960) Pollen analysis of the Burn of Benholm peat-bed, Kincardineshire, Scotland. *Societas Scientarum Fennica, Commentationes Biologicae*, **22**, 1–13.

Donner, J.J. (1979) The Early or Middle Devensian peat at the Burn of Benholm, Kincardineshire. *Scottish Journal of Geology*, **15**, 247–50.

Donner, J.J. and West, R.G. (1955) Ett drumlinsfält på ön Skye, Skottland. *Terra*, **67**, 45–8.

Dougal, J.W. (1928) Observations on the geology

of Lewis. *Transactions of the Edinburgh Geological Society*, **12**, 12–18.

Dougall, J. (1868) Sketch of the geology of the Falls of Clyde, the Mouse Valley, and Cartland Crags. *Transactions of the Geological Society of Glasgow*, **3**, 44–53.

Dreimanis, A. (1976) Tills: their origin and properties. In *Glacial Till. An Inter-disciplinary Study* (ed. R.F. Leggett). *The Royal Society of Canada Special Publications*, **12**, Royal Society of Canada, Ottawa, pp. 11–49.

Dreimanis, A. (1989) Tills: their genetic terminology and classification. In *Genetic Classification of Glacigenic Deposits* (eds R.P. Goldthwait and C.L. Matsch). A.A. Balkema, Rotterdam, pp. 17–83.

Drewry, D.J. (1986) *Glacial Geologic Processes*. Edward Arnold, London, 276 pp.

Dubois, A.D. (1984) On the climatic interpretation of the hydrogen isotope ratios in recent and fossil wood. *Bulletin de la Societé Géologique de Belgie*, **93**, 267–70.

Dubois, A.D. and Ferguson, D.K. (1985) The climatic history of pine in the Cairngorms based on radiocarbon dates and stable isotope analysis, with an account of the events leading up to its colonisation. *Review of Palaeobotany and Palynology*, **46**, 55–80.

Dubois, A.D. and Ferguson, D.K. (1988) Additional evidence for the climatic history of pine in the Cairngorms, Scotland, based on radiocarbon dates and tree ring D/H ratios. *Review of Palaeobotany and Palynology*, **54**, 181–5.

Duck, R.W. (1990) S.E.M. study of clastic fabrics preserved in calcareous concretions from the late-Devensian Errol Beds, Tayside. *Scottish Journal of Geology*, **26**, 33–9.

Duncan, W. (1794) Parish of Abernethy. In *The Statistical Account of Scotland*, Vol. 11 (ed. J. Sinclair). William Creech, Edinburgh, pp. 435–48.

Duplessy, J.-C., Delibrias, G., Turon, J.L., Pujol, C. and Duprat, J. (1981). Deglacial warming of the north-eastern Atlantic Ocean; correlation with the palaeoclimatic evolution of the European continent. *Palaeogeography, Palaeoclimatology, Palaeoecology*, **35**, 121–44.

Durno, S.E. (1956) Pollen analysis of peat deposits in Scotland. *Scottish Geographical Magazine*, **72**, 177–87.

Durno, S.E. (1957) Certain aspects of vegetational history in north-east Scotland. *Scottish Geographical Magazine*, **73**, 176–84.

Durno, S.E. (1958) Pollen analysis of peat deposits in eastern Scotland and Caithness. *Scottish Geographical Magazine*, **74**, 127–35.

Durno, S.E. (1959) Pollen analysis of peat deposits in the eastern Grampians. *Scottish Geographical Magazine*, **75**, 102–11.

Durno, S.E. (1961) Evidence regarding the rate of peat growth. *Journal of Ecology*, **49**, 347–51.

Dury, G.H. (1951) A 400-foot bench in south-eastern Warwickshire. *Proceedings of the Geologists' Association*, **62**, 167–73.

Dury, G.H. (1953) A glacial breach in the north-western Highlands. *Scottish Geographical Magazine*, **69**, 106–17.

Dyke, A.S. (1976) Tors and associated weathering phenomena, Somerset Island, District of Franklin. *Geological Survey of Canada, Paper*, **76–1B**, 209–16.

Dyke, A.S. (1983) Quaternary geology of Somerset Island, District of Franklin. *Geological Survey of Canada, Memoir*, **404**, 32 pp.

Eckford, R.J.A. (1952) Glacial phenomena in the West Linton–Dolphinton region. *Transactions of Edinburgh Geological Society*, **15**, 133–49.

Eckford, R.J.A. and Manson, W. (1927) Glacial phenomena around Loch Skene. *Proceedings of the Geologists' Association*, **38**, 508–10.

Eden, R.A., Carter, A.V.F. and McKeown, M.C. (1969) Submarine examination of Lower Carboniferous strata on inshore regions of the continental shelf of south-east Scotland. *Marine Geology*, **7**, 235–51.

Edmond, J.M. and Graham, J.D. (1977) Peterhead power station cooling water intake tunnel: an engineering case study. *Quarterly Journal of Engineering Geology*, **10**, 281–301.

Edwards, K.J. (1978) Palaeoenvironmental and archaeological investigations in the Howe of Cromar, Grampian region, Scotland. Unpublished PhD thesis, University of Aberdeen.

Edwards, K.J. (1979a) Earliest fossil evidence for *Koenigia islandica* – middle-Devensian interstadial pollen from Lewis, Scotland. *Journal of Biogeography*, **6**, 375–7.

Edwards, K.J. (1979b) Environmental impact in the prehistoric period. In *Early Man in the Scottish Landscape* (ed. L.M. Thoms). *Scottish Archaeological Forum,* **9**. Edinburgh University Press, Edinburgh, pp. 27–42.

Edwards, K.J. and Connell, E.R. (1981) Interglacial and interstadial sites in north-east Scotland. *Quaternary Newsletter*, **33**, 22–8.

Edwards, K.J. and and McIntosh, C.J. (1988) Improving the detection rate of cereal-type pollen grains from *Ulmus* decline and earlier deposits from Scotland. *Pollen et Spores*, **30**, 179–88.

Edwards, K.J. and Ralston, I. (1985) Postglacial hunter-gatherers and vegetational history in Scotland. *Proceedings of the Society of Antiquaries in Scotland (1984)*, **141**, 15–34.

Edwards, K.J. and Rowntree, K.M. (1980) Radiocarbon and palaeoenvironmental evidence for changing rates of erosion at a Flandrian stage site in Scotland. In *Timescales in Geomorphology* (eds R.A. Cullingford, D.A. Davidson and J. Lewin). John Wiley, Chichester, pp. 207–23.

Edwards, K.J. and Whittington, G. (1990) Palynological evidence for the growing of *Cannabis sativa* L. (hemp) in medieval and historical Scotland. *Transactions of the Institute of British Geographers*, NS **15**, 60–9.

Edwards, K.J., Caseldine, C.J. and Chester, D.K. (1976) Possible interstadial and interglacial pollen floras from Teindland, Scotland. *Nature*, **264**, 742–4.

Edwards, W. (1937) A Pleistocene strandline in the Vale of York. *Proceedings of the Yorkshire Geological Society*, **23**, 103–18.

Ehlers, J., Gibbard, P.L. and Rose, J. (1991) Glacial deposits of Britain and Europe: general overview. In *Glacial Deposits in Great Britain and Ireland* (eds J. Ehlers, P.L. Gibbard and J. Rose). A.A. Balkema, Rotterdam, pp. 493–501.

Eisbacher, G.H. (1979) Cliff collapse and rock avalanches (sturzstroms) in the Mackenzie Mountains, north-western Canada. *Canadian Geotechnical Journal*, **16**, 309–34.

Elder, S., McCall, R.J.S., Neaves, W.D. and Pringle, A.K. (1935) The drumlins of Glasgow. *Transactions of the Geological Society of Glasgow*, **19**, 285–7.

Ellis, J.B. (1975) Some deglacial landforms in Strathardle. *Scottish Field Studies Association Annual Report for 1974*, 27–36.

Elliot, C.J. (1984) Palynology and palaeoenvironments of the BGS Mains of Kilmaronock Borehole, Loch Lomond, Scotland. Unpublished MSc thesis, City of London Polytechnic and the Polytechnic of North London.

Elton, C. (1938) Notes on the ecological and natural history of Pabbay, and other islands in the Sound of Harris, Outer Hebrides. *Journal of Ecology*, **26**, 275–97.

Embleton, C. and King, C.A.M. (1968) *Glacial and Periglacial Geomorphology*. Edward Arnold, London, 608 pp.

Embleton, C. and King, C.A.M. (1975a) *Glacial Geomorphology*. Edward Arnold, London, 573 pp.

Embleton, C. and King, C.A.M. (1975b) *Periglacial Geomorphology*. Edward Arnold, London, 203 pp.

Emeleus, C.H. (1983) Tertiary igneous activity. In *Geology of Scotland*, 2nd edn (ed. G.Y. Craig). Scottish Academic Press, Edinburgh, pp. 357–97.

England, J. (1987) Glaciation and the evolution of the Canadian high arctic landscape. *Geology*, **15**, 419–24.

Erdtman, G. (1924) Studies in the micropalaeontology of postglacial deposits in northern Scotland and the Scotch Isles, with especial reference to the history of the woodlands. *Journal of the Linnaean Society (Botany)*, **46**, 449–504.

Erdtman, G. (1928) Studies in the postarctic history of the forests of north-western Europe. I. Investigations in the British Isles. *Geologiska Föreningens i Stockholm Förhandlingar*, **50**, 123–92.

Ergin, M., Harkness, D.D. and Walton, A. (1972) Glasgow University radiocarbon measurements V. *Radiocarbon*, **14**, 321–5.

Etheridge, R. (1876) Note on the fossils from the glacial deposits of the north-west coast of the Island of Lewis, Outer Hebrides. *Geological Magazine*, **3**, 552–5.

Evans, D.J.A. (1989) The nature of glaciotectonic structures and sediments at sub-polar glacier margins, north-west Ellesmere Island, Canada. *Geografiska Annaler*, **71A**, 113–23.

Evans, D.J.A. and Hansom, J.D. (1991) Scottish landform examples – 1. The Parallel Roads of Glen Roy. *Scottish Geographical Magazine*, **107**, 63–6.

Eyles, C.H., Eyles, N. and Miall, A.D. (1985) Models of glaciomarine sedimentation and their application to the interpretation of ancient glacial sequences. *Palaeogeography, Palaeoclimatology, Palaeoecology*, **51**, 15–84.

Eyles, N. (1979) Facies of supraglacial sedimentation on Icelandic and Alpine temperate glaciers. *Canadian Journal of Earth Sciences*, **16**, 1341–61.

Eyles, N. (1983) Modern Icelandic glaciers as depositional models for 'hummocky moraine' in the Scottish Highlands. In *Tills and Related Deposits* (eds E.B. Evenson, Ch. Schlüchter and J. Rabassa). A.A. Balkema, Rotterdam, pp. 47–59.

Eyles, N. and McCabe, A.M. (1989) The Late Devensian (<22,000 BP) Irish Sea basin: the sedimentary record of a collapsed ice-sheet margin. *Quaternary Science Reviews*, **8**, 307–51.

Eyles, N. and Sladen, J.A. (1981) Stratigraphy and geotechnical properties of weathered lodgement till in Northumberland, England. *Quarterly Journal of Engineering Geology*, **14**, 129–41.

Eyles, N., Sladen, J.A. and Gilroy, S. (1982) A depositional model for stratigraphic complexes and facies superimposition in lodgement tills. *Boreas*, **11**, 317–33.

Eyles, N., Eyles, C.H. and Miall, A.D. (1983) Lithofacies types and vertical profile models; an alternative approach to the description and environmental interpretation of glacial diamict and diamictite sequences. *Sedimentology*, **30**, 393–410.

Eyles, V.A. (1922) In *Memoirs of the Geological Survey. Summary of Progress of the Geological Survey of Great Britain and the Museum of Practical Geology for 1921*. HMSO, London, pp. 77–8.

Eyles, V.A. and Anderson, J.G.C. (1946) Brick clays of north-east Scotland. *Geological Survey of Britain Wartime Pamphlet*, No. 47, 89 pp.

Eyles, V.A., Simpson, J.B. and McGregor, A.G. (1949) *Geology of Central Ayrshire*, 2nd edn. Explanation of One-inch Sheet 14, Memoirs of the Geological Survey of Scotland. HMSO, Edinburgh, 160 pp.

Fahey, B.D. (1981) Origin and age of upland schist tors in central Otago, New Zealand. *New Zealand Journal of Geology and Geophysics*, **24**, 399–413.

Ferguson, R.I. (1981) Channel form and channel changes. In *British Rivers* (ed. J. Lewin). George Allen and Unwin, London, pp. 90–125.

Ferguson, R.I. and Werritty, A. (1983) Bar development and channel change in the gravelly River Feshie, Scotland. In *Modern and Ancient Fluvial Systems* (eds J. Collinson and J. Lewin). *Special Publication of the International Association of Sedimentologists*, **6**, 181–93.

Ferguson, W. (1850) Notice of the occurrence of chalk flints and Greensand fossils in Aberdeenshire. *Philosophical Magazine*, **37**, 430–8.

Ferguson, W. (1855) On the geological features of part of the district of Buchan, in Aberdeenshire, including notes on the occurrences of chalk-flints and Greensand. *Proceedings of the Philosophical Society of Glasgow*, **3**, 35–50.

Ferguson, W. (1857) Note on the chalk-flints and Greensand found in Aberdeenshire. *Quarterly Journal of the Geological Society of London*, **13**, 88–9.

Ferguson, W. (1877) On the occurrence of chalk flints and Greensand fossils in Aberdeenshire. *Transactions of the Edinburgh Geological Society*, **3**, 112–21.

Ferguson, W. (1893) On the occurrence of chalk flints and Greensand in the north-east district of Aberdeenshire. *Transactions of the Buchan Field Club*, **3**, 61–78.

Finlay, T.M. (1926) A töngsbergite boulder from the boulder-clay of Shetland. *Transactions of the Edinburgh Geological Society*, **12**, 180.

Finlay, T.M. (1930) The Old Red Sandstone of Shetland. *Transactions of the Royal Society of Edinburgh*, **56**, 671–94.

Firth, C.R. (1984) Raised shorelines and ice limits in the inner Moray Firth and Loch Ness areas, Scotland. Unpublished PhD thesis, Coventry (Lanchester) Polytechnic.

Firth, C.R. (1986) Isostatic depression during the Loch Lomond Stadial; preliminary evidence for the Great Glen, northern Scotland. *Quaternary Newsletter*, **48**, 1–9.

Firth, C.R. (1989a) Late Devensian raised shorelines and ice limits in the inner Moray Firth area, northern Scotland. *Boreas*, **18**, 5–21.

Firth, C.R. (1989b) A reappraisal of the Ardersier Readvance, inner Moray Firth. *Scottish Journal of Geology*, **25**, 249–61.

Firth, C.R. (1989c) Isostatic depression during the Loch Lomond Stadial (Younger Dryas): evidence from the inner Moray Firth, Scotland. *Geologiska Föreningens i Stockholm Förhandlingar*, **111**, 296–8.

Firth, C.R. (1990a) Late Devensian relative sea-level changes associated with the deglaciation of the Inverness Firth and Beauly Firth. In *Beauly to Nairn: Field Guide* (eds C.A. Auton, C.R. Firth and J.W. Merritt). Quaternary Research Association, Cambridge, pp. 5–9.

Firth, C.R. (1990b) An excursion to view the evidence of changes in relative sea-level during, and following, the deglaciation of the coastal lowlands bordering the northern shore of the Beauly Firth and the southern shore of the Inverness (Inner Moray) Firth. In *Beauly to Nairn: Field Guide* (eds C.A. Auton, C.R. Firth and J.W. Merritt). Quaternary Research Association, Cambridge, pp. 90–116.

Firth, C.R. and Haggart, B.A. (1989) Loch Lomond Stadial and Flandrian shorelines in the inner Moray Forth area, Scotland. *Journal of Quaternary Science*, **4**, 37–50.

Firth, C.R. and Haggart, B.A. (1990) The patterns of deglaciation and relative sea-level change around Muir of Ord and Beauly and the establishment of a Holocene sea-level curve for the Beauly Firth: an excursion guide. In *Beauly to Nairn: Field Guide* (eds C.A. Auton, C.R. Firth and J.W. Merritt). Quaternary Research Association, Cambridge, pp. 116–36.

Fisher, D.A., Reeh, N. and Langley, K. (1985) Objective reconstructions of the Late Wisconsin Laurentide ice sheet and the significance of deformable beds. *Géographie Physique et Quaternaire*, **39**, 229–38.

FitzPatrick, E.A. (1956) An indurated soil horizon formed by permafrost. *Soil Science*, **7**, 248–54.

FitzPatrick, E.A. (1958) An introduction to the periglacial geomorphology of Scotland. *Scottish Geographical Magazine*, **74**, 28–36.

FitzPatrick, E.A. (1963) Deeply weathered rock in Scotland, its occurrence, age and contribution to the soils. *Journal of Soil Science*, **14**, 33–43.

FitzPatrick, E.A. (1965) An interglacial soil at Teindland, Morayshire. *Nature*, **207**, 621–2.

FitzPatrick, E.A. (1969) Some aspects of soil evolution in north-east Scotland. *Soil Science*, **107**, 403–8.

FitzPatrick, E.A. (1972) The principal Tertiary and Pleistocene events in north-east Scotland. In *North-East Scotland Geographical Essays* (ed. C.M. Clapperton). Department of Geography, University of Aberdeen, Aberdeen, pp. 1–4.

FitzPatrick, E.A. (1975a) Particle size distribution and stone orientation patterns in some soils of north-east Scotland. In *Quaternary Studies in North-east Scotland* (ed. A.M.D. Gemmell). Department of Geography, University of Aberdeen, Aberdeen, pp. 49–60.

FitzPatrick, E.A. (1975b) Windy Hills. In *Aberdeen Field Excursion Guide*. Quaternary Research Association, Aberdeen, pp. 23–5.

FitzPatrick, E.A. (1987) Periglacial features in the soils of north-east Scotland. In *Periglacial Processes and Landforms in Britain and Ireland* (ed. J. Boardman). Cambridge University Press, Cambridge, pp. 153–62.

Fleet, H. (1938) Erosion surfaces in the Grampian Highlands of Scotland. Union Géographique Internationale – Commission Cartographie des surfaces d'Aplanissement Tertiaries, Rapport (for International Geographical Congress, Amsterdam 1938), pp. 91–4.

Fleming, J. (1847) Notes on the superficial strata of the neighbourhood of Edinburgh, concluded. *Proceedings of the Royal Society of Edinburgh*, **2**, 111–3.

Fleming, J. (1859) *The Lithology of Edinburgh* (Edited, with a Memoir, by the Rev. John Duns) William P. Kennedy, Edinburgh. Memoir pp. i–civ.; Lithology pp. 1–102.

Flenley, J.G. and Pearson, M.C. (1967) Pollen analysis of a peat from the island of Canna (Inner Hebrides). *New Phytologist*, **66**, 299–306.

Flett, J.S. (1898) On Scottish rocks containing orthite. *Geological Magazine*, **45**, 388–92.

Flett, J.S. and Read, H.H. (1921) Tertiary gravels of the Buchan district of Aberdeenshire. *Geological Magazine*, **58**, 215–25.

Flinn, D. (1964) Coastal and submarine features around the Shetland Islands. *Proceedings of the Geologists' Association*, **75**, 321–40.

Flinn, D. (1969) On the development of coastal profiles in the north of Scotland, Orkney and Shetland. *Scottish Journal of Geology*, **5**, 393–9.

Flinn, D. (1974) The coastline of Shetland. In *The Natural Environment of Shetland* (ed. R. Goodier). Nature Conservancy Council, Edinburgh, pp. 13–23.

Flinn, D. (1977) The erosion history of Shetland: a review. *Proceedings of the Geologists' Association*, **88**, 129–46.

Flinn, D. (1978a) The most recent glaciation of the Orkney–Shetland Channel and adjacent areas. *Scottish Journal of Geology*, **14**, 109–23.

Flinn, D. (1978b) The glaciation of the Outer Hebrides. *Geological Journal*, **13**, 195–9.

Flinn, D. (1980) The glaciation of the Outer Hebrides: reply. *Scottish Journal of Geology*, **16**, 85–6.

Flinn, D. (1981) A note on the glacial and late glacial history of Caithness. *Geological Journal*, **13**, 195–9.

Flinn, D. (1983) Glacial meltwater channels in the northern isles of Shetland. *Scottish Journal of Geology*, **19**, 311–20.

Flint, R.F. (1957) *Glacial and Pleistocene Geology*. John Wiley, New York and London, 553 pp.

Folk, R.L. and Patton, E.B. (1982) Buttressed

expansion of granite and development of grus in central Texas. *Zeitschrift für Geomorphologie*, N.F. **26**, 17–32.

Forbes, J.D. (1846) Notes on the topography and geology of the Cuchullin Hills in Skye, and on the traces of ancient glaciers which they present. *Edinburgh New Philosophical Journal*, **40**, 76–99.

Forsyth, I.H. and Chisholm, J.I. (1977) *The Geology of East Fife*. Explanation of the Fife portion of 'One-inch' Geological Sheet 41 and part of Sheet 49. Memoirs of the Geological Survey of Great Britain: Scotland. HMSO, Edinburgh, 284 pp.

Francis, E.H. (1975) Dunbar. In *The Geology of the Lothians and South East Scotland: An Excursion Guide* (eds G.Y. Craig and P. McL. D. Duff). Scottish Academic Press, Edinburgh, pp. 93–106.

Francis, E.H., Forsyth, I.H., Read, W.A. and Armstrong, M. (1970) *The Geology of the Stirling District*. Explanation of One-inch Geological Sheet 39. Memoirs of the Geological Survey of the Great Britain: Scotland. HMSO, Edinburgh, 357 pp.

Fraser, J. (1877) Report of the field excursion to the Nairn valley. *Transactions of the Inverness Scientific Society and Field Club*, **1**, 63–4.

Fraser, J. (1880) The recent formations and glacial phenomena of Strathnairn. *Transactions of the Inverness Scientific Society and Field Club*, **1**, 211–23. (Also in *Transactions of the Edinburgh Geological Society*, **4**, 55–66 (1881).)

Fraser, J. (1882a) The shell-bed at Clava. *Transactions of the Inverness Scientific Society and Field Club*, **2**, 169–76.

Fraser, J. (1882b) First notice of a post-Tertiary shell-bed, at Clava, in Nairnshire, indicating an arctic climate, and a sea-bed at a height of 500 feet. *Transactions of the Edinburgh Geological Society*, **4**, 136–42.

Fulton, R.J. (1989) *Quaternary Geology of Canada and Greenland*. Geological Survey of Canada, Geology of Canada, No.1 (Also Geological Society of America, The Geology of North America, V. K–1.) Geological Survey of Canada, Ottawa, 839 pp.

Futty, D.W. and Dry, F.T. (1977) *The Soils of the Country round Wick*. Sheets 110, 116 and part of 117. Memoirs of the Soil Survey of Great Britain: Scotland. HMSO, Edinburgh, 287 p.

Gailey, R.A. (1959) Glasgow University Expedition to North Rona. *Scottish Geographical Magazine*, **75**, 48–50.

Galloway, R.W. (1958) Periglacial phenomena in Scotland. Unpublished PhD thesis, University of Edinburgh.

Galloway, R.W. (1961a) Solifluction in Scotland. *Scottish Geographical Magazine*, **77**, 75–87.

Galloway, R.W. (1961b) Periglacial phenomena in Scotland. *Geografiska Annaler*, **43**, 348–53.

Galloway, R.W. (1961c) Ice wedges and involutions in Scotland. *Biuletyn Peryglacjalny*, **10**, 169–93.

Gaunt, G.D. (1970) A temporary section across the Escrick moraine at Wheldrake, East Yorkshire. *Journal of Earth Science, Leeds*, **8**, 163–70.

Gaunt, G.D. (1981) Quaternary history of the southern part of the Vale of York. In *The Quaternary in Britain* (eds J. Neale and J. Flenley). Pergamon Press, Oxford, pp. 82–97.

Gear, A.J. and Huntley, B (1991) Rapid changes in the range limits of Scots Pine 4000 years ago. *Science*, **251**, 544–7.

Geikie, A. (1863a) On the glacial drift of Scotland. *Transactions of the Geological Society of Glasgow*, **1**, 1–190.

Geikie, A. (1863b) *The Geology of Eastern Berwickshire*. Map 34. Memoirs of the Geological Survey of Great Britain (Scotland), HMSO, London, 58 pp.

Geikie, A. (1865) *The Scenery of Scotland Viewed in Connection with its Physical Geology*. Macmillan, London, 360 pp.

Geikie, A. (1874) *Explanation of Sheet 23. Lanarkshire: Central Districts*. Memoirs of the Geological Survey of Scotland. HMSO, Edinburgh, 107 pp.

Geikie, A. (1877) The glacial geology of Orkney and Shetland. *Nature*, **16**, 414–6.

Geikie, A. (1900) *The Geology of Central and Western Fife and Kinross*. Description of Sheet 40 and parts of Sheets 32 and 48. Memoirs of the Geological Survey of Scotland, HMSO, Glasgow, 284 pp.

Geikie, A. (1901) *The Scenery of Scotland Viewed in Connection with its Physical Geology*, 3rd edn. Macmillan, London, 540 pp.

Geikie, A. (1902) *The Geology of Eastern Fife*. Description of Sheet 41 and parts of Sheets 40, 48 and 49. Memoirs of the Geological Survey of Scotland, HMSO, Glasgow, 421 pp.

Geikie, A. (1903) *Textbook of Geology*, Vol. 1. Macmillan, London, 702 pp.

Geikie, A., Geikie, J. and Peach, B.N. (1869) *Ayrshire: Southern District*. Explanation of Sheet 14. Memoirs of the Geological Survey of Scotland, HMSO, Edinburgh, 27 pp.

Geikie, A., Peach, B.N., Jack, R.L., Skae, H. and Horne, J. (1871) *Explanation of Sheet 15, Dumfriesshire (north-west part); Lanarkshire (south part); Ayrshire (south-east part)*. Memoirs of the Geological Survey of Scotland, HMSO, Edinburgh, 43 pp.

Geikie, J. (1873) On the glacial phenomena of the Long Island or Outer Hebrides. First Paper. *Quarterly Journal of the Geological Society of London*, **29**, 532–45.

Geikie, J. (1874) *The Great Ice Age and its Relation to the Antiquity of Man*, 1st Edn. W. Isbister, London, 575 pp.

Geikie, J. (1877) *The Great Ice Age and its Relation to the Antiquity of Man*, 2nd edn. Daldy, Isbister and Co., London, 624 pp.

Geikie, J. (1878) On the glacial phenomena of the Long Island or Outer Hebrides. Second Paper. *Quarterly Journal of the Geological Society of London*, **34**, 819–70.

Geikie, J. (1881) *Prehistoric Europe. A Geological Sketch*. Edward Stanford, London, 592 pp.

Geikie, J. (1894) *The Great Ice Age and its Relation to the Antiquity of Man*, 3rd edn. Edward Stanford, London, 850 pp.

Gellatly, A.F., Gordon, J.E., Whalley, W.B. and Hansom, J.D. (1988) Thermal regime and geomorphology of plateau ice caps in northern Norway: observations and implications. *Geology*, **16**, 983–6.

Gemmell, A.M.D. (1973) The deglaciation of the Island of Arran, Scotland. *Transactions of the Institute of British Geographers*, **59**, 25–39.

Gemmell, A.M.D. (1975) The Kippet Hills. In *Aberdeen Field Excursion Guide*. Quaternary Research Association, Aberdeen, pp. 14–19.

Gemmell, A.M.D. and Kesel, R.H. (1979) Developments in the study of the Buchan flint deposits. In *Early Man in the Scottish Landscape* (ed. L.M. Thoms). *Scottish Archaeological Forum*, 9. Edinburgh University Press, Edinburgh, pp. 66–77.

Gemmell, A.M.D. and Kesel, R.H. (1982) The 'Pliocene' gravels of Buchan: a reappraisal: reply. *Scottish Journal of Geology*, **18**, 333–5.

Gemmell, A.M.D. and Ralston, I.B.M. (1984) Some recent discoveries of ice-wedge cast networks in north-east Scotland. *Scottish Journal of Geology*, **20**, 115–8.

Gemmell, A.M.D. and Ralston, I.B.M. (1985) Ice wedge polygons in north-east Scotland: a reply. *Scottish Journal of Geology*, **21**, 109–11.

George, T.N. (1958) The geology and geomorphology of the Glasgow district. In *The Glasgow Region* (eds R. Miller and J. Tivy). Handbook prepared for the meeting of the British Association, Glasgow. T and A. Constable, Edinburgh, pp. 17–61.

George, T.N. (1965) The geological growth of Scotland. In *The Geology of Scotland* (ed. G.Y. Craig). Oliver and Boyd, Edinburgh, pp. 1–48.

George, T.N. (1966) Geomorphic evolution in Hebridean Scotland. *Scottish Journal of Geology*, **2**, 1–34.

Gibb, A.W. (1905) On the occurrence of pebbles of white chalk in Aberdeenshire clay. *Report of the British Association for 1904*, 573.

Gibb, A.W. (1909) On the relation of the Don to the Avon at Inchrory, Banffshire. *Transactions of the Edinburgh Geological Society*, **9**, 227–9.

Gjessing, J. (1965) On plastic scouring and subglacial erosion. *Norsk Geografisk Tidsskrift*, **20**, 1–37.

Glentworth, R. and Muir, J.W. (1963) *The Soils of the Country round Aberdeen, Inverurie and Fraserburgh*. Sheets 77, 76 and 87/97. Memoirs of the Soil Survey of Great Britain, HMSO, Edinburgh, 371 pp.

Glentworth, R., Mitchell, W.A. and Mitchell, B.D. (1964) The red glacial drifts of north-east Scotland. *Clay Minerals Bulletin*, **5**, 373–81.

Godard, A. (1959) Contemporary periglacial phenomena in western Scotland. *Scottish Geographical Magazine*, **75**, 55.

Godard, A. (1961) L'efficacité de l'érosion glaciaire en Écosse du Nord. *Revue de Géomorphologie Dynamique*, **12**, 32–42.

Godard, A. (1965) *Recherches de Géomorphologie en Écosse du Nord-Ouest*. Université de Strasbourg, Publications de la Faculté des Lettres, Fondation Baulig, Tome 1, 701 pp.

Godard, A. (1989) Les vestiges des manteaux d'altération sur les socles des hautes latitudes: identification, signification. *Zeitschrift für Geomorphologie*, NF Supplementband, **72**, 1–20.

Godwin, H. (1943) Coastal peat beds of the British Isles and North Sea. *Journal of Ecology*, **31**, 199–247.

Godwin, H. (1975) *The History of the British Flora*. Cambridge University Press, Cambridge, 541 pp.

Godwin, H. and Switsur, V.R. (1966) Cambridge

University natural radiocarbon measurements VIII. *Radiocarbon*, **8**, 390–400.
Godwin, H. and Willis, E.H. (1962) Cambridge University natural radiocarbon measurements V. *Radiocarbon*, **4**, 52–70.
Godwin, H. and Willis, E.H. (1964) Cambridge University natural radiocarbon measurements VI. *Radiocarbon*, **6**, 116–37.
Godwin, H., Walker, D. and Willis, E.H. (1957) Radiocarbon dating and post-glacial vegetational history: Scaleby Moss. *Proceedings of the Royal Society of London*, **B 147**, 353–66.
Godwin, H., Willis, E.H. and Switsur, V.R. (1965) Cambridge University natural radiocarbon measurements VII. *Radiocarbon*, **7**, 205–12.
Goldthwait, R.P. (1976) Frost-sorted patterned ground: a review. *Quaternary Research*, **6**, 27–35.
Goodchild, J.G. (1896) Glacial furrows. *The Glacialists' Magazine*, **4**, 1–7.
Goodier, R. and Ball, D.F. (1975) Ward Hill, Hoy, Orkney: patterned ground features and their origins. In *The Natural Environment of Orkney* (ed. R. Goodier). Nature Conservancy Council, Edinburgh, pp. 47–56.
Goodlet, G.A. (1964) The kamiform deposits near Carstairs, Lanarkshire. *Bulletin of the Geological Survey of Great Britain*, **21**, 175–96.
Goodlet, G.A. (1970) Sands and gravels of the southern counties of Scotland. *Report of the Institute of Geological Sciences*, No. 70/4, 81 pp.
Gordon, A.D. and Birks, H.J.B. (1972) Numerical methods in Quaternary palaeoecology I: zonation of pollen diagrams. *New Phytologist*, **71**, 961–79.
Gordon, J.E. (1977) Morphometry of cirques in the Kintail–Affric–Cannich area of north-west Scotland. *Geografiska Annaler*, **59A**, 177–94.
Gordon, J.E. (1979) Reconstructed Pleistocene ice-sheet temperatures and glacial erosion in northern Scotland. *Journal of Glaciology*, **22**, 331–44.
Gordon, J.E. (1981) Ice-scoured topography and its relationships to bedrock structure and ice movement in parts of northern Scotland and West Greenland. *Geografiska Annaler*, **63A**, 55–65.
Gordon, J.E., Birnie, R.V. and Timmis, R. (1978) A major rockfall and debris slide on the Lyell Glacier, South Georgia. *Arctic and Alpine Research*, **10**, 49–60.
Gordon, S. (1943) Perpetual snowbeds of the Scottish hills. *The Field*, **181**, 88–9.
Graham, D.K. (1990) The fauna of the Clava Shelly Clay and Shelly Till. In *Beauly to Nairn: Field Guide* (eds C.A. Auton, C.R. Firth and J.W. Merritt). Quaternary Research Association, Cambridge, pp. 20–3.
Graham, D.K. and Gregory, D.M. (1981) A revision of C.F. Davidson's arctic fauna from Inchcoonans Claypit, Errol, held by the Museum and Art Gallery, Perth. *Scottish Journal of Geology*, **17**, 215–22.
Graham, D.K., Harland, R., Gregory, D.M., Long, D. and Morton A.C. (1990) The biostratigraphy and chronostratigraphy of BGS Borehole 78/4, North Minch. *Scottish Journal of Geology*, **26**, 65–75.
Gray, J.M. (1972) The Inter-, Late- and Post-glacial shorelines, and ice-limits of Lorn and eastern Mull. Unpublished PhD thesis, University of Edinburgh.
Gray, J.M. (1974a) The Main Rock Platform of the Firth of Lorn, western Scotland. *Transactions of the Institute of British Geographers*, **61**, 81–99.
Gray, J.M. (1974b) Lateglacial and postglacial shorelines in western Scotland. *Boreas*, **3**, 129–38.
Gray, J.M. (1975a) The Loch Lomond Readvance and contemporaneous sea-levels in Loch Etive and neighbouring areas of western Scotland. *Proceedings of the Geologists' Association*, **86**, 227–38.
Gray, J.M. (1975b) Lateglacial, *in situ* barnacles from Glen Cruitten Quarry, Oban. *Quaternary Newsletter*, **15**, 1–2.
Gray, J.M. (1978a) Low-level shore platforms in the south-west Scottish Highlands: altitude, age and correlation. *Transactions of the Institute of British Geographers*, NS, **3**, 151–64.
Gray, J.M. (1978b) Report of a short field meeting at Oban. *Quaternary Newsletter*, **26**, 14–6.
Gray, J.M. (1981) P-forms from the Isle of Mull. *Scottish Journal of Geology*, **17**, 39–47.
Gray, J.M. (1982a) The last glaciers (Loch Lomond Advance) in Snowdonia, North Wales. *Geological Journal*, **17**, 111–33.
Gray, J.M. (1982b) Unweathered, glaciated bedrock on an exposed lake bed in Wales. *Journal of Glaciology*, **28**, 483–97.
Gray, J.M. (1983) The measurement of shoreline altitudes in areas affected by glacio-isostasy, with particular reference to Scotland. In *Shorelines and Isostasy* (eds D.E. Smith and

A.G. Dawson). Academic Press, London, pp. 97–128.

Gray, J.M. (1984) A p-form site on the Isle of Islay, Scottish Inner Hebrides. *Quaternary Newsletter*, **42**, 17–20.

Gray, J.M. (1985) *Glacio-isostatic Shoreline Development in Scotland: An Overview*. Department of Geography and Earth Science, Queen Mary College, University of London. Occasional Paper, No. 24, 61 pp.

Gray, J.M. (1987) Age of the Main Rock Platform, western Scotland. INQUA XII Congress, Ottawa, Canada, 1987. *Programme and Abstracts*, 177.

Gray, J.M. (1989) Distribution and development of the Main Rock Platform, western Scotland: comment. *Scottish Journal of Geology*, **25**, 227–31.

Gray, J.M. (1991) Glaciofluvial landforms. In *Glacial Deposits in Great Britain and Ireland* (eds J. Ehlers, P.L. Gibbard and J. Rose). A.A. Balkema, Rotterdam, pp. 443–54.

Gray, J.M. and Brooks, C.L. (1972) The Loch Lomond Readvance moraines of Mull and Menteith. *Scottish Journal of Geology*, **8**, 95–103.

Gray, J.M. and Coxon, P. (1991) The Loch Lomond Stadial glaciation in Britain and Ireland. In *Glacial Deposits in Great Britain and Ireland* (eds J. Ehlers, P.L. Gibbard and J. Rose). A.A. Balkema, Rotterdam, pp. 89–105.

Gray, J.M. and Ivanovich, M. (1988) Age of the Main Rock Platform, western Scotland. *Palaeogeography, Palaeoclimatology, Palaeoecology*, **68**, 337–45.

Gray, J.M. and Lowe, J.J. (1977a) Introduction. In *Studies in the Scottish Lateglacial Environment* (eds J.M. Gray and J.J. Lowe). Pergamon Press, Oxford, pp. xi–xiii.

Gray, J.M. and Lowe, J.J. (1977b) The Scottish Lateglacial environment: a synthesis. In *Studies in the Scottish Lateglacial Environment* (eds J.M. Gray and J.J. Lowe). Pergamon Press, Oxford, pp. 163–81.

Gray, J.M. and Lowe, J.J. (1982) Problems in the interpretation of small-scale erosional forms on glaciated bedrock surfaces: examples from Snowdonia, North Wales. *Proceedings of the Geologists' Association*, **93**, 403–14.

Gray, J.M. and Sutherland, D.G. (1977) The 'Oban–Ford moraine': a reappraisal. In *Studies in the Scottish Lateglacial Environment* (eds J.M. Gray and J.J. Lowe). Pergamon Press, Oxford, pp. 33–44.

Green, F.H.W. (1968) Persistent snowbeds in the western Cairngorms. *Weather*, **23**, 206–8.

Gregory, D. (1980) Assemblages of the ostracod *Cytheropteron montrosiense* Brady, Crosskey and Robertson from offshore Devensian deposits in the Minch, west of Scotland. *Scottish Journal of Geology*, **16**, 281–7.

Gregory, J.W. (1912) The relations of kames and eskers. *Geographical Journal*, **40**, 169–75.

Gregory, J.W. (1913) The Polmont kame and on the classification of Scottish kames. *Transactions of the Geological Society of Glasgow*, **14**, 199–218.

Gregory, J.W. (1915a) The kames of Carstairs. *Scottish Geographical Magazine*, **31**, 465–76.

Gregory, J.W. (1915b) The geology of the Glasgow district. I. General account of the district. *Proceedings of the Geologists' Association*, **26**, 151–65.

Gregory, J.W. (1915c) The geology of the Glasgow district. XI. The kames of Carstairs. *Proceedings of the Geologists' Association*, **26**, 187–8.

Gregory, J.W. (1926) The Scottish kames and their evidence on the glaciation of Scotland. *Transactions of the Royal Society of Edinburgh*, **54**, 395–432.

Gregory, J.W. (1927) The moraines, boulder clay and glacial sequence of south-western Scotland. *Transactions of the Geological Society of Glasgow*, **17**, 354–76.

Gregory, J.W. (1928) The geology of Loch Lomond. *Transactions of the Geological Society of Glasgow*, **18**, 301–23.

Greig, D.C. (1981) Ice-wedge cast network in eastern Berwickshire. *Scottish Journal of Geology*, **17**, 119–22.

Gunson, A.R. (1975) The vegetation history of north-east Scotland. In *Quaternary Studies in North-east Scotland* (ed. A.M.D. Gemmell). Department of Geography, University of Aberdeen, Aberdeen, pp. 61–72.

Gustavson, T.C., Ashley, G.H. and Boothroyd, J.C. (1975) Depositional sequences in glaciolacustrine deltas. In *Glaciofluvial and Glaciolacustrine Sedimentation* (eds A.V. Jopling and B.C. McDonald). Society of Economic Palaeontologists and Mineralogists, Special Publication, No. 23, pp. 264–80.

Gwynne, D., Milner, C. and Hornung, M. (1974) The vegetation and soils of Hirta. In *Island Survivors: the Ecology of the Soay Sheep of St Kilda* (eds P.A. Jewell, C. Milner and J.M. Boyd). Athlone Press, London, pp. 36–7.

Haggart, B.A. (1978) A pollen and stratigraphical investigation into a peat deposit in St Michael's Wood near Leuchars, Fife. Unpublished MA dissertation, University of St Andrews.

Haggart, B.A. (1982) Flandrian sea-level changes in the Moray Firth area. Unpublished PhD thesis, University of Durham.

Haggart, B.A. (1986) Relative sea-level change in the Beauly Firth Scotland. *Boreas*, **15**, 191–207.

Haggart, B.A. (1987) Relative sea-level changes in the Moray Firth area, Scotland. In *Sea-level Changes* (eds M.J. Tooley and I. Shennan). Blackwell Scientific, Oxford, pp. 67–108.

Haggart, B.A. (1988a) A review of radiocarbon dates on peat and wood from Holocene coastal sedimentary sequences in Scotland. *Scottish Journal of Geology*, **24**, 125–44.

Haggart, B.A. (1988b) The stratigraphy, depositional environment and dating of a possible tidal surge deposit in the Beauly Firth area, north-east Scotland. *Palaeogeography, Palaeoclimatology, Palaeoecology*, **66**, 215–30.

Haggart, B.A. (1989) Variations in the pattern and rate of isostatic uplift indicated by a comparison of Holocene sea-level curves from Scotland. *Journal of Quaternary Science*, **4**, 67–76.

Hall, A.M. (1982) The 'Pliocene' gravels of Buchan: a reappraisal: discussion. *Scottish Journal of Geology*, **18**, 336–8.

Hall, A.M. (1983) Deep weathering and landform evolution in north-east scotland. Unpublished PhD thesis, University of St Andrews.

Hall, A.M. (1984a) *Buchan Field Guide*. Quaternary Research Association, Cambridge, 120 pp.

Hall, A.M. (1984b) Introduction. In *Buchan Field Guide* (ed. A.M. Hall). Quaternary Research Association, Cambridge, pp. 1–26.

Hall, A.M. (1984c) Central Buchan. In *Buchan Field Guide* (ed. A.M. Hall). Quaternary Research Association, Cambridge, pp. 27–45.

Hall, A.M. (1985) Cenozoic weathering covers in Buchan, Scotland and their significance. *Nature*, **315**, 392–5.

Hall, A.M. (1986) Deep weathering patterns in north-east Scotland and their geomorphological significance. *Zeitschrift für Geomorphologie*, NF **30**, 407–22.

Hall, A.M. (1987) Weathering and relief development in Buchan, Scotland. In *International Geomorphology 1986*, Part II, (ed. V. Gardiner). John Wiley, Chichester, pp. 991–1005.

Hall, A.M. (1989a) Pre-Late Devensian coastal rock platforms around Dunbar. *Scottish Journal of Geology*, **25**, 361–5.

Hall, A.M. (1989b) Location and excavation of the 'Indigo Boulder Clay' at Bellscamphie, Ellon, Grampian Region. Nature Conservancy Council, Peterborough, CSD Report No. 975, 26 pp.

Hall, A.M. (1991) Pre-Quaternary landscape evolution in the Scottish Highlands. *Transactions of the Royal Society of Edinburgh: Earth Sciences*, **82**, 1–26.

Hall, A.M. and Bent, A.J.A. (1990) The limits of the last British ice sheet in northern Scotland and the adjacent shelf. *Quaternary Newsletter*, **61**, 2–12.

Hall, A.M. and Connell, E.R. (1982) Recent excavations at the Greensand locality of Moreseat, Grampian Region. *Scottish Journal of Geology*, **18**, 291–6.

Hall, A.M. and Connell, E.R. (1986) A preliminary report on the Quaternary sediments at Leys gravel pit, Buchan, Scotland. *Quaternary Newsletter*, **48**, 17–28.

Hall, A.M. and Connell, E.R. (1991) The glacial deposits of Buchan, northeast Scotland. In *Glacial Deposits in Great Britian and Ireland* (eds J. Ehlers, P.L. Gibbard and J. Rose). A.A. Balkema, Rotterdam, pp. 129–36.

Hall, A.M. and Jarvis, J. (1989) A preliminary report on the Late Devensian glaciomarine deposits around St Fergus, Grampian Region. *Quaternary Newsletter*, **59**, 5–7.

Hall, A.M. and Mellor, A. (1988) The characteristics and significance of deep weathering in the Gaick area, Grampian Highlands, Scotland. *Geografiska Annaler*, **70A**, 309–14.

Hall, A.M. and Sugden, D.E. (1987) Limited modification of mid-latitude landscapes by ice sheets. *Earth Surface Processes and Landforms*, **12**, 531–42.

Hall, A.M. and Whittington, G. (1989) Late Devensian glaciation of southern Caithness. *Scottish Journal of Geology*, **25**, 307–24.

Hall, A.M., Mellor, A.M. and Wilson, M.J. (1989a) The clay mineralogy and age of deeply weathered rocks in north-east Scotland. *Zeitschrift für Geomorphologie*, NF, Supplementband, **72**, 97–108.

Hall, A.M., Jarvis, J. and Duck, R. (1989b) Survey of Quaternary deposits adjacent to Kirkhill Quarry. Nature Conservancy Council, Peterborough, CSD Report 924, 75 pp.

Hall, Sir J. (1815) On the revolutions of the Earth's surface. *Transactions of the Royal Society of Edinburgh*, **7**, 139–211.

Hallet, B. and Anderson, R.S. (1980) Detailed glacial geomorphology of a proglacial bedrock area at Castleguard Glacier, Alberta, Canada. *Zeitschrift für Gletscherkunde und Glazialgeologie*, **16**, 171–84.

Hansom, J.D. (1983) Shore-platform development in the South Shetland Islands, Antarctica. *Marine Geology*, **53**, 211–29.

Hansom, J.D. (1991) Holocene coastal development in the Dornoch Firth. In *Late Quaternary Coastal Evolution in the Inner Moray Firth. Field Guide* (eds C.R. Firth and B.A. Haggart). West London Press, London, pp. 45–54.

Harding, R.R., Merriman, R.J. and Nancarrow, P.H.A. (1984) St Kilda: an illustrated account of the geology. *Report of the British Geological Survey*, **16(7)**, 46 pp.

Harker, A. (1899a) Glaciated valleys in the Cuillins, Skye. *Geological Magazine*, **46**, 196–9.

Harker, A. (1899b) Notes on subaerial erosion in the Isle of Skye. *Geological Magazine*, **46**, 485–91.

Harker, A. (1901) Ice erosion in the Cuillin Hills, Skye. *Transactions of the Royal Society of Edinburgh*, **40**, 76–99.

Harker, A. (1908) *The Geology of the Small Isles of Inverness-shire*. Sheet 60. Memoirs of the Geological Survey of Scotland, HMSO, Glasgow, 210 pp.

Harkness, D.D. (1981) Scottish Universities Research and Reactor Centre, radiocarbon measurements IV. *Radiocarbon*, **23**, 252–304.

Harkness, D.D. (1983) The extent of natural ^{14}C deficiency in the coastal environment of the United Kingdom. In *^{14}C and Archaeology. Symposium held at Groningen, August 1981* (eds W.C. Mook and H.T. Waterbolk). PACT, **8**, 351–64.

Harkness, D.D. (1990) Radiocarbon dating of the Odhar Peat. In *Beauly to Nairn: Field Guide* (eds C.A. Auton, C.R. Firth and J.W. Merritt). Quaternary Research Association, Cambridge, pp. 69–70.

Harkness, D.D. and Wilson, H.W. (1979) Scottish Universities Research and Reactor Centre radiocarbon measurements III. *Radiocarbon*, **21**, 203–56.

Harris, A.L. and Peacock, J.D. (1969) Sand and gravel resources of the inner Moray Firth. *Report of the Institute of Geological Sciences*, **69/9**, 18 pp.

Harry, W.T. (1965) The form of the Cairngorm pluton. *Scottish Journal of Geology*, **1**, 1–8.

Harvey, A.M. (1986) Geomorphic effects of a 100 year storm in the Howgill Fells, northwest England. *Zeitschrift für Geomorphologie*, NF, **30**, 71–91.

Harvey, A.M., Oldfield, F., Baron, A.F. and Pearson, G.W. (1981) Lichens, soil development and the age of Holocene valley-floor landforms: Howgill Fells, Cumbria. *Geografiska Annaler*, **66A**, 353–66.

Hawksworth, D.L. (1970) Studies on the peat deposits of the island of Foula, Shetland. *Transactions and Proceedings of the Botanical Society of Edinburgh*, **40**, 576–91.

Haworth, E.Y. (1969) The diatoms of a sediment core from Blea Tarn, Langdale. *Journal of Ecology*, **57**, 429–41.

Haworth, E.Y. (1976) Two late-glacial (Late Devensian) diatom assemblage profiles from northern Scotland. *New Phytologist*, **77**, 227–56.

Haynes, V.M. (1968) The influence of glacial erosion and rock structure on corries in Scotland. *Geografiska Annaler*, **50A**, 221–34.

Haynes, V.M. (1969) The relative influence of rock properties and erosion processes in the production of glaciated landforms, with especial reference to corries in Scotland. Unpublished PhD thesis, University of Cambridge.

Haynes, V.M. (1977a) The modification of valley patterns by ice-sheet activity. *Geografiska Annaler*, **59A**, 195–207.

Haynes, V.M. (1977b) Landslip associated with glacier ice. *Scottish Journal of Geology*, **13**, 337–8.

Haynes, V.M. (1983) Scotland's landforms. In *Scotland. A New Study* (ed. C.M. Clapperton). David and Charles, Newton Abbot, pp. 28–63.

Hays, J.D., Imbrie, J. and Shackleton, N.J. (1976) Variations in the Earth's orbit: pacemaker of the ice ages. *Science*, **194**, 1121–32.

Heddle, M.F. (1880) The geognosy and mineralogy of Scotland. *Mineralogical Magazine*, **3**, 147–77.

Heijnis, H. (1990) Dating of the Odhar Peat at the Allt Odhar site by the Uranium series disequilibrium dating method. In *Beauly to Nairn: Field Guide* (eds C.A. Auton, C.R. Firth and J.W. Merritt). Quaternary Research Association, Cambridge, pp. 72–4.

Helland, A. (1879) Ueber die Vergletscherung

der Färöer, sowie der Shetland–Orkney–Inseln. *Zeitschrift der Deutschen Geologischen Gesellschaft*, **31**, 716–55.

Henderson, J. (1872) On Corstorphine Hill, near Edinburgh. *Transactions of the Edinburgh Geological Society*, **2**, 29–33.

Henderson, J. (1874) On some sections of boulder clay, peat and stratified beds, exposed in a quarry recently opened at Redhill, Slateford, near Edinburgh. *Transactions of the Edinburgh Geological Society*, **2**, 391–5.

Heslop Harrison, J.W. and Blackburn, K.B. (1946) The occurrence of a nut of *Trapa natans* L. in the Outer Hebrides, with some account of the peat bogs adjoining the loch in which the discovery was made. *New Phytologist*, **45**, 124–31.

Hibbert, F.A. and Switsur, R. (1976) Radiocarbon dating of Flandrian pollen zones in Wales and northern England. *New Phytologist*, **77**, 793–807.

Hibbert, F.A., Switsur, V.R. and West, R.G. (1971) Radiocarbon dating of the Flandrian pollen zones at Red Moss, Lancashire. *Proceedings of the Royal Society of London*, **B177**, 161–76.

Hibbert, S. (1831) On the direction of the diluvial wave in Shetland. *Edinburgh Journal of Science*, NS, **4**, 85–91.

Hickling, G. (1908) The Old Red Sandstone of Forfarshire. *Geological Magazine*, Decade 5, **5**, 396–408.

Hill, W. (1915) Chalk boulders from Aberdeen and fragments of chalk from the sea floor off the Scottish coast. *Proceedings of the Royal Society of Edinburgh*, **35**, 263–96.

Hills, R.C. (1969) Comparative weathering of granite and quarzite in a periglacial environment. *Geografiska Annaler*, **51A**, 46–7.

Hinxman, L.W. (1896) *Explanation of Sheet 75. West Aberdeenshire, Banffshire, parts of Elgin and Inverness*. Memoirs of the Geological Survey of Scotland, HMSO, Edinburgh, 48 pp.

Hinxman, L.W. (1901) The River Spey. *Scottish Geographical Magazine* **17**, 185–93.

Hinxman, L.W. and Anderson, E.M. (1915) *The Geology of Mid-Strathspey and Strathdearn, including the Country between Kingussie and Grantown*. Explanation of Sheet 74. Memoirs of the Geological Survey of Scotland, HMSO, Edinburgh, 97 pp.

Hinxman, L.W. and Wilson, J.S.G. (1902) *The Geology of Lower Strathspey*. Explanation of Sheet 85. Memoirs of the Geological Survey: Scotland, HMSO, Glasgow, 91 pp.

Hirons, K.R. and Edwards, K.J. (1986) Events at and around the first and second *Ulmus* decline: palaeoecological investigations in Co Tyrone, Northern Ireland. *New Phytologist*, **104**, 131–53.

Hirons, K.R. and Edwards, K.J. (1990) Pollen and related studies at Kinloch, Isle of Rhum, Scotland, with particular reference to possible early human impacts on vegetation. *New Phytologist*, **116**, 715–27.

Hodgson, D.M. (1982) Hummocky and fluted moraines in part of NW Scotland. Unpublished PhD thesis, University of Edinburgh.

Hodgson, D.M. (1986) A study of fluted moraines in the Torridon area, NW Scotland. *Journal of Quaternary Science*, **1**, 109–18.

Hodgson, D.M. (1987) Coire a'Cheud Chnoic. In *Wester Ross Field Guide* (eds C.K. Ballantyne and D.G. Sutherland). Quaternary Research Association, Cambridge, pp. 127–34.

Holden, W.G. (1977a) The glaciation of Central Ayrshire. Unpublished PhD thesis, University of Glasgow.

Holden, W.G. (1977b) Afton Lodge. In *Western Scotland I* (ed. R.J. Price). INQUA X Congress 1977. Guidebook for Excursion A12. Geo Abstracts, Norwich, pp. 12–3.

Holden, W.G. (1977c) Nith Bridge. In *Western Scotland I* (ed. R.J. Price). INQUA X Congress 1977, Guidebook for Excursion A12. Geo Abstracts, Norwich, pp. 13–5.

Holden, W.G. (1977d) Greenock Mains. In *Western Scotland I* (ed. R.J. Price). INQUA X Congress 1977. Guidebook for Excursion A12. Geo Abstracts, Norwich, pp. 15–6.

Holden, W.G. and Jardine, W.G. (1980) Greenock Mains and Nith Bridge. In *Field Guide to the Glasgow Region* (ed. W.G. Jardine). Quaternary Research Association, Cambridge, pp. 18–21.

Holdsworth, G. (1973) Ice deformation and moraine formation at the margin of an ice cap adjacent to a proglacial lake. In *Research in Polar and Alpine Geomorphology* (eds B.D. Fahey and R.D. Thompson). Geo Abstracts, Norwich, pp. 187–99.

Hollingworth, S.E. (1931) The glaciation of western Edenside and adjoining areas, and the drumlins of Edenside and the Solway basin. *Quarterly Journal of the Geological Society of London*, **87**, 281–359.

Holmes, G. (1984) Rock slope failure in parts of the Scottish Highlands. Unpublished PhD thesis, University of Edinburgh.

Holmund, P. (1991) Cirques at low altitudes need not necessarily have been cut by small glaciers. *Geografiska Annaler,* **73A**, 9–16.

Hoppe, G. (1965) Submarine peat in the Shetland Islands. *Geografiska Annaler,* **47A**, 195–203.

Hoppe, G. (1974) The glacial history of the Shetland Islands. *Institute of British Geographers. Special Publication*, **7**, 197–210.

Horne, J., Robertson, D., Jamieson, T.F., Fraser, J., Kendal, P.F. and Bell, D. (1894) The character of the high-level shell-bearing deposits at Clava, Chapelhall and other locations. *Report of the British Association for the Advancement of Science for 1893*, 483–514.

Horne, J., Robertson, D., Jamieson, T.F., Fraser, J., Kendall, P.F. and Bell, D. (1895) The character of the high-level shell-bearing deposits at Clava, Chapelhall, and other localities (Chapelhall section). *Report of the British Association for the Advancement of Science for 1894*, 307–15.

Horne, J., Robertson, D., Jamieson, T.F., Fraser, J., Kendall, P.F. and Bell, D. (1897) The character of the high-level shell-bearing deposits in Kintyre. *Report of the British Association for the Advancement of Science for 1896*, 378–99.

Horne, J. and Hinxman, L.W. (1914) *The Geology of the Country round Beauly and Inverness: including part of the Black Isle*. Explanation of Sheet 83. Memoirs of the Geological Survey of Scotland, HMSO, Edinburgh, 108 pp.

Horne, J. (1923) *The Geology of the Lower Findhorn and Lower Strath Nairn: including part of the Black Isle near Fortrose*. Sheet 84 and part of 94. Memoirs of the Geological Survey of Scotland, HMSO, Edinburgh, 128 pp.

Horsfield, B.R. (1983) The deglaciation pattern of the western Grampians of Scotland. Unpublished PhD thesis, University of East Anglia.

Howden, J.C. (1868) On the superficial deposits at the estuary of the South Esk. *Transactions of the Edinburgh Geological Society*, **1**, 138–50.

Howden, J.C. (1886) On the glacial deposits at Montrose. *Report of the British Association for the Advancement of Science for 1885*, 1040.

Howell, H.H. and Geikie, A. (1861) *The Geology of the Neighbourhood of Edinburgh*. Map 32. Memoirs of the Geological Survey of Great Britain (Scotland), HMSO, London, 151 pp.

Howell, H.H., Geikie, A. and Young, J. (1866) *The Geology of East Lothian, including parts of the Counties of Edinburgh and Berwick*. Maps 33, 34 and 41. Memoirs of the Geological Survey of Scotland, HMSO, London, 77 pp.

Hsü, K.J. (1975) Catastrophic debris streams (Sturzstroms) generated by rockfalls. *Bulletin of the Geological Society of America*, **86**, 129–40.

Hughes, T.J. (1987) Deluge II and the continent of doom: rising sea level and collapsing Antartic ice. *Boreas*, **16**, 89–100.

Hull, E. (1866) The raised beach of Cantyre. *Geological Magazine*, **3**, 5–10.

Hull, E. (1895) The glacial deposits of Aberdeenshire. *Geological Magazine*, Decade 4, **2**, 450–2.

Hulme, P.D. (1979) *Calliergon richardsonii* (Mitt.) Kindb. from the Late Devensian of Lang Lochs mire, Shetland. *Journal of Bryology*, **10**, 281.

Hulme, P.D. and Durno, S.E. (1980) A contribution to the phytogeography of Shetland. *New Phytologist*, **84**, 165–9.

Huntley, B. (1976) The past and present vegetation of the Morrone Birkwoods and Caenlochan National Nature Reserve. Unpublished PhD thesis, University of Cambridge.

Huntley, B. (1979) The past and present vegetation of the Caenlochan National Nature Reserve, Scotland. I. Present vegetation. *New Phytologist*, **83**, 215–84.

Huntley, B. (1981) The past and present vegetation of the Caenlochan National Nature Reserve, Scotland. II. Palaeoecological investigations. *New Phytologist*, **87**, 189–222.

Huntley, B. (1991) Historical lessons for the future. In *Scientific Management of Temperate Communities for Conservation* (eds I.F. Spellerberg, F.B. Goldsmith and M.G. Morris). British Ecological Society Symposium, No. 31. Blackwell, Oxford, pp. 473–503.

Huntley, B. and Birks, H.J.B. (1979a) The past and present vegetation of the Morrone Birkwoods National Nature Reserve, Scotland. I. A primary phytosociological survey. *Journal of Ecology*, **67**, 417–46.

Huntley, B. and Birks, H.J.B. (1979b) The past and present vegetation of the Morrone Birkwoods National Nature Reserve, Scotland. II. Woodland vegetation and soils. *Journal of Ecology*, **67**, 447–67.

Huntley, B. and Birks, H.J.B. (1983) *An Atlas of Past and Present Pollen Maps for Europe: 0–13,000 Years Ago*. Cambridge University Press, Cambridge, 667 pp.

Huntley, J.P. (in press) Palynological investigations at Freswick, Caithness. In *Excavations at a Norse Settlement Site in Northern Scotland* (eds C.D. Morris and C.E. Batey). Highland Regional Monograph, No. 1. Highland Regional Council, Inverness.

Hyde, W.T. and Peltier, W.R. (1985) Sensitivity experiments with a model of the ice age cycle: the response to harmonic forcing. *Journal of the Atmospheric Sciences*, **42**, 2170–88.

Imbrie, J. and Imbrie, K.P. (1979) *Ice Ages. Solving the Mystery*. Macmillan, London, 224 pp.

Imbrie, J., Hays, J.D., Martinson, D.G., McIntyre, A., Mix, A.C., Morley, J.J., Pisias, N.G., Prell, W.L. and Shackleton, N.J. (1984) The orbital theory of Pleistocene climate: support from revised chronology of the marine $\delta^{18}O$ record. In *Milankovitch and Climate. Understanding the Response to Astronomical Forcing* (eds A. Berger, J. Imbrie, J. Hays, G. Kukla and B. Saltzman). D. Reidel, Dordrecht, pp. 269–305.

Imrie, Lt.-Col. (1814) A geological account of the southern district of Stirlingshire, commonly called the Campsie Hills, with a few remarks relative to the two prevailing theories as to geology, and some examples given illustrative of these remarks. *Memoirs of the Wernerian Natural History Society*, **2**, 24–50.

Innes, J.L. (1982) Debris flow activity in the Scottish Highlands. Unpublished PhD thesis, University of Cambridge.

Innes, J.L. (1983a) Stratigraphic evidence of episodic talus accumulation on the Isle of Skye, Scotland. *Earth Surface Processes and Landforms*, **8**, 399–403.

Innes, J.L. (1983b) Lichenometric dating of debris flow deposits in the Scottish Highlands. *Earth Surface Processes and Landforms*, **8**, 579–88.

Innes, J.L. (1983c) Debris flows. *Progress in Physical Geography*, **7**, 469–501.

Innes, J.L. (1985) Magnitude–frequency relations of debris flows in north-west Europe. *Geografiska Annaler*, **67A**, 23–32.

Innes, J.L. (1989) Rapid mass movements in upland Britain: a review with particular reference to debris flows. *Studia Geomorphologica Carpatho–Balcanica*, **23**, 53–67.

Insch, E.V. (1976) The glacial geomorphology of part of the south-east Grampians and Strathmore. Unpublished MPhil thesis, University of Edinburgh.

Institute of Geological Sciences (1982) IGS Boreholes 1980. *Report of the Institute of Geological Sciences*, No. **81/11**, 12 pp.

International Geological Congress (1948) Guide to Excursion A19: Scotland (general tour), and Guide to Excursion C13: Edinburgh and St Andrews. International Geological Congress, London, 17 pp and 20 pp.

Irvine, D.R. (1872) *Wigtonshire – Mull of Galloway*. Explanation of Sheet 1. Memoirs of the Geological Survey of Scotland, HMSO, Edinburgh, 12 pp.

Iversen, J. (1944) *Viscum, Hedera* and *Ilex* as climatic indicators. *Geologiska Föreningens i Stockholm Förhandlingar*, **66**, 463–83.

Iversen, J. (1958) The bearing of glacial and interglacial epochs on the formation and extinction of plant taxa. *Uppsala Universitets Årbok*, **6**, 210–5.

Ives, J.D. (1958) Glacial geomorphology of the Torngat Mountains, northern Labrador. *Geographical Bulletin*, **12**, 47–75.

Ives, J.D. (1978) The maximum extent of the Laurentide ice sheet along the east coast of North America during the last glaciation. *Arctic*, **31**, 24–53.

Jack, R.L. (1875) Notes on a till or boulder clay with broken shells, in the lower valley of the River Endrick, near Loch Lomond, and its relation to certain other glacial deposits. *Transactions of the Geological Society of Glasgow*, **5**, 5–25.

Jahns, R.H. (1943) Sheet structure in granite: its origin and use as a measure of glacial erosion in New England. *Journal of Geology*, 51, 71–98.

James, H. (1874) *Notes on the Parallel Roads of Lochaber*. Ordnance Survey Office, Southampton, 7 pp.

Jamieson, T.F. (1858) On the Pleistocene deposits of Aberdeenshire. *Quarterly Journal of the Geological Society of London*, **14**, 509–32.

Jamieson, T.F. (1859) An outlier of Lias in Aberdeenshire. *Quarterly Journal of the Geological Society of London*, **15**, 131–3.

Jamieson, T.F. (1860a) On the occurrence of crag strata beneath the boulder clay in Aberdeenshire. *Quarterly Journal of the Geological Society of London*, **16**, 371–3.

Jamieson, T.F. (1860b) On the drift and rolled gravel of the north of Scotland. *Quarterly Journal of the Geological Society of London*, **16**, 347–71.

Jamieson, T.F. (1862) On the ice-worn rocks of

Scotland. *Quarterly Journal of the Geological Society of London*, **18**, 164–84.

Jamieson, T.F. (1863) On the Parallel Roads of Glen Roy, and their place in the history of the glacial period. *Quarterly Journal of the Geological Society of London*, **19**, 235–59.

Jamieson, T.F. (1865) On the history of the last geological changes in Scotland. *Quarterly Journal of the Geological Society of London*, **21**, 161–203.

Jamieson, T.F. (1866) On the glacial phenomena of Caithness. *Quarterly Journal of the Geological Society of London*, **11**, 261–81.

Jamieson, T.F. (1874) On the last stage of the glacial period in North Britain. *Quarterly Journal of the Geological Society of London*, **30**, 317–38.

Jamieson, T.F. (1882a) On the crag shells of Aberdeenshire and the gravel-beds containing them. *Quarterly Journal of the Geological Society of London*, **38**, 145–59.

Jamieson, T.F. (1882b) On the Red Clay of the Aberdeenshire coast and the direction of ice-movement in that quarter. *Quarterly Journal of the Geological Society of London*, **38**, 160–77.

Jamieson, T.F. (1892) Supplementary remarks on Glen Roy. *Quarterly Journal of the Geological Society of London*, **48**, 5–28.

Jamieson, T.F. (1905) Some changes of level in the glacial period. *Geological Magazine*, Ser. 5, **2**, 484–90.

Jamieson, T.F. (1906) The glacial period in Aberdeenshire and the southern border of the Moray Firth. *Quarterly Journal of the Geological Society of London*, **62**, 13–39.

Jamieson, T.F. (1908) A geologist on the Cairngorms. *Cairngorm Club Journal*, **5**, 82–8.

Jamieson, T.F. (1910) On the surface geology of Buchan. In *The Book of Buchan* (ed. J.F. Tocher). Aberdeen University Press, Aberdeen and P. Scrogie Ltd, Peterhead, pp. 19–25.

Jamieson, T.F., Jukes Browne, A.J. and Milne, J. (1898) Cretaceous fossils in Aberdeenshire. *Report of the British Association for the Advancement of Science for 1897*, 333–7.

Jansen, E. and Bjørklund, K.R. (1985) Surface ocean circulation in the Norwegian Sea 15,000 BP to present. *Boreas*, **14**, 243–57.

Jansen, E., Befring, S., Bugge, T., Eidvin, T., Holtedahl, H. and Sejrup, H.P. (1987) Large submarine slides on the Norwegian continental margin: sediments, transport and timing. *Marine Geology*, **78**, 77–107.

Jansen, J.H.F. (1976) Late Pleistocene and Holocene history of the northern North Sea, based on acoustic reflection records. *The Netherlands Journal of Sea Research*, **10**, 1–43.

Janssen, C.H. and Ten Hove, H.A. (1971) Some late-Holocene pollen diagrams from the Peel raised bogs (southern Netherlands). *Review of Palaeobotany and Palynology*, **11**, 7–53.

Jardine, W.G. (1962) Post-glacial sediments at Girvan, Ayrshire. *Transactions of the Geological Society of Glasgow*, **24**, 262–78.

Jardine, W.G. (1963) Pleistocene sediments at Girvan, Ayrshire. *Transactions of the Geological Society of Glasgow*, **24**, 4–16.

Jardine, W.G. (1964) Post-glacial sea-levels in south-west Scotland. *Scottish Geographical Magazine*, **80**, 5–11.

Jardine, W.G. (1967) Sediments of the Flandrian transgression in south-west Scotland: terminology and criteria for facies distinction. *Scottish Journal of Geology*, **3**, 221–6.

Jardine, W.G. (1968) The 'Perth' Readvance. *Scottish Journal of Geology*, **4**, 185–7.

Jardine, W.G. (1971) Form and age of late-Quaternary shore-lines and coastal deposits of south-west Scotland: critical data. *Quaternaria*, **14**, 103–14.

Jardine, W.G. (1973) The Quaternary geology of the Glasgow district. In *Excursion Guide to the Geology of the Glasgow District* (ed. B.J. Bluck). Geological Society of Glasgow, Glasgow, pp. 156–69.

Jardine, W.G. (1975) Chronology of Holocene marine transgression and regression in south-western Scotland. *Boreas*, **4**, 173–96.

Jardine, W.G. (1977) The Quaternary marine record in south-west Scotland and the Scottish Hebrides. In *The Quaternary History of the Irish Sea* (eds C. Kidson and M.J. Tooley). Seel House Press, Liverpool, pp. 99–118.

Jardine, W.G. (1978) Radiocarbon ages of raised-beach shells from Oronsay, Inner Hebrides, Scotland: a lesson in interpretation and deduction. *Boreas*, **7**, 183–96.

Jardine, W.G. (1980a) *Field Guide to the Glasgow Region*. Quaternary Research Association, Cambridge, 70 pp.

Jardine, W.G. (1980b) Holocene raised coastal sediments and former shorelines of Dumfriesshire and eastern Galloway. *Transactions of the Dumfriesshire and Galloway Natural History and Antiquarian Society*, 3rd Series, **55**, 1–59.

Jardine, W.G. (1981) Status and relationships of

the Loch Lomond Readvance and its stratigraphical correlatives. In *The Quaternary in Britain* (eds J. Neale and J. Flenley). Pergamon Press, Oxford, 168–73.

Jardine, W.G. (1982) Sea-level changes in Scotland during the last 18,000 years. *Proceedings of the Geologists' Association*, **93**, 25–41.

Jardine, W.G. (1986) The geological and geomorphological setting of the estuary and firth of Clyde. *Proceedings of the Royal Society of Edinburgh,* **90B**, 25–41.

Jardine, W.G. (1987) The Mesolithic coastal setting. In *Excavations on Oronsay: Prehistoric Human Ecology on a Small Island* (ed. P.A. Mellars). Edinburgh University Press, Edinburgh, pp. 25–51.

Jardine, W.G. and Dickson, J.H. (1980) Carstairs area. In *Field Guide to the Glasgow Region* (ed. W.G. Jardine). Quaternary Research Association, Cambridge, pp. 63–5.

Jardine, W.G. and Dickson, J.H. (1987) Significance of a recently-discovered interstadial site in western Scotland. INQUA XII Congress, Ottawa, Canada, 1987. Programme and Abstracts, p. 193.

Jardine, W.G. and Morrison, A. (1976) The archaeological significance of Holocene coastal deposits in south-western Scotland. In *Geoarchaeology. Earth Science and the Past* (eds D.A. Davidson and M.L. Shackley). Duckworth, London, pp. 175–95.

Jardine, W.G. and Morrison, A. (1980) Irvine area. In *Field Guide to the Glasgow Region* (ed. W.G. Jardine). Quaternary Research Association, Cambridge, pp. 13–19.

Jardine, W.G. and Peacock, J.D. (1973) Scotland. In *A Correlation of Quaternary Deposits in the British Isles* (eds G.F. Mitchell, L.F. Penny, F.W. Shotton and R.G. West). Geological Society of London, Special Report, No. 4, 53–9.

Jardine, W.G., Dickson, J.H., Haughton, P.D.W., Harkness, D.D., Bowen, D.Q. and Sykes, G.A. (1988) A late Middle Devensian interstadial site at Sourlie, near Irvine, Strathclyde. *Scottish Journal of Geology*, **24**, 288–95.

Jehu, T.J. and Craig, R.M. (1923a) The geology of the Outer Hebrides. Part I – The Barra Isles. *Transactions of the Royal Society of Edinburgh*, **53**, 419–41.

Jehu, T.J. and Craig, R.M. (1923b) The geology of the Outer Hebrides. Part II – South Uist and Eriskay. *Transactions of the Royal Society of Edinburgh*, **53**, 615–41.

Jehu, T.J. and Craig, R.M. (1926) The geology of the Outer Hebrides. Part III – North Uist and Benbecula. *Transactions of the Royal Society of Edinburgh*, **54**, 467–89.

Jehu, T.J. and Craig, R.M. (1927) The geology of the Outer Hebrides. Part IV – South Harris. *Transactions of the Royal Society of Edinburgh*, **55**, 457–88.

Jehu, T.J. and Craig, R.M. (1934) Geology of the Outer Hebrides. Part V – North Harris and Lewis. *Transactions of the Royal Society of Edinburgh*, **57**, 839–74.

Jenkins, A., Ashworth, P.J., Ferguson, R.I., Grieve, I.C., Rowling, P. and Stott, T.A. (1988) Slope failures on the Ochil Hills, Scotland, November 1984. *Earth Surface Processes and Landforms*, **13**, 69–76.

Jenkins, K.A. (1991) The origin of eskers and fluvioglacial features: an example from the area south of the Pentland Hills. Unpublished report, Department of Geology and Geophysics, University of Edinburgh, 36 pp.

Jessen, A. (1905) On the shell-bearing clay in Kintyre. *Transactions of the Edinburgh Geological Society*, **13**, 76–86.

Jessen, K. and Milthers, V. (1928) Stratigraphical and palaeontological studies of interglacial fresh-water deposits in Jutland and northwest Germany. *Danmarks Geologiske Undersøgelse*, Series II, **48**, 1–379.

Johansen, J. (1975) Pollen diagrams from the Shetland and Faroe Islands. *New Phytologist*, **75**, 369–87.

Johansen, J. (1978) The age of the introduction of *Plantago lanceolata* to the Shetland Islands. *Danmarks Geologiske Undersøgelse Årbok 1976*, 45–8.

Johansen, J. (1985) Studies in the vegetational history of the Faroe and Shetland Islands. *Annales Societatis Scientiarum Faeroensis Supplementum*, **11**, 117 pp.

John, B.S. (1973) Vistulian periglacial phenomena in south-west Wales. *Biuletyn Peryglacjalny*, **22**, 185–212.

Johnsen, S.J., Dansgaard, W., Clausen, H.B. and Langway, C.C. Jr. (1972) Oxygen isotope profiles through the Antarctic and Greenland ice sheets. *Nature*, **235**, 429–34.

Johnson, D.W. (1919) *Shore Processes and Shoreline Development.* John Wiley, New York, 584 pp.

Johnstone, G.S. and Mykura, W. (1989) *The*

Northern Highlands of Scotland, 4th edn. British Regional Geology. HMSO, London, 219 pp.

Jolly, W. (1868) On the evidences of glacier action in Galloway. *Transactions of the Edinburgh Geological Society*, **1**, 155–85.

Jolly, W. (1873) Note on the glaciers of Glen Spean and their relations to Glen Roy. *Transactions of the Edinburgh Geological Society*, **2**, 220–2.

Jolly, W. (1880a) The Parallel Roads of Lochaber. *Transactions of the Inverness Scientific Society and Field Club*, **1**, 287–92.

Jolly, W. (1880b) The Parallel Roads of Lochaber – the problem and its various solutions. *Nature*, **22**, 68–70.

Jolly, W. (1886a) The Parallel Roads of Lochaber; the problem, its conditions and solutions. *Transactions of the Geological Society of Glasgow*, **8**, 40–7.

Jolly, W. (1886b) The joint excursion of the Edinburgh and Glasgow Geological Societies to Ben Nevis and the Parallel Roads of Lochaber in July 1885. *Transactions of the Geological Society of Glasgow*, **8**, 72–105.

Jones, G.A. and Keigwin, L.D. (1988) Evidence from Fram Strait (78°N) for early deglaciation. *Nature*, **336**, 56–9.

Jones, V.J. (1987) A palaeoecological study of the post-glacial acidification of the Round Loch of Glenhead and its catchment. Unpublished PhD thesis, University of London.

Jones, V.J., Stevenson, A.C. and Battarbee, R.W. (1986) Lake acidification and the land-use hypothesis: a mid post-glacial analogue. *Nature*, **322**, 157–8.

Jones, V.J., Stevenson, A.C. and Battarbee, R.W. (1989) The acidification of lakes in Galloway, south west Scotland: a diatom and pollen study of the post-glacial history of the Round Loch of Glenhead. *Journal of Ecology*, **77**, 1–23.

Jones, V.J., Kreiser, A.M., Appleby P.G., Brodin, Y-W., Dayton, J., Natkanski, J., Richardson, N., Rippey, B., Sandøy, S. and Battarbee, R.W. (1990) The recent palaeolimnology of two sites with contrasting acid-deposition histories. *Philosophical Transactions of the Royal Society of London*, **B327**, 397–402.

Jouzel, J., Lorius, C., Petit, J.R., Genthon, C., Barkov, N.I., Kotlyakov, V.M. and Petrov, V.M. (1987) Vostok ice core: a continuous isotope temperature record over the last climatic cycle (160,000 years). *Nature*, **329**, 403–8.

Jouzel, J., Petit, J.R. and Raynaud, D. (1990) Palaeoclimatic information from ice cores: the Vostok records. *Transactions of the Royal Society of Edinburgh: Earth Sciences*, **81**, 349–55.

Kaitanen, V. (1969) A geographical study of the morphogenesis of northern Lapland. *Fennia*, **99(5)**, 85 pp.

Kaitanen, V. (1989) Relationships between ice-sheet dynamics and bedrock relief in dissected plateau areas in Finnish Lapland north of 69° latitude. *Geografiska Annaler*, **71A**, 1–15.

Kamb, B., Raymond, C.F., Harrison, W.D., Englehardt, H., Echelmeyer, K.A., Humphrey, N., Brugman, M.M. and Pfeffer, T. (1985) Glacier surge mechanism: 1982–1983 surge of Variegated Glacier, Alaska. *Science*, **227**, 469–79.

Karlsson, W., Vollsett, J., Bjørlykke, K. and Jørgensen, P. (1979) Changes in mineralogical composition of Tertiary sediments from North Sea wells. *Proceedings of the VIth International Clay Conference*, **27**, 281–9.

Keatinge, T.H. and Dickson, J.H. (1979) Mid-Flandrian changes in vegetation on Mainland Orkney. *New Phytologist*, **82**, 585–612.

Kelletat, D. (1970a) Rezante Periglazial Erscheinungen im Schottischen Hochland. Untersuchungen zu ihrer Verbreitung und Vergesellschaftung. *Gottinger Geographische Abhandlungen*, **51**, 67–140.

Kelletat, D. (1970b) Zum problem der Verbreitung, des Alters und der Bildungsdauer alter (inaktiver) Periglazialerscheinungen im Schottischen Hochland. *Zeitschrift für Geomorphologie*, NF, **14**, 510–19.

Kelletat, D. (1972) Zum Problem der Abgrenzung und ökologischen Differenzierung des Hochgebirges in Schottland. In *Geoecology of the High Mountain Regions of Eurasia* (ed. C. Troll). *Erdwissenschaftliche Forschung (Wiesbaden)*, **4**, 110–30.

Kemp, D.D. (1971) The stratigraphy and sub-carse morphology of an area on the northern side of the River Forth, between the Lake of Menteith and Kincardine-on-Forth. Unpublished PhD thesis, University of Edinburgh.

Kemp, D.D. (1976) Buried raised beaches on the northern side of the Forth Valley, central Scotland. *Scottish Geographical Magazine*, **92**, 120–8.

Kendall, P.F. (1902) A system of glacier-lakes in the Cleveland Hills. *Quarterly Journal of the Geological Society of London*, **58**, 471–571.

Kendall, P.F. and Bailey, E.B. (1908) The glaciation of East Lothian south of the Garleton Hills. *Transactions of the Royal Society of Edinburgh*, **46**, 1–31.

Kerney, M.P. and Cameron, R.A.D. (1979) *A Field Guide to the Land Snails of Britain and North-West Europe*. Collins, London, 288 pp.

Kerr, R.J. (1978) The nature and derivation of glacial till in part of the Tweed Basin. Unpublished PhD thesis, University of Edinburgh.

Kerr, W.B. (1982) Pleistocene ice movements in the Rhins of Galloway. *Transactions of Dumfriesshire and Galloway Natural History and Antiquarian Society*, **57**, 1–10.

Kerslake, P.D. (1982) Vegetational history of wooded islands in Scottish lochs. Unpublished PhD thesis, University of Cambridge.

Kesel, R.H. and Gemmell, A.M.D. (1981) The 'Pliocene' gravels of Buchan: a reappraisal. *Scottish Journal of Geology*, **17**, 185–203.

Kinahan, G.H. (1887) The terraces of the Great American Lakes and the Roads of Glenroy. *Transactions of the Edinburgh Geological Society*, **5**, 221–3.

King, R.B. (1968) Periglacial features in the Cairngorm Mountains. Unpublished PhD thesis, University of Edinburgh.

King, R.B. (1971a) Boulder polygons and stripes in the Cairngorm mountains, Scotland. *Journal of Glaciology*, **10**, 375–86.

King, R.B. (1971b) Vegetation destruction in the sub-alpine and alpine zones of the Cairngorm Mountains. *Scottish Geographical Magazine*, **87**, 103–15.

King, R.B. (1972) Lobes in the Cairngorm Mountains, Scotland. *Biuletyn Peryglacjalny*, **21**, 153–67.

Kinloch, B.B., Westfall, R.D. and Forrest, G.I. (1986) Caledonian Scots Pine: origins and genetic structure. *New Phytologist*, **104**, 703–29.

Kirby, R.P. (1966) The glacial geomorphology of the Esk Basin, Midlothian. Unpublished PhD thesis, University of Edinburgh.

Kirby, R.P. (1968) The ground moraines of Midlothian and East Lothian. *Scottish Journal of Geology*, **4**, 209–20.

Kirby, R.P. (1969a) Variation in glacial deposition in a sub-glacial environment: an example from Midlothian. *Scottish Journal of Geology*, **5**, 49–53.

Kirby, R.P. (1969b) Till fabric analyses from the Lothians, central Scotland. *Geografiska Annaler*, **51A**, 48–60.

Kirby, R.P. (1969c) Morphometric analysis of glaciofluvial terraces in the Esk Basin, Midlothian. *Transactions of the Institute of British Geographers*, **48**, 1–18.

Kirk, W. and Godwin, H. (1963) A late glacial site at Loch Droma, Ross and Cromarty. *Transactions of the Royal Society of Edinburgh,* **65**, 225–49.

Kirk, W., Rice, R.J. and Synge, F.M. (1966) Deglaciation and vertical displacement of shorelines in Wester and Easter Ross. *Transactions of the Institute of British Geographers*, **39**, 65–78.

Kirsop, J. (1882) *In* Proceedings of the Geological Society of Glasgow for session 1878–79. *Transactions of the Geological Society of Glasgow*, **6**, 291.

Koppi, A.J. (1977) Weathering of Tertiary gravels, a schist, and a metasediment in north-east Scotland. Unpublished PhD thesis, University of Aberdeen.

Koppi, A.J. and FitzPatrick, E.A. (1980) Weathering in Tertiary gravels in north-east Scotland. *Journal of Soil Science*, **31**, 525–32.

Kotarba, A. (1984) Slope features in areas of high relief in maritime climate (with the Isle of Rhum as example). *Studia Geomorphologica Carpatho–Balcanica*, **17**, 77–90.

Kotarba, A. (1987) Glacial cirques transformation under differentiated maritime climate. *Studia Geomorphologica Carpatho–Balcanica*, **21**, 77–92.

Kynaston, H. and Hill, J.B. (1908) *The Geology of the Country near Oban and Dalmally*. Explanation of Sheet 45. Memoirs of the Geological Survey of Scotland, HMSO, Glasgow, 184 pp.

Lacaille, A.D. (1954) *The Stone Age in Scotland*. Wellcome Historical Medical Museum, Oxford, 345 pp.

Laing, D. (1976) *The Soils of the Country round Perth, Arbroath and Dundee*. Sheets 48 and 49. Memoirs of the Soil Survey of Great Britain: Scotland, HMSO, Edinburgh, 328 pp.

Laing, S. (1877) Glacial geology of Orkney and Shetland. *Nature*, **16**, 418–9.

Lamb, H.H. (1977) *Climate. Present, Past and Future. Volume 2. Climatic History and the Future*. Methuen, London, 835 pp.

Lamb, H.H. (1982a) *Climate, History and the Modern World*. Methuen, London, 387 pp.

Lamb, H.H. (1982b) Reconstructing the course of

postglacial climate over the world. In *Climatic Change in Later Prehistory* (ed. A. Harding). Edinburgh University Press, Edinburgh, pp. 11–32.

Lambeck, K. (1991a) A model for Devensian and Flandrian glacial rebound and sea-level change in Scotland. In *Glacial Isostasy, Sea Level and Mantle Rheology* (eds R. Sabadini, K. Lambeck and E. Boschi). Kluwer, Dordrecht, 33–61.

Lambeck, K. (1991b) Glacial rebound and sea-level change in the British Isles. *Terra Nova*, **3**, 379–89.

Lamplugh, G.W. (1906) On British drifts and the interglacial problem. *Report of the British Association for the Advancement of Science for 1905*, 532–58.

Lamplugh, G.W. (1911) On the shelly moraine of the Sefstrom Glacier and other Spitsbergen phenomena illustrative of British glacial conditions. *Proceedings of the Yorkshire Geological Society*, **17**, 216–41.

Langmuir, E.C.D. (1970) Snow profiles in Scotland. *Weather*, **25**, 203–9.

Law, G.R. (1962) The sub-surface geology of the Bay of Nigg – Tullos area. Unpublished MSc thesis, University of Aberdeen.

Lawson, D.E. (1979) *Sedimentological analysis of the western terminus region of the Matanuska Glacier, Alaska*. United States Army Corps of Engineers Cold Regions Research and Engineering Laboratory, CRREL Report, 79–9, Hanover, New Hampshire, 112 pp.

Lawson, T.J. (1981a) First Scottish date from the last interglacial. *Scottish Journal of Geology*, **17**, 301–4.

Lawson, T.J. (1981b) The 1926–27 excavations of the Creag nan Uamh bone caves near Inchnadamph, Sutherland. *Proceedings of the Society of Antiquaries of Scotland*, **111**, 7–20.

Lawson, T.J. (1983) Quaternary geomorphology of the Assynt area, NW Scotland. Unpublished PhD thesis, University of Edinburgh.

Lawson, T.J. (1984) Reindeer in the Scottish Quaternary. *Quaternary Newsletter*, **42**, 1–7.

Lawson, T.J. (1986) Loch Lomond Advance glaciers in Assynt, Sutherland, and their palaeoclimatic implications. *Scottish Journal of Geology*, **22**, 289–98.

Lawson, T.J. (1990) Former ice movement in Assynt, Sutherland, as shown by the distribution of glacial erratics. *Scottish Journal of Geology*, **26**, 25–32.

Lawson, T.J. and Bonsall, J.C. (1986a) Early settlement in Scotland: the evidence from Reindeer Cave Assynt. *Quaternary Newsletter*, **49**, 1–7.

Lawson, T.J. and Bonsall, J.C. (1986b) The Palaeolithic in Scotland: a reconsideration of the evidence from Reindeer Cave, Assynt. In *The Palaeolithic of Britain and its Nearest Neighbours: Recent Trends* (ed. S.N. Colcutt). Department of Archaeology and Prehistory, University of Sheffield, Sheffield, pp. 85–9.

Laxton, J.L. (1984) The occurrence of possible Late-glacial estuarine deposits at levels above the carse clay west of Stirling. *Scottish Journal of Geology*, **20**, 107–14.

Laxton, J.L. and Nickless, E.F.P. (1980) The sand and gravel resources of the country around Lanark, Strathclyde Region. Description of 1:25,000 sheet NS 94 and part of NS 84. *Mineral Assessment Report, Institute of Geological Sciences*, No. 49, 144 pp.

Laxton, J.L. and Ross, D.L. (1983) The sand and gravel resources of the country west of Stirling, Central Region. Description of 1:25,000 sheet NS 69 and 79. *Mineral Assessment Report, Institute of Geological Sciences*, No. 131, 82 pp.

Le Coeur, C. (1988) Late Tertiary warping and erosion in western Scotland. *Geografiska Annaler*, **70A**, 361–7.

Le Coeur, C. (1989) La question des altérites profondes dans la région des Hébrides internes (Ecosse occidentale). *Zeitschrift für Geomorphologie*, NF, Supplementband, **72**, 109–24.

Lee, G.W. (1909) Palaeontological work. In *Memoirs of the Geological Survey. Summary of Progress of the Geological Survey of Great Britain and the Museum of Practical Geology for 1908*. HMSO, London, 75 pp.

Leftley, D.C. (1991) The Late Devensian development of the Strathrory river valley. Unpublished BSc dissertation, West London Institute of Higher Education.

Leverett, F. and Taylor, F.B. (1915) The Pleistocene of Indiana and Michigan and the history of the Great Lakes. *United States Geological Survey Monograph*, **53**, 529 pp.

Lewis, F.J. (1905) The plant remains in the Scottish peat mosses. Part I. The Scottish Southern Uplands. *Transactions of the Royal Society of Edinburgh*. **41**, 699–723.

Lewis, F.J. (1906) The plant remains in the Scottish peat mosses. Part II. The Scottish Highlands. *Transactions of the Royal Society of Edinburgh*, **45**, 335–60.

Lewis, F.J. (1907) The plant remains in the Scottish peat mosses. Part III. The Scottish Highlands and the Shetland Islands. *Transactions of the Royal Society of Edinburgh*, **46**, 33–70.

Lewis, F.J. (1911) The plant remains in the Scottish peat mosses. Part IV. The Scottish Highlands and Shetland, with an appendix on the Icelandic peat deposits. *Transactions of the Royal Society of Edinburgh*, **47**, 793–833.

Lewis, W.V. (1938) A meltwater hypothesis of cirque formation. *Geological Magazine*, **75**, 249–65.

Lewis, W.V. (1947) Valley steps and glacial valley erosion. *Transactions of the Institute of British Geographers*, **14**, 19–44.

Liestøl, O. (1956) Glacier dammed lakes in Norway. *Norsk Geografisk Tidsskrift*, **15**, 122–49.

Liestøl, O. (1961) Talus terraces in Arctic regions. *Norsk Polarinstitutt Årbok*, 102–5.

Lindner, L. and Marks, L. (1985) Types of debris slope accumulations and rock glaciers in south Spitsbergen. *Boreas*, **14**, 139–53.

Lindsay, R.A., Charman, D.J., Everingham, F., O'Reilly, R.M., Palmer, M.A., Rowell, T.A. and Stroud, D.A. (1988) *The Flow Country. The Peatlands of Caithness and Sutherland.* Nature Conservancy Council, Peterborough, 174 pp.

Linton, D.L. (1933) The 'Tinto Glacier' and some glacial features in Clydesdale. *Geological Magazine*, **70**, 549–54.

Linton, D.L. (1949a) Some Scottish river captures re-examined. *Scottish Geographical Magazine*, **65**, 123–32.

Linton, D.L. (1949b) Unglaciated areas in Scandinavia and Great Britain. *Irish Geography*, **2**, 25–33.

Linton, D.L. (1950a) The scenery of the Cairngorm Mountains. *Journal of the Manchester Geographical Society*, **55**, 45–9.

Linton, D.L. (1950b) Unglaciated enclaves in glaciated regions. *Journal of Glaciology*, **1**, 451–2.

Linton, D.L. (1950c) Discussion. *Comptes Rendus du Congrès International de Géographie, Lisbonne 1949*, Tome 2, pp. 298–300.

Linton, D.L. (1951a) Watershed breaching by ice in Scotland. *Transactions of the Institute of British Geographers*, **15**, 1–15.

Linton, D.L. (1951b) Problems of Scottish scenery. *Scottish Geographical Magazine*, **67**, 65–85.

Linton, D.L. (1952) The significance of tors in glaciated lands. In *Proceedings of the 17th International Geographical Congress, Washington DC, August 8–15, 1952*, pp. 354–7.

Linton, D.L. (1954) Some Scottish river captures re-examined. III. The beheading of the Don. *Scottish Geographical Magazine*, **70**, 64–78.

Linton, D.L. (1955) The problem of tors. *Geographical Journal*, **121**, 470–87.

Linton, D.L. (1957) Radiating valleys in glaciated lands. *Tijdschrift van het Koninklijk Nederlandsch Aardrijkskundig Genootschap*, **74**, 297–312.

Linton, D.L. (1959) Morphological contrasts between eastern and western Scotland. In *Geographical Essays in Memory of Alan G. Ogilvie* (eds R. Miller and J.W. Watson). Nelson, Edinburgh, pp. 16–45.

Linton, D.L. (1962) Glacial erosion on soft-rock outcrops in central Scotland. *Biuletyn Peryglacjalny*, **11**, 247–57.

Linton, D.L. (1963) The forms of glacial erosion. *Transactions of the Institute of British Geographers*, **33**, 1–28.

Linton, D.L. and Moisley, H.A. (1960) The origin of Loch Lomond. *Scottish Geographical Magazine*, **76**, 26–37.

Livingston, C. (1880) The valley of the Spean, with reference to the Lochaber Parallel Roads. *Transactions of the Inverness Scientific Society and Field Club*, **1**, 101–19.

Livingston, C. (1906) The Parallel Roads of Lochaber with relation to an ice-cap. *Transactions of the Geological Society of Glasgow*, **12**, 326–53.

Long, D. and Morton A.C. (1987) An ash fall within the Loch Lomond Stadial. *Journal of Quaternary Science*, **2**, 97–101.

Long, D. and Skinner, A.C. (1985) Glacial meltwater channels in the northern isles of Shetland. *Scottish Journal of Geology*, **21**, 222–4.

Long, D., Bent, A.J.A., Harland, R., Gregory, D.M., Graham, D.K. and Morton, A.C. (1986) Late Quaternary palaeontology, sedimentology and geochemistry of a vibrocore from the Witch Ground Basin, Central North Sea. *Marine Geology*, **73**, 109–23.

Long, D., Smith, D.E. and Dawson, A.G. (1989a) A Holocene tsunami deposit in eastern Scotland. *Journal of Quaternary Science*, **4**, 61–6.

Long, D., Dawson, A.G. and Smith, D.E. (1989b) Tsunami risk in north-western Europe: a Holocene example. *Terra Nova*, **1**, 532–7.

Loubere, P. and Moss, K. (1986) Late Pliocene climatic change and the onset of Northern

Hemisphere glaciation as recorded in the northeast Atlantic Ocean. *Bulletin of the Geological Society of America*, **97**, 818–28.

Lowe, J.J. (1977) Pollen analysis and radiocarbon dating of Lateglacial and Early Flandrian deposits in southern Perthshire. Unpublished PhD thesis, University of Edinburgh.

Lowe, J.J. (1978) Radiocarbon-dated Lateglacial and early Flandrian pollen profiles from the Teith Valley, Perthshire, Scotland. *Pollen et Spores*, **20**, 367–97.

Lowe, J.J. (1982a) Three Flandrian pollen profiles from the Teith Valley, Perthshire, Scotland. I. Vegetational history. *New Phytologist*, **90**, 355–70.

Lowe, J.J. (1982b) Three Flandrian pollen profiles from the Teith Valley, Perthshire, Scotland. II. Analysis of deteriorated pollen. *New Phytologist*, **90**, 371–85.

Lowe, J.J. (1984) A critical evaluation of pollen-stratigraphic investigations of pre-Late Devensian sites in Scotland. *Quaternary Science Reviews*, **3**, 405–32.

Lowe, J.J. (1991) Stratigraphic resolution and radiocarbon dating of Devensian Lateglacial sediments. In *Radiocarbon Dating: Recent Applications and Future Potential* (ed. J.J. Lowe). Quaternary Proceedings No. 1. Quaternary Research Association, Cambridge, pp. 19–25.

Lowe, J.J. and Cairns, P. (1989) Palynological investigations: biostratigraphy. In *Glen Roy Area: Field Guide* (eds J.D. Peacock and R. Cornish). Quaternary Research Association, Cambridge, pp. 10–11.

Lowe, J.J. and Cairns, P. (1991) New pollen-stratigraphic evidence for the deglaciation and lake drainage chronology of the Glen Roy–Glen Spean area. *Scottish Journal of Geology*, **27**, 41–56.

Lowe, J.J. and Walker, M.J.C. (1976) Radiocarbon dates and deglaciation of Rannoch Moor, Scotland. *Nature*, **246**, 632–3.

Lowe, J.J. and Walker, M.J.C. (1977) The reconstruction of the Lateglacial environment in the southern and eastern Grampian Highlands. In *Studies in the Scottish Lateglacial Environment* (eds J.M. Gray and J.J. Lowe). Pergamon Press, Oxford, pp. 101–18.

Lowe, J.J. and Walker, M.J.C. (1980) Problems associated with radiocarbon dating the close of the Lateglacial period in the Rannoch Moor area, Scotland. In *Studies in the Lateglacial of North-west Europe* (eds J.J. Lowe, J.M. Gray and J.E. Robinson). Pergamon Press, Oxford, pp. 123–38.

Lowe, J.J. and Walker, M.J.C. (1981) The early Postglacial environment of Scotland: evidence from a site near Tyndrum, Perthshire. *Boreas*, **10**, 281–94.

Lowe, J.J. and Walker, M.J.C. (1984) *Reconstructing Quaternary Environments*. Longman, Harlow, 389 pp.

Lowe, J.J. and Walker, M.J.C. (1986a) Lateglacial and early Flandrian environmental history of the Isle of Mull, Inner Hebrides, Scotland. *Transactions of the Royal Society of Edinburgh: Earth Sciences*, **77**, 1–20.

Lowe, J.J. and Walker, M.J.C. (1986b) Flandrian environmental history of the Isle of Mull, Scotland. II. Pollen analytical data from sites in western and northern Mull. *New Phytologist*, **103**, 417–36.

Lowe, J.J. and Walker, M.J.C. (1991) Vegetational history of the Isle of Skye: II. The Flandrian. In *The Quaternary of the Isle of Skye: Field Guide* (eds C.K. Ballantyne, D.I. Benn, J.J. Lowe and M.J.C. Walker). Quaternary Research Association, Cambridge, pp. 119–42.

Lowe, P.A., Duck, R.W. and McManus, J. (1991) A bathymetric reappraisal of Loch Muick, Aberdeenshire. *Scottish Geographical Magazine*, **107**, 110–5.

Lubbock, J. (1868) On the Parallel Roads of Glen Roy. *Quarterly Journal of the Geological Society of London*, **24**, 83–93.

Luckman, B.H. (1992) Debris flows and snow avalanche landforms in the Lairig Ghru, Cairngorm Mountains, Scotland. *Geografiska Annaler*, **74A**, 109–121.

Lucy, W.C. (1886) The terrace gravels of Achnasheen, Ross-shire. *Proceedings of the Cotteswold Naturalists' Field Club*, **8**, 118–20.

Lyell, C. (1841a) On the geological evidence of the former existence of glaciers in Forfarshire. *Proceedings of the Geological Society of London*, **3**, 337–45.

Lyell, C. (1841b) *Elements of Geology*, 2nd edn. John Murray, London, 2 Vols, 437 and 460 pp.

Lyell, C. (1863) *The Geological Evidences of the Antiquity of Man with Remarks on Theories of the Origin of Species by Variation*. John Murray, London, 520 pp.

McAdam, A.D. (1978) Sand and gravel resources of the Lothian Region of Scotland. *Report of the Institute of Geological Sciences*, No. **78/1**, 18 pp.

McAdam, A.D. and Tulloch, W. (1985) *Geology*

of the Haddington District. Memoir for 1:50,000 sheet 33W and part of sheet 41. British Geological Survey, HMSO, London, 99 pp.

McCall, J. and Goodlet, G.A. (1952) Indicator stones from the drift of south Midlothian and Peebles. *Transactions of the Edinburgh Geological Society*, **14**, 401–9.

McCallien, W.J. (1937a) Late-glacial and early post-glacial Scotland. *Proceedings of the Society of Antiquaries of Scotland*, **71**, 174–206.

McCallien, W.J. (1937b) Rhu (Row) Point – a readvance moraine. *Transactions of the Geological Society of Glasgow*, **19**, 385–9.

McCallien, W.J. (1938) *Geology of Glasgow and District*. Blackie, London and Glasgow, 190 pp.

McCann, S.B. (1961a) The raised beaches of western Scotland. Unpublished PhD thesis, University of Cambridge.

McCann, S.B. (1961b) Some supposed 'raised beach' deposits at Corran, Loch Linnhe and Loch Etive. *Geological Magazine*, **98**, 131–42.

McCann, S.B. (1964) The raised beaches of north-east Islay and western Jura, Argyll. *Transactions of the Institute of British Geographers*, **35**, 1–16.

McCann, S.B. (1966a) The limits of the Late-glacial Highland, or Loch Lomond, Readvance along the West Highland seaboard from Oban to Mallaig. *Scottish Journal of Geology*, **2**, 84–95.

McCann, S.B. (1966b) The main post-glacial raised shoreline of western Scotland from the Firth of Lorne to Loch Broom. *Transactions of the Institute of British Geographers*, **39**, 87–99.

McCann, S.B. (1968) Raised shore platforms in the Western Isles of Scotland. In *Geography at Aberystwyth. Essays Written on the Occasion of the Departmental Jubilee 1917–18 – 1967–68* (eds E.G. Bowen, H. Carter and J.A. Taylor). University of Wales Press, Cardiff, pp. 22–34.

McCann, S.B. and Richards, A. (1969) The coastal features of the island of Rhum in the Inner Hebrides. *Scottish Journal of Geology*, **5**, 15–25.

MacCulloch, J. (1817) On the Parallel Roads of Glen Roy. *Transactions of the Geological Society of London*, **4**, 314–92.

MacCulloch, J. (1819) *A Description of the Western Islands of Scotland, Including the Isle of Man: Comprising an Account of Their Geological Structure; With Remarks on Their Agriculture, Scenery, and Antiquities*, 3 Vols. Archibald Constable and Co., Edinburgh, 587, 589 and 91 pp.

MacDonald, K. (1881) Glacial drift in the Craggie Burn. *Transactions of the Inverness Scientific Society and Field Club*, **2**, 47–53.

MacDonald, K. (1903) The Parallel Roads of Glen Roy. *Transactions of the Inverness Scientific Society and Field Club*, **6**, 229–51.

MacDonald, K. and Fraser, J. (1881) Excursion to Craggie and Loch Moy. *Transactions of the Inverness Scientific Society and Field Club*, **2**, 99–114.

McEwen, L.J. (1986) River channel planform changes in upland Scotland, with specific reference to climatic fluctuations and land use changes over the past 250 years. Unpublished PhD thesis, University of St Andrews.

McEwen, L.J. and Werritty, A. (1988) The hydrology and long-term geomorphic significance of a flash flood in the Cairngorm Mountains, Scotland. *Catena*, **15**, 361–77.

Macfadzean, J. (1883) *The Parallel Roads of Glen Roy: Their Origin and Relation to the Glacial Period and the Deluge*. J. Menzies, Edinburgh, 149 pp.

McFarlane, J. (1858) Memorandum of shells and a deer's horn found in a cutting of the Forth and Clyde Junction Railway, Dumbartonshire. *Proceedings of the Royal Physical Society of Edinburgh*, **1**, 163–5.

MacGregor, A.R. (1973) *Fife and Angus Geology. An Excursion Guide*. Scottish Academic Press, Edinburgh, 281 pp.

McGregor, D.F.M. (1974) A quantitative analysis of some fluvioglacial deposits from east-central Scotland. Unpublished PhD thesis, University of Edinburgh.

Macgregor, M. (1927) The Carstairs district. *Proceedings of the Geologists Association*, **38**, 495–9.

Macgregor, M. and Ritchie, J. (1940) Early glacial remains of reindeer from the Glasgow district. *Proceedings of the Royal Society of Edinburgh*, **60**, 322–32.

Mackereth, F.J. (1965) Chemical investigations of lake sediments and their interpretation. *Proceedings of the Royal Society London,* **61**, 295–309.

Mackereth, F.J. (1966) Some chemical observations on post-glacial lake sediments. *Philosophical Transactions of the Royal Society of London*, **B250**, 165–213.

Mackenzie, G.S. (1848) An attempt to classify the phenomena in the glens of Lochaber with those of the Diluvium, or Drift, which covers the face of the country. *Edinburgh New Philosophical Journal*, **44**, 1–12.

Mackie, S.J. (1863) The Parallel Roads of Glen Roy. *Geologist*, **6**, 121–33.

Mackie, W. (1901) Some notes on the distribution of erratics over eastern Moray. *Transactions of the Edinburgh Geological Society*, **8**, 91–7.

Maclaren, C. (1828) Changes on the surface of the globe. *The Scotsman*, **12(918)**, 25 October, 683.

Maclaren, C. (1838) *A Sketch of the Geology of Fife and the Lothians Including Detailed Descriptions of Arthur's Seat and Pentland Hills*. Adam and Charles Black, Edinburgh, 235 pp.

Maclaren, C. (1839) Account of the Parallel Roads of Glen Roy, in Invernesshire. *Edinburgh New Philosophical Journal*, **27**, 395–402.

Maclaren, C. (1840) Discovery of the former existence of glaciers in Scotland especially in the Highlands, by Professor Agassiz. *The Scotsman*, **24(2165)**, 7 October, p. 3.

Maclaren, C. (1841) The glacial theory. No. 2. *The Scotsman*, **25(2190)**, 2 January, 2.

Maclaren, C. (1842a) The glacial theory of Prof Agassiz. *American Journal of Science and Arts*, **42**, 346–65.

Maclaren, C. (1842b) Striated rocks of Corstorphine Hill. *The Scotsman*, **26(2344)**, 25 June, 2.

Maclaren, C. (1845) Glaciers and icebergs in Scotland in ancient times. *The Scotsman*, **29(2685)**, 1 October, p. 2. (Also in *Edinburgh New Philosophical Journal*, **40**, 1846, 125–42; Cox and Nicol, ed. 1869, 103–19.)

Maclaren, C. (1846) Further evidence of the existence of glaciers in Scotland in ancient times. *The Scotsman*, **30(2798)** 31 October, p. 2. (Also in *Edinburgh New Philosophical Journal*, **42**, 1847, 25–37; Cox and Nicol, ed. 1869, 120–30.)

Maclaren, C. (1849) On grooved and striated rocks in the middle region of Scotland. *Edinburgh New Philosophical Journal*, **47**, 161–82.

Maclaren, C. (1866) *A Sketch of the Geology of Fife and the Lothians including Detailed Descriptions of Arthur's Seat and Pentland Hills*, 2nd edn. Adam and Charles Black, Edinburgh, 320 pp.

Maclean, A.F. (1991) The formation of valley-wall rock glaciers. Unpublished PhD thesis, University of St Andrews.

McLean, F. (1977) The glacial sediments of a part of east Aberdeenshire. Unpublished PhD thesis, University of Aberdeen.

McLellan, A.G. (1967a) The distribution, origin and use of sand and gravel deposits in central Lanarkshire. Unpublished PhD thesis, University of Glasgow.

McLellan, A.G. (1967b) The distribution of sand and gravel deposits in west central Scotland and some problems concerning their utilisation. Monograph, Department of Geography, University of Glasgow, 45 pp.

McLellan, A.G. (1969) The last glaciation and deglaciation of central Lanarkshire. *Scottish Journal of Geology*, **5**, 248–68.

Macleod, J. (1794) Parish of Harris. In *Statistical Account of Scotland*, Vol. 10 (ed. J. Sinclair). William Creech, Edinburgh, pp. 342–92.

McManus, J. (1972) Estuarine development and sediment distribution, with particular reference to the Tay. *Proceedings of the Royal Society of Edinburgh*, **71**, 97–113.

McMillan, A.A. and Aitken, A.M. (1981) The sand and gravel resources of the country west of Peterhead, Grampian Region. Description of 1:25,000 Sheet NK04 and parts of NJ94, 95 and NK05, 14 and 15. *Mineral Assessment Report, Institute of Geological Sciences,* No. 58, 99 pp.

McMillan, A.A. and Merritt, J.W. (1980) A reappraisal of the 'Tertiary' deposits of Buchan, Grampian Region. *Report of the Institute of Geological Sciences*, No. 80/1, 18–25.

MacPherson, J.B. (1978) Pollen chronology of the Glen Roy–Loch Laggan proglacial lake drainage. *Scottish Journal of Geology*, **14**, 125–39.

MacPherson, J.B. (1980) Environmental change during the Loch Lomond Stadial: evidence from a site in the upper Spey Valley, Scotland. In *Studies in the Lateglacial of North-west Europe* (eds J.J. Lowe, J.M. Gray and J.E. Robinson). Pergamon Press, Oxford, pp. 89–102.

McRae, F. (1845) Parish of North Uist. In *The New Statistical Account of Scotland*. Vol. 14. Blackwood and Sons, Edinburgh, pp. 159–81.

MacSween, M.D. (1959) Transhumance in North

Skye. *Scottish Geographical Magazine*, **75**, 75–88.

McVean, D.N. (1961) Flora and vegetation of the islands of St Kilda and North Rona in 1958. *Journal of Ecology*, **49**, 39–54.

McVean, D.N. (1963a) The ecology of Scots pine in the Scottish Highlands. *Journal of Ecology*, **18**, 671–86.

McVean, D.N. (1963b) Snow cover in the Cairngorms 1961–62. *Weather*, **18**, 339–42.

McVean, D.N. and Ratcliffe, D.A. (1962) *Plant Communities in the Scottish Highlands. A Study of Scottish Mountain, Moorland and Forest Vegetation*. HMSO, London, 445 pp.

Madgett, P.A. and Catt, J.A. (1978) Petrography, stratigraphy and weathering of Late Pleistocene tills in East Yorkshire, Lincolnshire and North Norfolk. *Proceedings of the Yorkshire Geological Society*, **42**, 55–108.

Maizels, J.K. (1976) A comparison of Pleistocene and present-day proglacial environments with particular reference to morphology and sedimentology. Unpublished PhD thesis, University of London.

Maizels, J.K. (1983a) Channel changes, palaeohydrology and deglaciation: evidence from some Lateglacial sandur deposits of north-east Scotland. *Quaternary Studies in Poland*, **4**, 171–87.

Maizels, J.K. (1983b) Proglacial channel systems: change and thresholds for change over long, intermediate and short time-scales. In *Modern and Ancient Fluvial Systems* (eds J.D. Collinson and J. Lewin). Blackwell Scientific, Oxford, International Association of Sedimentologists, Special Publication, **6**, 251–66.

Maizels, J.K. (1983c) Palaeovelocity and palaeodischarge determination for coarse gravel deposits. In *Background to Palaeohydrology* (ed. K.J. Gregory). John Wiley, Chichester, pp. 101–39.

Maizels, J.K. (1985) The physical background of the River Dee. In *The Biology and Management of the River Dee* (ed. D. Jenkins). Institute of Terrestrial Ecology, Huntingdon, pp. 7–22.

Maizels, J.K. (1986) Modelling of palaeohydrologic change during deglaciation. *Géographie Physique et Quaternaire*, **40**, 263–77.

Maizels, J.K. and Aitken, J.F. (1991) Palaeohydrological change during deglaciation in upland Britain: a case study from northeast Scotland. In *Temperate Palaeohydrology* (eds L. Starkel, K.J. Gregory and J.B. Thornes). John Wiley, Chichester, pp. 105–145.

Mangerud, J., Andersen, S.T., Berglund, B.G. and Donner, J.J. (1974) Quaternary stratigraphy of Norden, a proposal for terminology and classification. *Boreas*, **3**, 109–28.

Mangerud, J., Lie, S.E., Furnes, H., Kristiansen, I.L. and Lømo, L. (1984) A Younger Dryas ash bed in western Norway, and its possible correlations with tephra in cores from the Norwegian Sea and the North Atlantic. *Quaternary Research*, **21**, 85–104.

Manley, G. (1949) The snowline in Britain. *Geografiska Annaler*, **31**, 179–93.

Manley, G. (1971) Scotland's semi-permanent snows. *Weather*, **26**, 458–71.

Mannerfelt, C.M. (1945) Några glacialmorfologiska formelement. *Geografiska Annaler*, **27**, 1–239.

Mannerfelt, C.M. (1949) Marginal drainage channels as indicators of the gradients of Quaternary ice-caps. *Geografiska Annaler*, **31**, 194–9.

Mannion, A.M. (1978a) Late Quaternary deposits from Linton Loch, south-east Scotland. I. Absolute and relative pollen analyses of limnic sediments. *Journal of Biogeography*, **5**, 193–206.

Mannion, A.M. (1978b) Late Quaternary deposits from south-east Scotland II. The diatom assemblage of a marl core. *Journal of Biogeography*, **5**, 301–18.

Mannion, A.M. (1978c) Chemical analysis of the basal sediments from Linton Loch, south-east Scotland. *Chemosphere*, **3**, 291–6.

Mannion, A.M. (1978d) *A palaeogeographical study from south-east Scotland*. University of Reading, Department of Geography, Geographical Papers, No. 67, 42 pp.

Mannion, A.M. (1981a) The diatom assemblage of a marl core from Linton Loch. *Transactions of the Botanical Society of Edinburgh*, **43**, 263–70.

Mannion, A.M. (1981b) Chemical analyses of a marl core from south-east Scotland. *Chemosphere*, **10**, 495–504.

Mannion, A.M. (1982) Palynological evidence for lake-level changes during the Flandrian in Scotland. *Transactions of the Botanical Society of Edinburgh*, **44**, 13–8.

Marangunić, Č. and Bull, C. (1968) The landslide on Sherman Glacier. In *The Great Alaska Earthquake of 1964. 3. Hydrology*. NAS

Publication 1603. National Academy of Sciences, Washington, pp. 383–94.

Marthinussen, M. (1960) Coast and fjord area of Finnmark. In *Geology of Norway* (ed. O. Holdedahl). Norges Geologiske Undersøkelse, **208**, 416–29.

Martin, H.E. and Whalley, W.B. (1987) Rock glaciers. Part 1: rock glacier morphology, classification and distribution. *Progress in Physical Geography*, **11**, 260–82.

Martin, J. (1856) On the northern drift, as it is developed on the southern shore of the Moray Firth. *Edinburgh New Philosophical Journal*, New Series, **4**, 209–38.

Martin, J.H. (1981) Quaternary glaciofluvial deposits in central Scotland: Sedimentology and economic geology. Unpublished PhD thesis, University of Edinburgh.

Martin, M. (1884) A description of the Western Islands of Scotland *circa* 1695. T.D. Morison, Glasgow, 392 pp.

Martini, A. (1969) Sudetic tors formed under periglacial conditions. *Biuletyn Peryglacjalny*, **19**, 351–69.

Martinson, D.G., Pisias, N.G., Hays, J.D., Imbrie, J., Moore, T.C. and Shackleton, N.J. (1987) Age dating and the orbital theory of the ice ages: development of a high resolution 0 to 300,000-year chronostratigraphy. *Quaternary Research*, **27**, 1–27.

Mathews, W.H. (1973) Record of two jökulhlaups. *International Association for Scientific Hydrology Publication*, No. **95**, 99–110.

Matthews, J.A., Dawson, A.G. and Shakesby, R.A. (1986) Lake shoreline development, frost weathering and rock platform erosion in an alpine periglacial environment, Jotunheimen, southern Norway. *Boreas*, **15**, 33–50.

May, J. (1981) The glaciation and deglaciation of Upper Nithsdale and Annandale. Unpublished PhD thesis, University of Glasgow.

Mellor, A. and Wilson, M.J. (1989) Origin and significance of gibbsitic montane soils in Scotland, UK. *Arctic and Alpine Research*, **21**, 417–24.

Melvin, J. (1887) On the Parallel Roads of Lochaber. *Transactions of the Edinburgh Geological Society*, **5**, 268–74.

Menzies, J. (1976) The glacial geomorphology of Glasgow with particular reference to the drumlins. Unpublished PhD thesis, University of Glasgow.

Menzies, J. (1981) Investigations into the Quaternary deposits and bedrock topography of central Glasgow. *Scottish Journal of Geology*, **17**, 155–68.

Merritt, J.W. (1981) The sand and gravel resources of the country around Ellon, Grampian Region. Description of 1:25,000 resource sheets NJ93 with part of NJ82, 83 and 92, and NK03 and parts of NK02 and 13. *Mineral Assessment Report, Institute of Geological Sciences*, No. 76, 114 pp.

Merritt, J.W. (1990a) Evidence for events predating the maximum development of the last ice-sheet. In *Beauly to Nairn: Field Guide* (eds C.A. Auton, C.R. Firth and J.W. Merritt). Quaternary Research Association, Cambridge, pp. 1–3.

Merritt, J.W. (1990b) The lithostratigraphy at Clava and new evidence for the shell-bearing deposits being glacially-transported rafts. In *Beauly to Nairn: Field Guide* (eds C.A. Auton, C.R. Firth and J.W. Merritt). Quaternary Research Association, Cambridge, pp. 24–40.

Merritt, J.W. (1990c) The Allt Odhar interstadial site, Moy, Inverness-shire: lithostratigraphy. In *Beauly to Nairn: Field Guide* (eds C.A. Auton, C.R. Firth and J.W. Merritt). Quaternary Research Association, Cambridge, pp. 62–9.

Merritt, J.W. (in press) The high-level marine shell-bearing deposits of Clava, Inverness-shire, and their origin as glacial rafts. *Quaternary Science Reviews*, **11**.

Merritt, J.W. and Auton, C.A. (1990) The Dalcharn interglacial site, near Cawdor, Nairnshire: lithostratigraphy. In *Beauly to Nairn: Field Guide* (eds C.A. Auton, C.R. Firth and J.W. Merritt). Quaternary Research Association, Cambridge, pp. 41–54.

Merritt, J.W. and Laxton, J.L. (1982) The sand and gravel resources of the country around Callander and Dunblane, Central Region. Description of 1:25,000 sheet NN60 and 70. *Mineral Assessment Report, Institute of Geological Sciences*, No. 121, 100 pp.

Merritt, J.W. and McMillan, A.A. (1982) The 'Pliocene' gravels of Buchan: a reappraisal. *Scottish Journal of Geology*, **18**, 329–32.

Merritt, J.W., Coope, G.R., Taylor, B.J. and Walker, M.J.C. (1990) Late Devensian organic deposits beneath till in the Teith Valley, Perthshire. *Scottish Journal of Geology*, **26**, 15–24.

Metcalfe, G. (1950) The ecology of the Cairngorms. Part II. The mountain Callunetum. *Journal of Ecology*, **38**, 46–74.

Miall, A.D. (1978) Lithofacies types and vertical profile models in braided river deposits: a summary. In *Fluvial Sedimentology* (ed. A.D. Miall). Memoir of the Canadian Society of Petroleum Geologists, **5**, 597–604.

Miller, G.H., Jull, A.J.T., Linick, T., Sutherland, D., Sejrup, H.P., Brigham, J.K., Bowen, D.Q. and Mangerud, J. (1987) Racemization-derived late Devensian temperature reduction in Scotland. *Nature*, **326**, 593–5.

Miller, H. (1851) On peculiar scratched pebbles and fossil specimens from the boulder clay, and on chalk flints and oolitic fossils from the boulder clay in Caithness. *Report of the British Association for the Advancement of Science for 1850*, 93–6.

Miller, H. (1858) *The Cruise of the Betsy*; or, *A Summer Ramble Among the Fossiliferous Deposits of the Hebrides*. With *Rambles of a Geologist*; or, *Ten Thousand Miles Over the Fossiliferous Deposits of Scotland*. Thomas Constable and Co., Edinburgh, 486 pp.

Miller, H. (1859) *Sketch-book of Popular Geology: Being a Series of Lectures Delivered Before the Philosophical Institution of Edinburgh*. Thomas Constable and Co., Edinburgh, 358 pp.

Miller, H. (1864) *Edinburgh and Its Neighbourhood, Geological and Historical.* With *The Geology of the Bass Rock*. Adam and Charles Black, Edinburgh, 313pp and 24 pp.

Miller, H. (1884) On boulder-glaciation. *Proceedings of the Royal Physical Society of Edinburgh*, **8**, 156–89.

Miller, H. (1887) The Black Rock of Novar. *Transactions of the Inverness Scientific Society and Field Club*, **2**, 308–9.

Miller, M. (1987) Laminated sediments in Glen Roy, Inverness-shire. Unpublished MSc Thesis, City of London Polytechnic.

Miller, R., Common, R. and Galloway, R.W. (1954) Stone stripes and other surface features of Tinto Hill. *Geographical Journal*, **120**, 216–9.

Milne Home, D. (1840) On the Mid-Lothian and East-Lothian coal-fields. *Transactions of the Royal Society of Edinburgh*, **14**, 253–358.

Milne Home, D. (1846) Notice of polished and striated rocks recently discovered on Arthur Seat, and in some other places near Edinburgh. *Edinburgh New Philosophical Journal*, **41**, 206–8.

Milne Home, D. (1847a) On polished and striated rocks lately discovered on Arthur Seat, and other places near Edinburgh. *Edinburgh New Philosophical Journal*, **42**, 154–72.

Milne Home, D. (1847b) On the Parallel Roads of Lochaber with remarks on the change of relative levels of sea and land in Scotland. *Edinburgh New Philosophical Journal*, **43**, 339–64.

Milne Home, D. (1849) On the Parallel Roads of Lochaber, with remarks of the change of relative levels of sea and land in Scotland, and on the detrital deposits in that country. *Transactions of the Royal Society of Edinburgh*, **16**, 395–418.

Milne Home, D. (1871) *The Estuary of the Forth and Adjoining Districts Viewed Geologically*. Edmonston and Douglas, Edinburgh, 126 pp.

Milne Home, D. (1872a) Scheme for the conservation of remarkable boulders in Scotland, and for the indication of their positions on maps. *Proceedings of the Royal Society of Edinburgh*, **7**, 475–88.

Milne Home, D. (1872b) First report by the Committee on Boulders appointed by the Society. *Proceedings of the Royal Society of Edinburgh*, **7**, 703–51.

Milne Home, D. (1874a) Notice of a striated boulder lately found in a sand pit at Tynecastle, near Edinburgh. *Transactions of the Edinburgh Geological Society*, **2**, 347–50.

Milne Home, D. (1874b) Notice of a striated boulder found at Drylaw, near Linton, East Lothian. *Transactions of the Edinburgh Geological Society*, **2**, 350–1.

Milne Home, D. (1876) On the Parallel Roads of Lochaber. *Transactions of the Royal Society of Edinburgh*, **27**, 595–649.

Milne Home, D. (1877) Address by the President. *Transactions of the Edinburgh Geological Society*, **3**, 98–107.

Milne Home, D. (1878) Fourth report of Boulder Committee. *Proceedings of the Royal Society of Edinburgh*, **9**, 660–706.

Milne Home, D. (1879) Additional memoir on the Parallel Roads of Lochaber. *Transactions of the Royal Society of Edinburgh*, **28**, 93–117.

Milne Home, D. (1880a) Sixth report of the Boulder Committee. *Proceedings of the Royal Society of Edinburgh*, **10**, 577–635.

Milne Home, D. (1880b) Valedictory address. *Transactions of the Edinburgh Geological Society*, **3**, 357–64.

Milne Home, D. (1881a) On the glaciation of the Shetlands. *Geological Magazine*, **8**, 205–12.

Milne Home, D. (1881b) The glaciation of the Shetlands. *Geological Magazine*, Decade 2, **8**, 449–54.

Milne Home, D. (1881c) Notes from a diary in 1862 by D. Milne Home, LL.D, of some geological observations made (1) at points along the line of the Caledonian Railway from Carstairs southwards, (2) at Moffat, and (3) at Loch Skene. *Transactions of the Edinburgh Geological Society*, **4**, 69–74.

Milne Home, D. (1881d) Valedictory address for Session 1880–81. *Transactions of the Edinburgh Geological Society*, **4**, 104–15.

Milne Home, D. (1884) Tenth and final report of the Boulder Committee; with Appendix, containing an abstract of the information in the nine annual reports of the Committee; and a summary of the principal points apparently established by the information so received. *Proceedings of the Royal Society of Edinburgh*, **12**, 765–926.

Mitchell, G.F. (1948) Late-glacial deposits in Berwickshire. *New Phytologist*, **47**, 262–4.

Mitchell, G.F., Penny, L.F., Shotton, F.W. and West, R.G. (1973) A correlation of Quaternary deposits in the British Isles. *Geological Society of London, Special Report*, No. 4, 99 pp.

Mitchell, G.H. and Mykura, W. (1962) *The Geology of the Neighbourhood of Edinburgh*, 3rd edn. Explanation of One-inch Sheet 32. Memoirs of the Geological Survey of Scotland, HMSO, Edinburgh, 159 pp.

Moar, N.T. (1963) Pollen analysis of four samples from the River Annan, Dumfriesshire. *Transactions and Proceedings of Dumfriesshire and Galloway Natural History and Antiquarian Society*, **40**, 133–5.

Moar, N.T. (1964) The history of the Late Weichselian and Flandrian vegetation in Scotland. Unpublished PhD thesis, University of Cambridge.

Moar, N.T. (1969a) Two pollen diagrams from the Mainland, Orkney Islands. *New Phytologist*, **68**, 201–8.

Moar, N.T. (1969b) Late Weichselian and Flandrian pollen diagrams from south-west Scotland. *New Phytologist*, **68**, 433–67.

Moore, J.C. (1850) Notice of the occurrence of marine shells in the till. *Quarterly Journal of the Geological Society of London*, **6**, 388–9.

Moore, I.C. and Gribble, C.D. (1980) The suitability of aggregates from weathered Peterhead granites. *Quarterly Journal of Engineering Geology*, **13**, 305–13.

Moore, P.D. (1977) Stratigraphy and pollen analysis of Claish Moss, north-west Scotland: significance for the origin of surface pools and forest history. *Journal of Ecology*, **65**, 375–97.

Moran, S.R. (1971) Glaciotectonic structures in drift. In *Till. A Symposium* (ed. R.P. Goldthwait). Ohio State University Press, Ohio, pp. 127–48.

Moran, S.R., Clayton, L., Hooke, R. Le B., Fenton, M.M. and Andriashek, L.D. (1980) Glacier-bed landforms of the Prairie region of North America. *Journal of Glaciology*, **25**, 457–76.

Morrison, A. (1980) *Early Man in Britain and Ireland. An Introduction to Palaeolithic and Mesolithic Cultures*. Croom Helm, London, 209 pp.

Morrison, A. and Bonsall, C. (1989) The early post-glacial settlement of Scotland: a review. In *The Mesolithic in Europe* (ed. C. Bonsall). John Donald, Edinburgh, pp. 134–55.

Morrison, J., Smith, D.E., Cullingford, R.A. and Jones, R.L. (1981) The culmination of the main postglacial transgression in the Firth of Tay area, Scotland. *Proceedings of the Geologists' Association*, **92**, 197–209.

Morrison, W. (1888) Terraces at Achnasheen, Ross-shire. *Transactions of the Edinburgh Geological Society*, **5**, 275–9.

Mottershead, D.N. (1978) High altitude solifluction and post-glacial vegetation, Arkle, Sutherland. *Transactions of the Botanical Society of Edinburgh*, **43**, 17–24.

Munro, M. (1986) *Geology of the Country around Aberdeen*. Memoir of the British Geological Survey, 1:50,000 Sheet 77 (Scotland), HMSO, London, 124 pp.

Munthe, H. (1897) On the interglacial submergence of Great Britain. *Bulletin of the Geological Institute of Upsala*, **13**, 369–411.

Murdoch, W. (1975) The geomorphology and glacial deposits of the area around Aberdeen. In *Quaternary Studies in North East Scotland* (ed. A.M.D. Gemmell). Department of Geography, University of Aberdeen, Aberdeen, pp. 14–8.

Murdoch, W.M. (1977) The glaciation and deglaciation of south-east Aberdeenshire. Unpublished PhD thesis, University of Aberdeen.

Murray, J. and Pullar, L. (1910) *Bathymetrical Survey of the Scottish Freshwater Lochs*, 6 Vols. Challenger Office, Edinburgh.

Myers, A.S. (1872) On blocks and boulders lying loose on, and imbedded in, the rocks which form the seaboard of Benholm Parish, Kincardineshire, between high and low water-marks.

Transactions of the Edinburgh Geological Society, **2**, 141–4.

Mykura, W. (1976) *Orkney and Shetland*. British Regional Geology. HMSO, Edinburgh, 149 pp.

Mykura, W. and Phemister, J. (1976) *The Geology of Western Shetland*. Explanation of One-inch Geological Sheet Western Shetland; comprising Sheet 127 and parts of 125, 126 and 128. Memoirs of the Geological Survey of Great Britain: Scotland, HMSO, Edinburgh, 304 pp.

Mykura, W., Ross, D.L. and May, F. (1978) Sand and gravel resources of the Highland Region. *Report of the Institute of Geological Sciences*, **78/8**, 60 pp.

Nesje, A (1989) The geographical and altitudinal distribution of block fields in southern Norway and its significance to the Pleistocene ice sheets. *Zeitschrift für Geomorphologie*, NF, Supplementband, **72**, 41–53.

Nesje, A. and Sejrup, H.P. (1988) Late Weichselian/Devensian ice sheets in the North Sea and adjacent areas. *Boreas*, **17**, 371–84.

Newey, W.W. (1965a) Post-glacial vegetational and climatic changes in part of south-east Scotland as indicated by the pollen analysis and stratigraphy of some of its peat and lacustrine deposits. Unpublished PhD thesis, University of Edinburgh.

Newey, W.W. (1965b) Pollen analyses from south-east Scotland. *Transactions of the Botanical Society of Edinburgh*, **40**, 424–34.

Newey, W.W. (1966) Pollen analysis of sub-carse peats of the Forth Valley. *Transactions of the Institute of British Geographers*, **39**, 53–9.

Newey, W.W. (1970) Pollen analysis of Late-Weichselian deposits at Corstorphine, Edinburgh. *New Phytologist*, **69**, 1167–77.

Nichols, H. (1967) Vegetational change, shoreline displacement and the human factor in the late Quaternary history of south-west Scotland. *Transactions of the Royal Society of Edinburgh*, **67**, 145–87.

Nicol, J. (1844) *Guide to the Geology of Scotland*. Oliver and Boyd, Edinburgh, 272 pp.

Nicol, J. (1848) Observations on the recent formations in the vicinity of Edinburgh. *Quarterly Journal of the Geological Society of London*, **5**, 20–5.

Nicol, J. (1852) On the geology of the southern portion of the peninsula of Cantyre, Argyllshire. *Quarterly Journal of the Geological Society of London*, **8**, 406–25.

Nicol, J. (1869) On the origin of the Parallel Roads of Glen Roy. *Quarterly Journal of the Geological Society of London*, **25**, 282–91.

Nicol, J. (1872) How the Parallel Roads of Glen Roy were formed. *Quarterly Journal of the Geological Society of London*, **28**, 237–42.

Ockelmann, W.K. (1958) The zoology of east Greenland: marine lamellibranchiata. *Meddelelser om Grønland*, **122**, 1–256.

Odhner, N.H. (1915) Die Molluskenfauna des Eisfjordes. *Kungelige Svenska Vetenskap Akademiens Handlingar*, **54**, 1–274.

Oeschger, H. and Langway, C.C. Jr. (1989) *The Environmental Record in Glaciers and Ice Sheets*. John Wiley, Chichester, 401 pp.

Ogilvie, A.G. (1914) The physical geography of the entrance to the Inverness Firth. *Scottish Geographical Magazine*, **30**, 21–35.

Ogilvie, A.G. (1923) The physiography of the Moray Firth coast. *Transactions of the Royal Society of Edinburgh*, **53**, 377–404.

Ollier, C.D. (1969) *Weathering*. Oliver and Boyd, Edinburgh, 304 pp.

Olsson, I.U. (1979) A warning against radiocarbon dating of samples containing little carbon. *Boreas*, **8**, 203–7.

Olsson, I.U. (1986) Radiocarbon dating. In *Handbook of Holocene Palaeoecology and Palaeohydrology* (ed. B.E. Berglund). John Wiley, Chichester, pp. 273–312.

Omand, D. (1973) The glaciation of Caithness. Unpublished MSc thesis, University of Strathclyde.

O'Sullivan, P.E. (1970) The ecological history of the Forest of Abernethy, Inverness-shire. Unpublished PhD thesis, New University of Ulster.

O'Sullivan, P.E. (1973a) Pollen analysis of mor humus layers from a native Scots Pine ecosystem, interpreted with surface samples. *Oikos*, **24**, 259–72.

O'Sullivan, P.E. (1973b) Land-use changes in the Forest of Abernethy, Inverness-shire (1750–1900 AD). *Scottish Geographical Magazine*, **89**, 95–106.

O'Sullivan, P.E. (1974a) Two Flandrian pollen diagrams from the east-central Highlands of Scotland. *Pollen et Spores*, **16**, 33–57.

O'Sullivan, P.E. (1974b) Radiocarbon-dating and prehistoric forest clearance on Speyside (east-central Highlands of Scotland). *Proceedings of the Prehistoric Society*, **40**, 206–8.

O'Sullivan, P.E. (1975) Early and Middle-Flandrian pollen zonation in the Eastern Highlands of Scotland. *Boreas*, **4**, 197–207.

O'Sullivan, P.E. (1976) Pollen analysis and radiocarbon dating of a core from Loch Pityoulish, eastern Highlands of Scotland. *Journal of Biogeography*, **3**, 293–302.

O'Sullivan, P.E. (1977) Vegetation history and the native pinewoods. In *Native Pinewoods of Scotland* (eds R.G.H. Bunce and J.N.R. Jeffries). Institute of Terrestrial Ecology, Cambridge, pp. 60–9.

Otlet, R.L. and Walker, A.J. (1979) Harwell radiocarbon measurements III. *Radiocarbon*, **21**, 358–83.

Outcalt, S.I. and Benedict, J.B. (1965) Photo-interpretation of two types of rock glacier in the Colorado Front Range, USA. *Journal of Glaciology*, **5**, 849–56.

Overpeck, J.T., Peterson, L.C., Kipp, N., Imbrie, J. and Rind, D. (1989) Climate change in the circum-North Atlantic region during the last deglaciation. *Nature*, **338**, 553–7.

Page, N. (1972) On the age of the Hoxnian Interglacial. *Geological Journal*, **8**, 129–42.

Paine, A. (1982) Origin and development of blockfields in the Cairngorm Mountains. Unpublished BA dissertation, University of Cambridge.

Palmer, J. and Radley, J. (1961) Gritstone tors of the English Pennines. *Zeitschrift für Geomorphologie*, **5**, 37–52.

Panton, G.A. (1873) Note on a striated and water-worn cliff at Blackford Hill, near Edinburgh, and on a sandhill there. *Transactions of the Edinburgh Geological Society*, **2**, 238–42.

Panzer, W. (1928) Zur Oberflächengestalt der Ausseren Hebriden, Beobachtungen und Fragen. *Zeitschrift für Geomorphologie*, **3**, 167–203.

Paterson, I.B. (1974) The supposed Perth Readvance in the Perth district. *Scottish Journal of Geology*, **10**, 53–66.

Paterson, I.B. (1977) Sand and gravel resources of the Tayside region. *Report of the Institute of Geological Sciences*, **77/6**, 30 pp.

Paterson, I.B. (1981) The Quaternary geology of the Buddon Ness area of Tayside, Scotland. *Report of the Institute of Geological Sciences*, **81/1**, 9 pp.

Paterson, I.B., Armstrong, M. and Browne, M.A.E. (1981) Quaternary estuarine deposits in the Tay–Earn area, Scotland. *Report of the Institute of Geological Sciences*, **81/7**, 35 pp.

Paterson, I.B., Hall, I.H.S. and Stephenson, D. (1990) *Geology of the Greenock District*. Memoir of the British Geological Survey, 1:50,000 Geological Sheet 30W and part of Sheet 29E (Scotland), HMSO, London, 69 pp.

Paterson, W.S.B. (1981) *The Physics of Glaciers*. 2nd edn. Pergamon Press, Oxford, 380 pp.

Paul, M.A. (1983) The supraglacial landsystem. In *Glacial Geology. An Introduction for Engineers and Earth Scientists* (ed. N. Eyles). Oxford, Pergamon Press, 71–90.

Payne, A.J. and Sugden, D.E. (1990a) Climate and the initiation of maritime ice sheets. *Annals of Glaciology*, **14**, 232–7.

Payne, A.J. and Sugden, D.E. (1990b) Topography and ice sheet growth. *Earth Surface Processes and Landforms*, **15**, 625–39.

Peach, A.M. (1909) Boulder distribution from Lennoxtown, Scotland. *Geological Magazine*, **46**, 26–31.

Peach, B.N. and Horne, J. (1879) The glaciation of the Shetland Islands. *Quarterly Journal of the Geological Society of London*, **35**, 778–811.

Peach, B.N. and Horne, J. (1880) The glaciation of the Orkney Islands. *Quarterly Journal of the Geological Society of London*, **36**, 648–63.

Peach, B.N. and Horne, J. (1881a) The glaciation of the Shetland Isles. *Geological Magazine*, **8**, 65–9.

Peach, B.N. and Horne, J. (1881b) The glaciation of the Shetland Isles. *Geological Magazine*, **8**, 364–72.

Peach, B.N. and Horne, J. (1881c) The glaciation of Caithness. *Proceedings of the Royal Society of Edinburgh*, **6**, 316–52.

Peach, B.N. and Horne, J. (1893a) The ice-shed in the north-west Highlands during the maximum glaciation. *Report of the British Association for the Advancement of Science for 1892*, 720.

Peach, B.N. and Horne, J. (1893b) On a bone cave in the Cambrian limestone in Assynt, Sutherlandshire. *Report of the British Association for the Advancement of Science for 1892*, 720–1.

Peach, B.N. and Horne, J. (1893c) On the occurrence of shelly boulder clay in North Ronaldshay, Orkney. *Transactions of the Edinburgh Geological Society*, **6**, 309–13.

Peach, B.N. and Horne, J. (1910) The Scottish lochs in relation to the geological features of the country. In *Bathymetrical Survey of the Scottish Freshwater Lochs. 1.* (eds J. Murray and L. Pullar). Challenger Office, Edinburgh, pp. 439–513.

Peach, B.N. and Horne, J. (1917) The bone-cave

in the valley of Allt nan Uamh (Burn of the Caves), near Inchnadamph, Assynt, Sutherlandshire. *Proceedings of the Royal Society of Edinburgh*, **37**, 327–49.

Peach, B.N. and Horne, J. (1930) *Chapters on the Geology of Scotland*. Oxford University Press, London, pp. 1–21.

Peach, B.N., Clough, C.T., Hinxman, L.W., Wilson, J.S.G., Crampton, C.B., Maufe, H.B. and Bailey, E.B. (1910a) *The Geology of the Neighbourhood of Edinburgh*, 2nd edn. Explanation of Sheet 32, with part of 31. Memoirs of the Geological Survey of Scotland, HMSO, Edinburgh, 445 pp.

Peach, B.N., Horne, J., Woodward, H.B., Clough, C.T., Harker, A. and Wedd, C.B. (1910b) *The Geology of Glenelg, Lochalsh and south-east part of Skye*. Explanation of One-inch Map 71. Memoirs of the Geological Survey of Scotland, HMSO, Edinburgh, 206 pp.

Peach, B.N., Wilson, J.S.G., Hill, J.B., Bailey, E.B. and Graham, G.W. (1911) *The Geology of Knapdale, Jura and north Kintyre*. Explanation of Sheet 28, with parts of 27 and 29. Memoirs of the Geological Survey of Scotland, HMSO, Edinburgh, 149 pp.

Peach, B.N., Gunn, W., Clough, C.T., Hinxman, L.W., Crampton, C.B. and Anderson, E.M. (1912) *The Geology of Ben Wyvis, Carn Chuinneag, Inchbae and the Surrounding Country, including Garve, Evanton, Alness and Kincardine*. Explanation of Sheet 93. Memoirs of the Geological Survey of Scotland, HMSO, Edinburgh, 189 pp.

Peach, B.N., Horne, J., Gunn, W., Clough, C.T., Greenly, E., Hinxman, L.W., Cadell, H.M., Pocock, T.I. and Crampton, C.B. (1913a) *The Geology of the Fannich Mountains and the Country around Upper Loch Maree and Strath Broom*. Explanation of Sheet 92. Memoirs of the Geological Survey of Scotland, HMSO, Edinburgh, 127 pp.

Peach, B.N., Horne, J., Hinxman, L.W., Crampton, C.B., Anderson, E.M. and Carruthers, R.G. (1913b) *The Geology of Central Ross-shire*. Explanation of Sheet 82. Memoirs of the Geological Survey of Scotland, HMSO, Edinburgh, 114pp.

Peach, C.W. (1858) On the discovery of calcareous zoophytes in the boulder clay of Caithness, N.B. *Transactions of the Royal Physical Society of Edinburgh*, **1**, 18.

Peach, C.W. (1859) On the discovery of nullipores (calcareous plants) and sponges in the boulder clay of Caithness. *Transactions of the Royal Physical Society of Edinburgh*, **2**, 98–101.

Peach, C.W. (1860) On the chalk flints of the Island of Stroma and vicinity of John o' Groats, in the County of Caithness. *Proceedings of the Royal Physical Society of Edinburgh*, **2**, 159–61.

Peach, C.W. (1863a) On the fossils of the boulder-clay of Caithness, N.B. *Proceedings of the Royal Physical Society of Edinburgh*, **3**, 38–42.

Peach, C.W. (1863b) Further observations on the boulder clay of Caithness, with an additional list of fossils. *Proceedings of the Royal Physical Society of Edinburgh*, **3**, 396–403.

Peach, C.W. (1863c) On the fossils of the boulder-clay in Caithness. *Report of the British Association for the Advancement of Science for 1862*, 83–4.

Peach, C.W. (1865a) Traces of glacial drift in the Shetland Islands. *Report of the British Association for the Advancement of Science for 1864*, 59–61.

Peach, C.W. (1865b) Additional list of fossils from the boulder-clay of Caithness. *Report of the British Association for the Advancement of Science for 1864*, 61–3.

Peach, C.W. (1867) Further observations on, and additions to, the list of fossils found in the boulder-clay of Caithness, N.B. *Report of the British Association for the Advancement of Science for 1866*, 64–5.

Peacock, J.D. (1966) Note on the drift sequence near Portsoy, Banffshire. *Scottish Journal of Geology*, **2**, 35–7.

Peacock, J.D. (1967) West Highland morainic features aligned in the direction of ice flow. *Scottish Journal of Geology*, **3**, 372–3.

Peacock, J.D. (1970a) Some aspects of the glacial geology of west Inverness-shire. *Bulletin of the Geological Survey of Great Britain*, **33**, 43–56.

Peacock, J.D. (1970b) Glacial geology of the Lochy–Spean area. *Bulletin of the Geological Survey of Great Britain*, **31**, 185–98.

Peacock, J.D. (1971a) A re-interpretation of the coastal deposits of Banffshire and their place in the Late-glacial history of Scotland. *Bulletin of the Geological Survey of Great Britain*, **37**, 81–9.

Peacock, J.D. (1971b) Terminal features of the

Creran glacier of Loch Lomond Readvance age in western Benderloch, Argyll, and their significance in the late-glacial history of the Loch Linnhe area. *Scottish Journal of Geology*, **7**, 349–56.

Peacock, J.D. (1971c) Marine shell radiocarbon dates and the chronology of deglaciation in western Scotland. *Nature Physical Science*, **230**, 43–5.

Peacock, J.D. (1974a) Borehole evidence for late- and post-glacial events in the Cromarty Firth, Scotland. *Bulletin of the Geological Survey of Great Britain*, **8**, 55–67.

Peacock, J.D. (1974b) Islay shelly till with *Palliolum groenlandicum. Scottish Journal of Geology*, **10**, 159–60.

Peacock, J.D. (1975a) Palaeoclimatic significance of ice-movement directions of Loch Lomond Readvance glaciers in the Glen Moriston and Glen Affric areas, northern Scotland. *Bulletin of the Geological Survey of Great Britain*, **49**, 39–42.

Peacock, J.D. (1975b) Depositional environment of glacial deposits at Clava, north-east Scotland. *Bulletin of the Geological Survey of Great Britain*, **49**, 31–7.

Peacock, J.D. (1975c) Scottish late and post-glacial marine deposits. In *Quaternary Studies in North East Scotland* (ed. A.M.D. Gemmell). Department of Geography, University of Aberdeen, Aberdeen, pp. 45–8.

Peacock, J.D. (1975d) Landslip associated with glacier ice. *Scottish Journal of Geology*, **11**, 363–5.

Peacock, J.D. (1976) Quaternary features of Rhum, Inner Hebrides. *Quaternary Newsletter*, **20**, 1–4.

Peacock, J.D. (1977a) Subsurface deposits of Inverness and the inner Cromarty Firth. In *The Moray Firth Area Geological Studies* (ed. G. Gill). The Inverness Field Club, Inverness, pp. 103–4.

Peacock, J.D. (1977b) South Shian. In *Western Scotland I* (ed. R.J. Price). INQUA X Congress 1977. Guide book for excursion A12. Geo Abstracts, Norwich, 38.

Peacock, J.D. (1980a) Glaciation of the Outer Hebrides: a reply. *Scottish Journal of Geology*, **16**, 87–9.

Peacock, J.D. (1980b) An overlooked record of interglacial or interstadial sites in north-east Scotland. *Quaternary Newsletter*, **32**, 14–5.

Peacock, J.D. (1981a) Report and excursion guide – Lewis and Harris. *Quaternary Newsletter*, **35**, 45–54.

Peacock, J.D. (1981b) Scottish Late-glacial marine deposits and their environmental significance. In *The Quaternary in Britain* (eds J. Neale and J. Flenley). Pergamon Press, Oxford, pp. 222–36.

Peacock, J.D. (1983a) A model for Scottish interstadial marine palaeotemperature 13,000 to 11,000 BP. *Boreas*, **12**, 73–82.

Peacock, J.D. (1983b) Quaternary geology of the Inner Hebrides. *Proceedings of the Royal Society of Edinburgh*, **83B**, 83–9.

Peacock, J.D. (1984a) Quaternary geology of the Outer Hebrides. *Report of the British Geological Survey*, **16(2)**, 26 pp.

Peacock, J.D. (1984b) Errolston. In *Buchan Field Guide* (ed. A.M. Hall). Quaternary Research Association, Cambridge, pp. 108–9.

Peacock, J.D. (1985) Marine bivalve bores above present high tide level in a limestone cliff near Ord, southern Skye. *Scottish Journal of Geology*, **21**, 209–12.

Peacock, J.D. (1986) Alluvial fans and an outwash fan in upper Glen Roy, Lochaber. *Scottish Journal of Geology*, **22**, 347–66.

Peacock, J.D. (1987) A reassessment of the probable Loch Lomond Stade marine molluscan fauna at Garvel Park, Greenock. *Scottish Journal of Geology*, **23**, 93–103.

Peacock, J.D. (1989a) Survey of Quaternary landforms and deposits of Glen Roy, Glen Gloy and Glen Spean, Lochaber District, and recommendations for conservation. Nature Conservancy Council, Peterborough, CSD Report, No. 923, 59 pp.

Peacock, J.D. (1989b) Marine molluscs and Late Quaternary environmental studies with particular reference to the Late-glacial period in north-west Europe: a review. *Quaternary Science Reviews*, **8**, 179–92.

Peacock, J.D. (1991) Glacial deposits of the Hebridean region. In *Glacial Deposits in Great Britain and Ireland* (eds J. Ehlers, P.L. Gibbard and J. Rose). A.A. Balkema, Rotterdam, pp. 109–19.

Peacock, J.D. and Cornish, R. (1989) *Field Guide to the Glen Roy Area*. Quaternary Research Association, Cambridge, 69 pp.

Peacock, J.D. and Harkness, D.D. (1990) Radiocarbon ages and the full-glacial to Holocene transition in seas adjacent to Scotland and southern Scandinavia: a review. *Transactions of the Royal Society of Edinburgh: Earth Sciences*, **81**, 385–96.

Peacock, J.D. and Ross, D.L. (1978) Anomalous glacial erratics in the southern part of the Outer Hebrides. *Scottish Journal of Geology*, **14**, 262.

Peacock, J.D., Berridge, N.G., Harris, A.L. and May, F. (1968) *The Geology of the Elgin District*. Explanation of Sheet 95. Memoirs of the Geological Survey of Scotland, HMSO, Edinburgh, 165 pp.

Peacock, J.D., Austin, W.E.N., Selby, I., Graham, D.K., Harland, R. and Wilkinson, I.P. (1992) Late Devensian and Flandrian palaeoenvironmental changes on the Scottish continental shelf west of the Outer Hebrides. *Journal of Quaternary Science*, **7**, 145–61.

Peacock, J.D., Graham, D.K., Robinson, J.E. and Wilkinson, I. (1977) Evolution and chronology of Lateglacial marine environments at Lochgilphead, Scotland. In *Studies in the Scottish Lateglacial Environment* (eds J.M. Gray and J.J. Lowe). Pergamon Press, Oxford, pp. 89–100.

Peacock, J.D., Graham, D.K. and Wilkinson, I.P. (1978) Late-glacial and post-glacial marine environments at Ardyne, Scotland, and their significance in the interpretation of the history of the Clyde sea area. *Report of the Institute of Geological Sciences*, **78/17**, 25 pp.

Peacock, J.D., Graham, D.K. and Gregory, D.M. (1980) Late and post-glacial marine environments in part of the inner Cromarty Firth. *Report of the Institute of Geological Sciences*, **80/7**, 11 pp.

Peacock, J.D., Harkness, D.D., Housley, R.A., Little, J.A. and Paul, M.A. (1989) Radiocarbon ages for a glaciomarine bed associated with the maximum of the Loch Lomond Readvance in west Benderloch, Argyll. *Scottish Journal of Geology*, **25**, 69–79.

Pears, N.V. (1964) The present tree-line in the Cairngorm Mountains of Scotland and its relation to former tree-lines. Unpublished PhD thesis, University of London.

Pears, N.V. (1967) Present tree-lines of the Cairngorm Mountains, Scotland. *Journal of Ecology*, **55**, 815–29.

Pears, N.V. (1968) Post-glacial tree-lines of the Cairngorm Mountains, Scotland. *Transactions of the Botanical Society of Edinburgh*, **40**, 361–94.

Pears, N.V. (1970) Post-glacial tree-lines of the Cairngorm Mountains, Scotland: some modifications based on radiocarbon dating. *Transactions of the Botanical Society of Edinburgh*, **40**, 536–44.

Pears, N.V. (1972) Interpretation problems in the study of tree-line fluctuations. In *Research Papers in Forest Meteorology, an Aberystwyth Symposium* (ed. J.A. Taylor). University College of Wales, Aberystwyth, pp. 31–45.

Pears, N.V. (1975a) Radiocarbon dating of peat macrofossils in the Cairngorm Mountains. *Transactions of the Botanical Society of Edinburgh*, **42**, 255–60.

Pears, N.V. (1975b) The growth rate of hill peats in Scotland. *Geologiska Föreningens i Stockholm Förhandlingar*, **97**, 265–70.

Pears, N.V. (1988) Pine stumps, radiocarbon dates and stable isotope analysis in the Cairngorm Mountains: some observations. *Review of Palaeobotany and Palynology*, **54**, 175–80.

Peglar, S. (1979) A radiocarbon-dated pollen diagram from Loch of Winless, Caithness, north-east Scotland. *New Phytologist*, **82**, 245–63.

Pennant, T. (1771) *A Tour in Scotland 1769*. John Monk, Chester, 316 pp.

Pennington, W. (1975a) Climatic changes in Britain, as interpreted from lake sediments, between 15,000 and 10,000 years ago. In *Palaeolimnology of Lake Biwa and the Japanese Pleistocene*, Vol. 3 (ed. S. Horie). Kyoto University, Institute of Palaeolimnology, pp. 536–69.

Pennington, W. (1975b) A chronostratigraphic comparison of Late-Weichselian and Late-Devensian subdivisions, illustrated by two radiocarbon-dated profiles from western Britain. *Boreas*, **4**, 157–71.

Pennington, W. (1975c) An application of principal components analysis to the zonation of two Late-Devensian profiles. II. Interpretation of the numerical analyses in terms of Late-Devensian (Late-Weichselian) environmental history. *New Phytologist*, **75**, 441–53.

Pennington, W. (1977a) Lake sediments and the Lateglacial environment in northern Scotland. In *Studies in the Scottish Lateglacial Environment* (eds J.M. Gray and J.J. Lowe). Pergamon Press, Oxford, pp. 119–41.

Pennington, W. (1977b) The Late Devensian flora

and vegetation of Britain. *Philosophical Transactions of the Royal Society*, **B280**, 247–71.

Pennington, W. (1978) Quaternary geology. In *The Geology of the Lake District* (ed. F.M. Moseley). Yorkshire Geological Society, Leeds, Occasional Publication No. 3, pp. 207–25.

Pennington, W. (1980) Modern pollen samples from West Greenland and the interpretation of pollen data from the British late-glacial (Late Devensian). *New Phytologist*, **84**, 171–201.

Pennington, W. (1986) Lags in adjustment of vegetation to climate caused by the pace of soil development: evidence from Britain. *Vegetatio*, **67**, 105–18.

Pennington, W. and Sackin, M.J. (1975) An application of principal components analysis to the zonation of two Late-Devensian profiles. I. Numerical analysis. *New Phytologist*, **75**, 419–41.

Pennington, W., Haworth, E.Y., Bonny, A.P. and Lishman, J.P. (1972) Lake sediments in northern Scotland. *Philosophical Transactions of the Royal Society of London*, **B264**, 191–294.

Perry, I. and Moore, P.D. (1987) Dutch elm disease as an analogue to Neolithic elm decline. *Nature*, **326**, 72–3.

Petch, C.P. (1933) The vegetation of St Kilda. *Journal of Ecology*, **21**, 92–100.

Phemister, J. (1960) *Scotland: the northern Highlands*, 3rd edn. British Regional Geology. HMSO, Edinburgh, 104 pp.

Phemister, T.C. and Simpson, S. (1949) Pleistocene deep-weathering in north-east Scotland. *Nature*, **164**, 318–9.

Phillips, L. (1974) Vegetational history of the Ipswichian/Eemian interglacial in Britain and continental Europe. *New Phytologist*, **73**, 589–604.

Phillips, L. (1976) Pleistocene vegetational history and geology in Norfolk. *Philosophical Transactions of the Royal Society of London*, **B275**, 215–86.

Piggot, C.D. and Walters, S.M. (1954) On the interpretation of the discontinuous distribution shown by certain British species of open habitats. *Journal of Ecology*, **42**, 95–116.

Pilcher, J. (1991) Radiocarbon dating for the Quaternary scientist. In *Radiocarbon Dating: Recent Applications and Future Potential* (ed. J.J. Lowe). Quaternary Proceedings No. 1. Quaternary Research Association, Cambridge, pp. 27–33.

Poore, M.E.D. and Robertson, V.C. (1948) The vegetation of St Kilda in 1948. *Journal of Ecology*, **37**, 82–99.

Powell, R.D. (1983) Glacial-marine sedimentation processes and lithofacies of temperate tidewater glaciers, Glacier Bay, Alaska. In *Glacial-marine Sedimentation* (ed. B.F. Molnia). Plenum Press, New York, pp. 185–232.

Preece, R.C., Bennett, K.D. and Robinson, J.E. (1984) The biostratigraphy of an early Flandrian tufa at Inchrory, Glen Avon, Banffshire. *Scottish Journal of Geology*, **20**, 143–59.

Prell, W.L., Imbrie, J., Martinson, D.G., Morley, J.J., Pisias, N.G., Shackleton, N.J. and Streeter, H.F. (1986) Graphic correlation of oxygen isotope stratigraphy application to the late Quaternary. *Palaeoceanography*, **1**, 137–62.

Prestwich, J. (1838a) Observations on the Ichthyolites of Gamrie in Banffshire, and on the accompanying red conglomerates and sandstones. *Proceedings of the Geological Society*, **2**, 187–8.

Prestwich, J. (1838b) On some recent elevations of the coast of Banffshire; and on a deposit of clay, formerly considered to be Lias. *Proceedings of the Geological Society*, **2**, 545.

Prestwich, J. (1840) On the structure of the neighbourhood of Gamrie, Banffshire, particularly on the deposit containing Ichthyolites. *Transactions of the Geological Society*, 2nd Series, **5**, 139–48.

Prestwich, J. (1880) On the origin of the Parallel Roads of Lochaber and their bearing on other phenomena of the glacial period. *Philosophical Transactions of the Royal Society of London*, **170**, 663–726.

Price, R.J. (1960) Glacial meltwater channels in the upper Tweed drainage basin. *Geographical Journal*, **126**, 483–9.

Price, R.J. (1963a) A glacial meltwater drainage system in Peebles-shire, Scotland. *Scottish Geographical Magazine*, **79**, 133–41.

Price, R.J. (1963b) The glaciation of a part of Peeblesshire. *Transactions of the Edinburgh Geological Society*, **19**, 323–48.

Price, R.J. (1973) *Glacial and Fluvioglacial Landforms*. Oliver and Boyd, Edinburgh, 242 pp.

Price, R.J. (1975) The glaciation of west-central Scotland – a review. *Scottish Geographical Magazine*, **91**, 134–45.

Price, R.J. (1976) *Highland Landforms*. Highlands and Islands Development Board, Inverness, 109 pp.

Price, R.J. (1980) Geomorphological implications

of environmental changes during the last 30 000 years in central Scotland. *Zeitschrift für Geomorphologie*, NF, Supplementband, **36**, 74–83.

Price, R.J. (1983) *Scotland's Environment During the Last 30,000 years*. Scottish Academic Press, Edinburgh, 224 pp.

Price, R.J., Browne, M.A.E. and Jardine, W.G. (1980) Introduction. The Quaternary of the Glasgow region. In *Field Guide to the Glasgow Region* (ed. W.G. Jardine). Quaternary Research Association, Cambridge, pp. 3–9.

Pringle, J (1936) Ammonites from a transported mass of Jurassic clay at Plaidy, Aberdeenshire, with the description of a new species. *Transactions of the Edinburgh Geological Society*, **13**, 308–10.

Pye, K. and Paine, A.D.M. (1984) Nature and source of aeolian deposits near the summit of Ben Arkle, north-west Scotland. *Geologie en Mijnbouw*, **63**, pp. 13–18.

Rae, D.A. (1976) Aspects of glaciation in Orkney. Unpublished PhD thesis, University of Liverpool.

Ragg, J.M. (1960) *The Soils of the Country round Kelso and Lauder*. Sheets 25 and 26. Memoirs of the Soil Survey of Great Britain: Scotland, HMSO, Edinburgh, 201 pp.

Ragg, J.M. and Bibby, J.S. (1966) Frost weathering and solifluction products in southern Scotland. *Geografiska Annaler*, **48**, 12–23.

Ragg, J.M. and Futty, D.W. (1967) *The Soils of the Country round Haddington and Eyemouth*. Sheets 33, 34 and part of 41. Memoirs of the Soil Survey of Great Britain: Scotland. HMSO, Edinburgh, 310 pp.

Ramsay, A.C. (1878) *The Physical Geology and Geography of Great Britain*, 5th edn. Edward Stanford, London, 639 pp.

Rapson, S.C. (1985) Minimum age of corrie moraine ridges in the Cairngorm Mountains, Scotland. *Boreas*, **14**, 155–9.

Rapson, S.C. (1990) The age of the Cairngorm coire moraines. *Scottish Mountaineering Club Journal*, **34(181)**, 457–63.

Ratcliffe, D.A. (1977) *A Nature Conservation Review*, 2 vols. Cambridge University Press, Cambridge, 401pp and 320 pp.

Read, H.H. (1923) *The Geology of the Country round Banff, Huntly and Turriff.* Lower Banffshire and north-west Aberdeenshire. Explanation of Sheets 86 and 96. Memoirs of the Geological Survey of Scotland, HMSO, Edinburgh, 240 pp.

Read, H.H., Bremner, A., Campbell, R. and Gibb, A.W. (1923) Records of the occurrence of boulders of Norwegian rocks in Aberdeenshire and Banffshire. *Transactions of the Edinburgh Geological Society*, **11**, 230–1.

Reade, T.M. (1896) The present aspects of glacial geology. *Geological Magazine*, Decade 4, **3**, 542–51.

Reed, W.J. (1988) The vertical dimensions of the last ice sheet and Late Quaternary glacial events in northern Ross-shire. Unpublished PhD thesis, University of St Andrews.

Reid, J.R. (1969) Effects of a debris slide on 'Sioux Glacier', south-central Alaska. *Journal of Glaciology*, **8**, 353–67.

Renberg, I. and Hellberg, T. (1982) The pH history of lakes in south-western Sweden, as calculated from the subfossil flora of the sediments. *Ambio*, **11**, 30–3.

Rendell, H., Worsley, P., Green, F. and Parks, D. (1991) Thermoluminescence dating of the Chelford Interstadial. *Earth and Planetary Science Letters*, **103**, 182–9.

Renwick, J. (1895) Notes on an excursion to Glen Fruin, with a description of the moraine on its south side. *Transactions of the Geological Society of Glasgow*, **10**, 96–104.

Renwick, J. and Gregory, J.W. (1907) The Loch Lomond moraines. *Transactions of the Geological Society of Glasgow*, **13**, 45–55.

Rhind, D.W. (1965) Evidence of sea level changes along the coast north of Berwick. *Proceedings of the Berwickshire Naturalists' Club*, **37**, 10–15.

Rhind, D.W. (1972) The buried valley of the lower Tweed. *Transactions of the Natural History Society of Northumberland, Durham and Newcastle upon Tyne*, **4**, 159–64.

Rhind, W. (1836) *Excursions Illustrative of the Geology and Natural History of the Environs of Edinburgh*, 2nd edn. Maclachlan and Stewart, Edinburgh, 72 pp.

Rice, R.J. (1959) The glacial deposits of the Lunan and Brothock valleys in south-eastern Angus. *Transactions of Edinburgh Geological Society*, **17**, 241–59.

Rice, R.J. (1961) The glacial deposits at St Fort in north-eastern Fife: a re-examination. *Transactions of the Edinburgh Geological Society*, **18**, 113–23.

Rice, R.J. (1962) The morphology of the Angus coastal lowlands. *Scottish Geographical Magazine*, **78**, 5–14.

Richardson, R. (1877a) On phenomena of glaciation exhibited by the rocks of Corstorphine

Hill, near Edinburgh. *Transactions of the Edinburgh Geological Society*, **3**, 31–40.

Richardson, R. (1877b) Notice of glaciated rock surfaces (displaying corals) near Bathgate, recently quarried away. *Transactions of the Edinburgh Geological Society*, **3**, 108–9.

Richardson, R. (1882) On the discovery of arctic shells at high levels in Scotland. *Transactions of the Edinburgh Geological Society*, **4**, 179–200.

Richey, J.E. (1961) *Scotland: the Tertiary Volcanic Districts*, 3rd edn. British Regional Geology. HMSO, Edinburgh, 120 pp.

Richey, J.E. and Thomas, H.H. (1930) *The Geology of Ardnamurchan, North-west Mull and Coll.* A description of Sheet 51 and part of Sheet 52 of the Geological Map. Memoirs of the Geological Survey of Scotland, HMSO, Edinburgh, 393 pp.

Richey, J.E., Anderson, E.M., MacGregor, A.G., Bailey, E.B., Wilson, G.V., Burnett, G.A. and Eyles, V.A. (1930) *The Geology of North Ayrshire,* 2nd edn. Explanation of One-inch Sheet 22. Memoirs of the Geological Survey of Scotland, HMSO, Edinburgh, 417 pp.

Rind, D., Peteet, D. and Kukla, G. (1989) Can Milankovitch orbital variations initiate the growth of ice sheets in a general circulation model? *Journal of Geophysical Research*, **94**, 12,851–71.

Ringrose, P.S. (1987) Fault activity and palaeoseismicity during Quaternary time in Scotland. Unpublished PhD thesis, University of Strathclyde.

Ringrose, P.S. (1989a) Palaeoseismic(?) liquefaction event in Late Quaternary lake sediment at Glen Roy, Scotland. *Terra Nova*, **1**, 57–62.

Ringrose, P.S. (1989b) Recent fault movement and palaeoseismicity in western Scotland. *Tectonophysics,* **163**, 305–14.

Ringrose, P.S. (1989c) Evidence for palaeoseismicity: evidence from lake sediments. In *Glen Roy Area: Field Guide* (eds J.D. Peacock and R. Cornish). Quaternary Research Association, Cambridge, pp. 8–9.

Ritchie, J. (1928) The fauna of Scotland during the Ice Age. *Proceedings of the Royal Physical Society of Edinburgh*, **21**, 185–94.

Ritchie, W. (1966) The post-glacial rise in sea-level and coastal changes in the Uists. *Transactions of the Institute of British Geographers*, **39**, 79–86.

Ritchie, W. (1979) Machair development and chronology of the Uists and adjacent islands. *Proceedings of the Royal Society of Edinburgh*, **B77**, 107–22.

Ritchie, W. (1985) Inter-tidal and sub-tidal organic deposits and sea-level changes in the Uists, Outer Hebrides. *Scottish Journal of Geology*, **21**, 161–76.

Ritchie, W. (1986) Anomalous east coast machair in the Uists, Outer Hebrides. In *Essays for Professor R.E.H. Mellor* (eds W. Ritchie, J.C. Stone and A.S. Mather). Department of Geography, University of Aberdeen, Aberdeen, pp. 383–9.

Ritchie, W. and Mather, A. (1970) *The Beaches of Lewis and Harris.* Department of Geography, University of Aberdeen, Aberdeen, 113 pp.

Robertson, D. (1868) On clay beds of Ross Arden, on the banks of Loch Lomond. *Proceedings of the Natural History Society of Glasgow*, **1**, 92–3.

Robertson, D. (1875) Notes on the recent Ostracoda and Foraminifera of the Firth of Clyde, with some remarks on the distribution of Mollusca. *Transactions of the Geological Society of Glasgow*, **5**, 112–53.

Robertson, D. (1883) On the post-Tertiary beds of Garvel Park, Greenock. *Transactions of the Geological Society of Glasgow,* **7**, 1–37.

Robertson-Rintoul, M.S.E. (1986a) River planform, soil stratigraphy and the temporal and palaeoenvironmental significance of terraced valley fill deposits in upland Scotland. Unpublished PhD thesis, University of Hull.

Robertson-Rintoul, M.S.E. (1986b) A quantitative soil–stratigraphic approach to the correlation and dating of post-glacial river terraces in Glen Feshie, western Cairngorms. *Earth Surface Processes and Landforms*, **11**, 605–17.

Robin, G. de Q. (1983) *The Climatic Record in Polar Ice Sheets*. Cambridge University Press, Cambridge, 212 pp.

Robinson, D.E. (1981) The vegetational and land use history of the west of Arran, Scotland. Unpublished PhD thesis, University of Glasgow.

Robinson, D.E. (1983) Possible Mesolithic activity in the west of Arran. *Glasgow Archaeological Journal*, **10**, 1–6.

Robinson, D.E. (1987) Investigations into the Aukhorn peat mounds, Keiss, Caithness: pollen, plant macrofossil and charcoal analyses. *New Phytologist*, **106**, 185–200.

Robinson, D.E. and Dickson, J.H. (1988) Vegetational history and land use: a radiocarbon-dated pollen diagram from Machrie Moor,

Arran, Scotland. *New Phytologist*, **109**, 223–36.

Robinson, G., Peterson, J.A. and Anderson, P.M. (1971) Trend surface analysis of corrie altitudes in Scotland. *Scottish Geographical Magazine*, **87**, 142–6.

Robinson, M. (1977) Glacial limits, sea-level changes and vegetational development in part of Wester Ross. Unpublished PhD thesis, University of Edinburgh.

Robinson, M. (1982) Diatom analysis of early Flandrian lagoon sediments from East Lothian, Scotland. *Journal of Biogeography*, **9**, 207–22.

Robinson, M. (1987a) The Loch Lomond Readvance in Torridon and Applecross. In *Wester Ross Field Guide* (eds C.K. Ballantyne and D.G. Sutherland). Quaternary Research Association, Cambridge, pp. 123–7.

Robinson, M. (1987b) North-west Applecross peninsula. In *Wester Ross Field Guide* (eds C.K. Ballantyne and D.G. Sutherland). Quaternary Research Association, Cambridge, pp. 144–50.

Robinson, M. (1987c) Glasscnock (NG 8670 4610) and Druim Dubh (NC 8845 4720). In *Wester Ross Field Guide* (eds C.K. Ballantyne and D.G. Sutherland). Quaternary Research Association, Cambridge, 154–64.

Robinson, M. and Ballantyne, C.K. (1979) Evidence for a glacial readvance pre-dating the Loch Lomond Advance in Wester Ross. *Scottish Journal of Geology*, **15**, 271–7.

Rogers, H.D. (1862) On the origin of the Parallel Roads of Lochaber (Glen Roy), Scotland. *Proceedings of the Royal Institution of Great Britain*, **3**, 341–5.

Rolfe, W.D.I. (1966) Woolly rhinoceros from the Scottish Pleistocene. *Scottish Journal of Geology*, **2**, 253–8.

Romans, J.C.C. (1977) Stratigraphy of buried soil at Teindland Forest, Scotland. *Nature*, **268**, 622–3.

Romans, J.C.C. and Robertson, L. (1974) Some aspects of the genesis of alpine and upland soils in the British Isles. In *Soil Microscopy: Proceedings of the Fourth International Working-Meeting on Soil Micromorphology, Department of Geography, Queen's University, Kingston, Ontario, Canada, 27–31, August 1973* (ed. A.K. Rutherford). Limestone Press, Kingston, Ontario, pp. 498–510.

Romans, J.C.C., Stevens, J.H. and Robertson, L. (1966) Alpine soils of north-east Scotland. *Journal of Soil Science*, **17**, 184–99.

Rose, J. (1975) Raised beach gravels and ice wedge casts at Old Kilpatrick, near Glasgow. *Scottish Journal of Geology*, **11**, 15–21.

Rose, J. (1980a) Geilston. In *Glasgow Region Field Guide* (ed. W.G. Jardine). Quaternary Research Association, Glasgow, pp. 25–9.

Rose, J. (1980b) Ardmore Point. In *Glasgow Region Field Guide* (ed. W.G. Jardine). Quaternary Research Association, Glasgow, pp. 29–31.

Rose, J. (1980c) Rhu. In *Glasgow Region Field Guide* (ed. W.G. Jardine). Quaternary Research Association, Glasgow, pp. 31–7.

Rose, J. (1980d) The western side of Loch Lomond. In *Glasgow Region Field Guide* (ed. W.G. Jardine). Quaternary Research Association, Glasgow, pp. 37–9.

Rose, J. (1980e) Gartness. In *Glasgow Region Field Guide* (ed. W.G. Jardine). Quaternary Research Association, Glasgow, pp. 46–9.

Rose, J. (1980f) Ross Priory and the southern shore of Loch Lomond. In *Glasgow Region Field Guide* (ed. W.G. Jardine). Quaternary Research Association, Glasgow, pp. 49–51.

Rose, J. (1981) Field guide to the Quaternary geology of the south-eastern part of the Loch Lomond basin. *Proceedings of the Geological Society of Glasgow 1980–81*, 1–19.

Rose, J. (1987) Drumlins as part of a glacier bed continuum. In *Drumlin Symposium* (eds J. Menzies and J. Rose). Proceedings of the Drumlin Symposium, First International Conference on Geomorphology, Manchester, 16–18 September 1985. A.A. Balkema, Rotterdam, pp. 103–16.

Rose, J. (1989) Stadial type sections in the British Quaternary. In *Quaternary Type Sections: Imagination or Reality?* (eds J. Rose and C. Schlüchter). A.A. Balkema, Rotterdam, pp. 45–67.

Rose, J. and Letzer, J.M. (1975) Drumlin measurements: a test of the reliability of data derived from 1:25000 scale topographic maps. *Geological Magazine*, **112**, 361–71.

Rose, J. and Letzer, J.M. (1977) Superimposed drumlins. *Journal of Glaciology*, **18**, 471–80.

Rose, J. and McLellan, A.G. (1967) Landforms of eastern Perthshire. *Scottish Field Studies Association Annual Report for 1966*, 14–30.

Rose, J., Lowe, J.J. and Switsur, R. (1988) A radiocarbon date on plant detritus beneath till from the type area of the Loch Lomond Readvance. *Scottish Journal of Geology*, **24**, 113–24.

Ross, G. (1927) The superficial deposits in the Clyde Valley at Bonnington, 1½ miles south of Lanark. Geological Survey of Great Britain. *Summary of Progress of the Geological Survey of Great Britain and the Museum of Practical Geology for the Year of 1926.* HMSO, London, pp. 158–60.

Ross, T. (1796) Parish of Kilmanivaig. In *The Statistical Account of Scotland*, Vol. **17** (ed. Sir J. Sinclair). William Creech, Edinburgh, 543–50.

Röthlisberger, H. and Lang, H. (1987) Glacial hydrology. In *Glacio-fluvial Sediment Transfer. An Alpine Perspective* (eds A.M. Gurnell and M.J. Clark). John Wiley, Chichester, pp. 207–84.

Round, F.E. (1957) The late-glacial and post-glacial diatom succession in the Kentmere valley deposit. I. Introduction, methods and flora. *New phytologist*, **56**, 98–126.

Round, F.E. (1961) Diatoms from Estwaite. *New Phytologist*, **60**, 43–59.

Royal Commission On The Ancient And Historical Monuments Of Scotland (1980) *Argyll: an inventory of the ancient monuments. Vol. 3: Mull, Tiree, Coll and northern Argyll*: (excluding the early medieval and later monuments of Iona). HMSO, London, 281 pp.

Ruddiman, W.F. and Kutzbach, J.E. (1990) Late Cenozoic plateau uplift and climate change. *Transactions of the Royal Society of Edinburgh: Earth Sciences*, **81**, 301–14.

Ruddiman, W.F. and McIntyre, A. (1973) Time-transgressive deglacial retreat of polar waters from the North Atlantic. *Quaternary Research*, **3**, 117–30.

Ruddiman, W.F. and McIntyre, A. (1976) North-east Atlantic palaeoclimatic changes over the past 600,000 years. *Geological Society of America Memoir*, **145**, 111–46.

Ruddiman, W.F. and McIntyre, A. (1979) Warmth of the subpolar North Atlantic Ocean during Northern Hemisphere ice-sheet growth. *Science*, **204**, 173–5.

Ruddiman, W.F. and McIntyre, A. (1981a) The mode and mechanism of the last deglaciation: oceanic evidence. *Quaternary Research*, **16**, 125–34.

Ruddiman, W.F. and McIntyre, A. (1981b) The North Atlantic during the last deglaciation. *Palaeogeography, Palaeoclimatology, Palaeoecology*, **35**, 145–214.

Ruddiman, W.F. and Raymo, M.E. (1988) Northern Hemisphere climate régimes during the past 3Ma: possible tectonic connections. *Philosophical Transactions of the Royal Society of London*, **B318**, 411–30.

Ruddiman, W.F., Sancetta, C.D. and McIntyre, A. (1977) Glacial/interglacial response rate of subpolar North Atlantic waters to climatic change: the record in oceanic sediments. *Philosophical Transactions of the Royal Society of London*, **B280**, 119–42.

Ruddiman, W.F., McIntyre, A., Niebler-Hunt, V. and Durazzi, J.T. (1980) Oceanic evidence for the mechanism of rapid northern hemisphere glaciation. *Quaternary Research*, **13**, 33–64.

Ruddiman, W.F., Raymo, M.E. and McIntyre, A. (1986) Matuyama 41,000-year cycles: North Atlantic Ocean and Northern Hemisphere ice sheets. *Earth and Planetary Science Letters*, **80**, 117–29.

Ruddiman, W.F., Raymo, M.E., Martinson, D.G., Clement, B.M. and Backman, J. (1989) Pleistocene evolution: Northern Hemisphere ice sheets and North Atlantic Ocean. *Palaeoceanography*, **4**, 353–412.

Rudwick, M. (1962) Hutton and Werner compared George Greenough's geological tour of Scotland in 1805. *The British Journal for the History of Science*, **1**, 117–35.

Rudwick, M. (1974) Darwin and Glen Roy: a 'great failure' in scientific method? *Studies in the History and Philosophy of Science*, **5**, 97–185.

Russell, A.J. (1989) A comparison of two recent jökulhlaups from an ice-dammed lake, Søndre Strømfjord, West Greenland. *Journal of Glaciology*, **35**, 157–62.

Rust, B.R. (1978) Depositional models for braided alluvium. In *Fluvial Sedimentology* (ed. A.D. Miall). Memoir of the Canadian Society of Petroleum Geologists, **5**, 605–25.

Ryder, J.M. (1971) The stratigraphy and morphology of paraglacial alluvial fans in south-central British Columbia. *Canadian Journal of Earth Sciences*, *8*, 279–98.

Ryder, J.M. and Church, M. (1986) The Lillooet terraces of Fraser River: a palaeoenvironmental enquiry. *Canadian Journal of Earth Sciences*, **23**, 869–84.

Ryder, R.H. (1968) Geomorphological mapping of the Isle of Rhum. Unpublished MSc thesis, University of Glasgow.

Ryder, R.H. (1975) Rhum: geomorphology. 1:20,000 map published by the Department of Geography, University of Glasgow.

Ryder, R.H. and McCann, S.B. (1971) Periglacial

phenomena on the island of Rhum in the Inner Hebrides. *Scottish Journal of Geology*, **7**, 293–303.

Rymer, L. (1974) The palaeoecology and historical ecology of the Parish of North Knapdale, Argyllshire. Unpublished PhD thesis, University of Cambridge.

Rymer, L. (1977) A late-glacial and early post-glacial pollen diagram from Drimnagall, North Knapdale, Argyllshire. *New Phytologist*, **79**, 211–21.

Sale, C.J. (1970) Cirque distribution in Great Britain. A statistical analysis of variations in elevation, aspect and density. Unpublished MSc thesis, University College London.

Salter, J.W. (1857) On the Cretaceous fossils of Aberdeenshire. *Quarterly Journal of the Geological Society of London*, **13**, 83–7.

Saltzman, B. and Maasch, K.A. (1990) A first-order global model of late Cenozoic climatic change. *Transactions of the Royal Society of Edinburgh: Earth Sciences*, **81**, 315–25.

Samuelsson, G. (1910) Scottish peat mosses. A contribution to the knowledge of the late-Quaternary vegetation and climate of north-western Europe. *Bulletin of the Geological Institution of the University of Upsala*, **10**, 197–260.

Saunderson, H.C. (1977) The sliding bed facies in esker sands and gravels: a criterion for full-pipe (tunnel) flow? *Sedimentology*, **24**, 623–38.

Sauramo, M. (1923) Studies on the Quaternary varve sediments in southern Finland. *Bulletin of the Commission for Geology in Finland*, **60**, 1–164.

Saville, A. and Bridgland, D. (1992) Exploratory work at Den of Boddam, a flint extraction site on the Buchan Gravels near Peterhead, north-east Scotland. *Quaternary Newsletter*, **66**, 4–13.

Saxton, W.I. and Hopwood, A.T. (1919) On a Scandinavian erratic from the Orkneys. *Geological Magazine*, **56**, 273–4.

Sejrup, H.P., Aarseth, I., Ellingsen, K.L., Reither, E., Jansen, E., Løvlie, R., Bent, A., Brigham-Grette, J., Larsen, E. and Stoker, M. (1987) Quaternary stratigraphy of the Fladen area, central North Sea: a multidisciplinary study. *Journal of Quaternary Science*, **2**, 35–58.

Selby, I.C. (1987) Glaciated shorelines in Barra and Vatersay. *Quaternary Newsletter*, **53**, 16–22.

Selby, I.C. (1989) Quaternary geology of the Hebridean continental margin. Unpublished PhD thesis, University of Nottingham.

Shackleton, N.J. (1987) Oxygen isotopes, ice volume and sea level. *Quaternary Science Reviews*, **6**, 183–90.

Shackleton, N.J. and Pisias, N.G. (1985) Atmospheric carbon dioxide, orbital forcing and climate. In *The Carbon Cycle and Atmomospheric CO_2: Natural Variations Archean to Present* (eds E.T. Sundquist and W.S. Broecker). American Geophysical Union, Geophysical Monograph, **32**, 303–17.

Shackleton, N.J. and Opdyke, N.D. (1973) Oxygen isotope and palaeomagnetic stratigraphy of equatorial Pacific core V28–238: oxygen isotope temperatures and ice volumes on a 10^5 and 10^6 year scale. *Quaternary Research*, **3**, 39–55.

Shackleton, N.J., Hall, M.A., Line, J. and Cang Shuxi (1983) Carbon isotope data in core V19–30 confirm reduced carbon dioxide concentration in the ice age atmosphere. *Nature*, **306**, 319–22.

Shackleton, N.J., Backman, J., Zimmerman, H., Kent, D.V., Hall, M.A., Roberts, D.G., Schnitker, D., Baldauf, J.G., Despraires, A., Homrighausen, R., Huddlestun, P., Keene, J.B., Kaltenback, A.J., Krumslek, K.A.O., Morton, A.C., Murray, J.W. and Westberg-Smith, J. (1984) Oxygen isotope calibration of the onset of ice-rafting and history of glaciation in the North Atlantic region. *Nature*, **307**, 620–3.

Shackleton, N.J., Berger, A. and Peltier, W.R. (1990) An alternative astronomical calibration of the lower Pleistocene timescale based on ODP Site 677. *Transactions of the Royal Society of Edinburgh: Earth Sciences*, **81**, 251–61.

Shakesby, R.A. (1978) Dispersal of glacial erratics from Lennoxtown, Stirlingshire. *Scottish Journal of Geology*, **14**, 81–6.

Shakesby, R.A. (1979) The pattern of glacial dispersal and comminution of rock fragments and mineral grains from two point sources. *Zeitschrift für Gletscherkunde und Glazialgeologie*, **15**, 31–45.

Shakesby, R.A. (1981) The application of trend surface analysis to directional data. *Geological Magazine*, **118**, 39–48.

Shakesby, R.A. (1985) Geomorphological effects of jökulhlaups and ice-dammed lakes, Jotunheimen, Norway. *Norsk Geografisk Tidsskrift*, **39**, 1–16.

Shakesby, R.A. and Matthews, J.A. (1987) Frost

weathering and rock platform erosion on periglacial lake shorelines: a test of a hypothesis. *Boreas*, **67**, 197–203.
Sharp, M. (1985) Sedimentation and stratigraphy at Eyjabakkajökull – an Icelandic surging glacier. *Quaternary research*, **24**, 268–84.
Sharp, M., Dowdeswell, J.A. and Gemmell, J.C. (1989a) Reconstructing past glacier dynamics and erosion from glacial geomorphic evidence: Snowdon, North Wales. *Journal of Quaternary Science*, **4**, 115–30.
Sharp, M., Gemmell, J.C. and Tison, J.-L. (1989b) Structure and stability of the former subglacial drainage system of the Glacier de Tsanfleuron, Switzerland. *Earth Surface Processes and Landforms*, **14**, 119–34.
Sharpe, D.R. (1987) Glaciomarine fans built within and marginal to the Champlain Sea. In *Quaternary of the Ottawa Region and Guides for Day Excursions* (ed. R.J. Fulton). (XII INQUA Congress). National Research Council of Canada, pp. 63–74.
Sharpe, D.R. and Shaw, J. (1989) Erosion of bedrock by subglacial meltwater, Cantley, Quebec. *Geological Society of Amercia Bulletin*, **101**, 1011–20.
Shaw, J. (1988) Subglacial erosional marks, Wilton Creek, Ontario. *Canadian Journal of Earth Sciences*, **25**, 1256–67.
Shaw, J., Kvill, D. and Rains, B. (1989) Drumlins and catastrophic subglacial floods. *Sedimentary Geology*, **62**, 177–202.
Shaw, R. (1977) Periglacial features in part of the south-east Grampian Highlands of Scotland. Unpublished PhD thesis, University of Edinburgh.
Shennan, I. (1989) Holocene crustal movements and sea-level change in Great Britain. *Journal of Quaternary Science*, **4**, 77–89.
Shotton, F.W. (1953) The Pleistocene deposits of the area between Coventry, Rugby and Leamington and their bearing on the topographic development of the Midlands. *Philosophical Transactions of the Royal Society of London*, **B237**, 209–60.
Shotton, F.W. and Williams, R.E.G. (1971) Birmingham University radiocarbon dates V. *Radiocarbon*, **13**, 141–56.
Shotton, F.W. and Williams, R.E.G. (1973) Birmingham University radiocarbon dates VII. *Radiocarbon*, **15**, 51–468.
Shotton, F.W., Blundell, D.J. and Williams, R.E.G. (1968) Birmingham University radiocarbon dates II. *Radiocarbon*, **10**, 200–6.
Shotton, F.W., Blundell, D.J. and Williams, R.E.G. (1970) Birmingham University radiocarbon dates IV. *Radiocarbon*, **12**, 385–99.
Shreve, R.L. (1966) Sherman landslide, Alaska. *Science*, **154**, 1639–43.
Shreve, R.L. (1968a) The Blackhawk Landslide. *Geological Society of America Special Paper*, **108**, 47 pp.
Shreve, R.L. (1968b) Leakage and fluidization in air-layer lubricated avalanches. *Geological Society of America Bulletin*, **79**, 653–8.
Shreve, R.L. (1972) Movement of water in glaciers. *Journal of Glaciology*, **11**, 205–14.
Shreve, R.L. (1985a) Esker characteristics in terms of glacier physics, Katahdin esker system, Maine. *Geological Society of America Bulletin*, **96**, 639–46.
Shreve, R.L. (1985b) Late Wisconsin ice-surface profile calculated from esker paths and types, Katahdin Esker System, Maine. *Quaternary Research*, **23**, 27–37.
Siegenthaller, U. and Eicher, U. (1986) Stable oxygen and carbon isotope analyses. In *Handbook of Holocene Palaeoecology and Palaeohydrology* (ed. B.E. Berglund). John Wiley, Chichester, pp. 407–22.
Simpson, I.M. and West, R.G. (1958) On the stratigraphy and palaeobotany of a late-Pleistocene organic deposit at Chelford, Cheshire. *New Phytologist*, **57**, 239–50.
Simpson, J.B. (1928) In *Summary of Progress of the Geological Survey of Great Britain and the Museum of Practical Geology for the Year 1927*. Part 1. Geological Survey of Great Britain. HMSO, London, pp. 59–60.
Simpson, J.B. (1929) The valley glaciation of Loch Lomond. *Report of the British Association for the Advancement of Science for 1928*, 547–8.
Simpson, J.B. (1933) The late-glacial readvance moraines of the Highland border west of the River Tay. *Transactions of the Royal Society of Edinburgh*, **57**, 633–45.
Simpson, S. (1948) The glacial deposits of Tullos and Bay of Nigg, Aberdeen. *Transactions of the Royal Society of Edinburgh*, **61**, 687–97.
Simpson, S. (1955) A re-interpretation of the drifts of north-east Scotland. *Transactions of the Edinburgh Geological Society*, **16**, 189–99.
Sinclair, W. (1911) The relationship between the raised beach and the boulder clay of the west coast of Kintyre. *Transactions of the Geological Society of Glasgow*, **14**, 170.
Sissons, J.B. (1958a) The deglaciation of part of

East Lothian. *Transactions of the Institute of British Geographers*, **25**, 59–77.

Sissons, J.B. (1958b) Supposed ice-dammed lakes in Britain with particular reference to the Eddleston Valley, southern Scotland. *Geografiska Annaler*, **40**, 159–87.

Sissons, J.B. (1960) Some aspects of glacial drainage channels in Britain. Part I. *Scottish Geographical Magazine*, **76**, 131–46.

Sissons, J.B. (1961a) Some aspects of glacial drainage channels in Britain. Part II. *Scottish Geographical Magazine*, **77**, 15–36.

Sissons, J.B. (1961b) A subglacial drainage system by the Tinto Hills, Lanarkshire. *Transactions of the Edinburgh Geological Society*, **18**, 175–93.

Sissons, J.B. (1961c) The central and eastern parts of the Lammermuir–Stranraer moraine. *Geological Magazine*, **98**, 380–92.

Sissons, J.B. (1963a) The Perth Readvance in Central Scotland. Part I. *Scottish Geographical Magazine*, **79**, 151–63.

Sissons, J.B. (1963b) The glacial drainage system around Carlops, Peeblesshire. *Transactions of the Institute of British Geographers*, **32**, 95–111.

Sissons, J.B. (1964) The Perth Readvance in central Scotland. Part II. *Scottish Geographical Magazine*, **80**, 28–36.

Sissons, J.B. (1965) Quaternary. In *The Geology of Scotland* (ed. G.Y. Craig). Oliver and Boyd, Edinburgh, pp. 467–503.

Sissons, J.B. (1966) Relative sea-level changes between 10,300 and 8300 BP in part of the Carse of Stirling. *Transactions of the Institute of British Geographers*, **39**, 19–29.

Sissons, J.B. (1967a) *The Evolution of Scotland's Scenery*. Oliver and Boyd, Edinburgh, 259 pp.

Sissons, J.B. (1967b) Glacial stages and radiocarbon dates in Scotland. *Scottish Journal of Geology*, **3**, 375–81.

Sissons, J.B. (1969) Drift stratigraphy and buried morphological features in the Grangemouth–Falkirk–Airth area, central Scotland. *Transactions of the Institute of British Geographers*, **48**, 19–50.

Sissons, J.B. (1971) The geomorphology of central Edinburgh. *Scottish Geographical Magazine*, **87**, 185–96.

Sissons, J.B. (1972a) The last glaciers in part of the south-east Grampians. *Scottish Geographical Magazine*, **88**, 168–81.

Sissons, J.B. (1972b) Dislocation and non-uniform uplift of raised shorelines in the western part of the Forth Valley. *Transactions of the Institute of British Geographers*, **55**, 149–59.

Sissons, J.B. (1973a) Hypotheses of deglaciation in the eastern Grampians, Scotland. *Scottish Journal of Geology*, **9**, 96.

Sissons, J.B. (1973b) Delimiting the Loch Lomond Readvance in the eastern Grampians. *Scottish Geographical Magazine*, **89**, 138–9.

Sissons, J.B. (1974a) Glacial readvances in Scotland. In *Problems of the Deglaciation of Scotland* (eds C.J. Caseldine and W.A. Mitchell). Journal of St Andrews Geographers Special Publication, 1, 5–15.

Sissons, J.B. (1974b) A lateglacial ice-cap in the central Grampians. *Transactions of the Institute of British Geographers*, **62**, 95–114.

Sissons, J.B. (1974c) The Quaternary in Scotland: a review. *Scottish Journal of Geology*, **10**, 311–37.

Sissons, J.B. (1974d) Lateglacial marine erosion in Scotland. *Boreas*, **3**, 41–8.

Sissons, J.B. (1975a) A fossil rock glacier in Wester Ross. *Scottish Journal of Geology*, **11**, 83–6.

Sissons, J.B. (1975b) The Loch Lomond Readvance in the south-east Grampians. In *Quaternary Studies in North East Scotland* (ed. A.M.D. Gemmell). Department of Geography, University of Aberdeen, Aberdeen, pp. 23–9.

Sissons, J.B. (1975c) The geomorphology of East Lothian. In *The Geology of the Lothians and South East Scotland. An Excursion Guide* (eds G.Y. Craig and P.McL.D. Duff). Scottish Academic Press, Edinburgh, pp. 131–43.

Sissons, J.B. (1976a) Lateglacial marine erosion in south-east Scotland. *Scottish Geographical Magazine*, **92**, 17–29.

Sissons, J.B. (1976b) *The Geomorphology of the British Isles: Scotland*. Methuen, London, 150 pp.

Sissons, J.B. (1976c) A remarkable protalus rampart complex in Wester Ross. *Scottish Geographical Magazine*, **92**, 182–90.

Sissons, J.B. (1976d) A fossil rock glacier in Wester Ross. Reply to W.B. Whalley. *Scottish Journal of Geology*, **12**, 178–9.

Sissons, J.B. (1976e) The geomorphology of the upper Forth Valley. *Forth Valley Naturalist and Historian*, **1**, 5–20.

Sissons, J.B. (1977a) The Loch Lomond Readvance in the northern mainland of Scotland. In *Studies in the Scottish Lateglacial Environment* (eds J.M. Gray and J.J. Lowe). Pergamon Press, Oxford, pp. 45–59.

Sissons, J.B. (1977b) Former ice-dammed lakes in Glen Moriston, Inverness-shire, and their significance in upland Britain. *Transactions of the Institute of British Geographers*, NS **2**, 224–42.
Sissons, J.B. (1977c) The Loch Lomond Readvance in southern Skye and some palaeoclimatic implications. *Scottish Journal of Geology*, **13**, 23–36.
Sissons, J.B. (ed.) (1977d) *The Scottish Highlands*. INQUA X Congress 1977. Guidebook for Excursions A11 and C11. Geo Abstracts, Norwich, 51 pp.
Sissons, J.BL. (1977e) *Glen Roy National Nature Reserve: The Parallel Roads of Glen Roy*. Nature Conservancy Council, London, 12 pp.
Sissons, J.B. (1978) The Parallel Roads of Glen Roy and adjacent glens, Scotland. *Boreas* **7**, 229–44.
Sissons, J.B. (1979a) The later lakes and associated fluvial terraces of Glen Roy, Glen Spean and vicinity. *Transactions of the Institute of British Geographers*, NS **4**, 12–29.
Sissons, J.B. (1979b) The limit of the Loch Lomond Advance in Glen Roy and vicinity. *Scottish Journal of Geology*, **15**, 31–42.
Sissons, J.B. (1979c) Catastrophic lake drainage in Glen Spean and the Great Glen, Scotland. *Journal of the Geological Society of London*, **136**, 215–24.
Sissons, J.B. (1979d) Palaeoclimatic inferences from former glaciers in Scotland and the Lake District. *Nature*, **278**, 518–21.
Sissons, J.B. (1979e) The Loch Lomond Stadial in the British Isles. *Nature*, **280**, 199–203.
Sissons, J.B. (1979f) The Loch Lomond Advance in the Cairngorm Mountains. *Scottish Geographical Magazine*, **95**, 66–82.
Sissons, J.B. (1980a) The Loch Lomond Advance in the Lake District, northern England. *Transactions of the Royal Society of Edinburgh: Earth Sciences*, **71**, 13–27.
Sissons, J.B. (1980b) Palaeoclimatic inferences from Loch Lomond Advance glaciers. In *Studies in the Lateglacial of North-west Europe* (eds J.J. Lowe, J.M. Gray and J.E. Robinson). Pergamon Press, Oxford, pp. 31–43.
Sissons, J.B. (1980c) The glaciation of the Outer Hebrides (Letter). *Scottish Journal of Geology*, **16**, 81–4.
Sissons, J.B. (1981a) British shore platforms and ice-sheets. *Nature*, **291**, 473–5.
Sissons, J.B. (1981b) The last Scottish ice sheet: facts and speculative discussion. *Boreas*, **10**, 1–17.
Sissons, J.B. (1981c) Lateglacial marine erosion and a jökulhlaup deposit in the Beauly Firth. *Scottish Journal of Geology*, **17**, 7–19.
Sissons, J.B. (1981d) Ice-dammed lakes in Glen Roy and vicinity: a summary. In *The Quaternary in Britain* (eds J. Neale and J. Flenley). Pergamon Press, Oxford, pp. 174–83.
Sissons, J.B. (1982a) A former ice-dammed lake and associated glacier limits in the Achnasheen area, central Ross-shire. *Transactions of the Institute of British Geographers*, NS, **7**, 98–116.
Sissons, J.B. (1982b) The so-called high 'interglacial' rock shoreline of western Scotland. *Transactions of the Institute of British Geographers*, NS, **7**, 205–16.
Sissons, J.B. (1982c) Interstadial and last interglacial deposits covered by till in Scotland: a reply. *Boreas*, **11**, 123–4.
Sissons, J.B. (1983a) Shorelines and isostasy in Scotland. In *Shorelines and Isostasy* (eds D.E. Smith and A.G. Dawson). Academic Press, London, pp. 209–25.
Sissons, J.B. (1983b) Quaternary. In *Geology of Scotland*, 2nd edn. (ed. G.Y. Craig). Scottish Academic Press, Edinburgh, pp. 399–424.
Sissons, J.B. (1983c) The Quaternary geomorphology of the Inner Hebrides: a review and reassessment. *Proceedings of the Geologists' Association*, **94**, 165–75.
Sissons, J.B. and Brooks, C.L. (1971) Dating of early postglacial land and sea-level changes in the Western Forth Valley. *Nature Physical Science*, **234**, 124–7.
Sissons, J.B. and Cornish, R. (1982a) Rapid localized glacio-isostatic uplift at Glen Roy, Scotland. *Nature*, **297**, 213–4.
Sissons, J.B. and Cornish, R. (1982b) Differential glacio-isostatic uplift of coastal blocks at Glen Roy, Scotland. *Quaternary Research*, **18**, 268–88.
Sissons, J.B. and Cornish, R. (1983) Fluvial landforms associated with ice-dammed lake drainage in upper Glen Roy, Scotland. *Proceedings of the Geologists' Association*, **94**, 45–52.
Sissons, J.B. and Dawson, A.G. (1981) Former sea-levels and ice limits in part of Wester Ross, north-west Scotland. *Proceedings of the Geologists' Association*, **12**, 115–24.
Sissons, J.B. and Grant, A.J.H. (1972) The last

glaciers in the Lochnagar area, Aberdeenshire. *Scottish Journal of Geology*, **8**, 85–93.

Sissons, J.B. and Rhind, D.W. (1970) Drift stratigraphy and buried morphology beneath the Forth at Rosyth. *Scottish Journal of Geology*, **6**, 272–84.

Sissons, J.B. and Smith, D.E. (1965a) Raised shorelines associated with the Perth Readvance in the Forth Valley and their relation to glacial isostasy. *Transactions of the Royal Society of Edinburgh*, **66**, 143–68.

Sissons, J.B. and Smith, D.E. (1965b) Peat bogs in a post-glacial sea and a buried raised beach in the western part of the Carse of Stirling. *Scottish Journal of Geology*, **1**, 247–55.

Sissons, J.B. and Sutherland, D.G. (1976) Climatic inferences from former glaciers in the south-east Grampian Highlands, Scotland. *Journal of Glaciology*, **17**, 325–46.

Sissons, J.B. and Walker, M.J.C. (1974) Lateglacial site in the central Grampian Highlands. *Nature*, **249**, 822–4.

Sissons, J.B., Cullingford, R.A. and Smith, D.E. (1965) Some pre-carse valleys in the Forth and Tay basins. *Scottish Geographical Magazine*, **81**, 115–24.

Sissons, J.B., Smith, D.E. and Cullingford, R.A. (1966) Late-glacial and post-glacial shorelines in south-east Scotland. *Transactions of the Institute of British Geographers*, **39**, 9–18.

Sissons, J.B., Lowe, J.J., Thompson, K.S.R. and Walker, M.J.C. (1973) Loch Lomond Readvance in the Grampian Highlands of Scotland. *Nature Physical Science*, **244**, 75–7.

Small, A. and Smith, J.S. (1971) *The Strathpeffer and Inverness area*. British Landscapes Through Maps. Geographical Association, Sheffield, 25 pp.

Smiles, S. (1878) *Robert Dick, Baker, of Thurso, Geologist and Botanist*. John Murray, London, 436 pp.

Smith, A.G. (1984) Newferry and the Boreal–Atlantic transition. *New Phytologist*, **98**, 35–55.

Smith, A.G. and Cloutman, E.W. (1988) Reconstruction of Holocene vegetation history in three dimensions at Waun-Fignen-Felen, an upland site in south Wales. *Philosophical Transactions of the Royal Society of London*, **B322**, 159–219.

Smith, A.G. and Pilcher, J.R. (1973) Radiocarbon dates and vegetational history of the British Isles. *New Phytologist*, **72**, 903–14.

Smith, D.E. (1965) Late and Postglacial changes of shoreline on the north side of the Forth Valley and estuary. Unpublished PhD thesis, University of Edinburgh.

Smith, D.E. (1968) Post-glacial displaced shorelines in the surface of the carse clay on the north bank of the River Forth, in Scotland. *Zeitschift für Geomorphologie*, NF, **12**, 388–408.

Smith, D.E. (1984) Lower Ythan Valley: deglaciation and Flandrian relative sea level change. In *Buchan Field Guide* (ed. A.M. Hall). Quaternary Research Association, Cambridge, pp. 47–58.

Smith, D.E. (1986) Relative sea-level changes in the lower South Esk Valley. Excursion Guide for a meeting of IGCP-200 at the Burn, Edzell, May 1986, 15 pp.

Smith, D.E. and Cullingford, R.A. (1985) Flandrian relative sea-level changes in the Montrose Basin area. *Scottish Geographical Magazine*, **101**, 91–105.

Smith, D.E. and Dawson, A.G. (1990) Tsunami waves in the North Sea. *New Scientist*, **127** (1728), 46–9.

Smith, D.E., Cullingford, R.A. and Jones, R.L. (1977) Radiocarbon assay of duck bones from Late Devensian raised marine deposits near Montrose, Angus. *Quaternary Newsletter*, **23**, 5–6.

Smith, D.E., Morrison, J., Jones, R.L. and Cullingford, R.A. (1980) Dating the Main Postglacial Shoreline in the Montrose area, Scotland. In *Timescales in Geomorphology* (eds R.A. Cullingford, D.A. Davidson and J. Lewin). John Wiley, Chichester, pp. 225–45.

Smith, D.E., Cullingford, R.A. and Seymour, W.P. (1982) Flandrian relative sea-level changes in the Philorth Valley, north-east Scotland. *Transactions of the Institute of British Geographers*, NS **7**, 321–36.

Smith, D.E., Cullingford, R.A. and Brooks, C.L. (1983) Flandrian relative sea level changes in the Ythan Valley, north-east Scotland. *Earth Surface Processes and Landforms*, **8**, 423–38.

Smith, D.E., Cullingford, R.A. and Haggart, B.A. (1985a) A major coastal flood during the Holocene in eastern Scotland. *Eizeitalter und Gegenwart*, **35**, 109–18.

Smith, D.E., Dawson, A.G., Cullingford, R.A. and Harkness, D.D. (1985b) The stratigraphy of Flandrian relative sea-level changes at a site in Tayside, Scotland. *Earth Surface Processes and Landforms*, **10**, 17–25.

Smith, D.E., Sissons, J.B. and Cullingford, R.A.

(1969) Isobases for the Main Perth raised shoreline in south-east Scotland as determined by trend surface analysis. *Transactions of the Institute of British Geographers*, **46**, 45–52.

Smith, D.E., Thompson, K.S.R. and Kemp, D.D. (1978) The Late Devensian and Flandrian history of the Teith Valley, Scotland. *Boreas*, **7**, 97–107.

Smith, D.E., Turbayne, S.C., Dawson, A.G. and Hickey, K.R. (1991a) The temporal and spatial variability of major floods around European coasts. C.E.C. Contract Report EV4C0047UK-(H), Brussels, 129 pp.

Smith, D.E., Turbayne, S.C., Firth, C.R. and Brooks, C.L. (1991b) Creich. In *Late Quaternary Coastal Evolution in the Inner Moray Firth. Field Guide* (eds C.R. Firth and B.A. Haggart). West London Press, London, pp. 33–42.

Smith, J. (1838) On the last changes in the relative levels of the land and sea in the British Islands. *Edinburgh New Philosophical Journal*, **25**, 378–94.

Smith, J. (of Jordanhill) (1862) *Researches in Newer Pliocene and Post-Tertiary Geology*. John Gray, Glasgow, 191 pp.

Smith, J. (1850a) Note on the shells found in the till by Mr Cleghorn. *Quarterly Journal of the Geological Society of London*, **6**, 386.

Smith, J. (1850b) On the occurrence of marine shells in the stratified beds below the till. *Quarterly Journal of the Geological Society of London*, **6**, 386–8.

Smith, J. (1891) The great ice age in the Garnock Valley. *Transactions of the Geological Society of Glasgow*, **9**, 151–91.

Smith, J. (1896a) The great submergence: an interpretation of the Clava section, near Inverness, Scotland. *Geological Magazine*, Decade 4, **3**, 498–502.

Smith, J. (1896b) The geological position of the Irvine whale bed. *Transactions of the Geological Society of Glasgow*, **10**, 29–50.

Smith, J. (1896c) Discovery of interglacial shell-beds in Ayrshire. *Geological Magazine*, Decade 4, **3**, 286–7.

Smith, J. (1898) The drift or glacial deposits of Ayrshire. *Transactions of the Geological Society of Glasgow*, **11**, (supplement), 134 pp.

Smith, J. (1901) The drift or glacial formation of the Clyde drainage area. In *Fauna, Flora and Geology of the Clyde Area. Handbook on the Natural History of Glasgow and the West of Scotland* (eds G.F.S. Elliot, M. Laurie and J.B. Murdoch). British Association for the Advancement of Science, Glasgow, pp. 520–7.

Smith, J.A. (1858) Notice of the horn of a reindeer (*Cervus tarandus*, Linn), found in Dumbartonshire. *Proceedings of the Royal Physical Society of Edinburgh*, **1**, 247–49. (Also *Edinburgh New Philosophical Journal*, **6**, 165–7.)

Smith, J.A. (1871) Notice of remains of the reindeer, *Cervus tarandus*, found in Ross-shire, Sutherland and Caithness; with notes of its occurrence throughout Scotland. *Proceedings of the Society of Antiquaries of Scotland*, **8**, 186–222.

Smith, J.S. (1966) Morainic limits and their relationship to raised shorelines in the east Scotland Highlands. *Transactions of the Institute of British Geographers*, **39**, 61–4.

Smith, J.S. (1968) Shoreline evolution in the Moray Firth. Unpublished PhD thesis, University of Aberdeen.

Smith, J.S. (1977) The last glacial epoch around the Moray Firth. In *The Moray Firth Area Geological Studies* (ed. G. Gill). Inverness Field Club, Inverness, pp. 72–82.

Smith, S.M. (1971) Palaeoecology of post-glacial beaches in East Lothian. *Scottish Journal of Geology*, **8**, 31–49.

Smith, S.M. and Heppell, D. (1991) Checklist of British marine Mollusca. *National Museums of Scotland Information Series*, No. **11**, 114 pp.

Söderman, G., Kejonen, A. and Kujansuu, R. (1983) The riddle of the tors at Lauhavuori, western Finland. *Fennia*, **161**, 91–144.

Sollid, J.L., Andersen, S., Hamre, N., Kjeldsen, O., Salvigsen, O., Sturød, S., Tveitå, T. and Willhelmsen, A. (1973) Deglaciation of Finmark, North Norway. *Norsk Geografisk Tidsskrift*, **27**, 233–325.

Somervail, A. (1879) Observations on the higher summits of the Pentland Hills. *Transactions of the Edinburgh Geological Society*, **3**, 191–9.

Sparks, B.W. and West, R.G. (1972) *The Ice Age in Britain*. Methuen, London, 302 pp.

Spence, D.H.N. (1957) Studies on the vegetation of Shetland. 1. The serpentine debris vegetation in Unst. *Journal of Ecology*, **45**, 917–45.

Spence, D.H.N. (1974) Subarctic debris and scrub vegetation of Shetland. In *The Natural Environment of Shetland* (ed. R. Goodier). Nature Conservancy Council, Edinburgh, pp. 73–88.

Spencer, J.W. (1890) Ancient shores, boulder pavements, and high-level gravel deposits in the region of the Great Lakes. *Geological Society of America Bulletin*, **1**, 71–86.

Stark, J. (1902) The surface geology of the Falls of Clyde district. *Transactions of the Geological Society of Glasgow*, **12**, 52–7.

Statham, I. (1976a) A scree slope rockfall model. *Earth Surface Processes and Landforms*, **1**, 43–62.

Statham, I. (1976b) Debris flows on vegetated screes in the Black Mountain, Carmarthenshire. *Earth Surface Processes and Landforms*, **1**, 173–80.

Stephens, N. (1957) Some observations on the 'interglacial' platform and the early postglacial raised beach on the east coast of Ireland. *Proceedings of the Royal Irish Academy*, **58B**, 129–49.

Steven, H.M. and Carlisle, A. (1959) *The Native Pinewoods of Scotland*. Oliver and Boyd, Edinburgh, 368 pp.

Stevenson, A.C., Jones, V.J. and Battarbee, R.W. (1990) The cause of peat erosion: a palaeolimnological approach. *New Phytologist*, **116**, 727–35.

Stevenson, W. (1868) On Bedshiel 'Kaims', and their relations to similar deposits to eastward and westward. *History of the Berwickshire Naturalists' Club, 1863–1868*, 124–8.

Stewart, D.A. (1979) The Flandrian vegetational history of the Loch Lomond area. Unpublished PhD thesis, University of Glasgow.

Stewart, D.A. (1987) Further palynology and radiometric dating of Flandrian marine and freshwater sediments of Loch Lomond, Scotland. *INQUA XII Congress, Ottawa, Canada, 1987*. Programme and Abstracts, p. 269.

Stewart, D.A., Walker, A. and Dickson, J.H. (1984) Pollen diagrams from Dubh Lochan, near Loch Lomond. *New Phytologist*, **98**, 531–49.

Stewart, M. (1932) Notes on the geology of North Rona. *Geological Magazine*, **69**, 179–85.

Stewart, M. (1933) Notes on the geology of Sgula Sgeir and the Flannan Islands. *Geological Magazine*, **70**, 110–6.

Stoker, M.S. (1988) Pleistocene ice–proximal glaciomarine sediments in boreholes from the Hebrides shelf and Wyville–Thomson Ridge, NW UK Continental Shelf. *Scottish Journal of Geology*, **24**, 249–62.

Stoker, M.S. and Bent, A.J.A. (1985) Middle Pleistocene glacial and glaciomarine sediments in the west central North Sea. *Boreas*, **14**, 325–32.

Stoker, M.S. and Holmes, R. (1991) Submarine end-moraines as indicators of Pleistocene ice-limits off north-west Britain. *Journal of the Geological Society of London*, **148**, 431–4.

Stoker, M.S. and Bent, A.J.A. (1987) Lower Pleistocene deltaic and marine sediments in boreholes from the central North Sea. *Journal of Quaternary Science*, **2**, 87–96.

Stoker, M.S. and Graham, C. (1985) Pre-Late Weichselian submerged rock platforms off Stonehaven. *Scottish Journal of Geology*, **21**, 205–8.

Stoker, M.S., Long, D. and Fyfe, J.A. (1985) A revised Quaternary stratigraphy for the central North Sea. *British Geological Survey Report*, **17**(2), 35 pp.

Stone, J.C. (1959) A description of glacial retreat features in mid-Nithsdale. *Scottish Geographical Magazine*, **75**, 164–8.

Stone, K.H. (1963) Alaskan ice-dammed lakes. *Annals of the Association of American Geographers*, **53**, 332–49.

Straw, A. (1979) Eastern England. In *Eastern and Central England* (eds A. Straw and K.M. Clayton). Methuen, London, pp. 1–139.

Stroud, D.A., Reed, T.M., Pienkowski, M.W. and Lindsay, R.A. (1987) *Birds, Bogs and Forestry. The Peatlands of Caithness and Sutherland*. Nature Conservancy Council, Peterborough, 121 pp.

Stuart, A.J. (1982) *Pleistocene Vertebrates in the British Isles*. Longman, London, 212 pp.

Sturlurdottir, S.A. and Turner, J. (1985) The elm decline at Pawlaw Mire: an anthropogenic interpretation. *New Phytologist*, **99**, 323–9.

Sugden, D.E. (1965) Aspects of the glaciation of the Cairngorm Mountains. Unpublished DPhil thesis, University of Oxford.

Sugden, D.E. (1968) The selectivity of glacial erosion in the Cairngorm Mountains, Scotland. *Transactions of the Institute of British Geographers*, **45**, 79–92.

Sugden, D.E. (1969) The age and form of corries in the Cairngorms. *Scottish Geographical Magazine*, **85**, 34–46.

Sugden, D.E. (1970) Landforms of deglaciation in the Cairngorm Mountains. *Transactions of the Institute of British Geographers*, **51**, 201–19.

Sugden, D.E. (1971) The significance of periglacial activity on some Scottish mountains. *Geographial Journal*, **137**, 388–92.

Sugden, D.E. (1973a) Hypotheses of deglaciation in the eastern Grampians, Scotland. *Scottish Journal of Geology*, **9**, 94–5.

Sugden, D.E. (1973b) Delimiting Zone III glaciers

in the eastern Grampians. *Scottish Geographical Magazine*, **89**, 62–3.

Sugden, D.E. (1974a) Landscapes of glacial erosion in Greenland and their relationship to ice, topographic and bedrock conditions. *Institute of British Geographers Special Publication*, 7, 177–95.

Sugden, D.E. (1974b) Deglaciation of the Cairngorms and its wider implications. In *Problems of the Deglaciation of Scotland* (eds C.J. Caseldine and W.A. Mitchell). Journal of St Andrews Geographers, Special Publication **1**, 17–28.

Sugden, D.E. (1974c) Landforms. In *The Cairngorms. Their Natural History and Scenery* (eds D. Nethersole-Thompson and A. Watson). Collins, London, pp. 210–21.

Sugden, D.E. (1977) Did glaciers form in the Cairngorms in the 17th–19th centuries? *Cairngorm Club Journal*, **18**, 189–201.

Sugden, D.E. (1978) Glacial erosion by the Laurentide ice sheet. *Journal of Glaciology*, **20**, 367–91.

Sugden, D.E. (1980) The Loch Lomond Advance in the Cairngorms (a reply to J.B. Sissons). *Scottish Geographical Magazine*, **96**, 18–19.

Sugden, D.E. (1983) The Cairngorm tors and their significance. *Scottish Mountaineering Club Journal*, **32**, 327–34.

Sugden, D.E. (1987) The polar and glacial world. In *Horizons in Physical Geography* (eds M.J. Clark, K.J. Gregory and A.M. Gurnell). Macmillan, London, pp. 214–31.

Sugden, D.E. (1989) Modification of old land surfaces by ice sheets. *Zeitschrift für Geomorphologie*, NF, Supplementband **72**, 163–72.

Sugden, D.E. (1991) The stepped response of ice sheets to climatic change. In *Antartica and Global Climatic Change* (eds C.M. Harris and B. Stonehouse). Belhaven Press, London, pp. 107–14.

Sugden, D.E. and Clapperton, C.M. (1975) The deglaciation of upper Deeside and the Cairngorm Mountains. In *Quaternary Studies in North East Scotland* (ed. A.M.D. Gemmell). Department of Geography, University of Aberdeen, Aberdeen, pp. 30–8.

Sugden, D.E. and John, B.S. (1976) *Glaciers and Landscape*. Edward Arnold, London, 376 pp.

Sugden, D.E. and Ward, R. (1980) Mountains in the making. *Geographical Magazine*, **42**, 425–6.

Sugden, D.E. and Watts, S.H. (1977) Tors, felsenmeer, and glaciation in northern Cumberland Peninsula, Baffin Island. *Canadian Journal of Earth Sciences*, **14**, 2817–23.

Sugden, D.E., Clapperton, C.M. and Knight, P.G. (1985) A jökulhlaup near Søndre Strømfjord, West Greenland, and some effects on the ice-sheet margin. *Journal of Glaciology*, **31**, 366–8.

Sutherland, A. (1920) A boulder containing ammonites from Dunbeath, Caithness. *Transactions of the Edinburgh Geological Society*, **11**, 1.

Sutherland, D.G. (1980) Problems of radiocarbon dating deposits from newly deglaciated terrain: examples from the Scottish Lateglacial. In *Studies in the Lateglacial of Northwest Europe* (eds J.J. Lowe, J.M. Gray and J.E. Robinson). Pergamon Press, Oxford, 139–49.

Sutherland, D.G. (1981a) The high-level marine shell beds of Scotland and the build-up of the last Scottish ice sheet. *Boreas*, **10**, 247–54.

Sutherland, D.G. (1981b) The raised shorelines and deglaciation of the Loch Long/Loch Fyne area, western Scotland. Unpublished PhD thesis, University of Edinburgh.

Sutherland, D.G. (1984a) The Quaternary deposits and landforms of Scotland and the neighbouring shelves: a review. *Quaternary Science Reviews*, **3**, 157–254.

Sutherland, D.G. (1984b) The Late Quaternary sequence at Castle Hill, Gardenstown and King Edward. In *Buchan Field Guide* (ed. A.M. Hall). Quaternary Research Association, Cambridge, pp. 89–96.

Sutherland, D.G. (1984c) The submerged landforms of the St Kilda archipelago, western Scotland. *Marine Geology*, **58**, 435–42.

Sutherland, D.G. (1986) A review of Scottish marine shell radiocarbon dates, their standardisation and interpretation. *Scottish Journal of Geology*, **22**, 145–64.

Sutherland, D.G. (1987a) Achnasheen. In *Wester Ross Field Guide* (eds C.K. Ballantyne and D.G. Sutherland). Quaternary Research Association, Cambridge, pp. 65–70.

Sutherland, D.G. (1987b) Corrieshalloch Gorge. In *Wester Ross Field Guide* (eds C.K. Ballantyne and D.G. Sutherland). Quaternary Research Association, Cambridge, pp. 93–4.

Sutherland, D.G. (1987c) Loch Droma. In *Wester Ross Field Guide* (eds C.K. Ballantyne and D.G. Sutherland). Quaternary Research Association, Cambridge, pp. 94–7.

Sutherland, D.G. (1987d) Submerged rock platforms on the continental shelf west of Sula

Sgeir. *Scottish Journal of Geology*, **23**, 251–60.

Sutherland, D.G. (1991a) Late Devensian glacial deposits and glaciation in Scotland and the adjacent offshore region. In *Glacial Deposits in Great Britain and Ireland* (eds J. Ehlers, P.L. Gibbard and J. Rose). A.A. Balkema, Rotterdam, pp. 53–9.

Sutherland, D.G. (1991b) The glaciation of the Shetland and Orkney Islands. In *Glacial deposits in Great Britain and Ireland* (eds J. Ehlers, P.L. Gibbard and J. Rose). A.A. Balkema, Rotterdam, pp. 121–7.

Sutherland, D.G. and Walker, M.J.C. (1984) A late Devensian ice-free area and possible interglacial site on the Isle of Lewis, Scotland. *Nature*, **309**, 701–3.

Sutherland, D.G., Ballantyne, C.K. and Walker, M.J.C. (1982) A note on the Quaternary deposits and landforms of St Kilda. *Quaternary Newsletter*, **37**, 1–5.

Sutherland, D.G., Ballantyne, C.K. and Walker, M.J.C. (1984) Late Quaternary glaciation and environmental change on St Kilda, Scotland, and their palaeoclimatic significance. *Boreas*. **13**, 261–72.

Synge, F.M. (1956) The glaciation of north-east Scotland. *Scottish Geographical Magazine*, **72**, 129–43.

Synge, F.M. (1963) The Quaternary succession round Aberdeen, north-east Scotland. *Report on the 6th International Congress on the Quaternary, Warsaw, 1961. Vol. 3. Geomorphological Section*. International Union for Quaternary Research, Lodz, pp. 353–61.

Synge, F.M. (1966) The relationship of the raised strandlines and main end-moraines on the Isle of Mull and in the district of Lorn, Scotland. *Proceedings of the Geologists' Association*, **77**, 315–28.

Synge, F.M. (1977a) Records of sea levels during the Late Devensian. *Philosophical Transactions of the Royal Society of London*, **B280**, 211–28.

Synge, F.M. (1977b) Land and sea level change during the waning of the last regional ice sheet in the vicinity of Inverness. In *The Moray Firth Area Geological Studies* (ed. G. Gill). Inverness Field Club, Inverness, pp. 83–102.

Synge, F.M. (1980) A morphometric comparison of raised shorelines in Fennoscandia, Scotland and Ireland. *Geologiska Föreningens i Stockholm Förhandlingar*, **102**, 235–49.

Synge, F.M. and Smith, J.S. (1980) *A Field Guide to the Inverness Area*. Quaternary Research Association, Aberdeen, 24 pp.

Synge, F.M. and Stephens, N. (1966) Late- and post-glacial shorelines, and ice limits in Argyll and north-east Ulster. *Transactions of the Institute of British Geographers*, **39**, 101–25.

Tait, C. (1794) An account of the peat-mosses of Kincardine and Flanders in Perthshire. *Transactions of the Royal Society of Edinburgh*, **3**, 266–79.

Tait, D. (1908) On egg-shaped stones dredged from Wick harbour. *Transactions of Edinburgh Geological Society*, **9**, 135–6.

Tait, D. (1909) On the occurrence of Cretaceous fossils in Caithness. *Transactions of Edinburgh Geological Society*, **9**, 318–21.

Tait, D. (1912) On a large glacially transported mass of Lower Cretaceous rock at Leavad in the County of Caithness. *Transactions of Edinburgh Geological Society*, **10**, 1–9.

Tait, D. (1934) Excavations in the old lake deposits at Corstorphine during 1930–1932. *Transactions of the Edinburgh Geological Society*, **13**, 110–25.

Taylor, W. (1792) Parish of Rynd. In *The Statistical Account of Scotland* Vol. 4 (ed. J. Sinclair). William Creech, Edinburgh, pp. 178–84.

Ten Brink, N.W. (1975) Holocene history of the Greenland Ice Sheet based on radiocarbon-dated moraines in West Greenland. *Grønlands Geologiske Undersøgelse Bulletin*, **113**, 44 p.

Ten Brink, N.W. and Weidick, A. (1974) Greenland Ice Sheet history since the last glaciation. *Quaternary Research*, **4**, 429–40.

Terwindt, J.H.J. and Augustinus, P.G.E.F. (1985) Lateral and longitudinal successions in sedimentary structures in the Middle Mause esker, Scotland. *Sedimentary Geology*, **45**, 161–88.

Thomas, G.S.P. (1984) Sedimentation of a subaqueous esker-delta at Strabathie, Aberdeenshire. *Scottish Journal of Geology*, **20**, 9–20.

Thomas, G.S.P. and Connell, R.J. (1985) Iceberg drop, dump, and grounding structures from Pleistocene glacio-lacustrine sediments, Scotland. *Journal of Sedimentary Petrology*, **55**, 243–9.

Thompson, H.R. (1950) Some corries of north-west Sutherland. *Proceedings of the Geologists' Association*, **61**, 145–55.

Thompson, K.S.R. (1972) The last glaciers in western Perthshire. Unpublished PhD thesis, University of Edinburgh.

Thompson, R. and Morton, D.J. (1979) Magnetic susceptibility and particle-size distribution in recent sediments of the Loch Lomond drainage basin, Scotland. *Journal of Sedimentary Petrology*, **49**, 801–11.

Thomson, J. Jr. (1848) On the Parallel Roads of Lochaber. *Edinburgh New Philosophical Journal*, **45**, 49–61.

Thorp, P.W. (1981a) A trimline method for determining the upper limit of the Loch Lomond Advance glaciers: examples from the Loch Leven and Glencoe areas. *Scottish Journal of Geology*, **17**, 49–64.

Thorp, P.W. (1981b) An analysis of the spatial variability of glacial striae and friction cracks in part of the western Grampians of Scotland. *Quaternary Studies*, **1**, 71–94 (City of London Polytechnic/Polytechnic of North London).

Thorp, P.W. (1984) The glacial geomorphology of part of the western Grampians of Scotland with especial reference to the limits of the Loch Lomond Advance. Unpublished PhD thesis, City of London Polytechnic.

Thorp, P.W. (1986) A mountain icefield of Loch Lomond Stadial age, western Grampians, Scotland. *Boreas*, **15**, 83–97.

Thorp, P.W. (1987) Late Devensian ice sheet in the western Grampians, Scotland. *Journal of Quaternary Science*, **2**, 103–12.

Thorp, P.W. (1991a) The glaciation and glacial deposits of the western Grampians. In *Glacial Deposits in Great Britain and Ireland* (eds J. Ehlers, P.L. Gibbard and J. Rose). A.A. Balkema, Rotterdam, pp. 137–49.

Thorp, P.W. (1991b) Surface profiles and basal shear stresses of outlet glaciers from a Lateglacial mountain ice field in western Scotland. *Journal of Glaciology*, **37**, 77–88.

Ting, S. (1936) Beach ridges and other shore deposits in south-west Jura. *Scottish Geographical Magazine*, **52**, 182–7.

Ting, S. (1937) The coastal configuration of western Scotland. *Geografiska Annaler*, **19**, 62–83.

Tipping, R.M. (1984) Late Devensian and Early Flandrian vegetational history and deglacial chronology of western Argyll. Unpublished PhD thesis, City of London Polytechnic.

Tipping, R.M. (1985) Loch Lomond Stadial *Artemisia* pollen assemblages and Loch Lomond Readvance regional firn-line altitudes. *Quaternary Newsletter*, **46**, 1–11.

Tipping, R.M. (1986) A Late-Devensian pollen site in Cowal, south-west Scotland. *Scottish Journal of Geology*, **22**, 27–40.

Tipping, R.M. (1987) The prospects for establishing synchroneity in the early post-glacial pollen peak of *Juniperus* in the British Isles. *Boreas*, **16**, 155–63.

Tipping, R. (1988) The recognition of glacial retreat from palynological data: a review of recent work in the British Isles. *Journal of Quaternary Science*, **3**, 171–82.

Tipping, R.M. (1989a) Devensian Lateglacial vegetation history at Loch Barnluasgan, Argyllshire, western Scotland. *Journal of Biogeography*, **16**, 435–47.

Tipping, R.M. (1989b) Palynological evidence for the extent of the Loch Lomond Readvance in the Awe Valley and adjacent areas, SE Highlands. *Scottish Journal of Geology*, **25**, 325–37.

Tipping, R.M. (1991a) Climatic change in Scotland during the Devensian Late Glacial: the palynological record. In *The Late Glacial in North-west Europe: Human Adaption and Environmental Change at the End of the Pleistocene* (eds N. Barton, A.J. Roberts and D.A. Roe). Council for British Archaeology Research Report, **77**, 7–21.

Tipping, R.M. (1991b) The climatostratigraphic subdivision of the Devensian Lateglacial: evidence from a pollen site near Oban, western Scotland. *Journal of Biogeography*, **18**, 89–101.

Tivy, J. (1962) An investigation of certain slope deposits in the Lowther Hills, Southern Uplands of Scotland. *Transactions of the Institute of British Geographers*, **30**, 59–73.

Tolonen, M. (1980) Identification of fossil *Ulmus* pollen in sediments of Lake Lamminjärvi, S. Finland. *Annales Botanici Fennici*, **17**, 7–10.

Traverse, A. and Ginsberg, R.N. (1966) Palynology of the surface sediments of the Great Bahama Bank, as related to water movement and sedimentation. *Marine Geology*, **4**, 417–59.

Trenhaile, A.S. (1983) The development of shore platforms in high latitudes. In *Shorelines and Isostasy* (eds D.E. Smith and A.G. Dawson). Academic Press, London, pp. 77–93.

Trenhaile, A.S. and Mercan, D.W. (1984) Frost weathering and the saturation of coastal rocks. *Earth Surface Processes and Landforms*, **9**, 321–31.

Tufnell, L. (1971) Erosion by snow patches in the north Pennines. *Weather*, **26**, 492–8.

Tulloch, W. and Walton, H.S. (1958) *The Geology of the Midlothian Coalfield*. Memoirs of the Geological Survey of Scotland. HMSO, Edinburgh, 157 pp.

Turner, G.M. and Thompson, R. (1979) Behaviour of the Earth's magnetic field as recorded in the sediment of Loch Lomond. *Earth and Planetary Science Letters*, **42**, 412–26.

Turner, J. (1965) A contribution to the history of forest clearance. *Proceedings of the Royal Society*, **B161**, 343–54.

Turner, J. (1970) Post-Neolithic disturbance of British vegetation. In *Studies in the Vegetational History of the British Isles* (eds D. Walker and R.G. West). Cambridge University Press, Cambridge, pp. 97–116.

Turrill, W.B. (1927) The flora of St Kilda. *Report of the Botanical Society and Exchange Club of the British Isles*, **8**, 428–44.

Tyrrell, G.W. (1928) *The Geology of Arran*. Memoirs of the Geological Survey of Scotland, HMSO, Edinburgh, 292 pp.

Tyndall, J. (1879) The Parallel Roads of Glen Roy. *Proceedings of the Royal Institution of Great Britain*, **8**, 233–46.

Van der Veen, C.J. (1987) The West Antarctic Ice Sheet: the need to understand its dynamics. In *Dynamics of the West Antartic Ice Sheet* (eds C.J. van der Veen and J. Oerlemans). D. Reidel, Dordrecht, 1–16.

Van Zeist, W. and van der Spoel-Walvius, M.R. (1980) A palynological study of the late-glacial and the post-glacial in the Paris Basin. *Palaeohistoria*, **22**, 67–109.

Vasari, Y. (1970) The Late-glacial period in north-east Scotland. In *Probleme der Weichsel-Spätglazialen Vegetationsentwicklung in Mittel- und Nordeuropa* (ed. K.-D. Jager). Deutschen Akademie der Wissenschaften zu Berlin/International Union for Quaternary Research, Frankfurt/Oder, pp. 61–77.

Vasari, Y. (1977) Radiocarbon dating of the Lateglacial and early Flandrian vegetational succession in the Scottish Highlands and the Isle of Skye. In *Studies in the Scottish Lateglacial Environment* (eds J.M. Gray and J.J. Lowe). Pergamon Press, Oxford, 143–62.

Vasari, Y. and Vasari, A. (1968) Late- and post-glacial macrophytic vegetation in the lochs of northern Scotland. *Acta Botanica Fennica*, **80**, 120 pp.

Veyret, Y. and Coque-Delhuille, B. (1989) Les versants à banquettes de Ronas Hill. Essai de définition d'une province périglaciaire aux îles Shetland. *Hommes et Terres du Nord*, **3**, 171–8.

Vincent, P.J. and Lee, M.P. (1982) Snow patches on Farleton Fell, south-east Cumbria. *Geographical Journal*, **148**, 337–42.

Von Weymarn, J.A. (1974) Coastline development in Lewis and Harris, Outer Hebrides, with particular reference to the effects of glaciation. Unpublished PhD thesis, University of Aberdeen.

Von Weymarn, J. (1979) A new concept of glaciation in Lewis and Harris, Outer Hebrides. *Proceedings of the Royal Society of Edinburgh*, **B77**, 97–105.

Von Weymarn, J. and Edwards, K.J. (1973) Interstadial site on the Island of Lewis. *Nature*, **246**, 473–4.

Wager, L.R. (1953) The extent of glaciation in the island of St Kilda. *Geological Magazine*, **90**, 177–81.

Wahrhaftig, C. and Cox, A. (1959) Rock glaciers in the Alaska Range. *Bulletin of the Geological Society of America*, **70**, 383–436.

Wain-Hobson, T. (1981) Aspects of the glacial and post-glacial history of north-west Argyll. Unpublished PhD thesis, University of Edinburgh.

Walker, M.J.C. (1975a) Late Glacial and Early Postglacial environmental history of the central Grampian Highlands, Scotland. *Journal of Biogeography*, **2**, 265–84.

Walker, M.J.C. (1975b) Two Lateglacial pollen diagrams from the eastern Grampian Highlands of Scotland. *Pollen et Spores*, **17**, 67–92.

Walker, M.J.C. (1975c) A pollen diagram from the Pass of Drumochter, central Grampian Highlands, Scotland. *Transactions of the Botanical Society of Edinburgh*, **42**, 335–43.

Walker, M.J.C. (1977) Corrydon: a Lateglacial site from Glenshee, south-east Grampian Highlands, Scotland. *Pollen et Spores*, **19**, 391–406.

Walker, M.J.C. (1984a) A pollen diagram from St Kilda, Outer Hebrides, Scotland. *New Phytologist*, **97**, 99–113.

Walker, M.J.C. (1984b) Pollen analysis and Quaternary research in Scotland. *Quaternary Science Reviews*, **4**, 369–404.

Walker, M.J.C. (1990a) The Dalcharn interglacial site, near Cawdor, Nairnshire. Results of pollen analysis on the Dalcharn Biogenic Complex. In *Beauly to Nairn: Field Guide* (eds C.A. Auton, C.R. Firth and J.W. Merritt). Quaternary Research Association, Cambridge, pp. 54–7.

Walker, M.J.C. (1990b) The Allt Odhar interstadial site, Moy, Inverness-shire: results of pollen analysis. In *Beauly to Nairn: Field Guide* (eds C.A. Auton, C.R. Firth and J.W. Merritt). Quaternary Research Association, Cambridge, pp. 70–2.

Walker, M.J.C. and Harkness, D.D. (1990) Radiocarbon dating the Devensian Lateglacial in Britian: new evidence from Llanilid, South Wales. *Journal of Quaternary Science*, **5**, 135–44.

Walker, M.J.C. and Lowe, J.J. (1977) Postglacial environmental history of Rannoch Moor, Scotland. I. Three pollen diagrams from the Kingshouse area. *Journal of Biogeography*, **4**, 333–51.

Walker, M.J.C. and Lowe, J.J. (1979) Postglacial environmental history of Rannoch Moor, Scotland. II. Pollen diagrams and radiocarbon dates from the Rannoch Station and Corrour areas. *Journal of Biogeography*, **6**, 349–62.

Walker, M.J.C. and Lowe, J.J. (1980) Pollen analyses, radiocarbon dates and the deglaciation of Rannoch Moor, Scotland, following the Loch Lomond Advance. In *Timescales in Geomorphology* (eds R.A. Cullingford, D.A. Davidson and J. Lewin). John Wiley, Chichester, pp. 247–59.

Walker, M.J.C. and Lowe, J.J. (1981) Postglacial environmental history of Rannoch Moor, Scotland. III. Early- and mid-Flandrian pollen stratigraphic data from sites on western Rannoch Moor and near Fort William. *Journal of Biogeography*, **8**, 475–91.

Walker, M.J.C. and Lowe, J.J. (1982) Lateglacial and early Flandrian chronology of the Isle of Mull, Scotland. *Nature*, **296**, 558–61.

Walker, M.J.C. and Lowe, J.J. (1985) Flandrian environmental history of the Isle of Mull, Scotland. I. Pollen-stratigraphic evidence and radiocarbon dates from Glen More, south-central Mull. *New Phytologist*, **99**, 587–610.

Walker, M.J.C. and Lowe, J.J. (1987) Flandrian environmental history of the Isle of Mull, Scotland. III. A high-resolution pollen profile from Gribun, western Mull. *New Phytologist*, **106**, 333–47.

Walker, M.J.C. and Lowe, J.J. (1990) Reconstruction of the environmental history of the last glacial–interglacial transition: evidence from the Isle of Skye, Inner Hebrides, Scotland. *Quaternary Science Reviews*, **9**, 15–49.

Walker, M.J.C. and Lowe, J.J. (1991) Vegetational history of the Isle of Skye: I. The Late Devensian Lateglacial period (13–10 ka BP). In *The Quaternary of the Isle of Skye. Field Guide* (eds C.K. Ballantyne, D.I. Benn, J.J. Lowe and M.J.C. Walker). Quaternary Research Association, Cambridge, pp. 98–118.

Walker, M.J.C., Gray, J.M. and Lowe, J.J. (1985) *Isle of Mull, Inner Hebrides, Scotland*. Field Guide, Quaternary Research Association, Cambridge, 89pp.

Walker, M.J.C., Ballantyne, C.K., Lowe, J.J. and Sutherland, D.G. (1988) A reinterpretation of the Lateglacial environmental history of the Isle of Skye, Inner Hebrides, Scotland. *Journal of Quaternary Science*, **3**, 135–46.

Walker, M.J.C., Merritt, J.W., Auton, C.A., Coope, G.R., Field, M.H., Heijnis, H. and Taylor, B.J. (1992) Allt Odhar and Dalcharn: two pre-Late Devensian (Late Weichselian) sites in northern Scotland. *Journal of Quaternary Science*, **7**, 69–86.

Wallace, T.D. (1883) Shells in glacial clay at Fort George, Inverness-shire. *Transactions of the Edinburgh Geological Society*, **4**, 143–4.

Wallace, T.D. (1898) Geological notes on Strathdearn and the Aviemore railway. *Transactions of the Edinburgh Geological Society*, **7**, 416–9.

Wallace, T.D. (1901) Additional notes on the geology of Strathdearn and the adjoining districts of the Aviemore railway. *Transactions of the Edinburgh Geological Society*, **8**, 10–4.

Wallace, T.D. (1906) Glacial evidence in the Moray Firth area. *Transactions of the Inverness Scientific Society and Field Club*, **6**, 145–61.

Walther, M. (1984) Geomorphologische Untersuchungen zum Spätglazial und Früholozän in den Cuillin Hills (Insel Skye, Schottland). Unpublished PhD thesis, Free University of Berlin.

Walther, M. (1987) Ergebnisse geomorphologischer und palynologischer Untersuchungen zum Spätglazial und Frühholozän in den Cuillin Hills anf der Insel Skye (Schottland). *Eiszeitalter und Gegenwart*, **37**, 119–37.

Walton, K. (1959) Ancient elements in the coastline of north-east Scotland. In *Geographical Essays in Honour of Alan G. Ogilvie* (eds R. Miller and J.W. Watson). Nelson, London, pp. 93–109.

Walton, K. (1963) Geomorphology. In *The*

North-east of Scotland (eds A.C. O'Dell and J. Mackintosh). British Association, Aberdeen, pp. 16–32.
Walton, K. (1964) Aspects of the geomorphology of Scotland. In *Field Studies in the British Isles* (ed. J.A. Steers). Nelson, London, pp. 391–405.
Warburton, J. (1987) Characteristic ratios of width to depth-of-sorting for sorted stripes in the English Lake District. In *Periglacial Processes and Landforms in Britain and Ireland* (ed. J. Boardman). Cambridge University Press, Cambridge, pp. 163–71.
Ward, R.G. (1977) The Lochwinnoch Gap. In *Western Scotland I* (ed. R.J. Price). INQUA X Congress, 1977. Guidebook for Excursion A12. Geo Abstracts, Norwich, pp. 10–12.
Ward, R.G.W. (1980) Avalanche hazard in the Cairngorm Mountains, Scotland. *Journal of Glaciology*, **26**, 31–41.
Ward, R.G.W. (1981) Snow avalanches in Scotland with particular reference to the Cairngorm Mountains. Unpublished PhD thesis, University of Aberdeen.
Ward, R.G.W. (1984a) Avalanche prediction in Scotland: I. A survey of avalanche activity. *Applied Geography*, **4**, 91–108.
Ward, R.G.W. (1984b) Avalanche prediction in Scotland: II. Development of a predictive model. *Applied Geography*, **4**, 109–33.
Ward, R.G.W. (1985a) An estimate of avalanche frequency in Glen Feshie, Scotland, using tree rings. In *Palaeoenvironmental Investigations: Research Design, Methods and Data Analysis* (eds N.R.J. Fieller, D.D. Gilbertson and N.G.A. Ralph). British Archaeological Reports International Series, **258**, 237–44.
Ward, R.G.W. (1985b) Geomorphological evidence of avalanche activity in Scotland. *Geografiska Annaler*, **67A**, 247–56.
Ward, R.G.W., Langmuir, E.D.G. and Beattie, B. (1985) Snow profiles and avalanche activity in the Cairngorm Mountains, Scotland. *Journal of Glaciology*, **31**, 18–27.
Washburn, A.L. (1979) *Geocryology. A Survey of Periglacial Processes and Environments*. Edward Arnold, London, 406 pp.
Waters, R.S. (1954) Pseudo-bedding in the Dartmoor granite. *Transactions of the Royal Geological Society of Cornwall*, **18**, 456–62.
Waterston, C.D. (1960) City of Edinburgh. In *Edinburgh Geology. An Excursion Guide* (eds G.H. Mitchell, E.K. Walton and D. Grant). Oliver and Boyd, Edinburgh, pp. 10–27.
Watson, R.B. (1866) On the marine origin of the Parallel Roads of Glen Roy. *Quarterly Journal of the Geological Society*, **22**, 9–12.
Watson, J. (1985) Northern Scotland as an Atlantic–North Sea divide. *Journal of the Geological Society of London*, **142**, 221–43.
Watson, J.A. (1934) General history. In *History of the Edinburgh Geological Society* (hon. ed. R. Campbell). Edinburgh Geological Society, Edinburgh, pp. 5–15.
Watson, W.R. (1945) The glaciation of the Highland border district of south-eastern Perthshire. Part I: glacial retreat phenomena from the River Isla to the River Tay. *Transactions of the Edinburgh Geological Society*, **14**, 8–20.
Watt, A.S. and Jones, E.W. (1948) The ecology of the Cairngorms. Part I. The environment and the altitudinal zonation of the vegetation. *Journal of Ecology*, **36**, 283–304.
Watters, R.J. (1972) Slope stability in metamorphic rocks of the Scottish Highlands. Unpublished PhD thesis, Imperial College, University of London.
Watts, S.H. (1981) Bedrock weathering features in a portion of eastern High Arctic Canada: their nature and significance. *Annals of Glaciology*, **2**, 175–6.
Watts, S.H. (1983) Weathering processes and products under arid arctic conditions. *Geografiska Annaler*, **65A**, 85–98.
Watts, W.A. (1967) Interglacial deposits in Kildromin Townland, near Herbertstown, Co. Limerick. *Proceedings of the Royal Irish Academy*, **65B**, 339–48.
Watts, W.A. (1977) The Late Devensian vegetation of Ireland. *Philosophical Transactions of the Royal Society of London*, **B280**, 273–93.
Watts, W.A. (1980) Regional variation in the response of vegetation to Lateglacial climatic events in Europe. In *Studies in the Lateglacial of North West Europe* (eds J.J. Lowe, J.M. Gray and J.E. Robinson). Pergamon Press, Oxford, pp. 1–22.
Watts, W.A. (1985) Quaternary vegetation cycles. In *The Quaternary History of Ireland* (eds K.J. Edwards and W.P. Warren). Academic Press, London, pp. 155–85.
Webb, J.A. and Moore, P.D. (1982) The Late Devensian vegetational history of the Whitlaw Mosses, south-east Scotland. *New Phytologist*, **91**, 341–98.
Weertman, J. (1961) Mechanisms for the formation of inner moraines found near the edge of

cold ice caps and ice sheets. *Journal of Glaciology*, **3**, 965–78.

Welin, E., Engstrand, L. and Vaczy, S. (1975) Institute of Geological Sciences radiocarbon dates VI. *Radiocarbon*, **17**, 157–9.

Werritty, A. and Brazier, V. (1991) *The Geomorphology, Conservation and Management of the River Feshie SSSI*. Peterborough, Nature Conservancy Council Report, 73pp + figures, plates and appendices.

Werritty, A. and Ferguson, R.I. (1980) Pattern Changes in a Scottish braided river over 1, 30 and 200 years. In *Timescales in Geomorphology* (eds R.A. Cullingford, D.A. Davidson and J. Lewin). John Wiley, Chichester, pp. 247–59.

West, R.G. (1970) Pollen zones in the Pleistocene of Great Britain and their correlation. *New Phytologist*, **69**, 1179–83.

West, R.G. (1977) *Pleistocene Geology and Biology*, 2nd edn. Longman, London, 440 pp.

West, R.G. (1980) Pleistocene forest history of East Anglia. *New Phytologist*, **85**, 571–622.

Westoll, T.S. (1942) The corries of the Cairngorms. *Cairngorm Club Journal*, **83**, 216–24.

Whalley, W.B. (1974) *The mechanics of high-magnitude, low frequency rock failure and its importance in a mountainous area.* University of Reading, Department of Geography, Geographical Papers, No. **27**, 48 pp.

Whalley, W.B. (1976a) A fossil rock glacier in Wester Ross. *Scottish Journal of Geology*, **12**, 175–8.

Whalley, W.B. (1976b) Some aspects of the structure and development of earth pillars and corrugated lateral moraine surfaces. *Studia Geomorphologica Carpatho–Balcanica*, **10**, 49–62.

Whalley, W.B. and Martin, H.E. (1992) Rock glaciers: II models and mechanisms. *Progress in Physical Geography*, **16**, 127–86.

Whalley, W.B., Gordon, J.E. and Thompson, D.L. (1981) Periglacial features on the margins of a receding plateau ice cap, Lyngen, North Norway. *Journal of Glaciology*, **27**, 492–6.

Whalley, W.B., Douglas, G.R. and Jonsson, A. (1983) The magnitude and frequency of large rockslides in Iceland in the Postglacial. *Geografiska Annaler*, **65A**, 99–110.

White, I.D. and Mottershead, D.N. (1972) Past and present vegetation in relation to solifluction on Ben Arkle, Sutherland. *Transactions of the Botanical Society of Edinburgh*, **41**, 475–89.

White, S.E. (1976) Rock glaciers and block fields, review and new data. *Quaternary Research*, **6**, 77–97.

Whittington, G. (1990) The Dalcharn interglacial site, near Cawdor, Nairnshire. Results of pollen analysis on the Dalcharn Biogenic Member. In *Beauly to Nairn: Field Guide* (eds C.A. Auton, C.R. Firth and J.W. Merritt). Quaternary Research Association, Cambridge, pp. 57–9.

Whittington, G. and Edwards, K.J. (1990) The cultivation and utilisation of hemp in Scotland. *Scottish Geographical Magazine*, **106**, 167–73.

Whittington, G. and Gordon, A.D. (1987) The differentiation of the pollen of *Cannabis sativa* L. from that of *Humulus lupulus* L. *Pollen et Spores*, **29**, 111–20.

Whittington, G. and Ritchie, W. (1988) Flandrian environmental evolution on north-east Benbecula and southern Grimsay, Outer Hebrides, Scotland. *O'Dell Memorial Monograph (Department of Geography, University of Aberdeen).* No. **21**, 46 pp.

Whittington, G., Edwards, K.J. and Cundill, P.R. (1990) Palaeoenvironmental investigations at Black Loch in the Ochil Hills of Fife, Scotland. *O'Dell Memorial Monograph (Department of Geography, University of Aberdeen)*, No. **22**, 64 pp.

Whittington, G., Edwards, K.J. and Cundill, P.R. (1991a) Late- and post-glacial vegetational change at Black Loch, Fife, eastern Scotland – a multiple core approach. *New Phytologist*, **118**, 147–66.

Whittington, G., Edwards, K.J. and Caseldine, C.J. (1991b) Late- and post-glacial pollen-analytical and environmental data from a near-coastal site in north-east Fife, Scotland. *Review of Palaeobotany and Palynology*, **68**, 65–85.

Whittington, G., Edwards, K.J. and Cundill, P.R. (1991c) Palaeoecological investigations of multiple elm declines at a site in north Fife, Scotland. *Journal of Biogeography*, **18**, 71–87.

Whittow, J.B. (1977) *Geology and Scenery in Scotland.* Penguin Books, Harmondsworth, 362 pp.

Wilkins, D.A. (1984) The Flandrian woods of Lewis (Scotland). *Journal of Ecology*, **72**, 251–8.

Wilkinson, S.B. (1900) In *Summary of Progress of the Geological Survey of the United Kingdom for 1899*. Memoirs of the Geological Survey. HMSO, London, pp. 163–4.

Wilkinson, S.B. (1907) *The Geology of Islay,*

including Oronsay and Portions of Colonsay and Jura. Explanation of Sheets 19 and 27, with the western part of Sheet 20. Memoirs of the Geological Survey of Scotland, HMSO, Glasgow, 82 pp.

Williams, R.B.G. (1975) The British climate during the last glaciation: an interpretation based on periglacial phenomena. In *Ice Ages: Ancient and Modern* (eds A.E. Wright and F. Moseley). Seel House Press, Liverpool, pp. 95–120.

Williams, W. (1977) The Flandrian vegetational history of the Isle of Skye and the Morar Peninsula. Unpublished PhD thesis, University of Cambridge.

Williams, R.E.G. and Johnson, A.S. (1976) Birmingham University radiocarbon dates X. *Radiocarbon*, **18**, 249–67.

Wilson, G.V., Edwards, W., Knox, J., Jones, R.C.B. and Stephens, J.V. (1935) *The Geology of the Orkneys*. Memoirs of the Geological Survey of Scotland, HMSO, Edinburgh, 205 pp.

Wilson, J.S.G. (1886) *North-east Aberdeenshire, and Detached Portions of Banffshire*. Explanation of sheet 87. Memoirs of the Geological Survey of Scotland, HMSO, Edinburgh, 32 pp.

Wilson, J.S.G. (1900) Glen Spean and the Great Glen. *Summary of Progress of the Geological Survey of the United Kingdom for 1899*. Memoirs of the Geological Survey, HMSO, London, pp. 158–62.

Wilson, J.S.G. and Hinxman, L.W. (1890) *Central Aberdeenshire*. Explanation of Sheet 76. Memoirs of the Geological Survey of Scotland, HMSO, Edinburgh, 43 pp.

Wilson, M.J. (1985) The mineralogy and weathering history of Scottish soils. In *Geomorphology and Soils* (eds K.S. Richards, R.R. Arnett and S. Ellis). George Allen and Unwin, London, pp. 233–44.

Wilson, M.J. and Tait, J.M. (1977) Halloysite in some soils from north-east Scotland. *Clay Minerals*, **12**, 59–66.

Wilson, M.J., Bain, D.C. and McHardy, W.J. (1971) Clay mineral formation in a deeply-weathered boulder conglomerate in north-east Scotland. *Clays and Clay Minerals*, **19**, 345–52.

Wilson, M.J., Russell, J.D., Tait, J.M., Clark, D.R., Fraser, A.R. and Stephen, I. (1981) A swelling haematite/layer-silicate complex in weathered granite. *Clay Minerals*, **16**, 261–78.

Wilson, M.J., Russell, J.D., Tait, J.M., Clark, D.R. and Fraser, A.R. (1984) Macaulayite, a new mineral from north-east Scotland. *Mineralogical Magazine*, **48**, 127–9.

Wilson, P. (1989) Nature, origin and age of Holocene aeolian sand on Muckish Mountain, Co. Donegal, Ireland. *Boreas*, **18**, 159–68.

Wilson, P. (1990a) Characteristics and significance of protalus ramparts and fossil rock glaciers on Errigal Mountain, County Donegal. *Proceedings of the Royal Irish Academy*, **90B**, 1–21.

Wilson, P. (1990b) Morphology, sedimentological characteristics and origin of a fossil rock glacier on Muckish Mountain, north-west Ireland. *Geografiska Annaler*, **72A**, 237–47.

Wood, J.M. (1975) *Witchcraft and Superstitious Records in the South-Western District of Scotland*. E.P. Publishing, Wakefield, 355 pp.

Wood, W. (1887) *The East Neuk of Fife: its History and Antiquities*, 2nd edn. David Douglas, Edinburgh, 586 pp.

Worsley, P. (1991) Possible early Devensian glacial deposits in the British Isles. In *Glacial Deposits in Great Britain and Ireland* (eds J. Ehlers, P.L. Gibbard and J. Rose). A.A. Balkema, Rotterdam, pp. 47–51.

Wright, J. (1896) Boulder-clay, a marine deposit. *Transactions of the Geological Society of Glasgow*, **10**, 263–72.

Wright, W.B. (1911) On a preglacial shoreline in the Western Isles of Scotland. *Geological Magazine*, Decade 5, **8**, 97–109.

Wright, W.B. (1914) *The Quaternary Ice Age*. Macmillan, London, 464 pp.

Wright, W.B. (1928) The raised beaches of the British Isles. *First Report of the Commission on Pliocene and Pleistocene Terraces*. International Geographical Union, Oxford, pp. 99–106.

Wright, W.B. (1937) *The Quaternary Ice Age*, 2nd edn. Macmillan, London, 478 pp.

Yamasaki, F., Hamada, T. and Hamada, C. (1969) Riken natural radiocarbon measurements V. *Radiocarbon*, **11**, 451–62.

Yates, E.M. and Moseley, F. (1967) A contribution to the glacial geomorphology of the Cheshire Plain. *Transactions of the Institute of British Geographers*, **42**, 107–25.

Young, I. (1988) Caves of the Allt nan Uamh basin. In *The Limestone Caves of Scotland, Part 2. Caves of Assynt* (ed. T.J. Lawson). The Grampian Speleological Group Occasional Publication, **6**, 41–64.

Young, J. (1864) On the former existence of glaciers in the high grounds of the south of Scotland. *Quarterly Journal of the Geological Society of London*, **20**, 452–62.

References

Young, J. and Craig, R. (1869) Notes on the occurrence of seeds of freshwater plants and arctic shells, along with the remains of the mammoth and reindeer in beds under the boulder clay at Kilmaurs. *Transactions of the Geological Society of Glasgow*, **3**, 310–21.

Young, J.A.T. (1966) Analysis of glacial deposits near Fala, Midlothian. Unpublished PhD thesis, University of Edinburgh.

Young, J.A.T. (1969) Variations in till macrofabric over very short distances. *Bulletin of the Geological Society of America*, **80**, 2343–52.

Young, J.A.T. (1974) Ice wastage in Glenmore, upper Spey Valley, Inverness-shire. *Scottish Journal of Geology*, **10**, 147–57.

Young, J.A.T. (1975a) Ice wastage in Glen Feshie, Inverness-shire. *Scottish Geographical Magazine*, **91**, 91–101.

Young, J.A.T. (1975b) A re-interpretation of the deglaciation of Abernethy Forest, Inverness-shire. *Scottish Journal of Geology*, **11**, 193–205.

Young, J.A.T. (1976) The terraces of Glen Feshie, Inverness-shire. *Transactions of the Royal Society of Edinburgh*, **69**, 501–12.

Young, J.A.T. (1977a) Glacial geomorphology of the Aviemore–Loch Garten area, Strathspey, Inverness-shire. *Geography*, **62**, 25–34.

Young, J.A.T. (1977b) Glacial geomorphology of the Dulnain Valley, Inverness-shire. *Scottish Journal of Geology*, **13**, 59–74.

Young, J.A.T. (1978) The landforms of upper Strathspey. *Scottish Geographical Magazine*, **94**, 76–94.

Young, J.A.T. (1980) The fluvioglacial landforms of mid-Strathdearn, Inverness-shire. *Scottish Journal of Geology*, **16**, 209–20.

Zagwijn, W.H. (1975) Variations in climate as shown by pollen analysis in the Lower Pleistocene of Europe. In *Ice Ages: Ancient and Modern* (eds A.E. Wright and F. Moseley). Seel House Press, Liverpool, pp. 137–52.

Zagwijn, W.H. (1986) The Pleistocene of the Netherlands with special reference to glaciation and terrace formation. *Quaternary Science Reviews*, **5**, 341–5.

Zbinden, H., Andrée, M., Oeschger, H., Ammann, B., Lotter, A., Bonani, G. and Wölfli, W. (1989) Atmospheric radiocarbon at the end of the last glacial: an estimate based on AMS radiocarbon dates on terrestrial macrofossils from lake sediments. *Radiocarbon*, **31**, 795–804.

Zimmerman, H.B., Shackleton, N.J., Backman, J., Baldauf, J.G., Kaltenbach, A.J. and Morton, A.C. (1984) History of Plio-Pleistocene climate in the north-eastern Atlantic, Deep Sea Drilling Project Hole 552A. *Initial Reports of the Deep Sea Drilling Project*, **81**, 861–75.

Index

Page numbers appearing in **bold** refer to figures and page numbers appearing in *italic* type refer to tables.

Index

Index

Index

Index

Index

Index

Index